BIOGEOCHEMISTRY OF MARINE DISSOLVED ORGANIC MATTER

SECOND EDITION

BIOGEOCHEMISTRY OF MARINE DISSOLVED ORGANIC MATTER

SECOND EDITION

Edited by

DENNIS A. HANSELL
Rosenstiel School of Marine and Atmospheric Science
University of Miami
Miami, Florida

and

CRAIG A. CARLSON
Department of Ecology Evolution And Marine Biology
University of California
Santa Barbara, California

AMSTERDAM • BOSTON • HEIDELBERG • LONDON • NEW YORK • OXFORD
PARIS • SAN DIEGO • SAN FRANCISCO • SYDNEY • TOKYO
Academic Press is an imprint of Elsevier

Academic Press is an imprint of Elsevier
32 Jamestown Road, London NW1 7BY, UK
225 Wyman Street, Waltham, MA 02451, USA
525 B Street, Suite 1800, San Diego, CA 92101-4495, USA
The Boulevard, Langford Lane, Kidlington, Oxford OX5 1GB, UK

Notice
No responsibility is assumed by the publisher for any injury and/or damage to persons or property as a
matter of products liability, negligence or otherwise, or from any use or operation of any methods, products,
instructions or ideas contained in the material herein

Library of Congress Cataloging-in-Publication Data
Biogeochemistry of marine dissolved organic matter/edited by Dennis A. Hansell, Craig A.
Carlson. – Second edition.
     pages cm
  ISBN 978-0-12-405940-5 (hardback)
1. Seawater–Organic compound content.   2. Chemical oceanography.   3. Biogeochemistry.   I. Hansell,
Dennis A., editor.   II. Carlson, Craig A., editor.
  GC118.B56 2014
  551.46'6–dc23
                                                                                              2014033432

British Library Cataloguing in Publication Data
A catalogue record for this book is available from the British Library

ISBN: 978-0-12-405940-5

For information on all Academic Press publications
visit our web site at http://store.elsevier.com

Working together
to grow libraries in
developing countries

www.elsevier.com • www.bookaid.org

Dedication

For the support and balance that only family can provide we dedicate this book to our parents Paul and Rose Marie Hansell and David and Paula Carlson, our beloved spouses Paula and Alison and our children Allison and Rachel, and Matthew, Hayden and Sydney. The foundation they have provided us is, through this book, extended to coming generations of marine scientists.

Contents

List of Contributors

Rainer M.W. Amon Department of Marine Sciences and Oceanography, Texas A&M University at Galveston, Galveston, Texas, USA

Thomas R. Anderson National Oceanography Centre Southampton, Southampton, UK

Leif G. Anderson Department of Chemistry and Molecular Biology, University of Gothenburg, Gothenburg, Sweden

Sandra Arndt School of Geographical Sciences, University of Bristol, Bristol, UK

Steven R. Beaupré Geology and Geophysics, Woods Hole Oceanographic Institution, Woods Hole, Massachusetts, USA

Karin M. Björkman Daniel K. Inouye Center for Microbial Oceanography: Research and Education, Department of Oceanography, School of Ocean and Earth Science and Technology, University of Hawaii, Honolulu, Hawaii, USA

Deborah A. Bronk Virginia Institute of Marine Science, College of William & Mary, Virginia, USA

David J. Burdige Department of Ocean, Earth and Atmospheric Sciences, Old Dominion University, Noroflk, Virginia, USA

Craig A. Carlson Department of Ecology, Evolution and Marine Biology, University of California, Santa Barbara, California, USA

James R. Christian Fisheries and Oceans Canada, Canadian Centre for Climate Modelling and Analysis, Victoria, British Columbia, Canada

Thorsten Dittmar Research Group for Marine Geochemistry (ICBM-MPI Bridging Group), Institute for Chemistry and Biology of the Marine Environment (ICBM), University of Oldenburg, Oldenburg, Germany

Kevin J. Flynn Centre for Sustainable Aquatic Research, Swansea University, Swansea, UK

Dennis A. Hansell Rosenstiel School of Marine and Atmospheric Science, University of Miami, Miami, Florida, USA

David M. Karl Daniel K. Inouye Center for Microbial Oceanography: Research and Education, Department of Oceanography, School of Ocean and Earth Science and Technology, University of Hawaii, Honolulu, Hawaii, USA

David J. Kieber College of Environmental Science and Forestry, Department of Chemistry, State University of New York, Syracuse, New York, USA

Tomoko Komada Romberg Tiburon Center, San Francisco State University, Tiburon, California, USA

Caroline Leck Department of Meteorology, University of Stockholm, Stockholm, Sweden

Kenneth Mopper Department of Chemistry and Biochemistry, Old Dominion University, Norfolk, Virginia, USA

Norman B. Nelson Earth Research Institute, University of California Santa Barbara, California, USA

Mónica V. Orellana Polar Science Center, University of Washington/Institute for Systems Biology, Seattle, Washington, USA

Peter A. Raymond Yale School of Forestry and Environmental Studies, New Haven, Connecticut, USA

Daniel J. Repeta Department of Marine Chemistry and Geochemistry, Woods Hole Oceanographic Institution, Woods Hole, Massachusetts, USA

Andy Ridgwell School of Geographical Sciences, University of Bristol, Bristol, UK

Chiara Santinelli Istituto di Biofisica, Pisa, Italy

Rachel E. Sipler Virginia Institute of Marine Science, College of William & Mary, Virginia, USA

Robert G.M. Spencer Department of Earth, Ocean and Atmospheric Science, Florida State University, Tallahassee, Florida, USA

Colin A. Stedmon National Institute of Aquatic Resources, Technical University of Denmark, Charlottenlund, Denmark

Aron Stubbins Skidaway Institute of Oceanography, Department of Marine Sciences, University of Georgia, Savannah, Georgia, USA

Foreword

As one of Earth's largest exchangeable carbon reservoirs, similar in scale to atmospheric CO_2, the biogeochemical behavior of marine dissolved organic matter (DOM) has major significance for the carbon cycle, climate, and global habitability. As Ducklow (2002) wrote in the first edition of this book: "Oceanic DOM is now recognized as an important component of the biogeochemical system and possibly a barometer of global change." Accordingly, marine DOM research has undergone a renaissance, moving rapidly beyond issues of measurement methodology to critically important spatial-temporal mapping of DOC distribution in the global ocean (Hansell et al., 2009), and now poised to address the underlying regulatory mechanisms. Despite important strides, our understanding of the biogeochemical behavior of marine DOM is still in its exciting "early exponential growth phase." Fundamental questions of the nature and sources of DOM and mechanisms of its production, transformation, and respiration remain unanswered. So, there is much to do, and there are fresh ideas and powerful new study tools, as reflected in this book. Below, I formulate some problems that I suggest must be solved to understand DOM behavior and to predict the future biogeochemical state of the ocean. I hope that some young—and not so young—scientists will take on the challenge to solve them.

The lifetimes of DOM constituents range from minutes to millennia: some are mineralized rapidly by heterotrophic microbes (labile; LDOM); others less readily (semi-labile; SLDOM); while an incredible diversity of molecules, 10^{12}-10^{15} (Hedges, 2002) have accumulated over time to comprise the huge (~642 PgC) but intriguing refractory DOM pool. A challenge is to understand the biological and physicochemical forces that mediate and regulate the biogeochemical behavior of various DOM components.

Most oceanic DOM is ultimately derived from primary production, and owes some of its chemical complexity to it, as phytoplankton generate enormous molecular diversity at the expense of CO_2 and just a few inorganic and trace nutrients in order to serve their diverse adaptive needs. Biochemical processing of primary production by genetically diverse bacteria further adds to the chemical complexity of DOM. Intriguingly, most of the biomass generated by primary production is particulate (POM) yet on average about one-half—but a variable fraction—of primary production becomes DOM within the upper mixed layer, assessed conservatively as bacterial carbon demand. The POM-DOM transition is a critical step in the flow of reduced carbon in the global ocean and the capacity of the ecosystem to retain elements in the upper ocean for air-sea exchange; yet we currently lack knowledge of the underlying mechanisms or the means of direct quantification. There is extensive literature showing that multiple physiological, biochemical, and trophic interactions cause the release of DOM from the particulate phase—including living organisms. Sloppy feeding and exudation were long believed to be the major mechanisms, but with discoveries of new links in the food web there has also been recognition of additional mechanisms of DOM production. The list now includes microbial ectohydrolase activity, viral lysis of phytoplankton and heterotrophic microbes, cellular release of transparent exopolymer particles and other gel particles, programmed cell death, and microbe-microbe antagonism. It is probable that all mechanisms—both those listed and unlisted—cause the POM-DOM transition, but their relative quantitative significance varies in time and location. In view of the critical importance of the POM-DOM transition for predictive models of carbon flow and sequestration in the ocean, I stress the need to develop quantitative methods and a better mechanistic understanding of the production of marine DOM.

The utilization side of the marine DOM dynamics is deceptively simple, since essentially all DOM uptake is due to bacteria and Archaea, the dominant osmotrophic heterotrophs. The strength of this coupling is a critical variable in the regulation of DOM utilization and respiration, and subsequent air-sea exchange of CO_2. Tight coupling also prevents excursions in DOM concentration, so this regulation has major ecological and biogeochemical implications. How do microbes manage to biochemically couple so tightly with primary production; and what biochemical and behavioral (e.g., chemokinesis) mechanisms regulate the nature and strength of bacteria-organic matter coupling? What is the role of microbial genetic diversity and biogeochemical expressions in maintaining the strength of coupling?

Genomic predictions have provided powerful constraints. They tell us the molecular interactions among DOM molecules and microbes that are *possible*. However, predicting the biogeochemical dynamics of the complex DOM pool also requires ecophysiological and biochemical studies of DOM-bacteria interactions to determine: *what DOM transformations do take place, at what rates, by what biochemical mechanisms, subject to what regulatory forces and in what ecosystem context*. This is indeed a tall order; but the problem is critical to solve because of the central role of DOM-bacteria interactions in predicting the carbon cycle of the future ocean. Ocean acidification and warming are likely to affect the nature and rates of microbial production and transformation of DOM with potential influence on the carbon balance between the ocean and the atmosphere. Understanding what renders some of the DOM semi-labile or refractory will also require such mechanistic studies. This important research on dissolved phase carbon cycling and sequestration requires new methods, model systems, and concepts (e.g., Microbial Carbon Pump; Jiao et al., 2010) addressing in situ dynamics and interactions among microbes and (DOM) molecules.

Method refinement has been an important goal in marine DOM research. The field was transformed in the 1990s by the fundamental discovery that DOM was measurably dynamic, contrary to earlier thinking that the DOM pool was inert. This discovery "changed everything." Interestingly, chemists and microbiologists reached this conclusion by different paths. Chemists worked diligently to refine the DOC method (in spite of or perhaps because of initial setbacks; Hedges, 2002; Sugimura and Suzuki, 1988), achieving ~1 μM precision, sufficient to show DOC gradients over days to seasons and across locations and depths (Hansell et al., 2009). Marine microbiologists had been finding as early as the 1960s that ^{14}C labeled amino acids and sugars added to seawater as metabolic tracers were readily assimilated and respired with lifetimes of hours to days. Clearly, a DOM fraction represented by the radiotracers was highly dynamic. Remarkably, the seemingly opposite views of DOM lability coexisted for a decade. As it turned out, the microbiologists had been observing the behavior of a tiny but dynamic labile DOM (~1 μM) embedded within an ~40 to 70-fold larger recalcitrant DOM pool. Thus, the divergent impressions of the "shape of the elephant" were due to the enormous range of lifetimes of DOM components. This problem of the biogeochemical behaviors of labile and recalcitrant DOM, and implications for dissolved phase carbon sequestration, is an active research area, formalized as the Microbial Carbon Pump (Jiao et al., 2010).

While high precision measurements of DOC concentrations (Sharp, 2002) has transformed marine DOM research, further method refinement and new method development is a high priority. First, achieving 1-2 µM precision still requires a magic touch; we need a plug and play method that *even* a microbial oceanographer could use! Second, experimental studies of DOM uptake, respiration, and sequestration require yet higher analytical precision. An order of magnitude improvement may even enable measurements of bacterial utilization of semi-labile DOC (lifetime 1.5 years; Hansell, 2013) albeit requiring long incubations. By analogy, consider if the scientists studying ocean acidification could only measure seawater pH with precision of 0.1 units! (While Dennis Hansell was waiting for me to finally finish this Foreword, X-Prize worth $2M was announced for precise and user-friendly pH instrument: http://www.xprize. org/prize/wendy-schmidt-ocean-health-xprize.) Ultra-precise DOC, DON, and DOP methods will transform marine DOM biogeochemistry and climate predictions. Finally, we need a standardized method to measure microheterotrophic (bacteria + Archaea) respiration that does not significantly perturb the process being observed. These prokaryotes essentially monopolize DOM and their carbon growth efficiency is low, typically 10-30% (i.e., 70-90% of the labile DOM-C is respired by heterotrophic microbes). It is currently debated whether or not respiration and primary production are in balance or instead display spatial patterns of imbalance related to oligotrophic versus meso-/eutrophic systems (Ducklow and Doney, 2013).

The "DOM problem" has been studied for the better part of a century against significant methodological odds, yet significant advances have been made in the last 10-20 years that have enriched the field. Today, there is strong interdisciplinary convergence and integration, as marine chemists, organic geochemists, analytical chemists, microbiologists and molecular biologists, and modelers join to address big questions of carbon cycling and sequestration, climate predictions, and the biogeochemical state of the future ocean. This multifaceted pursuit has also led to the emergence of modern biogeochemistry as a distinct and dynamic field that is maturing rapidly and attracting students and postdocs as well as accomplished scientists from other fields. This advance of marine biogeochemistry as a discipline responsive to challenges posed by climate change may well be the most important development that has been stimulated by research on marine DOM.

Farooq Azam
Scripps Institution of Oceanography
University of California, San Diego

References

Ducklow, H.W., 2002. Foreword. In: Hansell, D.A., Carlson, C.A. (Eds.), Biogeochemistry of Marine Dissolved Organic Matter. Academic Press, San Diego, pp. xv–xix.

Ducklow, H.W., Doney, S.C., 2013. What is the metabolic state of the oligotrophic ocean? A debate. Ann. Rev. Mar. Sci. 5, 525–533.

Hansell, D.A., 2013. Recalcitrant dissolved organic carbon fractions. Ann. Rev. Mar. Sci. 5, 421–445.

Hansell, D.A., Carlson, C.A., Repeta, D.J., Schlitze, R., 2009. Dissolved organic matter in the ocean: new insights stimulated by a controversy. Oceanography 22, 52–61.

Hedges, J., 2002. Why dissolved organics matter. In: Hansell, D.A., Carlson, C.A. (Eds.), Biogeochemistry of Marine Dissolved Organic Matter. Academic Press, San Diego, pp. 1–33.

Jiao, N., Hernd, G.J., Hansell, D.A., Benne, R., Kattner, G., Wilhelm, S.W., et al., 2010. Microbial production of recalcitrant dissolved organic matter: long-term carbon storage in the global ocean. Nat. Rev. Microbiol. 8, 593–599.

Sharp, J.H., 2002. Analytical methods for total DOM pools. In: Hansell, D.A., Carlson, C.A. (Eds.), Biogeochemistry of Marine Dissolved Organic Matter. Academic Press, San Diego, pp. 35–58.

Sugimura, Y., Suzuk, Y., 1988. A high-temperature catalytic oxidation method for the determination of non-volatile dissolved organic carbon in seawater by direct injection of a liquid sample. Mar. Chem. 24, 105–131.

Preface

Efforts by the ocean science community to understand cycling of the major bioactive elements (C, N, P) in the ocean have experienced great success in the past two decades and continues today. Intensive focus on elemental cycling resulted from society's need to determine the role of the ocean in climate change, in turn requiring an understanding of its essential functions today, in the past, and in the future. The fundamentals of the biological processes involved in the transformations of the major elements have been identified. The next phase of research requires linking the biological processes to the very large oceanic reservoirs of the major elements, and identifying the sensitivities of those links. Establishing a linkage between processes and reservoirs falls into the discipline of *biogeochemistry*.

One of Earth's largest exchangeable reservoirs of carbon is dissolved organic matter (DOM) in the ocean. With a stock of 662 PgC, the pool approximates the amount of carbon resident in atmospheric CO_2. Prior to the 1990s, this major pool of carbon was primarily evaluated from a geochemical viewpoint; resolving the biochemical and isotopic composition of the pool was a central goal. With an enhanced biogeochemical perspective on DOM, the scientific questions began to rest broadly on the role of DOM in the oceanic C, N, and P cycles. Central questions were: can we accurately measure the concentrations of DOM in the ocean; what are the distributions of the dissolved organic C/N/P pools and what processes controls these distributions; what are the rates, biogeographical locations, and controls on elemental cycling through the pools; what are the biological and

physicochemical sources and sinks; what is the composition of DOM and how does that illuminate elemental cycling? Finally, do we understand DOM in elemental cycling well enough to accurately represent the processes in models of the modern, past, and future oceans?

In this edition, the progress of the last decade in answering these questions is reported and synthesized by key contributors to those advances. The book opens with a chapter by A. Ridgwell and S. Arndt, setting the paleoceanographic context for marine DOM. Study of the chemical and isotopic compositions of DOM has provided unique information on elemental cycling; both subjects have seen great growth over the past decade. These works are reviewed in chapters by D. Repeta and S. Beaupre, respectively. The biological cycling of the major elements (C, N, P) through DOM is reviewed in chapters by C. Carlson and D. Hansell (focusing on biological processes and carbon budgets), D. Karl and K. Björkman (focusing on P), and R. Sipler and D. Bronk (focusing on N). Particular emphasis is placed in these chapters on marine microbes as active agents in the processing of DOM. Photochemical reactivity of DOM, and implications for elemental cycling, is presented by K. Mopper, D. Kieber, and A. Stubbins. The contributions of optically active (chromophoric) DOM are covered by C. Stedmon and N. Nelson. The book continues with chapters covering DOM at ocean interfaces and in marginal seas (i.e., the rivers draining into the ocean, the sediments, the Arctic Ocean, and the Mediterranean Sea, in chapters by P. Raymond and R. Spencer, D. Burdige and T. Komada, L. Anderson and R. Amon, and C. Santinelli, respectively). Topics new to this edition include marine microgels, introduced by M. Orellana and C. Leck, and the long-term stability of marine DOM by T. Dittmar. The book closes with the advances of DOM in ecosystem and global circulation models by T. Anderson, J. Christian and K. Flynn.

Many in the ocean science community have developed a strong biogeochemical view of the ocean. This book serves as a tool to provide foundation for their forays into the biogeochemistry of marine organic matter. The book maintains a particular focus on DOM in elemental cycling, and therefore does not revisit the many, well-documented advances made in organic geochemistry during the previous decades. Attention is largely to the marine environment, with little coverage of the fresh water systems other than the important rivers and dynamics adding DOM to the coastal seas. The book is directed at professional ocean scientists and advanced students of biological and chemical oceanography.

Many individuals and organizations must be thanked for support of the science that provided content for this book, as well as to development of the book itself. The U.S. federal agencies supporting much of what has been reported here, including individual research by the chapter authors, are the National Science Foundation (NSF), the National Oceanographic and Atmospheric Administration (NOAA), and the National Aeronautics and Space Administration (NASA). The agency program managers most important in supporting research conducted by the editors are Don Rice, David Garrison, and Eric Itsweire at NSF, David Legler and Joel Levy at NOAA, and Diane Wickland and Paula Bontempi at NASA. We greatly appreciate their leadership and support, and hope they are proud of these accomplishments.

Dennis A. Hansell and Craig A. Carlson

Why Dissolved Organics Matter: DOC in Ancient Oceans and Past Climate Change

Andy Ridgwell, Sandra Arndt

School of Geographical Sciences, University of Bristol, Bristol, UK

CONTENTS

I OVERVIEW

The ocean and its underlying sediments are the largest sinks of CO_2 within the Earth system that are able to respond to changes in atmospheric CO_2 on both human-induced (anthropogenic)

and geologically relevant time scales. We need a complete understanding of their dynamics and strength of feedbacks with climate and other drivers of global change if we are to make confident projections regarding the full consequences of continued fossil fuel CO_2 emissions. To this end,

Biogeochemistry of Marine Dissolved Organic Matter,
http://dx.doi.org/10.1016/B978-0-12-405940-5.00001-7

1

in the past couple of decades, we have made rapid progress in elucidating the roles that basic physical (e.g. ocean circulation) and inorganic geochemical processes (e.g. gas exchange) play in regulating the uptake of CO_2 from the atmosphere. In contrast, the role and response of the "biological carbon pump"—the interplay of biological, geochemical, and physical processes that transfer carbon from the surface ocean where it is fixed by primary producers to the depths where it is either consumed or buried in the underlying sediments—is much less well understood. Even the sign of some of the main feedbacks involved and whether the response of the biological carbon pump will act to amplify or reduce future climate warming and ocean deoxygenation is somewhat uncertain.

Two research directions are providing new insights into the marine carbon cycle and how it responds to perturbation. The first revolves around ongoing efforts to understand the mechanistic operation of the biological carbon pump and, in particular, the cycling of carbon through dissolved organic carbon (DOC) in the ocean, as described in this book. The second is an increased appreciation of what might be learned from the geological record. Earth history is punctuated by a huge variety of transitions and perturbations in climate and global biogeochemical cycles, with some events exhibiting evidence for greenhouse warming and CO_2 release and hence potentially providing clues regarding future changes. The conjunction of these two developments has led to the idea that both the cycling of carbon through DOC and its reservoir size could have been fundamentally different in the past and that changes in that cycling may be mechanistically linked to major events in the geological record. The breath of speculation about how the marine carbon cycle may have operated during the Precambrian (prior to 541 Ma) and under very different conditions of oxygenation and ecosystem function from today also highlights the importance of first being able to ground our geologic interpretation in a full mechanistic understanding of the sources and sinks of DOC in the modern ocean.

This chapter provides an overview of how the proposed link between DOC and major global carbon cycle perturbations in the geological record arises. We start by presenting a brief summary of how the marine carbon cycle operates. We then introduce how the geological record can be interpreted, focusing on the ways in which the carbon isotopic signature of sedimentary rocks reflects past changes in global carbon cycling. We finish by critically assessing recent thinking regarding the potential role of DOC dynamics as a driver or amplifier of extreme climate events in the past as well as the potential future implications.

II MARINE CARBON CYCLING

Geological rock reservoirs dominate the global inventory of carbon on Earth (Figure 1.1). However, the response time for the formation or any substantive depletion of these reservoirs is counted in 10s, if not 100s, of millions of years, being largely governed by plate tectonics and major biological evolutionary innovations such as the advent and proliferation of calcifying plankton (Ridgwell, 2005). At ~38,000 PgC ($1 PgC = 10^{15} gC$), the present-day ocean dissolved inorganic carbon (DIC) reservoir is the next largest carbon store on Earth and is an order of magnitude larger than the likely extractable resources of fossil fuel carbon or the terrestrial biosphere (and soils). One way to influence atmospheric pCO_2 and climate via greenhouse warming is to create an imbalance in the inputs versus sinks of carbon to the ocean. However, the response time of the ocean plus atmosphere as a whole—calculated as its carbon inventory (Figure 1.1) divided by the rate of carbon throughput from weathering and mantle CO_2 outgassing (Figure 1.2b and c)—comes out to be of the order of 100,000 years (100 ky). Perturbations of the global carbon cycle that change the DIC inventory of the ocean as a whole in this way are hence only arguably relevant on geological time scales. Ocean alkalinity (and pH) influence the speciation of DIC (between $CO_{2(aq)}$,

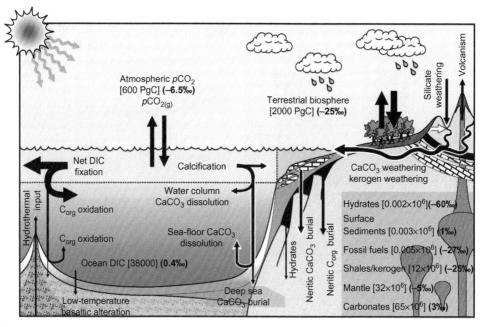

FIGURE 1.1 Illustration of the primary reservoirs and processes constituting the (natural) global carbon cycle. Approximate carbon inventories (brackets) and representative carbon isotopic compositions (bold and in parentheses) are shown for the main reservoirs and fluxes. *Adapted from Hönisch et al. (2012) with representative isotopic compositions from Maslin and Thomas (2003) and inventories from IPCC (2007).*

HCO_3^-, and CO_3^{2-}), but similar time scales apply if one wishes to change the alkalinity (ALK) inventory of the ocean as a whole.

In addition to changes in the total reservoir size, DIC (and alkalinity) can also be partitioned spatially within the ocean. This is important as the atmosphere only "sees" the surface ocean. The rapid exchange of CO_2 between ocean surface and atmosphere (Figure 1.2a) means that atmospheric pCO_2 and hence climate are highly responsive to changes in surface DIC which gives the ocean a unique importance to the global carbon cycle and climate system for reasons beyond its relative total carbon mass.

A A Tale of Three Ocean Carbon "Pumps"

The processes that control the partitioning of DIC between the surface ocean and ocean interior as well as its speciation at the surface are traditionally divided into three conceptual "pumps" in the ocean (e.g., Sarmiento and Gruber, 2006): (1) the "solubility" pump, (2) the "organic matter" ("organic carbon" or "soft tissue") pump, and (3) the "carbonate" (or "counter") pump, illustrated in Figure 1.2. In this conceptual framework, all three oceanic carbon pumps act primarily by vertically relocating carbon away from the surface, hence the concept of the "pumping" of carbon from the surface to depth. At the same time, the latter two also partition alkalinity (see Stumm and Morgan, 1981 or Zeebe and Wolf-Gladrow, 2001 for details) with depth, thus dictating the speciation of DIC (the balance between $CO_{2(aq)}$, HCO_3^-, CO_3^{2-}) to maintain overall charge neutrality. Modulating all three and hence influencing atmospheric pCO_2 are the large-scale patterns and rates of ocean circulation that

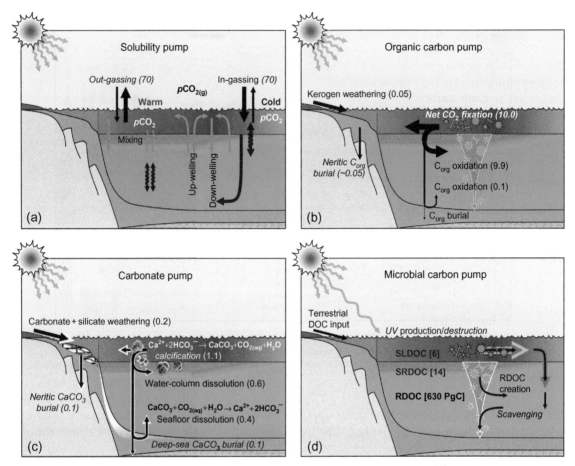

FIGURE 1.2   Schematic workings of the four components of the ocean's carbon pump. (a) The solubility. (b) The particulate organic carbon pump in the ocean. (c) The carbonate pump in the ocean. (d) The cycle of dissolved organic carbon in the ocean and the microbial carbon pump. Fluxes of DOC are broken down into semi-labile (light gray) and semi-refractory (dark gray), neither of which are transported far into the ocean interior. Fluxes of refractory DOC (RDOC) are indicated with black arrows. Reservoir inventories, in brackets, are from Hansell (2013). In all figures, carbon fluxes into the ocean are shown in normal font and sinks in italics. (a–c) These are quantified following Hönisch et al. (2012) in units of PgC year^{-1}.

control the return of DIC, ALK, and nutrients back to the surface.

The operation of the solubility pump in partitioning DIC between surface and the deep ocean is the simplest to conceptualize (Figure 1.2a). CO_2 is more soluble in colder compared to warmer seawater. As a consequence, carbon dissolution is enhanced in cold, high-latitude surface waters, where deep waters form. The rapid sinking of these dense,

DIC-rich surface waters and their redistribution in the deep ocean results in efficient pumping of carbon from the atmosphere to the deep ocean. Apart from temperature, the solubility of CO_2 is also affected by salinity. Thus, differences in salinity between major water masses and between surface and deep ocean will also make a contribution. Nevertheless, the contribution of salinity is generally minor compared to the effect of temperature.

The "organic carbon" ("organic matter" or "soft tissue") pump also acts to induce a vertical DIC gradient (Figure 1.2b). In this case, DIC is removed from solution as a result of photosynthesis and is fixed in the form of both particulate organic carbon (POC) and DOC. A proportion of POC escapes the intense recycling in the surface ocean, sinking vertically through the ocean interior where heterotrophic degradation results in a release of DIC. In general, the heterotrophic degradation of POC during sinking is highly efficient, with <1-6% of the POC export production reaching the seafloor, where of this flux, just 0.3% escape benthic degradation to be buried (Dunne et al., 2007). The further the POC sinks before the associated carbon is released back to the DIC pool, the more effective the partitioning between surface and deep ocean. Any reduction (or increase) in the efficiency of this degradation during sinking, perhaps as a result of changes in sinking speed or heterotrophic degradation rate, thus has the potential to affect pCO_2. The influence of the organic carbon pump is partially offset by ALK changes associated with nitrogen transformations, particularly from NO_3^- to organic N during photosynthesis and from organic N to NH_4^+ during heterotrophic degradation (Zeebe and Wolf-Gladrow, 2001), which at the surface has the effect of partitioning more DIC in the form of $CO_{2(aq)}$. In addition, the organic carbon pump controls nutrient cycling in the ocean. It creates nutrient-rich deep waters that are redistributed by global circulation and upwelling and thereby influences the spatial distribution of primary production in the ocean.

The third of the traditional carbon pumps, the carbonate pump, involves the biological precipitation of calcium (and to a lesser extent, magnesium) carbonate ($CaCO_3$) at the ocean surface. Perhaps as much as 50% of carbonate dissolves as it sinks (Feely et al., 2004). The remainder reaches the sediment surface where ~20% of the deposited carbonate escapes further dissolution to be buried (Feely et al., 2004), ultimately forming a new geological carbon reservoir (Figure 1.1). Although the carbonate pump also vertically redistributes carbon, the transfer of ALK dominates in terms of its influence on the speciation of dissolved carbon species and overall acts to increase, not decrease, surface ocean pCO_2.

By virtue of their influence on surface ocean pCO_2, all three of the pumps play an important role in the short-term carbon cycle. We define "short-term" here as the approximate mixing time scale of the ocean (nominally ~500-1000 years) and before the geochemical redistribution initially arising from any transient change in any of the C pumps will have been effectively homogenized by ocean mixing. On short time scales, the organic carbon pump dominates over the carbonate pump in terms of net influence on atmospheric pCO_2, simply as a result of its order-of-magnitude larger export flux. The organic matter pump also dominates over the solubility pump (Cameron et al., 2005).

All three pumps are also important components of the long-term carbon cycle, driving the accumulation, biogeochemical transformation, and burial of carbon in marine sediments, directly in the example of the two solid pumps and indirectly in controlling the saturation state (all three pumps) and oxygenation (organic carbon and solubility pumps) of bottom water. For instance, under conditions conducive to the preservation of biogenic calcium carbonate ($CaCO_3$)—primarily the shells of haptophyte algae (coccolithophores) and zooplankton (foraminifera)—sediments can become rich in $CaCO_3$, with large amounts of carbon locked away for millions of years in the form of chalk deposits. As mentioned earlier, the sediment preservation of organic matter is generally less efficient, with only a tiny fraction of the organic matter that is exported from the surface ocean ultimately buried. Nevertheless, organic carbon burial rates can vary significantly in space and time. Particular intervals of Earth history, for instance the Cretaceous (145.5-65.5My ago (Ma)) and Jurassic (201.3-145.5Ma), saw a widespread deposition of organic carbon rich (>1wt%) strata. The factors that may explain this spatial and temporal variability in organic carbon deposition and burial rates are a matter of debate. They include, but are not limited to, organic matter composition, terminal electron acceptor, and in particular oxygen availability, benthic community composition, microbial inhibition by specific metabolites, priming, physical protection, deposition rate, and macrobenthic activity (see review by Arndt et al., 2013). In addition, an enhanced deposition and burial of organic matter may promote the production of methane (CH_4) in the anoxic sub-seafloor. In conjunction with cold temperatures and/or high pressure, hydrates can form molecules of CH_4 trapped in a water-ice cage structure (Maslin et al., 2010). Methane hydrates are special among the geological reservoirs in that they also influence the global carbon cycle and climate on short time scales if the environmental conditions change sufficiently rapidly and substantially to destabilize the hydrate structure and release the CH_4 gas. An increase in ocean temperatures and depressurization by slumping and loss of the overlying sediment burden are possible drivers of significant and rapid CH_4 release (e.g., see Maslin et al., 2010).

The last glacial maximum (LGM; about 21,000 years ago) is a good example of the role that the ocean carbon pumps play in regulating atmospheric $p\mathrm{CO_2}$. At the time of the LGM, the concentration of CO_2 in the atmosphere was about 190 ppm compared to ~260 ppm at the start of the Holocene (11,700 years ago) and ~280 ppm just prior to the industrial revolution (Monnin et al., 2001). An enduring mystery is "why?", or in other words, in what way did the global carbon cycle function differently, resulting in lower atmospheric CO_2 during glacial periods (Kohfeld and Ridgwell, 2009). Aside from ideas revolving around lower glacial ocean temperatures and hence an increase in the strength of the solubility pump, a greater supply of eolian iron to surface ocean ecosystems and hence par-

tial relief of iron limitation in regions such as the Southern Ocean may have played a role (Watson et al., 2000). Iron deposition works its magic on atmospheric CO_2 via an increase in plankton productivity and POC export and hence a stronger organic matter pump in the ocean—increasing the partitioning of DIC away from the surface and to depth. The biological pump could have been more "efficient" as well as stronger. This is an important distinction to make. In terms of lowering atmospheric $p\mathrm{CO_2}$, it is not sufficient to simply export POC away from the surface more rapidly through increased biological export; the CO_2 that is eventually released through heterotrophic degradation needs to be kept away from the ocean surface and hence the atmosphere for as long as possible. For instance, carbon released from the degradation of POC in a more stratified glacial deep ocean (Adkins et al., 2002) would be less effectively mixed back towards the surface, while colder ocean temperatures could potentially have reduced the degradation rate of settling POC (e.g., Crill and Martens, 1987; Jørgensen and Sørensen, 1985; Matsumoto et al., 2007), pushing the average depth at which DIC is released to greater depths. Both processes would act to increase the efficiency of the biological carbon pump, even if a side effect was decreased export from the surface due to reduced nutrient return. That the glacial CO_2 question has yet to be settled, despite the existence of a wide variety of hypotheses (Kohfeld and Ridgwell, 2009), points to a still incomplete understanding of the marine carbon cycle and leaves the potential for new mechanisms and processes to be discovered.

B A Fourth Appears—The Microbial Carbon Pump

Appreciation of the "microbial carbon" pump (Figure 1.2d), a part of the biological pump involving the net production of DOC by microbial processing of organic matter, has increased rapidly over the past decade (Jiao et al., 2011).

It represents a further mechanism for repartitioning carbon among the non-geological (surficial) reservoirs. Alongside the production of POC and particulate inorganic carbon (PIC) by marine phytoplankton at the surface ocean, a variable but substantial fraction of primary production (30-50%) is released as DOC (e.g., Biddanda and Benner, 1997; Ducklow et al., 1995). DOC is also created by a variety of other activities in the surface ocean, including viral lysis, "sloppy feeding" by metazoan grazers, the egesta of protists and metazoa (e.g., Jiao et al., 2010), and the initial extracellular hydrolysis step involved in heterotrophic POC degradation. The most labile fractions of DOC are consumed rapidly and account for a mere 1% of the total ocean DOC inventory (Hansell, 2013). Only the more biologically recalcitrant DOC fractions last long enough to be transported into the deep ocean by ocean circulation. Even so, few of these newly produced, recalcitrant DOC fractions survive to reach depths of 1000 m (Hansell, 2013). The degradation product (DIC) will still continue to be advected into the ocean interior, however, helping partition DIC into the ocean interior.

At the highly refractory end of the reactivity spectrum, refractory DOC (RDOC) is removed only on multi-millennial time scales and hence on average survives multiple cycles of ocean turnover. Presumably, direct exudation during bacterial production, viral lysis of microorganisms or microbial necromass further contribute to the refractory pool. Microbial activity thus transfers, by the successive processing of DOC, organic carbon from low concentrations of reactive carbon to progressively higher concentrations of refractory carbon (hence the "pump" designation; Jiao et al., 2010). Despite its relatively low rate of production—perhaps only 0.043 PgC year^{-1} (Hansell, 2013) as compared to 1-10 PgC year^{-1} for PIC and POC export (Figure 1.1)—the most refractory DOC component accumulates in the ocean with estimated inventories of 630 ± 12 PgC for RDOC and >12 PgC for an ultra-refractory (e.g., black carbon) DOC

(Hansell, 2013). As such, RDOC constitutes a reservoir of similar size to the atmospheric burden of carbon present as CO_2 (Figure 1.1). Thus, although the existence of this refractory carbon pool has been known for decades (e.g., Hedges, 2002), its potential size and thus its potential significance for carbon sequestration and past and future climate change has only been recognized recently (e.g., Jiao et al., 2010).

In theory, any substantial change in the consumption (or degradation) rate of this DOC pool could exert a large effect on atmospheric pCO$_2$ and climate. A variety of environmental factors such as oxygen availability and ecosystem or temperature changes could potentially trigger substantial changes (see several other chapters in this book for insights on the processes that control the composition, reactivity, production, and removal of DOC in the ocean). In particular, the availability of oxygen often plays a central role in the discussion of substantial and large-scale changes in organic carbon consumption rates on climate-relevant time scales. Although the jury is still out on the direct effect of oxygen availability on heterotrophic organic carbon degradation, preservation of organic carbon tends to increase under anoxic conditions (e.g., Canfield et al., 1993). This increase could be the result of a thermodynamically limited degradation of refractory organic compounds in the absence of the powerful electron acceptor oxygen, the inability of anaerobic organisms to directly and completely degrade organic matter to CO_2, a decreased enzymatic activity, and/or a decreased availability of DOC adsorbed to mineral surfaces under anoxic conditions (e.g., see review by Arndt et al., 2013). Therefore, large-scale ocean redox changes could, in theory, result in a larger global ocean DOC inventory during periods of widespread ocean anoxia. This pool of DOC would then be rapidly oxidized to DIC once oxic conditions are re-established, releasing CO_2 to the atmosphere. This hence provides a new mechanism to alter the DIC inventory of the whole ocean and one that can operate with a response time much faster than the ~100 ky time scale driven by imbalances between weathering on land and sediment burial at the ocean floor. The possibility that such DOC reservoir variations occurred in the geological past is the subject of the remainder of this chapter.

III INTERPRETING THE GEOLOGICAL PAST

The strength and efficiency of the biological pump is controlled by a plethora of biological, geochemical, and physical factors, such as ecosystem composition, the reactivity and fate of organic matter, oxygen levels of the ocean's interior, and the cycling of nutrients (Hain et al., 2014). These factors interact with and respond to changes in climate. The resulting complexity makes it extremely challenging to make projections about the consequences of continuing global warming on the marine carbon cycle. The geological record helps here as Earth's history is punctuated by a huge variety of transitions and perturbations in global biogeochemical cycles and climate. These perturbations may exhibit evidence for greenhouse warming and CO_2 release (Hönisch et al., 2012), with some associated with major extinctions or biotic disruption. Such events potentially provide calibrations regarding, for instance, the response of marine ecosystems to ongoing global change (e.g., Gibbs et al., 2006, 2012), for which model-based projections are difficult.

A Carbon Isotopes as Proxies for Past Global Carbon Cycle Changes

To understand past carbon cycle feedbacks and biotic sensitivities, we need to know how key environmental conditions have changed, such as atmospheric CO_2 concentrations, surface temperatures (Dunkley Jones et al., 2013), ocean pH and carbonate saturation (Hönisch

et al., 2012), or depletion of the ocean's oxygen inventory (Keeling et al., 2010). However, reconstructing environmental variables is an imperfect science that becomes increasingly difficult as one goes further back in time. To about 800 ky ago (800 ka), ice cores drilled from the Antarctic ice cap reveal the composition of the ancient atmosphere (including, critically, pCO_2) encoded in bubbles (Etheridge et al., 1996; Lorius et al., 1993). Additional environmental variables, such as the fluxes of dust and aerosols to the ice sheet surface, are recorded as impurities in the ice itself (Lambert et al., 2012). Beyond this, we have no direct record of the concentration of CO_2 in the atmosphere and therefore have to rely on geological "proxies."

Proxies (here, entities or measures representing the value of something else) are rooted in measurements made of some physical property in a geological sample such as, for instance, the mass fraction of a particular solid component (e.g., $CaCO_3$) or a geochemical property, such as the ratio between different trace elements or isotopes of the same element (Wefer et al., 1999). The intent with proxies is to empirically (or in rare cases, theoretically) calibrate a physical or geochemical property against an environmental variable of interest. Such calibrations are based on a combination of laboratory manipulation experiments and correlations inherent in modern observations. For instance, the density of stomata—the physiologically controlled pores in the underside of leaves that regulate both CO_2 ingress and water vapor loss—tends to decrease with increasing ambient pCO_2 (Royer, 2001). The reason is that while plants require CO_2, they equally need to conserve water. Higher atmospheric pCO_2 simply means that fewer stomata are required by the plant in order to obtain the same diffusive flux of CO_2. Fewer stomata provide the benefit of reduced water loss. Hence, the density of stomata, which can be counted in fossil leaves, should inverse correlate with concentration of CO_2 in the atmosphere if the plant requires that water loss be minimized. However, for many other environmental variables no direct proxy may exist. Therefore, reconstruction of potential past environmental conditions and the testing of hypotheses often rely on more indirect proxies. The ratio of the stable isotopes of carbon recorded in sedimentary rocks has been particularly important in this respect and has been central to efforts in reconstructing past global carbon cycling.

Carbon has two stable isotopes with masses 12 and 13 and a mean global abundance ratio of ~99:1. By convention, the 12:13 ratio in a sample is written in the delta-notation:

$$\delta^{13}C_{sample} = \left(R_{sample}/R_{standard} - 1\right) \times 1000, \quad (1.1)$$

where R_{sample} is the $^{13}C/^{12}C$ ratio measured in the sample and $R_{standard}$ is the $^{13}C/^{12}C$ ratio in a substance of known composition (a standard). The sample isotopic composition ($\delta^{13}C_{sample}$) is scaled by a factor of 1000 and given in units of per mil (‰).

Isotope fractionation—a change in the ratio of 12:13 during conversion of one carbon-containing substance to a second—can occur under a number of circumstances (see Chapter 6). First, at the same temperature and hence same kinetic energy, the lighter isotope (^{12}C) has a higher velocity. During diffusion, the destination will become isotopically depleted (in ^{13}C) while the source of carbon will become enriched. Second, covalent bonds involving ^{13}C are stronger than those involving ^{12}C. Hence, the photosynthetic splitting of CO_2 and the breaking of bonds in organic molecules tends to occur to a greater extent for ^{12}C, again leaving the residual enriched in ^{13}C. The result of kinetic and particularly bond-breaking fractionation, as carbon moves between different carbon reservoirs and is transformed, is that the carbon reservoirs on Earth display a wide range of $^{12}C{:}^{13}C$ ratios. Typical $\delta^{13}C$ values for the major carbon reservoirs are illustrated in Figure 1.1. The importance of bond breaking in creating a distinctly

light (^{13}C-depleted) composition of organic matter is obvious, and the additional fermentation step during methanogenesis gives rise to ever more depleted methane carbon—that is, there is a fractionation during photosynthesis and carbon assimilation as inorganic carbon is turned into organic molecules, and then a second stronger biological fractionation as organic matter is broken down and partly converted into CH_4.

Because the various reservoirs are characterized by different values of $\delta^{13}C$, moving carbon from one to another will change not only the amount of carbon in the reservoirs but also potentially their $\delta^{13}C$. Repeated sampling of any one reservoir over time therefore enables input from (or loss to) a second reservoir to be inferred, even if changes in the size of either reservoir cannot be directly reconstructed. For instance, the isotopic composition of CO_2 in the atmosphere has declined from a pre-Industrial (year 1765) value of about −6‰ to a present-day value of about −8.3‰, reflecting the release of isotopically depleted fossil fuel carbon (ca. −27‰) to the atmosphere. Using this information together with a model representation of the exchange of carbon between atmosphere and ocean, it is possible to reconstruct the expected history of atmospheric pCO_2 increases since the industrial revolution consistent with the observed $\delta^{13}C$ changes. In this particular example, measurements of historical atmospheric pCO_2 change together with some information on the rate at which fossil fuel carbon has been burned and terrestrial biomass degraded are directly available and changes in pCO_2 do not have to be determined indirectly. However, in the absence of more direct measurements, $\delta^{13}C$ changes recorded in dated plant remains (the proxy), for example combined with models, would have allowed us to reconstruct the history of atmospheric pCO_2 and the associated time-varying emissions of fossil fuel CO_2 to the atmosphere.

B Reconstructing Past Steady-State Modes of Global Carbon Cycling

The geological record of carbon isotopic variability, illustrated in Figure 1.3, provides important insights into past carbon cycling. For the Phanerozoic (542 Ma to present; Figure 1.3a), the records are primarily derived from measurements made on the calcareous shells and skeletons of shallow dwelling and reef-forming organisms, supplemented by analyses on much smaller carbonate shells of planktonic and benthic foraminifera for the interval for which ancient oceanic crust and its overlying sediment burden still exists (the last ~180 My). Organic matter may also help reconstruct $\delta^{13}C$

trends, although marine carbonates ($CaCO_3$) have the advantage of being much more abundant in sedimentary rocks than organic matter.

Overall, changes in the global carbon cycle through the Phanerozoic (Figure 1.3a) are reflected in relatively slow, multimillion year transitions between higher and lower $\delta^{13}C$. These intervals of elevated (or depleted) $\delta^{13}C$ can last 10s to 100s of million years and are generally ascribed to changes in the balance of organic versus inorganic (carbonate) carbon burial. This is because the surficial carbon reservoirs have insufficient capacity (Figure 1.1) to accumulate (or release) carbon at rates anywhere near comparable to weathering or sediment burial (Figure 1.2)

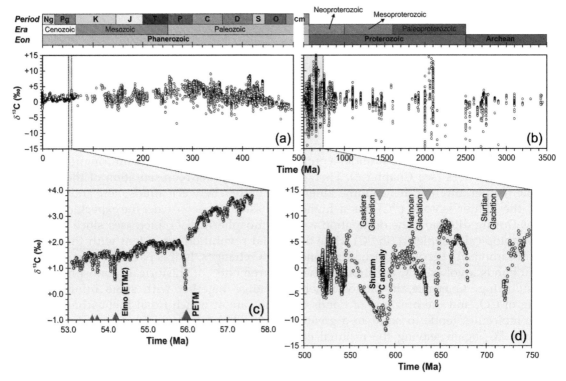

FIGURE 1.3   Carbon isotopic variability through Earth history. Shown are compilations of $\delta^{13}C$ measured in marine carbonates for the Phanerozoic (a) (Veizer et al., 1999) and for the Precambrian (b) (Shields and Veizer, 2002). (c) A more detailed benthic foraminiferal calcite $\delta^{13}C$ record spanning the Paleocene-Eocene boundary (Zachos et al., 2010). The timing of hyperthermal events are indicated by red triangles. (d) A more detailed record spanning the mid-through-late Neoproterozoic (Halverson et al., 2005; Macdondald et al., 2012). The approximate age of occurrence of glaciations are marked with inverted blue triangles. In the geological time scale at the top, the abbreviated Periods, from left to right are: Neogene ("Ng"), Paleogene ("Pg"), Cretaceous ("K"), Jurassic ("J"), Triassic ("T"), Permian ("P"), Carboniferous ("C"), Devonian ("D"), Silurian ("S"), Ordovician ("O"), and Cambrian ("Cm").

for such durations. Assuming steady state, that is, time scales much longer than the ca. 100 ky residence time of $\delta^{13}C$ in the modern surficial reservoirs, the required mass balance can be written in terms of the ratio between organic carbon ($F_{C_{org}}$) and total carbon ($F_{C_{org}} + F_{CaCO_3}$) burial fluxes:

$$\frac{F_{C_{org}}}{F_{C_{org}} + F_{CaCO_3}} = \frac{\delta^{13}C_{obs} - \delta^{13}C_{input}}{\Delta^{13}C_{C_{org} - CaCO_3}}, \quad (1.2)$$

where $\delta^{13}C_{input}$ is the isotopic signature of the average carbon input into the ocean (Figure 1.1), and $\Delta^{13}C_{C_{org} - CaCO_3}$ is the isotopic difference between the $\delta^{13}C$ of buried organic matter versus that of $CaCO_3$ (see, e.g., Kump and Arthur, 1999 for details). Because the system is expressed as a function of the observed $\delta^{13}C$ of the carbonate sediments ($\delta^{13}C_{obs}$), for approximately invariant rates of carbon inputs and burial, Eq. (1.2) reconstructs Earth's organic carbon burial history from the observed $\delta^{13}C$ record (Figure 1.3). (One also needs to assume appropriate values for the characteristic isotopic composition of carbon inputs and outputs ($\delta^{13}C_{input}$ and $\Delta^{13}C_{C_{org} - CaCO_3}$) as summarized in Figure 1.1.) In other words, $\delta^{13}C_{obs}$ can be used as a proxy for the global burial rate of organic matter, an important global carbon cycle flux that would otherwise be impossible to reconstruct.

As an example, the increasing $\delta^{13}C_{obs}$ trend associated with the Carboniferous (359-299 Ma) in Figure 1.3a is commonly interpreted as reflecting increasing organic matter burial in response to the evolution of lignin-forming plants. Lignin is a relatively recalcitrant terrestrial compound that could have caused a step increase in the efficiency of organic carbon preservation (Berner, 2004), for which the extensive coal measures of the Carboniferous (a series of coal strata characteristic of the period) are considered as supporting evidence. Proxies such as leaf stomata density suggest that atmospheric CO_2 declined to relatively low concentrations around this time

(Royer, 2006) and a long interval of glaciation occurred—both consistent with an increase in organic carbon burial that sequestered CO_2 from the ocean and atmosphere. The important lesson here is that large observed changes in $\delta^{13}C$ may indicate fundamental reorganizations of the global carbon cycle.

C Interpreting Transient Carbon Cycle Perturbations

Superimposed on the general underlying trends of gradually increasing and decreasing $\delta^{13}C$ are a wide variety of negative spikes and transients, illustrated in Figure 1.3c for the interval surrounding the Paleocene-Eocene boundary (ca. 56 Ma). Assuming that $\delta^{13}C$ transient changes reflect changes in the $\delta^{13}C$ of the seawater from which the carbonates are precipitated, rather than postdepositional diagenetic alteration, these negative spikes indicate that the ocean DIC and atmospheric CO_2 reservoirs were being augmented by carbon with a distinct isotopic signature. In a simple mass balance approach, which deliberately ignores the fluxes into and out of the system unlike the approach above (Eq. 1.2), we can write:

$$M_{final} \times \delta^{13}C_{M_{final}} \approx M_0 \times \delta^{13}C_{M_0} + \Delta M \times \delta^{13}C_{\Delta M}, (1.3a)$$

where the subscripts "0" and "final" represent the initial and final values of mass (M) and isotopic composition ($\delta^{13}C$) of the surficial carbon reservoir and ΔM and $\delta^{13}C_{\Delta M}$ are the magnitude of the added carbon and its isotopic composition, respectively. Equation (1.3a) simply states that the final mean isotopic composition of a carbon reservoir (here assumed to be surficial system of ocean + atmosphere + terrestrial biosphere) can be taken to be equal to the mean isotopic composition of the initial reservoir plus that of the added (or removed) carbon, all weighted by their respective masses. For a given isotopic signature of the new carbon, Eq. (1.3a) can be rearranged, substituting $M_{final} = M_0 + \Delta M$ (final carbon mass equals the initial mass plus carbon addition) and $\delta^{13}C_{M_{final}} = \delta^{13}C_{M_0} + \Delta\delta^{13}C$ (final isotopic composition equals the initial isotopic composition plus the recorded isotopic anomaly), to estimate the amount of carbon needed to explain a given magnitude of isotopic excursion ($\Delta\delta^{13}C$):

$$\Delta M = -\Delta\delta^{13}C \times \frac{M_0}{\delta^{13}C_{M_0} + \Delta\delta^{13}C - \delta^{13}C_{\Delta M}}. \quad (1.3b)$$

This relationship is illustrated in Figure 1.4 for assumed values of $\delta^{13}C_{M_0}$. An important caveat to this analysis is that the carbon addition (or removal) is assumed to be instantaneous. That is, that the excursion is not significantly modified as a result of dilution caused by the continual throughput of carbon from terrestrial weathering to marine sedimentation (Figure 1.2) while the new carbon is still being added (or removed).

The transient event that occurs at the boundary between the Paleocene and Eocene Epochs (Figure 1.3c) is a well-studied example of such a negative excursion event. The Paleocene (66-56 Ma) and Eocene (56-33.9 Ma) epochs are relatively warm intervals associated with elevated atmospheric pCO_2 compared to modern (Hönisch et al., 2012) and a peak in warmth towards the early-middle Eocene followed by a long-term cooling trend cumulating in the emergence of substantial ice on Antarctica in the Oligocene (33.9-23.0 Ma). Tectonically, these intervals saw the movement of India toward and then collision with the Asian plate and the disappearance of the remnant ancient Tethys Sea, together with the gradual opening in the late Eocene and Oligocene of seaways separating Antarctica from South America and Australia. The Paleocene-Eocene boundary itself is marked by a prominent extinction among deep-sea foraminifera (Thomas, 2007), associated with a pronounced surface warming characterized by a global mean temperature increase of about 5°C (Dunkley Jones et al., 2013)—known as the

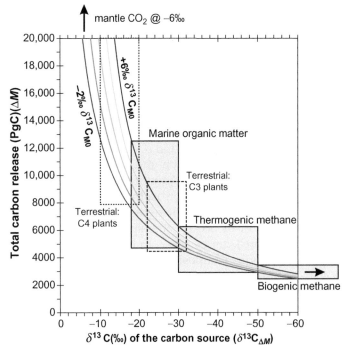

FIGURE 1.4 The relationship between carbon release, isotopic signature, and resulting excursion magnitude. Contours delineating the relationship between source magnitude and isotopic composition for a −4‰ carbon isotopic ($\delta^{13}C$) composition as given by Eq. (1.4). The assumed $\delta^{13}C$ of the carbon source is on the x-axis, with the resulting inferred mass of carbon (in PgC) on the y-axis. Contours are plotted for initial mean surficial carbon reservoir $\delta^{13}C$ values spanning the observed range of long-term Phanerozoic variability: −2‰ (pink), 0‰ (orange), 2‰ (green), 4‰ (light blue), and 6‰ (dark blue).

Paleocene-Eocene Thermal Maximum ("PETM") (Zachos et al., 2005).

A temporary negative shift in $\delta^{13}C$ associated with the PETM (Figure 1.3c) is recorded in all superficial carbon reservoirs, from terrestrial wood through marine algal compounds to biogenic carbonates (McInerney and Wing, 2011). Because excursions for this event are recorded in a variety of materials both on the land as well as in the ocean, we can be extremely confident that the $\delta^{13}C$ changes are not simply a diagenetic artifact. The magnitude of the carbon isotope excursion does vary somewhat among the different reservoirs, appearing amplified in organic matter closely connected to the atmosphere and damped in biogenic carbonates deposited in deep marine sedimentary settings, which complicates the interpretation (Sluijs and Dickens, 2012). But assuming that the event's "true" value is around −4‰ for instance (McInerney and Wing, 2011; Panchuk et al., 2008; Sluijs and Dickens, 2012), one can estimate the amount of carbon released using Figure 1.4 (Eq. 1.3b). The required carbon release depends upon the specific $\delta^{13}C$ of the assumed source of carbon in this calculation. Different sources of light carbon ranging from methane hydrates to deep-ocean DOC and with different characteristic $\delta^{13}C$ values have been suggested. Depending on the assumed carbon source and its isotopic value, different amounts of carbon release would thus be inferred (Panchuk et al., 2008). This is important information as knowing the amount of carbon released constrains the sensitivity of surface temperatures to CO_2 increase, known as "climate sensitivity" (PALAEOSENS Project Members, 2012). For instance, a popular hypothesis is that destabilization of methane hydrates, with a characteristic $\delta^{13}C$ value of −60‰ (Figure 1.1), was central to the PETM event (Dickens et al., 1995). Assuming an initial surficial carbon reservoir of ocean plus atmosphere plus terrestrial biosphere similar to modern (~41,000 PgC (38,000 + 600 + 2000)—Figure 1.1) and an initial ocean $\delta^{13}C$ of 2‰ (Figure 1.3c) requires a total carbon release from this source of 2800 PgC for an excursion of −4‰. Alternatively, a terrestrial organic carbon source at −22‰ (Panchuk et al., 2008) such as might be derived from (burning) peat deposits (Kurtz et al., 2003) or oxidizing organic matter in Antarctic permafrost (DeConto et al., 2012) would require 8200 PgC. An additional uncertainty is the assumed size of the past surficial carbon reservoir. While we have some proxy constraints on atmospheric pCO_2, we have to rely on marine geochemical models in making hindcasts for the ocean DIC inventory (e.g., Ridgwell, 2005). Models suggest a similar-to-modern inventory for the Paleocene and Eocene (Tyrrell and Zeebe, 2004), though much further back in the past it could have been substantially larger (Ridgwell, 2005).

D Ocean DOC and Ancient Carbon Cycling: An Example from the Paleocene and Eocene

Difficulties arise in interpreting the PETM because the amounts of carbon estimated using Eq. (1.3b) (and Figure 1.4) are often much larger than the size of the respective modern reservoirs (Figure 1.1). For a terrestrial carbon source of 8200 PgC, the buildup of vast peat deposits during the late Paleocene, followed by a rapid oxidation in a "global conflagration," would be required to explain the observations (Kurtz et al., 2003). For methane hydrates—the much warmer bottom water temperatures of the Paleocene and Eocene—11 °C in the deep ocean compared to ~2 °C today (Norris et al., 2013), would tend to substantially restrict the thickness and distribution of the hydrate stability zone and hence tend to decrease rather than increase the potential Paleocene hydrate reservoir (Buffett and Archer, 2004). The controversial nature of these source scenarios has stimulated thinking about the potential role of DOC in ancient oceans.

Motivated partly by the apparent symmetry in the decline and recovery in $\delta^{13}C$ across some of the smaller apparent hyperthermal events that occurred subsequent to the PETM (Figure 1.3c), Sexton et al. (2011) suggested that changes in the ocean (refractory) DOC reservoir might have played a key role. Specifically, they proposed that fluctuations in the oxygenation of the Eocene ocean may have substantially altered the volume of anoxic waters and hence enabled RDOC to accumulate. Oxidizing this pool, perhaps as a consequence of changes in ocean circulation and hence oxygenation of deep waters, would drive a decline in $\delta^{13}C$ recorded in marine carbonates. A subsequent replenishing of the reservoir would then drive the $\delta^{13}C$ of ocean DIC and hence carbonate carbon back to more positive values at the end of the perturbation. These smaller isotopic fluctuations are characterized by $\delta^{13}C$ excursions of only around −1‰ in magnitude compared to a PETM value of −4‰ and, therefore, require a smaller carbon input. Sexton et al. (2011) estimated that a periodic accumulation and subsequent oxidation of only ~1600 PgC explains the observations—a value almost identical to a result from Eq. (1.4) assuming a source isotopic signature of −25‰ (i.e., organic matter). An advantage of invoking ocean circulation changes and associated variations in the ocean RDOC inventory is that it provides a mechanism that explains why most of the hyperthermal events appear to be orbitally paced (Lunt et al., 2011). (In climate models, the seasonal insolation received at high latitudes can differ sufficiently between different orbital assumptions to drive large-scale changes in deep ocean ventilation in the South Atlantic—see Lunt et al., 2011.)

Problems with the Paleogene hyperthermal DOC hypothesis primarily involve the (unknown) sensitivity of RDOC degradation to ocean oxygenation. Assuming a modern production rate, an approximate doubling of the (modern) inventory to ~1600 PgC would require a slow-down of the degradation rate to just under half its current rate. If in a simple Gedankenexperiment, one assumes that the mean RDOC lifetime scales in inverse proportion to the oxic volume of the ocean, a doubling of the RDOC lifetime requires a reduction of the oxic ocean volume to half of that today, that is,

each molecule of RDOC would spend on average only half its time in oxygenated waters. Even these drastic assumptions would only result in an isotopic excursion of <0.5‰ in magnitude. Yet, there is no evidence for such pervasive ocean anoxia developing during the Eocene. Even during peak warmth of the PETM, only a few places in the deep ocean might have been anoxic (Panchuk, 2007). Suggestions that the deep ocean was not only anoxic but also stratified at depth (Sexton et al., 2011) has the advantage of creating an efficient deep-ocean trap for RDOM, generated deep in the water column or released from the sediments, but falls foul of similar paleoceanographic observational arguments against widespread seafloor anoxia. Also unanswered is the question of why the deep ocean would become stratified and deoxygenated prior to rather than during the transient warming and associated negative $\delta^{13}C$ excursion.

E Ocean DOC and Ancient Carbon Cycling: An Example from the Precambrian

A more enigmatic feature of the geological record of global carbon cycling is the extreme variability in $\delta^{13}C$ occurring during the Precambrian (prior to 541 Ma) (Figure 1.3b) and specifically during the Neoproterozoic (541-1000 Ma) (Figure 1.3d). The magnitude of this variation, with $\delta^{13}C$ reaching values as low as −10 to −15‰, is much harder to explain from a reservoir-change perspective using Eq. (1.3b) than is, for instance, the PETM. For example, prior to the evolution of land plants during the Ordovician (488-444 Ma), there would have been little if any organic carbon stored on land, removing one potential carbon source. This led Dan Rothman and colleagues (2003) to identify isotopically depleted oceanic DOC as a potentially key element in the Precambrian ocean. In their analysis, they posited that a sufficiently large DOC reservoir would allow relatively small changes in DOC inventory to exert a strong control on isotopic composition of oceanic DIC (+atmospheric pCO_2) and hence the measured $\delta^{13}C$ of marine carbonates. The absence of comparably extreme isotopic swings during the Phanerozoic (Figure 1.3a) is ascribed to a "terminal oxidation" event at the end of the Neoproterozoic (Rothman et al., 2003; Swanson-Hysell et al., 2010) and the emergence of a modern mode of marine carbon cycling characterized by a comparably small ocean DOC reservoir.

Building on this, Peltier et al. (2007) invoked oceanic DOC as part of a climate regulation feedback that might have helped regulate climate and prevent runaway "snowball" glaciation during the Neoproterozoic. Noting the temperature dependence of the solubility of oxygen in seawater, they suggested that cooling ocean temperatures, in increasing the degree of oxygenation and hence the rate of oxidation of DOC, would have the effect of increasing atmospheric pCO_2—a negative and stabilizing feedback on climate cooling. Analogous inferences were drawn by Swanson-Hysell et al. (2010) who proposed that the presence or absence of a massive oceanic DOC pool during intervals of the late Neoproterozoic could have dictated the duration of the intervals between glaciation via a strong negative feedback on cooling.

While mathematically elegant, there are a number of difficulties with a DOC-dominant picture of ancient carbon cycling as an explanation for observed variations in carbonate $\delta^{13}C$. Firstly, we lack direct or even indirect evidence for the existence of a massive pool of DOC in the ocean. Because in the Rothman et al. (2003) model this reservoir must be capable of buffering the isotopic composition of the system, such a massive pool was assumed to have been at least 10 times the size of the inorganic (ocean DIC + atmospheric pCO_2) reservoir. For a modern DIC + pCO_2 reservoir of 39,000 PgC, this equates to 390,000 PgC of DOC—more than 500 times larger than the modern reservoir. There are reasons why the magnitude of the DIC + pCO_2 reservoir inventory might have been still higher, hence requiring an even larger DOC inventory. For instance, model-based reconstructions

of long-term changes in ocean geochemistry suggest that at the beginning of the Cambrian (542 Ma), the oceanic DIC reservoir could have been around 4 times its modern value (a total of 152,000 PgC) (Ridgwell, 2005), with atmospheric pCO_2 as much as 20 times its modern value (Berner, 2004). This equates to a combined inorganic carbon reservoir of 164,000 PgC. The minimum DOC reservoir then becomes 1.6×10^6 PgC, equivalent to the concentration of a little over 1200 mg DOC per L of seawater (1.6×10^{21} gC in 1.3×10^{21} L of seawater) or ~0.1 mol kg^{-1}, which would make DOC the third most dominant dissolved species in the ocean after Cl$^-$ and Na$^+$. In comparison, even in the Black Sea—a restricted basin that becomes anoxic below about 150-m water depth and receives large dissolved organic matter loads via river input—observed DOC concentrations do not exceed 300 µmol kg^{-1} (0.0003 mol kg^{-1}) (Ducklow et al., 2007).

Despite the extremely high oceanic DOC concentrations required, there is no obvious mechanism or reasoning that precludes the buildup of such a reservoir given many millions of years. However, other considerations may help provide some (indirect) constraints on the dynamics of the ancient marine carbon cycle. The Shuram anomaly (Figure 1.3d), a feature occurring in multiple sections worldwide about 580 Ma and the largest carbon isotopic excursion in Earth history (Fike et al., 2006; Grotzinger et al., 2011), has been critically assessed in this context; that is, whether DOC oxidation could be involved or whether other considerations place important constraints. The Shuram anomaly is characterized by a $\delta^{13}C$ minimum of −12‰ and a total duration estimated at some tens of millions of years (Fike et al., 2006). Bristow and Kennedy (2008) tackled the feasibility of a DOC-based explanation by considering whether an oxidant reservoir sufficient to convert DOC to DIC could have existed during the Neoproterozoic. Making plausible assumptions regarding late Neoproterozoic atmospheric oxygen and ocean sulfate concentrations, they generated a −12‰

excursion in a geochemical box model. However, the simulated excursion lasted ~0.5 My—some 1-2 orders of magnitude shorter than existing estimates for the event. It does not help to invoke a faster rate of O_2 production to help oxidize the DOC because O_2 is generally thought to be made available by the burial of organic matter, that is, releasing oxygen associated with photosynthesis but preventing the organic matter produced from being reoxidized. Burying more organic matter would also drive oceanic DIC $\delta^{13}C$ to be heavy as per Eq. (1.2), opposing the influence of oxidized DOC driving $\delta^{13}C$ lighter. These arguments can also be extended to glacial-associated negative anomalies occurring slightly earlier in the Neoproterozoic (Figure 1.3d). Hence, in the Precambrian, the much-smaller-than-modern reservoirs of potential oxidants such as O_2 and SO_4^{2-} tend to argue against a dominant role for DOC in accounting for large and particularly long-lived carbon isotopic excursions.

IV IMPLICATIONS FOR FUTURE GLOBAL CHANGE?

As illustrated in Figure 1.3, Earth history contains a variety of perturbations of global climate and carbon cycling that are recorded in variations of geological proxies such as the carbon isotopic composition of sedimentary rocks. That many of these events have so far defied a simple or consensus explanation in terms of the source(s) of carbon behind the isotopic perturbation has stimulated interest in less "conventional" carbon reservoirs and processes. If the reservoir size and cycling of carbon through DOC in the ancient oceans does play, at least at times, a key role in the Earth system and particularly in modulating atmospheric pCO_2, it is tempting to speculate how the modern DOC cycle might respond to ongoing fossil fuel CO_2 emissions, with its attendant changes in climate. Of importance in the anthropogenic global change context is whether DOC might represent a positive (ex-

acerbating) or negative (damping) feedback on atmospheric pCO_2 and climate. After all, there is sufficient RDOC in the ocean such that if it were to be completely and instantaneously oxidized, it would deplete the global ocean dissolved oxygen inventory by about one third, driving substantial expansion of the intensity and extent of the oxygenation minimum zones of the ocean. Another view is that if all 630 PgC RDOC were to be oxidized to DIC and partitioned in the ~1:2 ratio characteristic of the equilibrium distribution of (fossil fuel) CO_2 between atmosphere and ocean (Archer et al., 1997), it would represent a potential 100 ppm increase in atmospheric CO_2. (It should be noted that achieving such an equilibrium distribution takes several thousand years to achieve.)

There may be DOC removal/oxidation mechanisms that we have not yet appreciated and it is possible that the 630 PgC of RDOC residing in the ocean today is much more susceptible to rapid decay than its mean radiocarbon age would suggest. But is there really any evidence in the geological record that this is the case? We suggest that explaining the geological record of carbon isotopic excursions in terms of RDOC inventory changes need not necessarily require new and potentially future-relevant mechanisms to be invoked. For instance, even for nominally rapid and abrupt geological events such as the hyperthermals of the Paleocene and Eocene, the onset of the event may have taken around 10 ky (McInerney and Wing, 2011). Taking a nominal lifetime (e-folding decay time) of RDOC of 16 ky (Hansell, 2013) implies that over an interval of 10 ky, simply ceasing production of RDOC at the start of the event would result in an approximate net CO_2 release of 300 PgC. For a marine DOC $\delta^{13}C$ source of ca. −25‰, this is sufficient, at steady state, to generate a −0.2‰ excursion in DIC $\delta^{13}C$ (Eqs. 1.3b and 1.3b, Figure 1.4). If prior to the PETM a 20-fold larger than modern RDOC inventory had accumulated, and again simply assuming that RDOC production ceases at the onset of the event, one could obtain an excursion of about $-4\permil$. Intermediate initial reservoir sizes could be combined with additional inputs from the terrestrial biosphere, marine hydrates, and/or enhanced volcanic outgassing to achieve a $-4\permil$ excursion overall. The net decay of such a RDOC reservoir would in this example play its role as a positive feedback, responding to an initial environmental change and amplifying it.

Although these are rather idealized and hypothetical scenarios, the point is that for negative $\delta^{13}C$ excursions recorded in the geological record to at least partly reflect DOC dynamics, new or novel removal/oxidation mechanisms are not needed per se. Instead, new mechanisms and understanding about the controls on DOC generation may be necessary. Essentially, this is the hypothesis of Tziperman et al. (2011), who invoked a decrease in the standing stocks of organic matter in the ocean as the trigger of the $\delta^{13}C$ excursion associated with Neoproterozoic glaciation (Figure 1.3d). Their assumed underlying driver was a biological innovation that led to the depletion of a preexisting pool of organic carbon in the ocean. Major evolutionary transitions cannot be invoked with the repeated isotopic fluctuations during the Paleocene and Eocene and hence exact parallels cannot be drawn with the Precambrian. However, biotic disruption is known to have occurred in association with some of the larger events such as the PETM (Hönisch et al., 2012; Norris et al., 2013) and an interval of suppression of RDOC production and hence net organic matter oxidation is feasible.

In the context of understanding past events, we may also be focusing on the "wrong" fraction of DOC. RDOC is an obvious candidate because the more labile fractions are rapidly degraded and in today's ocean comprise only a relatively small inventory—when combined, only 20 PgC compared to the RDOC inventory of 630 PgC (Hansell, 2013). However, it needs to be recognized that this is the situation for a well-oxygenated modern ocean. Analysis of the energetic potential from degrading different organic matter fractions (LaRowe and Van Cappellen, 2011) or lab incubations of sapropel organic matter (Moodley et al., 2005) illustrates that a range of molecular structures such as membrane-type compounds that are degraded by bacteria in a well-oxygenated ocean may be rendered effectively refractory under anoxic conditions. These compounds that today are classed as semi-refractory (SRDOC) might then accumulate in an anoxic (or euxinic) Precambrian ocean to create a massive DOC reservoir (Ridgwell, 2011). In this scenario, RDOC concentrations need not change at all. Variations in the degree of anoxia could then induce large changes in the SRDOC rather than RDOC inventory but with similar implications for the interpretation of recorded $\delta^{13}C$ and global carbon cycling (and climate). An advantage of considering SRDOC in global carbon cycle dynamics is that it is produced at an order of magnitude faster rate than RDOC (Hansell, 2013).

The multi-millennial geological time scale of many past global carbon cycle and climate events and perturbations is well aligned, coincidently or not, with the long lifetime of RDOC. However, the extraordinary persistence in the ocean of the most refractory DOC fraction does not lend itself to providing any substantive future feedback with future global change while the geological record has so far not provided any unambiguous evidence as to whether oxidation and removal has ever been appreciably any faster or proved to be sensitive to climate change. A promising way forward in this context would be to better understand the controls on and environmental sensitivities of production as well as consumption of RDOC today. Research into the fate of SRDOC in environments very different from those characterizing the modern ocean, perhaps the Black Sea, may also provide geologically relevant insights.

Acknowledgements

AR and SA acknowledge support from The Royal Society in the form of University Research Fellowship, and the Natural Environmental Research Council in the form of a Postdoctoral Fellowship.

References

Archer, D., Kheshgi, H., Maier-Riemer, E., 1997. Multiple timescales for neutralization of fossil fuel CO_2. Geophys. Res. Lett. 24, 405–408.

Arndt, S., Jørgensen, B.B., LaRowe, D.E., Middelburg, J.J., Pancost, R.D., Regnier, P., 2013. Quantifying the degradation of organic matter in marine sediments: a review and synthesis. Earth-Sci. Rev. 123, 53–86.

Adkins, J.F., McIntyre, K., Schrag, D.P., 2002. The salinity, temperature, and $\delta^{18}O$ of the glacial deep ocean. Science 298 (5599), 1769–1773. http://dx.doi.org/10.1126/science.1076252.

Berner, R.A., 2004. The Phanerozoic Carbon Cycle: CO_2 and O_2. Oxford University Press, Oxford, p. 160.

Biddanda, B., Benner, R., 1997. Carbon, nitrogen, and carbohydrate fluxes during the production of particulate and dissolved organic matter by marine phytoplankton. Limnol. Oceanogr. 42, 506–518.

Bristow, T., Kennedy, M.J., 2008. Carbon isotope excursions and the oxidant budget of the Ediacaran atmosphere and ocean. Geology 36, 863–866.

Buffett, B., Archer, D., 2004. Global inventory of methane clathrate: sensitivity to changes in the deep ocean. Earth Planet. Sci. Lett. 227, 185–199.

Canfield, D., Thamdrup, B., Hansen, J.W., 1993. The anaerobic degradation of organic matter in Danish coastal sediments. Geochim. Cosmochim. Acta 57, 3867–3883.

Cameron, D.R., Lenton, T.M., Ridgwell, A.J., Shepherd, J.G., Marsh, R., Yool, A., 2005. A factorial analysis of the marine carbon cycle and ocean circulation controls on atmospheric CO_2. Global Biogeochem. Cycles 19, http://dx.doi.org/10.1029/2005GB002489, GB4027.

Crill, P.M., Martens, C.S., 1987. Biogeochemical cycling in an organic rich coastal marine basin: temporal and spatial variations in sulfate reduction rates. Geochim. Cosmochim. Acta 51, 1175–1186.

DeConto, R.M., Galeotti, S., Pagani, M., Tracy, D., Schaefer, K., Zhang, T., et al., 2012. Past extreme warming events linked to massive carbon release from thawing permafrost. Nature 484 (7392), 87–91. http://dx.doi.org/10.1038/nature10929.

Dickens, G.R., O'Neil, J.R., Rea, D.K., Owen, R.M., 1995. Dissociation of oceanic methane hydrate as a cause of the carbon isotope excursion at the end of the Paleocene. Paleoceanography 10, 965–971. http://dx.doi.org/10.1029/95PA02087.

Ducklow, H.W., Quinby, H.L., Carlson, C.A., 1995. Bacterioplankton dynamics in the equatorial Pacific during the 1992 El Niño. Deep-Sea Res. II 42, 621–638.

Ducklow, H.W., Hansell, D.A., Morgan, J.A., 2007. Dissolved organic carbon and nitrogen in the Western Black Sea. Mar. Chem. 105, 140–150.

Dunkley Jones, T., Lunt, D.L., Schmidt, D.N., Ridgwell, A., Sluijs, A., Valdes, P.J., et al., 2013. Climate model and proxy data constraints on ocean warming across the Paleocene-Eocene Thermal Maximum. Earth-Sci. Rev. 125, 123–145.

Dunne, J., Sarmiento, J., Gnanadesikan, A., 2007. A synthesis of global particle export from the surface ocean and cycling through the ocean interior and on the seafloor. Global Biogeochem. Cycles 21, http://dx.doi.org/10.1029/2006GB002907, GB4006.

Etheridge, D.M., Steele, L.P., Langenfelds, R.L., Francey, R.J., 1996. Natural and anthropogenic changes in atmospheric CO_2 over the last 1000 years from air in Arctic ice and firn. J. Geophys. Res. 101, 4115–4128.

Feely, R.A., Sabine, C.L., Lee, K., Berelson, W., Kleypas, J., Fabry, V.J., et al., 2004. Impact of anthropogenic CO_2 on the $CaCO_3$ system in the oceans. Science 305, 362–366.

Fike, D.A., Grotzinger, J.P., Pratt, L.M., Summons, R.E., 2006. Oxidation of the Ediacaran ocean. Nature 444, 744–747. http://dx.doi.org/10.1038/nature05345.

Gibbs, S.J., Bralower, T.J., Bown, P.R., Zachos, J.C., Bybell, L., 2006. Shelf and open-ocean calcareous phytoplankton assemblages across the Paleocene-Eocene Thermal Maximum: implications for global productivity gradients. Geology 34, 233–236.

Gibbs, S.J., Bown, P.R., Murphy, B.H., Sluijs, A., Edgar, K.M., Pälike, H., et al., 2012. Scaled biotic disruption during early Eocene global warming events. Biogeosciences 9, 4679–4688.

Grotzinger, J., Fike, D., Fischer, W., 2011. Enigmatic origin of the largest known carbon isotope excursion in Earth's history. Nat. Geosci. 4, 285–292.

Hain, M.P., Sigman, D.M., Haug, G.H., 2014. 8.18 The Biological Pump in the Past. The Oceans and Marine Geochemistry, vol. 8, 2nd ed., Elsevier Ltd., pp. 485–508. http://dx.doi.org/10.1016/B978-0-08-095975-7.00618-5.

Halverson, G.P., Hoffman, P.F., Schrag, D.P., Maloof, A.C., Hugh, A., Rice, N., 2005. Toward a Neoproterozoic composite carbon-isotope record. GSA Bull. 117, 1181–1207. http://dx.doi.org/10.1130/B25630.1.

Hansell, D., 2013. Recalcitrant dissolved organic carbon fractions. Ann. Rev. Mar. Sci. 5, 421–445.

Hedges, J.I., 2002. Why dissolved organic matter? In: Hansell, D.A., Carlson, C.A. (Eds.), Biogeochemistry of Marine Dissolved Organic Matter. Academic Press, San Diego, California.

Hönisch, B., Ridgwell, A., Schmidt, D.N., Thomas, E., Gibbs, S.J., Sluijs, A., et al., 2012. The geological record of ocean acidification. Science 335, 1058–1063.

Jiao, N., Herndl, G.J., Hansell, D.A., Benner, R., Kattner, G., Wilhelm, S.W., et al., 2010. Microbial production of recalcitrant dissolved organic matter: long-term carbon storage in the global ocean. Nat. Rev. Microbiol. 8, 593–599.

Jiao, N., Azam, F., Sanders, S. (Eds.), 2011. Microbial Carbon Pump in the Ocean. Science/AAAS, Washington, DC, http://dx.doi.org/10.1126/science.opms.sb0001.

Jørgensen, B.B., Sørensen, J., 1985. Seasonal cycles of O_2, NO_3^- and SO_4^{2-} reduction in estuarine sediments: the significance of a nitrate reduction maximum in spring. Mar. Ecol. Prog. Ser. 24, 65–67.

Keeling, R.F., Körtzinger, A., Gruber, N., 2010. Ocean deoxygenation in a warming world. Ann. Rev. Mar. Sci. 2, 199–229. http://dx.doi.org/10.1146/annurev.marine.010908.163855.

Kohfeld, K.E., Ridgwell, A., 2009. Glacial-interglacial variability in atmospheric CO_2. In: Le Quere, C., Saltzman, E. (Eds.), Surface Ocean/Lower Atmosphere Processes. In: Geophysical Monograph Series, vol. 37. American Geophysical Union, Washington, DC.

Kump, L.R., Arthur, M.A., 1999. Interpreting carbon-isotope excursions: carbonates and organic matter. Chem. Geol. 161, 181–198.

Kurtz, A., Kump, L., Arthur, M., Zachos, J., Paytan, A., 2003. Early Cenozoic decoupling of the global carbon and sulfur cycles. Paleoceanography 18, 1090.

Lambert, F., Bigler, M., Steffensen, J.P., Hutterli, M., Fischer, H., 2012. Centenial mineral dust variability in high-resolution ice core data from Dome C. Antarctica Clim. Past 8, 609–623.

LaRowe, D.E., Van Cappellen, P., 2011. Degradation of natural organic matter: a thermodynamic analysis. Geochim. Cosmochim. Acta 75, 2030–2042.

Lorius, C., Jouzel, J., Raynaud, D., 1993. The ice core record: past archive of the climate and signpost to the future. In: Antarctica and Environmental Change. Oxford Science Publications, Oxford, UK, pp. 27–34.

Lunt, D., Ridgwell, A., Sluijs, A., Zachos, J., Hunter, S., Haywood, A., 2011. A model for orbital pacing of methane hydrate destabilization during the Palaeogene. Nat. Geosci. 4, 775–778.

Macdonald, F.A., et al., 2012. Early Neoproterozoic basin formation in the Yukon. In: Paul Hoffman Series, vol. 39, Geoscience, Canada, pp. 77–99.

Maslin, M.A., Thomas, E., 2003. Balancing the deglacial global carbon budget: the hydrate factor. QSR 15–17, 1729–1736.

Maslin, M., Owen, M., Betts, R., Day, S., Dunkley Jones, T., Ridgwell, A., 2010. Gas hydrates: past and future geohazard? Philos. Trans. R. Soc. A 368, 2369–2393. http://dx.doi.org/10.1098/rsta.2010.0065.

Matsumoto, K., Hashioka, T., Yamanaka, Y., 2007. Effect of temperature-dependent organic carbon decay on atmospheric pCO_2. J. Geophys. Res. 112, http://dx.doi.org/10.1029/2006JG000187, G02007.

McInerney, F.A., Wing, S.L., 2011. The Paleocene-Eocene thermal maximum: a perturbation of carbon cycle, climate, and biosphere with implications for the future. Annu. Rev. Earth Planet. Sci. 39, 489–516.

Monnin, E., Indermühle, A., Dällenbach, A., Flückiger, J., Stauffer, B., Stocker, T.F., et al., 2001. Atmospheric CO_2 concentrations over the last glacial termination. Science 291, 112–114.

Moodley, L., Middelburg, J.J., Herman, P.M.J., de Soetaert, K., Lange, G.J., 2005. Oxygenation and organic matter preservation in marine sediments: direct experimental evidence from ancient organic carbon-rich deposits. Geology 33, 889–892.

Norris, R.D., Kirtland Turner, S., Hull, P.M., Ridgwell, A., 2013. Marine ecosystem responses to Cenozoic global change. Science 341, 492–498.

Panchuk, K., 2007. Investigating the Paleocene/Eocene Carbon Cycle Perturbation: An Earth System Model Approach (Ph.D. dissertation). The Pennsylvania State University.

Panchuk, K., Ridgwell, A., Kump, L.R., 2008. Sedimentary response to Paleocene Eocene Thermal Maximum carbon release: a model-data comparison. Geology 36, 315–318.

Peltier, R.W., Liu, Y., Crowley, J.W., 2007. Snowball Earth prevention by dissolved organic carbon remineralization. Nature 450, 813–818.

Ridgwell, A., 2005. A Mid Mesozoic revolution in the regulation of ocean chemistry. Mar. Geol. 217, 339–357.

Ridgwell, A., 2011. Evolution of the ocean's "biological pump". Proc. Natl. Acad. Sci. U. S. A. 108, 16485–16486.

PALAEOSENS Project Members, Rohling, E.J., Sluijs, A., Dijkstra, H.A., Köhler, P., van de Wal, R.S.W., von der Heydt, A.S., et al., 2012. Making sense of palaeoclimate sensitivity. Nature 491, 683–691.

Rothman, D.H., Hayes, J.M., Summons, R.E., 2003. Dynamics of the Neoproterozoic carbon cycle. Proc. Natl. Acad. Sci. U. S. A. 100, 8124–8129.

Royer, D.L., 2001. Stomatal density and stomatal index as indicators of paleoatmospheric CO_2 concentration. Rev. Palaeobot. Palynol. 114.

Royer, D.L., 2006. CO_2-forced climate thresholds during the Phanerozoic. Geochim. Cosmochim. Acta 70, 5665–5675.

Sarmiento, J.L., Gruber, N., 2006. Ocean Biogeochemical Dynamics. Princeton University Press, Princeton, NJ, 526 pp.

Sexton, P.F., Norris, R.D., Wilson, P.A., Pälike, H., Westerhold, T., Röhl, U., et al., 2011. Eocene global warming events driven by ventilation of oceanic dissolved organic carbon. Nature 471, 349–352. http://dx.doi.org/10.1038/nature09826.

Shields, G., Veizer, J., 2002. Precambrian marine carbonate isotope database: version 1.1. Geochem. Geophys. Geosyst. 3, 108–113.

Sluijs, A., Dickens, G.R., 2012. Assessing offsets between the $\delta^{13}C$ of sedimentary components and the global exogenic carbon pool across early Paleogene carbon cycle perturbations. Global Biogeochem. Cycles 26, 1944–9224. http://dx.doi.org/10.1029/2011GB004224.

Solomon, S., Qin, D., Manning, M., Chen, Z., Marquis, M., Averyt, K.B., et al. (Eds.), 2007. Climate Change 2007: The physical science basis, contribution of working group I to the fourth assessment report of the intergovernmental panel on climate change. Cambridge University Press. ISBN 978-0-521-88009-1.

Stumm, W., Morgan, J.J., 1981. Aquatic Chemistry, second ed. Wiley-Interscience, New York, p. 780.

Swanson-Hysell, N.L., Rose, C.V., Calmet, C.C., Halverson, G.P., Hurtgen, M.T., Maloof, A.C., 2010. Cryogenian glaciation and the onset of carbon-isotope decoupling. Science 328, 608–611. http://dx.doi.org/10.1126/science.1184508.

Thomas, E., 2007. Cenozoic mass extinctions in the deep sea: what perturbs the largest habitat on Earth? Geological Society of America Special Papers 424, 1–23.

Tyrrell, T., Zeebe, R.E., 2004. History of carbonate ion concentration over the last 100 million years. Geochim. Cosmochim. Acta 68, 3521–3530.

Tziperman, E., Halevy, I., Johnston, D.T., Knoll, A.H., Schrag, D.P., 2011. Biologically induced initiation of Neoproterozoic snowball-Earth events. Proc. Natl. Acad. Sci. U. S. A. 108, 15091–15096.

Veizer, J., Ala, D., Azmy, K., Bruckschen, P., Buhl, D., Bruhn, F., et al., 1999. $^{87}Sr/^{86}Sr$, $\delta^{13}C$ and $\delta^{18}O$ evolution of Phanerozoic seawater. Chem. Geol. 161, 59–88. http://dx.doi.org/10.1016/S0009-2541(99)00081-9.

Watson, R.T., Noble, I.R., Bolin, B., Ravindramath, N.H., Verardo, D.J., Dokken, D.J. (Eds.), 2000. Land Use, Land-Use Change, and Forestry (a special report of the IPCC). Cambridge University Press, Cambridge, p. 377.

Wefer, G., Berger, W.H., Bijma, J., Fischer, G., 1999. Clues to ocean history: a brief overview of proxies. In: Fischer, G., Wefer, G. (Eds.), Use of Proxies in Paleoceanography: Examples from the South Atlantic. Springer, Berlin, pp. 1–68.

Zachos, J.C., Röhl, U., Schellenberg, S.A., Sluijs, A., Hodell, D.A., Kelly, D.C., et al., 2005. Rapid acidification of the ocean during the Paleocene-Eocene thermal maximum. Science 308 (5728), 1611–1615. http://dx.doi.org/10.1126/science.1109004.

Zachos, J.C., McCarren, H., Murphy, B., Rohl, U., Westerhold, T., 2010. Tempo and scale of late Paleocene and early Eocene carbon isotope cycles: implications for the origin of hyperthermals. Earth Planet. Sci. Lett. 299, 242–249.

Zeebe, R.E., Wolf-Gladrow, D., 2001. In: CO_2 in Seawater: Equilibrium, Kinetics, Isotopes. In: Elsevier Oceanography Series, vol. 65. Elsevier, Amsterdam, p. 346.

Chemical Characterization and Cycling of Dissolved Organic Matter

Daniel J. Repeta

Department of Marine Chemistry and Geochemistry, Woods Hole Oceanographic Institution,
Woods Hole, Massachusetts, USA

CONTENTS

Biogeochemistry of Marine Dissolved Organic Matter,
http://dx.doi.org/10.1016/B978-0-12-405940-5.00002-9

I INTRODUCTION

Each year, between 15 and 25 Pg of dissolved organic matter (DOM) are added to seawater by the activity of marine microbes, by atmospheric, fluvial, and groundwater transport of organic matter from the continents, and by the release of organic matter from the benthic boundary layer (Bauer and Bianchi, 2011; Burdige, 2007; Hansell, 2013; Jurado et al., 2008). Most DOM is immediately respired by marine microheterotrophs, oxidized by photochemical processes, or permanently buried in sediments. However, a significant fraction is stored in the water column where it interacts with a variety of biogeochemical cycles over timescales ranging from hours to millennia. Marine DOM stores nitrogen and phosphorus that would otherwise be immediately available to microbes in the upper water column, affects the bioavailability of essential trace metals, attenuates the penetration of UV and visible light in the euphotic zone, and sequesters an amount of carbon approximately equal to atmospheric carbon dioxide. All production, removal, and transformation processes leave an imprint on the composition of DOM and it is the potential that DOM composition can help to understand the sources and sinks of DOM and how it is cycled in the water column, as well as the interest in describing one of the largest reservoirs of organic matter on Earth, that drives much of the current research on DOM composition.

Composition refers to the broad suite of molecular characteristics that define DOM and includes levels of detail that range from simple elemental ratios of carbon, nitrogen, and phosphorus to stable and radio-isotopic content, to the stereochemistry of amino acids. How DOM is characterized, the extent to which the composition is known, therefore changes with the characteristics of interest. Global surveys of dissolved organic carbon (DOC), nitrogen (DON), and phosphorus (DOP) completed over the last decade now include $>10^4$ analyses and our knowledge of DOM elemental composition is therefore relatively comprehensive (Hansell et al., 2009, 2012). In contrast, our understanding of DOM molecular composition is often benchmarked as the inventory of dissolved compounds (simple sugars, amino acids, lipids, vitamins, pigments, etc.) that can be isolated from seawater. Viewed from this perspective, <10% of DOM has been characterized and it is fair to say that our knowledge of DOM has improved only marginally over the last decade and a half. However, it is often other features of composition, such as the distribution of major carbon, nitrogen, and phosphorus functional groups, the identity of major classes of compounds that contribute to labile and refractory fractions of DOM, or changes in the degree of oxidation, that are most informative to understanding the sources, sinks, and cycling of DOM. Viewed from this perspective, our understanding of DOM composition has advanced significantly to the point where between 60% and 70% of DOM has now been "characterized."

This chapter summarizes recent advances in our knowledge of DOM composition, with a particular emphasis on studies completed over the last decade that utilize high-field nuclear magnetic resonance, high-resolution mass spectrometry (MS), proteomics, and related immunochemical assays. A number of excellent reviews and workshop reports summarize the field of DOM composition up to the early 2000s (Aluwihare and Meador, 2008; Benner, 2002; McNichol and Aluwihare, 2007; Mopper et al., 2007), and other chapters in this book describe progress in DOM elemental composition (Chapter 3), isotopic composition (Chapter 6), dissolved organic nitrogen (Chapter 4) and phosphorus (Chapter 5), chromophoric DOM (Chapters 8 and 10), and DOM cycling in river and coastal systems (Chapter 11). These topics are integral to DOM composition and cycling and specific aspects are included in the discussion below, but the reader is referred to these chapters for comprehensive discussions on these topics.

In practical terms, all studies of DOM composition begin with sampling, which aside from providing the material used in chemical and spectral analyses also selectively defines the chemical fraction and spatial/temporal features of the DOM that is characterized. Because sampling is so integral to the interpretation of DOM composition, this chapter begins with a summary of sampling methods. From there, the chapter is organized into discussions of carbohydrates, proteins, and aliphatic organic matter, the major components of DOM that have been identified to date. Finally, the results from these studies are discussed within the broader perspective of how composition impacts the cycling of DOM in the water column.

II ISOLATION OF DOM FROM SEAWATER

DOM is operationally defined as the fraction of organic matter not retained by filtration. However, the specific choice of filter is determined by the preferences of the analyst and may be influenced by the subsequent suite of analyses that are to be performed. There is no universally agreed upon filter type or pore size that distinguishes dissolved and particulate phases. Many studies of marine particles have utilized Whatman GF/F glass fiber filters with a nominal pore size of 0.7μm due to the ease with which these filters can be cleaned and to their excellent flow characteristics, facilitating their adoption for DOM studies where large volume samples were often required. However, some marine bacteria and viruses are <0.7μm and pass through GF/F filters to be included in the "dissolved" fraction. Membranes with smaller pore sizes (0.1-0.2μm) effectively exclude bacteria, while much smaller pore sizes (10-15nm) are needed to exclude viruses. "Dissolved" is therefore a nonspecific term that can include bacteria-sized particles, colloidal organic matter, and truly dissolved species.

Although the concentrations of some organic compounds (amino acids, simple sugars, low

molecular weight (LMW) acids, ketones, and aldehydes) can be measured directly in seawater, >90% of DOM needs to be concentrated and isolated before spectral and further chemical characterization can proceed. Separation of DOM (~ 0.5-1mg/L) from salt (~35g/L) poses the major challenge for DOM isolation and no single approach recovers all DOM from seawater. Therefore, a number of different strategies have been developed that capitalize on either the larger molecular size or the lower polarity of DOM relative to sea salt. Advances in spectroscopic and spectrometric analyses have in many cases reduced sample requirements from mg to μg or even ng amounts of material and have mitigated some of the need to obtain DOM as a salt-free preparation. For example, samples for mass spectral analyses can be obtained from only a few liters of seawater, and proton NMR (^{1}H NMR) spectra have now been reported for unprocessed seawater (Dittmar et al., 2008a; Lam and Simpson, 2008). Presently, these approaches are limited to qualitative analyses of DOM; water suppression leads to biases in the determination of carbohydrate functional groups in ^{1}H NMR, while matrix effects and differences in ionization efficiency complicate the quantitative interpretation of mass spectra. As analytical techniques continue to advance, these limitations will be addressed, but for the immediate future, our ability to isolate only a portion of DOM will continue to limit our understanding of DOM composition.

A Isolation of Hydrophobic DOM by Solid-Phase Extraction

Simple passage of filtered seawater across a solid hydrophobic surface or mineral phase leads to the adsorption and concentration of DOM (Figure 2.1). Since the basis of the method is physisorptive attraction between DOM and the solid phase, the approach selectively concentrates the hydrophobic or surface-active fraction of DOM with distinct characteristics that are not representative of total DOM. For example, colored

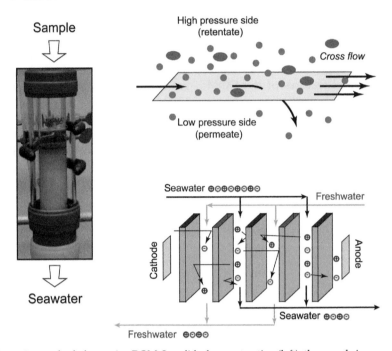

FIGURE 2.1   Extraction methods for marine DOM. In solid-phase extraction (left), the sample is passed through a column packed with an organic sorbent (polystyrene, octadecylsilyl coated silica gel, etc.). Physisorptive attraction between hydrophobic DOM and the sorbent leads to adsorption of DOM on the column, which in the photograph appears as a brown discoloration (due to the adsorption of colored dissolved organic matter). After the sample has been extracted, hydrophobic DOM is recovered by washing the column with methanol. In ultrafiltration (top, right), seawater is pressurized and passed through a filter with nanometer-sized pores (gray sheet in figure). Organic molecules with hydrodynamic diameters greater than the filter pore size (red ellipses) are retained, while salt, low molecular weight DOM, and water (blue circles) permeate through the filter. Once the sample has been concentrated to <1-2 L, residual salts are removed by serial dilution with ultra-pure water followed by filtration. In reverse osmosis/electrically assisted dialysis (RO/ED; bottom right), the sample is processed in serial ED and RO steps. In the ED step, the sample is passed through stacks of cation and anion exchange membranes that are under the influence of an applied electric potential and washed by alternating flows of sample (seawater, black arrows) and ultra-pure water (freshwater; gray arrows). Anions (chloride, sulfate, etc.) pass through the positively charged membrane (blue panels) toward the anode, while cations (sodium, potassium) pass through the negatively charged membranes (green panels) toward the cathode. This transfers salts from the sample into the freshwater feed, lowering the salinity of the sample. The sample is then concentrated by reverse osmosis and processed again by ED to further reduce the salt content of the sample. *Figure adapted from http://www.osmo-membrane.de.*

DOM is efficiently extracted by contact with polystyrene or octadecyl-silica (C-18) resins, but the absorption and fluorescent properties of the extracted samples are significantly different from the original seawater (Green and Blough, 1994). Early work on DOM solid-phase extraction (SPE) explored charcoal, freshly precipitated iron oxy-hydroxides, and synthetic hydrophobic resins as substrates; but over the past two decades commercially available octadecyl-bonded silica (C-18), cross-linked polystyrene (XAD-2, -4, and -16), and their derivatives (PPL), Isolute ENV, and polyacrylate (XAD-8)) have become the sorbents of choice for these studies (Mopper et al., 2007).

To maximize sample recovery, filtered samples are often acidified with hydrochloric acid to pH 2-2.5. At lower pH, most carboxylic acids and phenols are protonated, reducing their aqueous solubility and enhancing their adsorption, thereby leading to a higher recovery of carbon. The extraction efficiency will depend on the speed at which seawater is passed through the column and the amount of sorbent used per volume of seawater. A recent study used 1 g sorbent per 10-20 mg of total DOM at 20 column volumes per minute, but the effects of flow rate and sorbent/sample ratio have not been investigated in detail (Dittmar et al., 2008b). After collection, the sample is rinsed of salt using pH 2, ultra-pure water, and recovered by elution with methanol and/or methanolic or aqueous sodium or ammonium hydroxide. In performing SPE, the sample is subjected to large changes in pH, which have unknown effects on the sample composition.

The diversity of SPE products available to the analyst has led to a few comparisons of carbon recovery and the elemental, spectroscopic, and isotopic characteristics of SPE-DOM (Dittmar et al., 2008a; Macrellis et al., 2001). DOC and analyte recovery was highly dependent on the choice of solid phase. In one study, PPL (functionalized polystyrene/divinyl benzene) was found to be more efficient than C-18 (octadecylfunctionalized silica with 500Å pore size) in extracting DOC from surface seawater near the North Brazilian Shelf, adsorbing on average 62% of total DOC, with a lower C/N (20) and less depleted $\delta^{13}C$ value (−23.4‰) than recovered by C-18 (37%, 37 and −24.8‰, respectively). Given the large difference in recovery, C/N ratio, and isotope value, it would seem that PPL and C-18 extract different but overlapping fractions of DOM. However, the ^1H NMR spectra of the two samples were nearly identical. In this particular study, differences in extraction efficiency may have been accentuated by the presence of terrestrial organic matter sourced from the nearby continental margin. Extraction efficiencies for open-ocean seawater have not been widely reported, but are generally lower (10-20%).

B Isolation of High Molecular Weight DOM by Ultrafiltration

"Molecular" or "ultra-" filtration exploits the larger hydrodynamic diameter of high molecular weight DOM (HMWDOM) compared to most dissolved inorganic species to achieve a separation of organic matter from seawater. In this technique, hydrostatic pressure is applied across a semipermeable membrane perforated with very small pores, typically 1-15 nm in diameter. Salts, water, and organic matter of a hydrodynamic diameter smaller than the filter pore size pass through the membrane (as the permeate) while HMW organic matter is retained and concentrated in the original sample (as the retentate; Figure 2.1). Accumulation of organic matter on the filter surface rapidly leads to membrane polarization, which is reduced by applying a second (tangential or cross) flow perpendicular to the direction of filtration. The sample is concentrated to a fraction of its original volume, diluted with ultra-high purity, salt-free water, and filtered again. This "diafiltration" process is repeated until the salt content is reduced to a fraction of the organic matter concentration. Once this has been achieved, the remaining water is removed by lyophilization or rotary evaporation and the sample recovered.

Ultrafiltration concentrates hydrophilic HMWDOM. The amount and chemical characteristics of HMWDOM recovered is highly dependent on the membrane material (cellulose, polysulfone, etc.) and pore size, the strength of the cross flow, the ratio of original to final sample volume (concentration factor), and a number of other operational parameters. Plots of water volume filtered against DOC concentration show that ultrafiltration with a 1-nm pore-sized membrane typically retains ~30% of DOC, but losses associated with high concentration factors and diafiltration lead to far lower physical recoveries, typically 10-15% for final isolates that are 35-38% by weight carbon (Walker et al., 2011). Walker et al. (2011) applied a permeation model to ultrafiltration data where seawater was concentrated by factors of >1000-fold. HMW components of DOM behaved ideally, for example, even at high concentration factors they are efficiently retained by the ultrafiltration membrane. Hydrophobic, low molecular weight DOM (LMWDOM) that is nominally smaller than the membrane pore size, is also concentrated by ultrafiltration, probably due to membrane polarization and adsorption onto the membranes themselves. The amount of LMWDOM retained is highly variable, being sensitive to the membrane material, concentration factor, and perhaps other operational factors. A number of studies report differences in the recovery and chemical properties of HMWDOM. These differences could be due to the variable retention of LMW components. The selectivity inherent in ultrafiltration sampling, and the sensitivity of DOM recoveries to operation parameters that differ between systems and operators, suggests caution must be exercised when comparing data between studies.

C Isolation of DOM by Reverse Osmosis/Electrically Assisted Dialysis

A recent advance in DOM sampling is reverse osmosis/electrically assisted dialysis (RO/ED; Gurtler et al., 2008; Koprivnjak et al., 2009; Vetter et al., 2007) In this technique, the sample is desalted by an alternating series of positive and negative ion-exchange membranes under the influence of an electric potential (Figure 2.1). Anions pass through positively charged ion-exchange membranes toward the anode, while cations pass through negatively charged ion-exchange membranes toward the cathode. The resulting lower salinity sample is reduced in volume by reverse osmosis, then desalted a second time in a final ED phase. Recoveries of DOM range from 50% to >100% with an average carbon recovery of 76% ($n = 21$), similar to the 70% recovered when SPE and ultrafiltration are used in series (Simjouw et al., 2005). Final

preparations are ~25% salt by weight and have chemical characteristics of both hydrophobic organic matter isolated by SPE and hydrophilic DOM isolated by ultrafiltration.

III CHEMICAL CHARACTERIZATION OF DOM

DOM includes simple biochemicals (amino acids, simple sugars, vitamins, fatty acids), complex biopolymers (proteins, polysaccharides, lignins), and very complex degradation products of unknown origin that so far have defied full characterization (humic substances (HS), black carbon). The complexity and diversity of organic constituents in DOM have pushed the limits of the chemical and spectral techniques brought to bear on their characterization. High-resolution MS, high-field NMR, and new approaches to chemical degradation have significantly broadened and deepened our understanding of DOM composition, but have also highlighted the limits of even advanced spectral techniques.

Ultrafiltration, SPE, ED/RO, and direct chemical analysis of unfractionated seawater all access overlapping, but compositionally different fractions of DOM. NMR and chemical analysis of hydrolysis products have been most successful in characterizing polysaccharides and proteins in HMWDOM. High-resolution MS, 2D NMR, and chemical degradation approaches have found their most successful application in the characterization of hydrophobic DOM isolated by SPE. The following discussion is therefore organized around the characterization of carbohydrates, proteins, and hydrophobic HS that are the major components of DOM, highlighting the new approaches and methods that have come to the fore in the last decade.

A Polysaccharides in DOM

^{13}C NMR analysis of surface water HMWDOM gives a characteristic spectrum (Figure 2.2a) with

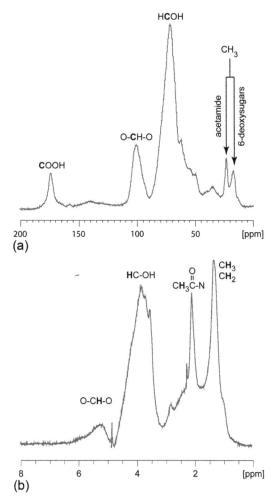

HCOH

CH₃

acetamide

6-deoxysugars

O-CH-O

COOH

200   150   100   50   [ppm]

(a)

HC-OH

CH₃C-N
O
‖

CH₃
CH₂

O-CH-O

8   6   4   2   [ppm]

(b)

FIGURE 2.2 Nuclear magnetic resonance spectra of HMWDOM collected by ultrafiltration of surface seawater collected from the North Pacific Ocean using a polysulfone membrane with 1-nm pore size (nominal 1 kDa molecular weight cutoff). (a) ^{13}C NMR spectra have major peaks from carboxyl (COOH, CON; ~175 ppm), aromatic and olefinic C (broad peak centered at ~140 ppm), anomeric (OCO; 100 ppm), O-alkyl C (HC–OH; 75 ppm), methine and substituted methylene C (~35-40 ppm), and two alkyl C peaks from acetamide (26 ppm), and 6-deoxysugars (20 ppm). (b)^{1}H NMR spectra show major peaks from anomeric protons (5.2 ppm), O-alkyl protons (3.5-4.5 ppm), acetamide methyl (2.0 ppm), and 6-deoxyl sugar methyl and methylene protons (1.3 ppm). These peaks sit atop a broad baseline between 0.9 and 4 ppm from substituted methine, methylene, and methyl protons from hydrophobic DOM.

major resonances assigned to carboxyl/amide carbon (~175 ppm; 5%), unsaturated C=C/aromatic carbon (broad peak centered at ~140 ppm; 5%), anomeric O–C–O (~100 ppm; 14%), O-alkyl C (~75 ppm; 56%), a broad peak centered at ~35-40 ppm (13%) from methylene and substituted methylene C, and two alkyl carbon peaks at 20 ppm (5%) and 26 ppm (5%). ^{1}H NMR spectra are similarly characteristic (Figure 2.2b), with major resonances from anomeric (5.2 ppm; HC(–O)$_2$), and O-alkyl (3.5-4.5 ppm; HO–CH) protons, as well as methyl carbons from acetamide (2.0 ppm; HN–C(O)CH$_3$) and deoxysugars (1.3 ppm; C(H$_2$)–CH$_3$). In addition, ^{1}H NMR spectra have a distinct signal at 2.7 ppm which is tentatively assigned to N-acetyl-N-methyl amino sugar (CH$_3$–NH–C(=O)CH$_3$) on the basis of chemical shift comparisons with literature values, but further corroborative evidence is lacking. The proton and carbon NMR spectra are complimentary, each showing a majority of functional groups can be assigned to carbohydrates. The positions of carbohydrate peaks change little between samples, but changes in their relative intensities are observed with depth (Benner et al., 1992; Hertkorn et al., 2006), sampling location (Aluwihare et al., 1997), and across salinity gradients (Abdulla et al., 2010a,b), supporting the idea of at least two major components to HMWDOM; a polysaccharide fraction referred to as acylated polysaccharide (APS; Aluwihare et al., 1997) or heteropolysacharide (HPS; Hertkorn et al., 2006; Abdulla et al., 2010a,b), which includes anomeric, O-alkyl, amide, and methyl carbon (from acetate and deoxy sugars), and a carboxylic acid-alkyl carbon-rich fraction referred to as carboxyl-rich aliphatic matter (CRAM) or carboxyl-rich compounds (CRC) reminiscent of aquatic HS.

The presence of distinct polysaccharide and CRAM/CRC fractions within HMWDOM can be demonstrated by passing HMWDOM samples through hydrophobic C-18 resin at low pH or through anionic exchange resin at neutral pH to selectively remove most of the CRAM/CRC fraction (Figure 2.3a; Panagiotopoulos et al.,

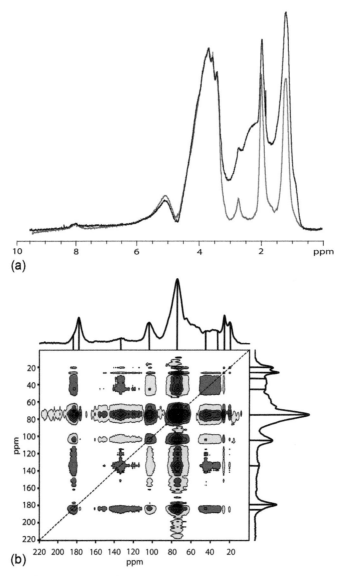

FIGURE 2.3   (a) Overlay of ^{1}H NMR spectra from North Pacific Ocean HMWDOM before (black trace) and after (red trace) passage through an anion exchange resin. Ultrafiltration concentrates from seawater high molecular weight, acylated polysaccharides (APS) as well as low molecular weight, hydrophobic humic substances. These two chemically distinct fractions of DOM can be separated by solid-phase extraction or ion-exchange chromatography, providing relatively pure fractions of acylated polysaccharide (APS) and carboxyl-rich aliphatic matter (CRAM). (b) Contour map of synchronous changes in peak intensity for ^{13}C NMR spectra of HMWDOM collected at estuarine and coastal marine sites. Positive correlations between functional groups appear in red as off-diagonal peaks. Negative correlations appear in blue, with the intensity of the colors scaled to the intensity of the correlation. *Figure courtesy of Drs. Hussain Abdulla and Patrick Hatcher. Samples were collected using a polysulfone membrane with a 1-nm pore size.*

2007) and through correlation spectroscopy (Figure 2.3b; Abdulla et al., 2010a,b; Abdulla et al., 2013) In correlation spectroscopy, the variable intensities of major resonances are correlated with a second variable (salinity, depth, location) to derive synchronous and nonsynchronous changes in spectral characteristics. Signal fluctuations that are synchronous indicate a common chemical constituent, while nonsynchronous signals indicate chemically distinct components. Correlation spectroscopy therefore identifies functional group relationships within the different fractions of DOM, but requires that different fractions of HWMDOM vary independently across changes in depth, location, salinity, etc. For example, Abdulla et al. (2013) used two-dimensional (2D) correlation analysis on a suite of samples from the Elizabeth River/Chesapeake Bay estuary to identify synchronous changes in ^{13}C NMR signal intensity. Strong correlations were observed between intense signals from O-alkyl carbon at 74 ppm, anomeric carbon at 103 ppm, amide carbon at 178 ppm, and two alkyl carbons at 20 and 26 ppm. These functional groups arise from the HMWDOM polysaccharide fraction. The carboxyl-rich component (CRAM/CRC) showed correlation between signals in the methylene and substituted carbon region (29-50 ppm), unsaturated C=C/aromatic region (115-160 ppm), carboxyl carbon at 183 ppm, and carbonyl (aldehyde and ketone) carbon at 190-200 ppm. The negative correlation between carbohydrate resonances at 110 and 74 ppm with aliphatic signals between 30 and 46 and unsaturated/aromatic signals centered at 130 ppm indicates that carbon functional groups in these regions of the spectrum are primarily associated with CRAM/CRC. HMWDOM polysaccharides have little to no aliphatic component other than carbon associated with acetate and deoxysugar carbon (Figure 2.2a).

Full characterization of any polysaccharide is challenging, and typically includes acid hydrolysis to determine simple sugar composition, methylation and reductive cleavage to determine branching, and spectroscopic or spectrometric characterization of partially degraded oligosaccharides to determine sequence. For HMWDOM, these approaches have proved to be only partially successful. Hydrolysis of HMWDOM using a wide variety of acids and hydrolysis conditions yields a characteristic suite of seven major neutral sugars including arabinose and xylose (pentoses), glucose, galactose, and mannose (hexoses), fucose and rhamnose, (6-deoxyhexoses) (Aluwihare et al., 1997; McCarthy et al., 1996; Panagiotopoulos and Sempéré, 2005; Sakugawa and Handa, 1985), and similar amounts of the amino sugars glucosamine and galactosamine (Aluwihare et al., 2002; Benner and Kaiser, 2003; Kaiser and Benner, 2000). Remarkably, the relative proportions of these sugars is largely conserved across samples collected in different ocean basins, at different times, and at different depths (Aluwihare et al., 1997; McCarthy et al., 1996; Sakugawa and Handa, 1985). However, small variations in the ratio of neutral sugars have been attributed to spatial/temporal changes in HMWDOM composition (Boon et al., 1998; Goldberg et al., 2009, 2010). Quantitatively, the seven neutral and two amino sugars represent only a minor fraction, between 10% and 20%, of the total HMWDOM carbohydrate. The composition of most HMWDOM polysaccharide, even to the level of simple sugar compliment, is therefore unknown.

Sugars in HMWDOM have also been characterized by direct temperature-resolved MS (DT-MS; Boon et al., 1998; Minor et al., 2001). In DT-MS, the sample is rapidly heated under vacuum. Thermal decomposition of the sample leads to the release of simple, volatile degradation products that are characteristic of the type of organic matter undergoing thermolysis. DT-MS of model compounds (proteins, lipids, carbohydrates, and nucleic acids) allow for the assignment of diagnostic masses for different compound classes. DT-MS spectra of HMWDOM display ions for hexoses, pentoses, and deoxysugars, that were attributed to N-acetyl-amino-, and monomethyl- and dimethy-, and methyl-deoxysugars. Ions from N-acetyl-aminosugars were often the most intense features of the mass spectra, consistent with the assignment of the large signal at 2 ppm in the ^{1}H NMR spectrum as $-C(=O)CH_3$ (Figure 2.2b). No ions for acetic acid were observed, and the study concluded that nitrogen, not oxygen, was the site of acetylation. Masses indicative of deoxy- and methyl sugars were also prominent, and these sugars make substantial contributions to the uncharacterized portion of HMWDOM. HMWDOM polysaccharides are introduced into the mass spectrometer by thermal desorption, which occurs in two distinct stages. The bimodal thermal evolution profile is indicative of at least two fractions of carbohydrate (Boon et al., 1998). Both fractions include the same suite of sugars, but the relative intensity of ions derived from furfural (from uronic acids or deoxysugars) was enhanced and the specificity of the spectra decreased in the high temperature fraction. No evidence was found in either fraction for extended homopolysaccharides: chitans, glycans, xylans, or arabinogalactans that are typical of many storage and some structural polysaccharides. DT-MS data suggest HMWDOM polysaccharides are very heterogenous, include a large fraction of deoxy- and N-acetyl aminosugars, and are highly branched and cross-linked.

Branching into two and three-dimensional polysaccharides can be assessed by selective permethylation of the polysaccharide to protect non-linked sites, followed by hydrolysis. Hydrolysis leaves methylated sites intact, but results in a suite of simple methylated sugars that can be identified and quantified by chromatography. Linkage analysis shows that of the sugars that can be recovered after acid hydrolysis (e.g., 10-20% of the total polysaccharide), 40% have only terminal linkages, 40% are linked without branching, and 20% are branched with one branch point. All of the seven major neutral sugars characteristic of HMWDOM (glucose, galactose, mannose, fucose, rhamnose, xylose, and arabinose) have terminal linkages, but only a few sugars are branched. Major linkage patterns for non-branched sugars include 1,3 and 1,4 linkages, with only small amounts of 1,2 branching (Aluwihare et al., 1997; Sakugawa and Handa, 1985). The degree of branching in HMWDOM polysaccharides is unusually high. Highly branched polysaccharides are typical of structural biopolymers found in microbial cell walls. The high degree of branching may impart some resistance to microbial degradation that persists and allows these polymers to accumulate as DOM.

The distinct peaks at 2.0 ppm in ^{1}H NMR and 26 ppm in ^{13}C NMR spectra (Figure 2.2a and b) are assigned to the methyl group of N-acetylated sugars (Aluwihare et al., 2005; Quan and Repeta, 2007). Correlation analysis (Figure 2.3b; Abdulla et al., 2010a,b) shows synchronous changes in resonances in amide carbon at 180 ppm, anomeric carbon, O-alkyl carbon, and methyl carbon at 26 and 20 ppm, in support of this assignment. Based on integration of the ^{13}C- and ^{1}H NMR spectra (Figure 2.2), acetate contributes 5-8% of the polysaccharide carbon, larger than any other molecular component of HMWDOM polysaccharide identified to date. On this basis, approximately one out of every four sugars in HMWDOM polysaccharide would have N-linked acetate. Aluwihare et al., 2005 inferred that N-acetyl amino sugars are incorporated into a family of related APS that includes major neutral sugars and amino sugars, but correlation analysis suggests a somewhat higher ratio (6:1) of neutral to amino sugars, and that there are at least two distinct fractions of neutral and amino polysaccharides. Correlation analysis distinguishes acylated amino sugar containing polysaccharide (APS) from (non-acylated) heteropolysaccharides on the basis of changes in the asymmetric amide stretching the IR spectrum at

$1660\,cm^{-1}$. In a suite of samples through a coastal estuary, Abdulla et al., 2010a found that as HMWDOM samples become increasing marine, with a relative increase from 50% to 70% in heteropolysaccharide content but little change in the acylated amino sugar containing polysaccharide component, suggesting that the major heteropolysaccharide and aminosugar components of HMWDOM behave independently. A more detailed characterization of HMWDOM and a better understanding of spatial/temporal changes in HMWDOM cycling will be needed to reconcile these two interpretations of polysaccharide composition.

The contribution of N-acetyl-aminosugars to HMWDOM polysaccharide has been quantified indirectly by monitoring the effects of mild acid hydrolysis on ^1H- and ^{15}N-NMR spectra (Aluwihare et al., 2005). ^{15}N NMR spectra of HMWDOM are characterized by a large peak from amide-N at 124 ppm and a smaller peak at 35 ppm from methyl- and amino-N (Aluwihare et al., 2005; McCarthy et al., 1997). The two common classes of biochemicals most likely to contribute to amide-N are proteins, polymers of amino acids linked through an amide functional group, and N-acetyl amino sugars such as chitin or peptidoglycan. Hydrolysis of either will convert amide-N to amino-N, however for proteins this results in depolymerization while in N-acetyl amino sugars it does not. If acetate occurs primarily as N-acetyl amino sugars, then the amount of amino acids plus acetic acid released by hydrolysis should equal the conversion of amide-N to amino-N measured by ^{15}N NMR. If conversion exceeds the sum of amino acids and acetic acid, then other biochemicals must contribute to amide-N. Alternatively, if the conversion is significantly less than the sum of amino acids and acetic acid, then a significant fraction of acetate is bound as (O-linked) esters.

Hydrolysis of surface (5-23 m) and midwater (600-1000 m) samples showed good agreement between the conversion of amide-N to amino-N and the production of amino acids and acetic acid (Figure 2.4). Between 97% and 116% of the new amino-N could be accounted for as the sum of amino acids and acetic acid. Of this, the majority (72-90%) was attributed to hydrolysis of N-acetyl amino sugars, while the balance (11-44%) was attributed to protein hydrolysis. Critical to this interpretation is the assumption that quantitative agreement between ^{15}N NMR and molecular level acetic acid analysis signifies that acetic acid is derived from N-acetyl amino sugars. The experiment does not show this directly, but given results from DT-MS that only acetamide is generated by HMWDOM thermolysis, fortuitous agreement between the two techniques seems unlikely.

Detailed characterization of methyl sugars followed their detection in DT-MS spectra by nearly a decade. Hydrolysis of HMWDOM yields an O-methylated sugar fraction that can be purified by chromatography and characterized in detail by ^1H NMR and MS (Panagiotopoulos et al., 2007, 2013). All seven major neutral sugars recovered by hydrolysis were found to have mono- and di-O-methylated homologues (Figure 2.5). In addition, many novel O-methyl sugars not yet identified in the nonmethylated fraction, including O-methyl heptose, O-methyl di-deoxyhexoses and yersiniose, a dideoxy-4-C-(1-hydroxyethyl)-D-xylo-hexose, occur in small amounts. The presence of methyl sugars does not impact the interpretation of linkage analysis, since methylation prevents linkage through a particular carbon, and the product in each case is a methyl sugar. O-methyl sugars are common in algal and bacterial structural polysaccharides, and as such have little biomarker value, but their occurrence in HMWDOM and the complexity of the O-methyl sugar mixture suggests a broad suite of structural polysaccharides may coexist in HMWDOM. Treatment of HMWDOM with periodate at high temperature yields methanol as a major oxidation product (Quan and Repeta, 2007). Assuming all this methanol is sourced from O-methyl sugars, O-methyl sugars contribute an important

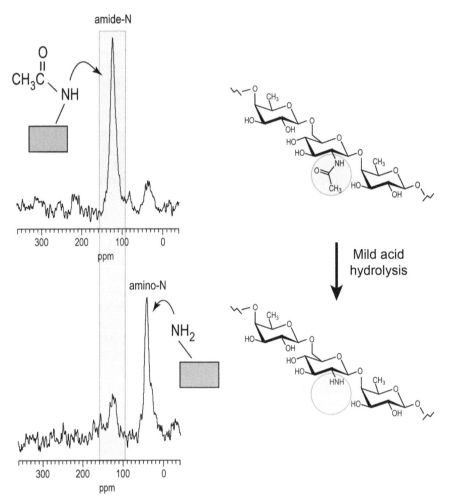

FIGURE 2.4   ^{15}N NMR spectra of HMWDOM (polysulfone membrane with 1-nm pore size) before (top) and after (bottom) acid hydrolysis. HMWDOM nitrogen occurs primarily as amide-N (124 pm) with smaller amounts of amine-N (35 ppm). Major biopolymers that incorporate amide-N are proteins and N-acetyl-amino sugars. The contribution of proteins and N-acetyl-amino sugars in HMWDOM can be assessed by combining ^{15}N NMR spectral data with measurements of acetic and amino acids released after acid hydrolysis. Treatment of HMWDOM with acid converts most amide-N into amine-N. For surface waters, companion chemical analyses show that N-acetyl amino sugars contribute 40-50% of amide-N while proteins contribute 8-13% of amide-N. *Figure adapted from Aluwihare and Meador (2008).*

fraction of HMWDOM, beyond the amount released by acid hydrolysis. Quantitative analysis of methanol after periodate oxidation could be useful in assessing the contribution of methyl sugars to non-hydrolyzable HMWDOM polysaccharide.

Attempts to further characterize HWMDOM polysaccharides have been met with limited success. For reasons that are still unknown, even aggressive hydrolysis by strong acid does not appear to depolymerize HMWDOM, and ~70% of the polysaccharide fraction remains

FIGURE 2.5 Representative structures of monosaccharides isolated from HMWDOM hydrolysis products. Treatment of HWMDOM with acid yields a suite of hexoses (mannose, galactose, glucose), a pentose (arabinose), 6-deoxyhexoses (fucose, rhamnose), hexosamines (glucosamine, galactosamine), and a large number of methylated hexoses (3-methyl rhamnose, etc.) as well as some novel deoxysugars (3-deoxygluose).

Arabinose

Fucose

Mannose

N-Acetyl glucosamine

3-Methyl rhamnose

3-deoxyglucosamine

relatively uncharacterized (Panagiotopoulos and Sempéré, 2005). Information on the non-hydrolyzable portion of HMWDOM polysaccharide can be obtained from 2D homo- and heteronuclear NMR correlation techniques (Figure 2.6). For example, the ^{1}H NMR correlation spectroscopy (COSY) spectrum of non-hydrolyzable HMWDOM polysaccharide shows strong cross peaks between methyl, H-6 protons (CH$_3$–), and carbohydrate H-5 (HC–OH) from 6-deoxysugars. Cross peaks are also observed between 2-3 and 3-4 ppm from 2-, 3-, or 4-deoxysugars (Figure 2.6). Deoxysugars are recovered from HMWDOM hydrolysis products only in low yields (~2% of total carbon), but the 2D NMR suggests a much higher contribution to HMWDOM carbohydrate.

Each class of sugar (hexose, pentose, deoxy-hexose, aminohexose, methylhexose) identified by DT-MS has a different molecular weight easily distinguished by MS, making this is an attractive approach for understanding the sequence and arrangement of sugars within DOM (Boon et al., 1998; Minor et al., 2001). However, a number of issues including the low ionization efficiency of carbohydrates relative to other classes of organic matter, incomplete removal of inorganic salts, and a molecular weight range that may easily exceed the mass range of most spectrometers (typically 2-4 kDa), need to be

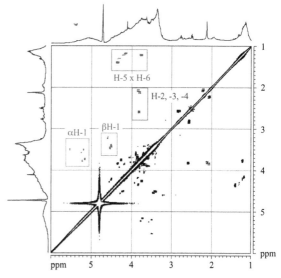

FIGURE 2.6 2D NMR homonuclear correlation spectroscopy (COSY) of HMWDOM carbohydrate from the North Pacific Ocean (Figure 2.2). In this example, strong cross peaks between H-5×H-6 (red) and H-2, H-3, and H-4 (blue) show the presence of 6-, 4-, 3-, and 2-deoxysugars in HMWDOM. Additional strong cross peaks are observed between αH-1×H-2 and βH-1×H-2 (violet), showing different stereochemical geometries within the polysaccharide.

overcome before MS can be applied to its full potential (Sakugawa and Handa, 1985; Schmidt et al., 2003). For example, high-resolution mass spectra for HMWDOM show no ions with H/C and O/C elemental ratios characteristic of poly-saccharides (~1.7-2.0 and ~0.8-1.0, respectively; Hertkorn et al., 2006). Numerous strategies have been developed in glycochemistry to enhance the MS detection of polysaccharides by conjugation with proteins or fluorescent tags that are amena-ble to ionization by electrospray and matrix as-sisted laser desorption ionization (ESI; MALDI). These approaches might prove promising for DOM polysaccharides, and could facilitate MS characterization. Separation of HMWDOM polysaccharides into defined fractions could also facilitate spectral characterization. Appropriate methods for size exclusion and ion chromatog-raphy for complex mixtures of oligo- and poly-saccharides in the molecular weight range of a few kDa and higher are limited, but have been applied to HMWDOM (Sakugawa and Handa, 1985). Recent advances in strong anion exchange chromatography at high pH also shows prom-ise as an approach for HMWDOM carbohydrate separations (Corradini et al., 2012).

Finally, some inferences as to the composi-tion of different HMWDOM polysaccharides can be drawn from changes in the temporal and spatial distribution of sugars between sam-ples. Multivariate analysis of DT-MS data from samples collected in the U.S. Mid Atlantic Bight between 35 and 43°N and the Gulf of Mexico find the same suite of ions from hexoses, pen-toses, deoxy-, N-acetyl-amino-, and methyl sug-ars in all samples, but the relative abundances changed with sample location (Boon et al., 1998). These differences were attributed to changes in the composition or relative abundance of different polysaccharides. The most striking differences were between sugars desorbed at low and high temperatures. The high tempera-ture fraction from all samples showed a remark-able similarity in composition, while the sugar distribution in the low temperature samples showed much more diversity, and some clus-tering by sampling site. The results suggest two distinct fractions of polysaccahrides coexist in coastal seawater, a fraction with highly similar composition that is ubiquitous at all sites, and a fraction of variable composition influenced by local production.

B Proteins and Amino Acids in DOM

Proteins account for up to 50% of the organic carbon and 80% of the organic nitrogen in ma-rine microbes. Grazing, release of extracellular enzymes, and viral lysis all introduce proteins and amino acids into seawater, and on this ba-sis alone, it is likely that proteins contribute to DOM. Analytical methods to identify and quan-tify dissolved proteins directly (proteomics) have only recently reached the point where they can be applied to DOM, and the next decade should see a rapid expansion in our understand-ing of the sources, cycling, and fate of proteins as these methods become more widely applied. However, studies of the distribution, stereo-chemistry, and isotopic values of amino acids re-leased after DOM hydrolysis have already made major contributions to our understanding of dis-solved protein cycling, and have set the stage for the future application of proteomics.

Sensitive methods for the detection of nano-molar concentrations of dissolved amino acids were first applied to seawater in the early 1960s and 1970s (Lee and Bada, 1975, 1977). Due to the low ambient concentrations of dissolved amino acids, and the potential for contamination by laboratory glassware and reagents, early studies gave variable results. Current analyses capitalize on the reaction of ortho-phthalaldehyde (OPA) with primary amines in basic, aqueous solutions to form fluorescent, hydrophobic products that can be retained and separated by high pressure liquid chromatography (Lindroth and Mopper, 1979; Mopper and Lindroth, 1982). The reac-tion proceeds rapidly and at high yields and has sub-nanomolar detection limits. Primary amines other than amino acids (ammonia, urea, etc.) also react with OPA and are included in the analysis. Some amino acids co-elute under some separation conditions, but with appropriate calibration the method provides good quantitative measurements of dissolved amino acids in seawater (Tada et al., 1998).

Dissolved amino acids are operationally classified by the methods used for sample processing. Dissolved "free" amino acids (DFAA) are measured by the direct reaction of OPA with seawater, and are thought to represent monomeric amino acids present in a sample. In order to minimize contamination during sample processing, some early studies did not use filtration to separate dissolved and particulate free amino acids (Mopper and Lindroth, 1982). However, subsequent studies typically include a filtration step (Fuhrman, 1987). Treatment of filtered seawater with acid hydrolyzes peptides, proteins, and glycoproteins and allows for the measurement of "total hydrolysable" or "total dissolved" amino acids (THAA; TDAA, respectively; Lee and Bada, 1975, 1977). The difference; THAA-DFAA represents dissolved combined amino acids (DCAA), the fraction of amino acids bound as proteins, peptides, and other amino acid polymers. Since DFAA are typically more than an order of magnitude less abundant than DCAA, recent studies have focused only on total hydrolyzable or total dissolved amino acids (McCarthy et al., 1996; Yamashita and Tanoue, 2003a).

Open-ocean profiles of DFAA in the equatorial Pacific and Sargasso Sea show very low values throughout the water column. Surface water concentrations (10-40 nM) are somewhat enriched relative to deep-sea samples, which have lower and more constant values (<10-20 nM). THAA distributions follow a pattern similar to DFAA, with higher and more variable concentrations in surface waters (200-450 nM) and lower and more stable values below the euphotic zone (100-200 nM) (Figure 2.7; Kaiser and Benner, 2009; Lee and Bada, 1975, 1977; McCarthy et al., 1996; Yamashita and Tanoue, 2003a) Semi-enclosed seas, coastal, and near-shore sites have higher concentrations of both DFAA and THAA.

Overall, spatial patterns of THAA concentrations reflect changes in DOC concentrations. THAA carbon to total DOC ratios (THAA-C/DOC) vary between sites and depth, but generally fall within 1-4% for surface waters and 0.4-0.8% at depths >1000 m (Kaiser and Benner, 2009; McCarthy et al., 1996; Yamashita and Tanoue, 2003a). THAA contribute a larger portion of DON (1.4-11%; Kaiser and Benner, 2009; McCarthy et al., 1996; Tada et al., 1998), and are the largest component of DON characterized to the molecular level. A decrease in the ratio of THAA-C to DOC has been observed between highly productive coastal waters and the open ocean along a line extending from Japan (Yamashita and Tanoue, 2003a). THAA carbon to DOC ratios fall from 4% in near-shore surface waters to 2% in offshore surface waters. Similar patterns have been observed in the Baltic Sea, Chesapeake Bay, Biscaye Bay (Florida), and in the Laptev Sea (Mopper and Lindroth, 1982). Nutrient and chlorophyll concentrations were quite high at all sites and higher rates of local primary production as well as more dynamic organic matter cycling closer to shore may result in relatively higher contributions of THAA to DOC.

Major amino acids include glycine, alanine, glutamic acid, serine, aspartic acid, arginine, and threonine, which typically contribute >90% of THAA (Figure 2.8). Other protein amino acids individually represent <5 mole% of THAA, although contributions from valine and leucine are sometimes in excess of this amount. Tryptophan decomposes under the acid and temperature conditions typically used for THAA analysis, and can be measured in samples only when alkaline conditions are used for hydrolysis. In open-ocean waters, tryptophan contributes ~1 mol% of THAA, although somewhat higher contributions were noted near shore (Yamashita and Tanoue, 2003a). The distribution of amino acids in THAA is significantly different than particulate organic matter (phytoplankton, suspended and sinking detrital material, Figure 2.8), with higher concentrations of aspartic acid and

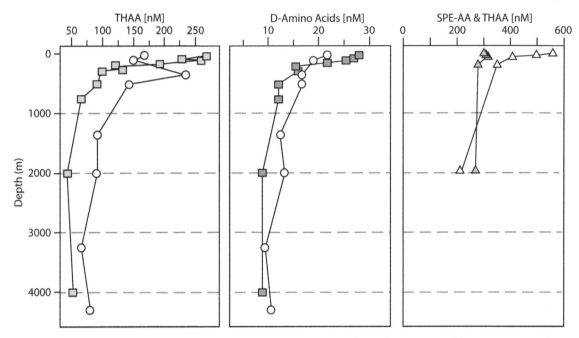

FIGURE 2.7   Distributions of total hydrolysable amino acids (THAA; left panel), D-amino acids (center panel), and SPE-amino acids and THAA (right panel). Data in the left and center panels are from the North Pacific (gray squares; Station ALOHA of the Hawaii Ocean Time-series (HOT) program) and North Atlantic (open circles; Bermuda Atlantic Time-series Study (BATS) program) Oceans. SPE-AA (filled triangles) and THAA (open triangles) data in the right panel are pooled data-sets from the high-latitude North Atlantic and Southern Oceans. Concentrations of THAA are high in the euphotic zone and fall through the mesopelagic zone. Concentrations stabilize in the deep ocean, at >1000 m. D-amino acids concentrations are also highest in the euphotic zone, decrease to about 1000 m and stabilize thereafter. SPE-AA are thought to represent amino acids and peptides that are part of refractory DOM, showing much less variation in concentration with depth. At 2000 m, nearly all THAA is recovered as SPE-AA. *Data on HOT and BATS are from Kaiser and Benner (2009). Data on SPE-AA and THAA in the right panel are from Hubberton et al. (1995).*

glycine, but lower concentrations of arginine, leucine, and isoleucine (Figure 2.8).

Amino acid distributions at open-ocean and coastal sites generally do not show large, systematic changes; however, subtle changes have been detected through correlative analysis. Using relative abundance as a measure of sample relatedness, Yamashita and Tanoue, 2003a compared changes in the distribution of amino acids across a north-south transect along 137° E from Ise Bay, Japan, into the northwest Pacific Ocean. Some amino acids (e.g., glycine and al-anine) are positively correlated in all samples, and on this basis, amino acids were grouped

into four categories. The positive correlations within a particular group indicates similarities in the biogeochemical cycling of the constitu-tive amino acids, while differences in the cor-relations between groups suggest divergence in THAA cycling due to differences in either the macromolecular form of the peptides or the way THAA are processed by microbial degradation. Amino acid distributions have also been analyzed by principal component analysis (PCA) to quan-titatively differentiate patterns of amino acid distribution. Cross correlation of the first and second PCA scores for each acid showed general groupings based on sample location and depth

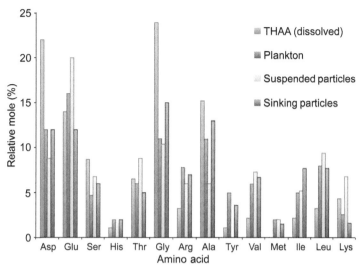

FIGURE 2.8   The relative distributions of individual amino acids within dissolved THAA, phytoplankton, suspended particulate matter, and sinking particles. Plankton, suspended, and sinking particle data are from the equatorial Pacific Ocean, while THAA is from Station Aloha, near Hawaii (Kaiser and Benner, 2009; Lee et al., 2000). Sinking particle data are from short-term deployments of floating sediment traps. Data from long-term deployments of moored traps have higher concentrations of glycine (Gly), similar to THAA in DOM. All sample types have the same suite of amino acids. The major difference between dissolved and particulate amino acid distributions are the higher relative abundance of aspartic acid and glycine in THAA. Amino acids are: Asp (aspartic acid), Glu (glutamic acid), Ser (serine), His (Histidine), Thr (Threonine), Gly (Glycine), Arg (Arginine), Ala (Alanine), Tyr (Tyrosine), Val (Valine), Met (Methionine), Ile (isoleucine), Leu (leucine), Lys (lysine).

(bay, coast, open ocean, surface, and deep water) and cross plots of factor coefficients grouped acids into four classes with the same make up as found for correlative analysis.

Amino acids occur in two enantiomeric forms (D- and L-amino acids). Proteins are made exclusively from L-amino acids, but many bacteria utilize D-amino acids in cellular regulatory functions and in bacterial cell-wall biopolymers (Cava et al., 2011). Enantiomeric amino acids can be chromatographically separated and quantified in THAA analysis, and the composition and distribution of D-amino acids have been used as a biomarker for the contribution of bacterial cell-wall material to DOM (Kaiser and Benner, 2006; Lee and Bada, 1977; McCarthy et al., 1998). Lee and Bada, 1977 reported enantiomeric (D:L) ratios of aspartic acid (0.07-0.44),

glutamic acid (0.04-0.07), alanine (0.05-0.14), valine (0.02-0.04), isoleucine (<0.01-0.22), and leucine (0.02-0.07). Subsequent studies have reported D:L ratios for serine of 0.1-1. McCarthy et al. (1998) found the HMWDOM fraction has higher D:L ratios for glutamic acid (0.08-0.15) and alanine (0.32-0.55) than THAA. These ratios are greater than D/L ratios in cellular organic matter and POM and are indicative of selective preservation of D-amino acid-containing bacterial structural polymers relative to proteins in THAA (Dittmar et al., 2001; Kawasaki and Benner, 2006; Kaiser and Benner, 2008; McCarthy et al., 1998; Nagata, 2003).

The high relative abundance of D-amino acids in THAA and HMWDOM has led to a series of investigations designed to assess the contribution of bacterial cell-wall material to DOM.

Bacteria are abundant at all depths in the ocean, and the selective preservation of D-amino acids indicated by high D/L ratios suggests that cell-wall biopolymers are more resistant to microbial degradation than proteins and are therefore likely to be persist in DOM (Calleja et al., 2013; Kaiser and Benner, 2008; McCarthy et al., 1996). Kaiser and Benner (2008) measured the abundance and distribution of D-amino acids and muramic acid (a constituent of peptidoglycan in bacterial cell walls) in marine heterotrophic and cyanobacteria in order to make a quantitative assessment of the peptidoglycan and other D-amino acid-containing bacterial cell-wall constituents to DOM. In purified peptidoglycan, the ratios of muramic acid to D-glutamic acid and D-alanine are ~1 and ~0.75, respectively. Ratios in marine POM and DOM above these values were interpreted to signify the presence of other (non-peptidoglycan) D-amino acid-containing constituents, while ratios below these values signify the preferential degradation of D-amino acids relative to muramic acid. Muramic acid was non-detectable (<1.2 nM) in filtered seawater, even though total D-amino acids concentrations ranged between 8 and 26 nM. These high relative concentrations suggest that D-amino acids in DOM are either derived from peptidoglycan degradation products or are constituents of bacterial polymers other than peptidoglycan.

Tanoue and colleagues pioneered the application of gel electrophoresis and N-terminal sequencing to characterize intact proteins in DOM (Tanoue, 1995; Tanoue et al., 1995, 1996; Yamada and Tanoue, 2003). Dissolved proteins can be concentrated and desalted using ultra-filtration (>10 kDa), and then partially purified by precipitation with cold trichloroacetic acid or methanol/chloroform/water (Powell and Timperman, 2005; Tanoue, 1995). Care must be exercised throughout the concentration steps to minimize contamination from lysed cells or bacterial growth, as well as from losses of proteins due to precipitation or adsorption onto filters and other surfaces. The insoluble protein pellet from trichloroacetic acid precipitation was washed of residual polysaccharides and lipids with ethanol and diethyl ether, and then analyzed by gel electrophoresis.

Electrophoretograms of dissolved proteins were compared to a suite of standard marker proteins separated on the gel electrophoresis plate by molecular size (Figure 2.9). Samples from the northcentral and southwest Pacific Ocean and the Gulf of Mexico show proteins with a wide range of sizes corresponding to molecular weights from <14 to >66 kDa (Powell et al., 2005; Tanoue, 1995; Tanoue et al., 1995, 1996). Major proteins were concentrated in 25-30 bands, some of which were present in all samples, while others changed with sample location and depth. Surface waters <50 m generally showed low numbers and concentrations of recognizable protein bands. The number and intensity of bands increased with depth (75-200 m); below 200 m, gels were heavily stained, with a high degree of background staining from unresolved proteins and perhaps HS. Proteins with putative molecular weights of 48 and 37-40 kDa were clearly visible at many of the stations irrespective of depth, while proteins designated as 66-63, 44, 41, 31-34, 26, 23, and 15 kDa varied with depth and sampling location, but were present in a number of samples. The strikingly similar patterns that appear in electrophoretograms of dissolved proteins led Tanoue et al. (1996) to conclude that the processes that transfer proteins from cellular material and allow for their accumulation in DOM are similar across broad expanses of the ocean.

The major protein band at ~48 kDa present in all samples was recovered from six samples collected between 45 and 462 m and the N-terminal sequence of 14-15 amino acids determined (Tanoue et al., 1995). Later analysis of these samples expanded the sequence to 24 amino acids (Yamada and Tanoue, 2003). All 48 kDa bands yielded the same N-terminal sequence of amino acids, indicating that all bands represented the same protein. The sequence shared

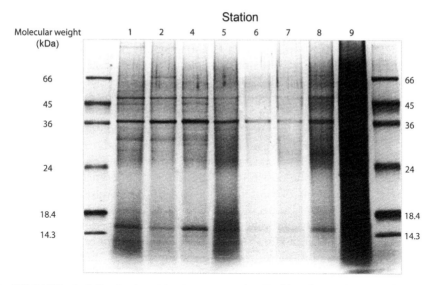

FIGURE 2.9 SDS-PAGE gel of dissolved proteins in seawater visualized by silver staining. Dark bands represent separated proteins. The left and right lanes are standard mixtures of known proteins used to calibrate the molecular separation of the gel. Sample proteins are from surface waters collected from the equator (Station 1) to 60° S (Stations 4, 5) along 120° E, then west across the Southern Ocean to ~30° E (stations 6-9). Note the similarity in protein bands between samples. Major protein bands appear in samples at 48, 37, and 15 kDa. *Figure used with kind permission from Dr. Eiichiro Tanoue.*

100% homology with porin-P and porin-O of the gram-negative bacterium *Pseudomonas aeruginosa* (Tanoue et al., 1995; Yamada and Tanoue, 2003). Porin-P and porin-O are membrane proteins expressed to facilitate cross membrane transport of small hydrophilic substrates, often under conditions of phosphate stress. The bacterial community composition at the sampling sites was not determined, but genomic analysis of bacterioplankton at Station ALOHA, one of the stations sampled by Tanoue, yielded 16s sequences of *Alteromonas, Vibrio,* and *Pseudomonas* spp., bacteria that are closely related to *P. aeruginosa*. Similar biosynthetic pathways between indigenous bacteria and *P. aeruginosa* might lead to porins of comparable homology. Recognizing that existing databases contain sequences from only a small number of proteins, Tanoue's data suggests that bacterial proteins such as porin-P contribute to DOM. Porin P has been shown to be resistant to proteases, which together with a potentially ubiquitous and abundant source,

may explain its appearance in all samples analyzed by gel electrophoresis in Tanoue's studies. Likewise a 40 kDa protein from the North Pacific had 100% homology with the family of outer membrane proteins (OmpAs) of *Acinetobacter* spp., however sequencing of the 30, 37, and 39 kDa proteins did not exhibit homology with any known proteins in searchable databases, and these proteins could not be identified

Suzuki et al. (1997) and Yamada et al. (2000) developed and applied an immunochemical assay against bovine serum albumin modified with the *N*-terminal 14 oligomer of *P. aeruginosa* porin P (α-48 DP N-14), and the whole outer membrane protein Omp35La from *Vibrio anguillarum*. Polyclonal antibodies developed against these antigens were used as sensitive screens for porin P and related proteins in DOM. Western blots of DOM proteins showed cross-reaction between the α-48 DP N-14 probe and the 39, 48, and 60 kDa bands and between the Omp35La probe and the 18, 34, and 70 kDa bands.

Cross-reaction of the α-48 DP N-14 probe and the 39 kDa protein is not fully understood, as a subsequent investigation showed this protein to be glycosylated, and therefore probably not a porin P homologue (Yamada and Tanoue, 2003). Cross-reactivity for the α-48 DP N-14 probe was also observed for natural populations of bacteria in all samples, although the number of bacteria cells that cross-reacted were 2-6 orders-of-magnitude less than enumerated by total bacterial cell counts. The results raise the possibility that the sources and composition of proteins dissolved in seawater might be highly specific.

Erdman degradation was used to partially sequence the 48 kDa protein in DOM, but HPLC-MS techniques offer the possibility of more comprehensive sequencing of dissolved proteins in environmental samples (Powell et al., 2005). In this approach, proteins purified by gel electrophoresis or capillary zone electrophoresis are partially digested by exposure to trypsin, a protease, and the peptides separated by HPLC and sequenced by MS. Fragment ion masses are used to reconstruct the peptide amino acid sequence that is then compared to sequence information stored in databases of known proteins. Using this approach, Powell et al. (2005) distinguished families of proteins, showing that proteins from both bacterial membranes (fatty acid synthetase, luminal binding protein) and enzymes (ribulose bisphosphate carboxylase, anthranilate synthethase) were present in DOM (Powell et al., 2005).

Bottom-up proteomics and amino acid D/L ratios both highlight the contribution of bacterial proteins to DOM. Another way to assess the sources, cycling, and distribution of dissolved proteins is to target abundant proteins using 2D gel electrophoresis and sensitive immunoassays such as ELISA (enzyme-linked immunosorbent assay) and MSIA (mass spectrometry immunoassay; Jones et al., 2004; Orellana and Hansell, 2013; Orellana et al., 2003). In their proteomic data, Powell et al. (2005) reported short peptide sequences suggestive of ribulose-1,5-bisphosphate carboxylate/

oxygenase (RuBisCo). RuBisCo is one of the most abundant proteins on Earth, and is essential in catalyzing carbon fixation in vascular plants, algae, photoautotrophic and chemo-autotrophic bacteria. RuBiscCo is abundant in particulate matter located in the euphotic zone, where grazing, viral lysis, and other processes transfer some particulate RuBisCO into DOM. Some of the DOM proteins visualized by electrophoresis and staining, particularly in the range of known RuBisCo subunit proteins of 55 and 13 kD, probably result from this cycling. Orellana and Hansell (2012) used a synthetic protein incorporating RuBisCo sequences to develop an immunoassay for anti-RbcL to measure RuBisCo concentrations in ~800 samples from the North Pacific Ocean. The large number of samples analyzed in this study, and sensitive detection limits (<1 ng RuBisCo L^{-1}), highlights the potential power of immunological approaches to track the distribution and cycling of biologically important proteins. They found RuBisCo in all samples from the surface to depths >4000 m. RuBisCo is synthesized in the euphotic zone, and the presence of high concentrations of RuBisCO (5-20 ng/L) throughout the deep ocean ties meso- and bathypelagic RuBisCO distributions to export production (a small amount of RuBisCO may be synthesized at depth from chemoautrophy, but this contribution is thought to be small). As large particles sink from the euphotic zone, microbial remineralization and physical disaggregation release smaller particles and inject DOM into underlying waters. High relative concentrations (15-20 ng L^{-1} between 2000 and 4000 m) of RuBisCO in the deep equatorial and subarctic Pacific coincide with higher carbon export in these regions. Low relative concentrations of RuBisCo (<10 ng L^{-1} between 2000 and 4000 m) below the northern and southern subtropical gyres were attributed to the low carbon export fluxes that characterize these regions. The coupling of deep water RuBisCo concentrations to surface processes implies a rapid turnover of RuBisCo related proteins, but spatial differences in deep RuBisCo concentrations also trace deep water mass flows. If correct, then the residence time of some deep RuBisCo is on the order of years to decades, a timescale typically associated with semi-labile DOM (Section IV.B).

C Humic Substances in Solid-Phase Extractable DOM (SPE-DOM)

Simple ^{1}H and ^{13}C NMR spectra along with elemental analysis of hydrophobic SPE-DOM made in the 1970s and 1980s showed SPE-DOM is distributed throughout the water column, is rich in COOH and aliphatic carbon, and has a COOH/ aliphatic carbon ratio of 1:4-5 (Hedges et al., 1988). Natural products with such a high ratio of COOH/alkyl carbon are rare in nature, and the SPE-DOM fraction is therefore thought to result from extensive transformations of marine lipids, carbohydrates, and proteins. Recent studies refer to this fraction of DOM as "carboxyl-rich aliphatic material" (CRAM), but it has alternatively been referred to as hydrophobic DOM, solid-phase extractable (SPE) DOM, carboxylate rich carbon (CRC), or marine HS. For simplicity, this fraction of DOM will hereafter be referred to as SPE-DOM, since all studies rely on isolation by adsorption onto a solid hydrophobic resin (XAD, C-18, PPL), and recent work has begun to distinguish distinct components within SPE-DOM and assign characteristic molecular features to specific terminology (CRAM, thermogenic DOM, etc.). However, in the literature, different terminologies are still in use. SPE-DOM as been compared to HS isolated from soils and freshwaters, but the marine version has a lower aromatic and olefinic content, a lower C/N ratio, and a carbon staple isotope (δ^{13}C) value of ~21-22‰, all of which suggested a autochthonous source (Druffel et al., 1992; Gagosian and Steurmer, 1977; Hedges et al., 1988). Although only a handful of measurements exist, SPE-DOM is highly depleted in radiocarbon (−310‰ to −587‰), and is considered to be a recalcitrant

fractory fraction of DOM resistant to microbial oxidation (Druffel et al., 1992).

1 Characterization of SPE-DOM by High-Field NMR

The processes that lead to the formation and removal of SPE-DOM in the ocean are not known, but high-field NMR and high-resolution MS have added unprecedented detail to our knowledge of its composition (Dittmar and Koch, 2006; Helms et al., 2013; Hertkorn et al., 2006, 2012). ^{1}H NMR spectra are characterized by broad peaks between 0.8 and 10 ppm with substantial signal overlap (Figure 2.10). Major signals have been assigned to aliphatic CH_3 and CH_2, carbonyl-rich aliphatics, particularly methine protons (($C)_2$–CH–COOH), methoxy protons (CH_3O), carbohydrate-derived methines (($C)_2$–CH–OH) and (O–CH–O)), and protons on sp^2 hybridized olefinic and aromatic carbon. When ^{1}H NMR spectra of samples collected in a depth profile from 5 to 5446 m were normalized to 100% total area (0-10.5 ppm), the

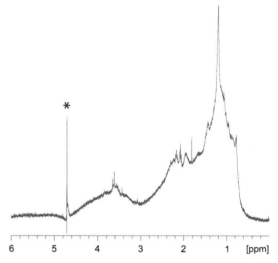

FIGURE 2.10 ^{1}H NMR of SPE-DOM from 900 m near Hawaii. NMR spectra of SPE-DOM differs from ^{1}H NMR spectra of HMWDOM by the absence of carbohydrates. Major resonances include methyl (H_3C-R; ~0.9-3.3 ppm), methylene (H_2CH$_2$-RR'; ~1.3-2 ppm), and methine (C-R,R',R''; ~1.4-4.5 ppm).

spectra showed nearly coinciding aliphatic terminal ($-CH_2-CH_3$; $\delta \sim 0.9$ ppm) methyl abundance, variable methylene ($-CH_2->4$ bonds from a heteroatom) abundance, and progressively increasing amounts of H associated with methylated, alicyclic rings with depth. Signals associated with polysaccharides ($N-C(O)-CH_3$; $H-C-OH$, $O-CH-O$, CH_3O-C) decreased with depth.

Although signal overlap limits molecular definition in 1D ^{1}H NMR spectra, 2D COSY spectra provide unprecedented definition of overlapping signals. About 4500 off-diagonal cross peaks have been observed for SPE-DOM collected in surface waters, of which ~75% were derived from sp^3-hybridized carbon ($HC_{sp3}-C_{sp3}H$, HC_{sp3} (O)$-C_{sp3}H$, and H_{sp3} (O)$-C_{sp3}$(O)H), and 25% from sp^2 hybridized carbon ($HC_{sp2}-C_{sp2}H$; $\delta_H > 5$ ppm; Hertkorn et al., 2012). Assuming an average spin system of ~3.5 protons, the COSY spectra of SPE-DOM suggests a mixture of at least several hundred distinct molecular species. With depth, the COSY spectra of SPE-DOM become progressively attenuated, however the position of major peaks remains the same. SPE-DOM composition is highly conserved with depth, but the attenuation of COSY cross peaks suggests subtle changes in composition that lead to faster transverse relaxation of NMR signals and/or increasing molecular diversity associated with DOM aging (Hertkorn et al., 2012).

^{13}C NMR spectra and 2D heteronuclear ^{1}H/^{13}C NMR spectra are consistent with these assignments, allowing for a better quantitative assessment of how carbon functional groups are distributed within SPE-DOM. ^{13}C NMR spectra also provide some unique insights not available through ^{1}H NMR alone. All ^{13}C NMR spectra show signals from carbonyl ($C=O$; 220-187 ppm), carboxyl (COX; where $X= -O, -N, -CH_3$; 187-167 ppm), aromatic $C-X$ (where $X=O$, N), and aromatic $C-H$ (167-145 and 145-108 ppm respectively), carbohydrate carbon (anomeric $O-CH-O$, 108-90 ppm and O-alkyl $HC-OH$, 90-47 ppm), and aliphatic carbon (47-0 ppm; Hertkorn et al., 2006, 2012) The relative amount of carboxylic acids and ketones increases from surface to deep water, and the amount of labile carbohydrate carbon declines, indicating a selective removal of carbohydrates and gradual oxidation of the non-carbohydrate fraction. Methine carbon associated with aliphatic branched functional groups (C_3-C-H) increased relative to aromatic and carbohydrate associated carbon from 46% in surface waters to 57% at 5446 m. The highly branched nature of marine SPE-DOM is also indicated by the relatively intense adsorption bands between 2970 and 2980 cm^{-1} (aliphatic methyl stretch) and 2944 cm^{-1} (aliphatic methyl stretch) in the infrared (Esteves et al., 2009). NMR and IR spectra indicate that aromatic carbon was not abundant at any depth (<5%), and some aromatic carbon associated with downfield NMR signals ($\delta_c = 164$ ppm) display a cross beak with aromatic protons ($\delta_H = 8.2$ ppm) that indicate nitrogen heterocycles (Esteves et al., 2009; Hertkorn et al., 2012). Only a minor fraction (15%) of methyl groups were purely aliphatic ($-CH_2-CH_3$), while a major fraction (70%) were shifted downfield to between 1 and 1.6 ppm. The downfield shift in most methyl groups suggests proximity (<3 C bonds away) to carboxyl carbon. The carboxyl carbon peak itself is Gaussian shaped and displays a sizable chemical shift range (6 ppm at half height) indicating high diversity and little preference or regularity in its chemical environment.

In summary, ^{1}H and ^{13}C NMR spectra show SPE-DOM is a highly complex, yet well defined mixture of molecular components, that includes carbohydrates and a carboxy-rich aliphatic (CRAM) fraction, along with a minor amount of extended aromatic and aromatic N-heterocycles. The CRAM fraction appears to become more highly branched with depth, either from transformations associated with aging, or from selective removal of less highly branched carbon. The major change in the distribution of carbon functional groups in SPE-DOM with depth results from the progressive loss of carbohydrate and the conservation of CRAM (Hertkorn et al., 2012).

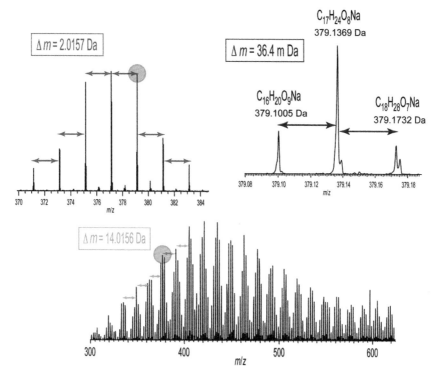

FIGURE 2.11   High-resolution mass spectrum (positive ion) from SPE-DOM collected at 250 m, Station Aloha, near Hawaii. The mass spectrum of the infused sample (bottom) shows a complex distribution of ions with reoccurring mass differences of $\Delta m = 14.0156$ due to methylene homologues (ΔCH_2). Expansion of the series centered at 377 Da (upper left) shows a second homologous series with $\Delta m = 2.0157$ (H_2). Further expansion (upper right) shows the high mass resolving power that can be achieved, which allows for the assignment of elemental formulae, and the distinction of formulae of the same nominal mass. Here ions that differ by the substitution of CH_4 (16.0313 Da) for O (15.9949) yield a mass difference of 36.4 mDa.

2 Characterization of SPE-DOM by High-Resolution MS

High-resolution mass spectra of SPE-DOM have been reported since the 1970s (Gagosian and Steurmer, 1977), but the introduction of fourier transform ion cyclotron resonance (FTICR) and more recently Orbitrap mass spectrometers, each capable of measuring ions at both very high mass accuracy and resolution, has changed our ability to characterize SPE-DOM at the molecular level (Kido Soule et al., 2010; Zubarev and Makarov, 2013). Coupled to electrospray ionization (ESI) these instruments provide the necessary resolution to distinguish the several thousand ions

with unique masses within DOM between 200 and 2000 Da (Figure 2.11). Due to the high mass resolution and accuracy (typically <1 ppm for ions <400 Da), elemental formulas for most ions can be assigned with a high level of confidence. Since the number of isomers for any given mass increases rapidly with molecular weight, each unique mass probably represents a mixture of different isomers. High-resolution MS therefore does not allow for the full identification of new compounds, however it has led to the discovery of important new compound classes, and it is proving to be a very powerful technique for detecting changes in DOM composition between samples. To fully exploit the large amount of data provided by high-resolution MS, analysts use a number of computational tools to extract information about DOM composition. Typically, elemental formulas with all possible combinations of atoms are calculated and matched to all ions to within a mass precision determined by the operational resolution of the particular instrument. Depending on the approach used 45-97% of observed masses can often be assigned molecular formulae. Elemental formulae with only C, H, and O dominate the molecular formulae of assigned masses, with fewer elemental formulae assigned to compounds having nitrogen (CHNO), sulfur (CHOS), and both nitrogen and sulfur (CHNOS).

In performing these calculations, some assumptions are needed. First, the type and likely number of atoms within a formula are inferred. For example, in a recent study of SPE-DOM in the southwest Atlantic Ocean, Flerus et al. (2012) constrained molecular formula calculations for masses between 200 and 600 Da within ^{12}C (0-∞), ^{13}C (0-1), ^{1}H (0-∞), ^{16}O (0-∞), ^{14}N (0-∞), ^{32}S (0-1). For positive ions ^{23}Na (0-1) is also allowed to account for sodiated DOM ions. The resulting mass list is filtered to remove ^{13}C isotopes (mass difference between peaks of 1.003 Da), and further filtered by making assumptions about nitrogen, the H/C ratio, and the number of double bonds (Koch et al., 2007). The power of high-resolution MS lies in its ability to distinguish between elemental formula with the same nominal masses. For example, the functional group $CH_3-CH-R-$ has the same nominal mass (16 Da) as the functional group $O=C-R-$, but differs in exact mass by 36.4 mDa (16.0313 vs. 15.9949), which is well within the resolution of high-field instruments. High-resolution MS data allow mass lists to be grouped into "pseudo-homologous" series that differ by a specified functionality (e.g., $-CH_2-$ or $-CO_2-$). The measured mass is converted to a "Kendrick mass" where each CH_2 unit is defined as 14.000 Da instead of its exact mass of 14.01565 Da. The difference between the exact

mass and Kendrick mass is then assessed as the Kendrick Mass Defect (KMD, where KMD = exact mass − Kendrick Mass). The KMD will be constant for compounds within a series that differ only by the number of $-CH_2-$ groups. The elemental formula of the lowest mass member within a series is determined, usually with a high degree of confidence, thereby determining the molecular formulas for all other members of the series. Analysis of mass data from a number of sites shows that most masses can be grouped into pseudo-homologous series with mass differences of 14.0156 (CH_2), 2.0157 (H_2; double bond series), and 0.0364 (replacement of CH_4 with oxygen).

Once elemental formulae have been determined, the data are reduced to their elemental H/C and O/C ratios and visualized in a van Krevelen plot (Figure 2.12). In a van Krevelen plot, the H/C and O/C ratio of each ion can be compared to the elemental ratios of likely biochemical precursors, to other samples grouped according to sample location or type, or queried with respect to differences in molecular weight, ionization mode, and other features that provide information on DOM composition and cycling. Figure 2.12 shows a van Krevelen plot of FTICR-MS from Station ALOHA of the Hawaii Ocean Time-series (HOT). The sample is typical of many open-ocean datasets, with the majority of compounds falling within a H/C range of 0.5-1.7 and an O/C ratio of 0.2-0.8 (Gonsior et al., 2011; Hertkorn et al., 2012; Koprivnjak et al., 2009; Kujawinski and Behn, 2006; Kujawinski et al., 2009). Surprisingly, this range of H/C and O/C falls outside the elemental ratios of most common proteins, carbohydrates, and lipids (Figure 2.12), indicating either SPE-DOM has been extensively reworked and altered from its biochemical precursors, or that SPE-DOM mass spectra target a fraction of carbon that is not abundant in cells. Intensity-weighted elemental data from a number of sites shows a narrower range of average values of between 1.3 and 1.4 for H/C and between 0.3 and 0.4 for O/C. Ratios may vary somewhat between different studies

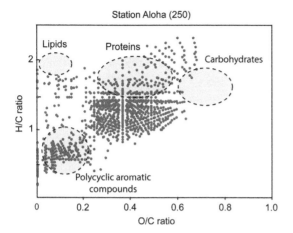

Station Aloha (250)

FIGURE 2.12 van Krevelen plot of data in Figure 2.11 from 250 m, Station ALOHA. Each dot represents an ion for which a unique molecular formula could be assigned from the exact mass. Here the elemental ratios H/C and O/C calculated from molecular formulae are plotted and compared with elemental H/C and O/C ratios from classes of known biochemical (lipids, proteins, and carbohydrates). Most SPE-DOM formulae plot outside the region of known biochemicals, perhaps due to extensive degradation and transformation of organic matter during DOM formation, or from unknown DOM precursors. A distinct group of ions with H/C < 1 and O/C < 0.2 appear in some SPE-DOM spectra; these have been assigned to polycyclic aromatic compounds (PCAs) of thermogenic origin (see text).

due to differences in sample handing or instrumental biases. Irrespective of these differences, all studies so far report fairly similar arrays of chemical formulae for SPE-DOM.

In one of the first reports of FTICR-MS data of SPE-DOM in the open ocean, Dittmar and Koch (2006) recognized a cluster of 244 ions with very low H/C and O/C ratios (0.5-0.9 and 0.1-0.25, respectively) and a high number of double bond equivalents (DBE; double bond equivalents, number or double bonds or rings). The high number of DBE and relatively LMWs (428-530 Da) that characterize these ions narrowly restricts the structural possibilities. Using conservative estimates of the number of DBE that can be assigned to oxygen, they postulated the low H/C, O/C group represents a class of condensed polycyclic

aromatic compounds with 5-8 rings along with different degrees of alkyl substitution and oxygen functionality. As there are no known biogenic precursors for compounds of this type, these compounds most likely originate from thermogenic processes such as terrestrial biomass burning, fossil fuel combustion, and/or reactions of organic matter in hydrothermal systems (Dittmar and Koch, 2006; Dittmar and Paeng, 2009). The discovery of thermogenic polycyclic aromatic compounds in DOM was a significant contribution to our understanding of DOM sources and cycling. Although the potential for a contribution of thermogenic black carbon to DOM had been recognized earlier from radiocarbon measurements, it was never characterized on the molecular level until the high resolution and specificity of FT-ICRMS was brought to bear on DOM characterization (Masiello and Druffel, 1998).

To quantify polycyclic aromatic carboxylic acids to DOM, an oxidative protocol designed to measure black carbon in soils was adapted to SPE-DOM (Dittmar, 2008). In the method, concentrated nitric acid at elevated temperature (170°C) oxidizes polycyclic aromatic compounds in SPE-DOM to benzene polycarboxylic acids (BPCAs; Figure 2.13), which are separated by HPLC and quantified by on-line spectroscopic detection. Analyses of marine SPE-DOM yields a suite of substituted BPCAs with a relatively high fraction of benzene penta- and hexacarboxylic acids consistent with highly condensed polycyclic aromatic precursors. The distribution of BPCA isomers varied little between samples, and it is inferred that molecular structures for SPE-DOM are therefore similar, irrespective of sample depth or location. BPCAs represent 1-3% of SPE-DOM. Subsequent work used the method to quantify total polycyclic aromatic compounds (PCAs) in a meridional section across the Southern Ocean at 30° E (Dittmar and Paeng, 2009). PCA concentrations ranged from 0.6-0.8 μM carbon or about 1-2% of total DOC. Both high and low values were measured in surface waters, implying inputs, and removal of

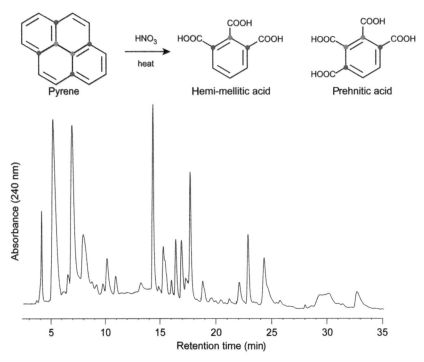

FIGURE 2.13   HPLC analysis of benzene polycarboxylic acids (BPCAs) derived from SPE-DOM. Treatment of SPE-DOM with nitric acid at high temperature degrades polycyclic aromatic compounds (top) into a series of benzoic acids that can be measured at high sensitivity by HPLC (bottom) and HPLC-MS. The sample was collected from 1100 m at Station ALOHA, near Hawaii. Benzene pentacarboxylic acid elutes at 5.1 min, while isomers of benzene tetracarboxylic and tricarboxylic acids elute in the regions 14-18 min and 22-26 min, respectively.

PCAs on relatively short timescales. Due to the remoteness of the Southern Ocean from significant river inputs, atmospheric deposition, and subsequent photochemical oxidation were considered to be the most likely sources and sinks for PCAs in this region. Radiocarbon in BPCAs from HMWDOM recovered by ultrafiltration was highly depleted (−880‰ to −918‰), with ages ranging from 17,000 to 20,100 ybp (Ziolkowski and Druffel, 2010). Concentrations were lower (90-330 nM) than reported by Dittmar and Paeng (2009), probably due to lower recoveries of hydrophobic LMWDOM by the ultrafiltration method. However, the results overall support the idea that PCAs represent a refractory DOM fraction. Further studies designed to determine PCA structures are needed to better describe the sources, cycling, and sinks of PCAs.

Studies that compare FT-ICRMS spectra on a simple presence/absence basis of individual ions report a remarkable uniformity in a large fraction of ions from samples collected from different locations and different depths in the ocean. Koch et al. (2005) report that of the 1580 chemical formulae recognized in a suite of samples collected in the Weddell Sea, ~30% were present in all samples. Only two formulae were found exclusively in surface waters, and only 79 formulae were found only in deep (>3500 m) water samples. Kujawinski et al. (2009) used statistical analysis of DOM collected in the surface, deep, and terrestrial-influenced coastal waters of the US Mid Atlantic Bight, also finding only a very small number of indicator species among >1000 formulae that could be attributed to exclusively surface marine (32 formulae) or exclusively terrestrial impacted (20 formulae) DOM. Finally, Flerus et al. (2012) found 54% of all masses were present in 90% of 137 samples collected in the eastern Atlantic Ocean, while 74% of mass were present in at least 100 samples. The uniformity in DOM molecular formulae was attributed to refractory DOM with common compositional features that persists over several millennia and is well mixed throughout the entire water column.

ESI-MS of uncharacterized, complex mixtures like SPE-DOM is inherently qualitative. Ionization is selective and subject to matrix effects that are not fully understood. Positive and negative modes yield highly overlapping but different sets of ions due to different ionization efficiencies for carboxylic acids and other functional groups. Only molecules that ionize are registered by the mass detector, and compound classes with low ionization efficiencies (e.g., carbohydrates) are therefore under-represented or can be absent from the mass spectrum (Hertkorn et al., 2006, 2012). However, with careful control of sample processing and analysis, recent studies are beginning to compare the intensity-weighted distribution of molecular formula within sample sets to identify spatial and temporal changes in SPE-DOM composition. Using this approach, Hertkorn et al. (2012) noted an increase in oxygen and a decrease in carbon content between surface (5m; 36% O; 50-52% C) and deep water (5446m; 42% O; 47% C) samples, which was attributed to progressive oxidation of DOM with age/depth, in agreement with NMR data that also shows a progressive increase in relative COOH and C=O % carbon with depth. However, Flerus et al. (2012) did not observe a shift in O/C ratios between surface and deep water samples collected on the same cruise. Hertkorn et al. (2012) noted that the abundance of formulae with CHO and CHNO increased with depth relative to formulae containing sulfur (CHOS and CHNOS), in contrast to Kujawinski et al. (2009) who reported an increase in both the number and intensity-weighted number of sulfur containing compounds between the surface (0m) and deep (1000m) Atlantic Ocean. Flerus et al.

(2012) used the intensity-weighted approach to distinguish masses that were relatively more abundant in surface waters from masses that were uniformly present (when corrected for total DOC) throughout the water column. They found that masses enriched in surface waters had a lower average mass (300 vs. 441 Da), and a lower range in DBE (2-11 vs. 7-14), again supported by NMR data suggesting a higher average degree of branching in deep water DOM (Flerus et al., 2012). The intensity-averaged molecular weight increased from 411 Da in surface waters to 417 Da in deep waters, suggesting at best a very small increase in molecular weight with water mass age (Flerus et al., 2012; Hertkorn et al., 2012).

IV LINKS BETWEEN DOM COMPOSITION AND CYCLING

The current paradigm of marine DOM cycling draws from a synthesis of rate measurements that span timescales from a few hours to several thousand years (Hansell, 2013; Chapter 3). On very short timescales, marine microbes produce and consume DOM that is "labile" or "reactive." Annually, a large flux of carbon passes through labile DOC, but at any given moment labile DOC represents only a small fraction (<0.2Pg) of the global DOC inventory. Over longer timescales of months to years, excess microbial carbon production and inputs of terrestrial carbon from atmospheric deposition, rivers, and groundwater leads to net accumulation of "semi-labile" or "semi-reactive" DOM in and immediately below the euphotic zone. Net accumulation of semi-labile DOM in the upper water column, and net removal in the mesopelagic ocean, give DOC profiles in temperate and tropical latitudes their characteristic shape, with high values in the surface and lower values at depth. Globally, semi-labile DOM contributes ~20GT C, or 3% of the marine DOC reservoir (Hansell et al., 2009, 2012). The annual flux of carbon through semi-labile DOM cannot be measured directly, and estimates of residence time span at least an order of magnitude, from a few months (the seasonal accumulation and export of semi-labile DOM from the euphotic zone) to several years (radiocarbon measurements of semi-labile DOM carbohydrates) (Hansell et al., 2009, 2012; Repeta and Aluwihare, 2006). This broad range of rates reflects differences in local production and consumption, differences in how rates are measured, how semi-labile DOM is defined, and the fraction of semi-labile DOM that is tracked. Lability changes with nutrient conditions, temperature, light, and consumer community structure.

In the deep ocean, DOC values are more stable and decrease only slowly along the path of abyssal circulation. (Hansell et al., 2012) Radiocarbon measurements show deep-sea DOC to be several thousand years old (Druffel et al., 1992; Williams and Druffel, 1987). Assuming production and removal of deep-sea DOM is a first order process, the turnover time of refractory DOC is ~16,000-30,000 years. The old radiocarbon age and the very slow net removal of deep-sea DOC imply that most marine DOM is "nonreactive" or "refractory." The distinction between labile, semi-labile, and refractory DOM is purely operational. Studies define these terms differently, depending on the nature of the study and the preferences of the investigators. Other nomenclatures exist in the literature which make finer classifications of DOM reactivity (Hansell, 2013; Hansell et al., 2012). However DOM lability is defined, it is inferred from the broad range of timescales over which DOM cycles that composition and lability are linked in some way. One goal of DOM chemical characterization is to identify characteristic features of labile, semi-labile, and refractory DOM, and understand how composition is linked to cycling.

A Composition and the Cycling of Labile DOM

Grazing, viral lysis of infected cells, and a host of routine cellular physiological processes act to release labile DOM into seawater (Chapter 3). Labile DOM is also produced from the photochemical oxidation of refractory organic matter as it upwells from the deep ocean into the euphotic zone (Kieber et al., 1990; Mopper et al., 1991). Photochemical degradation products include LMW organic acids, aldehydes, and ketones that are readily assimilated by marine microbes (Mopper and Stahovec, 1986). The annual carbon flux through labile DOM is measured as heterotrophic bacterial production, under the assumptions of steady state production and consumption and that microheterotrophs are the dominant sink for labile DOM. Measurements of bacterial production are imprecise and do not include organic matter consumed through light driven consumption by photoautotrophs and photochemical oxidation to CO_2, or losses due to the adsorption of labile DOM onto sinking particles. Current estimates of bacterial production show that annually, some ~20% of global primary production is released as labile DOM that is subsequently consumed through a "microbial loop" in which DOM is either respired or fixed again into microbial biomass (Figure 2.14). While most studies have focused on the consumption of DOM by heterotrophic bacteria and more recently archea, there is abundant evidence that photoautotrophs also assimilate simple organic acids and other LMW organic compounds. Some photoautrophs are auxotrophic for essential vitamins and some use siderophores produced by heterotophic bacteria to supplement their requirements for iron (Helliwell et al., 2011; Vraspir and Butler, 2009). The exchange of labile DOM between auto- and heterotrophic microbes therefore limits and shapes microbial production and marine biogeochemical cycling in a very fundamental way.

Labile DOM constituents include easily characterized, LMW biochemicals (simple amino acids, sugars, organic acids, ATP, vitamins, etc.) that are readily assimilated by marine microbes and simple biopolymers (proteins, unbranched homopolysaccharides, etc.) that can easily be

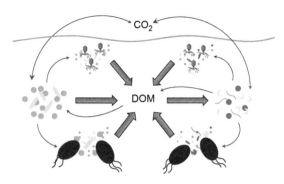

FIGURE 2.14 The microbial loop of labile DOM cycling. Carbon dioxide is fixed by photoautotrophs (yellow and green cells, left), which excrete about 10% of total fixed carbon as DOC. Grazing (bottom left, right) and cell lysis after viral infection (upper left, right) release additional labile DOM to be consumed by heterotrophic bacteria and archea (blue cells, right). The microbial loop consumes labile DOM, converting it to carbon dioxide and nutrients. Heterotrophs also release essential vitamins, siderophores, and other organic compounds that facilitate the growth of photoautotrophs. *Figure courtesy of Dr. Jamie Becker.*

hydrolyzed by extracellular hydrolytic enzymes. The high demand for labile organic substrates, nutrients, and metals by the microbial community keeps steady state concentrations of labile DOM constituents at nanomolar levels (Fuhrman, 1987; Kaiser and Benner, 2009; Skoog and Benner, 1997). As discussed in Section III.B, concentrations of dissolved free amino acids and THAA range from a few nanometer (for free amino acids) in the deep ocean, to 100s of nanometer (for THAA) in the surface open ocean. Radiocarbon tracer experiments show free amino acids are rapidly metabolized by heterotrophic bacteria, and the uptake and incorporation of the tritiated amino acid leucine is commonly used as a measure of bacterial production (Fuhrman, 1987). The lability of specific proteins and bacterial peptides that constitute THAA is unknown, but spatial and temporal changes in THAA concentrations along with experiments tracking extracellular peptidase activity by marine phytoplankton and bacteria show a significant fraction

of THAA are metabolized over short time scales (Amon et al., 2001). Dissolved simple sugars likewise occur at low nanometer concentrations in the euphotic zone and are taken up quickly by bacteria (Rich et al., 1996). Total hydrolyzable neutral and amino sugars concentrations range from >600 nM in open-ocean surface waters to 30-60 nM at depth. It is unclear what fraction of DOM polysaccharides are labile. Extracellular, dissolved polysaccharides are universally produced by marine algae and cyanobacteria when grown in laboratory pure culture (Aluwihare and Repeta, 1999; Aluwihare et al., 1997; Biddanda and Benner, 1997; Meon and Kirchman, 2001). A number of studies have explored carbohydrate degradation by inoculating filtered, spent culture media with heterotrophic bacteria and following simple sugar composition or DOM spectral characteristics over time. In these experiments homopolysaccharides are rapidly degraded by heterotrophic bacteria, but degradation is always selective; the composition of neutral sugars invariably changes over the course of the degradation experiments (Aluwihare et al., 1997; Meon and Kirchman, 2001). In the field, experiments using fluorescently labeled model compounds also show large changes in polysaccharide hydrolysis rates and substrate selectivity with location (Arnosti, 2011; Arnosti et al., 2011). Labile polysaccharides are readily produced and consumed during upper ocean carbon cycling, but steady state concentrations remain low as with other labile components of DOM due to tight coupling between production and consumption by marine microbes.

Although proteins and polysaccharides contribute the majority of carbon and nitrogen in living microbial biomass and marine particulate organic matter, microbes produce an enormous diversity of organic compounds as part to their metabolic systems. Untargeted characterization of DOM shows hundreds to thousands of different compounds can be recovered by SPE from pure cultures of marine photoautotrophs. Initial results suggest chemical diversity may track phylogenic diversity, such that the genomic variability in marine microbial communities may be matched by a similar chemical variability in labile DOM composition (Becker et al., 2014). Some labile DOM constituents may be generally available to the heterotrophic community, but others induce highly specific growth responses. For example, studies using defined radiolabeled substrates or labeled DOM produced in pure cultures of specific marine photoautotrophs show selective uptake by defined subclasses of marine heterotrophs (Cottrell and Kirchman, 2000; Sarmento and Gasol, 2012). DOM released by marine heterotrophs likewise enhances the growth of only certain marine autotrophs. A recent review highlights the interaction of marine bacteria and diatoms, while other studies have documented enhanced growth of the marine cyanobacteria *Prochlorococcus* in the presence of some heterotrophic bacteria (Amin et al., 2012; Sher et al., 2011). *Prochlorococcus*-heterotophic mutualism is highly strain specific, some strains of bacteria enhance growth of only some strains of *Prochlorococcus*.

In the examples above, the DOM constituents that lead to enhanced growth have not been identified, but B vitamins and siderophores are two examples of labile DOM nutrients that are known to impact microbial community production and diversity. Approximately half of all marine microalgae, including many harmful algal bloom forming species, as well as many marine bacterioplankton, are auxotrophic for vitamins B_1 and/or B_{12} (Helliwell et al., 2011; Tang et al., 2010). B-vitamins co-limit and shape marine microbial production, so knowledge of the distribution of B vitamins is important to assessments of nutrient limitation (Bertrand et al., 2011, 2012; Tang et al., 2010). Early measurements of B vitamin distributions relied on bioassays; but recently, direct analysis of B vitamins by mass spectrometry has been reported. Concentrations of B_{12} reach low nanometer levels in some coastal waters, but fall to <1 pM (detection limit) at open-ocean sites due to low bacterial production and high

uptake by B-vitamin auxotrophs (Sañudo-Wilhelmy et al., 2012).

Over 99% of dissolved iron in the ocean is complexed to organic ligands of unknown composition (Boye et al., 2001; Gledhill and Buck, 2012; Gledhill and van den Berg, 1994; Rue and Bruland, 1997). In many remote areas of the equatorial Pacific, northwest Pacific, and Southern Oceans iron concentrations fall to <100 pM, creating enormous selective pressures for microbes to develop efficient Fe uptake and utilization strategies (Barton et al., 2010; Boyd and Ellwood, 2010; Bragg et al., 2010; De Baar et al., 2005; Dutkiewicz et al., 2009, 2012; Follows et al., 2007; Jickells et al., 2005; Martin and Fitzwater, 1988; Martin et al., 1991; Miethke, 2013). Microbes that can access iron-organic ligand complexes have a distinct competitive advantage in iron-depleted regions of the ocean. One mechanism by which microbes can acquire iron is through the siderophores, strong iron-organic complexes that facilitate iron transport and uptake across the cell membrane (Vraspir and Butler, 2009). Like B vitamins, dissolved iron-organic complexes occur at extremely low concentrations that are difficult to measure. However, recent reports describe a number of siderophore and siderophore-like compounds in seawater at pM concentrations, and advances in analytical technologies should facilitate future measurements (Boiteau et al., 2013; Gledhill and Buck, 2012; Mawji et al., 2008, 2011; Velasquez et al., 2011). The distribution and cycling of vitamins and trace metal organic complexes are two developing areas of research that target specific, labile DOM constituents with measurable impacts on marine microbial communities. Studies of untargeted labile DOM composition and microbial interactions suggest that as yet unidentified labile DOM constituents may also influence an array of metabolic processes from cell-cell signaling (e.g., quorum sensing) to nutrient uptake. Understanding labile DOM composition is key to understanding many interactions between microbial communities, as well as many of the processes that limit and shape the microbial community production and diversity.

B Composition and the Cycling of Semi-Labile DOM

Traditionally, semi-labile DOM has been defined as the fraction of DOC that is present in surface waters but that is not at depths >1000m, the point where DOC concentration values begin to stabilize. This definition arises from two principal observations; (1) DOC concentrations in temperate and tropical surface waters are elevated relative to deep ocean values, and (2) radiocarbon models of DOC age use a mix of modern and aged components to explain upper ocean Δ^{14}C-DOC values (Druffel et al., 1992; Hansell et al., 2009, 2012). However, semi-labile DOM can also be defined on the basis of turnover time relative to labile and refractory DOM fractions. Viewed this way, semi-labile DOM is any organic matter that cycles on seasonal to decadal timescales. This definition includes DOM in the deep ocean (>1000m) that is in quasi-steady state with its production (via dissolution of sinking particles, in situ microbial production, etc.) and removal (Orellana and Hansell, 2013; Repeta and Aluwihare, 2006). The timescales over which semi-labile DOM cycles are too long to be captured experimentally in laboratory or field incubation experiments without significant bottle effects, but too short to be captured by natural abundance radiocarbon measurements. Linking semi-labile DOM constituents to cycling pathways is therefore difficult, and the processes that result in the net accumulation of semi-labile DOM in the surface ocean, and net degradation at depth, remain largely unexplained.

Marine algae and bacteria synthesize semi-labile DOM polysaccharides and proteins, which in laboratory degradation experiments can persist for several years (Fry et al., 1996; Jiao et al., 2010; Ogawa et al., 2001). HMWDOM carbohydrate amendments to seawater simulate rapid increases in bacterial carbon utilization by a succession of bacterial taxa, suggesting resource partitioning of semi-labile DOM between different species of bacteria (McCarren et al., 2010). Coordinated, cooperative interactions by a variety of different bacteria may be necessary to degrade the complex, highly branched APS that represents a large fraction of semi-labile DOM. The carbon respired by short-term incubation experiments of this type represents only a very small fraction (typically <1%) of the carbon amendment, and the link between short-term incubation results and semi-labile DOM degradation in situ is not clear. Longer incubations of nutrient amended surface waters show greater losses of DOC, but still fall far short of consuming a major fraction of the semi-labile DOC reservoir. In a series of experiments, Carlson and colleagues measured DOC respiration in a matrix of surface and 250m water treatments that had been 0.2µm filtered, amended with nutrients, and inoculated with either unfiltered (containing bacteria) surface or 250m water (Carlson et al., 2002, 2004). No significant DOC respiration and only a slight increase in bacteria cell numbers were observed in surface water inoculated with surface water bacteria. However, significant respiration (up to 10% of total DOC) and 3x increases in bacterial cell numbers were measured in treatments of surface water with 250m bacteria. The studies of Carlson and colleagues show that the bacterial community in surface waters has a limited capability to degrade polysaccharides and other components that contribute to semi-labile DOM, while mesopelagic bacterial communities have the requisite metabolic pathways to metabolize at least a fraction (up to 20%) of semi-labile DOM. The results of microbial degradation experiments have yet to be coupled to studies of semi-labile DOM composition. Characterizing the small amounts of semi-labile DOM that is consumed in these incubation will prove to be challenging, and developing improved experimental approaches to simulating semi-labile DOM degradation will be key to future progress in this area.

Metabolism of DOM by heterotrophic bacteria is selective and/or leads to subtle transformations of semi-labile DOM. A growing number of studies have measured differences in DOM composition between coastal and open-ocean DOM or between surface and deep water DOM and tied these differences to the "apparent diagenetic state" or potential reactivity of semilabile DOM. Goldberg and colleagues measured seasonal changes in DOC, total carbohydrates (TCHO), and dissolved combined neutral sugars (DCNS) as DOM accumulates in surface waters at the Bermuda Atlantic Time-series Study site (BATS; 31°40′N, 64°10′W) and spatially in surface waters across the North Atlantic subtropical gyre between 7 and 43°N (Goldberg et al., 2009, 2010). Values for DOC, THCO, and DCNS were highest during summertime stratification of the water column and in the mid-gyre (~24-27°N). Recently accumulated DOM had higher carbohydrate yields (%TCHO) and higher mol% galactose and mannose+xylose than DOM at depth, which was characterized by higher mol % glucose. As semi-labile DOM is processed by bacteria, it becomes less amenable to chemical characterization and the distribution of DCNS changes (Amon and Benner, 2003; Goldberg et al., 2009, 2010).

Patterns of amino acid and hexosamine distributions in particulate organic matter and sediments are also strongly imprinted with the combined effects of degradation and transformation. By comparing the distribution of amino acids in surface and subsurface sediments, and in fresh and recycled particulate organic matter, Dauwe proposed a degradation index that relates the composition of organic matter to lability (Dauwe et al., 1999). The degradation index assumes steady state deposition over time, so that the distribution of amino acids/hexosamines in contemporary surface sediments or fresh particulate matter is a good representation of the amino acid distribution of older sediments or more processed particulate organic matter at the time of synthesis. This concept has been applied

to studies of DOM cycling to provide a measure of DOM degradation. For semi-labile DOM, it is assumed that the distribution of amino acids/hexosamines in coastal surface waters is indicative of the initial product, which is transformed as waters move offshore or into the deep sea. Yamashita and Tanoue (2003a,b) observed decreases in the %THAA and increases in the relative amount of glycine and alanine in a suite of samples from near shore (fresh DOM) to offshore and greater depths (more recycled DOM), concluding that degradation had a major impact on the quantity and quality of THAA (Yamashita and Tanoue, 2003a). Likewise, Davis et al. (2009) measured the %THAA and degradation index in a suite of (organic matter) amended and unamended Arctic Ocean seawater samples, finding that %THAA changed rapidly in the days immediately following amendment, but stabilized and changed only slightly thereafter. These changes were accompanied by minor changes in the amino acid degradation index, which varied in the upper 200m of the water column but was invariant below this depth. These and other studies show that freshly produced organic matter has a relatively higher %THAA, which decreases on timescales of weeks to months, while the ratio of indicator amino acids can be used to track changes occurring over timescales of months to years.

Hubberten et al. (1994, 1995) used hydrophobic organic resins (XAD-2; a cross-linked polystyrene) to extract amino acids from the Greenland Sea and Southern Ocean. Seawater was filtered (Whatman GF/F), acidified to pH=2 with hydrochloric acid, then passed through the XAD column (Hubberten et al., 1994, 1995). Amino acids incorporated into hydrophobic substances are adsorbed onto the column, then recovered by sequentially rinsing with 0.2N NaOH and methanol. Each fraction was analyzed for amino acids. In all samples from <1000m, TDAA concentrations always exceeded amino acids in the XAD fractions (XAD-AA; Figure 2.7). However, the difference, the amount by which [TDAA] exceeded [XAD-AA], was greatest for surface waters and diminished with depth. Hubberten noted a linear correlation between chlorophyll-*a* [Chl-*a*] and THAA in surface waters. Extrapolation to [Chl-*a*]=0 yielded a positive intercept for [THAA] that was similar to values of XAD-AA. For depths below 100m, concentrations of THAA and XAD amino acids were approximately equal. Both observations led Hubberten et al. to suggest that there are at least two fractions of hydrolysable amino acids in seawater, a labile fraction that is largely hydrophilic and present in surface waters and a semi-labile or refractory component of about 150-250nM present throughout the water column. The distribution of amino acids in the XAD fraction was characterized and found to be similar to THAA, so no distinction of the two fractions could be made by amino acid distribution. No further characterization of XAD-AA was made; however proteins, peptides, and bacterial cell-wall material have some hydrophobic character, and can be retained on hydrophobic resins under low pH conditions. The amino acids recovered by Hubberten and colleagues could have included amino acids bound in HS, proteins/cell-wall biopolymers, and their degradation products. The decrease in the ratio THAA/XAD-AA with depth suggests some change in the macromolecular form of amino acids, a point corroborated by amino acid fluorescence data, which shows only tryptophan-like fluorophores in surface waters, but additional tyrosine-like fluorophores in deep water samples (Yamashita and Tanoue, 2003b, 2004).

Finally, DOM in filtered seawater spontaneously assembles into polymer gels, stable three-dimensional networks of DOM macromolecules that grow until they reach a size and density that allows them to sink or adhere to sinking particles; Chapter 9). Assembly is rapid and reversible, and gels are one mechanism by which semi-labile DOM can be removed from the surface ocean. As discussed in Section III.C, RuBisCo is produced in the euphotic zone, transported to the deep ocean by sinking particles, and released

by DOM-POM exchange. In regions of high particle flux such as the equatorial Pacific, RuBisCo can be detected well into the deep ocean where it persists as semi-labile DOM. Gels stain positively for carbohydrates, proteins, and lipids and these other constituents of polymer gels are likely carried into the deep ocean as well. Peptidases, proteases, and ATPase enzymes have been identified in polymer gels through proteomic analyses (Orellana et al., 2007). No similar analyses of polymer gel carbohydrates have been reported, but bubble formation and collapse concentrates polymer gels into sea surface foam, and chemical analyses of natural sea foams have been made (Orellana et al., 2011). NMR spectra of natural sea surface foam and foam produced after the microbial degradation of spent algal culture media have spectral features characteristic of APS that accumulate in surface water as semi-labile DOM carbohydrate. Combined neutral sugar distributions of semi-labile DOM and foam are also similar (Gogou and Repeta, 2010). ^{1}H NMR and ^{13}C NMR spectra of DOM collected by ultra-filtration, SPE, and ED/RO all show resonances assigned to semi-labile DOM carbohydrate as deep as 5200m, but further chemical and isotopic characterization is needed to link surface and deep reservoirs of semi-labile DOM (Benner et al., 1992; Helms et al., 2013; Quan and Repeta, 2007; Sannigrahi et al., 2005).

C Composition and the Cycling of Refractory DOM

By definition, refractory DOM does not degrade via the typical microbial and chemical processes that recycle labile and semi-labile DOM. Understanding why refractory DOM is so unique is therefore critical to a comprehensive description of the marine carbon cycle. The chemical composition of SPE-DOM is taken to be representative of refractory DOM. As discussed in Section III.C, SPE-DOM is a very complex mixture of at least several thousand different components with elemental

H/C and O/C ratios that lie outside the range of common lipids, proteins, and carbohydrates (Figure 2.12). SPE-DOM is therefore not thought to have a direct biological source, but the case for direct biological production and selective preservation of refractory organic matter has been made for sediments, and given the very small flux of carbon ($<0.04\,\mathrm{Pg\,year^{-1}}$) needed to support the radiocarbon age of refractory DOM, and the large annual flux of carbon through marine primary and secondary production, direct synthesis of refractory DOM cannot be discounted. The complex molecular features of SPE-DOM can be interpreted as the result of transformation of lipids, proteins, and carbohydrates that become scrambled in such a way as to make them difficult for marine microbes to assimilate or metabolize. Refractory DOM is characterized by a high proportion of carboxylate and fused alicyclic ring carbon that shares structural characteristics with polycyclic lipids (Hertkorn et al., 2006). The mechanisms that could convert simple lipids to refractory DOM are a matter of conjecture and more work is needed to assess both the feasibility of the formation pathway and its impact on microbial metabolism.

Although refractory DOM is not readily metabolized by marine microbes, a number of sinks that rely on physical/chemical processes have been identified. In a series of elegant experiments Mopper and coworkers detailed the photochemical oxidation of deep-sea colored DOM on exposure to light (Kieber et al., 1990; Mopper and Stahovec, 1986; Mopper et al., 1991). Photooxidation products include LMW, labile, organic acids, aldehydes, and ketones that are rapidly consumed by bacterial heterotrophs. Photochemical oxidation also decreases the adsorptive and fluorescent properties of DOM through the transformation or removal of unsaturated functional groups (Kieber et al., 1990; Weishaar et al., 2003). Subsequent studies have confirmed and expanded these observations and a photochemical sink and

transformation pathway for otherwise refractory DOM is now firmly established. However, photochemistry is confined to the upper portion of the euphotic zone and the amount of carbon potentially mineralized by this pathway is very limited. Photochemical oxidation of DOM probably leads to significant isotopic fractionation between products, but the isotopic value of DOM is similar to particulate organic matter, and it is unlikely that photochemical oxidation is the primary sink for refractory DOM.

Globally, the largest sinks for refractory DOM are in the deep sea (Hansell, 2013; Hansell and Carlson, 2013; Hansell et al., 2009, 2012). From an analysis of salinity and high precision DOC values in the deep Pacific Ocean, Hansell and Carlson (2013) were able to identify two regions of refractory DOC removal within the basin, a deep sink in the far North Pacific, and a mid depth sink in the tropical South Pacific. They estimated that these two sinks remove 7-29% of the 43 TG refractory DOC introduced into the deep global ocean by overturning circulation. Two processes, removal during hydrothermal circulation and adsorption onto sinking particles, have been identified as sinks for deep-sea refractory DOM. Hydrothermal fluids exiting permeable ocean crust at mid-ocean ridge crests and unsedimented/thinly sedimented ridge flanks have DOC concentrations that are lower by $20\text{-}25\,\mu M$ than deep seawater ($36\text{-}39\,\mu M$; typical vent fluid DOC concentrations are $11\text{-}19\,\mu M$; Lang et al., 2006; Lin et al., 2012). Global DOC losses in hydrothermal systems are low, $<0.0002\,\mathrm{PgC\,year^{-1}}$, but measurements are few and this may be an underestimate. The composition and lability of the $\sim\!20\,\mu M$ DOC that exits with hydrothermal fluids has not been studied, but radiocarbon values of the HMWDOC fraction range from $-772\%_0$ to $-835\%_0$, significantly depleted relative to overlying seawater (McCarthy et al., 2010). Other processes such as microbial oxidation and removal by self-assembling organic gels may be active and need further study.

Radiocarbon values of suspended particulate matter decrease by almost 100‰ with depth in the ocean (Druffel et al., 1992, 1998). Even at the low Stokes settling velocities calculated for μm sized particles (~1 m day^{-1}), suspended particulate matter settles quickly enough that no appreciable gradient in radiocarbon values due to aging should be measureable. To explain the gradient in suspended POC radiocarbon values, Druffel and colleagues postulated that radiocarbon depleted refractory DOM is adsorbed onto suspended particles. Adsorption forms the basis for SPE extraction of refractory DOM and it would be surprising if a similar process does not occur in the water column that is permeated with mineral surfaces and organic particles. A number of factors including complex POM dynamics in the deep ocean, resuspension and advection of fine-grained sediments with depleted radiocarbon values from continental margins, and adsorption of refractory DOM onto filters used to collect particles, potentially influence the DOM-POM adsorption model. However, based on Druffel's data, Hansell calculated that adsorption could remove ~0.05 PgC year^{-1} from the deep ocean, enough to sustain the observed deep-sea gradient in DOC concentration (Hansell et al., 2009). No analyses of particle-adsorbed, refractory DOM have been attempted. Given the recent progress in SPE-DOM characterization, and the ~500‰ separation in radiocarbon values for newly synthesized POM (+50‰) and refractory DOM (−400‰), it might be feasible to test the adsorption/removal hypothesis through a combination of structural and isotopic analyses.

V FUTURE RESEARCH

Over the past decade, our understanding of DOM composition and cycling has advanced in a number of areas. Reverse osmosis/electrodialysis now provides a larger fraction of DOM for study than either ultrafiltration or SPE alone. High-field, multidimensional NMR and high-resolution MS are providing unprecedented details of DOM composition, while new techniques of data integration and visualization are allowing marine chemists to couple different spectral datasets into a more integrated picture of DOM composition. This wealth of new information presents both opportunities and challenges to the DOM community. One challenge is to provide community-wide access to the very large amount of mass and NMR spectral data available through open-access databases. Presently, such data are often only summarized or provided as supplemental materials in published reports. A database of spectral information that includes details of the methodologies needs to be made available for interrogation by the wider scientific community if the full benefits of advanced methods of characterization are to be realized.

Much more effort needs to be placed on the validation of results by bringing different analyses to bear on common samples or at common study sites. It is unlikely and perhaps undesirable for sampling methods to converge at this time on one technique that provides the highest recovery of DOM. Each method has particular benefits of cost, speed, throughput, and selectivity for different fractions of DOM. A better understanding of the sampling overlap between techniques would however help to interpret similarities and differences in composition between SPE, ultrafiltration, and ED/RO samples. Most current studies focus on one or two methods of spectral characterization that are within the scope of expertise of particular analysts. Collaborative studies that integrate data from a suite of different spectral (MS, NMR, IR, etc.), chromatographic (HPLC, GC, etc.), elemental, and isotopic (δ^{13}C, Δ^{14}C, etc.) analyses would allow for more robust interpretation of complex datasets. For specific components of DOM that have biomarker or tracer potential (thermogenic DOM, APS), more comprehensive isotopic, and structural characterization, along with better methods for isolation and purification, would substantially improve existing models of DOM cycling.

Advances in technology have made in-depth studies of labile DOM composition and cycling feasible. Automated, high-resolution multidimensional liquid and gas chromatographic systems coupled with high-resolution MS and high-throughput microcapillary NMR offer the potential to better characterize the exometabolome of labile DOM. A large number of model photoautotrophs and heterotrophs representing major ecotypes of marine cyanobacteria, algae, and bacterioplankton are now available as pure cultures with sequenced genomes. Labile DOM characterization of these cultures at different growth stages would provide valuable links between genomic and exometabolomic composition, as well as microbial dynamics in natural systems. Paired cocultures studies further demonstrate that synergistic interactions between photoautotrophs and heterotrophs are probably common in the ocean. Little is known about the organic nutrients and substrates that stimulate these associations and this is a fruitful area for further study, particularly when paired with transcriptomic and proteomic datasets. There are only a few reports of the distribution of vitamins and other bioessential organic compounds (siderophores, quorum sensing compounds, etc.) in seawater. These compounds have a demonstrable impact on microbial ecosystems, but their production and uptake are only poorly understood. Finally, labile DOM composition needs to be linked to genomic studies already underway at time-series study sites and in global surveys of microbial populations and water column chemical properties. Labile DOM composition might provide an important new input to global models of microbial metabolism, diversity, and community structure.

^{13}C NMR spectra of DOM collected by RO/ED provides the most comprehensive view of semi-labile and refractory DOM composition currently available (Figure 2.15). These spectra show the presence of two major fractions, a

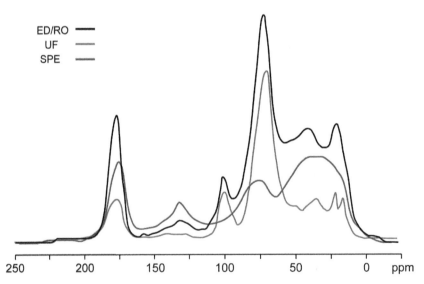

FIGURE 2.15 ^{13}C NMR of DOM recovered by RO/ED (black trace) superimposed on spectra of HMWDOM (ultrafiltration; gray trace) and SPE-DOM (red trace) from surface water. The spectra of HMWDOM and SPE-DOM are scaled to the RO/ED peak at 70 ppm. To a good approximation, the distribution of carbon functional groups in the RO/ED sample is a mixture of acylated polysaccharides (APS), which dominate the ^{13}C NMR spectra of HMWDOM, and hydrophobic humic substances/CRAM that dominate the ^{13}C NMR spectra of SPE-DOM. *Figure redrawn from Koprivnjak et al., 2009.*

semi-labile HMWDOM fraction with APS as the major component, and a refractory fraction with carboxy-rich aliphatic matter (CRAM) as the major component. The NMR spectral, elemental (C/N/P), and partial chemical (monosaccharide composition, linkage pattern) characteristics of APS are known. This level of detail has allowed for the detection of APS throughout the water column, but has not allowed for the identification of major sources of APS, or provided either an explanation for why APS accumulates in the upper ocean or a better estimate of carbon flux through semi-labile HMWDOM. The major sinks for APS have also not been identified. Better characterization of APS using MS and 2D NMR is an essential first step in addressing these questions, as are improved chemical techniques that can fully depolymerize APS for comprehensive monosaccharide and linkage analyses. A more detailed description of APS composition should allow for better distinction of APS sources in laboratory culture experiments which would help in the design and implementation of microbial degradation experiments, and experiments monitoring gel polymer formation as putative sinks for APS. Finally, there is evidence from the study of proteins in DOM for the transport of semi-labile DOM into the deep ocean. Determining the inventory of semi-labile DOM in the deep ocean, its radiocarbon value, and its sinks would provide important new insights into deep-sea carbon cycling and perhaps bathypelagic microbial ecology.

Understanding refractory DOM may be the most formidable challenge for DOM composition and cycling. High-resolution mass and high-field NMR spectral analyses have yielded unprecedented details into refractory DOM composition, but the results need to be independently verified by other techniques and made more quantitative. For example, the current view of SPE-DOM composition is of a very complex mixture of LMW carboxyl-rich aliphatic compounds. If correct, these compounds should

be amenable to separation and characterization by multidimensional gas chromatography-MS (GC-MS). 2D GC-MS has already proved to be a valuable tool in the characterization of petroleum, which is a mixture polycyclic alkanes and alkenes, some of which incorporate nitrogen and sulfur, similar in many respects to the proposed composition of refractory DOM.

Better integration of carbon isotope analyses with structural characterization is also needed. The isotopic composition of thermogenic DOM should provide important insights into the origin and cycling of polycyclic aromatic compounds in DOM. Only one such set of measurements have been made, and more are needed (Ziolkowski and Druffel, 2010). Finally, there are only a few reports of deep-sea DOM spectral characteristics. Given the large gradient in concentration and radiocarbon value through the deep ocean, it seems likely that aging and removal of DOM will impact DOM composition. A survey of DOM composition through the deep ocean may offer insights into the processes that cycle refractory DOM.

There is observational evidence for two deep ocean sinks of refractory DOM, removal during hydrothermal circulation of seawater through permeable crust and adsorption onto particles. Both sinks merit further study. The composition of DOM in vent fluids needs to be determined and compared with the isotopic and chemical composition of refractory DOM. Preliminary work suggests that hydrothermal circulation is not a large source or sink of refractory DOM, but further measurements are needed to verify this inference. Adsorption onto sinking particles has been proposed but not verified through experiments using molecular level analysis. Ultrafiltration and ED/RO or SPE allow for the concentration of deep-sea POM and DOM that could be used in to experimentally track adsorption. Given the high level of detail that can now be achieved using advanced analytical methods, it may also be feasible to isolate compounds with chemical and isotopic characteristics of refractory DOM from suspended particulate organic matter. Such studies would help to verify the adsorption hypothesis.

Finally, little is known about the role of deepsea microbes in DOM removal. Rates of microbial DOM cycling are thought to be very low and therefore difficult to simulate in laboratory experiments. However, using more sensitive and precise methods of DOM characterization that are now available, it may be possible to explore deep-sea microbial/refractory DOM interactions to provide new insights into this potentially important aspect of the marine carbon cycle.

Acknowledgments

The author has had the good luck to work with a large number of talented scientists over many years whose enthusiasm, ideas, and laboratory work have contributed to this chapter. In particular, this chapter has benefited from discussions with Lihini Aluwihare, Carl Johnson, Christos Panagiotopolous, Mar Nieto Cid, Aleka Gogou, Chiara Santinelli, Robert Chen, Jamie Becker, Rene Boiteau, and Chris Follett, and Tracy Quan. Hussain Abdulla, Eiichiro Tanoue provided many helpful comments and suggestions that significantly improved the manuscript. I would also like to thank the NSF Chemical Oceanography and National Science and Technology Center for Microbial Oceanography Research and Education (C-MORE) programs for their generous support over many years, and the Gordon and Betty Moore Foundation for their interest and support of microbial cycling and DOM through grants 1711 and 3298.

References

Abdulla, H.A.N., Minor, E.C., Hatcher, P.G., 2010a. Using two-dimensional correlations of ^{13}C NMR and FTIR to investigate changes in the chemical composition of dissolved organic matter along an estuarine transect. Environ. Sci. Technol. 44, 8044–8049.

Abdulla, H.A.N., Minor, E.C., Dias, R.F., Hatcher, P.G., 2010b. Changes in the compound classes of dissolved organic matter along an estuarine transect: a study using FTIR and ^{13}C NMR. Geochim. Cosmochim. Acta 74, 3815–3838.

Abdulla, H.A.N., Sleighter, R.L., Hatcher, P.G., 2013. Two dimensional correlation analysis of fourier transform ion cyclotron resonance mass spectra of dissolved organic matter: a new graphical analysis of trends. Anal. Chem. 85, 3895–3902.

Aluwihare, L.I., Meador, T., 2008. Chemical composition of marine dissolved organic nitrogen. In: Capone, D.G., Bronk, D.A., Mulholland, M., Carpenter, E. (Eds.), Nitrogen in the Marine Environment, Kluwer Press, Dordrecht.

Aluwihare, L.I., Repeta, D.J., 1999. A comparison of the chemical characteristics of oceanic and extracellualr DOM produced by marine algae. Mar. Ecol. Prog. Ser. 186, 105–117.

Aluwihare, L., Repeta, D., Chen, R., 1997. A major biopolymeric component to dissolved organic carbon in surface sea water. Nature 387, 166–169.

Aluwihare, L., Repeta, D., Chen, R., 2002. Chemical composition and cycling of dissolved organic matter in the Mid-Atlantic Bight. Deep Sea Res. Part II Top. Stud. Oceanogr. 49, 4421–4437.

Aluwihare, L.I., Repeta, D.J., Pantoja, S., Johnson, C.G., 2005. Two chemically distinct pools of organic nitrogen accumulate in the ocean. Science 308, 1007–1010.

Amin, S.A., Parker, M.S., Armbrust, E.V., 2012. Interactions between diatoms and bacteria. Microbiol. Mol. Biol. Rev. 76, 667–684.

Amon, R.M.W., Benner, R., 2003. Combined neutral sugars as indicators of the diagenetic state of dissolved organic matter in the Arctic Ocean. Deep Sea Res. Part I Oceanogr. Res. Pap. 50, 151–169.

Amon, R.M.W., Fitznar, H., Benner, R., 2001. Linkages among the bioreactivity, chemical composition, and diagenetic state of marine dissolved organic matter. Limnol. Oceanogr. 46, 287–297.

Arnosti, C., 2011. Microbial extracellular enzymes and the marine carbon cycle. Ann. Rev. Mar. Sci. 3, 401–425.

Arnosti, C., Steen, A.D., Ziervogel, K., Ghobrial, S., Jeffrey, W.H., 2011. Latitudinal gradients in degradation of marine dissolved organic carbon. PLoS One 6, e28900.

Barton, A.D., Dutkiewicz, S., Flierl, G., Bragg, J., Follows, M.J., 2010. Patterns of diversity in marine phytoplankton. Science 327, 1509–1511.

Bauer, J.E., Bianchi, T.S., 2011. Dissolved organic carbon cycling and transformation. In: Wolanski, E., McLusky, D.S. (Eds.), Treatise on Estuarine and Coastal Science. Academic Press, New York.

Becker, J.W., Berube, P.M., Follett, C.L., Waterbury, J.B., Chisholm, S.W., DeLong, E.F., et al., 2014. Closely related phytoplankton species produce similar suites of dissolved organic matter. Front. Microbiol. 5:111. doi:103398/fmicb.2014.00111.

Benner, R., 2002. Chemical composition and reactivity. In: Hansell, D.A., Carlson, C.A. (Eds.), Biogeochemistry of marine dissolved organic matter, Elsevier.

Benner, R., Kaiser, K., 2003. Abundance of amino sugars and peptidoglycan in marine particulate and dissolved organic matter. Limnol. Oceanogr. 48, 118–128.

Benner, R., Pakulski, J.D., McCarthy, M., Hedges, J.I., Hatcher, P.G., 1992. Bulk chemical characteristics of dissolved organic matter in the ocean. Science 255, 1561–1564.

Bertrand, E.M., Saito, M.A., Lee, P.A., Dunbar, R.B., Sedwick, P.N., Ditullio, G.R., 2011. Iron limitation of a springtime bacterial and phytoplankton community in the ross sea: implications for vitamin B_{12} nutrition. Front. Microbiol. 2, 160.

Bertrand, E.M., Allen, A.E., Dupont, C.L., Norden-Krichmar, T.M., Bai, J., Valas, R.E., 2012. Influence of cobalamin scarcity on diatom molecular physiology and identification of a cobalamin acquisition protein. Proc. Natl. Acad. Sci. U. S. A. 109, E1762–E1771.

Biddanda, B., Benner, R., 1997. Carbon, nitrogen, and carbohydrate fluxes during the production of particulate and dissolved organic matter by marine phytoplankton. Limnol. Oceanogr. 42, 506–518.

Boiteau, R., Fitzsimmons, J.N., Repeta, D.J., Boyle, E.A., 2013. A method for the characterization of iron ligands in seawater and marine cultures by HPLC-ICP-MS. Anal. Chem. 85, 4357–4362.

Boon, J.J., Klap, V.A., Eglinton, T.I., 1998. Molecular characterization of microgram amounts of oceanic colloidal organic matter by direct temperature-resolved ammonia chemical ionization mass spectrometry. Org. Geochem. 29, 1051–1061.

Boyd, P.W., Ellwood, M.J., 2010. The biogeochemical cycle of iron in the ocean. Nat. Geosci. 3, 675–682.

Boye, M., van den Berg, C.M.G., de Jong, J., Leach, H., Croot, P., de Baar, H.J.W., 2001. Organic complexation of iron in the Southern Ocean. Deep Sea Res. Part I Oceanogr. Res. Pap. 48, 1477–1497.

Bragg, J.G., Dutkiewicz, S., Jahn, O., Follows, M.J., Chisholm, S.W., 2010. Modeling selective pressures on phytoplankton in the global ocean. PLoS One 5, e9569.

Burdige, D.J., 2007. Preservation of organic matter in marine sediments: controls, mechanisms, and an imbalance in sediment organic carbon budgets? Chem. Rev. 107, 467–485.

Calleja, M.L., Batista, F., Peacock, M., Kudela, R., McCarthy, M.D., 2013. Changes in compound specific $\delta^{15}N$ amino acid signatures and d/l ratios in marine dissolved organic matter induced by heterotrophic bacterial reworking. Mar. Chem. 149, 32–44.

Carlson, C.A., Giovannoni, S., Hansell, D., Goldberg, S., Parsons, R., Otero, M., et al., 2002. Effect of nutrient amendments on bacterioplankton production, community structure, and DOC utilization in the northwestern Sargasso Sea. Aquat. Microb. Ecol. 30, 19–36.

Carlson, C.A., Giovannoni, S.J., Hansell, D.A., Goldberg, S.J., Parsons, R., Vergin, K., 2004. Interactions among dissolved organic carbon, microbial processes, and community structure in the mesopelagic zone of the northwestern Sargasso Sea. Limnol. Oceangr. 49, 1073–1083.

Cava, F., Lam, H., de Pedro, M., Waldor, M.K., 2011. Emerging knowledge of regulatory roles of D-amino acids in bacteria. Cell. Mol. Life Sci. 68, 817–831.

Corradini, C., Cavazza, A., Bignardi, C., 2012. High-performance anion-exchange chromatography coupled with pulsed electrochemical detection as a powerful tool to evaluate carbohydrates of food interest: principles and applications. Int. J. Carbohydr. Chem. 2012, 1–13.

Cottrell, M.T., Kirchman, D.L., 2000. Natural assemblages of marine proteobacteria and members of the Cytophaga-Flavobacter cluster consuming low- and high-molecular-weight dissolved organic matter. Appl. Environ. Microbiol. 66, 1692–1697.

Dauwe, B., Middelburg, J.J., Herman, P.M.J., Heip, C.H.R., 1999. Linking diagenetic alteration of amino acids and bulk organic matter reactivity. Limnol. Oceanogr. 44, 1809–1814.

Davis, J., Kaiser, K., Benner, R., 2009. Amino acid and amino sugar yields and compositions as indicators of dissolved organic matter diagenesis. Org. Geochem. 40, 343–352.

De Baar, H.J.W., Boyd, P.W., Coale, K.H., Landry, M.R., 2005. Synthesis of iron fertilization experiments: from the iron age in the age of enlightenment. J. Geophys. Res. 110, 1–24.

Dittmar, T., 2008. The molecular level determination of black carbon in marine dissolved organic matter. Org. Geochem. 39, 396–407.

Dittmar, T., Koch, B.P., 2006. Thermogenic organic matter dissolved in the abyssal ocean. Mar. Chem. 102, 208–217.

Dittmar, T., Paeng, J., 2009. A heat-induced molecular signature in marine dissolved organic matter. Nat. Geosci. 2, 175–179.

Dittmar, T., Fitznar, H., Kattner, G., 2001. Origin and biogeochemical cycling of organic nitrogen in the eastern Arctic Ocean as evident from D- and L-amino acids. Geochim. Cosmochim. Acta 65, 4103–4114.

Dittmar, T., Koch, B., Hertkorn, N., Kattner, G., 2008a. A simple and efficient method for the solid-phase extraction of dissolved organic matter (SPE-DOM) from seawater. Limnol. Oceanogr. Methods 6, 230–235.

Dittmar, T., Koch, B., Hertkorn, N., Kattner, G., 2008b. A simple and efficient method for the solid-phase extraction of dissolved organic matter (SPE-DOM) from seawater. Limnol. Oceanogr. Methods 6, 230–235.

Druffel, E.R.M., Williams, P.M., Bauer, J.E., Ertel, J.R., 1992. Cycling of dissolved and particulate organic matter in the open ocean. J. Geophys. Res. 97, 15639–15659.

Druffel, E.R.M., Griffin, S., Bauer, J.E., Wolgast, D., Wang, X.-C., 1998. Distribution of particulate organic carbon and radiocarbon in the water column from the upper slope to the abyssal NE Pacific Ocean. Deep Sea Res. Part II Top. Stud. Oceanogr. 45, 667–687.

Dutkiewicz, S., Follows, M.J., Bragg, J.G., 2009. Modeling the coupling of ocean ecology and biogeochemistry. Global Biogeochem. Cycles 23, 1–15.

Dutkiewicz, S., Ward, B.A., Monteiro, F., Follows, M.J., 2012. Interconnection of nitrogen fixers and iron in the Pacific Ocean: theory and numerical simulations. Global Biogeochem. Cycles 26, http://dx.doi.org/10.1029/2011GB004039, GB1012.

Esteves, V.I., Otero, M., Duarte, A.C., 2009. Comparative characterization of humic substances from the open ocean, estuarine water and fresh water. Org. Geochem. 40, 942–950.

Flerus, R., Lechtenfeld, O.J., Koch, B.P., McCallister, S.L., Schmitt-Kopplin, P., Benner, R., et al., 2012. A molecular perspective on the ageing of marine dissolved organic matter. Biogeosciences 9, 1935–1955.

Follows, M.J., Dutkiewicz, S., Grant, S., Chisholm, S.W., 2007. Emergent biogeography of microbial communities in a model ocean. Science 315, 1843–1846.

Fry, B., Hopkinson, C.S., Nolin, A., Norrman, B., Li, U., 1996. Long-term decomposition of DOC from experimental diatom blooms. Limnol. Oceanogr. 41, 1344–1347.

Fuhrman, J., 1987. Close coupling between release and uptake of dissolved free amino acids in seawater studied by an isotope dilution approach. Mar. Ecol. Prog. Ser. 37, 45–52.

Gagosian, R.B., Steurmer, D.H., 1977. The cycling of biogenic compounds and their dia- genetically transformed products in seawater. Mar. Chem. 5, 605–632.

Gledhill, M., Buck, K.N., 2012. The organic complexation of iron in the marine environment: a review. Front. Microbiol. 3, 1–17.

Gledhill, M., van den Berg, C.M.G., 1994. Determination of complexation of iron (III) with natural organic complexing ligands in seawater using cathodic stripping voltammetry. Mar. Chem. 47, 41–54.

Gogou, A., Repeta, D.J., 2010. Particulate-dissolved transformations as a sink for semi-labile dissolved organic matter: chemical characterization of high molecular weight dissolved and surface-active organic matter in seawater and in diatom cultures. Mar. Chem. 121, 215–223.

Goldberg, S.J., Carlson, C.A., Hansell, D.A., Nelson, N.B., Siegel, D.A., 2009. Temporal dynamics of dissolved combined neutral sugars and the quality of dissolved organic matter in the Northwestern Sargasso Sea. Deep Sea Res. Part I Oceanogr. Res. Pap. 56, 672–685.

Goldberg, S.J., Carlson, C.A., Bock, B., Nelson, N.B., Siegel, D.A., 2010. Meridional variability in dissolved organic matter stocks and diagenetic state within the euphotic and mesopelagic zone of the North Atlantic subtropical gyre. Mar. Chem. 119, 9–21.

Gonsior, M., Peake, B.M., Cooper, W.T., Podgorski, D.C., D'Andrilli, J., Dittmar, T., 2011. Characterization of dissolved organic matter across the Subtropical Convergence off the South Island, New Zealand. Mar. Chem. 123, 99–110.

Green, S.A., Blough, N.V., 1994. Optical absorption and fluorescence properties of chromophoric dissolved organic matter in natural waters. Limnol. Oceanogr. 39, 1903–1916.

Gurtler, B.K., Vetter, T.A., Perdue, E.M., Ingall, E., Koprivnjak, J.-F., Pfromm, P.H., 2008. Combining reverse osmosis and pulsed electrical current electrodialysis for improved recovery of dissolved organic matter from seawater. J. Membr. Sci. 323, 328–336.

Hansell, D.A., 2013. Recalcitrant dissolved organic carbon fractions. Ann. Rev. Mar. Sci. 5, 1–25.

Hansell, D.A., Carlson, C.A., 2013. Localized refractory dissolved organic sinks in the deep ocean. Global Biogeochem. Cycles 27, 705–710. http://dx.doi.org/10.1002/GBC.20067,2013.

Hansell, D.A., Carlson, C.A., Repeta, D.J., Schlitzer, R., 2009. Dissolved organic matter in the ocean. Oceanography 22, 202–211.

Hansell, D.A., Carlson, C.A., Schlitzer, R., 2012. Net removal of major marine dissolved organic carbon fractions in the subsurface ocean. Global Biogeochem. Cycles 26, GB1016. http://dx.doi.org/10.1029/2011GB004069.

Hedges, J.I., Hatcher, P.G., Ertel, J.R., Meyers-schulte, K.J., 1988. A comparison of dissolved humic substances from seawater with Amazon River counterparts by 13C-NMR spectrometry. Geochim. Cosmochim. Acta 56, 1753–1757.

Helliwell, K.E., Wheeler, G.L., Leptos, K.C., Goldstein, R.E., Smith, A.G., 2011. Insights into the evolution of vitamin B12 auxotrophy from sequenced algal genomes. Mol. Biol. Evol. 28, 2921–2933.

Helms, J.R., Mao, J., Chen, H., Perdue, E.M., Green, N., Hatcher, P.G., et al., 2013. Spectroscopic characterization of oceanic dissolved organic matter isolated by reverse osmosis coupled with electrodialysis: implications for oceanic carbon cycling. Limnol. Oceanogr. 53 (3), 955–969.

Hertkorn, N., Benner, R., Frommberger, M., Schmittkopplin, P., Witt, M., Kaiser, K., 2006. Characterization of a major refractory component of marine dissolved organic matter. Geochim. Cosmochim. Acta 70, 2990–3010.

Hertkorn, N., Harir, M., Koch, B.P., Michalke, B., Grill, P., Schmitt-Kopplin, P., 2012. High field NMR spectroscopy and FTICR mass spectrometry: powerful discovery tools for the molecular level characterization of marine dissolved organic matter from the South Atlantic Ocean. Biogeosci. Discuss. 9, 745–833.

Hubberten, U., Lara, R., Kattner, G., 1994. Amino acid composition of seawater and dissolved humic substances in the Greenland Sea. Mar. Chem. 45, 121–128.

Hubberten, U., Laral, R.J., Kattnerl, G., 1995. Refractory organic compounds in polar waters: relationship between humic substances and amino acids in the Arctic and Antarctic. J. Mar. Res. 53, 137–149.

Jiao, N., Herndl, G.J., Hansell, D.A., Benner, R., Kattner, G., Wilhelm, S.W., et al., 2010. Microbial production of recalcitrant dissolved organic matter: long-term carbon storage in the global ocean. Nat. Rev. Microbiol. 8, 593–599.

Jickells, T.D., An, Z.S., Andersen, K.K., Baker, A.R., Bergametti, G., Brooks, N., et al., 2005. Global iron connections between desert dust, ocean biogeochemistry, and climate. Science 308, 67.

Jones, V., Ruddell, C.J., Wainwright, G., Rees, H.H., Jaffé, R., Wolff, G.A., 2004. One-dimensional and two-dimensional polyacrylamide gel electrophoresis: a tool for protein characterisation in aquatic samples. Mar. Chem. 85, 63–73.

Jurado, E., Dachs, J., Duarte, C.M., Simó, R., 2008. Atmospheric deposition of organic and black carbon to the global oceans. Atmos. Environ. 42, 7931–7939.

Kaiser, K., Benner, R., 2000. Determination of amino sugars in environmental samples with high salt content by chromatography and pulsed amperometric detection. Anal. Chem. 72, 2566–2572.

Kaiser, K., Benner, R., 2008. Major bacterial contribution to the ocean reservoir of detrital organic carbon and nitrogen. Limnol. Oceanogr. 53, 99–112.

Kaiser, K., Benner, R., 2009. Biochemical composition and size distribution of organic matter at the Pacific and Atlantic time-series stations. Mar. Chem. 113, 63–77.

Kawasaki, N., Benner, R., 2006. Bacterial release of dissolved organic matter during cell growth and decline: molecular origin and composition. Limnol. Oceanogr. 51, 2170–2180.

Kido Soule, M.C., Longnecker, K., Giovannoni, S.J., Kujawinski, E.B., 2010. Impact of instrument and experiment parameters on reproducibility of ultrahigh resolution ESI FT-ICR mass spectra of natural organic matter. Org. Geochem. 41, 725–733.

Kieber, R.J., Zhou, X., Mopper, K., 1990. Formation of carbonyl compounds from UV-induced photodegradation of humic substances in natural waters: fate of riverine carbon in the sea. Limnol. Oceanogr. 35, 1503–1515.

Koch, B.P., Witt, M., Engbrodt, R., Dittmar, T., Kattner, G., 2005. Molecular formulae of marine and terrigenous dissolved organic matter detected by electrospray ionization Fourier transform ion cyclotron resonance mass spectrometry. Geochim. Cosmochim. Acta 69, 3299–3308.

Koch, B.P., Dittmar, T., Witt, M., Kattner, G., 2007. Fundamentals of molecular formula assignment to ultrahigh resolution mass data of natural organic matter. Anal. Chem. 79, 1758–1763.

Koprivnjak, J.-F., Pfromm, P.H., Ingall, E., Vetter, T.A., Schmitt-Kopplin, P., Hertkorn, N., et al., 2009. Chemical and spectroscopic characterization of marine dissolved organic matter isolated using coupled reverse osmosis–electrodialysis. Geochim. Cosmochim. Acta 73, 4215–4231.

Kujawinski, E.B., Behn, M.D., 2006. Automated analysis of electrospray ionization fourier transform ion cyclotron resonance mass spectra of natural organic matter. Anal. Chem. 78, 4363–4373.

Kujawinski, E.B., Longnecker, K., Blough, N.V., Vecchio, R., Del, Finlay L., Kitner, J.B., et al., 2009. Identification of possible source markers in marine dissolved organic matter using ultrahigh resolution mass spectrometry. Geochim. Cosmochim. Acta 73, 4384–4399.

Lam, B., Simpson, A.J., 2008. Direct (1)H NMR spectroscopy of dissolved organic matter in natural waters. Analyst 133, 263–269.

Lang, S.Q., Butterfield, D., Lilley, M.D., Paul Johnson, H., Hedges, J.I., 2006. Dissolved organic carbon in ridge-axis and ridge-flank hydrothermal systems. Geochim. Cosmochim. Acta 70, 3830–3842.

Lee, C., Bada, J.L., 1975. Amino acids in equatorial Pacific Ocean water. Earth Planet. Sci. Lett. 26, 61–68.

Lee, C., Bada, J.L., 1977. Dissolved amino acids in the equatorial Pacific, the Sargasso Sea, and Biscayne Bay. Limnol. Oceanogr. 22, 502–510.

Lee, C., Wakeham, S.G., Hedges, J.I., 2000. Composition and flux of particulate amino acids and chloropigments in equatorial Pacific seawater and sediments. Deep-Sea Res. I 47, 1535–1568.

Lin, H.-T., Cowen, J.P., Olson, E.J., Amend, J.P., Lilley, M.D., 2012. Inorganic chemistry, gas compositions and dissolved organic carbon in fluids from sedimented young basaltic crust on the Juan de Fuca Ridge flanks. Geochim. Cosmochim. Acta 85, 213–227.

Lindroth, P., Mopper, K., 1979. High performance liquid chromatographic determination of subpicomole amounts of amino acids by precolumn fluorescence derivatization with o-phthaldialdehyde. Anal. Chem. 51, 1667–1674.

Macrellis, H.M., Trick, C.G., Rue, E.L., Smith, G., Bruland, K.W., 2001. Collection and detection of natural iron-binding ligands from seawater. Mar. Chem. 76, 175–187.

Martin, J.H., Fitzwater, S.E., 1988. Iron deficiency limits phytoplankton growth in the north-east Pacific subarctic. Nature 371, 1223–1229.

Martin, J.H., Gordon, R.M., Fitzwater, S.E., 1991. The case for iron. Limnol. Oceanogr. 36, 1793–1802.

Masiello, C.A., Druffel, E.R.M., 1998. Black carbon in deep-sea sediments. Science 280, 1911–1913.

Mawji, E., Gledhill, M., Milton, J.A., Tarran, G.A., Ussher, S., Thompson, A., Wolff, G.A., Worsfold, P.J., Achterberg, E.P., 2008. Hydroxamate Siderophores: Occurrence and importance in the Atlantic Ocean. Environ, Sci. Technol. 42, 8675–8680.

Mawji, E., Gledhill, M., Milton, J.A., Zubkov, M.V., Thompson, A., Wolff, G.A., et al., 2011. Production of siderophore type chelates in Atlantic Ocean waters enriched with different carbon and nitrogen sources. Mar. Chem. 124, 90–99.

McCarren, J., Becker, J.W., Repeta, D.J., Shi, Y., Young, C.R., Malmstrom, R.R., et al., 2010. Microbial community transcriptomes reveal microbes and metabolic pathways associated with dissolved organic matter turnover in the sea. Proc. Natl. Acad. Sci. U. S. A. 107, 16420–16427.

McCarthy, M., Hedges, J., Benner, R., 1996. Major biochemical composition of dissolved high molecular weight organic matter in seawater. Mar. Chem. 55, 281–297.

McCarthy, M.D., Pratum, T., Hedges, J.I., Benner, R., 1997. Chemical composition of dissolved organic nitrogen in the ocean. Nature 390, 150–154.

McCarthy, M., Hedges, J., Benner, R., 1998. Major bacterial contribution to marine dissolved organic nitrogen. Science 281, 231–234.

McCarthy, M.D., Beaupré, S.R., Walker, B.D., Voparil, I., Guilderson, T.P., Druffel, E.R.M., 2010. Chemosynthetic origin of ^{14}C-depleted dissolved organic matter in a ridge-flank hydrothermal system. Nat. Geosci. 4, 32–36.

McNichol, A.P., Aluwihare, L.I., 2007. The power of radiocarbon in biogeochemical studies of the marine carbon cycle: insights from studies of dissolved and particulate organic carbon (DOC and POC). Chem. Rev. 107, 443–466.

Meon, B., Kirchman, D.L., 2001. Dynamics and molecular composition of dissolved organic material during experimental phytoplankton blooms. Mar. Chem. 75, 185–199.

Miethke, M., 2013. Molecular strategies of microbial iron assimilation: from high-affinity complexes to cofactor assembly systems. Metallomics 5, 15–28.

Minor, E.C., Boon, J.J., Harvey, H.R., Mannino, A., 2001. Estuarine organic matter composition as probed by direct temperature-resolved mass spectrometry and traditional geochemical techniques. Geochim. Cosmochim. Acta 65, 2819–2834.

Mopper, K., Lindroth, P., 1982. Diel and depth variations in dissolved free amino acids and ammonium in the Baltic Sea determined by shipboard HPLC analysis. Limnol. Oceanogr. 27, 336–347.

Mopper, K., Stahovec, W.L., 1986. Sources and sinks of low molecular weight organic carbonyl compounds in seawater. Mar. Chem. 19, 305–321.

Mopper, K., Zhou, X., Kieber, R.J., Kieber, D.J., Sikorski, R.J., Jones, R.D., 1991. Photochemical degradation of dissolved organic carbon and its impact on the ocean carbon cycle. Nature 353, 60–62.

Mopper, K., Stubbins, A., Ritchie, J.D., Bialk, H.M., Hatcher, P.G., 2007. Advanced instrumental approaches for characterization of marine dissolved organic matter: extraction techniques, mass spectrometry, and nuclear magnetic resonance spectroscopy. Chem. Rev. 107, 419–442.

Nagata, T., 2003. Microbial degradation of peptidoglycan in seawater. Limnol. Oceanogr. 48, 745–754.

Ogawa, H., Amagai, Y., Koike, I., Kaiser, K., Benner, R., 2001. Production of refractory dissolved organic matter by bacteria. Science 292, 917–920.

Orellana, M.V., Hansell, D., 2012. Ribulose-1,5-bisphosphate carboxylase/oxygenase (RubisCO): a long lived protein in the deep ocean. Limnol. Oceanogr. 57, 826–834.

Orellana, M.V., Lessard, E.J., Dycus, E., Chin, W.-C., Foy, M.S., Verdugo, P., 2003. Tracing the source and fate of biopolymers in seawater: application of an immunological technique. Mar. Chem. 83, 89–99.

Orellana, M.V., Petersen, T.W., Diercks, A.H., Donohoe, S., Verdugo, P., van den Engh, G., 2007. Marine microgels: optical and proteomic fingerprints. Mar. Chem. 105, 229–239.

Orellana, M.V., Matrai, P.A., Leck, C., Rauschenberg, C.D., Lee, A.M., 2011. Marine microgels as a source of cloud condensation nuclei in the high Arctic. Proc. Natl. Acad. Sci. U. S. A. 108, 13612–13617.

Panagiotopoulos, C., Sempéré, R., 2005. Analytical methods for the determination of sugars in marine samples: a historical perspective and future directions. Limnol. Oceanogr. Methods 3, 419–454.

Panagiotopoulos, C., Repeta, D.J., Johnson, C.G., 2007. Characterization of methyl sugars, 3-deoxysugars and methyl deoxysugars in marine high molecular weight dissolved organic matter. Org. Geochem. 38, 884–896.

Panagiotopoulos, C., Repeta, D.J., Mathieu, L., Rontani, J.-F., Sempéré, R., 2013. Molecular level characterization of methyl sugars in marine high molecular weight dissolved organic matter. Mar. Chem. 154, 34–45.

Powell, M.J., Timperman, A.T., 2005. Quantitative analysis of protein recovery from dilute, large volume samples by tangential flow ultrafiltration. J. Membr. Sci. 252, 227–236.

Powell, M.J., Sutton, J.N., Del Castillo, C.E., Timperman, A.T., 2005. Marine proteomics: generation of sequence tags for dissolved proteins in seawater using tandem mass spectrometry. Mar. Chem. 95, 183–198.

Quan, T.M., Repeta, D.J., 2007. Periodate oxidation of marine high molecular weight dissolved organic matter: evidence for a major contribution from 6-deoxy- and methyl sugars. Mar. Chem. 105, 183–193.

Repeta, D.J., Aluwihare, L.I., 2006. Radiocarbon analysis of neutral sugars in high-molecular-weight dissolved organic carbon: implications for organic carbon cycling. Limnol. Oceanogr. 51, 1045–1053.

Rich, J.H., Ducklow, H.W., Kirchman, D.L., 1996. Concentrations and uptake of neutral monosaccharides along 140 °W in the equatorial Pacific: contribution of glucose to heterotrophic bacterial activity and the DOM flux. Limnol. Oceanogr, 595–604.

Rue, E.L., Bruland, K.W., 1997. The role of organic complexation on ambient iron chemistry in the equatorial Pacific Ocean and the response of a mesoscale iron addition experiment. Limnol. Oceanogr. 42, 901–910.

Sakugawa, H., Handa, N., 1985. Isolation and chemical characterization of dissolved and particulate polysaccharides in Mikawa Bay. Geochim. Cosmochim. Acta 49, 1185–1193.

Sannigrahi, P., Ingall, E.D., Benner, R., 2005. Cycling of dissolved and particulate organic matter at station Aloha: insights from 13C NMR spectroscopy coupled with elemental, isotopic and molecular analyses. Deep Sea Res. Part I Oceanogr. Res. Pap. 52, 1429–1444.

Sañudo-Wilhelmy, S., Cutter, L.S., Durazo, R., Smail, E., Gómez-Consarnau, L., Webb, E., 2012. Multiple B-vitamin depletion in large areas of the coastal ocean. Proc. Natl. Acad. Sci. U. S. A. 109, 14041–14045.

Sarmento, H., Gasol, J.M., 2012. Use of phytoplankton-derived dissolved organic carbon by different types of bacterioplankton. Environ. Microbiol. 14, 2348–2360.

Schmidt, A., Karas, M., Dülcks, T., 2003. Effect of different solution flow rates on analyte ion signals in nano-ESI MS, or: when does ESI turn into nano-ESI? J. Am. Soc. Mass Spectrom. 14, 492–500.

Sher, D., Thompson, J.W., Kashtan, N., Croal, L., Chisholm, S.W., 2011. Response of Prochlorococcus ecotypes to co-culture with diverse marine bacteria. ISME J. 5, 1125–1132.

Simjouw, J.-P., Minor, E.C., Mopper, K., 2005. Isolation and characterization of estuarine dissolved organic matter: comparison of ultrafiltration and C18 solid-phase extraction techniques. Mar. Chem. 96, 219–235.

Skoog, A., Benner, R., 1997. Aldoses in various size fractions of marine organic matter: implications for carbon cycling. Limnol. Oceanogr. 42, 1803–1813.

Suzuki, S., Kogure, K., Tanoue, E., 1997. Immunochemical detection of dissolved proteins and their source bacteria in marine environments. Mar. Ecol. Prog. Ser. 158, 1–9.

Tada, K., Tada, M., Maita, Y., 1998. Dissolved free amino acids in coastal seawater using a modified fluorometric method. J. Oceanogr. 54, 313–321.

Tang, Y.Z., Koch, F., Gobler, C.J., 2010. Most harmful algal bloom species are vitamin B1 and B12 auxotrophs. Proc. Natl. Acad. Sci. U. S. A. 107, 20756–20761.

Tanoue, E., 1995. Detection of dissolved protein molecules in oceanic waters. Mar. Chem. 51, 239–252.

Tanoue, E., Nishiyama, S., Kamo, M., Tsugita, A., 1995. Bacterial membranes: possible source of a major dissolved protein in seawater. Geochim. Cosmochim. Acta 59, 2643–2648.

Tanoue, E., Masao, I., Takahashi, M., 1996. Discrete dissolved and particulate proteins in oceanic waters. Limnol. Oceanogr. 41, 1334–1343.

Velasquez, I., Nunn, B.L., Ibisanmi, E., Goodlett, D.R., Hunter, K.A., Sander, S.G., 2011. Detection of hydroxamate siderophores in coastal and Sub-Antarctic waters off the South Eastern Coast of New Zealand. Mar. Chem. 126, 97–107.

Vetter, T., Perdue, E., Ingall, E., Koprivnjak, J., Pfromm, P., 2007. Combining reverse osmosis and electrodialysis for more complete recovery of dissolved organic matter from seawater. Sep. Purif. Technol. 56, 383–387.

Vraspir, J.M., Butler, A., 2009. Chemistry of marine ligands and siderophores. Ann. Rev. Mar. Sci. 1, 43–63.

Walker, B.D., Beaupré, S.R., Guilderson, T.P., Druffel, E.R.M., McCarthy, M.D., 2011. Large-volume ultrafiltration for the study of radiocarbon signatures and size vs. age relationships in marine dissolved organic matter. Geochim. Cosmochim. Acta 75, 5187–5202.

Weishaar, J.L., Aiken, G.R., Bergamaschi, B.A., Fram, M.S., Fujii, R., Fujii, R., et al., 2003. Evaluation of specific ultraviolet absorbance as an indicator of the chemical composition and reactivity of dissolved organic carbon. Environ. Sci. Technol. 37, 4702–4708.

Williams, P.M., Druffel, E.R.M., 1987. Radiocarbon in dissolved organic matter in the central North Pacific Ocean. Nature 330, 246–248.

Yamada, N., Tanoue, E., 2003. Detection and partial characterization of dissolved glycoproteins in oceanic waters. Limnol. Oceanogr. 48, 1037–1048.

Yamada, N., Suzuki, S., Tanoue, E., 2000. Detection of Vibrio (Listonella) anguillarum porin homologue proteins and their source bacteria from coastal seawater. J. Oceanogr. 56, 583–590.

Yamashita, Y., Tanoue, E., 2003a. Distribution and alteration of amino acids in bulk DOM along a transect from bay to oceanic waters. Mar. Chem. 82, 145–160.

Yamashita, Y., Tanoue, E., 2003b. Chemical characterization of protein-like fluorophores in DOM in relation to aromatic amino acids. Mar. Chem. 82, 255–271.

Yamashita, Y., Tanoue, E., 2004. Chemical characteristics of amino acid-containing dissolved organic matter in seawater. Org. Geochem. 35, 679–692.

Ziolkowski, L.A., Druffel, E.R.M., 2010. Aged black carbon identified in marine dissolved organic carbon. Geophys. Res. Lett. 37, 4–7.

Zubarev, R.A., Makarov, A., 2013. Orbitrap mass spectrometry. Anal. Chem. 85, 5288–5296.

3

DOM Sources, Sinks, Reactivity, and Budgets

Craig A. Carlson*, Dennis A. Hansell†

*Department of Ecology, Evolution and Marine Biology, University of California, Santa Barbara, California, USA

†Rosenstiel School of Marine and Atmospheric Science, University of Miami, Miami, Florida, USA

C O N T E N T S

I INTRODUCTION

Dissolved organic matter (DOM) represents one of Earth's largest exchangeable reservoirs of organic material. It is operationally defined as organic matter that passes through glass fiber filters with a nominal pore size of 0.7 μm (e.g., Whatman type GF/F) or, less frequently, 0.2-μm pore-size plastic filters (typically track-etched polycarbonate). At ~662 ± 32 Pg (10^{15} g) C, dissolved organic carbon (DOC) exceeds the inventory of organic particles in the oceans by 200-fold, making it one of the largest bioreactive pools of carbon in the ocean (Hansell and Carlson, 1998a; Hansell et al., 2009; Williams and Druffel, 1987), second only to dissolved inorganic carbon (DIC) (38,100 Pg C; Sarmiento and Gruber, 2006). The ocean inventory of DOC is comparable to the mass of inorganic C in the atmosphere (Eppley et al., 1987; Fasham et al., 2001; MacKenzie, 1981). Net oceanic uptake of CO_2 is ~1.9 Pg C year^{-1} (Sarmiento and Gruber, 2006), thus small perturbations in the processes regulating DOC production and removal affect both ocean/atmosphere CO_2 balance and organic carbon export (Carlson et al., 1994; Copin-Montgut and Avril, 1993; Ducklow et al., 1995; Hansell

and Carlson, 2001) and its storage in the ocean interior (Hansell et al., 2009).

Only in the past two decades has the marine geochemistry community developed the analytical skill (Sharp et al., 2002) necessary for reliable description and quantification of DOM and its contributions to the ecology and biogeochemistry of the ocean's water column. Prior to the 2000s, the marine DOM community suffered from a paucity of high quality data to adequately describe the DOC distribution, inventories, and fluxes in the global ocean. In the past decade, time-series programs together with basin scale programs such as the US Climate Variability and Predictability (CLIVAR) Repeat Hydrography project have revealed temporal and spatial variability of DOC in unprecedented detail (Figure 3.1). Biological processes (largely microbial) together with ocean physics shape the observed patterns.

Since the first edition of this book, published a decade ago, there have been numerous advances toward understanding DOM dynamics in the ocean. The combination of large data sets (Figure 3.1) with physical and biogeochemical tracers has helped researchers assess the inventories and turnover rates of various fractions of marine DOM. New technologies that further

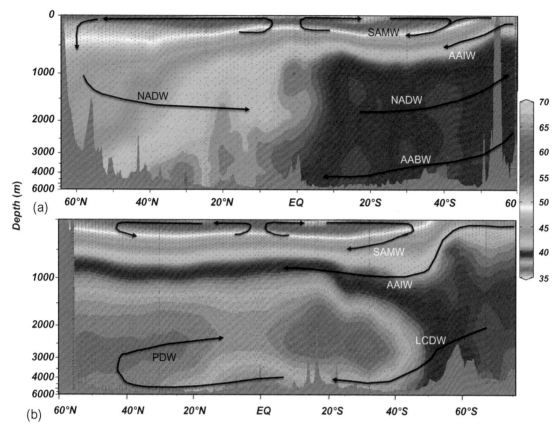

FIGURE 3.1   Distribution of DOC (µmol kg⁻¹) along meridional sections in the Atlantic (a) (CLIVAR line A16) and Pacific basins (b) (CLIVAR line P16). DOC production and consumption processes in combination with ocean circulation, and factors that control DOC reactivity, shape the distribution of DOC in the global ocean. This figure depicts accumulation of DOC in the surface waters, redistribution in the surface layer by wind driven circulation and export into the ocean interior via mode and deep water formation. DOC exported with North Atlantic Deep Water (NADW) formation flows to the south where a portion is removed by microbial and abiotic removal processes. In the Pacific, relatively DOC-enriched waters enters the interior of the South Pacific with northward flowing Antarctic Intermediate Water (AAIW) and Lower Circumpolar Water (LCDW), with slow DOC removal in the North Pacific. The return flow moves DOC-depleted waters to the south at intermediate depths within the Pacific Deep Water (PDW). The arrows represent simplified circulation of water masses, including NADW, Subantarctic Mode Water (SAMW), AAIW, Antarctic Bottom Water (AABW), LCDW, and PDW. *Data employed are available at http://ushydro.ucsd.edu.*

characterize the myriad of compounds comprising the DOM pool (see Chapter 2), together with new microbial ecological studies and the integration of "omic" approaches, have advanced our understanding of the biological mechanisms that act upon the pool.

This chapter reviews the many production and consumption processes that control the fluxes, accumulation, and inventory of DOM in the ocean. We (1) review the mechanisms

of DOM production, (2) review the processes of DOM removal, (3) review the factors and processes that lead to DOM accumulation, (4) examine DOM fractions in terms of biological lability and ecological and biogeochemical significance, and (5) describe our current understanding of the global ocean DOC budget. The conceptual model of the various DOM production and removal processes given in Figure 3.2 serves as a guide for the first two sections of this review.

II DOM PRODUCTION PROCESSES

The euphotic zone is the principal site of organic matter production in the open ocean. Net production of DOM results from the temporal and spatial uncoupling of biological production from biotic and abiotic removal processes (Figure 3.2). The magnitude and quality of DOM production varies considerably, being controlled by several biological, chemical, and physical parameters. While DOM production is ultimately constrained by the magnitude of primary production (PP), there are complementary mechanisms responsible for DOM production including (a) extracellular release (ER) by phytoplankton, (b) grazer-mediated release and excretion, (c) release via cell lysis (both viral and bacterial), (d) solubilization of detrital and sinking particles, and (e) release from prokaryotes.

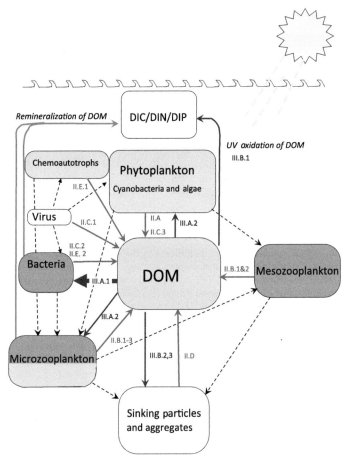

FIGURE 3.2　DOM production and consumption processes in marine systems. Dashed lines represent food web interactions and solid arrows represent DOM production (red) or consumption (blue) processes. Green shade represents autotrophs, red shade represents heterotrophs, and blue shade represents DOM. The large blue arrow is the dominant biotic removal, uptake by heterotrophic prokaryotes. While these production and removal processes have been identified in the field and in experiments, their individual contributions remain largely unquantified. Roman numerals and letters refer to the section in the chapter where the process is discussed.

A Extracellular Phytoplankton Production

Phytoplankton are commonly single-celled photoautotrophic microbes that have the complex cellular machinery required to synthesize their required macromolecules from simple inorganic nutrients. One might think that the photosynthetically generated organic compounds would be retained by the cell solely for biosynthesis. However, environmental factors such as irradiance, nutrient availability, pH, temperature, and phytoplankton community structure can lead to uncoupling of photosynthesis from cell growth, resulting in ER of photosynthate (Myklestad, 2000). The ER of polypeptides (Fogg, 1952) and carbohydrates (Guillard and Wangersky, 1958; Lewin, 1956) were first identified in algal cultures over five decades ago. Since then, a voluminous literature has arisen in which the physiological, ecological, and biogeochemical significance of ER continues to grow, though often in conflicting directions. For example, Duursma (1963) hypothesized that dead or dying cells were the major sources of phytoplankton-derived DOM in the ocean, yet Fogg (1966) argued that ER was an attribute of healthy cells with as much as 5-35% of fixed carbon being released immediately as DOM.

Early understanding of ER was hampered by methodological concerns (Sharp, 1977); but by the early 1980s careful physiological studies (Mague et al., 1980), further development of theory (Berman-Frank and Dubinsky, 1999; Bjørnsen, 1988; Fogg, 1983; Wood and Van Valen, 1990), as well as laboratory and field evidence (Baines and Pace, 1991; Marañón et al., 2005; Nagata, 2000; Teira et al., 2003; see Table 3.2 in Carlson, 2002) indicated that ER was indeed a "normal function" of healthy photoautotrophic growth. Despite the large and growing literature, the magnitude, importance, and mechanisms responsible for ER remain debated and difficult to elucidate. There are several reviews that describe the rates and potential mechanisms of phytoplankton ER in marine systems (Baines and Pace, 1991; Berman-Frank and Dubinsky, 1999; Fogg, 1983; Myklestad, 2000; Nagata, 2000; Williams, 1990; Wood and Van Valen, 1990) as well as predictive models that assess the extracellular production of DOM by several phytoplankton species (Flynn et al., 2008); thus, the topic of phytoplankton ER is only briefly discussed here.

1 Extracellular Release Models

Two models have been proposed to explain ER by photoautotrophs. They are the *overflow model* (Fogg, 1966, 1983; Williams, 1990) and the *passive diffusion model* (Bjørnsen, 1988; Fogg, 1966).

a OVERFLOW MODEL

Photosynthesis is largely regulated by irradiance while cellular growth is constrained by the availability of inorganic nutrients. Because light and nutrient availability are frequently uncoupled, a cell's photosynthate may be produced faster than it can be incorporated (Fogg, 1966). Healthy phytoplankton cells may use ER as an adaptive strategy to dissipate energy when nutrient availability is uncoupled from periods of excess light (Wood and Van Valen, 1990). It is argued that the release of biochemical reductants serves two roles: (1) as an alternative to photorespiration that protects the cell from photochemical damage and (2) it allows the cell's photosynthetic machinery to continue, thus reducing the lag time necessary to shift the photosynthate into anabolism once nutrients become available (Wood and Van Valen, 1990). The active release of photosynthate represents the carbon "overflow" mechanism, which strikes a balance between carbon demand for anabolism and photosynthetic assimilation; thus, ER serves as a functionally neutral process (Berman-Frank and Dubinsky, 1999).

According to the *overflow model*, DOM exudation should (1) correlate to the photosynthetic rate (Baines and Pace, 1991), (2) be absent at night, (3) be comprised of both low molecular weight (LMW < 1000 Da) and high molecular weight (HMW > 1000 Da) DOM (Fogg, 1966; Williams, 1990), and (4) be carbon rich (Sambrotto et al., 1993). The correlation between ER and PP represents an "income tax" (Bjørnsen, 1988), with light intensity and nutrient availability controlling the degree to which the *overflow model* functions.

The ecological consequence of overflow is that it provides an energy and carbon source to some heterotrophic bacterioplankton capable of growing and attaining nutrients at concentrations too low for phytoplankton to acquire. One argument suggests that the algal-bacterial interaction stimulates bacterial remineralization, enhancing micro-zones of elevated nutrients that benefit phytoplankton (Berman-Frank and Dubinsky, 1999). In contrast, a paradoxical situation exists whereby nutrient-stressed phytoplankton stimulate bacterial production (BP), thus increasing competition for nutrients (Bratbak and Thingstad, 1985). Mitra et al. (2013) proposed an interesting scenario in which active release of DOM by mixotrophs, protist capable of phagotrophy and phototrophy, support the nutrient demand for their photosynthesis through the "farming" of bacterioplankton for the nitrogen and phosphorous. In this model, the DOM released by mixotrophic protists stimulate the growth of heterotrophic bacterioplankton that in turn successfully compete for dissolved macronutrients (i.e., N or P) at low concentrations to build bacterial biomass. The mixotroph then consumes the bacteria, which serve as a direct conduit of N and P to support mixotrophic primary production that would otherwise be unavailable to the protist at such low concentrations.

b PASSIVE DIFFUSION MODE

Bjørnsen (1988) reasoned that if active release of DOC by phytoplankton enhances nutrient competition with heterotrophic bacterioplankton and exacerbates nutrient limitation, then ER must be unintended and passive. The *passive diffusion model* describes the maintenance of a concentration gradient of LMW photosynthate across phytoplankton cell membranes. LMW compounds such as dissolved free neutral sugars (DFNS) and dissolved free amino acids (DFAA) are present at millimolar concentrations intracellularly but at nanomolar concentrations extracellularly, thus creating an extreme gradient-favoring passive diffusion of such compounds across the cell membrane (Bjørnsen, 1988; Fogg, 1966). According to the model, DOM exudation should (1) be of LMW compounds (both carbon and nitrogen), (2) continue at night (perhaps at a lower rate), (3) be correlated to phytoplankton biomass rather than with PP, and (4) be relatively greater per cell in small phytoplankton with higher surface-to-volume ratios (Bjørnsen, 1988; Kiørboe, 1993; Williams, 1990). The subsequent uptake of LMW DOM by bacterioplankton maintains the gradient, eliciting further diffusion from phytoplankton cells (Bjørnsen, 1988; Bratbak and Thingstad, 1985). Accordingly, ER would be more highly correlated to photosynthetic biomass than to photosynthetic rate, constituting a "property tax" comparable to ~5% of phytoplankton biomass per day (Bjørnsen, 1988).

c MODEL COMPARISON

Whether ER of natural phytoplankton populations is best represented by the *overflow* or the *passive diffusion* model is debatable. Observations of ER at night (Berman and Kaplan, 1984; Mague et al., 1980) and the release of LMW compounds (Lee and Rhee, 1997; Mague et al., 1980; Møller-Jensen, 1983; Søndergaard and Schierup, 1981) support the *passive diffusion model*. But the molecular composition of ER can be distinct from the soluble cellular fractions, indicating that ER is not solely due to cell leakage or lysis (Myklestad, 1974). HMW DOM release (Biddanda and Benner, 1997; Guillard and Hellebust, 1971; Lancelot, 1983, 1984; Lignell, 1990), absence of ER at night (Veldhuis and Admiraal, 1985), and enhanced

ER during nutrient limitation (Goldman et al., 1992; Smith et al., 1998; Wood and Van Valen, 1990) support the *overflow model*.

The mechanisms controlling ER are still poorly understood. It is likely that these models are not mutually exclusive and that both models are correct, with one dominating the other under the right environmental conditions (light and nutrient field) and community structure (cell size and phylogenetic composition).

2 *Experimental and Field Observations*

a USING RADIOISOTOPIC TRACERS

Measuring the uptake of ^{14}C-bicarbonate by phytoplankton and then tracing the fraction that accumulates as DO^{14}C is a common method used to measure ER (Fogg, 1966). A useful term used by microbial ecologists is percent extracellular release (PER), which normalizes DO^{14}C release (ER) to total NPP (particulate plus dissolved NPP). Because other food web processes also produce DOM (Figure 3.2), PER has been redefined as the fraction of recently produced photosynthate directly released or leaked from intact phytoplankton cells rather than from other food web interactions (Teira et al., 2001).

In practice, this technique is straightforward; however, many procedural artifacts can result in either over- or underestimates of ER and PER. Overloading of cells on a filter, rupturing cells during filtration, and mishandling of sample (Goldman and Dennett, 1985; Sharp, 1977) or incomplete removal of DI^{14}C from the filtrate (Sharp, 1977) produce artificially high ER rates. Alternatively, if the incubation time with ^{14}C-bicarbonate is too short, such that the intracellular pools of organic metabolites do not reach isotopic equilibrium, then ER will be underestimated if a constant tracer release model is used (Lancelot, 1979; Smith, 1982). The contemporaneous uptake and remineralization of DO^{14}C by heterotrophs leads to underestimates of ER as well (Lancelot, 1979; Wiebe and Smith,

1977). Finally, filter types used to separate PO^{14}C from DO^{14}C affects estimates of PER. Adsorption of DO^{14}C to glass fiber filters overestimates PO^{14}C production by 30-100% compared to polycarbonate or mixed ester filters (Karl et al., 1998; Marañón et al., 2004; Maske and Garcia-Mendoza, 1994; Morán et al., 1999), thus underestimating ER rates.

In a review of culture experiments, Nagata (2000) reported PER values averaged 5% (typical range 2-10%) for a variety of marine phytoplankton isolates growing in exponential phase. Numerous studies have employed the DO^{14}C approach to examine ER and PER of natural assemblages, yet there remains considerable uncertainty regarding the magnitude and variability of PER, its relationship to PP, light intensity, nutrients and cell size, and the most appropriate model to explain ER across productivity gradients (see Tables II and III in Carlson, 2002).

Here, we summarize representative field studies examining trends in ER and PER over large spatial and productivity gradients. These studies employed a linear regression model of log ER versus log PP to assess variability of ER and PER (Table 3.1). Baines and Pace (1991) synthesized 17 studies from lacustrine, estuarine, and marine sites, observing that 78% of log ER variability could be explained by log total PP. A slope of nearly 1 in that study suggests DOC production is a constant fraction over a large range in PP. Similarly, Marañón et al. (2005) found that euphotic zone integrated rates of ER were highly correlated to PP with a mean PER of 20% for stations that spanned productive coastal upwelling systems (Ría de Vigo) to the moderately oligotrophic Celtic Sea. Teira et al. (2001), however, extended the trophic range even further to include several high-productivity upwelling systems as well highly oligotrophic sites in North Atlantic subtropical gyre, finding the slope of log ER versus log PP to be <1, indicating that ER was not a constant fraction of total PP across this

TABLE 3.1 Findings of Field Studies Assessing Total Primary Production (PP), Extracellular Release (ER) and Percent Extracellular Release (PER) across Productivity Gradients

Experimental Site	PER	PP Range	Comments	Reference
Lacustrine and estuarine systems (17 sites)	~13%	$0.3–109\,mg\,m^{-3}\,h^{-1}$	• Compared 225 volumetric analyses • ER was a relatively constant fraction of PP • Slope of Model II linear regression model between log ER versus log total PP was ~1.08 for all data; PER was relatively constant over large gradient in PP • High R^2 (0.78) of linear regression of log ER versus total log PP indicates physiological processes of the phytoplankton are dominant sources of ER • Used GF/F filters	Baines and Pace (1991)
Coastal upwelling (Ría de Vigo to moderately oligotrophic Celtic Sea)	~22%	$0.2–2\,mg\,m^{-3}\,h^{-1}$ in surface waters Mean EZ integrated PP $27–48\,mg\,m^{-2}\,h^{-1}$ [a]	• Compared 10 profiles of ER and PP • Integrated ER was a relatively constant fraction of integrated PP • Slope of Model II linear regression model between integrated log ER variability versus log total PP was ~0.96 for all data; PER was relatively constant over large gradient in PP • High R^2 (0.9) of linear regression of log ER versus log PP indicates physiological processes of the phytoplankton are dominant source of ER • Attributed higher PER compared to Baines and Pace (above) to the use of 0.2 polycarbonate instead of GF/F filters	Marañón et al. (2005)
Benguela, Mauritania, and Spain's coastal upwelling systems to highly oligotrophic North Atlantic system	4-9% in productive systems >35% in oligotrophic system	$<0.01–3\,mg\,m^{-3}\,h^{-1}$ in surface waters EZ integrated PP $2–72\,mg\,C\,m^{-2}\,h^{-1}$	• Compared 17 profiles of ER and PP • Integrated ER was not a constant fraction of integrated PP across entire data set • Slope of Model II linear regression model between integrated log ER variability versus log total PP was ~0.65 for all data: PER demonstrated inverse relationship with PP • Lower R^2 (0.67) of linear regression of log ER versus log PP indicates that other food web interactions contributed to ER • Used GF/F filters	Teira et al. (2001)

[a]*Daily rates presented in Marañón et al. (2005) were divided by a photo period of 12 h to approximate hourly rates. EZ refers to euphotic zone.*

large-productivity gradient. They reported PER to be low and relatively constant (4-9%) in high-productivity upwelling systems but high and variable (>35%) in oligotrophic systems (Table 3.1). High and variable PER in low-productivity systems is consistent with other studies showing a greater fraction of PP released as ER (high PER) with decreasing PP (Fogg, 1983; Mague et al., 1980; Morán et al., 2002). PER values >35% have been observed in other oligotrophic systems (Alonso-Sáez et al., 2008; Karl et al., 1998; López-Sandoval et al., 2011; Teira et al., 2003); but in most cases, it is not that the ER rate is enhanced but that ER remains relatively constant against low PP.

Finally, the production of DO^{14}C may not be attributable solely to phytoplankton ER but could include other trophic interactions within the incubation bottles (i.e., grazing, lysis, etc.; see below). The coefficient of determination from the linear regression of log ER versus log PP quantifies the relative contribution of physiological versus food web interactions to ER. Baines and Pace (1991) and Marañón et al. (2004,2005) accounted for 70-80% of ER variability from PP, arguing that physiological processes of the phytoplankton cells rather than food web interactions were the dominant forcing of ER during short ^{14}C incubations. In contrast, Teira et al. (2001, 2003) accounted for <40% of the variability of ER from PP in oligotrophic systems, suggesting that food web interactions were strongly involved in the release of dissolved compounds. These contrasting results indicate that the contribution of recent photosynthate to ER may vary across trophic gradients, influencing the degree to which phytoplankton physiology influences the composition of DOM in eutrophic and oligotrophic systems.

b MICROCOSM, MESOCOSM, AND FIELD OBSERVATIONS

Studies of PER provide important information at the physiological level of the cell but it is the magnitude of ER, and not PER, that is most informative at the ecosystem level (Berman-Frank and Dubinsky, 1999). While the ^{14}C tracer method is a sensitive assay, it only provides information about photosynthate fixed over the time of the incubation (generally hours) and only reflects ER of carbon. Measuring changes in DOM concentrations and constituents within microcosms (Obernosterer and Herndl, 1995; Smith et al., 1998; Wetz and Wheeler, 2003, 2007), mesocosms (Børsheim et al., 2005; Conan et al., 2007; Grossart et al., 2007; Rochelle-Newall et al., 1999), and time-series studies (Carlson and Hansell, 2003; Carlson et al., 1998; Halewood et al., 2012; Sintes et al., 2010; Williams, 1995) provides further information regarding partitioning of organic matter between dissolved and particulate phases, the chemical character of the resulting DOM, and how those dynamics are related to underlying factors such as nutrient stress and physiological state of phytoplankton populations.

Annual net DOM production (or accumulation) is most evident in ocean regions that experience phytoplankton blooms (i.e., net DOM production is associated with positive net community production). The quantity and quality of DOM produced during these bloom events varies considerably and is likely controlled by a number of biological, chemical, and physical factors. For example, one order of magnitude less DOC (mmol m^{-2}) accumulated during a bloom in the eutrophic Ross Sea compared to oligotrophic Sargasso Sea blooms despite four- to fivefold greater PP in the Ross Sea (Carlson et al., 1998). For the *Phaeocystis antarctica*-dominated bloom in the Ross Sea, 89% (~1 mol C m^{-2}) of the accumulated total organic carbon (TOC) was present as POC with the remaining 11% (~0.1 mol C m^{-2}) as DOC. In contrast, a mean of 86% (0.75-1.0 mol m^{-2}) of TOC accumulated as DOC during the 1992, 1993, and 1995 picoplankton-dominated blooms in the Sargasso Sea, with as little as 14% (0.08-0.29 mol C m^{-2}) accumulating as POC. These data indicate that net DOC production is not a constant fraction of PP, as is often presumed in pelagic ecosystem models, but that nutrient status and phytoplankton community structure are important controls of organic matter partitioning.

The inorganic nutrient field may be an important factor controlling quantity and quality of bloom-produced DOM and, in turn, its lability (Carlson and Hansell, 2003; Ogawa et al., 1999; Williams, 1995). Nutrient-limiting conditions alone do not necessarily lead to large-scale net DOM production. For example, during summer stratified, nutrient-depleted periods (July-October) in the Sargasso Sea no further DOC accumulation is resolved in the surface 40 m despite continued regenerative primary production (Carlson et al., 2002; Hansell and Carlson, 2001). Rather it may be that the rapid transition from nutrient-repleted to nutrient-depleted conditions leads to net DOC production. This assertion can be demonstrated in culture experiments (Figure 3.3; where unhealthy phytoplankton are retained within the incubation vessel) but is more difficult to observe directly in natural populations. The results in Figure 3.3 are consistent with several other culture and mesocosm studies (Conan et al., 2007; Goldman et al., 1992; Smith et al., 1998; Wetz and Wheeler, 2003). There are observations of C-rich DOM accumulation following nutrient depletion in natural field populations (Carlson et al., 1998; Halewood et al., 2012; Hopkinson and Vallino, 2005; Sintes et al., 2010; Williams, 1995), but such observations at the time nutrients depletion are lacking.

The type of nutrient limitation (whether N or P) can affect the quantity and quality of the DOM produced. Phosphorous depletion in micro- and mesocosms results in greater DOC ER compared to N limitation (Conan et al., 2007; Magaletti et al., 2004; Obernosterer and Herndl, 1995). Mesocosms under P-limitation can lead to the production of DOM that is more resistant to microbial degradation (Kragh and Sondergaard, 2009). Karl et al. (1998) suggested

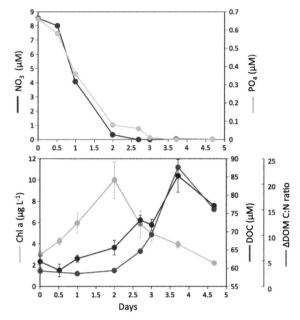

FIGURE 3.3   Results of a Bloom-in-a-Bottle experiment conducted in the Santa Barbara Channel in April 2006. Surface water was mixed 50:50 with nutrient enriched 75 m water and incubated in 50 L carboys within a flowing on deck incubator. During nutrient replete conditions net DOC production was relatively minor (<5 μM) and the ΔDOM was nitrogen rich. After inorganic N and P were depleted DOC concentrations increased by 20 μM C and the ΔDOM became C-rich with C:N ratio of 15-25. *C. Carlson previously unpublished data.*

that as the N:P ratio of the remaining nutrients within ocean systems increases, more DOC accumulation might be expected. However, questions remain on whether nutrient (N, P, Fe) limitation leads to production and accumulation of poor quality DOM directly or if it instead limits bacterial growth and thus consumption of DOM (Cotner et al., 1997; Obernosterer et al., 2003; Puddu et al., 2006; Radić et al., 2006; Thingstad et al., 1997).

c ER QUALITY AND TRANSPARENT EXOPOLYMER PARTICLES

Carbohydrates account for up to 80% of phytoplankton ER (Ittekkot et al., 1981; Lancelot, 1984;

Myklestad, 2000). Monosaccharide:polysaccharide ratios are generally <0.5, indicating that HMW polymeric compounds are a dominant fraction of ER (Myklestad, 2000). These ER products are distinct from the intracellular composition of phytoplankton cells (Myklestad and Haug, 1972) and a population's carbohydrate composition varies significantly between phytoplankton species (Myklestad, 1974; Percival et al., 1980). Peptides and amino acids are also observed in ER (Mague et al., 1980; Martin-Jezequel et al., 1988) but the C:N ratio of ER is generally C-rich relative to the canonical Redfield ratios.

Carbohydrates, particularly acidic polysaccharides (Fogg, 1983; Hellebust, 1974; Myklestad, 2000), are important precursors for the formation of transparent exopolymer particles (TEP) (Engel et al., 2004; Passow, 2002a). TEP are sticky particles sharing some characteristics with microgels (see Chapter 9). They form abiotically from dissolved precursors that aggregate into transparent sheets enriched with polysaccharides. Nutrient limitation can accelerate TEP production by some diatoms (Gärdes et al., 2012; Passow, 2002a). Some bacterioplankton lineages have been implicated in TEP production either by directly releasing precursors (Passow, 2002a; Stoderegger and Herndl, 1999) or by altering the dissolved precursor via enzymatic activity, thus forming TEP (Gärdes et al., 2012; Passow, 2002a). TEP has important biogeochemical implications in that it can aggregate senescing phytoplankton, enhancing the sinking flux of particulate organic matter (POM) (Passow, 2002b). TEP formation also has implications for estimates of ER and PER. If dissolved precursors aggregate to TEP caught on filters as POC, then TEP formation causes an underestimate of the true ER and PER (Wetz and Wheeler, 2003, 2007). Wetz and Wheeler (2007) determined that when the production of TEP was included in estimates of ER, PER of some diatoms (e.g., *C. closterium*) increased from <10% to as much as 80%.

Accounting for TEP production during phytoplankton blooms is important to ER and PER assessments.

B Grazer-Induced DOM Production

Johannes and Webb (1965) were among the first to measure the release of dissolved amino acids from mesozooplankton, estimating that ~25% of PP in the Sargasso Sea was released as DOC by zooplankton grazing. Jumars et al. (1989) argued that the principal pathway of DOM from phytoplankton to bacteria was via the byproducts of zooplankton ingestion and digestion. The four main processes by which zooplankton release DOM are direct excretory release, egestion (release and dissolution of unassimilated material), breakage of large prey during handling and feeding (sloppy feeding), and leaching from fecal pellets (Figure 3.2).

1 Herbivory

Marine phytoplankton fix ~50 Pg C year^{-1} as net PP (Chavez et al., 2011) that in turn supports most of the heterotrophy within the ocean. However, with the exception of phytoplankton blooms very little of this production accumulates as biomass, with nearly all being turned over by mortality processes such as metazoan and protistan grazing and viral infection (Strom, 2008). A fraction of plankton mortality is released as DOM, much of which supports BP (see Section III). Unlike bacterivory (see below), which essentially recycles DOM, herbivory represents a source of new DOM to the water column (Nagata and Kirchman, 1992a).

a MESOZOOPLANKTON

Mesozooplankton remove ~10-40% of daily PP in coastal and open ocean environments (Calbet, 2001). In weakly stratified or upwelling systems, mesozooplankton (e.g., copepods) form the "classical" ocean food web link, passing energy and nutrients from primary producers to higher trophic levels such as fish (Cushing, 1989). Lampert (1978) first demonstrated that as zooplankton graze on phytoplankton they damage the prey cell, releasing DOC via "sloppy feeding." Tracer experiments (^{14}C) confirmed that when labeled diatoms were fed to copepods up to 27% of the original phytoplankton radiocarbon appeared as DO^{14}C within 48 h (Copping and Lorenzen, 1980). DOM production by copepods is positively correlated to the size of the phytoplankton (Lampert, 1978; Møller, 2005; Møller and Nielsen, 2001) rather than prey density (Møller, 2007). Møller et al. (2003) demonstrated that when large diatoms or dinoflagellates were prey, 50-70% of grazed carbon was released as DOC via "sloppy feeding." As prey size decreases such that the predator-to-prey size ratio (based on equivalent spherical diameter) rises above a specific threshold (32-55 depending on the study), then prey are ingested whole, minimizing the DOM spillover associated with "sloppy feeding" (Bjørnsen, 1988; Møller, 2005, 2007; Saba et al., 2011; Figure 3.4). For example, in stratified oligotrophic systems where the phytoplankton are small, <16% of ingested prey biomass is released as DOM (Strom et al.,

1997). Direct excretion of DFAA and urea can be an important source of dissolved organic nitrogen (DON) (Vincent et al., 2007), ranging from 7% to 80% of total dissolved N release, at times exceeding inorganic N excretion (Steinberg and Saba, 2008; see Chapter 4).

Jumars et al. (1989) reasoned, based on digestion theory, that copepods would produce fecal pellets rich in unabsorbed digestive products when food concentrations where high. They hypothesized that labile DOM would rapidly diffuse from fecal pellets within minutes of release, representing a major source of fresh DOM. Urban-rich (1999) found that ~50% of fresh ^{14}C-labeled fecal pellet carbon was removed either by remineralization or released as DOC, but over hours to days rather than minutes. Other studies report that DOC and DON (i.e., urea) leached from copepod fecal pellets represent a minor percentage of ingested POC (0-6%) (Møller et al., 2003; Saba et al., 2011; Strom et al., 1997) and PON (6%) (Saba et al., 2011). The variability in the magnitude and rate of DOM release from fecal pellets may be related to the quality of food consumed by zooplankton. Other factors such as the destruction of fecal pellets by mesozooplankton coprophagy (form of sloppy feeding) may be necessary to maximize DOM release from fecal pellets (Lampitt et al., 1990; Strom et al., 1997).

b MICROZOOPLANKTON

Photosynthetic bacteria (Chisholm, 1992) and picoeukaryotes (Countway and Caron, 2006; Worden and Not, 2008; i.e., picophytoplankton 0.2-2 μm) typically comprise the majority of phytoplankton in open ocean systems, but they are too small to be effectively grazed by mesozooplankton (Calbet and Landry, 1999). Protistan microzooplankton (<200 μm), such as ciliates and large dinoflagellates, are the dominant herbivores in oceanic systems (Sherr and Sherr, 2002), removing on average 67% of daily PP (Calbet and Landry, 2004). Direct excretion of assimilated organic matter and egestion of

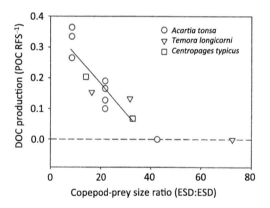

FIGURE 3.4 DOC production by sloppy feeding as a fraction of POC removed from suspension (RFS) as a function of copepod-prey size ratio based on estimated spherical diameters of each. DOC production decreased as the prey size decreased and copepod-prey ratio increased. *Adapted from Møller (2007) Copyright 2007 by the Association for the Sciences of Limnology and Oceanography, Inc.*

unassimilated DOM (i.e., undigested colloidal material and enzymes) are important mechanisms by which protozoan grazers release DOM (Nagata, 2000). Up to 20-40% of ingested algal biomass can be released as DOC when microzooplankton grazing pressure dominates phytoplankton mortality (Nagata, 2000). DOC release in the presence of grazing can be four- to sixfold greater than that attributed to direct phytoplankton release (Strom et al., 1997). Marine protozoa release 1-22% of ingested PON as DFAA (Nagata, 2000; Steinberg and Saba, 2008), thus a significant source of DON as well.

2 Omnivory and Carnivory

Zooplankton feeding strategies are not limited to herbivory, but also include omnivory and carnivory (Steinberg and Saba, 2008). Copepods such as *A. tonsa* derive as much as half of their diet from heterotrophic ciliates and dinoflagellates (Saba et al., 2009). Food source affects the relative organic and inorganic nutrient release rates and the stoichiometry of the resulting DOM. For example, copepods fed carnivorously had greater DOC and NH_4^+ release rates relative to those fed mixed diets, but released more urea when fed omnivorously (Saba et al., 2009). Herbivorous mesozooplankton may be a greater source of urea compared to their carnivorous counter parts because their phytoplankton prey have greater levels of arginine, a precursor to urea (Bidigare, 1983).

Omnivorous gelatinous zooplankton (i.e., ctenophores, pteropods, salps, scyphomedusae) assimilate large quantities of primary and secondary production in coastal as well as open ocean systems and can release significant amounts of C, N, and P as DOM (Condon et al., 2010, 2011; Hansson and Norrman, 1995; Pitt et al., 2009; Steinberg and Saba, 2008). Gelatinous zooplankton release dissolved and colloidal material by excretion, mucus production, and leakage of digestive enzymes to the DOM pools. Protein comprises up to ~70% of medusa and ctenophore biomass (Pitt et al., 2009), allowing these taxa to release 16-41% of their total dissolved nitrogen exudates as DON (i.e., primary amines, DFAA, and glycoproteins; Steinberg and Saba, 2008; Pitt et al., 2009). However, due to the release of large quantities of colloidal mucus the DOM released by gelatinous zooplankton is generally C-rich (DOC:DON = 29:1) (Condon et al., 2010, 2011). C-enriched DOM appears to favor opportunistic (copiotrophic) bacterioplankton (e.g., members of the *Gammaproteobacteria*) that utilize labile DOM rapidly but inefficiently. Condon et al. (2011) suggested that the flux of DOM from the "jelly shunt" enhances respiration and could affect the bacterioplankton community structure and subsequent biogeochemical cycling in coastal systems with large blooms of gelatinous zooplankton.

3 Bacterivory

Bacterioplankton cell densities are kept within a narrow range (10^5-10^6 cells L^{-1}) in surface waters (Ducklow and Carlson, 1992) despite large gradients in PP (bottom-up control), indicating that bacterivory and viral lysis are important top-down controls of BP. Ciliates (10-200 μm; Taylor et al., 1985) and smaller heterotrophic and mixotrophic nanoflagellates (1-10 μm) are major bacteriovores (Jürgens et al., 2008; Sherr et al., 2007; Strom, 2000) that can dominate bacterioplankton mortality in low productivity systems (Strom, 2000). Phagotrophic protists (<5 μm) can consume 54% and 75% of the daily cyanobacteria and heterotrophic bacterioplankton production, respectively (Caron et al., 1991; Sherr and Sherr, 2002; Strom, 2000).

Egestion and excretion by bacteriovores are significant sources of DOM in the open sea, controlling DOM composition through the production of both labile and recalcitrant compounds (Kujawinski, 2011; Nagata et al., 2000). Alteration of organic matter takes place within the protozoan digestive vacuole where prey are vigorously attacked by a suite of hydrolytic enzymes. Some subcellular components of the bacterial prey, being resistant to digestion (e.g., cell membranes), are egested along with the HMW digestive enzymes (Nagata and Kirchman, 1992b, 1999; Nagata et al., 2000) and colloidal material (Tranvik and Bertilsson, 2001). It has been estimated that 10-30% of DOM released via bacterivory is comprised of phospholipids colloids with aqueous center called liposome-like "picopellets" that are resistant to bacterial degradation (Nagata and Kirchman, 1999). Enhanced production of aliphatic compounds has been demonstrated in in the presence of phagotrophic nanoflagellates (Kujawinski et al., 2004) indicating that microzooplankton grazing may affect the recalcitrant nature of newly produced DOM. However, other studies have suggested that while protozoan grazers enhance DOC flux the majority of the released DOM is labile LMW material readily consumed by heterotrophic bacterioplankton, thus enhancing DOM remineralization (Taylor et al., 1985). Gruber et al. (2006) characterized DOM produced in the presence of bacteriovores by high-resolution mass spectrometry, finding it to be labile and remineralized rapidly, suggesting that bacteriovores played little role in adding to the complexity of accumulated DOM. Further study is required to resolve the role of bacterivory in transforming the bioavailability of DOM.

4 Biogeochemical Significance

Zooplankton grazing and DOM production has biogeochemical as well as ecological significance. Bacterivory increases the flux of inorganic nutrients to phytoplankton and bacterioplankton and subsequently reduces the export flux of organic particles (Michaels and Silver, 1988; Nagata and Kirchman, 1992b). Some species of mesozooplankton contribute to vertical export of carbon and nitrogen by consuming organic particles in the surface at night and respiring it as CO_2 and releasing ammonium below the epipelagic zone during the day (Longhurst and Harrison, 1989). Despite the relatively low absolute DOM production rates in the oligotrophic open ocean, migrating zooplankton can

excrete as much as 2-10% body $C d^{-1}$ and 1-6% of body $N d^{-1}$ as bulk DOC and DON, respectively (Steinberg et al., 2000, 2002). This mode of DOM flux can be significant relative to other types of POM flux in stratified oligotrophic systems. In the Sargasso Sea, DOC flux by migrating zooplankton can be comparable to 14% and 25% of POC flux at 150 and 300 m, respectively, and DON flux via migrating zooplankton comparable to 60% and >100% of PON flux at those depths, respectively (Steinberg et al., 2000, 2002).

C DOM Production via Cell Lysis

1 Viral Lysis and the Viral Shunt

Virioplankton are ubiquitous in the marine environment, with the majority believed to be phage or viruses that infect prokaryotic organisms (Breitbart, 2012; Suttle, 2007; Wommack and Colwell, 2000). Marine phage control microbial abundance and production, they are important agents of genetic exchange that can affect the microbial community structure (Breitbart, 2012; Suttle, 2007) and they are major biogeochemical agents that impact the cycling of micro and macro nutrients (Middelboe, 2008). There are ~10^{30} phage in the ocean, infecting prokaryotic cells at a rate of 10^{23} infections per second and removing 5-40% of prokaryotic standing stock on a daily basis (Middelboe, 2008; Suttle, 1994, 2007). Their contribution to bacterioplankton mortality is similar to that imposed by protozoan grazers (Fuhrman and Noble, 1995). Viruses also infect eukaryotic cells (Bratbak et al., 1992; Gobler et al., 1997; Suttle, 1994) and influence the termination of phytoplankton blooms (Brussaard, 2005). Thus, marine virioplankton are considered the most abundant predatory agents in the sea (Breitbart, 2012; Suttle, 2007). The production of viral progeny through lytic infection causes the death and lysis of the host, resulting in the release of DOM lysate. Coined the "viral shunt," this process effectively converts POM to DOM. An estimated 25% of annual ocean PP flows through the viral shunt, producing ~3-20 Pg of DOC each year (Wilhelm and Suttle, 1999).

a BIOGEOCHEMICAL SIGNIFICANCE

Viral lysate is composed of labile, dissolved deoxyribonucleic acids (DNA) (Holmfeldt et al., 2010), DFAAs and dissolved combined amino acids (DCAA) (Middelboe and Jorgensen, 2006), carbohydrates (Weinbauer and Peduzzi, 1995), as well as more recalcitrant cellular debris and colloidal particles (Shibata et al., 1997). Many of the components are utilized quickly, meeting 10-40% of bacterial carbon demand (BCD) (Middelboe, 2008). Using radioactively labeled lysate, Noble and Fuhrman (1999) demonstrated that much of it turned over on time scales of a day. Studies with bacterial isolates have shown that only 28% of viral lysate is converted to non-infected bacterial biomass (Middelboe et al., 2003). In contrast to protozoan grazing that passes a considerable fraction of microbial organic matter to the "classical food web," the viral shunt diverts organic matter toward microbially mediated recycling processes. Bacterial recycling of viral lysate represents a semi-closed trophic loop that stimulates metabolism of noninfected bacteria ("viral priming"), with the net effect of inorganic nutrient regeneration and increased DOC flux through bacterioplankton (Bonilla-Findji et al., 2009; Fuhrman, 1999; Middelboe and Lyck, 2002; Suttle, 2007; Weinbauer et al., 2011; Wilhelm and Suttle, 1999). Together, these observations indicate that viruses are a major influence in biogeochemical cycling in the global ocean.

2 Bacterial Lysis

Bacterially induced lysis of phytoplankton (Imai et al., 1993; Stewart and Brown, 1969) and other bacteria (Martin, 2002; Zusheng et al., 1998) has received relatively little attention as a potential contributor to DOM production. Stewart and Brown (1969) first observed lysis of cultured algae and bacteria by *Cytophaga* sp. isolated from sewage. Later, Imai et al. (1993) found that gliding

Cytophaga sp. isolated from a marine system had a rather large prey range, capable of lysing at least ten marine phytoplankton species. Long and Azam (2001) report that approximately half of bacterial isolates studied demonstrate antagonistic activity toward other prokaryotes through the production of antibiotics, enzymes, or through predatory behavior. This activity is especially true for organisms with a particle-attached lifestyle. Bacterially induced lysis by gliding bacteria may help regulate the colonization of macroaggregates (Imai et al., 1993). The production and release of hydrolytic enzymes by bacteria in close proximity (i.e., in biofilms or on marine particles) can result in nonspecific degradation and lysis of neighboring bacteria (Lin and McBride, 1996). For example, the production of peptidoglycan hydrolase by one group of bacteria can lyse surrounding dissimilar bacteria (Kadurugamuwa and Beveridge, 1996; Zusheng et al., 1998). Predatory bacteria like the gammaproteobacterium *Bdellovibrio* can burrow into the periplasm of prey bacteria, consume the labile components, and lyse their prey, thus releasing cytosol into its environment (Martin, 2002). The importance of bacterially induced lysis to DOM production remains unquantified.

3 Allelopathy

Allelopathic interactions occur when one microbial species releases biochemicals that actively inhibit competitors (Wolfe, 2000). The field of microbial chemical ecology is largely understudied (Wolfe, 2000) but allelopathic interactions may be an important source of direct and indirect production and release of DOM by microbes. Algicidal bacteria have been implicated in the regulation of phytoplankton blooms and release of DOM but further work is necessary (see review by Mayali and Azam, 2004). Protistan phytoplankton release specific organic compounds such as DFAAs (i.e., valine, cysteine, proline, alanine, and serine) and dimethyl-sulfoniopropionate (DMSP) to deter protozoan grazing and potentially promote harmful algal blooms of phytoplankton (Strom, 2008; Weissbach et al., 2010; Wolfe, 2000). Although the concentrations of individual allelopathic compounds are measured at nanomolar levels, the secondary effect of these "toxins" may impact DOM production by the affected species. For example, Tillmann (2003) demonstrated that the harmful algal bloom species *Prymnesiou parvum* (a mixotrophic flagellate) released the toxin prymesin, which can exert lytic effects on its dinoflagellate predator. Weissbach et al. (2010) showed that compounds extracted from the dinoflagellate *Alexandrium* sp. enhanced both lysis of the prymnesiophyte *Phaeocystis globosa* and the flux of DOM to bacterioplankton as a result of allelopathic interaction. "Allelopathic" is a relative term describing a specific form of extracellularly released DOM: compounds perceived as allelopathic to a targeted competitor may serve as neutral or even labile DOM resources to a minority or majority of bacterioplankton consumers.

D Solubilization of Particles

Particle aggregates range from submicron colloidal material (Wells, 2002; Wells and Goldberg, 1993) to macroscopic marine snow (Alldredge and Silver, 1988) enriched in organic and inorganic nutrients relative to the ambient seawater (Simon et al., 2002). The dominant biopolymers include carbohydrates, proteins, and lipids (Grossart and Ploug, 2001; Kiørboe and Jackson, 2001; Kiørboe et al., 2003; Simon et al., 2002; Skoog and Benner, 1997), with larger aggregates having lower C:N ratios (Alldredge, 1998; Mueller-Niklas et al., 1994) compared to smaller suspended POM or DOM. Aggregates can harbor orders of magnitude higher densities of "attached" bacteria per unit volume compared to the surrounding water (Alldredge and Gotschalk, 1990; Smith et al., 1995); these microbial assemblages are distinct from the free-living bacterioplankton (DeLong et al., 1993; Moeseneder et al., 2001; Rath et al., 1998).

Bacterial extracellular enzymes (exo- or ectoenzymes) are of fundamental importance in the hydrolysis of polymeric POM (Karner and Herndl, 1992; Simon et al., 2002; Smith et al., 1992, 1995) and DOM (Arnosti, 2011; Arnosti et al., 2005a; Christian and Karl, 1995; Hoppe, 1991). The production of homoserine lactone by attached microbial specialists may play an important role in quorum sensing and hydrolysis of sinking particles (Hmelo et al., 2011). Attached microbial assemblages express proteases, phosphatases, β-glucosidase, and chitinase in unique proportions (Arnosti, 2011; Karner and Herndl, 1992; Smith et al., 1992) and at greater activity rates compared to those produced by free-living cells in the surrounding water. The transformation of POM to DOM via hydrolysis is referred to as solubilization; this process can be uncoupled from substrate uptake and remineralization rates of attached bacteria (Smith et al., 1992). In the Southern California Bight, solubilization of POM to DCAA was uncoupled from bacterial uptake, with 50-98% of the released DCAA escaping rapid utilization by attached bacteria (Smith et al., 1992). This uncoupled solubilization can result in the release of a nutrient- and DOM-rich plume that is 10- to 100-fold greater in volume than the aggregate itself (Kiørboe and Jackson, 2001; Kiørboe et al., 2001; Figure 3.5). Particle attached microbial specialists are important in transforming POM to DOM or CO_2, thereby attenuating sinking particle flux as well as potentially providing substrate for free-living microbial plankton (Cho and Azam, 1988; Kiørboe and Jackson, 2001). These plume microenvironments, known as "hot spots," create heterogeneity in the ocean interior, further stimulating remineralization by free-living microbes (Azam, 1998; Azam and Long, 2001; Kiørboe and Jackson, 2001; Reinthaler et al., 2006).

Growth of particle attached microbial specialists on aggregates has several biogeochemical implications. First, increased hydrolysis and remineralization of sinking aggregates attenuates the sinking flux at a shallower depth,

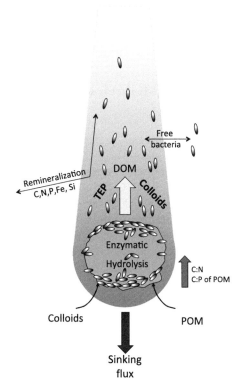

Schematic representation of a sinking organic aggregate functioning as an "enzymatic reactor." Attached bacteria express high levels of hydrolytic extracellular enzymes that mediate the production of DOM, colloidal material, and an inorganic plume by solubilizing sinking POM. These plumes form nutrient-rich "hot spots" that attract and fuel free-living bacterioplankton throughout the water column. *Adapted from Smith et al. (1992) and Azam and Malfatti (2007).*

potentially reducing passive carbon flux. Second, the release of HMW polymers via hydrolysis may produce slowly degraded DOM. Third, differential solubilization of organic nitrogen and phosphorus relative to carbon (Grossart and Ploug, 2001; Simon et al., 2002; Smith et al., 1995) can lead to qualitative transformation of sinking aggregates as well as retention of bioavailable nitrogen and phosphorus in the euphotic zone, thereby increasing the C:N and C:P of sinking material.

E Prokaryote Production of DOM

1 Chemoautotrophy

In addition to photoautotrophy within the thin photic layer of the ocean, members of the *Bacteria* and *Archaea* domains can oxidize reduced inorganic compounds to provide the reduction potential required to fix CO_2 toward the formation of organic matter (i.e., chemoautotrophy (aka chemolithotrophy)). Chemoautotrophic processes have been identified in hydrothermal vent systems (Jannasch and Mottl, 1985) and in hypoxic and anoxic regions of the ocean (Taylor et al., 2001; Walsh et al., 2009), but recent work reveals chemoautotrophy is prevalent throughout the oxygenated deep water column as well (Hansman et al., 2009; Ingalls et al., 2006). Chemoautotrophic rates of CO_2 fixation supported by nitrification within the global dark ocean are conservatively estimated at ~0.11 Pg C year^{-1} for the mesopelagic and bathypelagic realms (Middelburg, 2011). The discovery of the *Thaumarchaeota* (originally marine group I *Crenarchaeota*) *Nitrosopumilus maritimus* led many to assume that CO_2 fixation in the dark ocean was fueled by the oxidation of ammonia (Herndl et al., 2008; Middelburg, 2011; Reinthaler et al., 2010) introduced with sinking particles or vertically migrating organisms. Indeed genomic studies have identified ammonia monooxygenase-A genes in the majority of mesopelagic *Archaea* (Swan et al., 2011). Reduced N may not be the only reductant used by deep chemoautotrophs. Swan et al. (2011) reported the potential roles of reduced sulfur, CO, and methane. Consistent with this finding, Reinthaler et al. (2010) reported DIC uptake in the dark ocean can reach 15-53% of the export production in some systems, exceeding rates supported by exported N by factors of 20-50 (i.e., the Middelburg analysis). Sulfur oxidation genes have been observed in genomes of mesopelagic bacterioplankton that also contain ribulose-1,5-bisphosphate carboxylase-oxygenase (RuBisCO), a key enzyme mediating the Calvin-Benson-Bassham cycle (Swan et al., 2011). Members of various bacterioplankton clades (e.g., *Deltaproteobacteria* SAR324, *Gammaproteobacteria* ARCTI96 BD-19, and some *Oceanospirillales*) contain RuBisCO genes; experimental assays have confirmed CO_2 fixation by members of the SAR324 clade (Swan et al., 2011). Chemoautotrophy represents a potentially important autochthonous source of new labile organic matter in a dark ocean that is otherwise dominated by refractory organic matter (Middelburg, 2011). However, the main energy source for chemoautotrophy remains a topic of investigation and the bioavailability of DOM released remains unknown.

2 Chemoheterotrophy

Extracellular release of DOM via overflow metabolism results from an uncoupling of catabolic and anabolic processes when bacteria are grown on energy- and nutrient-rich substrates (Carlson et al., 2007; Liu, 1998). The extent to which energy dissipation via overflow metabolism occurs in natural systems is unknown but many cultivated marine bacteria produce copious amounts of HMW mucous exopolymers under favorable growth conditions (Decho, 1990). The heterotrophic bacterioplankton *Lentisphaera araneosa*, when grown in isolation, produces copious amounts of TEP, significantly altering the viscosity of their growth media (Cho et al., 2004). These secretions maintain a stable microenvironment, contain extracellular enzymes, as well as sequester and concentrate nutrients. Stoderegger and Herndl (1998, 1999) reported a bacterial capsular material release rate of 0.36 fmol C cell^{-1} d^{-1}; however, not all bacteria produce capsular material or exopolymers. Heterotrophic bacterioplankton also directly release DOM in the form of hydrolytic enzymes (Arnosti, 2011), siderophores for trace metal acquisition (Vraspir and Butler, 2009) and quorum-sensing compounds (i.e., acyl homoserine lactones; Hmelo et al., 2011). Using high-resolution mass spectrometry, Kujawinski et al. (2009) demonstrated the release of organic metabolites specific to lineages

of bacterioplankton. The contribution of chemoautotrophic and heterotrophic release of DOM in the oceans remains unquantified at this time.

III DOM REMOVAL PROCESSES

Despite a range of 30-8543 mg C m^{-2} d^{-1} (285-fold range in variability) for depth integrated PP (Behrenfeld and Falkowski, 1997), DOM concentrations are maintained in a remarkably narrow range of 34-80 µmol kg^{-1} C throughout the world ocean (Hansell et al., 2009; Figure 3.1). Given that DOM production can be an important fraction of PP, this striking offset between the range in PP rates and DOC concentrations indicates that several biotic and abiotic DOM removal processes are at work (Figure 3.2).

A Biotic Consumption of DOM

1 Prokaryotes

A major metabolic strategy in the contemporary aerobic ocean is chemoheterotrophy where heterotrophic prokaryotes use organic matter as both a carbon source and an electron donor. Heterotrophic prokaryotes (*Bacteria* and *Archaea*) represent the most abundant organisms in the sea, with ~1.2×10^{29} cells in the oceanic water column (Whitman et al., 1998). Most of the literature regarding heterotrophic prokaryotes have focused on bacterioplankton but recent studies have demonstrated that members of the domain *Archaea* can comprise ~20-40% of total prokaryotic cells in the mesopelagic and bathypelagic realms (Church et al., 2003; Herndl et al., 2005; Karner et al., 2001; Ouverney and Fuhrman, 2000). Some members of *Archaea* are capable of heterotrophic uptake of DOM (Ouverney and Fuhrman, 2000; Reinthaler et al., 2006; Teira et al., 2004, 2006), accounting for 10-80% of total heterotrophic prokaryotic production in the deep (dark) ocean. For ease of comparison and in keeping with the current literature, the term "bacterioplankton" is used here to describe heterotrophic prokaryotes in terms of their abundance and activity hereafter in this chapter.

Bacterioplankton comprise the majority of living surface area in the global ocean (Pomeroy et al., 2007). Membrane surfaces are important sites where the interactions between biology and ocean chemistry (geochemistry) occur. Ectoenzymes on these surfaces and in periplasmic spaces carry out the majority of nutrient and carbon transformations in the ocean. Prokaryotes are limited to the uptake of LMW (~600 Da; Dalton = atomic mass) compounds through their cell membranes via permeases, thus they must hydrolyze HMW polymeric organic matter to LMW fragments prior to transport (Amon and Benner, 1994; Arnosti, 2011; Arnosti et al., 2005b; Christian and Karl, 1995; Hoppe et al., 2002). The large active living surface, together with a small cell size (typically 0.1-0.8 µm in diameter), afford bacterioplankton with a high surface area-to-volume ratio, providing a high affinity and competitive advantage for substrates at low concentrations (Pomeroy et al., 2007).

Bacterioplankton are the principal consumers of DOM in the ocean. Their ecological role within the microbial loop facilitates the transformation of DOM to POM (Azam et al., 1983; Pomeroy et al., 2007) or the remineralization of DOM to its inorganic constituents (Ducklow et al., 1986; Goldman and Dennett, 2000). There are numerous biotic (viral, grazing) and abiotic (nutrient limitation, UV exposure, temperature, pH) controls on heterotrophic bacterioplankton that affect their biogeochemical roles in nutrient cycling and in the transformation of organic matter to more recalcitrant states (Benner and Herndl, 2011; Cowie and Hedges, 1994; Goldberg et al., 2011; Jiao et al., 2010; Skoog and Benner, 1997). See the excellent review by Church (2008) regarding the numerous resource controls on bacterioplankton dynamics.

Incorporation of radioactive tracers (i.e., ^{3}H-thymidine or ^{3}H-leucine) into bacterioplankton biomass (BB) is a common approach to estimate net BP (Ducklow, 2000) or net flux of DOC into BB. Many studies have employed these measurements and the factors needed to convert the incorporation rates to carbon units (see reviews by Ducklow and Carlson, 1992; Ducklow, 2000). In a cross system analysis covering large spatial and temporal ranges, Cole et al. (1988) concluded that net BP averaged 31% of local PP. Similarly, Ducklow (1999) concluded that net BP generally comprised <20% of local PP (Figure 3.6). While the rate of net BP appears modest compared to local PP, it is the gross flux of DOC or the BCD needed to support net BP that is most important in constraining energy and carbon flux through the microbial food web. As there is no direct measure of gross DOM flux, it must be derived from proxies including net BP, bacterioplankton respiration (BR), and bacterioplankton growth efficiency (BGE).

a BACTERIAL GROWTH EFFICIENCY

The amount of BB produced per unit of DOC consumed (BGE) is a basic property that determines the biogeochemical and ecological role of bacterioplankton. At the cellular level, BGE is a measure of the coupling between anabolic (biosynthetic; energy requiring) and

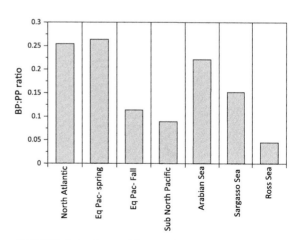

FIGURE 3.6 Ratio of bacterial production (BP) to primary production (PP) for various oceanic systems. All sample sites were occupied as part of the Joint Global Ocean Flux Study (JGOFS). *Figure adapted from Ducklow (1999).*

catabolic (metabolic breakdown; energy yielding) reactions (del Giorgio and Cole, 2000; del Giorgio and Williams, 2005; Carlson et al., 2007). Through respiration, chemoheterotrophs generate an electrochemical potential when electrons flow from reduced compounds through the electron transport system (ETS) to an electron acceptor (O_2 in most of the oceanic water column). That energy is collected, converted, and stored in the high-energy phosphate bonds of adenosine-5'-triphosphate (ATP). Subsequent hydrolysis of ATP releases the energy to drive anabolic processes; however, cells also expend energy independently of biomass synthesis, such as on maintenance metabolism (Russell and Cook, 1995). Diverting energy into these alternate pathways ultimately decouples anabolism from catabolism, with the ratio of anabolism:catabolism setting the BGE of a cell (Figure 3.7). Maintenance metabolism keeps cellular entropy low by repairing cellular damage and maximizing the functional integrity of the cell. Cells divert energy into maintenance metabolism to keep membranes and the ETS energized and functional, thereby keeping cells in a "physiologically active" state and thus poised to take advantage of ephemeral but favorable conditions such as patchiness of labile DOM and inorganic nutrients (growth response). Consequently, cells appear to sacrifice BGE to optimize their growth response (Carlson et al., 2007; del Giorgio and Cole, 2000; Westernhoff et al., 1983).

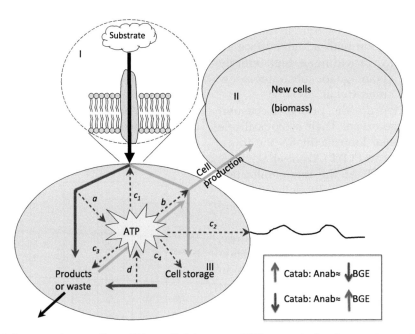

FIGURE 3.7   Substrate and energy flow within a cell. Substrates <600 Da are actively taken up via membrane proteins (I). As the substrate enters the cell it either enters catabolic (blue lines) or anabolic pathways (green lines). Monomers for anabolism can come preformed from the environment or as products of catabolism. The red-hashed lines represent the flow of energy to and from these metabolic pathways. Energy is conserved via substrate catabolism and ATP is produced at a rate a; in the absence of exogenous organic substrates the cell can yield ATP at rate d by catabolizing storage material (endogenous substrates). As ATP is hydrolyzed, energy is released and utilized at rate b to drive anabolic processes such as production of new cells (growth; II) and production cell storage products (III). Energy is also utilized at various rates to support processes that are independent of anabolism. This maintenance energy is used at rates c_1 to activate uptake systems, c_2 to fuel cell motility, c_3 to actively eliminate waste and c_4 repair cellular machinery. *Adapted from del Giorgio and Cole (1998) and Carlson et al. (2007).*

At the population level there are numerous factors affecting BGE, including temperature (Apple et al., 2006; Rivkin and Legendre, 2001), inorganic nutrient limitation (Goldman and Dennett, 2000; Goldman et al., 1987; Obernosterer et al., 2003; Tortell et al., 1999), energetic quality (i.e., how oxidized an organic substrate is; del Giorgio and Cole, 1998), viral interaction (Fuhrman, 1999; Middelboe et al., 1996, 2003), and diversity of heterotrophic prokaryotes (Reinthaler and Herndl, 2005; Table 3.2). Community BGE is driven by an integration of numerous variables rather than any single environmental condition, thus patterns of BGE versus single environmental factors are often conflicting (Table 3.2). Carlson et al. (2007) introduced a conceptual model demonstrating the relationship between BGE and environmental stressors or "hostility," where environmental "hostility" summarized a combination of resource limitation (i.e., organic matter or nutrients) with various levels of physical or chemical stressors (i.e., temperature, UV, pH). The more "hostile" the environment, the more energy diverted to maintenance metabolism, thus decreasing the BGE. The theoretical maximum yield of biomass per unit ATP (i.e., 32 g biomass dry weight per mole ATP) roughly corresponds to a BGE of 88% (Russell and Cook, 1995; del Giorgio and Cole, 2000); however, the BGE in natural systems is generally <50%, with a tendency to increase from <10% in oligotrophic systems to >20% in eutrophic coastal systems (see reviews by del Giorgio and Cole (2000)) and Carlson (2002)).

Measuring BGE in natural systems challenges microbial ecologists due to the shortcomings of current experimental designs (del Giorgio and Cole, 1998; Robinson and Williams, 2005). BGE can be expressed by the formula

$$BGE = BP/(BP + BR), \qquad (3.1)$$

where BR is bacterioplankton respiration. BGE can also be estimated empirically by growing natural assemblages of bacterioplankton on ambient DOM in a closed system and directly measuring changes in BB relative to DOC consumption. Here, BGE is expressed as:

$$BGE = \Delta BB/\Delta DOC, \qquad (3.2)$$

where ΔBB is change in BB and ΔDOC is the change in DOC concentration, from the initiation of an experiment through the beginning of stationary phase (Carlson et al., 1999, 2004; Halewood et al., 2012; Nelson et al., 2013).

The advantages, disadvantages, and associated problems of short- and long-term incubations in determining BGE are described by del Giorgio and Cole (1998). Short-term incubations should be more ecologically relevant; however, it is often not possible to measure changes in natural substrates on time scales of hours. The use of radiolabeled substrates (Williams, 1970, 1981) allows detection of substrate changes with high sensitivity and minimizes incubation time, but model compounds do not necessarily reflect ambient DOM quality. Seawater culture techniques, which separate bacterioplankton from grazers, allow assessment of bacterial dynamics and resolution of concentration changes of natural substrates or oxidation products (Biddanda et al., 1994; Bjørnsen, 1986; Carlson and Ducklow, 1996; Carlson et al., 1999, 2004; Cherrier et al., 1996; Coffin et al., 1993; Halewood et al., 2012; Hansell et al., 1995; Kähler et al., 1997; Nelson et al., 2013). However, these approaches require long incubations that introduce bottle effects such as increasing impacts by grazers, wall growth, and prokaryote community structure shifts. Additionally, grazer dilution experiments do not reduce top down control by viruses. Finally, the reduction in grazing activity can also affect bacterial physiology because nutrient recycling can be reduced (Christian and Karl, 1995). As a result, seawater cultures may indicate bacterial dynamics that are not entirely representative of in situ dynamics, giving lower BGE with longer incubations. However, del Giorgio and Cole (1998) did not find a systematic effect of incubation duration on BGE.

TABLE 3.2 Effects of Evironmental Parameters on Bacterial Growth Efficiency (BGE) in Natural Bacterioplankton Populations

Env. Factor	Description	Reference
Trophic status	Strong positive relationship between BGE and NPP (BGE <5% to >60%)	del Giorgio and Cole (1998)
	Strong positive relationship between BGE and Chl *a* (BGE <10% to >40%)	Biddanda et al. (2001)
	Strong positive relationship between BGE and bacterial growth rate (proxy for C availability) along North Pacific inshore–offshore transect; BGE <5% in oligotrophic system to >20% in coastal eutrophic system	del Giorgio et al. (2011)
Temperature *Large range*	In cross-system analysis over temperature range of 0-30°C there was apparent inverse relationship between BGE and temperature (BGE <5% to >60%)	Rivkin and Legendre (2001)
	With a given system and over large temperature ranges (i.e., 0-30°C for Maryland estuaries and 0-22°C for coastal North Sea) there appears to be an inverse relationship between BGE and temperature (BGE <5% to >60%)	Apple et al. (2006), Sintes et al. (2010)
Small range	Over an annual cycle in the North Sea there was no apparent relationship between BGE and temperature at 10-20°C (BGE <5% to >30%)	Reinthaler and Herndl (2005)
	In the Ross Sea, BGE showed systematic increase over the course of a phytoplankton bloom, with BGE increasing from 3% to 38% despite narrow temperature range of 0-2°C	Carlson and Hansell (2003)
Substrate stoichiometry	The substrate C:N molar ratio has inverse relationship with BGE in natural complex media and defined media at substrate C:N ratios <10	del Giorgio and Cole (1998), Goldman et al. (1987)
	Natural populations of bacterioplankton using multiple substrate sources (organic and inorganic), with substrate C:N ratios as great as 30, do not reveal systematic relationships between BGE and substrate C:N ratio; bulk substrate C:N does not reflect bioavailable fraction; other nutrients may be limiting	Goldman and Dennett (2000)
Quality of substrate	Inverse relationship between BGE and hydrolytic enzyme activity suggests hydrolysis of polymeric compounds requires energy and reduces BGE	Middelboe and Søndergaard (1993)
	Utilization of relatively oxidized substrates yields low BGE despite high supply and presence of inorganic nutrients due to low substrate C and energy content	del Giorgio and Cole (1998), Linton and Stephenson (1978), Vallino et al. (1996), Apple and del Giorgio (2007)
Light and proteorhodopsin	Positive relationship between biomass yield (BGE) and light exposure in marine *Flavobacteria*	Gómez-Consarnau et al. (2007)
	No relationship between biomass yield and light exposure with the proteorhodopsin-containing alphaproteobacteria *Pelagibacter ubique*	Giovannoni et al. (2005)
Viral impact	Viral infection lyses bacteria to form DOM; increased shunt of bacterial biomass to DOM elevates bacterial respiration and lowers BGE of bacterioplankton population	Fuhrman (1999), Motegi et al. (2009), Bonilla-Findji et al. (2009)
	BGE of non-infected populations of bacterioplankton decreases due to energy requirements to degrade polymeric DOM lysates, including cell membrane, cell wall, and colloidal material	Middelboe et al. (1996), Middelboe and Lyck (2002)
Microbial community structure and richness	BGE in the southern North Sea is negatively related to bacterioplankton richness (i.e., BGE decreases as number of bacterial OTUs increase)	Reinthaler et al. (2005)

b BACTERIAL CARBON DEMAND

It is possible to approximate BCD by combining measurements of BP with estimates of BGE, such that,

$$BCD = BP/BGE. \qquad (3.3)$$

Given this relationship, low BGE's have important implications for the biogeochemistry of aquatic systems. For example, if under high environmental hostility bacterioplankton maximize their metabolic flexibility by partitioning more energy into maintenance metabolism, then BGE will be low. A lower BGE requires a larger flux DOC to support an observed net BP. Thus, BCD can represent a significant fraction of local PP even if net BP:PP is low. The lower BGE determined for ambient bacterioplankton grown on natural substrates can result in calculated BCD exceeding phytoplankton production (del Giorgio and Cole, 2000). Figure 3.8 demonstrates the relationship between integrated BCD and integrated PP for a variety of oceanic sites. When PP is $>1\,g\,m^{-2}\,d^{-1}$, local PP and the resulting flux of DOC from food web processes is sufficient to support BCD. But with $PP < 1\,g\,m^{-2}\,d^{-1}$, BCD is similar to or greater than local PP, indicating a mismatch between demand and production of DOC. One must use caution when interpreting BCD data because considerable variability in BGE exists within a given system (Carlson and Ducklow, 1996; Carlson et al., 1999; Eichinger et al., 2006; Halewood et al., 2012; Jahnke and Craven, 1995). In addition, conversion factors to translate BP to carbon units have an enormous range (see Ducklow and Carlson, 1992), making BP and hence BCD relatively unconstrained (Anderson and Ducklow, 2001). It should also be noted that these BCD to PP comparisons do not include measures of DOM production, which can amount to 10-50% of particulate PP (see Section II), thus leading to overestimates of the BCD:PP ratio.

Direct measures of BR could better constrain BCD, expressed as:

$$BCD = BP + BR. \qquad (3.4)$$

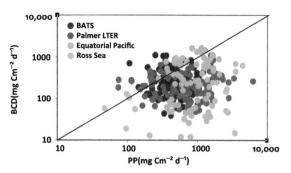

FIGURE 3.8 Integrated bacterial carbon demand (BCD) and integrated primary production (PP) within the euphotic zone of representative ocean sites. BGE of 0.1 was used to estimate BCD. The black line is the 1:1 line. Data points that lie above this line indicate that BCD was greater than local primary production at the time of sample collection. Note that PP is particulate primary production only and does not include ER; thus BCD:total PP is an overestimate. The BATS data represent monthly values from 1991 to 2003 ($n=155$; see Steinberg et al., 2001 for details; data available at http://bats.bbsr.edu/). Paired BP and PP from the Equatorial Pacific ($n=16$) and Ross Sea, Antarctica ($n=77$) were calculated according to Ducklow (1999) (data are available at http://usjgofs.whoi.edu/jg/dir/jgofs/). All data from the Palmer Peninsula, Antarctica ($n=112$) are courtesy of H. Ducklow and the Palmer LTER program. *Adapted from Carlson et al. (2007).*

BR is most commonly assessed by changes in O_2 (del Giorgio and Williams, 2005) or DIC (Hansell et al., 1995) during short-term (12-24 h) *in vitro* dark incubations. However, despite the importance of the BR, there is a paucity of direct measurements in the open ocean (<500 as of 2005; del Giorgio and Williams, 2005). Technological challenges, such as physical separation of bacterioplankton from other heterotrophic plankton and phytoplankton, the limit of detection of the analytical techniques, as well as other contamination and procedural errors (see review by Robinson, 2008) account for this paucity of BR data. Robinson (2008) reports the median (mean ± standard deviation) BR to be 0.5 (1.3 ± 2.3; $n=105$) and 2.7 (7.1 ± 12.2; $n=315$) $mmol\,C\,m^{-3}\,d^{-1}$ for ocean and coastal systems, respectively. To properly constrain BCD and the flux of labile DOC, further development regarding direct (accurate) measurements of BR are necessary (del Giorgio and Williams, 2005; Jahnke and Craven, 1995; Robinson, 2008; Williams et al., 2004).

c PHOTOHETEROTROPHY

Photoheterotrophy and the diversity of organisms that are capable of this process are discussed in excellent reviews (Béjà and Suzuki, 2008; Yurkov and Csotonyi, 2008; Zubkov, 2009) and will be considered only briefly here. Historically, microbial ecologists have considered DOM as the primary carbon and energy source fueling heterotrophic bacterioplankton in the oceanic euphotic zone. However, recent work has shown that several groups of bacterioplankton can use sunlight as an alternative energy source while at the same time consuming DOM to meet carbon and supplemental energy requirements (Béjà and Suzuki, 2008; Béjà et al., 2000, 2001; Karl, 2002; Kolber et al., 2001). The three basic photoheterotroph groups include facultative photoheterotrophs (i.e., cyanobacteria; Church et al., 2004; Michelou et al., 2007; Zubkov and Tarran, 2005), bacteriochlorophyll-based aerobic anoxygenic photoheterotrophic (AAnP) bacteria (Kolber et al., 2000, 2001), and bacteria that use the pigment proteorhodopsin (Béjà and Suzuki, 2008; Béjà et al., 2001).

The oxygenic photosynthetic *Prochlorococcus*'s genome has revealed that some ecotypes have oligopeptide and sugar transporters, indicating the potential for a partially heterotrophic lifestyle (Rocap et al., 2003). *Prochlorococcus* and *Synechococcus* enhance uptake of DFAA, urea, and DMSP during daylight in oligotrophic systems (Béjà and Suzuki, 2008; Church et al., 2004; Malmstrom et al., 2005b; Michelou et al., 2007). Thus, facultative heterotrophy must be more energetically favorable than complete reliance upon de novo synthesis of organic matter. AAnP, thought to comprise 2-11% of upper ocean bacterioplankton abundance (Schwalbach and Fuhrman, 2005), use bacteriochlorophyll to harvest light energy but consume DOM to meet ~80% of their energetic demands (Yurkov and Csotonyi, 2008). Numerous phylogenetic groups of heterotrophic bacterioplankton contain genes for the synthesis of proteorhodopsin, a photoactive protein that drives a proton pump for ATP synthesis (Béjà and Suzuki, 2008; de la Torre et al., 2003). Some groups of proteorhodopsin-containing bacterioplankton (i.e., members of the *Flavobacteria*) increase production rates in the presence of light (Gómez-Consarnau et al., 2007, 2010), while the role of proteorhodopsin is less obvious in cell growth for members of the abundant *alphaproteobacteria* SAR11 (Giovannoni et al., 2005). If photoheterotrophy leads to the generation of ATP, then more of the DOM taken up by cells can be assimilated into biomass, thus increasing the organism's BGE. The role of photoheterotrophs and its impacts on DOM dynamics remains a fascinating but currently unquantified area of research.

2 Eukaryotes

Several eukaryotes utilize labile DOM compounds directly. Colloidal DOM, with molecular weights from 55 to 2000 kDa and composed of carbohydrates and proteins, can be consumed by heterotrophic flagellates (First and Hollibaugh, 2009; Sherr and Sherr, 1988; Tranvik et al., 1993). Photoheterotrophic microalgae take up organic material via phagotrophy or osmotrophy to supplement their autotrophic metabolism under light or nutrient limiting conditions (Caron, 2000; Flynn et al., 2013; Rivkin and Putt, 1987). Michelou et al. (2007) demonstrated that phototrophic picoeukaryotes were responsible for measurable uptake of the amino acids but that their uptake rates were much lower compared to other prokaryotes, accounting for <3% of amino acid assimilation. Marine choanoflagellates and invertebrate larvae (i.e., urchins and bivalves) can utilize dissolved monosaccharides and DFAAs as a nutritional source (Manahan and Richardson, 1983; Marchant and Scott, 1993) while the isotopic composition of eel leptocephali indicate that DOM could serve as a food source (Otake et al., 1993). However, the magnitude of DOM removal by marine eukaryotes is likely to be small relative to heterotrophic prokaryotes.

B Abiotic Removal Processes

1 Phototransformation

UV excitation from sunlight impacts DOM removal both directly and indirectly. As refractory DOC (RDOC) is returned to the ocean surface via mixing a portion of it is photo-oxidized to CO_2 and to a lesser extent CO (Mopper and Kieber, 2002; Mopper et al., 1991; Moran and Zepp, 1997; Stubbins et al., 2006, 2012). Anderson and Williams (1999) estimated photochemical oxidation of RDOC to be ~65 mmol m^{-2} year^{-1}. Absorption of UV light by chromophores can also transform HMW DOM to biologically available LMW carbonyl compounds (Anderson and Williams, 1999; Benner and Biddanda, 1998; Kieber et al., 1989; Mopper et al., 1991; Moran and Zepp, 1997), facilitating the uptake and remineralization of DOM by heterotrophic bacterioplankton. Cherrier et al. (1999) showed that surface bacterioplankton DNA were depleted in $\Delta^{14}C$ relative to modern DIC, proposing the phototransformation of RDOC ($\Delta^{14}C$ depleted) to labile DOC and subsequent uptake by bacterioplankton into DNA. Photochemical processes also release labile N and P compounds such as ammonium, amino acids, and phosphate that further fuel microbial production (Moran and Zepp, 2000). Thus, photolysis may represent a significant sink of DOM; however, this process is ultimately limited to the surface ocean where UV light is present. See Chapter 8 for more on DOM photochemistry.

Removal of recalcitrant forms of DOC are also observed in the ocean interior where UV does not penetrate (Carlson et al., 2010; Hansell et al., 2009, 2012), thus other abiotic removal processes must be considered.

2 Sorption of DOM onto Particles

Sorption of DOM onto surfaces such as sinking particles has been proposed as an abiotic removal mechanism (Druffel and Williams, 1990; Druffel et al., 1996, 1998). Druffel et al. (1996) observed a mean $\Delta^{14}C$ reduction of 93‰ for POC suspended in the deep ocean compared to that suspended in the upper layer, indicating that deep ocean POC was significantly older (depleted $\Delta^{14}C$ values). If true, ~14% of the carbon on suspended POC in the deep Pacific could be a result of sorption of old DOC (Druffel and Williams, 1990) onto sinking aggregates. DOC removal by this process could account for ~1.4-2.8 nmol C kg^{-1} year^{-1} (Hansell et al., 2009) but further quantification of this process is required.

3 Condensation of Marine Microgels

HMW DOM spontaneously assembles into chemical and physical gels that are interconnected tangles of covalently cross-linked biopolymers called self-assembled microgels (SAGs) (Verdugo et al., 2012). Under subtle changes in physical (i.e., temperature) or chemical (i.e., pH) conditions, these swollen, neutrally buoyant SAGs can collapse into dense particles that can sink. The mass transfer from DOM to SAGs to dense sinking particles via abiotic gel formation represents an unquantified DOM removal process (see Chapter 9).

4 Hydrothermal Circulation

Circulation of fluids through porous basalts at high-temperature hydrothermal ridge-flanks can alter the chemistry of the flowing water. The outflow from these ridge-flank systems has been shown to have low DOC concentrations (i.e., 8-27 µmol L^{-1}) compared to surrounding off-axis water (Lang et al., 2006). Lang and colleagues estimated that circulation through such high-temperature vent systems could remove ~0.7-1.4×10^{10} g C year^{-1}, a relatively minor DOC sink compared to others discussed above.

IV DOM ACCUMULATION

As described in Section II, there are numerous DOM production mechanisms (Figure 3.2). While much of the newly produced DOM is rapidly processed by heterotrophic microbes (Section III), a portion resists or escapes microbial degradation to accumulate and persist. In this review, we refer to DOM that resists rapid microbial degradation and accumulates as recalcitrant DOM. The accumulation of DOC up to $80 \mu mol kg^{-1}$ in the oligotrophic subtropical gyres (Figure 3.1; Carlson et al., 2010; Hansell et al., 2009) is unambiguous evidence for the production of recalcitrant DOM.

There are a number of fractions of recalcitrant DOM that turn over on numerous time scales, detailed below (see Section V). Why does DOM accumulate? In the open ocean the net production of DOC is ultimately due to the decoupling of biological production and consumption processes (i.e., net community production; Hansell and Carlson, 1998b). While newly produced DOM may be biologically degradable, hydrolyzing it to a utilizable form may not be metabolically cost effective to heterotrophic bacterial under specific environmental conditions; thus, resulting in DOM accumulation (Floodgate (1995). Abiotic and biotic processes can release or transform DOM to molecular forms that are not available and shielded from rapid microbial degradation due to the intrinsic stability of the DOM (described in Sections IV.A and IV.B) or escape detection due to low concentrations of individual compounds (described in Section IV.C), thus allowing it to accumulate. See Chapter 7 for further discussion of long-lived DOM.

A Abiotic Formation of Biologically Recalcitrant DOM

Abiotic processes can lead to the restructuring of recognizable compounds into complex macromolecules that are not resolved by traditional (bio)chemical analysis and that impede biological degradation. The cross-linking polymerization of LMW DOM (condensation reactions catalyzed by light and metal complexation; Harvey et al., 1983) to compounds like melanodins (Maillard, 1912), the polymerization of LMW labile material (e.g., proteins; Hedges, 1988) such as the binding of monomers to macromolecular DOM (Carlson et al., 1985), and adsorption of labile DOC to colloids (Kirchman et al., 1989; Nagata and Kirchman, 1996) have been proposed as mechanisms that abiotically transform DOM to a form too complex for enzymatic degradation. DOM can spontaneously assemble into gels that under certain conditions (i.e., high temperature or low pH) may undergo abrupt condensation (Orellana and Hansell, 2012), immobilizing organic matter and associated metals and nutrients within an impermeable and recalcitrant organic submicron gel (Chin et al., 1998; Verdugo et al., 2012).

The effects of UV exposure on DOM are complex, yielding both labile components that stimulate heterotrophic microbes (Benner and Biddanda, 1998; Moran and Zepp, 1997; Mostofa et al., 2013) and recalcitrant DOM byproducts (Benner and Biddanda, 1998; Keil and Kirchman, 1994; Tranvik and Bertilsson, 2001; Tranvik and Kokalj, 1998). Benner and Biddanda (1998) found that exposure of euphotic zone DOM to UV irradiation reduced BP by 75% while exposure of deep DOM (150-1000 m) to UV enhanced BP by 40%. They concluded that the chemical composition of DOM dictates whether phototransformation produces bioavailable or bioresistant compounds. Recent high-resolution mass spectrometry confirms that UV exposure alters the molecular composition of DOM (Gonsior et al., 2009; Kujawinski et al., 2009). This alteration may impart resistance to microbial degradation, however UV can induce DNA damage to members of the bacterioplankton community (Meador et al., 2009), impeding DOM uptake.

The production of reactive oxygen species (i.e., singlet oxygen) during photolysis has been proposed as a mechanism that forms partially oxidized DOM, further reducing its bioavailability (Baltar et al., 2013; Cory et al., 2010; Scully et al., 2003). The incomplete combustion of biomass and fossil fuels produces thermogenic black carbon that becomes long-lived DOC after its introduction to the ocean (Dittmar and Paeng, 2009; Ziolkowski and Druffel, 2010).

B Biotic Formation of Recalcitrant DOM

1 Microbial Carbon Pump

The production of recalcitrant DOM compounds via heterotrophic microbial processes has been coined the "microbial carbon pump" (MCP) (Jiao et al., 2010, 2011). According to this conceptual model, a fraction of labile DOM metabolized by microbes is shunted to a form that resists further microbial degradation, being stored in the recalcitrant DOC pool for months to millennia. Unlike the biological carbon pump, which relies on vertical flux of organic matter to store carbon in the deep ocean, the MCP operates independently of depth, sequestering carbon in a recalcitrant form from the ocean surface to the deep sea.

Early experiments provided evidence consistent with the MCP conceptual model. Where heterotrophic microbes were provided labile model compounds as their sole carbon source (i.e., glucose or leucine), recalcitrant DOM accumulated in the growth media, resisting further bacterial remineralization for months (Brophy and Carlson, 1989; Heissenberger and Herndl, 1994) to a year (Ogawa et al., 2001). Heissenberger and Herndl (1994) showed that some fraction of the original ^{14}C-leucine growth substrate was transformed into recalcitrant HMW (>50,000 Da) DOM in bacterial batch culture experiments. Similarly, Tranvik (1993) found that ~3% of the initial glucose amended to seawater remineralization experiments was transformed to humic-like DOM within the first week of incubation.

Recently, high-resolution mass spectrometry applied to samples drawn from microbial remineralization experiments has characterized the chemical transformation of DOM via heterotrophic microbial processes. Gruber et al. (2006) reported that after 36 days of incubation 29% of the carbon initially amended as glucose remained as a resistant form of DOM. By comparing remineralization experiments with and without protistan grazers the same authors concluded that bacterioplankton, rather than grazers, were the primary shapers of complex recalcitrant DOM.

a DIRECT SOURCE VIA THE MCP

Biochemical characterization of DOM reveals that specific bacterial biomarkers, such as membrane proteins (i.e., porin P) (Tanoue et al., 1995), lipid components of lipopolysaccharides (Wakeham et al., 2003) and D-enantiomers of amino acids, unique to bacterial peptidoglycan (glucosamine and galactosamine; Kawasaki and Benner, 2006; McCarthy et al., 1998), are distributed throughout water column. Based on measurements of D-amino acids in the DOM pool, ~25% (165 Pg) of recalcitrant DOC is of bacterial origin (Benner and Herndl, 2011; Kaiser and Benner, 2008). These studies provide evidence that a significant portion the DOM pool originates from heterotrophic bacterioplankton.

How is bacterial DOM produced? It is likely that a significant fraction of the bacteria-derived DOM is released during cell death as bacterioplankton are lysed by viruses or egested during protistan bacterivory. Mitra et al. (2013) suggest that mixotrophs may be a source of complex recalcitrant DOM of bacterial origin as the protists egest undigested bacterial compounds. Viral lysis of prokaryotic organisms can yield structural lipids, D-amino acids, and diaminopimelic acid (DAPA) (biomarkers of peptidoglycan) as well as other colloidal material and cellular debris that are resistant to rapid microbial degradation (Middelboe and Jorgensen, 2006; Shibata et al., 1997).

Extracellular release of recalcitrant DOM by bacteria can occur as cells grow, elongate, and divide. The cell's peptidases cleave peptide bonds to release dipeptides and monopeptides from the cell. During exponential growth D-alanine can accumulate to relatively high concentrations as it is cleaved from peptidoglycan during transpeptidation and released from the cell. D-alanine is released in relatively high concentration when peptidoglycan bonds were cleaved for cell division. Kawasaki and Benner (2006) determined bacterial PER of DOM was 14-31% of bacterial growth in exponential phase, some of which persists as recalcitrant DOM. While some studies have demonstrated slow mineralization of peptidoglycan (Nagata et al., 2003), others have observed rapid turnover of the cell wall material (Middelboe and Jorgensen, 2006).

b MICROBIAL TRANSFORMATION

Prokaryotic ectoenzymes target specific chemical bonds within organic matter polymers, with POM and HMW DOM becoming transformed into other components as a result of this interaction. It is hypothesized that over time the most labile organic matter is consumed, leaving behind more recalcitrant components. Amon and Benner (1996) first proposed the size-reactivity continuum model, which suggests that as organic matter is decomposed it become less bioreactive and smaller in size, thus bioreactivity of organic matter decreases as follows:

$$POM \rightarrow HMW\,DOM \rightarrow LMW\,DOM,$$

where each size fraction consists of a continuum of composition, reactivities, and diagenetic states (Amon and Benner, 1996). The hypothesis is that through the production of hydrolytic enzymes and the preferential removal of compounds such as dissolved aldoses, amino sugars, and amino acids, prokaryotes diagenetically alter and transform organic matter to recalcitrant material (Amon and Benner, 1996; Cowie and Hedges, 1994; Goldberg et al., 2011; Skoog and Benner, 1997).

The diagenetic transformation can be assessed by comparing the relative contribution of specific compounds like dissolved combined neutral sugars (DCNS) (Cowie and Hedges, 1994; Goldberg et al., 2009; Skoog and Benner, 1997) or DCAA (Davis and Benner, 2007; Kaiser and Benner, 2009) relative to total DOC. Lower DCNS and DCAA yields as well as compositional shifts of individual sugars or amino acids (i.e., molar fractions) within marine sediments (Dauwe et al., 1999) or in the water column indicate the transformation of organic matter to more diagenetically altered material over depth and through time (Amon and Benner, 2003; Benner, 2002; Davis et al., 2009; Goldberg et al., 2010, 2011; Kaiser and Benner, 2012; Skoog and Benner, 1997).

c TIME SCALES OF DOM PERSISTENCE

Evidence of microbial transformation of organic matter and accumulation of bacterial biomarkers in the bulk DOM pool supports the MCP conceptual model yet details regarding the rates of recalcitrant DOM production, removal, and the contribution to annual carbon sequestrations via this process remain unknown at this point. Caution must be used when estimates of RDOC production rates are based on relatively short-term experiments (e.g., 6-12 months). For example, remineralization experiments conducted over months yield estimated RDOC production rates of 0.5-0.6 GtC year^{-1} (Brophy and Carlson, 1989); however, without following the fate of the experimentally produced RDOC those high rates probably do not apply to the natural ocean system. In fact, estimated removal rates based on mass balance of the RDOC pool (see below) may provide a better constraint on RDOC production. Based on these slower removal rates, the global RDOC production via the MCP has been estimated at 0.008-0.023 Pg C year^{-1} (Benner and Herndl, 2011), up to half the total RDOC export of 0.043 Pg C year^{-1} (Hansell et al., 2012). If true, then C sequestration via the MCP would be an order of magnitude less than the

C sequestered in the deep sea via the biological pump (i.e., POM flux >2000 m is 0.7 Pg C year^{-1}; Henson et al., 2012).

DOM that is persistent at one geographical location or depth can be become bioavailable at another (Carlson et al., 2011). This variability is most obvious when one considers that of the 1.9 Pg C of DOC that resists microbial degradation within and exported annually from the epipelagic only 0.2 Pg C survives to depths >500 m (Carlson et al., 2010; Hansell et al., 2009). The remineralization of exported DOC accounts for up to half the oxygen utilized within the mesopelagic zone (Doval and Hansell, 2000); even though the exported DOM was persistent in the surface waters it became bioavailable once transported to depth.

Field and experimental studies in the Bermuda Atlantic Time-series Study (BATS) site have shown that the seasonally accumulated surface DOC (Figure 3.9) that is resistant to microbial degradation by surface water bacterioplankton assemblage becomes bioavailable and removed on timescales of weeks once exported to the mesopelagic (Carlson et al., 1994, 2004; Hansell and Carlson, 2001; Figure 3.9). During or shortly following the annual convective mixing and DOC export there is a coincident increase in bacterioplankton abundance within the upper mesopelagic (140-250 m) at BATS (Carlson et al., 2009; Morris et al., 2005). Tracking changes in bacterioplankton community structure via 16S rRNA gene markers also reveal a marked shift in the mesopelagic bacterioplankton community structure during or shortly following mixing with taxa such as SAR11 (ecotype II) (Figure 3.10), marine *Actinobacteria*, SAR202, and OCS116 increasing their relative contribution to the total mesopelagic community following convective mixing (Carlson et al., 2009; Morris et al., 2005; Treusch et al., 2009; Vergin et al., 2013). These observations suggest that microbial community structure and their associated metabolic capabilities play an important role in controlling the utilization of recalcitrant DOM.

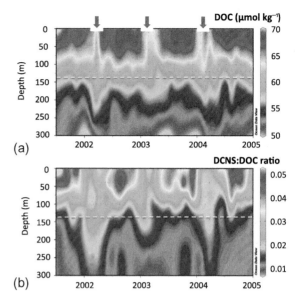

(a)

(b)

FIGURE 3.9 Time-series of DOC concentrations (a) and DCNS:DOC ratios (b) in the upper 300 m at the Bermuda Atlantic Time-series Study (BATS) site. DOC accumulates in the surface waters, persisting until deep convective overturn (red arrows indicate approximate timing of deep mixing). Upon mixing, surface layer DOC is exported to the mesopelagic zone (the zones are separated by dashed white line). Upon restratification a portion of the exported DOC is removed in the mesopelagic zone and the remaining DOC character, as indicated by the DCNS:DOC ratio, becomes diagenetically altered. *Data reformatted from Goldberg et al. (2009).*

2 Limitation of Microbial Growth and DOM Accumulation

Factors such as nutrient limitation (Church, 2008; Church et al., 2000; Obernosterer et al., 2003; Thingstad et al., 1997), microbial community structure and specific metabolic repertoire (Carlson et al., 2004; DeLong et al., 2006; Elifantz et al., 2007; Nelson and Carlson, 2012; Treusch et al., 2009), pressure (Tamburini et al., 2002, 2003), and temperature (Church, 2008; Ducklow and Shiah, 1993; Shiah and Ducklow, 1994) control where and when recalcitrant DOM becomes bioavailable.

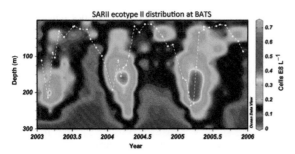

FIGURE 3.10 SAR11 subclade II cell densities (E8 L⁻¹) in the surface 300m from 2003 through 2005 at the BATS site. The white dashed line represents the mixed layer depth and is used to examine distribution patterns in the context of mixing and stratification. The data demonstrate a response of the mesopelagic subclade of the bacterioplankton SAR11 during or shortly following deep mixing. This response is presumed in part to result from the delivery of surface-derived organic matter (i.e., vertical DOC flux) into the mesopelagic during mixing. A portion of the DOC that persisted in the surface waters appears to become available to the organisms in the mesopelagic zone, triggering biomass production. *Adapted from Carlson et al. (2009).*

Given that accumulated DOM is recalcitrant, the growth of heterotrophic bacterioplankton should be initially limited by the availability of labile DOM (Carlson et al., 1996, 2002; Cherrier et al., 1996; Church, 2008; Church et al., 2000; Kirchman, 1990; Kirchman et al., 1990; Williams, 2000). This view has been challenged by the "malfunctioning microbial loop" hypothesis (Thingstad et al., 1997). This hypothesis states that competition for limiting inorganic macro or micro nutrients (Zweifel et al., 1993; Cotner et al., 1997; Thingstad et al., 1998; Church et al., 2000; Obernosterer et al., 2003; Mills et al., 2008) and grazing pressure (Thingstad and Lignell, 1997; Zweifel, 1999) reduce the bacterioplankton growth rate, biomass, and carbon demand to levels that allow accumulation of biodegradable DOC during biologically productive seasons. The results of experiments that directly assess inorganic nutrient limitation on bioavailability of DOM consumption are conflicting (see review by Church, 2008) and require further study.

Low temperatures may inhibit BP (Pomeroy et al., 1991; Shiah and Ducklow, 1994) and so foster DOM accumulation (Zweifel, 1999) but in the Ross Sea, Antarctica there is little evidence to support such temperature regulation (Carlson and Hansell, 2003; Ducklow et al., 2001). Temperature regulation may be more important in systems that demonstrate large seasonal temperature ranges. High pressure in the deep sea may affect bioavailability of DOM. For example, deep-sea piezophilic microorganisms display metabolic capabilities allowing them to thrive under cold, high-pressure conditions (Tamburini et al., 2003). Some piezophiles are capable of degrading complex organic matter (i.e., chitin, cellulose) by activating specific metabolic pathways under pressure (i.e., 28MPa) but turn those pathways off at atmospheric pressure (Lauro and Bartlett, 2007; Vezzi et al., 2005).

3 Eukaryote Source of Recalcitrant DOM

Eukaryotes produce recalcitrant DOM either directly or via food web interactions. The direct release of unmodified compounds such as acyl heteropolysaccharide (APS) from marine diatoms and haptophytes (Aluwihare and Repeta, 1999) leads to the accumulation of this metabolically resistant and dominant polysaccharide in the surface ocean (Aluwihare et al., 1997, 2002). The inverse relationship between hydrophobic humic DOM production and diatom biomass degradation (Lara and Thomas, 1995) indicate that cellular components of eukaryotic cells (i.e., cell wall material) may be a source of recalcitrant marine DOM. Cultures of the chrysophyte *Aurecoccus anophagefferens*, when incubated with active viruses, released 19% of the total cellular C content as DOC but its bioavailability to bacterioplankton was less than for DOC released from uninfected cultures (Gobler et al., 1997). The production of liposome-like colloids during microzooplankton grazing has been also been proposed as a mechanism of recalcitrant DOM production by eukaryotes (Nagata, 2000; Nagata and Kirchman, 1992a, 1999). These studies highlight the importance of eukaryotic processes in the production of recalcitrant DOM.

C Neutral Molecules and Preservation

The mechanisms discussed above require that DOM be physically, chemically, or biologically altered to a molecular structure that appears to "shield" the labile components from biological oxidation (Borch and Kirchman, 1999). This shielding is referred to as the "intrinsic stability hypothesis" (see Chapter 7). However, the molecular structure that can explain long-term stability has largely escaped detection (see Chapter 7). Dittmar and Kattner (2003) found that 40% of the deep Arctic DOC was composed of small aliphatic neutral molecules, a chemical structure different than previously described (Benner et al., 1992; McCarthy et al., 1996). Most molecules identified to date by ultra high-resolution mass spectrometry contain high oxygen content, suggesting they should be susceptible to prokaryotic uptake, but many of these compounds are also highly polar carboxyl-rich alicyclic molecules that require highly specific microbial uptake systems (Kattner et al., 2011). The "molecular diversity hypothesis" (see Chapter 7) posits that the long-term stability of DOM is not necessarily a result of a resistant chemical structure but that the accumulated DOM is comprised of a high diversity (perhaps millions) of individual compounds where any individual compound is found at extremely low substrate concentration (Chapter 7). According to this hypothesis, slow DOM decomposition is a function of low individual substrate concentration (Kattner et al., 2011). Factors that could lead to long-term preservation of highly diluted individual compounds include: concentrations that are far below the chemoreceptive threshold of prokaryotes (Kattner et al., 2011), low encounter rate between substrate and bacteria via molecular diffusion (Stocker, 2012) and a greater energy demand to acquire a substrate than the energy gained from its acquisition (thermodynamic inhibition; La Rowe et al., 2012).

D Biogeochemical Implications of Organic Matter Partitioning into Recalcitrant DOM

Biotic and abiotic processes that produce recalcitrant DOM create vertical DOM gradients. The DOM fractions that accumulate in excess of mesopelagic or bathypelagic concentrations can contribute significantly to the biological pump provided they escape microbial degradation in the surface waters until export via deep convective overturn (Amon et al., 2003; Carlson et al., 1994, 2010; Copin-Montgut and Avril, 1993; Hansell and Carlson, 2001; Hansell et al., 2009) or subduction along isopycnal surfaces into the ocean interior (Doval and Hansell, 2000; Hansell et al., 2002).

Understanding the mechanisms that control the partitioning of organic matter production between POM and DOM in marine systems is important because the two pathways have vastly different effects on the efficiency and magnitude of the biological pump. New POM production dominates the export of organic matter, accounting for ~80% of the carbon reaching the deep sea (Hansell, 2002; Hopkinson and Vallino, 2005). Upwelling or vertical mixing brings nutrients to surface waters in roughly Redfield proportions of 106C:16N:1P, and the formation of POM consumes nutrients at those ratios. Export and subsequent remineralization of that POM returns the C, N, and P to depth in amounts equal to the original vertical supply with no net flux of carbon to depth (Michaels et al., 2001).

When phytoplankton populations become nutrient depleted, carbon overproduction can proceed (Sambrotto et al., 1993), in which the photoautotrophic populations continue to fix C as an energy dissipation mechanism, releasing it as C-rich DOM in excess of Redfield stoichiometry (Carlson et al., 1998; Conan et al., 2007; Hopkinson and Vallino, 2005; Wetz and Wheeler,

2007; Figure 3.3). This mechanism produces "new" organic carbon in the sense that it allows organic matter production by phytoplankton in excess of the CO_2 supplied to the surface waters by vertical entrainment. Thus, the dynamics of DOM production can operate outside of the balanced Redfield production cycle, yielding a C-rich organic pool. The C:N:P ratios of bulk DOM range from 478:29:1 in a Pacific oceanic gyre (Church et al., 2002) to a mean of 778:54:1 in the Atlantic, with a mean stoichiometry of 199:20:1 for seasonally produced DOM in coastal and offshore regions of the North Atlantic and North Pacific central gyre (Hopkinson and Vallino, 2005; Figure 3.11).

The departure from Redfield stoichiometry means that for every new N and P atom introduced in the surface water, potentially more C

can be stored in DOM. If the C-rich DOM persists then the carbon can either be sequestered as DOM (i.e., the MCP) or it can support efficient C export through vertical mixing of DOM (Hopkinson and Vallino, 2005). Geochemical models estimate that ~20% of the annual net community production (1.9 Pg C) is exported to depths >100 m each year (Hansell et al., 2009; Figure 3.11). DOM can be a highly efficient C export mechanism with global implications for the operation of the biological pump if it is mixed deeply and advected horizontally at depth or along isopycnals to regions that escape ventilation in the subsequent season(s). An example of this is demonstrated in Figure 3.1a where relatively high DOC concentrations in NADW at high latitudes move south in the North Atlantic where it persists or is degraded at depth rather

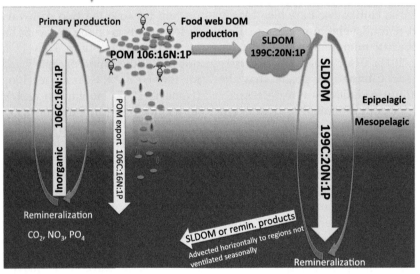

FIGURE 3.11   Scenario where mixing of DOM, due to its stoichiometry, contributes to net carbon export from the surface waters. Mixing leads to the entrainment of mesopelagic nutrients into the surface water at Redfield stoichiometry. POM is subsequently formed and exported at a ratio of [106]C:[16]N:1P. However, SLDOM is produced at a C-rich stoichiometry, with a mean of [199]C:[20]N:[1]P (Hopkinson and Vallino, 2005). The recently produced DOM that survives until deep mixing can result in an efficient DOM export C-rich organic matter from the epipelagic zone especially if mixing is deep enough or advected horizontally at depth to regions that do not ventilate in subsequent season.

than being mixed back to the surface. As stated by Hopkinson and Vallino (2005) *"if estimates of DOC export are correct we urgently need to understand the mechanisms that control the stoichiometry of DOM production, export and remineralization so that predictions of the response to climate and CO$_2$ changes can be made."*

V DOM REACTIVITY

"It is important to avoid the microbiological heresy that asserts that bacteria can defy the laws of chemistry and break down everything! The decomposition of some large molecules requires considerable energy, and the energy balance must in the end be to the organism's advantage" **Floodgate (1995).**

The bulk DOM pool is comprised of a myriad of compounds (Hertkorn et al., 2006; Koch et al., 2008; Kujawinski, 2011) that exhibit a spectrum of reactivities and turnover times (see Chapters 2 and 7; Hansell, 2013). At one extreme, the most labile DOM compounds turn over on time scales of minutes to weeks (Amon and Benner, 1996; Carlson et al., 1999; Cherrier and Bauer, 2004; Fuhrman, 1987; Halewood et al., 2012; Keil and Kirchman, 1999; Rich et al., 1996), while at the other extreme long-lived DOC compounds resist microbial degradation, turning over on timescales of centuries to millennia (Bauer, 2002; Williams and Druffel, 1987). Using DOM remineralization experiments, Ogura (1972) characterized an additional pool of DOC that degrades at intermediate rates with turnover time scales of months to years. Later, field observations of vertical DOC distributions in combination with radiocarbon ages of DOC (Bauer et al., 1992; Druffel et al., 1992) led to the conceptual partitioning of DOC into the labile (LDOC), refractory (RDOC), and "semi-labile" DOC (SLDOC) fractions. According to this partitioning, SLDOC represents a fraction of intermediate reactivity that resists rapid microbial degradation, accumulates in the surface ocean in excess of deep water concentrations and becomes available for vertical and horizontal export but is eventually remineralized on time scales of months to years (Carlson and Ducklow, 1995; Kirchman et al., 1993; Ogawa and Tanoue, 2003).

Although this conceptual model regarding the broad pools of lability has been useful in providing insight about DOC dynamics and assessing the importance/contribution of the various fractions to short-term biological processes versus longer biogeochemical processes, it has also been hampered by a lack of rigorous quantitative description. Since the first edition of this book and review of this subject (Carlson, 2002), there have been extensive field campaigns greatly increasing the number of global ocean DOC data (now >50,000 measurements; see Hansell et al., 2009). These coupled measures of bulk DOC, DOC characterization, water mass age tracers, and other biogeochemical variables (Carlson et al., 2010; Goldberg et al., 2011; Hansell et al., 2009, 2012) have greatly improved our understanding of bulk DOC fractions (Hansell, 2013). Here, we briefly describe five major fractions of DOC, their inventories and associated removal rates (Figure 3.12).

A Biologically Labile DOM

The LDOC pool is biologically significant in that it is rapidly consumed, supporting the metabolic energy and nutrient demands of heterotrophic prokaryotes, and turning over on time scales of hours to days. BCD (described above) best represents the flux of LDOC, with greater than 50% of net PP flowing through the labile DOM pool on a daily basis (Ducklow, 1999; Williams, 2000). There are numerous proposed sources of labile DOM within the marine microbial food web (Figure 3.2) and a global LDOC production rate of ~15-25 Pg C year^{-1} in the photic zone (Hansell, 2013). However, adequate quantification of any given production mechanism remains lacking.

The LDOC fraction is comprised of monomers and oligomers that include DFNS

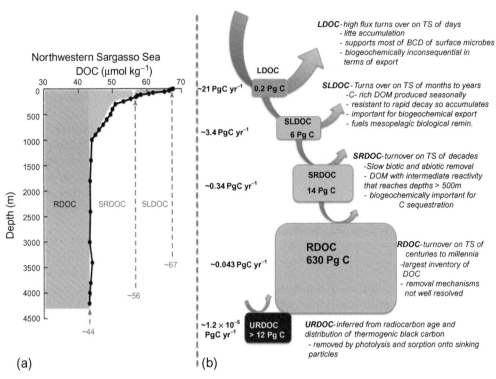

FIGURE 3.12 (a) DOC fractions, defined by reactivity, in the oceanic water column. Vertical distribution of the RDOC, SRDOC, and SLDOC pool (shaded areas) and the estimated range of DOC concentrations observed within those pools observed in stratified oligotrophic waters such as that of the Northwestern Sargasso Sea. (b) Estimated inventories (in box) and removal rates (figure to left of box) of each fraction of DOM. TS is time scale. *Adapted from Hansell (2013).*

(Kirchman et al., 2001; Rich and Kirchman, 1994; Rich et al., 1996), DFAA (Ferguson and Sunda, 1984; Meon and Amon, 2004; Simon and Rosenstock, 2007), organic sulfur compounds (i.e., DMSP) (Kiene et al., 2000; Malmstrom et al., 2004a), organic phosphorous compounds (i.e., ATP, lipids; see Chapter 5), as well as hydrolysable HMW compounds (Amon and Benner, 1994, 1996; Benner, 2002). Radioactive tracers of model compounds together with observed consumption of specific compounds (assessed by change in concentration) are used to estimate the flux and contribution to bacterioplankton metabolic demand. The flux of labile DOM meets from just a few % of the daily C and N demand to as much as 100% of the metabolic needs of natural bacterioplankton assemblages, depending on trophic state of the system (see reviews by Kirchman, 2000; Nagata, 2008). The high flux and tight coupling of production and consumption maintains these compounds at nanomolar (Ferguson and Sunda, 1984; Keil and Kirchman, 1999; Rich et al., 1996; Simon and Rosenstock, 2007; Skoog et al., 1999) to a few micromolar concentration in the open ocean (Carlson, 2002; Nagata, 2008), and an estimated inventory of <2 Pg C (Hansell, 2013). The rapid remineralization of LDOC is important in that most of the resulting inorganic constituents are retained within the euphotic zone; however, its contribution to the biological pump, redistribution of nutrients into the ocean interior and sequestration of carbon is biogeochemically inconsequential.

Tracing model compounds can be informative; however, the majority of labile DOM remains uncharacterized or quantified and bulk DOC concentration alone does not provide reliable insight regarding biological lability. Seawater culture experiments, combined with direct measurements of BP and DOM consumption (Figure 3.13; Carlson et al., 1996, 2004; Cherrier et al., 1996; Letscher et al., 2013) or respiratory gases (Carlson et al., 1999; Coffin et al., 1993; del Giorgio and Davis, 2002; Hansell et al., 1995; Kähler et al., 1997; Reinthaler et al., 2005) are used to assess the bioavailable fractions of naturally occurring uncharacterized DOM on timescales of days to weeks (Figure 3.12). The proportion of labile DOM varies spatially with an apparent gradient of decreasing concentrations of labile DOM from coastal to oceanic systems (del Giorgio and Davis, 2002). In some oligotrophic systems there is no measurable surplus of labile DOM (on the μM scale) available to microbes on time scales of hours to days (Carlson et al., 2002, 2004; Cherrier et al., 1996; Figure 3.13b).

Spectroscopic technologies such as nuclear magnetic resonance (NMR) and electrospray ionization Fourier transform ion cyclotron resonance mass spectrometry (ESI FT-ICR-MS) are elucidate the composition of DOM components (Kujawinski, 2011; see Chapter 2). NMR, for example, provides information on the functional groups in bulk DOM (Benner, 2002). ESI FT-ICR-MS resolves molecular markers (m/z values) in "fingerprinting" DOM compositional diversity (Kujawinski et al., 2009). These tools in combination with microbial experiments provide useful insight about the exometabolome (Kujawinski, 2011), how microbes utilize, produce, or transform the most labile components of the DOM pool.

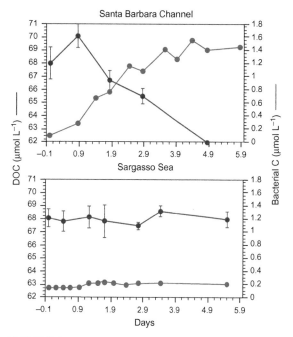

FIGURE 3.13 Bacterial carbon production (as change in biomass; red circles) and DOC utilization (change in concentrations; blue circles) in seawater culture remineralization experiments conducted in Santa Barbara Channel (SBC), California (a), and the northwestern Sargasso Sea (b). The ecosystems had similar initial DOC concentrations, however DOC removal was only observed in the SBC experiment. At the time the Sargasso Sea experiment was conducted (summer 1992), in situ DOC production and consumption processes were tightly coupled, such that accumulation of labile DOC was <1 μM C; limited availability of labile DOC prevented rapid microbial growth and DOC degradation. These results indicate that DOC concentration alone is a poor proxy for its bioreactivity. Details of experimental design are described in Carlson et al. (2004). *Figure adapted from Carlson and Ducklow (1996); previously unpublished data from the SBC (C. Carlson).*

B Biologically Semi-labile and Semi-refractory DOM

DOM fractions with intermediate reactivity (i.e., turnover rates greater than months but less than centuries (time scale of ocean mixing)), and observable as the DOC concentrations in excess of deeper water concentrations are important in vertical export via mixing. These materials were originally proposed as single SLDOM pool, however greater spatial resolution of bulk DOC (Hansell et al., 2009) together with better estimates of removal rates of various fractions (Abell et al., 2000; Carlson et al., 2010; Hansell and Carlson, 2001; Santinelli et al., 2010) have led to further partitioning the pools of intermediate reactivity into SLDOC and "semi-refractory" (SRDOC) fractions (Figure 3.12; Hansell et al., 2012). Partitioning between SLDOC and SRDOC is constrained by the observed removal rates of these pools.

1 SLDOC

SLDOC is characterized as DOC that accumulates and is resistant to decay in the surface waters but, following export into the upper mesopelagic, is removed over months to years (Hansell et al., 2012). The global inventory of SLDOC is maintained at ~6±2 PgC, with an estimated production rate of ~3.4 PgC year^{-1} (Figure 3.12; Hansell, 2013; Hansell et al., 2012). The export of SLDOC to depths >100m can exceed estimates of local POC flux in some oceanic systems (Carlson et al., 1994; Copin-Montgut and Avril, 1993; Hansell and Carlson, 2001). The global export rate of SLDOC is estimated at 1.5 PgC year^{-1} (Hansell et al., 2012), making it the most important DOM fraction contributing to carbon export out of the epipelagic zone. However the majority of this exported DOM is remineralized within the upper mesopelagic zone (100-500m) and the resulting inorganic by-products may be returned to the air/sea interface if subsequent seasonal ventilation is deep enough (Carlson et al., 1994; Doval and Hansell, 2000; Hansell and Carlson, 2001). Thus, while SLDOC may be a significant component of local export flux (depths > 100m) it plays only a limited role in long-term carbon sequestration (Carlson, 2002; Hansell et al., 2012).

LMW and HMW polymeric compounds of varying lability (Amon and Benner, 1994,

1996; Benner, 2002; Ogawa and Tanoue, 2003), including DCNS (Cowie and Hedges, 1994; Goldberg et al., 2009; Skoog and Benner, 1997), DCAA (Davis and Benner, 2007; Kaiser and Benner, 2009), N-acetyl amino polysaccharides (N-AAPs) (Aluwihare et al., 2005), and acyl heteropolysaccharides (APS) (Aluwihare and Repeta, 1999), comprise a portion of the SLDOC pool. Prior to bacterial uptake these polymeric compounds must first be hydrolyzed to monomers via extracellular enzymes (Arnosti, 2011; Billen and Fontigny, 1987). Thus, the temporal uncoupling between enzyme production and bacterial uptake can result in transient SLDOC accumulation. The concentration of specific compounds relative to TOC (yield) provides insight into patterns of OM (Cowie and Hedges, 1994; Dauwe et al., 1999) and DOM diagenesis (Amon and Benner, 2003; Davis et al., 2009; Goldberg et al., 2010, 2011; Kaiser and Benner, 2012; Skoog and Benner, 1997), with lower yields indicating more diagenetically altered DOM. Systematic changes in individual monomers within the DCNS or DCAA pools (i.e., molar fractions) proceed along diagenetic pathways revealing preferential removal of some monomeric subunits while others become enriched (glucose and glucosamine) as DOM becomes altered (Goldberg et al., 2011; Kaiser and Benner, 2012).

The variability of accumulated DOC is most pronounced above the seasonal pycnocline, with this variability largely attributed to dynamics of the SLDOC pool (Figure 3.1). As described above (Section IV.B.1.c), the BATS site reveals the seasonal patterns of SLDOC accumulation, export (Carlson et al., 1994; Hansell and Carlson, 2001) and remineralization that coincide with regular patterns of diagenetic alteration of DOC within the euphotic and mesopelagic zones (Goldberg et al., 2009; Figure 3.9). Shortly after thermal stratification, the SLDOC concentrations increase by 3-10 µmol kg^{-1} C within the euphotic zone, and the pool becomes enhanced in DCNS and enriched in mannose and galactose (Goldberg et al., 2009). Despite a relatively high DCNS yield, the seasonally accumulated DOC remains largely resistant to rapid microbial degradation by the surface microbial consortium (Carlson et al., 2004). During winter convective overturn a portion of the surfaced accumulated SLDOC is exported into the mesopelagic zone (Carlson et al., 1994; Hansell and Carlson, 2001), also exhibiting a high DCNS yield at depth (Goldberg et al., 2009; Figure 3.9b). The exported SLDOC represents an allochthonous pool of DOC supplementing the metabolic needs of the mesopelagic heterotrophic bacterioplankton (Anderson and Williams, 1998; Carlson et al., 2004; Hansell and Carlson, 2001; Luo et al., 2010). At BATS, ~3-6 μmol C kg^{-1} of DOC are remineralized after restratification (Figure 3.9) by distinct mesopelagic microbial assemblages (see Section IV.B.1.c; Carlson et al., 2004; Morris et al., 2005; Treusch et al., 2009) on time scales of weeks. Coincident with SLDOC removal the character of the DOM becomes altered with a reduction in the DCNS yield (Figure 3.9b) and preferential removal of mannose and galactose (Goldberg et al., 2009, 2011). Similarly, Abell et al. (2000) reported removal rates of SLDOC to be 2-9 μmol kg^{-1} year^{-1} in the North Pacific.

2 SRDOC

The fraction of DOC that turns over on times scales of decades was differentiated from the SLDOC and later termed SRDOC (Hansell et al., 2012). SRDOC becomes evident in the vertical DOC gradients observed in deeper portions of the mesopelagic zones (>500-1000 m) of ocean regions that exhibit a permanent pycnocline. These gradients can also extend into the bathypelagic realm at high latitudes where deep water formation results in significant DOC export (i.e., North Atlantic and Mediterranean Sea; Amon et al., 2003; Carlson et al., 2010; Santinelli et al., 2010). By examining concomitant changes in DOC concentrations and chlorofluorocarbon (CFC)-ages within deep water masses of the North

Atlantic, Carlson et al. (2010) estimated DOC removal rates of ≤1 μmol kg^{-1} year^{-1}. The inventory of SRDOC is ~14 ± 2 Pg C globally with an annual production and subsequent export rate of ~0.34 Pg C year^{-1} (Figure 3.12; Hansell et al., 2012). Thus, SRDOC is considered a minor contributor in terms of annual vertical carbon export; however if a large percentage of SRDOC is remineralized below the permanent pycnocline then the respiratory products are returned to the surface slowly (i.e., decades to centuries), making SRDOC potentially important in C sequestration (Hansell et al., 2012). DOC removal accounts for 5-20% of apparent oxygen utilization (AOU) (Arístegui et al., 2003; Carlson et al., 2010; Doval and Hansell, 2000), though we cannot rule out abiotic removal mechanisms such as scavenging by sinking particles (Druffel et al., 1998) or gel aggregation and sedimentation (Verdugo et al., 2004); thus the correlation between DOC and AOU concentrations needs further evaluation.

C Biologically Refractory and Ultra-refractory DOM Pools

The refractory pools of DOM include the refractory (RDOC) and ultra-refractory (URDOC) fractions. Together these fractions are operationally defined as the DOM that remains after LDOM, SLDOM, and SRDOC are removed. The combined RDOC + URDOC pools are most clearly resolved in deep waters >1000 m where the vertical DOC gradients are minimal or absent (i.e., mean DOC concentrations ≤42 μmol kg^{-1}). The global inventory of these combined pools is estimated at 642 Pg C (Hansell et al., 2012). Together the RDOC and URDOC pools have an apparent mean age of 4000-6000 year in the North Atlantic and the central North Pacific Oceans, respectively (Bauer et al., 1992; Druffel et al., 1989; Williams and Druffel, 1987). Based on mass balance calculations and the fact that DO^{14}C ages are greater than ocean mixing times, Druffel et al. (1992) suggested that the most refractory components of the bulk DOM pool were distributed uniformly throughout the water column, representing ~60% of surface DOC in thermally stratified systems (Carlson, 2002; Carlson and Ducklow, 1995; Cherrier et al., 1996; Hansell, 2013; Ogawa and Tanoue, 2003; Figure 3.12a).

1 URDOC

Given the existing DOC data sets, differentiating between URDOC from RDOC fractions is not possible from direct observations of bulk DOC change. The existence of URDOC must be inferred from the contribution of polycyclic aromatic hydrocarbons, also known as thermogenic black carbon (Dittmar and Paeng, 2009). The inventory of black carbon ranges from 2% to 22% of bulk DOC or ~13-145 Pg C (Figure 3.12; Dittmar, 2008; Hansell, 2013; Ziolkowski and Druffel, 2010). Thermogenic black carbon has radiocarbon ages much greater than ambient bulk deep DOC, is assumed to be inert on time scales of ocean mixing, and may reside in the DOC pool for 2500-13,900 ^{14}C-year prior to sedimentation (Ziolkowski and Druffel, 2010). The primary source of black carbon is the burning of terrestrial forests and fossil fuels, which is produced at a rate of ~1.2×10^{-5} Pg year^{-1} (Hansell, 2013). Photolysis (Stubbins et al., 2012) and sorption onto sinking particles or sediments (Dittmar and Paeng, 2009) are considered the primary sinks for thermogenic black carbon.

2 RDOC

By difference, RDOC represents the largest fraction of accumulated DOC, comprising a global inventory up to 630 ± 32 Pg C (Figure 3.12; Hansell, 2013). It is dominated by LMW DOM (Amon and Benner, 1996; Benner, 2002; Kattner et al., 2011; Chapter 7), a surprising finding given that bacterioplankton typically take up LMW compounds. However, while some LMW monomers are highly labile, the vast majority of LMW compounds are either diagenetically altered and resistant to microbial remineralization (Amon and Benner, 1994, 1996; Benner, 2002) or

the concentrations of individual compounds are sufficiently low enough to avoid detection by microbial uptake systems (i.e., molecular diversity hypothesis; above). Regardless, the sum of individual LMW compounds comprises a large unavailable pool (see Chapter 7).

While RDOC is considered biologically refractory (Barber, 1968), the deep ocean DOC range is 34-50 μmol C kg^{-1}, with the gradient developing over a single global cycle of abyssal circulation (Hansell and Carlson, 1998a; Hansell et al., 2009). The concurrent decrease in DOC with an increase in Δ^{14}C age suggests that a portion of the RDOC is removed on the time scales of centuries within the ocean interior. Despite the potential to sequester carbon for long periods of time, the estimated RDOC production rate of ~0.043 Pg year^{-1} indicates that its annual contribution is modest. However, even minor changes in the sink mechanism can have significant implication on longer climatic timescales. Peltier et al. (2007) proposed that elevated rates of RDOC remineralization in the Neoproterozoic increased the atmospheric CO_2 reservoir, enhancing the greenhouse warming of the Earth's surface and preventing global scale glaciation. Similar roles for DOC in past climates have been proposed (Rothman et al., 2003; Sexton et al., 2011). See this chapter for further review of this topic.

What are the processes that affect modern day RDOC removal? The rate constants required to account for continuous removal are on the order of 10^{-3}-10^{-4} year^{-1}, too slow for continuous microbial decomposition to be evident (Anderson and Williams, 1999; Williams, 2000). Anderson and Williams (1999) proposed a coupled biological-photochemical model in which deep DOC is biologically refractory but photochemically reactive. Once exposed to surface UV irradiation, some RDOC is removed via photooxidation or broken into labile compounds and removed via microbial remineralization (Kieber et al., 1989; Mopper and Kieber, 2002; Mopper et al., 1991). This mechanism may account for a significant portion of RDOC loss (Moran and Zepp, 1997) but it is restricted to the surface ocean and cannot account for the deep DOC gradients (Carlson et al., 2010; Hansell et al., 2009, 2012). Williams (2000) proposed a third mode of deep DOC decomposition, in which attached bacteria that are associated with sinking particles generate short bursts of microbially mediated DOM removal. Further research is needed to fully elucidate refractory DOM removal mechanisms and the associated rates.

VI THE PRIMING EFFECT

Sinks for recalcitrant DOC exist in the ocean but the mechanisms are unknown. Soil scientists have long recognized that inputs of labile organic matter to soil can lead to increased remineralization of recalcitrant organic matter. This phenomenon is referred to as the priming effect (Bianchi, 2011; Guenet et al., 2010). The controlling factors responsible for the priming effect remain unresolved but some combination of three proposed mechanisms, reviewed by Guenet et al. (2010), may explain the removal of recalcitrant marine DOM. The mechanisms include:

Co-metabolism—In this mechanism, oxidation of recalcitrant organic matter is an inadvertent consequence of enhanced microbial activity stimulated by labile organic matter. As such, labile organic matter supplies energy that stimulates growth and activates the production of hydrolytic enzymes capable of degrading recalcitrant forms of organic matter (Horvath, 1972).

Net mutualism between two microbial communities—In this mechanism, there are two distinct communities of heterotrophic microbes (i.e., labile organic matter consumers and recalcitrant organic matter consumers). Upon addition of labile organic matter, the byproducts of the decomposed labile organic matter provide energy that activates the production of hydrolytic enzymes by a second community of microbes specializing in the decomposition of recalcitrant organic

matter. Thus, remineralization of recalcitrant organic matter is a result of energy redistribution between two functionally distinct populations of microbes (Fontaine and Barot, 2005; Fontaine et al., 2003; Guenet et al., 2010).

Alternate metabolisms of single community—Here a homogenous population of heterotrophic microbes becomes capable of utilizing recalcitrant compounds when labile organic matter provides the energy necessary for the purposeful synthesis of recalcitrant organic matter degrading enzymes (Guenet et al., 2010).

The priming effect has been discussed largely in the context of the remineralization of recalcitrant terrigenous dissolved organic carbon (tDOC) as it flows into aquatic systems (Bianchi, 2011; Guenet et al., 2010). Approximately 0.3 Pg C of tDOC is discharged to the global ocean annually; however, little of this material can be identified in the open ocean based on chemical biomarkers or stable isotope data (Bianchi, 2011). The removal of recalcitrant tDOC can be attributed to direct photooxidation (Mopper et al., 1991), photolytic conversion from recalcitrant to labile DOM and subsequent uptake by heterotrophic bacteria (Benner and Kaiser, 2011), or microbial priming (Bianchi, 2011; Guenet et al., 2010). There is evidence that the priming effect exists in the open ocean as well. Microcosm experiments in the oligotrophic Sargasso Sea showed that the addition of glucose, ammonia, and phosphorous stimulated bacterioplankton to reduce DOC concentrations to below that observed in unamended control treatments conducted over the course of weeks to months (Carlson et al., 2002). Stimulation of SLDOC removal was also observed in experiments where surface accumulated DOC was inoculated with mesopelagic water (Carlson et al., 2004; Letscher et al., 2013). Cherrier et al. (1999) showed that the nucleic acids of oceanic bacterioplankton were depleted in ^{14}C with respect to modern surface inorganic carbon. They proposed co-metabolism as one potential mechanism to explain the depleted ^{14}C bacterial signature, suggesting that some bacterioplankton were capable of utilizing older recalcitrant DOC while they used modern labile DOC to support their metabolism. Recent observations in the deep Pacific Ocean showed large departures of RDOC distribution from conservative mixing, suggesting a relatively rapid removal of RDOC in some realms of the ocean's interior (Hansell and Carlson, 2013); priming was proposed as mechanism to explain this phenomenon.

VII MICROBIAL COMMUNITY STRUCTURE AND DOM UTILIZATION

In addition to the availability of macro and trace nutrients, the structure of the microbial community plays an important role in regulating the accumulation and subsequent remineralization of recalcitrant DOM. Cultivation-independent approaches (i.e., 16S rRNA fingerprinting, multiplex pyrosequencing, metagenomics) reveal that the taxonomic diversity of prokaryotes within the oceans is incredibly high (DeLong et al., 2006; Giovannoni, 2005; Giovannoni and Stingl, 2005), although there remains uncertainty as to how much of this diversity is functionally significant (DeLong et al., 2006; Rappe and Giovannoni, 2003; Venter et al., 2004). Field studies demonstrate free-living (unattached) microbial populations that are vertically structured, particularly between the euphotic and aphotic regions of the oceanic water column (Carlson et al., 2009; DeLong et al., 2006; Field et al., 1997; Giovannoni et al., 1996; Gordon and Giovannoni, 1996; Moreira et al., 2006; Morris et al., 2005; Treusch et al., 2009). The mechanisms responsible for variation in microbial communities with depth are not well understood, but presumably the stratification results from niche partitioning among microbial specialists along with the vertical gradients in nutrients and energy availability (DeLong et al., 2006; Field et al., 1997; Giovannoni et al., 1996; Gordon and Giovannoni, 1996).

DOM quality is also an important determinant of the microbial communities developing in experimental and field studies. Adding DOM of varying quantity and quality to natural microbial assemblages yields significant differential responses among the major and minor groups of bacterioplankton (Alonso-Sáez et al., 2012; Arnosti et al., 2005a; Carlson et al., 2002, 2004; Cottrell and Kirchman, 2000; Landa et al., 2013; Malmstrom et al., 2005a; McCarren et al., 2010; Nelson and Carlson, 2012; Nelson et al., 2013; Sarmento et al., 2013; Steen et al., 2010). Figure 3.14 provides an example where distinct bacterioplankton community structures develop between treatments of varying DOM character, from model compounds to complex algal-derived DOM.

Genomic and transcriptomic approaches reveal the potential of marine bacterioplankton to utilize different DOM targets (Giovannoni, 2005; Kujawinski, 2011; McCarren et al., 2010; Poretsky et al., 2010). Profiles of metagenomes identified an enhanced number of putative polysaccharide degradation genes from deep microbial populations relative to that found at the surface (DeLong et al., 2006). These data suggest that the deep populations may be better adapted to utilize more recalcitrant polysaccharides (Kujawinski, 2011).

As described above, there are numerous examples of correlative relationships between DOM quality and microbial community structure and metabolic potential. However, microbial oceanographers must continue to develop ways of detecting and tracking specific compounds into specific microbes to better understand when, where, and why specific compounds become available for heterotrophic uptake. Some progress has been made recently with microautoradiography (Cottrell and Kirchman, 2000; Malmstrom et al., 2004b; Ouverney and Fuhrman, 1999). Malmstrom and colleagues used microautoradiography coupled with fluorescent in situ hybridization (Micro-FISH) to trace the uptake of DMSP into members of *Synechococcus*,

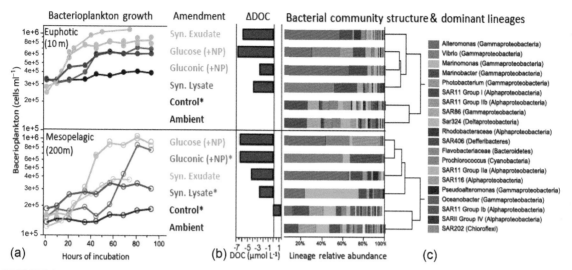

FIGURE 3.14   Sargasso Sea bacterioplankton abundance (a), DOC utilization (b), and variability in bacterioplankton community structure (c) in response to different model and algal-derived organic substrates. Solid circles and solid lines represent response in bacterioplankton cell density, measured by flow cytometry in euphotic zone incubations; open circles represent response in mesopelagic incubations. All organic amendments were made at a final concentration of ~10 μM C, including *Synechococcus* exudate and lysate, glucose (plus 1 μM NH$_4$Cl, 0.1 μM K$_2$HPO$_4$), and gluconic acid (plus 1 μM NH$_4$Cl, 0.1 μM K$_2$HPO$_4$). ΔDOC is change in DOC concentration from time 0 to 50h (euphotic zone treatments) or 70h (mesopelagic treatments). The response of the microbial community was assessed at 50 or 70h using Stable Isotope Probing (SIP) with ^{13}C labeled substrates. Color bars represent the relative contribution of various bacterioplankton lineages. The hierarchical cluster dendrograms show community similarity among intact ambient and culture end-point communities according to treatment and depth. *Data reformatted from Nelson and Carlson (2012).*

SAR11, *Roseobacter* and *Cytophaga*-like clades (Malmstrom et al., 2004a,2004b, 2005b). Stable isotope probing (SIP) is another promising approach in which the incorporation of ^{13}C or ^{15}N alters the buoyant density of macromolecules (i.e., DNA, RNA, phospholipids), allowing them to be separated on a cesium chloride density gradient (Neufeld et al., 2007). SIP coupled with 16S rDNA pyrosequencing in dark seawater cultures allows one to track the incorporation of model compounds or complex labeled organic matter (i.e., algal extracts) into specific microbial communities without the a priori selection of targeted phylotypes (Figure 3.14; Nelson and Carlson, 2012). The use of secondary-ion mass spectrometry (nanoSIMS) provides promise of tracking labeled elements within substrate into cells (Sheik et al.,

2013). Tracing the production and disappearance of specific labeled compounds via FT-ICR MS is an exciting twist for using SIP to identify microbial DOM interactions; it represents a promising research direction (Kujawinski, 2011).

VIII DOC IN THE OCEAN CARBON BUDGET

Above we discuss in detail the production, consumption, and reactivity of DOC in the global ocean. In this section a budget for DOC in the cycling of ocean carbon is proposed as shown in Figure 3.15. Both internal carbon cycling (autochthonous) and transport to the ocean (allochthonous) are included. The ocean inventories of

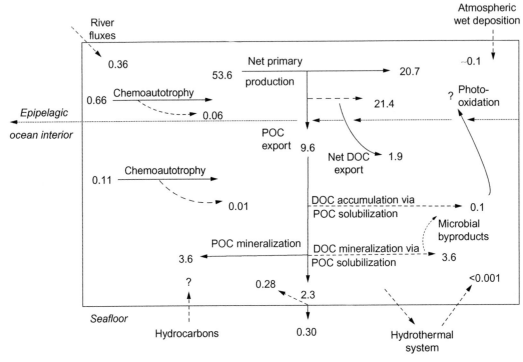

FIGURE 3.15   The global ocean organic carbon budget (PgC year^{-1}) including both particulate (solid line) and dissolved (dashed line) matter. The ocean carbon inventories of these pools are assumed to be at steady state; all input fluxes (value at origin of arrow) are balanced by removal (value at end of arrow). Fluxes that have very high uncertainty are noted by a question mark (?).

the pool are assumed to be held at steady state; inputs (sources) are balanced by removal (sinks), though over widely ranging time scales. Sinks are described above; inputs are considered here.

A Autochthonous Sources

1 Epipelagic

Net primary (PP) and particle export production in the global ocean were estimated by Dunne et al. (2007) at 53.6 ± 7.4 and $9.6 \pm 3.6\,\mathrm{PgC\,year^{-1}}$, respectively. LDOC production in the epipelagic is taken to be 30-50% of PP (Williams, 2000), or $21.4 \pm 5.3\,\mathrm{PgC\,year^{-1}}$. Most of this material (~91%) is mineralized in the epipelagic, supporting the microbial loop there. The balance of DOC produced in the epipelagic has a lifetime

of adequate duration to be exported to the upper mesopelagic at a rate of $1.9\,\mathrm{PgC\,year^{-1}}$ (Hansell et al., 2009), with DOC contributing 16.5% of total export from the euphotic zone. Mass balance requires the amount of NPP left for grazing in the epipelagic to be $20.7 \pm 9.5\,\mathrm{PgC\,year^{-1}}$ (i.e., epipelagic grazing = PP-DOC consumed-DOC exported-POC exported).

2 Ocean Interior

Sinking POC and DOC exported with overturning circulation enter the ocean interior at 9.6 ± 3.6 and $1.9\,\mathrm{PgC\,year^{-1}}$, respectively. At steady state, the exported DOC is fully removed at depth, though the variable lifetimes of the exported fractions dictate where and when removal occurs in the water column (Hansell,

2013; Hansell et al., 2012). One fate of POC exported from the surface ocean is accumulation at the seafloor at $2.3 \pm 0.9 \, \text{Pg C year}^{-1}$, of which $0.30 \pm 0.15 \, \text{Pg C year}^{-1}$ is buried (Hedges and Keil, 1995), $0.28 \pm 0.14 \, \text{Pg C year}^{-1}$ is released to the water column as porewater DOC (Dunne et al., 2007), and the balance is mineralized ($1.83 \pm 0.20 \, \text{Pg C year}^{-1}$). The sedimentary DOC flux to the water column (Dunne et al., 2007) is based on Papadimitriou et al. (2002), who reported sedimentary DOC effluxes at ~14% and ~36% of POC influx to the seafloor at <2000 and >2000 m, respectively. The Dunne et al. (2007) sedimentary DOC efflux of $0.18 \pm 0.07 \, \text{Pg C year}^{-1}$ for depths <2000 m is equivalent to that by Burdige et al. (1999), lending confidence to the global estimate. See Chapter 12 for more on sedimentary DOC fluxes.

Given the imbalance in fluxes of POC leaving the epipelagic and that reaching the seafloor, $7.3 \, \text{Pg C year}^{-1}$ must be consumed or transformed in the water column. The fate of this material, whether grazed by metazoans, consumed in solubilized form by prokaryotes (i.e., as DOC), or accumulated as suspended POC or as DOC is not adequately known. Heterotrophic C demand in the ocean interior is commonly reported to exceed the export flux supporting it (Burd et al., 2010), so our understanding of the system is far from complete. Steinberg et al. (2008) investigated the fate of exported POC in the ocean's twilight zone. Mesopelagic (150-1000 m) BCD exceeded zooplankton C demand by up to threefold in the subtropical Pacific (22 °45'N, 158 °W), while bacteria and zooplankton required relatively equal amounts of exported POC in the subarctic Pacific (47 °N, 160 °E). Given that most global POC export occurs in high productivity systems, which are better represented by the subarctic than the subtropics, the carbon budget in Figure 3.15 allocates half of the available POC export to metazoans (POC mineralization by grazing) and half to heterotrophic prokaryotes (DOC mineralization via POC solubilization), each at $3.6 \, \text{Pg C year}^{-1}$.

An unknown (but likely small) fraction of the exported POC accumulates in the water column as a recalcitrant, humic-like DOC (Hansell, 2013). Some of this material accumulates as a function of AOU in the water column (Swan et al., 2009; Yamashita and Tanoue, 2008), apparently a byproduct of POC mineralization, with the relative roles of metazoans versus prokaryotes unknown. Yamashita and Tanoue (2008) estimated the in situ production rates of fluorescent DOM in the ocean interior to be $0.022\text{-}0.051 \, \text{Pg C year}^{-1}$. There are non-optically active fractions of DOC accumulating as well, such as proteins (Orellana and Hansell, 2012). While both fluorescent and non-optically active forms of accumulated DOC leave signals in the water column (Orellana and Hansell, 2012; Swan et al., 2009; Yamashita and Tanoue, 2008), the global rates of release are presumably very small compared to the POC being transformed. Assuming that the optically and non-optically active fractions accumulate at similar rates, a global accumulation rate of this processed DOC is taken to be $\sim 0.1 \, \text{Pg C year}^{-1}$.

a DEEP CHEMOAUTOTROPHY

There is uncertainty in the magnitude of chemoautotrophy occurring in the dark reaches of the deep ocean. Middelburg (2011) estimated C fixation rates by nitrifying bacteria of $0.66 \, \text{Pg C year}^{-1}$ for the upper ocean and $0.11 \, \text{Pg C year}^{-1}$ in the dark ocean, respectively. In Figure 3.15, 10% of these fluxes have been allocated to DOC. Chemoautotrophy apparently results in the formation of a labile form of DOC, as its accumulation is not evident; DOC in the deepest reaches of the ocean largely behaves conservatively over great distances, with neither addition nor removal evident at the micromolar level (Hansell and Carlson, 2013). As described above other reductants such as sulfur, CO, and methane may fuel chemoautotrophs (Swan et al., 2011), supporting higher rates compared to those based on ammonia oxidation alone. The lower rates suggested by Middelburg are employed in Figure 3.15.

B Allochthonous Sources

Three important external sources of DOC to the ocean are the atmosphere, rivers (see Chapter 11), and hydrocarbon seeps at the seafloor. Precipitation delivers $0.3\,Pg\,C\,year^{-1}$ organic matter to the earth surface (Willey et al., 2006), with $0.1\,Pg\,C\,year^{-1}$ falling to the ocean (Willey et al., 2000). Aitkenhead and McDowell (2000) estimated riverine DOC flux at $0.36\,Pg\,C\,year^{-1}$, consistent with Meybeck (1981) and similar to the $0.41\,Pg\,C\,year^{-1}$ by Schlesinger and Melack (1981). Much of this material is likely oxidized while still on the continental shelves (Alling et al., 2008; Benner and Opsahl, 2001; Hansell et al., 2004; Holmes et al., 2008; Letscher et al., 2011), but a portion survives for transport to the deep ocean (Benner et al., 2005; Yamashita and Tanoue, 2008). The input and accumulation of DOC in the global ocean via seafloor hydrocarbon seeps is unknown, as is its input from submarine groundwater discharge.

IX SUMMARY

In this chapter, we have outlined our present understanding of the DOM production and removal processes, characteristics of the DOM fractions defined by lability, factors that lead to DOM accumulation, and the role of DOC in the oceanic carbon budget.

1. DOM production mechanisms are a normal function of food web processes, including photoautotrophic and chemoautotrophic releases, zooplankton associated processes (i.e., grazing and excretion), viral- and bacterial-induced release, solubilization of particles, and heterotrophic prokaryote DOM production. Numerous autochthonous DOM production mechanisms exist, but the contribution of any given process varies depending upon environmental and food web controls and the contribution of specific production mechanisms remains largely unquantified.

2. Microbial heterotrophic activity largely shapes the distribution and dynamics of DOC within the ocean's surface and interior by altering DOM chemical character as well as affecting the magnitude and rate of DOM removal. In pelagic marine ecosystems, heterotrophic *Bacteria*, and *Archaea* transform the vast reservoir of biochemically heterogeneous DOM, altering its molecular structure with enzymatic activities at the cell surface, oxidizing it to CO_2, or transforming it to microbial biomass. Prokaryotic oxidation of DOM is considered the main sink for recently produced DOM; however, the role of UV oxidation is an important removal process especially for refractory DOM. Sorption of DOM onto sinking particles is a potential DOM removal mechanism within the oceans interior. Work continues toward trying to identify and quantify processes that remove refractory DOM.

3. The bulk DOM pool consists of a myriad of compounds that span a great range of biological reactivities, ranging from material that turns over on timescales of minutes to millennia. Recent observational data in combination with water mass age tracers and ocean circulation models have helped to resolve inventories and flux of DOM within the labile, semi-labile, semi-refractory, refractory, and ultra-refractory fractions. While LDOC is extremely important in fueling the microbial food web, its rapid turnover makes it inconsequential to carbon export and sequestration. The SLDOC pool, with turnover rates of months to years, contributes the most to export; however, the majority of this pool is remineralized at depths too shallow to result in long-term carbon storage. The SRDOC contribution to annual export is about an order of magnitude less than the SLDOC pool but its longer

turnover rates results in deeper export and greater contribution to long term storage of DOC compared to the SLDOC pool. The combined RDOC and URDOC turns over on centennial to millennial timescales and represents the vast majority of the accumulated DOC, yet the dynamics of these pools are not well understood because they are under-constrained by observations.

4. Accumulation of DOM results from the uncoupling of DOM production and removal processes. The production of biologically resistant compounds occurs in part via physical processes such as condensation reactions or phototransformations. Unmodified recalcitrant components of DOM, formed by direct biosynthesis, have been identified for both phytoplankton and bacterioplankton. Evidence of DOM with prokaryotic biomarkers lends support to the MCP concept, a process that differs from the biological carbon pump in that it contributes to the sequestration of carbon throughout the water column. The degree to which the MCP shunts carbon to long-term storage as organic matter, versus brief storage associated with export and rapid, deep remineralization, remains unknown; however the rates of carbon sequestration appear to be $<0.023\,PgC\,year^{-1}$. The quantitative importance of the MCP is a topic of investigation. In contrast to the ideas that DOM accumulates because it is transformed to compounds that "shield" it from enzymatic attack, the "neutral reactivity theory" proposes that individual "available" compounds are kept below detection threshold and accumulate offers intriguing new ideas regarding DOM accumulation.

5. A central challenge in DOM biogeochemistry is understanding what makes DOM persist at one geographical location or depth and become biologically available at another? Factors such as DOC chemical composition, surrounding inorganic macro and micro nutrient fields, microbial community structure, and associated expression of enzymes and uptake pathways, likely play roles in controlling the availability of DOM to heterotrophic prokaryotes. Integration of "omic" technologies that reveal microbial community structure and metabolic strategies with new DOM chemical characterization approaches provide the promise of linking specific microbial groups with specific DOM compounds.

6. Improvement in the quality and number of ocean DOC measurements continues to enhance our estimates of DOC inventory and fluxes in the ocean and account for the production and removal terms indicating $1.9\,PgC$ as DOC exported from the epipelagic every year. We expect our understanding of the dynamics of this important carbon reservoir to advance with the introduction of new data, technologies, and modeling efforts.

Acknowledgments

We thank the U.S. National Science Foundation and the National Atmospheric and Oceanographic Administration for continued supporting our research over the past two decades. This chapter benefited greatly from discussions with numerous colleagues including but not limited to Craig Nelson, Steve Giovannoni, Stuart Goldberg, Mark Brzezinski, David Siegel, Chuck Hopkinson, Norm Nelson, and Hugh Ducklow. We are indebted to the numerous students, post doctoral scientists, and technicians who have contributed to our research programs over the years, including but not limited to Elisa Halewood, Meredith Myers, Maverick Carey, Rachel Parsons, Robert Morris, Emma Wear, Anna James, Wenhao Chen, Charles Farmer, Susan Becker, Robert Letscher, Nicholas Huynh, Jeremy Mathis, Qian Li, Lauren Zamora, Andrew Margolin, Sarah Bercovici, Meredith Jennings, and Paula Hansell.

References

Abell, J., Emerson, S., Renaud, P., 2000. Distribution of TOP, TON, TOC in the North Pacific subtropical gyre: implications for nutrient supply in the surface ocean and remineralization in the upper thermocline. J. Mar. Res. 58, 203–222.

Aitkenhead, J., McDowell, W., 2000. Soil C:N ratio as a pre-dictor of annual riverine DOC flux at local and global scales. Global Biogeochem. Cycles 14, 127–138.

Alldredge, A., 1998. The carbon, nitrogen and mass content of marine snow as a function of aggregate size. Deep-Sea Res. I 45, 529–541.

Alldredge, A.L., Gotschalk, C.C., 1990. The relative contri-bution of marine snow of different origins to biological processes in coastal waters. Cont. Shelf Res. 10, 41–58.

Alldredge, A.L., Silver, M.W., 1988. Characteristics, dynam-ics and significance of marine snow. Prog. Oceanogr. 20, 41–82.

Alling, V., Humborg, C., Mörth, C.-M., Rahm, L., Pollehne, F., 2008. Tracing terrestrial organic matter by dS and dC signatures in a subarctic estuary. Limnol. Oceanogr. 53, 2594–2602.

Alonso-Sáez, L., Vázquez-Domínguez, E., Cardelús, C., Pinhassi, J., Sala, M.M., Lekunberri, I., et al., 2008. Factors controlling the year-round variability in carbon flux through bacteria in a coastal marine system. Ecosystems 11, 397–409.

Alonso-Sáez, L., Sánchez, O., Gasol, J.M., 2012. Bacterial up-take of low molecular weight organics in the subtropical Atlantic: are major phylogenetic groups functionally dif-ferent? Limnol. Oceanogr. 57, 798.

Aluwihare, L.I., Repeta, D.J., 1999. A comparison of the chemical characteristics of oceanic DOM and extracellu-lar DOM produced by marine algae. Mar. Ecol. Prog. Ser. 186, 105–117.

Aluwihare, L.I., Repeta, D.J., Chen, R.F., 1997. A major bio-polymeric component to dissolved organic carbon in sur-face sea water. Nature 387, 166–169.

Aluwihare, L.I., Repeta, D.J., Chen, R.F., 2002. Chemical composition and cycling of dissolved organic matter in the Mid-Atlantic Bight. Deep-Sea Res. II 49, 4421–4437.

Aluwihare, L., Repeta, D.J., Pantoja, S., Johnson, C.G., 2005. Two chemically distinct pools of organic nitrogen accu-mulate in the ocean. Science 308, 1007–1010.

Amon, R.M.W., Benner, R., 1994. Rapid cycling of high-molecular-weight dissolved organic matter in the ocean. Nature 369, 549–552.

Amon, R.M.W., Benner, R., 1996. Bacterial utilization of dif-ferent size classes of dissolved organic matter. Limnol. Oceanogr. 41, 41–51.

Amon, R.M.W., Benner, R., 2003. Combined neutral sug-ars as indicators of the diagenic state of dissolved organic matter in the Arctic Ocean. Deep-Sea Res. I 50, 151–169.

Amon, R.M.W., Budus, G., Meon, B., 2003. Dissolved or-ganic carbon distribution an origin in the Nordic Seas: exchanges with the Arctic Ocean and North Atlantic. J. Geophys. Res. 108, 3221–3238.

Anderson, T.R., Ducklow, H.W., 2001. Microbial loop carbon cycling in ocean environments studied using a simple steady-state model. Aquat. Microb. Ecol. 26, 37–49.

Anderson, T.R., Williams, P.J.L.B., 1998. Modelling the sea-sonal cycle of dissolved organic carbon at Station E1 in the English Channel. Estuar. Coast. Shelf Sci. 46, 93–109.

Anderson, T.R., Williams, P.J.L., 1999. A one-dimensional model of dissolved organic carbon cycling in the water column incorporating combined biological-photochemical decomposition. Global Biogeochem. Cycles 13, 337–349.

Apple, J.K., Del Giorgio, P.A., 2007. Organic substrate qual-ity as the link between bacterioplankton carbon demand and growth efficiency in a temperate salt-marsh estuary. ISME J. 1(8), 729–742.

Apple, J.K., del Giorgio, P., Kemp, W., 2006. Temperature reg-ulation of bacterial production, respiration, and growth efficiency in a temperate salt-marsh estuary. Aquat. Microb. Ecol. 43, 243–254.

Arístegui, J., Agusti, S., Duarte, C.M., 2003. Respiration in the dark ocean. Geophys. Res. Lett. 30, 1041.

Arnosti, C., 2011. Microbial extracellular enzymes and the marine carbon cycle. Ann. Rev. Mar. Sci. 3, 401–425.

Arnosti, C., Durkin, S., Jeffrey, W., 2005a. Patterns of extra-cellular enzymatic activities among pelagic microbial communities: implications for cycling of dissolved or-ganic carbon. Aquat. Microb. Ecol. 38, 135–145.

Arnosti, C., Durkin, S., Jeffrey, W.H., 2005b. Patterns of extra-cellular enzyme activities among pelagic marine micro-bial communities: implications for cycling of dissolved organic carbon. Aquat. Microb. Ecol. 38, 135–145.

Azam, F., 1998. Microbial control of oceanic carbon flux: the plot thickens. Science 280, 694–696.

Azam, F., Malfatti, F., 2007. Microbial structuring of marine ecosystems. Nat. Rev. Microbiol. 5, 782–791.

Azam, F., Long, R.A., 2001. Sea snow microcosms. Nature 414, 495–498.

Azam, F., Fenchel, T., Field, J.G., Gray, J.S., Meyer-Reil, L.A., Thingstad, F., 1983. The ecological role of water-column microbes in the sea. Mar. Ecol. Prog. Ser. 10, 257–263.

Baines, S.B., Pace, M.L., 1991. The production of dissolved organic matter by phytoplankton and its importance to bacteria: patterns across marine and freshwater systems. Limnol. Oceanogr. 36, 1078–1090.

Baltar, F., Reinthaler, T., Herndl, G.J., Pinhassi, J., 2013. Major effect of hydrogen peroxide on bacterioplankton metabo-lism in the Northeast Atlantic. PLoS One 8, e61051.

Barber, R.T., 1968. Dissolved organic carbon from deep water resists microbial oxidation. Nature 220, 274–275.

Bauer, J.E., 2002. Carbon isotopic compositioin of DOM. In: Hansell, D.A., Carlson, C.A. (Eds.), Biogeochemistry of Marine Dissolved Organic Matter. Academic Press, San Diego, pp. 405–453.

Bauer, J.E., Williams, P.M., Druffel, E.R.M., 1992. [14]C activity of dissolved organic carbon fractions in the north-central Pacific and Sargasso Sea. Nature 357, 667–670.

Behrenfeld, M.J., Falkowski, P.G., 1997. Photosynthetic rates derived from satellite-based chlorophyll concentration. Limnol. Oceanogr. 42, 1–20.

Béjà, O., Suzuki, M.T., 2008. Photoheterotrophic marine prokaryotes. In: Kirchman, D.L. (Ed.), Microbial Ecology of the Oceans, Second ed. Wiley-Blackwell, New Jersey, pp. 131–157.

Béjà, O., Aravind, L., Koonin, E.V., Suzuki, M.T., Hadd, A., Nguyen, L.P., et al., 2000. Bacterial rhodopsin: evidence for a new type of phototrophy in the sea. Science 289, 1902–1906.

Béjà, O., Spudich, E.N., Spudich, J.L., Leclerc, M., DeLong, E.F., 2001. Proteorhodopsin phototrophy in the ocean. Nature 411, 786–789.

Benner, R.H., 2002. Composition and reactivity. In: Hansell, D.A., Carlson, C.A. (Eds.), Biogeochemistry of Marine Dissolved Organic Matter. Academic Press, San Diego, pp. 59–90.

Benner, R., Biddanda, B., 1998. Photochemical transformation of surface and deep marine dissolved organic matter: effects on bacterial growth. Limnol. Oceanogr. 43, 1373–1378.

Benner, R., Herndl, G., 2011. Bacterially derived dissolved organic matter in the microbial carbon pump. In: Jiao, N., Azam, F., Sanders, S. (Eds.), Microbial Carbon Pump in the Ocean. Science/AAAS, Washington, DC, pp. 46–48.

Benner, R., Kaiser, K., 2011. Biological and photochemical transformations of amino acids and lignin phenols in riverine dissolved organic matter. Biogeochemistry 102, 209–222.

Benner, R., Opsahl, S., 2001. Molecular indicators of the sources and transformations of dissolved organic matter in the Mississippi river plume. Org. Geochem. 32, 597–611.

Benner, R., Pakulski, J.D., McCarthy, M., Hedges, J.I., Hatcher, P.G., 1992. Bulk chemical characteristics of dissolved organic matter in the ocean. Science 255, 1561–1564.

Benner, R., Louchouarn, P., Amon, R.M., 2005. Terrigenous dissolved organic matter in the Arctic Ocean and its transport to surface and deep waters of the North Atlantic. Global Biogeochem. Cycles 19, GB2025.

Berman, T., Kaplan, B., 1984. Diffusion chamber studies of carbon flux from living algae to heterotrophic bacteria. Hydrobiologia 108, 127–134.

Berman-Frank, I., Dubinsky, Z., 1999. Balanced growth in aquatic plants: myth or reality? Phytoplankton use the imbalance between carbon assimilation and biomass production to their strategic advantage. BioScience 49, 29–37.

Bianchi, T.S., 2011. The role of terrestrially derived organic carbon in the coastal ocean: a changing paradigm and the priming effect. Proc. Natl. Acad. Sci. 108, 19473–19481.

Biddanda, B., Benner, R., 1997. Carbon, nitrogen and carbohydrate fluxes during the production of particulate and dissolved organic matter by marine phytoplankton. Limnol. Oceanogr. 42, 506–518.

Biddanda, B., Ogdahl, M., Cotner, J., 2001. Dominance of bacterial metabolism in oligotrophic relative to eutrophic waters. Limnol. Oceanogr. 46, 730–739.

Biddanda, B., Opsahl, S., Benner, R., 1994. Plankton respiration and carbon flux through bacterioplankton on the Louisiana shelf. Limnol. Oceanogr. 39, 1259–1275.

Bidigare, R.R., 1983. Nitrogen excretion by marine zooplankton. In: Carpenter, E.J., Capone, D.G. (Eds.), Nitrogen in the Marine Environment. Academic Press, New York, pp. 385–409.

Billen, G., Fontigny, A., 1987. Dynamics of a Phaeocystis-dominated spring bloom in Belgian coastal waters II. Bacterioplankton dynamics. Mar. Ecol. Prog. Ser. 37, 249–257.

Bjørnsen, P.K., 1986. Bacterioplankton growth yield in continuous seawater cultures. Mar. Ecol. Prog. Ser. 30, 191–196.

Bjørnsen, P.K., 1988. Phytoplankton exudation of organic matter: why do healthy cells do it ? Linmol. Oceanogr. 33, 151–154.

Bonilla-Findji, O., Herndl, G.J., Gattuso, J.-P., Weinbauer, M.G., 2009. Viral and flagellate control of prokaryotic production and community structure in offshore Mediterranean waters. Appl. Environ. Microbiol. 75, 4801–4812.

Borch, N.H., Kirchman, D.L., 1999. Protection of protein from bacterial degradation by submicron particles. Aquat. Microb. Ecol. 16, 265–272.

Børsheim, K.Y., Vadstein, O., Myklestad, S.M., Reinertsen, H., Kirkvold, S., Olsen, Y., 2005. Photosynthetic algal production, accumulation and release of phytoplankton storage carbohydrates and bacterial production in a gradient in daily nutrient supply. J. Plank. Res. 27, 743–755.

Bratbak, G., Thingstad, T.F., 1985. Phytoplankton-bacteria interactions: an apparent paradox? Analysis of a model system with both competition and commensalism. Mar. Ecol. Prog. Ser. 25, 23–30.

Bratbak, G., Egge, J.K., Heldal, M., 1992. Viral mortality of the marine alga Emiliana huxleyi (Haptophyceae) and termination of algal bloom. Mar. Ecol. Prog. Ser. 83, 273–280.

Breitbart, M., 2012. Marine viruses: truth or dare. Ann. Rev. Mar. Sci. 4, 425–448.

Brophy, J.E., Carlson, D.J., 1989. Production of biologically refractory dissolved organic carbon by natural seawater microbial populations. Deep-Sea Res. I 36, 497–507.

Brussaard, C.P., 2005. Viral control of phytoplankton populations—a review. J. Eukaryot. Microbiol. 51, 125–138.

Burd, A.B., Hansell, D.A., Steinberg, D.K., Anderson, T.R., Arístegui, J., Baltar, F., et al., 2010. Assessing the apparent imbalance between geochemical and biochemical indicators of meso- and bathypelagic biological activity: what the @$#! is wrong with present calculations of carbon budgets? Deep-Sea Res. II 57, 1519–1536.

Burdige, D.J., Berelson, W.M., Coale, K.H., McManus, J., Johnson, K.S., 1999. Fluxes of dissolved organic carbon from California continental margin sediments. Geochim. Cosmochim. Acta 63, 1507–1515.

Calbet, A., 2001. Mesozooplankton grazing effect on primary production: a global comparative analysis in marine ecosystems. Limnol. Oceanogr. 46, 1824–1830.

Calbet, A., Landry, M.R., 1999. Mesozooplankton influences on the microbial food web: direct and indirect trophic interactions in the oligotrophic open ocean. Limnol. Oceanogr. 44, 1370–1380.

Calbet, A., Landry, M.R., 2004. Phytoplankton growth, microzooplankton grazing, and carbon cycling in marine systems. Limnol. Oceanogr. 49, 51–57.

Carlson, C.A., 2002. Production and removal processes. In: Hansell, D.A., Carlson, C.A. (Eds.), Biogeochemistry of Marine Dissolved Organic Matter. Academic Press, San Diego, pp. 91–151.

Carlson, C.A., Ducklow, H.W., 1995. Dissolved organic carbon in the upper ocean of the central Equatorial Pacific, 1992: daily and finescale vertical variations. Deep-Sea Res. II 42, 639–656.

Carlson, C.A., Ducklow, H.W., 1996. Growth of bacterioplankton and consumption of dissolved organic carbon in the Sargasso Sea. Aquat. Microb. Ecol. 10, 69–85.

Carlson, C.A., Hansell, D.A., 2003. The contribution of dissolved organic carbon and nitrogen to biogeochemistry of the Ross Sea. In: DiTullio, G., Dunbar, R. (Eds.), Biogeochemical Cycles in the Ross Sea. AGU Press, Washington DC, pp. 123–142.

Carlson, D.J., Mayer, L.M., Brann, M.L., Mague, T.H., 1985. Binding of monomeric organic compounds to macromolecular dissolved organic matter in seawater. Mar. Chem. 16, 141–153.

Carlson, C.A., Ducklow, H.W., Michaels, A.F., 1994. Annual flux of dissolved organic carbon from the euphotic zone in the northwestern Sargasso Sea. Nature 371, 405–408.

Carlson, C.A., Ducklow, H.W., Sleeter, T.D., 1996. Stocks and dynamics of bacterioplankton in the Northwestern Sargasso Sea. Deep-Sea Res. II 43, 491–515.

Carlson, C.A., Ducklow, H.W., Hansell, D.A., Smith, W.O., 1998. Organic carbon partitioning during spring phytoplankton blooms in the Ross Sea Polynya and the Sargasso Sea. Limnol. Oceanogr. 43, 375–386.

Carlson, C.A., Bates, N.R., Ducklow, H.W., Hansell, D.A., 1999. Estimation of bacterial respiration and growth efficiency in the Ross Sea Antarctica. Aquat. Microb. Ecol. 19, 229–244.

Carlson, C.A., Giovannoni, S.J., Hansell, D.A., Goldberg, S.J., Parsons, R., Otero, M.P., et al., 2002. The effect of nutrient amendments on bacterioplankton growth, DOC utilization, and community structure in the northwestern Sargasso Sea. Aquat. Microb. Ecol. 30, 19–36.

Carlson, C.A., Giovannoni, S.J., Hansell, D.A., Goldberg, S.J., Parsons, R., Vergin, K., 2004. Interactions among dissolved organic carbon, microbial processes, and community structure in the mesopelagic zone of the northwestern Sargasso Sea. Limnol. Oceanogr. 49, 1073–1083.

Carlson, C.A., del Giorgio, P.A., Herndl, G.J., 2007. Microbes and the dissipation of energy and respiration: from cells to cosystems. Oceanography 20, 89–100.

Carlson, C.A., Morris, R., Parsons, R., Treusch, A.H., Giovannoni, S.J., Vergin, K., 2009. Seasonal dynamics of SAR11 populations in the euphotic and mesopelagic zones of the northwestern Sargasso Sea. ISME J. 3, 283–295.

Carlson, C.A., Hansell, D.A., Nelson, N.B., Siegel, D.A., Smethie, W.M., Khatiwala, S., et al., 2010. Dissolved organic carbon export and subsequent remineralization in the mesopelagic and bathypelagic realms of the North Atlantic basin. Deep-Sea Res. II 57, 1433–1445.

Carlson, C.A., Hansell, D.A., Tamburini, C., 2011. DOC persistence and its fate after export within the ocean interior. In: Jiao, N., Azam, F., Sanders, S. (Eds.), Microbial Carbon Pump in the Ocean. Science/AAAS, Washington DC, pp. 57–59.

Caron, D.A., 2000. Symbiosis and mixotropy among pelagic microorganisms. In: Kirchman, D.L. (Ed.), Microbial Ecology of the Oceans. Wiley-Liss, Inc., New York, pp. 495–523.

Caron, D.A., Lim, E.L., Miceli, G., Waterbury, J.B., Valois, F.W., 1991. Grazing and utilization of chroococcoid cyanobacteria and heterotrophic bacteria by protozoa in laboratory cultures and a coastal plankton community. Mar. Ecol. Prog. Ser. 76, 205–217.

Chavez, F.P., Messié, M., Pennington, J.T., 2011. Marine Primary production in relation to climate variability and change. Ann. Rev. Mar. Sci. 3, 227–260.

Cherrier, J., Bauer, J.E., 2004. Bacterial utilization of transient plankton-derived dissolved organic carbon and nitrogen inputs in surface ocean waters. Aquat. Microb. Ecol. 35, 229–241.

Cherrier, J., Bauer, J.E., Druffel, E.R.M., 1996. Utilization and turnover of labile dissolved organic matter by bacterial heterotrophs in eastern North Pacific surface waters. Mar. Ecol. Prog. Ser. 139, 267–279.

Cherrier, J., Bauer, J.E., Druffel, E.R.M., Coffin, R.B., Chanton, J.P., 1999. Radiocarbon in marine bacteria: evidence for the ages of assimilated carbon. Limnol. Oceanogr. 44, 730–736.

Chin, W.-C., Orellana, M.V., Verdugo, P., 1998. Spontaneous assembly of marine dissolved organic matter into polymer gels. Nature 391, 568–572.

Chisholm, S.W., 1992. Phytoplankton size. In: Primary Productivity and Biogeochemical Cycles in the Sea. Springer US, New York, pp. 213–237.

Cho, B.C., Azam, F., 1988. Major role of bacteria in biogeochemical fluxes in the ocean's interior. Nature 332, 441–443.

Cho, J.C., Vergin, K.L., Morris, R.M., Giovannoni, S.J., 2004. Lentisphaera araneosa gen. nov., sp. nov, a transparent exopolymer producing marine bacterium, and the description of a novel bacterial phylum Lentisphaerae. Environ. Microbiol. 6, 611–621.

Christian, J.R., Karl, D.M., 1995. Bacterial ectoenzymes in marine waters: activity ratios and temperature responses in three oceanographic provinces. Limnol. Oceanogr. 40, 1042–1049.

Church, M.J., 2008. Resource control of bacterial dynamics in the sea. In: Kirchman, D.L. (Ed.), Microbial Ecology of the Oceans. second ed.. Wiley-Liss, New York, pp. 335–382.

Church, M.J., Hutchins, D.A., Ducklow, H.W., 2000. Limitation of bacterial growth by dissolved organic matter and iron in the Southern Ocean. Appl. Environ. Microbiol. 66, 455–466.

Church, M.J., Ducklow, H.W., Karl, D.M., 2002. Multiyear increases in dissolved organic matter inventories at station ALOHA in the North Pacific Subtropical Gyre. Limnol. Oceanogr. 47, 1–10.

Church, M.J., DeLong, E.F., Ducklow, H.W., Karner, M.B., Preston, C.M., Karl, D.M., 2003. Abundance and distribution of planktonic Archaea and Bacteria in the waters west of the Antarctic Peninsula. Limnol. Oceanogr. 48, 1893–1902.

Church, M.J., Ducklow, H.W., Karl, D.M., 2004. Light dependence of [3H] leucine incorporation in the oligotrophic North Pacific Ocean. Appl. Environ. Microbiol. 70, 4079–4087.

Coffin, R.B., Connolly, J.P., Harris, P.S., 1993. Availability of dissolved organic carbon to bacterioplankton examined by oxygen utilization. Mar. Ecol. Prog. Ser. 101, 9–22.

Cole, J.J., Finlay, S., Pace, M.L., 1988. Bacterial production in fresh and saltwater ecosystems: a cross-system overview. Mar. Ecol. Prog. Ser. 43, 1–10.

Conan, P., Sondergaard, M., Kragh, T., Thingstad, F., Pujo-Pay, M., Williams, P.J.L.B., et al., 2007. Partitioning of organic production in marine plankton communities: the effects of inorganic nutrient ratios and community composition on new dissolved organic matter. Limnol. Oceanogr. 52, 753–765.

Condon, R.H., Steinberg, D.K., Bronk, D.A., 2010. Production of dissolved organic matter and inorganic nutrients by gelatinous zooplankton in the York River estuary, Chesapeake Bay. J. Plank. Res. 32, 153–170.

Condon, R.H., Steinberg, D.K., del Giorgio, P.A., Bouvier, T.C., Bronk, D.A., Graham, W.M., et al., 2011. Jellyfish blooms result in a major microbial respiratory sink of carbon in marine systems. Proc. Natl. Acad. Sci. U. S. A. 108, 10225–10230.

Copin-Montgut, G., Avril, B., 1993. Vertical distribution and temporal variation of dissolved organic carbon in the North-Western Mediterranean Sea. Deep-Sea Res. 40, 1963–1972.

Copping, A.E., Lorenzen, C.J., 1980. Carbon budget of marine phytoplankton-herbivore system with carbon-14 as tracer. Limnol. Oceanogr. 25, 873–882.

Cory, R.M., McNeill, K., Cotner, J.P., Amado, A., Purcell, J.M., Marshall, A.G., 2010. Singlet oxygen in the coupled photochemical and biochemical oxidation of dissolved organic matter. Environ. Sci. Technol. 44, 3683–3689.

Cotner, J.B., Ammerman, J.W., Peele, E.R., Bentzen, E., 1997. Phosphorus-limited bacterioplankton growth in the Sargasso Sea. Aquat. Microb. Ecol. 13, 141–149.

Cottrell, M.T., Kirchman, D.L., 2000. Natural assemblages of marine proteobacteria and members of the Cytophaga-Flovobacter cluster consuming low and high molecular weight dissolved organic matter. Appl. Environ. Microbiol. 66, 1692–1697.

Countway, P.D., Caron, D.A., 2006. Abundance and distribution of Ostreococcus sp. in the San Pedro Channel, California, as revealed by quantitative PCR. Appl. Environ. Microbiol. 72, 2496–2506.

Cowie, G.L., Hedges, J.I., 1994. Biochemical indicators of diagenetic alteration in natural organic matter mixtures. Nature 369, 304–307.

Cushing, D., 1989. A difference in structure between ecosystems in strongly stratified waters and in those that are only weakly stratified. J. Plank. Res. 11, 1–13.

Dauwe, B., Middelburg, J., Herman, P., Heip, C., 1999. Linking diagenetic alterations of amino acids and bulk organic matter reactivity. Limnol. Oceanogr. 43, 1809–1814.

Davis, J., Benner, R., 2007. Quantitative estimates of labile and semi-labile dissolved organic carbon in the western Arctic Ocean: a molecular approach. Limnol. Oceanogr. 52, 2434–2444.

Davis, J., Kaiser, K., Benner, R., 2009. Amino acid and amino sugar yields and compositions as indicators of dissolved organic matter diagenesis. Org. Geochem. 40, 343–352.

de la Torre, J.R., Christianson, L.M., Beja, O., Suzuki, M.T., Karl, D.M., Heidelberg, J., et al., 2003. Proteorhodopsin genes are distributed among divergent marine bacterial taxa. Proc. Natl. Acad. Sci. U. S. A. 100, 12830–12835.

Decho, A.W., 1990. Microbial exopolymer secretions in ocean environments: their role(s) in food webs and marine processes. Oceanogr. Mar. Biol. Annu. Rev. 28, 73–153.

del Giorgio, P.A., Cole, J.J., 2000. Bacterial energetics and growth efficiency. In: Kirchman, D.L. (Ed.), Microbial Ecology of the Oceans. Wiley-Liss, Inc, New York, pp. 289–325.

del Giorgio, P.A., Condon, R., Bouvier, T., Longnecker, K., Bouvier, C., Sherr, E., Gasol, J.M., 2011. Coherent patterns in bacterial growth, growth efficiency, and leucine metabolism along a northeastern Pacific inshore-offshore transect. Limnol. Oceanogr. 56, 1–16.

del Giorgio, P., Davis, J., 2002. Patterns in dissolved organic matter lability and consumption across aquatic ecosystems. In: Aquatic Ecosystems: Interactivity of Dissolved Organic Matter. Academic Press, San Diego, CA, pp. 399–424.

del Giorgio, P.A., Williams, P.J.L.B., 2005. The global significance of respiration in aquatic ecosystems: from single cells to the biosphere. In: Giorgio, P.A.D., Williams, P.J.L.B. (Eds.), Respiration in Aquatic Ecosystems. Oxford, New York, pp. 267–303.

del Giorgio, P., Cole, J.J., 1998. Bacterial growth efficiency in natural aquatic systems. Annu. Rev. Ecol. Syst. 29, 503–541.

DeLong, E.F., Franks, D.G., Alldredge, A.L., 1993. Phylogenetic diversity of aggregate-attached vs. free-living marine bacterial assemblages. Limnol. Oceanogr. 38, 924–934.

DeLong, E.F., Preston, C.M., Mincer, T., Rich, V., Hallam, S.J., Frigaard, N.U., et al., 2006. Community genomics among stratified microbial assemblages in the ocean's interior. Science 311, 496–503.

Dittmar, T., 2008. The molecular level determination of black carbon in marine dissolved organic matter. Org. Geochem. 39, 396–407.

Dittmar, T., Kattner, G., 2003. Recalcitrant dissolved organic matter in the ocean: major contribution of small amphiphilics. Mar. Chem. 82, 115–123.

Dittmar, T., Paeng, J., 2009. A heat-induced molecular signature in marine dissolved organic matter. Nat. Geosci. 2, 175–179.

Doval, M.D., Hansell, D.A., 2000. Organic carbon and apparent oxygen utilization in the western South Pacific and central Indian Ocean. Mar. Chem. 68, 249–264.

Druffel, E.R.M., Williams, P.M., 1990. Identification of deep marine source of particulate organic carbon using bomb 14C. Nature 347, 172–174.

Druffel, E.R.M., Williams, P.M., Robertson, K., Griffin, S., Jull, A.J.T., Donahue, D., et al., 1989. Radiocarbon in dissolved organic and inorganic carbon from the central north Pacific. Radiocarbon 31, 523–532.

Druffel, E.R.M., Williams, P.M., Bauer, J.E., Ertel, J.R., 1992. Cycling of dissolved and particulate organic matter in the open ocean. J. Geophys. Res. 97, 15,639–15,659.

Druffel, E.R.M., Bauer, J.E., Williams, P.M., Griffin, S., Wolgast, D., 1996. Seasonal variability of particulate organic radiocarbon in the northeast Pacific Ocean. J. Geophys. Res. 101, 20543–20552.

Druffel, E.R.M., Griffin, S., Bauer, J.E., Wolgast, D.M., Wang, X.-C., 1998. Distribution of particulate organic carbon and radiocarbon in the water column from the upper slope to the abyssal NE Pacific Ocean. Deep-Sea Res. II 45, 667–687.

Ducklow, H.W., 1999. The bacterial component of the oceanic euphotic zone. FEMS Microbiol. Ecol. 30, 1–10.

Ducklow, H., 2000. Bacterial production and biomass in the ocean. In: Kirchman, D.L. (Ed.), Microbial Ecology of the Oceans. Wiley-Liss, Inc., New York, pp. 85–120.

Ducklow, H.W., Carlson, C.A., 1992. Oceanic bacterial production. In: Marshall, K.C. (Ed.), Advances in Microbial Ecology. Plenum Press, New York, pp. 113–181.

Ducklow, H.W., Shiah, F., 1993. Bacterial production in estuaries. In: Ford, T. (Ed.), Aquatic Microbiology: An Ecological Approach. Blackwell Science, Cambridge, MA, pp. 261–288.

Ducklow, H.W., Purdie, D.A., Williams, P.J.L., Davies, J.M., 1986. Bacterioplankton: a sink for carbon in a coastal marine plankton community. Science 232, 865–867.

Ducklow, H.W., Carlson, C.A., Bates, N.R., Knap, A.H., Michaels, A.F., 1995. Dissolved organic carbon as a component of the biological pump in the North Atlantic Ocean. Phil. Trans. R. Soc. A 348, 161–167.

Ducklow, H.W., Carlson, C.A., Church, M., Kirchman, D.L., Smith, D.C., Steward, G., 2001. The seasonal development of the bacterioplankton bloom in the Ross Sea, Antarctica, 1994–1997. Deep-Sea Res. II 48, 4199–4221.

Dunne, J.P., Sarmiento, J.L., Gnanadesikan, A., 2007. A synthesis of global particle export from the surface ocean and cycling through the ocean interior and on the seafloor. Global Biogeochem. Cycles 21, GB4006.

Duursma, E.K., 1963. The production of dissolved organic matter in the sea, as related to the primary gross production of organic matter. Neth. J. Sea Res. 2, 85–94.

Eichinger, M., Poggiale, J.C., Van Wambeke, F., Lefevre, D., Sempere, R., 2006. Modelling DOC assimilation and bacterial growth efficiency in biodegradation experiments: a case study in the Northeast Atlantic Ocean. Aquat. Microb. Ecol. 43, 139–151.

Elifantz, H., Dittell, A.I., Cottrell, M.T., Kirchman, D.L., 2007. Dissolved organic matter assimilation by heterotrophic bacterial groups in the western Arctic Ocean. Aquat. Microb. Ecol. 50, 39–49.

Engel, A., Thoms, S., Riebesell, U., Rochelle-Newall, E., Zondervan, I., 2004. Polysaccharide aggregation as a potential sink of marine dissolved organic carbon. Nature 428, 929–932.

Eppley, R.W., Stewart, E., Abbott, M.R., Owen, R.W., 1987. Estimating ocean production from satellited-derived chlorophyll: insights from the Eastropac data set. Oceanol. Acta SP, 109–113.

Fasham, M.J.R., Balino, B.M., Bowles, M.C., Anderson, R., Archer, D., Bathmann, U., et al., 2001. A new vision of ocean biogeochemistry after a decade of the Joint Global Ocean Flux Study (JGOFS). Ambio 10, 4–31.

Ferguson, R.L., Sunda, W.G., 1984. Utilization of amino acids by planktonic marine bacteria: importance of clean technique and low substrate additions. Linmol. Oceanogr. 29, 258–274.

Field, K.G., Gordon, D., Wright, T., Rapp, M., Urbach, E., Vergin, K., et al., 1997. Diversity and depth-specific distribution of SAR 11 cluster rRNA genes from marine planktonic bacteria. Appl. Environ. Microbiol. 63, 63–70.

First, M.R., Hollibaugh, J.T., 2009. The model high molecular weight DOC compound, dextran, is ingested by the benthic ciliate Uronema marinum but does not supplement ciliate growth. Aquat. Microb. Ecol. 57, 79.

Floodgate, G.D., 1995. Some environmental aspects of hydrocarbon bacteriology. Aquat. Microb. Ecol. 9, 3–11.

Flynn, K.J., 2008. The importance of the form of the quota curve and control of non-limiting nutrient transport in phytoplankton models. J. Plank. Res. 30, 423–438.

Flynn, K.J., Stoecker, D.K., Mitra, A., Raven, J.A., Gilbert, P.M., Hansen, P.J., et al., 2013. Misuse of the phytoplankton-zooplankton dichotomy: the need to assign organisms as mixotrophs within plankton functional types. J. Plankton Res. 35, 3–11.

Fogg, G., 1952. The production of extracellular nitrogenous substances by a blue-green alga. Proc. R. Soc. Lond. B 139, 372–397.

Fogg, G.E., 1966. The extracellular products of algae. Oceanogr. Mar. Biol. Annu. Rev. 4, 195–212.

Fogg, G.E., 1983. The ecological significance of extracellular products of phytoplankton. Bot. Mar. 26, 3–14.

Fontaine, S., Barot, S., 2005. Size and functional diversity of microbe populations control plant persistence and long-term soil carbon accumulation. Ecol. Lett. 8, 1075–1087.

Fontaine, S., Mariotti, A., Abbadie, L., 2003. The priming effect of organic matter: a question of microbial competition? Soil Biol. Biochem. 35, 837–843.

Fuhrman, J., 1987. Close coupling between release and uptake of dissolved free amino acids in seawater studied by an isotope dilution approach. Mar. Ecol. Prog. Ser. 37, 45–52.

Fuhrman, J.A., 1999. Marine viruses and their biogeochemical and ecological effects. Nature 399, 541–548.

Fuhrman, J.A., Noble, R.T., 1995. Viruses and protist cause similar bacterial mortality in coastal seawater. Limnol. Oceanogr. 40, 1236–1242.

Gärdes, A., Ramaye, Y., Grossart, H.-P., Passow, U., Ullrich, M.S., 2012. Effects of *Marinobacter adhaerens* HP15 on polymer exudation by *Thalassiosira weissflogii* at different N:P ratios. Mar. Ecol. Prog. Ser. 461, 1.

Giovannoni, S., 2005. The shape of microbial diversity. Environ. Microbiol. 7, 476.

Giovannoni, S.J., Stingl, U., 2005. Molecular diversity and ecology of microbial plankton. Nature 437, 343–348.

Giovannoni, S.J., Rapp, M.S., Vergin, K., Adair, N., 1996. 16S rRNA genes reveal stratified open ocean bacterioplankton populations related to the green non-sulfur bacteria phylum. Proc. Natl. Acad. Sci. U. S. A. 93, 7979–7984.

Giovannoni, S.J., Bibbs, L., Cho, J.C., Stapels, M.D., Desiderio, R., Vergin, K.L., et al., 2005. Proteorhodopsin in the ubiquitous marine bacterium SAR11. Nature 438, 82–85.

Gobler, C.J., Hutchins, D.A., Fisher, N.S., Cosper, E.M., Saudo-Wilhelmy, S.A., 1997. Release and bioavailability of C, N, P, Se and Fe following viral lysis of marine chrysophyte. Limnol. Oceanogr. 42, 1492–1504.

Goldberg, S., Carlson, C., Hansell, D., Nelson, N., Siegel, D., 2009. Temporal dynamics of dissolved combined neutral sugars and the quality of dissolved organic matter in the Northwestern Sargasso Sea. Deep-Sea Res. I 56, 672–685.

Goldberg, S., Carlson, C.A., Brock, B., Nelson, N.B., Siegel, D.A., 2010. Meridional variability in dissolved organic matter stocks and diagenetic state within the euphotic and mesopelagic zone of the North Atlantic subtropical gyre. Mar. Chem. 56, 9–21.

Goldberg, S., Carlson, C., Brzezinski, M., Nelson, N., Siegel, D., 2011. Systematic removal of neutral sugars within dissolved organic matter across ocean basins. Geophys. Res. Lett. 38, http://dx.doi.org/10.1029/2011GL048620, L17606.

Goldman, J.C., Dennett, M.R., 1985. Susceptibility of some marine phytoplankton species to cell breakage during filtration and post-filtration rinsing. J. Exp. Mar. Biol. Ecol. 86, 47–58.

Goldman, J.C., Dennett, M.R., 2000. Growth of marine bacteria in batch and continuous culture under carbon and nitrogen limitation. Limnol. Oceanogr. 45, 789–800.

Goldman, J., Caron, D.A., Dennett, M.R., 1987. Regulation of gross growth efficiency and ammonium regeneration in bacteria by substrate C:N ratio. Linmol. Oceanogr. 32, 1239–1252.

Goldman, J.C., Hansell, D.A., Dennett, M.R., 1992. Chemical characterization of three large oceanic diatoms: potential impact on water column chemistry. Mar. Ecol. Prog. Ser. 88, 257–270.

Gómez-Consarnau, L., González, J.M., Coll-Lladó, M., Gourdon, P., Pascher, T., Neutze, R., Pedrós-Alió, C., Pinhassi, J., 2007. Light stimulates growth of proteorhodopsin-containing marine Flavobacteria. Nature 445(7124), 210–213.

Gómez-Consarnau, L., Akram, N., Lindell, K., Pedersen, A., Neutze, R., Milton, D.L., et al., 2010. Proteorhodopsin phototrophy promotes survival of marine bacteria during starvation. PLoS Biol. 8, e1000358.

Gonsior, M., Peake, B.M., Cooper, W.T., Podgorski, D., D'Andrilli, J., Cooper, W.J., 2009. Photochemically induced changes in dissolved organic matter identified by ultrahigh resolution Fourier transform ion cyclotron resonance mass spectrometry. Environ. Sci. Technol. 43, 698–703.

Gordon, D., Giovannoni, S.J., 1996. Detection of stratified microbial populations related to Chlorobium and Fibrobacter species in the Atlantic and Pacific oceans. Appl. Environ. Microbiol. 62, 1171–1177.

Grossart, H.P., Ploug, H., 2001. Microbial degradation of organic carbon and nitrogen on diatom aggregates. Limnol. Oceanogr. 46, 267–277.

Grossart, H.-P., Engel, A., Arnosti, C., De La Rocha, C.L., Murray, A.E., Passow, U., 2007. Microbial dynamics in autotrophic and heterotrophic seawater mesocosms III. Organic matter fluxes. Aquat. Microb. Ecol. 49, 143–156.

Gruber, D.F., Simjouw, J.-P., Seitzinger, S.P., Taghon, G.L., 2006. Dynamics and characterization of refractory dissolved organic matter produced by a pure bacterial culture in an experimental predator-prey system. Appl. Environ. Microbiol. 72, 4184–4191.

Guenet, B., Danger, M., Abbadie, L., Lacroix, G., 2010. Priming effect: bridging the gap between terrestrial and aquatic ecology. Ecology 91, 2850–2861.

Guillard, R.R.L., Hellebust, J.A., 1971. Growth and the production of extracellular substances by two strains of *Phaeocystis pouchetii*. J. Phycol. 7, 330–338.

Guillard, R.R.L., Wangersky, P.J., 1958. The production of extracellular carbohydrates by some marine flagellates. Limnol. Oceanogr. 3, 449–454.

Halewood, E.R., Carlson, C.A., Brzezinski, M.A., Reed, D.C., Goodman, J., 2012. Annual cycle of organic matter partitioning and its availability to bacteria across the Santa Barbara Channel continental shelf. Aquat. Microb. Ecol. 67, 189–209.

Hansell, D.A., 2002. DOC in the global ocean carbon cycle. In: Hansell, D.A., Carlson, C.A. (Eds.), Biogeochemistry of Marine Dissolved Organic Matter. Academic Press, San Diego, pp. 685–716.

Hansell, D.A., 2013. Recalcitrant dissolved organic carbon fractions. Ann. Rev. Mar. Sci. 5, 421–445.

Hansell, D.A., Carlson, C.A., 1998a. Deep ocean gradients in dissolved organic carbon concentrations. Nature 395, 263–266.

Hansell, D.A., Carlson, C.A., 1998b. Net community production of dissolved organic carbon. Global Biogeochem. Cycles 12, 443–453.

Hansell, D.A., Carlson, C.A., 2001. Biogeochemistry of total organic carbon and nitrogen in the Sargasso Sea: control by convective overturn. Deep-Sea Res. II 48, 1649–1667.

Hansell, D.A., Carlson, C.A., 2013. Localized refractory dissolved organic carbon sinks in the deep ocean. Global Biogeochem. Cycles 27, 705–710.

Hansell, D.A., Bates, N.R., Gundersen, K., 1995. Mineralization of dissolved organic carbon in the Sargasso Sea. Mar. Chem. 51, 201–212.

Hansell, D.A., Carlson, C.A., Suzuki, Y., 2002. Dissolved organic carbon export with North Pacific intermediate water formation. Global Biogeochem. Cycles 16, 1007.

Hansell, D.A., Kadko, D., Bates, N.R., 2004. Degradation of terrigenous dissolved organic carbon in the western Arctic Ocean. Science 304, 858–861.

Hansell, D.A., Carlson, C.A., Repeta, D.J., Shlitzer, R., 2009. Dissolved organic matter in the ocean: a controversy stimulates new insights. Oceanography 22, 202–211.

Hansell, D.A., Carlson, C.A., Schlitzer, R., 2012. Net removal of major marine dissolved organic carbon fractions in the subsurface ocean. Global Biogeochem. Cycles 26, http://dx.doi.org/10.1029/2011GB004069, GB1016.

Hansman, R.L., Griffin, S., Watson, J.T., Druffel, E.R., Ingalls, A.E., Pearson, A., et al., 2009. The radiocarbon signature of microorganisms in the mesopelagic ocean. Proc. Natl. Acad. Sci. U. S. A. 106, 6513–6518.

Hansson, L., Norrman, B., 1995. Release of dissolved organic carbon (DOC) by the scyphozoan jellyfish Aurelia aurita and its potential influence on the production of planktic bacteria. Mar. Biol. 121, 527–532.

Harvey, G.R., Boran, D.A., Chesal, L.A., Tokar, J.M., 1983. The structure of marine fulvic and humic acids. Mar. Chem. 12, 119–132.

Hedges, J.I., 1988. Polymerization of humic substances in natural environments. In: Frimmel, F.H., Christman, R.F. (Eds.), Humic Substances and Their Role in the Environment. Wiley, Chichester, pp. 45–48.

Hedges, J.I., Keil, R.G., 1995. Sedimentary organic matter preservation: an assessment and speculative synthesis. Mar. Chem. 49, 81–115.

Heissenberger, A., Herndl, G.H., 1994. Formation of high molecular weight material by free-living marine bacteria. Mar. Ecol. Prog. Ser. 111, 129.

Hellebust, J.A., 1974. Extracellular products. Bot. Monogr. 10, 838–863.

Henson, S.A., Sanders, R., Madsen, E., 2012. Global patterns in efficiency of particulate organic carbon export and transfer to the deep ocean. Global Biogeochem. Cycles 26, GB1028.

Herndl, G.J., Reinthaler, T., Teira, E., van Aken, H., Veth, C., Pernthaler, A., et al., 2005. Contribution of Archaea to total prokaryotic production in the deep Atlantic Ocean. Appl. Environ. Microbiol. 71, 2303–2309.

Herndl, G.J., Agogué, H., Baltar, F., Reinthaler, T., Sintes, E., Varela, M.M., 2008. Regulation of aquatic microbial processes: the'microbial loop'of the sunlit surface waters and the dark ocean dissected. Aquat. Microb. Ecol. 53, 59.

Hertkorn, N., Benner, R., Frommberger, M., Schmitt-Kopplin, P., Witt, M., Kaiser, K., et al., 2006. Characterization of a major refractory component of marine dissolved organic matter. Geochim. Cosmochim. Acta 70, 2990–3010.

Hmelo, L.R., Mincer, T.J., Van Mooy, B.A.S., 2011. Possible influence of bacterial quorum sensing on the hydrolysis of sinking particulate organic carbon in marine environments. Environ. Microbiol. Rep. 3, 682–688.

Holmes, R.M., McClelland, J.W., Raymond, P.A., Frazer, B.B., Peterson, B.J., Stieglitz, M., 2008. Lability of DOC transported by Alaskan rivers to the Arctic Ocean. Geophys. Res. Lett. 35, L03402.

Holmfeldt, K., Titelman, J., Riemann, L., 2010. Virus production and lysate recycling in different Sub-basins of the Northern Baltic Sea. Microb. Ecol. 60, 572–580.

Hopkinson, C.S., Vallino, J.J., 2005. Efficient export of carbon to the deep ocean through dissolved organic matter. Nature 433, 142–145.

Hoppe, H.G., 1991. Microbial extracellular enzyme activity: a new key parameter in aquatic ecology. In: Chrost, R.J. (Ed.), Microbial Enzymes in Aquatic Environments. Springer-Verlag, New York, pp. 60–83.

Hoppe, H.G., Arnosti, C., Herndl, G.J., 2002. Ecological significance of bacterial enzymes in the marine environment. In: Burns, R.G., Dick, R.P. (Eds.), Ezymes in the Environment: Activity, Ecology and Applications. Marcel Dekker, New York, pp. 73–108.

Horvath, R.S., 1972. Microbial co-metabolism and the degradation of organic compounds in nature. Bacteriol. Rev. 36, 146.

Imai, I., Ishida, Y., Hata, Y., 1993. Killing of marine phytoplankton by a gliding bacterium *Cytophaga* sp., isolated from the coastal sea of Japan. Mar. Biol. 116, 527–532.

Ingalls, A.E., Shah, S.R., Hansman, R.L., Aluwihare, L.I., Santos, G.M., Druffel, E.R.M., et al., 2006. Quantifying archaeal community autotrophy in the mesopelagic ocean using natural radiocarbon. Proc. Natl. Acad. Sci. U. S. A. 103, 6442–6447.

Ittekkot, V., Brockmann, U., Michaelis, W., Degens, E.T., 1981. Dissolved free and combined carbohydrates during a phytoplankton bloom in the northern North Sea. Mar. Ecol. Prog. Ser. 4, 299–305.

Jahnke, R., Craven, D.B., 1995. Quantifying the role of heterotrophic bacteria in the carbon cycle: a need for respiration rate measurements. Limnol. Oceanogr. 40, 436–441.

Jannasch, H.W., Mottl, M.J., 1985. Geomicrobiology of deep-sea hydrothermal vents. Science 229, 717–725.

Jiao, N., Herndl, G.J., Hansell, D.A., Benner, R., Kattner, G., Wilhelm, S.W., et al., 2010. Microbial production of recalcitrant dissolved organic matter: long-term carbon storage in the global ocean. Nature 8, 593–599.

Jiao, N., Herndl, G.J., Hansell, D.A., Benner, R., Kattner, G., Wilhelm, S.W., et al., 2011. The microbial carbon pump and the oceanic recalcitrant dissolved organic matter pool. Nat. Rev. Microbiol. 9, 555.

Johannes, R., Webb, K., 1965. Release of dissolved amino acids by marine zooplankton. Science 150, 76–77.

Jumars, P.A., Penry, D.L., Baross, J.A., Perry, M.J., Frost, B.W., 1989. Closing the microbial loop: dissolved carbon pathway to heterotrophic bacteria from incomplete ingestion, digestion and absorption in animals. Deep-Sea Res. 36, 483–495.

Jürgens, K., Massana, R., Kirchman, D., 2008. Protist grazing on marine bacterioplankon. In: Kirchman, D. (Ed.), Microbial Ecology of the Oceans. Wiley-Liss, New York, pp. 383–442.

Kadurugamuwa, J.L., Beveridge, T.J., 1996. Bacteriolytic effect of membrane vesicles from *Pseudomonas aeruginosa* on other bacteria including pathogens: conceptually new antibiotics. J. Bact. 178, 2767–2774.

Kähler, P., Bjornsen, P.K., Lochte, K., Antia, A., 1997. Dissolved organic matter and its utilization by bacteria during spring in the Southern Ocean. Deep-Sea Res. II 44, 341–353.

Kaiser, K., Benner, R., 2008. Major bacterial contribution to the ocean reservoir of detrital organic carbon and nitrogen (vol 53, pg. 99, 2008). Limnol. Oceanogr. 53, 1192.

Kaiser, K., Benner, R., 2009. Biochemical composition and size distribution of organic matter at the Pacific and Atlantic time-series stations. Mar. Chem. 113, 63–77.

Kaiser, K., Benner, R., 2012. Organic matter transformations in the upper mesopelagic zone of the North Pacific:

chemical composition and linkages to microbial community structure. J. Geophys. Res.-Oceans 117, 1023.

Karl, D., 2002. Hidden in a sea of microbes. Nature 415, 590–591.

Karl, D.M., Hebel, D.V., Björkman, K., Letelier, R.M., 1998. The role of dissolved organic matter release in the productivity of the oligotrophic North Pacific Ocean. Limnol. Oceanogr. 43, 1270–1286.

Karner, M., Herndl, G.J., 1992. Extracellular enzymatic activity and secondary production in free-living and marine-snow-associated bacteria. Mar. Biol. 113, 341–347.

Karner, M.B., DeLong, E.F., Karl, D.M., 2001. Archaeal dominance in the mesopelagic zone of the Pacific Ocean. Nature 409, 507–510.

Kattner, G., Simon, M., Koch, B., 2011. Molecular characterization of dissolved organic matter and constraints for prokaryotic utilization. In: Jiao, N., Azam, F., Sanders, S. (Eds.), Microbial Carbon Pump in the Ocean. Science/AAAS, Washington, DC, pp. 60–61.

Kawasaki, N., Benner, R., 2006. Bacterial release of dissolved organic matter during cell growth and decline: molecular origin and composition. Limnol. Oceanogr. 51, 2170–2180.

Keil, R.G., Kirchman, D.L., 1994. Abiotic transformation of labile protein to refractory protein in sea water. Mar. Chem. 45, 187–196.

Keil, R.G., Kirchman, D.L., 1999. Utilization of dissolved protein and amino acids in the northern Sargasso Sea. Aquat. Microb. Ecol. 18, 293–300.

Kieber, D.J., McDaniel, J., Mopper, K., 1989. Photochemical source of biological substrates in sea water: implications for carbon cycling. Nature 341, 637–639.

Kiene, R.P., Linn, L.J., Bruton, J.A., 2000. New and important roles for DMSP in marine microbial communities. J. Sea Res. 43, 209–224.

Kiørboe, T., 1993. Turbulence, phytoplankton cell size, and the structure of pelagic food webs. Adv. Mar. Biol. 29, 72.

Kiørboe, T., Jackson, G.A., 2001. Marine snow, organic solute plumes and optimal chemosensory behavior of bacteria. Limnol. Oceanogr. 46, 1309–1318.

Kiørboe, T., Ploug, H., Thygesen, U.H., 2001. Fluid motion and solute distribution around sinking aggregates I. Small scales fluxes and heterogeneity of nutrients in the pelagic environment. Mar. Ecol. Prog. Ser. 211, 1–13.

Kiørboe, T., Tang, K., Grossart, H.-P., Ploug, H., 2003. Dynamics of microbial communities on marine snow aggregates: colonization, growth, detachment, and grazing mortality of attached bacteria. Appl. Environ. Microbiol. 69, 3036–3047.

Kirchman, D.L., 1990. Limitation of bacterial growth by dissolved organic matter in the subarctic Pacific. Mar. Ecol. Prog. Ser. 62, 47–54.

Kirchman, D.L., 2000. Uptake and regeneration of inorganic nutrients by marine heterotrophic bacteria. In: Kirchman,

D.L. (Ed.), Microbial Ecology of the Oceans. Wiley-Liss, Inc., New York, pp. 261–288.

Kirchman, D.L., Henry, D.L., Dexter, S.C., 1989. Adsorption of protein to surfaces in seawater. Mar. Chem. 27, 201–217.

Kirchman, D., Keil, R., Wheeler, P.A., 1990. Carbon limitation of ammonium uptake by heterotrophic bacteria in the subarctic Pacific. Linmol. Oceanogr. 35, 1267–1278.

Kirchman, D.L., Lancelot, C., Fasham, M., Legendre, L., Radach, G., Scott, M., 1993. Dissolved organic matter in biogeochemical models of the ocean. In: Evans, G.T., Fasham, M.J.R. (Eds.), Towards a Model of Ocean Biogeochemical Processes. Springer-Verlag, Berlin, pp. 209–225.

Kirchman, D.L., Meon, B., Ducklow, H.W., Carlson, C.A., Hansell, D.A., Steward, G.F., 2001. Glucose fluxes and concentrations of dissolved combined neutral sugars (polysaccharides) in the Ross Sea and Polar Front Zone Antarctica. Deep-Sea Res. II 48, 4179–4197.

Koch, B.P., Ludwichowski, K.-U., Kattner, G., Dittmar, T., Witt, M., 2008. Advanced characterization of marine dissolved organic matter by combining reversed-phase liquid chromatography and FT-ICR-MS. Mar. Chem. 111, 233–241.

Kolber, Z.S., Van Dover, C., Niederman, R., Falkowski, P., 2000. Bacterial photosynthesis in surface waters of the open ocean. Nature 407, 177–179.

Kolber, Z., Plumley, F.G., Lang, A.S., Beatty, J.T., Blankenship, R.E., Vandover, C.L., et al., 2001. Contribution of aerobic photoheterotrophic bacteria to carbon cycle in the ocean. Science 292, 2492–2495.

Kragh, T., Sondergaard, M., 2009. Production and decomposition of new DOC by marine plankton communities: carbohydrates, refractory components and nutrient limitation. Biogeochemistry 96, 177–187.

Kujawinski, E.B., 2011. The impact of microbial metabolism on marine dissolved organic matter: insights from analytical chemistry and microbiology. Ann. Rev. Mar. Sci. 3, 567–599.

Kujawinski, E.B., Del Vecchio, R., Blough, N.V., Klein, G.C., Marshall, A.G., 2004. Probing molecular-level transformations of dissolved organic matter: insights on photochemical degradation and protozoan modification of DOM from electrospray ionization Fourier transform ion cyclotron resonance mass spectrometry. Mar. Chem. 92, 23–37.

Kujawinski, E.B., Longnecker, K., Blough, N.V., Del Vecchio, R., Finlay, L., Kitner, J.B., et al., 2009. Identification of possible source markers in marine dissolved organic matter using ultrahigh resolution mass spectrometry. Geochim. Cosmochim. Acta 73, 4384–4399.

La Rowe, D.E., Dale, A.W., Amend, J.P., Van Cappellen, P., 2012. Thermodynamic limitations on microbially catalyzed reaction rates. Geochim. Cosmochim. Acta 90, 96–109.

Lampert, W., 1978. Release of dissolved organic carbon by grazing zooplankton. Limnol. Oceanogr. 23, 831–834.

Lampitt, R.S., Noji, T., Bodugen, B.v., 1990. What happens to zooplankton fecal pellets? Implication for material flux. Mar. Biol. 104, 15–23.

Lancelot, C., 1979. Gross excretion rates of natural marine phytoplankton and heterotrophic uptake of excreted products in the southern North Sea, as determined by short-term kinetics. Mar. Ecol. Prog. Ser. 1, 179–186.

Lancelot, C., 1983. Factors affecting phytoplankton extracellular release in the Southern Bight of the North Sea. Mar. Ecol. Prog. Ser. 12, 115–121.

Lancelot, C., 1984. Extracellular release of small and large molecules by phytoplankton in the Southern Bight of the North Sea. Estuar. Coast. Shelf Sci. 18, 65–77.

Landa, M., Cottrell, M.T., Kirchman, D.L., Blain, S., Obernosterer, I., 2013. Changes in bacterial diversity in response to dissolved organic matter supply in a continuous culture experiment. Aquat. Microb. Ecol. 69, 157–168.

Lang, S.Q., Butterfield, D.A., Lilley, M.D., Paul Johnson, H., Hedges, J.I., 2006. Dissolved organic carbon in ridge-axis and ridge-flank hydrothermal systems. Geochim. Cosmochim. Acta 70, 3830–3842.

Lara, R.J., Thomas, D.N., 1995. Formation of recalcitrant organic matter: humification dynamics of algal derived dissolved organic carbon and its hydrophobic fractions. Mar. Chem. 51, 193–199.

Lauro, F., Bartlett, D., 2007. Prokaryotic lifestyles in deep sea habitats. Extremophiles 12, 15–25.

Lee, D.Y., Rhee, G.Y., 1997. Kinetics of cell death in the cyanobacterium Anabaena flos-auae and the production of dissolved organic carbon. J. Phycol. 33, 991–998.

Letscher, R.T., Hansell, D.A., Carlson, C.A., Lumpkin, R., Knapp, A.N., 2013. Dissolved organic nitrogen in the global surface ocean: distribution and fate. Global Biogeochem. Cy. 27, 141–153.

Letscher, R.T., Hansell, D.A., Kadko, D., 2011. Rapid removal of terrigenous dissolved organic carbon over the Eurasian shelves of the Arctic Ocean. Mar. Chem. 123, 78–87.

Lewin, R.A., 1956. Extracellular polysaccharides of green algae. Can. J. Microbiol. 2, 665–672.

Lignell, R., 1990. Excretion of organic carbon by phytoplankton: its relation to algal biomass, primary productivity and bacterial secondary productivity in the Baltic Sea. Mar. Ecol. Prog. Ser. 68, 85–99.

Lin, D., McBride, M.J., 1996. Development of techniques for the genetic manipulation of the gliding bacteria Lysobacter enzymogenes and Lysobacter brunescens. Can. J. Microbiol. 42, 896–902.

Linton, J.D., Stephenson, R.J., 1978. A preliminary study on growth yields in relation to the carbon and energy content of various organic growth substrates. FEMS Microbiol. Lett. 3(2), 95–98.

Liu, Y., 1998. Energy uncoupling in microbial growth under substrate-sufficient conditions. Appl. Microbiol. Biotechnol. 49, 500–505.

Long, R.A., Azam, F., 2001. Microscale patchiness of bacterioplankton assemblage richness in seawater. Aquat. Microb. Ecol. 26, 103–113.

Longhurst, A.R., Harrison, W.G., 1989. The biological pump: profiles of plankton production and consumption in the upper ocean. Prog. Oceanogr. 22, 47–123.

López-Sandoval, D., Fernández, A., Maranón, E., 2011. Dissolved and particulate primary production along a longitudinal gradient in the Mediterranean Sea. Biogeosciences 8, 815–825.

Luo, Y.-W., Friedrichs, M.A.M., Doney, S.C., Church, M.J., Ducklow, H.W., 2010. Oceanic heterotrophic bacterial nutrition by semilabile DOM as revealed by data assimilative modeling. Aquat. Microb. Ecol. 60, 273–278.

MacKenzie, F.T., 1981. Global carbon cycle: some minor sinks for CO_2. In: Likens, G.E., MacKenzie, F.T., Richey, J.E., Sedell, J.R., Turekian, K.K. (Eds.), Flux of Organic Carbon by Rivers to the Ocean. U.S. Department of Energy, Washington, DC, pp. 360–384.

Magaletti, E., Urbani, R., Sist, P., Ferrari, C.R., Cicero, A.M., 2004. Abundance and chemical characterization of extracellular carbohydrates released by the marine diatom Cylindrotheca fusiformis under N-and P-limitation. Eur. J. Phycol. 39, 133–142.

Mague, T.H., Freberg, E., Hughes, D.J., Morris, I., 1980. Extracellular release of carbon by marine phytoplankton; a physiological approach. Linmol. Oceanogr. 25, 262–279.

Maillard, L.C., 1912. Action des acides amines sur les sucres; formation des melanoidines par voie methodique. Acad. Sci. Compt. Rend. 154, 66–68.

Malmstrom, R.R., Kiene, R.P., Cottrell, M.T., Kirchman, D.L., 2004a. Contribution of SAR11 bacteria to dissolved dimethylsulfoniopropionate and amino acid uptake in the North Atlantic ocean. Appl. Environ. Microbiol. 70, 4129–4135.

Malmstrom, R.R., Kiene, R.P., Kirchman, D.L., 2004b. Identification and enumeration of bacteria assimilating dimethylsulfoniopropionate (DMSP) in the North Atlantic and Gulf of Mexico. Limnol. Oceanogr. 49, 597–606.

Malmstrom, R.R., Cottrell, M.T., Elifantz, H., Kirchman, D.L., 2005a. Biomass production and assimilation of dissolved organic matter by SAR11 bacteria in the Northwest Atlantic Ocean. Appl. Environ. Microbiol. 71, 2979–2986.

Malmstrom, R.R., Kiene, R.P., Vila, M., Kirchman, D.L., 2005b. Dimethylsulfoniopropionate (DMSP) assimilation by Synechococcus in the Gulf of Mexico and northwest Atlantic Ocean. Limnol. Oceanogr. 50, 1924–1931.

Manahan, D.T., Richardson, K., 1983. Competition studies on the uptake of dissolved organic nutrients by bivalve larvae (Mytilus edulis) and marine bacteria. Mar. Biol. 75, 241–247.

Maranón, E., Cermeno, P., Fernandez, E., Rodriguez, J., Zabala, L., 2004. Significance and mechanisms of photosynthetic production of dissolved organic carbon in a coastal eutrophic ecosystem. Limnol. Oceanogr. 49, 1652–1666.

Maranón, E., Cermeno, P., Perez, V., 2005. Continuity in the photosynthetic production of dissolved organic carbon from eutrophic to oligotrophic waters. Mar. Ecol. Prog. Ser. 299, 7–17.

Marchant, H.J., Scott, F.J., 1993. Uptake of sub-micrometer particles and dissolved organic material by Antarctic choanoflagellates. Mar. Ecol. Prog. Ser. 92, 59–64.

Martin, M.O., 2002. Predatory prokaryotes: an emerging research opportunity. J. Mol. Microbiol. Tech. 4, 467–478.

Martin-Jezequel, V., Poulet, S., Harris, R., Moal, J., Samain, J., 1988. Interspecific and intraspecific composition and variation of free amino acids in marine phytoplankton. Mar. Ecol. Prog. Ser. 44, 303–313.

Maske, H., Garcia-Mendoza, E., 1994. Adsorption of dissolved organic matter to the inorganic filter substrate and its implications for 14C uptake measurements. Appl. Environ. Microbiol. 60, 3887–3889.

Mayali, X., Azam, F., 2004. Algicidal bacteria in the sea and their impact on algal blooms. J. Eukaryot. Microbiol. 51, 139–144.

McCarren, J., Becker, J.W., Repeta, D.J., Shi, Y., Young, C.R., Malmstrom, R.R., et al., 2010. Microbial community transcriptomes reveal microbes and metabolic pathways associated with dissolved organic matter turnover in the sea. Proc. Natl. Acad. Sci. U. S. A. 107, 16420–16427.

McCarthy, M., Hedges, J., Benner, R., 1996. Major biochemical composition of dissolved high molecular weight organic matter in seawater. Mar. Chem. 55, 281–297.

McCarthy, M.D., Hedges, J.I., Benner, R., 1998. Major bacterial contribution to marine dissolved organic nitrogen. Science 281, 231–234.

Meador, J.A., Baldwin, A.J., Catala, P., Jeffrey, W.H., Joux, F., Moss, J.A., et al., 2009. Sunlight-induced DNA Damage in Marine Micro-organisms Collected Along a Latitudinal Gradient from 70° N to 68° S. Photochem. Photobiol. 85, 412–420.

Meon, B., Amon, R.M., 2004. Heterotrophic bacterial activity and fluxes of dissolved free amino acids and glucose in the Arctic rivers Ob, Yenisei and the adjacent Kara Sea. Aquat. Microb. Ecol. 37, 121–135.

Meybeck, M., 1981. River transport of organic carbon to the ocean. NAS-NRC Carbon Dioxide Effects Res. and Assessment Program: Flux of Org. Carbon by Rivers to the Ocean, pp. 219–269 (SEE N 81-30674 21-45).

Michaels, A.F., Silver, M.W., 1988. Primary production, sinking fluxes and the microbial food web. Deep-Sea Res. 35, 473–490.

Michaels, A.F., Karl, D.M., Capone, D.G., 2001. Element stoichiometry, new production and nitrogen fixation. Oceanography 14, 68–77.

Michelou, V.K., Cottrell, M.T., Kirchman, D.L., 2007. Light-stimulated bacterial production and amino acid assimilation by cyanobacteria and other microbes in the North Atlantic Ocean. Appl. Environ. Microbiol. 73, 5539–5546.

Middelboe, M., 2008. Microbial disease in the sea: effects of viruses on carbon and nutrient cycling. In: Ostfeld, R.S., Keesing, F., Eviner, V.T. (Eds.), Infectious Disease Ecology: Effects of Ecosystems on Disease and of Disease on Ecosystems. Princeton University Press, Princeton, pp. 242–259.

Middelboe, M., Jorgensen, N.O.G., 2006. Viral lysis of bacteria: an important source of dissolved amino acids and cell wall compounds. J. Mar. Biol. Assoc. U. K. 86, 605–612.

Middelboe, M., Lyck, P.G., 2002. Regeneration of dissolved organic matter by viral lysis in marine microbial communities. Aquat. Microb. Ecol. 27, 187–194.

Middelboe, M., Jorgensen, N., Kroer, N., 1996. Effects of viruses on nutrient turnover and growth efficiency of noninfected marine bacterioplankton. Appl. Environ. Microbiol. 62, 1991–1997.

Middelboe, M., Riemann, L., Steward, G.F., Hansen, V., Nybroe, O., 2003. Virus-induced transfer of organic carbon between marine bacteria in a model community. Aquat. Microb. Ecol. 33, 1–10.

Middelboe, M., Søndergaard, M. , 1993. Bacterioplankton growth yield: seasonal variations and coupling to substrate lability and β-glucosidase activity. Appl. Environ. Microbiol. 59, 3916–3921.

Middelburg, J.J., 2011. Chemoautotrophy in the ocean. Geophys. Res. Lett. 38, L24604.

Mills, M.M., Moore, C., Langlois, R., Milne, A., Achterberg, E., Nachtigall, K., et al., 2008. Nitrogen and phosphorus co-limitation of bacterial productivity and growth in the oligotrophic subtropical North Atlantic. Limnol. Oceanogr. 53, 824–834.

Mitra, A., Flynn, K.J., Burkholder, J., Berge, T., Calbet, A., Raven, J.A., et al., 2013. The role of mixotrophic protists in the biological carbon pump. Biogeosciences Discussions 10, 13535–13562.

Moeseneder, M.M., Winter, C., Herndl, G.J., 2001. Horizontal and vertical complexity of attached and free-living bacteria of the Eastern Mediterranean Sea, determined by 16S rDNA and 16S rRNA fingerprints. Limnol. Oceanogr. 46, 95–107.

Møller, E.F., 2005. Sloppy feeding in marine copepods: prey-size-dependent production of dissolved organic carbon. J. Plank. Res. 27, 27–35.

Møller, E.F., 2007. Production of dissolved organic carbon by sloppy feeding in the copepods "Acartia tonsa," "Centropages typicus," and "Temora longicornis". Limnol. Oceanogr. 52, 79–84.

Møller, E.F., Nielsen, T.G., 2001. Production of bacterial substrate by marine copepods: effect of phytoplankton biomass and cell size. J. Plank. Res. 23, 527–536.

Møller, E.F., Thor, P., Nielsen, T.G., 2003. Production of DOC by Calanus finmarchicus, C. glacialis and C. hyperboreus through sloppy feeding and leakage from fecal pellets. Mar. Ecol. Prog. Ser. 262, 185–191.

Møller-Jensen, L., 1983. Phytoplankton release of extracellular organic carbon, molecular weight composition, and bacterial assimilation. Mar. Ecol. Prog. Ser. 11, 39–48.

Motegi, C., Nagata, T.. Miki, T.. Weinbauer, M.G.. Legendre, L.. Rassoulzadegan, F., 2009. Viral control of bacterial growth efficiency in marine pelagic environments. Limnol. and Oceanogr. 54(6), 1901.

Mopper, K., Kieber, D.J., 2002. Photochemistry and cycling of carbon, sulfur, nitrogen and phosphorus. In: Hansell, D.A., Carlson, C.A. (Eds.), Biogeochemistry of Marine Dissolved Organic Matter. Academic Press, San Diego.

Mopper, K., Zhou, X.L., Kieber, R.J., Kieber, D.J., Sikorski, R.J., Jones, R.D., 1991. Photochemical degradation of dissolved organic carbon and its impact on the oceanic carbon cycle. Nature 353, 60–62.

Moran, M.A., Zepp, R.G., 1997. Role of photoreactions in the formation of biologically labile compounds from dissolved organic matter. Limnol. Oceanogr. 42, 1307–1316.

Moran, M.A., Zepp, R.G., 2000. UV radiation effects on microbes and microbial processes. In: Kirchman, D.L., (Ed.), Microbial Ecology of the Oceans. Wiley-Liss Inc., New York pp. 201–228.

Morán, X.A.G., Gasol, J.M., Arin, L., Estrada, M., 1999. A comparison between glass fiber and membrane filters for the estimation of phytoplankton POC and DOC production. Mar. Ecol. Prog. Ser. 187, 31–41.

Morán, X., Estrada, M., Gasol, J., Pedros-Alio, C., 2002. Dissolved primary production and the strength of phytoplankton–bacterioplankton coupling in contrasting marine regions. Microb. Ecol. 44, 217–223.

Moreira, D., Rodriguez-Valera, F., Lopez-Garcia, P., 2006. Metagenomic analysis of mesopelagic Antarctic plankton reveal a novel deltaproteobacteial group. Microbiology 152, 505–517.

Morris, R.M., Vergin, K.L., Cho, J.C., Rappe, M.S., Carlson, C.A., Giovannoni, S.J., 2005. Temporal and spatial response of bacterioplankton lineages to annual convective overturn at the Bermuda Atlantic Time-series Study site. Limnol. Oceanogr. 50, 1687–1696.

Mostofa, K.M., Liu, C.-q, Minakata, D., Wu, F., Vione, D., Mottaleb, M.A., et al., 2013. Photoinduced and microbial degradation of dissolved organic matter in natural waters. In: Photobiogeochemistry of Organic Matter. Springer, New York, pp. 273–364.

Mueller-Niklas, G., Schuster, S., Kaltenboeck, E., Herndl, G.J., 1994. Organic content and bacterial metabolism in amorphous aggregations of the northern Adriatic Sea. Limnol. Oceanogr. 39, 58–68.

Myklestad, S., 1974. Production of carbohydrates by marine planktonic diatoms. I. Comparison of nine different species in culture. J. Exp. Mar. Biol. Ecol. 15, 261–274.

Myklestad, S., 2000. Dissolved organic carbon from phytoplankton. Mar. Chem. 111–148.

Myklestad, S., Haug, A., 1972. Production of carbohydrates by the marine diatom *Chaetoceros affinis* var. *willei* (Gran) Hustedt. I. Effect of the concentration of nutrients in the culture medium. J. Exp. Mar. Biol. Ecol. 9, 125–136.

Nagata, T., 2000. Production mechanisms of dissolved organic matter. In: Kirchman, D.L. (Ed.), Microbial Ecology of the Oceans. Wiley-Liss, New York, pp. 121–152.

Nagata, T., 2008. Organic matter—bacteria interactions in seawater. In: Kirchman, D.L. (Ed.), Microbial Ecology of the Oceans. second ed.. Wiley-Liss, New York, pp. 207–241.

Nagata, T., Kirchman, D.L., 1992a. Release of dissolved organic matter by heterotrophic protozoa: implications for microbial foodwebs. Arch. Hydrobiol. 35, 99–109.

Nagata, T., Kirchman, D.L., 1992b. Release of macromolecular organic complexes by heterotrophic marine flagellates. Mar. Ecol. Prog. Ser. 83, 233–240.

Nagata, T., Kirchman, D.L., 1996. Bacterial degradation of protein adsorbed to model submicron particles in seawater. Mar. Ecol. Prog. Ser. 132, 241–248.

Nagata, T., Kirchman, D., 1999. Bacterial mortality: a pathway for the formation of refractory DOM. New Frontiers in Microbial Ecology: Proceedings of the Eighth International Symposium on Microbial Ecology. Atlantic Canada Society for Microbial Ecology, Halifax.

Nagata, T., Fukuda, H., Fukuda, R., Koike, I., 2000. Bacterioplankton distribution and production in deep Pacific waters: large-scale geographic variations and possible coupling with sinking particle fluxes. Limnol. Oceanogr. 45, 426–435.

Nagata, T., Meon, B., Kirchman, D., 2003. Microbial degradation of peptidoglycan in seawater. Limnol. Oceanogr. 48, 745–754.

Nelson, C., Carlson, C., 2012. Tracking differential incorporation of dissolved organic carbon types among diverse lineages of Sargasso Sea bacterioplankton. Environ. Microbiol. 14, 1500–1516.

Nelson, C.E., Goldberg, S.J., Kelly, L.W., Haas, A.F., Smith, J.E., Rohwer, F., et al., 2013. Coral and macroalgal exudates vary in neutral sugar composition and differentially enrich reef bacterioplankton lineages. ISME J. 7, 962–979.

Neufeld, J.D., Vohra, J., Dumont, M.G., Leuders, T., Mandfield, M., Friedrich, M.W., et al., 2007. DNA stable-isotope probing. Nat. Protoc. 2, 860–866.

Noble, R.T., Fuhrman, J.A., 1999. Breakdown and microbial uptake of marine viruses and other lysis products. Aquat. Microb. Ecol. 20, 1–11.

Obernosterer, I., Herndl, G.J., 1995. Phytoplankton extracellular release and bacterial growth: dependence on the inorganic N:P ratio. Mar. Ecol. Prog. Ser. 116, 247–257.

Obernosterer, I., Kawasaki, N., Benner, R., 2003. P-limitation of respiration in the Sargasso Sea and uncoupling of bacteria from P-regeneration in size-fractionation experiments. Aquat. Microb. Ecol. 32, 229–237.

Ogawa, H., Tanoue, E., 2003. Dissolved organic matter in oceanic waters. J. Oceanogr. 59, 129–147.

Ogawa, H., Fukuda, R., Koike, I., 1999. Vertical distribution of dissolved organic carbon and nitrogen in the Southern Ocean. Deep-Sea Res. I 46, 1809–1826.

Ogawa, H., Amagai, Y., Koike, I., Kaiser, K., Benner, R., 2001. Production of refractory dissolved organic matter by bacteria. Science 292, 917–920.

Ogura, N., 1972. Rate and extent of decomposition of dissolved organic matter in the surface water. Mar. Biol. 13, 89–93.

Orellana, M.V., Hansell, D.A., 2012. Ribulose-1, 5-bisphosphate carboxylase/oxygenase (RuBisCO): a long-lived protein in the deep ocean. Limnol. Oceanogr. 57, 826.

Otake, T., Nogami, K., Maruyama, K., 1993. Dissolved and particulate organic matter as possible food sources for eel leptocephali. Mar. Ecol. Prog. Ser. 92, 27–34.

Ouverney, C.C., Fuhrman, J.A., 1999. Combined Microautoradiography-16S rRNA probe technique for determination of radioisotope uptake by specific microbial cell types in situ. Appl. Environ. Microbiol. 65, 1746–1752.

Ouverney, C.C., Fuhrman, J.A., 2000. Marine planktonic Archaea take up amino acids. Appl. Environ. Microbiol. 66, 4829.

Papadimitriou, S., Kennedy, H., Bentaleb, I., Thomas, D., 2002. Dissolved organic carbon in sediments from the eastern North Atlantic. Mar. Chem. 79, 37–47.

Passow, U., 2002a. Production of transparent exopolymer particles (TEP) by phyto-and bacterioplankton. Mar. Ecol. Prog. Ser. 236, 1–12.

Passow, U., 2002b. Transparent exopolymer particles (TEP) in aquatic environments. Prog. Oceanogr. 55, 287–333.

Peltier, W.R., Liu, Y., Crowley, J.W., 2007. Snowball Earth prevention by dissolved organic carbon remineralization. Nature 450, 813–818.

Percival, E., Anisur Rahman, M., Weigel, H., 1980. Chemistry of the polysaccharides of the diatom *Coscinodiscus nobilis*. Phytochemistry 19, 809–811.

Pitt, K.A., Welsh, D.T., Condon, R.H., 2009. Influence of jellyfish blooms on carbon, nitrogen and phosphorus cycling and plankton production. Hydrobiologia 616, 133–149.

Pomeroy, L.R., Wiebe, W.J., Deibel, D., Thompson, R.J., Rowe, G.R., Pakulski, J.D., 1991. Bacterial response to temperature and substrate concentrations during the Newfoundland spring bloom. Mar. Ecol. Prog. Ser. 75, 143–159.

Pomeroy, L.R., Williams, P.J.I., Azam, F., Hobbie, J.E., 2007. The microbial loop. Oceanography 20, 28–33.

Poretsky, R.S., Sun, S., Mou, X., Moran, M.A., 2010. Transporter genes expressed by coastal bacterioplankton in response to dissolved organic carbon. Environ. Microbiol. 12, 616–627.

Puddu, A., Zoppini, A., Fazi, S., Rosati, M., Amalfitano, S., Magaletti, E., 2006. Bacterial uptake of DOM released from P-limited phytoplankton. FEMS Microbiol. Ecol. 46, 257–268.

Radić, T., Ivančić, I., Fuks, D., Radić, J., 2006. Marine bacterioplankton production of polysaccharidic and proteinaceous particles under different nutrient regimes. FEMS Microbiol. Ecol. 58, 333–342.

Rappe, M.S., Giovannoni, S.J., 2003. The uncultured microbial majority. Annu. Rev. Microbiol. 57, 369–394.

Rath, J., Wu, K.Y., Herndl, G.J., DeLong, E.F., 1998. High phylogenetic diversity in a marine-snow-associated bacterial assemblage. Aquat. Microb. Ecol. 14, 261–269.

Reinthaler, T., Herndl, G.J., 2005. Seasonal dynamics of bacterial growth efficiencies in relation to phytoplankton in the southern North Sea. Aquat. Microb. Ecol. 39, 7–16.

Reinthaler, T., Winter, C., Herndl, G.J., 2005. Relationship between bacterioplankton richness, respiration, and production in the southern North Sea. Appl. Environ. Microbiol. 71, 2260–2266.

Reinthaler, T., van Aken, H., Veth, C., Arístegui, J., Robinson, C., Williams, P., et al., 2006. Prokaryotic respiration and production in the meso- and bathypelagic realm of the eastern and western North Atlantic basin. Limnol. Oceanogr. 51, 1262–1273.

Reinthaler, T., van Aken, H.M., Herndl, G.J., 2010. Major contribution of autotrophy to microbial carbon cycling in the deep North Atlantic's interior. Deep-Sea Res. II 57, 1572–1580.

Rich, J.H., Kirchman, D.L., 1994. Differential uptake of monosaccharides by heterotrophic bacteria and the effects of ammonium in the Equatorial Pacific. Eos Trans. Am. Geophys. Union 75, 49.

Rich, J.H., Ducklow, H.W., Kirchman, D.L., 1996. Concentration and uptake of neutral monosaccharides along 140 W in the equatorial Pacific: contribution of glucose to heterotrophic bacterial activity and the DOM flux. Limnol. Oceanogr. 41, 595–604.

Rivkin, R.B., Legendre, L., 2001. Biogenic carbon cycling in the upper ocean: effects of microbial respiration. Science 291, 2398–2400.

Rivkin, R.B., Putt, M., 1987. Heterotrophy and photoheterotrophy by Antarctic microalgae: light-dependent incorporation of amino acids and glucose. J. Phycol. 23, 442–452.

Robinson, C., 2008. Heterotrophic bacterial respiration. In: Kirchman, D.L. (Ed.), Microbial Ecology of the Oceans. second ed. John Wiley & Sons, Inc, Hoboken, NJ, pp. 299–334.

Robinson, C., Williams,, P.J.I.B., 2005. Respiration and its measurement in surface marine waters. In: delGiorgio, P.A., Williams, P.J.l.B (Eds.), Respiration in Aquatic Ecosystems. Oxford University Press, New York, pp. 147–180.

Rocap, G., Larimer, F.W., Lamerdin, J., Malfatti, S., Chain, P., Ahlgren, N.A., et al., 2003. Genome divergence in two Prochlorococcus ecotypes reflects oceanic niche differentiation. Nature 424, 1042–1047.

Rochelle-Newall, E.J., Fisher, T.R., Fan, C., Glibert, P.M., 1999. Dynamics of chromophoric dissolved organic matter and dissolved organic carbon in experimental mesocosms. Int. J. Remote Sens. 20, 627–641.

Rothman, D.H., Hayes, J.M., Summons, R.E., 2003. Dynamics of the Neoproterozoic carbon cycle. Proc. Natl. Acad. Sci. U. S. A. 100, 8124–8129.

Russell, J.B., Cook, G.M., 1995. Energetics of bacterial growth: balance of anabolic and catabolic reactions. Microbiol. Rev. 59, 48–62.

Saba, G.K., Steinberg, D.K., Bronk, D.A., 2009. Effects of diet on release of dissolved organic and inorganic nutrients by the copepod Acartia tonsa. Mar. Ecol. Prog. Ser. 386, 147–161.

Saba, G.K., Steinberg, D.K., Bronk, D.A., 2011. The relative importance of sloppy feeding, excretion, and fecal pellet leaching in the release of dissolved carbon and nitrogen by Acartia tonsa copepods. J. Exp. Mar. Biol. Ecol. 404, 47–56.

Sambrotto, R.N., Savidge, G., Robinson, C., Boyd, P., Takahashi, T., Karl, D.M., et al., 1993. Elevated consumption of carbon relative to nitrogen in the surface ocean. Nature 363, 248–250.

Santinelli, C., Nannicini, L., Seritti, A., 2010. DOC dynamics in the meso and bathypelagic layers of the Mediterranean Sea. Deep-Sea Res. II 57, 1446–1459.

Sarmento, H., Romera-Castillo, C., Lindh, M., Pinhassi, J., Sala, M.M., Gasol, J.M., et al., 2013. Phytoplankton species-specific release of dissolved free amino acids and their selective consumption by bacteria. Linmol. Oceanogr. 58, 1123–1135.

Sarmiento, J.L., Gruber, N., 2006. Ocean Biogeochemical Dynamics. Princeton University Press, Princeton, NJ.

Schlesinger, W.H., Melack, J.M., 1981. Transport of organic carbon in the world's rivers. Tellus 33, 172–187.

Schwalbach, M.S., Fuhrman, J.A., 2005. Wide-ranging abundances of aerobic anoxygenic phototrophic bacteria in the world ocean revealed by epifluorescence microscopy and quantitative PCR. Limnol. Oceanogr. 50, 620–628.

Scully, N.M., Cooper, W.J., Tranvik, L.J., 2003. Photochemical effects on microbial activity in natural waters: the interaction of reactive oxygen species and dissolved organic matter. FEMS Microbiol. Ecol. 46, 353–357.

Sexton, P.F., Norris, R.D., Wilson, P.A., Pälike, H., Westerhold, T., Röhl, U., et al., 2011. Eocene global warming events driven by ventilation of oceanic dissolved organic carbon. Nature 471, 349–352.

Sharp, J.H., 1977. Excretion of organic matter by marine phytoplankton: do healthy cells do it? Limnol. Oceanogr. 22, 381–399.

Sharp, J.H., Carlson, C.A., Peltzer, E.T., Castle-Ward, D.M., Savidge, K.B., Rinker, K.R., 2002. Final dissolved organic carbon broad community intercalibration and preliminary use of DOC reference materials. Mar. Chem. 77, 239–253.

Sheik, A.R., Brussaard, C.P., Lavik, G., Lam, P., Musat, N., Krupke, A., et al., 2013. Responses of the coastal bacterial community to viral infection of the algae Phaeocystis globosa. ISME J. 8, 212–225.

Sherr, E., Sherr, B., 1988. Roles of microbes in pelagic food webs: a revised concept. Linmol. Oceanogr. 33, 1225–1257.

Sherr, E.B., Sherr, B.F., 2002. Significance of predation by protists in aquatic microbial food webs. Antonie van Leeuwenhoek 81, 293–308.

Sherr, B.F., Sherr, E.B., Caron, D.A., Vaulot, D., Worden, A.Z., 2007. Oceanic protists. Oceanography 20, 130–134.

Shiah, F.K., Ducklow, H.W., 1994. Temperature regulation of heterotrophic bacterioplankton abundance, production, and specific growth rate in Chesapeake Bay. Limnol. Oceanogr. 39, 1243–1258.

Shibata, A., Kogure, K., Koike, I., Ohwada, K., 1997. Formation of submicron colloidal particles from marine bacteria by viral infection. Mar. Ecol. Prog. Ser. 155, 303–307.

Simon, M., Rosenstock, B., 2007. Different coupling of dissolved amino acid, protein, and carbohydrate turnover to heterotrophic picoplankton production in the Southern Ocean in austral summer and fall. Limnol. Oceanogr. 52, 85.

Simon, M., Grossart, H.-P., Schweitzer, B., Ploug, H., 2002. Microbial ecology of organic aggregates in aquatic ecosystems. Aquat. Microb. Ecol. 28, 175–211.

Sintes, E., Stoderegger, K., Parada, V., Herndl, G.J., 2010. Seasonal dynamics of dissolved organic matter and microbial activity in the coastal North Sea. Aquat. Microb. Ecol. 60, 85–95.

Skoog, A., Benner, R., 1997. Aldoses in various size fractions of marine organic matter: implications for carbon cycling. Limnol. Oceanogr. 42, 1803–1813.

Skoog, A., Biddanda, B., Benner, R., 1999. Bacterial utilization of dissolved glucose in the upper water column or the Gulf of Mexico. Limnol. Oceanogr. 44, 1625–1633.

Smith, R.E., 1982. The estimation of phytoplankton production and excretion by carbon-14. Mar. Biol. Lett. 3, 325–334.

Smith, D., Simon, M., Alldredge, A.L., Azam, F., 1992. Intense hydrolytic enzyme activity on marine aggregates and implications for rapid particle dissolution. Nature 359, 139–142.

Smith, D.C., Steward, G.F., Long, R.A., Azam, F., 1995. Bacterial mediation of carbon fluxed during a diatom bloom in a mesocosm. Deep-Sea Res. II 42, 75–98.

Smith, W.O., Carlson, C.A., Ducklow, H.W., Hansell, D.A., 1998. Growth dynamics of Phaeocystis antarctic-dominated plankton assemblages from the Ross Sea. Mar. Ecol. Prog. Ser. 168, 229–244.

Søndergaard, M., Schierup, H.H., 1981. Release of extracellular organic carbon during a diatom bloom in Lake Moss: molecular weight fractionation. Freshwater Biol. 12, 313–320.

Steen, A.D., Ziervogel, K., Arnosti, C., 2010. Comparison of multivariate microbial datasets with the Shannon index: an example of using enzyme activity from diverse marine environments. Org. Geochem. 41, 1019–1021.

Steinberg, D., Saba, G., 2008. Nitrogen consumption and metabolism in marine zooplankton. In: Capone, D., Bronk, D., Mulholland, M., Carpenter, E. (Eds.), Nitrogen in the Marine Environment. second ed.. Elsevier, New York, pp. 1135–1196.

Steinberg, D.K., Carlson, C.A., Bates, N.R., Johnson, R.J., Michaels, A.F., Knap, A.H., 2001. Overview of the U.S. JGOFS Bermuda Atlantic Time-series Study (BATS): a decade-scale look at ocean biology and biogeochemistry. Deep-Sea Res. II 48, 1405–1447.

Steinberg, D.K., Carlson, C.A., Bates, N.R., Goldthwait, S.A., Madin, L.P., Michaels, A.F., 2000. Zooplankton vertical migration and the active transport of dissolved organic and inorganic carbon in the Sargasso Sea. Deep-Sea Res. I 47, 137–158.

Steinberg, D.K., Goldthwait, S.A., Hansell, D.A., 2002. Zooplankton vertical migration and the active transport of dissolved organic and inorganic nitrogen in the Sargasso Sea. Deep-Sea Res. I 49, 1445–1461.

Steinberg, D.K., Cope, J.S., Wilson, S.E., Kobari, T., 2008. A comparison of mesopelagic mesozooplankton community structure in the subtropical and subarctic North Pacific Ocean. Deep-Sea Res. II 55, 1615–1635.

Stewart, J.R., Brown, R.M., 1969. Cytophaga that kills or lyses algae. Science 164, 1523–1524.

Stocker, R., 2012. Marine microbes see a sea of gradients. Science 338, 628–633.

Stoderegger, K., Herndl, G.J., 1998. Production and release of bacterial capsular material and its subsequent utilization by marine bacterioplankton. Limnol. Oceanogr. 43, 877–884.

Stoderegger, K.E., Herndl, G.J., 1999. Production of exopolymer particles by marine bacterioplankton under contrasting turbulence conditions. Mar. Ecol. Prog. Ser. 189, 9–16.

Strom, S.L., 2000. Bacerivory interactions between bacteria and their grazers. In: Kirchman, D.L. (Ed.), Microbial Ecology of the Ocean. Wiley-Liss, Inc, New York, pp. 351–386.

Strom, S.L., 2008. Microbial ecology of ocean biogeochemistry: a community perspective. Science 320, 1043–1045.

Strom, S.L., Benner, R., Ziegler, S., Dagg, M.J., 1997. Planktonic grazers are a potentially important source of marine dissolved organic carbon. Limnol. Oceanogr. 42, 1364–1374.

Stubbins, A., Uher, G., Law, C.S., Mopper, K., Robinson, C., Upstill-Goddard, R.C., 2006. Open-ocean carbon monoxide photoproduction. Deep-Sea Res. II 53, 1695–1705.

Stubbins, A., Niggemann, J., Dittmar, T., 2012. Photo-lability of deep ocean dissolved black carbon. Biogeosciences 9, 1661–1670.

Suttle, C.A., 1994. The significance of viruses to mortality in aquatic microbial communities. Microb. Ecol. 28, 237–243.

Suttle, C.A., 2007. Marine viruses—major players in the global ecosystem. Nat. Rev. Microbiol. 5, 801–812.

Swan, C., Siegel, D., Nelson, N., Carlson, C., Nasir, E., 2009. Biogeochemical and hydrographic controls on chromophoric dissolved organic matter distribution in the Pacific Ocean. Deep-Sea Res. I 56, 2175–2192.

Swan, B.K., Martinez-Garcia, M., Preston, C.M., Sczyrba, A., Woyke, T., Lamy, D., et al., 2011. Potential for chemolithoautotrophy among ubiquitous bacteria lineages in the dark ocean. Science 333, 1296–1300.

Tamburini, C., Garcin, J., Ragot, M., Bianchi, A., 2002. Biopolymer hydrolysis and bacterial production under ambient hydrostatic pressure through a 2000 m water column in the NW Mediterranean. Deep-Sea Res. II 49, 2109–2123.

Tamburini, C., Garcin, J., Bianchi, A., 2003. Role of deep-sea bacteria in organic matter mineralization and adaptation to hydrostatic pressure conditions in the NW Mediterranean Sea. Aquat. Microb. Ecol. 32, 209–218.

Tanoue, E., Nishiyama, S., Kamo, M., Tsugita, A., 1995. Bacterial membranes: possible source of a major dissolved protein in seawater. Geochim. Cosmochim. Acta 59, 2643–2648.

Taylor, G.T., Iturriaga, R., Sullivan, C.W., 1985. Interactions of bacterivorous grazers and heterotrophic bacteria with dissolved organic matter. Mar. Ecol. Prog. Ser. 23, 129–141.

Taylor, G.T., Iabichella, M., Ho, T.-Y., Scranton, M.I., Thunell, R.C., Muller-Karger, F., et al., 2001. Chemoautotrophy in the redox transition zone of the Cariaco Basin: a significant midwater source of organic carbon production. Limnol. Oceanogr. 46, 148–163.

Teira, E., Pazo, M.J., Serret, P., Fernandez, E., 2001. Dissolved organic carbon production by microbial populations in the Atlantic Ocean. Limnol. Oceanogr. 46, 1370–1377.

Teira, E., Pazo, M.J., Quevedo, M., Fuentes, M.V., Niell, F.X., Fernandez, E., 2003. Rates of dissolved organic carbon production and bacterial activity in the eastern North Atlantic Subtropical Gyre during summer. Mar. Ecol. Prog. Ser. 249, 53–67.

Teira, E., Reinthaler, T., Pernthaler, A., Pernthaler, J., Herndl, G.J., 2004. Hybridization and microautoradiography to detect substrate utilization by Bacteria and Archaea in the deep ocean. Appl. Environ. Microbiol. 70, 4411–4414.

Teira, E., Lebaron, P., van Aken, H., Herndl, G.J., 2006. Distribution and activity of Bacteria and Archaea in the deep water masses of the North Atlantic. Limnol. Oceanogr. 51, 2131–2144.

Thingstad, T.F., Lignell, R., 1997. Theoretical models for the control of bacterial growth rate, abundance, diversity and carbon demand. Aquat. Microb. Ecol. 13, 19–27.

Thingstad, T.F., Hagstrom, A., Rassoulzadegan, F., 1997. Accumulation of degradable DOC in surface waters: is it caused by a malfunctioning microbial loop? Limnol. Oceanogr. 42, 398–404.

Thingstad, T.F., Zweifel, U.L., Rassoulzadegan, F., 1998. P limitation of heterotrophic bacteria and phytoplankton in the northwest Mediterranean. Limnol. Oceanogr. 43, 88–94.

Tillmann, U., 2003. Kill and eat your predator: a winning strategy of the planktonic flagellate Prymnesium parvum. Aquat. Microb. Ecol. 32, 73–84.

Tortell, P.D., Maldonado, M.T., Granger, J., Price, N.M., 1999. Marine bacteria and biogeochemical cycling of iron in the oceans. FEMS Microbiol. Ecol. 29, 1–11.

Tranvik, L.J., 1993. Microbial transformation of labile dissolved organic matter into humic-like matter in seawater. FEMS Microbiol. Ecol. 12, 177–183.

Tranvik, L.J., Bertilsson, S., 2001. Contrasting effects of solar UV radiation on dissolved organic sources for bacterial growth. Ecol. Lett. 4, 458–463.

Tranvik, L., Kokalj, S., 1998. Decreased biodegradability of algal DOC due to interactive effects of UV radiation and humic matter. Aquat. Microb. Ecol. 14, 301–307.

Tranvik, L.J., Sherr, E., Sherr, B.F., 1993. Uptake and utilization of 'colloidal DOM' by heterotrophic flagellates in seawater. Mar. Ecol. Prog. Ser. 92, 301–309.

Treusch, A.H., Vergin, K.L., Finlay, L.A., Donatz, M.G., Burton, R.M., Carlson, C.A., et al., 2009. Seasonality and vertical structure of microbial communities in an ocean gyre. ISME J. 3, 1148–1163.

Urban-Rich, J., 1999. Release of dissolved organic carbon from copepod fecal pellets in the Greeland Sea. J. Exp. Mar. Biol. Ecol. 232, 107–124.

Veldhuis, M.J.W., Admiraal, W., 1985. Transfer of photosynthetic products in gelatinous colonies of Phaeocystis pouchetii (Haptophyceae) and its effect on the measurement of excretion rate. Mar. Ecol. Prog. Ser. 26, 301–304.

Vallino, J.J., Hopkinson, C.S., Hobbie, J.E., 1996. Modeling bacterial utilization of dissolved organic matter: Optimization replaces Monod growth kinetics. Limnol. Oceanogr. 41, 1591–1609.

Venter, J.C., Remington, K., Heidelberg, J.F., Halpern, A.L., Rusch, D., Eisen, J.A., et al., 2004. Environmental genome shotgun sequencing of the Sargasso Sea. Science 304, 66–72.

Verdugo, P., Alldredge, A.L., Azam, F., Kirchman, D.L., Passow, U., Santschi, P.H., 2004. The oceanic gel phase: a bridge in the DOM-POM continuum. Mar. Chem. 92, 67–85.

Verdugo, P., Carlson, C., Giovannoni, S., 2012. Marine microgels. Ann. Rev. Mar. Sci. 4, 375–400.

Vergin, K.L., Beszteri, B., Monier, A., Thrash, J.C., Temperton, B., Treusch, A.H., et al., 2013. High-resolution SAR11 ecotype dynamics at the Bermuda Atlantic Time-series Study site by phylogenetic placement of pyrosequences. ISME J. 7, 1322–1332.

Vezzi, A., Campanaro, S., D'Angelo, M., Simonato, F., Vitulo, N., Lauro, F.M., et al., 2005. Life at depth: *Photobacterium profundum* genome sequence and expression analysis. Science 307, 1459–1461.

Vincent, D., Slawyk, G., L'Helguen, S., Sarthou, G., Gallinari, M., Seuront, L., et al., 2007. Net and gross incorporation of nitrogen by marine copepods fed on ^{15}N-labelled diatoms: methodology and trophic studies. J. Exp. Mar. Biol. Ecol. 352, 295–305.

Vraspir, J.M., Butler, A., 2009. Chemistry of marine ligands and siderophores. Ann. Rev. Mar. Sci. 1, 43.

Wakeham, S.G., Pease, T.K., Benner, R., 2003. Hydroxy fatty acids in marine dissolved organic matter as indicators of bacterial membrane material. Org. Geochem. 34, 857–868.

Walsh, D.A., Zaikova, E., Howes, C.G., Song, Y.C., Wright, J.J., Tringe, S.G., et al., 2009. Metagenome of a versatile chemolithoautotroph from expanding oceanic dead zones. Science 326, 578–582.

Weinbauer, M.G., Peduzzi, P., 1995. Effect of virus-rich high molecular weight concentrates of seawater on the dynamics of dissolved amino acids and carbohydrates. Mar. Ecol. Prog. Ser. 127, 245–253.

Weinbauer, M.G., Bonilla-Findji, O., Chan, A.M., Dolan, J.R., Short, S.M., Simek, K., et al., 2011. Synechococcus growth in the ocean may depend on the lysis of heterotrophic bacteria. J. Plank. Res. 33, 1465–1476.

Weissbach, A., Tillmann, U., Legrand, C., 2010. Allelopathic potential of the dinoflagellate *Alexandrium tamarense* on marine microbial communities. Harmful Algae 10, 9–18.

Wells, M.L., 2002. Marine colloids and trace metals. In: Biogeochemistry of Marine Dissolved Organic Matter. Elsevier Science, USA, pp. 367–404.

Wells, M.L., Goldberg, E.D., 1993. Colloid aggregation in seawater. Mar. Chem. 41, 353–358.

Westernhoff, H.V., Hellingwerf, K.J., Dam, K.V., 1983. Thermodyanic efficiency of microbial growth is low but optimal for maximal growth rate. Proc. Natl. Acad. Sci. U. S. A. 80, 305–309.

Wetz, M.S., Wheeler, P.A., 2003. Production and partitioning of organic matter during simulated phytoplankton blooms. Limnol. Oceanogr. 48, 1808–1817.

Wetz, M.S., Wheeler, P.A., 2007. Release of dissolved organic matter by coastal diatoms. Limnol. Oceanogr. 52, 798–807.

Whitman, W.B., Coleman, D.C., Wiebe, W.J., 1998. Prokaryotes: the unseen majority. Proc. Natl. Acad. Sci. U. S. A. 95, 6578–6583.

Wiebe, W.J., Smith, D.F., 1977. Direct measurement of dissolved organic carbon release by phytoplankton and incorporation by microheterotrophs. Mar. Biol. 42, 213–223.

Wilhelm, S.W., Suttle, C.A., 1999. Viruses and nutrient cycles in the sea. BioScience 49, 781–788.

Willey, J.D., Kieber, R.J., Eyman, M.S., Avery Jr., G.B., 2000. Rainwater dissolved organic carbon: concentrations and global flux. Global Biogeochem. Cycles 14, 139–148.

Willey, J.D., Kieber, R.J., Avery, G.B., 2006. Changing chemical composition of precipitation in Wilmington, North Carolina, USA: implications for the continental USA. Environ. Sci. Technol. 40, 5675–5680.

Williams, P.J.L.B., 1970. Heterotrophic utilization of dissolved organic compound in the sea. I. Size distribution of population and relationship between respiration and incorporation of growth substrates. J. Mar. Biol. Assoc. U.K. 50, 859–870.

Williams, P.J.L., 1981. Incorporation of microheterotrophic processes into the classical paradigm of the planktonic food web. Kiel. Meeresforsch. 5, 1–28.

Williams, P.J.L.B., 1990. The importance of losses during microbial growth: commentary on the physiology, measurement and ecology of the release of dissolved organic material. Mar. Microb. Food Webs 4, 175–206.

Williams, P.J.L., 1995. Evidence for the seasonal accumulation of carbon-rich dissolved organic material, its scale in comparison with changes in particulate material and the consequential effect on net C/N assimilation ratios. Mar. Chem. 51, 17–29.

Williams, P.J.l, 2000. Heterotrophic bacteria and the dynamics of dissolved organic material. In: Kirchman, D.L. (Ed.), Microbial Ecology of the Oceans. Wiley-Liss, New York, pp. 153–200.

Williams, P.M., Druffel, E.R.M., 1987. Radiocarbon in dissolved organic matter in the central North Pacific Ocean. Nature 330, 246–248.

Williams, P.J.L., Morris, P.J., Karl, D.M., 2004. Net communtiy production and metabolic balance at the oligotrophic ocean site, station ALOHA. Deep-Sea Res. I 51, 1563–1578.

Wolfe, G.V., 2000. The chemical defense ecology of marine unicellular plankton: constraints, mechanisms, and impacts. Biol. Bull. 198, 225–244.

Wommack, K.E., Colwell, R.R., 2000. Virioplankton: viruses in aquatic ecosystems. Microbiol. Mol. Biol. Rev. 64, 69–114.

Wood, A., Van Valen, L., 1990. Paradox lost? On the release of energy-rich compounds by phytoplankton. Mar. Microb. Food Webs 4, 103–116.

Worden, A.Z., Not, F., 2008. Ecology and diversity of picoeukaryotes. In: Kirchman, D.L. (Ed.), Microbial Ecology of the Oceans. second ed.. Wiley-Liss, New York, pp. 159–205.

Yamashita, Y., Tanoue, E., 2008. Production of bio-refractory fluorescent dissolved organic matter in the ocean interior. Nat. Geosci. 1, 579–582.

Yurkov, V., Csotonyi, J.T., 2008. New light on aerobic anoxygenic phototrophs. In: The Purple Phototrophic Bacteria. Springer, New York, pp. 31–55.

Ziolkowski, L., Druffel, E., 2010. Aged black carbon identified in marine dissolved organic carbon. Geophys. Res. Lett. 37, L16601.

Zubkov, M.V., 2009. Photoheterotrophy in marine prokaryotes. J. Plank. Res. 31, 933–938.

Zubkov, M.V., Tarran, G.A., 2005. Amino acid uptake of Prochlorococcus spp. in surface waters across the South Atlantic Subtropical Front. Aquat. Microb. Ecol. 40, 241–249.

Zusheng, L., Clarke, A.J., Beveridge, R.J., 1998. Gram-negative bacteria produce membrane vesicles which are capable of killing other bacteria. J. Bact. 180, 5478–5483.

Zweifel, U.L., 1999. Factors controlling accumulation of labile dissolved organic carbon in the Gulf of Riga. Estuar. Coast. Shelf Sci. 48, 357–370.

Zweifel, U.L., Norrman, B., Hagstrm, K., 1993. Consumption of dissolved organic carbon by marine bacteria and demand for inorganic nutrients. Mar. Ecol. Prog. Ser. 101, 23–32.

CHAPTER

4

Dynamics of Dissolved Organic Nitrogen

Rachel E. Sipler, Deborah A. Bronk

Virginia Institute of Marine Science, College of William & Mary, Virginia, USA

CONTENTS

Biogeochemistry of Marine Dissolved Organic Matter,
http://dx.doi.org/10.1016/B978-0-12-405940-5.00004-2

I INTRODUCTION

Dissolved organic nitrogen (DON) is that subset of the dissolved organic carbon (DOC) pool that also contains N. The DON pool is still where the action is—a chemical drive thru where microorganisms get N, C, and energy. Research into dissolved organic matter (DOM), like this book, is still a bit like aged detritus—very C-rich. Though we set out to cover all things DON, if only briefly, page limitations, the explosion of relevant literature over the last decade, and time prevented us from being all-inclusive. Here, we focus on work published largely after 2001 and topics not included in the earlier review (Bronk, 2002) such as new information from the growing field of high-resolution chemical characterization. This means that much of the chapter in the earlier edition is still relevant and the two chapters together should be viewed as a unit; in particular many older citations were cut from this edition to save space. We further direct the reader to other reviews that cover DON in general (Antia et al., 1991), DON uptake (Berman and Bronk, 2003; Bronk et al., 2007; Mulholland and Lomas, 2008), DON in cultures

(Bronk and Flynn, 2006), DON chemical composition (Aluwihare and Meador, 2008), analytical methods (McCarthy and Bronk, 2008; Worsfold et al., 2008), and N regeneration, including DON (Bronk and Steinberg, 2008).

II DON CONCENTRATIONS IN AQUATIC ENVIRONMENTS

The measurement of DON concentrations has become much more common over the last decade. Here, we briefly discuss methods to measure concentrations of DON, followed by a survey of DON concentrations in aquatic environments. We focus on studies published in the last decade but include data from Bronk (2002) in our calculation of concentration means across systems (Table 4.1). We also provide a discussion of how DON varies vertically and seasonally in the ocean.

A Methods to Measure DON Concentrations

A reliable, high precision method for quantifying DON concentrations is required to study

TABLE 4.1 Total Dissolved Nitrogen (TDN) and Dissolved Organic Nitrogen (DON) Concentrations in the Literature

Location	Sampling Date	Sample Depth (m)	TDN Concentration (μmol N L^{-1})	DON Concentration (μmol N L^{-1})	DON:TDN (%)	C:N DOM Pool	Method	Reference
OCEANIC—SURFACE								
Bronk (2002) mean (n =27)			9.6 ± 4.1	5.6 ± 2.5	70.5 ± 24.7	14.0 ± 2.9		Bronk (2002)
Inside Cyclone Opal—lee of Hawaiian Islands	Mar 2005	Upper 110	4.9^g	4.0^g	82.3		HTO	Mahaffey et al. (2008)
Outside Cyclone Opal—lee of Hawaiian Islands	Mar 2005	Upper 110	4.1^g	3.9^g	94.8		HTO	Mahaffey et al. (2008)
Sargaso Sea, BATS	Apr 2009	0	2.2	1.8	81.9 ± 0.0	3.0 ± 0.0	HTO	Baer (unpublished data)
Sargaso Sea, BATS	Apr 2009	120	0.7	0.3	45.9 ± 0.0	14.0 ± 0.0	HTO	Baer (unpublished data)
North Sea	1995-2005	Surface	8.4 ± 3.1	5.6 ± 1.4	67.2 ± 4.7	13.6 ± 1.3	PO	Van Engeland et al. (2010)
Tropical Atlantic (10S-10N)	Aug & Oct 2003, Feb 2005, Apr 2010	Upper 50	5.2 ± 1.5	4.8 ± 0.4	92.2	15.4	HTO	Letscher et al. (2013a)
Eastern Subtropical North Atlantic (10–40N 10-50W)	Jan-Feb 1998, Jul 2003	Upper 50	4.6 ± 0.6	4.5 ± 0.5	97.7	16.0	HTO	Letscher et al. (2013a)
Western Subtropical North Atlantic (10–40N 50-80W)	Feb 1998, Oct-Nov 2003	Upper 50	4.9 ± 0.9	4.7 ± 0.6	95.8	15.3	HTO	Letscher et al. (2013a)
Subpolar North Atlantic (40-66N)	Jun-Sept 2003	Upper 50	7.1 ± 2.8	4.5 ± 0.6	63.7	14.0	HTO	Letscher et al. (2013a)
Eastern Subtropical South Atlantic (10–40S 20E-10W)	Mar-Apr 2010	Upper 50	4.8 ± 0.6	4.7 ± 0.5	97.9	14.7	HTO	Letscher et al. (2013a)
Western Subtropical South Atlantic (10–40S 10-50W)	Feb 2005	Upper 50	4.4 ± 0.4	4.4 ± 0.4	100.0	16.2	HTO	Letscher et al. (2013a)
Equatorial Pacific (10S-10N)	Feb 2006, Dec-Jan 2007/8	Upper 50	9.7 ± 4.5	4.4 ± 0.6	45.2	15.1	HTO	Letscher et al. 2013a
Eastern Subtropical North Pacific (10–40N 85-140W)	Aug 2004, Dec 2007, May 2011	Upper 50	9.7 ± 7.2	5.1 ± 1.0	52.6	13.3	HTO	Letscher et al. (2013a)

(Continued)

TABLE 4.1 Total dissolved nitrogen (TDN) and dissolved organic nitrogen (DON) concentrations in the literature—cont'd

Location	Sampling Date	Sample Depth (m)	TDN Concentration (μmol N L⁻¹)	DON Concentration (μmol N L⁻¹)	DON:TDN (%)	C:N DOM Pool	Method	Reference
Western Subtropical North Pacific (10-40N 120E-140W)	Jun-Aug 2004, Feb-Mar 2006	Upper 50	4.4 ± 0.5	4.3 ± 0.4	97.7	15.8	HTO	Letscher et al. (2013a)
NE Subpolar Pacific (40-60N 125-160W)	Mar 2006, May 2011	Upper 50	15.6 ± 5.7	4.4 ± 0.5	28.3	12.9	HTO	Letscher et al. (2013a)
Eastern Subtropical South Pacific (10-40S 80-140W)	Jan 2008	Upper 50	6.6 ± 3.3	4.3 ± 0.5	65.6	15.7	HTO	Letscher et al. (2013a)
Bay of Bengal (0-23N 78-100E)	Oct 1995, Apr 2007	Upper 50	5.5 ± 0.5	5.4 ± 0.4	98.1	13.6	HTO	Letscher et al. (2013a)
Arabian Sea (0-25N 50-75E)	Mar-Apr 1995	Upper 50	7.2 ± 1.7	6.0 ± 0.6	84.3	13.5	HTO	Letscher et al. (2013a)
South Indian (0-40S 20-120E)	Oct 2005, Mar-Apr 2007, Feb 2008, Mar-May 2009	Upper 50	5.0 ± 1.4	4.5 ± 0.5	89.8	15.2	HTO	Letscher et al. (2013a)
Sargasso Sea, Bermuda	Feb 2001	100		3.8 ± 0.0[a,b]			PO	Knapp et al. (2005)
Sargasso Sea, Bermuda	July 2000	60		4.7 ± 0.4[a,b]			PO	Knapp et al. (2005)
Sargasso Sea, Bermuda	July 2000	Surface		4.3 ± 0.0[a,b]			PO	Knapp et al. (2005)
Sargasso Sea, Bermuda	Jan 2001	100		4.1 ± 0.1[a,b]			PO	Knapp et al. (2005)
East Japan Sea	May 2007	<100	9.3 ± 0.8	4.4 ± 1.3	47.7	17.0 ± 3.0	HTO	Kim and Kim (2013)
Subarctic North East Pacific	Winter 1995-1998	<100	11.0 to 17.3	4.6 to 5.1[b]			PO	Wong et al. (2002)
Subarctic North East Pacific	Spring 1995-1998	<90	8.5 to 14.9	4.7 to 6.6[b]			PO	Wong et al. (2002)
Subarctic North East Pacific	Summer 1995-1998	<55	4.8 to 10.7	4.8 to 6.3[b]			PO	Wong et al. (2002)
Subarctic North East Pacific	Feb 1995-Feb 1998	Surface		4.0 to 6.0[b]			PO	Wong et al. (2002)
40°N-50°N Eastern Atlantic	Spring 2000-Fall 2005	Upper 100		4.3 to 6.3[a]			UV	Torres-Valdés et al. (2009)
30°N-40°N Atlantic	Spring 2000-Fall 2005	Upper 100		3.6 to 4.8[a]			UV	Torres-Valdés et al. (2009)
20°N-30°N Atlantic	Spring 2000-Fall 2005	Upper 100		4.9 to 5.8[a]			UV	Torres-Valdés et al. (2009)

Location	Date	Depth		Concentration	%	Method	Reference
10°N-20°N Eastern Atlantic	Spring 2000-Fall 2005	Upper 100		4.9 to 5.4[a]		UV	Torres-Valdés et al. (2009)
10°N-10°S Equatorial Atlantic	Spring 2000-Fall 2005	Upper 100		3.6 to 5.1[a]		UV	Torres-Valdés et al. (2009)
10°S-20°S Eastern Atlantic	Spring 2000-Fall 2005	Upper 100		4.8 to 5.3[a]		UV	Torres-Valdés et al. (2009)
20°S -30°S Atlantic	Spring 2000-Fall 2005	Upper 100		3.6 to 5.2[a]		UV	Torres-Valdés et al. (2009)
30°S-40°S Atlantic	Spring 2000-Fall 2005	Upper 100		4.2 to 4.8[a]		UV	Torres-Valdés et al. (2009)
40°S-50°S Eastern Atlantic	Spring 2000-Fall 2005	Upper 100	~5 to ~8	3.9 ± 1.9[a]		UV	Torres-Valdés et al. (2009)
13.51°S-32.33°W Atlantic	Apr-May 2000	<180		5.6 ± 1.0[c,f]	>60[f]	UV[h]	Mahaffey et al. (2004)
~35°S, ~49°W Atlantic	Apr-May 2000	<250		~4 to ~5[f]	90 to 95[f]	UV[h]	Mahaffey et al. (2004)
~20-25°S, ~35°W Atlantic	Apr-May 2000	<250		~3 to ~3.8[f]	>95[f]	UV[h]	Mahaffey et al. (2004)
~15°S, ~32°W Atlantic	April-May 2000	<250		~5 to ~6[f]	~95[f]	UV[h]	Mahaffey et al. (2004)
~10°S, ~30°W Atlantic	Apr-May 2000	<250		~5 to ~6[f]	>97[f]	UV[h]	Mahaffey et al. (2004)
~8°S, ~29°W Atlantic	Apr-May 2000	<250		~4.5 to ~5.5[f]	>95[f]	UV[h]	Mahaffey et al. (2004)
Equatorial Atlantic ~25°W	Apr-May 2000	<250		~4.75 to ~5.75[f]	~90 to 95[f]	UV[h]	Mahaffey et al. (2004)
~5°N, ~24°W Atlantic	Apr-May 2000	<250		~6 to ~7[f]	>97[f]	UV[h]	Mahaffey et al. (2004)
~10°N, ~22°W Atlantic	Apr-May 2000	<250		~5.5 to ~6.3[f]	>97[f]	UV[h]	Mahaffey et al. (2004)
~22°N, ~21°W Atlantic	Apr-May 2000	<250		~4 to ~4.8[f]	>95[f]	UV[h]	Mahaffey et al. (2004)
~35°N, ~20°W Atlantic	Apr-May 2000	<250		~5.8 to ~6.5[f]	~90 to 95[f]	UV[h]	Mahaffey et al. (2004)
~38°N, ~20°W Atlantic	Apr-May 2000	<250		~3.5 to ~4.5[f]	~70 to 75[f]	UV[h]	Mahaffey et al. (2004)
~47°N, ~20°W Atlantic	Apr-May 2000	<250		~5 to ~7[f]	~55[f]	UV[h]	Mahaffey et al. (2004)
~50°N, ~20°W Atlantic	Apr-May 2000	<250		~6.5 to ~7[f]	~55[f]	UV[h]	Mahaffey et al. (2004)
Western North Atlantic	Sept-Oct 2000	60		3.9 ± 1.1[c]		PO	Varela et al. (2005)
North Atlantic Subtropical Gyre	Sept-Oct 2000	110		3.5 ± 0.2[c]		PO	Varela et al. (2005)
Eastern Coastal Canary Islands	Sept-Oct 2000	80		6.9 ± 1.6[c]		PO	Varela et al. (2005)
Eastern Tropical Atlantic	Sept-Oct 2000	80		5.3 ± 0.4[c]		PO	Varela et al. (2005)
South Atlantic Gyre	Sept-Oct 2000	140		5.2 ± 0.4[c]		PO	Varela et al. (2005)
Brasil Coastal Current	Sept-Oct 2000	90		4.5 ± 0.9[c]		PO	Varela et al. (2005)

(Continued)

TABLE 4.1 Total dissolved nitrogen (TDN) and dissolved organic nitrogen (DON) concentrations in the literature—cont'd

Location	Sampling Date	Sample Depth (m)	TDN Concentration (μmol N L^{-1})	DON Concentration (μmol N L^{-1})	DON:TDN (%)	C:N DOM Pool	Method	Reference
17.737° S, 3.126° E Angola Basin, Southern Atlantic	Nov 2008	5		~5		~14	HTO	Hertkorn et al. (2013)
17.737° S, 3.126° E Angola Basin, Southern Atlantic	Nov 2008	48		~7		~11	HTO	Hertkorn et al. (2013)
	*Mean ± std		8.0 ± 3.9	5.1 ± 1.7	76.6 ± 22.3	14.0 ± 2.9		
OCEANIC—DEEP								
Bronk (2002) mean (n = 21)			31.7 ± 11.5	3.9 ± 2.3	19.6 ± 29.4	16.3 ± 1.5		Bronk (2002)
Sargaso Sea, BATS	Apr 2009	400	5.6	0.4	7.3 ± 0.0	10.1 ± 0.0	HTO	Baer (unpublished data)
Sargaso Sea, BATS	Apr 2009	800	20.9	1.1	5.3 ± 0.0	3.4 ± 0.0	HTO	Baer (unpublished data)
40°N-50°N Atlantic	Fall 2004	200-300		~4 to 5[a]	~50 to >90[a]		UV	Torres-Valdés et al. (2009)
30°N-40°N Atlantic	Fall 2004	200-300		~3.5 to 4[a]	60 to >90[a]		UV	Torres-Valdés et al. (2009)
20°N-30°N Atlantic	Fall 2004	200-300		~3.8 to 4[a]	75 to >90[a]		UV	Torres-Valdés et al. (2009)
10°N-20°N Atlantic	Fall 2004	200-300		~3 to 4[a]	50 to >90[a]		UV	Torres-Valdés et al. (2009)
10°N-10°S Equatorial Atlantic	Fall 2004	200-300		~2.5 to ~4[a]	<10 to ~30[a]		UV	Torres-Valdés et al. (2009)
10°S-20°S Atlantic	Fall 2004	200-300		~2.8 to ~4.5[a]	<25 to >90[a]		UV	Torres-Valdés et al. (2009)
20°S-30°S Atlantic	Fall 2004	200-300		<3 to ~4.3[a]	~20 to >90[a]		UV	Torres-Valdés et al. (2009)
30°S-40°S Atlantic	Fall 2004	200-300		<3 to 4[a]	~50 to ~70[a]		UV	Torres-Valdés et al. (2009)
40°N-50°N Atlantic	Spring 2000, 2003, 2004	200-300		~2 to 4[a]	<25 to 40[a]		UV	Torres-Valdés et al. (2009)
30°N-40°N Atlantic	Spring 2000, 2003, 2004	200-300		~2 to 4[a]	~10 to ~35[a]		UV	Torres-Valdés et al. (2009)
20°N-30°N Atlantic	Spring 2000, 2003, 2004	200-300		~1 to ~2.8[a]	<25 to ~40[a]		UV	Torres-Valdés et al. (2009)
10°N-20°N Atlantic	Spring 2000, 2003, 2004	200-300		2 to ~4[a]	~10 to 25[a]		UV	Torres-Valdés et al. (2009)
10°N-10°S Equatorial Atlantic	Spring 2000, 2003, 2004	200-300		~1 to ~3.5[a]	~10 to ~25[a]		UV	Torres-Valdés et al. (2009)
10°S-20°S Atlantic	Spring 2000, 2003, 2004	200-300		~0.5 to ~3.5[a]	~10 to 50[a]		UV	Torres-Valdés et al. (2009)
20°S-30°S Atlantic	Spring 2000, 2003, 2004	200-300		~0.5 to ~5[a]	~40 to ~75[a]		UV	Torres-Valdés et al. (2009)

Location	Date	Depth					Method	Reference
13.51°S–32.33°W Atlantic	Apr–May 2000	>500	~27	2.2 ± 0.3[f]	~7 to 8[f]		UV[h]	Mahaffey et al. (2004)
17.737°S, 3.126°E Angola Basin, Southern Atlantic	Nov 2008	200		~11 to 12		~6	HTO	Hertkorn et al. (2013)
17.737°S, 3.126°E Angola Basin, Southern Atlantic	Nov 2008	5446		~5		~9	HTO	Hertkorn et al. (2013)
*Mean ± std			27.9 ± 12.6	3.6 ± 2.2	33.7 ± 28.1	14.0 ± 4.5		
COASTAL/CONTINENTAL SHELF								
Bronk (2002) mean (n = 14)			16.4 ± 9.8	10.8 ± 7.9	70.7 ± 21.7	17.9 ± 4.4	PO	Bronk (2002)
Arkona Sea	Aug 2005	Surface	18.5	10.2	55.1		PO	Stedmon et al. (2007)
Kotka	Aug 2005	Surface	18.9	18.0	95.2		PO	Stedmon et al. (2007)
Oulu	Sept 2005	Surface	12.5	7.8	62.4		PO	Stedmon et al. (2007)
North Sea	1995–2005	Surface	39.0 ± 20.2	11.3 ± 3.0	30.3 ± 7.6	11.2 ± 0.4	PO—Koroleff 1983	Van Engeland et al. (2010)
North Sea—Pre-bloom	Jan–Mar 2010	1	15.4	5.8	39		HTO	Johnson et al. (2013a)
North Sea—Bloom/transition	Apr 2010	1	11.4	7.9	74		HTO	Johnson et al. (2013a)
North Sea—Post-bloom	May–Sept 2010	1	9.5	8.4	90		HTO	Johnson et al. (2013a)
Eastern Gulf of Mexico	Oct 2007	Surface	15.7 ± 1.1	13.6 ± 1.4	86.6 ± 1.4	31.8 ± 1.5	HTO	Sipler et al. (2013)
Coastal Gulf of Mexico	Oct 2007, 2008, 2009	Surface	13.4 ± 1.2	12.8 ± 1.4	94.9 ± 3.2	14.5 ± 1.4	PO	Bronk et al. (in press)
Off-shore Gulf of Mexico	Oct 2007, 2008, 2009	Surface	6.8 ± 0.4	6.2 ± 0.5	91.4 ± 3.2	20.8 ± 8.3	PO	Bronk et al. (in press)
West Florida Shelf	Oct 2008	Surface	16.5	16.2 ± 0.3	98.2		PO	Wawrik et al. (2009)
Jeju Island, Southern Sea Korea	Oct 2010, Jan 2011, June 2011	Surface		25.0 ± 15.0			HTO	Kim et al. (2013)
Horsens Fjord	Sept 2004–Jul 2005	1		12.0 to 35.0		11.4 ± 3.8	HTO	Lønborg and Søndergaard (2009)
Darss Sill	Sept 2004–Jul 2005	1		17.0 to 36.0		10.6 ± 1.7	HTO	Lønborg and Søndergaard (2009)
Northeastern shelf of the Gulf of Cádiz (SW Iberian Peninsula)	June 2006	3		4.8 ± 1.5[b]		20.0 ± 8.0[d]	HTO	Ribas-Ribas et al. (2011)
Northeastern shelf of the Gulf of Cádiz (SW Iberian Peninsula)	Nov 2006	3		6.2 ± 2.3[b]		13.0 ± 7.0[d]	HTO	Ribas-Ribas et al. (2011)

(Continued)

TABLE 4.1 Total dissolved nitrogen (TDN) and dissolved organic nitrogen (DON) concentrations in the literature—cont'd

Location	Sampling Date	Sample Depth (m)	TDN Concentration (µmol N L^{-1})	DON Concentration (µmol N L^{-1})	DON:TDN (%)	C:N DOM Pool	Method	Reference
Northeastern shelf of the Gulf of Cádiz (SW Iberian Peninsula)	Feb 2007	3		4.2 ± 2.8[b]		30.0 ± 26.0[d]	HTO	Ribas-Ribas et al. (2011)
Northeastern shelf of the Gulf of Cádiz (SW Iberian Peninsula)	May 2007	3		8.6 ± 3.1[b]		14.0 ± 8.0[d]	HTO	Ribas-Ribas et al. (2011)
Mid-Atlantic Bight (NJ)	Jul 2002	Surface	8.0 ± 1.8	8.0 ± 1.8	99.8 ± 0.3		PO	Bradley et al. (2010b)
Mid-Atlantic Bight (NJ)	Jul 2002	Bottom (14)	12.1 ± 1.8	5.4 ± 1.0	46.8 ± 19.5		PO	Bradley et al. (2010b)
Eastern Coastal Canary Islands	Sept-Oct 2000	80		6.9 ± 1.6[c]			PO	Varela et al. (2005)
Brasil Coastal Current	Sept-Oct 2000	90		4.5 ± 0.9[c]			PO	Varela et al. (2005)
Southern California Bight-nearshore (5.6 km from shore)	Oct 1992	10		6.7 ± 0.1			UV	Bronk and Ward (2005)
Oregon continental shelf	Apr 2005	Surface		12.0		13.0	PO	Wetz et al. (2008)
Oregon continental shelf	Aug 2005	Surface		10.0		10.5	PO	Wetz et al. (2008)
Oregon continental shelf	Sept 2005	Surface		8.2		7.1	PO	Wetz et al. (2008)
Gulf of Riga, Baltic Sea	May 1999	Surface		16 to 34			PO	Berg et al. (2003a)
Gulf of Riga, Baltic Sea	Jun 1999	Surface		12 to 23			PO	Berg et al. (2003a)
Gulf of Riga, Baltic Sea	Jul 1999	Surface		22 to 30			PO	Berg et al. (2003a)
Trondheimsfjord, Norway	Aug 2003	Surface		15.5 to 20.7			HTO	Davidson et al. (2007)
Raunefjord, Norway	Mar 2003	Surface	11	3.9	35		PO	Sanderson et al. (2008)
Raunefjord, Norway	Apr 2005	Surface	6	5.7	95		PO	Bradley et al. (2010c)
Baltic Sea—Gulf of Riga	May 1996	Upper 20	20 to 74		65 to 100		HTO	Berg et al. (2001)

Location	Date	Upper 20	*Mean ± std				HTO	Reference
Baltic Sea—Gulf of Riga	Jul 1996		18 to 42; 16.9 ± 10.6	11.4 ± 7.3	90 to 100; 73.3 ± 23.2	16.4 ± 7.0	HTO	Berg et al. (2001)
ESTUARINE								
Bronk (2002) mean (n = 14)			38.5 ± 37.0	25.9 ± 21.6	68.9 ± 22.4	17.0 ± 12.1		Bronk (2002)
Chesapeake Bay (north)	Aug 2010	0	21.6 ± 0.1	17.4 ± 0.2	80.7 ± 1.2		PO	Baer (unpublished data)
Chesapeake Bay (mid-Bay)	Aug 2010	0	17.9 ± 0.2	17.5 ± 0.2	97.8 ± 1.5		PO	Baer (unpublished data)
Chesapeake Bay (south)	Aug 2010	0	11.7 ± 0.3	11.4 ± 0.3	97.7 ± 3.1		PO	Baer (unpublished data)
Chesapeake Bay	Aug-Sept 2004	0	14.0 to 53.0	12.0 to 15.0			PO	Bradley et al. (2010a)
York River	Aug 2009	0	21.6 ± 0.1	20.5 ± 0.1	94.9		PO	Killberg-Thoreson et al. (2010)
York River	Summer (June-Aug) 2010 & 2011	0	17.0 ± 1.5	16.5 ± 1.4	96.9 ± 12.1	17.9 ± 34.7	PO	Baer 2013
York River	Fall (Sept-Nov) 2010 & 2011	0	25.8 ± 5.7	16.0 ± 2.0	62.1 ± 25.2	19.5 ± 17.4	PO	Baer 2013
York River	Winter (Dec-Feb) 2010 & 2011	0	21.5 ± 4.5	16.2 ± 1.1	75.3 ± 21.9	17.3 ± 12.6	PO	Baer 2013
York River	Spring (Mar-May) 2010 & 2011	0	21.2 ± 8.4	20.4 ± 8.1	96.2 ± 55.9	17.0 ± 51.7	PO	Baer 2013
James River/Chesapeake Bay	Aug 2008	0	11.3 to 28.8	10.7 to 21.3			PO	Bronk et al. (2010)
Charlotte Harbor, FL	Oct 2007, 2008, & 2009	Surface	19.7 ± 3.7	18.2 ± 3.9	92.2 ± 5.1	16.6 ± 1.9	PO	Bronk et al. (in press)
Rowley estuary, Massachusetts	Summer 2000	Estuary water column		~10 to 50.0			PO	Tobias et al. (2003)
Plym River Estuary, SE England	June 2002-Sept 2003	0.2	32.1 ± 27.1		36 ± 17	7.2 ± 6.5	HTO	Badr et al. (2008)
Yealm River Estuary, SE England	Feb 2002-Sept 2003	0.2	24.5 ± 27.7		38 ± 22	10.1 ± 10.3	HTO	Badr et al. (2008)
Juan de Fuca and Georgia Straits	Winter 1994-1997	<50		3.0 to 7.6[b]			PO	Wong et al. (2002)
Juan de Fuca and Georgia Straits	Spring 1994-1998	<50		5.3 to 12.4[b]			PO	Wong et al. (2002)
Juan de Fuca and Georgia Straits	Summer 1994-1999	<50		9.1 to 14.5[b]			PO	Wong et al. (2002)

(Continued)

4. DYNAMICS OF DISSOLVED ORGANIC NITROGEN

TABLE 4.1 Total dissolved nitrogen (TDN) and dissolved organic nitrogen (DON) concentrations in the literature—cont'd

Location	Sampling Date	Sample Depth (m)	TDN Concentration (µmol N L⁻¹)	DON Concentration (µmol N L⁻¹)	DON:TDN (%)	C:N DOM Pool	Method	Reference
Juan de Fuca and Georgia Straits	Fall 1994-2000	<50		3.9 to 7.3[b]			PO	Wong et al. (2002)
Black Sea—Northwest Shelf	May-Jun 2001	Surface		15		15 to 19	HTO	Ducklow et al. (2007)
Black Sea—Western Gyre	May-June 2001	Surface		11		15 to 19	HTO	Ducklow et al. (2007)
Randers Fjord	April 2001	Surface & bottom		20.0 to 57.0	17 to 92		PO	Veuger et al. (2004)
Randers Fjord	August 2001	Surface & bottom		11.0 to 68.0	22 to 69		PO	Veuger et al. (2004)
	*Mean ± std		30.1 ± 27.9	21.6 ± 15.9	71.8 ± 23.0	16.0 ± 7.4		
SELECTED RIVERS								
Bronk (2002) mean ($n = 13$, representing 43 rivers)			61.2 ± 32.2	34.7 ± 20.7	60.1 ± 23.5	25.7 ± 12.5		Bronk (2002)
Ob River	Jun 2003-Nov 2006	Surface	31.1 ± 1.7	18.4 ± 1.0	59.2	39.8	HTO	Letscher et al. (2013b); Cooper et al. (2008)
Yenisey River	Jun 2003-Nov 2006	Surface	18.3 ± 1.0	12.5 ± 0.7	68.3	36.3	HTO	Letscher et al. (2013b); Cooper et al. (2008)
Lena River	Jun 2003-Nov 2006	Surface	20.1 ± 1.1	16.6 ± 0.9	82.6	45.9	HTO	Letscher et al. (2013b); Cooper et al. (2008)
Kolyma River	Jun 2003-Nov 2006	Surface	16.1 ± 0.9	10.9 ± 0.6	67.7	41.1	HTO	Letscher et al. (2013b); Cooper et al. (2008)
Mackenzie River	Jun 2003-Nov 2006	Surface	14.4 ± 0.8	7.4 ± 0.4	51.4	49.2	HTO	Letscher et al. (2013b); Cooper et al. (2008)
Yukon River	Jun 2003-Nov 2006	Surface	23.0 ± 1.2	16.1 ± 0.9	70.0	24.1	HTO	Letscher et al. (2013b); Cooper et al. (2008)
Bass River	Jul 1998	Surface	6.4 ± 0.1	3.7 ± 0.1	58.0 ± 1.0	87.0 ± 5.0	HTO	Wiegner et al. (2006)
Delaware River	Jul 1998	Surface	66.0 ± 1.2	7.8 ± 0.7	12.0 ± 1.0	26.0 ± 1.0	HTO	Wiegner et al. (2006)
Hudson River	Aug 1998	Surface	41.5 ± 1.9	11.9 ± 2.0	28.0 ± 4.0	28.0 ± 5.0	HTO	Wiegner et al. (2006)
Altamaha River	Jul 1998	Surface	30.7 ± 2.0	12.8 ± 2.0	35.0 ± 0.0	32.0 ± 4.0	HTO	Wiegner et al. (2006)
Savannah River	Jul 1998	Surface	42.6 ± 1.6	8.4 ± 1.8	20.0 ± 4.0	31.0 ± 7.0	HTO	Wiegner et al. (2006)
Pocomoke River	Aug 1998	Surface	37.1 ± 0.3	34.7 ± 0.3	94.0 ± 0.0	22.0 ± 1.0	HTO	Wiegner et al. (2006)
Choptank River	Aug 1998	Surface	31.3 ± 2.7	26.1 ± 2.9	83.0 ± 2.0	15.0 ± 2.0	HTO	Wiegner et al. (2006)

Location	Date	Depth					Method	Reference
Peconic River	Jul 1998	Surface	25.9 ± 0.8	15.3 ± 0.6	59.0 ± 1.0	32.0 ± 2.0	HTO	Wiegner et al. (2006)
ARCTIC								
Bronk (2002) mean (n = 3)		*Mean ± std	37.4 ± 24.6	23.8 ± 18.1	57.7 ± 23.7	32.5 ± 16.3	HTO	Bronk (2002)
Western Arctic	May-Jun 2002	Upper 100		4.7 ± 1.0		14.8 ± 1.3	HTO	Davis and Benner (2005)
Western Arctic	Jul-Aug 2002	Upper 100		3.9 ± 0.2[e]		18.2 ± 0.6	HTO	Davis and Benner (2005)
Western Arctic	May-Aug 2002	Surface		4.3 ± 0.1[e]		17.9 ± 0.8	HTO	Davis and Benner (2005)
Western Arctic	May-Aug 2002	41-299		4.4 ± 0.1		17.5[d]	HTO	Davis and Benner (2005)
Western Arctic	May-Aug 2002	201-1000		3.5 ± 0.2		20.0	HTO	Davis and Benner (2005)
Western Arctic	May-Aug 2002	1000		2.4 ± 0.5		23.8	HTO	Davis and Benner (2005)
				1.7 ± 0.1		29.4	HTO	Davis and Benner (2005)
Chukchi Sea (Barrow, AK)	Jan 2011 & 2012	Surface	13.5 ± 2.1	3.9 ± 1.0	28.7 ± 30.0	21.8 ± 26.8	HTO	Baer 2013
Chukchi Sea (Barrow, AK)	Apr 2010 & 2011	Surface	12.3 ± 3.6	4.8 ± 0.5	39.0 ± 31.3	15.9 ± 12.9	HTO	Baer 2013
Chukchi Sea (Barrow, AK)	Aug 2010 & 2011	Surface	8.2 ± 2.1	7.4 ± 2.1	90.2 ± 38.2	14.7 ± 31.6	HTO	Baer 2013
Near Shore, Laptev Sea Arctic	Jun-Jul 1994	Surface	10.3	7.4 ± 1.3	71.8	20.9	PO	Dittmar et al. (2001)
Laptev Sea, Arctic	Jul-Sept 1995	0-30	9.2	6.2 ± 0.4	67.4	20.2	PO	Dittmar et al. (2001)
Laptev Sea, Arctic	Jul-Sept 1995	30-200	12.6	4.5 ± 0.3	35.7	18.2	PO	Dittmar et al. (2001)
Laptev Sea, Arctic	Jul-Sept 1995	200-500	14.5	3.6 ± 0.4	24.8	17.2	PO	Dittmar et al. (2001)
Laptev Sea, Arctic	Jul-Sept 1995	>500	17.3	3.4 ± 0.3	19.7	20.0	PO	Dittmar et al. (2001)
Chukchi Sea (66-74N 150-180W)	May-Jun 2002	Upper 30	9.9	4.0 ± 0.7	40.4	18.4	HTO	Letscher et al. (2013b)
Chukchi Sea (66-74N 150-180W)	Jul-Aug 2002, Sept 2009	Upper 30	7.0 to 9.5	6.0 to 8.0	85.0	13.2	HTO	Letscher et al. (2013b)
Chukchi Borderland (74-79N 170E-160W)	Sept 2008	Upper 30	5.5	5.0	90.9	13.4	HTO	Letscher et al. (2013b)
Beaufort Sea (68-71N 130-140W)	Aug 2008	Upper 30	5.2	5.0	96.2	20.1	HTO	Letscher et al. (2013b)
Canada Basin (72-82N 120-170W)	Aug 2002, Aug-Sep 2008	Upper 30	3.3 to 4.3	3.0 to 4.0	92.0	15.2	HTO	Letscher et al. (2013b)
East Siberian Sea (70-74N 170-180E)	Sept 2009	Upper 30	6.2 to 7.2	6.0 to 7.0	100.0	15.8	HTO	Letscher et al. (2013b)

(Continued)

TABLE 4.1   Total dissolved nitrogen (TDN) and dissolved organic nitrogen (DON) concentrations in the literature—cont'd

Location	Sampling Date	Sample Depth (m)	TDN Concentration (µmol N L⁻¹)	DON Concentration (µmol N L⁻¹)	DON:TDN (%)	C:N DOM Pool	Method	Reference
Makarov Basin (79-82N 140-180E)	Sept 2008	Upper 30	6.8 to 7.9	6.0 to 7.0	88.5	18.8	HTO	Letscher et al. (2013b)
Amundsen/Nansen Basin (78-82N 100-140E)	Sept-Oct 2008	Upper 30	6.5	5.0	76.9	18.3	HTO	Letscher et al. (2013b)
*Mean±std			9.4 ± 3.8	4.7 ± 1.5	65.5 ± 28.9	18.1 ± 3.6		
SOUTHERN OCEAN								
Bronk (2002) mean (n = 11)			32.2 ± 4.9	4.4 ± 1.8	13.8 ± 3.8	12.1 ± 2.9		Bronk (2002)
Southern Ocean >40S	Jan-Feb 2005, Feb-Mar 2007, Feb-Apr 2008, Mar 2010, Mar-Apr 2011	Upper 50	25.9 ± 6.8	4.2 ± 0.8	16.2	11.4	HTO	Letscher et al. (2013a)
*Mean±std			31.6 ± 5.0	4.4 ± 1.8	14.0 ± 3.6	12.0 ± 2.8		

Data were taken from text, tables, estimated from graphs, or obtained from the authors and are presented as the mean ± standard deviation. The means presented in the first edition (Bronk, 2002) are presented at the start of most sections. All data from the earlier edition (Bronk, 2002) and the more recent data presented here were included in the calculated mean ± standard deviation presented at the end of each section. When a range is given, the average of that range was used in the calculation of the mean ± standard deviation. Methods used were high temperature oxidation (HTO), persulfate oxidation (PO), and ultraviolet oxidation (UV). We note that this table is far from complete (e.g., means were not presented for some studies, units were unclear, etc.).

Means for each region include all data from Bronk (2002) as well as new data listed in table.

[a] *Samples were not filtered so data are for TN.*

[b] *Ammonium is included in the TON concentration.*

[c] *Standard error reported, not standard deviation.*

[d] *DOC:DON, actual C:N values not presented.*

[e] *DON does not include urea (urea was treated as DIN).*

[f] *DON = TDN − NO₃.*

[g] *Calculated from integrated value throughout the upper 110 m.*

[h] *UV as described in Sanders and Jickells (2000).*

any aspect of DON cycling. The measurement of DON concentrations is still problematic, and there is no single accepted method for DON analysis. Concentrations of DON are determined as the difference between total dissolved N (TDN) concentration and the sum of the dissolved inorganic nitrogen (DIN) forms ammonium (NH_4^+), nitrate (NO_3^-), and nitrite (NO_2^-). There are currently three methods commonly used to measure TDN concentrations in aqueous samples—persulfate oxidation (Menzel and Vaccaro, 1964), ultraviolet oxidation (Armstrong et al., 1966), and high temperature oxidation (Sharp et al., 2004; Walsh 1989). DON concentrations are estimated by difference; therefore, they have the combined analytical error and uncertainty of three separate analyses: TDN, NH_4^+, and combined NO_3^-/NO_2^-. This is especially problematic for samples with proportionally high DIN concentrations, such as those from the deep-ocean, high nutrient low chlorophyll zones, and eutrophic systems. A number of method comparisons have been performed (Bronk et al., 2000; Sharp, 2002; Sharp et al., 2004; Torres-Valdés et al., 2009; Walsh, 1989; Worsfold et al., 2008); however, no method has emerged as clearly superior. Over the past decade, the most common method for determining TDN (and DON) has been high temperature combustion, likely because this method analyzes TDN in tandem with DOC thus optimizing throughput. For analytically challenging open ocean samples, however, the analyses are frequently decoupled.

A note about filtration—DON samples are generally filtered, hence the "D" for dissolved. The filters used commonly range from 0.2 μm to glass fiber (GF/F) filters, with a nominal pore size of 0.7 μm. However, many open ocean studies do not filter samples and so measure total organic N (TON), rather than DON. This is especially true for samples collected below the nitracline or deeper than 100-200 m. Researchers working in these waters remove the filtration step because the particulate N (PN) pool is generally so small (<10% of TON) and the risk of contamination and cell breakage during filtration is so great (e.g., Abell et al., 2000). We note

that DON samples analyzed by the U.S. Global Repeat Hydrography program discussed below are filtered using in-line pre-combusted GF/F filters attached directly to the Niskin bottle.

To conclude this brief review of methods we make a plea for specificity in units. Though we have stopped using our favorite unit, the much-maligned μg-at NL^{-1} (Williams, 2004), for the purpose of DON work we encourage the use of $\mu mol N L^{-1}$, rather than μM to avoid confusion, particularly in studies that include organic substrates that contain more than one N in the molecule (e.g., urea). We also suggest that the use of $\mu mol L^{-1}$, without the N, leaves too much to the imagination of the reader. Lack of specificity in units is one reason some papers were not included in the tables.

B Global Distributions and Fate

Our understanding of global DON distributions has been greatly enhanced by the U.S. Global Ocean Carbon Repeat Hydrography program (http://ushydro.ucsd.edu/), which measures DON throughout the water column every ~60 nautical miles along transects (Letscher et al., 2013a). The mean global DON concentration in surface waters from this survey is $4.4 \pm 0.5 \,\mu mol N L^{-1}$ with a range of 2-7 (Letscher et al., 2013a). In the ocean, regions that are either adjacent to or immediately downstream of eastern boundaries have DON concentrations >5 $\mu mol N L^{-1}$ (Figure 4.1). These elevated DON concentrations are also observed in upwelling zones in equatorial waters and in the monsoon-driven Arabian Sea indicating that there is a net source of DON resulting from enhanced NO_3^--supported biological production within these systems (Letscher et al., 2013a). Poleward and westward of the upwelling regions, concentrations of DON are lower, so it is assumed that the subtropical gyres are sinks for DON due to biological consumption. In the Southern Ocean, concentrations are similar to other surface ocean waters though more variable, which could be the result of the high NO_3^- concentrations and associated analytical uncertainty.

FIGURE 4.1   Dissolved organic nitrogen [DON] (μmol N Kg^{-1}; colored dots) and nitrate [NO$_3^-$](μmol N Kg^{-1}; isolines) concentrations in the surface ocean (10m depth). *From Letscher et al. (2013a).*

Concentrations of DON and TON decrease with depth, coincident with an increase in NO$_3^-$ (Karl et al., 2001; Torres-Valdés et al., 2009), with deep-ocean values generally ranging between 2 and 3 μmol N L^{-1} (Abell et al., 2000; Hansell and Carlson, 2001; Torres-Valdés et al., 2009). The number of DON concentration measurements in the deep ocean is limited due to an inability of existing methods to resolve low DON concentrations with high NO$_3^-$ concentrations present. Though DON samples are collected from deep water as part of the US CLIVAR Repeat Hydrography program, they are currently being archived in hope of a better method. In addition to vertical structure through the water column, there is also evidence that a sea-surface microlayer exists at the interface between the ocean and atmosphere and that this is a zone where concentrations of DON are elevated, particularly in the form of dissolved free amino acids (DFAA) (Reinthaler et al., 2008).

DON research in the Atlantic has received the most attention over the last decade and some patterns have emerged. In the tropics, DON concentrations are highest in the upper 50m, but elevated concentrations extend to 100m in the subtropical gyres (Torres-Valdés et al., 2009). The concentrations in the North Atlantic are slightly lower than those in the South Atlantic (3-5 versus 4-5 μmol N L^{-1}; Torres-Valdés et al., 2009). DON contributes >90% of the TDN in these surface waters, but the percentage of NO$_3^-$ contribution to TDN increases with depth (Torres-Valdés et al., 2009). There are generally higher DON concentrations in the western side of the Atlantic basin (>5 μmol N L^{-1}) compared to the east (~4.5 μmol N L^{-1}; Letscher et al., 2013a; Roussenov et al., 2006), although Torres-Valdés et al. (2009) notes that this trend is weak. The Atlantic basin also contains a number of hot spots associated with upwelling areas at 18°N off North Africa, at 5°S within the equatorial Atlantic, and along 24.5°N in the western Atlantic (Torres-Valdés et al., 2009). Two modeling studies investigated the contribution of DON to export production in the Atlantic. Though the models were of different resolution, they both found that ~40% or greater of PN export could be supported by semi-labile DON inputs in the southern and eastern sides of the North Atlantic subtropical gyre (Roussenov et al., 2006; Torres-Valdés et al., 2009).

The interior of the oligotrophic subtropical gyres may receive allochthonous input of DON from net production occurring at the gyre margins. Concentrations of NO_3^- in euphotic, upwelling regions create a net accumulation of DON that is able to evade rapid remineralization by microbes, allowing it to be transported horizontally to gyre interiors. As a result, inverse relationships between concentrations of DON/TON and NO_3^- are commonly observed (reviewed in Bronk, 2002; Torres-Valdés et al., 2009). Theoretically, DON advected into the subtropical gyre could be considered a source of "new" or export production as long as a fraction of it was remineralized in the subtropical gyre. In addition to horizontal transport of DON, cross gyre transport of NO_3^--rich water from neighboring upwelling regions may occur via Ekman drift or geostrophic eddies resulting in DON production within the interior of the subtropical gyre (Williams and Follows, 1998). A study by Letscher et al. (2013a) found that the surface bacterioplankton community in the Florida Straight did not readily utilize DON. On the other hand, the same DON was rapidly mineralized by bacterioplankton from the upper mesopelagic region. Similar trends have also been observed for DOC (Carlson et al., 2004). Letscher et al. (2013a) demonstrated that DON is primarily removed through vertical mixing and subsequent remineralization by microbes below the mixed layer.

C Cross System Comparison

In general, the lowest mean concentrations of DON are found in the deep-ocean and the highest mean concentrations are found in rivers (Table 4.1). The consistent sampling and measurement protocols used by the CLIVAR US Repeat Hydrography program yield a robust estimate of surface ocean DON concentrations of $4.4 \pm 0.5\,\mu mol\,N\,L^{-1}$ (Letscher et al., 2013a), which is slightly lower than the $5.1 \pm 1.7\,\mu mol\,N\,L^{-1}$ estimated from literature values presented in Table 4.1. Deep-ocean DON concentrations are lower with most measurements in the 2-5 $\mu mol\,N\,L^{-1}$ range with a mean of $3.6 \pm 2.2\,\mu mol\,N\,L^{-1}$ (Table 4.1). Concentrations tend to be progressively higher in coastal, then estuarine, and then river waters, with corresponding increases in the range of variability (Figure 4.2). For example, DON concentrations can vary dramatically in estuaries in response to river flow (Agedah et al., 2009). In all environments, except the deep-ocean and high nutrient low chlorophyll Southern Ocean, ~58-77% of the bulk of the TDN pool is comprised of DON (Table 4.1). The high NO_3^- concentrations in the deep-ocean and Southern Ocean surface waters, result in ~34% and 14% of the TDN pool being composed of DON in these environments, respectively. The C:N ratio of DOM increases from 14 in the surface ocean along a decreasing salinity gradient to a maximum mean of 32 in riverine systems (Table 4.1).

D Seasonal Variations

As DON has increasingly been measured at time-series sites, much has been learned about its seasonal pattern. There are areas, generally more oligotrophic in nature, where no seasonal pattern is indicated including the Santa Monica Basin (Hansell et al., 1993) and the Hawaiian Ocean Time-series site in the North Pacific gyre (Karl et al., 2001). There was also no seasonal pattern observed at the Bermuda Atlantic Time Series (BATS) site in the Sargasso Sea with respect to DON concentrations (Hansell and Carlson, 2001) or DON isotopic composition (Knapp et al., 2005). In contrast, some studies suggest that DON increases in late spring and summer, including work in the Gulf of Mexico (López-Veneroni and Cifuentes, 1994), Chesapeake Bay (Bronk et al., 1998), and North Inlet, SC (Lewitus et al., 2000).

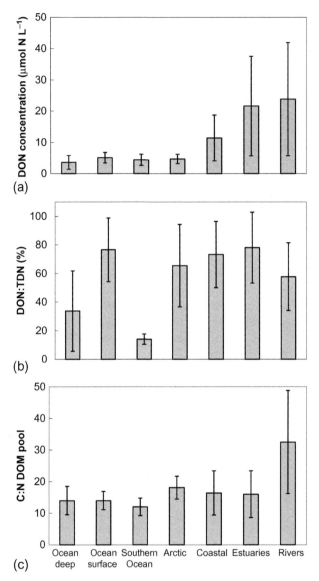

FIGURE 4.2 Means and standard deviations for data presented in Table 4.1 including (a) dissolved organic nitrogen (DON) concentrations, (b) DON: total dissolved nitrogen (TDN) ratios as a percentage, and (c) the carbon to nitrogen (C:N) ratio of the dissolved organic matter (DOM). Means and standard deviations were calculated using data from all studies presented in Table 1 of Bronk (2002) and Table 4.1 in this chapter.

Butler et al. (1979) conducted an 11-year study of DON concentrations in the English Channel and documented a steady increase in DON concentrations from January through August, and then a steady decline from August to December.

A time-series study from 1995 to 2005 in the North Sea documented concentrations that averaged $11\,\mu mol\,N\,L^{-1}$ at coastal areas and $5\,\mu mol\,N\,L^{-1}$ at ocean sites (Van Engeland et al., 2010). Despite the lack of robust inter-annual variability, the coastal region had consistently higher DON concentrations in the spring and summer compared to that observed in the autumn and winter. Overall, tropical and subtropical open ocean sites appear generally less influenced by seasonal variability compared to estuarine, coastal and high latitude sites.

The greatest seasonal differences in DON concentration are generally observed in high latitude coastal regions influenced by riverine discharge of terrestrially derived DON. This seasonality is particularly pronounced in the Arctic Ocean where ~67% of TDN entering the Arctic Ocean via rivers is DON (Holmes et al., 2012). The contribution of DON increases from 32% in winter (Nov-April), when the tundra is mostly frozen, to 74% in spring (May-June) during the initial thaw (i.e., the freshet), and reaches 80% of the TDN pool entering the Arctic during summer (July-October; Holmes et al., 2012). Inter-annual or decadal climatic changes may also impact DON concentrations. For example, changes in DON concentrations in the subarctic northeast Pacific (British Columbia, Canada) were larger in non-El Niño years than El Niño years (Wong et al., 2002).

III COMPOSITION OF THE DON POOL

The DON pool is frequently treated like a black box made up of a heterogeneous mixture of compounds. From the standpoint of short term (hours) measurements of DON uptake or regeneration, the DON box contains two pools—a large refractory pool composed of compounds that may persist for years and a smaller labile pool that turns over on time scales of seconds to hours and that often barely accumulates to measurable concentrations. The processes that control the smaller labile pool in the surface ocean are the focus of Sections IV and V below. From the standpoint of global distributions, however, the black box of DON holds at minimum three

pools—the two already mentioned and a third semi-labile pool that varies depending on the season. Though this simplified view of DON composition can be useful, in reality there are many factors, such as abiotic transformation and microbial community composition and associated metabolic capacity and demand, that will define which compounds occupy which pool and when. For example, compounds that may be refractory in the surface waters may be highly labile to mesopelagic bacteria after vertical transport (described in Section II).

Our knowledge of the chemical composition of the DON pool has increased within the last decade and is poised for rapid increase in the future. This section reviews our current understanding of the chemical composition of the pool and individual organic substrates. This is followed by a discussion of methods used to interrogate the large uncharacterized fraction of DON and what we have learned from them. We end with suggested research priorities for the future. Due to space limitations DON in lakes, streams, and ground water, with some exceptions, are not included.

A Chemical Composition— Characterizable DON

Several classes of compounds have been identified or classified within the DON pool, including urea, DFAA, dissolved combined amino acids (DCAA), nucleic acids and operationally defined structural families including humic and fulvic substances. The remainder of the DON pool is a mixture of mostly unidentified compounds. In this section, we briefly describe individual organic compounds including measurement techniques and a summary of concentrations in the environment. For a more detailed review of the quantification and characterization of aquatic DON see McCarthy and Bronk (2008) and Worsfold et al. (2008) and the more general review of DOM characterization by Repeta in this volume.

1 Urea

When it comes to chemical composition, all DON is DOC, but not all DOC is DON. Urea, $CO(NH_2)_2$, is one compound, however, where the distinction has been blurred historically. Some definitions of an organic compound require a C–H or C–C bond—urea has neither. Further complicating the issue, urea is produced industrially via the Wöhler process purely from inorganic salts (potassium cyanate and ammonium sulfate). As a result, some studies, particularly older ones, consider urea as an inorganic form of N. This practice needs to be considered when looking at historical data on DON concentrations (Bronk, 2002). The simplest definition of an organic compound, however, is one that contains C. We like to keep it simple with DON so we put urea squarely down on the side of organic and place it in the DON pool. From an oceanographic perspective, treating urea as DON also makes sense because it is naturally produced through organismal excretion and organic matter decomposition in the environment. We also suggest that using the simplest definition of a DON compound will be particularly important in the future. To use any other definition could set the field up for real problems when analytical tools are developed that can routinely determine the presence of C–H or C–C bonds in N compounds.

There are two methods commonly used to measure urea concentrations in aquatic systems—the direct colorimetric measurement of urea using diacetyl monoxime (Price and Harrison, 1987) and the urease method, which involves enzymatic hydrolysis of urea to CO_2 and ammonia (McCarthy, 1970). Revilla et al. (2005) did a method comparison and found that the direct monoxime method had greater accuracy and varied less with salinity.

In general, urea concentrations in open ocean systems tend to be low with a mean of $<1.0 \, \mu mol \, N \, L^{-1}$ (Table 4.2). While urea typically accounts for a minor percentage of the overall DON pool (6–19%; Table 4.2), its impact on aquatic systems may be increasing. Urea concentrations are higher and more variable in coastal systems, with concentrations as high as $25 \, \mu mol \, N \, L^{-1}$ in near coastal and estuarine waters (e.g., Glibert et al., 2005, 2006; Kudela et al., 2008; Lomas et al., 2002). The industrial production of urea has increased 100-fold over the last 40 years and at least a proportion of this urea is reaching our estuaries and coastal oceans primarily via fertilizer runoff (Glibert et al., 2006). Glibert et al. (2006) suggest that increases in urea concentration are linked to a number of ecological impacts including an increase in harmful algal blooms (see Section V.D.2).

2 Amino Acids

Amino acids are of interest to chemists because their composition and enantiomeric properties can give insight into production and diagenetic processes. Biologists are interested in amino acids because they are highly labile forms of organic N that have traditionally been thought to fuel bacterial productivity in the ocean (see Section V.D.3). Amino acids occur as DFAA and DCAA with total hydolyzable amino acids (THAA) being the sum of the two (detailed discussion in Aluwihare and Meador, 2008). The two groups in THAA differ in that DFAA are simple amino acids, while DCAA are amino acids combined with other structures such as humics, polypeptides, or proteins (Thurman, 1985).

Dissolved primary amines (DPAs) are distinguished from DFAAs and DCAAs based on a difference in functionality. Although DFAAs and DPAs both contain an amine group ($-NH_2$), DPAs have an akyl functional group and DFAAs have a carboxyl group and side chain (Wedyan, 2008). DPAs typically form when amino acids are broken down and are most commonly measured fluorometrically (Parsons et al., 1984) using o-phthaldialdehyde (OPA) and mercaptoethanol as a reagent. However, a study by Aminot and Kérouel (2006) suggested using 3-mercaptopropionic acid in place of mercaptoethanol to produce more reliable results.

TABLE 4.2 Literature Values of Concentrations of Some Dissolved Organic Compounds Including Urea, Dissolved Free Amino Acids/Dissolved Primary Amines (Dfaa/Dpa), and Dissolved Combined Amino Acids (dcaa) In Aquatic Systems

Location	Sampling Date	Sampling Depth (m)	Compound Concentration (μmol N L^{-1})	%DON Pool	Method	Reference
UREA—OCEANIC SURFACE						
Bronk (2002) mean (n=2)			4.04 ± 6.46			Bronk (2002)
Sargasso Sea, BATS	Apr 2009	0	0.07 ± 0.00	3.8 ± 1.5	Monoxime	Baer (unpublished data)
Sargasso Sea, BATS	Apr 2009	120	0.08 ± 0.01	22.1 ± 10.7	Monoxime	Baer (unpublished data)
Western North Atlantic	Sept-Oct 2000	60	0.32 ± 0.20 [a]	~8.2	Goeyens et al. (1998)	Varela et al. (2005)
North Atlantic Subtropical Gyre	Sept-Oct 2000	110	0.15 ± 0.03[a]	~4.2	Goeyens et al. (1998)	Varela et al. (2005)
Eastern Tropical Atlantic	Sept-Oct 2000	80	0.43 ± 0.06[a]	~8	Goeyens et al. (1998)	Varela et al. (2005)
South Atlantic Gyre	Sept-Oct 2000	140	0.24 ± 0.06[a]	~4.5	Goeyens et al. (1998)	Varela et al. (2005)
Ross Sea—Station O	Jan 1997	3.9	0.20		Monoxime	Cochlan and Bronk (2003)
Ross Sea—Station O	Dec 1997	4.9	0.00		Monoxime	Cochlan and Bronk (2003)
Ross Sea—Station Emperor	Feb 1997	5.3	0.13		Monoxime	Cochlan and Bronk (2003)
Ross Sea—Station Pack Ice	Feb 1997	4.1	0.38		Monoxime	Cochlan and Bronk (2003)
Ross Sea—Station Orca	Nov 1997	14.7	0.00		Monoxime	Cochlan and Bronk (2003)
Ross Sea—Station E	Dec 1997	4.7	0.12		Monoxime	Cochlan and Bronk (2003)
*Mean ± std			**0.95 ± 2.92**	**8.5 ± 6.9**		
UREA—OCEANIC DEEP						
Sargasso Sea, BATS	Apr 2009	400	0.06 ± 0.02	14.7 ± 39.3	Monoxime	Baer (unpublished data)

(Continued)

TABLE 4.2   Literature Values of Concentrations of Some Dissolved Organic Compounds Including Urea, Dissolved Free Amino Acids/Dissolved Primary Amines (Dfaa/Dpa), and Dissolved Combined Amino Acids (dcaa) In Aquatic Systems—cont'd

Location	Sampling Date	Sampling Depth (m)	Compound Concentration ($\mu mol\,N\,L^{-1}$)	%DON Pool	Method	Reference
Sargasso Sea, BATS	Apr 2009	800	0.06 ± 0.00	5.4 ± 1.7	Monoxime	Baer (unpublished data)
		*Mean ± std	0.06 ± 0.00	10.0 ± 6.6		
UREA—UPWELLING						
Bronk (2002) mean ($n=2$)			0.09 ± 0.07			Bronk (2002)
		*Mean ± std	0.09 ± 0.07			
UREA—COASTAL/CONTINENTAL SHELF						
Bronk (2002) mean ($n=4$)			1.09 ± 0.55			Bronk (2002)
Ría de Vigo, NW Iberia	Sept 2006	Surface	0.28 to 2.29		Monoxime	Seeyave et al. (2013)
Ría de Vigo, NW Iberia	Jun 2007	Surface	0.10 to 0.41		Monoxime	Seeyave et al. (2013)
Florida Keys > 4 km off shore	Summer 2009	Surface	0.50 ± 0.10		Monoxime	Crandall and Teece (2012)
Florida Keys sea grass bed (Key Largo)	Summer 2010	1-4	1.90 ± 0.20		Monoxime	Crandall and Teece (2012)
Florida Keys areas of schooling fish	Summer 2010	1-4	1.80 ± 0.50		Monoxime	Crandall and Teece (2012)
Florida Keys coral formations	Summer 2010	1-4	1.10 ± 0.20		Monoxime	Crandall and Teece (2012)
Florida Keys bottom near sediment	Summer 2010	4-80	1.10 ± 0.30		Monoxime	Crandall and Teece (2012)
West Florida Shelf	Oct 2008	Surface	0.33 ± 0.07		Monoxime	Wawrik et al. (2009)
Mid-Atlantic Bight (NJ)	Jul 2002	1	2.06 ± 0.45	26.2 ± 6.2	Monoxime	Bradley et al. (2010b)
Mid-Atlantic Bight (NJ)	Jul 2002	14	1.81 ± 0.11	34.6 ± 5.7	Monoxime	Bradley et al. (2010b)
Eastern Coastal Canary Islands	Sept-Oct 2000	80	0.37 ± 0.04[a]	~5.3	Goeyens et al. (1998)	Varela et al. (2005)
Brazil Coastal Current	Sept-Oct 2000	90	0.46 ± 0.12[a]	~10.2	Goeyens et al. (1998)	Varela et al. (2005)

	Date	Depth	*Mean ± std		Method	Reference
Monterey Bay, CA	Aug 2006	Surface	0.64		Monoxime	Kudela et al. (2008)
Monterey Bay, CA	Sept 2006	Surface	0.42		Monoxime	Kudela et al. (2008)
Oslofjord, Norway	1980	0-16	0.10 to 10.00		Monoxime	Kristiansen (1983)[b]
Raunefjord, Norway	Mar 2003	Surface	1.20		Monoxime	Sanderson et al. (2008)
Raunefjord, Norway	Apr 2005	Surface	0.45		Monoxime	Bradley et al. (2010c)
Baltic Sea	May-Oct 1999	Upper 5	0.09 to 6.91		Monoxime	Stepanauskas et al. (2002)[b]
Baltic Sea—Gulf of Riga	May 1996	Upper 20	<0.7		Parsons et al. (1984)	Berg et al. (2001)
Baltic Sea—Gulf of Riga	Jul 1996	Upper 20	0.40		Parsons et al. (1984)	Berg et al. (2001)
		*Mean ± std	**1.24 ± 1.12**	**19.1 ± 13.7**		
UREA—ESTUARINE						
Bronk (2002) mean (n = 6)			0.38 ± 0.21	4.3 ± 3.1		Bronk (2002)
Pocomoke River (MD)	May 1999		0.00 ± 0.00		Urease	Mulholland et al. (2003)
Pocomoke River (MD)	Aug 1999		0.19 ± 0.14		Urease	Mulholland et al. (2003)
Pocomoke River (MD)	May 2000		0.35 ± 0.34		Urease	Mulholland et al. (2003)
Pocomoke River (MD)	Aug 2000		0.36 ± 0.04		Urease	Mulholland et al. (2003)
Neuse River Estuary	Jan-Dec 2001	0.5	1.50 to 4.50		Monoxime	Twomey et al. (2005)
Chesapeake Bay (north)	Aug 2010	0	0.40 ± 0.00	2.3 ± 1.4	Monoxime	Baer (unpublished data)
Chesapeake Bay (mid-Bay)	Aug 2010	0	0.21 ± 0.01	1.2 ± 5.6	Monoxime	Baer (unpublished data)
Chesapeake Bay (south)	Aug 2010	0	0.17 ± 0.01	1.5 ± 7.7	Monoxime	Baer (unpublished data)
Chesapeake Bay	Aug-Sept 2004	0	0.50 to 1.00		Monoxime	Bradley et al. (2010a)
York River	Aug 2009	0	0.10 ± 0.10		Monoxime	Killberg-Thoreson et al. (2010)
York River	Summer (Jun-Aug) 2010 & 2011	0	0.33 ± 0.06	1.9 ± 21.1	Monoxime	Baer 2013
York River	Fall (Sept-Nov) 2010 & 2011	0	0.67 ± 0.28	2.6 ± 47.3	Monoxime	Baer 2013

(Continued)

TABLE 4.2   Literature Values of Concentrations of Some Dissolved Organic Compounds Including Urea, Dissolved Free Amino Acids/Dissolved Primary Amines (Dfaa/Dpa), and Dissolved Combined Amino Acids (dcaa) In Aquatic Systems—cont'd

Location	Sampling Date	Sampling Depth (m)	Compound Concentration (µmol N L^{-1})	%DON Pool	Method	Reference
York River	Winter (Dec-Feb) 2010 & 2011	0	0.45 ± 0.19	2.1 ± 46.1	Monoxime	Baer 2013
York River	Spring (Mar-May) 2010 & 2011	0	0.37 ± 0.16	1.7 ± 58.1	Monoxime	Baer 2013
James River/Chesapeake Bay	Aug 2008	0	0.10 to 0.99		Monoxime	Bronk et al. (2010)
Randers Fjord	Apr 2001	Surface & bottom	0.80 to 3.20		Colorometric	Veuger et al. (2004)
Randers Fjord	Aug 2001	Surface & bottom	1.20 to 3.90		Colorometric	Veuger et al. (2004)
Moreton Bay, Queensland, Australia	Sept 1997	Surface		2.0 to 65.0	Urease	Glibert et al. (2006)
Moreton Bay, Queensland, Australia	Feb 1998	Surface		2.0 to 45.0	Urease	Glibert et al. (2006)
Moreton Bay, Queensland, Australia	Jul 1998	Surface		0.0 to 18.0	Urease	Glibert et al. (2006)
Middle River—Chesapeake Bay Tributary	2000-2002	Surface	0.25 to 1.25		Urease	Glibert et al. (2005)
Manokin and Kings Creek—Chesapeake Bay Tributary	1998-2002	Surface	0.75 to 1.90		Urease	Glibert et al. (2005)
Transquacking and Chicamacomico Rivers—Chesapeake Bay Tributary	1998-2002	SurfacE	1.00 to 2.40		Urease	Glibert et al. (2005)
Pocomoke River—Chesapeake Bay Tributary	1998-2002	Surface	0.45 to 0.90		Urease	Glibert et al. (2005)

Location	Date	Depth	Concentration	Mean ± std	Method	Reference
Sounds—Chesapeake Bay	1998–2002	Surface	0.25 to 0.75		Urease	Glibert et al. (2005)
Coastal Bays—Chesapeake Bay	1998–2002	Surface	0.60 to 2.60		Urease	Glibert et al. (2005)
Great South Bay, New York	Summer 1979 & Spring 1980	Surface	0.60 to 9.40		Monoxime	Kaufman et al. (1983)[b]
Mankyung and Dongjin River Estuary, Korea	Nov 1992, Feb & Jun 1993	Surface	0.60 to 4.30		Monoxime	Cho et al. (1996)[b]
Chesapeake Bay, mainstream	1972–1973, 1988–1994, 1996–1998	Surface & bottom	<0.01 to 8.16		Urease	Lomas et al. (2002)[b]
Florida Bay	Nov 2002	Surface	0.36 to 1.70		Urease	Glibert et al. (2004)[b]
Coastal Bays, Maryland	Apr–Oct 1998–2002	Surface	<0.01 to 14.40		Urease	Glibert et al. (2005)[b]
Kings Creek, Chesapeake Bay, Maryland	Apr–Oct 1998–2002	Surface	0.30 to 24.20		Urease	Glibert et al. (2005)[b]
Chicamicomico R., Chesapeake Bay, Maryland	Apr–Oct 1998–2002	Surface	1.00 to 23.40		Urease	Glibert et al. (2005)[b]
Knysna Estuary, South Africa	Jun 2000–Feb 2001	Surface	0.40 to 5.80		Grasshoff et al. (1983)	Switzer (unpublished data)[b]
		*Mean ± std	1.85 ± 2.95	6.9 ± 9.7		
UREA—RIVERINE						
Bronk (2002) mean (n = 2)			1.68 ± 1.58	6.4 ± 5.7		Bronk (2002)
Savannah River, Georgia	Mar 1971	Upper 10	0.59 to 8.89		Monoxime	Remsen et al. (1972)[b]
Ogeechee River, Georgia	Mar 1971	0	1.26 to 4.89		Monoxime	Remsen et al. (1972)[b]
		*Mean ± std	2.79 ± 1.72	6.4 ± 5.7		
UREA—ARCTIC						
Bronk (2002) mean (n = 2)			3.86 ± 3.84			Bronk (2002)
Chukchi Sea (Barrow, AK)	Jan 2011 & 2012	Surface	0.38 ± 0.44	9.7 ± 119.4	Monoxime	Baer et al. (submitted)

(*Continued*)

TABLE 4.2   Literature Values of Concentrations of Some Dissolved Organic Compounds Including Urea, Dissolved Free Amino Acids/Dissolved Primary Amines (Dfaa/Dpa), and Dissolved Combined Amino Acids (dcaa) In Aquatic Systems—cont'd

Location	Sampling Date	Sampling Depth (m)	Compound Concentration (µmol N L⁻¹)	%DON Pool	Method	Reference
Chukchi Sea (Barrow, AK)	Apr 2010 & 2011	Surface	0.21 ± 0.10	4.4 ± 49.3	Monoxime	Baer et al. (submitted)
Chukchi Sea (Barrow, AK)	Aug 2010 & 2011	Surface	0.66 ± 0.35	8.9 ± 59.7	Monoxime	Baer et al. (submitted)
		*Mean ± std	**1.79 ± 2.70**	**7.7 ± 2.9**		
DFAA/DPA—OCEANIC SURFACE						
Bronk (2002) mean ($n=5$)			0.13 ± 0.21	11.0		Bronk (2002)
Sargasso Sea, BATS	Apr 2009	0	0.07 ± 0.00	4.1 ± 0.8	F	Baer (unpublished data)
Sargasso Sea, BATS	Apr 2009	120	0.08 ± 0.00	22.5 ± 2.0	F	Baer (unpublished data)
		*Mean ± std	**0.11 ± 0.17**	**12.5 ± 9.3**		
DFAA/DPA—OCEANIC DEEP						
Bronk (2002) mean ($n=2$)			0.14 ± 0.20	8.0		Bronk (2002)
35°S, 49°W to 48°S, 20°W Atlantic	Apr–May 2000	>250	0.01 to 0.05	0.2 to 0.6	F	Mahaffey et al. (2004)
Sargasso Sea, BATS	Apr 2009	400	0.06 ± 0.01	14.0 ± 22.0	F	Baer (unpublished data)
Sargasso Sea, BATS	Apr 2009	800	0.04 ± 0.00	3.9 ± 8.8	F	Baer (unpublished data)
		*Mean ± std	**0.08 ± 0.11**	**6.6 ± 5.9**		
DFAA/DPA—UPWELLING						
Bronk (2002) mean ($n=1$)			0.20			Bronk (2002)
		*Mean ± std	**0.20**			
DFAA/DPA—COASTAL/CONTINENTAL SHELF						
Bronk (2002) mean ($n=5$)			0.12 ± 0.09	3.4 ± 2.1		Bronk (2002)
Mid-Atlantic Bight (NJ)	Jul 2002	1	0.26 ± 0.24	3.0 ± 2.1	HPLC	Bradley et al. (2010b)
Mid-Atlantic Bight (NJ)	Jul 2002	14	0.17 ± 0.10	3.1 ± 1.3	HPLC	Bradley et al. (2010b)

West Florida Shelf	Oct 2008	Surface	0.85 ± 0.22		F	Wawrik et al. (2009)
Raunefjord, Norway	Mar 2003	Surface	0.20		F	Sanderson et al. (2008)
Raunefjord, Norway	Apr 2005	Surface	0.25		F	Bradley et al. (2010c)
Jiaozhou Bay, China	Jun 2007	Surface	0.49 to 4.48		HPLC/F	Yanping et al. (2009)
Gulf of Riga, Baltic Sea	May 1999	Surface	0.11 to 0.34		HPLC/F	Berg et al. (2003a)
Gulf of Riga, Baltic Sea	Jun 1999	Surface	0.07 to 0.29		HPLC/F	Berg et al. (2003a)
Gulf of Riga, Baltic Sea	Jul 1999	Surface	0.10 to 0.17		HPLC/F	Berg et al. (2003a)
Gulf of Riga, Baltic Sea	May 1996	Upper 20	<0.80		HPLC	Berg et al. (2001)
Gulf of Riga, Baltic Sea	Jul 1996	Upper 20	~0.10		HPLC	Berg et al. (2001)
		*Mean ± std	**0.39 ± 0.61**	**3.2 ± 1.2**		
DFAA/DPA—ESTUARINE						
Bronk (2002) mean (n = 4)			0.42 ± 0.09	1.7		Bronk (2002)
Pocomoke River (MD)	May 1999	<0.5	0.20 ± 0.15		HPLC/F	Mulholland et al. (2003)
Pocomoke River (MD)	Aug 1999	<0.5	0.24 ± 0.20		HPLC/F	Mulholland et al. (2003)
Pocomoke River (MD)	May 2000	<0.5	0.71 ± 0.57		HPLC/F	Mulholland et al. (2003)
Pocomoke River (MD)	Aug 2000	<0.5	0.44 ± 0.27		HPLC/F	Mulholland et al. (2003)
Chesapeake Bay (north)	Aug 2010	0	0.17 ± 0.02	1.0 ± 10.5	F	Baer (unpublished data)
Chesapeake Bay (mid-Bay)	Aug 2010	0	0.20 ± 0.01	1.1 ± 3.7	F	Baer (unpublished data)
Chesapeake Bay (south)	Aug 2010	0	0.15 ± 0.00	1.3 ± 3.1	F	Baer (unpublished data)
Chesapeake Bay	Aug-Sept 2004	0	<0.25		F	Bradley et al. (2010a)
York River	Summer (Jun-Aug) 2010 & 2011	0	0.30 ± 0.04	1.8 ± 15.8	F	Baer 2013
York River	Fall (Sept-Nov) 2010 & 2011	0	0.28 ± 0.11	1.1 ± 45.3	F	Baer 2013

(Continued)

TABLE 4.2　Literature Values of Concentrations of Some Dissolved Organic Compounds Including Urea, Dissolved Free Amino Acids/Dissolved Primary Amines (Dfaa/Dpa), and Dissolved Combined Amino Acids (dcaa) In Aquatic Systems—cont'd

Location	Sampling Date	Sampling Depth (m)	Compound Concentration (μmol N L^{-1})	%DON Pool	Method	Reference
York River	Winter (Dec-Feb) 2010 & 2011	0	0.21 ± 0.02	1.0 ± 22.9	F	Baer 2013
York River	Spring (Mar-May) 2010 & 2011	0	0.35 ± 0.13	1.7 ± 54.3	F	Baer 2013
James River/Chesapeake Bay	Aug 2008	0	0.18 to 0.28		F	Bronk et al. (2010)
Chesapeake Bay /James River	Aug 2008	Surface	0.15 to 0.60		HPLC	Liu et al. (2010)
Randers Fjord	Apr 2001	Surface & bottom	0.00 to 0.65		HPLC	Veuger et al. (2004)
Randers Fjord	Aug 2001	Surface & bottom	0.02 to 1.40		HPLC	Veuger et al. (2004)
		*Mean ± std	0.34 ± 0.16	1.3 ± 0.3		
DFAA/DPA—RIVERS						
Bronk (2002) mean (n=2)			0.30 ± 0.00	1.2 ± 0.1		Bronk (2002)
		*Mean ± std	0.30 ± 0.00	1.2 ± 0.1		
DFAA/DPA—ARCTIC						
Bronk (2002) mean (n=1)			0.22			Bronk (2002)
Chukchi Sea (Barrow, AK)	Jan 2011 & 2012	Surface	0.13 ± 0.07	3.3 ± 62.6	F	Baer et al. (submitted)
Chukchi Sea (Barrow, AK)	Apr 2010 & 2011	Surface	0.11 ± 0.05	2.3 ± 43.8	F	Baer et al. (submitted)
Chukchi Sea (Barrow, AK)	Aug 2010 & 2011	Surface	0.21 ± 0.07	2.8 ± 45.2	F	Baer et al. (submitted)
Ob River	Aug-Sept 2001	Surface	0.01		HPCL/F	Meon and Amon (2004)

Yenisei River	Aug–Sept 2001	Surface	0.02 to 0.04		HPCL/F	Meon and Amon (2004)
Arctic Estuaries	Aug–Sept 2001	Surface	0.01 to 0.02		HPCL/F	Meon and Amon (2004)
Arctic Estuaries	Aug–Sept 2001	Pycnocline	0.01 to 0.05		HPCL/F	Meon and Amon (2004)
Arctic Estuaries	Aug–Sept 2001	Above bottom	0.01 to 0.01		HPCL/F	Meon and Amon (2004)
Open Kara Sea	Aug–Sept 2001	Surface	0.05 ± 0.05		HPCL/F	Meon and Amon (2004)
Open Kara Sea	Aug–Sept 2001	Pycnocline	0.02 ± 0.02		HPCL/F	Meon and Amon (2004)
Open Kara Sea	Aug–Sept 2001	Above bottom	0.02 ± 0.01		HPCL/F	Meon and Amon (2004)
		*Mean ± std	**0.07 ± 0.08**	**2.8 ± 0.5**		
DFAA/DPA—SOUTHERN OCEAN						
Bronk (2002) mean ($n = 2$)			0.36 ± 0.12	11.5 ± 0.7		Bronk (2002)
		*Mean ± std	**0.36 ± 0.12**	**11.5 ± 0.7**		
DCAA—OCEANIC SURFACE						
Bronk (2002) mean ($n = 2$)			0.57 ± 0.19			Bronk (2002)
		*Mean ± std	**0.57 ± 0.19**			
DCAA—OCEANIC DEEP						
Bronk (2002) mean ($n = 1$)			0.35			Bronk (2002)
		*Mean ± std	**0.35**			
DCAA—COASTAL/CONTINENTAL SHELF						
Bronk (2002) mean ($n = 2$)			1.51 ± 1.23	5.3 ± 4.1		Bronk (2002)
		*Mean ± std	**1.51 ± 1.23**	**5.3 ± 4.1**		
DCAA—ESTUARINE						
Bronk (2002) mean ($n = 1$)			3.74	12.6		Bronk (2002)

(Continued)

TABLE 4.2 Literature Values of Concentrations of Some Dissolved Organic Compounds Including Urea, Dissolved Free Amino Acids/Dissolved Primary Amines (Dfaa/Dpa), and Dissolved Combined Amino Acids (dcaa) In Aquatic Systems—cont'd

Location	Sampling Date	Sampling Depth (m)	Compound Concentration ($\mu mol\,N\,L^{-1}$)	%DON Pool	Method	Reference
Pocomoke River (MD)	May 1999	<0.5	2.10 ± 0.06		HPLC/F	Mulholland et al. (2003)
Pocomoke River (MD)	Aug 1999	<0.5	2.25 ± 1.43		HPLC/F	Mulholland et al. (2003)
Pocomoke River (MD)	May 2000	<0.5	2.25 ± 0.70		HPLC/F	Mulholland et al. (2003)
Pocomoke River (MD)	Aug 2000	<0.5	2.99 ± 0.17		HPLC/F	Mulholland et al. (2003)
Chesapeake Bay/James River	Aug 2008	Surface	0.80 to 2.20		VPH- HPLC	Liu et al. (2010)
Randers Fjord	Apr 2001	Surface & bottom	0.90 to 4.50		HPLC	Veuger et al. (2004)
Randers Fjord	Aug 2001	Surface & bottom	1.20 to 2.90		HPLC	Veuger et al. (2004)
		*Mean ± std	2.45 ± 0.68	12.6		
DCAA—RIVERS						
Bronk (2002) mean (n = 2)			1.86 ± 0.50	7.0 ± 1.8		Bronk (2002)
		*Mean ± std	1.86 ± 0.50	7.0 ± 1.8		

Data were taken from text, tables, estimated from graphs, or obtained from the authors and are presented as the mean ± standard deviation. The means presented in the first edition of this chapter (Bronk, 2002) are presented at the start of most sections. All data from the earlier edition (Bronk, 2002) and the more recent data presented here were included in the calculated mean ± standard deviation presented at the end of each section. When a range is given, the average of that range was used in the calculation of the mean ± standard deviation for that system. Methods used were high-pressure liquid chromatography (HPLC), vapor phase HPLC (VP-HPLC), and fluorometry (F).

*Means for each region include all data from Bronk (2002) as well as new data listed in table.

[a]Standard error reported, not standard deviation.

[b]Data from Gilbert et al. (2006).

In the literature, concentrations of DFAA are often reported as DPA because in waters with low NH_4^+ concentrations, the two measurements are generally equal (Kirchman et al., 1989). Like DFAAs, concentrations of DPA tend to be very low due to the close coupling between uptake and release (Fuhrman, 1990), however, this is not always the case. For example, extremely high levels of DPA (up to $17 \mu mol\,NL^{-1}$) were measured during a dinoflagellate bloom in Chesapeake Bay (Sellner and Nealley, 1997).

Concentrations of DFAA are measured using high-pressure liquid chromatography (HPLC; Mopper and Lindroth, 1982). DFAA concentrations range from 0.001 to $1.4 \mu mol\,NL^{-1}$ in the studies surveyed here and represents 1.2-12.5% of the total DON pool (Table 4.2). Concentrations of DFAA vary with season, depth, location, and time of day (Bronk et al., in press; Fuhrman, 1987, 1990; Mopper and Lindroth, 1982; Williams and Poulet, 1986). Sources of DFAA include primary producers, many of which have large intracellular pools of amino acids that can be released via a number of processes (see Section IV.D.3), viral lysis of bacteria (Middelboe and Jørgensen, 2006), release from micro- and macrozooplankton (reviewed in Steinberg and Saba, 2008), and excretory products of jellyfish and ctenophores (Pitt et al., 2009). Laboratory culture experiments indicate that intra-and extracellular amino acids can vary as a function of growth phase and thus can be very important indicators of cellular metabolic state in phytoplankton (reviewed in Bronk and Flynn, 2006).

In the few studies available, concentrations of DCAA range from 0.57 to $2.99 \mu mol\,NL^{-1}$ and represent 5.3-12.6% of the total DON pool (Table 4.2). Concentrations of DCAA are typically measured by HPLC after acid hydrolysis with 6N HCl at 110°C for 20-24h (Parsons et al., 1984). Tsugita et al. (1987) introduced a quicker vapor phase hydrolysis technique, which produced DCAA concentrations that were 1.5 ± 0.4 times higher than the traditional hydrolysis method. The chemical structure of DCAA is largely unknown but can include a suite of compounds where amino acids are polymerized to each other (i.e., proteins and oligopeptides) or where amino acids are bound to humic or fulvic materials or adsorbed to clays or other materials (reviewed in Bronk, 2002). Between 30% and 100% of DCAAs are in the HMW fraction of the DON pool (Aluwihare and Meador, 2008). Concentrations of THAA (DCAA+DFAA) range from 0.15 to $0.65 \mu mol\,NL^{-1}$ in marine surface waters (Dittmar et al., 2001; Hubberten et al., 1995; Yamashita and Tanoue, 2003) with concentrations declining by roughly a factor of three in waters below the euphotic zone (Hubberten et al., 1995; Lee and Bada, 1975; Yamashita and Tanoue, 2003).

Beyond concentration, the composition and D/L enantiomeric ratio of amino acids can be a useful indicator of the source and diagenic state of the amino acids and DON pool (e.g., Calleja et al., 2013; Kaiser and Benner, 2008; McCarthy et al., 1998). The D enantiomers, abundant in bacterial cell walls, are indicative of refractory components. In contrast, L enantiomers are indicative of freshly produced amino acids (Calleja et al., 2013; Kaiser and Benner, 2008) such that as DOM is degraded by bacteria, the D/L enantiomeric ratio increases.

Similarly the amino acid specific $\delta^{15}N$ composition has also been used to determine the source and trophic transfer history of DON (e.g., McCarthy et al., 2007, 2013; McClelland and Montoya, 2002; McClelland et al., 2003). Several studies have also used the very short lived $\delta^{13}N$ to investigate amino acid cycling in bacteria and protists and data indicate that strong fractionation only occurs in a small number of amino acids (McCarthy et al., 2004; Ziegler and Fogel, 2003). Amino acid specific $\delta^{15}N$ signatures may also be useful in determining the source (e.g., prokaryotes versus eukaryotes) of marine DON (McCarthy et al., 2013). Although all phytoplankton have a similar $\delta^{15}N$ pattern, an offset of the $\delta^{15}N$ of glutamine plus glutamic acid may help to determine if observed DON was produced by cyanobacteria or eukaryotic phytoplankton.

3 Humic and Fulvic Substances

The study of humic substances seems to be declining among marine chemists as more isolation approaches become available allowing other fractions to be interrogated. We kept this section in this edition, however, because humics are turning out to be surprisingly important biologically (see Section V.D.4). Humics are comprised of three operationally defined structural families including fulvic acids, humic acids and humins. Fulvic acids are defined as hydrophilic acids soluble in water at any pH, range in size between 500 and 2000 Da, and can be isolated using XAD-4 resin. Humic acids are also operationally defined and tend to be larger with sizes >2000 Da; they form precipitates at pH < 2 but are soluble at higher pHs and are isolated using DAX-8 (formerly XAD-8) resins (Peuravuori et al., 2002). Humins are insoluble in water at any pH and are thus the least common fraction in marine waters (Aiken, 1985). In addition to resins, reverse osmosis (Serkiz and Perdue, 1990) has also been used to produce humic and fulvic standards that are now commonly used as reference material for ultra-high resolution Fourier transform ion cyclotron resonance mass spectrometry (FT-ICR MS) studies investigating the composition of marine DOM (e.g., Koch et al., 2005). Several FT-ICR MS studies have investigated humic substances from freshwater systems in great detail, characterizing the majority of the fulvic pool at the molecular level and providing insight into the distribution of structural types (e.g., Kujawinski, 2002; Kujawinski et al., 2002a, 2002b; Sleighter and Hatcher, 2007; Stenson, 2008; Stenson et al., 2002).

In seawater, humic substances typically make up 10-20% of the DOM pool, while hydrophilic fulvic acids contribute 50% or more (reviewed in Hessen and Tranvik, 1998). Humic substances isolated from natural waters originate in either a terrestrial or a marine environment. Humics from both environments are colored organic acids that have similar metal complexing capabilities and redox functions (Harvey and Boran, 1985).

Humic substances of terrestrial origin are more aromatic and have a higher C:N ratio compared to marine humics, which are more aliphatic in nature (Stuermer et al., 1978). The building blocks of marine humics are believed to be biosynthetic compounds such as amino acids, sugars, amino-sugars, and fatty acids (Gagosian and Lee, 1981). The general consensus is that both marine and terrestrial humic substances arise from microbial degradation of plant material, but the exact mechanisms of humification remain unknown (Hedges, 1988).

Collating a table of humic N concentrations proved to be frustrating. Researchers routinely report concentrations of DOC in humic substances or elemental analyses of C:N ratios. In most cases, however, both pieces of information are not presented for a given sample so that the concentration of DON cannot be estimated. Bronk (2002) presents concentrations of humic substances that range from 0.4 to 12.3 μmol N L^{-1}. The C:N of aquatic humic substances isolated with XAD resin can be highly variable but ranges from 18 to 30:1 for humic acids and 45 to 55:1 for fulvic acids (See, 2003; See and Bronk, 2005; Thurman, 1985).

4 Other Organic Compounds

There has been a suite of other organic compounds identified in seawater including nucleic acids, purines and pyrimidines, pteridines, methylamines, and creatine (see review by Antia et al., 1991). The nucleic acids include dissolved deoxyribonucleic acids (D-DNA) and dissolved ribonucleic acids (D-RNA). There are a number of methods used to measure nucleic acid concentrations in seawater (reviewed in Karl and Bailiff, 1989) as well as a centrifugal concentration method for D-DNA quantification (Brum et al., 2004). Bacteria produce D-DNA during growth (Paul et al., 1987, 1990), through protozoan grazing on bacteria (Alonso et al., 2000; Ishii et al., 1998; Kawabata et al., 1998; Turk et al., 1992), and by viral lysis (Alonso et al., 2000, Corinaldesi et al., 2007, Riemann et al.,

2009; Weinbauer et al., 1993). Bacteria use extracellular nucleases to hydrolyze D-DNA and then transport the low molecular weight (LMW) compounds through membrane transport proteins (Paul et al., 1987). RNA concentrations within a cell are dependent on the growth rate of the cell, while DNA concentrations are not nearly as variable. In general, D-DNA concentrations are highest in near-shore surface waters with concentrations decreasing with depth and distance from shore (reviewed by Karl and Bailiff, 1989). In Tokyo Bay, concentrations of both D-DNA and D-RNA decrease from the head to the mouth, and vertical profiles have surface maxima with lower concentrations deeper in the water column (Sakano and Kamatani, 1992). Vertical profiles of D-DNA at Station ALOHA in the North Pacific subtropical gyre are also higher in the upper water column and decrease with depth (Brum, 2005). Concentrations of particulate DNA and RNA are greater than the dissolved forms, and both D-DNA and total nucleic acids (D-DNA + RNA) concentrations are correlated with dissolved organic phosphorus (DOP) concentrations. D-DNA has also been measured within Arctic sea ice and concentrations in the ice were higher than the underlying seawater (Collins and Deming, 2011).

Purines and pyrimidines are N-heterocyclic, aromatic compounds that contain five and two atoms of N, respectively. They are present in DNA and RNA and are components of coenzymes. Adenine and guanine are the dominant purine bases found in nucleic acids, and thymine, cytosine and uracil are the major pyrimidine bases. The range of concentrations reported in Antia et al. (1991) is 4×10^{-5} to $12.6 \mu mol N L^{-1}$. Significant differences in lability and turnover rates of purines and pyrimidines have been documented with purines being catabolized by bacteria more quickly than the relatively less labile pyrimidines (Berg and Jørgensen, 2006). The catabolism of purines may also be important in the production of urea in the marine environment (Berg and Jørgensen 2006).

Methylamines come in mono-, di-, and trimethyl forms that are primary, secondary, and tertiary methylated homologues of ammonia. In the Arabian Sea, concentrations of methylamines are low in oligotrophic open ocean regions and higher in productive coastal areas influenced by upwelling (Gibb et al., 1999). Concentrations of the different forms generally follow the pattern of mono- > di- > tri-methylamines (Table 2 in Gibb et al., 1999). Overall, methylamines contribute <1% of the measured DON pool, and their distributions appear to be biologically controlled with maxima in the mono- and di-forms associated with diatom abundance and microzooplankton grazing.

B Chemical Composition—Opening the Black Box of Uncharacterized DON

The marine DON pool is often referred to as a black box due to its chemical complexity, our lack of insight into the factors controlling its cycling, and current analytical limitations for molecular level characterization. As a result of these challenges, the majority of compounds that comprise the DON pool have not been identified. The issues of isolation and characterization are particularly important because of the debate over the total number of marine DON compounds. It remains unknown whether or not the majority of unidentified DON compounds are quantitatively significant in nature or whether they are produced via matrix effects associated with current hydrolysis methods. The compound level analysis needed to end this debate requires a method to concentrate but not alter DON and an instrument with very high sensitivity that allows for the quantification of individual compounds in a salt matrix. With an average salinity of 35 and an estimated DON concentration of $5 \mu mol N L^{-1}$, the salts in seawater outnumber DON 500,000:1. These salts contribute to our current analytical limitations by clogging channels, reacting with the DOM and solvents, and eroding analytical columns.

Here we describe methods for DON isolation and characterization and briefly review chemical characteristics of the DON pool.

1 DON Isolation Methods

DON isolation continues to be a difficult undertaking and there is still no one universally agreed upon method. All currently available methods are biased in some way, and few studies directly investigate the retention of DON using the various approaches (Aluwihare and Meador, 2008; Dittmar et al., 2008; Worsfold et al., 2008). Here we briefly describe several of the most commonly used isolation methods including size fractionation via ultrafiltration, solid phase extraction, and early yet promising descriptions of tandem reverse osmosis and electro dialysis.

Ultrafiltration (UF) separates DON by size in the same way that we use GF/Fs to separate particulates from the dissolved fractions of the TDN pool. Depending upon sample volume there are two approaches. Tangential flow UF is generally used when concentrating large volumes (liters) often needed for open ocean investigations. Stirred cells hold smaller volumes (~100-400 mL) and are useful in regions with higher DON concentrations like coastal waters and estuaries. Although UF is an effective way to reproducibly separate the DOM into smaller pools, only the high molecular weight (HMW) fractions are retained. The isolated HMW fraction most commonly includes compounds larger than 1 nm or with molecular weights >1000 Da and represents 20-40% of the DON pool. The LMW fraction, those compounds <1000 Da including urea, free amino acids, and up to 80% of the marine DON pool (Kaiser and Benner, 2009), are not retained and thus not concentrated or desalted. The loss of the LMW compounds is problematic because substrates available for direct assimilation via transporters by microorganisms are in this fraction. Therefore, incubations using HMW ultrafiltered DON (UDON) do not assess the lability of a large part of the DON pool but do provide insight into the composition and lability of

the arguably more complex, larger size fraction (reviewed in Aluwihare and Meador, 2008).

Recently, radiocarbon dating has shown that LMW DOM compounds are much older and have thus been assumed to be more refractory than HMW compounds (e.g., Amon and Benner, 1994; Walker et al., 2014). Following this reasoning, one way to study the labile DOM pool is to target the HMW fraction. While intriguing, the approach of focusing on the HMW fraction will miss compounds like urea and free amino acids that are known to be highly labile and reside clearly in the LMW DON pool. If studying DON cycling, focusing on UDOM may be particularly misleading in systems where DOC and DON cycling appears to be decoupled (e.g., Abell et al., 2000, 2005; Church et al., 2002; Knapp et al., 2005). We suggest that assuming that all LWM DON is refractory is a significant misstep in efforts to better understand DON composition and cycling.

Solid phase extraction (SPE) isolates DOM based on its chemical composition rather than size. Dittmar et al. (2008) compared the DOM retention of different SPE sorbents including silica based C8, C18, C18OH, and C18EWP and styrene divinyl benzene based PPL and ENV sorbents. C18 and PPL sorbents are the most commonly used of this group but both unfortunately selectively omit some LMW ionic compounds and large biomolecules like proteins. Several of these sorbents including C18s and PPLs also appear to discriminate against some N-containing moieties. For example, Sleighter and Hatcher (2008) showed that 9.4% of compounds within whole water samples from the Dismal Swamp contained N but that percentage dropped to 1.4% after that same sample had been C18 extracted. Results from the South Atlantic indicate that the N extraction efficiency of samples isolated with PPLs ranged between 10.2% and 28.3% depending on depth, with deep-ocean efficiencies being 2-3 times lower than surface water extractions (Hertkorn et al., 2013). Conversely, the DOC extraction efficiencies were much more stable ranging between 37% and 44% with no discernible trend with depth. This illustrates the need for studies to report both the N and C extraction efficiencies instead of assuming the maximum possible retention for C alone. The discrimination of DON by SPEs is also indicated by the high C:N ratios (21 to 42) of the isolated DOM compared to the C:N (10 to 17) of the initial seawater (De Jesus, 2008; Dittmar et al., 2008; Hertkorn et al., 2013). Although neither method is ideal, PPLs appeared to be more efficient at isolating DON from seawater compared to C18s based on the C:N ratio of the extracted DOM, however, these results may be site specific.

Although some SPEs may be better at retaining labile compounds like urea and free amino acids, neither is retained by C18 or PPLs (Sipler, unpublished data). The fact that urea and free amino acids are not retained on C18 or PPLs is actually fortuitous because these compounds can be quantified separately. As a result the C18 or PPL DON fractions could be added to total urea and amino acid (e.g., DFAA or DPA) concentrations to provide a more complete view of the overall DON pool.

Although there are clearly a number of caveats (e.g., low DON extraction efficiencies) to SPE there are also several benefits to the approach. In theory, if a compound can stick to a particular SPE, it will stick and if it cannot, it will pass through the column as filtrate. The use of SPE is reasonably easy, can be performed in the field without large power requirements, requires relatively small sample volumes (mLs to Ls depending on the source), requires short extraction times compared to UF, has a high degree of reproducibility, and new promising SPE resins are continually being developed.

Since UF and SPE do not necessarily retain the same compounds, Simjouw et al. (2005) and Dittmar et al. (2008) both suggest combining UF and SPE to increase the overall DOM recovery. Although studies combing UF and SPE are not prevalent, the combination of UF and C18 in series has successfully retained 66-81% of DON

collected during a red tide bloom in the Gulf of Mexico (Sipler et al., 2013) and 40-56% of DON from samples collected in Florida Bay (Sipler, 2009). Therefore, based on current isolation methods, the combination of multiple isolation approaches will provide the greatest recovery of DON. We direct the reader to several studies that compare or review isolation methods for a more thorough discussion (Aluwihare and Meador, 2008; Dittmar et al., 2008; Mopper et al., 2007; Simjouw et al., 2005; Tfaily et al., 2012).

Another relatively new approach to isolating DOM is reverse osmosis (RO) and electro dialysis (ED; Koprivnjak et al., 2009; Vetter et al., 2007). Like many isolation methods, RO/ED appears to work well for freshwater samples but are still not fully vetted when it comes to saline samples and even less so when specifically considering DON. Published DOC recoveries range between 61% and 95% (Gurtler et al., 2008; Koprivnjak et al., 2009; Vetter et al., 2007). Unfortunately DON recovery was not measured directly in any of these studies, however, Koprivnjak et al. (2009) estimated that 75% of DON was recovered based on assumptions of the original DON concentration and measured C:N ratio of the final RO/ED DOM. This RO/ED method has also been used in the study of DOP and shown to recover 54-84% of the DOP pool (e.g., Young and Ingall, 2010). There are a few drawbacks to current RO/ED methods including the sample processing time and volume (6 h to process 200 L; Vetter et al., 2007), cost of the equipment that is not currently commercially available, and electrical demands, which may be an issue for ship-based manipulations. However, if this method proves to be effective in removing salts while retaining ~75% of the marine DON pool, then all efforts will be well worth it.

2 DON Characterization Methods

There are a number of characterization methods available but the maximum proportion of DON recovered and degree to which DON has been or can be chemically characterized is ultimately limited by the recovery of DON from the specific sample and isolation method used. Similar to isolation methods, no single chemical characterization method has emerged as the clear winner when characterizing DOM(N). Several different instruments and techniques have been used to characterize DON including nuclear magnetic resonance (NMR) spectroscopy (e.g., Aluwihare et al., 2005), Fourier transform ion cyclotron resonance mass spectrometry (FT-ICR MS; e.g., Kujawinski et al., 2004), and natural abundance mass spectrometry (Knapp et al., 2005). Studies using a combination of several different analytical methods have been the most informative. For example, identification of specific compound classes or functional groups via NMR provided information regarding the molecular formulas assigned through FT-ICR MS. ^{15}N-NMR has been used to detect a number of DON compounds and functional groups including amine, amide and N-heterocycle groups (Worsfold et al., 2008). Solid-state NMR analyses estimate that <15% of the total N in the deep-ocean is heterocyclic N (Aluwihare and Meador, 2008). FT-ICR MS, on the other hand, has the benefit of ultra-high resolution that, with the proper inlet system, allows complete non-fragmented compounds to be detected and, through sub ppm resolution, molecular formulas to be assigned. FT-ICR is non-quantitative, however, and does not supply structural information about the individual compounds without additional analyses (Mopper et al., 2007).

Electrospray ionization (ESI) and atmospheric pressure photoionization (APPI) are common inlet systems associated with FT-ICR MS. Both are soft ionization inlet systems and are less likely to fragment compounds compared to other common MS techniques including gas chromatography (GC). It is the inlet system that allows FT-ICR MS to detect whole compounds and can be run in either the positive or negative ionization mode. Samples detected in the positive ionization mode contain more basic functional groups and those

detected in the negative mode tend to be more acidic. Compounds containing a mixture of functional groups may be detected in both modes. A recent comparison of ESI and APPI inlet systems coupled to an FT-ICR MS showed that APPI preferentially ionizes more DON compounds than ESI (Podgorski et al., 2012). However, this comparison was made between positive ionization mode APPI and negative mode ESI. More DON compounds are detected in the positive ionization mode than the negative ionization mode (Seitzinger et al., 2005) so that is not necessarily a direct comparison. However, APPI does have the benefit of increased ionization of DON compounds and reduced adduct formation (i.e., when two or more elements combine) compared to positive ionization mode ESI. D'Andrilli et al. (2010) suggest analyzing samples using both inlet systems to attain the most complete suite of compounds.

3 Chemical Characteristics of Marine DON

At the time of the last edition, chemical characterization of DON focused largely on the UDOM fraction. The field has moved far beyond that with the cutting edge research combining the metabolism of organisms and the compounds they create, destroy or transform (reviewed in Kujawinski, 2011). From a chemical perspective, a number of studies have used NMR, FT-ICR MS or both to investigate the chemical composition of marine DOM, including DON. NMR has been used on DON containing samples collected from the South Atlantic, Angola coast (Hertkorn et al., 2013), Florida Bay (Maie et al., 2006), the Gulf Stream (Mao et al., 2012), Ogeechee River Mouth (Mao et al., 2012), and an estuarine to marine transect from the Chesapeake Bay to the coastal Atlantic (Abdulla et al., 2010), to name a few. FT-ICR MS has also been used to chemically characterize DON from the subtropical convergence along the South Island of New Zealand (Gonsior et al., 2011), eastern Atlantic (Flerus et al., 2012), Angola Basin in the South Atlantic (Hertkorn et al., 2013; Reemtsma et al., 2008), estuarine to marine transect from the Chesapeake

Bay to the coastal Atlantic (Sleighter and Hatcher, 2008), Gulf of Mexico (Podgorski et al., 2012; Sipler et al., 2013), Ochlockonee River, FL (Podgorski et al., 2012), Pacific Ocean (Hertkorn et al., 2006), and the Weddell Sea (D'Andrilli et al., 2010). It is important to note that most of the chemical characterization methods require DOM isolation, and all of the data reviewed in the remainder of this section were isolated using UF, SPE or RO/ED prior to analysis. Here we describe three studies that used a combination of characterization methods to gain insight into DON composition.

High-field NMR spectroscopy and FT-ICR MS were combined to investigate the composition of DOM over depth, including samples from surface (5 m), the fluorescence maximum (48 m), the upper mesopelagic (200 m), and the abyssopelagic (5446 m) collected in the Angola Basin of the South Atlantic (Hertkorn et al., 2013). All samples were PPL extracted prior to analysis and although the N recovery decreased with depth, the C:N ratio of the extracted DOM stayed relatively stable (24.7 ± 2.4 to 27.0 ± 0.9). The number of compounds containing C, hydrogen, oxygen and N (CHON) increased with depth but those CHON compounds that also contained sulfur (CHONS) decreased with depth. Aliphatic CHON compounds were enriched at the surface and six member N-heterocycles, similar to pyridines, were detected at all depths (Hertkorn et al., 2013).

Pyridines were detected in UDON samples from Florida Bay using a combination of ^{15}N cross-polarization magnetic angle spinning (CPMAS) NMR and X-ray photoelectron spectroscopy (Maie et al., 2006). Pyridinic N, peptide bond N and primary amine N were 18.5%, 74.2%, and 7.3% of the DON pool, respectively. Maie et al. (2006) speculated that the source of the pyridinic-N was from wild fires, common in the Everglade region. They also suggested that the high proportions ($27 \pm 4\%$ UDON) and relative proportions of individual hydrolysable amino acids (HAA) were due to biological

production within the system and are not associated with significant diagenic processing as would be expected if HAAs had been transported downstream.

Gonsior et al. (2011) investigated the DOM composition across the subtropical convergence zone off the coast of New Zealand using FT-ICR MS and excitation emission matrix fluorescence (EEM) spectroscopy. Summer and winter seawater samples were collected from three locations of increasing distance from the coast and extracted using C18s. The number of DON compounds detected did not change dramatically between seasons (spring and summer). Approximately 100 compounds detected at the near shore sites were detected in the positive mode, which is better for detecting N-containing compounds, but half as many (~50) were detected in the off-shore site, which suggests that the number of DON compounds were removed with distance from the coast. Approximately 70% of all N-containing compounds detected at the offshore site were also detected at the near-shore sites but none of these formulas were found within a nearby freshwater system (Gonsior et al., 2011). The same N-containing molecular formulas were assigned in both winter and spring samples indicating homogeneity within the DON pool or at least the fraction of DON retained by C18. Sleighter and Hatcher (2008) found the opposite to be true for the Chesapeake Bay region, where the number of DON compounds increased with distance from land, and the offshore DON site was more diverse than near shore.

C Concentration and Composition of the DON Pool: Research Priorities

The last decade has seen an explosion in the number of DON measurements and the research community has made real progress on two of the recommendations offered in the first edition—the need for large, high quality data sets of DON distributions and looking beyond the

HMW UDOM pool with respect to composition. We now have a global picture for DON concentration in surface waters and a suite of analytical approaches that are beginning to shed light on the entire DON pool. Although much has been accomplished, we still have quite a ways to go on the number one recommendation from the first edition—the development of a high sensitivity method that directly measures concentrations of DON. As long as a DON concentration is determined by taking the difference between TDN and DIN measurements, analytical errors will confound our ability to resolve small but significant changes in DON. These limitations currently prevent the research community from peering into the ocean interior with respect to DON distribution and its associated dynamics. Due to the large amount of inorganic salt in seawater, DON isolation methods that recover all or most DON are also still desperately needed for more complete DON characterization. As this research area progresses it is important that DON recoveries be reported for each study rather than assumed based on earlier work or inferred from data on DOC retention. We should also not ignore any fraction of the DON pool. Efforts to develop or combine current analytical techniques to chemically characterize DON at the compound level should be a priority over the next decade, particularly as it relates to DON production and consumption.

IV SOURCES OF DON TO THE WATER COLUMN

DON in the ocean can originate from two general source types—autochthonous, which produce DON within the system, in this case the water column, and allochthonous, which supply DON from outside the system. Each source will be reviewed, followed by methods to measure rates of DON release, an updated summary of rates of DON release across environments, and suggested research priorities for the future.

We note that when rate data are available for a specific source, such as DON release from N_2 fixers, they are included in the discussion of the source. We cover *in situ* rates that cannot be linked to a specific source, and likely are a function of multiple sources, in the later section on DON release rates in the literature.

A Autochthonous Sources

DON can be produced within the water column by a number of organisms—phytoplankton, N_2 fixers, bacteria, micro- and macrozooplankton, and viruses (Figure 4.3). There are a number of additional sources of DON, many of which are more important to shallower coastal systems, which will not be considered here due to space limitations. These include release from macro organismal excretion, direct release from macroalgae, detrital particle release via dissolution, and release from sediments via digenesis (all reviewed in Bronk and Steinberg, 2008); also see the sediment review by Burdige and Komada in this volume. We note that photochemical processes have been shown to result in the release of amino acids from organic matter including DON and so is considered in the next section on sinks for DON.

1 Phytoplankton

Phytoplankton can be a source of DON either through active release or exudation, through the passive diffusion of metabolites through cell membranes, and trophic mediated release where phytoplankton are acted upon by zooplankton (i.e., sloppy feeding) or viruses. With respect to direct release there are two models proposed—the overflow model and the passive diffusion model (Fogg, 1966), both of which are discussed in some detail in Carlson (2002), Chapter 3 in this volume, and Bronk (2002) and so are only briefly considered here.

In the overflow model, active release occurs when excess photosynthate accumulates due to nutrient limitation, which is termed exudation.

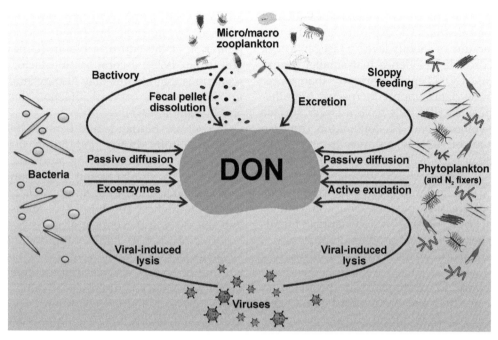

FIGURE 4.3   Conceptual diagram of processes involved in *in situ* dissolved organic nitrogen (DON) release in aquatic systems.

The requirement of nutrient limitation, however, makes it problematic as a mechanism for DON release. There are a number of other direct release processes including release due to osmotic changes, release of reduced inorganic or organic N due to excess light, and autolysis, also known as programmed cell death, which occurs when an organism produces enzymes that destroy its own cell membranes (e.g., Llabrés and Agusti, 2006, see below). Another form of DON release that has still received little attention is protoplasm lysis during spermatogenesis (i.e., the creation of sperm when the cell enters a sexual reproduction phase), which can be brought on by N starvation (Collos, 1986; Sakshaug and Holm-Hansen, 1977). With respect to osmotic stress, organisms are generally adapted to a relatively narrow salinity range. If cells encounter conditions outside of that range some can synthesize and accumulate or release

osmoprotective compounds, including some N-containing compounds such as glycine betaine (Keller et al., 1999; Ning et al., 2002). In response to light stress, some phytoplankton shunt photons into pathways that reduce oxidized N compounds resulting in the release of LMW organic compounds, such as amino acids and inorganic N (NH_4^+ or NO_2^-), as part of a protective futile cycle (e.g., Lomas and Glibert, 1999).

In the passive diffusion model, DON is released from the cell during the diffusion of LMW compounds through the cell membrane from highly concentrated intracellular pools to very dilute extracellular pools. In this scenario bacteria act as ectoparasites by taking up extracellular DOM, which maintains the large concentration gradient between intracellular and extracellular pools (Bjørnsen, 1988). This passive release could be influenced by any process that affects the integrity of cell membranes such as physiological stress, UV exposure, and changes in temperature or salinity levels. Hasegawa et al. (2000a) provided evidence that smaller plankton release DO^{15}N more efficiently than larger ones. These data support Bjørnsen's hypothesis, because smaller cells have a larger surface area to volume ratio. Phytoplankton can also be a source of DON when other trophic levels act on them such as lysis due to viral infection (Brussaard, 2004) or during zooplankton sloppy feeding (i.e., when grazers do not consume a cell whole but break it apart releasing intracellular contents; Dagg, 1974). Also see Chapter 3 for additional details of DOM production processes.

Culture work in the 1960s and 1970s was the foundation of what we know about direct DON release in phytoplankton (reviewed in Antia et al., 1991 and Bronk and Flynn, 2006). Based on a survey of culture studies, an average of $19.3 \pm 16.2\%$ of gross N uptake is released as DON (see Table 1 in Bronk and Flynn, 2006). This value is generally lower than the mean percent release observed in field studies, which suggests that trophic level interactions (e.g., grazing) are important. Culture work has also shown that percent release is highest immediately after cells are transferred from one set of environmental conditions to another (Collos et al., 1992; Slawyk and Raimbault, 1995). The question is whether this is an artifact of the cells being stressed during the transfer or a natural response of phytoplankton to a changing environment. If it is the latter, then the percentage of extracellular release is likely much higher in the turbulent natural environment than in the more static laboratory studies.

Culture studies are also very useful to look at release at different stages in the cells cycle—lag phase, exponential, or stationary. The highest rates of DCAA release occur during stationary phase but are low during active growth (Flynn and Berry, 1999). This same pattern is seen in nature where it is common to measure low rates of release in actively growing cells early in a phytoplankton bloom, but high rates of release are observed when cells are physiologically stressed by nutrient limitation and/or light inhibition at the end of a bloom (Larsson and Hagström, 1979); a complicating factor in this scenario is that blooms may also decline due to grazing pressure or viral infection (Jenkinson and Biddanda, 1995).

In contrast, Biddanda and Benner (1997) performed culture studies and found that most DOM was produced during nutrient replete conditions. A similar result was found in a study of batch cultures of marine *Synechococcus* where the highest absolute DON release rates, and release as a percentage of gross N uptake, occurred during N sufficient growth (Bronk, 1999). DON release rates were also higher when cultures of *Scenedesmus quadricauda* were grown under N-replete conditions, relative to cultures grown under N-limited conditions, again suggesting that release may be suppressed under nutrient limited growth so that cells can meet their N quota (Nagao and Miyazaki, 2002).

One caveat is that the DON release rates that are generally measured are net rates, particularly under N-limited conditions. The degree and type of nutrient limitation can affect if DON is released directly or taken up and, if released, how long it will persist in the water column (e.g., Arrigo, 2005). Recently released DON that is reincorporated can result in an underestimate of the true DON release rate. Organic N can be used by many phytoplankters to fulfill their nutritional requirements making reincorporation likely, particularly when cells are N-limited (reviewed in Bronk et al., 2007).

Another source of released DON is programmed cell death (PCD), also known as autocatalytic cell death, in which an endogenous biochemical pathway leads to changes in cell morphology and ultimately cell dissolution (reviewed in Bidle and Falkowski, 2004). Cells contain proteases that perform functions such as metabolic regulation through enzyme degradation and protein breakdown. PCD can occur in response to advanced age, environmental stress such as nutrient deprivation, exposure to bright light or extended darkness (e.g., Berges and Falkowski, 1998; Llabrés and Agusti, 2006) or salinity or oxidation stress (Ning et al., 2002). A number of studies provide evidence that PCD occurs in cultures (Berges and Falkowski, 1998; Berman-Frank et al., 2004; Brussaard et al., 1997; Ning et al., 2002; Segovia et al., 2003; Vardi et al., 1999); a brief review of these studies is given in Bronk and Steinberg (2008).

There is little doubt that PCD occurs but there is considerable debate as to its frequency. At the heart of the debate are the methodological challenges one confronts when trying to quantify PCD from other processes that may be occurring simultaneously, particularly viral infection (Agusti and Duarte, 2002; Agusti et al., 1998; Brussaard et al., 1995; Riegman et al., 2002; van Boekel et al., 1992; Vardi et al., 2009). One reason why quantifying the role of PCD in DON release is important is that different mechanisms of cell lysis likely produce DON of differing lability. For example, a cell that lyses due to viral infection has likely had host cell N shunted towards viral production. This is in contrast to a cell that lyses due to PCD, which results in DOM and enzymes released into the environment such that DON released from cells that died from PCD may be more bioavailable to bacterioplankton compared to compounds released during lytic viral infection.

In the ocean, cells with large intracellular vacuoles are likely to be of particular importance in dissolved N release. As noted earlier, a ^{15}N study by Hasegawa et al. (2000a) showed more DON release by smaller phytoplankton cells than by larger cells. In upwelling regions of the central Atlantic where phytoplankton >2 μm dominate, the percent of total N uptake released as DON was <30% (Varela et al., 2005). In contrast, the percent of total N released as DON was often >50% in oligotrophic regions dominated by pico-phytoplankton, although it was highly variable (Varela et al., 2005). In another study, low rates of DON release were measured where small

cells (<2 μm) dominated total biomass and primary production (Varela et al., 2003). However, the highest DON release rates occurred when smaller cells contributed more to total biomass than to primary production. Flagellates were the primary small cells, and it is likely that they were heterotrophs due to the minor contribution this size class made to primary production. Still, direct release, enhanced by the flagellates' small surface to volume ratio, and egestion are mechanisms that would have contributed to DON release.

The colonial, mat-forming diatom *Rhizosolenia* is one genera of algae that has received considerable attention (e.g., Villareal et al., 1993, 2011). *Rhizosolenia*, and other large cells including *Pyrocystis* spp. and *Ethmodiscus*, are able to persist in oligotrophic regions because they are capable of vertically migrating into the nitracline to fill their large internal vacuoles with NO_3^- (Singler and Villareal, 2005; Villareal et al., 1999). Cellular carbohydrate levels may influence the buoyancy of the mats, with negatively buoyant mats observed to have higher C:N and carbohydrate: protein ratios (Villareal et al., 1996). Once the mats are at the surface, NO_3^-, NO_2^-, NH_4^+ and presumably DON may be released by the cells either directly or when grazed during sloppy feeding. Vertically migrating mats from the oligotrophic North Pacific have been shown to release 27.1 ± 8.3 and 3.6 ± 1.9 nmol N $(\mu g\, Chl a)^{-1} h^{-1}$ as NO_3^- and NH_4^+, respectively (Singler and Villareal, 2005). Pilskaln et al. (2005) estimated the upward flux of N by large and small mats to be ~50 μmol N m^{-2} day^{-1} in the North Pacific subtropical gyre, which is significant when compared to an estimate of turbulent NO_3^- flux of 200 μmol N m^{-2} day^{-1} (Richardson et al., 1998).

2 N₂ Fixers

N_2 fixers can be a significant source of new N to tropical and subtropical marine systems in which they occur (Karl et al., 2002). Traditionally the non-heterocystous colonial cyanobacterium, *Trichodesmium*, has been considered the primary

N_2 fixer in the ocean. Over the past decade, however, single celled N_2 fixers were discovered in the subtropical Pacific whose importance is still being evaluated (Montoya et al., 2004; Zehr et al., 2001).

Trichodesmium is known to contribute to N fluxes in marine systems both directly, through the release of amino acids, DON, and NH_4^+, and indirectly through regeneration of DIN and DON by bacteria and grazers living in association with the colonies (Mulholland, 2007; Mulholland et al., in press; Sheridan et al., 2002). *Trichodesmium* in the Caribbean Sea and Atlantic Ocean can release up to half of the recently fixed N_2 as DON during growth (Glibert and Bronk, 1994), primarily as DFAA (Capone et al., 1994). In the Gulf of Mexico, *Trichodesmium* released 40-51% of recently fixed N with approximately one quarter to one half of this release as NH_4^+ and DON presumably making up the difference (Mulholland et al., 2006). Rates of N_2 fixation and DON release produced from recently fixed N were measured in <10 µm and >10 µm size fractions (pre-fractionated prior to the $^{15}N_2$ gas addition) in surface waters along a transect across the Atlantic at ~24.5° N (Benavides et al., 2013b). DON release rates ranged from 0.001 to ~0.09 nmol N L^{-1} h^{-1} and there was no significant difference between the size fractions. In the <10 µm and >10 µm fractions ~23 and ~14% of gross N_2 fixation was released as DON respectively (Benavides et al., 2013b). There was no clear trend in DON release with longitude along the transect, in contrast to rates of N_2 fixation, which were higher at the western stations. There were also no significant day night differences in rates of N_2 fixation or DON release.

Elevated DON concentrations observed within *Trichodesmium* blooms in the subtropical North Pacific at station ALOHA (Karl et al., 1992; Letelier and Karl, 1996) provides additional evidence of DON release by *Trichodesmium*, which is likely attributed to enhanced grazing and/or viral infection within the blooms. *Macrosetella gracilis* is one of the few known

grazers of *Trichodesmium* that can release NH_4^+ and likely DON during sloppy feeding and excretion (O'Neil and Roman, 1992; O'Neil et al., 1996). A notable characteristic of *M. gracilis* is that they do not make solid fecal pellets when they ingest *Trichodesmium* such that much of the N they release is dissolved in form. With respect to viruses, populations of *Trichodesmium* in the Caribbean Sea and cultured populations of *Trichodesmium* NIBB1067 have been shown to have lysogenic phages that can result in cell lysis and DOM release (Ohki, 1999). In the tropical North Pacific an estimated 0.3-6.5% of *Trichodesmium* trichomes were lysed by viruses each day (Hewson et al., 2004). In addition, an estimated 0.32-15 nmol N colony^{-1} h^{-1} is released by bacterial peptidase and β-glucosamidase activity associated with *Trichodesmium* colonies in the subtropical Atlantic Ocean (Nausch, 1996).

Though studies are limited, N_2 fixers in culture tend to release a smaller percentage of their recently fixed N as DON. The unicellular diazotroph *Cyanothece* released <1% of the total N_2 fixed (Benavides et al., 2013a) and *Trichodesmium* IMS101 released ~8% (Mulholland et al., 2004). These low DON release percentages in culture support the importance of grazers, and perhaps viruses, as mediators of release. In a field study that included single celled N_2 fixers, Benavides et al. (2011) used the difference between gross N_2 fixation, measured by acetylene reduction, and net N_2 fixation, measured by $^{15}N_2$ uptake, to estimate DON release in the subtropical northeast Atlantic. A range of 41-76% of gross N_2 fixed was lost as DON in their study. However, because of the apparent inclusion of NH_4^+ regeneration in their DON calculations their rates are likely overestimates (Mulholland et al. 2004; see Section IV.C below).

3 Bacteria

Bacteria contribute to DON release via active release of enzymes, passive diffusion through cell membranes, remineralization of particulate organic matter and trophic mediated release when affected by bacterivory or viral infection (see review by Azam and Malfatti, 2007). Carlsson and Granéli (1993) and Carlsson et al. (1993) provided evidence that DON is utilized by phytoplankton indirectly through bacterial incorporation of the DON, and then release of NH_4^+ during bacterivory. Bacteria are also known to release DON through the mineralization of organic aggregates (Smith et al., 1992). In another study, Jørgensen et al. (1999) documented urea release in bioassays in the Gulf of Riga. Therkildsen et al. (1997) measured urea production in cultures of two marine bacteria, and found that the highest accumulation occurred during the growth deceleration phase and at the beginning of stationary phase. They argued that internal RNA provides the precursors for the release of urea. During growth, the intracellular RNA pool is dynamic with some portion of it being degraded to urea and NH_4^+, particularly during exponential growth (Mason and Engli, 1993). Urea is also produced as a result of use of nucleotide bases (pyrimidines, xanthine, hypoxanthine, adenine, and guanine) as demonstrated in dark bioassays with water from the coast of Denmark (Berg and Jørgensen, 2006). Like Therkildsen et al. (1997) they found that urea production was highest during growth deceleration. They also found that urea was produced at a faster rate in treatments containing guanine, hypoxanthine and xanthine compared to the other nucleotide bases (Berg and Jørgensen, 2006).

Bacteria are capable of releasing amino acids and there is an abundant literature in biotechnology related to bacterial amino acid and protein secretion (e.g., Harwood and Cranenburgh, 2008) but in marine systems much less is known. Kawasaki and Benner (2006) measured the release of D-enatiomers of amino acids by bacteria collected from several different aquatic environments and found that D-alanine, D-aspartic acid, and D-glutamic acid were primarily released during cell growth and cell decline. Specifically, D-alanine was released in relatively high concentration during exponential growth when peptidoglycan bonds were cleaved for cell division. More recently, Kaiser and Benner (2008) measured bacterial release of DON using D-amino acids and muramic acid as biomarkers and found that D-amino acids are derived from numerous macromolecules in addition to peptidoglycan. This study also suggested that bacterial organic matter accounts for ~50% of DON in the ocean.

Bacterial release of DON varies by region suggesting that different processes control DON release. Varela et al. (2005) measured DON release at numerous sites across the Atlantic Ocean and found that the highest rates of DON release occurred in oligotrophic waters. Though advances have been made in the past decade to understand bacterial DON release, the exact composition of the DON that is released is still relatively unknown (Kujawinski, 2011).

4 Micro- and Macrozooplankton

Zooplankton contribute to DON release. Microzooplankton, such as flagellates and ciliates, release DON via excretion and egestion. Macrozooplankton, such as copepods, can produce DON via dissolution of fecal pellets, excretion, and sloppy feeding (Figure 4.4; reviewed in Bronk and Steinberg, 2008 and Steinberg and Saba, 2008) though most of the work has focused on DOC. The processes of grazing and bacterivory are known to be important in transforming and partitioning C and N within planktonic systems (reviewed in Nagata, 2000; Strom, 2000). These processes can result in DON release via sloppy feeding (Møller, 2007; Saba et al., 2011a) or bacterivory by microzooplankton where undigested material is egested resulting in release of dissolved and colloidal intracellular materials. DON release can also result from excretion, where DON compounds are released as waste (Conover and Gustavson, 1999; Miller and Glibert, 1998), or via diffusion away from fecal pellets or fecal pellet dissolution (Jumars et al., 1989; Saba et al., 2011a). In addition to

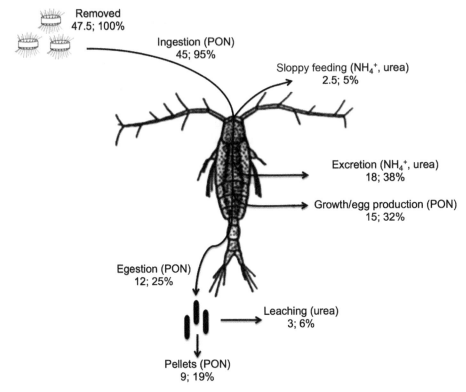

FIGURE 4.4   Nitrogen flow through the copepod *Acartia tonsa* during feeding on the diatom *Thalassiosira weissflogii*. The first value shown is the calculated average rate of release or assimilation (ng N individual^{-1}h^{-1}) from which the percentage of nitrogen removed from suspension was calculated (second value). *From Saba et al. (2011a), which was modified from Møller et al. (2003) and Steinberg and Saba (2008).*

being a source of DON, zooplankton that migrate vertically can contribute to active flux of DON through the water column. A significant proportion of the TDN excreted by zooplankton that migrate to the mesopelagic is excreted as DON. In the equatorial Pacific, 46-60% of TDN excreted by vertical migrators was as DON in a high nutrient low chlorophyll region and an oligotrophic region (Le Borgne and Rodier, 1997). In the Sargasso Sea 32% of the TDN excreted by migrators was as DON (Steinberg et al., 2002).

Microzooplankton tend to ingest whole cells, such that the release of DOM during egestion is relatively minor, although some workers have found that microzooplankton enhance DOM fluxes as well as influence the composition of the DOM pool (Flynn and Davidson, 1993; Strom et al., 1997). In a number of studies reviewed in Carlson (2002), 9±6% (n=5) of the N ingested by a number of microzooplankton species is released as DFAA or DFAA and DCAA. Ferrier-Pagès et al. (1998) measured release of DPA from two flagellates (*Strombidium sulcatum*, ciliate, and *Pseudobodo* sp., aplastidic flagellate) grazing on bacteria in culture. They found high rates of DPA release during the exponential phase, when ingestion rates were at a maximum, but lower rates during the other growth phases. In another bacterivory study, Alonso et al. (2000) found rates of up to 281 μg DNA L^{-1}h^{-1} in bacterial cultures with the ciliate *Uronema* sp. and the flagellate *Pseudobodo* sp. In the Adriatic Sea, concentrations of D-DNA and nanoflagellates were found to co-vary during a study of diel dynamics, and D-DNA release rates increase sixfold when the nanoflagellate, *Ochromonas* sp., was present (Turk et al., 1992).

Sloppy feeding by macrozooplankton is responsible for the generation of significant amounts of labile DOM (Dagg, 1974; Urban-Rich, 1999). Vincent et al. (2007) used ^{15}N tracer techniques to quantify DON release from two diatoms to two copepods. They found that direct excretion from the zooplankton was the largest source of release (64-79% of total N ingested) with the remainder coming from the diatoms directly via direct release or sloppy feeding. Saba et al. (2011a) used a similar approach and found that the amount of DOC, NH_4^+, and urea released by *Acartia tonsa* after feeding on *Thalassiosira weissflogii* varied by mode of release (Figure 4.4). DOC production and NH_4^+ release was greatest by excretion and accounted for 80% of total DOC release and 93% of total NH_4^+ release, respectively. In comparison, sloppy feeding accounted for 20% of total DOC release and 7% of total NH_4^+ release. Of total N measured (NH_4^+ plus urea), urea from fecal pellet leaching accounted for 62% while sloppy feeding accounted for 25%. Urea accounted for 20% of total NH_4^+ plus urea released from copepods.

Macrozooplankton release $13 \pm 12\%$ ($n = 11$) of their body N as DON, often measured as DFAA, urea, or DCAA (Carlson, 2002). In a number of other studies, $25 \pm 12\%$ ($n = 6$) of the TDN released is organic in form (Carlson, 2002). The relative coupling of C and N during this release, however, is poorly known. Daly et al. (1999) suggested that C and N release by macrozooplankton results in quantitatively different fluxes of DOC and DON, which ultimately results in non-Redfieldian, C-rich ratios within those pools. It is important to note that the stoichiometric ratio of what is released may be very different from the ambient DOM pool, both in

the surface and even more so deeper in the water column. The "microbial carbon pump" has been proposed as a mechanism by which bacteria transform labile DOM to a C-rich recalcitrant state that persists. The selective removal of N and phosphorus ultimately results in recalcitrant DOM with a very high DOC:DON (or DOP) (e.g., Azam and Malfatti, 2007; Jiao et al., 2010 and associated correspondence).

Copepod diet may play an important role in determining the quantity and composition of regenerated C and N available to the microbial loop. Saba et al. (2009) found that *Acartia tonsa* release rates of DOC and NH_4^+ were highest on a carnivore diet and lowest on an omnivore diet. Urea was 32-59% of total N released on an omnivore diet, much higher than on a carnivore diet. Additionally, grazing and the subsequent form and rate of N regeneration can be affected by the presence of HABs. In one study, DON release by zooplankton was higher when they fed on toxic algal cultures suggesting that exposure to karlotoxin may have increased membrane permeability (Saba et al., 2011b).

It has been hypothesized that the majority of DOM diffuses away from fecal pellets within minutes of its release, assuming that the peritrophic membrane surrounding the pellet is permeable (Jumars et al., 1989). The mucus coating of some fecal pellets, however, may restrict this avenue of exchange, and some studies suggest that fecal pellets must be broken up mechanically before a substantial amount of DOM is released (Lampitt et al., 1990; Strom et al., 1997). In terms of DOC, leaching from fecal pellets increases when copepods are fed higher food concentrations (Møller et al., 2003). Copepods feeding on dinoflagellates also generate fecal pellets that release more DOC than copepods fed on diatoms (Thor et al., 2003). Urban-Rich (1999) measured abiotic and biotic release of C from copepod fecal pellets using ^{14}C tracers in the Greenland Sea, and found that 86% of the DOC is utilized or leached from fecal pellets in the first 6 h. Roy and Poulet (1990) found a rapid decrease in copepod fecal pellet total DFAA concentration within the first 3-5 days. In addition to passive leaching, DOC and DFAA can be rapidly released from fecal pellets due to bacterial degradation processes and hydrolysis of proteins (Roy and Poulet, 1990; Urban-Rich, 1999). Shorter-term incubations were not done, however, that could address the issue of time-scale raised by Jumars et al. (1989).

Roy et al. (1989) investigated the release of DFAA during sloppy feeding by two copepod species fed a large diatom, known to have large intracellular DFAA pools. Although the presence of particulate debris indicated that sloppy feeding occurred, total DFAA concentrations and extracellular composition were not significantly different after copepod grazing. This suggests that the DFAAs released were rapidly taken up by bacteria or other cells or were adsorbed onto debris produced by the copepods. Fuhrman (1987) found that combinations of copepods and microplankton had a much higher release rate of DFAA than either alone. He also found that release of DFAA due to sloppy feeding by copepods or excretion can be similar in magnitude to direct release by microzooplankton.

In estuarine mesocosms, rates of urea and DPA release from the copepod, *Acartia tonsa*, tend to be higher during early morning and early evening and are of the same order of magnitude as NH_4^+ regeneration (Miller and Glibert, 1998). Micrograzers are important agents of DON release and bacteria rapidly consume 58-103% of the recently released DON in Japanese coastal waters (Hasegawa et al., 2000b). The magnitude of DON release is ~40% of the rate of NH_4^+ regeneration in the Southern California Bight and Monterey Bay (Bronk and Ward, 1999) and 59% of NH_4^+ regeneration in Japanese coastal waters (Hasegawa et al., 2000b). A tight positive correlation between rates of DON release and NH_4^+ regeneration ($p < 0.001$) has also been observed, suggesting that the same process is likely responsible for both DON and NH_4^+ regeneration. For example, Bronk and Steinberg (2008) plotted

56 pairs of total DON release and NH_4^+ regeneration and observed a slope of 0.237 ($r^2 = 0.51$).

Gelatinous zooplankton directly release DON via excretion or leaching from fecal pellets or via sloppy feeding. The DOM released from jellyfish can be both colloidal and dissolved. Condon et al. (2011) found that gelatinous zooplankton in Chesapeake Bay released large quantities of extremely labile DOM with a mean C:N ratio of 25.6 ± 31.6. The labile C-rich DOM represents a substrate that supports bacterial respiration rather than production thus limiting the passage of organic matter to higher trophic levels.

5 Viruses

Viruses have been found to infect both phytoplankton and bacteria and recently have been identified in a freshwater zooplankton (Frischer et al., 2010). Phage, viruses that infect bacteria and cyanobacteria, are common but the number of identified viruses that infect eukaryotic phytoplankton is relatively few (Bidle and Falkowski, 2004; Brussaard, 2004). Viruses can affect DON release in three ways. First, when cells burst in the final stage of lytic infection they release the dissolved cytosol. While estimates of viral-mediated mortality exhibit a wide range, studies indicate that ~20-30% of heterotrophic bacteria production is lysed daily due to viral infection (Suttle, 2005). In contrast, an estimated ~10-40% of phytoplankton production is lysed daily due to viral infection with extreme events that account for 100% mortality (reviewed in Brussaard, 2004). Second, because of their small size (20-200 nm; Brussaard, 2004), viruses are part of the DON pool. Virus concentrations in aquatic systems range from 10^9 to 10^{11} viral particles L^{-1} (Suttle, 2005; Wommack and Colwell, 2000). Due to their high abundance, viruses contribute a large fraction of dissolved DNA in seawater; at station ALOHA in the North Pacific subtropical gyre, viruses contributed 49-63% of the total dissolved DNA pool (Brum, 2005). Third, viruses can exert important selective controls over microbial community structure infecting some plankton groups but not others and thus yielding microbial communities that are more or less susceptible to grazing or sloppy feeding, for example (reviewed in Breitbart, 2012; Rohwer and Thurmer, 2009; Suttle, 2005; Wommack and Colwell, 2000).

Lysis of phytoplankton has also been observed in the northwestern Mediterranean Sea where lysis rates were ~50% of gross phytoplankton growth in surface waters but only 7% at the deep chlorophyll maximum (Agusti et al., 1998). The current understanding is that when grazing rates and sinking rates are low, lytic infection and viral lysis rates can be high. When cells are not grazed or do not rapidly sink, the enhanced cell density of a population increases the encounter rate and potential for viral infection. Another study in the Mediterranean found a strong correlation between phytoplankton lysis rates and temperature where cell lysis was highest in summer and lowest in winter (Agusti and Duarte, 2000). There may be reasons to speculate the exact magnitude of measured lysis rates, however, the general patterns discussed below appear robust.

Numerous studies have been conducted to examine the effects of viral lysis on the flux and composition of the DOM pool (e.g., Middelboe and Lyck, 2002; Miki and Yamamura, 2005).

In mesocosm experiments, it was found that phage lysis products stimulate bacterial growth and metabolic activity, indicating that viral lysis can contribute to the "viral loop" in which a portion of available nutrients cycle between the DOM pool and bacterial production without passing to higher trophic levels (Middelboe et al., 1996). The ability of viruses to change DOM composition to recalcitrant material has also been demonstrated. Middelboe and Jorgensen (2006) used a model system of the marine bacterium (*Cellulophaga* sp.) and a virus specific to it to quantify the release of DFAA, DCAA, and compounds derived from bacterial cell walls following viral lysis. The study showed that ~83% of the peptidoglycan-derived D-DFAA released

was late in the incubations, representing a potential source of recalcitrant DOM.

Viral infection can also impact DON release in the surface ocean via their effect on sinking rates of phytoplankton. In one study, viral infection of *Heterosigma akashiwo* resulted in cells having higher sinking rates (Lawrence and Suttle, 2004). If this occurs in areas with a shallow mixed layer, the higher sinking rates could cause cells to sink deeper in the water column such that the N released upon lysis would no longer be available in the upper water column to support productivity (Lawrence and Suttle, 2004).

As an interesting twist on phytoplankton DOM release, Murray (1995) used a modeling approach to suggest that DOM exudation can be a cost-effective, indirect means of reducing viral infection in phytoplankton. The reasoning is that DOM exudation supports bacterial populations that can remove viruses, which the bacteria contact with much greater frequency because of their small size. If DOM exudation leads to a reduction in phytoplankton losses from infection greater than the cost of the exudation, then DOM exudation is a cost effective strategy.

Fungal infection also has the potential to affect DON release. Most work to date in this area has been conducted with freshwater phytoplankton species (e.g., Holfeld, 1998). Just as with viruses, fungal infections have been implicated in bloom crashes and in controlling the composition of the phytoplankton community (Park et al., 2004). One difference between viral and fungal infection is the size of the phytoplankton at risk. A number of studies show viral infection of smaller phytoplankton (chrysophytes and cyanobacteria) but not of the larger dinoflagellates and diatoms (Brussaard, 2004). In contrast, larger cells are more often infected by eukaryotic parasites, such as fungi (Park et al., 2004).

Zooplankton are also at risk of infection, both viral (Frischer et al., 2010) and in the form of epibiotic associations (reviewed in Carman and Dobbs, 1997). Negative effects associated with infection can include the development of lesions and loss of DOM (Gomez-Gutierrez et al., 2003; Park et al., 2004).

B Allochthonous Sources

Allochthonous sources of DON include rivers, ground water, and atmospheric inputs. We briefly discuss each below.

1 Rivers

Several comprehensive studies have been completed to quantify the flux of riverine N delivered to the ocean. A review of the literature indicates that riverine DON concentrations average $23.8 \pm 18.1 \, \mu mol \, N \, L^{-1}$, comprise $57.7 \pm 23.7\%$ of the TDN pool, with a C:N ratio of 32.5 ± 16.3 (Table 4.1). Seitzinger and Harrison (2008) estimated that $23.81 \, Tg \, N \, year^{-1}$ (TN, includes DON, DIN, and PN) is delivered to the coastal ocean by the world's largest 25 rivers, $5.02 \, Tg \, N \, year^{-1}$ of which is DON. Wiegner et al. (2006) found that DON accounted for 8-94% of the TDN pool for nine rivers in the eastern United States. A study of the Mississippi-Atchafalaya River Basin revealed that 24% of the annual total N flux was DON (Goolsby et al., 2001). Lobbes et al. (2000) studied 12 Russian Rivers to determine the biogeochemical composition of organic matter including DON that enters the Arctic Ocean. DON concentrations were similar to PON, ranging from 6.7 to $17.9 \, \mu mol \, N \, L^{-1}$ (mean of 11.5). The study did not calculate export rates of DON specifically, but they did calculate that these rivers were responsible for 38% of the total DOC discharge of the Arctic Rivers. These studies provide evidence that DON can be a substantial fraction of total N in rivers and should be included in N loading budgets to coastal waters.

The fate, bioavailabilty, and concentration of riverine DON once it reaches the ocean is still unclear and likely varies between rivers (reviewed in Dagg et al., 2004; see also riverine DOM review by Raymond and Spencer in this volume). Organic matter found in rivers is typically derived from plant material, and its composition reflects the properties of the soil, which is strongly influenced by the vegetation and climate of the region (Lobbes et al., 2000). Pellerin et al. (2004) looked at how land usage impacts the amount of DON present in rivers. Over 100 rivers in the northeastern United States were analyzed to determine what factors control DON concentrations in rivers and streams. The study found that the percentage of wetlands that surround a watershed can explain 60% of the variability of DON concentrations, and that watersheds with abundant wetlands have higher DON concentrations and DON:TDN ratios. Wiegner et al. (2006) established that ~23% of riverine DON was bioavailable to heterotrophic bacteria in rivers in the eastern United States, suggesting that a significant fraction of riverine DON is available once it reaches the coast.

Changes to hydrological cycles may influence the delivery of DON to the coast. The Pan-Arctic River Transport of Nutrients, Organic Matter, and Suspended Sediments (PARTNERS) project was initiated in the early 2000s to measure the biogeochemistry of the largest Arctic Rivers. Holmes et al. (2012) estimated DON flux data from the PARTNERS project and conclude that previous estimates for some Arctic Rivers (Gordeev et al., 1996) may have been two to three times too high. Tank et al. (2012) used the flux data from Holmes et al. (2012) to determine the distribution and availability of riverine N to the Arctic Ocean and discovered that riverine DON is available for uptake far into the open ocean because it is remineralized slowly. They estimate that riverine N supports $0.5-1.5 \, Tmol \, C \, year^{-1}$ of primary production that occurs in the Arctic Ocean; however, this represents only a small fraction of total production in the entire Arctic. The study concluded that rates of riverine DON regeneration in nearshore N-limited regions lead to highly localized rates of photosynthesis and account for a large portion of coastal production.

There has recently been a fair bit of research focused on DON in streams. Using data from

850 river stations, Scott et al. (2007) calculated long-term mean-annual and interannual loads of DON and found that DON was the dominant N pool within rivers throughout most of the United States in both pristine streams and in streams with high anthropogenic impacts. Johnson et al. (2013b) measured DON production in 36 head-water streams 24 h after a ^{15}N-NO$_3^-$ tracer addition and detected DON production in 15 of the streams. They also found that streams in areas with a higher percentage of the land modified for agricultural or urban use were more likely to have measurable DON production. Overall, production of DON was a median of 8% of total NO$_3^-$ uptake after 24 h and was positively correlated with ecosystem respiration. This study confirms that N cycling processes in headwater streams can quickly convert NO$_3^-$ to DON with rates similar in magnitude to stream denitrification and nitrification.

2 Groundwater

Submarine groundwater discharge (SGD) studies have mostly focused on NO$_3^-$, largely due to the legal regulations limiting the amount of NO$_3^-$ in drinking water. The relatively few studies that have looked at the reduced forms of N in groundwater find that DON is a significant component. Kroeger et al. (2007) found that N release via groundwater into Tampa Bay, FL is mostly in the reduced forms of NH$_4^+$ and DON. Another study in the northeastern Gulf of Mexico estimate that 27.7% of the groundwater-derived total N-load was in the form of DON, and that land-derived DON makes up ~52% of the total submarine groundwater discharge for the Gulf of Mexico (Santos et al., 2009). Concentrations of DON in aquifers on the east coast of the United States range from 0 to 107 µmol N L^{-1} with means for most studies in the 10-20 µmol N L^{-1} range (Joye et al. 2006).

Groundwater can be a significant source of DON to the ocean. The flux of SGD from a volcanic island in Hwasun Bay, Jeju, Korea, was estimated to be 1.3×10^5 mol DON day^{-1},

a value two times greater than DON released from bottom sediment diffusion and on par with DON discharge from major rivers (Kim et al., 2013). The total SGD DON concentrations (50 ± 27 µmol N L^{-1}) approximated those observed in Tampa Bay (Kroeger et al., 2007). This work demonstrated that DON from SGD mixes conservatively with seawater indicating that DON SGD may be transported farther offshore than DIN forms like NO$_3^-$.

In a large study of DON release via groundwater, measurements were made in 10 watersheds on Cape Cod, MA (Kroeger et al., 2006). They found that rates of both TDN and DON export increased as population within the watershed increased, and that DON was often the main form of N exported. These results are important because DON in groundwater is frequently assumed to be natural, while NO$_3^-$ is due to anthropogenic modification. This study provides evidence that a significant portion of N introduced to watersheds via human activity is exported in groundwater as DON.

3 Atmospheric Deposition

Atmospheric deposition is another source of DON to marine systems (Cornell et al., 1995; for more complete reviews on organic N in the atmosphere see Cape et al. (2011), Cornell (2011), and Jickells et al. (2013)). The source of atmospheric deposition can be local and there are often gradients in concentration from land to sea with evidence for long-range atmospheric transport (Cornell et al., 2003). Several recent studies have shown that water soluble organic N (WSON), once thought to be a minor contributor, becomes DON if it enters the ocean and can account for ~30% of the total fixed N deposition (Duce et al., 2008). Organic N concentrations in rainwater collected in marine locations are ~5 µmol N L^{-1} (Cornell, 2011). Marine/coastal atmospheric organic N is comprised of a complex mixture of compounds including urea (e.g., Cornell et al., 2001; Mace et al., 2003), amines (e.g., Facchini et al., 2008), amino acids and peptides

(Kieber et al., 2005; Mace et al., 2003; Matsumoto and Uematsu, 2005), amides (Hawkins and Russell, 2010), nitro-polycyclic aromatic hydrocarbons (PAHs; Tsapakis and Stephanou, 2007), humic-like substances (Graber and Rudich, 2006), and yet unidentified compounds (Altieri et al., 2009, 2012). DFAA can account for as much as 50% of WSON (Mace et al., 2003). Glycine is the most abundant amino acid in marine WSON accounting for ~40-60% of the DFAA or DCAA pools (Mandalakis et al., 2011; Matsumoto and Uematsu, 2005). Atmospheric urea concentrations are more variable and represent ~2% to ~50% of the WSON (Cornell et al., 2001; Mace et al., 2003). Both amino acids and urea are bioavailable to marine phytoplankton and thus, would directly support marine phytoplankton growth. Approximately 20-75% of the atmospheric organic N is bioavailable to coastal or estuarine microbial communities on relatively short (hours to days) time scales (Peierls and Paerl, 1997; Seitzinger and Sanders, 1999).

Dry deposition may contribute more WSON than wet (rain) deposition. A study of atmospheric DON deposited in the Eastern Mediterranean showed that annual dry deposition concentrations were over three times greater (17.4 mmol Nm^{-2}) than wet deposition (~4.8 mmol Nm^{-2}) concentrations (Violaki et al., 2010). This difference is also reflected in the percent contribution of DON to the overall pool with 22.7% being deposited as DON during the wet season compared to 38.6% during the dry season (Violaki et al., 2010). They also suggested that of the new production stimulated by atmospheric deposition 20-30% could be attributed to atmospheric deposition of DON (Violaki et al., 2010).

Continental atmospheric organic N concentrations are higher (7.2-117 µmol N L^{-1}; Cornell, 2011) and appear to be much more complex at the compound level (Altieri et al., 2009) than more remote marine locations like Bermuda (Altieri et al., 2012). There also appear to be seasonal patterns. For example WSON concentrations over the East China Sea and western North Pacific are much higher (54 ± 36 nmol N L^{-1}) in the fall than in the spring (16 ± 19 nmol N L^{-1}; Nakamura et al., 2006). Spatially, atmospheric DON concentrations are greatest in air masses that have recently crossed over landmasses, but this complexity may reflect, in part, anthropogenic atmospheric inputs. A global three-dimensional chemistry transport model simulation of the present-day atmospheric flux of organic N to the global ocean suggests that anthropogenic N sources accounts for ~40% of the oceanic organic N deposition, while oceanic sources only account for ~33% (Kanakidou et al., 2012). This and similar studies (Duce et al., 2008) show that atmospheric deposition is an important and potentially increasing (Cornell, 2011) contributor to the ocean N budget, which can support up to 3% of new biological production in marine environments (Duce et al., 2008).

Furthermore, aerosols originating in biologically productive marine systems appear to be enriched in organic N, suggesting that atmospheric exchange is a potential sink for marine DON (Miyazaki et al., 2011). No clear trend of increasing DON concentrations in rainwater was observed in a review of DON concentrations over nearly a century, though the authors raise the issue of potential problems with data comparability over the years (Cornell et al., 2003).

C Methods to Estimate Rates of Autochthonous DON Release

Rates of DON release are measured using two general approaches—monitoring changes in concentrations through time or the use of ^{15}N tracer techniques (see McCarthy and Bronk, 2008). In the first approach, the measure of DON or individual organic compounds over time provides a net DON release rate. In the second approach, ^{15}N tracers are used to assess the gross DON release rate. Using the tracer approach, the DON must be separated from the inorganic N substrate added at the start of the experiment (NH_4^+, NO_3^-, or NO_2^-). Currently, three general ways to isolate DON have been used in tracer studies—wet chemistry, ion retardation, and dialysis. Since the last edition of the book there has been a depressing lack of progress in new methods to isolate DON in this regard. As such the earlier edition of this chapter (Bronk, 2002) will still serve as a useful introduction to the approaches and problems. Here we give a brief review and describe a few new approaches that have been more recently applied.

Wet chemical isolation is still the method most commonly used in which NH_4^+ is removed by raising the pH slightly, thus effecting a change from soluble protonated NH_4^+, to the more volatile NH_3 via diffusion in a heated oven (Slawyk and Raimbault, 1995) or vacuum distillation (Bronk and Ward, 1999). Nitrate in the sample is converted to NH_3 with DeVarda's alloy, with the NH_3 again removed through volatilization (Bronk and Ward, 1999; Slawyk and Raimbault, 1995). There is a thorough discussion of the sequential diffusion approach and the subtleties of the calculations in Raimbault and Garcia (2008). All of these techniques run the risk of hydrolyzing labile DON because the basic conditions used remove NH_4^+ and NO_3^-/NO_2^- from solution (Bronk and Ward, 2000). This same approach is actually used as a way of removing labile DON before isolation of NO_3^- (Sigman et al., 1997). If this approach is applied, analysts and reviewers should insist that qualifying statements be made about the likelihood that the rates measured are underestimates due to the hydrolysis of labeled DON to NH_4^+.

Sadly, there is still no reliable source for ion retardation resin (BioRad AG 11 A8). The resin quantitatively removes salts including NH_4^+, NO_3^-, and NO_2^- allowing DON to be isolated in the eluate (Bronk and Glibert, 1991, 1993a and b; Hu and Smith, 1998; Nagao and Miyazaki, 1999). One can make AG 11 A8 resin by chemically altering another resin (Dowex anion exchange resin, BioRad AG1-X8; Hatch et al., 1957). However, our own experience trying to make reliable resin has been disappointing in most cases. Dialysis has also been used to isolate DON but remains challenging in saltwater. An approach for freshwater samples uses rotary evaporation to preconcentrate the DON followed by dialysis (100 Da cutoff) to remove DIN (Feuerstein et al., 1997).

Benavides et al. (2013a and b) used a mass balance approach to estimate DON release. They measured the ^{15}N enrichment of the NH_4^+ pool following solid phase extraction (described in Bronk et al., in press) and the ^{15}N enrichment of the NO_3^-/NO_2^- pool using the denitrifier method (Sigman et al., 2001). The TDN pool was persulfate oxidized to NO_3^- and then again analyzed with the denitrifier method. The final ^{15}N enrichment of the DON pool was estimated by difference. Mulholland et al. (2004) estimated N release from N_2 fixers by taking the difference between gross N_2 fixation, measured using acetylene reduction, and net N_2 fixation, measured using $^{15}N_2$ uptake. The resulting N release rate is not specific to DON release, however, as it would also include N lost via NH_4^+ regeneration.

Measurements of urea release rates are still relatively rare. Uptake and regeneration rates of urea appear to be tightly coupled such that trying to estimate urea regeneration rates by measuring changes in urea concentration is problematic. That leaves tracer approaches but these are labor intensive. Hansell and Goering (1989) used a dual ^{15}N and ^{14}C approach to measure urea regeneration. The principle of isotope dilution can also be used to estimate regeneration of urea (Slawyk et al., 1990). In this approach, urea is first converted to NH_4^+ using urease. The NH_4^+ is then isolated and the analysis proceeds as a regular NH_4^+ regeneration analysis (reviewed in McCarthy and Bronk, 2008).

D Literature Values of DON Release Rates in Aquatic Environments

DON release rates can vary widely within and between systems (Table 4.3). In most cases, rates were measured using whole water samples, and the specific processes that produced the measured rates are unknown.

TABLE 4.3　Literature Values of Rates of DON Release and Ratios of DON Release to Gross N Uptake, Expressed as a Eercent, Determined Using ^{15}N Tracer Techniques (^{15}N) and Wet Chemistry (WC).

Location	Date	Substrate	DON release rate (nmol N l^{-1}h^{-1})	DON release: Gross N Uptake (%)	Method	Reference
OCEANIC—NH$_4^+$						
Bronk (2002) mean ($n=4$)		NH$_4^+$	9.5 ± 5.5	42.9 ± 10.3		Bronk (2002)
Atlantic—off Spain—oceanic	Apr 1997-Oct 1999[a]	NH$_4^+$	14.6 ± 22.6		^{15}N-WC	Varela et al. (2003)
Atlantic—off Spain—shelf break	Apr 1997-Oct 1999[a]	NH$_4^+$	2.2 ± 2.3		^{15}N-WC	Varela et al. (2003)
Atlantic—productive (chl> 0.25 µg L^{-1})	Aug 1998-Oct 2000	NH$_4^+$	1.4 ± 0.3	11.3	^{15}N-WC	Varela et al. (2006)
Atlantic—oligotrophic (chl <0.25 µg L^{-1})	Aug 1998-Oct 2000	NH$_4^+$	8.9 ± 2.6	14.4	^{15}N-WC	Varela et al. (2006)
North Atlantic Drift Providence	Sept-Oct 2000	NH$_4^+$	0.7 ± 0.2[b]	20.7 ± 11.0[c]	^{15}N-WC	Varela et al. (2005)
North Atlantic Subtropical Gyre Providence	Sept-Oct 2000	NH$_4^+$	0.2 ± 0.0[b]	24.7 ± 4.8[c]	^{15}N-WC	Varela et al. (2005)
East Canary Coastal Providence	Sept-Oct 2000	NH$_4^+$	0.2 ± 0.1[b]	15.8 ± 7.4[c]	^{15}N-WC	Varela et al. (2005)
Eastern Tropical Atlantic Providence	Sept-Oct 2000	NH$_4^+$	0.2 ± 0.0[b]	17.4 ± 4.1[c]	^{15}N-WC	Varela et al. (2005)
South Atlantic Subtropical Gyre Providence	Sept-Oct 2000	NH$_4^+$	0.3 ± 0.0[b]	22.7 ± 5.4[c]	^{15}N-WC	Varela et al. (2005)
Brazil Current Coastal Providence	Sept-Oct 2000	NH$_4^+$	0.1 ± 0.0[b]	14.2 ± 3.7[c]	^{15}N-WC	Varela et al. (2005)
Atlantic—North East	Feb-Mar 2001	NH$_4^+$	0.5 ± 0.6[d]			Fernández and Raimbault (2007)
Atlantic—North East	Mar-May 2001	NH$_4^+$	4.0 ± 5.7[d]			Fernández and Raimbault (2007)
Southern California Bight	Oct 1992	NH$_4^+$	0.1 to 38.5	0.5 to 89.0	^{15}N-WC	Bronk and Ward (2005)
Southern California Bight	Apr 1994	NH$_4^+$	0.8 to 8.8	6.4 to 63.2	^{15}N-WC	Bronk and Ward (2005)
Marquesas archipeligo, South Pacific	Oct 2004	NH$_4^+$	0.3 ± 0.3	3.7 ± 2.0[h]	^{15}N-WC	Raimbault and Garcia (2008)
HNCL, South Pacific	Oct 2004	NH$_4^+$	0.1 ± 0.0	1.3 ± 0.6[h]	^{15}N-WC	Raimbault and Garcia (2008)
South Pacific gyre—oligotrophic	Nov & Dec 2004	NH$_4^+$	0.5 ± 1.0	20.1 ± 7.4[h]	^{15}N-WC	Raimbault and Garcia (2008)
South Pacific gyre—Eastern boundary	Nov 2004	NH$_4^+$	1.6 ± 2.1	12.0 ± 7.4[h]	^{15}N-WC	Raimbault and Garcia (2008)
Chilean upwelling	Jul & Dec 2004	NH$_4^+$	2.0 ± 1.1	15.6 ± 14.1[h]	^{15}N-WC	Raimbault and Garcia (2008)
		*Mean ± std	4.3 ± 5.8	23.4 ± 14.8		

OCEANIC—NO₃⁻

Bronk (2002) mean ($n=5$)		NO_3^-	19.2 ± 28.8	41.8 ± 26.1		Bronk (2002)
North Atlantic Drift Providence	Sept-Oct 2000	NO_3^-	0.7 ± 0.3[b]	51.3 ± 30.1[c]	¹⁵N-WC	Varela et al. (2005)
North Atlantic Subtropical Gyre Providence	Sept-Oct 2000	NO_3^-	0.3 ± 0.1[b]	60.6 ± 18.7[c]	¹⁵N-WC	Varela et al. (2005)
East Canary Coastal Providence	Sept-Oct 2000	NO_3^-	10.8 ± 7.5[b]	31.3 ± 24.9[c]	¹⁵N-WC	Varela et al. (2005)
Eastern Tropical Atlantic Providence	Sept-Oct 2000	NO_3^-	4.3 ± 1.6[b]	47.6 ± 23.6[c]	¹⁵N-WC	Varela et al. (2005)
South Atlantic Subtropical Gyre Providence	Sept-Oct 2000	NO_3^-	0.6 ± 0.3[b]	39.5 ± 25.4[c]	¹⁵N-WC	Varela et al. (2005)
Brazil Current Coastal Providence	Sept-Oct 2000	NO_3^-	0.5 ± 0.3[b]	31.1 ± 21.1[c]	¹⁵N-WC	Varela et al. (2005)
Southern California Bight	Oct 1992	NO_3^-	1.1 to 29.2	26.3 to 98.5	¹⁵N-WC	Bronk and Ward (2005)
Southern California Bight	Apr 1994	NO_3^-	0.3 to 34.2	3.0 to 92.7	¹⁵N-WC	Bronk and Ward (2005)
Marquesas archipeligo, South Pacific	Oct 2004	NO_3^-	0.2 ± 0.2	11.3 ± 3.6[h]	¹⁵N-WC	Raimbault and Garcia (2008)
HNCL, South Pacific	Oct 2004	NO_3^-	0.1 ± 0.1	10.8 ± 6.1[h]	¹⁵N-WC	Raimbault and Garcia (2008)
South Pacific gyre—oligotrophic	Nov & Dec 2004	NO_3^-	0.0 ± 0.0	19.3 ± 10.1[h]	¹⁵N-WC	Raimbault and Garcia (2008)
South Pacific gyre—Eastern boundary	Nov 2004	NO_3^-	0.1 ± 0.1	11.3 ± 8.0[h]	¹⁵N-WC	Raimbault and Garcia (2008)
Chilean upwelling	Jul & Dec 2004	NO_3^-	1.1 ± 1.4	12.4 ± 7.1[h]	¹⁵N-WC	Raimbault and Garcia (2008)
*Mean ± std			6.5 ± 16.4	35.9 ± 20.9		

COASTAL—NH₄⁺

Bronk (2002) mean ($n=11$)		NH_4^+	75.6 ± 78.7	31.5 ± 26.0		Bronk (2002)
Southern California Bight	Oct 1992	NH_4^+	16.4 to 31.5	18.5 to 71.7	¹⁵N-WC	Bronk and Ward (2005)
Atlantic—off Spain—mid-shelf	Apr 1997-Oct 1999[a]	NH_4^+	6.1 ± 9.5		¹⁵N-WC	Varela et al. (2003)
Atlantic—off Spain—coastal	Apr 1997-Oct 1999[a]	NH_4^+	18.1 ± 26.6		¹⁵N-WC	Varela et al. (2003)

(Continued)

TABLE 4.3   Literature Values of Rates of DON Release and Ratios of DON Release to Gross N Uptake, Expressed as a Percent, Determined Using 15N Tracer Techniques (15N) and Wet Chemistry (WC)—cont'd

Location	Date	Substrate	DON release rate (nmol N l⁻¹ h⁻¹)	DON release: Gross N Uptake (%)	Method	Reference
Atlantic Shelf—high productivity region	Aug 1998-Oct 2000	NH_4^+	12.9 ± 2.3	18.2	^{15}N-WC	Varela et al. (2006)
Atlantic Shelf—low productivity region	Aug 1998-Oct 2000	NH_4^+	10.5 ± 3.1	23.6	^{15}N-WC	Varela et al. (2006)
Northwest Spain	Oct 1998-Sept 1999	NH_4^+	4.4 to 64.4	6.5 to 62.7	^{15}N-WC	Bode et al. (2004)
		*Mean ± std	50.8 ± 65.3	31.2 ± 22.4		
COASTAL—NO_3^-						
Bronk (2002) mean (*n* = 7)		NO_3^-	8.4 ± 6.4	48.7 ± 24.3		Bronk (2002)
Southern California Bight	Oct 1992	NO_3^-	0.3 to 1.4	5.0 to 35.6	^{15}N-WC	Bronk and Ward (2005)
		*Mean ± std	7.4 ± 6.5	45.1 ± 24.7		
ESTUARINE—NH_4^+						
Bronk (2002) mean (*n* = 6)		NH_4^+	73.7 ± 60.1	25.5 ± 6.7		Bronk (2002)
		*Mean ± std	73.7 ± 60.1	25.5 ± 6.7		
ESTUARINE—NO_3^-						
Bronk (2002) mean (*n* = 1)		NO_3^-	60.6	11.0		Bronk (2002)
		*Mean ± std	60.6	11.0		
		*Mean ± std for NH_4^+ uptake	29.1 ± 50.4	26.5 ± 17.1		
		*Mean ± std for NO_3^- uptake	8.8 ± 17.2	37.7 ± 22.3		
		*Mean ± std for oceanic sytems	5.3 ± 11.5	29.5 ± 18.9		
		*Mean ± std for coastal sytems	35.0 ± 55.7	36.3 ± 23.7		

*Mean ± std for estuatine sytems	71.8 ± 55.1	23.4 ± 8.2
*Mean ± std all measurements	21.3 ± 41.9	31.1 ± 20.0

[e]Calculated using raw data from manuscript and equations in Slawyk et al. (1998).

[f]<10 μm size fraction

[g]<202 μm size fraction

[i]Calculated from uptake and percent release data.

[j]Calculated with data from the <90 μm fraction so grazing is likely reduced or absent.

[k]Release mediated by grazers.

[l]Combined data for <10 μm and <202 μm.

*Means for each region include all data from Bronk (2002) as well as new data listed in table.

[m]Rates were measured in April 1997, August and September 1998, and October 1999.

[b]Rates were calculated by taking the mean euphotic zone integrated rate and dividing by the mean integration depth.

[c]Percentage was calculated using gross nitrogen uptake rates.

[d]Hourly rates calculated by dividing by daily rates by 24.

[e]Hourly rates calculated by dividing by daily rates by 24. All rates within euphotic zone.

[p]Data provided by authors. Hourly rates calculated by dividing by daily rates by 24.

Data were taken from text, tables, estimated from graphs, or obtained from the authors and are presented as the mean ± standard deviation. The means presented in the first edition of this chaper (Bronk, 2002) are presented at the start of each section. All data from the earlier edition (Bronk, 2002) and the more recent data presented here were included in the calculated mean ± standard deviation presented at the end of each section. When a range is given, the average of that range was used in the calculation of the mean ± standard deviation for that system.

1 Bulk DON

Estimates of DON release in the ocean have increased significantly since the last edition. Using ^{15}N tracer techniques, rates of DON release can be measured after NH_4^+ or NO_3^- incorporation; see review by Bronk and Steinberg (2008) for a more detailed analysis. Though the data are variable, overall rates of DON release resulting from NH_4^+ uptake were higher ($29\,nmol\,N\,L^{-1}h^{-1}$) than those resulting from NO_3^- ($9\,nmol\,N\,L^{-1}h^{-1}$) though the percentage of NH_4^+ uptake released as DON was lower (27%) compared to when NO_3^- was the substrate (38%; Table 4.3; Figure 4.5). This is consist with results from a north and south transect through the central Atlantic, where rates of DON release, resulting from NH_4^+ uptake, were higher in shelf waters and lower in offshore oceanic waters (Varela et al., 2006). At open oceanic sites, however, rates of DON release resulting

from NH_4^+ uptake were greater for oligotrophic sites (defined as areas with $<0.25\,\mu g\,Chl\,L^{-1}$) relative to productive sites (14% vs. 11%, respectively; Varela et al., 2006). Similarly in coastal sites, the percent of NH_4^+ released as DON was higher in low productivity waters relative to high productivity waters (23% vs. 18%, respectively; Varela et al., 2006). In contrast, however, other data collected along a coastal to ocean continuum of stations had higher rates of DON release at offshore stations relative to coastal (Varela et al., 2003).

The average rate of DON release in marine systems, from NH_4^+ or NO_3^- uptake, is $21\,nmol\,N\,L^{-1}h^{-1}$ (Table 4.3). Though highly variable, DON release rates tend to be lowest at oceanic sites (mean of $5\,nmol\,N\,L^{-1}h^{-1}$), intermediate in coastal areas (mean of $35\,nmol\,N\,L^{-1}h^{-1}$) and highest at estuarine sites (mean of $72\,nmol\,N\,L^{-1}h^{-1}$). An average of 31%

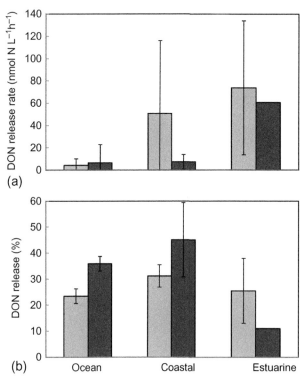

(a)

(b)   Ocean   Coastal   Estuarine

FIGURE 4.5 Means and standard deviations for data presented in Table 4.3 including (a) dissolved organic nitrogen (DON) release rates and (b) DON release as a percentage of gross nitrogen uptake. Means and standard deviations were calculated using data from all studies presented in Table 4 of Bronk (2002) and Table 4.3 in this chapter.

of gross N uptake is released as DON (Table 4.3). Gross N uptake is the amount of N taken into cells regardless of its ultimate fate, be it production of PN or DON release (Bronk et al., 1994). This is in contrast to the traditionally measured net N uptake that only includes production of PN (Bronk et al., 1994). DON release, as a percentage of gross N uptake, appears to be similar in oceanic and coastal environments with $29 \pm 19\%$ and $36 \pm 24\%$ of gross N uptake released as DON, respectively (Table 4.3). We are unaware of new DON release studies from estuarine environments. The data available suggest that release rates tend to be higher in estuarine environments, $72 \, \text{nmol} \, N \, L^{-1} h^{-1}$ with a mean of 23% of gross N being lost as DON (Table 4.3).

Vertical profiles of DON release rates also vary. In the Gulf of Lions, DON release was generally higher in surface waters and lower at depth (Diaz and Raimbault, 2000). DON production, resulting from NH_4^+ and NO_3^- uptake, was measured along a transect across the South Pacific gyre and then compared to rates measured in the upwelling region off Chile and in the waters around the Marquesas Island (Raimbault and Garcia, 2008). DON release due to NH_4^+ uptake was greatest in upwelling regions but still significant at more oligotrophic sites (50-100 versus 5-$20 \, \text{nmol} \, N \, L^{-1} day^{-1}$); DON release due to NO_3^- uptake showed similar patterns. The more oligotrophic sites tended to have the highest percent of recently fixed N released as DON (>20%), while percentages in other regions were <15%.

In contrast, the percentage of N released as DON increased with depth in Monterey Bay, suggesting that deeper in the water column, a smaller percentage of the N taken up was incorporated into sinking particles (Bronk and Ward, 1999). In this study there was also seasonality with regard to the primary fate of N uptake with particle production dominating in spring and DON production dominating in autumn (Bronk and Ward, 1999). These data suggest that the DON pool acts as an intermediate between DIN assimilation and

the net formation of particles for export and will thus affect C flow in Monterey Bay.

In the Choptank River, a subestuary of Chesapeake Bay, rates of total DON release were significantly higher in a <202 µm fraction relative to the <1.2 µm plankton, likely due to feeding processes associated with the larger size fraction (Bronk and Glibert, 1993b). Several lines of evidence indicate that active grazing was responsible for DON release including a low ratio of LMW DON to total DON release in the <202 µm fraction, increased DON release rates at night, and a doubling of phaeopigment concentrations during the 36h experiment. In contrast, rates of LMW DON release in the <1.2 µm plankton were not significantly different from rates of total DON release, and rates of DON release decreased by over 95% in the dark, indicating that passive release from autotrophs was a more important release process. DON release also often appears to decrease at night, although the difference is not statistically significant (Bronk et al., 1998).

2 Urea

Urea has been shown to be excreted by cryptomonads, herbivorous marine zooplankton, oceanic and lake microzooplankton, bivalve molluscs, marine and freshwater teleost fish, and freshwater crabs (Antia et al., 1991). Berg and Jørgensen (2006) found that bacterial catabolism of purines, and, to a lesser extent, pyrimidines was an important source of urea. In the Bering Sea, *in situ* production of urea was approximately equal to the consumption of urea (Hansell and Goering, 1989). In the central channel of Chesapeake Bay, rates of urea regeneration were generally less than rates of NH_4^+ regeneration, but, at the highest rates measured, urea regeneration contributed up to 100% of the phytoplankton N requirement (Bronk et al., 1998). Lomas et al. (2002) reviewed urea regeneration rates in Chesapeake Bay and found that mean Bay-wide surface urea regeneration rates were highest but most variable during the fall. In the Southern California Bight, urea decomposition was significantly deeper in the water column, particularly at the base of the euphotic zone, and the activity was primarily in the bacterial size fraction (Cho and Azam, 1995). These data suggest that urea was an important intermediate between sinking particles and release of NH_4^+ mediated by bacteria in the mesopelagic. Urea is also an important part of the N cycle of coral reefs with large variability in concentrations in and around the reef likely due to release from sediments, sea grasses, and the coral formations themselves (Crandall and Teece, 2012).

Size fractionated urea regeneration was measured using isotope dilution (see Section IV.C) in the western English Channel over an annual cycle (L'Helguen et al., 2005). Rates of regeneration were lowest in winter and highest in summer with 51% of the regeneration due to nanoplankton and 36% by microplankton. Uptake of urea was approximately equal to rates of urea regeneration, which indicates that *in situ* processes were sufficient to support measured rates of urea uptake (L'Helguen et al., 2005).

3 Amino Acids and Other Organics

DCAA and DFAA are other important organic release products. In cultures enriched with DIN, extracellular DFAA often accumulate with the highest concentrations generally present during stationary phase (Myklestad et al., 1989). Diatoms showed the greatest rates of DFAA excretion during exponential growth (Myklestad et al., 1989 and references therein). The types of amino acids released from *Chaetoceros affinis* (diatom) changes during exponential growth relative to stationary growth (Myklestad et al., 1989). In cultures of *Phaeodactylum tricornutum*, DFAA accumulate extracellularly, particularly glycine, threonine, and serine, which are all important in cellular respiration or are components of cell walls that are likely resistant to decomposition (Marsot et al., 1991).

Amino acid enantiomeric ratios have also been used to infer DOM release by capitalizing on the fact that eukaryotic organisms release the L enantiomer exclusively. The DCAA in refractory DOM generally has a low L/D ratio of 3 to 4 (Calleja et al., 2013; McCarthy et al., 1998; see Section III.A.2). Significant phytoplankton release, however, can increase the L/D ratio up to 8, making the ratio a useful indicator of new DOM production.

A recent study investigated the response of natural bacterial communities to the addition of culture extracts and the subsequent production of DFAA by four axenic phytoplankton cultures (Sarmento et al., 2013). All cultures were grown under the same conditions, but no two cultures produced the same total concentration or proportional composition of amino acids. Prasinophyte *Micromonas pusilla* released the most DFAA of any of the species tested, and the dominant amino acid was L-alanine while histidine was the dominant DFAA product of *Skeletonema* and *Prorocentrum*. The DFAA consumption by bacteria was proportional to availability, however, the dominant bacterial "species" was distinct for each algal source. Sarmento et al. (2013) observed a tight inverse relationship between DFAA removal and bacterial abundance but could not rule out the possibility that the algal produced DOM as a whole influenced the bacterial communities. A similar trend was observed in the Gulf of Mexico where DON filtrate from *Trichodesmium* cultures caused a change in the abundance of red tide *Karenia brevis* cells as well as changes in the bacterial community (Sipler et al., 2013).

Mesocosm studies with and without nutrient enrichment revealed enhanced DCAA concentrations in the enriched mesocosm but not in the unamended control (Meon and Kirchman, 2001). These observations suggested that degradation processes, rather than production processes, exerted greater control over composition of the DCAA pool.

Another organic pool is the nucleic acid fraction. In the North Pacific subtropical gyre, Brum (2005) observed that only 11-35% of the production of enzymatically hydrolysable dissolved

DNA was due to lysis of bacteria by viruses. This suggests that other processes, such as bacterial exudation, bacterial autolysis, and bacterivory, contributed to the D-DNA flux. Production of D-DNA from viruses was lowest at the surface (5 m), relatively constant within the upper 45 m, and greatest at the deepest station sampled (75 m; Brum, 2005).

E Sources of DON: Research Priorities

A quick non-destructive means of isolating DON in quantities sufficient for mass spectrometric and other chemical analyses was listed as a Holy Grail of marine N research in the first edition. Sadly the quest continues. Such a method could be used to isolate DON in tracer experiments to allow researchers the ability to measure rates of DON release accurately and efficiently. Though there has been considerable work investigating the processes for bulk DOC (see Chapter 3), studies that consider bulk DON in parallel or exclusively are still relatively scarce, likely due to methodological limitations. There are still many open questions related to the mechanisms that produce DON, especially the impact of PCD, the role of viruses and the pathways involving micro- and macrozooplankton (i.e., sloppy feeding versus excretion versus fecal pellet dissolution). One fruitful area for future research is defining how cellular metabolism and community composition control the rate of DON production and the type of DON produced. In this case we already have the tools in hand to make real progress.

V SINKS FOR DON

On a global scale sinks for DON in surface waters include biological uptake, chemical transformations, and vertical mixing; the latter was discussed in Section II. Here we discuss DON bioavailability, the methods used to measure it, mechanisms and transformations that can affect it, rates of DON uptake in the environment and future research directions.

A DON Bioavailability

As discussed in Section III, concentrations of DON in the ocean are generally highest in surface waters and then decrease with depth. From the standpoint of lability, the largest pool within the DON box is composed of truly refractory components that persist for timescales of years to millennia. The semi-labile fraction includes compounds such as proteins, DCAA, and amino polysaccharides, which turn over on seasonal to annual time scales. The smallest pool in the DON box contains the highly labile forms such as urea, DFAA, and nucleic acids (Bronk et al., 2007). These labile compounds turn over on timescales of minutes for amino acids (e.g., Fuhrman, 1987) to days for urea (e.g., Bronk et al., 1998) and DNA (e.g., Jørgensen et al., 1993). The bulk of research on DON availability has focused on this labile fraction. Recent work, however, has also shown that even HMW compounds such as humic substances, historically considered to be highly refractory, can be used by organisms on time scales of hours (i.e., See et al., 2006).

DON uptake in marine systems has been historically attributed to heterotrophic bacteria production (Figure 4.6). Over the last decades, however, there is a greater appreciation that autotrophic microbes as well as heterotrophic bacteria can utilize similar pools of inorganic and organic N substrates in the ocean (reviewed in Bronk et al., 2007; Mulholland and Lomas, 2008). Now that we know that both autotrophs and heterotrophs use DON the question becomes who is doing what and when? There are a number of very basic reasons why this question is important. First, autotrophs take up CO_2 as a C source while heterotrophic microbes consume DOM and release CO_2. Knowing which group is using different N forms thus becomes important in terms of the global C budget. Second, oxygenic phytoplankton produce

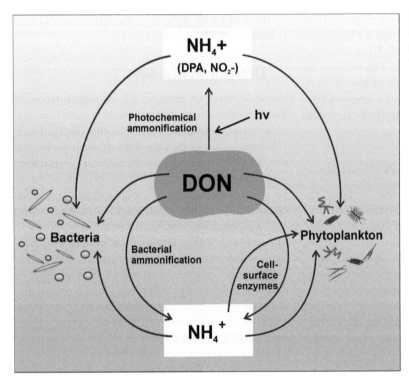

FIGURE 4.6 Conceptual diagram of processes involved in dissolved organic nitrogen (DON) utilization in aquatic systems.

oxygen while heterotrophic bacteria consume it. Knowing how N inputs are partitioned between the two groups can inform estimates of extent and susceptibility to hypoxia and anoxia. Third, most phytoplankton are bigger than bacteria. If we know what N inputs are supporting phytoplankton we know what inputs are more likely to contribute to higher trophic levels. Fourth, the f-ratio, which is the ratio of new to new plus regenerated primary production, is a parameter that estimates the magnitude of export production in open ocean systems (Eppley and Peterson, 1979). Traditionally the f-ratio has been calculated by dividing new production ($^{15}NO_3^-$ uptake) by new plus regenerated ($^{15}NH_4^+$ plus $^{15}NO_3^-$ uptake; e.g., Harrison et al., 1987). This classic approach, however, does not include uptake of regenerated DON by phytoplankton. In addition, uptake measurements have historically included variable fractions of bacterial uptake (Kirchman et al., 1992). Measuring N uptake

of inorganic and organic N by phytoplankton and bacteria independently would yield a much clearer, more useful picture. Fifth, some phytoplankton are toxic or form harmful blooms. Defining which substrates support HAB species is valuable from a management standpoint. The list could go on.

B Methods to Estimate Rates of DON Uptake

Two issues are important when considering DON uptake—what is the rate of uptake and what groups are responsible for, or contribute to, the uptake rate. With respect to measuring DON uptake rates, the very nature of DON is problematic because it is composed of a large number of compounds of unknown composition (Aluwihare and Meador, 2008). The two general approaches that are used are bioassays and tracers (reviewed in McCarthy and Bronk, 2008).

1 Measuring Uptake Rates

Bioassays are most commonly used to study the bioavailability of bulk DON (e.g., Sipler et al., 2013) or during additions of undefined organic compounds (e.g., Bronk et al., 2010). Bioassays are also useful in studies of recalcitrant DON, such as humics (e.g., Carlsson et al., 1995). One drawback of bioassays is that they often involve the addition of a DON concentrate or individual compounds at elevated concentrations to increase the likelihood of seeing a signal. A similar approach with the potential to circumvent this problem is the use of chemostats (e.g., Kisand et al., 2008) or ecostats (Hutchins et al., 2003). Another drawback of bioassays is that they only measure net changes such that uptake rates of a given substrate may be masked if regeneration rates are high. The use of tracers can avoid this problem, and a number of commercially available stable or radioactive tracers have been used including ^{15}N, ^{14}C, ^3H, and ^{35}S (e.g., Fuhrman, 1987; Hartmann et al., 2009; Lomas et al., 2002; Zubkov and Tarran, 2005). One advantage of using stable isotope tracers over radioactive tracers is that they allow for the simultaneous measurement of uptake and regeneration using the principle of isotope dilution (reviewed in McCarthy and Bronk, 2008).

In terms of organic N tracers, the most common are urea and amino acids, which are commercially available in single- (^{15}N) and dual-labeled (^{15}N and ^{13}C) forms. Other organic tracers include individual amino acids, amino acid mixtures, or combined amino acids produced in algal cultures, purines, pyrimidines, adenosine triphosphate (ATP), whole DNA, acetamide, and creatine, among others. In addition, ^{15}N-labeled DON has also been produced *in situ* by growing plankton on $^{15}NH_4^+$ or $^{15}NO_3^-$ (Bronk and Glibert, 1993a; Veuger et al., 2004), from *Trichodesmium* grown on $^{15}N_2$ gas (Bronk et al., 2004), by growing kelp on $^{15}NH_4^+$ (Hyndes et al., 2012), or by adding ^{15}N-labeled NH_4^+ to the sediment where *Spartina alterniflora* is grown to produce labeled humic substances (See and Bronk, 2005; See et al., 2006).

2 Determining Which Organisms Contribute to Uptake

The second issue is defining what organisms are responsible for, or contributing to, the rates measured; approaches include quantitative and non-quantitative methods. To quantify heterotrophic from autotrophic N uptake, it is necessary to separate the two groups. The most common method is size fractionation, where different fractions are separated using filters of varying pore size (e.g., Rodrigues and Williams, 2002). Glass fiber filters are the most common in ^{15}N studies because they are inexpensive, can be precombusted to remove N, and are compatible with mass spectrometry. Size fractionation is the simplest approach, but can produce inaccurate or imprecise rates of N uptake by phytoplankton due to the size overlap between bacteria and phytoplankton and variations in the size of the particle collected due to clogging. Studies have shown that GF/F filters retain up to 93% (40-75% on average) of the bacterial community depending on the system (Berg et al., 2001; Gasol and Morán, 1999; Lee and Fuhrman, 1987; Lee et al., 1995). Metabolic inhibitors such as cycloheximide and streptomycin, have also been used to block the function of ribosomes in eukaryotes and prokaryotes, respectively (e.g., Middelburg and Nieuwenhuize, 2000a, 2000b; Tungaraza et al., 2003a; Veuger et al., 2004). Inhibitors, however, can lack specificity and run the risk of incomplete inhibition of the target organism (Lee et al., 1992; Oremland and Capone, 1988). Dark bioassays have also been used to distinguish between phytoplankton and bacterial N uptake under the premise that phytoplankton go into suspended animation when in the dark (e.g., Glibert et al., 2004; Gobler et al., 2002; Joint et al., 2002), but these results can be difficult to interpret.

3 Flow Cytometric Sorting

More recently flow cytometric sorting has been used to physically separate phytoplankton from bacteria at the end of tracer experiments (reviewed in Lomas et al., 2011). Flow cytometry was introduced into oceanography to quickly estimate the abundance of pico- and nanophytoplankton in the ocean (e.g., Olson et al., 1985; Yentsch et al., 1983). The use of flow cytometry has expanded into a number of applications in aquatic microbial ecology, including the routine analysis of bacteria and virus abundance (Collier, 2004; see reviews in Wang et al., 2010; Lomas et al., 2011). Key to using flow cytometry to measure rates was the development of cell sorting, which allows researchers to isolate cells based on unique optical properties. Cells can be sorted based on the presence or absence of pigments, or labeled using dyes (Lomas et al., 2011). At its simplest, chlorophyll-containing phytoplankton can be sorted at the end of traditional ^{15}N tracer incubations to directly measure rates of phytoplankton N uptake. This approach was pioneered by Lipschultz (1995), who used flow cytometric sorting to separate phytoplankton from bacteria and detritus so he could measure phytoplankton-specific uptake rates of ^{15}N-labeled NH_4^+ and NO_3^-.

4 Molecular Approaches

There are a large suite of molecular approaches that focus on the detection of biological macromolecules. Early molecular studies in this area mainly focused on the detection of specific phylogenetic markers such as the 16S small subunit ribosomal RNA gene, which allows the assessment of microbial diversity (e.g., Giovannoni et al. 1990). Similarly, functional genes can often be used to assess whether natural communities contain organisms with the capacity to perform metabolic activities of biogeochemical significance such as nitrate assimilation (Allen et al., 2001). A bewildering array of molecular technologies have since been developed to specifically enumerate and assess the distribution and activity of microbial populations, often greatly influencing our perspective of global biogeochemical transformations including the N cycle (e.g., Zehr and Ward, 2002). In their latest incarnation,

molecular tools have come to oceanography by way of the -omics revolution. This is an area of rapid progress and great promise, which is facilitated by the extraordinarily rapid advances in nucleic acid sequencing technologies, computer science and bioinformatics.

At a high level -omics approaches fall into three categories based on the biological macromolecule that is analyzed—genomics, transcriptomics, and proteomics. Genomics and metagenomics focus at the level of the genome and provide information on the metabolic potential of an organism or microbial community, respectively. Transcriptomics takes it a step further and focuses on mRNA transcripts to define the biochemical and metabolic processes an organism may regulate up or down (McCarren et al., 2010). Gene transcription is a highly regulated process and specific mRNA transcripts are generally not made unless microbes sense a need to synthesize the requisite enzymes. However, because of post-transcriptional regulation not all transcripts are translated and there can be an uncoupling between the transcriptome and the proteome. Proteomics leverages genomic data for mass spectrometric analysis of peptide fragments. Post-transcriptional mechanisms often greatly affect the composition of the proteome, particularly in eukaryotes, so the analysis focuses on the final gene products in the form of proteins and the biochemical underpinnings of protein regulation (Keller and Hettich, 2009).

The shear number of reviews on molecular approaches published in recent years is a testament to its importance to the future of oceanography and its rapid evolution. We will let the experts do the work and direct readers to reviews on the history and general applications of molecular and genomic techniques to oceanography (Bowler et al., 2009; Doney et al., 2004; Sowell et al., 2009), and the application of these approaches to the study of DOM (McCarren et al., 2010; Solomon et al., 2010), viruses (Rohwer and Thurmer, 2009), bacteria (DeLong, 2009; Morris et al., 2010), phytoplankton in general (Rynearson and

Palenik, 2011), diatoms in specific (Armbrust, 2009), harmful algae (McLean, 2013), and general environmental assessments (Velhoen et al., 2012). Central to the accelerating progress in this area are the increasing number of bacterial and phytoplankton genomes that have been sequenced including *Synechococcus* (Palenik et al., 2003), *Prochlorococcus* (Dufresne et al., 2003; Rocap et al., 2003), the diatoms *Thalassiosira pseudonana* (Armbrust et al., 2004) and *Phaeodactylum tricornutum* (Bowler et al., 2008), the prasinophyte *Ostreococcus* (Palenik et al., 2007), the picoeukaryote *Micromonas* (Worden et al., 2009), the coccolithophore *Emiliania huxleyi* (Read et al., 2013), among others. These data can inform field research by suggesting fruitful directions to pursue based on, for example, the presence or absence of specific substrate transporters (e.g., Allen et al., 2011). Combining molecular tools with biogeochemical tracers can be a powerful approach to determining which groups are active and the rates of their activity. Here we focus on one approach where we have some direct experience—stable isotope probing (SIP).

5 Stable Isotope Probing

A key limitation in biogeochemical studies is linking uptake of substrates with the organisms responsible for that uptake. Enter nucleic acid based SIP (Radajewski et al., 2000; see reviews in Chen and Murrell, 2010; Wawrik, 2013), an approach in which ^{13}C- or ^{15}N-labeled substrates are tracked into the DNA of organisms that actively assimilate labeled substrate. Uptake of the labeled compounds results in the incorporation of the heavy isotope into DNA, which has a small but significant density difference from the more abundant natural isotope (e.g., Buckley et al., 2007; Gallagher et al., 2005). After purification of the DNA, it is suspended in a concentrated solution of cesium chloride (CsCl) and the DNA molecules of different densities can be separated by ultracentrifugation and fractionation. The distinct communities that incorporate the substrate labeled with the heavy isotope are then interrogated by demonstrating

density shifts in some organisms using molecular ecology methods such as qPCR (quantitative PCR, Radajewski et al., 2000), gene probing and sequence analysis (e.g., Gallagher et al., 2005; Nelson and Carlson, 2012). The SIP approach is also being applied at the protein level (reviewed in Jehmlich et al., 2010; Seifert et al., 2012).

SIP has emerged as a powerful tool for the investigation of C and N transformations in microbial communities (see reviews by Chen and Murrell, 2010; Dumont and Murrell, 2005; Friedrich, 2006; Uhlík et al., 2009; Wawrik, 2013), but has thus far not been extensively used to investigate marine DON. The majority of ^{15}N-SIP work has been done with $^{15}N_2$ uptake in soils. In Section V.D, we review the relatively few studies that have used SIP in the marine environment to demonstrate the promise of the technique (e.g., Wawrik et al., 2009).

CsCl gradient-based SIP has several drawbacks, however. Most importantly, its application requires that relatively high concentrations of labeled substrate be added to achieve effective density shifts (Buckley et al., 2007; Dumont and Murrell, 2005). In the case of N this typically requires additions above background levels thereby limiting its utility in some environmental settings. Additionally, in some instances relatively long incubations are required to ensure adequate incorporation into DNA. During long incubations, there is an increased likelihood of crossfeeding, which occurs when a substrate added in one form is transformed to another form (e.g., nitrification converts NH_4^+ to NO_3^-) during the incubation, potentially confounding the results (Neufeld et al., 2005).

A promising new strategy for SIP that attempts to overcome these limitations is Chip-SIP (Mayali et al., 2011). In a Chip-SIP experiment, a custom microarray is generated from molecular sequence data and hybridizations are performed with nucleic acids that are recovered from incubations with tracer-level additions of ^{13}C and/or ^{15}N substrates. Subsequently, the array is scanned with secondary ion mass spectrometer imaging (i.e., nanoSIMS; Finzi-Hart et al., 2009; Ploug et al., 2010), which can identify the locations of spots with ^{13}C and ^{15}N enrichment. By correlating the phylogenetic identity of probe spots with enrichment, the technique gives a highly sensitive identification of the organisms actively involved in substrate assimilation, even with complex mixtures in the natural environment (Mayali et al., 2011; Mayali et al., 2012). Compared to traditional SIP it allows for tracer level additions of labeled substrate, which limits the potential for cross-feeding and stimulation. It is conducted using short time-scale incubations by targeting mRNA, which turns over rapidly. Additionally, dually labeled substrates (^{13}C and ^{15}N) can be used because each label can be followed separately. While Chip-SIP is relatively new and limited by the availability of nanoSIMS instruments, it holds great promise in deciphering the specific communities involved in marine DON uptake.

C Mechanisms that Contribute to DON Bioavailability

Organic compounds can be large and the composition of the DON pool diverse. As a result, there are likely a number of mechanisms that are responsible for the uptake rates of DON we measure. Here we briefly review enzymatic decomposition, pino- and phagocytosis, photochemical-mediated uptake, salinity-mediated uptake, and ammonification.

1 Enzymatic Decomposition

One important variable relevant to DON uptake is the molecular size of the substrate under consideration. For LMW substrates like urea and some amino acids, uptake likely occurs by active transport driven by a sodium ion pump (reviewed in Mulholland and Lomas, 2008). Theoretically it could also occur through facilitated diffusion but it is unlikely that the concentrations outside the cell would be high enough

(mM) to create a concentration gradient into the cell. For compounds greater than ~1 kDa, direct uptake cannot occur and phytoplankton must use other mechanisms often involving enzymes (reviewed in Berges and Mulholland, 2008; Lewitus and Subba Rao, 2006). Proteolytic enzymes break down HMW compounds into their LMW constituent molecules, which are more accessible to the cell. Proteolytic enzymes occur both intracellularly, attached to the outside of the cell, or extracellularly, released from the cell into the environment (Berges and Mulholland, 2008). Bacterial degradation of DON followed by phytoplankton uptake of the released compounds has been demonstrated by a number of researchers (Antia et al., 1991; Berman et al., 1999; Ietswaart et al., 1994; Lisa et al., 1995; Palenik and Henson, 1997). When the decomposition results in NH_4^+ production, the process is known as ammonification. Direct uptake of DON, without bacterial mediation, was considered to be relatively minor, but this view is changing.

Recent studies have shown that phytoplankton have proteolytic processes including amino acid oxidation and peptide hydrosis (Mulholland et al., 2003). The importance of amino acid oxidases was first introduced to oceanography by Palenik and Morel (1991). These enzymes cleave free amino acids and primary amines to produce extracellular H_2O_2, α-keto acids or aldehydes, and NH_4^+, which can then be taken up by the cell. Fluorescently labeled amino acids, including LYA-lysine, have been used to quantify the pathway (Mulholland et al., 1998; Pantoja and Lee, 1994). A number of species of phytoplankton are known to possess these enzymes (Mulholland et al., 2003; Palenik and Morel, 1991). Autotrophic cell surface enzymes occur across a range of systems facilitating an estimated 20% of the DON utilization in coastal and oceanic environments (Mulholland et al., 1998). One open question is whether these enzymes are capable of cleaving the N from a wide range of DON substrates, which would allow phytoplankton to "farm" N from these compounds while leaving the remainder of the compound external to the cell (Berg et al., 2003b; Bronk et al., 2007; Stoecker and Gustafson, 2003).

In peptide hydrolysis, peptide bonds of proteins and polypeptides are severed producing smaller peptides and amino acids. Leucine aminopeptidase (LAP) is one class of proteolytic enzymes able to hydrolyze peptide bonds (Langheinrich, 1995). In addition to making their own enzymes, phytoplankton may also benefit from those produced by other cells. Bacteria can release proteolytic enzymes directly or through cell lysis or bacterivory (Berges and Mulholland, 2008).

2 Pinocytosis

Pinocytosis is also a mechanism that some plankton may use to take up DON (reviewed in Glibert and Legrand, 2006; Flynn et al., 2013). In pinocytosis, cells ingest dissolved macromolecules from the medium outside the cell by expanding a vesicle from the plasma membrane, engulfing the molecules and then pulling them inside the cell, creating a vacuole. Pinocytosis has been seen in dinoflagellates, euglenoids, and chlorophytes, including *Alexandrium catenella*, *Amphidinium carterae* and *Prorocentrum micans* (reviewed in Bronk, 2002).

One challenge is detecting when pinocytosis is occurring. In a study of humic uptake, incubations were conducted using dual-labeled humic substances (^{15}N and ^{13}C; See et al., 2006). They assumed that if pino- or phagocytosis were used, as opposed to enzymatic breakdown by phytoplankton or bacteria, to take up the humics, the ^{15}N and ^{13}C would be taken up in the same stoichiometric ratio as the labeled humics. No pinocytosis or phagocytosis was observed in cultures of *Synechococcus* sp., *Amphidinium carterae*, and *Thalassiosira cf. miniscula*, which are coastal phytoplankton strains known to take up humic-N (See et al., 2006).

3 Photochemistry

Another mechanism of DON uptake is photo-oxidation of an organic substrate followed by uptake of the photoproducts (reviewed in Kitidis and Uher, 2008; see Chapter 8 in this volume). The first rates of photochemical release of N were published by Bushaw et al. (1996), and numerous subsequent studies have shown that photochemical reactions occur when DOM is exposed to sunlight, resulting in the production of carbon monoxide, carbon dioxide, carbonyl compounds, among others (reviewed in Grzybowski and Tranvik, 2008; Tarr et al., 2001).

Many of the studies of photochemical N release have been done in fresh or brackish water environments. Though a number of studies have observed photoproduction of NH_4^+, DFAA, DCAA, DPA, and NO_2^- (Aarnos et al., 2012; reviewed in Bronk, 2002), other studies either can not resolve release or measure extremely low rates of release (Bertilsson et al., 1999; Koopmans and Bronk, 2002). Humic substances are likely important substrates for photoproduction because their aromaticity and color allow them to absorb UV light, making them more photochemically reactive than other classes of marine DOM. Furthermore, an estimated 50-75% of the N associated with humic substances exists as DFAA, amino sugars, and other N-rich compounds that are likely sources of the labile N forms produced photochemically (Rice, 1982; Stevenson, 1994; Thurman, 1985). A review of NH_4^+ photoproduction rates indicates a range from 0 to $220 \, nmol \, L^{-1} h^{-1}$ in estuarine systems and $0.7\text{-}46 \, nmol \, L^{-1} h^{-1}$ in marine systems (Kitidis and Uher, 2008). Photoproduction was shown in the range of ~6-29 $nmol \, NL^{-1} h^{-1}$ for urea and 0-41 $nmol \, NL^{-1} h^{-1}$ for amino acids (Kitidis and Uher, 2008). Stedmon et al. (2007) observed rates of photoproduction of NH_4^+ in the Baltic as high as $237 \, \mu mol \, N \, m^{-2} d^{-1}$, which is on the same scale as the rate of atmospheric deposition for the region. In the Gulf of Mexico, rates of NH_4^+ photoproduction were ~10% of *in situ* NH_4^+

regeneration, while rates of amino acid photo-production were low and frequently below detection, though they could provide 2-12% of the amino acid uptake measured in parallel (Bronk et al., in press).

4 Salinity-mediated Uptake

One mechanism for accessing N in DON that has received little attention is salinity-mediated release, where organic compounds act as a transport mechanism. Humic substances have cation exchange sites that allow them to retain and exchange ions from the surrounding waters (Baalousha et al., 2006). Ammonium can be adsorbed to these exchange sites on humics in estuarine and coastal areas (Gardner et al., 1991; Rosenfeld, 1979). As the humics are transported into more saline waters, the NH_4^+ adsorbed to the humics can be replaced with salt ions thereby releasing NH_4^+ into the surrounding water (Gardner et al., 1998; See, 2003; See and Bronk, 2005). Termed the "humic shuttle," this may be an important mechanism that delivers labile N to more N-limited mesohaline regions in estuaries and the coastal ocean (See, 2003; Bronk et al., 2007). Along salinity gradients in the York, Altamaha, and Satilla Rivers, on the U.S. east coast, it was demonstrated that a linear release of NH_4^+ from humics occurs as salinity increased (See, 2003; See and Bronk, 2005). This phenomenon was also observed with organics in wastewater treatment plant effluent, which releases NH_4^+ when the effluent enters receiving waters (Bronk et al., 2010).

D Literature Values of DON Uptake in Aquatic Environments

Here we review uptake rates of bulk DON (i.e., the total DON pool), urea, DFAA and DCAA, humic substances, and other DON compounds. Additional reviews to consult include Bronk et al. (2007) and Mulholland and Lomas (2008).

1 Bulk DON

Most of the work conducted on bulk DON utilization has been in freshwater systems using a bioassay approach. This work suggests that 12-72% of the DON pool in those systems is bioavailable on the order of days to weeks. In the Delaware and Hudson Rivers, 40-72% of the DON was shown to be consumed during 10-15 day dark bioassays, and DON consumption resulted in both an increase in PN and the release of DIN (Seitzinger and Sanders, 1997). These data suggest that the bioavailable DON can be utilized within estuaries with residence times on the order of weeks to months. In systems where residence times are shorter, riverine DON will be a source of bioavailable N to coastal waters.

Stepanauskas et al. (1999a) measured the concentration and bioavailability of three MW DOM fractions in samples collected seasonally in Swedish wetlands. They found that bioavailable DON was higher in seawater than in freshwater, and that bioavailability did not correlate with the C:N ratio of the DOM. In additional studies in wetlands, AMPase activity was twofold higher in seawater, relative to freshwater, indicating that hydrolysis and turnover of terrestrial DON increased when it entered the coastal ocean (Stepanauskas et al., 1999b). In two streams in Sweden, 19-55% of the bulk DON was bioavailable in short-term bioassays (Stepanauskas et al., 2000), though only 5-18% of the DON was identified as urea, DCAA or DFAA. Bioassay experiments conducted in the Gulf of Riga showed an average of 77% (8-136%) of the bacterial N biomass accumulation resulted from the uptake of DFAA and DCAA, and 13% of the DON was bioavailable over the course of 7-8 days (Jørgensen et al., 1999). In another study, as much as 40% of the DON in water samples from rivers and estuaries along a continuum of anthropogenic modification was consumed during a 6-day dark incubation and up to 80% of the TDN used was organic (Wiegner et al., 2006). Bulk DON produced in cultures of *Trichodesmium* was also found to be highly labile to a microbial community in the Gulf of Mexico dominated by *Karenia brevis* (Sipler et al., 2013). Chemical characterization of the DON indicated that the DON compounds consumed were highly reduced with nearly half having molar ratios suggestive of lipids or proteins.

Lønborg and Álvarez-Salgado (2012) reviewed published DOM incubation studies and found that $35 \pm 13\%$ of total DON in coastal oceans was bioavailable on timescales ranging from 2.5 to 300 days (median = 135.8 days); this was in contrast to $22 \pm 12\%$ for DOC and $70 \pm 18\%$ for DOP. The DON taken up was also more N rich, with C:N ratios of 8.8 ± 4.4, compared to 15.6 ± 4.6 for refractory DON (Lønborg and Álvarez-Salgado, 2012).

2 Urea

Measurements of urea uptake are becoming more common providing robust literature in a number of systems. Mean uptake rates of urea in the literature were lowest in the Arctic, then continually increased in ocean, upwelling, Southern Ocean, coastal, estuaries, to a high in rivers (Table 4.4). Across all systems, urea contributed a mean of $27.7 \pm 19.0\%$ of measured uptake.

Phytoplankton and bacteria must split the urea molecule prior to use, a process catalyzed by the enzyme urease, producing two molecules of NH_4^+ (Solomon et al., 2010). In bacteria, urease activity appears to be controlled by multiple mechanisms, including the availability of other sources of labile N substrates and the availability of extracellular urea (reviewed in Jørgensen, 2006). In general, phytoplankton were believed to be the primary users of urea in marine systems and molecular studies have shown that the genetic machinery for urea use is widely available (Allen et al., 2011; Collier et al., 2009). Recent studies, however, have called this belief into question, particularly in estuaries. Jørgensen (2006) conducted a series

of bioassay experiments focused on bacterial uptake at four sites. He found that urea uptake was highly variable, that urea uptake rates were typically lower than uptake rates of NH_4^+ or DFAA, but that urea could provide up to 44% of the bacterial N demand. Lomas et al. (2002) reviewed urea uptake rates for over a decade in Chesapeake Bay and found that urea is consistently an important N source for the plankton community, and that the highest mean Bay-wide rates are observed during the summer. Hansell and Goering (1989) found that urea uptake rates, based on urea disappearance, were an average of 140% greater than those based on rates of N accumulation in the Bering Sea, thus illustrating the close coupling between urea uptake and urea regeneration. Because urea regeneration was prevalent in their samples, the isotope dilution correction increased measured uptake rates by an average of 54%. Bradley et al. (2010a) used a flow cytometric approach to measure uptake and demonstrated that phytoplankton at the mouth of Chesapeake Bay utilized more urea and DFAA as N sources while bacteria relied more on DFAA and NH_4^+. Over the entire study area (main axis of Chesapeake Bay) urea accounted for $10 \pm 9\%$ of the total N uptake and DFAA accounted for $9 \pm 7\%$.

Urea has also been shown to be important in more marine systems. Sanderson et al. (2008) found urea uptake rates to account for the majority of total N uptake in both phytoplankton and bacterial size classes during a *Phaeocystis* bloom within a Raunefjord, Norway mesocosm study. Similarly Bradley and colleagues also found that urea contributed 10-25% of total N uptake by both phytoplankton and bacteria in mesocosm studies in Raunefjord, Norway (Bradley et al., 2010c) as well as in surface waters at LEO-15 in the Mid-Atlantic Bight where urea served as the dominant N form utilized (Bradley et al., 2010b). Uptake of urea in the coastal Chukchi Sea was determined to be very low under ice conditions (Baer et al., submitted), however, rates were much higher in the

4. DYNAMICS OF DISSOLVED ORGANIC NITROGEN

TABLE 4.4 Literature Values of Uptake Rates for Some Individual Organic Nitrogen Compounds and the Bulk DON Pool

Location	Sampling Date	Sampling Depth (m)	Fraction	Uptake Rate (nmol N l^{-1}h^{-1}-%)			Total N Uptake (%)			Substrates Included in Total N Uptake	Method	Reference
UREA—OCEANIC												
Bronk (2002) mean (n = 4)				20.1	±	25.7	15.7	±	4.7			Bronk (2002)
North Atlantic	Sept-Oct 2000	100, 7, & 1% Io	WW	0.1	to	4.4	28.0	to	52.0	N4, N3, U	^{15}N	Varela et al. (2005)
Equatorial Atlantic	Sept-Oct 2000	100, 7, & 1% Io	WW	0.0	to	23.8	55			N4, N3, U	^{15}N	Varela et al. (2005)
South Atlantic	Sept-Oct 2000	100, 7, & 1% Io	WW	0.9	to	8.7	65			N4, N3, U	^{15}N	Varela et al. (2005)
Sargasso Sea	May 1982	60, 22, & 3 & Io	<153 µm	0.2	to	4.1					^{15}N	Glibert et al. (1988)
NW Atlantic—Gulf Stream	May 1982	60, 22, & 3 & Io	<153 µm	0.2	to	0.5					^{15}N	Glibert et al. (1988)
NE Pacific (includes Ocean Station P)	winter-Mar 1993 & Feb 1994	Euphotic zone	WW	0.2	to	1.0					^{15}N	Varela and Harrison (1999)
NE Pacific (includes Ocean Station P)	spring-May 1993 & 1994	Euphotic zone	WW	0.6	to	8.5					^{15}N	Varela and Harrison (1999)
NE Pacific (includes Ocean Station P)	summer-Sept 1992 & 1994	Euphotic zone	WW	0.2	to	2.0					^{15}N	Varela and Harrison (1999)
Sub Arctic Pacific	May-Oct 1987 & 1998	Surface	WW	0.5	to	9.2	10.0	to	43.0	N4, N3, U	^{15}N	Wheeler and Kokkinakis (1990)
N Pacific	Nov-Dec 1982	Surface	WW	0.4	to	1.1					^{15}N	Kanda et al. (1985)
Tropical N Pacific	Dec 1982-Jan 1983	Surface	WW	0.9	to	4.0					^{15}N	Kanda et al. (1985)
Subtropical N Pacific	Jan-Feb 1983	Surface	WW	0.9	to	1.8					^{15}N	Kanda et al. (1985)

Region	Date	Depth	Fraction	Rate	Mean ± SD	Substrate	Method	Reference
NE Arabian Sea	Apr 2006	Euphotic zone	WW	1.6 to 3.8			^{15}N	Gandhi et al. (2010)
Arabian Sea—Spring Intermonsoon	Mar-Apr 1995	Euphotic zone	WW	0.0 to 20.0			^{15}N	Sambrotto (2001)
Arabian Sea—SW Monsoon	Jul-Aug 1995	Euphotic zone	WW	0.0 to 13.0			^{15}N	Sambrotto (2001)
Western Indian Ocean -NE Monsoon	Nov-Dec 1994	Euphotic zone	WW	0.0 to 3.0			^{15}N	Watts and Owens (1999)
Gulf of Mexico	Summer/Fall 2001-2002	Surface	WW	3.9 to 1950[h]			^{15}N	Mulholland et al. (2006)
Gulf of Mexico—Offshore	Oct 2007-2009	Surface	WW	39.4 ± 29.7	12.9 ± 5.1	N4, N3, U, AA	^{15}N	Bronk et al. (in press)
*Mean ± std				52.5 ± 206.9	29.1 ± 19.8			
UREA—UPWELLING								
Bronk (2002) mean ($n=2$)				2.6 ± 3.3	16.8 ± 9.7			Bronk (2002)
NW Africa Upwelling	Sept-Oct 2000	100, 7, & 1% Io	WW	0.00 to 19.90	19	N4, N3, U	^{15}N	Varela et al. (2005)
NW Iberian upwelling	Aug 1998	Upper 40	WW	3.96 to 7.92			^{15}N	Joint et al. (2001)
Namibian Upwelling	Sept-Oct 1985	Euphotic zone	WW	1.68 to 9.31			^{15}N	Probyn (1985)
Benguela Upwelling	Dec 1983	Surface	<212 μm	0.00 to 80.00			^{15}N	Probyn (1985)
Benguela Upwelling	Dec 1983	Surface	<10 μm	0.00 to 30.00			^{15}N	Probyn (1985)
Benguela Upwelling	Dec 1983	Surface	<1 μm	0.00 to 12.00			^{15}N	Probyn (1985)
North Sea	Mar-June	10 & 40	WW	0.0 to 19			^{15}N	Tungaraza et al. (2003b)[a]

(Continued)

TABLE 4.4   Literature Values of Uptake Rates for Some Individual Organic Nitrogen Compounds and the Bulk DON Pool—cont'd

Location	Sampling Date	Sampling Depth (m)	Fraction	Uptake Rate (nmol N l⁻¹ h⁻¹ %)			Total N Uptake (%)	Substrates Included in Total N Uptake	Method	Reference
Westland, New Zealand	Winter-Jul	Upper 10	<200 μm	0.1	to	1			^{15}N	Chang et al., 1992[a]
Westland, New Zealand	Winter-Jul	Upper 10	<20 μm	0.1	to	0.5			^{15}N	Chang et al., 1992[a]
Westland, New Zealand	Winter-Jul	Upper 10	<2 μm	0.01	to	0.25			^{15}N	Chang et al., 1992[a]
Westland, New Zealand	Summer-Feb	10	<200 μm	0.6	to	13.2			^{15}N	Chang et al., 1995[a]
Westland, New Zealand	Summer-Feb	10	<20 μm	0.6	to	9.9			^{15}N	Chang et al., 1995[a]
Westland, New Zealand	Summer-Feb	10	<2 μm	0.4	to	2.6			^{15}N	Chang et al., 1995[a]
Oregon Coast	Summer-low NO_3^-	15	WW	2.0	to	69			^{15}N	Kokkinakis and Wheeler, 1987[a]
Oregon Coast	Summer-high NO_3^-	15	WW	9.0	to	59			^{15}N	Kokkinakis and Wheeler, 1987[a]
Oregon Coast	Summer	10-15	10-200 μm	0.00	to	42			^{15}N	Kokkinakis and Wheeler, 1988[a]
Oregon Coast	Summer	10-15	<10 μm	2.00	to	30			^{15}N	Kokkinakis and Wheeler, 1988[a]
Westland, New Zealand	Winter	Euphotic zone	<200 μm	0.25	to	0.62			^{15}N	Chang et al. (1992)[a]
Westland, New Zealand	Winter	Euphotic zone	<20 μm	0.17	to	0.35			^{15}N	Chang et al. (1992)[a]
Westland, New Zealand	Winter	Euphotic zone	<2 μm	0.07	to	0.13			^{15}N	Chang et al. (1992)[a]
			*Mean ± std	10.0	±	12.2	17.5 ± 7.0			

UREA—COASTAL

Bronk (2002) mean (n=4)				118.5	±	152.8	30.6	±	14.8		Bronk (2002)
Coastal S Atlantic—Brazil	Sept-Oct 2000	100, 7, & 1% Io	WW	0.30	to	18.30	80		N4, N3, U	[15]N	Varela et al. (2005)
Gulf of Mexico—Coastal	Oct 2007-2009	Surface	WW	40.2	±	16.3	10.6	±	3.0	N4, N3, U, AA	[15]N	Bronk et al. (in press)
West Florida Shelf	Oct 2008	Surface	WW	13.0	±	4.0				[15]N	Wawrik et al. (2009)
Oslofjord, Norway	Annual	1 m	WW	0	to	3250				[15]N	Kristiansen (1983)[a]
Norwegian Fjords	Spring	Upper 15 m	WW	2.1	to	35				[15]N	Fernández et al.1996[a]
Raunefjord, Norway	Mar 2003	Surface	>8 μm	4.5	to	106.8	28.0	to	73.0	N4, N3, U, AA	[15]N	Sanderson et al. (2008)
Raunefjord, Norway	Mar 2003	Surface	0.2-8 μm	3.3	to	17.6	39.0	to	64.0	N4, N3, U, AA	[15]N	Sanderson et al. (2008)
Raunefjord, Norway	Apr 2005	Surface	Phyto	2	to	28	19.0	±	10.0	N4, N3, U, AA	[15]N	Bradley et al. (2010c)
Raunefjord, Norway	Apr 2005	surface	Bacteria	2	to	16	24.0	±	8.0	N4, N3, U, AA	[15]N	Bradley et al. (2010c)
Skagerrak, Sweden	Annual	50% LD	WW	0.6	to	16.7				[15]N	Pettersson (1991)[a]
Skagerrak, Sweden	Spring	Above, in and below pycnocline	WW	0.9	to	17.5				[15]N	Pettersson and Sahlsten (1990)[a]
Skagerrak, Sweden—coastal	Spring	Above, in and below pycnocline	WW	3	to	10.7				[15]N	Rosenberg et al. (1990)[a]
Skagerrak, Sweden—open	Spring	Above, in and below pycnocline	WW	4.4	to	9				[15]N	Rosenberg et al. (1990)[a]

(Continued)

TABLE 4.4 Literature Values of Uptake Rates for Some Individual Organic Nitrogen Compounds and the Bulk DON Pool—cont'd

Location	Sampling Date	Sampling Depth (m)	Fraction	Uptake Rate (nmol N l⁻¹ h⁻¹%)			Total N Uptake (%)	Substrates Included in Total N Uptake	Method	Reference
Western English Channel	Annual	50% LD	<200 μm	2	to	15			¹⁵N	L'Helguen et al. (1996)[a]
Humber Estuary Plume, England	Spring/summer	5	WW	5.0	to	40			¹⁵N	Shaw et al. (1998)[a]
Humber Estuary Plume, England	Spring/Summer	5	WW	6.1	to	22			¹⁵N	Shaw et al. (1998)[a]
Belgian Coastal waters	Feb-Jun	Surface	WW	0.2	to	17			¹⁵N	Tungaraza et al. (2003a)[a]
Baltic Sea—Gulf of Riga	Spring-May	Upper mixed layer	WW	27	to	42			¹⁵N	Berg et al. (2001)[a]
Baltic Sea—Gulf of Riga	Summer-Jul	Upper mixed layer	WW	40	to	220			¹⁵N	Berg et al. (2001)[a]
Baltic Sea—Gulf of Riga[b]	Spring-May	Surface	<200 μm	1.7	to	22.1			¹⁵N	Berg et al., 2003a[a]
Baltic Sea—Gulf of Riga[b]	Spring-Jun	Surface	<200 μm	0.15	to	4.2			¹⁵N	Berg et al. (2003a)[a]
Baltic Sea—Gulf of Riga[b]	Summer-Jul	Surface	<200 μm	4.7	to	51.6			¹⁵N	Berg et al. (2003a)[a]
SW Finland[c]	Summer	3-4	WW	37	to	94			¹⁵N	Tamminen and Irmisch (1996)[a]
Chesapeake Bay plume, USA	Winter	Surface	WW	8	to	72			¹⁵N	Glibert and Garside (1992)[a]
Chesapeake Bay plume, USA	Spring	Surface	WW	5	to	140			¹⁵N	Glibert and Garside (1992)[a]
Chesapeake Bay plume, USA	Summer	Surface	WW	35	to	650			¹⁵N	Glibert and Garside (1992)[a]
Chesapeake Bay plume, USA	Spring	Surface	WW	0	to	267			¹⁵N	Mulholland and Lomas (2008)[a]

Location	Season	Depth	Filter	Value				Tracer	Reference
Chesapeake Bay plume, USA	Summer	Surface	WW	2.3	to	179		^{15}N	Mulholland and Lomas, 2008[a]
Chesapeake Bay plume, USA	Fall	Surface	WW	24.1	to	54.5		^{15}N	Mulholland and Lomas (2008)[a]
Mid-Atlantic shelf, USA	Spring	Surface	WW	0.04	to	25.5		^{15}N	Mulholland and Lomas (2008)[a]
Mid-Atlantic shelf, USA	Summer	Surface	WW	2.3	to	248		^{15}N	Mulholland and, Lomas, (2008)[a]
Mid-Atlantic Bight, USA	Jul 2002	Surface	WW	600	to	1900		^{15}N	Bradley et al. (2010b)
Mid-Atlantic Bight, USA	Jul 2002	Bottom (~15 m)	WW	5	to	42		^{15}N	Bradley et al. (2010b)
Strait of Georgia, Canada	Summer	50% LD-stratified	WW	40	to	60		^{15}N	Price et al. (1985)[a]
Strait of Georgia, Canada	Summer	50% LD-frontal	WW	100	to	150		^{15}N	Price et al. (1985)[a]
Southern California Bight[d]	Summer	Euphotic zone	<183 µm	0	to	18.8		^{15}N	McCarthy (1972)[a]
Eastern Agulhas Bank	Summer-Jan	Euphotic zone	WW	3.2	to	11.1		^{15}N	Probyn et al. (1995)[a]
Westland, New Zealand	Winter	10	<200 µm	1.8	to	6.5		^{15}N	Chang et al. (1989)[a]
Westland, New Zealand	Winter	10	<20 µm	1.1	to	5.6		^{15}N	Chang et al. (1989)[a]
Westland, New Zealand	Winter	10	<2 µm	0.7	to	3.1		^{15}N	Chang et al. (1989)[a]
			*Mean ± std	112.7	±	304.5	36.4 ± 22.3		

UREA—ESTUARINE

Location	Season	Depth	Filter	Value				Tracer	Reference
Bronk (2002) mean (n = 8)				102.6	±	89.0	23.4 ± 21.8		Bronk (2002)

(Continued)

TABLE 4.4 Literature Values of Uptake Rates for Some Individual Organic Nitrogen Compounds and the Bulk DON Pool—cont'd

Location	Sampling Date	Sampling Depth (m)	Fraction	Uptake Rate (nmol N l^{-1}h^{-1}%)	Total N Uptake (%)	Substrates Included in Total N Uptake	Method	Reference
Gulf of Mexico—Estuarine	Oct 2007-2009	Surface	WW	39.7 ± 18.1	12.3 ± 3.6	N4, N3, U, AA	^{15}N	Bronk et al. (in press)
Cheaspeake Bay, USA	Aug-Sept 2004	Surface	WW	2 to 147			^{15}N	Bradley et al. (2010a)
Bellport Bay (Great South Bay), NY, USA	Annual	Surface	WW	~20 to 270			^{15}N	Carpenter and Dunham (1985)[a]
Long Island Sound, NY, USA	Spring	Surface	WW	0.3 to 15.7			^{15}N	Mulholland and Lomas, 2008[a]
Great South Bay, Long Island Sound, USA	Annual	Surface	WW	<50 to 900			^{15}N	Kaufman et al. (1983)[a]
Great South Bay, Long Island Sound, USA	Annual	Surface	WW	20 to 270			^{15}N	Carpenter and Dunham (1985)[a]
Chincoteague Bay, VA, USA	Apr-Sept	Surface	WW	70 to 4690			^{15}N	Mulholland and Lomas (2008)[a]
Hog Island Bay, VA, USA	Spring	Surface	WW	2 to 36.9			^{15}N	Mulholland and Lomas (2008)[a]
Neuse River Estuary, USA	Annual	Surface	WW	<100 to 3750			^{15}N	Twomey et al. (2005)[a]
Chesapeake Bay, USA	Winter	Surface	WW	7 to 70			^{15}N	Glibert et al. (1992)[a]
Chesapeake Bay, USA	Spring	Surface	WW	8 to 550			^{15}N	Glibert et al. (1992)[a]
Chesapeake Bay, USA	Summer	Surface	WW	35 to 660			^{15}N	Glibert et al. (1992)[a]
Chesapeake Bay, USA	Spring	1	WW	63 to 105			^{15}N	Glibert et al. (1995)[a]

Location	Season	Depth	Type	Low	to	High	Rate range	label	Reference
Randers Fjord, Denmark	April	Surface	WW	4	to	43		15N	Veuger et al. (2004)[a]
Randers Fjord, Denmark	August	Surface	WW	2	to	130		15N	Veuger et al. (2004)[a]
Scheldt estuary, Netherlands[e]	Annual[f]	2	WW	0	to	148		15N	Andersson et al. (2006)[a]
Thames estuary, England	Winter-Feb	Surface	WW	0	to	7		15N	
Humber Estuary, England	Spring/summer	5	WW	0	to	30		15N	Shaw et al. (1998)[a]
Hong Kong waters	Jul 2008	Euphotic zone	WW				4.0 to 59.0 N4, N3, U	15N	Xu et al. (2012)
Moreton Bay, Queensland, Australia	Sept 1997	Surface	WW				6.0 to 40.0 N4, N3, U	15N	Glibert et al. (2006)
Moreton Bay, Queensland, Australia	Feb 1998	Surface	WW				20.0 to 29.0 N4, N3, U	15N	Glibert et al. (2006)
Moreton Bay, Queensland, Australia	Jul 1998	Surface	WW				10.0 to 15.0 N4, N3, U	15N	Glibert et al. (2006)
*Mean ± std				267.8	±	570.5	22.2 ± 16.3		

UREA—RIVERINE

Location	Season	Depth	Type	Low	to	High	label	Reference
Lafayette River (Ches. Bay), USA	Summer	Surface	WW	0	to	200	15N	Mulholland and Lomas (2008)[a]
Lafayette River (Ches. Bay), USA	Spring	Surface	WW	90	to	120	15N	Mulholland and Lomas (2008)[a]
York River (Ches. Bay), USA	Spring	Surface	WW	221	to	7662	15N	Mulholland and Lomas (2008)[a]
York River (Ches. Bay), USA	Summer	Surface	WW	202	to	3652	15N	Mulholland and Lomas (2008)[a]
*Mean ± std				1518.4	±	1830.1		

(Continued)

TABLE 4.4   Literature Values of Uptake Rates for Some Individual Organic Nitrogen Compounds and the Bulk DON Pool—cont'd

Location	Sampling Date	Sampling Depth (m)	Fraction	Uptake Rate (nmol N l⁻¹ h⁻¹%)			Total N Uptake (%)			Substrates Included in Total N Uptake	Method	Reference
UREA—ARCTIC												
Arctic Ocean—Northeast Water Polynya	May-Jul 1993	Surface	WW	0.7	to	2.1					^{15}N	Smith et al. (1997)
Arctic Ocean—North Water (Northern Baffin Bay)	Sept 1999	Surface	WW	0.2	to	3.7	16 (max %)			N4, N3, U	^{15}N	Fouilland et al. (2007)
Barents Sea	Winter—1984-1988	Surface (0.25-15)	WW		<0.1						^{15}N	Kristiansen et al. (1994)
Barents Sea	Spring—Apr-Jun 1984-1988	Euphotic	WW	0.7	to	3.8					^{15}N	Kristiansen et al. (1994)
Arctic—Coastal Chukchi Sea	Jan 2011 & 2012	1-2			0.01						^{15}N	Baer et al. (submitted)
Arctic—Coastal Chukchi Sea	Apr 2010 & 2011	4-8		0.02	to	0.18					^{15}N	Baer et al. (submitted)
Arctic—Coastal Chukchi Sea	Aug 2010 & 2011	2-8		1.34	to	4.92					^{15}N	Baer et al. (submitted)
*Mean ± std				1.3	±	1.2	16					
UREA—SOUTHERN OCEAN												
Bronk (2002) mean (n=2)				3.1	±	3.3	13.7	±	1.0			Bronk (2002)
Southern Ocean—N of Antarctic Peninsula	Nov-Dec 1992	2	WW	0.4	to	59.0					^{15}N	Bury et al. (1995)

Location	Date	Depth	Sample	Value 1			Value 2			Tracer	Reference
Southern Ocean—N of Antarctic Peninsula	Nov-Dec 1992	Under ice	WW			1.3				¹⁵N	Bury et al. (1995)
Southern Ocean—N of Antarctic Peninsula	Nov-Dec 1992	Ice	ice	0.5	to	1.0				¹⁵N	Bury et al. (1995)
Southern Ocean—Atlantic Sector	Mar-Apr 1984	Surface	<200 µm	22.0	to	277.0				¹⁵N	Probyn and Painting (1985)
Southern Ocean—Atlantic Sector	Mar-Apr 1984	Surface	<15 µm	16.0	to	153.0				¹⁵N	Probyn and Painting (1985)
Southern Ocean—Atlantic Sector	Mar-Apr 1984	Surface	<1 µm	12.0	to	129.0				¹⁵N	Probyn and Painting (1985)
Southern Ocean—Pacific Sector	Nov-Dec 2001	Euphotic zone	WW	0.0	to	1.8				¹⁵N	Savoye et al. (2004)
*Mean ±std				**38.1**	**±**	**52.8**	**13.7**	**±**	**1.0**		
DCAA—OCEANIC											
Bronk (2002) mean (n = 1)						37.6			46.3		Bronk (2002)
*Mean ±std						**37.6**			**46.3**		
DCAA—COASTAL											
Bronk (2002) mean (n = 3)				31.5	±	24.0	34.9	±	21.2		Bronk (2002)
*Mean ±std				**31.5**	**±**	**24.0**	**34.9**	**±**	**21.2**		
DFAA/DPA—OCEANIC											
Bronk (2002) mean (n = 1)						5.0			6.2		Bronk (2002)

(Continued)

TABLE 4.4 Literature Values of Uptake Rates for Some Individual Organic Nitrogen Compounds and the Bulk DON Pool—cont'd

Location	Sampling Date	Sampling Depth (m)	Fraction	Uptake Rate (nmol N l^{-1} h^{-1}%)			Total N Uptake (%)			Substrates Included in Total N Uptake	Method	Reference
Gulf of Mexico	Summer/Fall 2001-2002	Surface	WW	1.1	to	246					^{15}N	Mulholland et al. (2006)
Gulf of Mexico—Offshore	Oct 2007-2009	Surface	WW	6.3	±	3.7	2.5	±	1.2	N4, N3, U, AA	^{15}N	Bronk et al. (in press)
	*Mean ± std			45.0	±	68.1	4.4	±	2.6			
DFAA/DPA—COASTAL												
Bronk (2002) mean (n=6)				20.7	±	19.3	17.8	±	29.5			Bronk (2002)
Gulf of Mexico—coastal	Oct 2007-2009	Surface	WW	25.3	±	11.8	6.8	±	2.0	N4, N3, U, AA	^{15}N	Bronk et al. (in press)
West Florida Shelf	Oct 2008	Surface	WW	55.0	±	1.0					^{15}N	Wawrik et al. (2009)
Baltic Sea—Gulf of Riga	Spring-May	Upper mixed layer	WW	50	to	90					^{15}N	Berg et al. (2001)[a]
Baltic Sea—Gulf of Riga	Summer-Jul	Upper mixed layer	WW	25	to	60					^{15}N	Berg et al. (2001)[a]
Baltic Sea—Gulf of Riga[b]	Spring-May	Surface	<200 μm	1.4	to	33.1					^{15}N	Berg et al. (2003a)[a]
Baltic Sea—Gulf of Riga[b]	Spring-Jun	Surface	<200 μm	0.2	to	2.3					^{15}N	Berg et al. (2003a)[a]
Baltic Sea—Gulf of Riga[b]	Summer-Jul	Surface	<200 μm	1.3	to	13.8					^{15}N	Berg et al. (2003a)[a]
Chesapeake Bay plume, USA	Spring	Surface	WW	0	to	111					^{15}N	Mulholland and Lomas (2008)[a]
Chesapeake Bay plume, USA	Summer	Surface	WW	3.2	to	126					^{15}N	Mulholland and Lomas (2008)[a]
Chesapeake Bay plume, USA	Fall	Surface	WW	9.7	to	13.0					^{15}N	Mulholland and Lomas (2008)[a]

Location	Season/Date	Depth	Type	Value 1	Value 2	Method	Compounds	Reference
Mid-Atlantic shelf, USA	Spring	Surface	WW	0.04 to 30.6		[15]N		Mulholland and Lomas (2008)[a]
Mid-Atlantic shelf, USA	Summer	Surface	WW	3.1 to 60		[15]N		Mulholland and Lomas, 2008[a]
Mid-Atlantic Bight, USA	Jul 2002	Surface	WW	20 to 200		[15]N		Bradley et al. (2010b)
Mid-Atlantic Bight, USA	Jul 2002	Bottom (~15)	WW	12 to 20		[15]N		Bradley et al. (2010b)
Raunefjord, Norway	Mar 2003	Surface	>8 μm	0.5 to 7.5	2 to 12	[15]N	N4, N3, U, AA	Sanderson et al. (2008)
Raunefjord, Norway	Mar 2003	Surface	0.2-8 μm	0.8 to 6.6	15 to 38	[15]N	N4, N3, U, AA	Sanderson et al. (2008)
Raunefjord, Norway	Apr 2005	Surface	Phyto	0.1 to 0.8	2 to 1	[15]N	N4, N3, U, AA	Bradley et al. (2010c)
Raunefjord, Norway	Apr 2005	Surface	Bacteria	0.75 to 4.3	11 to 9	[15]N	N4, N3, U, AA	Bradley et al. (2010c)
	*Mean ± std			27.4 ± 28.1	14.2 ± 21.0			
DFAA/ DPA—ESTUARINE								
Bronk (2002) mean (n = 6)				34.1 ± 34.4	40.0 ± 23.6			Bronk (2002)
Gulf of Mexico—Estuarine	Oct 2007-2009	Surface	WW	36.7 ± 26.1	12.6 ± 10.6	[15]N	N4, N3, U, AA	Bronk et al. (in press)
Cheaspeake Bay, USA	Aug-Sept 2004	Surface	WW	9 to 86		[15]N		Bradley et al. (2010a)
Long Island Sound, NY, USA	Spring	Surface	WW	0.4 to 5.7		[15]N		Mulholland and Lomas (2008)[a]
Chincoteague Bay, VA, USA	Apr-Sept	surface	WW	34 to 290		[15]N		Mulholland and Lomas (2008)[a]
Hog Island Bay, VA, USA	Spring	Surface	WW	0 to 156		[15]N		Mulholland and Lomas (2008)[a]
Randers Fjord, Denmark	April	Surface	WW	1 to 12		[15]N		Veuger et al. (2004)[a]

(Continued)

TABLE 4.4   Literature Values of Uptake Rates for Some Individual Organic Nitrogen Compounds and the Bulk DON Pool—cont'd

Location	Sampling Date	Sampling Depth (m)	Fraction	Uptake Rate (nmol N l⁻¹ h⁻¹%)			Total N Uptake (%)	Substrates Included in Total N Uptake	Method	Reference
Randers Fjord, Denmark	August	Surface	WW	4	to	180			[15]N	Veuger et al. (2004)[a]
Scheldt estuary, Netherlands[e]	Annual[f]	2	WW	0	to	110			[15]N	Andersson et al. (2006)[a]
Thames estuary, England	Winter-Feb	Surface	WW	15	to	75			[15]N	
			*Mean ± std	48.7	±	43.8	33.2 ± 23.6			
DFAA/ DPA—RIVERINE										
Bronk (2002) mean (n = 1)					4.5		2.8			Bronk (2002)
York River (Ches. Bay), USA	Spring	Surface	WW	36	to	303			[15]N	Mulholland and Lomas (2008)[a]
York River (Ches. Bay), USA	Summer	Surface	WW	3.9	to	343			[15]N	Mulholland and Lomas (2008)[a]
			*Mean ± std	115.8	±	96.4	2.8			
DFAA/DPA—ARCTIC										
Bronk (2002) mean (n = 1)					2.8					
Arctic—Coastal Chukchi Sea	Jan 2011 & 2012	1-2		0.11	to	0.34			[15]N	Baer et al. (submitted)
Arctic—Coastal Chukchi Sea	Apr 2010 & 2011	4-8		0.09	to	1.27			[15]N	Baer et al. (submitted)
Arctic— Coastal Chukchi Sea	Aug 2010 & 2011	2-8		1.42	to	3.8			[15]N	Baer et al. (submitted)
			*Mean ± std	1.6	±	1.3				

PURINES AND PYRIMINDINES—COASTAL

Location	Date	Depth		Value			Compound		Type	Reference
Øresund coast, Denmark	Apr-Aug 1998	Surface	B	129			Adenine		CC	Berg and Jørgensen (2006)
Øresund coast, Denmark	April-August 1998	Surface	B	264			Guanine		CC	Berg and Jørgensen (2006)
Øresund coast, Denmark	Apr-Aug 1998	Surface	B	210			Hypoxan-thine		CC	Berg and Jørgensen (2006)
Øresund coast, Denmark	Apr-Aug 1998	Surface	B	207			Xanthine		CC	Berg and Jørgensen (2006)
Øresund coast, Denmark	Apr-Aug 1998	Surface	B	24			Cytosine		CC	Berg and Jørgensen (2006)
Øresund coast, Denmark	Apr-Aug 1998	Surface	B	7			Thymine		CC	Berg and Jørgensen (2006)
Øresund coast, Denmark	Apr-Aug 1998	Surface	B	30			uracil		CC	Berg and Jørgensen (2006)
***Mean ± std**				**124.4**	**±**	**105.2**				

BULK DON—OCEANIC

Location				Value						Reference
Bronk (2002) mean (n = 2)				43.8	±	30.1				Bronk (2002)
***Mean ± std**				**43.8**	**±**	**30.1**				

BULK DON—COASTAL

Location	Date	Depth	Fraction	Value						Reference
Bronk (2002) mean (n = 1)				31.4						Bronk (2002)
Baltic Sea—Gulf of Riga[b]	Spring-May	Surface	<200 µm	0.7	to	3.8[g]				Berg et al. (2003a)[a]
Baltic Sea—Gulf of Riga[b]	Spring-Jun	Surface	<200 µm	0.2	to	1.7[g]				Berg et al. (2003a)[a]

(Continued)

4. DYNAMICS OF DISSOLVED ORGANIC NITROGEN

TABLE 4.4 Literature Values of Uptake Rates for Some Individual Organic Nitrogen Compounds and the Bulk DON Pool—cont'd

Location	Sampling Date	Sampling Depth (m)	Fraction	Uptake Rate (nmol N l⁻¹ h⁻¹%)			Total N Uptake (%)		Substrates Included in Total N Uptake	Method	Reference
Baltic Sea—Gulf of Riga[b]	Summer-Jul	Surface	<200 μm	0.9	to	8.3[g]					Berg et al. (2003a)[a]
			*Mean ± std	9.8	±	14.5					
BULK DON—ESTUARINE											
Bronk (2002) mean (n=3)				212.8	±	186.5	41.6 ±	19.7			Bronk (2002)
Randers Fjord, Denmark	April	Surface	WW	<5	to	60					Veuger et al. (2004)[a]
Randers Fjord, Denmark	August	Surface	WW	130	to	430					Veuger et al. (2004)[a]
Plym and Yealm River Estuary, SE England	Jun 2002-Sept 2003	0.2		3300	±	2300				CC	Badr et al. (2008)
			*Mean ± std	708.5	±	1277.8	41.6 ±	19.7			

Data were taken from text, tables, estimated from graphs, or obtained from the authors and are presented as the mean ± standard deviation. The mean presented in the first edition of this chapter (Bronk, 2002) is presented at the start of each section. All data from the earlier edition (Bronk, 2002) and the more recent data presented here were included in the calculated mean ± standard deviation presented at the end of each section. Note that some older papers are included here because they have been broken out into separate environments not covered in Bronk (2002). When a range is given, the average of that range was used in the calculation of the mean ± standard deviation for that system. Methods used include ¹⁵N tracer techniques (¹⁵N) or changes in concentration over time (CC). Substrates include ammonium (N4), nitrate (N3), urea (U), and amino acids (AA).

[a]Means for each region include all data from Bronk (2002) as well as new data listed in table.

[b]Unpublished data included in Mulholland and Lomas (2008).

[c]From Mulholland and Lomas (2008)—Absolute rates were obtained by multiplying specific rates by total nitrogen biomass. Total nitrogen biomass was calculated from C biomass using a C:N of 6.6.

[d]From Mulholland and Lomas (2008)—Rates were obtained using ¹⁴C tracer.

[e]From Mulholland and Lomas (2008)—Areal rates reported were averaged over 1-100% light level; this depth varied from 60 to 100 m so volumetric rates were calculated using an average euphotic depth of 80 m; daily rates reported were based onbserved daily variation in uptake, the authors state these were similar to those observed by Dugdale et al. (1992), and so hourly rates were calculated using a 18 h uptake period for urea.

[e]From Mulholland and Lomas (2008)—Excluded freshwater station.

[f]From Mulholland and Lomas (2008)—Annual = annual range of values.

[g]From Mulholland and Lomas (2008)—Rates are for adenine uptake.

[h]Data not included in mean.

summer. Urea uptake accounted for 4-59% of total N uptake in Hong Kong waters (Xu et al., 2012), and Yuan et al. (2012) found that 20-60% of total N uptake in western Hong Kong waters could be attributed to combined urea and AA. In 2000, Varela et al. (2005) determined that urea accounted for 40%, 55%, and 65% of the total N uptake in the North, Equatorial and South Atlantic, respectively. The lowest percent contribution (20%) of urea to total N uptake was measured in the NW Africa upwelling area and the highest urea contribution of 80% was measured along the coastal region of Brazil. Results of another recent study investigating the impact of upwelling and downwelling conditions on N uptake rates showed that urea uptake rates were nearly ten times greater in spring during downwelling conditions near the Ría de Vigo on the NW Iberian Peninsula (Seeyave et al., 2013). In northern Baffin Bay, Fouilland et al. (2007) found that uptake of urea by phytoplankton account for 58-95% of the total urea uptake. A study in the Southern Ocean in 2001 showed that although there was urea uptake, uptake was dominated by either NH_4^+ or NO_3^- (Savoye et al., 2004). Urea can also contribute to N nutrition through cyanate, which is one of the by-products of urea breakdown. Studies in the Red Sea show that cyanate could be taken up directly or via breakdown to NH_4^+ and CO_2 (Kamennaya et al., 2008). Natural abundance data also indicates that cyanate contributes to *Prochlorococcus* nutrition but was not important to *Synechococcus*.

Using SIP to investigate uptake of DIN and DON by phytoplankton in the Gulf of Mexico, Wawrik et al. (2009) were able to demonstrate that both *Synechococcus* and diatoms actively incorporated urea, while only the former was involved in the uptake of amino acids. Investigation of 16S rDNA in the Arctic Ocean has revealed that urea uptake was not found in bacterial or archaeal populations during summer, but both populations incorporated urea during the winter (Cooper et al., submitted). Furthermore, this same study was able to

identify marine group I Crenarchaea, SAR11, SAR324, and *Oceanospirillum* as the clades responsible for most of the uptake, potentially supporting the hypothesis of urea-fueled nitrification by polar microorganisms (Alonso-Sáez et al., 2012).

Flow cytometric sorting was used in a study of N uptake along the main axis of Chesapeake Bay in the summer. Results indicate that bacterial retention on the GF/F filters, used for most uptake measurements, result in an overestimate of phytoplankton-specific uptake of NH_4^+, urea, and DFAA by 61%, 53%, and 135%, respectively (Bradley et al., 2010a) and an underestimate of 58% for NO_3^- uptake. Underestimates (and overestimates) of this size will affect such things as estimates of new or regenerated production and *f*-ratio estimation, which has implications for global C models and will affect estimates of potential drawdown of anthropogenic CO_2 by the oceans.

HABs represent useful tools in the study of DON, particularly urea, use by autotrophs. Severe HAB blooms are high in biomass and can be virtual monocultures of a given species. It has been suggested that the ability, or lack thereof, to use DON could be an important driver of phytoplankton community composition (Paerl, 1997). If DIN supply is low, phytoplankton able to use DON may have a competitive advantage. A number of harmful algae species can use DON for N nutrition and there is some evidence they prefer it to inorganic N substrates (e.g., Berg et al., 1997; Herndon and Cochlan, 2007; Lomas et al., 1996). For example, dinoflagellates appear to prefer organic N (Dyhrman and Anderson, 2003; Fan et al., 2003; Glibert and Terlizzi, 1999; Palenik and Morel, 1991). Urea was an important N source during a bloom of the dinoflagellate *Prorocentrum minimum* in the Neuse River estuary (Fan et al., 2003), an important source of N to the red tide alga, *Lingulodinium polyedrum* (Kudela and Cochlan, 2000) and was shown to be an important contribution to the N nutrition of the brown tide *Aureococcus anophagefferens* (Berg et al., 1997; Gobler and Sañudo-Wilhelmy,

2001). A more detailed discussion of early HAB work can be found in Bronk (2002). In contrast, the toxigenic diatom *Pseudo-nitzschia australis* had lower growth rates when grown on urea, relative to NH_4^+ or NO_3^-. In addition, domoic acid production rates were two to three times higher when cells were grown on urea relative to NO_3^- or NH_4^+ (Howard et al., 2007).

3 Amino Acids

There are still few rates of DCAA uptake in the literature (Table 4.4). Measurements of DFAA uptake are more common but still relatively few in number as well. The mean uptake rates of DFAA measured in ocean, coastal, and estuarine environments are similar in magnitude (27.4-48.7 nmol N $L^{-1} h^{-1}$) but the percent contribution of DFAA to total measured uptake appears to increase from ocean, to coastal, to estuarine sites (Table 4.4). Mean DFAA uptake rates were highest in rivers but the variability is large.

Bacteria are generally considered the primary users of DFAA and DCAA. In a number of studies DFAA and DCAA supplied ~50% of the bacterial N demand in estuarine and coastal systems (Keil and Kirchman 1991, 1993; Middelboe et al., 1995). In a salt marsh phytoplankton community, addition of organic N, including glycine, glutamic acid, and an amino acid mixture, enhanced phytoplankton growth (Lewitus et al., 2000). The physiological response of the phytoplankton community to organic N additions, in the presence and absence of antibiotics, suggests that the stimulation caused by organic N additions results directly from uptake of the organic substrates and indirectly through bacterial decomposition. Glibert et al. (2004) found that in central Florida Bay, urea and AA uptake rates contributed substantially to the total N uptake. A recent study by Van Engeland et al. (2013) found that specific uptake rates of amino acids by microbes were inversely related to the structural complexity of the compounds with glycine having the highest rates followed by L-leucine and L-phenylalanine. Studies in the Mediterranean Sea and the Pacific Ocean near California, indicated that ~60% of the archaea exhibit measurable DFAA uptake at nanomolar levels (Ouverney and Fuhrman, 2000).

There is increasing recognition that the utilization of DFAA and DCAA may be affected by abiotic reactions. Though we reviewed mechanisms that can make DON available, glucosylation and adsorption processes appear to be important in making labile compounds more refractory. Rates of protein utilization decrease when the protein is adsorbed to submicron particles (Nagata and Kirchman, 1996). This is potentially a very important mechanism because the surface area of colloids in the surface ocean likely exceeds that of bacteria (Schuster et al., 1998). Accordingly, a given amino acid released from a phytoplankton cell is much more likely to come into contact with colloidal material, rendering it less biologically available, than to come into direct contact with a bacterial cell. These studies suggest that competition between abiotic adsorption onto colloids and bacterial uptake can have large implications for the cycling of DOM, particularly small labile moieties such as amino acids. Schuster et al. (1998) estimated that ~11-55% of the DFAA detectable by HPLC may be adsorbed to colloidal DOM in oceanic surface waters. Borch and Kirchman (1999) demonstrated that natural bacterial populations could degrade ~92% of dissolved unprotected proteins in 72-90 h. Protein adsorbed to or present within liposomes, which were designed to mimic protein that is adsorbed or trapped within particles similar to those produced by protists, however, had substantially lower degradation rates. The fecal pellets of some flagellates are believed to be similar in structure to liposomes (Nagata and Kirchman, 1992), and viral lysis can also produce liposome-like structures (Shibata et al., 1997). Reduction in the degradation rates of organics associated with liposome-like structures may explain the presence of membrane proteins in the deep ocean DOM pool (McCarthy et al., 1998; Tanoue et al., 1996). However, it has also been shown that adsorption of DFAA can make refractory organics more bioavailable. Adsorption of DFAA to dextran and phytoplankton-derived colloidal DOM, for example, results in approximately three times more efficient utilization of dextran or colloidal DOM by marine bacteria when compared to dextran or DOM without adsorbed DFAA (Schuster et al., 1998).

Plasma membrane amino acid transporters found in the genome of the marine diatom, *Thalassiosira pseudonana* (Armbrust et al., 2004) indicate that the potential for AA uptake exists, despite the low concentrations of amino acids in the environment. This finding is supported by early autoradiography studies that measured diatom uptake of ^{14}C-labeled amino acids (Wheeler et al., 1974, 1977). Baer et al. (submitted), found amino acids supported the majority of the N nutritional requirements in the coastal Chukchi Sea during ice cover. Uptake rates of AA by the bacterial size class were often greater than that of NH_4^+ in a bloom of *Phaeocystis* within mesocosm studies in Raunefjord, Norway (Sanderson et al., 2008). Bradley et al. (2010c) showed that although DFAA uptake was negligible in phytoplankton it accounted for 11% of total N utilized by bacteria. A similar approach was used with ^{35}S-labeled methionine to estimate DFAA uptake by *Prochlorococcus* and *Synechococcus* in the South Atlantic (Zubkov and Tarran, 2005).

The first study utilizing Chip-SIP found that heterotrophic bacterial uptake of amino acids was not ubiquitous, with 6 out of 52 active taxa not incorporating ^{15}N amino acids (Mayali et al., 2011). This result casts doubt on the use of bacterial production estimates, which are largely based on uptake of leucine or thymidine. Following up on that work, Mayali et al. (2013) found further evidence of taxon-specific uptake of amino acids in an estuarine system. Additionally, they discovered that *Roseobacters* incorporated relatively more N and *Bacteroidetes* relatively more C from an amino acid mixture, potentially because they are selectively incorporating specific components of the mixture.

4 Humic Substances

Humic substances are produced from the decay of plant material, can have either terrestrial or marine (e.g., phytoplankton) origins, and constitute a large reservoir of organic C and N in aquatic systems (Bronk, 2002). Humic acids also contribute greatly to the fluorescence signal of terrestrial chromophoric (colored) DOM (CDOM; Stedmon et al., 2003) and can transport metals and nutrients to the coast where they are released (see Section V.C.4).

More recent studies have shown that DOM, including humics, are directly available to some bacteria (Coates et al., 2002; Cottrell and Kirchman, 2000; Rosenstock et al., 2005) and phytoplankton (Heil, 2005; See et al., 2006). They have also been shown to induce changes in eukaryotic and prokaryotic microbial community composition (Alonso-Sáez et al., 2009; Fagerberg et al., 2010; Sipler et al., unpublished data). These observations suggest that humic substances play a complex, biogeochemically active role in coastal ecosystems (Bronk et al., 2007; See and Bronk, 2005; Steinberg et al., 2004).

Experiments that were amended with natural humic substances, isolated from river water, revealed that phytoplankton growth and biomass formation was stimulated (Carlsson et al., 1993). The literature suggests that the N associated with humic substances can be removed via one of three mechanisms—through microbial activity (Müller-Wegener, 1988), via excision by phytoplankton cell-surface enzymes (Palenik and Morel, 1990a, 1990b; see Section V.C.1), or through photodegradation to LMW compounds by exposure to UV radiation (Geller, 1986; Kieber et al., 1990; Mopper et al., 1991).

Humic substances have been implicated as a potential source of C and N to the toxic dinoflagellate *Alexandrium catenella* (Doblin et al., 2000), and growth of another toxic dinoflagellate *Alexandrium tamarense* was shown to increase when exposed to humic substances (Gagnon et al., 2005). Uptake of humic-N into phytoplankton biomass was also measured directly using

[15]N-labeled humic substances produced in the laboratory (See and Bronk, 2005). In this experiment, non-axenic cultures of 17 recently isolated estuarine and coastal phytoplankton strains took up ^{15}N-labeled humic-N (See et al., 2006), however, elevated rates of humic-N uptake were not sustained over long periods of time suggesting that a finite pool of labile N was associated with these compounds (See et al., 2006). Two of the cultures examined were also available in axenic form. No uptake of ^{15}N-labeled humic-N was detected in the axenic cultures suggesting that at least with these two cultures, bacterial remineralization was required to make the humic-N bioavailable.

5 Other Organic Compounds

Additional studies that measure uptake of other organic N compounds such as purines, pyrimidines, and amines show that though phytoplankton and bacteria can utilize these compounds, the uptake rates are generally quite low (reviewed in Antia et al., 1991 and Bronk et al., 2007). Berg and Jørgensen (2006) found that bacterial catabolism of purines and, to a lesser extent, pyrimidines represents an important source of urea that can then be used by both phytoplankton and bacteria. Some purines and pteridines are primary excretory products that are the end products of N catabolism and their breakdown can be a source of ammonia and urea (reviewed in Antia et al., 1991). They can also serve as C and N sources for yeast, bacteria, archaea, protozoa, and phytoplankton (reviewed in Berg and Jørgensen, 2006). There is still a debate as to whether D-DNA is actually used as a source of N for bacteria; D-DNA is ~16% N and so it has the potential to be a N source. Paul et al. (1988) found evidence that D-DNA was used as a source of nucleic acids for bacteria and that it is also degraded to provide phosphate needed by the cell. Jørgensen et al. (1993) measured uptake rates of DFAA, DCAA, and D-DNA in seawater cultures, and found that D-DNA was used primarily as a source of N. When DFAA,

DCAA, and D-DNA were combined, they provided 14-49% of the net bacterial N uptake measured in that study. Using turnover times of unidentified HMW DON, estimated with $\delta^{15}N$ data, DON concentrations, and rates of primary production, Benner et al. (1997) estimated that DON remineralization could support 30-50% of daily phytoplankton N demand in the equatorial Pacific region.

Flow cytometric sorting was used to quantify *Aureococcus*-specific uptake during a bloom in Chincoteague Bay, MD. Results indicated that *A. anophagefferens* and bacteria both took up NH_4^+, NO_3^-, urea, and leucine, but that uptake of organic compounds by *A. anophagefferens* far exceeded uptake by bacteria at all stages of the bloom and for all compounds (Boneillo and Mulholland, 2014). Bronk and Glibert (1993a) used ^{15}N-labeled DON produced *in situ* in Chesapeake Bay and found that during the decline of the spring bloom, uptake rates of DON are higher than uptake rates of NH_4^+ and NO_3^-. In August, rates of DON uptake are again higher than uptake rates of NO_3^- though not higher than NH_4^+. Similar results were observed in Randers Fjord where uptake rates of ^{15}N-labeled algal-derived DON were similar in magnitude to NH_4^+ uptake in April and greater than the combined rates of NH_4^+, NO_3^-, urea, and DFAA in August (Veuger et al., 2004). The magnitude of DON rates indicates that compounds other than urea and DFAA were used. Uptake rates of a suite of organic N substrates, including urea, glutamate and DON produced using ^{15}N-labeled N_2 gas added to *Trichodesmium* collected in the Gulf of Mexico, were found to be significant in waters with high concentrations of *Karenia brevis* (Bronk et al., 2004). A number of other studies have also demonstrated the importance of release of recently fixed N to supporting growth of other organisms in the field including diatoms (Chen et al., 2011), the harmful prymnesiophyte *Karenia brevis* in the Gulf of Mexico (Bronk et al., 2004), picoplankton in the Baltic Sea (Ohlendieck et al., 2000) and the southwest Pacific (Garcia et al., 2007) as well as in cultures (Agawin et al., 2007). In a study in the Baltic, Ohlendieck et al. (2000) measured the accumulation of N recently fixed by two other N_2 fixers, *Aphanizomenon* and *Nodularia*. In two experiments in July 1995 and July 1996, they found that $7.7 \pm 2.1\%$ and $6.7 \pm 2.1\%$ of the recently fixed ^{15}N was present in the picoplankton, indicating organic release and subsequent reincorporation.

E Sinks for DON: Research Priorities

The challenge of the coming decade will be to harness the explosion in molecular data to point biogeochemical studies in new and exciting directions. Omics tools provide the potential for opening the black boxes of diversity and function for phytoplankton and bacteria. After years of catching everything on a GF/F filter, tools like flow cytometric sorting, stable isotope probing, and nanoSIMS provide unprecedented ability to discriminate between autotrophic and heterotrophic uptake, between individual species, and even between individual cells. Though we can now look at substrate utilization in even finer detail we should not lose focus of the big picture that the planet is changing and we need to understand the effects of those changes on the large-scale movement of C, N, P, and other elements. The classic tools of organic matter analysis and tracer-based rate measurements still have much to teach us.

VI SUMMARY

There has been remarkable progress in our understanding of DON in the ocean since the publication of the last edition. A single chapter can no longer do the topic justice and so we have inserted references to other reviews throughout. Here we tried to cover the basics, direct the reader to the previous edition, and focus on results published since the last edition.

A DON Concentrations in Aquatic Environments

One of the greatest accomplishments in the last decade is the production of a global map of surface ocean DON concentrations (Letscher et al., 2013a), which shows very low concentrations of DON in the central gyres and higher concentrations near land and in areas of upwelling (Figure 4.1). DON research in the Atlantic over the past decade has revealed that an estimated 40% of PN export could be supported by DON in regions of the North subtropical gyre. Vertical mixing is increasingly recognized as an important control on DON distributions. DON that is recalcitrant in surface waters can be removed by microbes after transport to the ocean interior. Vertical profiles of DON generally show a surface enrichment, and DON concentrations tend to be inversely correlated with NO_3^- concentrations as depth increases, and concentrations of DON and NO_3^- are also often inversely correlated over time in surface waters. Cross system analyses demonstrates that DON generally accounts for the largest percentage of the TDN pool (~70%), with exceptions being the deep-ocean and Southern Ocean (Figure 4.2). Concentrations of DON are generally lowest in the deep ocean, and then increase in the surface ocean through the coastal ocean, estuaries and rivers with C:N ratios increasing in parallel with concentrations (Table 4.1).

B Composition of the DON Pool

Much progress has been made in defining DON composition since the last edition. In the most general sense, the generic DON pool appears to have the following characteristics. Identifiable LMW compounds such as urea, DFAA, and DCAA make up ~5-10% of the total DON pool each (Table 4.2). Roughly 30% of the pool is HMW (>1 kDa), and of that HMW fraction, ~20-30% is comprised of hydrolyzable amino acids with the remainder being amide in form. Humic substances would be included in the HMW fraction though there are relatively few data on humic-N concentrations. This leaves a substantial fraction of the pool yet to be identified. Though there are several promising techniques to isolate and characterize DON, there is still no technique that can isolate the entire DON pool, which is a barrier to more rapid progress in this area.

C Sources of DON to the Water Column

There are two main source types for DON—autochthonous (produced within the water column) and allochthonous (from outside of the system). This review primarily focuses on autochthonous water column processes (Figure 4.3). Sources of DON include phytoplankton and N_2 fixers (via passive diffusion, active release, sloppy feeding, and viral lysis), bacteria (via passive diffusion, release of exoenzymes, bacterivory, and viral lysis), micro- and macrozooplankton (via fecal pellet dissolution and excretion; Figure 4.4), and viruses (via viral lysis and control on microbial community composition). The largest source of DON is phytoplankton, with sloppy feeding on phytoplankton by zooplankton likely being the most important mechanism of DON release. In terms of release from zooplankton directly, direct release via excretion or egestion dominates. Allochthonous sources of DON to the water column include rivers, groundwater, and atmospheric deposition. Rates of DON release were measured by monitoring changes in concentration, which produce net rates, or by using ^{15}N as a tracer, which more closely approximate gross rates. Over all rates of DON release are lowest at oceanic sites, intermediate at coastal sites, and highest in estuaries (Table 4.3). On average ~30% of inorganic N uptake is released as DON (Figure 4.5). In terms of allochthonous sources, research over the last decade suggests that DON contributes ~60% of TDN in rivers, is a significant component in groundwater, and contributes ~30% of total fixed N in atmospheric deposition with the fractions likely to increase in all three sources in the future due to anthropogenic alterations.

D Sinks for DON

This review focuses on uptake of DON by heterotrophs and autotrophs and the mechanisms that can contribute to DON bioavailability (Figure 4.6). Bioassays and ^{15}N tracer techniques are the methods used to measure rates of DON uptake. Over the past decade, our ability to partition uptake by autotrophs versus heterotrophs and into individual species has been enabled by advances in the use of flow cytometric sorting, molecular and genomic techniques, and SIP. Though heterotrophs have been traditionally considered the primary users of DON, research in the past decade has continued to provide evidence that DON can be an important source of N for phytoplankton. A number of mechanisms can contribute to DON uptake including enzymatic decomposition, pinocytosis, photochemistry, and salinity effects. Uptake rates of bulk DON and individual organic N compounds vary widely across systems and even within systems (Table 4.4). The work summarized here indicates that greater than a third of the bulk DON pool is bioavailable on the order of days to weeks. Urea uptake measurements have become much more common with mean rates increasing from the ocean, through the coastal zone into estuaries and rivers. Amino acid uptake measurements are still relatively scarce. In the case of DFAA, available data show they contribute an increasing fraction of measured N uptake as one moves from the ocean, through the coast and into estuaries. Evidence is also accumulating that suggests that humics can be important sources of N in some systems.

Acknowledgments

We thank S. Baer, Q. Roberts, M. Sanderson, and J. Spackeen for help preparing the manuscript, Boris Wawrik for help with the molecular section, Mike Lomas and an anonymous reviewer for helpful reviews, and C. Carlson and D. Hansell for their patience. This work was supported by the National Science Foundation (OCE-0960806 and ANT-1043635 to DAB). This paper is Contribution No. 3375 from the Virginia Institute of Marine Science, College of William & Mary.

References

Aarnos, H., Ylöstalo, P., Vähätalo, A.V., 2012. Seasonal phototransformation of dissolved organic matter to ammonium, dissolved inorganic carbon, and labile substrates supporting bacterial biomass across the Baltic Sea. J. Geophys. Res. 117, doi:10.1029/2010JG001633, G01004.

Abdulla, H.A.N., Minor, E.C., Dias, R.F., Hatcher, P.G., 2010. Changes in the compound classes of dissolved organic matter along an estuarine transect: a study using FTIR and ^{13}C NMR. Geochim. Cosmochim. Acta 74, 3815–3838.

Abell, J., Emerson, S., Renaud, P., 2000. Distributions of TOP, TON, and TOC in the North Pacific subtropical gyre: implications for nutrient supply in the surface ocean and remineralization in the upper thermocline. J. Mar. Res. 58, 203–222.

Abell, J., Emerson, S., Keil, R.G., 2005. Using preformed nitrate to infer decadal changes in DOM remineralization in the subtropical North Pacific. Global Biogeochem. Cycles 19, GB1008. doi:10.1029/2004, GB002285.

Agawin, N.S.R., Rabouille, S., Veldhuis, M.J.W., Servatius, L., Hol, S., van Overzee, H.M.J., et al., 2007. Competition and facilitation between unicellular nitrogen-fixing cyanobacteria and non-nitrogen-fixing phytoplankton species. Limnol. Oceanogr. 52, 2233–2248.

Agedah, E.C., Binalaiyifa, H.E., Ball, A.S., Nedwell, D.B., 2009. Sources, turnover and bioavailability of dissolved organic nitrogen (DON) in the Colne estuary, UK. Mar. Ecol. Prog. Ser. 382, 23–33.

Agusti, S., Duarte, C.M., 2000. Strong seasonality in phytoplankton cell lysis in the NW Mediterranean littoral. Limnol. Oceanogr. 45, 940–947.

Agusti, S., Duarte, C.M., 2002. Addressing uncertainties in the assessment of phytoplankton lysis rates in the sea. Limnol. Oceanogr. 47, 921–924.

Agusti, S., Satta, M.P., Mura, M.P., Benavent, E., 1998. Dissolved esterase activity as a tracer of phytoplankton lysis: evidence of high phytoplankton lysis rates in the northeastern Mediterranean. Limnol. Oceanogr. 43, 1836–1849.

Aiken, G.R., 1985. Isolation and concentration techniques for aquatic humic substances. In: Aiken, G.R., McKnight, D.M., Wershaw, R.L., MacCarthy, P. (Eds.), Humic Substances in Soil Sediment and Water. Wiley-Interscience, New York (USA).

Allen, A.E., Booth, M.G., Frischer, M.E., Verity, P.G., Zehr, J.P., Zani, S., et al., 2001. Diversity and detection of nitrate assimilation genes in marine bacteria. Appl. Environ. Microbiol. 67, 5343–5348.

Allen, A.E., Dupont, C.L., Obornik, M., Horak, A., Nunes-Nesi, A., McCrow, J.P., et al., 2011. Evolution and metabolic significance of the urea cycle in photosynthetic diatoms. Nature 473, 203–207.

Alonso, M.C., Rodriguez, V., Rodriguez, J., Borrego, J.J., 2000. Role of ciliates, flagellates and bacteriophages on the mortality of marine bacteria and on dissolved-DNA concentrations in laboratory experimental systems. J. Exp. Mar. Biol. Ecol. 244, 239–252.

Alonso-Sáez, L., Unanue, M., Latatu, A., Azua, I., Ayo, B., Artolozaga, I., et al., 2009. Changes in marine prokaryotic community induced by varying types of dissolved organic matter and subsequent grazing pressure. J. Plankton Res. 31, 1373–1383.

Alonso-Sáez, L., Waller, A.S., Mende, D.R., Bakker, K., Farnelid, H., Yager, P.L., et al., 2012. Role for urea in nitrification by polar marine Archaea. Proc. Natl. Acad. Sci. 109, 17989–17994.

Altieri, K.E., Turpin, B.J., Seitzinger, S.P., 2009. Composition of dissolved organic nitrogen in continental precipitation investigated by ultra-high resolution FT-ICR mass spectrometry. Environ. Sci. Technol. 43, 6950–6955.

Altieri, K., Hastings, M., Peters, A., Sigman, D., 2012. Molecular characterization of water soluble organic nitrogen in marine rainwater by ultra-high resolution electrospray ionization mass spectrometry. Atmos. Chem. Phys. 12, 3557–3571.

Aluwihare, L.I., Meador, T., 2008. Chemical composition of marine dissolved organic nitrogen. In: Capone, D.G., Bronk, D.A., Mulholland, M., Carpenter, E.J. (Eds.), Nitrogen in the Marine Environment. Elsevier, Amsterdam.

Aluwihare, L.I., Repeta, D.J., Pantoja, S., Johnson, C.G., 2005. Two chemically distinct pools of organic nitrogen accumulate in the ocean. Science 308, 1007–1010.

Aminot, A., Kérouel, R., 2006. The determination of total dissolved free primary amines in seawater: critical factors, optimized procedure and artifact correction. Mar. Chem. 98, 223–240.

Amon, R.M.W., Benner, R., 1994. Rapid cycling of high-molecular-weight dissolved organic matter in the ocean. Nature 369, 549–552.

Andersson, M.G.I., van Rijswijk, P., Middelburg, J.J., 2006. Uptake of dissolved inorganic nitrogen, urea and amino acids in the Scheldt estuary: comparison of organic carbon and nitrogen uptake. Aquat. Microb. Ecol. 44, 303–315.

Antia, N.J., Harrison, P.J., Oliveira, L., 1991. Phycological reviews: the role of dissolved organic nitrogen in phytoplankton nutrition, cell biology, and ecology. Phycologia 30, 1–89.

Armbrust, E.V., 2009. The life of diatoms in the world's oceans. Nature 459, 185–192.

Armbrust, E.V., Berges, J.A., Bowler, C., Green, B.R., Martinez, D., Putnam, N.H., et al., 2004. The genome of the diatom *Thalassiosira pseudonana*: ecology, evolution, and metabolism. Science 306, 79–86.

Armstrong, F.A.J., Williams, P.M., Strickland, J.D.H., 1966. Photo-oxidation of organic matter in sea water by ultra-violet radiation, analytical and other applications. Nature 211, 481–483.

Arrigo, K.R., 2005. Marine microorganisms and global nutrient cycles. Nature 437, 249–355.

Azam, F., Malfatti, F., 2007. Microbial structuring of marine ecosystems. Nat. Rev. Microbiol. 5, 782–791.

Baalousha, M., Motelica-Heino, M., Le Coustumer, P., 2006. Conformation and size of humic substances: effects of major cation concentration and type, pH, salinity and residence time. Colloids Surf. 272, 48–55.

Badr, E.A., Tappin, A.D., Achterberg, E.P., 2008. Distributions and seasonal variability of dissolved organic nitrogen in two estuaries in SW England. Mar. Chem. 110, 153–164.

Baer, S.E., Sipler, R.E., Roberts, Q., Yager, P.L., Frischer, M.E., Bronk, D.A., Submitted. Seasonal nitrogen uptake and regeneration in the western coastal Arctic.

Baer, S.E., 2013. Seasonal Nitrogen Uptake and Regeneration in the Water Column and Sea-Ice of the Western Coastal Arctic. PhD thesis. Virginia Institute of Marine Science, College of William & Mary, Williamsburg, VA, 187 p.

Benavides, M., Agawin, N.S.R., Arístegui, J., Ferriol, P., Stal, L.J., 2011. Nitrogen fixation by *Trichodesmium* and small diazotrophs in the subtropical northeast Atlantic. Aquat. Microb. Ecol. 65, 43–53.

Benavides, M., Agawin, N.S.R., Arístegui, J., Peene, J., Stal, L.J., 2013a. Dissolved organic nitrogen and carbon release by a marine unicellular diazotrophic cyanobacterium. Aquat. Microb. Ecol. 69, 69–80.

Benavides, M., Bronk, D.A., Agawin, N.S.R., Pérez-Hernández, M.D., Hernández-Guerra, A., Arístegui, J., et al., 2013b. Longitudinal variability of size-fractionated N_2 fixation and DON release rates along 24.5°N in the subtropical North Atlantic. J. Geophys. Res.-Ocean 118, 3406–3415.

Benner, R., Biddanda, B., Black, B., McCarthy, M., 1997. Abundance, size distribution, and stable carbon and nitrogen isotopic compositions of marine organic matter isolated by tangential-flow ultrafiltration. Mar. Chem. 57, 243–263.

Berg, G.M., Jørgensen, N.O.G., 2006. Purine and pyrimidine metabolism by estuarine bacteria. Aquat. Microb. Ecol. 42, 215–226.

Berg, G.M., Glibert, P.M., Lomas, M.W., Burford, M.A., 1997. Organic nitrogen uptake and growth by the chrysophyte *Aureococcus anophagefferens* during a brown tide event. Mar. Biol. 129, 377–387.

Berg, G.M., Glibert, P.M., Jørgensen, N.O.G., Balode, M., Purina, I., 2001. Variability in inorganic and organic nitrogen uptake associated with riverine nutrient input in the Gulf of Riga, Baltic Sea. Estuaries 24, 204–214.

Berg, G.M., Balode, M., Purina, I., Bekere, S., Bechemin, C., Maestrini, S.Y., et al., 2003a. Plankton community composition in relation to availability and uptake of oxidized and reduced nitrogen. Aquat. Microb. Ecol. 30, 263–274.

Berg, G.M., Repeta, D.J., La Roche, J., 2003b. The role of the picoeukaryote *Aureococcus anophagefferens* in cycling of marine high-molecular weight dissolved organic nitrogen. Limnol. Oceanogr. 48, 1825–1830.

Berges, J.A., Falkowski, P.G., 1998. Physiological stress and cell death in marine phytoplankton: induction of proteases in response to nitrogen or light limitation. Limnol. Oceanogr. 43 (1), 129–135.

Berges, J., Mulholland, M., 2008. Enzymes and cellular N cycling. In: Capone, D.G., Bronk, D.A., Mulholland, M., Carpenter, E.J. (Eds.), Nitrogen in the Marine Environment. Elsevier, San Diego.

Berman, T., Bronk, D.A., 2003. Dissolved organic nitrogen: a dynamic participant in aquatic ecosystems. Aquat. Microb. Ecol. 31, 279–305.

Berman, T., Bechemin, C., Maestrini, S.Y., 1999. Release of ammonium and urea from dissolved organic nitrogen in aquatic ecosystems. Aquat. Microb. Ecol. 16, 295–302.

Berman-Frank, I., Bidle, K.D., Haramaty, L., Falkowski, P.G., 2004. The demise of the marine cyanobacterium, *Trichodesmium* spp., via an autocatalyzed cell death pathway. Limnol. Oceanogr. 49, 997–1005.

Bertilsson, S., Stepanauskas, R., Cuadros-Hansson, R., Granéli, W., Wikner, J., Tranvik, L., et al., 1999. Photochemically induced changes in bioavailable carbon and nitrogen pools in a boreal watershed. Aquat. Microb. Ecol. 19, 47–56.

Biddanda, B., Benner, R., 1997. Carbon, nitrogen and carbohydrate fluxes during the production of particulate and dissolved organic matter by marine plankton. Limnol. Oceanogr. 42, 506–518.

Bidle, K.D., Falkowski, P.G., 2004. Cell death in planktonic, photosynthetic microorganisms. Nat. Rev. 2, 643–655.

Bjørnsen, P.K., 1988. Phytoplankton exudation of organic matter: why do healthy cells do it? Limnol. Oceanogr. 33, 151–155.

Bode, A., Varela, M., Teira, E., Fernánadez, J.A., Gonzalez, N., Varela, M., et al., 2004. Planktonic carbon and nitrogen cycling off northwest Spain: variations in production of particulate and dissolved organic pools. Aquat. Microb. Ecol. 37, 95–107.

Boneillo, G., Mulholland, M., 2014. Interannual variability influences brown tide (*Aureococcus anophagefferens*) blooms in coastal embayments. Estuar. Coast. 37, 147–163.

Borch, N.H., Kirchman, D.L., 1999. Protection of protein from bacterial degradation by submicron particles. Aquat. Microb. Ecol. 16, 265–272.

Bowler, C., Allen, A.E., Badger, J.H., Grimwood, J., Jabbari, K., Kuo, A., et al., 2008. The *Phaeodactylum* genome reveals the evolutionary history of diatom genomes. Nature 456, 239–244.

Bowler, C., Karl, D.M., Colwell, R.R., 2009. Microbial oceanography in a sea of opportunity. Nature 459, 180–184.

Bradley, P., Lomas, M., Bronk, D.A., 2010a. Inorganic and organic nitrogen use by phytoplankton along Chesapeake Bay, measured using a flow cytometric sorting approach. Estuar. Coast. 33, 971–984.

Bradley, P., Sanderson, M.P., Frischer, M.E., Brofft, J., Booth, M.G., Kerkhof, L.J., et al., 2010b. Inorganic and organic nitrogen uptake by phytoplankton and heterotrophic bacteria in the stratified Mid-Atlantic Bight. Estuar. Coast. Shelf Sci. 88, 429–441.

Bradley, P., Sanderson, M.P., Nejstgaard, J.C., Sazhin, A.F., Killberg, L.M., Verity, P.G., et al., 2010c. Nitrogen uptake by phytoplankton and bacteria during an induced *Phaeocystis pouchetii* bloom, measured using size fractionation and flow cytometric sorting. Aquat. Microb. Ecol. 61, 89–104.

Breitbart, M., 2012. Marine viruses: truth or dare. Ann. Rev. Mar. Sci. 4, 425–448.

Bronk, D.A., 1999. Rates of NH_4^+ uptake, intracellular transformation, and dissolved organic nitrogen release in two clones of marine *Synechococcus* spp. J. Plankton Res. 21, 1337–1353.

Bronk, D.A., 2002. Dynamics of DON. In: Hansell, D.A., Carlson, C.A. (Eds.), Biogeochemistry of Marine Dissolved Organic Matter. Academic Press, San Diego.

Bronk, D.A., Flynn, K.J., 2006. Algal cultures as a tool to study the cycling of dissolved organic nitrogen. In: Rao, D.V.S. (Ed.), Algal Cultures, Analogues of Blooms and Applications. Oxford & IBH Publishing Co. Pvt. Ltd., New Delhi.

Bronk, D.A., Glibert, P.M., 1991. A ^{15}N tracer method for the measurement of dissolved organic nitrogen release by phytoplankton. Mar. Ecol. Prog. Ser. 77, 171–182.

Bronk, D.A., Glibert, P.M., 1993a. Application of a ^{15}N tracer method to the study of dissolved organic nitrogen uptake during spring and summer in Chesapeake Bay. Mar. Biol. 115, 501–508.

Bronk, D.A., Glibert, P.M., 1993b. Contrasting patterns of dissolved organic nitrogen release by two size fractions of estuarine plankton during a period of rapid NH_4^+ consumption and NO_2^- production. Mar. Ecol. Prog. Ser. 96, 291–299.

Bronk, D.A., Steinberg, D.K., 2008. Nitrogen regeneration. In: Capone, D.G., Bronk, D.A., Carpenter, E.J., Mulholland, M.R. (Eds.), Nitrogen in the Marine Environment, second ed. Elsevier Inc. Burlington, MA.

Bronk, D.A., Ward, B.B., 1999. Gross and net nitrogen uptake and DON release in the euphotic zone of Monterey Bay, California. Limnol. Oceanogr. 44, 573–585.

Bronk, D.A., Ward, B.B., 2000. Magnitude of DON release relative to gross nitrogen uptake in marine systems. Limnol. Oceanogr. 45, 1879–1883.

Bronk, D.A., Ward, B.B., 2005. Inorganic and organic nitrogen cycling in the Southern California Bight. Deep Sea Res. Part I: Oceanogr. Res. Pap. 52, 2285–2300.

Bronk, D.A., Glibert, P.M., Ward, B.B., 1994. Nitrogen uptake, dissolved organic nitrogen release, and new production. Science 265, 1843–1846.

Bronk, D.A., Glibert, P.M., Malone, T.C., Banahan, S., Sahlsten, E., 1998. Inorganic and organic nitrogen cycling in Chesapeake Bay: autotrophic versus heterotrophic processes and relationships to carbon flux. Aquat. Microb. Ecol. 15, 177–189.

Bronk, D.A., Lomas, M.W., Glibert, P.M., Schukert, K.J., Sanderson, M.P., 2000. Total dissolved nitrogen analysis: comparisons between the persulfate, UV and high temperature oxidation methods. Mar. Chem. 69, 163–178.

Bronk, D.A., Sanderson, M.P., Mulholland, M.R., Heil, C.A., O'Neil, J.M., 2004. Organic and inorganic nitrogen uptake kinetics in field populations dominated by *Karenia brevis*. In: Steidinger, K., Vargo, G.A., Heil, C.A. (Eds.), Harmful Algae, 2002. Florida Fish and Wildlife Conservation Commission, Florida Institute of Oceanography and Intergovernmental Oceanographic Commission of UNESCO, St. Petersburg, FL, pp. 80–82.

Bronk, D.A., See, J.H., Bradley, P., Killberg, L., 2007. DON as a source of bioavailable nitrogen for phytoplankton. Biogeosciences 4, 283–296.

Bronk, D.A., Roberts, Q., Sanderson, M.P., Canuel, E., Hatcher, P.G., Mesfioui, R., et al., 2010. Effluent Organic Nitrogen (EON): bioavailability and photochemical and salinity-mediated release. Environ. Sci. Technol. 44, 5830–5835.

Bronk, D.A., Killberg-Thoreson, L., Sipler, R.E., Mulholland, M.R., Roberts, Q.N., Bernhardt, P.W., et al., in press. Nitrogen uptake and regeneration (ammonium regeneration, nitrification, and photoproduction) in waters of the west Florida shelf prone to blooms of Karenia brevis. Harmful Algae.

Brum, J.R., 2005. Concentration, production and turnover of viruses and dissolved DNA pools at Stn ALOHA, North Pacific Subtropical Gyre. Aquat. Microb. Ecol. 41, 103–113.

Brum, J.R., Steward, G.F., Karl, D.M., 2004. A novel method for the measurement of dissolved deoxyribonucleic acid in seawater. Limnol. Oceanogr. Methods 2, 248–255.

Brussaard, C.P.D., 2004. Viral control of phytoplankton populations—a review. J. Eukaryotic Microbiol. 51, 125–138.

Brussaard, C.P.D., Riegman, R., Noordeloos, A.A.M., Cadee, G.C., Witte, H., Kop, A.J., et al., 1995. Effects of grazing, sedimentation and phytoplankton cell lysis on the structure of a coastal pelagic food web. Mar. Ecol. Prog. Ser. 123, 259–271.

Brussaard, C.P.D., Noordeloos, A.A.M., Riegman, R., 1997. Autolysis kinetics of the marine diatom *Ditylum brightwellii* (Bacillariophyceae) under nitrogen and phosphorus limitation and starvation. J. Phycol. 33, 980–987.

Buckley, D.H., Huangyutitham, V., Hsu, S.-F., Nelson, T.A., 2007. Stable isotope probing with ^{15}N achieved by disentangling the effects of genome G+C content and isotope enrichment on DNA density. Appl. Environ. Microbiol. 73, 3189–3195.

Bury, S.J., Owens, N.J.P., Preston, T., 1995. ^{13}C and ^{15}N uptake by phytoplankton in the marginal ice zone of the Bellingshausen Sea. Deep-Sea Res. II. 42, 1225–1252.

Bushaw, K.L., Zepp, R.G., Tarr, M.A., Schulz-Jander, D., Bourbonniere, R.A., Hodson, R., et al., 1996. Photochemical release of biologically labile nitrogen from dissolved organic matter. Nature 381, 404–407.

Butler, E.I., Knox, S., Liddicoat, M.I., 1979. The relationship between inorganic and organic nutrients in seawater. J. Mar. Biol. Assoc. U.K 59, 239–250.

C Fernández, I., Raimbault, P., 2007. Nitrogen regeneration in the NE Atlantic Ocean and its impact on seasonal new, regenerated and export production. Mar. Ecol. Prog. Ser. 337, 79–92.

Calleja, M.L., Batista, F., Peacock, M., Kudela, R., McCarthy, M.D., 2013. Changes in compound specific $\delta^{15}N$ amino acid signatures and D/L ratios in marine dissolved organic matter induced by heterotrophic bacterial reworking. Mar. Chem. 149, 32–44.

Cape, J.N., Cornell, S.E., Jickells, T.D., Nemitz, E., 2011. Organic nitrogen in the atmosphere—where does it come from? a review of sources and methods. Atmos. Res. 102, 30–48.

Capone, D., Ferrier, M., Carpenter, E., 1994. Amino acid cycling in colonies of the planktonic marine cyanobacterium *Trichodesmium thiebautii*. Appl. Environ. Microbiol. 60, 3989–3995.

Carlson, C.A., 2002. Production and removal processes. In: Hansell, D.A., Carlson, C.A. (Eds.), Biogeochemistry of Marine Dissolved Organic Matter. Academic Press, Amsterdam.

Carlson, C.A., Giovannoni, S.J., Hansell, D.A., Goldberg, S.J., Parsons, R., Vergin, K., et al., 2004. Interactions among dissolved organic carbon, microbial processes, and community structure in the mesopelagic zone of the northwestern Sargasso Sea. Limnol. Oceanogr. 49, 1073–1083.

Carlsson, P., Granéli, E., 1993. Availability of humic bound nitrogen for coastal phytoplankton. Estuar. Coast. Shelf Sci. 36, 433–447.

Carlsson, P., Segatto, A.Z., Granéli, E., 1993. Nitrogen bound to humic matter of terrestrial origin—a nitrogen pool for coastal phytoplankton? Mar. Ecol. Prog. Ser. 97, 105–116.

Carlsson, P., Granéli, E., Tester, P., Boni, L., 1995. Influences of riverine humic substances on bacteria, protozoa, phytoplankton, and copepods in a coastal plankton community. Mar. Ecol. Prog. Ser. 127, 213–221.

Carman, K.R., Dobbs, F.C., 1997. Epibiotic microorganisms on copepods and other marine crustaceans. Microsc. Res. Tech. 37, 116–135.

Carpenter, E.J., Dunham, S., 1985. Nitrogenous nutrient uptake, primary production, and species composition of phytoplankton in the Carmans River estuary, Long Island, New York. Limnol. Oceanogr. 30(3), 513–526.

Chang, F.H., Vincent, W.F., Woods, P.H., 1989. Nitrogen assimilation by three size fractions of the winter phytoplankton off Westland, New Zealand. New Zealand. J. Mar. Freshwat. Res. 23, 491–505.

Chang, F.H., Vincent, W.F., Woods, P.H., 1992. Nitrogen-utilization by size-fractionated phytoplankton assemblages associated with an upwelling even off Westland, New Zealand. New Zealand. J. Mar. Freshwat. Res. 26, 287–301.

Chang, F.H., Bradford-Grieve, J.M., Vincent, W.F., Woods, P.H., 1995. Nitrogen uptake by summer size-fractionated phytoplankton assemblages in the Westland, New Zealand, upwelling system, New Zealand. J. Mar. Freshwat. Res. 29, 147–161.

Chen, Y., Murrell, J.C., 2010. When metagenomics meets stable-isotope probing: progress and perspectives. Trends Microbiol. 18, 157–163.

Chen, Y.L., Tuo, S., Chen, H., 2011. Co-occurrence and transfer of fixed nitrogen from *Trichodesmium* spp. to diatoms in the low-latitude Kuroshio Current in the NW Pacific. Mar. Ecol. Prog. Ser. 421, 25–38.

Cho, B.C., Azam, F., 1995. Urea decomposition by bacteria in the Southern California Bight and its implications for the mesopelagic nitrogen cycle. Mar. Ecol. Prog. Ser. 122, 21–26.

Cho, B., Park, M., Shim, J., Azam, F., 1996. Significance of bacteria in urea dynamics in coastal surface waters. Mar. Ecol. Prog. Ser. 142, 19–26.

Church, M.J., Ducklow, H.W., Karl, D.M., 2002. Multiyear increases in dissolved organic matter inventories at Station ALOHA in the North Pacific Subtropical Gyre. Limnol. Oceanogr. 47, 1–10.

Coates, J.D., Cole, K.A., Chakraborty, R., O'Connor, S.M., Achenbach, L.A., 2002. Diversity and ubiquity of bacteria capable of utilizing humic substances as electron donors for anaerobic respiration. Appl. Environ. Microbiol. 68, 2445.

Cochlan, W.P., Bronk, D.A., 2003. Effects of ammonium on nitrate utilization in the Ross Sea, Antarctica: implications for f-ratio estimates. In: Ditullio, G.R., Dunbar, R.B. (Eds.), Biogeochemistry of the Ross Sea. American Geophysical Union, Washington, D. C..

Collier, J.L., 2004. Flow cytometry and the single compound in plankton ecology. J. Phycol. 40, 805–807.

Collier, J.L., Baker, K.M., Bell, S.L., 2009. Diversity of urea-degrading microorganisms in open-ocean and estuarine planktonic communities. Environ. Microbiol. 11, 3118–3131.

Collins, R.E., Deming, J.W., 2011. Abundant dissolved genetic material in Arctic sea ice Part I: extracellular DNA. Polar Biol. 34, 1819–1830.

Collos, Y., 1986. Time-lag algal growth dynamics: biological constraints on primary production in aquatic environments. Mar. Ecol. Prog. Ser. 33, 193–206.

Collos, Y., Dohler, G., Biermann, I., 1992. Production of dissolved organic nitrogen during uptake of nitrate by *Synedra planctonica*: implications for estimating new production in the oceans. J. Plankton Res. 14, 1025–1029.

Condon, R.H., Steinberg, D.K., del Giorgio, P.A., Bouvier, T.C., Bronk, D.A., Graham, W.M., Ducklow, H.W., 2011. Jellyfish blooms result in a major microbial respiratory sink of carbon in marine systems. Proc. Natl. Acad. Sci. U. S. A. 108 (25), 10225–10230.

Conover, R.J., Gustavson, K.R., 1999. Sources of urea in arctic seas: zooplankton metabolism. Mar. Ecol. Prog. Ser. 179, 41–54.

Cooper, L.W., McClelland, J.W., Holmes, R.M., Raymond, P.A., Gibson, J.J., Guay, C.K., et al., 2008. Flow-weighted values of runoff tracers ($\delta^{18}O$, DOC, Ba, alkalinity) from the six largest Arctic rivers. Geophys. Res. Lett. 35, doi:10.1029/2008GL035007.

Cooper, J.T., Wawrik, B., Baer, S.E., Connelly, T.L., Bronk, D.A., Submitted. Analysis of ammonium, nitrate, and urea uptake by marine pelagic Bacteria and Archaea during the Arctic summer and winter seasons via stable isotope probing.

Corinaldesi, C., Dell' Anno, A., Danovaro, R., 2007. Viral infection plays a key role in extracellular DNA dynamics in marine anoxic systems. Limnol. Oceanogr. 52, 508–516.

Cornell, S.E., 2011. Atmospheric nitrogen deposition: revisiting the question of the importance of the organic component. Environ. Pollut. 159, 2214–2222.

Cornell, S., Rendell, A., Jickells, T., 1995. Atmospheric inputs of dissolved organic nitrogen to the oceans. Nature 376, 243–246.

Cornell, S.E., Mace, K., Coeppicus, S., Duce, R., Huebert, R., Jickells, T., et al., 2001. Organic nitrogen in Hawaiian rain and aerosol. J. Geophys. Res. 106, 7973–7983.

Cornell, S.E., Jickells, T.D., Cape, J.N., Rowland, A.P., Duce, R.A., 2003. Organic nitrogen deposition on land and coastal environments: a review of methods and data. Atmos. Environ. 37, 2173–2191.

Cottrell, M., Kirchman, D.L., 2000. Natural assemblages of marine proteobacteria and members of the *Cytophaga-Flavobacter* cluster consuming low-and high-molecular-weight dissolved organic matter. Appl. Environ. Microbiol. 66, 1692–1697.

Crandall, J., Teece, M., 2012. Urea is a dynamic pool of bioavailable nitrogen in coral reefs. Coral Reefs 31, 207–214.

Dagg, M.J., 1974. Loss of prey body contents during feeding by an aquatic predator. Ecology 55, 9903–9906.

Dagg, M., Benner, R., Lohrenz, S., Lawrence, D., 2004. Transformation of dissolved and particulate materials on continental shelves influenced by large rivers: plume processes. Cont. Shelf Res. 24, 833–858.

Daly, K.L., Wallace, D.W.R., Smith Jr., W.O., Skoog, A., Lara, R., Gosselin, M., et al., 1999. Non-Redfield carbon and nitrogen cycling in the Arctic: effects of ecosystem structure and dynamics. J. Geophys. Res. 104, 3185–3199.

D'Andrilli, J., Dittmar, T., Koch, B.P., Purcell, J.M., Marshall, A.G., Cooper, W.T., et al., 2010. Comprehensive characterization of marine dissolved organic matter by Fourier transform ion cyclotron resonance mass spectrometry with electrospray and atmospheric pressure photoionization. Rapid Commun. Mass Spectrom. 24, 643–650.

Davidson, K., Gilpin, L.C., Hart, M.C., Fouilland, E., Mitchell, E., Calleja, I.Á., et al., 2007. The influence of the balance of inorganic and organic nitrogen on the trophic dynamics of microbial food webs. Limnol. Oceanogr. 52, 2147–2163.

Davis, J., Benner, R., 2005. Seasonal trends in the abundance, composition and bioavailability of particulate and dissolved organic matter in the Chukchi/Beaufort Seas and western Canada Basin. Deep Sea Res. Part II: Top. Stud. Oceanogr. 52, 3396–3410.

De Jesus, R.P., 2008. Natural Abundance Radiocarbon Studies of Dissolved Organic Carbon (DOC) in the Marine Environment. ProQuest. San Diego, CA.

DeLong, E., 2009. The microbial ocean from genomes to biomes. Nature 459, 200–206.

Diaz, F., Raimbault, P., 2000. Nitrogen regeneration and dissolved organic nitrogen release during spring in a NW Mediterranean coastal zone (Gulf of Lions): implications for the estimation of new production. Mar. Ecol. Prog. Ser. 197, 51–65.

Dittmar, T., Fitznar, H.P., Kattner, G., 2001. Origin and biogeochemical cycling of organic nitrogen in the eastern Arctic Ocean as evident from D- and L-amino acids. Geochim. Cosmochim. Acta 65, 4103–4114.

Dittmar, T., Koch, B., Hertkorn, N., Kattner, G., 2008. A simple and efficient method for the solid-phase extraction of dissolved organic matter (SPE-DOM) from seawater. Limnol. Oceanogr. Methods 6, 230–235.

Doblin, M., Legrand, C., Carlsson, P., Hummert, C., Granéli, E., Hellegraeff, G., 2000. Uptake of humic substances by the toxic dinoflagellate *Alexandrium catenella*. In: Harmful Algal Blooms, pp. 336–339. Proceedings of the 9th International Conference - Harmful Algal Blooms 2000, Hobart, Australia.

Doney, S.C., Abbott, M.R., Cullen, J.J., Karl, D.M., Rothstein, L., 2004. From genes to ecosystems: the ocean's new frontier. Front. Ecol. Environ. 2, 457–466.

Duce, R.A., LaRoche, J., Altieri, K., Arrigo, K.R., Baker, A.R., Capone, D.G., et al., 2008. Impacts of atmospheric antropogenic nitrogen on the open ocean. Science 320, 893–897.

Ducklow, H.W., Hansell, D.A., Morgan, J.A., 2007. Dissolved organic carbon and nitrogen in the Western Black Sea. Mar. Chem. 105, 140–150.

Dufresne, A., Salanoubat, M., Partensky, F., Artiguenave, F., Axmann, I.M., Barbe, V., et al., 2003. Genome sequencing of the cyanobacterium *Prochlorococcus marinus* SS120, a nearly minimal oxyphototrophic genome. Proc. Natl. Acad. Sci. 100, 10020–10025.

Dumont, M.G., Murrell, J.C., 2005. Stable isotope probing—linking microbial identity to function. Nat. Rev. Microbiol. 3, 499–504.

Dyhrman, S.T., Anderson, D.M., 2003. Urease activity in cultures and field populations of the toxic dinoflagellate, *Alexandrium*. Limnol. Oceanogr. 48, 647–655.

Eppley, R.W., Peterson, B.J., 1979. Particulate organic matter flux and planktonic new production in the deep ocean. Nature 282, 677–680.

Facchini, M.C., Decesari, S., Rinaldi, M., Carbone, C., Finessi, E., Mircea, M., et al., 2008. Important source of marine secondary organic aerosol from biogenic amines. Environ. Sci. Tech. 42, 9116–9121.

Fagerberg, T., Jephson, T., Carllson, P., 2010. Molecular size of riverine dissolved organic matter influences coastal and phytoplankton communities. Mar. Ecol. Prog. Ser. 409, 17–25.

Fan, C., Glibert, P., Burkholder, J., 2003. Characterization of the affinity for nitrogen, uptake kinetics, and environmental relationships for *Prorocentrum minimum* in natural blooms and laboratory cultures. Harmful Algae 2, 283–299.

Fernández, E., Marañón, E., Harbour, D.S., Kristiansen, S., Heimdal, B.R., 1996. Patterns of carbon and nitrogen uptake during blooms of *Emiliania huxleyi* in two Norwegian fjords. J. Plankton Res. 18, 2349–2366.

Ferrier-Pagès, C., Karner, M., Rassoulzadegan, F., 1998. Release of dissolved amino acids by flagellates and ciliates grazing on bacteria. Oceanol. Acta 21, 485–494.

Feuerstein, T., Ostrom, P., Ostrom, N., 1997. Isotope biogeochemistry of dissolved organic nitrogen: a new technique and application. Org. Geochem. 27, 363–370.

Finzi-Hart, J.A., Pett-Ridge, J., Weber, P.K., Popa, R., Fallon, S.J., Gunderson, T., et al., 2009. Fixation and fate of C and N in the cyanobacterium *Trichodesmium* using nanometer-scale secondary ion mass spectrometry. Proc. Natl. Acad. 106, 6345–6350.

Flerus, R., Lechtenfeld, O., Koch, B.P., McCallister, S., Schmitt-Kopplin, P., Benner, R., et al., 2012. A molecular perspective on the ageing of marine dissolved organic matter. Biogeosciences 9, 1935–1955.

Flynn, K.J., Berry, L.S., 1999. The loss of organic nitrogen during marine primary production may be significantly overestimated when using ^{15}N substrates. Proc. R. Soc. Lond. B 266, 641–647.

Flynn, K.J., Davidson, K., 1993. Predator-prey interactions between *Isochrysis galbana* and *Oxyrrhis marina* II. Prelease of non-protein amines and faeces during predation of Isochrysis. J. Plankton Res. 15, 893–905.

Flynn, K.J., Stoecker, D.K., Mitra, A., Raven, J.A., Glibert, P.M., Hansen, P.J., et al., 2013. Misuse of the phytoplankton-zooplankton dichotomy: the need to assign organisms as

mixotrophs within plankton functional types. J. Plankton Res. 35, 3–11.

Fogg, G.E., 1966. The extracellular products of algae. Oceanogr. Mar. Biol. Ann. Rev. 4, 195–212.

Fouilland, E., Gosselin, M., Rivkin, R.B., Vasseur, C., Mostajir, B., 2007. Nitrogen uptake by heterotrophic bacteria and phytoplankton in Arctic surface waters. J. Plankton Res. 29, 369–376.

Friedrich, M.W., 2006. Stable-isotope probing of DNA: insights into the function of uncultivated microorganisms from isotopically labeled metagenomes. Curr. Opin. Biotechnol. 17, 59–66.

Frischer, M.G., Allen, M.J., Wilson, W.H., Suttle, C.A., 2010. Giant virus with a remarkable complement of genes infects marine zooplankton. Proc. Natl. Acad. 107, 19508–19513. doi:10.1073/pnas.1007615107.

Fuhrman, J., 1987. Close coupling between release and uptake of dissolved free amino acids in seawater studied by an isotope dilution approach. Mar. Ecol. Prog. Ser. 37, 45–52.

Fuhrman, J., 1990. Dissolved free amino acid cycling in an estuarine outflow plume. Mar. Ecol. Prog. Ser. 66, 197–203.

Gagnon, R., Levasseur, M., Weise, A.M., Fauchot, J., 2005. Growth stimulation of *Alexandrium tamarense* (Dinophyceae) by humic substances from the Manicouagan River (eastern Canada). J. Phycol. 41, 489–497.

Gagosian, R.B., Lee, C., 1981. Processes controlling the distribution of biogenic organic compounds in seawater. In: Duursma, E.K., Dawson, R. (Eds.), Marine Organic Chemistry: Evolution, Composition, Interactions and Chemistry of Organic Matter in Seawater. Elsevier, Amsterdam.

Gallagher, E., McGuinness, L., Phelps, C., Young, L.Y., Kerkhof, L.J., 2005. ^{13}C-Carrier DNA shortens the incubation time needed to detect benzoate-utilizing denitrifying bacteria by stable-isotope probing. Appl. Environ. Microbiol. 71, 5192–5196.

Gandhi, N., Ramesh, R., Srivastava, R., Sheshshayee, M.S., Dwivedi, R.M., Raman, M., 2010. Nitrogen uptake rates during spring in the NE Arabian Sea. Int. J. Oceanogr. 2010, 1–10.

Garcia, N., Raimbault, P., Sandroni, V., 2007. Seasonal nitrogen fixation and primary production in the Southwest Pacific: nanoplankton diazotrophy and transfer of nitrogen to picoplankton organisms. Mar. Ecol. Prog. Ser. 343, 25–33.

Gardner, W.S., Seitzinger, S.P., Malczyk, J.M., 1991. The effects of sea salts on the forms of nitrogen released from estuarine and freshwater sediments: does ion pairing affect ammonium flux? Estuaries 14, 157–166.

Gardner, W.S., Cavaletto, J.F., Bootsma, H.A., Lavrentyev, P.J., Tanvone, F., 1998. Nitrogen cycling rates and light effects in tropical Lake Maracaibo, Venezuela. Limnol. Oceanogr. 43, 1814–1825.

Gasol, J.M., Morán, X.A.G., 1999. Effects of filtration on bacterial activity and picoplankton community structure as assessed by flow cytometry. Aquat. Microb. Ecol. 16, 251–264.

Geller, A., 1986. Comparison of mechanisms enhancing biodegradability of refractory lake water constituents. Limnol. Oceanogr. 31, 755–764.

Gibb, S.W., Mantoura, R.F.C., Liss, P.S., Barlow, R.G., 1999. Distributions and biogeochemistries of methylamines and ammonium in the Arabian Sea. Deep-Sea Res. II. 46, 593–615.

Giovannoni, S.J., Britschgi, T.B., Moyer, C.L., Field, K.G., 1990. Genetic diversity in Sargasso Sea bacterioplankton. Nature 345, 60–63.

Glibert, P.M., Bronk, D.A., 1994. Release of dissolved organic nitrogen by marine diazotrophic cyanobacteria Trichodesmium spp. Appl. Environ. Microbiol. 36, 3996–4000.

Glibert, P.M., Garside, C., 1992. Diel variability in nitrogenous nutrient uptake by phytoplankton in the Chesapeake Bay. J. Plankton Res. 14, 271–288.

Glibert, P.M., Legrand, C., 2006. The diverse nutrient strategies of harmful algae: focus on osmotropy. In: Graneli, E., Turner, J.T. (Eds.), Ecology of Harmful Algae. Ecological Studies, Vol. 189. Springer-Verlag, Berlin.

Glibert, P.M., Terlizzi, D.E., 1999. Cooccurrence of elevated urea levels and dinoflagellate blooms in temperate estuarine aquaculture ponds. Appl. Environ. Microbiol. 65, 5594–5596.

Glibert, P.M., Dennett, M.R., Caron, D.A., 1988. Nitrogen uptake and ammonium regeneration by pelagic microplankton and marine snow from the North Atlantic. J. Mar. Res. 46, 837–852.

Glibert, P.M., Miller, C.A., Garside, C., Roman, M.R., McManus, G.B., 1992. NH_4^+ regeneration and grazing: interdependent processes in size-fractionation $^{15}NH_4^+$ experiments. Mar. Ecol. Prog. Ser. 82, 65–74.

Glibert, P.M., Conley, D.J., Fisher, T.R., Harding, J., Lawrence, W., Malone, T.C., 1995. Dynamics of the 1990 winter/spring bloom in Chesapeake Bay. Mar. Ecol. Prog. Ser. 122, 27–43.

Glibert, P.M., Heil, C.A., Hollander, D., Revilla, M., Hoare, A., Alexander, J., et al., 2004. Evidence for dissolved organic nitrogen and phosphorus uptake during a cyanobacterial bloom in Florida Bay. Mar. Ecol. Prog. Ser. 280, 73–83.

Glibert, P., Trice, T.M., Michael, B., Lane, L., 2005. Urea in the tributaries of the Chesapeake and coastal bays of Maryland. Water Air Soil Poll. 160, 229–243.

Glibert, P.M., Harrison, J., Heil, C., Seitzinger, S., 2006. Escalating worldwide use of urea—a global change contributing to coastal eutrophication. Biogeochemistry 77, 441–463.

Gobler, C., Sañudo-Wilhelmy, S.A., 2001. Temporal variability of groundwater seepage and brown tide blooms in a Long Island embayment. Mar. Ecol. Prog. Ser. 217, 299–309.

Gobler, C.J., Renaghan, M.J., Buck, N.J., 2002. Impacts of nutrients and grazing mortality on the abundance of Aureococcus anaphagefferens during a New York brown tide bloom. Limnol. Oceanogr. 47, 129–141.

Goeyens, L., Kindermans, N., Abu Yusuf, M., Elskens, M., 1998. A room temperature procedure for the manual determination of urea in seawater. Estuar. Coast. Shelf Sci. 47 (4), 415–418.

Gomez-Gutierrez, J., Peterson, W.T., De Robertis, A., Brodeur, R.D., 2003. Mass mortality of krill caused by parasitoid ciliates. Nature 301, 339.

Gonsior, M., Peake, B.M., Cooper, W.T., Podgorski, D.C., D'Andrilli, J., Dittmar, T., et al., 2011. Characterization of dissolved organic matter across the Subtropical Convergence off the South Island, New Zealand. Mar. Chem. 123, 99–110.

Goolsby, D.A., Battaglin, W.A., Aulenbach, B.T., Hooper, R.P., 2001. Nitrogen input to the Gulf of Mexico. J. Environ. Qual. 30, 329–336.

Gordeev, V.V., Martin, J.M., Sidorov, I.S., Sidorova, M.V., 1996. A reassessment of the Euarasian river input of water, sediment, major elements, and nutrients to the Arctic Ocean. Am. J. Sci. 296, 664–691.

Graber, E.R., Rudich, Y., 2006. Atmospheric HULIS: how humic-like are they? A comprehensive and critical review. Atmos. Chem. Phys. 6, 729–753.

Grzybowski, W., Tranvik, L., 2008. Phototransformations of dissolved organic nitrogen. In: Capone, D.G., Bronk, D.A., Mulholland, M., Carpenter, E.J. (Eds.), Nitrogen in the Marine Environment.

Gurtler, B.K., Vetter, T.A., Perdue, E.M., Ingall, E., Koprivnjak, J.F., Pfromm, P.H., et al., 2008. Combining reverse osmosis and pulsed electrical current electrodialysis for improved recovery of dissolved organic matter from seawater. J. Membr. Sci. 323, 328–336.

Hansell, D.A., Carlson, C.A., 2001. Biogeochemistry of total organic carbon and nitrogen in the Sargasso Sea: control by convective overturn. Deep Sea Res. Part II: Top. Stud. Oceanogr. 48, 1649–1667.

Hansell, D.A., Goering, J.J., 1989. A method for estimating uptake and production rates for urea in seawater using ^{14}C urea and ^{15}N urea. J. Fisheries Aquat. Sci. 46, 198–202.

Hansell, D.A., Williams, P.M., Ward, B.B., 1993. Measurements of DOC and DON in the Southern California Bight using oxidation by high temperature combustion. Deep Sea Res. Part I: Oceanogr. Res. Pap. 40, 219–234.

Harrison, W.G., Platt, T., Lewis, M.R., 1987. f-Ratio and its relationship to ambient nitrate concentrations in coastal waters. J. Plankton Res. 9, 235–248.

Hartmann, M., Zubkov, M.V., Martin, A.P., Scalan, D.J., Burkill, P.H., 2009. Assessing amino acid uptake by phototrophic nanoflagellates in nonaxenic cultures using flow cytometric sorting. FEMS Microbiol. Lett. 298, 166–173.

Harvey, G.R., Boran, D.A., 1985. Geochemistry of humic substances in seawater. In: Aiken, G.R., Mcknight, D.M., Wershaw, R.L., MacCarthy, P. (Eds.), Humic Substances in Soil Sediment and Water. John Wiley and Sons, New York.

Harwood, C.R., Cranenburgh, R., 2008. Bacillus protein secretion: an unfolding story. Trends Microbiol. 16 (2), 73–79.

Hasegawa, T., Koike, I., Mukai, H., 2000a. Release of dissolved organic nitrogen by size-fractionated natural planktonic assemblages in coastal waters. Mar. Ecol. Prog. Ser. 198, 43–49.

Hasegawa, T., Koike, I., Mukai, H., 2000b. Estimation of dissolved organic nitrogen release by micrograzers in natural plankton assemblages. Plankton Biol. Ecol. 47, 23–30.

Hatch, M.J., Dillon, J.A., Smith, H.B., 1957. Preparation and use of snake-cage polyelectrolytes. Ind. Eng. Chem. 49, 1812–1818.

Hawkins, L.N., Russell, L.M., 2010. Polysaccharides, proteins, and phytoplankton fragments: four chemically distinct types of marine primary organic aerosol classified by single particle spectromicroscopy. Adv. Meteorol. 2010, 1–14 .

Hedges, J.I., 1988. Polymerization of humic substances in natural environments. In: Frimmel, F.H., Christman, R.F. (Eds.), Humic Substances and Their Role in the Environment. John Wiley and Sons Limited, New York.

Hedges, J.I., Blanchette, R.A., Weliky, K., Devol, A.H., 1988. Effects of fungal degradation on the CuO oxidation products of lignin: a controlled laboratory study. Geochim. Cosmochim. Acta 52, 2717–2726.

Heil, C., 2005. Influence of humic, fulvic and hydrophilic acids on the growth, photosynthisis and respiration of the dinoflagellate *Prorocentrum minimum* (Pavillard) Schiller. Harmful Algae 4, 603–618.

Herndon, J., Cochlan, W.P., 2007. Nitrogen utilization by the raphidophyte *Heterosigma akashiwo*: growth and uptake kinetics in laboratory cultures. Harmful Algae 6, 260–270.

Hertkorn, N., Benner, R., Frommberger, M., Schmitt-Kopplin, P., Witt, M., Kaiser, K., et al., 2006. Characterization of a major refractory component of marine dissolved organic matter. Geochim. Cosmochim. Acta 70, 2990–3010.

Hertkorn, N., Harir, M., Koch, B., Michalke, B., Schmitt-Kopplin, P., 2013. High-field NMR spectroscopy and FTICR mass spectrometry: powerful discovery tools for the molecular level characterization of marine dissolved organic matter. Biogeosciences 10, 1583–1624.

Hessen, D.O., Tranvik, L.J., 1998. Aquatic Humic Substance. Ecology and Biogeochemistry. Springer, Berlin.

Hewson, I., Govil, S.R., Capone, D.G., Carpenter, E.J., Fuhrman, J.A., 2004. Evidence of *Trichodesmium* viral lysis and potential significance for biogeochemical cycling in the oligotrophic ocean. Aquat. Microb. Ecol. 36, 1–8.

Holfeld, H., 1998. Fungal infections of the phytoplankton: seasonality, minimal host density, and specificity in a mesotrophic lake. New Phytol. 138, 507–517.

Holmes, R., McClelland, J., Peterson, B., Tank, S., Bulygina, E., Eglinton, T., et al., 2012. Seasonal and annual fluxes of nutrients and organic matter from large rivers to the arctic ocean and surrounding seas. Estuar. Coast. 35, 369–382.

Howard, M.D.A., Cochlan, W.P., Ladizinsky, N., Kudela, R.M., 2007. Nitrogenous preference of toxigenic *Pseudo-nitzschia australis* (Bacillariophyceae) from field and laboratory experiments. Harmful Algae 6, 206–217.

Hu, S.H., Smith, W.O., 1998. The effects of irradiance on nitrate uptake and dissolved organic nitrogen release by phytoplankton in the Ross Sea. Continental Shelf Res. 18, 971–990.

Hubberten, U., Lara, R.J., Kattner, G., 1995. Refractory organic compounds in polar waters: relationship between humic substances and amino acids in the Arctic and Antarctic. J. Mar. Res. 53, 137–149.

Hutchins, D.A., Pustizzi, F., Hare, C.E., DiTullio, G.R., 2003. A shipboard natural community continuous culture system for ecologically relevant low-level nutrient enrichment experiments. Limnol. Oceanogr. Methods 1, 82–91.

Hyndes, G.A., Lavery, P.S., Doropoulos, C., 2012. Dual processes for cross-boundary subsidies: incorporation of nutrients from reef-derived kelp into a seagrass ecosystem. Mar. Ecol. Prog. Ser. 445, 97–107.

Ietswaart, T., Schneider, P.J., Prins, R.A., 1994. Utilization of organic nitrogen sources by two phytoplankton species and a bacterial isolate in pure and mixed cultures. Appl. Environ. Microbiol. 60, 1554–1560.

Ishii, N., Kawabata, Z., Nakano, S., Min, M., Takata, R., 1998. Microbial interactions responsible for dissolved DNA production in a hypereutrophic pond. Hydrobiologia 380, 67–76.

Jemlich, N., Schmidt, F., Taubert, M., Seifert, J., Bastida, F., von Bergen, M., et al., 2010. Protein-based stable isotope probing. Nat. Protocols 5, 1957–1966.

Jenkinson, I.R., Biddanda, B.A., 1995. Bulk-phase viscoelastic properties of seawater: relationship with plankton components. J. Plankton Res. 17, 2251–2274.

Jiao, N., Herndl, G.J., Hansell, D.A., Benner, R., Kattner, G., Wilhelm, S.W., et al., 2010. Microbial production of recalcitrant dissolved organic matter: long term carbon storage in the global ocean. Nat. Rev. Microbiol. 8, 593–599.

Jickells, T., Baker, A.R., Cape, J.N., Cornell, S.E., Nemitz, E., 2013. The cycling of organic nitrogen through the atmosphere. Phil. Trans. R. Soc. Lond. B Biol. Sci. 368, .

Johnson, M.T., Greenwood, N., Sivyer, D.B., Thomson, M., Reeve, A., Weston, K., et al., 2013a. Characterizing the seasonal cycle of dissolved organic nitrogen using Cefas SmartBuoy high-resolution time-series samples from the southern North Sea. Biogeochemistry 113, 23–36.

Johnson, L.T., Tank, J.L., Hall, R.O., Mulholland Jr., P.J., Hamilton, S.K., Balett, H.M., et al., 2013b. Quantifying the production of dissolved organic nitrogen in headwater streams using ^{15}N tracer additions. Limnol. Oceanogr. 58, 1271–1285.

Joint, R., Rees, A.P., Woodward, E.M.S., 2001. Primary production and nutrient assimilation in the Iberian upwelling in August 1998. Prog. Oceanogr. 51, 303–320.

Joint, I., Henriksen, P., Fonnes, G.A., Bourne, D., Thingstad, T.F., Riemann, B., et al., 2002. Competition for inorganic nutrients between phytoplankton and bacterioplankton in nutrient manipulated mesocosms. Aquat. Microb. Ecol. 29, 145–159.

Jørgensen, N.O.G., 2006. Uptake of urea by estuarine bacteria. Aquat. Microb. Ecol. 42, 227–242.

Jørgensen, N.O.G., Kroer, N., Coffin, R.B., Yang, X.-H., Lee, C., 1993. Dissolved free amino acids, combined amino acids, and DNA as sources of carbon and nitrogen to marine bacteria. Mar. Ecol. Prog. Ser. 98, 135–148.

Jørgensen, N.O.G., Kroer, N., Coffin, R.B., Hoch, M.P., 1999. Relations between bacterial nitrogen metabolism and growth efficiency in an estuarine and an open-water ecosystem. Aquat. Microb. Ecol. 18, 247–261.

Joye, S.B., Bronk, D.A., Koopmans, D.J., Moore, W.S., Joye, S.B., Bronk, D.A., et al., 2006. Evaluating the potential importance of groundwater-derived carbon, nitrogen, and phosphorus inputs to South Carolina and Georgia coastal ecosystems. In: Kleppel, G.S., De Voe, R., Rawson, M.L. (Eds.), Changing Land Use Patters in the Coastal Zone. Springer, New York, NY, pp. 139–178.

Jumars, P.A., Penry, D.L., Baross, J.A., Perry, M.J., Frost, B.W., 1989. Closing the microbial loop: dissolved carbon pathway to heterotrophic bacteria from incomplete ingestion, digestion, and absorption in animals. Deep-Sea Res. 36, 483–495.

Kaiser, K., Benner, R., 2008. Major bacterial contribution to the ocean reservoir of detrital organic carbon and nitrogen. Limnol. Oceanogr. 53, 99–113.

Kaiser, K., Benner, R., 2009. Biochemical composition and size distribution of organic matter at the Pacific and Atlantic time-series stations. Mar. Chem. 113, 63–77.

Kamennaya, N.A., Chernihovsky, M., Post, A.F., 2008. The cyanate utilization capacity of marine unicellular Cyanobacteria. Limnol. Oceanogr. 53, 2485–2494.

Kanakidou, M., Duce, R.A., Prospero, J.M., Baker, A.R., Benitez-Nelson, C., Dentener, F.J., et al., 2012. Atmospheric fluxes of organic N and P to the global ocean. Global Biogeochem. Cycles 26, .

Kanda, J., Saino, T., Hattori, A., 1985. Nitrogen uptake by natural populations of phytoplankton and primary production in the Pacific Ocean: regional variability of uptake capacity. Limnol. Oceanogr. 30, 987–999.

Karl, D.M., Bailiff, M.D., 1989. The measurement and distribution of dissolved nucleic acids in aquatic environments. Limnol. Oceanogr. 34, 543–558.

Karl, D.M., Letelier, R., Hebel, D.V., Bird, D.F., Winn, C.D., 1992. Trichodesmium blooms and new nitrogen in the North Pacific Gyre. In: Carpenter, E.J., Capone, D.G., Reuter, J.G. (Eds.), Marine Pelagic Cyanobacteria: Trichodesmium and Other Diazotrophs. Kluwer Academic Pub, Dordrecht, The Netherlands.

Karl, D.M., Bjorkman, K.M., Dore, J.E., Fujieki, L., Hebel, D.V., Houlihan, T., et al., 2001. Ecological nitrogen-to-phosphorus stoichiometry at station ALOHA. Deep-Sea Res. II. 48, 1529–1566.

Karl, D., Michaels, A., Capone, D., Carpenter, E., Letelier, R., Lipschultz, F., et al., 2002. Dinitrogen fixation in the world's oceans. Biogeochemistry 57/58, 47–98.

Kaufman, Z.G., Lively, J.S., Carpenter, E.J., 1983. Uptake of nitrogenous nutrients by phytoplankton in a barrier island estuary: Great South Bay, New York. Estuar. Coast. Shelf Sci. 17, 483–493.

Kawabata, Z., Ishii, N., Nasu, M., Min, M., 1998. Dissolved DNA produced through a prey — predator relationship in a species-defined aquatic microcosm. Hydrobiologia 387, 71–76.

Kawasaki, N., Benner, R., 2006. Bacterial release of dissolved organic matter during cell growth and decline: molecular origin and composition. Limnol. Oceanogr. 51, 2170–2180.

Keil, R.G., Kirchman, D.L., 1991. Contribution of dissolved free amino acids and ammonium to the nitrogen requirements of heterotrophic bacterioplankton. Mar. Ecol. Prog. Ser. 73, 1–10.

Keil, R.G., Kirchman, D.L., 1993. Dissolved combined amino acids: chemical form and utilization by marine bacteria. Limnol. Oceanogr. 38, 1256–1270.

Keller, M., Hettich, R., 2009. Environmental proteomics: a paradigm shift in characterizing microbial activities at the molecular level. Microbiol. Mol. Biol. Rev. 73, 62–70.

Keller, M.D., Kiene, R.P., Matrai, P.A., Bellows, W.K., 1999. Production of glycine betaine and dimethylsulfoniopropionate in marine phytoplankton. II. N-limited chemostat cultures. Mar. Biol. 135, 249–257.

Kieber, R.J., Zhou, X., Mopper, K., 1990. Formation of carbonyl compounds from UV-induced photodegration of humic substances in natural waters: fate of riverine carbon in the sea. Limnol. Oceanogr. 35, 1503–1515.

Kieber, R., Long, M., Willey, J., 2005. Factors influencing nitrogen speciation in coastal rainwater. J. Atmos. Chem. 52, 81–99.

Killberg-Thoreson, L., Reay, W.G., Roberts, Q.N., Sanderson, M.P., Bronk, D.A., 2010. Seasonal nitrogen uptake dynamics in the York River, Virginia. In: Proceedings from the 2010 AGU Ocean Sciences Meeting. American

Geophysical Union, 2000 Florida Ave., N. W. Washington DC 20009 USA, 2010.

Kim, T.-H., Kim, G., 2013. Factors controlling the C:N:P stoichiometry of dissolved organic matter in the N-limited, cyanobacteria-dominated East/Japan Sea. J. Mar. Syst. 115–116, 1–9.

Kim, T.-H., Kwon, E., Kim, I., Lee, S.-A., Kim, G., 2013. Dissolved organic matter in the subterranean estuary of a volcanic island, Jeju: importance of dissolved organic nitrogen fluxes to the ocean. J. Sea Res. 78, 18–24.

Kirchman, D.L., Keil, P.G., Wheeler, P.A., 1989. The effect of amino acids on ammonium utilization and regeneration by heterotrophic bacteria in the subarctic Pacific. Deep-Sea Res. 36, 1763–1776.

Kirchman, D.L., Moss, J., Keil, R.G., 1992. Nitrate uptake by heterotrophic bacteria: does it change the f-ratio? Arch. Hydrobiol. Beih. 37, 129–138.

Kisand, V., Rocker, D., Simon, M., 2008. Significant decomposition of riverine humic-rich DOC by marine but not estuarine bacteria assessed in sequential chemostat experiments. Aquat. Microb. Ecol. 53, 151–160.

Kitidis, V., Uher, G., 2008. Photochemiccal mineralization of dissolved organic nitrogen. In: Mertens, L.P. (Ed.), Biological Oceanography Research Trends. Nova Science Publishers, pp. 131–156.

Knapp, A.N., Sigman, D.M., Lipschultz, F., 2005. N isotopic composition of dissolved organic nitrogen and nitrate at the Bermuda Atlantic Time-series Study site. Global Biogeochem. Cycles 19, doi:10.1029/2004gb002320 GB1018.

Koch, B.P., Witt, M., Engbrodt, R., Dittmar, T., Kattner, G., 2005. Molecular formulae of marine and terrigenous dissolved organic matter detected by electrospray ionization Fourier transform ion cyclotron resonance mass spectrometry. Geochim. Cosmochim. Acta 69, 3299–3308.

Kokkinakis, S.A., Wheeler, P., 1987. Nitrogen uptake and phytoplankton growth in coastal upwelling regions. Limnol. Oceanogr. 32, 1112–1123.

Kokkinakis, S.A., Wheeler, P.A., 1988. Uptake of ammonium and urea in the northeast Pacific: comparison between netplankton and nanoplankton. Mar. Ecol. Prog. Ser. 43, 113–124.

Koopmans, D.J., Bronk, D.A., 2002. Photochemical production of inorganic nitrogen from dissolved organic nitrogen in waters of two estuaries and adjacent surficial groundwaters. Aquat. Microb. Ecol. 26, 295–304.

Koprivnjak, J.-F., Pfromm, P., Ingall, E., Vetter, T., Schmitt-Kopplin, P., Hertkorn, N., et al., 2009. Chemical and spectroscopic characterization of marine dissolved organic matter isolated using coupled reverse osmosis–electrodialysis. Geochim. Cosmochim. Acta 73, 4215–4231.

Koroleff, F., 1983. Simultaneous oxidation of nitrogen and phosphorus compounds by persulfate. Meth. Seawater Anal. 2, 205–206.

Kristiansen, S., 1983. Urea as a nitrogen source for the phytoplankton in the Oslofjord. Mar. Biol. 74, 17–24.

Kristiansen, S., Farbrot, T., Wheeler, P.A., 1994. Nitrogen cycling in the Barents Sea—seasonal dynamics of new and regenerated production in the marginal ice zone. Limnol. Oceanogr. 39, 1630–1642.

Kroeger, K.D., Cole, M.L., Valiela, I., 2006. Groundwater-transported dissolved organic nitrogen exports from coastal watersheds. Limnol. Oceanogr. 51, 2248–2261.

Kroeger, K.D., Swarzenski, P.W., Greenwood, W.J., Reich, C., 2007. Submarine groundwater discharge to Tampa Bay: nutrient fluxes and biogeochemistry of the coastal aquifer. Mar. Chem. 104, 85–97.

Kudela, R.W., Cochlan, W.P., 2000. Nitrogen and carbon uptake kinetics and the influence of irradiance for a red tide bloom off southern California. Aquat. Microb. Ecol. 21, 31–47.

Kudela, R.M., Lane, J.Q., Cochlan, W.P., 2008. The potential role of anthropogenically derived nitrogen in the growth of harmful algae in California, USA. Harmful Algae 8, 103–110.

Kujawinski, E.B., 2002. Electrospray ionization Fourier transform ion cyclotron resonance mass spectrometry (ESI FT-ICR MS): characterization of complex environmental mixtures. Environ. Forensics 3, 207–216.

Kujawinski, E.B., 2011. The impact of microbial metabolism on marine dissolved organic matter. Ann. Rev. Mar. Sci. 3, 567–599.

Kujawinski, E.B., Freitas, M.A., Zang, X., Hatcher, P.G., Green-Church, K.B., Jones, R.B., et al., 2002a. The application of electrospray ionization mass spectrometry (ESI MS) to the structural characterization of natural organic matter. Organic Geochem. 33, 171–180.

Kujawinski, E.B., Hatcher, P.G., Freitas, M.A., 2002b. High-resolution Fourier transform ion cyclotron resonance mass spectrometry of humic and fulvic acids: improvements and comparisons. Anal. Chem. 74, 413–419.

Kujawinski, E.B., Del Vecchio, R., Blough, N.V., Klein, G.C., Marshall, A.G., 2004. Probing molecular-level transformations of dissolved organic matter: insights on photochemical degradation and protozoan modification of DOM from electrospray ionization Fourier transform ion cyclotron resonance mass spectrometry. Mar. Chem. 92, 23–37.

Lampitt, R.S., Noji, T., von Bodungen, B., 1990. What happens to zooplankton faecal pellets? Implications for material flux. Mar. Biol. 104, 15–23.

Langheinrich, U., 1995. Plasm membrane-associated aminopeptidase activities of Chlamydomonas reinhardtii and their biochemical characterization. Biochim. Biophys. Acta 12491, 45–57.

Larsson, U., Hagström, Å., 1979. Phytoplankton exudate release as an energy source for the growth of pelagic bacteria. Mar. Biol. 52, 199–206.

Lawrence, J.E., Suttle, C.A., 2004. Effect of viral infection on sinking rates of *Heterosigma akashiwo* and its implications for bloom termination. Aquat. Microb. Ecol. 37, 1–7.

Le Borgne, R., Rodier, M., 1997. Net zooplankton and the biological pump: a comparison between the oligotrophic and mesotrophic equatorial Pacific. Deep Sea Res. II. 44, 2003–2023.

Lee, C., Bada, J., 1975. Amino acids in Equatorial Pacific ocean water. Earth Planet. Sci. Lett. 26, 61–68.

Lee, S., Fuhrman, J.A., 1987. Relationship between biovolume and biomass of naturally derived bacterioplankton. Appl. Environ. Microbiol. 53, 1298–1303.

Lee, C., Hedges, J.I., Wakeman, S.G., Zhu, N., 1992. Effectiveness of various treatments in retarding microbial activity in sediment trap material and their effects on the collection of swimmers. Limnol. Oceanogr. 37, 117–130.

Lee, S., Kang, Y.-C., Fuhrman, J.A., 1995. Imperfect retention of natural bacterioplankton cells by glass fiber filters. Mar. Ecol. Prog. Ser. 119, 285–290.

Letelier, R.M., Karl, D.M., 1996. Role of *Trichodesmium* spp. in the productivity of the subtropical North Pacific Ocean. Mar. Ecol. Prog. Ser. 133, 263–273.

Letscher, R.T., Hansell, D.A., Carlson, C.A., Lumpkin, R., Knapp, A.N., 2013a. Dissolved organic nitrogen in the global surface ocean: distribution and fate. Global Biogeochem. Cycles 27, 141–153.

Letscher, R.T., Hansell, D.A., Kadko, D., Bates, N.R., 2013b. Dissolved organic nitrogen dynamics in the Arctic Ocean. Mar. Chem. 148, 1–9.

Lewitus, A.J., Subba Rao, D.V., 2006. Osmotrophy in marine microalgae. In: Algal Cultures, Analogues of Blooms and Applications, vol. 1. pp. 343–383.

Lewitus, A.J., Koepfler, E.T., Pigg, R.J., 2000. Use of dissolved organic nitrogen by a salt marsh phytoplankton bloom community. Arch. Hydrobiol. Spec. Issues Adv. Limnol. 55, 441–456.

L'Helguen, S., Madec, C., Le Corre, P., 1996. Nitrogen uptake in permanently well-mixed temperature coastal waters. Estuar. Coast. Shelf Sci. 42, 803–818.

L'Helguen, S., Slawyk, G., Le Corre, P., 2005. Seasonal patterns of urea regeneration by size-fractioned microheterotrophs in well-mixed temperate coastal waters. J. Plankton Res. 27 (3), 263–270.

Lipschultz, F., 1995. Nitrogen-specific uptake rates of marine phytoplankton isolated from natural populations of particles by flow cytometry. Mar. Ecol. Prog. Ser. 123, 245–258.

Lisa, T., Piedras, P., Cardenas, J., Pineda, M., 1995. Utilization of adenine and quanine as nitrogen sources by *Chlamydomonas reinhardtii* cells. Plant Cell Environ. 18, 583–588.

Liu, Z., Kobiela, M.E., McKee, G.A., Tang, T., Lee, C., Mulholland, M.R., et al., 2010. The effect of chemical structure on the hydrolysis of tetrapeptides along a river-to-ocean transect: AVFA and SWGA. Mar. Chem. 119, 108–120.

Llabrés, M., Agusti, S., 2006. Picophytoplankton cell death induced by UV radiation: evidence for oceanic Atlantic communities. Limnol. Oceanogr. 51, 21–29.

Lobbes, J.M., Fitznar, H.P., Kattner, G., 2000. Biogeochemical characteristics of dissolved and particulate organic matter in Russian rivers entering the Arctic Ocean. Geochim. Cosmochim. Acta 64, 2973–2983.

Lomas, M.W., Glibert, P.M., 1999. Temperature regulation of nitrate uptake: a novel hypothesis about nitrate uptake and reduction in cool-water diatoms. Limnol. Oceanogr. 44, 556–572.

Lomas, M.W., Glibert, P.M., Berg, G.M., Burford, M., 1996. Characterization of nitrogen uptake by natural populations of *Aureococcus anaphagefferens* (Chrysophyceae) as a function of incubation duration, substrate concentration, light, and temperature. J. Phycol. 32, 907–916.

Lomas, M.W., Trice, T.M., Glibert, P.M., Bronk, D.A., McCarthy, J.J., 2002. Temporal and spatial dynamics of urea concentrations in Chesapeake Bay: biological versus physical forcing. Estuaries 25, 469–482.

Lomas, M.W., Bronk, D.A., van den Engh, G., 2011. Use of flow cytometry to measure biogeochemical rates and processes in the ocean. Ann. Rev. Mar. Sci. 3, 537–566.

Lønborg, C., Álvarez-Salgado, X.A., 2012. Recycling versus export of bioavailable dissolved organic matter in the coastal ocean and efficiency of the continental shelf pump. Global Biogeochem. Cycles 26, .

Lønborg, C., Søndergaard, M., 2009. Microbial availability and degradation of dissolved organic carbon and nitrogen in two coastal areas. Estuar. Coast. Shelf Sci. 81, 513–520.

López-Veneroni, D., Cifuentes, L.A., 1994. Transport of dissolved organic nitrogen in Mississippi River Plume and Texas-Louisiana continental shelf near-shore waters. Estuaries 17, 796–808.

Mace, K.A., Artaxo, P., Duce, R.A., 2003. Water-soluble organic nitrogen in Amazon Basin aerosols during the dry (biomass burning) and wet seasons. J. Geophys. Res. Atmos. 108, 4512.

Mahaffey, C., Williams, R.G., Wolff, G.A., Anderson, W.T., 2004. Physical supply of nitrogen to phytoplankton in the Atlantic Ocean. Global Biogeochem. Cycles 18, .

Mahaffey, C., Benitez-Nelson, C.R., Bidigare, R.R., Rii, Y., Karl, D.M., 2008. Nitrogen dynamics within a wind-driven eddy. Deep Sea Res. Part II: Top. Stud. Oceanogr. 55, 1398–1411.

Maie, N., Parish, K.J., Watanabe, A., Knicker, H., Benner, R., Abe, T., et al., 2006. Chemical characteristics of dissolved organic nitrogen in an oligotrophic subtropical coastal ecosystem. Geochim. Cosmochim. Acta 70, 4491–4506.

Mandalakis, M., Apostolaki, M., Tziaras, T., Polymenakou, P., Stephanou, E.G., 2011. Free and combined amino acids in marine background atmospheric aerosols over the Eastern Mediterranean. Atmos. Environ. 45, 1003–1009.

Mao, J., Kong, X., Schmidt-Rohr, K., Pignatello, J.J., Perdue, E.M., 2012. Advanced solid-state NMR characterization of marine dissolved organic matter isolated using the coupled reverse osmosis/electrodialysis method. Environ. Sci. Tech. 46, 5806–5814.

Marsot, P., Cembella, A.D., Colombo, J.C., 1991. Intracellular and extracellular amino acid pools of the marine diatom *Phaeodactylum tricornutum* (Bacillariophyceae) grown on unenriched seawater in high cell density dialysis culture. J. Phycol. 27, 478–491.

Martinez, J., Smith, D.C., Steward, G.F., Azam, F., 1996. Variability in ectohydrolytic enzyme activities of pelagic marine bacteria and its significance for substrate processing in the sea. Aquat. Microb. Ecol. 10, 223–230.

Mason, C.A., Engli, T., 1993. Dynamics of microbial growth in the decelerating and stationary phase of batch culture. In: Kjelleberg, S. (Ed.), Starvation in Bacteria. Plenum Press, New York.

Matsumoto, K., Uematsu, M., 2005. Free amino acids in marine aerosols over the western North Pacific Ocean. Atmos. Environ. 39, 2163–2170.

Mayali, X., Weber, P.K., Brodie, E.L., Mabery, S., Hoeprich, P.D., Pett-Ridge, J., et al., 2011. High-throughput isotopic analysis of RNA microarrays to quantify microbial resource use. ISME J. 6, 1210–1221.

Mayali, X., Weber, P.K., Pett-Ridge, J., 2012. High-throughput isotopic analysis of RNA microarrays to quantify microbial resource use. ISME J. 6, 1210–1221.

Mayali, X., Weber, P.K., Pett-Ridge, J., 2013. Taxon-specific C/N relative use efficiency for amino acids in an estuarine community. FEMS Microbiol. Ecol. 83, 402–412.

McCarren, J., Becker, J.W., Repeta, D.J., Shi, Y., Young, C.R., Malmstrom, R.R., Chisholm, S.W., DeLong, E.F., 2010. Microbial community transcriptomes reveal microbes and metabolic pathways associated with dissolved organic matter turnover in the sea. Proc. Natl. Acad. Sci. U. S. A. 107 (38), 16420–16427.

McCarthy, J.J., 1970. A urease method for urea in seawater. Limnol. Oceanogr. 15, 309–313.

McCarthy, J.J., 1972. The uptake of urea by natural populations of marine phytoplankton. Limnol. Oceanogr. 17, 738–748.

McCarthy, M.D., Bronk, D.A., 2008. Analytical methods for the study of nitrogen. In: Capone, D.G., Bronk, D.A., Mulholland, M., Carpenter, E.J. (Eds.), Nitrogen in the Marine Environment. Elsevier, San Diego.

McCarthy, M.D., Hedges, J.I., Benner, R., 1998. Major bacterial contribution to marine dissolved organic nitrogen. Science 281, 231–234.

McCarthy, M.D., Benner, R., Lee, C., Hedges, J.I., Fogel, M.L., 2004. Amino acid carbon isotopic fractionation patterns in oceanic dissolved organic matter: an unaltered photoautotrophic source for dissolved organic nitrogen in the ocean? Mar. Chem. 92, 123–134.

McCarthy, M.D., Benner, R., Lee, C., Fogel, M.L., 2007. Amino acid nitrogen isotopic fractionation patterns as indicators of heterotrophy in plankton, particulate, and dissolved organic matter. Geochim. Cosmochim. Acta 71, 4727–4744.

McCarthy, M.D., Lehman, J., Kudela, R., 2013. Compound-specific amino acid δ^{15}N patterns in marine algae: tracer potential for cyanobacterial vs. eukaryotic organic nitrogen sources in the ocean. Geochim. Cosmochim. Acta 103, 104–120.

McClelland, J.W., Montoya, J.P., 2002. Trophic relationships and the nitrogen isotopic composition of amino acids. Ecology 83, 2173–2180.

McClelland, J.W., Holl, C.M., Montoya, J.P., 2003. Relating low δ^{15}N values of zooplankton to N_2-fixation in the tropical North Atlantic: insights provided by stable isotope ratios of amino acids. Deep-Sea Res. Part I. 50, 849–861.

McLean, T.I., 2013. "Eco-omics": a review of the application of genomics, transcriptomics, and proteomics for the study of the ecology of harmful algae. Microb. Ecol. 65, 901–915.

Menzel, D.W., Vaccaro, R.F., 1964. The measurement of dissolved organic and particulate carbon in seawater. Limnol. Oceanogr. 9, 138–142.

Meon, B., Amon, R.M.W., 2004. Heterotrophic bacterial activity and fluxes of dissolved free amino acids and glucose in the Arctic rivers Ob Yenisei and the adjacent Kara Sea. Aquat. Microb. Ecol. 37, 121–135.

Meon, B., Kirchman, D.L., 2001. Dynamics and molecular composition of dissolved organic material during experimental phytoplankton blooms. Mar. Chem. 75, 185–199.

Middelboe, M., Jorgensen, N.O.G., 2006. Viral lysis of bacteria: an important source of dissolved amino acids and cell wall compounds. J. Mar. Biol. Assoc. U.K 86, 605–612.

Middelboe, M., Lyck, P.G., 2002. Regeneration of dissolved organic matter by viral lysis in marine microbial communities. Aquat. Microb. Ecol. 27.

Middelboe, M., Sondergaard, M., Letarte, Y., Borch, N.H., 1995. Attached and free-living bacteria: production and polymer hydrolysis during a diatom bloom. Microb. Ecol. 29, 231–248.

Middelboe, M., Jørgensen, N.O.G., Kroer, N., 1996. Effects of viruses on nutrient turnover and growth efficiency of noninfected marine bacterioplankton. Appl. Environ. Microbiol. 62, 1991–1997.

Middelburg, J.J., Nieuwenhuize, J., 2000a. Uptake of dissolved inorganic nitrogen in turbid, tidal estuaries. Mar. Ecol. Prog. Ser. 192, 79–88.

Middelburg, J.J., Nieuwenhuize, J., 2000b. Nitrogen uptake by heterotrophic bacteria and phytoplankton in the nitrate-rich Thames estuary. Mar. Ecol. Prog. Ser. 203, 13–21.

Miki, T., Yamamura, N., 2005. Intraguild predation reduces bacterial species richness and loosens the viral loop in aquatic systems: 'kill the killer of the winner' hypothesis. Aquat. Microb. Ecol. 40, 1–12.

Miller, C.A., Glibert, P.M., 1998. Nitrogen excretion by the calanoid copepod *Acartia tonsa*: results from mesocosm experiments. J. Plankton Res. 20, 1767–1780.

Miyazaki, Y., Kawamura, K., Jung, J., Furutani, H., Uematsu, M., 2011. Latitudinal distributions of organic nitrogen and organic carbon in marine aerosols over the western North Pacific. Atmos. Chem. Phys. 11, 3037–3049.

Møller, E.F., 2007. Production of dissolved organic carbon by sloppy feeding in the copepods *Acartia tonsa*, *Centropages typicus*, and *Temora longicornis*. Limnol. Oceanogr. 52, 79–84.

Møller, E.F., Thor, P., Nielsen, T.G., 2003. Production of DOC by *Clanus finmarchicus* C. glacialis and C. hyperboreus through sloppy feeding and leakage from fecal pellets. Mar. Ecol. Prog. Ser. 262.

Montoya, J.P., Holl, C.M., Zehr, J.P., Hansen, A., Villareal, T.A., Capone, D.G., et al., 2004. High rates of N_2 fixation by unicellular diazotrophs in the oligotrophic Pacific Ocean. Nature 430, 1027–1031.

Mopper, K., Lindroth, P., 1982. Diel and depth variations in dissolved free amino acids and ammonium in the Baltic Sea determined by shipboard HPLC analysis. Limnol. Oceanogr. 27, 336–347.

Mopper, K., Zhou, X., Kieber, R.J., Kieber, D.J., Sikorski, R.J., Jones, R.D., et al., 1991. Photochemical degradation of dissolved organic carbon and its impact on the oceanic carbon cycle. Nature 353, 60.

Mopper, K., Stubbins, A., Ritchie, J.D., Bialk, H.M., Hatcher, P.G., 2007. Advanced instrumental approaches for characterization of marine dissolved organic matter: extraction techniques, mass spectrometry, and nuclear magnetic resonance spectroscopy. Chem. Rev. 107, 419–442.

Morris, R.M., Nunn, B.L., Frazar, C., Goodlett, D.R., Ting, Y.S., Rocap, G., et al., 2010. Comparative metaproteomics reveals ocean-scale shifts in microbial nutrient utilization and energy transduction. ISME J. 4, 673–685.

Mulholland, M.R., 2007. The fate of nitrogen fixed by diazotrophs in the ocean. Biogeosciences 4, 37–51.

Mulholland, M., Lomas, M.W., 2008. Nitrogen uptake and assimilation. In: Capone, D.G., Bronk, D.A., Mulholland, M., Carpenter, E.J. (Eds.), Nitrogen in the Marine Environment, second ed. Elsevier Inc. Burlington, MA.

Mulholland, M.R., Glibert, P.M., Berg, G.M., Van Heukelem, L., Pantoja, S., Lee, C., et al., 1998. Extracellular amino acid oxidation by microplankton: a cross-system comparison. Aquat. Microb. Ecol. 15, 141–152.

Mulholland, M.R., Lee, C., Glibert, P.M., 2003. Extracellular enzyme activity and uptake of carbon and nitrogen along an estuarine salinity and nutrient gradient. Mar. Ecol. Prog. Ser. 258, 3–17.

Mulholland, M.R., Bronk, D.A., Capone, D.G., 2004. Dinitrogen fixation and release of ammonium and dissolved organic nitrogen by *Trichodesmium* IMS101. Aquat. Microb. Ecol. 37, 85–94.

Mulholland, M.R., Bernhardt, P., Heil, C.A., Bronk, D.A., O'Neil, J.M., 2006. Nitrogen fixation and release of fixed nitrogen in the Gulf of Mexico. Limnol. Oceanogr. 51, 1762–1776.

Mulholland, M.R., Bernhardt, P., Ozmon, I., Procise, L. A., O'Neil, J.M., Heil, C.A., et al., in press. Contributions of N_2 fixation to N inputs supporting Karenia brevis blooms in the Gulf of Mexico. Harmful Algae.

Müller-Wegener, U., 1988. Interaction of humic substances with biota. In: Frimmel, F.H., Christman, R.F. (Eds.), Humic Substances and Their Role in the Environment. John Wiley and Sons Limited, New York.

Murray, A.G., 1995. Phytoplankton exudation: exploitation of the microbial loop as a defense against algal viruses. J. Plankton Res. 17, 1079–1094.

Myklestad, S., Holm-Hansen, O., Varum, K.M., Volcani, B.E., 1989. Rate of release of extracellular amino acids and carbohydrates from the marine diatom *Chaetoceros affinis*. J. Plankton Res. 11, 763–773.

Nagao, F., Miyazaki, T., 1999. A modified ^{15}N tracer method and new calculation for estimating release of dissolved organic nitrogen by freshwater planktonic algae. Aquat. Microb. Ecol. 16, 309–314.

Nagao, F., Miyazaki, T., 2002. Release of dissolved organic nitrogen from *Scenedesmus quaricauda* (Chlorophyta) and *Microcystis novcekii* (Cyanobacteria). Aquat. Microb. Ecol. 27, 275–284.

Nagata, T., 2000. Production mechanisms of dissolved organic matter. In: Kirchman, D.L. (Ed.), Microbial Ecology of the Oceans. Wiley, New York.

Nagata, T., Kirchman, D.L., 1992. Release of macromolecular organic complexes by heterotrophic marine flagellates. Mar. Ecol. Prog. Ser. 83, 233–240.

Nagata, T., Kirchman, D.L., 1996. Bacterial degradation of protein adsorbed to model submicron particles in seawater. Mar. Ecol. Prog. Ser. 132, 241–248.

Nakamura, T., Ogawa, H., Maripi, D.K., Uematsu, M., 2006. Contribution of water soluble organic nitrogen to total nitrogen in marine aerosols over the East China Sea and western North Pacific. Atmos. Environ. 40, 7259–7264.

Nausch, M., 1996. Microbial activities on *Trichodesmium* colonies. Mar. Ecol. Prog. Ser. 141, 173–181.

Nelson, C., Carlson, C., 2012. Tracking differential incorporation of dissolved organic carbon types among diverse lineages of Sargasso Sea bacterioplankton. Environ. Microbiol. 14, 1500–1516.

Neufeld, J.D., Dumont, M.G., Vohra, J., Murrell, J.C., 2005. Methodological considerations for the use of stable isotope probing in microbial ecology. Microb. Ecol. 53, 435–442.

Ning, S.B., Guo, H.L., Wang, L., Song, Y.C., 2002. Salt stress induces programmed cell death in prokaryotic organism *Anabaena*. J. Appl. Microbiol. 93, 15–28.

Ohki, K., 1999. A possible role of temperate phage in the regulation of *Trichodesmium* biomass. In: Charpy, L., Larkum, A.W.D. (Eds.), Marine Cyanobacteria. Bulletin de l'Institute Oceanographique, Monaco.

Ohlendieck, U., Stuhr, A., Siegmund, H., 2000. Nitrogen fixation by diazotrophic cyanobacteria in the Baltic Sea and transfer of the newly fixed nitrogen to picoplankton organisms. J. Mar. Syst. 25, 213–219.

Olson, R.J., Valout, D., Chisholm, S.W., 1985. Marine phytoplankton distributions measured using shipboard flow cytometry. Deep-Sea Res. 32, 1273–1280.

O'Neil, J.M., Roman, M.R., 1992. Grazers and associated organisms of *Trichodesmium*. In: Carpenter, E.J., Capone, D.G., Reuter, J.G. (Eds.), Marine Pelagic Cyanobacteria: *Trichodesmium* and Other Diazotrophs. Kluwer Academic Press, Dordrecht.

O'Neil, J.M., Metzler, P.M., Glibert, P.M., 1996. Ingestion of $^{15}N_2$-labeled *Trichodesmium* sp. and ammonium regeneration by the harpacticoid copepod *Macrosetella gracilis*. Mar. Biol. 125, 89–96.

Oremland, D.S., Capone, D., 1988. Use of specific inhibitors in biogeochemistry and microbial ecology. Adv. Microb. Ecol. 10, 285–383.

Ouverney, C.C., Fuhrman, J.A., 2000. Marine planktonic archaea take up amino acids. Appl. Environ. Microbiol. 66, 4829–4833.

Paerl, H.W., 1997. Coastal eutrophication and harmful algal blooms: importance of atmospheric deposition and groundwater as "new" nitrogen and other nutrient sources. Limnol. Oceanogr. 42, 1154–1165.

Palenik, B., Henson, S.E., 1997. The use of amides and other organic nitrogen sources by the phytoplankton *Emiliana huxleyi*. Limnol. Oceanogr. 42, 1544–1551.

Palenik, B., Morel, F.M.M., 1990a. Amino acid utilization by marine phytoplankton: a novel mechanism. Limnol. Oceanogr. 35, 260–269.

Palenik, B., Morel, F.M.M., 1990b. Comparison of cell-surface L-amino acid oxidases from several marine phytoplankton. Mar. Ecol. Prog. Ser. 59, 195–201.

Palenik, B., Morel, F.M.M., 1991. Amine oxidases of marine phytoplankton. Appl. Environ. Microbiol. 57, 2440–2443.

Palenik, B., Brahamsha, B., Larimer, F.W., Land, M., Hauser, L., Chain, P., et al., 2003. The genome of a motile marine *Synechococcus*. Nature 424, 1037–1042.

Palenik, B., Grimwood, J., Aerts, A., Rouzé, P., Salamov, A., Putnam, N., et al., 2007. The tiny eukaryote *Ostreococcus* provides genomic insights into the paradox of plankton speciation. Proc. Natl. Acad. Sci. 104 (18), 7705–7710.

Pantoja, S., Lee, C., 1994. Cell-surface oxidation of amino acids in seawater. Limnol. Oceanogr. 39, 1718–1726.

Park, M.G., Yih, W., Coats, D.W., 2004. Parasites and phytoplankton, with special emphasis on dinoflagellate infections. J. Eukaryotic Microb. 51, 145–155.

Parsons, T.R., Maita, Y., Lalli, C.M., 1984. A Manual of Chemical and Biological Methods for Seawater Analysis. Pergamon, Oxford.

Paul, J.H., Jeffrey, W.H., DeFlaun, M.F., 1987. Dynamics of extracellular DNA in the marine environment. Appl. Environ. Microbiol. 53, 170–179.

Paul, J.H., DeFlaun, M.F., Jeffrey, W.H., David, A.W., 1988. Seasonal and diel variability in dissolved DNA and in microbial biomass and activity in a subtropical estuary. Appl. Environ. Microbiol. 54, 718–727.

Paul, J.H., Jeffrey, W.H., Cannon, J.P., 1990. Production of dissolved DNA, RNA, and protein by microbial populations in a Florida reservoir. Appl. Environ. Microbiol. 56, 2957–2962.

Peierls, B.L., Paerl, H.W., 1997. Bioavailability of atmospheric organic nitrogen deposition to coastal phytoplankton. Limnol. Oceanogr. 42, 1819–1823.

Pellerin, B.A., Wollheim, W.M., Hopkinson, C.S., McDowell, W.H., Williams, M.R., Vörösmarty, C.J., et al., 2004. Role of wetlands and developed land use on dissolved organic nitrogen concentrations and DON/TDN in northeastern U.S. rivers and streams. Limnol. Oceanogr. 49, 910–918.

Pettersson, K., 1991. Seasonal uptake of carbon and nitrogen and intracellular storage of nitrate in planktonic organisms in the Skagerrak. J. Exp. Mar. Biol. Ecol. 151, 121–137.

Pettersson, K., Sahlsten, E., 1990. Diel patterns of combined nitrogen uptake and intracellular storage of nitrate by phytoplankton in the open Skagerrak. J. Exp. Mar. Biol. Ecol. 138, 167–182.

Peuravuori, J., Lehtonen, T., Pihlaja, K., 2002. Sorption of aquatic humic matter by DAX-8 and XAD-8 resins: comparative study using pyrolysis gas chromatography. Anal. Chim. Acta 471, 219–226.

Pilskaln, C.H., Villareal, T.A., Dennett, M., Darkangelo-Wood, C., Meadows, G., 2005. High concentrations of marine snow and diatom algal mats in the North Pacific Subtropical Gyre: implications for carbon and nitrogen cycles in the oligotrophic ocean. Deep Sea Res. Part I: Oceanogr. Res. Pap. 52, 2315–2332.

Pitt, K.A., Welsh, D.T., Condon, R.H., 2009. Influence of jellyfish blooms on carbon, nitrogen and phosphorus cycling and plankton production. Hydrobiologia 616, 133–149.

Ploug, H., Musat, N., Adam, B., Moraru, C.L., Lavik, G., Vagner, T., et al., 2010. Carbon and nitrogen fluxes associated with the cyanobacterium *Aphanizomenon* sp. in the Baltic Sea. ISME J. 4, 1215–1223.

Podgorski, D.C., McKenna, A.M., Rodgers, R.P., Marshall, A.G., Cooper, W.T., 2012. Selective ionization of dissolved organic nitrogen by positive ion atmospheric pressure photoionization coupled with Fourier transform ion cyclotron resonance mass spectrometry. Anal. Chem. 84, 5085–5090.

Price, N., Harrison, P., 1987. Comparison of methods for the analysis of dissolved urea in seawater. Mar. Biol. 94, 307–317.

Price, N.M., Cochlan, W.P., Harrison, P.J., 1985. Time course of uptake of inorganic and organic nitrogen by phytoplankton in the Strait of Georgia: comparison of frontal and stratified communities. Mar. Ecol. Prog. Ser. 27, 39–53.

Probyn, T.A., 1985. Nitrogen uptake by size-fractionted phytoplankton populations in the southern Benguela upwelling systems. Mar. Ecol. Prog. Ser. 22, 249–258.

Probyn, T.A., 1988. Nitrogen utilization by phytoplankton in the Namibian upwelling region during and austral spring. Deep Sea Res. Part A, Oceanogr. Res. Pap. 38, 1387–1404.

Probyn, T.A., Painting, S.J., 1985. Nitrogen uptake by size-fractionated phytoplankton populations in Antarctic surface waters. Limnol. Oceanogr. 30, 1327–1332.

Probyn, T., Mitchell-Innes, B., Searson, S., 1995. Primary productivity and nitrogen uptake in the subsurface chlorophyll maximum on the Eastern Agulhas Bank. Continental Shelf Res. 15, 1903–1920.

Radajewski, S., Ineson, P., Parekh, N.R., Murrell, J.C., 2000. Stable-isotope probing as a tool in microbial ecology. Nature 403, 646–649.

Raimbault, P., Garcia, N., 2008. Evidence for efficient regenerated production and dinitrogen fixation in nitrogen-deficient waters of the South Pacific Ocean: impact on new and export production estimates. Biogeosciences 5, 323–338.

Read, B.A., Kegel, J., Klute, M.J., Kuo, A., Lefebvre, S.C., Maumus, F., et al., 2013. Pan genome of the phytoplankton *Emiliania* underpins its global distribution. Emiliania huxleyi Annotation Consortium, Nature, 499, 209–213.

Reemtsma, T., These, A., Linscheid, M., Leenheer, J., Spitzy, A., 2008. Molecular and structural characterization of dissolved organic matter from the deep ocean by FTICR-MS, including hydrophilic nitrogenous organic molecules. Environ. Sci. Tech. 42, 1430–1437.

Reinthaler, T., Sintes, E., Herndl, G.J., 2008. Dissolved organic matter and bacterial production and respiration in the sea-surface microlayer of the open Atlantic and the western Mediterranean Sea. Limnol. Oceanogr. 53, 122–136.

Remsen, C.C., Carpenter, E.J., Schroeder, B.W., 1972. Competition for urea among estuarine microorganisms. Ecology 53 (5), 921–926.

Revilla, M., Alexander, J., Glibert, P.M., 2005. Urea analysis in coastal waters: comparison of enzymatic and direct methods. Limnol. Oceanogr. Methods 3, 290–299.

Ribas-Ribas, M., Gómez-Parra, A., Forja, J.M., 2011. Spatiotemporal variability of the dissolved organic carbon and nitrogen in a coastal area affected by river input: the north eastern shelf of the Gulf of Cádiz (SW Iberian Peninsula). Mar. Chem. 126, 295–308.

Rice, D.L., 1982. The detritus nitrogen problem: new observations and perspectives from organic geochemistry. Mar. Ecol. Prog. Ser. 9, 153–162.

Richardson, T.L., Cullen, J.J., Kelley, D.E., Lewis, M.R., 1998. Potential contribution of vertically migrating *Rhizoselenia* to nutrient cycling and new production in the open ocean. J. Plankton Res. 20, 219–241.

Riegman, R., van Bleijswijk, J.D.L., Brussaard, C.P.D., 2002. The use of dissolved esterase activity as a tracer of phytoplankton lysis. Limnol. Oceanogr. 47, 916–920.

Riemann, L., Holmfeldt, K., Titelman, J., 2009. Importance of viral lysis and dissolved DNA for bacterioplankton activity in a P-limited estuary, Northern Baltic Sea. Microb. Ecol. 57, 286–294.

Rocap, G., Larimer, F., Lamerdin, J., Malfatti, S., Chain, P., Ahlgren, N., et al., 2003. Genome divergence in two Prochlorococcus ecotypes reflects oceanic niche differentiation. Nature 424, 1042–1047.

Rodrigues, R.M.N.V., Williams, P.J.B., 2002. Inorganic nitrogen assimilation by picoplankton and whole plankton in a coastal ecosystem. Limnol. Oceanogr. 47, 1608–1616.

Rohwer, F., Thurmer, R.V., 2009. Viruses manipulate the marine environment. Nature 459, 207–212.

Rosenberg, R., Dahl, E., Edler, L., Fryberg, L., Granéli, E., Granéli, W., et al., 1990. Pelagic nutrient and energy transfer during spring in the open and coastal Skagerrak. Mar. Ecol. Prog. Ser. 61, 215–231.

Rosenfeld, J.K., 1979. Ammonium adsorption in nearshore anoxic sediments. Limnol. Oceanogr. 24, 356–364.

Rosenstock, B., Zwisler, W., Simon, M., 2005. Bacterial Consumption of Humic and Non-Humic Low and High Molecular Weight DOM and the Effect of Solar Irradiation on the Turnover of Labile DOM in the Southern Ocean. Microb. Ecol. 50, 90–101.

Roussenov, V., Williams, R.G., Mahaffey, C., Wolff, G.A., 2006. Does the transport of dissolved organic nutrients affect export production in the Atlantic Ocean? Global Biogeochem. Cycles 20, GB3002. doi:10.1029/2005GB002510.

Roy, S., Poulet, S.A., 1990. Laboratory study of the chemical composition of aging copepod fecal material. J. Exp. Mar. Biol. Ecol. 135, 3–18.

Roy, S., Harris, R.P., Poulet, S.A., 1989. Inefficient feeding by *Calanus helgolandicus* and *Temora longicornis* on *Coscinodiscus wailesii* : quantitative estimation using chlorophyll-type pigments and effects on dissolved free amino acids. Mar. Ecol. Prog. Ser. 52, 145–153.

Rynearson, T.A., Palenik, B., 2011. Learning to read the oceans: genomics of marine phytoplankton. Advances in Marine Biology. 60, 1–39.

Saba, G.K., Steinberg, D.K., Bronk, D.A., 2009. Effects of diet on release of dissolved organic and inorganic nutrients by the copepod *Acartia tonsa*. Mar. Ecol. Prog. Ser. 386, 147–161.

Saba, G.K., Steinberg, D.K., Bronk, D.A., 2011a. The relative importance of sloppy feeding, excretion, and fecal pellet leaching in the release of dissolved carbon and nitrogen by *Acartia tonsa* copepods. J. Exp. Mar. Biol. Ecol. 404, 47–56.

Saba, G.K., Steinberg, D.K., Bronk, D.A., Place, A.R., 2011b. The effects of harmful algae species and food concentration on zooplankton grazer production of dissolved organic matter and inorganic nutrients. Harmful Algae 10, 291–303.

Sakano, S., Kamatani, A., 1992. Determination of dissolved nucleic acids in seawater by the fluorescence dye, ethidium bromide. Mar. Chem. 37, 239–255.

Sakshaug, E., Holm-Hansen, O., 1977. Chemical composition of *Skeletonema costatus* (Grev.) *Cleve* and *Pavlova* (Monochrysis) *lutheri* (Droop) Green as a function of nitrate-, phosphate-, and iron-limited growth. J. Exp. Mar. Biol. Ecol. 29, 1–34.

Sambrotto, R.N., 2001. Nitrogen production in the northern Arabian Sea during the Spring Intermonsoon and Southwest Monsoon seasons. Deep Sea Res. II 48, 1173–1198.

Sanderson, M.P., Bronk, D.A., Nejstgaard, J.C., Verity, P.G., Sazhin, A.F., Frischer, M.E., 2008. Phytoplankton and bacterial uptake of inorganic and organic nitrogen during an induced bloom of *Phaeocystis pouchetii*. Aquat. Microb. Ecol. 51, 153–168.

Santos, I.R., Burnett, W.C., Dittmar, T., Suryaputra, I.G., Chanton, J., 2009. Tidal pumping drives nutrient and dissolved organic matter dynamics in a Gulf of Mexico subterranean estuary. Geochim. Cosmochim. Acta 73, 1325–1339.

Sarmento, H., Romera-Castillo, C., Lindh, M., Pinhassi, J., Sala, M.M., et al., 2013. Phytoplankton species-specific release of dissolved free amino acids and their selective consumption by bacteria. Limnol. Oceanogr. 58, 1123–1135.

Savoye, N., DeHairs, F., Elskens, M., Cardinal, D., Kopczynska, E.E., Trull, T.W., et al., 2004. Regional variation of spring N-uptake and new production in the Southern Ocean. Geophys. Res. Lett. 31, doi:10.1029/2003GL018946.

Schuster, S., Arrieta, J.M., Herndl, G.J., 1998. Adsorption of dissolved free amino acids on colloidal DOM enhances colloidal DOM utilization but reduces amino acid uptake by orders of magnitude in marine bacterioplankton. Mar. Ecol. Prog. Ser. 166, 99–108.

Scott, D., Harvey, J., Alexander, R., Schwarz, G., 2007. Dominance of organic nitrogen from headwater streams to large rivers across the conterminous United States. Global Biogeochem. Cycles 21, doi:10.1029/2006GB002730, GB1003.

See, J.H., 2003. Availability of Humic Nitrogen to Phytoplankton (Ph.D.). The College of William and Mary.

See, J.H., Bronk, D.A., 2005. Changes in C:N ratios and chemical structures of estuarine humic substances during aging. Mar. Chem. 97, 334–346.

See, J.H., Bronk, D.A., Lewitus, A.J., 2006. Uptake of *Spartina*-derived humic nitrogen by estuarine phytoplankton in nonaxenic and axenic culture. Limnol. Oceanogr. 51, 2290–2299.

Seeyave, S., Probyn, T., Álvarez-Salgado, X.A., Figueiras, F.G., Purdie, D.A., Barton, E.D., et al., 2013. Nitrogen uptake of phytoplankton assemblages under contrasting upwelling and downwelling conditions: the Ría de Vigo, NW Iberia. Estuar. Coast. Shelf Sci. 124, 1–12.

Segovia, M., Haramaty, L., Berges, J.A., Falkowski, P.G., 2003. Cell death in the unicellular chlorophyte *Dunaliella tertiolecta*: a hypothesis on the evolution of apoptosis in higher plants and metazoans. Plant Physiol. 132, 99–105.

Seifert, J., Taubert, M., Jehmlich, N., Schmidt, F., Volker, U., Vogt, C., et al., 2012. Protein-based stable isotope probing (protein-SIP) in functional metaproteomics. Mass Spect. Rev. 31, 683–697.

Seitzinger, S.P., Harrison, J.A., 2008. Land-based nitrogen sources and their delivery to coastal systems. In: Nitrogen in the Marine Environment, second ed. Elsevier, Amsterdam, pp. 469–510.

Seitzinger, S., Sanders, R., 1997. Contribution of dissolved organic nitrogen from rivers to estuarine eutrophication. Mar. Ecol. Prog. Ser. 159, 1–12.

Seitzinger, S., Sanders, R., 1999. Atmospheric inputs of dissolved organic nitrogen stimulate estuarine bacteria and phytoplankton. Limnol. Oceanogr. 44, 721–730.

Seitzinger, S.P., Harrison, J.A., Dumont, E., Beusen, A.H.W., Bouwman, A.F., 2005. Sources and delivery of carbon, nitrogen, and phosphorus to the coastal zone: an overview of Global Nutrient Export from Wastersheds (NEWS) models and their application. Global Biogeochem. Cycles 19, GB4S01. doi:10.1029/2005GB002606.

Sellner, K.G., Nealley, E.W., 1997. Diel fluctuations in dissolved free amino acids and monosaccharides in Chesapeake Bay dinoflagellate blooms. Mar. Chem. 56, 193–200.

Serkiz, S.M., Perdue, E.M., 1990. Isolation of dissolved organic matter from the Suwannee river using reverse osmosis. Water Res. 24, 911–916.

Sharp, J.H., 2002. Analytical methods for dissolved organic carbon, nitrogen, and phosphorus. In: Hansell, D.A., Carlson, C.A. (Eds.), Biogeochemistry of Marine Dissolved Organic Matter. Academic Press, San Diego.

Sharp, J.H., Beauregard, A.Y., Burdige, D., Cauwet, G., Curless, S.E., Lauck, R., et al., 2004. A direct instrument comparison for measurement of total dissolved nitrogen in seawater. Mar. Chem. 84, 181–193.

Shaw, P.J., Chapron, C., Purdie, D.A., Rees, A.P., 1998. Impacts of phytoplankton activity on dissolved nitrogen fluxes in the tidal reaches and estuary of the Tweed, UK. Mar. Pollut. Bull. 37, 280–294.

Sheridan, C.C., Steinberg, D.K., Kling, G.W., 2002. The microbial and metazoan community associated with colonies of Trichodesmium spp.: a quantitative survey. J. Plankton Res. 24, 913–922.

Shibata, A., Kogure, K., Koike, I., Ohwada, K., 1997. Formation of submicron colloidal particles from marine bacteria by viral infection. Mar. Ecol. Prog. Ser. 155, 303–307.

Sigman, S.D., Altabet, M.A., Michner, R., McCorkle, D.C., Fry, B., Holmes, R.M., et al., 1997. Natural abundance-level measurement of the nitrogen isotopic composition of oceanic nitrate: an adaptation of the ammonia diffusion method. Mar. Chem. 57, 227–242.

Sigman, D.M., Casciotti, K.L., Andreani, M., Barford, C., Galanter, M., Bohlke, J.K., et al., 2001. A bacterial method for the nitrogen isotopic analysis of nitrate in seawater and freshwater. Anal. Chem. 73, 4145–4153.

Simjouw, J.-P., Minor, E.C., Mopper, K., 2005. Isolation and characterization of estuarine dissolved organic matter: comparison of ultrafiltration and C18 solid-phase extractions techniques. Mar. Chem. 96, 219–235.

Singler, H., Villareal, T.A., 2005. Nitrogen inputs into the euphotic zone by vertically migrating Rhizosolenia mats. J. Plankton Res. 27, 545–556.

Sipler, R.E., 2009. The Role of Dissolved Organic Matter in Structuring Microbial Community Composition. Rutgers University-Graduate School, New Brunswick.

Sipler, R.E., Bronk, D.A., Seitzinger, S.P., Lauck, R.J., McGuinness, L.R., Kirkpatrick, G.J., et al., 2013. Trichodesmium-derived dissolved organic matter is a source of nitrogen capable of supporting the growth of toxic red tide Karenia brevis. Mar. Ecol. Prog. Ser. 483, 31–45.

Slawyk, G., Raimbault, P., 1995. Simple procedure for simultaneous recovery of dissolved inorganic and organic nitrogen in 15N-tracer experiments and improving the isotopic mass balance. Mar. Ecol. Prog. Ser. 124, 289–299.

Slawyk, G., Raimbault, P., L'Helguen, S., 1990. Recovery of urea nitrogen from seawater for measurement of 15N abundance in urea regeneration studies, using the isotope-dilution approach. Mar. Chem. 30, 343–362.

Sleighter, R.L., Hatcher, P.G., 2007. The application of electrospray ionization coupled to ultrahigh resolution mass spectrometry for the molecular characterization of natural organic matter. J. Mass Spect. 42, 559–574.

Sleighter, R.L., Hatcher, P.G., 2008. Molecular characterization of dissolved organic matter (DOM) along a river to ocean transect of the lower Chesapeake Bay by ultrahigh resolution electrospray ionization Fourier transform ion cyclotron resonance mass spectrometry. Mar. Chem. 110, 140–152.

Smith, D.C., Simon, M., Alldredge, A.L., Azam, F., 1992. Intense hydrolytic enzyme activity on marine aggregates and implications for rapid particle dissolution. Nature 359, 139–141.

Smith, J., O., W., Gosselin, M., Legendre, R., Wallace, D.W.R., Daly, K.L., et al., 1997. New production in the Northeast Water Polynya: 1993. J. Mar. Sys. Sys. 10, 199–209.

Solomon, C.M., Collier, J.L., Berg, G.M., Glibert, P.M., 2010. Role of urea in microbial metabolism in aquatic systems: a biochemical and molecular review. Aquat. Microb. Ecol. 59, 67–88.

Sowell, S.M., Wilhelm, L.J., Norbeck, A.D., Lipton, M.S., Nicora, C.D., Barofsky, D.F., et al., 2009. Transport functions dominate the SAR11 metaproteome at low-nutrient extremes in the Sargasso Sea. ISME J. 3, 93–105.

Stedmon, C.A., Markager, S., Bro, R., 2003. Tracing dissolved organic matter in aquatic environments using a new approach to fluorescence spectroscopy. Mar. Chem. 82, 239–254.

Stedmon, C.A., Markager, S., Tranvik, L.J., Kronberg, L., Slätis, T., Martinsen, W., et al., 2007. Photochemical production of ammonium and transformation of dissolved organic matter in the Baltic Sea. Mar. Chem. 104, 227–240.

Steinberg, D.K., Saba, G.K., 2008. Nitrogen consumption and metabolism in marine zooplankton. In: Capone, D.G., Bronk, D.A., Mulholland, M., Carpenter, E.J. (Eds.), Nitrogen in the Marine Environment, second ed. Elsevier Inc. Burlington, MA.

Steinberg, D.K., Goldthwait, S.A., Hansell, D.A., 2002. Zooplankton vertical migration and the active transport of dissolved organic and inorganic nitrogen in the Sargasso Sea. Deep Sea Res. 49, 1445–1461.

Steinberg, D.K., Nelson, N.B., Carlson, C.A., Prusak, A.C., 2004. Production of chromophoric dissolved organic matter (CDOM) in the open ocean by zooplankton and the colonial cyanobacterium Trichodesmium spp. Mar. Ecol. Prog. Ser. 267, 45–64.

Stenson, A.C., 2008. Reverse-phase chromatography fractionation tailored to mass spectral Characterization of humic substances. Environ. Sci. Tech. 42, 2060–2065.

Stenson, A.C., Landing, W.M., Marshall, A.G., Cooper, W.T., 2002. Ionization and fragmentation of humic substances in electrospray ionization fourier transform-ion cyclotron resonance mass spectrometry. Anal. Chem. 74, 4397–4409.

Stepanauskas, R., Edling, H., Tranvik, L.J., 1999a. Differential dissolved organic nitrogen availability and bacterial aminopeptidase activity in limnic and marine waters. Microb. Ecol. 38, 264–272.

Stepanauskas, R., Leonardson, L., Tranvik, L.J., 1999b. Bioavailability of wetland-derived DON to freshwater and marine bacterioplankton. Limnol. Oceanogr. 44, 1477–1485.

Stepanauskas, R., Laudon, H., Jørgensen, N.O.G., 2000. High DON bioavailability in boreal streams during a spring flood. Limnol. Oceanogr. 45, 1298–1307.

Stepanauskas, R., Jorgensen, N.O.G., Eigaard, O.R., Zvikas, A., Tranvik, L.J., Leonardson, L., et al., 2002. Summer inputs of riverine nutrients to the Baltic Sea: bioavailability and eutrophication relevance. Ecolo. Monographs 72, 579–597.

Stevenson, F.J., 1994. Humus Chemistry. Wiley, New York.

Stoecker, D.K., Gustafson, D.E., 2003. Cell-surface proteolytic activity of photosynthetic dinoflagellates. Aquat. Microb. Ecol. 30, 175–183.

Strom, S.L., 2000. Bacterivory: interactions between bacteria and their grazers. In: Kirchman, D.L. (Ed.), Microbial Ecology of the Oceans. Wiley-Liss, Inc., New York.

Strom, S.L., Benner, R., Ziegler, S., Dagg, M.J., 1997. Planktonic grazers are a potentially important source of marine dissolved organic carbon. Limnol. Oceanogr. 42, 1364–1374.

Stuermer, D.H., Peters, K.E., Kaplan, I.R., 1978. Source indicators of humic substances and proto-kerogen: stable isotope ratios, elemental compositions, and electron spin resonance spectra. Geochim. Cosmochim. Acta 42, 989–997.

Suttle, C.A., 2005. Viruses in the sea. Nature 437, 356–361.

Tamminen, T., Irmisch, A., 1996. Urea uptake kinetics of a midsummer planktonic community on the SW coast of Finland. Mar. Ecol. Prog. Ser. 130, 201–211.

Tank, S.E., Manizza, M., Holmes, R.M., McClelland, J.W., Peterson, B.J., 2012. The processing and impact of dissolved riverine nitrogen in the Arctic Ocean. Estuar. Coast. 35, 401–415.

Tanoue, E., Ishii, M., Midorikawa, T., 1996. Discrete dissolved and particulate proteins in oceanic waters. Limnol. Oceanogr. 41, 1334–1343.

Tarr, M.A., Wang, W., Bianchi, T.S., Engelhaupt, E., 2001. Mechanisms of ammonia and amino acid photoproduction from aquatic humic and colloidal matter. Water Res. 35, 3688–3696.

Tfaily, M.M., Hodgkins, S., Podgorski, D.C., Chanton, J.P., Cooper, W.T., 2012. Comparison of dialysis and solid-phase extraction for isolation and concentration of dissolved organic matter prior to Fourier transform ion cyclotron resonance mass spectrometry. Anal. Bioanal. Chem. 404, 447–457.

Therkildsen, M., Isaksen, M.F., Lonstein, B.A., 1997. Urea production by the marine bacteria *Delaya venusta* and *Pseudomonas stutzeri* grown in a minimal medium. Aquat. Microb. Ecol. 13, 213–217.

Thor, P., Dam, H.G., Rogers, D.R., 2003. Fate of organic carbon released from decomposing copepod fecal pellets in relation to bacterial production and ectoenzymatic activity. Aquat. Microb. Ecol. 33, 279–288.

Thurman, E.M., 1985. Organic Geochemistry of Natural Waters. Niyhoff/Junk, Boston.

Tobias, C.R., Cieri, M., Peterson, B.J., Deegan, L.A., Vallino, J., Hughes, J.E., et al., 2003. Processing watershed-derived nitrogen in a well-flushed New England estuary. Limnol. Oceanogr. 48, 1766–1778.

Torres-Valdés, S., Roussenov, V.M., Sanders, R., Reynolds, S., Pan, X., Mather, R., et al., 2009. Distribution of dissolved organic nutrients and their effect on export production over the Atlantic Ocean. Global Biogeochem. Cycles 23, GB4019.

Tsapakis, M., Stephanou, E.G., 2007. Diurnal cycle of PAHs, nitro-PAHs, and oxy-PAHs in a high oxidation capacity marine background atmosphere. Environ. Sci. Technol. 41, 8011–8017.

Tsugita, A., Uchida, T., Mewes, H.W., Atake, T., 1987. A rapid vapor-phase acid (hydrochloric acid and trifluoroacetic acid) hydrolysis for peptide and protein. J. Biochem. 102, 1593–1597.

Tungaraza, C., Brion, N., Rousseau, V., Baeyens, W., Goeyens, L., 2003a. Influence of bacterial activities on nitrogen uptake rates determined by the application of antibiotics. Oceanologia. 45, 473–489.

Tungaraza, C., Rousseau, V., Brion, N., Lancelot, C., Gichuki, J., Baeyens, W., et al., 2003b. Contrasting nitrogen uptake by diatom and *Phaeocystis*-dominated phytoplankton assemblages in the North Sea. J. Exp. Mar. Biol. Ecol. 292, 19–41.

Turk, V., Rehnstam, A.-S., Lundberg, E., Hagström, Å., 1992. Release of bacterial DNA by marine nanoflagellates, an intermediate step in phosphorus regeneration. Appl. Environ. Microbiol. 58, 3744–3750.

Twomey, L.J., Piehler, M.F., Paerl, H.W., 2005. Phytoplankton uptake of ammonium, nitrate, and urea in the Neuse River Estuary, NC, USA. Hydrobiologia 533, 123–134.

Uhlík, O., Jecná, K., Leigh, M.B., Macková, M., Macek, T., 2009. DNA-based stable isotope probing: a link between community structure and function. Sci. Total Environ. 407, 3611–3619.

Urban-Rich, J., 1999. Release of dissolved organic carbon from copepod fecal pellets in the Greenland Sea. J. Exp. Mar. Biol. Ecol. 232, 107–124.

van Boekel, W.H.M., Hansen, F.C., Riegman, R., Bak, R.P.M., 1992. Lysis-induced decline of a *Phaeocystis* bloom and coupling with the microbial foodweb. Mar. Ecol. Prog. Ser. 81, 269–276.

Van Engeland, T., Soetaert, K., Knuijt, A., Laane, R.W.P.M., Middelburg, J.J., 2010. Dissolved organic nitrogen dynamics in the North Sea: a time series analysis (1995–2005). Estuar. Coast. Shelf Sci. 89, 31–42.

Van Engeland, T., Bouma, T.J., Morris, E.P., Brun, F.G., Peralta, G., Lara, M., et al., 2013. Dissolved organic matter uptake in a temperate seagrass ecosystem. Mar. Ecol. Prog. Ser. 478, 87–100.

Vardi, A., Berman-Frank, I., Rozenberg, T., Hadas, O., Kaplan, A., Levine, A., et al., 1999. Programmed cell death of the dinoflagellate *Peridinium gatunense* is mediated by CO_2 limitation and oxidative stress. Curr. Biol. 9, 1061–1064.

Vardi, A., Van Mooy, B.A., Fredricks, H.F., Popendorf, K.J., Ossolinski, J.E., Haramaty, L., et al., 2009. Viral glycosphingolipids induce lytic infection and cell death in marine phytoplankton. Science 326 (5954), 861–865.

Varela, D.E., Harrison, P.J., 1999. Effect of ammonium on nitrate utilization by Emiliania huxleyi, a coccolithophore from the oceanic northeastern Pacific. Mar. Ecol. Prog. Ser. 186, 67–74.

Varela, M.M., Barquero, S., Bode, A., Fernádez, E., Gonzalez, N., Teira, E., et al., 2003. Microplanktonic regeneration of ammonium and dissolved organic nitrogen in the upwelling area of the NW of Spain: relationships with dissolved organic carbon production and phytoplankton size structure. J. Plankton Res. 25, 719–736.

Varela, M.M., Bode, A., Fernández, E., Gónzalez, N., Kitidis, V., Varela, M., et al., 2005. Nitrogen uptake and dissolved organic nitrogen release in planktonic communities characterized by phytoplankton size structure in the central North Atlantic. Deep Sea Res. Part I: Oceanogr Res. Pap. 52, 1637–1661.

Varela, M.M., Bode, A., Morán, X.A.G., Valencia, J., 2006. Dissolved organic nitrogen release and bacterial activity in the upper layers of the Atlantic Ocean. Microb. Ecol. 51, 487–500.

Velhoen, N., Ikonomou, M.G., Helbing, C.A., 2012. Molecular profiling of marine fauna: integration of omics with the environmental assessment of the world's oceans. Ecotoxicol. Environ. Saf. 76, 23–38.

Vetter, T.A., Perdue, E.M., Ingall, E., Koprivnjak, J.F., Pfromm, P.H., 2007. Combining reverse osmosis and electrodialysis for more complete recovery of dissolved organic matter from seawater. Sep. Purif. Technol. 56, 383–387.

Veuger, B., Middelburg, J.J., Boschker, H.T.S., Nieuwenhuize, J., van Rijswijk, P., Rocchelle-Newall, E.J., et al., 2004. Microbial uptake of dissolved organic and inorganic nitrogen in Randers Fjord. Estuar. Coast. Shelf Sci. 61, 507–515.

Villareal, T.A., Altabet, M.A., Culver-Rymsza, K., 1993. Nitrogen transport by vertically migrating diatom mats in the North Pacific Ocean. Nature 363, 709–712.

Villareal, T., Woods, S., Moore, J., Culver-Rymsza, K., 1996. Vertical migration of Rhizosolenia mats and their significance to NO_3^- fluxes in the central North Pacific gyre. J. Plankton Res. 18, 1103–1121.

Villareal, T.A., Pilskaln, C., Brzezinski, M., Lipschultz, F., Dennett, M., Gardner, G.B., et al., 1999. Upward transport of oceanic nitrate by migrating diatom mats. Nature 397, 423–425.

Villareal, T.A., Adornato, L., Wilson, C., Schoenbaechler, C.A., 2011. Summer blooms of diatom-diazotroph assemblages and surface chlorophyll in the North Pacific gyre: a disconnect. J. Geophys. Res. 116, doi:10.1029/2010JC006268, C03001.

Vincent, D., Slawyk, G., L'Helguen, S., Sarthou, G., Gallinari, M., Seuront, L., Sautour, B., Ragueneau, O., 2007. Net and gross incorporation of nitrogen by marine copepods fed on 15N-labeled diatoms: methodology and trophic studies. J. Exp. Mar. Biol. Ecol. 352 (2), 295–305.

Violaki, K., Zarbas, P., Mihalopoulos, N., 2010. Long-term measurements of dissolved organic nitrogen (DON) in atmospheric deposition in the Eastern Mediterranean: fluxes, origin and biogeochemical implications. Mar. Chem. 120, 179–186.

Walker, B.D., Guilderson, T.P., Okimura, K.M., Peacock, M.B., McCarthy, M.D., 2014. Radiocarbon signatures and size–age–composition relationships of major organic matter pools within a unique California upwelling system. Geochim. Cosmochim. Acta 126, 1–17.

Walsh, T.W., 1989. Total dissolved nitrogen in seawater: a new high temperature combustion method and a comparison with photo-oxidation. Mar. Chem. 29, 295–311.

Wang, Y., Hammes, F., De Roy, K., Verstraete, W., Boon, N., 2010. Past, present and future applications of flow cytometry in aquatic microbiology. Trends Biotechnol. 28, 416–424.

Ward, B.B., Bronk, D.A., 2001. Net nitrogen uptake and DON release in surface waters: importance of trophic interactions implied from size fractionation experiments. Mar. Ecol. Prog. Ser. 219, 11–24.

Watts, L.J., Owens, N.J.P., 1999. Nitrogen assimilation and the f-ratio in the northwestern Indian Ocean during and intermonsoon period. Deep Sea Res. II 46, 725–743.

Wawrik, B., 2013. Stable isotope probing the N cycle: current applications and future directions. In: Marco, D. (Ed.), Metagenomics of the Microbial Nitrogen Cycle: Theory, Methods and Applications. Horizon Scientific Press.

Wawrik, B., Callaghan, A.V., Bronk, D.A., 2009. Use of inorganic and organic nitrogen by Synechococcus spp. and diatoms on the West Florida Shelf as measured using stable isotope probing. Appl. Environ. Microbiol. 75, 6662–6670.

Wedyan, M.D.A., 2008. Characteristics of amino acids in the atmospheric and marine environment. World Appl. Sci. J. 3, 454–469.

Weinbauer, M.G., Peduzzi, P., 1995. Effect of virus-rich high molecular weight concentrates of seawater on the dynamics of dissolved amino acids and carbohydrates. Mar. Ecol. Prog. Ser. 127, 245–253.

Weinbauer, M.G., Fuks, D., Peduzzi, P., 1993. Distribution of viruses and dissolved DNA along a coastal trophic gradient in the Northern Adriatic Sea. Appl. Environ. Microbiol. 59, 4074–4082.

Wetz, M.S., Hales, B., Wheeler, P.A., 2008. Degradation of phytoplankton-derived organic matter: implications for carbon and nitrogen biogeochemistry in coastal ecosystems. Estuar. Coast. Shelf Sci. 77, 422–432.

Wheeler, P.A., Kokkinakis, S.A., 1990. Ammonium recycling limits nitrate use in the oceanic subarctic Pacific. Limnol. Oceanogr. 35, 1267–1278.

Wheeler, P.A., North, B.B., Stephens, G.C., 1974. Amino acid uptake by marine phytoplankters. Limnol. Oceanogr. 19, 249–259.

Wheeler, P.A., North, B., Littler, M., Stephens, G., 1977. Uptake of glycine by natural phytoplankton communities. Limnol. Oceanogr. 22, 900–910.

Wiegner, T.N., Seitzinger, S.P., Glibert, P.M., Bronk, D.A., 2006. Bioavailability of dissolved organic nitrogen and carbon from nine rivers in the eastern United States. Aquat. Microb. Ecol. 43, 277–287.

Williams, P.J.L.B., 2004. Meters, kilograms, seconds, but no bomb units. A zero tolerance approach to units. (With apologies to Lynne Truss). Limnol. Oceanogr. Bull. 13, 29–32.

Williams, R.G., Follows, M.J., 1998. The Ekman transfer of nutrients and maintenance of new production over the North Atlantic. Deep Sea Res. 45, 461–489.

Williams, R., Poulet, S.A., 1986. Relationships between the zooplankton, phytoplankton, particulate matter, and dissolved free amino acids in the Celtic Sea. Mar. Biol. 90, 279–284.

Wommack, K.E., Colwell, R.R., 2000. Virioplankton: viruses in aquatic ecosystems. Microbiol. Mol. Biol. Rev. 64, 69–114.

Wong, C.S., Waser, N.A.D., Whitney, F.A., Johnson, W.K., Page, J.S., 2002. Time-series study of the biogeochemistry of the North East subarctic Pacific: reconciliation of the Corg/N remineralization and uptake ratios with the Redfield ratios. Deep Sea Res. II 49, 5717–5738.

Worden, A.Z., Lee, J.H., Mock, T., Rouzé, P., Simmons, M.P., Aerts, A.L., et al., 2009. Green evolution and dynamic adaptations revealed by genomes of the marine picoeukaryotes Micromonas. Science 324 (5924), 268–272.

Worsfold, P.J., Monbet, P., Tappin, A.D., Fitzsimons, M.F., Stiles, D.A., McKelvie, I.D., et al., 2008. Characterisation and quantification of organic phosphorus and organic nitrogen components in aquatic systems: a review. Anal. Chim. Acta 624, 37–58.

Xu, J., Glibert, P.M., Liu, H., Yin, K., Yuan, X., Chen, M., et al., 2012. Nitrogen sources and rates of phytoplankton uptake in different regions of Hong Kong waters in summer. Estuar. Coast. 35, 559–571.

Yamashita, Y., Tanoue, E., 2003. Distribution and alteration of amino acids in bulk DOM along a transect from bay to oceanic waters. Mar. Chem. 82, 145–160.

Yanping, Z., Guipeng, Y., Yan, C., 2009. Chemical characterization and composition of dissolved organic matter in Jiaozhou Bay. Chinese J. Oceanol. Limnol. 27, 851–858.

Yentsch, C.M., Horan, P.K., Muirhead, K., et al., 1983. Flow cytometry and cell sorting: a technique for analysis and sorting of aquatic particles. Limnol. Oceanogr. 28, 1275–1280.

Young, C.L., Ingall, E.D., 2010. Marine dissolved organic phosphorus compostion: insights from samples recovered using combined electrodialysis/reverse osmosis. Aquat. Geochem. 16, 563–574.

Yuan, X., Glibert, P.M., Xu, J., Liu, H., Chen, M., Liu, H., et al., 2012. Inorganic and organic nitrogen uptake by phytoplankton and bacteria in Hong Kong waters. Estuar. Coast. 35, 325–334.

Zehr, J.P., Ward, B.B., 2002. Nitrogen cycling in the ocean: new perspectives on processes and paradigms. Appl. Environ. Microbiol. 68, 1015–1024.

Zehr, J.P., Waterbury, J.B., Turner, P.J., Montoya, J.P., Amoregie, E., Steward, G.F., et al., 2001. Unicellular cyanobacteria fix N_2 in the subtropical North Pacific Ocean. Nature 412, 635–638.

Ziegler, S., Fogel, M.L., 2003. Seasonal and diel relationships between the isotopic compositions of dissolved and particulate organic matter in freshwater ecosystems. Biogeochemistry 64, 25–52.

Zubkov, M.V., Tarran, G.A., 2005. Amino acid uptake of Prochlorococcus spp. in surface waters across the South Atlantic Subtropical Front. Aquat. Microb. Ecol. 40, 241–249.

5

Dynamics of Dissolved Organic Phosphorus

David M. Karl, Karin M. Björkman

Daniel K. Inouye Center for Microbial Oceanography: Research and Education, Department of Oceanography,
School of Ocean and Earth Science and Technology, University of Hawaii, Honolulu, Hawaii, USA

CONTENTS

Biogeochemistry of Marine Dissolved Organic Matter,
http://dx.doi.org/10.1016/B978-0-12-405940-5.00005-4

I INTRODUCTION

Phosphorus (P) is an essential macronutrient for all living organisms; life is truly built around P (deDuve, 1991). In the sea, P exists in both dissolved and particulate pools with inorganic and organic forms. The uptake, remineralization, and physical and biological exchanges among these various pools are the essential components of the marine P-cycle (Figure 5.1). Compared to the much more comprehensive investigations of carbon (C) and nitrogen (N) dynamics in the sea, P pool inventories, and fluxes are less well documented though no less important.

During cell growth, P is incorporated into a broad spectrum of organic compounds with vital functions including structure, metabolism, and regulation. In time, selected P-containing organic compounds are lost from the cells to the surrounding environment by combined exudation and excretion processes. When cells turn over, whether by death/autolysis, grazing, parasitism, or viral infection, there is an enhanced release of intracellular P-containing compounds as both dissolved and particulate organic matter (DOM and POM, respectively). In this broad view, dissolved organic P (DOP) is simply the intermediate between inorganic P (Pi) uptake and Pi regeneration (Figure 5.1). For this and other ecological and analytical interdependencies of Pi and DOP, it is impossible

to isolate DOP from the remainder of the marine P cycle. It is also imperative to emphasize that the production and cycling of P-containing compounds are inextricably linked to C and N dynamics by virtue of the fact that marine DOM and POM pools include many compounds that contain both C and P (e.g., phospholipids, sugar phosphates, and selected phosphonates) or C, N, and P (e.g., nucleotides, nucleic acids, and selected vitamins and phosphonates; see Figures 5.2 and 5.3 and Table 5.1). It is, therefore, inappropriate to consider DOP as separate from dissolved organic C (DOC) and dissolved organic N (DON), or to view the P cycle in biogeochemical isolation.

This review will take a holistic approach to the marine P-cycle with an emphasis on the production and turnover of P-containing and N-P-containing DOM (i.e., DOC-P and DOC-N-P, hereafter collectively referred to as DOP). By design, this chapter will focus on the pelagic environment, especially the open sea. Investigation of the marine sedimentary P-cycle is further complicated by the presence of numerous poorly defined P reservoirs (e.g., Föllmi, 1996; Ruttenberg, 2014), which precludes a straightforward determination of P inventories and fluxes. While the majority of P-cycle processes occur throughout the world's oceans, net DOM/POM production is enhanced in the euphotic zone (e.g., the upper 0-100 m of the water

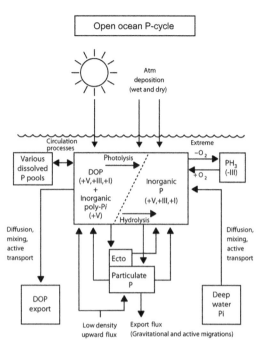

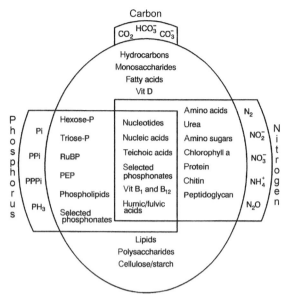

FIGURE 5.1   The open ocean P-cycle showing the various sources and sinks of inorganic and organic P, including biotic and abiotic interconversions. The large rectangle in the center represents the upper water column TDP pool comprised of Pi, inorganic polyphosphate and a broad spectrum of largely uncharacterized DOP compounds. Ectoenzymatic activity (Ecto) is critical for microbial assimilation of selected TDP compounds. Particulate P, which includes all viable microorganisms, sustains the P-cycle by assimilating and regenerating Pi, producing and hydrolyzing selected non-Pi P, especially DOP compounds, and by supporting net particulate matter production and export. The exported DOP and particulate P are eventually remineralized to Pi in the deep sea. Atmospheric deposition, horizontal transport and the upward flux of low-density organic P compounds are generally poorly constrained processes in most marine habitats. The numbers in parentheses represent P valence states indicating an active reduction-oxidation of P in the open sea. Phosphine (PH_3), shown at the right, is the most reduced form of P in the biosphere and is generally negligible except under very unusual, highly reduced conditions. *Redrawn from Karl (2014).*

FIGURE 5.2   Bar and shield representation of dissolved matter in seawater showing the intersection of C, N and P compound classes. For example, dissolved P can exist in a variety of inorganic P forms (outside portion of the shield) or as DOC-P and DOC-N-P compounds. Likewise, C and N have both unique and intersecting pools. Compound symbols include: Pi = orthophosphate, PPi = pyrophosphate (pyro-Pi), PPPi = inorganic polyphosphate (poly-Pi), PH_3 = phosphine, RuBP = Ribulose-1,5-bisphosphoric acid, PEP = phospho(enol) pyruvic acid, N_2 = nitrogen, NO_2^- = nitrite, NO_3^- = nitrate, NH_4^+ = ammonium and N_2O = nitrous oxide.

column) while net remineralization of DOM/POM generally occurs at greater water depths. This vertical stratification of the marine P-cycle (as well as C- and N-cycles) is an important factor that ultimately controls the distributions

and abundances of microbial biomass and rates of global ocean biomass production. Further, the stratification of these cycles greatly impacts the sources, sinks, and, most likely, the chemical composition of marine DOM.

This chapter is an update of Karl and Björkman (2002), which appeared in the first edition of this volume. We will present selected information on DOP formation, distribution, and turnover in the sea building upon several recent, authoritative reviews by Benitez-Nelson (2000), Canfield et al. (2005), Paytan and McLaughlin (2007), and Karl (2007b, 2014) on various aspects of the marine P-cycle, as well as nearly one century of field and laboratory research on this subject. The "Organic Phosphorus Workshop" held in July 2003 in Ascona, Switzerland, and later published

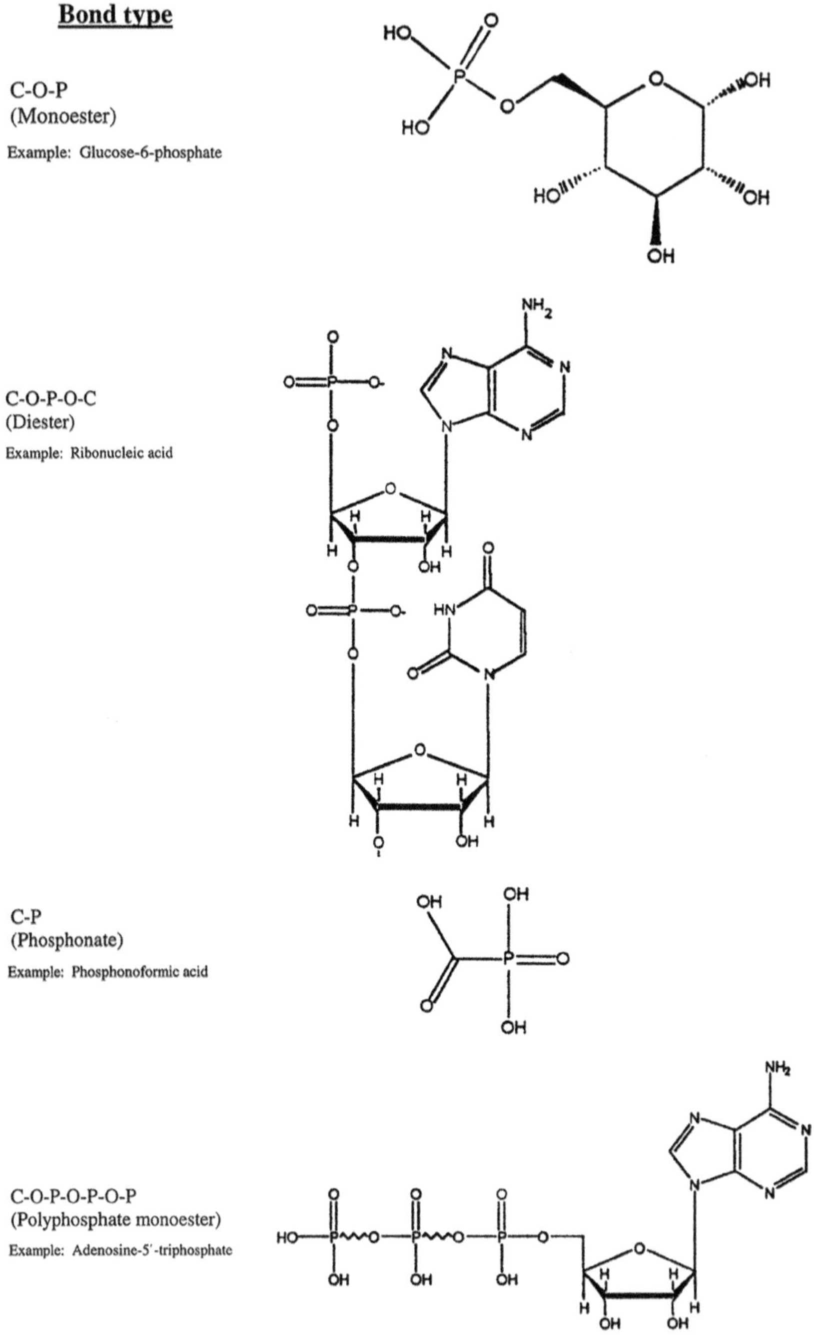

Bond type

C-O-P
(Monoester)

Example: Glucose-6-phosphate

C-O-P-O-C
(Diester)

Example: Ribonucleic acid

C-P
(Phosphonate)

Example: Phosphonoformic acid

C-O-P-O-P-O-P
(Polyphosphate monoester)

Example: Adenosine-5'-triphosphate

FIGURE 5.3   Selected structures of representative DOP pool compound classes with specific examples. Not shown here are the less well-known classes such as phosphoramidate (N-P bonded) or phosphorothionate (S-P bonded) compound that could also be present in cells and in seawater.

TABLE 5.1   Selected Organic P Compounds Either Known to be or Likely to be Present in Seawater

Compound	Chemical Formula (Molecular Weight)	P (% by Weight)	Molar C:N:P
MONOPHOSPHATE ESTERS			
Ribose-5-phosphoric acid (R-5-P)	$C_5H_{11}O_8P$ (230.12)	13.5	5:-:1
Phospho(enol)pyruvic acid (PEP)	$C_3H_5O_6P$ (168)	18.5	3:-:1
Glyceraldehyde 3-phosphoric acid (G-3-P)	$C_3H_7O_6P$ (170.1)	18.2	3:-:1
Glycerophosphoric acid (Gly-3-P)	$C_3H_9O_6P$ (172.1)	18.0	3:-:1
Creatine phosphoric acid (CP)	$C_4H_{10}N_3O_5P$ (211.1)	14.7	4:3:1
Glucose-6-phosphoric acid (Glu-6-P)	$C_6H_{13}O_9P$ (260.14)	11.9	6:-:1
Ribulose-1,5-bisphosphoric acid (RuBP)	$C_5H_6O_{11}P_2$ (304)	20.4	2.5:-:1
Fructose-1,6-diphosphoric acid (F-1,6-DP)	$C_6H_{14}O_{12}P_2$ (340.1)	18.2	3:-:1
Phosphoserine (PS)	$C_3H_8NO_6P$ (185.1)	16.7	3:1:1
NUCLEOTIDES AND DERIVATIVES			
Adenosine 5'-triphosphoric acid (ATP)	$C_{10}H_{16}N_5O_{13}P_3$ (507.2)	18.3	3.3:1.7:1
Uridylic acid (UMP)	$C_9H_{13}N_2O_9P$ (324.19)	9.6	9:2:1
Uridine diphosphate glucose (UDPG)	$C_{15}H_{24}N_2O_{17}P_2$ (566.3)	10.9	7.5:1:1
Guanosine 5'-diphosphate 3'-diphosphate or "magic spot" (ppGpp)	$C_{10}H_{17}N_5O_{17}P_4$ (603)	20.6	2.5:1.25:1
Pyridoxal 5-monophosphoric acid (PyMP)	$C_8H_{10}NO_6P$ (247.2)	12.5	8:1:1
Nicotinamide adenine dinucleotide phosphate (NADP)	$C_{22}H_{28}N_2O_{14}N_6P_2$ (662)	9.4	11:3:1
Ribonucleic acid (RNA)	Variable	~9.2%	~9.5:4:1
Deoxyribonucleic acid (DNA)	Variable	~9.5%	~10:4:1
Inositohexaphosphoric acid or phytic acid (PA)	$C_6H_{18}O_{24}P_6$ (660.1)	28.2	1:-:1
VITAMINS			
Thiamine pyrophosphate (Vit B_1)	$C_{12}H_{19}N_4O_7P_2S$ (425)	14.6	6:2:1
Riboflavine 5'-phosphate (Vit B_2-P)	$C_{17}H_{21}N_4O_9P$ (456.3)	6.8	17:4:1
Cyanocobalamin (Vit B_{12})	$C_{63}H_{88}CoN_{14}O_{14}P$ (1355.42)	2.3	63:14:1
PHOSPHONATES			
Methylphosphonic acid (MPn)	CH_5O_3P (96)	32.3	1:-:1
Phosphonoformic acid (FPn)	CO_5PH_3 (126)	24.6	1:-:1
2-aminoethylphosphonic acid (2-AEPn)	$C_2H_8NO_4P$ (141)	22.0	2:1:1

(Continued)

TABLE 5.1   Selected Organic P Compounds Either Known to be or Likely to be Present in Seawater—cont'd

Compound	Chemical Formula (Molecular Weight)	P (% by Weight)	Molar C:N:P
OTHER COMPOUNDS//COMPOUND CLASSES			
Marine fulvic acid[a] (MFA)	Variable	0.4-0.8	80-100:-:1
Marine humic acid[a] (MHA)	Variable	0.1-0.2	>300:-:1
Phospholipids (PL)	Variable	≤ 4	~40:1:1
Malathion (Mal)	$C_9H_{16}O_5PS$ (267)	11.6	9:-:1
"Redfield" plankton	Variable	1-3	106:16:1

[a]Marine FA and HA are operationally defined fractions, thus their composition may vary with source. These values are from Nissenbaum (1979)

as the proceedings (Turner et al., 2005) provided useful updates on the chemical characterization of P in environmental samples, including DOP in marine ecosystems and new information on the dynamics of organic P. For reasons already mentioned, it is impossible to discuss DOP in any useful ecological framework without also considering other DOM/POM pools and related biogeochemical processes. Although dissolved inorganic P concentrations (typically reported as soluble reactive phosphorus or SRP) are routinely measured in physical, chemical, and biological studies of the marine environment, estimates of total P (i.e., the sum of reactive and nonreactive forms of dissolved inorganic and organic P, also called total dissolved P or TDP) are rare, despite the existence, for over 50 years, of reliable analytical methods. Although TDP was included as a core measurement during the International Geophysical Year (IGY) Atlantic Basin hydrographic survey of 1957-1958 (McGill, 1963), none of the "modern" oceanographic sampling programs, including Geochemical Ocean Sections (GEOSECS) and World Ocean Circulation Experiment (WOCE) and Climate Variability (CLIVAR) Repeat Hydrography program included TDP as a core measurement. Even the Joint Global Ocean Flux Study (JGOFS) program, which sponsored regional-scale field studies of ocean biogeochemistry, mostly ignored P-cycle processes. Consequently, the extant database of high-quality paired SRP and TDP in the world's oceans is relatively sparse in comparison with the global coverage of SRP.

II TERMS, DEFINITIONS, AND CONCENTRATION UNITS

The total P (TP) fraction in seawater is divided, unequally, among particulate P (PP) and TDP fractions (TP=PP+TDP) with both fractions containing inorganic and organic P derivatives. The TDP pool greatly exceeds the PP pool in most open ocean marine environments, but it is the biogenic PP pool (i.e., cells or living biomass) that ultimately produces and remineralizes DOP, thereby sustaining the marine P-cycle. Recently, Cai and Guo (2009) reported the presence of colloidal organic P (COP) and colloidal inorganic P using an ultrafiltration permeation method for samples collected from several aquatic environments. For the northern Gulf of Mexico, COP concentrations ranged from 30 to 60 nM and were >50% of the total "DOP" (Cai and Guo, 2009). It is conceivable that the chemical composition, uptake pathways, and dynamics of COP may be distinct from truly dissolved DOP substrates.

The inorganic forms of P consist mostly of orthophosphoric acid (in seawater at 33% salinity, 20 °C and pH 8.0 as 1% $H_2PO_4^-$ / 87% HPO_4^{2-} / 12% PO_4^{3-};

Kester and Pytkowicz, 1967), pyrophosphate ($P_2O_7^{4-}$; hereafter abbreviated pyro-Pi), and other condensed cyclic (metaphosphate) and linear (polyphosphate) polymers of various molecular weights (hereafter abbreviated poly-Pi). The condensed phosphates can exist in the dissolved, colloidal, and particulate matter fractions of seawater, whereas Pi and pyro-Pi are mostly contained in the truly dissolved fraction or within intracellular pools. Although technically inorganic, both pyro-Pi and poly-Pi are produced biologically with specific metabolic functions, so they should also be considered biomolecules. Recently, the presence of reduced Pi in the form of phosphite (+III valence state) or hypophosphite (+I valence state) have been detected or hypothesized to occur in nature (Hanrahan et al., 2005; White and Metcalf, 2007) and may also be present in marine ecosystems (Karl, 2014). Of these various inorganic forms, only Pi is quantitatively detected by the standard molybdenum blue assay procedure (see Section V.C for more information on reaction specificity). Therefore, the measurements of pyro-Pi, poly-Pi, and reduced inorganic P (+I and +III valence states) require sample hydrolysis to yield reactive Pi, and are analytically part of the TDP pool. However, as DOP commonly is defined as TDP–SRP, these Pi compounds will often be included as part of the DOP pool.

The organic P fractions include primarily monomeric and polymeric esters (C-O-P bonded compounds with P valence of either +III or +V), phosphonates (C-P bonded compounds with P valence generally < +V), and organic condensed phosphates (Table 5.1 and Figure 5.3). Among the ester-linked DOP compounds, both phosphomonoesters and phosphodiesters are present (Table 5.1). Each compound has unique chemical and physical properties, with characteristic phosphohydrolytic enzyme susceptibility. Numerous compound classes (e.g., nucleotides, nucleic acids, phospholipids, phosphoproteins, sugar phosphates, phosphoamides, vitamins) have been detected in seawater and these will be discussed in subsequent sections. Oxidative destruction of the associated organic matter is generally required to convert organic P to reactive Pi, although certain compound classes are partially hydrolyzed during Pi analysis and thus may contribute to an overestimation of the true Pi concentration. For this reason, the standard molybdenum blue assay measures an operationally defined pool, SRP, and the difference between TDP (i.e., equal to SRP following sample hydrolysis) and the initial SRP value has been termed the soluble nonreactive P (SNP) pool. Although SRP is often equated to Pi, in reality, SRP only sets an upper constraint on Pi. Depending upon oxidation/hydrolysis conditions that are used for analysis, the SNP pool includes organic P, pyro-Pi, and poly-Pi as well as partially reduced Pi molecules (e.g., phosphite) that do not react with the standard SRP assay. Consequently, SNP concentration is technically not equal to DOP due to the two independent conditions: SRP $\geq$ Pi and SNP $\geq$ DOP. This may have important analytical and ecological implications as discussed in subsequent sections.

P in seawater can also be characterized by origin (e.g., biogenic or lithogenic) or by physical characteristics (e.g., molecular weight or photolytic lability). Because many different forms can be used as P sources for marine microorganisms, albeit at variable rates and efficiencies, the most ecologically relevant fraction is the biologically available P (BAP) pool. Ideally, BAP consisting of both Pi plus the biolabile fraction of the SNP pool should be measured to constrain oceanic P cycle fluxes, but routine analytical methods do not exist. It might be argued, a priori, that SRP measurements by the Murphy and Riley (1962) procedure place a lower bound on BAP, because both Pi and acid-labile DOP must be biologically available; however, this may not always be the case. Pi contained in colloidal associations or adsorbed to nanoparticles would assay as part of the SRP pool but might be unavailable for uptake under ambient conditions. In all likelihood, only microbioassay analysis can provide an accurate estimate of BAP (see Section VIII.F for details). Suffice it to say, we are still lacking a comprehensive chemical description of dissolved P in seawater (see Section VII).

The measurement of TDP is also operationally defined; typically a high-intensity ultraviolet (UV) photooxidation (Armstrong et al., 1966) or high temperature wet chemical oxidation (Menzel and Corwin, 1965) or a combined (Ridal and Moore, 1990) pretreatment is used to convert SNP to Pi for subsequent analysis by the standard molybdenum blue assay. However, it is well known that certain P-containing compounds (e.g., poly-Pi, nucleotide di-, and triphosphates) are not quantitatively recovered by standard UV photooxidation procedures; neither method quantitatively recovers P from all phosphonate compounds (Monaghan and Ruttenberg, 1999; Thomson-Bulldis and Karl, 1998). As emphasized previously, there is no a priori relationship between these operationally defined pools and the more ecologically relevant BAP pool. Although several SNP compound classes have been reported to exist in seawater, including poly-Pi (Solórzano and Strickland, 1968), nucleotides (Azam and Hodson, 1977; Nawrocki and Karl, 1989), nucleic acids (DeFlaun et al., 1986; Karl and Bailiff, 1989), and monophosphate esters (Strickland and Solórzano, 1966), the SNP pool in seawater remains largely uncharacterized.

The earliest reports of Pi and TDP in seawater, prior to approximately 1930, all reported P as milligrams of P pentoxide (P_2O_5) per cubic meter of seawater (e.g., Atkins, 1923). Ironically, the chemical form P_2O_5 decomposes in water; the correct form should be P_4O_{10} (Olson, 1967). Despite a logical recommendation by Atkins (1925), "to convert the conventional P_2O_5 values into the more rational values for the phosphate ion the factor 1.338, or very approximately 4/3, may be used to multiply the former," the P_2O_5 equivalence reporting practice continued. In 1933, Cooper (1933) made another plea for the importance of consistency in reporting dissolved nutrient and other elemental data. He suggested the gram atom (or submultiple thereof, e.g., milligram atom, microgram atom) of the element under investigation, per cubic meter, as the most useful and meaningful concentration unit. This would provide for the direct comparison with other elements, and a relatively straightforward calculation of bioelemental atomic stoichiometry (i.e., C:N:P:Si) for dissolved or particulate matter. Atomic, molecular, and ionic ratios would all be numerically identical. Cooper (1933) went on to state, "it is felt that such a radical change in the method of reporting results, before being put into service, requires the concurrence of the majority of oceanographical chemists, as uniformity in practice above all else is desirable." The contemporary community of scholars did not immediately accept this bold suggestion, and even at the present time, there is no uniformity of reporting dissolved and particulate matter P concentrations.

A variety of units, all interchangeable, have been used to report DOP in seawater. In preparing this review, we have converted all of the reported concentration data to ng-at P L^{-1} (nM-P) or µg-at P L^{-1} (µM P) as appropriate. For organic P pools, this refers to P only; so 507 ng L^{-1} of dissolved adenosine 5′-triphosphate (ATP), for example, would be equal to 1 nM ATP, but 3 nM-P because each mole of ATP contains three P atoms. This practice of reporting DOP in P molar equivalents is absolutely necessary because the exact chemical composition remains largely unknown. For quantitative measurements of polymeric compounds such as DNA, RNA, and lipid-P we also report the assumptions that we made regarding the mole percentage of P in the specific polymeric compound. Sometimes molality (mol kg^{-1} of seawater) rather than molarity (mol L^{-1} of seawater) is used so that one does not have to calculate changes in volume that occur due to variations in temperature, pressure, or salinity but, for the purposes of this review, we will consider these differences to be negligible.

III THE EARLY YEARS OF PELAGIC MARINE P-CYCLE RESEARCH (1884-1955)

Several pathfinding scientific studies, especially those conducted during the first half of the twentieth century, provided a sound foundation for contemporary investigations of the marine P-cycle. The creation of the Marine Biological Association of the United Kingdom in 1884 and dedication of their marine laboratory at Plymouth in 1888 and the creation of the Marine Biological Laboratory at Woods Hole, MA (USA) in 1888 are especially noteworthy because of the major impact these two research centers have had, and continue to have, on the field of marine ecology and biogeochemistry. In 1903, working out of the Plymouth laboratory, Donald J. Matthews began a systematic study of the oceanographic features of the English Channel. His time-series research program that was later continued by Atkins, Cooper, and Harvey led to a comprehensive understanding of the fundamental links between nutrients, phytoplankton, and fish production in the sea. Matthews (1916, 1917) is also credited with making the first reliable estimations of phosphate in seawater, and with the discovery of oceanic DOP. The colorimetric method that he selected was based on the Pouget and Chouchak reagent (sodium molybdate/strychnine sulfate/nitric acid), which yielded a yellow colored product, the intensity of which was proportional to the amount of phosphate in the water sample. To increase sensitivity, Matthews (1917) first had to concentrate the dissolved phosphate by coprecipitation using either ammonia or a mixture of ammonia and an iron salt. The former, a predecessor to the modern "MAGIC" technique (Karl and Tien, 1992), removed phosphate by adsorption onto magnesium hydroxide, $Mg(OH)_2$, and the latter as ferric phosphate and ferric hydroxide.

Using this laborious but robust method for samples collected near Knap Buoy in the English Channel, Matthews (1916, 1917) made two very important observations: (1) the concentration of phosphate in seawater was approximately $0.85 \mu M$ in December 1915, decreasing systematically to a minimum of $<0.1 \mu M$ between late April and late May, increasing again in January 1917 to a similar winter value, and (2) TDP, measured as phosphate following sample oxidation by potassium permanganate, exceeded the initial concentration of phosphate in the untreated sample by up to a factor of 2-3 fold. In other words, Matthews documented for the first time the seasonal dynamics of phosphate during the vernal blooming of phytoplankton and, more significant to the topic of this review, the presence of DOP in coastal seawaters. While he was very careful to emphasize that the reported DOP values were not necessarily quantitative, the organic component was highest when the phosphate was at a minimum suggesting that DOP formation may be coupled to particulate matter production. He concluded this key discovery paper by stating that "the nature and origin of this organic P is, of course, quite unknown" (Matthews, 1917). The importance of Matthew's research cannot be overstated; however, today, nearly a century after his pioneering contributions, we still lack a comprehensive understanding of DOP dynamics in seawater.

Shortly after the completion of these initial studies, Matthews was called into military service and joined the Hydrographic Office of the Navy where he remained following the end of World War I. A reorganization of research programs at the Plymouth Marine Laboratory led to the formation of a Department of General Physiology and the addition of W. R. G. Atkins and H. W. Harvey to the scientific staff (Southward and Roberts, 1987). Along with L. H. N. Cooper, F. S. Russell, and other staff chemists and plankton biologists at the Plymouth Laboratory, they reestablished in 1921 the monthly time-series sampling program at several key locations in the English Channel. The phosphate detection system used earlier by Matthews had been replaced by the more sensitive Denigès (1921) method.

The "molybdenum blue" method employed ammonium molybdate, sulfuric acid, and stannous chloride (Atkins and Wilson, 1926) that developed an intense blue color in the presence of phosphate. This method, or slight variations thereof, continues to be the method of choice for contemporary studies of the marine P-cycle (see Section V.C). While the seasonal phosphate concentration dynamics using the Denigès-Atkins method revealed trends that were similar to the results published a decade earlier by Matthews, the time-series measurements documented significant interannual variations in both the date of initiation of the spring phytoplankton bloom and, therefore, in the net rate of phosphate uptake (Atkins, 1928). There was also significant interannual variation in total phosphate consumed during the spring period, a value that was subsequently related to annual variations in the potential production of fish. That is to say, knowledge of the rate of phosphate consumption provided a lower limit constraint on annual phytoplankton production. The seasonal net consumption of phosphate was highly correlated to removal of both silicate and carbon dioxide, suggesting that diatoms were largely responsible for organic matter production in the English Channel ecosystem (Atkins, 1930). However, by comparison to this deliberate focus on net plankton production for fisheries management, studies of organic matter decomposition and phosphate remineralization were generally ignored.

Despite these field successes at Plymouth during the 1920s, research on the nature of the "enigmatic" DOP pools, discovered earlier by Matthews, was placed on hold. Even worse, Atkins considered the observed increase in phosphate following permanganate oxidation to be an analytical artifact due to the presence of arsenite in seawater and concluded "much of what was formerly considered to be P in organic combination, in seawater, is in reality arsenic" (Atkins and Wilson, 1927). This inappropriate conclusion from one of the intellectual giants of those times was sufficient to preclude further investigations

of DOP at Plymouth or elsewhere, for at least a decade, or more.

H. W. Harvey and L. H. N. Cooper, two of Atkins' contemporaries at the Plymouth Marine Laboratory who shared common interests in nutrient and plankton dynamics, began systematic studies of the coupled N and P cycles, including nutrient remineralization. It was reasoned that the inorganic nutrients assimilated into organic compounds by marine plankton must eventually be recycled back to nitrate and phosphate by the combined processes of digestion, decay, and chemical hydrolysis. Consequently, both DOP and PP must be considered integral components of the marine P-cycle. At about this same time, studies on the role of marine bacteria as agents of organic matter decomposition were getting underway at several independent marine laboratories worldwide (e.g., Renn, 1937; Waksman and Renn, 1936; ZoBell and Grant, 1943). Furthermore, considerations of coupled particle sinking and decomposition (Seiwell and Seiwell, 1938) provided the incentive for investigations of the role of biological processes in the distributions of nonconservative properties (e.g., nutrients, oxygen, and carbon dioxide) as a function of water depth and distance from landmasses. Included in these investigations were studies of the release of phosphate during dark storage of water samples with and without added plankton (Cooper, 1935; Gill, 1927) or specific DOP compounds (Harvey, 1940). It was concluded that much of the total P in the sea, especially in the surface waters, was tied up in physiologically important classes of living and nonliving particulate matter, which upon the initial period of decomposition were released as DOP. Only after microbial decomposition, was the P released back as free phosphate. A renewed appreciation for the role of DOP in the marine P-cycle came from these laboratory experiments including a focused effort on field measurements. Independent, but similar, studies on the N:P stoichiometry of plankton production and decomposition conducted by Redfield et al. (1937) and Cooper (1938) also accelerated during the 1930s leading, eventually, to an ecumenical theory of nutrient dynamics in the sea. It should be noted, however, that all field data on SRP and TDP concentrations measured using the Denigès-Atkins method and collected prior to about 1940 had an inadvertent analytical error caused by salt interference that required a correction factor of 1.35 (Cooper, 1939a). This correction, when applied to the data collected for the N:P of plankton and water samples collected off Plymouth, U.K. resulted in a revised N:P molar stoichiometry of 15:1, rather than the 20:1 that had been reported previously; this revised ratio was termed the "Cooper ratio" (Cooper, 1938, 1939b).

By the early 1940s, the fundamental role of DOP in the marine P-cycle was firmly established (Atkins, 1930; Cooper, 1938; Kreps, 1934; Newcombe and Brust, 1940; Redfield et al., 1937). The sustained time-series investigations of the English Channel provided evidence for a seasonally variable pool of DOP, with maximum concentrations observed in phase with the height of the phytoplankton bloom in late spring to early summer (Armstrong and Harvey, 1950; Harvey, 1950; Figure 5.4). Field studies conducted in the epipelagic waters of the Gulf of Maine (Redfield et al., 1937) and in Chesapeake Bay (Newcombe and Brust, 1940) revealed similar results.

Confirmation of the presence of a significant pool of DOP in seawaters from diverse habitats further stimulated research to ascertain the sources and sinks of these potentially diverse, but biologically relevant compound classes. Until this time, bacteria had been considered to be the principal agents of DOP remineralization to Pi, the preferred substrate for phytoplankton growth. However, careful laboratory experiments conducted by Chu (1946) documented the ability of selected bacteria-free phytoplankton cultures to utilize DOP, thereby providing a novel, alternative pathway in the marine P-cycle. Presumably, these microorganisms would be

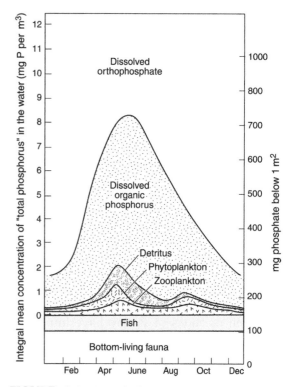

FIGURE 5.4 Annual changes in various living and nonliving P pools and total inventories for the samples collected in the English Channel. The annual dominance of TDP (>80% of total P) is evident. The most notable seasonal change in the P inventory is the shift from Pi dominance in winter to DOP dominance in late spring-summer. *Redrawn with permission from Harvey (1955).*

selected for during the summer months when Pi concentrations were low and DOP/Pi ratios were high, a scenario that could promote a seasonal succession of phytoplankton species in certain habitats.

It seems appropriate to end this section on "The early years of marine P-cycle research" with the publication of Harvey's seminal monograph "The Chemistry and Fertility of Sea Waters" (Harvey, 1955). While his field observations concentrated mainly on the English Channel, the conceptual framework presented in this now classic volume received worldwide attention and provided the incentive for a large portion of the DOP research which followed during the next half century.

IV THE PELAGIC MARINE P-CYCLE: KEY POOLS AND PROCESSES

Compared to the more complex cycles of C, N, and S that are characterized and sustained by redox transformations, the marine P-cycle at first appeared rather simple with a redox invariant pentavalent state (+V) as PO_4^{3-}, whether as free orthophosphate or as P incorporated into organic compounds. However, this conceptual view of a redox neutral P-cycle is no longer tenable (Hanrahan et al., 2005; White and Metcalf, 2007), and suggests that microbial oxidation-reduction reactions of P may have important bioenergetic and ecological consequences in the sea that are analogous to the other major bioelement cycles (Karl, 2014).

Cellular P metabolism in the marine environment is complex. The transfer of phosphoryl groups is a fundamental characteristic of intermediary metabolism and is, therefore, crucial for life. Numerous enzymes share the ability to catalyze phosphoryl group transfer including phosphatases, phosphokinases, phosphomutases, nucleotidases, nucleases, phosphodiesterases, phospholipases, and nucleotidyl transferases and cyclases (Knowles, 1980). New enzymes and pathways are still being discovered, for example, with the metabolism of phosphonates (Martinez et al., 2010; McSorley et al., 2012; Quinn et al., 2007). A surprise discovery by Yang and Metcalf (2004) was that *Escherichia coli* alkaline phosphatase (APase), one of the most well-studied phosphohydrolases, also catalyzed the oxidation of phosphite (+III) to phosphate (+V) and molecular hydrogen. This represented an unprecedented reaction in both P and H biochemistry. Of these enzyme classes, the phosphatases (mono- and diesterases), nucleotidases

and nucleases have been most frequently studied in the marine environment (Figure 5.5). The depolymerization reactions converting high molecular weight (HMW) DOP to intermediate and low molecular weight (IMW and LMW, respectively) are probably slow relative to the hydrolysis and direct uptake of monomeric DOP. Consequently, HMW-DOP turnover is probably the "bottleneck" in marine P-cycle dynamics.

Marine microorganisms can assimilate three separate classes of P compounds: (1) inorganic P, including Pi, phosphite, pyro-P, and poly-P, (2) ester-linked DOP compounds, including both oxidized (+V) and reduced (<+V) derivatives, and (3) phosphonates in a variety of P valence states. Typically, Pi is the preferred substrate, thus, during Pi-sufficient growth conditions, the synthesis of specific enzymes for phosphoester and phosphonate utilization are usually repressed (see Section IX.D). Most microorganisms synthesize one or more phosphohydrolytic enzymes in order to degrade selected DOP compounds. However, hydrolysis prior to transport is necessary for those DOP compounds that cannot be transported directly into the cell. In bacteria, DOP hydrolysis usually occurs at the outer cell membrane or in the periplasmic space, and under certain conditions, enzymes are exported from the cell for hydrolytic activity in the surrounding medium. This may be one source of the reported presence of "dissolved" enzymatic activity in the sea. Pi that is released upon hydrolysis is often transported into the cell and assimilated into new biomolecules for metabolism or cell growth. However, depending upon the DOP substrate in question, the Pi-free organic molecule may be the preferred target substrate and the newly released Pi remains behind in the medium. By example, the hydrolysis of glucose-6-P (glu-6-P) most likely provides Pi, whereas the hydrolysis of adenosine 3′, 5′-monophosphate (AMP) could provide a purine base for nucleic acid synthesis with the Pi remaining in the medium (Ammerman and Azam, 1985; Björkman and Karl, 1994; Tamminen, 1989). Consequently, the composition of the DOP pool

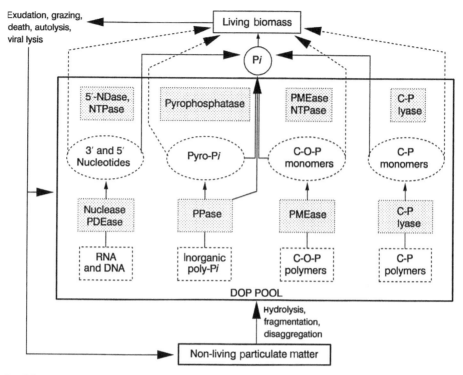

FIGURE 5.5 Schematic presentation of the role of selected enzymes (both dissolved and cell/organism-associated) in DOP pool dynamics. The production of detrital P, including both particulate and IMW/HMW dissolved inorganic and organic P pools provides key substrates (open boxes) for the specific enzymes (shaded boxes). The continued supply of monomeric compounds (<1 kDa; open circles) fuels additional enzymatic processes that lead to DOP or Pi assimilation by living biomass. Because most DOP compounds start as nonliving particulate and HMW cellular constituents but are assimilated as monomeric compounds, there is a degradation 'bottleneck' leading to the continued production of bioavailable DOP substrates in marine ecosystems. Abbreviations: C-O-P = ester-linked P compounds; C-P = phosphonates; PPase = inorganic polyphosphatase; PMEase = phosphomonoesterase (including but not limited to APase); 5NDase = 5′ nucleotidase; NTPase = nucleotide triphosphatase; PDEase = phosphodiesterase.

as well as the biosynthetic needs of the microbial assemblage will largely determine whether cellular phosphohydrolase activity is a mechanism for net Pi regeneration or net Pi sequestration. In either case, however, the enzymatic activity represents a net sink for DOP compounds and effectively sustains DOP pool turnover.

The combined processes of excretion, exudation, death, and autolysis (including viral lysis) provide a constant flux into the DOP pool, and the combined processes of hydrolysis (including ectoenzymatic activity, chemical hydrolysis, and

photolysis) and active microbial uptake comprise the major DOP sinks. The ability to transport and metabolize selected DOP compounds appears to be enhanced during conditions when the preferred growth substrate, Pi, is present at limiting concentrations. The regulation of this switch from Pi to DOP assimilation is under complex metabolic control. DOP turnover is inextricably linked to Pi supply, and this generally results in a negative correlation between Pi and DOP concentrations in the global ocean (see Section VI). Thus, DOP production and DOP utilization may be both spatially and temporally decoupled in the marine environment.

In coastal regions, there can be both point source (e.g., outfalls, rivers, ground waters) and nonpoint source (e.g., runoff) inputs of both inorganic and organic P; the chemical composition of the latter may be fundamentally distinct from the autochthonous DOP pool. Atmospheric deposition of P is another potential source term, but these fluxes, especially for DOP, are poorly constrained at present. Atmospheric inputs are also likely to be regionally variable with the largest fluxes immediately adjacent to (i.e., downwind) major continental, industrial, or volcanogenic sources. Active volcanoes may be unique in that they can catalyze the formation of gaseous P (P_4O_{10}) by the intense heating of basaltic rocks (Yamagata et al., 1991). The gaseous byproducts condense in the volcanogenic steam plume and precipitate as pyro-Pi and Pi, and possibly partially reduced P molecules.

Because there is no significant gas phase in the oceanic P-cycle, the primary net removal mechanism for P in the open ocean is via gravitational settling of particulate matter, downward diffusion, and advection and, for selected regions of the world's ocean, horizontal transport. However, for most open ocean habitats where concentration gradients are weak or nonexistent, horizontal advection represents both a source and sink with a near zero net impact on the P budget. Over sufficiently long time scales (decades to centuries), there is a balance between the sources and sinks, which leads to relatively time-invariant pool inventories and fluxes. However, over seasonal-to-decadal time scales, both the concentrations and fluxes can vary in response to changes in local and regional physical forcing and global climate variability. These perturbations can also lead to changes in the relative proportions of Pi to DOP, or DOP to TDP in selected habitats. The North Pacific Subtropical Gyre (NPSG) case study, presented later in this review (see Section VII.B), will focus on the P-cycle in a climate-forced marine ecosystem.

The use of enzymatic biomarkers to assess the P status of natural microbial assemblages has been extensively employed in microbiological oceanography. Constitutive enzymes (those found in cells regardless of nutrient status), inducible enzymes, (those produced by an organism when exposed to a specific substrate), and repressible enzymes, (those synthesized when the concentration of a specific repressor becomes very low) are the three main classes of enzymes used to efficiently regulate the cell's biodegradation potential. Consequently, the presence or absence of a specific enzyme can be used as an ecological indicator of metabolic readiness (see Section IX.D).

The importance of P in microbial metabolism and, therefore, in ecosystem processes is best demonstrated by the impressive ecophysiological response of bacteria to Pi-stress or Pi-limitation. Under such conditions, substantial and coordinated changes occur that prepare the cells for competition and survival, including the synthesis of more efficient and specific Pi-capture systems, and the ability to utilize a broader range of potential P-containing organic substrates, including phosphonates. Enzymatic hydrolysis of the phosphonate C-P bond requires either phosphonatase or C-P lyase. These independent enzymes are distinguishable by their substrate specificities (Wanner, 1993); both pathways are stimulated during periods of Pi-limitation (Metcalf and Wanner, 1991).

Another group of enzymes that are an integral part of the Pi-limitation stimulon are phosphatases, a general term used to refer to an enzyme that catalyzes the hydrolysis of esters and anhydrides of phosphoric acid. The most common is a class of alkaline phosphomonoesterases, also called APase (EC 3.1.3.1), which can potentially hydrolyze a broad spectrum of DOP compounds optimally at seawater pH. APase is a relatively nonspecific enzyme that releases Pi from a variety of phosphomonoesters, including di-, tri-, and polyphosphate (e.g., nucleotides) organic derivatives. The ecological advantage is obvious; selected DOP compound classes could serve as reliable growth substrates during periods of Pi depletion. APase enzymes are diverse with variable substrate specificities, physical and kinetic properties, and metal-ion requirements. Even within a single species, there may be multiple forms of APase (DePrada et al., 1996). Some, but not all, APases also possess mononucleotidase activity. Although APase activity has been reported for numerous marine habitats, most ecological studies lack a detailed description of the enzymes under consideration.

The relative ecophysiological roles of APase versus 5' nucleotidase (e.g., 5NDase), ATPase, phosphonate lyase, and other free and cell-associated enzymes will undoubtedly vary with the DOP pool composition. Together, these enzymes act primarily on LMW, monomeric substrates and undoubtedly help to keep the ambient pools of LMW DOP at low (10^{-9} M, or less) levels. We hypothesize that the DOP pool turnover is controlled largely by the production rate of IMW and LMW substrates from the hydrolysis of HMW-DOP, rather than by LMW DOP utilization. This would lead to an enhanced ecological role for those enzymes that are capable of hydrolyzing the HMW-DOP, especially nucleases (including DNase and RNase), phosphodiesterases (exonuclease), and proteases (Figure 5.5). P-cycle closure would ultimately require the coordinated suite of enzymes. If this model is correct, then the MW spectrum of DOP in the marine environment would be skewed towards HMW (>10 kDa). At present, there are few open ocean data available to test this ecological prediction.

V SAMPLING, INCUBATION, STORAGE, AND ANALYTICAL CONSIDERATIONS

A Sampling

The reliability of any field-collected data set is determined by the methods used for sampling and analysis. Although this chapter focuses on the ecological implications of the field data themselves, it is important to comment briefly on sampling, experimental design for P flux estimations, and laboratory analysis of the respective P pools.

There are several general concerns that should be mentioned in regard to the methods reviewed in this chapter. First and foremost, P contamination is a major potential problem. Furthermore, sample contamination by toxic trace metals during sampling and subsampling is also a concern for P flux studies that require sample incubation. It is recommended that all sampling gear, as well as storage and incubation bottles, be thoroughly acid-cleaned (1 M or ~10% HCl) and distilled water rinsed before use, and sample water rinsed before collection of the target seawater.

Sampling is one of the most important, but often overlooked, aspects of oceanography. Because of the ease with which a seawater or sediment sample is obtained it is tacitly assumed that sampling is a straightforward and simple task; in reality, it is not (Karl and Dore, 2001). Questions of time and space (Dickey, 1991; Karl, 2010), minimum size and number of samples, sample replication, contamination and post-collection treatment of the primary samples are all relevant. Most of our conceptual views of the marine environment and, therefore, the basis for our sampling protocols focus on the vertical structure of the marine environment, despite the fact that the ocean is clearly a "horizontal" habitat (e.g., horizontal-to-vertical scale of the North Pacific Ocean is >1000:1; Karl and Dore, 2001). In the open sea, there is a well-defined vertical zonation of biological communities based on light (in the near surface) and other physical and chemical properties at depths greater than approximately 200 m. In designing a sampling program for P-cycle research, it is essential to consider this identifiable zonation as well as the source and other unique characteristics of each of the unique water masses.

When investigating P cycling rates, for example by using radioisotope techniques (see Section V.E), several precautions must be taken in order to ensure that the rates measured during the post-collection incubation procedure are representative of those occurring in nature. First and foremost, the initial sample must be collected with great care so as to minimize chemical and microbiological contamination. Furthermore, exposure of viable microorganisms to environmental conditions that are substantially different from those at the collection site should be avoided so as to minimize any deleterious effects ranging from short-term transitions in metabolism to death.

The investigator should be aware of at least three separate areas where variability can be introduced into field measurements: replication at the level of sampling (i.e., multiple water samples collected from a common depth), replication at the level of subsampling (i.e., multiple subsamples from a single sample), and analytical replication (i.e., multiple analyses of a single sample extract). Because of the heterogeneous distribution of microbial communities in nature, and therefore of DOP production, variance between sampling bottles is generally the largest source of error. Replication is most meaningful when performed at the highest level, i.e., multiple samples of water from a given environment (Kirchman et al., 1982). It has also been demonstrated that the overall variance and the precision with which the sample variance can be estimated are functions of the procedure used to subsample the initial sample collection (Venrick, 1971).

B Sample Processing, Preservation, and Storage

Early investigations made no attempt to preserve seawater samples prior to analysis for Pi or TDP, even during prolonged storage (weeks to months). The regeneration of Pi with time would have resulted in systematic overestimations of Pi and underestimations of DOP, without greatly affecting TDP. Since that time, there has been a great deal of discussion and some acrimonious debate on the issue of seawater sample processing, preservation, and storage prior to analysis for SRP and TDP (Dore et al., 1996).

The first important consideration is whether or not to filter the water sample prior to preservation and long-term storage. The answer to this question is generally site-specific and will depend entirely on the objectives of the study and on the relative proportions of particulate and dissolved P. In most open ocean ecosystems and in all subeuphotic zone habitats, PP rarely exceeds 1-5% of the total P in the sample so filtration may not be necessary. However, when sample filtration is desirable, then the choice of filter matrix (polycarbonate, cellulose acetate, glass fibers, silver, aluminum oxide), porosity, pressure differential employed, and volume-to-area (i.e., filter loading) are equally relevant concerns. Filtration can also increase SRP and DOP concentrations by cell leakage or breakage (Pilson and Betzer, 1973).

No matter what filter is selected, there will be some limitation. For example, glass fiber filters (Whatman GF/F or equivalent) that are routinely used in oceanography have two potential problems: (1) the porosity (0.7 µm nominal) precludes a unique separation of small particles and colloids from the truly dissolved P pools and (2) in certain habitats, there is a significant adsorption of dissolved matter onto the GF/F filter matrix. This latter problem is a special concern in oligotrophic ocean environments where DOP concentrations exceed PP by 1-2 orders of magnitude. Furthermore, the ability of poly-Pi to quantitatively bind to powdered silica glass (Ault-Riché et al., 1998) suggests that dissolved poly-Pi in seawater may be underestimated if water samples are first filtered through silica glass fiber filters. The advantage of glass fiber filters is that they can be combusted at 450 °C and acid-cleaned before use to thoroughly remove any P contamination. They also have favorable flow characteristics and high loading capacity which are important for the measurements of certain P pools.

If filtration is not employed then PP must also be measured independently to provide the most accurate estimate of DOP (i.e., DOP = TP − (Pi + PP)). The use of independent analytical procedures, one for TP (e.g., UV photooxidation) and another for PP (e.g., high temperature ashing followed by acid hydrolysis) has the potential for large systematic errors if, for example, particulate poly-Pi is present. To the extent possible, the oxidation/hydrolysis methods used for TDP and PP should be matched.

For decades, marine chemists have tested the suitability of various preservation techniques, considering the potential effects of storage container, temperature, chemical additions, and radiation on samples of different water types (Gilmartin, 1967; Maher and Woo, 1998; Murphy and Riley, 1956). The considerable body of literature presents varied and often contradictory opinions on the effectiveness of various preservation methods. The most extreme position is that all Pi, SRP, and TDP measurements must be conducted in the field on fresh sample materials. This, of course, is not always possible and sometimes not even desirable even when it is possible. A study by Dore et al. (1996) has reevaluated several long-standing criticisms of the seawater sample storage problem. In their hands, immediate freezing ($-20\,°C$) of the sample in a high-density polyethylene bottle, stored upright in the dark, provides a simple and suitable method for storage of seawater for periods up to 1 year for subsequent analysis of SRP and DOP. Alternatively, preservation with mercuric chloride followed by storage at $4\,°C$ in the dark (Kattner, 1999), and acidification to pH 1 followed by storage at 4-5 °C (Monaghan and Ruttenberg, 1999) have also been used.

C Detection of Pi and P-Containing Compounds in Seawater

The analysis of dissolved and PP-containing compounds in seawater is neither simple nor straightforward. Olson (1967) has created a glossary for P-containing compounds that includes 75 potential forms of inorganic and organic P that might be found in nature. Strickland and Parsons (1972) have defined eight of the most relevant operational classes of P compounds based on reactivity with the acidic molybdate reagents, ease of hydrolysis and particle size. These range from "inorganic, soluble and reactive," presumably Pi, through "enzyme hydrolyzable phosphate" (Pi released following treatment with APase), to "inorganic, particulate, and unreactive" (presumably P-containing minerals). Some of the operationally defined pools have no convenient analytical method of determination while others can only be estimated as the difference between two operational classes with partially overlapping specificity. Only a very few specific P-containing compounds or compound classes can be readily detected at the low concentrations typically found in seawater (see Section VIII). Our inability to completely characterize these various dissolved and particulate pools currently limits further progress towards a comprehensive understanding of the marine P cycle.

1 Analysis of Pi

The quantitative estimation of DOP (as well as TP, TDP, and PP) relies upon the measurement of Pi, both before and after sample oxidation/hydrolysis (see below). Consequently, the precision and accuracy of DOP pool estimation is tied directly to the specificity and reliability of Pi analysis. Although Pi can be measured by any of a number of analytical techniques (Boltz, 1972), quantitative analysis of Pi in seawater has traditionally relied upon the formation of a 12-molybdophosphoric acid (12-MPA) complex and its subsequent reduction to yield a highly colored blue solution, the extinction of which is measured by absorption spectrophotometry (Atkins, 1923; Denigès 1920, 1921; Fiske and Subbarow, 1925; Murphy and Riley, 1958, 1962; Osmond, 1887). Ironically, molybdate enhances the hydrolysis of selected organic-P compounds and pyro-Pi (Weil-Malherbe and Green, 1951), so the molybdenum blue protocol appears prepositioned for Pi overestimation in natural seawater samples. However, without knowledge of the precise chemical composition of seawater DOP, we cannot predict the magnitude of this interference.

The measurement of Pi (i.e., SRP) has a very rich and diverse history that, unfortunately, cannot be fully chronicled here. Over the years, numerous improvements have been introduced to the basic method so that substantial variability now exists in the conditions used for color development, final reduction of the 12-MPA complex, and the treatment of potentially interfering compounds; Armstrong (1965), Olson (1967) and Broberg and Petterson (1988) provide comprehensive historical accounts of these changes. In 1962, the single "mixed reagent" (sulfuric acid, ammonium molybdate, ascorbic acid, potassium antimonyl tartrate) was introduced and this is the protocol most commonly employed today (Murphy and Riley, 1962). One critical difference between the methods is the reducing agent used; ascorbic acid in the Murphy-Riley method versus stannous chloride by many other researchers, and the use of antimony (+III) as a reaction catalyst. In addition to being able to combine all required reagents into a single dose addition of "mixed reagent" to the sample, color development is rapid and relatively stable. Furthermore, the salt error, which for the use of stannous chloride is significant (e.g., $\geq 15\%$), is $<1\%$ (Murphy and Riley, 1962). Both methods, however, are affected by arsenate (+V) reactivity and by the acid-catalyzed hydrolysis of labile organic-P compounds. The consequences of these potential interferences for quantitative determinations of Pi, TDP and, therefore, DOP are discussed below.

Although the stepwise chemical procedure for Pi determination is straightforward and fully amenable to automated analysis, there are many analytical and conceptual complexities inherent in measuring and interpreting Pi concentrations in seawater (Tarapchak, 1983). The SRP pool measured by the standard Murphy-Riley procedure is not necessarily equivalent to the concentration of Pi, but may also include non-Pi P-containing compounds that are hydrolyzed under the acidic reaction conditions (Figure 5.6). Rigler (1956, 1973) was the first to demonstrate that the conventional method of SRP determination in aquatic environments can result in a serious overestimation of Pi, especially when the SNP concentration equals or exceeds the Pi pool. Since then, many investigators have struggled with this problem, and it is still a challenge to obtain a reliable measurement of the Pi concentration in seawater. Drummond and Maher (1995) described an adaptation of the Murphy and Riley (1962) procedure that yields full phosphoantimonyl molybdic acid color development in less than 1 min. Although they did not comment further on this, it is conceivable that this rapid reaction rate might provide a more accurate estimate of Pi by eliminating or at least reducing the time that is available for DOP hydrolysis, similar to the 6 second assay (Chamberlain and Shapiro, 1969). Significant differences between SRP and Pi have been observed for nearly every natural aquatic ecosystem where more rigorous and specific methods of Pi analysis have been employed (Jones and Spencer, 1963; Karl and Tien, 1997; Kuenzler and Ketchum, 1962; Pettersson, 1979; Rigler, 1968; Thomson-Bulldis and Karl, 1998). Even different SRP methods return unequal estimates of "Pi" when tested with common seawater samples (Karl and Tien, 1997). Accurate determination of Pi is absolutely essential for the application of artificial ^{32}Pi or ^{33}Pi tracer studies if mass flux estimation (Pi uptake or regeneration rates, DOP

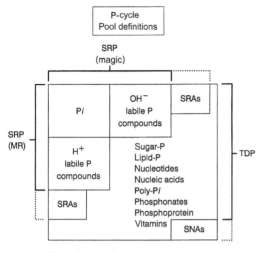

Pi = orthophosphate

SRP (MR) = soluble reactive P (Murphy & Riley)

SRP (MAGIC) = soluble reactive P (by MAGIC)

SRAs = soluble reactive As

SNAs = soluble non-reactive As [TAs-SRAs]

TDP = total dissolved P

SNP = soluble non-reactive P [TDP-SRP]

Non Pi-P = non Pi P [TDP-Pi]

FIGURE 5.6 Schematic presentation of the chemically diverse pool of P-containing compounds in the marine environment. Shown on the left is the subset of compounds that are typically detected using the standard Murphy-Riley molybdenum blue spectrophotometric SRP assay including Pi and certain acid (H^+) labile P compounds. The dashed line indicates that the water sample can be pre-treated to remove the inhibitory effects of SRAs, but this method is not always employed. Shown at the top is the spectrum of target compounds using the MAGIC-based SRP analysis method and, at right, the TDP procedure which theoretically measures all P-containing compounds. Depending upon the TDP method selected, however, linear polyphosphates or phosphonates may not be quantitatively recovered. Various operational definitions, given below the figure, are derived from these analyses but none of these chemical techniques yields an accurate estimate of the BAP pool, which is the pool of greatest interest in studies of the microbially driven P-cycle.

production rates) is the experimental objective, and for accurate estimation of SNP and DOP, because these pools are difference estimations (i.e., TDP–SRP) (see Section V.E). Any overestimation of Pi (e.g., by partial DOP hydrolysis) will result in an equimolar underestimation of

DOP, and thus will greatly impact the calculated Pi/DOP ratio. Although the use of malachite green oxalate for Pi measurements in seawater may reduce, or even eliminate DOP hydrolysis (Fernández et al., 1985), this method has not been widely tested under field conditions.

According to Strickland and Parsons (1972), the limit of Pi detection using the standard Murphy-Riley method is approximately 0.03 μM Pi and the precision at the 0.3 μM Pi level is $\pm 0.02/n^{0.5}$ μM, where n is number of determinations (e.g., if triplicate samples are prepared, then precision at 0.3 μM Pi level is 0.012, or 3.8%). For measurements of DOP, which rely upon independent estimates of SRP and TDP, both the detection limit and the precision are degraded. The accuracies of Pi, TDP, and DOP estimations, on the other hand, are not easily determined because reliable, certifiable SNP standards do not exist. Furthermore, the Pi concentration in many oligotrophic oceanic habitats is below this reported detection limit of 0.03 μM P, so alternate, high-sensitivity Pi detection methods are sometimes required.

Long pathlength spectrophotometry, including liquid wave-guide "total reflection" capillary cells (up to tens of m in length), has been used to enhance detection sensitivity (Fuwa et al., 1984). Ormaza-González and Statham (1991) reported a Pi detection limit of 1 nM with a relative standard deviation of 6% with a 0.6 m Pyrex® capillary system. Teflon AF-2400 capillary tubing (ID = 280 μm, length 4.5 m) has also been used for seawater absorbance spectroscopy applications (Waterbury et al., 1997). More recently, Zhang and Chi (2002) used a commercially available 2 m, 500 μm ID quartz capillary liquid core waveguide flow cell that was coated with Teflon-1600 to measure Pi concentrations as low as 0.5 nM, with a precision of 2% at a Pi concentration of 10 nM. Alternative detection systems, including luminol chemiluminescence emission (luminol oxidation by 12-MPA complex) have also been devised (Yaqoob et al., 2004). Patey et al. (2008) present an excellent summary of high-sensitivity (detection limit ≤ 1 nM) methods for Pi detection in seawater.

The sensitivity of colorimetric Pi analysis has been enhanced by thermal lensing or the use of solvents or C18 solid-phase sorbent (e.g., Liang et al., 2006) to extract and concentrate the 12-MPA dye, but none of these methods improves the signal:noise (blank) ratio for Pi detection. Karl and Tien (1992) devised the MAGnesium Induced Coprecipitation (MAGIC) method for Pi analysis that improves both the detection limit and assay precision by concentrating the Pi from seawater prior to the addition of the reagents, thereby enhancing the signal-to-noise ratio. This method has been used to reliably detect subnanomolar concentrations of Pi in seawater (Wu et al., 2000). MAGIC has also been combined with a novel luminol chemiluminescence technique to provide a convenient, alternative subnanomolar Pi detection system (Zui and Birks, 2000).

The MAGIC technique employs an alkaline solution for preconcentration of Pi, rather than an acidic solution. The advantage of this technique is the reduction of potential interference from the hydrolysis of acid-labile DOP compounds during processing if they are not coprecipitated (Karl and Tien, 1992). However, it is conceivable that seawater also contains base-labile DOP that might also be included in the MAGIC-Pi assay (e.g., phosphoproteins). The pH excursion from ambient seawater required for $Mg(OH)_2$ formation is only about 1-1.5 pH units ($Mg(OH)_2$ buffers seawater at pH 9.4), thus, Pi isolation from non-Pi SRP and SNP compounds is possible (Thomson-Bulldis and Karl, 1998). The MAGIC procedure has also been used to separate Pi from DOP, thereby providing a method for the direct determination of DOP (technically, non-Pi TDP) in seawater samples with high Pi/DOP ratios (Thomson-Bulldis and Karl, 1998). Consequently, this improved method enhances the reliability of both Pi and DOP determinations in seawater. In their analysis of high-sensitivity Pi methods, Patey et al.

(2008) conclude that MAGIC is labor intensive and not amenable to automation compared to other analytical approaches. However, it is the only method devised to date that improves the sample signal-to-noise ratio.

Hudson et al. (2000) devised a steady-state radio-bioassay to measure Pi in aquatic habitats. Application of this novel method to a variety of oligotrophic lakes indicated that the biologically active Pi pool was much lower than the chemically measured SRP pool, usually by more than an order of magnitude, and in some cases the active Pi pool was as low as 27 pM (Hudson et al., 2000). Moutin et al. (2002) used the Hudson et al. (2000) method to estimate bioavailable Pi turnover time and as an estimate of the pool size along a W to E transect in the central Mediterranean Sea. They reported extremely low concentrations of bioavailable Pi, many subnanomolar, with turnover times in the range of 1-3 h. The presence of a diverse microbial assemblage despite what must be intense competition for Pi, indicates that they must possess very high-affinity transport systems, and living very close to the theoretical diffusion limit for uptake. Indeed, the success of *Synechococcus* in the oligotrophic Mediterranean Sea has been tied to its ability to acquire Pi at both low (<5 nM) and high (5-25 nM) Pi concentrations (Moutin et al., 2002).

2 Analysis of TDP

The measurement of SNP (and DOP) in seawater requires paired measurements of SRP and TDP, the latter following pretreatment to effect a quantitative conversion of all inorganic and organic, nonreactive P to SRP. The key to successful and accurate SNP (and DOP) estimation is complete oxidation and hydrolysis of the combined P. Methods for the breakdown of P-containing organic matter can be classified as: (1) dry combustion with or without subsequent acid hydrolysis of poly-Pi, (2) wet combustion with permanganate, persulfate, or perchloric acid as the oxidant, (3) UV photooxidation, or

(4) alkali fusion with sodium carbonate or sodium nitrate, followed by acid digestion. Of these, the first three method classes have been used extensively to estimate marine DOP. Several comprehensive reviews of environmental organic-P measurements have appeared (Maher and Woo, 1998; Robards et al., 1994; Worsfold et al., 2008), so only a few of the many analytical concerns will be discussed here.

The earliest investigators relied on permanganate oxidation (Matthews, 1917), Kjeldahl digestion (Jones and Perkins, 1923), fuming sulfuric acid in a nitric/hydrochloric acid mixture (Juday et al., 1928), or fuming sulfuric acid plus hydrogen peroxide (Redfield et al., 1937). However, it was Harvey (1948) that devised a simpler, safer, and generally more convenient method for DOP analysis. His method required only the addition of sulfuric acid (0.28N final concentration) to a whole or filtered seawater sample followed by autoclaving at 135-140 °C for 5-6 h. In his hands, this method quantitatively converted nucleic acids, phosphoprotein, and phosphate esters to Pi (Harvey, 1948). An additional advantage was that this procedure could be performed at sea for near real-time estimation of DOP (technically SNP) concentrations, if necessary or desirable. Unfortunately, only a few years later Pratt (1950) reported that TP concentrations obtained using the Harvey (1948) method were frequently less than the sum of SRP plus PP, so he concluded that this method was unsatisfactory for seawater analysis; no alternative method was presented.

Hansen and Robinson (1953) proposed a method based on initial sample oxidation with perchloric acid, followed by treatment with concentrated hydrochloric acid. Several advantages at that time included a lower blank (compared to sulfuric acid) and more rapid oxidation of organic matter. However, like the earlier methods, the Hansen and Robinson method required time-consuming heating to evaporate the seawater during sample oxidation, not to mention the inherent hazards with the use of perchloric acid.

In 1964, Menzel and Vaccaro showed the complete oxidation of C from diverse organic compounds using persulfate as an oxidizing agent and a year later an adaptation of this method was described for the quantitative determination of TP and TDP in seawater (Menzel and Corwin, 1965). The method requires only the addition of potassium persulfate (0.7% $K_2S_2O_8$ final concentration) to a seawater sample followed by autoclaving at 120 °C for 30 min, or boiling (100 °C) for 1 h. Microwave digestion (450 W, 10 min) has also been used (Woo and Maher, 1995). During the procedure, decomposition of persulfate to sulfuric acid results in a drop in pH to about 1.5-1.8, with hydrogen peroxide as another key byproduct. This highly oxidizing, hot acidic environment ensured the complete oxidation of organic matter. Tests with standard P-containing compounds and P mass balance reconciliation with diatom cultures confirmed the efficacy of this procedure (Menzel and Corwin, 1965). It should be mentioned that inorganic poly-Pi, if present, would be hydrolyzed to Pi by the Menzel and Corwin (1965) method so its presence would result in a stoichiometric overestimation of DOP if the assumption, TDP–SRP = DOP, is made. Also, any reduced inorganic P (e.g., phosphite) present in the sample would be oxidized to SRP, leading to an overestimation of DOP.

Various chemical oxidation methods require evaporation of seawater. However, only the magnesium sulfate-hydrochloric acid hydrolysis (Solórzano and Sharp, 1980) and the magnesium nitrate oxidation methods (Cembella et al., 1986) have been employed for DOP estimation. These straightforward, efficient, but somewhat laborious methods involve drying a small volume (10 ml) sample with exogenous magnesium salt, followed by high temperature (450-500 °C) combustion for 2 h. This decomposes the organic matter and converts some of the DOP to Pi. The residue is then treated with hot (80 °C) hydrochloric acid to completely hydrolyze any poly-Pi that may be present to reactive Pi. The Cembella et al. (1986) magnesium nitrate method targeted hydrolysis of the extremely stable phosphonate compound class, and was shown to be effective.

Armstrong et al. (1966) devised a fundamentally different approach to the same end; namely, photochemical combustion of organic matter by UV radiation. This method, which employs quartz reaction tubes, a 1200-W mercury arc lamp (Hanovia #189A, or equivalent) and the addition of only hydrogen peroxide as a source of oxygen, provides an efficient and rapid (1h) means for the conversion of selected DOP compounds to Pi. Golimowski and Golimowska (1996) have published a comprehensive review of the probable mechanisms of organic matter destruction during UV treatment, including characteristics of the various UV lamps, effects of pH and choice of oxidant(s). Several advantages include low (effectively, zero) reagent blank and, therefore, higher precision, and the ability to measure both DON (by oxidation to nitrate plus nitrite) and DOP in a single sample. Complete hydrolysis of the phosphate ester (Gly-3-P, ribose-5-P) and phosphonate compounds (2-AEP) to Pi was observed following an irradiation period of approximately 1h at 60-80 °C (note: the sample tubes were positioned 7cm from the UV lamp with an incident actinic energy of 200-250mW cm^{-2}; Armstrong et al., 1966). However, aging lamps, hot spots/cold spots in the irradiation apparatus, and the need for adequate temperature control, plague UV photooxidation methods. The use of internal standards and adequate reference materials is highly recommended (see Kérouel and Aminot, 1996).

One unique characteristic of the UV photo-oxidation method, which may be construed as either an analytical advantage or as a disadvantage, is the inability to degrade inorganic and organic polyphosphates. As mentioned above, the persulfate oxidation procedure will overestimate DOP in the presence of inorganic poly-Pi. In contrast, inorganic poly-Pi will not interfere in the UV method but DOP will be underestimated

if organic poly-Pi compounds, such as ATP, are present. To date, the reactivity of reduced inorganic or organic compounds (+I to +III valence) to UV photooxidation remains unknown. Armstrong and Tibbitts (1968), using a lower intensity 380-W UV mercury arc lamp to decrease the hydrolysis reaction rates, suggested that organic poly-Pi esters initially released inorganic poly-Pi that was slowly hydrolyzed with time. They also suggested that TDP concentration, including poly-Pi, could be determined by including a post-irradiation acid hydrolysis step, a method that could theoretically be used to infer relative contributions of monoester-linked P versus poly-Pi. Yanagi et al. (1992) later developed this approach of differential UV-lability into a quantitative assay for the partial characterization of DOP pool (see Section VIII.D). More recently an online 40-W UV reactor coupled to flow injection analysis has been described for quantitative detection of the DOP, including condensed poly-Pi compounds (Tue-Ngeun et al., 2005).

TDP has also been measured in seawater using hydride generation and gas chromatography (Hashimoto et al., 1987). This method has several advantages over colorimetry, in particular, improved detection limit and reaction specificity. More importantly, the analytical basis for this method (reduction of all forms of phosphate to phosphine gas by borohydride reduction) is fundamentally different from the colorimetric procedures. They reported excellent quantitative correspondence between TP and TDP concentrations measured by the P hydride method and persulfate digestion for samples collected from 0-9600 m in the Japan Trench.

With this spectrum of potentially available methods for the measurement of TDP (i.e., SNP and, therefore, DOP) it is only natural that individual researchers might endeavor to compare two or more of these methods before adopting the most reliable one for the seawater samples under consideration. It is conceivable, even likely, that the efficiencies of these methods are site-specific due to regional variations in the chemical composition of the ambient SNP pools. It is important to mention that in presenting the DOP concentration data, later in this review, we make no corrections for variable DOP recovery based on the method selected because, quite frankly, there is no way to estimate this unless multiple TDP protocols were employed—and this situation is rare.

Four comprehensive TDP methods comparisons are worthy of mention. Ridal and Moore (1990) compared the persulfate oxidation and UV-irradiation methods against a new method that relies on a sequential UV-persulfate technique. This stepwise UV-acid treatment follows the analytical recommendation made originally by Armstrong et al. (1966) for the quantitative recovery of inorganic and organic poly-Pi compounds. Their results for samples collected from a variety of coastal and open ocean marine habitats suggested that, primarily in open ocean ecosystems, the combined method returned values 1.25-1.50 times higher than either method separately (Ridal and Moore, 1990). A subsequent study was conducted in the northeast subarctic Pacific Ocean (Ridal and Moore, 1992). For this habitat, the UV method returned an average concentration that was only $71 \pm 9\%$ and the persulfate method $83 \pm 9\%$ of the "DOP" concentration estimated by the combined UV-persulfate technique. They concluded that the "standard" methods of analysis used for most marine studies should be considered minimum estimates of the ambient DOP concentrations; however, without additional information on the presence of inorganic poly-Pi it is not known which DOP estimation is the most accurate. We note that recent advances in methodology now allow for the unique identification and quantification of poly-Pi (Diaz and Ingall, 2010; Martin and van Mooy, 2013). However, to date, few have employed these novel techniques in marine environments, none of which addressed the dissolved poly-Pi pool (Martin and van Mooy, 2013). Nevertheless, a cross-habitat/regional poly-Pi comparison should be of great value.

Nedashkovskiy et al. (1995) compared the wet persulfate oxidation and the dry magnesium nitrate combustion methods using seawater collected off Vladivostok and in the Northwestern Bering Sea. For samples containing low concentrations of organic-P ($\leq 0.5\,\mu M$) the two methods were indistinguishable. However in productive Bering Sea waters, where organic-P exceeded $0.5\,\mu M$, the mean difference approached 20% in favor of the dry combustion method.

Ormaza-González and Statham (1996) compared five independent methods of TDP analysis. The highest concentrations were obtained by the magnesium nitrate oxidation method while the lowest values were obtained by the UV photooxidation technique. These data suggest that inorganic and organic poly-Pi compounds, not detected by the UV method, may comprise a significant percentage of the TDP pool in the North Sea waters used for this comparison.

Monaghan and Ruttenberg (1999) compared the dry combustion method, using two separate oxidants (magnesium nitrate and magnesium sulfate), to the wet chemical persulfate oxidation method. This comprehensive study tested the recovery of Pi from a variety of organic-P compounds as well as the estimation of TDP from continental shelf seawaters collected off California. Although quantitative recovery of Pi from selected phospholipids and phosphonates was only achieved by dry combustion, TDP concentrations for natural seawater samples were comparable for magnesium sulfate oxidation and persulfate, and were on average 7% lower for the nitrate oxidation method. This indicates a negligible presence in these natural samples, of the known organic-P compounds that are poorly recovered by persulfate oxidation, namely phospholipids and phosphonates (Monaghan and Ruttenberg, 1999).

Finally, because many marine biogeochemists seek fundamental information on the coupled N and P pool dynamics in seawater, several methods have been devised to optimize the measurements of DON and DOP or DOC/DON/DOP in a single chemical oxidation treatment. As mentioned above, UV photooxidation can be used to measure both compound classes, though the optimal irradiation time for DOP is 1-2h compared to 24h for DON (Walsh, 1989). A recent comparison of DON estimation using UV photooxidation, persulfate wet chemical oxidation and high temperature combustion oxidation indicated higher variability and lower recoveries for the UV method (Bronk et al., 2000); however, their inability to recover NH_4^+ as NO_3^- casts some doubt on the efficiency of their UV photooxidation procedures.

The fundamental problem with simultaneous wet chemical oxidation of DON and DOP is that organic N requires an alkaline medium while organic P requires an acidic medium for optimum performance. This problem was solved by F. Koroleff, as reported by Valderrama (1981), using a mixture of potassium persulfate, boric acid, and NaOH which provides for a time-temperature dependent evolution from alkaline (pH 9.7) to acidic (pH 5-6) conditions during the reaction period. A modification of this procedure that provides for the simultaneous determinations of DOC, DON, and DOP has also been published (Raimbault et al., 1999).

D Analytical Interferences in SRP and TDP Estimation

Of the various potential interferences on the accuracy of Pi and TDP estimation, two in particular deserve mention. The first is the aforementioned potential overestimation of Pi by "reactive" organic-P compounds (Figure 5.6). Their ubiquity in marine ecosystems worldwide, especially surface waters, will contribute to the measurement of SRP, and systematically underestimate SNP and DOP. TDP is unaffected by their presence. An equally insidious analytical interference derives from the nearly ubiquitous presence of arsenic (As) in seawater. Arsenate (AsO_4^{3-}), the most oxidized form of As (+V), reacts with the SRP reagents and forms a

blue-colored complex of an equivalent molar absorptivity to that of Pi, while arsenite (AsO_3^{3-}, +III) does not react at all. Dissolved AsO_4^{3-} can be reduced to AsO_3^{3-} with thiosulfate in acidic medium in the presence of excess metabisulfate (Johnson, 1971; von Schouwenburg and Walinga, 1967) to prevent it from interfering with SRP estimation (Figure 5.6). The concentration of soluble reactive As (SRAs) sometimes exceeds the concentration of Pi so attention to this potential analytical problem is imperative. The addition of thiosulfate is required to eliminate AsO_4^{3-} interference. As a result of this step, the time for full color development is prolonged and this procedure is not routinely employed in most automated SRP analyses (Downes, 1978), even though AsO_4^{3-} may still interfere. For accurate TDP determinations, it is also necessary to correct for AsO_4^{3-} that is produced during sample oxidation as a result of the conversion of SNAs to SRAs (e.g., oxidation of organic-As), though this is seldom, if ever, done. Also, the presence of AsO_3^{3-} (+III) would overestimate TDP and DOP.

More than just an analytical nuisance, As has an important biogeochemical cycle that has many features in common with the marine P-cycle. In fact, because of the variable redox states commonly found in seawater (+III, +V) and the role marine microorganisms have in As oxidation, reduction, and methylation, we can anticipate an active As cycle in most marine habitats. Furthermore, AsO_4^{3-} is a well-studied analog of Pi, and it is transported and incorporated by many of the same enzyme systems. It is also a well-recognized uncoupler of oxidative- and photo-phosphorylation and, hence, a metabolic poison (Benson, 1984; Francesconi and Edmonds, 1993). When the ambient pool of AsO_4^{3-} is high, relative to Pi, as occurs in many oligotrophic marine habitats (Karl and Tien, 1997) there should be a selection for microorganisms with either high-specificity Pi transport systems or high-capacity detoxification mechanisms, or both. The former would lead to a preponderance of AsO_4^{3-} and the latter to a preponderance of AsO_3^{3-} and DOAs. As mentioned above, both SRP (Pi) and TDP measurements need to be cognizant of potential interference from all possible forms of As.

E Use of Isotopic Tracers in P-cycle Research

The use of stable and radioisotopic tracers to monitor and quantify the rates of microbial growth, metabolism and biogeochemical cycling of key elements and compounds has revolutionized the field of microbiological oceanography. For P-cycle research, there are two major categories: (1) the use of naturally occurring, cosmogenic radioactive isotopes and (2) the use of exogenously supplied radioactive isotopes. There are two radioisotopic tracers for P; ^{32}P (E_{max} = 1.71 MeV, ½ life = 14.3 days) and ^{33}P (E_{max} = 0.25 MeV, ½ life = 25.3 days) which both exhibit β^- particle decay. The detection and quantification of the cosmogenic radiotracers ^{32}P and ^{33}P (Lal and Lee, 1988; Lal et al., 1988) are most useful for long-term (day to week) whole ecosystem studies. Applications in the Sargasso Sea, the Gulf of Maine, and the NPSG have demonstrated the efficacy of using natural cosmogenic $^{32}P/^{33}P$ radioisotopes in studies of the marine P cycle (Benitez-Nelson and Buesseler, 1999; Benitez-Nelson and Karl, 2002; Waser et al., 1996). The use of exogenously supplied ^{32}P- and ^{33}P-labeled inorganic and organic compounds is best suited for short-term (hour to day) studies of metabolic pathways, nutrient fluxes, and organic tissue labeling patterns. Several whole lake (pond) ecosystem studies (Hutchinson and Bowen, 1947, 1950; Rigler, 1956) and at least one marine reef flat ^{32}Pi experiment (Atkinson and Smith, 1987) have been conducted, but direct tracer release has not yet been used in either open ocean patch studies or in mesocosm enclosures.

Although P has only a single stable isotope, ^{31}P, the oxygen atoms that are in association with both inorganic and organic P pools contain three isotopes: ^{16}O, ^{17}O, and ^{18}O. These could, in theory, assist in a quantitative study of the marine P-cycle (Longinelli et al., 1976), although no comprehensive study of oxygen isotopes in DOP has yet been published. Phosphate oxygen is tightly bound to P such that under ambient conditions in the sea, exchange of oxygen between Pi and the surrounding water is negligible (Blake et al., 1997). However, it has been hypothesized that biological cycling of P would act to isotopically equilibrate the phosphate oxygen with ambient water (Colman et al., 2005). Liang and Blake (2009) showed isotopic fractionation that was dependent on the DOP compound, and the mediating enzyme involved (e.g., APase vs. 5NDase). More recently, the $\delta^{18}O$ of Pi have been used to assess the utilization of DOP in the Sargasso Sea (McLaughlin et al., 2013). Consequently, time and space measurements in the $\delta^{18}O$ of Pi and DOP could thus provide invaluable information on P biogeochemistry.

The use of exogenous radioisotopic tracers has become routine for many P-cycle investigations. Often this is the only approach that is sensitive and specific enough to measure the sometimes low fluxes of P that occur in open ocean ecosystems. The details of selected individual methods are discussed elsewhere; however, there are several general considerations regarding the use of $^{32}P/^{33}P$ radioactive isotope tracers in studies of microbial ecology that merit attention. These include: (1) the overall reliability of the added element (or compound) as a tracer, including an evaluation of the site of labeling, its uniqueness, and stability during cellular metabolism and biosynthesis, (2) isotope discrimination factors, (3) the partitioning of the added tracer into existing exogenous and internal pools of identical atoms, molecules, or compounds, (4) the importance of measuring the specific activity of the incorporated tracer, and (5) the design and implementation of experimental procedures and proper kinetic analysis of the resulting data. The underlying assumption of these methods is that the subsequent incubation conditions do not alter the in situ rates of compound uptake, metabolism or biosynthesis. This assumption is usually difficult to verify (Karl and Dore, 2001).

A very important but often overlooked principle in the use of radioisotopic tracers in marine ecological studies is the evaluation of the specific activity (radioactivity per unit mass) of the added, incorporated, or metabolized element, molecule or compound during the incubation/labeling period. The ideal tracer is one that can be added without perturbing the steady-state concentration of the ecosystem as a whole. In ecological studies, an accurate assessment of the specific activity during the incubation/labeling period is further complicated by the dilution of the added tracer with exogenous pools present in the environment and by endogenous pools present in living microbial cells. Without a reliable measurement of the extent of dilution prior to incorporation, tracer uptake data by themselves are of limited use in quantitative microbial ecology (see Section IX). Furthermore, isotope specific activities may change over the course of the labeling period due to the combined effects of depletion (uptake) of the added tracer or isotope dilution by a constant regeneration of the exogenous pools (assuming steady-state conditions). In fact, Pi regeneration rates have been estimated in environmental samples by measuring the extent of isotope dilution during short-term sample incubation periods (Harrison, 1983).

VI DOP IN THE SEA: VARIATIONS IN SPACE

Biogeochemical cycles of C, N, and P in the sea are ultimately sustained by solar energy via the process of photosynthesis. Consequently, DOP production is highly correlated with the primary formation of organic matter in the euphotic zone. The C:N:P molar stoichiometries of POM and DOM pools are key ecological parameters in the sea and will therefore also be considered briefly in this review. However, the primary

focus will be on total DOP, which likely contains a broad spectrum of compounds of variable C:P and C:N:P stoichiometry (Figure 5.2, Table 5.1).

The instantaneous concentration of DOP in seawater is expected to vary geographically, with depth in the water column and, for a given location, with time. Consequently, the ambient DOP pool must be viewed as a transient that is largely controlled by the balance between local DOP production and DOP removal processes. The physical, chemical, and biological influences on these key ecosystem processes will be discussed later, as will information on the chemical characterization of the heterogeneous DOP pool. In this section, we present the geographical and depth distributions of oceanic DOP to establish generalized concentration patterns and broad correlations with other relevant parameters, including Pi. We have employed two separate data sources: (1) the U.S. National Oceanic and Atmospheric Administration—National Ocean Data Center's (NOAA-NODC) online oceanographic profile database (http://www.nodc.noaa.gov/cgi-bin/JOPI/jopi) and (2) our own DOP database collated from the archival literature which includes many published profiles that are not included in the NOAA-NODC database.

The NOAA-NODC global ocean search acquired all SRP/TDP data entries; the full extracted data set included $n = 250,694$ paired measurements. We then screened these data and removed all entries where the reported SRP exceeded the reported TDP (i.e., "negative" DOP); this decreased the number of measurements to $n = 233,118$ data pairs. This was our primary Global Ocean DOP data set. We also subsampled the Global Ocean DOP data set to obtain a secondary database for stations where the depth of the water column was $\geq 200\,m$ (i.e., the pelagic marine environment which is the stated focus of this review). This otherwise unedited Global Open Ocean DOP (or GOOD) database, which includes $n = 139,747$ measurement pairs is available along with our enhanced DOP summary upon request of the senior author. The GOOD database includes measurements from all major ocean basins but has several large data gaps, most notably the Eastern North Pacific Ocean, the South Pacific Ocean, and the Southern Ocean (Figure 5.7). Nevertheless, the number of SRP/TDP measurements (n = 139,747 pairs) greatly exceeded our initial expectation, especially considering the relatively few open ocean DOP profiles that have been published in the refereed literature. A notable exception, that we highlight here, is the extensive survey of the North and South Atlantic Ocean basins conducted as part of the IGY. During this 2-year (1957-1958) investigation, McGill (1963, 1964) compiled what amounts to the most comprehensive study of oceanic DOP yet attempted. This must be considered the exception to the otherwise sparse open ocean database.

A Regional and Depth Variations in DOP

There are several general features of the global ocean DOP distributions. First, concentration versus depth profiles in the open ocean almost exclusively reveal elevated DOP in the upper 0-100 m of the water column. For example, the 12-year climatology of SRP and DOP at Station ALOHA (22°45'N, 158°W) documents a characteristic inverse depth relationship with high concentrations of DOP in the surface water, decreasing with water depth, and vice versa for SRP concentrations (Figure 5.8). The only possible exception to this general pattern might be for high latitude habitats in winter where deep vertical mixing and low solar irradiance preclude contemporaneous near-surface DOP production via primary production.

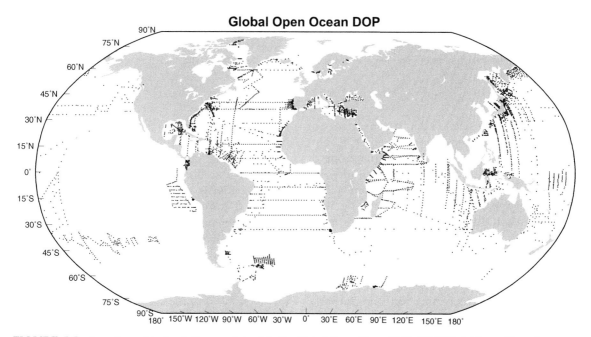

FIGURE 5.7 Inventory of hydrostations contained in the Global Open Ocean DOP (GOOD) database. These data were obtained from the National Oceanic and Atmospheric Administration—National Oceanic Data Center (NOAA-NODC) via their online World Wide Web-based search and were edited by us, as described in the text. The fully edited GOOD database is available from the senior author upon request.

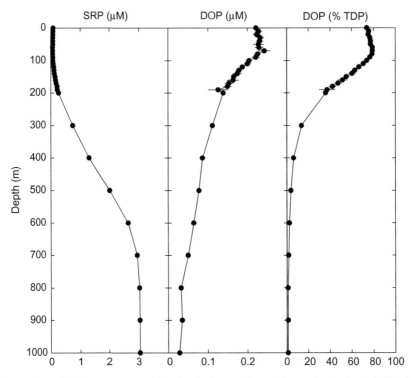

FIGURE 5.8   SRP, DOP, and DOP as a percent of TDP for the Hawaii Ocean Time-series Station ALOHA (22° 45'N, 158° W). These data are mean values and 95% confidence intervals based on samples collected during 110 research cruises between Oct. 1988 and Dec. 1999.

Despite this predictable depth dependence of total DOP in the open sea, individual DOP compounds or compound classes can have one of three fundamentally different distributions as a function of depth in the water column: (1) local enrichments near the sea surface with decreasing concentrations beneath the euphotic zone (similar to total DOP), (2) near-surface depletion with increasing concentrations beneath the euphotic zone, or (3) constant concentration with depth. These depth distributions are a result of net production/consumption and import/export processes (see Section VII).

At the subtropical location of Station ALOHA, DOP in the upper 100 m of the water column averages 0.20-0.22 μM or approximately 70-80% of the TDP pool. Below 100 m, DOP decreases with

increasing water depth to values <0.05 μM; DOP concentrations at depths ≥ 300 m are consistently <10% of the corresponding TDP, indicating a deep water dominance by SRP (Figure 5.8). Similar patterns are also observed for the GOOD data set (Figure 5.9), with the exception of a generally lower percentage of DOP in near surface waters (i.e., 50-60% of the TDP pool compared to 70-80% for the subtropical North Pacific; Figure 5.8). For all oceanic DOP profiles, however, both the absolute DOP concentration and the DOP/TDP ratio vary systematically, and therefore predictably, with water depth.

It is well known that the concentration of SRP, especially in subthermocline (>600 m) waters increases as the age of the water mass also increases (Levitus et al., 1993). This is a result of the

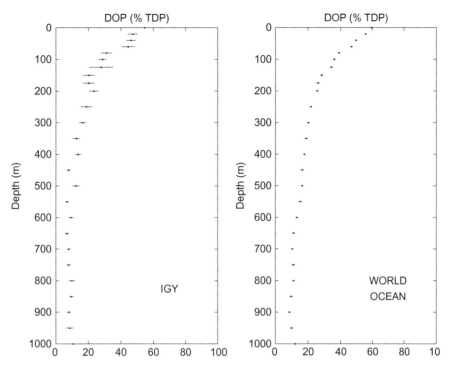

FIGURE 5.9 Vertical profiles of DOP, expressed as percent of TDP, for the entire Global Open Ocean DOP (GOOD) database and International Geophysical Year (IGY) subsampled data (see text for more details). The data are presented as mean values and 95% confidence intervals. The data were depth binned, as shown, prior to the determinations of the summary statistics. For the IGY profile, n ranged from 32 (125 m) to 594 (surface) with a median of $n = 130$. For the GOOD profile, n ranged from 805 (850 m) to 12,234 (25 m) with a median of $n = 2700$.

long-term net remineralization of organic matter. These spatial patterns are clearly evident in the IGY data set (McGill, 1963, 1964). A comparison of two zonal sections, one at 40°N and the other at 24.25°S in the Atlantic Ocean documents the following: (1) a contrast between relatively low TDP/low SRP waters at 40°N compared to high TDP/high SRP in the northward flowing Antarctic Bottom Water in the west (>4000 m) and Antarctic Intermediate Water (~1000 m) seen at 24.25°S, (2) nearly homogeneous concentrations of all forms of P in the relatively "young, well mixed" North Atlantic compared to the South Atlantic, especially for the subeuphotic zone waters, and (3) a general increase in surface water DOP concentrations in the South

Atlantic basin especially in the near-surface water (McGill, 1963, 1964; McGill et al., 1964). There also appears to be a significant basin-scale east-to-west gradient for DOP, at least at 40°N, with higher DOP concentrations in the western North Atlantic, and minimum DOP concentrations in the deep central waters of both basins. These regional variations are superimposed on the general global DOP distributions described previously, and are probably related to circulation patterns and processes. A similar Atlantic Ocean intra-basin gradient in DOP was reported by Vidal et al. (1999) along a transect from the Canary Islands to Argentina (22°N to 31°S). Euphotic zone (0-200 m) DOP concentrations in the western portion of the basin were approximately two to three times greater than they were in the eastern portion of the Atlantic basin (0.2 to 0.3 μM vs. <0.1 μM; Vidal et al., 1999). Roussenov et al. (2006) used nutrient observations and a simplified cycling and transport model to hypothesize a northward transport of DOP in the North Atlantic Ocean from the tropics into the subtropical gyre. Based on model calculations, the input of DOP (locally equivalent to "new" P) is 12 mmol P m^{-2}year^{-1}, which is sufficient to support approximately 50% of P lost in sinking particulate matter (Roussenov et al., 2006). In a separate study, Lomas et al. (2010) supported this hypothetical flux of DOP into the North Atlantic subtropical gyre via transport by subtropical mode water. Based on a comprehensive 5-year data set collected at the Bermuda Atlantic Time-series Site (BATS) that included multiple P pools and their dynamics, the authors concluded that imported DOP supports approximately 25% of annual primary production and essentially all of the P exported as sinking particles. Furthermore, analysis of a longer time-series of DOP concentrations at BATS indicates that the more recently obtained (2004-2008) data were 30-50% lower than those obtained a decade earlier. They hypothesized that the ecosystem is not in long-term steady-state, and presented Pi concentration data from the approximately 500 m depth horizon that showed a 1.7 ± 0.5 nmol kg^{-1}year^{-1} increase in Pi over the past two decades. Increased DOP flux or more efficient utilization in surface waters, or both, apparently led to an enhanced export flux and remineralization in subeuphotic zone waters (Lomas et al., 2010).

If the DOP:TDP ratio of the global ocean remains more or less constant for a given water depth, as the GOOD data set implies (Figure 5.9), then one might predict a systematic inter-ocean basin increase in DOP concentration with highest values in the North Pacific and lowest concentrations in the North Atlantic resulting from the ocean's "conveyor belt"-like circulation patterns. Accordingly, the North Atlantic Ocean is

the origin and the North Pacific Ocean is the terminus of the global transport system. These inter-basin differences in the concentration of DOP, and perhaps in the C:N:P stoichiometry of the DOM pool as well, are anticipated, especially if remineralization processes and long-term accumulation of recalcitrant DOM (R-DOM) is an important process. Regional variations within a given ocean basin are also possible.

Unfortunately, subeuphotic zone DOP measurements, especially where DOP is ≤10% of the TDP pool, are not very reliable given the nature of the paired analyses of SRP and TDP and estimation of DOP by difference. For this reason, several investigators acknowledge the uncertainty of the deep water DOP pool estimations despite reporting them. The global compilation, with individual estimates based on fairly large sample sizes ($n > 200$) indicates: (1) a detectable pool of DOP at all ocean depths and (2) a decreasing DOP percentage with increasing depth. These features would be consistent with the time-dependent remineralization of labile and semi-labile DOP. While the global open ocean DOP data sets are poorly positioned to evaluate this biogeochemical prediction rigorously, owing to the inherent inaccuracies of DOP estimation in subeuphotic zone waters (see Section IV.B), there does appear to be an increase in DOP along the circulation trajectory (Figure 5.10). The global pattern of increasing SRP in deep waters has been previously observed, but we believe this is the first evidence for inter-basin variations in near-surface DOP.

This rather robust inverse relationship between SRP and DOP in the surface ocean is a manifestation of Pi uptake, net biomass production, and the rapid remineralization of POM. The higher the ratio of recycled-to-total production, the higher the turnover rate of POM and DOM. These processes lead to the eventual, local accumulation of semi-labile and refractory DOM compounds, resulting in the broad geographical distributions observed for the global DOP data set. While it is not explicitly demonstrated here,

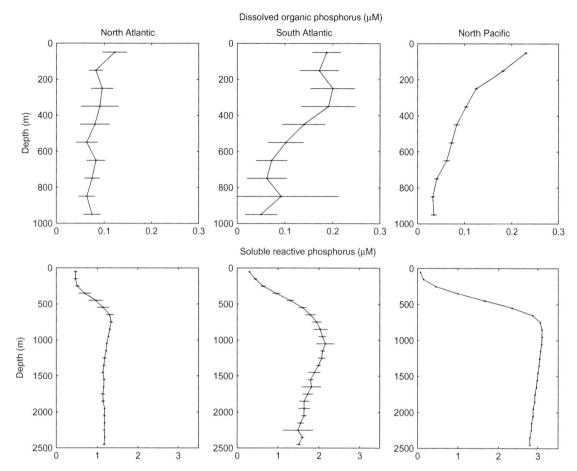

FIGURE 5.10 Changes in SRP and DOP pools along the global ocean circulation trajectory from the North Atlantic to the North Pacific Oceans. North and South Atlantic measurements are from the NOAA-NODC Global Ocean DOP data set, and the North Pacific measurements are from Station ALOHA (22°45′N, 158°W). Values shown are means and 95% confidence intervals for each respective data set.

there is likely to be a corresponding gradient in the chemical composition of DOP with a higher percentage of labile phosphate esters in high latitude regions and, for a given location, in near surface waters.

In spite of these broad generalized patterns of DOP in the world ocean, local and regional variations are also evident. DOP is enriched (>0.25 μM) in most coastal and estuarine habitats that have been investigated (Table 5.2). Upwelling-induced organic matter production

and coupled DOP production, combined with coastal runoff and locally restricted flushing can all contribute to both local and regional elevations in surface DOP.

B DOP Concentrations in the Deep Sea

The precision of DOP estimation is eroded when SRP is ≥ 90% of the TDP pool, for example at all ocean depths greater than approximately 500 m in the global open ocean (Figure 5.9), as well

TABLE 5.2　Selected Marine Dissolved Organic Phosphorus (DOP) Concentrations and DOP as Percentage of Total Dissolved Phosphorus (% of TDP) Reported from a Variety of Geographic Locations.

Location	Depth (m)	Method[a]	DOP[b] (µM)	% of TDP	Reference
Coastal/Estuarine					
Suruga Bay, Japan	0-1000	UV	0.11-0.29		Yanagi et al. (1992)
Prydz Bay, Antarctica	0-200	UV	0.05-0.90		Yanagi et al. (1992)
Florida Bay, USA	Surface	DA-AH	0.37 ± 0.02 range 0.04-2.03 ($n = 183$)		Fourqurean et al. (1993)
Sandfjord, Norway	Surface 10	PO	0.224-0.264 0.318-0.337	98-98 90-95	Thingstad et al. (1993)
Mamala Bay, USA	Surface	PO	0.12-0.17	37-43	Björkman and Karl (1994)
Chesapeake Bay, USA	Surface	PO-AA	0.2-0.6		Conley et al. (1995)
Bay of Aarhus, Denmark	3 11 16	PO	0.48 ± 0.01 0.63 ± 0.03 0.98 ± 0.05		Thingstad et al. (1996)
Apalachicola Bay, USA	Surface	PO	0.1-1.1		Mortizavi et al. (2000)
Tokyo Bay, Japan Corpus Christy Bay, USA	Surface	DA-AH	0.50-0.67 0.49 ± 0.06	70 50	Suzumura and Ingall (2001)
Gotland Basin (57°19′20N, 20°03′00E)	1-15	PO-AA	0.16-0.23		Nausch et al. (2004)
York River Estuary, USA	Surface	PO	0.28-0.42	~60	McCallister et al. (2005)
Baltic Sea	1-2	PO-AA	0.20-0.32		Nausch and Nausch (2006)
Mississippi Bight Gulf of Mexico	Surface 2-500	PO	0.17 0.053-0.078	~80 3-85	Cai and Guo (2009)
Mediterranean Sea	8	PO-AA	0.04 to 0.06	25 to >50	Tanaka et al. (2011)
NE Mediterranean Sea (43°5′N, 6°0′E)	3	PO	BLD-0.329		Bogé et al. (2012)
Mississippi Sound Gulf of Mexico	Surface	PO	0.31-0.67	75-90	Lin et al. (2012)
Continental shelf					
Southern California Bight (offshore site)	5 20	UV	0.21 0.22	51 42	Ammerman and Azam (1985)
Santa Catalina Basin (33°18.5′N, 118°40′W)	0-50 75-1300	UV	0.27 to 0.42~0.15		Holm-Hansen et al. (1966)
NE Pacific off Mexico (15°40′N, 107°30′W)	10 25 75	PO	0.126 0.171 0.120	66 70 56	Orrett and Karl (1987)

TABLE 5.2   Selected Marine Dissolved Organic Phosphorus (DOP) Concentrations and DOP as Percentage of Total Dissolved Phosphorus (% of TDP) Reported from a Variety of Geographic Locations.—cont'd

Location	Depth (m)	Method[a]	DOP[b](μM)	% of TDP	Reference
North Atlantic Scotian Shelf	10 50 90	UV-PO	0.14 0.39 0.33	29 35 3	Ridal and Moore (1990)
NE continental shelf slope (George Banks)	Surface 200 400-800	UV	0.17 0.06 0.03		Hopkinson et al. (1997)
Southern NW shelf, Australia	<25m	PO	0.06-0.19	33-86	Furnas and Mitchell (1999)
Eel River Shelf, N. California (summer values)	3 10 35	DA-AH	0.40 0.45 0.18	82 76 11	Monaghan and Ruttenberg (1999)
Southern Ocean (61°05′S, 54°8′W)	10 50 150	UV	0.20 0.16 0.06		Sanders and Jickells (2000)
Bering Sea, Chukchi Sea	0-140	PO	BDL-0.36	0-75	Lin et al. (2012)
Open ocean					
North Pacific Subtropical Gyre (~28°30′N, 155°30′W)	Surface 100 900 5000	UV	0.27 0.21 0.12 0.22		Williams et al. (1980)
North Pacific Subtropical Gyre (21°22′N, 155°W)	0-200	UV	0.18		Jackson and Williams (1985)
North Pacific Subtropical Gyre (21°22′N, 158°14′W)	0-100 900	UV	0.15-0.20 0.03	57-60 1	Smith et al. (1986)
North Pacific Subtropical Gyre	0-50	UV	0.10-0.35		Abell et al. (2000)
North Pacific Subtropical Gyre (22°45′N, 158°W)	Surface	PO	0.140-0.285	66-90	Björkman et al. (2000)
Gulf Stream (41°10′N, 64°33′W)	20 75 125 3600	UV-PO	0.10 0.07 0.01 BDL[c]		Ridal and Moore (1990)
NW Mediterranean Sea (42°30′N, 4°22′E)	Surface 125 400	PO-AA	0.13 0.09 BDL	95 50 0	Raimbault et al. (1999)
Sargasso Sea (26°N, 70°W to 31°40′N, 64°10′W)	Surface	UV	0.1-0.5	~95-100	Cavender-Bares et al. (2001)
Sargasso Sea (31°67′N, 64°17′ to 26°1′N, 70°W)	Surface	UV	0.074 ± 0.042	~99	Wu et al. (2000)

(Continued)

TABLE 5.2   Selected Marine Dissolved Organic Phosphorus (DOP) Concentrations and DOP as Percentage of Total Dissolved Phosphorus (% of TDP) Reported from a Variety of Geographic Locations.—cont'd

Location	Depth (m)	Method[a]	DOP[b](μM)	% of TDP	Reference
Southern Ocean (57°35′S, 57°W)	Surface 300 500	UV	0.16 0.10 BDL		Sanders and Jickells (2000)
Atlantic Ocean (transect; 27°13′N to 36°21′S)	5 DCM	PO	BLD-0.19 BLD-0.17		Cañellas et al. (2000)
N. Pacific Ocean (22°45′N, 158°00′W)	5-175	PO	0.172 ± 0.022 to 0.279 ± 0.011	~45-90	Björkman and Karl (2003)
NE Atlantic, Bay of Biscay (46°45.0′N, 5°59.6′W)	Surface 200-600 800-1500 3000-4000		0.07 0.03-0.05 0.02 0.01-0.02		Aminot and Kérouel (2004)
Atlantic Ocean (transect; 48°N to 35°S)	Mixed layer	UV	0.07-0.43	67-100	Mahaffey et al. (2004)
NE Mediterranean Sea	0-2600	UV	0.023-0.087		Krom et al. (2005)
Atlantic Ocean (transect; 40°N to 40°S)	25	UV	~0.030-0.310		Mather et al. (2008)
Sargasso Sea (32°50′N, 64°10′W)	Surface 200 500	PO	0.045-0.060 0.065-0.090 0.006-0.036	~80 ~43	Lomas et al. (2010)
N. Pacific Ocean (30°N, 145-150°W, 24°N, 156°W)	5-45	PO-AA	0.246 ± 0.038 0.251 ± 0.030		Watkins-Brandt et al. (2011)
N. Pacific Ocean (22°45′N, 158°00′W)	110 200 300	UV	0.22 0.15 0.17	35-72	Kaiser and Benner (2012)
Bering Sea, Beaufort Sea (55°56′N, 173°11′E,)	0-3800	PO	BLD-0.27	~10	Lin et al. (2012)

[a]Method abbreviations are: UV = ultraviolet photooxidation, PO = wet persulfate oxidation, PO-AA = persulfate oxidation under alkaline-acid conditions, DA-AH = dry ashing, acid hydrolysis, AH-PO = ash hydrolysis and persulfate oxidation, UV-PO = ultraviolet photooxidation-persulfate oxidation.

[b]Data are mean concentrations, mean ± 1 S.D. or ranges as shown.

[c]BDL = below detection limit which for the standard paired SRP/TDP assay is ≤ 20-30 nM.

as in many high latitude surface waters. Small relative errors in SRP and TDP determinations translate into large errors in the calculation of DOP. Ketchum et al. (1955) presented a comprehensive assessment of the analytical and statistical considerations for samples collected in the equatorial Atlantic Ocean. Using the method of Harvey (1948), they analyzed paired Pi and TDP for more than 1000 seawater samples and concluded that unless the difference (i.e., TDP-Pi) exceeded 10% of the TDP value, that the DOP (technically, SNP) estimate could not be considered to be significantly different from zero. For their analyses, 95% of the surface water samples contained significant DOP, decreasing through the region of the phosphate-cline where

P*i* increases and DOP decreases with increasing water depth. At depths greater than 1000 m there was no measurable (statistically significant) DOP present; only 13 out of 259 deep water samples (5%) gave positive differences that exceeded 1 standard deviation (Ketchum et al., 1955). It could not be determined whether DOP was present at concentrations below the analytical detection limit, or whether DOP was truly absent. Consequently, few reliable data sets exist for DOP concentrations in the deep mesopelagic and abyssopelagic zones (>1000 m). This is quite unfortunate because the poorly understood, stepwise conversion of PP and DOP to P*i* is a key metabolic process in these regions.

This analytical uncertainty, for better or for worse, has not prevented ocean researchers from sampling, analyzing, and reporting subeuphotic zone DOP concentrations. The caution we urge here is to be cognizant of the statistical constraints on the methodologies employed, as they will clearly affect the ecological interpretations of the data obtained. Examination of these data sets documents a fairly broad range in mesopelagic and deep sea DOP concentrations that cannot be easily reconciled with any known oceanic processes. A difference of just 20-50 nM DOP between these determinations, when scaled to dimensions of the deep sea, creates or eliminates a DOP pool that becomes significant for global ocean P budgets. It is imperative that we obtain reliable deep water DOP measurements if we ever hope to understand subeuphotic zone DOP dynamics or marine P cycle as a whole.

In theory, one might anticipate a small, but finite DOP pool that would represent a balance between local DOP production and utilization processes. The supply of DOP to depth depends to a large extent on the nature of organic matter export processes (e.g., downward diffusion and mixing of DOM vs. gravitational settling of POM) and on the pathways and coupling between export and remineralization mechanisms. However, only the process of gravitational settling of particulate matter can export significant amounts of "fresh" organic materials to subthermocline (>500 m) ocean depths. During the 1-2-month-long journey to the seabed in the open ocean, these exported particles are disaggregated, hydrolyzed, and otherwise reworked with a continuous, and sometimes variable, loss of organic C, N, and P. Much of this organic matter is remineralized at depth which accounts for the generally increasing concentrations of dissolved inorganic nutrients (see Figures 5.8 and 5.10). For open ocean habitats, the flux of organic-P from the euphotic zone (150 m reference depth) is approximately 5 mmol P m^{-2} year^{-1}. This statistical population of sinking particles is attenuated nearly an order of magnitude by the 1000 m depth horizon, and nearly two orders of magnitude at the seabed (5000 m). This attrition of organic-P from the sinking particulate pool as a predictable function of depth is the primary starting material for the suspended particulate and DOM pools beneath the euphotic zone. Only when a sufficient number of determinations are available, can the statistical significance of DOP in the deep sea be assured.

Repeated observations of SRP and TDP in a section between Montauk Point, N.Y., and Bermuda during the period 1958-1961 have demonstrated the appearance, at depth, of a low-salinity subarctic intermediate water mass that covaries with DOP concentration (McGill et al., 1964). Samples collected at hydrostation "S" south of Bermuda demonstrated statistically significant time-varying concentrations of DOP (i.e., TDP–SRP; Figure 5.11). It was concluded that these were advective rather than in situ features, and provided a constraint on the interpretation of single-point, Eularian design time-series experiments. It is possible that these variations in the DOP concentrations of deep North Atlantic Ocean waters were real; McGill et al. (1964) previously reported time-dependent variations of DOP from <0.04 to >0.15 μM for waters below 1500 m at hydrostation "S" in the Sargasso Sea near Bermuda

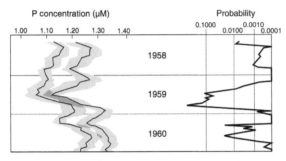

FIGURE 5.11   Mean SRP (shaded, left) and TDP (shaded, right) concentrations, and respective 95% confidence intervals, for deep water (>1500 m) samples collected at hydrostation "S" south of Bermuda (32°10'N, 64°30'W) for the period 1958-1960. Also shown, on the right, are the probabilities that the analyses for SRP and TDP are identical (e.g., a low probability indicates a statistically significant amount of DOP). The TDP analyses were performed according to the method of Harvey (1948). *Redrawn with permission from McGill et al. (1964).*

(Figure 5.11). Using a mass balance approach, they estimated a net in situ DOP consumption rate of 0.1 μM year^{-1}, so it is conceivable that interannual variations may exist even in the deep sea. Significant seasonal and interannual variations in particulate matter export at the 1000 m reference level on the order of ±50% of the long-term climatological mean are not unexpected, even in the subtropical gyres of the world ocean. This could easily lead to the changes observed by McGill et al. (1964) for samples collected at hydrostation "S". Even if these "temporal" patterns can be reconciled by spatial variability at this site, one would still need to explain the cause(s) of spatial variability in deep water DOP.

Thomson-Bulldis and Karl (1998) devised a novel method that can provide more accurate and reliable DOP measurements in the presence of a large SRP pool; deep sea DOP estimation is a perfect application for the modified MAGIC method. This method provides a separation of SRP from most DOP compounds prior to the direct measurement of DOP following oxidation–hydrolysis to Pi. Application to deep Pacific Ocean seawater has indicated an

abyssopelagic DOP concentration of <40 nM; values that would have been undetectable by the conventional TDP–SRP difference methodology (Figure 5.12). For comparison, we present two other "credible" published profiles for the North Pacific Ocean obtained using different methods for TDP but both based upon the more traditional DOP estimation by difference between SRP and TDP. While there does appear to be a detectable DOP pool in the deep waters

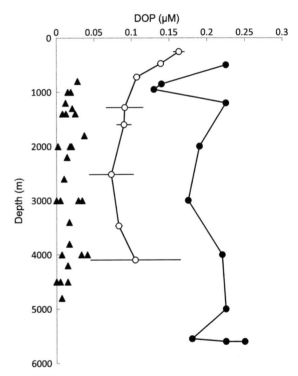

FIGURE 5.12   DOP concentration versus depth profiles at several stations in the North Pacific Ocean. Symbols: (●)—data from Williams et al. (1980) for samples collected in June 1977 at 28°N, 155°W and analyzed using the method of Armstrong et al. (1966); (○)—data from Loh and Bauer (2000), presented as mean ± 1 standard deviation, for samples collected in Jun. 1995 at 34°50'N, 123°W and analyzed using the method of Solórzano and Sharp (1980); (▲)—data from Thomson-Bulldis and Karl (1998) and K. Björkman, unpublished for samples collected in Dec. 1996, Jan. and Feb. 1997 and Nov. 2000 at Station ALOHA (22°45'N, 158°W) and analyzed using the MAGIC-persulfate oxidation method.

of the North Pacific it is uncertain whether it is <40 nM as the MAGIC method implies, or more than twice that amount (Figure 5.12).

An important ecological prediction of this depth-dependent delivery of organic-P to the subeuphotic zone waters is the potential selection against DOP utilization by the relatively high and depth-dependent ambient concentrations of Pi. All else being equal, one would predict a relative "preservation" of DOP in the deep sea due to preferential assimilation of Pi. However, if DOP is utilized for reasons other than Pi acquisition (i.e., as a biosynthetic precursor for nucleic acids) then there may be a simultaneous utilization of both Pi and DOP in these otherwise Pi-sufficient deep sea habitats. In fact, the case for preferential DOP utilization has been suggested by the discovery of APase activity in the deep ocean (Koike and Nagata, 1997; Hoppe and Ullrich, 1999; see Section IX.D). However, to our knowledge this important aspect of microbial ecophysiology has not been systematically investigated.

Jiao et al. (2010) have recently hypothesized that microorganisms are responsible for the production of recalcitrant organic matter through long-term, repetitive processing of the reactive DOM pool. In their microbial C pump (MCP) model, the resultant R-DOM would be extremely P-poor, with a molar C:P stoichiometry of ~3500:1 (Jiao et al., 2010). However, there are few data from the deep sea to examine the molecular structure, C:P stoichiometry and bioavailability of deep sea DOM/DOP, or to otherwise support or reject this MCP hypothesis. Flerus et al. (2012) used Fourier transform ion cyclotron resonance mass spectrometry (FT-ICR MS) with electrospray ionization to characterize and age DOM from 137 different sites in the eastern Atlantic Ocean. They confirm a previously proposed DOM degradation continuum of microbially mediated refractory DOM (RDOM) formation. However, given the "young" maximum ages of ~10,000 years for individual compound classes, relative to the age of the ocean, there must be a removal mechanism which is yet to be discovered for RDOM, including RDOP if that pool does exist.

C Stoichiometry of Dissolved and Particulate Matter Pools

All known organisms contain a nearly identical suite of biomolecules with common structural and metabolic functions; this biochemical uniformitarianism serves to constrain the bulk elemental composition of life. In a seminal paper, Redfield et al. (1963) summarized much of the earlier research on C, N, and P stoichiometry of DOM and POM pools in the sea and combined these data sets into an important unifying concept that has served as the basis for many subsequent field and modeling studies in oceanic biogeochemistry. As Paul Falkowski has so eloquently stated "the elemental stoichiometry that we call the Redfield ratio is a result of nested processes that have a molecular foundation but are coupled to biogeochemical processes on large spatial and long temporal scales" (Falkowski, 2000). These biogeochemical characteristics of oceanic habitats involve both macro- (e.g., N, P) and trace nutrient (e.g., iron) limitation, including DOP pool bioavailability.

Despite this perceived uniformity, it is well known that the chemical composition of living organisms can vary considerably as a function of growth rate, energy (including light) availability, ambient nutrient (including both major and trace elements) concentrations, and bioelemental ratios. For example, under conditions of saturating light and limiting N, certain photoautotrophic organisms can store C as lipid or carbohydrate, thereby increasing their C:N and C:P ratios. Likewise, if P is present in excess of cellular demands, it can be taken up and stored as poly-Pi, causing a decrease in the bulk C:P and N:P ratios. Conversely, when the bioavailable N:P ratio is greater than what is present in "average" organic matter (i.e., >16N:1P by atoms) selected groups of microorganisms can exhibit a metabolic "P-sparing" effect and produce biomass with C:P and N:P ratios significantly greater than the hypothesized Redfield ratios of 106:1 and 16:1, respectively. Based on theoretical biochemical arguments, Raven (1994) predicted that the C:N:P ratio in microorganisms is non-scalable with decreasing cell size. He concluded that the C:P ratio in small (0.5 μm) eukaryotes might be 30-50% lower (C:P=64:1) than in cyanobacteria of the same size due to the higher DNA content of eukaryotes. Consequently, there does not appear to be a robust constraint on ecological C:N:P stoichiometry in the marine environment.

Loh and Bauer (2000) assembled one of the most complete biogeochemical data sets, including particulate C, N, P as well as dissolved C, N, P pool measurements. They used their data from the Eastern North Pacific Ocean to test the hypothesis of preferential remineralization of N relative to P, and of N and P relative to C. Based on C:P and N:P ratios, they concluded that organic P is preferentially remineralized over organic C and N resulting in increasing C:P and N:P ratios of the DOM pool with increasing water depth. Our inability to provide a comprehensive inventory of DON and DOP compounds has promoted a stoichiometric assessment of the DOM pool based on separate measurements of DOC, DON, and DOP. However, this does little to provide a biochemical characterization of the representative compound classes dissolved in seawater.

Cavender-Bares et al. (2001) measured near-surface (3 m) dissolved inorganic and organic N and P distributions along a >2500 km transect (sampled in March 1998). Their study area included the eastern North Atlantic subtropical gyre, the Gulf Stream, and temperate coastal shelf habitats. Throughout the subtropics, the total dissolved N (TDN) and TDP pools were dominated by organic components (Figure 5.13). SRP was very low throughout the southern portion of the transect with concentrations between 1-10 nM and occasionally below 1 nM (Cavender-Bares et al., 2001). Only north of the Gulf Stream did the dissolved inorganic

constituents represent a significant proportion of the total dissolved nutrient pools. Furthermore, the DON and DOP pools were relatively invariant over a broad geographical range (26-37°N) with mean concentrations of 6.3 μM N and 0.12-0.13 μM P, respectively (Figure 5.13). The mean molar N:P ratio of the DOM pool averaged 50:1, a value that is three times greater than the canonical Redfield stoichiometry of 16N:1P. North of the Gulf Stream the molar ratio TDN:TDP pool dropped to 16:1 even though the N:P ratio of the DOM pool remained >25 (Figure 5.13). The accumulation of N, relative to P, in the marine DOM pool within the subtropical gyre is likely caused by the addition of "extra" N by the metabolic activities of N_2 fixing microorganisms that are selected for during periods of low (<16N:1P) dissolved inorganic nutrient availability, especially south of 31°N (Cavender-Bares et al., 2001). Further implications of this ecological selection

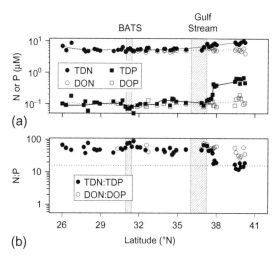

FIGURE 5.13   (a) Total N and P (TDN and TDP) and organic N and P (DON and DOP), all in μM, and (b) molar N:P stoichiometry of the respective pools for samples collected along a North Atlantic Ocean transect in Mar. 1998. Vertical striped bars indicate the approximate location of the Bermuda Atlantic Time-series Study (BATS) and the Gulf Stream. The dotted line in the bottom plot shows the Redfield N:P ratio (16:1) as a point of reference. *Reprinted with permission from Cavender-Bares et al. (2001).*

process on related aspects of the marine P-cycle are discussed in the following section.

Martiny et al. (2013) have reported strong latitudinal gradients in the C-N-P stoichiometry of both marine plankton and total suspended POM in ocean basins worldwide. The global average C:P and N:P ratios were 23-88% higher than the canonical Redfield ratio of 106:1 and 16:1, respectively, with highest values in tropical and subtropical ecosystems. They hypothesized that selection for N_2 fixation in low fixed N regions could lead to P-stress and to P-depleted POM (Martiny et al., 2013), and possibly DOM. Sustained N_2 fixation requires a continuous supply of iron so N_2 fixation and P-cycle dynamics are also inextricably tied to atmospheric delivery of iron. Indeed, the high concentrations of Pi (>100nM), long turnover time (>6months) and low rates of N_2 fixation in the South Pacific gyre have been attributed to N_2 fixation limitation by iron (Moutin et al., 2008). Research conducted at Station ALOHA in the NPSG has shown decadal-scale variability in both POM (Hebel and Karl, 2001) and DOM (Church et al., 2002) pool stoichiometry, possibly in response to variability in habitat and microbial population structure (Karl, 1999; Karl et al., 2001a; see Section VII.B).

VII DOP IN THE SEA: VARIATIONS IN TIME

Long-term ecological studies are predicated on the straightforward assertion that certain processes, such as succession and climate variability, are long-term processes and must be studied as such (Strayer et al., 1986). Indeed, there are many examples in the scientific literature where interpretations from short-term ecological studies are at odds with data sets collected over much longer time scales. Because it is difficult to observe slow or abrupt environmental changes directly, much less to understand the fundamental cause and effect relations of these changes,

Magnuson (1990) has coined the term "the invisible present" to refer to these complex ecological interactions. Most of what we know about the marine DOP pool is locked up in the invisible present and opaque past. As data accumulate in a long-term ecological context, new phenomena will become apparent and new understanding will be derived.

Our presentation, above, of the compiled GOOD database implies that DOP pools are constant in time for a given location. This is, most likely, an inaccurate assumption. Because the generally inverse relationship between subeuphotic zone SRP and DOP concentrations is a manifestation of plankton growth in the surface ocean, one might predict a significant but opposite seasonal cycle in the concentrations of SRP and DOP, with DOP maxima following the vernal bloom in temperate ocean habitats (Butler et al., 1979; Harvey, 1955; Strickland and Austin, 1960). During the Research on Antarctic Coastal Ecosystems and Rates (RACER) program, a detailed study of Marguerite Bay from open water to near the fast ice-edge revealed a systematic and coherent shift from a SRP-dominated to a DOP-dominated euphotic zone (Figure 5.14). This spatial mosaic was a result of an ice-edge induced bloom of phytoplankton and the temporal uncoupling of DOP production and DOP utilization processes (Karl et al., 1992). Similar processes are likely to occur wherever and whenever net planktonic biomass is produced. These seasonally phased near-surface ocean production processes should also drive a seasonal cycle in subeuphotic zone processes via a coupled particulate matter production-export cycle. However, for reasons already presented above, we currently lack the analytical tools to observe small changes in DOP within habitats where the DOP concentrations are $\leq 10\%$ of the corresponding TDP pools.

During the past century, there have been several systematic time-series studies of the marine P cycle; we shall present two case studies in this section as well as the result of an ecosystem scale

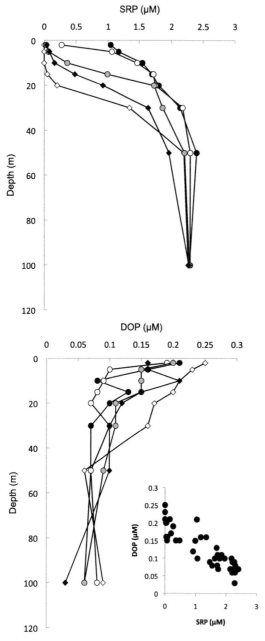

P*i* enrichment experiment. The first case study is from coastal waters of the English Channel that began in 1916 with the pioneering research of Matthews, Atkins, Cooper, Harvey, and others at the Plymouth Marine Laboratory (see Section III). The second, the Hawaii Ocean Time-series (HOT) study of the subtropical North Pacific Ocean, began in October 1988 and continues to the present. Both research programs have revealed significant time-dependent and climate-driven changes in P biogeochemistry and in the potential role of DOP in ocean productivity. The Eastern Mediterranean Sea P*i* release experiment was conducted to test several aspects of the marine P-cycle in a well-characterized, P*i*-stressed habitat.

A English Channel

The nearly continuous 60-year data set (1923-1987) for P*i* and nitrate concentrations at station E1 in the English Channel has been summarized and interpreted by Joint et al. (1997) and Jordan and Joint (1998). Winter maxima in the concentrations of P*i* varied considerably with significant interannual and, especially, interdecadal frequencies. Independent analyses of these same data had previously suggested a temporal trend in the chemistry and biology of the English Channel waters beginning in the 1930s, one that was broadly related to North Atlantic climate variability (Russell et al., 1971; Southward, 1980). During the period 1924-1929, wintertime surface water P*i* concentrations averaged 0.67 μM compared to a mean concentration of 0.48 μM during 1931-1938. After 1969, P*i* returned to the 1920s value of 0.62 μM. Coincident with these decadal-scale alterations in wintertime P*i*, there was a significant change in fisheries including the disappearance of herring after 1930. It was hypothesized that these ecosystem processes were linked to climate changes, and to the resultant effects on water movements and associated planktonic assemblages. This has become known as the "Russell Cycle" hypothesis (Karl 2010).

FIGURE 5.14 SRP and DOP pools for water samples collected along an offshore (● and ○) -to-ice edge (◇) transect in Marguerite Bay, south of Adelaide Island on the western Antarctic Peninsula (near 68°S, 68°W). The inset at bottom right shows the negative correlation between SRP and DOP for this region.

However, after careful reanalysis, even this heroic data collection effort appears to be inadequate for a rigorous statistical test of this hypothesis (Joint et al., 1997). Changes (improvements) in P*i* measurement protocols, uncertain retrospective "corrections" and the World War II sampling gap all contribute uncertainties to this otherwise incredibly rich data set. From this careful time-series analysis (Joint et al., 1997), one might wonder about the ecological significance of any single P*i* or DOP profile collected in the expeditionary mode of most oceanographic investigations!

Irrespective of the temporal trends in P*i* concentrations, the N:P stoichiometry of the dissolved inorganic nutrient pool (reported as the nitrate:P*i* molar ratio) varies seasonally with significant deviations from the Redfield ratio of 16N:1P (Jordan and Joint, 1998). The most intriguing result was the significant shift in the frequency distribution from a N:P ratio >6 during winter and fall to a N:P ratio <6 during summer (Jordan and Joint, 1998). These relatively low inorganic N:P ratios in summer, in the absence of known water column denitrification appeared enigmatic, and suggested a decoupling of nitrate and P*i* regeneration from the organic matter formed during the vernal bloom of phytoplankton.

Based on the pioneering research of Harvey and others we would have anticipated a fairly high net rate of DOP (and DON) production during the summertime period (see Figure 5.4). If DOP was recycled more rapidly than DON, or if DOP was assimilated directly in preference of P*i*, this could account for the shift in dissolved inorganic nutrient ratios in this habitat. A comprehensive 11-year study (1969-1977) of nutrient dynamics in the English Channel, which included measurements of both the dissolved inorganic and total dissolved nutrient pools (and, therefore, DON:DOP as well) clearly reveals the inverse correlations between nitrate/DON and between P*i*/DOP during the year (Butler et al., 1979; Figure 5.15). Whereas the inorganic N:P

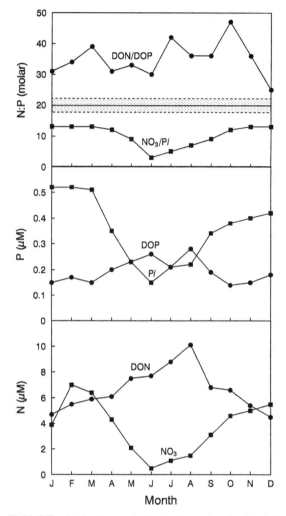

FIGURE 5.15 Seasonal fluctuations in dissolved P (P*i* and DOP) and N (nitrate (NO$_3$) and DON) pools in waters of the English Channel, and corresponding N:P molar stoichiometries of DON:DOP and NO$_3$:P*i* fractions. The shaded region in the top panel is the mean ± 1 standard deviation for the water column integrated TDN:TDP pool. These data are the monthly climatologies for an 11-year study. *Redrawn from Butler et al. (1979).*

ratios varied from a minimum of 3 in summer to a maximum of 13 in winter, the organic N:P ratios varied from a maximum of 42 in summer to a minimum of 25 in winter; TDN:TDP was "buffered" at a value of 19.83 ± 2.25. This ratio is believed to reflect the stoichiometric balance of plankton production and export processes (Butler et al., 1979).

In late 1987, the western English Channel time-series program was terminated due to lack of government funding. However, several programs related to nutrient and plankton dynamics have recently begun (Aiken et al., 2004; Pinkerton et al., 2003; Rodriguez et al., 2000).

B North Pacific Subtropical Gyre

The subtropical gyres of the world ocean are extensive, coherent regions that occupy approximately 40% of the surface of the Earth. With a surface area of nearly $2 \times 10^7 \, \text{km}^2$, the NPSG is the largest of these regions and, therefore, Earth's largest contiguous biome. Once thought to be homogeneous and static habitats, there is increasing evidence that mid-latitude gyres exhibit substantial physical and biological variability on several timescales from months to decades.

There is long-standing interest and substantive debate over the nature of nutrient control of primary production in the NPSG and in the world's ocean as a whole. Codispoti (1989) has summarized the key scientific issues, specifically the balance between rates of N_2 fixation and denitrification and the bioavailability of P. Extended periods of fixed N-limitation should select for N_2-fixing microorganisms and force the ecosystem towards P-limitation; thus, P would be the ultimate production rate-limiting macronutrient. However, this rather simple conceptual model assumes that N_2-fixing microorganisms can effectively compete for P and other required trace elements (especially iron) that are required for the activity of nitrogenase, the enzyme responsible for the reduction of N_2 to ammonium. Other environmental factors likely to influence N_2 fixation are temperature, turbulence, and dissolved oxygen concentration; oxygen inhibits nitrogenase activity (Karl et al., 2002). The input of new N into the surface ocean can decouple the otherwise linked regional C-N-P cycles, leading to an altered P-deficient stoichiometry in DOM and POM pools and a pulsed, net sequestration of atmospheric carbon dioxide. Consequently, the two most crucial environmental controls on N_2 fixation are the bioavailabilities of iron and P, including DOP (Karl, 2007b; Karl et al., 2002).

Pioneering research in the NPSG conducted by Perry (1972, 1976) suggested that P might control microbial growth in the near-surface waters. Several physiological parameters, including high particulate C:P ratios and high biomass-normalized rates of APase, were indicative of P-deficiency. Nevertheless, analytical and intellectual assets at this time were directed elsewhere. The Plankton Rate Processes in Oligotrophic Oceans Study (PRPOOS) had a deliberate focus on gross and net rates of primary production in the NPSG, but not on the nutrient controls thereof. In fact, PRPOOS completed its field work on marine production just a few years before the prokaryotic microorganism *Prochlorococcus* sp. was discovered as the dominant photoautotrophic component of these open ocean habitats (Chisholm et al., 1988). One might legitimately wonder, "What else didn't we know at that time?" The VERtical Transport and EXchange (VERTEX) program conducted extensive studies of coupled primary production and particulate matter export in the NPSG, but again lacked an explicit focus on the ecophysiological controls thereof. The Asian Dust Inputs to Oligotrophic Seas (ADIOS) project focused on the eolian deposition of trace elements, including iron, but did not systematically evaluate the inter-relationships between iron deposition, N_2 fixation, and P pool dynamics. In fact, none of these three major biogeochemical programs included core measurements of DOP. At that time, the unifying concept of new and regenerated production (sensu Dugdale and Goering, 1967), based strictly on nitrate and ammonium pool dynamics respectively, reigned as biogeochemical dogma. There is presently compelling evidence to suggest that this N-centric view of new and export production in the NPSG is in need of revision to accommodate both N_2 fixation and P biodynamics (Karl, 1999, 2000).

Since October 1988, a comprehensive suite of ocean measurements including P pool and P flux estimations has been obtained at the oligotrophic Station ALOHA in the NPSG. Core measurements were selected to provide a data set to evaluate existing C-N-P biogeochemical models and, if necessary, to improve them. The emergent data set from Station ALOHA is unique, robust, and rich with previously undocumented phenomena (Karl, 2007b, 2014). During the initial investigation period it became evident that the NPSG P-cycle was unexpectedly complex. SRP pool dynamics were characterized by both high-frequency (weeks to months) and lower-frequency (years) changes that included aperiodic DOP pool inflation (up to 50%) above the longer-term mean. The nearly quantitative shift from dissolved "inorganic" to dissolved "organic" P suggested an accumulation of biorefractory materials, a condition that could be a manifestation of enhanced N_2 fixation and a switch from N-limitation to P-limitation as first suggested by Karl et al. (1995). At the end of 1994, it was difficult to predict how long these features could persist without fundamentally disrupting new and export production processes in the NPSG. Continued measurements of SRP and DOP have documented further decreases in the SRP inventory to ~2.2 mmol P m^{-2} by December 1997, but without a concomitant increase in DOP (Church et al., 2002; Karl et al., 2001b). Apparently, the initial DOP pool inflation was only a temporary perturbation in the NPSG P-cycle dynamics. Enhanced net utilization of semi-labile DOP would be predicted under sustained conditions of P-limitation (Karl, 2007a). A shift in bioavailable P, from Pi to DOP could also promote species selection, thereby affecting overall community structure and key ecological processes.

The stoichiometry of the TDN:TDP pools in the upper 0-100 m of the water column (mostly dominated by DON and DOP) also displayed variability on monthly, seasonal, and interannual time scales. For example, monthly observations revealed occasional high-frequency changes in the N:P ratio of the total DOM between consecutive cruises (e.g., during spring 1989, spring 1997-1999; Figure 5.16). These features were characterized by decreases in the N:P ratio from values that were significantly higher than the Redfield ratio to values approximating it (Figure 5.16). When observed, these events always coincided with pulsed inputs of inorganic nutrients as detected by elevated nitrate plus nitrite (N+N) inventories (Karl et al., 2001b). In certain years (e.g., 1991, 1993, 1996), these nutrient injections were either absent or, more likely, missed by the relatively coarse monthly frequency of our sampling program. We also observed several time periods of sustained, systematic change in the N:P ratio; e.g., Jan 1991 to July 1992 where the N:P decreased from >20 to values equivalent to the Redfield ratio, followed by an approximately 18-month period during which the N:P ratio slowly increased back to values approaching 25:1 (Figure 5.16). These features resulted in significant interannual variations in the TDN:TDP ratio (Figure 5.16), with 1993 standing out as a year with an anomalously high mean TDN:TDP ratio of 22.8 (S.D. = 1.8). The mean TDN:TDP ratio for the complete 9-year data set was 19.6 (S.D. = 2.6), well above the 16N:1P Redfield ratio.

Major differences were also observed for the molar N:P stoichiometries of the dissolved inorganic nutrient pools (i.e., N+N:SRP) versus the total dissolved nutrient pools (i.e., TDN:TDP; Figure 5.17). The greatest differences were observed in the upper 0-400 m of the water column (and, especially in the upper 0-100 m) where dissolved organic nutrients are present as significant fractions of the TDN and TDP pools. Whereas the dissolved inorganic N:P ratios in the upper water column were significantly lower than the Redfield ratio of 16N:1P, the N:P stoichiometry of the total dissolved pool (inorganic

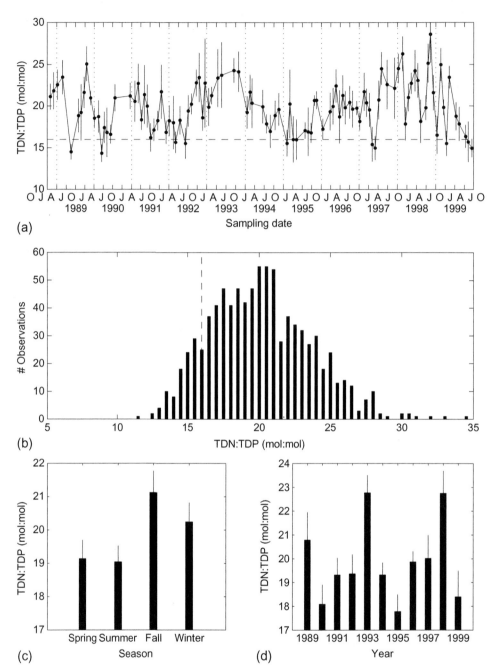

FIGURE 5.16 Nitrogen-to-phosphorus (N:P) ratios for the total dissolved matter pools in the upper 0-100 m of the water column at Station ALOHA during a decade-long observation period. (a) TDN:TDP versus sampling date. For each cruise, the mean value ± 1 standard deviation is presented. As a point for reference, the horizontal dashed line is the Redfield ratio of 16N:1P. (b) Frequency histogram of TDN:TDP values for the data set. As a point for reference, the vertical dashed line is the Redfield ratio of 16N:1P. (c) Seasonal variability in TDN:TDP at Station ALOHA. Seasons are: Spring = Mar.-May, Summer = Jun.-Aug., Fall = Sep.-Nov., Winter = Dec.-Feb. The values presented are the mean ± 1 standard deviation for each data set. (d) Interannual variability in TDN:TDP at Station ALOHA. The values presented are the mean ± 1 standard deviation for each data set. *Redrawn from Karl et al. (2001b), with additional data from the Hawaii Ocean Time-series Data Organization and Graphical System (HOT-DOGS).*

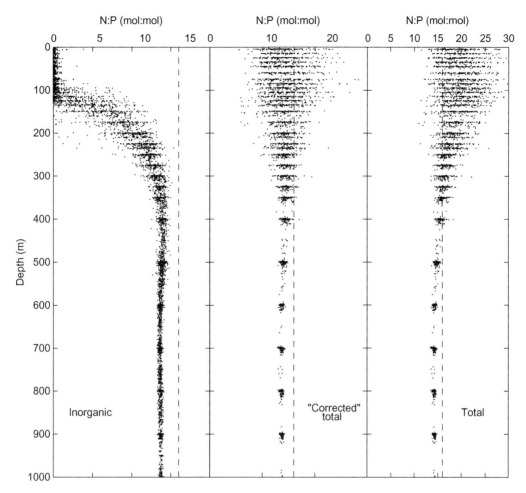

FIGURE 5.17   Nitrogen-to-phosphorus (N:P) ratios versus water depth for samples collected at Station ALOHA during the period Oct. 1988 to Dec. 1997. (Left) Molar N:P ratios for dissolved inorganic pools calculated as nitrate plus nitrite (N + N): soluble reactive phosphorus (SRP). (Center) Molar N:P ratios for the "corrected" total dissolved matter pools (see text for details). (Right) Molar N:P ratios for total dissolved matter pools, including both inorganic and organic compounds, calculated as total dissolved nitrogen (TDN): total dissolved phosphorus (TDP). As a point for reference, the vertical dashed line in each graph is the Redfield molar ratio of 16N:1P. *Redrawn from Karl et al. (2001b), with additional data as in Figure 5.16.*

plus organic) was significantly greater than the Redfield ratio by as much as 50% (Figure 5.17). Furthermore, there were systematic changes in the N:P stoichiometry as a function of water depth; inorganic N:P increased towards a ratio of approximately 14, while total N:P decreased towards the same value (Figure 5.17). In both

data sets, the greatest rate of change in N:P with depth was in the 100-400 m region of the water column.

The relatively high TDN:TDP ratios in the near-surface waters are consistent with the hypothesis that P, not N, is the (or one of several) production rate-limiting nutrient(s) in this ecosystem. This conclusion assumes that the TDN and TDP pools are fully bioavailable (see Jackson and Williams, 1985; Smith, 1984). However, more recent research on DOM suggests that near-surface pools are comprised of at least two components: one that is locally produced and consumed during microbial metabolism (the labile pool), and one that may be more refractory (Karl, 2007a). Although it is impossible to quantify these subcomponents using existing analytical techniques (and in reality there may be a continuum of bioavailabilities) for the sake of the present discussion we will assume that the mean deep water (>600 m) DON and DOP pools are refractory. If TDN and TDP are corrected for these non-labile components, the depth profile of N:P ratios assumes the characteristic "T-shape" (Fanning, 1992), but for a fundamentally different reason than the original author suggested. Rather than being a consequence of analytical uncertainties at low surface ocean concentrations, we hypothesize that the T-shaped profile for the corrected TDN:TDP ratios at Station ALOHA is a manifestation of an alternation between periods of N-limitation (left-hand side of the T) and periods of P-limitation (right-hand side of the T). Using cosmogenic ^{33}P and ^{32}P inventories and a non steady-state model, Benitez-Nelson and Karl (2002) documented time-varying (weeks to months) labeling of the DOP pool in surface waters at Station ALOHA. They hypothesized the presence of multiple DOP subpools, each with a different $^{33}P/^{32}P$ concentration, activity ratio, bioavailability, and residence time. These various pools, possibly unique compound classes, can buffer Pi starvation during periods of high productivity and export.

N_2 fixation is one of two major microbiological processes (the other being denitrification) that can significantly influence oceanic N:P stoichiometry on global scales. Several lines of evidence from Station ALOHA suggest that N_2 fixation is an important contemporary source of new N for the pelagic ecosystem of the NPSG. In addition to the observed secular changes in SRP inventories and the N:P ratios already discussed, other independent measurements include: (1) *Trichodesmium* population abundances and estimates of their potential rates of biological N_2 fixation, (2) seasonal variations in the natural ^{15}N abundances of particulate matter exported to the deep sea and collected in bottom-moored sediment traps, and (3) increases in the DON pools during the period of increased rates of N_2 fixation (Karl et al., 1997). Since 2003, N_2 fixation has been measured at Station ALOHA using state-of-the-art $^{15}N_2$ tracer methods (Montoya et al., 1996). Then a high-profile report claimed that N_2 fixation rates had been systematically underestimated using this method (Mohr et al., 2010). Global oceanic rates of N_2 fixation are probably at least twice as high (Grosskopf et al., 2012). This discrepancy has also been confirmed at Station ALOHA (Wilson et al., 2012) so all evidence suggests that the NPSG is becoming more P-stressed and that DOP may become increasingly more important in microbial metabolism (Karl, 2007a). However, over the past years (2011-2012), the Pi pool in the surface waters (0-100 m) at Station ALOHA have more than doubled, returning to 1988 levels (Karl, 2014). Whether this is a result of decadal-scale climate–ocean interactions (Karl, 1999), predicted oscillations in N-controlled versus P-controlled ecosystem dynamics (Karl, 2002), iron limitation of N_2 fixation (Moutin et al., 2008), or the result of combined or independent processes is not known at the present time (Karl, 2014).

At the beginning of the HOT program in 1988, biogeochemical processes in the gyre were thought to be well understood. New and export production were limited by the supply of nitrate from below the euphotic zone, and rates of primary production were thought to be largely supported by locally regenerated nitrogen. However, the contemporary view recognizes the gyre as a very different ecosystem (Karl, 1999; Karl et al., 2001a). Based on multiple decade-scale data sets (1988-2010), we hypothesize that there has been a fundamental shift from N-limitation to P-limitation (Karl and Tien, 1997; Karl et al., 1995). The ecological consequences of this hypothesized N_2 fixation-forced Pi-limitation, especially on DOP pool dynamics, is presented elsewhere (Karl et al., 2001b; see also Figure 5.18). Suffice it to say that enhanced Pi cycling rates, shifts in the chemical composition of the DOP pool, and microbial biodiversity changes are all relevant features of these decade-scale ecosystem processes. The fundamental role of nutrient dynamics in biogeochemical processes and ecosystem modeling demands that we have a comprehensive, mechanistic understanding of inventories and fluxes. Although the present ongoing ocean time-series study at Station ALOHA has certainly not resolved all of

these important matters, it does provide an unprecedented data set to begin the next phase of hypothesis testing (Karl, 2014).

C Eastern Mediterranean Sea

The Eastern Mediterranean Sea is one of the most oligotrophic regions on Earth, at least with respect to the distributions of suspended particulate and dissolved matter (Berman et al., 1985). Comprehensive field investigations conducted over the past few decades have suggested that P limits biological production in this region (Krom et al., 1991). In May 2002, a multi-disciplinary team of scientists conducted a large-scale Pi-fertilization experiment of an approximately 16-km^2 patch of open sea at 33.3°N, 32.3°E;

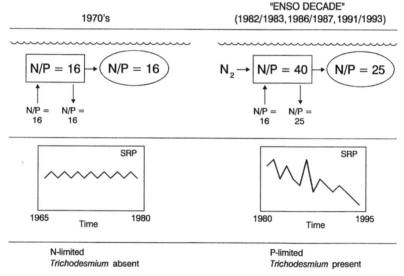

FIGURE 5.18 Schematic presentation of the NPSG alternating ecosystem state hypothesis. This cartoon depicts the contrasting N and P nutrient cycles during periods of low rates of N_2 fixation (e.g., 1970s) and enhanced rates of N_2 fixation (1980-present). It is believed that the increased frequency and duration of the El Niño-Southern Oscillation (ENSO) cycle since the early 1980s is a major cause of the N_2 fixation rate enhancement (see Karl, 1999; Karl et al., 2001a). The small rectangles and ovals at the top of each panel represent the average N:P ratios in particulate and dissolved matter, respectively, and the upward and downward arrows are the N:P stoichiometry of imported (mostly dissolved) and exported (mostly particulate) matter. N_2 fixation (on right) decouples the N:P stoichiometry of the NPSG ecosystem. The center panels depict the inventories of SRP during both phases of the cycle showing a secular decrease in SRP following the selection and growth of N_2-fixing microorganisms, such as *Trichodesmium*. Many of these predictions have been confirmed during the ongoing study at Station ALOHA (see text).

the Cycling of P in the Eastern Mediterranean (CYCLOPS) project (Thingstad et al., 2005).

Following the addition of a diluted mixture of phosphoric acid to increase the Pi concentration by two orders of magnitude to ~110 nM (SF$_6$ was also added) in this P-limited habitat, there was a 40% decrease (rather than the predicted 40-fold increase) of chlorophyll in the microbial assemblage compared to control regions outside the Pi-enriched patch. The chlorophyll "hole" developed over approximately 5 days, before returning to background levels after a period of about 1 week. Rates of primary production and phytoplankton growth after Pi addition were also reduced (Thingstad et al., 2005). In contrast to the response observed for the photoautotrophic assemblage, the addition of Pi stimulated "bacterial" production (as measured by ^{14}C-leucine incorporation) and resulted in the accumulation of PP, presumably a result of the net growth of the chemoorganoheterotrophic assemblage. Rates of N$_2$ fixation also increased following Pi addition (Rees et al., 2006), but the microorganisms responsible were not identified.

This unique study provided an opportunity to follow P-cycle dynamics including, but not limited to, Pi uptake, DOP turnover, DOM and POM elemental stoichiometry. Prior to the Pi pulse, DOP concentrations (reported as UV-oxidizable DOP or DOP$_{UV}$; Krom et al., 2005b) were extremely low (~50 nM) consistent with the hypothesized P-limited nature of this ecosystem. Furthermore, there was little change in DOP$_{UV}$ with depth (0-2000 m), which suggested that this might be the residual, non-labile fraction (Krom et al., 2005b). The DON:DOP molar ratio was ~100:1 and the PON:POP ratio was 27-32:1, both significantly elevated relative to the Redfield stoichiometry of 16:1 (Krom et al., 2005a,b).

Additional details of this elegant Pi-addition field experiment—the first of its kind ever conducted in the open sea—have appeared in a special volume of Deep-Sea Research (Krom, 2005). The unprecedented results obtained during the

CYCLOPS project required a new conceptual framework with much more complex and possibly non-linear trophic interactions. Because this was a "one-off" experiment, it is not clear whether the results obtained are reproducible, whether the concentration of exogenous P is a critical variable (e.g., Do the results scale on P loading?), or whether the results obtained are unique to hyperoligotrophic ecosystems. Perhaps more importantly, this field experiment demonstrates the ever-present possibility that manipulation of ocean ecosystems can yield entirely unexpected results—a message that is central to the present debate regarding the intentional fertilization of the ocean for C offsets (Chisholm et al., 2001).

VIII DOP POOL CHARACTERIZATION

A major analytical challenge in DOP pool characterization is the detection of individual compounds typically present at pM to nM concentrations dissolved in seawater medium containing approximately 35 g l^{-1} of inorganic salts. Preconcentration and separation using ion exchange resins, ion exclusion, or similar chromatographic procedures or even lyophilization that have proven useful for the characterization of DOP in soil extracts and freshwater habitats (e.g., Espinosa et al., 1999; Hino, 1989; Minear, 1972; Nanny et al., 1995) are generally not applicable for the analysis of marine DOP. However, remarkable progress has been made during the past decade largely due to advances in methodology for organic matter isolation and analysis (Hertkorn et al., 2013; Kujawinski et al., 2009; Nebbioso and Piccolo, 2013; Repeta et al., 2002), but most of these methods have focused on DOC and DON-N, rather than on DOC-P or DOC-N-P.

Because abiotic synthesis of organic P is not likely to occur in the marine environment, both the presence of a detectable DOP pool, as well as its molecular weight spectrum and chemical composition are dependent upon biological, mostly microbiological, processes. If marine DOP is derived from living organisms, as it ultimately must be, then the molecular spectrum of P in living cells or in marine particulate matter should be a first order inventory of DOP sources. The macromolecular composition (by weight %) of an "average" bacterial cell is: protein 52%, polysaccharide 17%, RNA 16%, lipid 9.4%, DNA 3.2%, other <3% (Stouthamer, 1977). Of these compound classes, only the nucleic acids and, to a lesser extent, lipids are P-rich. Correll (1965) measured the percent distribution of P in natural particulate matter from 11 stations in the subantarctic and antarctic waters, and reported a predominance of RNA-P (15-74%) and lipid-P (3-29%) with trace amounts (2-15%) of acid soluble organic-P; these results are consistent with expected subcellular pool distributions. Miyata and Hattori (1986) also investigated the composition of natural plankton populations (i.e., PP) in Tokyo Bay via differential chemical extraction. Their results indicated that nucleic acid-P and lipid-P were the dominant forms of organic-P, together accounting for about 60-70% of the total. A large cellular pool of Pi was also detected.

As Waksman and Carey (1935) correctly noted many years ago, the DOM pool in seawater is in a "state of dynamic equilibrium" in which residues and waste products are the source terms and bacterial activity is the sink. The ambient chemical composition of DOM and individual DOP compound concentrations, therefore, reflect the net balance between production and utilization. Approximately 80 years ago, Krough (1934) first estimated that the amount of DOM present in the sea was nearly 300 times greater than that contained in all living organisms. He believed that this large reservoir of organic matter, including DOP, must represent "waste products" that cannot be recovered and may even be slowly accumulating. Furthermore, partial degradation, including photo- and chemical alteration processes, may potentially create DOP

compounds that are not present in the organisms themselves. Consequently, the chemical composition of DOP in a given habitat can vary significantly as a result of either preferential production or preferential utilization of selected organic-P compounds or compound classes.

The marine DOP pool is also expected to vary considerably, ranging from truly dissolved monomeric compounds through polymeric (IMW and HMW) compound classes, to small (<0.2 μm) particles and colloids. Colloidal organic matter is extremely abundant in the marine environment (Koike et al., 1990; Wells and Goldberg, 1991), but remains poorly characterized. The cumulative high surface-to-volume ratio of marine colloids may facilitate complexation, aggregation, or adsorption interactions with truly dissolved compounds.

The technique of cross-flow filtration (CFF) has been used to isolate and concentrate colloidal marine materials (e.g., Buesseler et al., 1996; Cai and Guo 2009; Lin et al., 2012; Whitehouse et al., 1990) for subsequent analysis. In the absence of selective adsorption or contamination, TP should behave conservatively in CFF systems. However, in practice, large deviations are observed including an enigmatic production of colloidal-P during sample processing (Bauer et al., 1996). A similar production of colloidal-P during ultrafiltration processing of freshwater samples has also been reported (Nanny and Minear, 1997), suggesting that some caution is advised in the interpretation of colloidal-P data sets. For the purposes of this review and except where otherwise noted, colloids are included in the "dissolved" matter fraction.

Marine DOP can be characterized by a number of independent techniques including, for example, direct chemical analysis of selected molecules or compound classes, the use of specific hydrolytic enzymes or partial photochemical degradation, chromatographic and molecular weight fractionation, and ^{31}P-NMR. The combined use of molecular weight separation and specific chemical class characterization by, for example, ^{31}P-NMR or enzymatic hydrolysis, is beginning to provide a glimpse of complexity of the DOP pool. The direct measurements of specific P-containing organic compounds are the most useful data sets, from an ecological and biogeochemical perspective, because their fluxes can be traced to specific sources and sinks. However, the few studies that have been conducted have identified only a minority fraction of the ambient DOP pool, leaving the majority uncharacterized.

Below, we present several independent approaches to DOP pool characterization, beginning with the least specific method of physical separation by molecular weight/size and ending with specific methods of single compound or compound class characterization. Selected marine DOP data sets will also be presented for each major application. We end this section with some speculation on the possible chemical composition of the uncharacterized majority of the DOP pool.

A Molecular Weight Characterization of the DOP Pool

Various techniques have been used to fractionate DOP by molecular weight, size, ionic charge, or the ability to adsorb onto specific resins (e.g., nonionic XAD). When combined with chemical detection systems these procedures can be used to characterize the total DOP pool. However, it is uncertain how much ecological information can be obtained by these separation techniques alone because DOP bioavailability is probably not directly correlated with these physical properties. Although the HMW organic fraction has often been considered to be more biorefractory than LMW matter, metabolic evidence has shown just the opposite for selected marine habitats (Amon and Benner, 1996). Currently, it is impossible to determine how representative this HMW-DOP is of either the total DOP or the more ecologically relevant BAP pool.

A basic problem in studies of DOP pool characterization is the need for preconcentration of selected compounds prior to analysis. A widely used method in freshwater habitats, but less so in marine ecosystems, is gel chromatography, usually employing Sephadex® gels. This provides for the separation of HMW-P (>5 kDa), including colloids, in the "void volume" from IMW (>400 Da but <5 kDa) and LMW (<400 Da) fractions (Broberg and Persson, 1988).

Matsuda and Maruyama (1985) used Sephadex G-25 gel chromatography to characterize the DOP pool in coastal waters of Tokyo Bay. Several model DOP compounds were co-analyzed to establish the overall efficacy of this method; HMW-DOP compounds were contained in the void volume, LMW monophosphate esters (e.g., Gly-3-P) eluted prior to Pi and nucleotides eluted after Pi. Prior to separation, the TDP was first concentrated 30- to 35-fold by rotary evaporation and salt exclusion, a method which recovered only approximately 50% of the total P. Separate HMW and LMW DOP pools were observed in all seawater samples that were analyzed. Both fractions decreased with increasing water depth suggesting that biochemical lability was independent of molecular weight. In contrast, DOP compounds of IMW were relatively constant and, perhaps, biochemically stable (Matsuda and Maruyama, 1985). The IMW DOP fraction also absorbed UV light, consistent with nucleotide bases; no further chemical characterization was given and no ecological implications were discussed.

The method of tangential flow ultrafiltration has been used to isolate and concentrate DOM for subsequent chemical analysis (e.g., see Benner et al., 1992). However, this method is very selective for the IMW and HMW (generally >1 kDa) compounds, which for many aquatic ecosystems may be ≤35% of the total organic matter pool. In a series of carefully controlled experiments, Nanny et al. (1994) evaluated the behavior of ultrafiltration and reverse osmosis (RO) for concentration and molecular size fractionation of SNP in freshwater lakes. They evaluated both exogenously supplied standard compounds and the effects of increasing ionic strength by NaCl addition. Their results demonstrated that membrane type, including pore size, ionic strength, specific DOP test compound and volume concentration factor were all key variables that affected model compound—and presumably natural DOP compound—recovery. Another potential problem with the interpretation of molecular weight separations is the real possibility of LMW, monomeric DOP compound or Pi association with HMW materials. While the exact mechanism is not well understood, adsorption, hydrogen bonding, metal bridging, and even Maillard reaction are all distinct possibilities (Carlson et al., 1985). A novel and significant seawater application of tangential flow ultrafiltration for HMW-DOP pool concentration, coupled with ^{31}P-NMR molecular characterization (Clark et al., 1998) is presented in a subsequent section of this review (see Section VIII.C).

Llewelyn et al. (2002) described a novel procedure for pre-concentrating HMW-DOP using CFF and barium acetate precipitation followed by analysis electrospray ionization and FT-ICR MS characterization. Using water samples collected from the Florida Everglades, they detected a range of molecular fragments in the mass range 241-626 Da, from the initial >1000 Da sample, indicating a LMW bias. Unfortunately, even this sophisticated analytical approach was unable to significantly advance our understanding of the biogeochemistry of marine DOP (Cooper et al., 2005). P X-ray absorption near-edge spectroscopy can be used to characterize both inorganic and organic P species on particles and cells, but to date has not been employed to characterize marine DOP (Brandes et al., 2007; Ingall et al., 2011).

B DOP Pool Characterization by Enzymatic Reactivity

There are at least three separate applications for the use of specific enzymes in ecological studies of the marine P-cycle: (1) the presence and relative activities of specific enzymes in natural microbial assemblages can be used as physiological indicators of Pi-stress, deficiency, or other ecological processes, (2) in vitro or in vivo measurements of the rates of specific enzymes in natural assemblages of microorganisms can provide relevant information on DOP turnover rates, and (3) exogenous additions of specific enzymes to whole, filtered or partially purified seawater samples can be used to help characterize the DOP pool composition and to determine its potential bioavailability. The first two applications are discussed in a subsequent section; the third application is discussed below.

In theory, selected DOP compounds or compound classes could be estimated by measurements of Pi accumulation in cell-free seawater samples following a timed incubation with an exogenous purified enzyme or multiple enzyme cocktail. This is a straightforward and versatile approach. For example, Herbes et al. (1975) used three separate enzyme treatments to characterize both the LMW- and HMW-DOP fractions from a variety of aquatic habitats. This method could also be used to characterize nascent DOP that is produced during ^{32}Pi or ^{33}Pi tracer addition experiments, but to our knowledge this has not yet been attempted. The specificity of the enzyme(s) used will establish the specificity of DOP compound or compound class analysis.

In 1966, Strickland and Solórzano described a quantitative assay for total dissolved phosphomonoester P (PME-P) using exogenous APase from E. coli. An adaptation of this method using immobilized E. coli APase has also been described (Shan et al., 1994). This assay monitors the appearance of SRP during timed incubations following APase addition, relative to controls without exogenous enzyme. Conversely, the in situ activity of APase can be estimated by adding excess substrates followed by timed measurements of Pi or other byproducts (see Section IX.D). Significant PME-P concentrations ranging from 0.05 to 0.45 μM P were detected in coastal California seawater samples (Strickland and Solórzano, 1966). APase can also release Pi from pyro-Pi and poly-Pi (Rivkin and Swift, 1980), so technically Pi increase from APase treatment only provides an upper constraint on PME-P concentration in seawater. However, independent estimations of pyro-Pi and poly-Pi concentrations could be made to establish limits on this potential source of interference. Technically, though, this should be termed the APase-hydrolyzable P (APHP) pool, rather than PME-P.

Taft et al. (1977) applied this assay to samples collected in Chesapeake Bay over a full year. PME-P/APHP was >10% of the total DOP pool in 25 of 61 samples (but typically <20%); PME-P/APHP concentrations co-varied with DOP, and both peaked in late summer (Taft et al., 1977). Kobori and Taga (1979) measured PME-P/APHP concentrations, APase activity, bacterial cell number, % of isolates that were APase positive, and DOP in a variety of Japan coastal habitats. Ambient concentrations of PME-P/APHP ranged from 19-50% of the total DOP but, in general, appeared to be scavenged to low concentrations, presumably because the APase-containing bacteria that dominated these habitats readily use PME-P/APHP. At depths greater than 100 m in Sagami Bay, PME-P/APHP was undetectable in the environments that were studied.

A similar approach to detect other components of the marine DOP pool has employed DNase and/or RNase treatments of isolated dissolved nucleic acids to estimate the specific enzyme hydrolyzable components of these macromolecular fractions by measurements of DNA or RNA before and after specific enzyme treatments (DeFlaun et al., 1987; Siuda and Chróst, 2000). Most, but not all, of the extracellular nucleic acid pool appears to be nuclease-hydrolyzable, which implies that it is readily available to selected microorganisms. Direct estimates of ^{3}H-DNA turnover (Paul et al., 1987) have also supported this conclusion.

The "residual" nucleic acid fraction that is not degraded by specific nuclease treatments could be adsorbed to clay particles or otherwise chemically or physically altered. These semi-labile or truly refractory nucleic acid fractions could accumulate over time.

McKelvie et al. (1995) devised a method using immobilized phytase that was designed to selectively release Pi from phytic acid and related inositol phosphates. However, in practice the phytase system was less specific than anticipated so the defined pool was termed "phytase hydrolysable P" or PHP (McKelvie et al., 1995). PHP was compared to SRP, and for a variety of aquatic environments including several estuarine but no open ocean habitats. PHP was a significant percentage of the "apparent DOP." However, potential problems with the inability to separate Pi from labile DOP in the standard SRP determination could lead to an underestimation of DOP, and overestimation of the percentage of PHP by the methods employed. This analytical limitation is not unique to the phytase assay. In any case, most of the DOP pool in the environments studied appeared to be enzyme hydrolyzable.

Suzumura et al. (1998) used tangential flow ultrafiltration techniques to isolate LMW and HMW-SNP fractions in Tokyo Bay, Japan. These pools were then characterized using two phosphohydrolytic enzymes: APase (as above) and phosphodiesterase (PDEase; EC 3.1.4.1). The former enzyme alone will release Pi from a variety of PME, and the combination of the two enzymes will release Pi from nucleic acids. The molecular weight spectrum revealed a dominance of LMW-SNP (54-76% of the bulk SNP); no further characterization of this fraction was reported. Enzymatic characterization of the HMW-SNP fractions revealed the presence of both phosphomonoesters and diesters, with diester compounds in greatest abundance. However, on average, the majority of the isolated HMW-SNP was resistant to the enzyme treatments (Suzumura et al., 1998). Pretreatment with chloroform converted a portion of HMW-SNP to an enzymatically labile form and, on this basis, the authors suggested the presence of hydrophobic P-containing compounds, perhaps phospholipids. Presumably, this latter pool of enzymatically resistant compounds is also unavailable for microbial decomposition without prior hydrolysis or other diagenetic alteration.

C DOP Pool Characterization by ^{31}P-NMR

Nuclear magnetic resonance (NMR) spectroscopy has proven to be a powerful analytical tool for the molecular characterization of marine DOM. The abundant, naturally occurring isotope of P, ^{31}P, has a magnetic moment that is detectable by dipole resonance in an applied magnetic field. Although resonance is observed that is due solely to the P atom, chemical shifts due to the electron shells of the other atoms with which P is associated produce the diagnostic NMR spectra. Specifically, mono- and di-P esters are readily distinguishable from phosphonates and can be identified in bulk DOP concentrates. Both solution and solid-state magic angle spinning (MAS) ^{31}P-NMR techniques have been employed in marine ecosystems (Ingall et al., 1990). A major limitation with this method is the detection limit; marine DOP must be concentrated several thousand-fold prior to analysis. During the concentration process, DOP can be selectively eliminated (e.g., by molecular weight) or chemically altered. Detection of Pi in the HMW fraction indicated either post-concentration DOP hydrolysis or desorption of Pi from HMW organic/inorganic matter, or both (Nanny and Minear, 1997). Ironically, the NMR measurement technique itself is "non-destructive."

Nanny and Minear (1997) have combined ^{31}P-NMR with selective chemical hydrolysis and other reaction techniques to further characterize the major compound classes of HMW-DOP isolated from aquatic ecosystems. For analysis

of Crystal Lake, they detected phosphonates in the HMW-DOP, but not in the IMW or LMW fractions.

Using tangential flow ultrafiltration and solid-state cross-polarized MAS ^{31}P-NMR, Clark et al. (1998, 1999) demonstrated that approximately 75% of the HMW-DOP collected at a station in the South Pacific Ocean (12°S, 134°W) was comprised of ester-linked P compounds and the remainder (~25%) were phosphonates (Figure 5.19). Both classes could potentially include a broad spectrum of individual compounds with independent sources and sinks and variable residence times. Clark et al. (1998) also noted a significant shift in the bulk C:N:P elemental composition of the isolated HMW-DOM with depth (C:N:P molar ratios of 247:15:1, 321:19:1 and 539:30:1 for surface, 375 and 4000 m, respectively), indicating a selective remineralization of P from seawater DOM.

The relatively high proportion of phosphonates was unexpected and interpreted to be a result of the selective retention of phosphonates

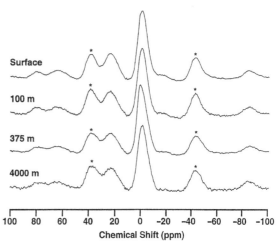

FIGURE 5.19 ^{31}P-NMR spectra of HMW-DOP from several reference depths in the Pacific Ocean (12°S, 135°W). The peak at 0 ppm is indicative of phosphate esters, and the peak at 25 ppm denotes phosphonates. Asterisks indicate spinning sidebands, an artifact of magic angle spinning. *Reprinted with permission from Clark et al. (1998).*

over time, based on the presumption that these compounds are not readily degraded in the sea. Equally intriguing and enigmatic is the fact that despite a six-fold decrease in HMW-DOP concentration from 90 nM in the surface to 15 nM at a depth of 4000 m, the ^{31}P-NMR spectra show P-esters and phosphonates at approximately identical proportions indicating a coupled utilization. In the absence of a deep water source of P-esters, the hypothesized higher rates of P-ester biodegradation should have resulted in an increasingly lower proportion of P-esters with depth in the water column; this is contrary to what was observed. From the information presented, one is led to conclude: (1) both compound classes are ubiquitous in the marine environment including the South Pacific abyss, (2) both P-esters and phosphonates have similar, probably slow, turnover times in the open ocean, and (3) both compound classes are used in proportion to their ambient concentrations. What is still unresolved from the extant data set is the primary source of the high phosphonate to P-ester DOP at all water depths.

Kolowith (née Clark) et al. (2001) extended their South Pacific Ocean HMW-DOP/^{31}P-NMR measurement program to a station in the North Atlantic subtropical gyre (32°N, 64°W), three stations in the North Pacific Ocean (10°N, 140°W; 18°N, 134°W; 22°N, 130°W) and two stations in the North Sea. The cumulative result from all 16 individual samples shows a remarkable uniformity in composition regardless of water depth or distance from shore; P-esters and phosphonates are present in similar proportions in the HMW-DOP throughout the world ocean (Kolowith et al., 2001). For selected sampling sites simultaneous collections of PP (0.1-60 μm fraction) were also obtained, and these were subjected to the same ^{31}P-NMR analysis. Only P-ester compounds were detected in the particulate matter concentrates re-emphasizing an apparent, relative enrichment of phosphonates in the DOP pools (Kolowith et al., 2001).

More recently, a novel coupled reverse osmosis–electrodialysis (RO/ED) method has been employed for the isolation and partial purification of marine DOM (Gurtler et al., 2008; Koprivnjak et al., 2009; Mao et al., 2012; Vetter et al., 2007). Young and Ingall (2010) used RO/ED to recover both LMW and HMW-DOM and ^{31}P-NMR to characterize the concentrated materials. They employed this method to a variety of coastal and offshore environments. In general, their results indicated a greater proportion of poly-P and P-ester and a lower proportion of phosphonates than were observed using tangential flow ultrafiltration. Although the results are not directly comparable, the authors suggested that poly-P may be enriched in the LMW- and phosphonates in the HMW-DOP fractions.

Finally, Sannigrahi et al. (2006) employed ^{13}C and ^{31}P NMR spectroscopy to characterize ultrafiltered DOM (UDOM) and suspended POM pools from the surface to the deep abyss at Station ALOHA in the North Pacfic Subtropical Gyre. The average molar C:P and N:P ratios of the UDOM were 211:1 and 13.9:1, respectively. The stoichiometry of UDOM was relatively constant with depth even though the absolute concentration of DOP decreased by approximately 70% from the surface to 4000 m (Sannigrahi et al., 2006). Because the C:P ratios of total DOM are up to an order of magnitude higher (Thomson-Bulldis and Karl, 1998), the ultrafiltered fraction (>1000 Da) must be significantly enriched in P relative to the LMW fraction. Significant compositional differences between UDOM and POM indicated that UDOM probably does not originate from POM, and that aggregation of UDOM probably does not produce a significant fraction of POM (Sannigrahi et al., 2006). Furthermore, a large fraction of the P in UDOM appears to be associated with carbohydrates, amino acids, and lipids, and the P functional groups in UDOM among the various molecular classes are relatively constant throughout the entire water column (Sannigrahi et al., 2006).

D DOP Pool Characterization by Partial Photochemical Oxidation

Karl and Yanagi (1997) used continuous-flow UV photo-decomposition to provide a partial characterization of SNP in the subtropical North Pacific Ocean. TDP was reproducibly subdivided into three chemically distinct pools: SRP (presumably dominated by Pi), UV-labile SNP (P$_{UV-L}$; containing primarily monophosphate esters) and UV-stable SNP (P$_{UV-S}$; containing primarily nucleotide di- and triphosphates, nucleic acids, and other compounds that are resistant to the low intensity UV treatment that was developed for the purpose of organic-P pool characterization). Field application of these procedures to samples collected at Station ALOHA (22°45′N, 158°W) during the period September 1991 to March 1992 revealed the presence of all three operationally defined pools with an upper water column (0-100 m) average of 23% SRP, 26% P$_{UV-S}$ and 51% P$_{UV-L}$ (Karl and Yanagi, 1997). However, the P$_{UV-S}$ pool did vary nearly three-fold (4.38 mmol P m^{-2} in December to 11.07 mmol P m^{-2} in March; Figure 5.20), suggesting variable production or consumption rates. Depth profiles also revealed near-surface enrichments in the P$_{UV-L}$ pool suggesting a direct coupling with photosynthetic processes (Karl and Yanagi, 1997).

E Direct Measurement of DOP Compounds

Despite recent progress in TDP pool characterization, individual compound analyses are difficult and therefore rare. Only a few organic-P compounds or compound classes have been measured in seawater. These individual compounds probably have multiple sources and sinks, and variable residence times in the marine environment. For example, some compounds are probably excreted from living cells (e.g., vitamins and c-AMP), while others are most likely produced only following cell death or lysis (e.g.,

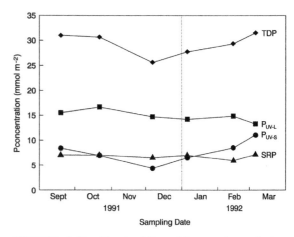

FIGURE 5.20 Dissolved P pool inventories at Station ALOHA (22°45′N, 158°W) for the period Sep. 1991-Mar. 1992. Data shown are the 0-100 m depth-integrated P concentrations (mmol m^{-2}) for each of the pools identified: SRP (▲), P$_{UV-L}$ (■), P$_{UV-S}$ (●) and TDP (◆); see text for operational definitions. *Redrawn from Karl and Yanagi (1997).*

nucleic acids). Unfortunately, no comprehensive study has been conducted to attempt a P mass balance for a given water sample, but it appears that no more than half, and probably less, of the marine DOP pool has been chemically characterized. This uncertainty in SNP pool composition is a major impediment in our attempts to quantify P fluxes in the marine environment.

1 Nucleic Acids

None of the five major nucleic acid bases contain P; only the nucleotide derivatives and nucleic acid polymers thereof are part of the DOP pool (Table 5.1). Pioneering research efforts to measure DNA in the sea began in the late 1960s with the quantitative laboratory and field studies of Holm-Hansen and colleagues (Holm-Hansen, 1969; Holm-Hansen et al., 1968). During these initial investigations, it was established that a large proportion of particulate DNA was associated with nonliving organic matter (Holm-Hansen, 1969). Subsequent studies confirmed the presence of a large pool of detrital DNA, both in particulate (Karl and Winn, 1984) and in dissolved fractions (DeFlaun et al., 1986). Paul et al. (1990) have also detected the intact gene for the enzyme ribulosebisphosphate carboxylase/oxygenase (rbcL) as part of the dissolved DNA (D-DNA) pool indicating that phytoplankton must be considered a probable source for extracellular DNA. Dissolved RNA (D-RNA) has also been detected in seawater (Karl and Bailiff, 1989; Sakano and Kamatani, 1992). Compared to the volume of research focused on D-DNA very little is presently known about D-RNA, even though it is sometimes the larger of the two pools in seawater and in cells. If viral lysis is a major control on bacterial and phytoplankton populations in the marine environment (Suttle, 2007), and there is still active debate on this matter, then the production of dissolved nucleic acids in seawater may ultimately be controlled by the viral lytic cycle.

The measurement of dissolved nucleic acids requires isolation either by adsorption onto barium sulfate (Pillai and Ganguly, 1972) or hydroxyapatite (Hicks and Riley, 1980), or by precipitation using ethanol (DeFlaun et al., 1986), cetyltrimethylammonium bromide (Karl and Bailiff, 1989), or polyethylene glycol (Maruyama et al., 1993), followed by colorimetric or fluorometric dye detection. Depending on the fluorescent dye selected, either single-stranded DNA or single- plus double-stranded DNA is measured (Sakano and Kamatani, 1992). Some DNA may be bound to histone or similar protein, and in this state may be inaccessible to nucleases or to the specific dyes commonly used to quantify nucleic acids. Dissolved nucleic acid can also be measured, indirectly, by high-performance liquid chromatography (HPLC) analysis of the free nucleic acid bases released following polymer hydrolysis (Breter et al., 1977). There are unique advantages and disadvantages of each method (Siuda and Güde, 1996), but a detailed discussion is beyond the scope of this review.

Research conducted over the past decade has indicated that extracellular DNA has at least three forms: (1) naked, free DNA, (2) DNase resistant naked DNA possibly adsorbed onto small particles or contained within colloids, and (3) protein-DNA complexes, perhaps virus particles. Though less carefully described, D-RNA should have a similar distribution spectrum. Initially, certain researchers thought that much of the dissolved nucleic acid was essentially virus particles (see Wommack and Colwell, 2000 for a historical account); this important controversy cannot be resolved here. Nevertheless, at least three lines of independent evidence argue against this: (1) the co-occurrence of large concentrations of D-RNA (RNA-containing marine viruses are rare; Steward et al., 1992), (2) the broad D-DNA molecular weight spectrum (<0.1 to >36 k base pairs; DeFlaun et al., 1987), and (3) direct measurements of virus particles and quantitative estimations of their potential contributions to the D-DNA pool. Virus particles, if present, would contribute to the "non-DNase digestible D-DNA" that has been measured for many marine habitats (e.g., Maruyama et al., 1993).

During an extensive, time-series investigation in the North Adriatic Sea, Weinbauer et al. (1993) reported that virus particles averaged 17.1% (range, 0.7-88.3%) of the measured D-DNA. Using the method of vortex flow filtration, Paul et al. (1991) demonstrated that viral DNA averaged only 3.7% (range 0.9-12.3%) of the D-DNA for a variety of aquatic habitats of different trophic states. Jiang and Paul (1995) used differential centrifugation to separate seawater D-DNA into truly soluble and bound (viral particles, colloids, adsorbed D-DNA) forms; D-DNA pool averaged 50% soluble, 8-15% viral, and 35-42% other bound D-DNA. Kingdom probing of the isolated D-DNA using 16S rRNA-targeted oligonucleotide probes (universal, eubacterial, and eukaryotic) indicated that D-DNA was a complex domain mixture. However, these results also confirmed a relatively low viral particle contribution to D-DNA in the variety of marine habitats investigated (Jiang and Paul, 1995). More recently, Brum et al. (2004) devised a method to distinguish between free, enzymatically hydrolysable DNA (ehD-DNA), DNA within viruses, and uncharacterized, bound DNA. Application of this method to water samples collected at Station ALOHA demonstrated that ~50% of the total D-DNA was ehD-DNA and that this pool turned over 3-10 times faster than DNA within viruses (Brum, 2005). The probable fate of free virus particles is to be consumed by protozoan grazers, degraded by cell-associated or free enzymes, or adsorbed onto sinking particles. Consequently, virus particles are like all other nonliving organic-P pools in the sea from an ecological perspective.

The distribution of D-DNA follows the general pattern of DOM in the marine environment; namely, highest concentrations in coastal waters, decreasing with distance from shore and with depth in the water column (DeFlaun et al., 1987; Figure 5.21). For samples collected at an oligotrophic North Pacific station, Karl and Bailiff (1989) reported higher concentrations of D-DNA than particulate DNA (P-DNA), and D-RNA to D-DNA ratios ranging from 3 to 10, similar to particulate RNA:DNA ratios found in growing microorganisms. Although no DOP data were presented, the P content of the D-RNA plus D-DNA fractions could have accounted for approximately 30-40 nM DOP in the euphotic zone of their oligotrophic North Pacific station; DOP in this habitat is typically 250-400 nM (Orrett and Karl, 1987). For a series of stations in the English Channel, Hicks and Riley (1980) reported that P contained in the dissolved nucleic acid fraction (RNA plus DNA) accounted for 27-49% of the total DOP. A similar estimation for samples collected in Tokyo and Sagami Bays yielded a mean dissolved nucleic acid-P of 12.9% of the total DOP (Sakano and Kamatani, 1992). The latter authors also documented significant temporal changes in the concentrations of both D-RNA and D-DNA and especially in the D-RNA/D-DNA ratio (Figure 5.22). In another study of D-DNA dynamics in Tampa Bay,

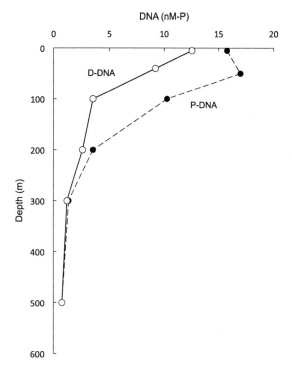

FIGURE 5.21 Vertical distributions of D-DNA (○, solid line) and P-DNA (●, dashed line) for samples collected in the Gulf of Mexico at 26°30′N, 85°W. *Redrawn with permission from DeFlaun et al. (1987).* DNA-P was calculated from the reported concentrations (µg l−1), assuming a P content of 9.5% by weight.

Florida both seasonal and diel concentration variations were observed (Jiang and Paul, 1994; Paul et al., 1988). The mechanism for the significant diel periodicity was not identified, and it is likely to be a result of the balance between production and utilization processes. D-DNA-P averaged 6.6% of the total DOP, but because D-RNA was not measured, this value should be considered a lower constraint on the contribution of dissolved nucleic acids to marine DOP.

A significant research effort has been invested to ascertain extracellular nucleic acid production and utilization processes. Although selected bacterial species produce extracellular DNA during growth, no comparable experimental data are available for D-RNA production.

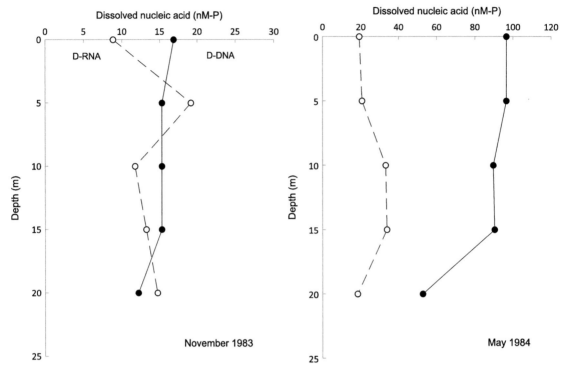

FIGURE 5.22   Vertical distributions of D-DNA (●, solid line) and D-RNA (○, dashed line) for samples collected in Tokyo Bay (35°08′N, 139°50′E) during Nov. 1983 (left) and May 1984 (right). *Redrawn from Sakano and Kamatani (1992).* Nucleic acid-P was calculated from the reported RNA/DNA concentrations (µg l−1), assuming P contents of 9.2% and 9.5% by weight, respectively, for RNA and DNA.

Paul et al. (1987) investigated the dynamics of D-DNA (i.e., production and turnover rates) in subtropical coastal and oceanic environments. Daily D-DNA production by heterotrophic bacteria (using ^{3}H-thymidine) was ~5% of the ambient pool. Active consumption of radiolabeled *E. coli* DNA by cell-associated and extracellular nucleases was also demonstrated. The utilization of D-DNA appears to involve hydrolysis by nonspecific cell-associated nucleases, uptake of individual nucleic acid bases and salvage pathway biosynthesis back into cellular nucleic acids. Likewise, viral RNA seeded into filtered-sterilized coastal seawater was stable for approximately 1 month compared to only 2 days in paired unfiltered samples (Tsai et al., 1995), suggesting a rapid turnover of D-RNA.

Novitsky (1986) also examined the degradation of ^{3}H-labeled detrital DNA and RNA in sedimentary marine ecosystems. He found that: (1) RNA was degraded faster than DNA, (2) both nucleic acids were ultimately degraded to the same extent and (3) both dissimilation to ^{3}H$_2$O and assimilation via salvage pathways into new RNA and DNA were evident.

The production of extracellular nucleic acids may also be a manifestation of microzooplankton grazing processes or viral-induced cell lysis. The concentrations of D-DNA and nanoflagellates were found to co-vary at a station in the Adriatic Sea (Turk et al., 1992); quantitative estimates indicated that most of the ingested DNA was subsequently released into the environment. The turnover of D-DNA was greater in P-limited than in N-limited habitats (Lorenz and Wackernagel, 1994), suggesting that DNA, a P-enriched macromolecule, may be an important source of P for microbial assemblages.

The presence of extracellular nucleic acids and their rapid turnover have important ecological implications. First, RNA and DNA are N- and P-enriched components of the seawater DOM pool (Table 5.1), which could provide nutrients for autotrophic and heterotrophic microorganisms. Second, dissolved nucleic acids could provide a supply of purine and pyrimidine bases for nucleic acid biosynthesis. This would spare the cell the energy that would otherwise be required for de novo synthesis of these invaluable precursors. Third, and perhaps most importantly, free DNA could effect genetic transformation under ecologically permissive conditions. Natural transformation by extracellular DNA is a potential mechanism for lateral gene exchange in natural aquatic habitats. Prokaryotes are the only organisms known to actively take up DNA and recombine it into their genomes, a process called natural transformation (Redfield et al., 1997), and may have evolved it as a means for bacteria to adapt to changing environments. The presence of gene-sized DNA fragments in seawater prompted DeFlaun and Paul (1989) to look for natural transformation among marine microbial assemblages and, later, Paul et al. (1991) were the first to document transformation in seawater samples under ambient environmental conditions. The D-DNA pool may, therefore, represent the "community genome" (DeFlaun and Paul, 1989). The ecological implications of horizontal gene transfer are profound; we believe that this is an important contemporary area of research, poised for rapid progress in the next decade.

2 ATP and Related Nucleotides

ATP is a biologically labile but chemically stable compound, with key functions in cellular energetics, metabolism, and biosynthesis. Its inherent chemical stability predicts that there might exist a detectable dissolved ATP (D-ATP) pool in seawater that would exist as a transient between ATP production and utilization. Azam and Hodson (1977) were the first to report the presence of D-ATP in seawater and since that time it has been detected in all marine and freshwater environments where measurements have been attempted (e.g., Figure 5.23, left).

The quantitative determination of D-ATP requires isolation and partial purification (e.g., H_2SO_4 extraction, adsorption onto activated charcoal, desalting, elution into an ammoniacal-ethanol solution and vacuum evaporation; Hodson et al., 1976) prior to detection by the firefly luciferin-luciferase bioluminescence assay. The recovery of D-ATP by this rather labor-intensive procedure is typically around 30% (Karl and Holm-Hansen, 1978; McGrath and Sullivan, 1981) although higher recovery has also been reported (Hodson et al., 1976). Other adenine and non-adenine nucleotides are also co-adsorbed and co-concentrated by the charcoal method. The firefly assay confers a high-sensitivity and substrate specificity. Only ATP and, to a lesser extent, guanosine triphosphate (GTP) and uridine triphosphate react with crude luciferase enzyme preparations to yield light; however, the presence and relative concentrations of these various substrates can be determined by the kinetics of luciferase light emission (Karl, 1978). Björkman and Karl (2001) have devised an alternative D-ATP concentration technique that is based upon coprecipitation with magnesium hydroxide (i.e., the MAGIC technique; Karl and Tien, 1992), followed by firefly bioluminescence. This simple and straightforward technique also routinely yields a high recovery of ATP ($\geq 90\%$) and, if necessary or desired, can yield quantitative recovery. Non-adenine nucleotide triphosphates (e.g., GTP) are also quantitatively coprecipitated. A HPLC technique has also been described (Admiraal and Veldhuis, 1987), but the relatively poor detection limit precludes its use without some preconcentration method. The advantage of HPLC is that all nucleotides and related derivatives can be separated and quantified from a single run.

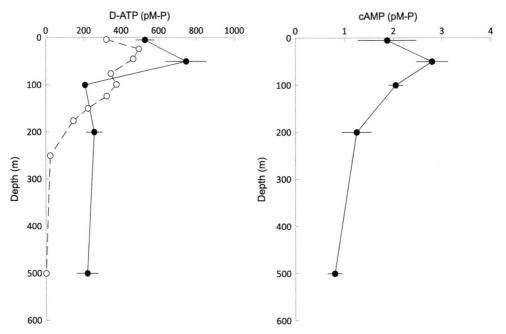

FIGURE 5.23   Dissolved ATP and c-AMP concentration versus depth profiles for samples collected in the North Pacific Ocean. (Left) D-ATP concentrations: (●, solid line) at 32°42′N, 117°23′W in the San Diego Trough *(redrawn with permission from Azam and Hodson, 1977)* and (○, dashed line) at Station ALOHA 22°45′N, 158°W (Björkman and Karl, 2001) versus depth. (Right) c-AMP versus depth at a station 41 km offshore in the Southern California Bight. *Redrawn from Ammerman and Azam (1981).*

D-ATP is typically highest in surface coastal waters, decreasing substantially with distance from shore and with depth (Azam and Hodson, 1977; Hodson et al., 1981; McGrath and Sullivan, 1981; Figure 5.23, left); however, ambient near-surface ocean concentrations are not static. For example, the ambient D-ATP pool in the Southern Ocean varied considerably during and following the spring bloom of phytoplankton, with a peak abundance of >2.5 nmol L^{-1} in Dec-Jan and minimum concentrations (<0.004 nmol L^{-1}) in March (Nawrocki and Karl, 1989). At the height of the bloom, D-ATP was approximately 60% of the corresponding particulate ATP (P-ATP), but by March P-ATP was dominant pool.

D-ATP typically exceeds P-ATP within the subeuphotic zone waters off Southern California (500 m depth), sometimes by an order of magnitude (Azam and Hodson, 1977). However, more recent and, perhaps, more reliable MAGIC-based estimates of subeuphotic zone D-ATP concentrations from the subtropical North Pacific Ocean revealed a much steeper concentration versus depth gradient and much lower deep water concentrations than previously suggested (Björkman and Karl, 2001); nevertheless, D-ATP was still detected at a depth of 500 m (Figure 5.23, left). This presence of D-ATP in deep seawater argues for a local source, a long residence time, or both. While ATP is chemically stable in seawater, with a half-life of approximately 8 years at 21 °C (Hulett, 1970) it is microbiologically labile with ambient pool turnover times of hours to days in near-surface temperate waters (Azam and Hodson, 1977; Hodson et al., 1981; McGrath and Sullivan, 1981). D-ATP is used almost exclusively for biosynthesis and less than 2% of the ^{14}C-ATP tracer was respired

(Azam and Hodson, 1977). Marine microorganisms can and do assimilate D-ATP, in part, for P and, in part, for purine salvage and nucleic acid synthesis. Consequently, D-ATP concentration measurements coupled with exogenous radiolabeled ($^{32}P/^{33}P$, ^3H, ^{14}C) D-ATP uptake measurements can be used to estimate D-ATP and mass flux through marine microbial assemblages (Azam and Hodson, 1977). D-ATP is one of the few organic-P compounds that can be measured at ambient pool concentrations and can be tracked through the microbial food web with specific radioisotopic tracers. In this regard, and especially considering its fundamental role in cellular bioenergetics and biosynthesis, ATP might be considered a "model" compound in studies of the marine P cycle.

Application of the D-ATP MAGIC method to water samples collected from Station ALOHA showed that D-ATP comprised >65% of the total ATP pool (dissolved plus particulate) in summer and ~50% in winter. The calculated net production rates of D-ATP varied from 40 to 150 pM day^{-1}, with a pool turnover of 1-2 days (Björkman and Karl, 2005).

Because ATP is the primary energy currency in all living cells, the presence of D-ATP, often at concentrations exceeding P-ATP, may seem enigmatic. Why would cells leak or otherwise lose such a vital intracellular constituent? The potential mechanisms of D-ATP production, including exudation, inefficient grazing, cell death, or viral lysis, are not well constrained. It is possible that the presence of non-ATP nucleotides, measured along with D-ATP, might help identify the major sources. For example, if the grazing of healthy actively growing cells is the main source for D-ATP in seawater, then one might also expect to observe other nucleotide triphosphate compounds that are vital for biosynthesis (e.g., GTP; in which case there would be a D-GTP:D:ATP of 0.5-1.0 indicative of growing microbial cells; Karl 1978), and a high ratio of ATP relative to total adenylates (sum of ATP + ADP + AMP). If, on the other hand, D-ATP

was a result of viral lysis of debilitated microbial cells or programmed death and autolysis, then non-ATP adenine nucleotides (e.g., ADP and AMP) would be expected to dominate the total extracellular nucleotide pool.

Nawrocki and Karl (1989) compared the pattern of intracellular and extracellular adenine nucleotides (reported as the adenylate energy charge; $EC_A = (ATP + \frac{1}{2} ADP)/(ATP + ADP + AMP)$) for samples collected in Bransfield Strait, Antarctica as a direct test of the D-ATP source hypothesis. During the early phase of the spring bloom (Dec), the particulate and dissolved EC_A ratios were indistinguishable (D-EC_A/P-EC_A = 1.01 ± 0.06) at all stations. These results suggested that grazing or excretion could be the primary source of D-ATP in this habitat. A calculation revealed that a loss of approximately 2-3% of the total ATP production by the cells present in the water column would be needed to account for the measured ATP flux. Björkman and Karl (2001) have reported elevated D-GTP concentrations and high D-GTP/D-ATP ratios (approximately 1:1) in near-surface waters (0-100 m) of the subtropical North Pacific gyre. These field results are consistent with a grazing/excretion source.

3 Cyclic AMP

Adenosine 3',5'-cyclic monophosphate (c-AMP) is a ubiquitous cellular compound that is required for the synthesis of a number of inducible catabolic proteins, bacterial flagellum synthesis, catabolite gene activation and other key functions (Botsford and Harman, 1992). It is found in both prokaryotes and eukaryotes, including marine bacteria, phytoplankton and, probably, Archaea. Intracellular concentrations of c-AMP are carefully regulated by enzymatic synthesis and degradation, as well as by active excretion from the cell.

In 1981, Ammerman and Azam documented the existence of c-AMP in coastal seawater. The dissolved c-AMP was extracted from seawater and concentrated using the $H_2SO_4^-$ charcoal adsorption technique used previously by Hodson et al. (1976) for D-ATP measurements. The isolated c-AMP was quantified by a commercially available radioimmunoassay. At one offshore station in the Southern California Bight, c-AMP was maximal (2-3 pM-P) in the near-surface water, decreasing to <1 pM-P at 500 m (see Figure 5.23, right). In near-shore seawaters, the concentration of dissolved c-AMP was generally higher and varied approximately four-fold with time of day, suggesting a dynamic production and consumption cycle; concentrations peaked at sunset (Ammerman and Azam, 1981).

Although c-AMP is stable in filter-sterilized (0.2 μm) surface seawater, it is rapidly assimilated by marine microorganisms, especially bacteria (Ammerman and Azam, 1981). Uptake of c-AMP is facilitated by a high-specificity, high-affinity (K_m <10 pM) transport system (Ammerman and Azam, 1982). Evidence from well-designed dual labeled (c-AM^{32}P/(2,8-^{3}H)-c-AMP) experiments indicated that c-AMP was assimilated intact and that it remained in the cytoplasm rather than being incorporated into biomolecules. This is fully consistent with its role as a regulatory molecule. More recently, Dunlap et al. (1992) reported growth of the marine luminous bacterium, *Vibrio fischeri*, on c-AMP as a sole source of C, energy, N and P (also see Callahan et al., 1995). Growth was dependent on a high specific activity, narrow substrate specificity c-AMP phosphodiesterase located in the periplasm of the cell. Ammerman and Azam (1982) suggested that uptake and release of c-AMP may control intracellular levels and, hence, metabolism for microbes in the marine environment. In their ecosystem model, starved or otherwise metabolically debilitated bacterial cells would take up dissolved c-AMP from seawater to supplement or replace de novo synthesis of c-AMP from cellular ATP; new catabolic enzymatic pathways would be produced and alternate ambient substrates metabolized (Ammerman and Azam, 1987). To our knowledge, this intriguing model has not yet been evaluated under field conditions.

4 Lipids

Lipids are major constituents of all living cells. As a class of biomolecules, lipids have enormous structural diversity that, in part, reflects phylogenetic diversity. For example, whereas bacteria contain phospholipids containing ester-linked fatty acids, the membrane phospholipids in *Archaea* have ether-linked isoprenoid side chains (Tornabene and Langworthy, 1979). Several categories of phospholipids (PL) have been detected in the marine DOP pool (Suzumura, 2005). Despite their billing, certain PL molecules contain as much N as they do P (Table 5.1), and neither N nor P are in very high relative abundance compared to C (molar C:P ratio is ~40:1).

Phospholipids are ubiquitous in Bacteria and Eukarya and present to a lesser extent in *Archaea*; they function primarily as structural components of the membrane. The PL content of cells can be quite large. According to Dobbs and Findlay (1993), PL-P in marine bacteria can account for approximately 15-20% of total cellular P—despite the relatively low percent P content—but the exact amount will vary with nutrient status and growth rate. The turnover time of phospholipids following cell death is rapid (2-10 days), at least in aerobic sedimentary habitats, but lipid turnover has not been investigated in the more dilute, pelagic ecosystem.

Phospholipids are easily separated from other DOP compounds because of their polar nature; although soluble in organic solvents, they have an associated hydrophilic group at the P-bonded end. Phospholipids can be separated and detected by HPLC or by thin-layer chromatography coupled with flame ionization detection (TLC-FID, Chromarod-Iatroscan system; Parrish and Ackman, 1985). Several different types of phospholipids are present but, most often, "total PL" is reported. However, total PL can be further fractionated by altering hydrolysis conditions into diacyl phospholipids, plasmalogens (lipids containing vinyl ether) and phosphonolipids. The inherent stability of the C-P bond in organic phosphonate compounds has been employed, analytically, to separate phosphate ester-linked lipid compounds from phosphonolipids. Using a stepwise treatment with hydrochloric acid (6 M, 120 °C, 2 h) to cleave phosphodiesters, followed by APase to cleave monophosphate esters, only phosphonolipids are expected to remain (Snyder and Law, 1970).

Phosphonolipids are produced by eukaryotes, but degraded by prokaryotes (Kittredge and Roberts 1969; Kononova and Nesmeyanova, 2002). Certain membrane phosphonolipid derivatives of phosphorylethanolamine (e.g., 2-AEP; Table 5.1), detected in relatively high concentrations in certain protozoans (Kennedy and Thompson, 1970), may increase the stability of the surface membranes by resistance to degradation. This would also predict a longer residence time in DOM and could lead to an accumulation relative to more labile DOP compounds. They may have an important physiological and ecological role, especially under low nutrient conditions.

Liu et al. (1998) used CFF to concentrate the colloidal fraction of DOM (>10 kDa) from phytoplankton cultures and surface seawater collected in Conception Bay, Newfoundland. As expected, colloidal lipids were always present but, on average, lipids were higher in the <10 kDa fraction. Relative to total lipids, the percentage of PL was nearly twice as high in the colloids than in the truly dissolved fraction (Liu et al., 1998). This indicates that colloidal-P may be chemically distinct from truly dissolved compounds.

Suzumura and Ingall (2001) compared the PL content of DOM and POM collected in two contrasting coastal marine environments. Samples collected from both Tokyo Bay and Corpus Christi Bay, had particulate PL that ranged from 31 to 300 nM-P, corresponding to 3 to 14% of the total PP. By contrast, dissolved PL concentrations were low, ranging from 0.7 to 6 nM-P, corresponding to <1% of total DOP. The authors concluded that PL was a minor component of DOP in their study regions. In a subsequent study, Suzumura and Ingall (2004) investigated

the dynamics of "hydrophobic" P (H-P; primarily PL) at several stations in the North Pacific Ocean. In addition to measurements of dissolved and particulate H-P, they also conducted laboratory experiments that tracked changes in H-P during particulate matter remineralization. At Station ALOHA where the most extensive sampling was performed, dissolved H-P ranged from 4 to 10 nM-P and represented approximately <3% in surface waters (0-150 m), increasing to 17.6% at 4000 m (Suzumura and Ingall, 2004). The percentage of H-P in suspended particulate matter showed a similar depth trend. During a 166-day laboratory-based degradation experiment, approximately 83% and 50% of the initial PP (0.30 μM) and DOP (0.49 μM) was degraded respectively, and for both pools the percentage of H-P increased significantly (Suzumura and Ingall, 2004).

Perhaps, the most comprehensive study of PL in the marine environment was that conducted in Bedford Basin, Canada by C. Parrish and colleagues (Parrish, 1986, 1987; Parrish and Wangersky, 1987). The total lipid fraction was extracted with cold dichloromethane, followed by Chromarod-Iatroscan system detection. Repeat analyses of the major classes of both particulate and dissolved lipids leading up to and following the vernal primary production maximum revealed complex dynamics especially for the dissolved lipid classes (Parrish, 1987). Their results for near-surface waters (~5 m) collected during the 1982 spring bloom revealed a phytoplankton community with peak chl a values of >30 μg l^{-1} on Julian day 86; nitrate decreased throughout the bloom period (Figure 5.24). Dissolved PL concentration was high (equivalent to ~50-65 nM-P) prior to the spring bloom, but decreased substantially during the bloom. In contrast, the particulate PL concentration was relatively low (<5 nM-P) before and during the initial stages of the bloom, increasing to ~20 nM-P at the height of the bloom (Figure 5.24). Dissolved PL consistently accounted for 15-20% of the total dissolved lipid content of the water

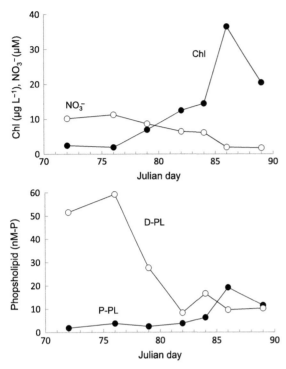

FIGURE 5.24 Phospholipid (PL) dynamics during the initiation of the vernal bloom of phytoplankton in Bedford Basin, Canada during 1982. Shown on the top graph is chlorophyll a ($\mu g\, l^{-1}$; ●) and nitrate concentration (μM; ○) as a function of time shown in Julian day. Shown on the bottom graph are concentrations of dissolved PL-P (nM-P; ○) and particulate PL-P (nM-P; ●). *Redrawn from Parrish (1987)*. PL (μM-P) was estimated from data on PL ($\mu g\, l-1$) assuming that PL is 4% P, by weight.

column. Similar results, notably the decreased concentrations of dissolved PL at the height of the bloom and a large increase in dissolved PL during the demise phase, were also observed at this location in 1984 (Parrish, 1987). It was suggested that phytoplankton in this coastal marine habitat can use PL-Pi as a P source for growth. A post-bloom resupply of dissolved PL, hypothesized to be a result of intense grazing pressure, was also observed (Parrish, 1987).

5 Vitamins

Vitamins are a class of otherwise unrelated organic compounds that are needed in trace

amounts for the normal functioning of all organisms. Thiamine pyro-Pi (the biologically active form of vitamin B$_1$), pyridoxal phosphate (the biologically active form of vitamin B$_6$) and cobalamin (vitamin B$_{12}$), have been detected in seawater where they contribute to the DOP pool. Recently, there has been a renewed interest in "vitamin ecology" that derives from two independent lines of research. The first involved analytical improvements that enabled direct determinations of B vitamins in seawater using C$_{18}$ solid-phase extraction/concentration and quantification by reverse phase HPLC with subpicomolar detection limits for vitamins B$_1$ and B$_{12}$ (Okbamichael and Sañudo-Wilhelmy, 2004, 2005). The second area of research involves physiological studies of novel model systems combined with genome sequencing that supports B vitamin auxotrophy for the most abundant heterotrophic bacterium in the sea (e.g., SAR 11 clade; Giovannoni et al., 2005; Tripp et al., 2008). It is unlikely that vitamins comprise a significant percentage of the total DOP pool, but there is no question that they are important from an ecological perspective, and in some regions may control rates of primary production (Sañudo-Wilhelmy et al., 2012).

Vitamin B$_{12}$, in particular, is an obligate growth factor for many marine phytoplankton, so net growth of these organisms requires vitamin B$_{12}$ uptake from seawater. Vitamin B$_{12}$ is synthesized exclusively by microorganisms, predominantly bacteria, so there may be an obligate vitamin syntrophy within the marine microbial world (Provasoli and Pintner, 1953). Metabolic relationships between algae and heterotrophic bacteria may effectively exchange utilizable organic matter for vitamin B$_{12}$, thereby maintaining a viable population of B$_{12}$-requiring phytoplankton during periods of vitamin deficient growth (Croft et al., 2005; Haines and Guillard, 1974), as might occur in low nutrient, subtropical gyres. Bonnet et al. (2010) have recently reported excretion of vitamin B$_{12}$ by axenic cultures of the marine cyanobacteria *Crocosphaera* and *Synechococcus*. Extracellular release of vitamin B$_{12}$ was much higher for the diazotroph *Crocosphaera*, and was comparable during growth on either NH_4^+ or N_2. Production continued even when the medium was amended with 400 pM vitamin B_{12}, but ceased when the external concentration was increased to 800 pM (Bonnet et al., 2010). This threshold value (>400 pM) exceeds vitamin B_{12} values measured in the Northeast Pacific Ocean by 1-2 orders of magnitude (Sañudo-Wilhelmy et al., 2012) suggesting that these cyanobacteria are probably net vitamin producers under in situ conditions.

Vitamins B_1 and B_{12} were historically measured using bioassay procedures based on the incorporation of ^{14}C-bicarbonate during growth of specific, vitamin-requiring phytoplankton (Carlucci and Silbernagel, 1966a,b). Application of these extremely sensitive (e.g., detection limits for vitamin $B_{12} < 0.05$ ng of total vitamin per liter which equates to 4×10^{-14} moles of P) bioassays revealed low concentrations (sometimes undetectable) in the surface ocean, especially in open ocean oligotrophic ecosystems. Higher concentrations were detected at depth (Carlucci and Silbernagel, 1966c). Menzel and Spaeth (1962) documented a seasonal cycle in vitamin B_{12} concentrations in the Sargasso Sea, with lowest concentrations during the spring-summer period. They also showed that vitamin B_{12} requiring phytoplankton species (mostly diatoms) were absent during periods of undetectable vitamin B_{12} concentration, suggesting a key ecological selection pressure.

Sañudo-Wilhelmy et al. (2012) have reported multiple B vitamin depletion over a large area of the Northeast Pacific Ocean. More importantly, they showed large spatial variability in vitamin concentrations both horizontally and vertically (Figure 5.25). For example, three selected B_{12} versus water column depth profiles showed three distinct patterns: (1) low, but measurable concentrations in the surface and increasing B_{12} concentrations with depth, (2) near-surface ocean B_{12} maximum, with lower concentrations at depth, and (3) undetectable surface ocean B_{12} concentrations (<30 pM) with variable subsurface (>100 m) concentrations (Figure 5.25, Top). These complex patterns were hypothesized to be the result of phytoplankton community structure and productivity as well as water mass characteristics (Sañudo-Wilhelmy et al., 2012). In certain regions, B_{12} deficiency may limit solar energy capture in the sea.

6 Inorganic Poly-P*i* and Pyro-P*i*

Inorganic poly-P*i* are linear polymers of P*i*, joined by phosphoanhydride bonds identical to high-energy P-bonds of ATP (Kulaev and Vagabov, 1983). The poly-P*i* chain length can vary considerably from $n = 2$ (pyro-P*i*) to $n \geq 1000$. This diversity of form has important implications both for their analysis in seawater and for their potential ecological role. While these important cell-derived compounds are obviously not part of the DOP pool per se, they do contribute to the TDP pool and, therefore, to the usual operational definition of "DOP" (i.e., [TDP]-[SRP]). Furthermore, both pyro-P*i* and poly-P*i* can be formed by enzymatic or photolytic hydrolysis of selected organic-P compounds, so as a courtesy we will include them here as probable constituents of the marine DOP pool.

Isolation and partial purification of poly-P*i* is difficult because of the variable molecular weight (chain length) and variable chemical properties. For example, laboratory studies have identified at least three poly-P*i* fractions: (1) acid soluble, (2) base soluble, and (3) other (Clark et al., 1988). Quantitative estimation is usually by difference measurement based on the relative photochemical stability of poly-P*i*. This difference approach is inherently unreliable, especially when total poly-P*i* is $\leq 20\%$ of TDP. Furthermore, it is impossible to separate inorganic poly-P*i* from organic poly-P*i* by these methods.

Solórzano and Strickland (1968) detected poly-P*i* in both coastal and open ocean ecosystems. They employed a sequential UV photooxidation followed by hydrochloric acid hydrolysis.

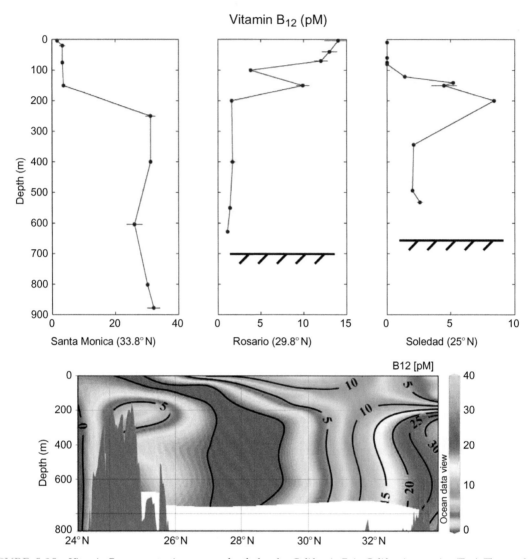

FIGURE 5.25 Vitamin B$_{12}$ concentration versus depth for the California-Baja California margin. (Top) Three selected depth profiles from left: Santa Monica 33.84°N, 119.03°W, Rosario 29.80°N, 116.08°W, and Soledad 25.22°N, 112.71°W. (Bottom) Vitamin B$_{12}$ contour plot for the entire region showing the locations of the three selected stations.

The UV-stable, acid-labile Pi was taken as a measure of the sum of organic plus inorganic poly-Pi. Concentrations in surface waters ranged from 200 nM-P (25% of SNP) in a red-tide off Newport Beach to undetectable levels (<50 nM-P) in the East-Central Pacific Ocean.

Isotope dilution has also been proposed as a method for pyro-Pi and poly-Pi estimation in samples containing other potentially interfering P compounds (Quimby et al., 1954) but, to our knowledge, this method has not been applied to seawater samples. More recently, poly-Pi has been detected and quantified using a firefly luciferase-based assay for ATP following incubation with exogenous additions of the enzyme polyphosphate kinase and the substrate adenosine 5'-diphosphate (ADP) (Ault-Riché et al., 1998). To our knowledge this sensitive and specific assay for inorganic poly-Pi has not yet been applied to seawater.

Poly-Pi is ubiquitous in every cell in nature where it participates in multiple, key functions (Kornberg et al., 1999). Pyro-Pi, like poly-Pi and the γ- and β-P groups of ATP, is a "high-energy" P compound that is important in cellular bioenergetics. Some researchers believe that pyro-Pi was the evolutionary precursor to ATP (e.g., Lipmann, 1965). Extracellular pyro-Pi can also serve as the sole source of energy for the growth of certain aquatic, anaerobic bacteria (Liu et al., 1982; Varma et al., 1983). Furthermore, the poly-Pi polymer has a clear osmotic advantage over free Pi (i.e., it is osmotically inert), and may also act as a metal-ion chelator or high-capacity buffering system (Kornberg et al., 1999). Perhaps, one of the most intriguing potential roles for poly-Pi in marine microorganisms is its role in enabling competence for bacterial transformation by facilitating the cross-membrane transport of extracellular DNA (Reusch and Sadoff, 1988), and perhaps other small and large DOP compounds. Poly-Pi, the "molecular fossil," appears to have recently come back to life (Kornberg and Fraley, 2000).

The most important function of poly-Pi, however, appears to be regulation of gene expression, especially as it relates to nutritional stresses including P- and N-limitation. When microorganisms are placed into a Pi-deficient medium they are induced to take up large amounts of Pi when it becomes available again. The Pi is converted into poly-Pi and stored until needed for macromolecular biosynthesis (Jensen and Sicko, 1974). In *E. coli*, poly-Pi transients, accounting for changes of ≥100-fold in periods of minutes to hours, have been observed during shifts from nutrient-replete to minimal growth media (Ault-

Riché et al., 1998). Cells deficient in poly-Pi are noncompetitive during periods of nutritional stress, whether acute or chronic (Kornberg et al., 1999). Rapid uptake and storage of Pi would be a key survival strategy for growth in a fluctuating nutrient environment. There is also an intracellular transient accumulation of poly-Pi at the onset of Pi depletion. This process, termed "poly-Pi overplus" (Voelz et al., 1966), is fundamentally distinct from "luxury uptake" of Pi which also results in poly-Pi formation and storage but does not require Pi depletion. Of the two processes, the poly-Pi overplus phenomenon is probably most important in the marine environment and especially so in the open ocean. For example, if near-surface ocean microbes are exposed to alternating periods of high and low Pi, as they probably are (Karl, 1999), then this could lead to a poly-Pi overplus response and intracellular sequestration of P as poly-Pi. Consequently, the current view that poly-Pi would not be expected to exist in low-nutrient open-ocean seawaters may be incorrect; a focused research effort on this topic should be undertaken.

F Biologically Available P

Regardless of the rigor and precision with which P-containing compound pools are measured, the ecological significance of these analytical determinations will be incomplete until reliable estimates of the BAP pool are routinely available. In addition to Pi, which is generally the preferred substrate for microorganisms, the P contained in a variety of polymeric inorganic compounds, in monomeric and polymeric organic compounds and in selected P-containing minerals is available to some or all microorganisms. However, some microorganisms may prefer ester-linked P sources to free orthophosphate (Cotner and Wetzel, 1992; Tarapchak and Moll, 1990). Nevertheless, the bioavailability of most organic-P pools depends on ambient Pi pool concentrations and on the expression of specific enzymes that control transport, salvage, and substrate hydrolysis. Because many of these enzymes are induced by low Pi, bioavailability may be a variable, time- and habitat condition-dependent parameter, rather than an easily predicted or measured metric. Assessment of the BAP may also depend on the time scale of consideration; for example, substrates that appear to be recalcitrant on short-time scales (e.g., <1 day) may fuel longer-term (annual to decadal) microbial metabolism, especially as the ambient Pi pool becomes limiting.

Theoretically, only a well-designed bioassay procedure can ever be expected to provide reliable data on bioavailable P; chemical assay systems, regardless of design, cannot replicate the metabolism of a living cell. One of the first P bioassay systems, described by Stewart et al. (1970), used P-starved cells of the N_2-fixing cyanobacterium *Anabaena*, and a response based on acetylene reduction (N_2 fixation) to detect bioavailable P in freshwater ecosystems. Björkman and Karl (1994) have assessed the short-term bioavailability of 7 organic and 2 inorganic P compounds to natural marine microbial communities by measuring the sparing effect on the uptake of exogenous [32]Pi relative to unamended (negative control) and Pi-supplemented (positive control) samples, following a procedure described by Berman (1988) for freshwater habitats. These short-term bioavailability surveys indicated that there were marked differences among the compounds tested, but confirmed a strong metabolic preference for Pi. Of the compounds tested, nucleotides (ATP, UDP, GDP) were the most available organic-P compounds, followed by poly-Pi and Gly-3-P. Hexose-P compounds had "bioavailability factors" that were consistently <5% those of equimolar additions of Pi (Björkman and Karl, 1994).

Karl and Bossard (1985) estimated BAP using a γ-P labeling technique that they developed for ATP pool turnover. Because the intracellular ATP pool turns over rapidly, the specific radioactivity in the γ-P position of the P-ATP pool will reach equilibrium with the available pool of extracellular P (i.e., the BAP pool). Bossard and Karl (1986)

successfully applied this ^{32}P labeling method to a variety of marine and freshwater habitats and reported that BAP was greater than or nearly equal to measured SRP. More recently, Björkman et al. (2000) used the γ-P-ATP labeling method to estimate the BAP pool at two stations in the oligotrophic North Pacific Ocean. In their study BAP was generally greater than Pi or SRP but, as in Bossard and Karl (1986), less than the TDP pool. The BAP:TDP molar ratio averaged 0.24, and generally exceeded the Pi:TDP and SRP:TDP ratios, sometimes by more than a factor of two. These results suggest that some fraction of the ambient DOP has a bioavailability factor equal to, or greater than, Pi in these open ocean habitats.

At Station ALOHA, Björkman and Karl (2003) estimated the BAP pool using the ATP pool isotope dilution method of Karl and Bossard (1985). By this analysis, the BAP pool in the upper 0-175 m of the water column consistently exceeded the measured Pi pool by factors of 1.4 to 2.8 and amounted to approximately 7-15% of the DOP pool (Björkman and Karl, 2003).

Nausch and Nausch (2006) estimated the fraction of DOP that was bioavailable by adding C and N to natural assemblages of Baltic Sea microorganisms to induce Pi depletion and DOP utilization. At their site BAP ranged from 55-65% of the total DOP in May, when DOP concentrations were at an annual maximum (0.3-0.5 μM), to much lower proportions in late summer when DOP concentrations were typically ~0.2 μM (Nausch and Nausch, 2006). They hypothesized the existence of a "refractory" DOP pool (0.14 to 0.21 μM) that was present throughout the year. However, it is possible that even the "refractory" DOP pool could be utilized over longer periods of time (their experiments lasted 4-6 days), or following the selection for substrate responsive microorganisms (Karl, 2007a).

G DOP: The "Majority" View

Although no comprehensive characterization has been attempted using all possible DOP compound or compound class-specific detection capabilities simultaneously, it is probable that less than 50% of the total DOP, and perhaps much less, can be identified. That leaves greater than half of the DOP uncharacterized by existing protocols. What is this "majority DOP?"

Seawater DOM is marine in origin (Hedges et al., 2000; Lee and Wakeham, 1988); however, it appears to be diagenetically altered and chemically distinct from the source materials. This may account, in large part, for our inability to characterize it. Hedges et al. (2000) have reviewed the status of what they term the molecularly uncharacterized component of nonliving organic matter, or simply "MUC." It is generally believed that partial degradation leads to the production of refractory materials that selectively accumulate in seawater. The relatively old mean ^{14}C age of the DOM pool (of which DOP is a subcomponent) even in the surface ocean implies that much of it may be turned over on time scales of centuries to millennia. However, if we assume that the deep water RDOM (~40 μM C, mean age of 6000 years) is also present in surface waters where it represents about 35-50% of the total pool, then the remainder of the surface pool must be zero age (i.e., modern) to reconcile the measured mean surface DOM age estimate of 1500 years. These age estimates may or may not also apply to the P-containing DOM pool because these fractions have never been isolated for direct age determination. The fact is that we simply do not know how the age of the DOM pool scales on mass, class, or other characteristics. Information regarding the age of specific components of the DOP pool would, in our view, be invaluable from an ecological perspective.

There are two general processes potentially leading to the formation of MUC (Hedges et al., 2000): (1) heteropolycondensation reactions whereby small reactive organic intermediates combine to from larger and more complex organic compounds (also known as humification) and (2) differential/partial biodegradation of HMW organic matter. The starting materials for the latter process would likely be complex compounds like bacterial peptidoglycan-rich cell walls, DNA-histones, or viral particles. While there appears to be an ongoing paradigm shift from process (1) to process (2), few data currently exist for DOP compounds.

Humification, the process whereby organic materials that are relatively resistant to microbial and free-enzyme mineralization are polymerized, condensed, and otherwise reworked into a diverse suite of compounds called "humic substances," is widespread in nature (Rashid, 1985). Based on their relative solubilities in acidic or basic solutions, humic substances are further subdivided into humic acids (HA, insoluble in acid, soluble in base), fulvic acids (FA, acid and base soluble), and humins (insoluble in both acid and base). Marine humic substances also have variable molecular weights, ranging from LMW to colloidal-sized particles, and are isotopically and chemically unique from soil humus. They can adsorb, complex, and chelate most trace metals, and probably adsorb Pi as well. This would be in addition to the actual P content of the compounds themselves.

Dispersed HA and FA compounds may account for some measurable fraction of the marine DOP pool, but the amount remains uncertain. According to Nissenbaum (1979), sedimentary marine HA and FA fractions contain 0.1-0.2% and 0.4-0.8% P by weight, respectively, with corresponding bulk C:P ratios ranging from 300-400 in HA to <100 in FA (Table 5.1). The P-enriched FA fraction is thought to be more recent. Nissenbaum (1979) suggests the following diagentic scheme: plankton → DOM → FA → HA → kerogen. The fact that FA compounds are enriched in P (relative to C) compared to the average C and P contents of plankton (106C:1P) seems enigmatic. This P-enrichment of FA is especially interesting considering the fact that the mean C:P molar ratio of DOM in surface and deep seawater indicates a P depletion (C:P$_{surface}$ ~200-400, C:P$_{deep}$ >800) relative to living biomass. Even many labile DOP compounds have C:P ratios larger than the quoted value of <100 for marine FA (Nissenbaum, 1979). From these analyses we must conclude that the FA fraction of seawater is a probable, but ill-constrained, P trap. Mopper and Schultz (1993) have identified a "humic-type" signature as one of two main fluorescence components of the DOM pool at Station ALOHA. While quantitative information on HA/FA abundance is lacking, it is clear that humic substances are present throughout the water column in the global ocean. More careful field work certainly needs to be conducted to substantiate this claim.

Brophy and Carlson (1989) have documented the microbiological formation of biorefractory HMW-DOM from bioreactive LMW precursors, and suggested that this might be a pathway for local accumulations of organic matter. In similar experiments, labile protein added to seawater and allowed to "age" for 40 days was degraded four-fold more slowly than the non-aged protein (Keil and Kirchman, 1994). The chemical modification was determined to result from abiotic, organic-organic interaction, which the authors conclude may be a necessary first step in the formation of biorefractory compounds.

It is probable that a fraction of the marine DOP pool is also derived from cell wall-associated compound classes. McCarthy et al. (1997) suggested that degradation-resistant biomolecules, rather than abiotically-produced heterocyclic geopolymers, dominate the marine DOM pool. This refractory DOM pool appears to contain a large percentage of bacterial peptidoglycan (McCarthy et al., 1998). Lipopolysaccharide (LPS) is a component of the cell wall of gram-negative bacteria including some cyanobacteria (Mayer et al., 1985; Wilkinson, 1996). It is localized in the center portion of the outer membrane where it contributes to cell integrity. Although LPS is rapidly degraded following cell death and lysis, dissolved LPS (D-LPS) has been detected in aquatic environments and may represent >90% of the total LPS in certain habitats

(Karl and Dobbs, 1998). Spontaneous release of LPS has been documented for many bacteria and appears to occur during normal growth (Cadieux et al., 1983). D-LPS may also accumulate as a consequence of protozoan grazing activities or viral lysis. The MCP model is consistent with these observations of a biologically produced R-DOM pool.

Phosphorylation of LPS is a probable mechanism of metabolic regulation in the sea (Ray et al., 1994). A striking feature is the presence of a high level of P (2-6%, by weight) that exists as either ethanolamine-P, ethanolamine-pyro-Pi, protein-P or as Pi (Mühlradt et al., 1977). A majority of marine bacteria are gram-negative and thus may contain LPS-P. Currently it is not known how much P is contained in the D-LPS fraction in seawater (the assay for D-LPS does not measure P content directly), and the P content of individual forms of LPS varies considerably. As discussed above, LPS phosphorylation may serve as a regulatory mechanism in bacteria, so the LPS-P content of seawater could be variable in both space and time. We hypothesize that both LPS and the gram-positive bacterial cell wall analog compounds teichoic and techuronic acids (which contain Gly-3-P and ribotol-P polymers) may be present as components of the DOP pool because bacterial cell walls are such a significant percentage of total bacterial cell biomass. This prediction, however, awaits direct experimental evaluation.

The coordination and regulation of cellular metabolism are key ecological processes. Bacteria, for example, impose regulatory mechanisms on most catabolic and biosynthetic processes to ensure that needs are met, but not exceeded (Saier et al., 1995). Large portions of the bacterial and archaeal genomes appear to be used for regulation, so it is conceivable that regulatory molecules are constantly produced and used by marine microorganisms. These compounds could, in theory, be selectively retained in the DOM pool following exudation or cell death.

The phosphorylation and dephosphorylation of proteins is an important metabolic process in living organisms that may be a form of signal transduction and gene regulation (Stock et al., 1989). Both exogenous and endogenous cues can elicit specific adaptive responses that optimize metabolism and maximize survival. The low-temperature dependence of the protein phosphorylation process in certain bacteria (Ray et al., 1994) provides a potential mechanism for a geographical variability in DOP production. The phosphoenolpyruvate:sugar phosphotransferase system (PTS) in bacteria is one of the best characterized examples of phosphoprotein regulation (Saier, 1989); numerous other examples also exist (Bourret et al., 1991). The PTS is used for sugar transport, phosphorylation, and chemoreception; it has been detected in marine bacteria (Hodson and Azam, 1979).

Synthesis of polysaccharides (including LPS) from monosaccharides generally relies on a biosynthetic pathway that involves a nucleoside-sugar intermediate (e.g., UDP-glucose, CDP-fructose). To our knowledge these phosphorylated intermediates have not been reported in seawater, though they undoubtedly exist largely for the same reasons as dissolved nucleotides. In addition, a family of nucleotide derivatives, including guanosine-3′-diphosphate-5′-diphosphate (ppGpp) has been implicated in the growth rate-dependent regulation of ribosomal RNA and protein synthesis since its discovery more than four decades ago (Cashel and Gallant, 1969). These regulatory processes have unknown but probably key ecological roles and may be present in variable intracellular and extracellular concentrations in seawater. To our knowledge, they have not yet been detected in seawater.

A novel group of phosphoinositol derivatives has recently been reported in very high concentrations (50-350 mM) in the cytoplasm of certain hyperthermophilic bacteria and *Archaea* (Martins et al., 1997; Ramakrishnan et al., 1997; Scholz et al., 1992). Initially, it was thought that they served as thermoprotectors, but it now appears that they have other, yet undiscovered, functions—perhaps osmolytic (Chen and Roberts, 1998). Inositols are stable to hot hydrochloric acid hydrolysis (6 M, 110 °C, 48 h), so it is possible that they have not been quantitatively measured by use of wet chemical oxidation techniques. Inositol isomers have also been detected in marine sediments (Suzumura and Kamatani, 1995; White and Miller, 1976), but not yet in the water column. Another phosphorylated compound, 2,3-diphosphoglycerate, has been shown to be a novel phosphagen in selected *Archaea* (Kanodia and Roberts, 1983; Matussek et al., 1998). The recent discovery of high concentrations of planktonic *Archaea* in seawater (DeLong, 1992; DeLong et al., 1994; Fuhrman et al., 1992; Karner et al., 2001) demands that we also consider them as a potential source for marine DOP.

Finally, several organic-P compounds have commercial applications; for example, triphenyl phosphate is a plasticizer and tricresyl phosphate is an additive in gasoline. In addition, organophosphorus pesticides (parathion, malathion, ethion, thimet) are in use worldwide. While most of these are eventually degraded, some compounds persist longer than others, especially when present in low concentrations. Atmospheric, riverine, or outfall introduction into the sea may arrest the degradation processes via substrate dilution. This could lead to the selective accumulation of biochemically refractory or otherwise "unwanted" DOP compounds, in spite of low relative production or delivery rates.

In summary, a complete molecular characterization of marine DOP has not yet been possible. This situation may be due to a lack of effort as much as it is due to a lack of methods. When one considers the combined probable contribution from dissolved nucleic acids, dissolved phospholipids, dissolved proteins, LMW intermediates (including nucleotides, regulatory molecules, and osmolytes) and HA and FA fractions, budget reconciliation may be possible in the near future.

IX DOP PRODUCTION, UTILIZATION, AND REMINERALIZATION

DOP production in the sea typically begins with biological Pi uptake and incorporation into one of the many intracellular P-containing compounds in the cell. For this reason, the environmental controls on rates of Pi uptake and pathways of Pi assimilation into organic-P compounds are critical to our understanding of DOP dynamics. Because Pi concentrations in the surface ocean rarely exceed 3 μM, one might anticipate the universal presence of the high-specificity, high-sensitivity Pi transport system for many microorganisms in the sea.

Most of the Pi that is assimilated by microorganisms is locally regenerated back to Pi to sustain in situ production processes. In the open ocean, typically less than 10% of the total P that is assimilated is exported from the system. As for Pi uptake, accurate Pi regeneration or DOP production estimation requires information on the ambient pool of BAP in the habitat under investigation. When combined with independent methods of DOP pool characterization, the radiotracer experiments described below can also be used to quantify the fluxes of specific DOP compounds or compound classes (e.g., nucleotides). A "pulse-chase" experimental design (i.e., pre-incubation with ^{32}Pi or ^{33}Pi radiotracer followed by the addition of a ten-fold excess of ^{31}Pi) could be employed to follow both net DOP production, during the labeling phase, and DO^{32}P turnover in the post-chase treatments. Dual labeled ^{33}Pi/^{32}Pi pulsed experiments can also be conducted to trace DOP pool dynamics and to estimate DOP pool residence times by models that take advantage of the differential half-lives of these independent radiotracers. Finally, the addition of exogenous, ^{32}P (or ^{33}P)-radiolabeled DOP compounds (e.g., Glu-6-P, ATP, Gly-3-P) can provide information on the turnover time of individual pools and potentially (if combined with direct measurements of ambient pool concentra-tions) on mass fluxes (production and utilization rates) through selected organic-P pools. Size fractionation of uptake, the use of taxon-specific fluorescent probes combined with microautoradiography and flow cytometric cell sorting can be used to identify the substrate-reactive groups. However, our present inability to measure most DOP compounds (a notable exception is ATP, as discussed above); limits the field application of these experimental approaches.

A DOP Production and Remineralization

The use of ^{32}P (or ^{33}P) as a tracer for Pi uptake, DOP production, and Pi regeneration has been used extensively in oceanography to measure the growth and metabolic activities of algal and bacterial assemblages (Atkinson, 1987; Harrison et al., 1977; Perry and Eppley, 1981; Rigler, 1956; Sorokin, 1985; Sorokin and Vyshkvartsev, 1974; Taft et al., 1975; Watt and Hayes, 1963). Certain P radiotracer field experiments have employed size fractionation treatments (Harrison et al., 1977), metabolic inhibitors (Krempin et al., 1981) or multiple labeled substrates (Cuhel et al., 1983) to separate algal and heterotrophic bacterial activities. Typically, the radiotracer is added as carrier free ^{32}Pi (or ^{33}Pi) in order to minimize perturbations that may be caused by Pi pool enrichment.

The use of exogenous radiotracers to measure DOP production during timed incubation experiments generally requires a protocol to separate unused or regenerated radiolabeled Pi from the recently produced radiolabeled DOP compounds. This is an analytical challenge in many field studies because most of the radioisotope is generally present as Pi and because of the potential diversity of the individual DOP compounds that are produced. Smith et al. (1985) employed an isobutanol extraction technique to separate the ^{33}Pi (SRP)-molybdenum blue complex from DO^{33}P in solution. To the extent that SRP exceeds Pi, or if any of the DO^{33}P compounds are soluble in isobutanol, then the solvent extraction method will underestimate the net DO^{33}P production rate. Orrett and Karl (1987) compared the isobutanol technique to the much simpler "Bochner and Ames" (1982) technique; the latter involved a procaine-tungstate-tetraethylammonium dependent selective precipitation of ^{32}Pi leaving DO^{32}P in solution. The mean DO^{32}P production rate estimates from an open ocean field application of these two methods showed no significant differences (Orrett and Karl, 1987). More recently, Björkman et al. (2000) have employed the MAGIC technique, described earlier, as an effective means to separate ^{32}Pi from DO^{32}P in sample filtrates.

Once experimental data are available on ^{32}P/^{33}P-labeled DOP accumulation rates (expressed as Bq per volume per time) conversion to the more ecologically meaningful absolute P fluxes requires the assumption/application of a specific P assimilation model to convert P radioactivity to P mass. The key to providing a reliable mass flux is, again, the accurate estimation of the BAP pool. Orrett and Karl (1987) reviewed four independent models that have been used for seawater: (1) SRP (or Pi) model, (2) TDP model, (3) RNA model, and (4) ATP model. Each model differs in the assumptions used for P uptake and assimilation. For example, the SRP model assumes that the exogenous precursor ^{32}Pi/^{33}Pi mixes completely with the SRP pool prior to uptake and incorporation. Consequently, the radiolabeled DOP that is produced during the incubation would have a specific radioactivity (radioactive per unit mass) equivalent to that calculated for the SRP pool. However, in most open ocean environments where DOP > Pi, it may be more reasonable to acknowledge the possible role of DOP in microbial metabolism. The TDP model, therefore, assumes that both Pi and DOP are bioavailable during these relatively short (<1 day) incubation experiments. In our opinion, this assumption may also be incorrect and would lead to an underestimation of the true precursor specific radioactivity. The remaining two metabolic models rely on direct measurements of the specific radioactivity in two independent subcellular constituent pools that are expected to reflect the isotope dilution of the added precursor by all bioavailable extracellular and intracellular P pools (Orrett and Karl, 1987). In the RNA model, the specific radioactivity of nascent RNA is measured directly (using double-labeled ^{3}H-adenine and ^{32}Pi incubations; Karl, 1981) and this value is taken to represent an average for the entire DOP pool. The final method, the ATP model, is analogous to the RNA model but assumes that the terminal phosphate group of ATP (the γ-P pool rather than the α-P-pool as in the RNA model) accurately reflects isotope dilution. Two key advantages of the ATP model are the rapid turnover of cellular ATP which leads to rapid isotopic equilibration, and the compelling evidence that the γ-P position of ATP is very likely the precursor for the biosynthesis of most P-containing organic compounds (see Section VIII.F).

Hudson and Taylor (1996) devised and applied a novel method for the direct measurement of DOP production in planktonic communities. Their procedure requires a pre-incubation with high specific activity ^{33}Pi for 17-76 h to label the metabolically active organisms. A subsequent pulse of non-radioactive Pi competitively inhibited further ^{33}Pi uptake and provided a time-zero starting point for the timed appearance of ^{33}P-DOP from the combined processes of excretion, exudation, grazing, and cell lysis. The addition of the ^{31}Pi pulse is assumed to have no effect on DOP loss by the community of microorganisms. The DOP production (moles P L^{-1} day^{-1}) can be estimated from the mean specific radioactivity of the PP pool at the beginning of the second incubation. To our knowledge, this method has not yet been applied to marine ecosystems.

Watt and Hayes (1963) and Johannes (1964) were among the first to demonstrate contemporaneous Pi uptake and DOP production in near-surface planktonic assemblages using ^{32}Pi radiotracer incubations. Johannes (1964) also documented a coupled production of DOP by diatoms and DOP uptake by bacteria. It was also shown that mesozooplankton produced substantial amounts of DOP. According to Satomi and Pomeroy (1965) zooplankton, in general, have low oxygen consumption to Pi release ratios, which suggests that they do not completely oxidize their food. This is consistent with their observations of high DOP release rates.

Smith et al. (1985) conducted ^{33}Pi uptake experiments in coastal and offshore Hawaiian waters. They detected net ^{33}P-DOP production in all of their incubations, and used both 3-component (Pi→PP→DOP) and 4-component (Pi→PP$_1$/ PP$_2$→DOP) steady-state models to evaluate the observed P pools and fluxes. Their results were consistent with a rapid and coupled production and utilization of DOP in these marine habitats.

Dolan et al. (1995) examined coupled Pi uptake and passage through a simplified planktonic food web as defined by specific particulate matter size-fractions in Villefranche Bay, France. Pi uptake was dominated (>50%) by the smallest size-fraction (0.2-1 μm), presumably auto- and heterotrophic bacteria. Pi turnover times were rapid - less than a few hours for most of the 3-month observation period. Release of incorporated ^{32}P from various particulate size-fractions was investigated by incubating with ^{32}Pi for a 3-h period, followed by the addition of an excess of unlabeled AMP (100 μM). The addition of AMP, they reasoned, would partially inhibit the assimilation of recently produced DO^{32}P compound and provide a more accurate estimate of gross DOP fluxes. The measured rates of DO^{32}P release in these experiments were low, generally ≤1% of the corresponding particulate ^{32}P activity per hour (Dolan et al., 1995). However, when the concentration of oligotrich ciliates (predators of microorganisms in the 1-6 μm size class) was artificially increased, there was a significant transfer of ^{32}P from particulate to dissolved pools. These field results confirmed the role of protozoan grazing in nutrient cycling processes (Andersen et al., 1986), including the Pi→PP→DOP→Pi pathway.

Thingstad et al. (1993) conducted a comprehensive study of microbial transformations of P in P-limited Sandsfjord, western Norway, including the coupling of Pi uptake, DOP production, specific DOP compound hydrolysis and enzymatic hydrolysis. They focused on the production and turnover of nucleotides, and used ATP as a "model" compound. DOP/Pi concentration ratios in this habitat varied considerably but were generally between 10-100:1; Pi (reported as SRP) was ≤5 nM at selected stations. An averaged flow model was devised to best accommodate the measurements and their underlying assumptions. The most striking result was the presence of a large DOP pool (~200-250 nM) that was dominated (>99% of total DOP pool) by polymeric compounds (presumably RNA and DNA). These polymeric compounds turned over very slowly compared to the relatively small but rapidly assimilated nucleotide pool. In this regard, their results are consistent with the nuclease/phosphodiesterase "bottleneck" hypothesis discussed in a previous section of this review (see Section IV, Figure 5.5).

Løvdal et al. (2007) conducted a large mesocosm experiment in the Baltic Sea to examine the turnover of P from ^{33}P-labeled DNA and ^{33}P-labeled ATP compared to Pi uptake. They also investigated the partitioning of P into different microbial taxa following the induction of a nutrient-enhanced (N+P added at 16:1 molar ratio) phytoplankton bloom and various post-bloom nutrient amendments. During the post-bloom experimental phase, replicate 50 m^3 mesocosms received either: (1) (N+P) at 16:1 molar ratio, (2) (N+P) at 80:1 molar ratio or (3) N only at the higher loading as in (2), plus glucose. Forced P-limitation led to depletion of Pi (<1 nM) and to the rapid and efficient utilization of both LMW- and HMW-DOP (Løvdal et al., 2007). While Pi was still the preferred substrate, the rapid turnover times measured for ATP (~5 min) and DNA (~1.5 h) in the N plus glucose treatment, and the ability of all size-fractions (1-0.2 μm, 10-1 μm, >10 μm) to retain ^{33}P indicated the ability of both bacteria and phytoplankton to compete for all sources of P that were tested (Løvdal et al., 2007).

Orrett and Karl (1987) reported 0-100 m depth-integrated DOP production rates ranging from 0.3-0.8 mmol P m^{-2} day^{-1} (TDP specific activity model) and 0.5-1.2 mmol P m^{-2} day^{-1} (RNA specific activity model) for water samples collected in the NPSG. They reasoned that these DOP production rates could be further extrapolated to organic C, if the mean C:P molar ratio was either known or correctly assumed. An upper bound on C:P was taken as the whole cell C:P (106C:1P; Redfield et al., 1963), although it is unlikely that DOP compounds are, on average, this C rich (see Table 5.1). They assigned a value of 3C:1P as the theoretical lower bound which is identical to the nucleotide triphosphate pool. It is equally unlikely that the DOP pool would be that C poor, relative to P. A ratio of 9.5C:1P, the approximate value for RNA, was taken as the most reasonable estimate; the true C:P ratio for DOP is likely to be closer to the lower bound than to the upper bound. The extrapolated rate of DOC production, 24 mmol C m^{-2} day^{-1}, was about 50% of net primary production for this region (Karl et al., 1996, 1998). Because this estimate is based on accumulation (net production) of DOP during the incubation period and, therefore, does not include contemporaneous DO^{32}P production and DO^{32}P utilization, gross DOP fluxes will be even larger. These results suggest an important role for DOP in microbial loop processes in these low nutrient, open ocean habitats. At steady-state, DOP production and DOP remineralization rates would be in balance. Consequently, given the DOP pool size and estimates of DOP turnover rates, these organic pools are likely to serve as an important source of P, as well as C and N, for microorganisms. If the compound C:P ratio is less than the whole cell C:P ratio, or if C is derived from additional or alternative sources, then Pi is likely to be released into the medium. These coupled processes most likely sustain the marine P-cycle in the euphotic zone of the sea.

Björkman et al. (2000) measured coupled rates of Pi uptake and DOP production at several stations in the NPSG using ^{32}Pi as a tracer. Pi uptake rates varied from 3-8.2 nM Pi day^{-1}; Pi pool turnover time was 2-40 days. Net DOP production (i.e., accumulation) was 10-40% of the net Pi uptake. The estimated turnover time for the entire DOP pool, assuming compositional singularity with nascent DOP, was 60-300 days. In all likelihood, the recently produced materials are assimilated much more rapidly than this simple calculation would suggest. Pi regeneration from selected, exogenously added DOP compounds was rapid and efficient; highest rates of Pi release were observed for nucleotides (Björkman et al., 2000).

Mather et al. (2008) conducted a basin-wide study of P cycling in the Atlantic Ocean. A comparison of the iron-replete, Pi-stressed North Atlantic Subtropical Gyre (NASG) with the iron-poor, Pi-replete southern gyre revealed a distinct asymmetry in DOP dynamics which they hypothesized was a consequence of enhanced N$_2$ fixation in the nutrient stressed NASG. Indirect estimates of DOP utilization based on ambient DOP concentrations and in situ APase activity, led them to conclude that DOP pool turnover in the NASG was 5.5 ± 2.3 months compared to >10 years in the southern gyre. This DOP flux is sufficient to provide up to 30% of the boreal primary production in the NASG (Mather et al., 2008).

Although coupled Pi uptake and DOP production is well documented in a variety of marine ecosystems, the actual mechanisms of DOP production remain elusive. Admiraal and Werner (1983) investigated the production of DOP by two coastal marine diatom species in laboratory culture. In addition to total DOP production rates, they concentrated a fraction of the DOP pool, using Sephadex® G-10 chromatography, and documented partial reabsorption of the isolated DOP compounds by the same two species during Pi-limited growth. The inadvertent diffusive loss of LMW compounds or the active excretion of both LMW and HMW compounds are both feasible; the list of specific compounds that are liberated by growing algae is very large (Fogg, 1966; Hellebust, 1974). Alternatively, DOP release could result from cell autolysis, predator grazing, or viral lysis. Each separate pathway might be expected to produce a different spectrum of compounds. Most of the research conducted to date has focused on extracellular production of DOC, not DOP, but suffice it to say that DOP production by healthy microorganisms is probably a universal phenomenon.

More recently, White et al. (2012) assessed the lability and environmental controls of remineralization of a range of model DOP compounds in coastal and open ocean habitats. Their results supported previous findings of relatively rapid remineralization of exogenous C-O-P (glucose-6-P, AMP and poly-P), and relatively slow remineralization for the C-P substrate (White et al., 2012). The decay rates were higher for coastal waters, and increased with increasing temperature in both coastal and open ocean experiments. The DOP remineralization rates exceeded bacterial P demand so these pathways likely provide a local source of recycled Pi for the microbial assemblage.

B Direct Utilization of DOP

DOP compounds in seawater consist of both labile and refractory compounds. The labile DOP pool includes both transportable and non-transportable organic compounds, either of which can serve as P sources for microbial growth. The outer membrane of gram-negative bacteria allows the transport of molecules up to about 600 Da (Weiss et al., 1991). Therefore, many DOP compounds can, and probably are, taken up directly without the need for prior hydrolytic alteration. For example, Gly-3-P and AMP can be assimilated intact by certain bacteria (Ruby et al., 1985; Wanner, 1993), whereas larger DOP compounds must be enzymatically hydrolyzed, either at the cell surface (or in the

periplasmic space for bacteria) or in the surrounding medium prior to assimilation. The Pi released is then available for assimilation and biosynthesis. Vitamin B$_{12}$ (MW = 1355 Da) is a notable exception since many marine microbes assimilate it intact.

The ability of an organism to grow on one or multiple DOP substrates as the sole source of cellular P can be traced to one of two independent properties: the presence of cell membrane or periplasmic bound enzymes that catalyze the DOP compound dephosphorylation and thereby enhance Pi availability or the presence of a DOP compound- or compound class-specific uptake system. Both pathways are present in marine microorganisms; growth of both prokaryotic and eukaryotic microorganisms on a variety of different DOP compounds is well documented (Antia et al., 1991; Cembella et al., 1984a,b; Kuenzler, 1965; van Boekel, 1991; White et al., 2010).

The phosphate regulon (Pho) of *Escherichia coli* includes at least 31 genes arranged in eight separate operons (Wanner, 1993). These genes are co-regulated by environmental Pi concentrations and, working together, facilitate P assimilation. During Pi-limitation, the synthesis of selected Pho enzymes are up-regulated including, but not limited to: (1) a high-affinity, high-specificity periplasmic permease (Pst) to enhance Pi assimilation capacity, (2) a periplasmic APase to facilitate monophosphate ester-linked DOP hydrolysis, and (3) specific enzymes to facilitate the uptake and hydrolysis of phosphonates. Expression of the Pi regulon has been reported for marine bacteria (McCarter and Silverman, 1987), so the much more extensively studied *E. coli* may be an adequate model at least for hypothesis testing in Pi-stressed marine habitats. The Pho regulon also controls the transport of selected, intact DOP compounds into bacterial cells. Several proteins of the outer membrane of many bacteria (termed "porins") are involved in the formation of aqueous pores through which small hydrophilic molecules (<600 Da, including

Pi and selected DOP molecules) can pass through the membrane. Pi concentration regulates the synthesis of these proteins; Pi starvation enhances their biosynthesis (Saier, 2000; Tommassen and Lugtenberg, 1981). Tanoue (1995) has reported that porins, specifically porin-P (a protein that is synthesized during Pi-limitation) may be a major component of the total dissolved protein pool in seawater. Using sodium dodecylsulfate-polyacrylamide gel electrophoresis, Suzuki et al. (1997) detected several distinct bands with molecular weights 14.3-66 kDa. One frequently observed band (MW = 48 kDa) was identified as bacterial porin-P. Consequently, it appears that porins may be an ecologically significant pathway of Pi and DOP assimilation in low nutrient, open ocean habitats. In addition, several recent studies have found evidence of presence and expression of various Pho genes in marine microbes such as SAR11, cyanobacteria, and diazotrophs (Dyhrman and Haley, 2006; Sowell et al., 2009).

Laboratory studies conducted with *E. coli* have documented independent and fairly specific transport systems for Gly-3-P and Glu-6-P that are derepressed during Pi-limitation. Although both systems also co-transport Pi, they are low-affinity in this regard, especially when compared to the Pst system that is also active during periods of Pi-stress. These specific DOP transport systems appear to be one example of an anion-exchange mechanism in bacteria (Maloney et al., 1990). The uptake of the DOP compound is linked with the export of Pi and may, therefore, integrate into a vital chemiosmotic circuit or H$^+$ pump; in effect, these DOP transport systems are Pi-linked antiporters. The two most potentially significant functions of these anion-exchange mechanisms are (Maloney et al., 1990): (1) to catalyze the heterologous exchange of Pi and a selected sugar phosphate (e.g., Pi:Gly-3-P and Pi:Glu-6-P) and (2) to exchange intracellular AsO$_4^{3-}$ for either Pi or for a sugar phosphate. The first function, the asymmetric exchange of Pi for a sugar phosphate, may be part of the cell's mechanism to balance the supply of C and P. Because many DOP compounds have excess P relative to C, compared to whole cell C:P stoichiometry (e.g., the molar C:P ratios for Gly-3-P and Glu-6-P are 3 and 6, respectively, compared to a Redfield stoichiometry of 106; also see Table 5.1) this would provide an efficient system to achieve a physiological C-to-P balance. This control mechanism might be especially important during periods of low growth rate when C demands for maintenance energy generation are high, but biosynthetic demands for P are low, as in the mesopelagic or abyssopelagic zones. The second function, the exchange of AsO$_4^{3-}$ for Pi or the exchange of AsO$_4^{3-}$ for a sugar phosphate may be one of several strategies for As detoxification, especially in open ocean environments where the As/Pi concentration ratio exceeds 10 (Karl and Tien, 1997).

Another potential advantage of the direct uptake of DOP, especially nucleotide monophosphates, is the use of these molecules as biosynthetic precursors. This conservation of pre-existing phosphate bonds has significant implications for cellular energetics and for the growth efficiency of microorganisms (Rittenberg and Hespell, 1975).

The availability of water insoluble or hydrophobic DOP compounds, like membrane phospholipids, may require additional, specific enzymes for assimilation. Lemke et al. (1995) evaluated the role of cell surface hydrophobicity, a measurable property of all microbial cells, on the relative utilization rates of hydrophobic and hydrophilic P compounds. Hydrophilic bacteria grew rapidly on Pi and Gly-3-P, but could not assimilate the hydrophobic substrate, phosphatidic acid (PA) or membrane phospholipids (Lemke et al., 1995). Conversely, bacteria with hydrophobic cell surfaces efficiently utilized PA and lipid-P. These cell-specific metabolic capabilities could lead to DOP resource partitioning among otherwise competing microheterotrophs, or to species selection and succession following production or exhaustion of one or more key DOP substrates.

An interesting and potentially important study of the effect of fluid motion on the utilization of selected LMW DOM (not explicitly DOP compounds) by heterotrophic bacteria revealed that uptake rate was enhanced by advective flow, but only at low subsaturating DOM concentrations (Logan and Kirchman, 1991). One prediction of their results is that particle-bound bacteria, especially those sinking through the water column, might be more important for deep water DOP remineralization than the solitary microorganisms suspended in the water column.

Finally, competition between bacteria (chemoheterotrophs) and phytoplankton for Pi and DOP in aquatic environments has been studied extensively (e.g., Rhee, 1972). These investigations have examined uptake affinities, cell quotas, storage capabilities, and other ecophysiological parameters under both P-sufficient and P-limited growth conditions. Despite some contradictory field results, most investigations revealed a tight metabolic coupling between the producer and consumer species and explicit nutrient resource-based competition, especially at limiting concentrations of Pi. Less well documented is the ability to switch from Pi-based to DOP-based metabolism or the sequential versus simultaneous utilization of two or more P-containing substrates. The bacterial-phytoplankton dichotomy disappears in open ocean environments because most of the "phytoplankton" are bacteria (e.g., the picophytoplankton assemblage). Most low nutrient environments are dominated by prokaryotes, e.g., *Prochlorococcus* or *Synechococcus*, and may even be supported by mixotrophic growth (e.g., simultaneous utilization of photoautotrophic and chemoheterotrophic metabolic pathways). In any event, all prokaryotes and eukaryotes function as P-traps and ultimately must compete with each other.

C The Methylphosphonate "Cycle"

Among the various reduced DOP compounds in the sea, methylphosphonate (MPn) has recently received great attention since its metabolism has been linked to aerobic production of methane, a potent greenhouse gas (Karl et al., 2008). This discovery provides an explanation for the presence of excess methane in oxygenated surface ocean waters, the so-called "marine methane enigma" (Kiene, 1991), assuming there is a natural source of MPn. Subsequent laboratory-based experiments showed that *Trichodesmium* could use MPn as their sole source of P, and that methane was quantitatively produced during MPn metabolism (Beversdorf et al., 2010; White et al., 2010). Similar results were obtained for the growth of other marine microorganisms, including the abundant and ubiquitous heterotroph, SAR 11 (Carini, 2013), so the ability to metabolize MPn appears to be common.

Watkins-Brandt et al. (2011) assessed the short-term (24h) responses of inorganic C assimilation and N₂ fixation to the addition of either Pi or MPn at several stations in the NPSG. Their results indicated that Pi and MPn contributed equally to the observed stimulation (relative to unamended control treatment) of both C and N fixation, indicating P-limitation of both processes at the time of their investigation (Jul.-Aug. 2008; Watkins-Brandt et al., 2011). Furthermore, both Pi and MPn stimulated C fixation rates beyond those attributable to N₂-fixing microorganisms, indicating that MPn appears to be readily bioavailable to a much broader spectrum of the microbial assemblage (Watkins-Brandt et al., 2011).

Recently, Metcalf et al. (2012) discovered a novel MPn biosynthesis pathway, and the presence of MPn synthase-encoding genes in the marine metagenomic sequence database. In addition the catalytic mechanism for aerobic formation of methane by bacteria has been reported (Kamat et al., 2013). There now appears to be little doubt that MPn cycling contributes to the near-surface ocean methane supersaturation observed worldwide. However, the distribution of MPn synthesizing microorganisms, the physiological role(s) of MPn, and the ambient MPn concentrations and fluxes in the sea all remain as key research foci.

D Enzymes as P-cycle Facilitators

Because Pi starvation causes a significant increase in APase activity it has been suggested that the detection of APase in field-collected samples may be an ecophysiological indicator of Pi-stress. However, the literature on this topic is confusing and, at times, contradictory. For example, it has also been shown that starvation for nucleic acid bases will induce APase synthesis even under conditions of excess Pi (Wilkins, 1972). Consequently, there may be alternative ecological interpretations for the presence of elevated APase in natural assemblages of microorganisms.

APase activity can be measured by colorimetry (using para-nitrophenyl phosphate as the substrate), fluorometry (using either o-methyl-fluorescein phosphate or 4-methylumbelliferyl phosphate as the substrate) or firefly bioluminescence (using ATP as the substrate). Substrate selection determines to a large extent measurement sensitivity and assay specificity. For example, para-nitrophenyl phosphate and ATP are both hydrolyzed by APase and the related enzyme nucleotidase, so these assays are not necessarily specific for APase unless combined with other sample treatments (e.g., Pi additions; Karl and Craven, 1980). Furthermore, the use of artificial substrates has the potential to overestimate the in situ rates of enzymatic activity (Admiraal and Veldhuis, 1987; Cembella et al., 1984a). Only ATP is likely to be a native substrate in natural samples and only the use of ATP provides a detection system that can provide in situ rate estimates. A novel insoluble fluorogenic substrate for APase, termed enzyme labeled fluorescence (ELF), has been used to detect the presence of the enzyme in single microbial cells (Dyhrman and Palenik, 1999; González-Gil et al., 1998). The ELF assay does not provide quantitative estimates of APase activity but does provide enzyme localization at the single cell level. Cell detection is by either epifluorescence microscopy or laser-based flow cytometry.

Laboratory studies of APase in pure cultures of marine bacteria (Hassan and Pratt, 1977) reported that APase was completely repressed, partially repressed or not repressed at all by the presence of 50 mM Pi in the growth medium. This variable response suggested the presence of different APase isozymes some of which appear to be constitutive. In any case, the relatively high Pi concentrations used in this and other laboratory studies of the Pi repression of APase activity is 10^4 to 10^6 times higher than Pi concentrations in most surface ocean waters (open ocean 1-100 nM Pi, coastal ocean <1 μM) so even "relatively high" seawater concentrations must be considered Pi-depleted for the purposes of the microbial APase derepression response. An ecological prediction is that APase should be ubiquitous in seawater.

Morita and Howe (1957) were the first to report APase activity in cultures of marine bacteria. Their interests centered on the effects of ambient hydrostatic pressure on the potential efficiency of Pi regeneration in the deep sea. Wai et al. (1960) published the first report of APase activity in seawater from the coastal waters near Taiwan. They detected and quantified APase activity by several procedures including the measurement of Ca^{2+} increase following a timed incubation with fish bone powder (calcium phosphate) and by an increase in Pi following a timed incubation with Gly-3-P. APase activity was ubiquitous, but varied with sample location (Wai et al., 1960).

The model organism *Trichodesmium*, a well-studied diazotroph, can use a variety of DOP compounds as its sole source of P (Mulholland et al., 2002; White et al., 2010). It also likely supplements its P requirements with DOP under field conditions, including both phosphate esters and phosphonates (Dyhrman et al., 2002, 2006; Sohm and Capone, 2006; Stihl et al., 2001; Yentsch et al., 1972).

All living organisms in the sea, from bacteria to marine mammals, can produce APase and other phosphatases, thus it is difficult to determine the exact source of the enzyme activity in most environmental samples. However, from an ecological perspective it is critical to know which organisms are expressing APase activity under a particular set of environmental conditions. APase activity in the dissolved fraction is also common (Reichardt et al., 1967). Most investigators measure APase activity in either whole water (dissolved plus particulate activity) or filtered (or filter size-fractionated) subsamples. For example, in Toulon Bay (France), APase in the >90 μm fraction accounted for more than 80% of the total activity whereas in other coastal regions of the Mediterranean APase was restricted to the picoplankton (0.25-5 μm) fraction (Gambin et al., 1999). Boon (1994) has devised a very clever technique to distinguish between prokaryotic and eukaryotic APase activities. His method relies on a differential inhibition by various physical (e.g., thermal deactivation) and chemical (e.g., Zn^{2+} and Cu^{2+} ion inhibition) treatments.

Since the first field report of APase activity in seawater by Wai et al. (1960), numerous studies have been conducted using a variety of increasingly more sensitive and specific assay systems. APase synthesis is derepressed following Pi-limitation and is not induced by the presence of suitable substrates, thus in vivo enzyme activity under saturated substrate concentrations is not equivalent to P flux. Theoretically, in situ rates of DOP hydrolysis can be estimated only if the concentration of exogenous substrate is comparable to in situ specific DOP compound concentrations. Consequently, most APase and other enzyme activity measurements in seawater must be considered "potential" activities.

Perry (1972) was the first to document APase activity in an oligotrophic, open ocean ecosystem. The APase activity in the NPSG was greatest for samples collected in the upper 60 m of the water column and decreased with increasing water depth and ambient Pi concentration. Microbial assemblages lacking detectable APase activity at time of sample collection produced APase within 1-2 days when incubated in bottles; the addition of 5 μM Pi to water samples immediately repressed APase activity (Perry, 1972). Li et al. (1998) measured significant dissolved (<0.2 μm) and particulate APase activities in the Gulf of Aqaba, northern Red Sea. Dissolved APase activities ranged from 40-70% of the total APase activity; most of the particulate APase activity was attributed to the picoplankton fraction. Cell-free activities were stable in the dark at 4 °C for extended periods (up to 40 days) and could lead to variable dissolved:particulate ratios, for example as a consequence of transient periods of Pi-limitation. In contrast to this open ocean study, Taga and Kobori (1978) reported a positive, not negative, relationship between APase activity and Pi concentrations for samples collected in Tokyo Bay. As mentioned previously, even the highest Pi concentrations found in marine ecosystems (i.e., 3-4 μM and generally much lower) are low relative to the Pi concentrations required to repress APase activity in laboratory cultures of bacteria and eukaryotic algae. Consequently, the positive relationship between APase activity and Pi is probably controlled more by seston biomass than by per cell APase activities. In this regard, Smith et al. (1992) reported intense APase activity associated with marine aggregates compared to surrounding seawater; volume-normalized enhancements were 10^3- to 10^4-fold. The interstitial fluids of large-dimension suspended and sinking particles may have chemical compositions that are different from the surrounding bulk fluid environments. This could select for, or against, specific enzyme activities.

There are at least two ecologically relevant post-synthesis controls on in situ APase activities in seawater: trace metal concentrations and UV-B radiation. Reuter (1983) demonstrated that APase activities in phytoplankton cultures, cell-free enzyme preparations and field-collected samples are inhibited by free copper ions at concentrations comparable to those found in the marine environment (cupric ion activities of 10^{-9} to 10^{-12} M). This trace metal effect could interfere with or totally block the direct utilization of selected DOP compounds by natural microbial assemblages and therefore alter marine P-cycle dynamics. More recently, Garde and Gustavson (1999) have documented a UV-B radiation (280-320 nm) sensitivity of APase activity in the marine environment. A major implication of this work is that photo-degradation of APase activity may limit a cell's ability to obtain Pi from the ambient DOP pool, thereby exacerbating the effects of Pi-limitation in well-lighted, near-surface habitats. This UV-B control of APase activity would probably be more important in clear, open ocean ecosystems than in coastal habitats. In more productive coastal shelf and estuarine habitats there may be a positive photolytic effect that derives from the light-dependent release of APase previously bound (and therefore deactivated) by humic substances (Boavida, 2000; Boavida and Wetzel, 1998). This photodegradation/photoactivation dichotomy may also affect other enzymes in the marine P-cycle and, therefore, influence DOP turnover rates.

Finally, Hoppe and Ullrich (1999) have presented a comprehensive assessment of APase activities not only in the euphotic zone but to abyssal ocean depths in the Indian Ocean. Contrary to expectation, the measured total APase activities generally increased with increasing depth despite decreased bacterial biomass, increased ambient concentrations of Pi and a decrease in the activities of other hydrolytic enzymes. Elevated APase activities in the deep ocean (>1000 m) had previously been reported for particulate matter samples collected in the central North Pacific Ocean (Koike and Nagata, 1997), and may be a general feature of marine ecosystems. Hoppe and Ullrich (1999) hypothesized that the elevated deep water APase activities may be a manifestation of an enhanced C acquisition

system involving bioavailable DOP compounds. The elevated APase would locally regenerate Pi but, more importantly, would capture the DOP compound C skeleton, which could be respired to provide energy for cell maintenance. If this intriguing deep sea metabolic model is supported by future field research it would be an important lesson regarding our general ignorance of sub-euphotic zone ocean processes. Regardless, the report of elevated cell-specific APase activities in Pi-sufficient deep sea habitats casts doubt on past interpretations of environmental APase activity as being an indicator of Pi-stress only (Karl, 2014).

The development and application of novel molecular techniques, in particular metagenomics, has resulted in a renewed interest in APase form and function. Recently, Luo et al. (2009) identified several forms of APase (PhoA, PhoD, PhoX) in the marine metagenomics dataset. Of the 3733 bacterial APase sequences surveyed, 41% were cytoplasmic suggesting that transport and intracellular hydrolysis of LMW DOP may be an important mechanism for P acquisition in the open sea. Each of these APase gene families has unique substrate specificity, metal co-factors and kinetics. Subsequent investigation supported the major ecological role for PhoX, compared to the classical PhoA type of APase (Luo et al., 2011; Sebastian and Ammerman, 2009), in a broad range of marine environments. Since previous APase assay methods were designed to detect primarily periplasmic and outer membrane proteins, and extracellular APase, it is unclear whether the quantitative marine APase database is biased against the more abundant intracellular proteins. At the very least, we probably need to revise our current conceptual understanding (Dyhrman et al., 2007) to accommodate these new discoveries (White, 2009).

APase is just one of several potential enzymatic facilitators of DOP turnover in the sea (Figure 5.5); we have discussed it at length here as a "case study" because it has been measured extensively in the sea. The enzyme 5NDase is a membrane-bound protein found in most bacteria

(Bengis-Garber and Kushner, 1982) and in eukaryotic algae (Flynn et al., 1986); it is responsible for the hydrolysis of extracellular nucleotides (Figure 5.5). In nature the substrate specificity of 5NDase overlaps with other enzymes; for example, ATP is also hydrolyzed by APase as well as by more specific ATPases (EC 3.6.1.4) found in many cells. A major difference seems to be in the regulation of enzyme production, especially the effects of Pi-limitation. In contrast to APase, the activity and synthesis of 5NDase in *E. coli* are not inhibited by Pi and in this respect resemble inorganic pyrophosphatase (Neu, 1967). 5NDase may play a key role in nucleotide pool turnover and biosynthetic salvage pathways. Ammerman and Azam (1985) were the first to detect 5NDase in natural marine microbial assemblages and since that time numerous reports have appeared. Ironically, the assay system most frequently employed releases ^{32}Pi from γ^{32}P-ATP. This would integrate the activities of several different classes of phosphohydrolytic enzymes (5NDase, APase, ATPase), thereby overestimating specific 5NDase activity. The addition of 100 μM of Pi to the assay mixture would improve the accuracy of 5NDase detection in collected samples (Cotner and Wetzel, 1991). From an ecological perspective, however, it is not so important which enzyme class is active but rather what the total rate of catalysis is under in situ conditions.

Siuda and Güde (1994) found that ATP was hydrolyzed much more rapidly by 5NDase than by APase, whereas the rates of AMP and ADP were comparable. If total extracellular nucleotide production in the ocean is controlled by cell exudation or grazing activities, then ATP may be a large portion of the pool and 5NDase activity would be important. If, on the other hand, hydrolysis of polymeric DNA and RNA supplies most of the dissolved nucleotides in seawater then both APase and 5NDase would be important to the cells (Figure 5.5). In fact, Ammerman (1991) concludes that the major function of 5NDase may be to assist in the coupled re-cycling of RNA and DNA by hydrolyzing the ribo- and deoxyribonucleotides produced by nuclease enzyme activity. As discussed previously, the "bottleneck" in the DOP cycle may be nuclease, rather than 5NDase activity (see Section IV, Figure 5.5).

Nucleases and related enzymes catalyze the breakdown of nucleic acids by hydrolysis of the phosphodiester bonds. Some nucleases are specific for RNA (RNases); others act only on DNA (DNases), while still others are nonspecific. Furthermore, nucleases can be classified by mode of substrate attack; polynucleotides can be hydrolyzed at many interior locations in the polymer (endonucleases) or stepwise from one end of the chain (exonucleases). Additionally, the exonuclease attack can be directed from either the $3' \rightarrow 5'$ or the $5' \rightarrow 3'$. Therefore the hydrolysis products from nuclease activity in seawater can result in a mixture of oligonucleotides as well as 3' and 5' mononucleotides (Figure 5.5). This is a potentially important point, since 5NDase is specific for 5'-nucleotides and will not hydrolyse 3' compounds (Bengis-Garber and Kushner, 1981). APase, on the other hand, will hydrolyze both 3'- and 5'-nucleotides (Figure 5.5). Nuclease activity is also important for control of intracellular RNA concentrations, especially during C and energy starvation. In laboratory studies with *E. coli*, the RNA degradation products—including both 3' and 5'-nucleotides and oligonucleotides—appear in the medium (Cohen and Kaplan, 1977). Consequently this may be an important source for DOP production under certain non-growth conditions (Novitsky and Morita, 1977).

E Taxon-specific DOP Uptake

Field measurements of DOP uptake are generally made using ^{32}P/^{33}P labeled substrates and time course incubations. Typically, phosphomonoesters (e.g., glucose-6-P, glycerol-P) or nucleotide triphosphates (e.g., ATP) have been employed, and the use of pre- or post-incubation size fractionation has been used to gain a better identity of the substrate responsive assemblages. The use of model substrates in field experiments designed to study DOP dynamics is analogous to the use of laboratory-reared microorganisms to understand the physiological or genetic response to controlled growth conditions. Both approaches are important, but neither is able to capture the diversity and complexity of natural marine habitats and their microbial assemblages.

Alonso-Sáez and Gasol (2007) used microautoradiography combined with catalyzed reporter deposition fluorescence in situ hybridization (CARDFISH) to analyze the patterns of glucose, amino acid and ATP uptake by specific bacterial groups in the NW Mediterranean Sea over an annual cycle. While the different target groups (*Alpha-proteobacteria*, *Gamma-proteobacteria*, and *Bacteroidetes*) showed considerable variability in their assimilation of glucose and amino acids with some essentially nonresponsive to the substrate at certain times of the year, ATP was actively and universally assimilated (Alonso-Sáez and Gasol, 2007). They attributed this result to strong, year-round P-limitation at their study site.

Orchard et al. (2010) compared Pi and ATP uptake for *Trichodesmium* colonies collected from the upper 20 m of the Sargasso Sea. Their results indicated that *Trichodesmium* actively assimilated ATP and, based on uptake rates and kinetics and on alkaline phosphatase measurements, concluded that DOP (specifically P-esters) can exceed 25% of the P requirement in this region.

Recently, taxon-specific, quantative uptake of specific DOP substrates has been achieved by combining radioisotopic cell labeling with flow cytometric cell sorting and analysis. This approach is an improvement over the use of size fractionation, which cannot distinguish heterotrophic bacteria from picocyanobacteria due to overlapping size spectra, or previous methods that detected active cells using microautoradiography. Cell sorting also provides the opportunity to test hypotheses regarding relative substrate preference, P resource partitioning and competition among co-occurring microbial taxa. To date, the most commonly identified taxa in field studies include *Prochlorococcus*,

Synechococcus, and "non-pigmented" (assumed to be heterotrophic) bacteria; the latter group is sometimes further subdivided by the intensity of DNA-staining into low- and high-nucleic acid (LNA and HNA, respectively) subpopulations. Within each sorted group or taxon, there is likely to be cell-to-cell variability so the sorted populations record only the mean condition for the aggregate assemblage, which typically ranges from 10^3-10^6 cells per sample depending upon the design of the experiment.

Michelou et al. (2011) compared per cell, per surface area and per cellular P-quota uptake of both ^{32}Pi and AT^{32}P for microorganisms in the western Sargasso Sea. Their results indicated that the average Pi and ATP uptake rates per cell were 50- and 80-fold higher, respectively, for *Synechococcus* than for either *Prochlorococcus* or bacteria. The same was true if uptake was normalized to cellular P-quota (Michelou et al., 2011). However, despite their minimal per cell (and per cellular P-quota) uptake capacity, bacteria as a group dominated (>90%) total Pi and ATP uptake due to the overwhelming numerical abundance of this microbial group. Comparison of Pi and ATP uptake patterns for the four major groups of phytoplankton (*Prochlorococcus*, *Synechococcus*, picoeukaryotes, and nanoeukaryotes) in the Sargasso Sea indicated that all four groups actively assimilated ATP, especially in surface waters where Pi concentrations were <7 nM and often <1 nM (Casey et al., 2009). It was concluded that more than 50% of the P demand in these Pi-stressed habitats might have been derived from ATP.

Björkman et al. (2012) conducted similar taxon-specific uptake experiments in the NPSG. They reported that the dominant photoautotroph, *Prochlorococcus*, competes equally with bacteria for ^{32}Pi uptake on a seawater volume basis (Björkman et al., 2012). However, on a per cell basis, *Prochlorococcus* was three times more efficient (20 amol cell^{-1} day^{-1}) for Pi assimilation at ambient (~50 nM) concentrations. The kinetic response of both *Prochlorococcus* and bacteria to the addition of exogenous Pi (10-500 nM) was small, indicating that internal Pi pools for both groups were nearly saturated at the time of this study (Björkman et al., 2012). Their model DOP substrate, AT^{32}P (labeled in the γ-position), displayed more rapid pool total turnover times than were measured for Pi and had a greater kinetic response with V_{max} values at saturating ATP concentrations (>20 nM) exceeding those at ambient ATP concentrations by more than 50-fold (Björkman et al., 2012). A follow-on study investigated the light dependence of ^{32}Pi and γ-AT^{32}P uptake. Incubation of samples at in situ light levels led to significant increases in ^{32}Pi uptake for *Prochlorococcus* compared to dark incubations, with mean L:D uptake ratios of 2.5-fold; there was no significant impact on ^{32}Pi uptake by bacteria (Duhamel et al., 2012). The uptake of γ-AT^{32}P by *Prochlorococcus* was also higher in the light than in the dark, but again there was no effect for bacteria (Duhamel et al., 2012).

It remains to be shown whether inorganic and organic P substrates are assimilated simultaneously by the same cell or ecotype, or if these resources are partitioned among separate taxa or subpopulations of the total microbial assemblage (Karl, 2014). Of importance to the marine microbial P-cycle is the relative bioavailability of DOP compounds which likely leads to preferential uptake of monophosphate esters, nucleotides, vitamins, and alkylphosphonates, and the accumulation of polymeric DOP. The turnover of HMW-DOP is probably the flux "bottleneck" in the microbial P-cycle, but at the same time may represent a P buffer or surplus for possible use when more bioavailable supplies are exhausted (Karl, 2007a,b). Ultimately, most (essentially all) DOP is consumed because deep sea DOP is <10% of the surface concentration.

F DOP Interactions with Light and Suspended Minerals

The marine DOP pool is known to react with UV light; in fact, this is one of the many techniques used to characterize (Karl and Yanagi, 1997) and quantify it (Armstrong et al., 1966). Less is known about solar-induced DOP photolysis in the sea. Francko and Heath (1982) reported the existence of UV-sensitive P (SNP which released SRP following timed exposure to natural UV-irradiation) in lake ecosystems and it is probable that similar compound classes also exist in the marine environment (see Kieber et al., 1989). One intriguing implication of photochemical alteration of DOM is the possible co-occurrence of iron photoreduction (reduction of hydrous iron oxides) and Pi desorption from colloidal particles (Tate et al., 1995). If similar actinic effects occur in the marine P-cycle, one might anticipate that they would be restricted to the near-surface waters where light fluxes are maximal. Near sea-surface enrichments of SRP have recently been reported for samples collected in the North Pacific Ocean (Haury and Shulenberger, 1998; Haury et al., 1994; Karl and Tien, 1997), which could in principle arise via natural DOP photolysis (although other explanations are also possible). The near-surface accumulation of lipid-rich, positively buoyant colloidal, or particulate material may be important in this hypothesized pathway (i.e., organic matter + light → SRP).

McCallister et al. (2005) investigated the effects of sunlight on the decomposition of DOM in the York River estuary. Although their results varied with location in the estuary and with season, in general there was no abiotic production of NH$_4^+$ or Pi or loss of DOC during a 9-hr exposure of filtered water samples contained in quartz tubes to natural sunlight (UV-B plus UV-A exposure ranging from 801 to 910 W m^{-2} depending on experiment). However, incubation of light-exposed samples with a 1% (vol/vol) inoculum of water from the same site showed an enhanced bacterial utilization of DOC, but not of DON or DOP, over an incubation period of up to 28 days relative to dark (non-irradiated) controls (McCallister et al., 2005). No molecular characterization of the DOM pools was attempted but it appears from these results that DOC-N,

DOC-P, and DOC-N-P may be less photolabile than organic compounds that do not contain N and P.

Finally, metal-ion catalysis of phosphate ester hydrolysis in aqueous solutions is well documented (Dixon et al., 1982). However, the generally low concentrations of transition metal ions in open ocean habitats may preclude this process. More important in these habitats, phosphate ester hydrolysis may be facilitated by suspended minerals, including amorphous iron and manganese hydroxides (Baldwin et al., 1995). This could lead to an ecologically significant coupling between the atmospheric delivery of bioessential trace metals and the pulsed, abiotic release of Pi from semi-labile or biorefractory DOP.

X CONCLUSIONS AND PROSPECTUS

The biogeochemical cycle of P (as well as C and N) in the sea is sustained by energy supplied to the surface ocean by sunlight. Photosynthetic production of organic matter fuels a complex series of trophic interactions and organic matter export (both as DOM and POM) from the euphotic zone that ultimately sustains life throughout the world ocean.

In pelagic marine ecosystems, the supply rates of both N and P exert primary controls on ecosystem productivity (Smith et al., 1986). Over long time scales, P rather than N is probably the biomass- and production rate-limiting nutrient in the global ocean (Codispoti, 1989; Redfield, 1958), due to P removal in shallow-water, carbonate-dominated ecosystems and due to the role of bacterial N_2 fixation as a mechanism for relieving the ecosystem of fixed N-limitation. Consequently, field studies of P cycling in the epipelagic zone are of fundamental importance for our understanding of microbial processes in oligotrophic oceanic habitats. Nevertheless, P-cycle investigations conducted beyond the continental shelf are rare, especially by comparison to the relatively large body of knowledge on N cycling in oceanic habitats. This situation is due, in part, to technical limitations in the ability to obtain precise and accurate determinations of low concentrations of dissolved P.

Research conducted during the past few decades has uncovered several novel aspects of the marine P-cycle (Karl, 2014). For example, we now recognize that P exists in a variety of valence states in nature, but we are just beginning to understand the reduction-oxidation pathways and their relationships to cellular bioenergetics, biogeochemistry, and ecology. Furthermore, recent development and application of –omics technology (e.g., genomics, transcriptomics, proteomics, and metabolomics) have opened new windows of opportunity, improving our general understanding of the dynamics and regulation of the marine microbial P-cycle. Finally, research conducted at two key open ocean time-series stations (BATS and HOT) have revealed complex, non steady-state microbial P dynamics that are dependent on large-scale climate variations and inextricably linked to global ocean C- and N-cycles. By any measure, the ocean is an undersampled habitat so we should anticipate additional novel discoveries as new field-based research is conducted.

A more complete chemical characterization of the DOM pools in seawater would be a most welcomed addition to contemporary studies of the marine P-cycle. Likewise, a better understanding of the ecosystem processes that are responsible for extracellular accumulation of relevant biomolecules such as ATP, GTP, and nucleic acids will help to constrain DOP sources, sinks, and fluxes. Finally, it is imperative that future field studies recognize the P-cycle as an integral component of microbiological oceanography and marine biogeochemistry, along with the better studied though no less relevant C- and N-cycles. As McGill (1963) remarked nearly four decades ago, "The organic portion of the P cycle will undoubtedly receive much closer scrutiny in the future as its importance becomes more evident to biologists and oceanographers." While prophetic at that time, we still have some unfinished business. We look forward to new pathfinding contributions on marine DOP pool dynamics in the next decade.

Acknowledgments

We thank the volume co-editors, D. Hansell and C. Carlson for the invitation to prepare this review on marine DOP and, especially, all of the scientists who provided the original ideas, novel methodologies and oceanic data sets that we summarize herein. The NOAA-NODC online database was indispensable for the preparation of this review, as were the many scientists and technicians—both past and present— who have contributed to the HOT core measurement program. L. Lum and L. Fujieki provided invaluable assistance in the preparation of text and figures. Generous support for our investigations of the marine P-cycle has been provided by the National Science Foundation and the Gordon and Betty Moore Foundation.

References

Abell, J., Emerson, S., Renaud, P., 2000. Distribution of TOP, TON, and TOC in the North Pacific subtropical gyre: implications for nutrient supply in the surface ocean and remineralization in the upper thermocline. J. Mar. Res. 58, 203–222.

Admiraal, W., Veldhuis, M.J.W., 1987. Determination of nucleosides and nucleotides in seawater by HPLC; application to phosphatase activity in cultures of the alga *Phaeocystis pouchetii*. Mar. Ecol. Prog. Ser. 36, 277–285.

Admiraal, W., Werner, D., 1983. Utilization of limiting concentrations of ortho-phosphate and production of extracellular organic phosphates in cultures of marine diatoms. J. Plankton Res. 5, 495–513.

Aiken, J., Fishwick, J., Moore, G., Pemberton, K., 2004. The annual cycle of phytoplankton photosynthetic quantum efficiency, pigment composition and optical properties in the western English Channel. J. Mar. Biol. Assoc. UK 84, 301–313.

Alonso-Sáez, L., Gasol, J.M., 2007. Seasonal variations in the contributions of different bacterial groups to the uptake of low-molecular-weight compounds in northwestern Mediterranean coastal waters. Appl. Environ. Microbiol. 73, 3528–3535.

Aminot, A., Kérouel, R., 2004. Dissolved organic carbon, nitrogen and phosphorus in the N-E Atlantic and the N-W Mediterranean with particular reference to non-refractory fractions and degradation. Deep-Sea Res. 51, 1975–1999.

Ammerman, J.W., 1991. Role of ecto-phosphohydrolases in phosphorus regeneration in estuarine and coastal ecosystems. In: Chróst, R.J. (Ed.), Microbial Enzymes in Aquatic Environments. Springer-Verlag, New York, pp. 165–186.

Ammerman, J.W., Azam, F., 1981. Dissolved cyclic adenosine monophosphate (cAMP) in the sea and uptake of cAMP by marine bacteria. Mar. Ecol. Prog. Ser. 5, 85–89.

Ammerman, J.W., Azam, F., 1982. Uptake of cyclic AMP by natural populations of marine bacteria. Appl. Environ. Microbiol. 43, 869–876.

Ammerman, J.W., Azam, F., 1985. Bacterial 5′-nucleotidase in aquatic ecosystems: a novel mechanism of phosphorus regeneration. Science 227, 1338–1340.

Ammerman, J.W., Azam, F., 1987. Characteristics of cyclic AMP transport by marine bacteria. Appl. Environ. Microbiol. 53, 2963–2966.

Amon, R.M.W., Benner, R., 1996. Bacterial utilization of different size classes of dissolved organic matter. Limnol. Oceanogr. 41, 41–51.

Andersen, O.K., Goldman, J.C., Caron, D.A., Dennett, M.R., 1986. Nutrient cycling in a microflagellate food chain: III Phosphorus dynamics. Mar. Ecol. Prog. Ser. 31, 47–55.

Antia, N.J., Harrison, P.J., Oliveira, L., 1991. The role of dissolved organic nitrogen in phytoplankton nutrition, cell biology and ecology. Phycologia 30, 1–89.

Armstrong, F.A.J., Harvey, H.W., 1950. The cycle of phosphorus in the waters of the English Channel. J. Mar. Biol. Assoc. UK 29, 145–162.

Armstrong, F.A.J., 1965. Phosphorus. In: Riley, J.P., Skirrow, G. (Eds.), Chemical Oceanography. Academic Press, New York, pp. 323–365.

Armstrong, F.A.J., Tibbitts, S., 1968. Photochemical combustion of organic matter in sea water, for nitrogen, phosphorus and carbon determination. J. Mar. Biol. Assoc. UK 48, 143–152.

Armstrong, F.A., Williams, P.M., Strickland, J.D.H., 1966. Photo-oxidation of organic matter in seawater by ultraviolet radiation, analytical and other applications. Nature 211, 481–483.

Atkins, W.R.G., 1923. The phosphate content of fresh and salt waters in its relationship to the growth of the algal plankton. J. Mar. Biol. Assoc. UK 13, 119–150.

Atkins, W.R.G., 1925. Seasonal changes in the phosphate content of sea water in relation to the growth of the algal plankton during 1923 and 1924. J. Mar. Biol. Assoc. UK 13, 700–720.

Atkins, W.R.G., 1928. Seasonal variations in the phosphate and silicate content of sea water during 1926 and 1927 in relation to the phytoplankton crop. J. Mar. Biol. Assoc. UK 15, 191–205.

Atkins, W.R.G., 1930. Seasonal variations in the phosphate and silicate content of sea-water in relation to the phytoplankton crop. Part V. November 1927 to April 1929, compared with earlier years from 1923. J. Mar. Biol. Assoc. UK 16, 821–852.

Atkins, W.R.G., Wilson, E.G., 1926. The colorimetric estimation of minute amounts of compounds of silicon, of phosphorus, and of arsenic. Biochem. J. 20, 1223–1228.

Atkins, W.R.G., Wilson, E.G., 1927. The phosphorus and arsenic compounds of sea-water. J. Mar. Biol. Assoc. UK 14, 609–614.

Atkinson, M.J., 1987. Rates of phosphate uptake by coral reef flat communities. Limnol. Oceanogr. 32, 426–435.

Atkinson, M.J., Smith, D.F., 1987. Slow uptake of ^{32}P over a barrier reef flat. Limnol. Oceanogr. 32, 436–441.

Ault-Riché, D., Fraley, C.D., Tzeng, C.-M., Kornberg, A., 1998. Novel assay reveals multiple pathways regulating stress-induced accumulations of inorganic polyphosphate in Escherichia coli. J. Bacteriol. 18, 1841–1847.

Azam, F., Hodson, R.E., 1977. Dissolved ATP in the sea and its utilisation by marine bacteria. Nature 267, 696–698.

Baldwin, D.S., Beattie, J.K., Coleman, L.M., Jones, D.R., 1995. Phosphate ester hydrolysis facilitated by mineral phases. Environ. Sci. Tech. 29, 1706–1709.

Bauer, J.E., Ruttenberg, K.C., Wolgast, D.M., Monaghan, E., Schrope, M.K., 1996. Cross-flow filtration of dissolved and colloidal nitrogen and phosphorus in seawater: results from an intercomparison study. Mar. Chem. 55, 33–52.

Bengis-Garber, C., Kushner, D.J., 1981. Purification and properties of 5′-nucleotidase from the membrane of Vibrio costicola, a moderately halophilic bacterium. J. Bacteriol. 146, 24–32.

Bengis-Garber, C., Kushner, D.J., 1982. Role of membrane-bound 5′-nucleotidase in nucleotide uptake by the moderate halophile Vibrio costicola. J. Bacteriol. 149, 808–815.

Benitez-Nelson, C.R., 2000. The biogeochemical cycling of phosphorus in marine systems. Earth Sci. Rev. 51, 109–135.

Benitez-Nelson, C.R., Buesseler, K.O., 1999. Variability of inorganic and organic phosphorus turnover rates in the coastal ocean. Nature 398, 502–505.

Benitez-Nelson, C.R., Karl, D.M., 2002. Phosphorus cycling in the North Pacific Subtropical Gyre using cosmogenic ^{32}P and ^{33}P. Limnol. Oceanogr. 47, 762–770.

Benner, R., Pakulski, J.D., McCarthy, M., Hedges, J.I., Hatcher, P.G., 1992. Bulk chemical characteristics of dissolved organic matter in the ocean. Science 255, 1561–1564.

Benson, A.A., 1984. Phytoplankton solved the arsenate-phosphate problem. In: Holm-Hansen, O., Bolis, L., Gilles, R. (Eds.), Marine Phytoplankton and Productivity. Springer-Verlag, Berlin, pp. 55–59.

Berman, T., 1988. Differential uptake of orthophosphate and organic phosphorus substrates by bacteria and algae in Lake Kinneret. J. Plankton Res. 10, 1239–1249.

Berman, T., Walline, P.D., Schneller, A., Rothenberg, J., Townsend, D.W., 1985. Secchi disk depth record: a claim for the eastern Mediterranean. Limnol. Oceanogr. 30, 447–448.

Beversdorf, L.J., White, A.E., Björkman, K.M., Letelier, R.M., Karl, D.M., 2010. Phosphonate metabolism of Trichodesmium IMS101 and the production of greenhouse gases. Limnol. Oceanogr. 55, 1768–1778.

Björkman, K., Karl, D.M., 1994. Bioavailability of inorganic and organic phosphorus compounds to natural assemblages of microorganisms in Hawaiian coastal waters. Mar. Ecol. Prog. Ser. 111, 265–273.

Björkman, K., Karl, D.M., 2001. A novel method for the measurement of dissolved adenosine and guanosine triphosphate in aquatic habitats: applications to marine microbial ecology. J. Microbiol. Meth. 47, 159–167.

Björkman, K., Karl, D.M., 2003. Bioavailability of dissolved organic phosphorus in the euphotic zone at Station ALOHA, North Pacific Subtropical Gyre. Limnol. Oceanogr. 48, 1049–1057.

Björkman, K., Karl, D.M., 2005. Presence of dissolved nucleotides in the North Pacific subtropical gyre and their role in cycling of dissolved organic phosphorus. Aquat. Microb. Ecol. 39, 193–203.

Björkman, K., Thomson-Bulldis, A.L., Karl, D.M., 2000. Phosphorus dynamics in the North Pacific subtropical gyre. Aquat. Microb. Ecol. 22, 185–198.

Björkman, K., Duhamel, S., Karl, D.M., 2012. Microbial group specific uptake kinetics of inorganic phosphate and adenosine-5′-triphosphate (ATP) in the North Pacific Subtropical Gyre. Frontiers Aquat. Microbiol. 3, doi: 10.3389/fmicb.2012.00189.

Blake, R.E., O'Neil, J.R., Garcia, G.A., 1997. Oxygen isotope systematics of microbially mediated reactions of phosphate I.: Degradation of organophosphorus compounds. Geochim. Cosmochim. Acta 61, 4411–4422.

Boavida, M.-J., 2000. Phosphatases in phosphorus cycling: a new direction for research on an old subject. Arch. Hydrobiol. Spec. Issues Advanc. Limnol. 55, 433–440.

Boavida, M.-J., Wetzel, R.G., 1998. Inhibition of phosphatase activity by dissolved humic substances and hydrolytic reactivation by natural ultraviolet light. Freshwater Biol. 40, 285–293.

Bochner, B.R., Ames, B.N., 1982. Selective precipitation of orthophosphate from mixtures containing labile phosphorylated metabolites. Anal. Biochem. 122, 100–107.

Bogé, G., Lespilette, M., Jamet, D., Jarnet, J.L., 2012. Role of seawater DIP and DOP in controlling bulk alkaline phosphatase activity in the N.W. Mediterranean Sea (Toulon, France). Mar. Poll. Bull. 64, 1989–1996.

Boltz, D.F., 1972. Total phosphorus. In: Halmann, M. (Ed.), Analytical Chemistry of Phosphorus Compounds. Wiley-Interscience, New York, pp. 9–65.

Bonnet, S., Webb, E.A., Panzeca, C., Karl, D.M., Capone, D.G., Sañudo-Wilhelmy, S.A., 2010. Vitamin B_{12} excretion by cultures of the marine cyanobacteria *Crocosphaera* and *Synechococcus*. Limnol. Oceanogr. 55, 1959–1964.

Boon, P.I., 1994. Discrimination of algal and bacterial alkaline phosphatases with a differential-inhibition technique. Aust. J. Mar. Freshwater Res. 45, 83–107.

Bossard, P., Karl, D.M., 1986. The direct measurement of ATP and adenine nucleotide pool turnover in microorganisms: a new method for environmental assessment of metabolism, energy flux and phosphorus dynamics. J. Plankton Res. 8, 1–13.

Botsford, J.L., Harman, J.G., 1992. Cyclic AMP in prokaryotes. Microbiol. Rev. 56, 100–122.

Bourret, R.B., Borkovich, K.A., Simon, M.I., 1991. Signal transduction pathways involving protein phosphorylation in prokaryotes. Ann. Rev. Biochem. 60, 401–441.

Brandes, J., Ingall, E., Paterson, D., 2007. Characterization of minerals and organic phosphorus species in marine sediments using soft X-ray fluorescence spectomicroscopy. Mar. Chem. 103, 250–265.

Breter, H.-J., Kurelec, B., Müller, W.E.G., Zahn, R.K., 1977. Thymine content of sea water as a measure of biosynthetic potential. Mar. Biol. 40, 1–8.

Broberg, O., Persson, G., 1988. Particulate and dissolved phosphorus forms in freshwater: composition and analysis. Hydrobiologia 170, 61–90.

Broberg, O., Petterson, K., 1988. Analytical determination of orthophosphate in water. Hydrobiologia 170, 45–59.

Bronk, D.A., Lomas, M.W., Glibert, P.M., Schukert, K.J., Sanderson, M.P., 2000. Total dissolved nitrogen analysis: comparisons between the persulfate, UV and high temperature oxidation methods. Mar. Chem. 69, 163–178.

Brophy, J.E., Carlson, D.J., 1989. Production of biologically refractory dissolved organic carbon by natural seawater microbial populations. Deep-Sea Res. 36, 497–507.

Brum, J.R., 2005. Concentration, production, and turnover of viruses and dissolved DNA pools at Station ALOHA, North Pacific Subtropical Gyre. Aquat. Microb. Ecol. 41, 103–113.

Brum, J.R., Steward, G.F., Karl, D.M., 2004. A novel method for the measurement of dissolved deoxyribonucleic acid in seawater. Limnol. Oceanogr. Meth. 2, 248–255.

Buesseler, K.B., Bauer, J.B., Chen, R.F., Eglinton, T.I., Gustafsson, O., Landing, W., et al., 1996. An intercomparison of cross-flow filtration techniques used for sampling marine colloids: overview and organic carbon results. Mar. Chem. 55, 1–31.

Butler, E.I., Knox, S., Liddicoat, M.I., 1979. The relationship between inorganic and organic nutrients in sea water. J. Mar. Biol. Assoc. UK 59, 239–250.

Cadieux, J.E., Kuzio, J., Milazzo, F.H., Kropinski, A.M., 1983. Spontaneous release of lipopolysaccharide by *Pseudomonas aeruginosa*. J. Bacteriol. 155, 817–825.

Cai, Y., Guo, L., 2009. Abundance and variation of colloidal organic phosphorus in riverine, estuarine, and coastal waters in the northern Gulf of Mexico. Limnol. Oceanogr. 54, 1393–1402.

Callahan, S.M., Cornell, N.W., Dunlap, P.V., 1995. Purification and properties of periplasmic 3′:5′-cyclic nucleotide phosphodiesterase. J. Biol. Chem. 270, 17627–17632.

Cañellas, M., Agustí, S., Duarte, C., 2000. Latitudinal variability in phosphate uptake in the Central Atlantic. Mar. Ecol. Prog. Ser. 194, 283–294.

Canfield, D.E., Kristensen, E., Thamdrup, B., 2005. The phosphorus cycle. Adv. Mar. Biol. 48, 419–440.

Carini, P.J., 2013. Genome-enabled investigation of the minimal growth requirements and phosphate metabolism for *Pelagibacter* marine bacteria. PhD dissertation. Oregon State University, 196 pp.

Carlson, D.J., Mayer, L.M., Brann, M.L., Mague, T.H., 1985. Binding of monomeric organic compounds to macromolecular dissolved organic matter in seawater. Mar. Chem. 16, 141–153.

Carlucci, A.F., Silbernagel, S.B., 1966a. Bioassay of seawater. I. A ^{14}C-uptake method for the determination of concentrations of vitamin B_{12} in seawater. Can. J. Microbiol. 12, 175–183.

Carlucci, A.F., Silbernagel, S.B., 1966b. Bioassay of seawater. II. Methods for the determination of concentrations of dissolved vitamin B_1 in seawater. Can. J. Microbiol. 12, 1079–1089.

Carlucci, A.F., Silbernagel, S.B., 1966c. Determination of vitamins in seawater. In: Proceedings of an I.B.P.-Symposium held in Amsterdam and Nieuwersluis 10-16 October 1966. Koninlijke Nederlandse Akademie van Wetenschappen, Amsterdam, pp. 239–244.

Casey, J., Lomas, M.W., Michelou, V., Orchard, E.D., Dyhrman, S.T., Ammerman, J.W., et al., 2009. Phytoplankton taxon-specific orthophosphate (P*i*) and ATP uptake in the northwestern Atlantic subtropical gyre. Aquat. Microb. Ecol. 58, 31–44.

Cashel, M., Gallant, J., 1969. Two compounds implicated in the function of the RC gene of *Escherichia coli*. Nature 221, 838–841.

Cavender-Bares, K.K., Karl, D.M., Chisholm, S.W., 2001. Nutrient gradients in the western North Atlantic Ocean: relationship to microbial community structure, and comparison to patterns in the Pacific Ocean. Deep-Sea Res. 48, 2373–2395.

Cembella, A.D., Antia, N.J., Harrison, P.J., 1984a. The utilization of inorganic and organic phosphorous compounds as nutrients by eukaryotic microalgae: a multidisciplinary perspective: Part 1. CRC Crit. Rev. Microbiol. 10, 317–391.

Cembella, A.D., Antia, N.J., Harrison, P.J., 1984b. The utilization of inorganic and organic phosphorous compounds as nutrients by eukaryotic microalgae: a multidisciplinary perspective. Part 2. CRC Crit. Rev. Microbiol. 11, 13–81.

Cembella, A.D., Antia, N.J., Taylor, F.J.R., 1986. The determination of total phosphorus in seawater by nitrate oxidation of the organic component. Water Res. 20, 1197–1199.

Chamberlain, W., Shapiro, J., 1969. On the biological significance of phosphate analysis; comparison of standard and new methods with a bioassay. Limnol. Oceanogr. 14, 921–927.

Chen, L., Roberts, M.F., 1998. Cloning and expression of the inositol monophosphatase gene from *Methanococcus jannaschii* and characterization of the enzyme. Appl. Environ. Microbiol. 64, 2609–2615.

Chisholm, S.W., Olson, R.J., Zettler, E.R., et al., 1988. A novel free-living prochlorophyte abundant in the oceanic euphotic zone. Nature 334, 340–343.

Chisholm, S.W., Falkowski, P.G., Cullen, J.J., 2001. Discrediting ocean fertilization. Science 294, 309–310.

Chu, S.P., 1946. The utilization of organic phosphorus by phytoplankton. J. Mar. Biol. Assoc. UK 26, 285–295.

Church, M.J., Ducklow, H.W., Karl, D.M., 2002. Multiyear increases in dissolved organic matter inventories at Station ALOHA in the North Pacific Subtropical Gyre. Limnol. Oceanogr. 47, 1–10.

Clark, J.E., Beegen, H., Wood, H.G., 1988. Isolation of intact chains of polyphosphate from "*Propionibacterium shermanii*" grown on glucose or lactate. J. Bacteriol. 168, 1212–1219.

Clark, L.L., Ingall, E.D., Benner, R., 1998. Marine phosphorus is selectively remineralized. Nature 393, 426.

Clark, L.L., Ingall, E.D., Benner, R., 1999. Marine organic phosphorus cycling: novel insights from nuclear magnetic resonance. Am. J. Sci. 299, 724–737.

Codispoti, L.A., 1989. Phosphorus *vs.* nitrogen limitation of new and export production. In: Berger, W.H., Smetacek, V.S., Wefer, G. (Eds.), Productivity of the Ocean: Present and Past. John Wiley and Sons, New York, pp. 377–394.

Cohen, L., Kaplan, R., 1977. Accumulation of nucleotides by starved *Escherichia coli* cells as a probe for the involvement of ribonucleases in ribonucleic acid degradation. J. Bacteriol. 129, 651–657.

Colman, A.S., Blake, R.E., Karl, D.M., Fogel, M.L., Turekian, K.K., 2005. Marine phosphate oxygen isotopes and organic matter remineralization in the oceans. Proc. Natl. Acad. Sci. U. S. A. 102, 13023–13028.

Conley, D.J., Smith, W.M., Cornwell, J.C., Fisher, T.R., 1995. Transformation of particle bound phosphorus at the land-sea interface. Estuar. Coast. Shelf Sci. 40, 161–176.

Cooper, L.H.N., 1933. Chemical constituents of biological importance in the English Channel. Part I. November 1930 to January 1932. Phosphate, silicate, nitrate, nitrite, ammonia. J. Mar. Biol. Assoc. UK 18, 677–728.

Cooper, L.H.N., 1935. The rate of liberation of phosphate in sea water by the breakdown of plankton organisms. J. Mar. Biol. Assoc. UK 20, 197–202.

Cooper, L.H.N., 1938. On the ratio of nitrogen to phosphorus in the sea. J. Mar. Biol. Assoc. UK 22, 177–182.

Cooper, L.H.N., 1939a. Redefinition of the anomaly of the nitrate-phosphate ratio. J. Mar. Biol. Assoc. UK 23, 179.

Cooper, L.H.N., 1939b. Salt error in determinations of phosphate in sea water. J. Mar. Biol. Assoc. UK 23, 171–178.

Cooper, W.T., Llewelyn, J.M., Bennett, G.L., Salters, V.J.M., 2005. Mass spectrometry of natural organic phosphorus. Talanta 66, 348–358.

Correll, D.L., 1965. Pelagic phosphorus metabolism in antarctic waters. Limnol. Oceanogr. 10, 364–370.

Cotner, J.B., Wetzel, R.G., 1991. 5'-Nucleotidase activity and inhibition in a eutrophic and an oligotrophic lake. Appl. Environ. Microbiol. 57, 1306–1312.

Cotner, J.B., Wetzel, R.G., 1992. Uptake of dissolved inorganic and organic phosphorus compounds by phytoplankton and bacterioplankton. Limnol. Oceanogr. 37, 232–243.

Croft, M.T., Lawrence, A.D., Raux-Deery, E., Warren, M.J., Smith, A.G., 2005. Algae acquire vitamin B_{12} through a symbiotic relationship with bacteria. Nature 438, 90–93.

Cuhel, R.L., Jannasch, H.W., Taylor, C.D., Lean, D.R.S., 1983. Microbial growth and macromolecular synthesis in the northwestern Atlantic Ocean. Limnol. Oceanogr. 28, 1–18.

deDuve, C., 1991. Blueprint for a Cell: The Nature and Origin of Life. Neil Patterson Publishers, Burlington, NC.

DeFlaun, M.F., Paul, J.H., 1989. Detection of exogenous gene sequences in dissolved DNA from aquatic environments. Microb. Ecol. 18, 21–28.

DeFlaun, M.F., Paul, J.H., Davis, D., 1986. Simplified method for dissolved DNA determination in aquatic environments. Appl. Environ. Microbiol. 52, 654–659.

DeFlaun, M.F., Paul, J.H., Jeffrey, W.H., 1987. Distribution and molecular weight of dissolved DNA in subtropical estuarine and oceanic environments. Mar. Ecol. Prog. Ser. 38, 65–73.

DeLong, E.F., 1992. Archaea in coastal marine environments. Proc. Natl. Acad. Sci. U. S. A. 89, 5685–5689.

DeLong, E.F., Wu, K.Y., Prezelin, B.B., Jovine, R.V.M., 1994. High abundance of archaea in antarctic marine picoplankton. Nature 371, 695–697.

Denigès, G., 1920. Réaction de coloration extrêmement sensible des phosphates et des arséniates. Ses applications. C. R. Acad. Des. Sci. 171, 802–804.

Denigès, G., 1921. Détermination quantitative des plus faibles quantités de phosphates dans les produits biologiques par la méthode céruléomolybdique. C. R. Soc. Biol. 84, 875–84, 877.

DePrada, P., Loveland-Curtze, J., Brenchley, J.E., 1996. Production of two extracellular alkaline phosphatases by a psychrophilic Arthrobacter strain. Appl. Environ. Microbiol. 62, 3732–3738.

Diaz, J.M., Ingall, E.D., 2010. Fluorometric quantification of natural inorganic polyphosphate. Environ. Sci. Technol. 44, 4665–4671.

Dickey, T.D., 1991. The emergence of concurrent high-resolution physical and bio-optical measurements in the upper ocean and their applications. Rev. Geophys. 29, 383–413.

Dixon, N.E., Jackson, W.G., Marty, W., Sargeson, A.M., 1982. Base hydrolysis of pentaaminecobalt (III) complexes of urea, dimethyl sulfoxide, and trimethyl phosphate. Inorg. Chem. 21, 688–697.

Dobbs, F.C., Findlay, R.H., 1993. Analysis of microbial lipids to determine biomass and detect the response of sedimentary microorganisms to disturbance. In: Kemp, P.F., Sherr, B.F., Sherr, E.B., Cole, J.J. (Eds.), Handbook of Methods in Aquatic Microbial Ecology. Lewis Publishers, Boca Raton, FL, pp. 347–358.

Dolan, J.R., Thingstad, T.F., Rassoulzadegan, F., 1995. Phosphate transfer between microbial size-fractions in Villefranche Bay (N. W. Mediterranean Sea), France in autumn 1992. Ophelia 41, 71–85.

Dore, J.E., Houlihan, T., Hebel, D.V., Tien, G., Tupas, L., Karl, D.M., 1996. Freezing as a method of sample preservation for the analysis of dissolved inorganic nutrients in seawater. Mar. Chem. 53, 173–185.

Downes, M.T., 1978. An automated determination of low reactive phosphorus concentrations in natural waters in the presence of arsenic, silicon and mercuric chloride. Water Res. 12, 743–745.

Drummond, L., Maher, W., 1995. Determination of phosphorus in aqueous solution *via* formation of the phosphoantimonylmolybdenum blue complex: re-examination of optimum conditions for the analysis of phosphate. Anal. Chim. Acta 302, 69–74.

Dugdale, R.C., Goering, J.J., 1967. Uptake of new and regenerated forms of nitrogen in primary productivity. Limnol. Oceanogr. 12, 196–206.

Duhamel, S., Björkman, K.M., Karl, D.M., 2012. Light dependence of phosphorus uptake by microorganisms in the North and South Pacific Ocean. Aquat. Microb. Ecol. 67, 225–238.

Dunlap, P.V., Mueller, U., Lisa, T.A., Lundberg, K.S., 1992. Growth of the marine luminous bacterium *Vibrio fischeri* on 3′:5′-cyclic AMP: correlation with a periplasmic 3′:5′-cyclic AMP phosphodiesterase. J. Gen. Microbiol. 138, 115–123.

Dyhrman, S.T., Haley, S.T., 2006. Phosphorus scavenging in the unicellular marine diazotroph *Crocosphaera watsonii*. Appl. Environ. Microbiol. 72, 1452–1458.

Dyhrman, S.T., Palenik, B., 1999. Phosphate stress in cultures and field populations of the dinoflagellate *Prorocentrum minimum* detected by a single-cell alkaline phosphatase assay. Appl. Environ. Microbiol. 65, 3205–3212.

Dyhrman, S.T., Webb, E., Anderson, D.M., Moffett, I., Waterbury, J., 2002. Cell specific detection of phosphorus stress in *Trichodesmium* from the Western North Atlantic. Limnol. Oceanogr. 47, 1823–1836.

Dyhrman, S.T., Chappell, P.D., Haley, S.T., Moffett, J.W., Orchard, E.D., Waterbury, J.B., et al., 2006. Phosphonate utilization by the globally important marine diazotroph *Trichodesmium*. Nature 439, 68–71.

Dyhrman, S.T., Ammerman, J.W., Van Mooy, B.A.S., 2007. Microbes and the marine phosphorus cycle. Oceanography 20, 110–116.

Espinosa, M., Turner, B.L., Haygarth, P.M., 1999. Preconcentration and separation of trace phosphorus compounds in soil leachate. J. Environ. Qual. 28, 1497–1504.

Falkowski, P.G., 2000. Rationalizing elemental ratios in unicellular algae. J. Phycol. 36, 3–6.

Fanning, K.A., 1992. Nutrient provinces in the sea: concentration ratios, reaction rate ratios, and ideal covariation. J. Geophys. Res. 97, 5693–5712.

Fernández, J.A., Niell, F.X., Lucena, J., 1985. A rapid and sensitive automated determination of phosphate in natural waters. Limnol. Oceanogr. 30, 227–230.

Fiske, C.H., Subbarow, Y., 1925. The colorimetric determination of phosphorus. J. Biol. Chem. 66, 375–400.

Flerus, R., Lechtenfeld, O.J., Koch, B.P., McCallister, S.L., Schmitt-Kopplin, P., Benner, R., et al., 2012. A molecular perspective on the ageing of marine dissolved organic matter. Biogeosciences 9, 1935–1955.

Flynn, K.J., Öpik, H., Syrett, P.J., 1986. Localization of the alkaline phosphatase and 5′-nucleotidase activities of the diatom *Phaeodactylum tricornutum*. J. Gen. Microbiol. 132, 289–298.

Fogg, G.E., 1966. The extracellular products of algae. Oceanogr. Mar. Biol. Ann. Rev. 4, 195–212.

Föllmi, K.B., 1996. The phosphorus cycle, phosphogenesis and marine phosphate-rich deposits. Earth Sci. Rev. 40, 55–124.

Fourqurean, J.W., Jones, R.D., Zieman, J.C., 1993. Processes influencing water column nutrient characteristics and phosphorus limitation of phytoplankton biomass in Florida Bay, FL, USA: inferences from spatial distributions. Estuar. Coast. Shelf Sci. 36, 295–314.

Francesconi, K.A., Edmonds, J.S., 1993. Arsenic in the sea. Oceanogr. Mar. Biol. Ann. Rev. 31, 111–151.

Francko, D.A., Heath, R.T., 1982. UV-sensitive complex phosphorus: association with dissolved humic material and iron in a bog lake. Limnol. Oceanogr. 27, 564–569.

Fuhrman, J.A., McCallum, K., Davis, A.A., 1992. Novel major archaebacterial group from marine plankton. Nature 356, 148–149.

Furnas, M.J., Mitchell, A.W., 1999. Wintertime carbon and nitrogen fluxes on Australia's Northwest shelf. Estuar. Coast. Shelf Sci. 49, 165–175.

Fuwa, K., Lei, W., Fujiwara, K., 1984. Colorimetry with a total-reflection long-capillary cell. Anal. Chem. 56, 1640–1644.

Gambin, F., Bogé, G., Jamet, D., 1999. Alkaline phosphatase in a littoral Mediterranean marine ecosystem: role of the main plankton size classes. Mar. Environ. Res. 47, 441–456.

Garde, K., Gustavson, K., 1999. The impact of UV-B radiation on alkaline phosphatase activity in phosphorus-depleted marine ecosystems. J. Exp. Mar. Biol. Ecol. 238, 93–105.

Gill, R., 1927. The influence of plankton on the phosphate content of stored sea-water. J. Mar. Biol. Assoc. UK 14, 1057–1065.

Gilmartin, M., 1967. Changes in inorganic phosphate concentration occurring during seawater sample storage. Limnol. Oceanogr. 12, 325–328.

Giovannoni, S.J., Tripp, H.J., Givan, S., Podar, M., Vergin, K.L., Baptista, D., et al., 2005. Genome streamlining in a cosmopolitan oceanic bacterium. Science 309, 1242–1245.

Golimowski, J., Golimowska, K., 1996. UV-photooxidation as pretreatment step in inorganic analysis of environmental samples. Anal. Chim. Acta 325, 111–133.

González-Gil, S., Keafer, B.A., Jovine, R.V.M., Aguilera, A., Lu, S., Anderson, D.M., 1998. Detection and quantification of alkaline phosphatase in single cells of phosphorus-starved marine phytoplankton. Mar. Ecol. Prog. Ser. 164, 21–35.

Grosskopf, T., Mohr, W., Baustian, T., Schunck, H., Gill, D., Kuypers, M.M.M., et al., 2012. Doubling of marine dinitrogen-fixation rates based on direct measurements. Nature 488, 361–364.

Gurtler, B.K., Vetter, T.A., Perdue, E.M., Ingall, E., Koprivnjak, J.-F., Pfromm, P.H., 2008. Combining reverse osmosis and pulsed electrical current electrodialysis for improved recovery of dissolved organic matter from seawater. J. Membr. Sci. 323, 328–336.

Haines, K.C., Guillard, R.R.L., 1974. Growth of vitamin B_{12}-requiring marine diatoms in mixed laboratory cultures with vitamin B_{12}-producing marine bacteria. J. Phycol. 10, 245–252.

Hanrahan, G., Salmassi, T.M., Khachikian, C.S., Foster, K.L., 2005. Reduced inorganic phosphorus in the natural environment: significance, speciation and determination. Talanta 66, 435–444.

Hansen, A.L., Robinson, R.J., 1953. The determination of organic phosphorus in sea water with perchloric acid oxidation. J. Mar. Res. 12, 31–42.

Harrison, W.G., 1983. Uptake and recycling of soluble reactive phosphorus by marine microplankton. Mar. Ecol. Prog. Ser. 10, 127–135.

Harrison, W.G., Azam, F., Renger, E.H., Eppley, R.W., 1977. Some experiments on phosphate assimilation by coastal marine plankton. Mar. Biol. 40, 9–18.

Harvey, H.W., 1940. Nitrogen and phosphorus required for the growth of phytoplankton. J. Mar. Biol. Assoc. UK 24, 115–123.

Harvey, H.W., 1948. The estimation of phosphate and of total phosphorus in sea waters. J. Mar. Biol. Assoc. UK 27, 337–359.

Harvey, H.W., 1950. On the production of living matter in the sea off Plymouth. J. Mar. Biol. Assoc. UK 29, 97.

Harvey, H.W., 1955. The Chemistry and Fertility of Sea Waters. Cambridge University Press, London.

Hashimoto, S., Fujiwara, K., Fuwa, K., 1987. Determination of phosphorus in natural water using hydride generation and gas chromatography. Limnol. Oceanogr. 32, 729–735.

Hassan, H.M., Pratt, D., 1977. Biochemical and physiological properties of alkaline phosphatases in five isolates of marine bacteria. J. Bacteriol. 129, 1607–1612.

Haury, L., Shulenberger, E., 1998. Surface nutrient enrichment in the California Current off Southern California: description and possible causes. Deep-Sea Res. II 45, 1577–1601.

Haury, L.R., Fey, C.L., Shulenberger, E., 1994. Surface enrichment of inorganic nutrients in the North Pacific Ocean. Deep-Sea Res. 41, 1191–1205.

Hebel, D.V., Karl, D.M., 2001. Seasonal, interannual and decadal variations in particulate matter concentrations and composition in the subtropical North Pacific Ocean. Deep-Sea Res. II 48, 1669–1696.

Hedges, J.I., Eglinton, G., Hatcher, P.G., Kirchman, D.L., Arnosti, C., Derenne, S., et al., 2000. The molecularly-uncharacterized component of nonliving organic matter in natural environments. Org. Chem. 31, 945–958.

Hellebust, J.A., 1974. Extracellular products. Bot. Monogr. 10, 838–863.

Herbes, S.E., Allen, H.E., Mancy, K.H., 1975. Enzymatic characterization of soluble organic phosphorus in lake water. Science 187, 432–434.

Hertkorn, N., Harir, M., Koch, B.P., Michalke, B., Schmitt-Kopplin, P., 2013. High field NMR spectroscopy and FTICR mass spectrometry: powerful discovery tools for the molecular level characterization of marine dissolved organic matter. Biogeosciences 10, 1583–1624.

Hicks, E., Riley, J.P., 1980. The determination of dissolved total nucleic acids in natural waters including sea water. Anal. Chim. Acta 116, 137–144.

Hino, S., 1989. Characterization of orthophosphate release from dissolved organic phosphorus by gel filtration and several hydrolytic enzymes. Hydrobiologia 174, 49–55.

Hodson, R.E., Azam, F., 1979. Occurrence and characterization of a phosphoenolpyruvate: glucose phosphotransferase system in a marine bacterium, *Serratia marinorubra*. Appl. Environ. Microbiol. 38, 1086–1091.

Hodson, R.E., Holm-Hansen, O., Azam, F., 1976. Improved methodology for ATP determination in marine environments. Mar. Biol. 34, 143–149.

Hodson, R.E., Maccubbin, A.E., Pomeroy, L.R., 1981. Dissolved adenosine triphosphate utilization by free-living and attached bacterioplankton. Mar. Biol. 64, 43–51.

Holm-Hansen, O., 1969. Determination of microbial biomass in ocean profiles. Limnol. Oceanogr. 14, 740–747.

Holm-Hansen, O., Strickland, J.D.H., Williams, P.M., 1966. A detailed analysis of biologically important substances in a profile off southern California. Limnol. Oceanogr. 11, 548–561.

Holm-Hansen, O., Sutcliffe Jr., W.H., Sharp, J., 1968. Measurement of deoxyribonucleic acid in the ocean and its ecological significance. Limnol. Oceanogr. 13, 507–514.

Hopkinson Jr., C.S., Fry, B., Nolin, A.L., 1997. Stoichiometry of dissolved organic matter dynamics on the continental shelf of the northeastern U.S.A. Cont. Shelf Res. 17, 473–489.

Hoppe, H.-G., Ullrich, S., 1999. Profiles of ectoenzymes in the Indian Ocean: phenomena of phosphatase activity in the mesopelagic zone. Aquat. Microb. Ecol. 19, 139–148.

Hudson, J.J., Taylor, W.D., 1996. Measuring regeneration of dissolved phosphorus in planktonic communities. Limnol. Oceanogr. 41, 1560–1565.

Hudson, J.J., Taylor, W.D., Schindler, D.W., 2000. Phosphate concentration in lakes. Nature 406, 54–56.

Hulett, H.R., 1970. Non-enzymatic hydrolysis of adenosine phosphates. Nature 225, 1248–1249.

Hutchinson, G.E., Bowen, V.T., 1947. A direct demonstration of the phosphorus cycle in a small lake. Proc. Natl. Acad. Sci. U. S. A. 33, 148–153.

Hutchinson, G.E., Bowen, V.T., 1950. Limnological studies in Connecticut. IX. A quantitative radiochemical study of the phosphorus cycle in Linsley Pond. Ecology 31, 194–203.

Ingall, E.D., Schroeder, P.A., Berner, R.A., 1990. The nature of organic phosphorus in marine sediments: new insights from ^{31}P NMR. Geochim. Cosmochim. Acta 54, 2617–2620.

Ingall, D.D., Brandes, J.A., Diaz, J.M., deJonge, M.D., Paterson, D., McNulti, I., et al., 2011. Phosphorus K-edge XANES spectroscopy of mineral standards. J. Synchrotron Radiat. 18, 189–197.

Jackson, G.A., Williams, P.M., 1985. Importance of dissolved organic nitrogen and phosphorus to biological nutrient cycling. Deep-Sea Res. 32, 223–235.

Jensen, T.E., Sicko, L.M., 1974. Phosphate metabolism in blue-green algae. I. Fine structure of the "polyphosphate overplus" phenomenon in Plectonema boryanum. Can. J. Microbiol. 20, 1235–1239.

Jiang, S.C., Paul, J.H., 1994. Seasonal and diel abundance of viruses and occurrence of lysogeny/bacteriocinogeny in the marine environment. Mar. Ecol. Prog. Ser. 104, 163–172.

Jiang, S.C., Paul, J.H., 1995. Viral contribution to dissolved DNA in the marine environment as determined by differential centrifugation and kingdom probing. Appl. Environ. Microbiol. 61, 317–325.

Jiao, N., Herndl, G.J., Hansell, D.A., Benner, R., Kattner, G., Wilhelm, S.W., et al., 2010. Microbial production of recalcitrant dissolved organic matter: long-term carbon storage in the global ocean. Nat. Rev. Microbiol. 8, 593–598.

Johannes, R.E., 1964. Uptake and release of dissolved organic phosphorus by representatives of a coastal marine ecosystem. Limnol. Oceanogr. 9, 224–234.

Johnson, D.L., 1971. Simultaneous determination of arsenate and phosphate in natural waters. Environ. Sci. Tech. 5, 411–414.

Joint, I., Jordan, M.B., Carr, M.R., 1997. Is phosphate part of the Russell Cycle? J. Mar. Biol. Assoc. UK 77, 625–633.

Jones, W., Perkins, M.E., 1923. The gravimetric determination of organic phosphorus. J. Biol. Chem. 55, 343–351.

Jones, P.G.W., Spencer, C.P., 1963. Comparison of several methods of determining inorganic phosphate in seawater. J. Mar. Biol. Assoc. UK 43, 251–273.

Jordan, M.B., Joint, I., 1998. Seasonal variation in nitrate: phosphate ratios in the English Channel 1923-1987. Estuar. Coast. Shelf Sci. 46, 157–164.

Juday, C., Birge, E.A., Kemmerer, G., Robinson, R.J., 1928. Phosphorus content of lake waters of northeastern Wisconsin. Trans. Wis. Acad. Sci. Arts Lett. 23, 233–248.

Kaiser, K., Benner, R., 2012. Organic matter transformations in the upper mesopelagic zone of the North Pacific: chemical composition and linkages to microbial community structure. J. Geophys. Res. 117, doi:10.1029/2011JC007141, C01023.

Kamat, S.S., Williams, H.J., Dangott, L.J., Chakrabarti, M., Raushel, F.M., 2013. The catalytic mechanism for aerobic formation of methane by bacteria. Nature 497, 132–136.

Kanodia, S., Roberts, M.F., 1983. Methanophosphagen: unique cyclic pyrophosphate isolated from Methanobacterium thermoautotrophicum. Proc. Natl. Acad. Sci. U. S. A. 80, 5217–5221.

Karl, D.M., 1978. Occurrence and ecological significance of GTP in the ocean and in microbial cells. Appl. Environ. Microbiol. 36, 349–355.

Karl, D.M., 1981. Simultaneous rates of ribonucleic acid and deoxyribonucleic acid syntheses for estimating growth and cell division of aquatic microbial communities. Appl. Environ. Microbiol. 42, 802–810.

Karl, D.M., 1999. A sea of change: biogeochemical variability in the North Pacific subtropical gyre. Ecosystems 2, 181–214.

Karl, D.M., 2000. A new source of 'new' nitrogen in the sea. Trends Microbiol. 8, 301.

Karl, D.M., 2002. Nutrient dynamics in the deep blue sea. Trends Microbiol. 10, 410–418.

Karl, D.M., 2007a. Microbial oceanography: paradigms, processes and promise. Nat. Rev. Microbiol. 5, 759–769.

Karl, D.M., 2007b. The marine phosphorus cycle. In: Hurst, C.J., Crawford, R.L., Garland, J.L., Lipson, D.A., Mills,

A.L., Stetzenbach, L.D. (Eds.), Manual of Environmental Microbiology. Third ed.. American Society of Microbiology, Washington, DC, pp. 523–539.

Karl, D.M., 2010. Oceanic ecosystem time-series programs: ten lessons learned. Oceanography 23, 104–125.

Karl, D.M., 2014. Microbially mediated transformations of phosphorus in the sea: new views of an old cycle. Ann. Rev. Mar. Sci. 6, 279–337.

Karl, D.M., Bailiff, M.D., 1989. The measurement and distribution of dissolved nucleic acids in aquatic environments. Limnol. Oceanogr. 34, 543–558.

Karl, D.M., Björkman, K.M., 2002. Dynamics of DOP. In: Hansell, D., Carlson, C. (Eds.), Biogeochemistry of Marine Dissolved Organic Matter. Academic Press, San Diego, pp. 249–366.

Karl, D.M., Bossard, P., 1985. Measurement and significance of ATP and adenine nucleotide pool turnover in microbial cells and environmental samples. J. Microbiol. Meth. 3, 125–139.

Karl, D.M., Craven, D.B., 1980. Effects of alkaline phosphatase activity on nucleotide measurements in aquatic microbial communities. Appl. Environ. Microbiol. 40, 549–561.

Karl, D.M., Dobbs, F.C., 1998. Molecular approaches to microbial biomass estimation in the sea. In: Cooksey, K.E. (Ed.), Molecular Approaches to the Study of the Ocean. Chapman & Hall, London, pp. 29–89.

Karl, D.M., Dore, J.E., 2001. Microbial ecology at sea: sampling, subsampling and incubation considerations. In: Paul, J.H. (Ed.), Methods in Marine Microbiology, Academic Press, pp. 13–39.

Karl, D.M., Holm-Hansen, O., 1978. Methodology and measurement of adenylate energy charge ratios in environmental samples. Mar. Biol. 48, 185–197.

Karl, D.M., Tien, G., 1992. MAGIC: a sensitive and precise method for measuring dissolved phosphorus in aquatic environments. Limnol. Oceanogr. 37, 105–116.

Karl, D.M., Tien, G., 1997. Temporal variability in dissolved phosphorus concentrations in the subtropical North Pacific Ocean. Mar. Chem. 56, 77–96.

Karl, D.M., Winn, C.D., 1984. Adenine metabolism and nucleic acid synthesis: applications to microbiological oceanography. In: Hobbie, J.E., Williams, P.J.leB. (Eds.), Heterotrophic Activity in the Sea. Plenum Publishing Corp, New York, pp. 197–215.

Karl, D.M., Yanagi, K., 1997. Partial characterization of the dissolved organic phosphorus pool in the oligotrophic North Pacific Ocean. Limnol. Oceanogr. 42, 1398–1405.

Karl, D.M., Amos, A., Holm-Hansen, O., Huntley, M.E., Vernet, M., 1992. RACER: the Marguerite Bay ice-edge reconnaissance. Antarct. J. US 27, 175–177.

Karl, D.M., Letelier, R., Hebel, D., Tupas, L., Dore, J., Christian, J., et al., 1995. Ecosystem changes in the North Pacific subtropical gyre attributed to the 1991-92 El Niño. Nature 373, 230–234.

Karl, D.M., Christian, J.R., Dore, J.E., Hebel, D.V., Letelier, R.M., Tupas, L.M., et al., 1996. Seasonal and interannual variability in primary production and particle flux at Station ALOHA. Deep-Sea Res. II 43, 539–568.

Karl, D., Letelier, R., Tupas, L., Dore, J., Christian, J., Hebel, D., 1997. The role of nitrogen fixation in biogeochemical cycling in the subtropical North Pacific Ocean. Nature 388, 533–538.

Karl, D.M., Hebel, D.V., Björkman, K., Letelier, R.M., 1998. The role of dissolved organic matter release in the productivity of the oligotrophic North Pacific Ocean. Limnol. Oceanogr. 43, 1270–1286.

Karl, D.M., Bidigare, R.R., Letelier, R.M., 2001a. Long-term changes in plankton community structure and productivity in the North Pacific Subtropical Gyre: the domain shift hypothesis. Deep-Sea Res. II 48, 1449–1470.

Karl, D.M., Björkman, K.M., Dore, J.E., Fujieki, L., Hebel, D.V., Houlihan, T., et al., 2001b. Ecological nitrogen-to-phosphorus stoichiometry at Station ALOHA. Deep-Sea Res. II 48, 1529–1566.

Karl, D., Michaels, A., Bergman, B., Capone, D., Carpenter, E., Letelier, R., et al., 2002. Dinitrogen fixation in the world's oceans. Biogeochemistry 57 (58), 47–98.

Karl, D.M., Beversdorf, L., Björkman, K.M., Church, M.J., Martinez, A., DeLong, E.F., 2008. Aerobic production of methane in the sea. Nat. Geosci. 1, 473–478.

Karner, M.B., DeLong, E.F., Karl, D.M., 2001. Archaeal dominance in the mesopelagic zone of the Pacific Ocean. Nature 409, 507–510.

Kattner, G., 1999. Storage of dissolved inorganic nutrients in seawater: poisoning with mercuric chloride. Mar. Chem. 67, 61–66.

Keil, R.G., Kirchman, D.L., 1994. Abiotic transformation of labile protein to refractory protein in sea water. Mar. Chem. 45, 187–196.

Kennedy, K.E., Thompson Jr., G.A., 1970. Phosphonolipids: localization in surface membranes of *Tetrahymena*. Science 168, 989–991.

Kérouel, R., Aminot, A., 1996. Model compounds for the determination of organic and total phosphorus dissolved in natural waters. Anal. Chim. Acta 318, 385–390.

Kester, D.R., Pytkowicz, R.M., 1967. Determination of the apparent dissociation constants of phosphoric acid in seawater. Limnol. Oceanogr. 12, 243–252.

Ketchum, B.H., Corwin, N., Keen, D.J., 1955. The significance of organic phosphorus determinations in ocean waters. Deep-Sea Res. 2, 172–181.

Kieber, D.J., McDaniel, J., Mopper, K., 1989. Photochemical source of biological substrates in sea water: implications for carbon cycling. Nature 341, 637–639.

Kiene, R.P., 1991. Production and consumption of methane in aquatic systems. In: Rogers, J.E., Whitman, W.B. (Eds.), Microbial Production and Consumption of Greenhouse

Gases: Methane, Nitrogen Oxides, and Halomethanes. ASM, Washington, DC, pp. 111–146.

Kirchman, D.L., Sigda, J., Kapuscinski, R., Mitchell, R., 1982. Statistical analysis of the direct count method for enumerating bacteria. Appl. Environ. Microbiol. 43, 376–382.

Kittredge, J.S., Roberts, E., 1969. A carbon-phosphorus bond in nature. Science 164, 37–42.

Knowles, J.R., 1980. Enzyme-catalyzed phosphoryl transfer reactions. Ann. Rev. Biochem. 49, 877–919.

Kobori, H., Taga, N., 1979. Phosphatase activity and its role in the mineralization of organic phosphorus in coastal sea water. J. Exp. Mar. Biol. Ecol. 36, 23–39.

Koike, I., Nagata, T., 1997. High potential activity of extracellular alkaline phosphatase in deep waters of the central Pacific. Deep-Sea Res. II 44, 2283–2294.

Koike, I.S., Hara, S., Terauchi, K., Kogue, K., 1990. Role of sub-micron particles in the ocean. Nature 345, 242–244.

Kolowith, L.C., Ingall, E.D., Benner, R., 2001. Composition and cycling of marine organic phosphorus. Limnol. Oceanogr. 46, 309–320.

Kononova, S.V., Nesmeyanova, M.A., 2002. Phosphonates and their degradation by microorganisms. Biochemistry 67, 184–195.

Koprivnjak, J.-F., Pfromm, P.H., Ingall, E., Vetter, T.A., Schmitt-Kopplin, P., Hertkorn, N., et al., 2009. Chemical and spectroscopic characterization of marine dissolved organic matter isolated using coupled reverse osmosis-electrodialysis. Geochim. Cosmochim. Acta 73, 4215–4231.

Kornberg, A., Fraley, C.D., 2000. Inorganic polyphosphate: a molecular fossil come to life. ASM News 66, 275–280.

Kornberg, A., Rao, N.N., Ault-Riché, D., 1999. Inorganic polyphosphate: a molecule of many functions. Ann. Rev. Biochem. 68, 89–125.

Krempin, D.W., McGrath, S.M., SooHoo, J.B., Sullivan, C.W., 1981. Orthophosphate uptake by phytoplankton and bacterioplankton from the Los Angeles Harbor and southern California coastal waters. Mar. Biol. 64, 23–33.

Kreps, E., 1934. Organic catalysts or enzymes in sea water. James Johnstone Memorial Volume. University of Liverpool Press, Liverpool, pp. 193.

Krom, M.D., 2005. Preface: CYCLOPS dedicated volume. Deep-Sea Res. II 52, 2877–2878.

Krom, M.D., Kress, N., Brenner, S., Gordon, L.I., 1991. Phosphorus limitation of primary productivity in the eastern Mediterranean Sea. Limnol. Oceanogr. 36, 424–432.

Krom, M.D., Thingstad, T.F., Brenner, S., Carbo, P., Drakopoulos, P., Fileman, T.W., et al., 2005a. Summary and overview of the CYCLOPS P addition Lagrangian experiment in the Eastern Mediterranean. Deep-Sea Res. II 52, 3090–3108.

Krom, M.D., Woodward, E.M.S., Herut, B., Kress, N., Carbo, P., Mantoura, R.F.C., et al., 2005b. Nutrient cycling in the south east Levantine basin of the eastern Mediterranean: results from a phosphorus starved system. Deep-Sea Res. II 52, 2879–2896.

Krough, A., 1934. Conditions of life in the ocean. Ecol. Monogr. 4, 421–429.

Kuenzler, E.J., 1965. Glucose-6-phosphate utilization by marine algae. J. Phycol. 1, 156–164.

Kuenzler, E.J., Ketchum, B.H., 1962. Rate of phosphorus uptake by Phaeodactylum tricornutum. Biol. Bull. 123, 134–145.

Kujawinski, E.G., Longnecker, K., Blough, N.V., Del Vecchio, R., Finlay, L., Kitner, J.B., et al., 2009. Identification of possible source markers in marine dissolved organic matter using ultrahigh resolution mass spectrometry. Geochim. Cosmochim. Acta 73, 4384–4399.

Kulaev, I.S., Vagabov, V.M., 1983. Polyphosphate metabolism in micro-organisms. Adv. Microb. Physiol. 24, 83–171.

Lal, D., Lee, T., 1988. Cosmogenic ^{32}P and ^{33}P used as tracers to study phosphorus recycling in the ocean. Nature 333, 752–754.

Lal, D., Chung, Y., Platt, T., Lee, T., 1988. Twin cosmogenic radiotracer studies of phosphorus cycling and chemical fluxes in the upper ocean. Limnol. Oceanogr. 33, 1559–1567.

Lee, C., Wakeham, S.G., 1988. Organic matter in seawater: Biogeochemical processes. In: Riley, J.P. (Ed.), Chemical Oceanography, Vol. 9. Academic Press, New York, pp. 1–51.

Lemke, M.J., Churchill, P.F., Wetzel, R.G., 1995. Effect of substrate and cell surface hydrophobicity on phosphate utilization in bacteria. Appl. Environ. Microbiol. 61, 913–919.

Levitus, S., Conkright, M.E., Reid, J.L., Najjar, R.G., Mantyla, A., 1993. Distribution of nitrate, phosphate and silicate in the world oceans. Prog. Oceanogr. 31, 245–273.

Li, H., Veldhuis, M.J.W., Post, A.F., 1998. Alkaline phosphatase activities among planktonic communities in the northern Red Sea. Mar. Ecol. Prog. Ser. 173, 107–115.

Liang, Y., Blake, R.E., 2009. Compound- and enzyme-specific phosphodiester hydrolysis mechanisms revealed by δ^{18}O of dissolved inorganic phosphate: implications for marine P cycling. Geochim. Cosmochim. Acta 73, 3782–3794.

Liang, Y., Yuan, D., Li, Q., Lin, Q., 2006. Flow injection analysis of ultratrace orthophosphate in seawater with solid-phase enrichment and luminal chemiluminescence detection. Anal. Chim. Acta 571, 184–190.

Lin, P., Guo, L., Chen, M., Tong, J., Lin, F., 2012. The distribution and chemical speciation of dissolved and particulate phosphorus in the Bering Sea and the Chukchi-Beaufort Seas. Deep-Sea Res. II 81–84, 79–94.

Lipmann, F., 1965. Projecting backward from the present stage of evolution of biosynthesis. In: Fox, S.W. (Ed.), The Origins of Prebiological Systems. Academic Press, New York, pp. 259–280.

Liu, C.-L., Hart, N., Peck Jr., H.D., 1982. Inorganic pyrophosphate: energy source for sulfate-reducing bacteria of the genus *Desulfotomaculum*. Science 217, 363–364.

Liu, Q., Parrish, C.C., Helleur, R., 1998. Lipid class and carbohydrate concentrations in marine colloids. Mar. Chem. 60, 177–188.

Llewelyn, J.M., Landing, W.M., Marshall, A.G., Cooper, W.T., 2002. Electrospray ionization Fourier transform ion cyclotron resonance mass spectrometry of dissolved organic phosphorus species in a treatment wetland after selective isolation and concentration. Anal. Chem. 74, 600–606.

Logan, B.E., Kirchman, D.L., 1991. Uptake of dissolved organics by marine bacteria as a function of fluid motion. Mar. Biol. 111, 175–181.

Loh, A.N., Bauer, J.E., 2000. Distribution, partitioning and fluxes of dissolved and particulate organic C, N and P in the eastern North Pacific and Southern Oceans. Deep-Sea Res. 47, 2287–2316.

Lomas, M.W., Burke, A.L., Lomas, D.A., Bell, D.W., Shen, C., Dyhrman, S.T., et al., 2010. Sargasso Sea phosphorus biogeochemistry: an important role for dissolved organic phosphorus (DOP). Biogeosciences 7, 695–710.

Longinelli, A., Bartelloni, M., Cortecci, G., 1976. The isotopic cycle of oceanic phosphate I. Earth Planet Sci. Lett. 32, 389–392.

Lorenz, M.G., Wackernagel, W., 1994. Bacterial gene transfer by natural genetic transformation in the environment. Microbiol. Rev. 58, 563–602.

Løvdal, T., Tanaka, T., Thingstad, T.F., 2007. Algal-bacterial competition for phosphorus from dissolved DNA, ATP, and orthophosphate in a mesocosm experiment. Limnol. Oceanogr. 52, 1407–1419.

Luo, H., Benner, R., Long, R.A., Hu, J., 2009. Subcellular localization of marine bacterial alkaline phosphatases. Proc. Natl. Acad. Sci. U. S. A. 106, 21219–21223.

Luo, H., Zhange, H., Long, R.A., Benner, R., 2011. Depth distributions of alkaline phosphatase and phosphonate utilization genes in the North Pacific Subtropical Gyre. Aquat. Microb. Ecol. 62, 61–69.

Magnuson, J.J., 1990. Long-term ecological research and the invisible present. BioScience 40, 495–501.

Mahaffey, C., Williams, R.G., Wolff, G.A., Anderson, W.T., 2004. Physical supply of nitrogen to phytoplankton in the Atlantic Ocean. Global Biogeochem. Cycles 18, GB1034.

Maher, W., Woo, L., 1998. Procedures for the storage and digestion of natural waters for the determination of filterable reactive phosphorus, total filterable phosphorus and total phosphorus. Anal. Chim. Acta 375, 5–47.

Maloney, P.C., Ambudkar, S.V., Anantharam, V., Sonna, L.A., Varadhachary, A., 1990. Anion-exchange mechanisms in bacteria. Microbiol. Rev. 54, 1–17.

Mao, J.D., Kong, X.Q., Schmidt-Rohr, K., Pignatello, J.J., Perdue, E.M., 2012. Advanced solid-state NMR characterization of marine dissolved organic matter isolated using the coupled reverse osmosis/electrodialysis method. Environ. Sci. Technol. 46, 5806–5814.

Martin, P., van Mooy, B.A.S., 2013. Fluorometric quantification of polyphosphate in environmental plankton samples: extraction protocols, matrix effects, and nucleic acid interference. Appl. Environ. Microbiol. 79, 273–281.

Martinez, A., Tyson, G.W., DeLong, E.F., 2010. Widespread known and novel phosphonate utilization pathways in marine bacteria revealed by functional screening and metagenomic analyses. Environ. Microbiol. 12, 222–238.

Martins, L.O., Huber, R., Huber, H., Stetter, K.O., da Costa, M.S., Santos, H., 1997. Organic solutes in hyperthermophilic *Archaea*. Appl. Environ. Microbiol. 63, 896–902.

Martiny, A.C., Pham, C.T.A., Primeau, F.W., Vrugt, J.A., Moore, J.K., Levin, S.A., et al., 2013. Strong latitudinal patterns in the elemental ratios of marine plankton and organic matter. Nat. Geosci. 6, 279–283.

Maruyama, A., Oda, M., Higashihara, T., 1993. Abundance of virus-sized non-DNase-digestible (coated DNA) in eutrophic seawater. Appl. Environ. Microbiol. 59, 712–717.

Mather, R.L., Reynolds, S.E., Wolff, G.A., Williams, R.G., Torres-Valdes, S., Woodward, E.M.S., et al., 2008. Phosphorus cycling in the North and South Atlantic Ocean subtropical gyres. Nat. Geosci. 1, 439–443.

Matsuda, O., Maruyama, A., 1985. Gel chromatographic characterization of dissolved organic phosphorus in eutrophic seawater during a phytoplankton bloom. Bull. Plankton Soc. Jpn 32, 91–99.

Matthews, D.J., 1916. On the amount of phosphoric acid in the seawater off Plymouth Sound. J. Mar. Biol. Assoc. UK 11, 122–130.

Matthews, D.J., 1917. On the amount of phosphoric acid in the sea-water off Plymouth Sound II. J. Mar. Biol. Assoc. UK 11, 251–257.

Matussek, K., Moritz, P., Brunner, N., Eckerskorn, C., Hensel, R., 1998. Cloning, sequencing, and expression of the gene encoding cyclic 2,3-diphosphoglycerate synthetase, the key enzyme of cyclic 2,3-diphosphoglycerate metabolism in *Methanothermus fervidus*. J. Bacteriol. 180, 5997–6004.

Mayer, H., Tharanathan, R.N., Weckesser, J., 1985. Analysis of lipopolysaccharides of gram-negative bacteria. In: Gottschalk, G. (Ed.), Methods in Microbiology, Vol. 18. Academic Press, London, pp. 157–207.

McCallister, S.L., Bauer, J.E., Kelly, J., Dhcklow, H.W., 2005. Effects of sunlight on decomposition of estuarine dissolved organic C, N and P and bacterial metabolism. Aquat. Microb. Ecol. 40, 25–35.

McCarter, L.L., Silverman, M., 1987. Phosphate regulation of gene expression in *Vibrio parahaemolyticus*. J. Bacteriol. 169, 3441–3449.

McCarthy, M., Pratum, T., Hedges, J., Benner, R., 1997. Chemical composition of dissolved organic nitrogen in the ocean. Nature 390, 150–154.

McCarthy, M., Hedges, J., Benner, R., 1998. Major bacterial contribution to marine dissolved organic nitrogen. Science 281, 231–234.

McGill, D.A., 1963. The distribution of phosphorus and oxygen in the Atlantic Ocean. PhD dissertation, Woods Hole Oceanographic Institution, Woods Hole, Massachusetts.

McGill, D.A., 1964. The distribution of phosphorus and oxygen in the Atlantic ocean, as observed during the I.G.Y., 1957-1958. Prog. Oceanogr. 2, 129–211.

McGill, D.A., Corwin, N., Ketchum, B.H., 1964. Organic phosphorus in the deep water of the western North Atlantic. Limnol. Oceanogr. 9, 27–34.

McGrath, S.M., Sullivan, C.W., 1981. Community metabolism of adenylates by microheterotrophs from the Los Angeles and Southern California coastal waters. Mar. Biol. 62, 217–226.

McKelvie, I.D., Hart, B.T., Cardwell, T.J., Cattrall, R.W., 1995. Use of immobilized 3-phytase and flow injection for the determination of phosphorus species in natural waters. Anal. Chim. Acta 316, 277–289.

McLaughlin, K., Sohm, J.A., Cutter, G.A., Lomas, M.W., Paytan, A., 2013. Phosphorus cycling in the Sargasso Sea: investigation using the oxygen isotopic composition of phosphate, enzyme-labeled fluorescence, and turnover times. Global Biogeochem. Cycles 27, 375–378.

McSorley, F.R., Wyatt, P., Martinez, A., DeLong, E.F., Hove-Jensen, B., Zechel, D.L., 2012. PhnY and PhnZ comprise a new oxidative pathway for enzymatic cleavage of a carbon-phosphorus bond. J. Am. Chem. Soc. 134, 8364–8367.

Menzel, D.W., Corwin, N., 1965. The measurement of total phosphorus in seawater based on the liberation of organically bound fractions by persulfate oxidation. Limnol. Oceanogr. 10, 280–282.

Menzel, D.W., Spaeth, J.P., 1962. Occurrence of vitamin B$_{12}$ in the Sargasso Sea. Limnol. Oceanogr. 7, 151–154.

Menzel, D.W., Vaccaro, R.F., 1964. The measurement of dissolved organic and particulate carbon in seawater. Limnol. Oceanogr. 9, 138–142.

Metcalf, W.W., Wanner, B.L., 1991. Involvement of the *Escherichia coli phn* (psiD) gene cluster in assimilation of phosphorus in the form of phosphonates, phosphite, Pi esters and Pi. J. Bacteriol. 173, 587–600.

Metcalf, W.W., Griffin, B.M., Cicchillo, R.M., Gao, J., Janga, S.C., Cooke, H.A., et al., 2012. Synthesis of methylphosphonic acid by marine microbes: a source for methane in the aerobic ocean. Science 337, 1104–1107.

Michelou, V.K., Lomas, M.W., Kirchman, D.L., 2011. Phosphate and adenosine-5'-triphosphate uptake by cyanobacteria and heterotrophic bacteria in the Sargasso Sea. Limnol. Oceanogr. 56, 323–332.

Minear, R.A., 1972. Characterization of naturally occurring dissolved organophosphorus compounds. Environ. Sci. Tech. 6, 431–437.

Miyata, K., Hattori, A., 1986. A simple fractionation method for determination of phosphorus components in phytoplankton: application to natural populations of phytoplankton in summer surface waters of Tokyo Bay. J. Oceanogr. Soc. Japan 42, 255–265.

Mohr, W., Grosskopf, T., Wallace, D.W.R., LaRoche, J., 2010. Methodological underestimation of oceanic nitrogen fixation rates. PLoS ONE 5, e12583. doi:10.1371/journal.pone.0012583.

Monaghan, E.J., Ruttenberg, K.C., 1999. Dissolved organic phosphorus in the coastal ocean: reassessment of available methods and seasonal phosphorus profiles from the Eel River Shelf. Limnol. Oceanogr. 44, 1702–1714.

Montoya, J.P., Voss, M., Kähler, P., Capone, D.G., 1996. A simple, high-precision, high-sensitivity tracer assay for N$_2$ fixation. Appl. Environ. Microbiol. 62, 986–993.

Mopper, K., Schultz, C.A., 1993. Fluorescence as a possible tool for studying the nature and water column distribution of DOC components. Mar. Chem. 41, 229–238.

Moran, S.B., Buesseler, K.O., 1992. Short residence time of colloids in the upper ocean estimated from ^{238}U-^{234}Th disequilibria. Nature 359, 221–223.

Morita, R.Y., Howe, R.A., 1957. Phosphatase activity by marine bacteria under hydrostatic pressure. Deep-Sea Res. 4, 254–258.

Mortazavi, B., Iverson, R.L., Landing, W.M., Huang, W., 2000. Phosphorus budget of Apalachicola Bay: a river-dominated estuary in the northeastern Gulf of Mexico. Mar. Ecol. Prog. Ser. 198, 33–42.

Moutin, T., Thingstad, T.F., Van Wambeke, F., Marie, D., Claustre, H., 2002. Does competition for nanomolar phosphate supply explain the predominance of the cyanobacterium *Synechococcus*? Limnol. Oceanogr. 47, 1562–1567.

Moutin, T., Karl, D.M., Duhamel, S., Rimmelin, P., Raimbault, P., Van Mooy, B.A.S., et al., 2008. Phosphate availability and the ultimate control of new nitrogen input by nitrogen fixation in the tropical Pacific Ocean. Biogeosciences 5, 95–109.

Mühlradt, P.F., Wray, V., Lehmann, V., 1977. A ^{31}P-nuclear-magnetic-resonance study of the phosphate groups in lipopolysaccharide and lipid A from *Salmonella*. Eur. J. Biochem. 81, 193–203.

Mulholland, M.R., Floge, S., Carpenter, E.J., Capone, D.G., 2002. Phosphorus dynamics in cultures and natural populations of *Trichodesmium* spp. Mar. Ecol. Prog. Ser. 239, 45–55.

Murphy, J., Riley, J.P., 1956. The storage of sea-water samples for the determination of dissolved inorganic phosphates. Anal. Chim. Acta 14, 818–819.

Murphy, J., Riley, J.P., 1958. A single-solution method for the determination of soluble phosphate in seawater. J. Mar. Biol. Assoc. UK 37, 9–14.

Murphy, J., Riley, J.P., 1962. A modified single solution method for the determination of phosphate in natural waters. Anal. Chim. Acta 27, 31–36.

Nanny, M.A., Minear, R.A., 1997. Characterization of soluble unreactive phosphorus using ^{31}P nuclear magnetic resonance spectroscopy. Mar. Geol. 139, 77–94.

Nanny, M.A., Kim, S., Gadomski, J.E., Minear, R.A., 1994. Aquatic soluble unreactive phosphorus: concentration by ultrafiltration and reverse osmosis membranes. Water Res. 28, 1355–1365.

Nanny, M.A., Kim, S., Minear, R.A., 1995. Aquatic soluble unreactive phosphorus: HPLC studies on concentrated water samples. Water Res. 29, 2138–2148.

Nausch, M., Nausch, G., 2006. Bioavailability of dissolved organic phosphorus in the Baltic Sea. Mar. Ecol. Prog. Ser. 321, 9–17.

Nausch, M., Nausch, G., Wasmund, N., 2004. Phosphorus during the transition from nitrogen to phosphate limitation in the central Baltic Sea. Mar. Ecol. Prog. Ser. 266, 15–25.

Nawrocki, M.P., Karl, D.M., 1989. Dissolved ATP turnover in the Bransfield Strait, Antarctica during a spring bloom. Mar. Ecol. Prog. Ser. 57, 35–44.

Nebbioso, A., Piccolo, A., 2013. Molecular characterization of dissolved organic matter (DOM): a critical review. Anal. Bioanal. Chem. 405, 109–124.

Nedashkovskiy, A.P., Khokhlov, D.A., Salikova, N.N., Sapozhnikov, V.V., 1995. A comparison of two methods for total phosphorus determination in seawater (persulfate digestion by Koroleff-Valderrama and digestion with magnesium nitrate). Oceanology 34, 498–502.

Neu, H.C., 1967. The 5′-nucleotidase of Escherichia coli. J. Biol. Chem. 242, 3896–3904.

Newcombe, C.L., Brust, H.F., 1940. Variations in the phosphorus content of estuarine waters of the Chesapeake Bay near Solomons Island, Maryland. J. Mar. Res. 33, 76–88.

Nissenbaum, A., 1979. Phosphorus in marine and non-marine humic substances. Geochim. Cosmochim. Acta 43, 1973–1978.

Novitsky, J.A., 1986. Degradation of dead microbial biomass in a marine sediment. Appl. Environ. Microbiol. 52, 504–509.

Novitsky, J.A., Morita, R.Y., 1977. Survival of a psychrophilic marine vibrio under long-term nutrient starvation. Appl. Environ. Microbiol. 33, 635–641.

Okbamichael, M., Sañudo-Wilhelmy, S.A., 2004. A new method for the determination of vitamin B$_{12}$ in seawater. Anal. Chim. Acta 517, 33–38.

Okbamichael, M., Sañudo-Wilhelmy, S.A., 2005. Direct determination of vitamin B$_1$ in seawater by solid-phase extraction and high-performance liquid chromatography quantification. Limnol. Oceanogr. Meth. 3, 241–246.

Olson, S., 1967. Recent trends in the determination of orthophosphate in water. In: Golterman, H.L., Clymo, R.S. (Eds.), "Chemical Environment in the Aquatic Habitat", Proceedings of an I.B.P.-symposium held in Amsterdam and Nieuwersluis 10-16 October 1966. N.V. Noord-Hollandsche Uitgevers Maatschappij, Amsterdam, pp. 63–105.

Orchard, E.D., Ammerman, J.W., Lomas, M.W., Dyhrman, S.T., 2010. Dissolved inorganic and organic phosphorus uptake in Trichodesmium and the microbial community: the importance of phosphorus ester in the Sargasso Sea. Limnol. Oceanogr. 55, 1390–1399.

Ormaza-González, F.I., Statham, P.J., 1991. Determination of dissolved inorganic phosphorus in natural waters at nanomolar concentrations using a long capillary cell detector. Anal. Chim. Acta 244, 63–70.

Ormaza-González, F.I., Statham, P.J., 1996. A comparison of methods for the determination of dissolved and particulate phosphorus in natural waters. Water Res. 30, 2739–2747.

Orrett, K., Karl, D.M., 1987. Dissolved organic phosphorus production in surface seawaters. Limnol. Oceanogr. 32, 383–395.

Osmond, M.F., 1887. Sur une réaction pouvant servir au dosage colorimétrique du phosphore dans les foutes, les aciers, etc. Bull. Soc. Chim. Paris 47, 745–749.

Parrish, C.C., 1986. Dissolved and particulate lipid classes in the aquatic environment. PhD dissertation. Dalhousie University.

Parrish, C.C., 1987. Time series of particulate and dissolved lipid classes during spring phytoplankton blooms in Bedford Basin, a marine inlet. Mar. Ecol. Prog. Ser. 35, 129–139.

Parrish, C.C., Ackman, R.G., 1985. Calibration of the Iatroscan-Chromarod system for marine lipid class analysis. Lipids 20, 521–530.

Parrish, C.C., Wangersky, P.J., 1987. Particulate and dissolved lipid classes in cultures of Phaeodactylum tricornutum grown in cage culture turbidostats with a range of nitrogen supply rates. Mar. Ecol. Prog. Ser. 35, 119–128.

Patey, M.D., Rijkenberg, M.J.A., Statham, P.J., Stinchcombe, M.C., Achterberg, E.P., Mowlem, M., 2008. Determination of nitrate and phosphate in seawater at nanomolar concentrations. Trends Anal. Chem. 27, 169–182.

Paul, J.H., Jeffrey, W.H., DeFlaun, M.F., 1987. Dynamics of extracellular DNA in the marine environment. Appl. Environ. Microbiol. 53, 170–179.

Paul, J.H., DeFlaun, M.F., Jeffrey, W.H., David, A.W., 1988. Seasonal and diel variability in dissolved DNA and in microbial biomass and activity in a subtropical estuary. Appl. Environ. Microbiol. 54, 718–727.

Paul, J.H., Cazares, L., Thurmond, J., 1990. Amplification of the rbcL gene from dissolved and particulate DNA from aquatic environments. Appl. Environ. Microbiol. 56, 1963–1966.

Paul, J.H., Frischer, M.E., Thurmond, J.M., 1991. Gene transfer in marine water column and sediment microcosms by natural plasmid transformation. Appl. Environ. Microbiol. 57, 1509–1515.

Paytan, A., McLaughlin, K., 2007. The oceanic phosphorus cycle. Chem. Rev. 107, 563–576.

Perry, M.J., 1972. Alkaline phosphatase activity in subtropical Central North Pacific waters using a sensitive fluorometric method. Mar. Biol. 15, 113–119.

Perry, M.J., 1976. Phosphate utilization by an oceanic diatom in phosphorus-limited chemostat culture and in the oligotrophic waters of the central North Pacific. Limnol. Oceanogr. 21, 88–107.

Perry, M.J., Eppley, R.W., 1981. Phosphate uptake by phytoplankton in the central North Pacific Ocean. Deep-Sea Res. 28A, 39–49.

Pettersson, K., 1979. Enzymatic determination of orthophosphate in natural waters. Int. Rev. Hydrobiol. 64, 585–607.

Pillai, T.N.V., Ganguly, A.K., 1972. Nucleic acids in the dissolved constituents of sea-water. J. Mar. Biol. Assoc. India 14, 384–390.

Pilson, M.E.Q., Betzer, S.B., 1973. Phosphorus flux across a coral reef. Ecology 54, 581–588.

Pinkerton, M.H., Lavender, S.J., Aiken, J., 2003. Validation of SeaWiFS ocean color satellite data using a moored databuoy. J. Geophys. Res. 108 (C5), 3133. doi:10.1029/2002JC001337.

Pratt, D.M., 1950. Experimental study of the phosphorus cycle in fertilized salt water. J. Mar. Res. 9, 29–54.

Provasoli, L., Pintner, I.J., 1953. Ecological implications of *in vitro* nutritional requirements of algal flagellates. Ann. N.Y. Acad. Sci. 56, 839–851.

Quimby, O.T., Mabis, A.J., Lampe, H.W., 1954. Determination of triphosphate and pyrophosphate by isotope dilution. Anal. Chem. 26, 661–667.

Quinn, J.P., Kulakova, A.N., Cooley, N.A., McGrath, J.W., 2007. New ways to break an old bond: the bacterial carbon-phosphorus hydrolases and their role in biogeochemical phosphorus cycling. Environ. Microbiol. 9, 2392–2400.

Raimbault, P., Pouvesle, W., Diaz, F., Garcia, N., Sempéré, R., 1999. Wet-oxidation and automated colorimetry for simultaneous determination of organic carbon, nitrogen and phosphorus dissolved in seawater. Mar. Chem. 66, 161–169.

Ramakrishnan, V., Verhagen, M.F.J.M., Adams, M.W.W., 1997. Characterization of di-*myo*-inositol-1,1'-phosphate in the hyperthermophilic bacterium *Thermotoga maritime*. Appl. Environ. Microbiol. 63, 347–350.

Rashid, M.A., 1985. Geochemistry of Marine Humic Compounds. Springer-Verlag.

Raven, J.A., 1994. Why are there no picoplankton O_2 evolvers with volumes less than $10^{-19}\,m^3$. J. Plankton Res. 16, 565–580.

Ray, M.K., Kumar, G.S., Shivaji, S., 1994. Phosphorylation of lipopolysaccharides in the Antarctic psychrotroph *Pseudomonas syringae*: a possible role in temperature adaptation. J. Bacteriol. 176, 4243–4249.

Redfield, A.C., 1958. The biological control of chemical factors in the environment. Am. Sci. 46, 205–222.

Redfield, A.C., Smith, H.P., Ketchum, B., 1937. The cycle of organic phosphorus in the Gulf of Maine. Biol. Bull. 123, 421–443.

Redfield, A.C., Ketchum, B.H., Richards, F.A., 1963. The influence of organisms on the composition of sea water. In: Hill, M.N. (Ed.), The Sea, vol. 2. Interscience, New York, pp. 26–77.

Redfield, R.J., Schrag, M.R., Dean, A.M., 1997. The evolution of bacterial transformation: sex with poor relations. Genetics 146, 27–38.

Rees, A.P., Law, C.S., Woodward, E.M.S., 2006. High rates of nitrogen fixation during an *in-situ* phosphate release experiment in the Eastern Mediterranean Sea. Geophys. Res. Lett. 33, doi:10.1029/2006GL025791, L10607.

Reichardt, W., Overbeck, J., Steubing, L., 1967. Free dissolved enzymes in lake waters. Nature 216, 1345–1347.

Renn, C.E., 1937. Bacteria and the phosphorus cycle in the sea. Biol. Bull. 122, 190–195.

Repeta, D.J., Quan, T.M., Aluwihare, L.I., Chen, R.F., 2002. Chemical characterization of high molecular weight dissolved organic matter in fresh and marine waters. Geochim. Cosmochim. Acta 66, 955–962.

Reusch, R.N., Sadoff, H.L., 1988. Putative structure and functions of a poly-β-hydroxybutyrate/calcium polyphosphate channel in bacterial plasma membranes. Proc. Natl. Acad. Sci. U. S. A. 85, 4176–4180.

Reuter Jr., J.G., 1983. Alkaline phosphatase inhibition by copper: implications to phosphorus nutrition and use as a biochemical marker of toxicity. Limnol. Oceanogr. 28, 743–748.

Rhee, G.-Y., 1972. Competition between an alga and an aquatic bacterium for phosphate. Limnol. Oceanogr. 4, 505–514.

Ridal, J.J., Moore, R.M., 1990. A re-examination of the measurement of dissolved organic phosphorus in seawater. Mar. Chem. 29, 19–31.

Ridal, J.J., Moore, R.M., 1992. Dissolved organic phosphorus concentrations in the northeast subarctic Pacific Ocean. Limnol. Oceanogr. 37, 1067–1075.

Rigler, F.H., 1956. A tracer study of the phosphorus cycle in lake water. Ecology 37, 550–562.

Rigler, F.H., 1968. Further observations inconsistent with the hypothesis that the molybdenum blue method measures orthophosphate in lake water. Limnol. Oceanogr. 13, 7–13.

Rigler, F.H., 1973. A dynamic view of the phosphorus cycle in lakes. In: Griffith, E.J., et al. (Eds.), Environmental Phosphorus Handbook. Wiley-Interscience, New York, pp. 539–572.

Rittenberg, S.C., Hespell, R.B., 1975. Energy efficiency of intraperiplasmic growth of *Bdellovibrio bacteriovorus*. J. Bacteriol. 121, 1158–1165.

Rivkin, R.B., Swift, E., 1980. Characterization of alkaline phosphatase and organic phosphorus utilization in the oceanic dinoflagellate *Pyrocystis noctiluca*. Mar. Biol. 61, 1–8.

Robards, K., McKelvie, I.D., Benson, R.L., Worsfold, P.J., Blundell, N.J., Casey, H., 1994. Determination of carbon, phosphorus, nitrogen and silicon species in waters. Anal. Chim. Acta 287, 147–190.

Rodriguez, F., Fernandez, E., Head, R.N., Harbour, D.S., Bratbak, G., Heldal, M., et al., 2000. Temporal variability of viruses, bacteria, phytoplankton and zooplankton in the western English Channel off Plymouth. J. Mar. Biol. Assoc. UK 80, 575–586.

Roussenov, V., Williams, R.G., Mahaffey, C., Wolff, G.A., 2006. Does the transport of dissolved organic nutrients affect export production in the Atlantic Ocean? Global Biogeochem. Cycles 20, doi:10.1029/2005GB002510, GB3002.

Ruby, E.G., McCabe, J.B., Barke, J.I., 1985. Uptake of intact nucleoside monophosphates by *Bdellovibrio bacteriovorus* 109J. J. Bacteriol. 163, 1087–1094.

Russell, F.S., Southward, A.J., Boalch, G.T., Butler, E.I., 1971. Changes in biological conditions in the English Channel off Plymouth during the last half century. Nature 234, 468–470.

Ruttenberg, K.C., 2014. The global phosphorus cycle. In: Karl, D.M., Schlesinger, W.H. (Eds.), *Biogeochemistry vol. 8 Treatise on Geochemistry*. 2nd ed.. Elsevier, Oxford, UK.

Saier Jr., M.H., 1989. Protein phosphorylation and allosteric control of inducer exclusion and catabolite repression by the bacterial phosphoenolpyruvate: sugar phosphotransferase system. Microbiol. Rev. 53, 109–120.

Saier Jr., M.H., 2000. A functional-phylogenetic classification system for transmembrane solute transporters. Microbiol. Mol. Biol. Rev. 64, 354–411.

Saier Jr., M.H., Chauvaux, S., Deutscher, J., Reizer, J., Ye, J.-J., 1995. Protein phosphorylation and regulation of carbon metabolism in Gram-negative *versus* Gram-positive bacteria. TIBS 20, 267–271.

Sakano, S., Kamatani, A., 1992. Determination of dissolved nucleic acids in seawater by the fluorescence dye, ethidium bromide. Mar. Chem. 37, 239–255.

Sanders, R., Jickells, T., 2000. Total organic nutrients in the Drake Passage. Deep-Sea Res. 47, 997–1014.

Sannigrahi, P., Ingall, E.D., Benner, R., 2006. Nature and dynamics of phosphorus-containing components of marine dissolved and particulate organic matter. Geochim. Cosmochim. Acta 70, 5868–5882.

Sañudo-Wilhelmy, S.A., Cutter, L.S., Durazoc, R., Smaila, E.A., Gómez-Consarnau, L., Webb, E.A., et al., 2012. Multiple B-vitamin depletion in large areas of the coastal ocean. Proc. Natl. Acad. Sci. U. S. A. 109, 14041–14045.

Satomi, M., Pomeroy, L.R., 1965. Respiration and phosphorus excretion in some marine populations. Ecology 46, 877–881.

Scholz, S., Sonnenbichler, J., Schäfer, W., Hensel, R., 1992. Di-*myo*-inositol-1,1′-phosphate: a new inositol phosphate isolated from *Pyrococcus woesei*. FEBS Lett. 306, 239–242.

Sebastian, M., Ammerman, J.W., 2009. The alkaline phosphatase PhoX is more widely distributed in marine bacteria than the classical PhoA. ISME J. 3, 563–572.

Seiwell, H.R., Seiwell, G.E., 1938. The sinking of decomposing plankton in sea water and its relationship to oxygen consumption and phosphorus liberation. Proc. Am. Phil. Soc. 78, 465–481.

Shan, Y., McKelvie, I.D., Hart, B.T., 1994. Determination of alkaline phosphatase-hydrolyzable phosphorus in natural water systems by enzymatic flow injection. Limnol. Oceanogr. 39, 1993–2000.

Siuda, W., Chróst, R.J., 2000. Concentration and susceptibility of dissolved DNA for enzyme degradation in lake water—some methodological remarks. Aquat. Microb. Ecol. 21, 195–201.

Siuda, W., Güde, H., 1994. A comparative study on 5′-nucleotidase (5′-nase) and alkaline phosphatase (APA) activities in two lakes. Arch. Hydrobiol. 131, 211–229.

Siuda, W., Güde, H., 1996. Determination of dissolved deoxyribonucleic acid concentration in lake water. Aquat. Microb. Ecol. 11, 193–202.

Smith, S.V., 1984. Phosphorus *versus* nitrogen limitation in the marine environment. Limnol. Oceanogr. 29, 1149–1160.

Smith, R.E.H., Harrison, W.G., Harris, L., 1985. Phosphorus exchange in marine microplankton communities near Hawaii. Mar. Biol. 86, 75–84.

Smith, S.V., Kimmerer, W.J., Walsh, T.W., 1986. Vertical flux and biogeochemical turnover regulate nutrient limitation of net organic production in the North Pacific Gyre. Limnol. Oceanogr. 31, 161–167.

Smith, D.C., Simon, M., Alldredge, A.L., Azam, F., 1992. Intense hydrolytic enzyme activity on marine aggregates and implications for rapid particle dissolution. Nature 359, 139–142.

Snyder, W.R., Law, J.H., 1970. A quantitative determination of phosphonate phosphorus in naturally occurring aminophosphonates. Lipids 5, 800–802.

Sohm, J.A., Capone, D.G., 2006. Phosphorus dynamics of the tropical and subtropical north Atlantic: *Trichodesmium* spp. *versus* bulk plankton. Mar. Ecol. Prog. Ser. 317, 21–28.

Solórzano, L., Sharp, J.H., 1980. Determination of total dissolved phosphorus and particulate phosphorus in natural waters. Limnol. Oceanogr. 25, 754–758.

Solórzano, L., Strickland, J.D.H., 1968. Polyphosphate in seawater. Limnol. Oceanogr. 13, 515–518.

Sorokin, Y.I., 1985. Phosphorus metabolism in planktonic communities of the eastern tropical Pacific Ocean. Mar. Ecol. Prog. Ser. 27, 87–97.

Sorokin, Y.I., Vyshkvartsev, D.I., 1974. Consumption of mineral phosphate by a planktonic community in tropical waters. Oceanology 14, 552–556.

Southward, A.J., 1980. The Western English Channel—an inconstant ecosystem? Nature 285, 361–366.

Southward, A.J., Roberts, E.K., 1987. One hundred years of marine research at Plymouth. J. Mar. Biol. Assoc. UK 67, 465–506.

Sowell, S.M., Wilhelm, L.J., Norbeck, A.D., Lipton, M.S., Nicora, C.D., Barofsky, D.F., et al., 2009. Transport functions dominate the SAR11 metaproteome at low-nutrient extremes in the Sargasso Sea. ISME J. 3, 93–105.

Steward, G.F., Wikner, J., Cochlan, W.P., Smith, D.C., Azam, F., 1992. Estimation of virus production in the sea: II. Field results. Mar. Microb. Food Webs 6, 79–90.

Stewart, W.D.P., Fitzgerald, G.P., Burris, R.H., 1970. Acetylene reduction assay for determination of phosphorus availability in Wisconsin lakes. Proc. Natl. Acad. Sci. U. S. A. 66, 1104–1111.

Stihl, A., Sommer, U., Post, A.F., 2001. Alkaline phosphatase activities among populations of the colony-forming diazotrophic cyanobacterium *Trichodesmium* spp. (cyanobacteria) in the Red Sea. J. Phycol. 37, 310–317.

Stock, J.B., Ninfa, A.J., Stock, A.M., 1989. Protein phosphorylation and regulation of adaptive responses in bacteria. Microbiol. Rev. 53, 450–490.

Stouthamer, A.H., 1977. Energetic aspects of the growth of micro-organisms. Symp. Soc. Gen. Microbiol. 27, 285–315.

Strayer, D., Glitzenstein, J.S., Jones, C.G., Kolasoi, J., Likens, G.E., McDonnell, M.J., et al., 1986. Long-term ecological studies: an illustrated account of their design, operation, and importance to ecology. Occasional Publication of the Institute of Ecosystem Studies, No. 2. Millbrook, New York.

Strickland, J.D.H., Austin, K.H., 1960. On the forms, balances and cycle of phosphorus observed in the coastal and oceanic waters of the northeastern Pacific. J. Fish. Res. Bd. Canada 17, 337–345.

Strickland, J.D.H., Parsons, T.R., 1972. A Practical Handbook of Seawater Analysis. Fisheries Research Board of Canada.

Strickland, J.D.H., Solórzano, L., 1966. Determination of monoesterase hydrolysable phosphate and phosphomonoesterase activity in sea water. In: Barnes, H. (Ed.), Some Contemporary Studies in Marine Science. George Allen and Unwin Ltd., London, pp. 665–674.

Suttle, C.A., 2007. Marine viruses—major players in the global ecosystem. Nat. Rev. Microbiol. 5, 801–812.

Suzuki, Y., Kogure, K., Tanoue, E., 1997. Immunochemical detection of dissolved proteins and their source bacteria in marine environments. Mar. Ecol. Prog. Ser. 158, 1–9.

Suzumura, M., 2005. Phospholipids in marine environments: a review. Talanta 66, 422–434.

Suzumura, M., Ingall, E.D., 2001. Concentrations of lipid phosphorus and its abundance in dissolved and particulate organic phosphorus in coastal seawater. Mar. Chem. 75, 141–149.

Suzumura, M., Ingall, E.D., 2004. Distribution and dynamics of various forms of phosphorus in seawater: insights from field observations in the Pacific Ocean and a laboratory experiment. Deep-Sea Res. I 51, 1113–1130.

Suzumura, M., Kamatani, A., 1995. Origin and distribution of inositol hexaphosphate in estuarine and coastal sediments. Limnol. Oceanogr. 40, 1254–1261.

Suzumura, M., Ishikawa, K., Ogawa, H., 1998. Characterization of dissolved organic phosphorus in coastal seawater using ultrafiltration and phosphohydrolytic enzymes. Limnol. Oceanogr. 43, 1553–1564.

Taft, J.L., Taylor, W.R., McCarthy, J.J., 1975. Uptake and release of phosphorus by phytoplankton in the Chesapeake Bay Estuary, USA. Mar. Biol. 33, 21–32.

Taft, J.L., Loftus, M.E., Taylor, W.R., 1977. Phosphate uptake from phosphomonoesters by phytoplankton in the Chesapeake Bay. Limnol. Oceanogr. 22, 1012–1021.

Taga, N., Kobori, H., 1978. Phosphatase activity in eutrophic Tokyo Bay. Mar. Biol. 49, 223–229.

Tamminen, T., 1989. Dissolved organic phosphorus regeneration by bacterioplankton: 5′-nucleotidase activity and subsequent phosphate uptake in a mesocosm experiment. Mar. Ecol. Prog. Ser. 58, 89–100.

Tanaka, T., Thingstad, T.F., Christaki, U., Colombet, J., Cornet-Barthaux, V., Courties, C., et al., 2011. Lack of P-limitation of phytoplankton and heterotrophic prokaryotes in surface waters of three anticyclonic eddies in the stratified Mediterranean Sea. Biogeosciences 8, 525–538.

Tanoue, E., 1995. Detection of dissolved protein molecules in oceanic waters. Mar. Chem. 51, 239–252.

Tarapchak, S.J., 1983. Soluble reactive phosphorus measurements in lake water: evidence for molybdate-enhanced hydrolysis. J. Environ. Qual. 12, 105–108.

Tarapchak, S.J., Moll, R.A., 1990. Phosphorus sources for phytoplankton and bacteria in Lake Michigan. J. Plankton Res. 12, 743–758.

Tate, C.M., Broshears, R.E., McKnight, D.M., 1995. Phosphate dynamics in an acidic mountain stream: interactions involving algal uptake, sorption by iron oxide, and photoreduction. Limnol. Oceanogr. 40, 938–946.

Thingstad, T.F., Skjoldal, E.F., Bohne, R.A., 1993. Phosphorus cycling and algal-bacterial competition in Sandsfjord, western Norway. Mar. Ecol. Prog. Ser. 99, 239–259.

Thingstad, T.F., Riemann, B., Havskum, H., Garde, K., 1996. Incorporation rates and biomass content of C and P in phytoplankton and bacteria in the Bay of Aarhus (Denmark) June 1992. J. Plankton Res. 18, 97–121.

Thingstad, T.F., Krom, M.D., Mantoura, R.F.C., Flaten, G.A.F., Groom, S., Herut, B., et al., 2005. Nature of phosphorus limitation in the ultraoligotrophic eastern Mediterranean. Science 309, 1068–1071.

Thomson-Bulldis, A., Karl, D.M., 1998. Application of a novel method for phosphorus determinations in the oligotrophic North Pacific Ocean. Limnol. Oceanogr. 43, 1565–1577.

Tommassen, J., Lugtenberg, B., 1981. Localization of phoE, the structural gene for outer membrane protein e in Escherichia coli K-12. J. Bacteriol. 147, 118–123.

Tornabene, T.G., Langworthy, T.A., 1979. Diphytanyl glycerol and dibiphytanyl glycerol ether lipids of methanogenic archaebacteria. Science 203, 51–53.

Tripp, H.J., Kitner, J.B., Schwalbach, M.S., Dacey, J.W., Wilhelm, L.J., Giovannoni, S.J., 2008. SAR11 marine bacteria require exogenous reduced sulphur for growth. Nature 452, 741–744.

Tsai, Y.-L., Tran, B., Palmer, C.J., 1995. Analysis of viral RNA persistence in seawater by reverse transcriptase-PCR. Appl. Environ. Microbiol. 61, 363–366.

Tue-Ngeun, O., Ellis, P., McKelvie, I., Worsfold, P., Jakmunee, J., Grudpan, K., 2005. Determination of dissolved reactive phosphorus (DRP) and dissolved organic phosphorus (DOP) in natural waters by the use of rapid sequenced reagent injection flow analysis. Talanta 66, 453–460.

Turk, V., Rehnstam, A.-S., Lundberg, E., Hagström, Å., 1992. Release of bacterial DNA by marine nanoflagellates, an intermediate step in phosphorus regeneration. Appl. Environ. Microbiol. 58, 3744–3750.

Turner, B.L., Frossard, E., Baldwin, D.S. (Eds.), 2005. Organic Phosphorus in the Environment. CABI Publishing, Cambridge, MA, pp. 399.

Valderrama, J.C., 1981. The simultaneous analysis of total nitrogen and total phosphorus in natural waters. Mar. Chem. 10, 109–122.

van Boekel, W.H.M., 1991. Ability of Phaeocystis sp. to grow on organic phosphates: direct measurement and prediction with the use of an inhibition constant. J. Plankton Res. 13, 959–970.

Varma, A.K., Rigsby, W., Jordan, D.C., 1983. A new inorganic pyrophosphate utilizing bacterium from a stagnant lake. Can. J. Microbiol. 29, 1470–1474.

Venrick, E.L., 1971. The statistics of subsampling. Limnol. Oceanogr. 16, 811–818.

Vetter, T.A., Perdue, E.M., Ingall, E., Koprivnjak, J.-F., Pfromm, P.H., 2007. Combining reverse osmosis and electrodialysis for more complete recovery of dissolved organic matter from seawater. Sep. Pur. Technol. 56, 383–387.

Vidal, M., Duarte, C.M., Agustí, S., 1999. Dissolved organic nitrogen and phosphorus pools and fluxes in the central Atlantic Ocean. Limnol. Oceanogr. 44, 106–115.

Voelz, H., Voelz, U., Ortigoza, R.O., 1966. The polyphosphate overplus phenomenon in Myxococcus xanthus and its influence on the architecture of the cell. Arch. Mikrobiol. 53, 371–388.

von Schouwenburg, J.C., Walinga, I., 1967. The rapid determination of phosphorus in the presence of arsenic, silicon and germanium. Anal. Chim. Acta 37, 269–271.

Wai, N., Hung, T.-C., Lu, Y.-H., 1960. Alkaline phosphatase in Taiwan sea water. Bull. Inst. Chem. Acad. Sin. 3, 1–10.

Waksman, S.A., Carey, C.L., 1935. Decomposition of organic matter in sea water by bacteria. II. Influence of addition of organic substances upon bacterial activities. J. Bacteriol. 29, 545–561.

Waksman, S.A., Renn, C.E., 1936. Decomposition of organic matter in the sea by bacteria. Biol. Bull. 70, 472.

Walsh, T.W., 1989. Total dissolved nitrogen in seawater: a new-high-temperature combustion method and a comparison with photo-oxidation. Mar. Chem. 26, 295–311.

Wanner, B.L., 1993. Gene regulation by phosphate in enteric bacteria. J. Cell. Biochem. 51, 47–54.

Waser, N.A.D., Bacon, M.P., Michaels, A.F., 1996. Natural activities of ^{32}P and ^{33}P and the $^{33}P/^{32}P$ ratio in suspended particulate matter and plankton in the Sargasso Sea. Deep-Sea Res. II 43, 421–436.

Waterbury, R.D., Yao, W., Byrne, R.H., 1997. Long pathlength absorbance spectroscopy: trace analysis of Fe(II) using a 4.5 m liquid core waveguide. Anal. Chim. Acta 357, 99–102.

Watkins-Brandt, K.S., Letelier, R.M., Spitz, Y.H., Church, M.J., Böttjer, D., White, A.E., 2011. Addition of inorganic or organic phosphorus enhances nitrogen and carbon fixation in the oligotrophic North Pacific. Mar. Ecol. Prog. Ser. 432, 17–29.

Watt, W.D., Hayes, F.R., 1963. Tracer study of the phosphorus cycle in sea water. Limnol. Oceanogr. 8, 276–285.

Weil-Malherbe, H., Green, R.H., 1951. The catalytic effect of molybdate on the hydrolysis of organic phosphate bonds. Biochem. J. 49, 286–292.

Weinbauer, M.G., Fuks, D., Peduzzi, P., 1993. Distribution of viruses and dissolved DNA along a coastal trophic gradient in the Northern Adriatic Sea. Appl. Environ. Microbiol. 59, 4074–4082.

Weiss, M.S., Abele, U., Weckesser, J., Welte, W., Schiltz, E., Schulz, G.E., 1991. Molecular architecture and electrostatic properties of a bacterial porin. Science 254, 1627–1630.

Wells, M.L., Goldberg, E.D., 1991. Occurrence of small colloids in seawater. Nature 353, 342–344.

White, A.E., 2009. New insights into bacterial acquisition of phosphorus in the surface ocean. Proc. Natl. Acad. Sci. U. S. A. 106, 21013–21014.

White, A.K., Metcalf, W.W., 2007. Microbial metabolism of reduced phosphorus compounds. Annu. Rev. Microbiol. 61, 379–400.

White, R.H., Miller, S.L., 1976. Inositol isomers: occurrence in marine sediments. Science 193, 885–886.

White, A.E., Karl, D.M., Björkman, K.M., Beversdorf, L.J., Letelier, R.M., 2010. Production of organic matter by Trichodesmium IMS101 as a function of phosphorus source. Limnol. Oceanogr. 55, 1755–1767.

White, A.E., Watkins-Brandt, K.S., Engle, M.A., Burkhardt, B., Paytan, A., 2012. Characterization of the rate and temperature sensitivities of bacterial remineralization of dissolved organic phosphorus compounds by natural populations. Front. Microbiol. 3, 1–13.

Whitehouse, B.G., Yeats, P.A., Strain, P.M., 1990. Cross-flow filtration of colloids from aquatic environments. Limnol. Oceanogr. 35, 1368–1375.

Wilkins, A.S., 1972. Physiological factors in the regulation of alkaline phosphatase synthesis in Escherichia coli. J. Bacteriol. 110, 616–623.

Wilkinson, S.G., 1996. Bacterial lipopolysaccharides—themes and variations. Prog. Lipid Res. 35, 283–343.

Williams, P.M., Carlucci, A.F., Olson, R., 1980. A deep profile of some biologically important properties in the central North Pacific gyre. Oceanol. Acta 3, 471–476.

Wilson, S.T., Böttjer, D., Church, M.J., Karl, D.M., 2012. Comparative assessment of nitrogen fixation methodologies conducted in the oligotrophic North Pacific Ocean. Appl. Environ. Microbiol. 78, 6516–6523.

Wommack, K.E., Colwell, R.R., 2000. Virioplankton: viruses in aquatic ecosystems. Microbiol. Mol. Biol. Rev. 64, 69–114.

Woo, L., Maher, W., 1995. Determination of phosphorus in turbid waters using alkaline potassium peroxodisulphate digestion. Anal. Chim. Acta 315, 123–135.

Worsfold, P.J., Monbet, P., Tappin, A.D., Fitzsimons, M.F., Stiles, D.A., McKelvie, I.D., 2008. Characterisation and quantification of organic phosphorus and organic nitrogen components in aquatic systems: a review. Anal. Chim. Acta 624, 37–58.

Wu, J., Sunda, W., Boyle, E.A., Karl, D.M., 2000. Phosphate depletion in the western North Atlantic Ocean. Science 289, 759–762.

Yamagata, Y., Watanabe, H., Saitoh, M., Namba, T., 1991. Volcanic production of polyphosphates and its relevance to prebiotic evolution. Nature 352, 516–519.

Yanagi, K., Yasuda, M., Fukui, F., 1992. Reexamination of the fractionation of total dissolved phosphorus in seawater using a modified UV-irradiation procedure, and its application to samples from Suruga Bay and Antarctic Ocean. J. Oceanogr. 48, 267–281.

Yang, K., Metcalf, W.W., 2004. A new activity for an old enzyme: Escherichia coli bacterial alkaline phosphatase is a phosphite-dependent hydrogenase. Proc. Natl. Acad. Sci. U. S. A. 101, 7919–7924.

Yaqoob, M., Nabi, A., Worsfold, P.J., 2004. Determination of nanomolar concentrations of phosphate in freshwaters using flow injection with luminal chemiluminescence detection. Anal. Chim. Acta 510, 213–218.

Yentsch, C.M., Yentsch, C.S., Perras, J.P., 1972. Alkaline phosphatase activity in the tropical marine blue-green alga, Oscillatoria erythraea ("Trichodesmium"). Limnol. Oceanogr. 17, 772–774.

Young, C.L., Ingall, E.D., 2010. Marine dissolved organic phosphorus composition: insights from samples recovered using combined electrodialysis/reverse osmosis. Aquat. Geochem. 16, 563–574.

Zhang, J.-Z., Chi, J., 2002. Automated analysis of nanomolar concentrations of phosphate in natural waters with liquid waveguide. Environ. Sci. Technol. 36, 1048–1053.

ZoBell, C.E., Grant, C.W., 1943. Bacterial utilization of low concentrations of organic matter. J. Bacteriol. 45, 555–564.

Zui, O.V., Birks, J.W., 2000. Trace analysis of phosphorus in water by sorption preconcentration and luminol chemiluminescence. Anal. Chem. 72, 1699–1703.

CHAPTER

6

The Carbon Isotopic Composition of Marine DOC

Steven R. Beaupré

Geology and Geophysics, Woods Hole Oceanographic Institution, Woods Hole, Massachusetts, USA

CONTENTS

I INTRODUCTION

Marine dissolved organic matter (DOM) is loosely defined as the collection of small molecules ($<ca.$ 1 μm) in seawater that are built upon skeletons of carbon atoms. Despite this elemental limitation, the bonding properties of carbon combined with associated heteroatoms (e.g., H, O, N, P, S) enable the natural formation of countless unique molecules. In the oceans, they

Biogeochemistry of Marine Dissolved Organic Matter,
http://dx.doi.org/10.1016/B978-0-12-405940-5.00006-6

are primarily derived from living organisms by photosynthetic reduction of dissolved inorganic carbon (DIC) and subsequently transformed through progressive trophic levels and abiotic processes. The variety of compounds in the ensuing mixtures is both the potential key and principle challenge to understanding their biogeochemistry. Thus, for more than 100 years, scientists have been studying molecular structures, spatiotemporal distributions, and the timescales over which DOM constituents are processed in their respective environments (Hansell and Carlson, 2002; Natterer, 1892). This work has been augmented by examining the isotopic composition of carbon atoms in DOM (i.e., dissolved organic carbon, DOC) because isotopes serve as powerful tracers of both mass and time. Carbon isotope analyses of bulk DOC (i.e., the whole mixture of molecules) began in earnest in the 1960s and then progressed toward size class, compound class, and compound specific isotope analyses in the 1990s. This chapter provides a brief introduction to these canonical isotope analyses, more recent developments, and their associated interpretations.

II CARBON ISOTOPE GEOCHEMISTRY PRIMER

Of the 15 known isotopes of carbon that range in atomic mass number from 8 to 22 (Lide, 2007), only ^{12}C and ^{13}C are stable. Radiocarbon (^{14}C) is the longest lived of the radioactive carbon isotopes, undergoing beta (β) decay with a half-life of 5730 ± 40 years (Godwin, 1962). The half-lives of the remaining radioisotopes (Lide, 2007) are <20.3 min (i.e., ^{11}C) and preclude observations of their natural abundances in the sea. Consequently, only ^{12}C, ^{13}C, and ^{14}C have found widespread application in the study of marine DOM.

On average, ^{12}C, ^{13}C, and ^{14}C comprise 98.89%, 1.11%, and $<\sim 10^{-10}$% of Earth's carbon inventory, respectively. Indeed, a fundamental challenge

to carbon isotope geochemistry lies in successfully measuring isotopic abundances that span as much as 15 orders of magnitude in natural samples. This daunting ratio led Melvin Calvin and other pioneering carbon isotope scientists to proclaim, "It is doubtful that any tracer use can be made of naturally occurring C^{14} [sic]" (Calvin et al., 1949). Shortly afterward, however, Willard Libby summarized the principles of his ^{14}C dating method (Libby, 1952) and won the 1960 Nobel Prize in Chemistry. The first measurements of marine DOC ^{13}C and ^{14}C abundances followed less than a decade later (Williams, 1968; Williams et al., 1969). The ultimate goal of this work was not to make seemingly impossible measurements, but rather to discover how our planet works. This required an understanding of the fundamental processes that create slight deviations in ^{12}C, ^{13}C, and ^{14}C abundances from their global averages, as summarized below and explained in detail elsewhere (Criss, 1999; Faure, 1986; Hoefs, 1997).

A Carbon-13 and Stable Isotope Systematics

Compared to ^{12}C, a ^{13}C atom requires slightly more energy to break the chemical bonds that hold it in a molecule because it has the additional mass of one neutron. Likewise, molecules containing ^{13}C atoms require slightly more energy to undergo physical transformations than identical molecules constructed with ^{12}C atoms alone. The ensuing differences in associated properties such as molar masses, densities, melting points, vapor pressures, enthalpies of formation, chemical rate constants, equilibrium constants, etc. are known as "isotope effects." These effects alter the ratio of ^{13}C to ^{12}C abundances (i.e., $^{13}C/^{12}C$) in both the products and residual molecules of substances undergoing physical or chemical transformations. It is this natural partitioning, or "isotopic fractionation," that ultimately leads to the small but detectable variations in $^{13}C/^{12}C$ ratios throughout the environment (Criss, 1999; Hoefs, 1997). The extent of fractionation is characteristic of the chemical, physical, and biological processes at play, and it is sensitive to the conditions under which they occur. Therefore, the isotopic composition of DOC depends on the isotopic composition of its carbon source, the mechanisms by which that carbon was transformed, and the environments in which the transformations took place. In short, the isotopic composition of DOC is a collective record of its history.

For consistency and comparability, $^{13}C/^{12}C$ ratios (R) of samples are commonly reported as $\delta^{13}C$ values, which represent permil (i.e., parts per thousand, ‰) deviations from the $^{13}C/^{12}C$ ratio of an internationally accepted standard known as "PDB" (the Pee Dee Belemnite standard).

$$\delta^{13}C = \left(\frac{R_{sample}}{R_{standard}} - 1 \right) 1000\text{‰} \qquad (6.1)$$

A strong fractionation in preference for ^{12}C over ^{13}C during photosynthesis and a comparatively weak fractionation during respiration (<ca. 1‰, Shaffer et al., 1999) creates marine DOC that is significantly depleted in ^{13}C (ca. −21‰) relative to the DIC from which it is derived (~0‰) (Boutton, 1991). The extent of fractionation during photosynthesis, for example, depends on environmental factors such as temperature and CO_2 partial pressure (Degens et al., 1968b; Rau et al., 1992). It also depends upon biological factors ranging in scale from the enzymatic reduction of CO_2 (O'Leary, 1981, 1988) to variations in community composition and structure (Falkowski, 1991; Rau et al., 1982). Therefore, the $\delta^{13}C$ values should record the provenance of each compound in the DOC pool while simultaneously illuminating the significance of each of these controls.

All DOC isotope ratios, however, represent the average isotope ratio of all of the constituent molecules. This can be represented succinctly by the approximate equations for mass balance (Eqs. 6.2 and 6.3), in which the bulk isotope ratio

(R_{bulk}) is equal to the sum of the isotope ratios of each individual compound (R_i, for $i=1$ to n compounds) weighted by their respective mole fractions of carbon atoms (X_i, Eq. (6.4), where α_i is the stoichiometric ratio of carbon atoms in a given compound; e.g., for glucose ($C_6H_{12}O_6$), $\alpha_{glucose} = 6$ mole C/mole glucose).

$$R_{bulk} = \sum_{i=1}^{n} X_i R_i \qquad (6.2)$$

$$1 = \sum_{i=1}^{n} X_i \qquad (6.3)$$

$$X_i = \frac{\alpha_i \text{moles}_i}{\sum_{i=1}^{n} \alpha_i \text{moles}_i} \qquad (6.4)$$

The mole fraction is conveniently based upon concentration measurements in which all isotopes are simultaneously quantified as carbon atoms (i.e., $C = {}^{12}C + {}^{13}C + {}^{14}C$). In a rigorous derivation, Eq. (6.2) should be written in terms of fractional isotopic abundances (e.g., $^{13}C/(^{12}C + {}^{13}C + {}^{14}C)$) rather than simple isotope ratios (e.g., $R = {}^{13}C/^{12}C$) to obtain a true equality. However, the small natural abundances of ^{13}C and ^{14}C compared to ^{12}C permit the use of $^{13}C/^{12}C$ ratios, $\delta^{13}C$ values, or even $\Delta^{14}C$ values (see below) for R with negligible error for most environmental samples (i.e., $^{13}C/(^{12}C + {}^{13}C + {}^{14}C) \approx {}^{13}C/^{12}C$). As such, Eq. (6.2) is a powerful description of the isotopic composition, and hence, biogeochemistry of DOC. It can be used to identify components of the DOC pool or to trace their origin based on their isotopic signatures. If the isotopic signatures of individual components are sufficiently well known, then mass balance relationships can also be used to quantify their proportions in the mixture (i.e. mole fractions). For example, Williams and Gordon (1970) concluded in a pioneering study that terrestrial DOC (as represented by the Amazon River with $\delta^{13}C = -28.5 \pm 0.2$‰) delivered to the sea over geological time must comprise a small proportion of marine DOC

($\delta^{13}C = -24.4$ to $-22.0‰$) because their characteristic $\delta^{13}C$ values are disparate. Similar mass balance analyses have proven useful in coastal regions where terrestrial (-30 to $-25‰$) and marine DOC $\delta^{13}C$ values are sufficiently different for robust calculations (Boutton, 1991). In most cases, however, unambiguous attribution of bulk $\delta^{13}C$ variations to unique carbon sources in the marine environment has been hindered by significant overlap in the ranges of their individual $\delta^{13}C$ values. Therefore, the majority of this chapter will focus on the ^{14}C content of marine DOC.

B Carbon-14

Carbon-14 is radioactive and decays with a 5730 ± 40 year half-life (Godwin, 1962) that is much shorter than the 4.55 billion year age of the Earth (Dalrymple, 1991). Therefore, ^{14}C owes its existence on Earth to continuous natural production, primarily by the reaction between cosmic-ray generated neutrons and ^{14}N in the stratosphere (Anderson et al., 1947; Grosse, 1934; Libby, 1955). The newly created ^{14}C atoms eventually combine with oxygen to form CO_2, mix down into the lower atmosphere and surface ocean, and assimilate into organic matter via primary production (Schlesinger, 1997). If we assume the ^{14}C production rate was constant throughout recent history and that it mixes throughout Earth's principle carbon reservoirs much faster than its rate of radioactive decay, then it *should* have a constant global abundance. Following these same assumptions, the tissues of living organisms should also maintain a steady-state ^{14}C abundance. If the organisms cease to exchange carbon with the environment after dying, then the abundance of ^{14}C atoms in their remains will decrease over time due to radioactive decay. Thus, the minimum length of time since an organism manufactured an organic molecule can be calculated from the decrease in its ^{14}C abundance relative to the steady-state abundance of ^{14}C in the environment, i.e., assigned a "conventional radiocarbon date" (Libby, 1955). In this manner, the 4000-6000-year conventional ^{14}C ages of DOC in the deep N. Atlantic and N. Pacific illuminate the persistence of these molecules in the global carbon cycle (Bauer et al., 1992; Druffel et al., 1992; Williams and Druffel, 1987). However, this seemingly straightforward application of ^{14}C dating is nuanced with complications that, interestingly, allow more information to be extracted from DOC ^{14}C measurements than time alone.

The most fundamental complication to ^{14}C analyses is that isotopic fractionation can alter $^{14}C/^{12}C$ ratios independently of radioactive decay. Therefore, by convention, all ^{14}C data are corrected for fractionation based on independent measurements of ^{14}C, ^{13}C, and ^{12}C, assuming that ^{14}C is fractionated exactly twice as much as ^{13}C (Stuiver and Polach, 1977). Without this correction, for example, the loss of ^{14}C atoms due to fractionation would make a typical organic molecule (e.g., $\delta^{13}C \approx -20‰$, with an analogous $\delta^{14}C$ value changing by $\delta^{13}C \times 2 = -40‰$) appear to instantaneously and erroneously age *ca.* 330 ^{14}C years *during* its photosynthesis from surface DIC ($\delta^{13}C \approx \delta^{14}C \approx 0‰$). With this correction, however, molecules of DOC inherit the same ages as their carbon source (e.g., DIC) at the time of synthesis, and then continue to age as radioactive decay of ^{14}C proceeds. That is, ^{14}C is a powerful tracer of time because this correction ensures that variations in conventional ^{14}C ages are *not* due to isotopic fractionation.

Correcting for isotopic fractionation ensures that variations in ^{14}C ages ultimately derive from radioactive decay and a suite of processes that can be broadly categorized as mixing between isotopically unique sources of carbon (Eq. 6.2). Unfortunately, direct application of the linear isotopic mass balance equations to conventional ^{14}C ages is fundamentally complicated by radioactive decay, i.e., a logarithmic relationship between ages and ^{14}C abundances. Therefore, the

^{14}C abundances of geochemical samples, such as DOC, are more commonly reported as Δ^{14}C values (approximated by Eq. 6.5), from which conventional ^{14}C ages can be calculated (Eq. 6.6) (Stuiver and Polach, 1977).

$$\Delta^{14}C = \delta^{14}C - 2\left(\delta^{13}C + 25\right)\left(1 + \frac{\delta^{14}C}{1000}\right) \quad (6.5)$$

$$^{14}C \text{ age} = -\left(\frac{5568 \text{ years}}{\ln 2}\right)\ln\left(1 + \frac{\Delta^{14}C}{1000}\right) \quad (6.6)$$

Δ^{14}C values are analogous to δ^{13}C values (Eq. 6.1) in that they represent permil (‰) deviations in the ^{14}C/^{12}C ratios of samples relative to a standard, but differ from δ^{13}C values in two important respects. First, Δ^{14}C values are reported relative to a ^{14}C standard that, by definition, does not change with time (i.e., 95% of the ^{14}C abundance in Oxalic Acid-I (NIST SRM 4990B) as measured in 1950). Second, they are corrected for isotopic fractionation via δ^{13}C values, assuming that ^{14}C fractionates exactly twice as much as ^{13}C, and then placed on a scale that is relative to the postulated mean δ^{13}C value of terrestrial wood (−25‰). An additional correction may be applied (not shown in Eq. 6.5) for radioactive decay if significant time (e.g., >1 year) has passed between a ^{14}C measurement and either sample collection (e.g., DOC sampling) or tissue synthesis (e.g., $CaCO_3$ accretion in coral bands of known age). These attributes define Δ^{14}C values as the fractionation-corrected deviations in ^{14}C abundances from the preindustrial atmosphere at specified moments in history. Like conventional ^{14}C ages, DOC Δ^{14}C values are inherited from their original carbon source (e.g., DIC) and subsequently decrease as radioactive decay proceeds. Unlike ^{14}C ages, Δ^{14}C values can be used in mass balance mixing analyses (Eqs. 6.2–6.4) and are thus more appropriate for discussing the remaining complications to ^{14}C analyses of marine DOC.

First, the global carbon cycle is not perfectly well mixed with respect to ^{14}C. This leads to abundances that vary among the principle carbon reservoirs due to differences in their residence times and the persistence of radioactive decay. For example, the majority of ^{14}C enters the pool of marine DIC (DIC = $[CO_2]$ + $[H_2CO_3]$ + $[HCO_3^-]$ + $[CO_3^{2-}]$) in the surface ocean by air-sea gas exchange with atmospheric CO_2. It then experiences net radioactive decay during transport through the ocean's interior, resulting in very low DIC Δ^{14}C values at depth. The ^{14}C-depleted deep water continually shoals and mixes with recently dissolved CO_2 to maintain surface ocean DIC Δ^{14}C values that are lower than the values of atmospheric CO_2 (Figure 6.1a). The comparatively lower Δ^{14}C values and associated older ^{14}C ages of surface DIC are known as the marine "reservoir effect" and "reservoir age," respectively, and they vary with upwelling strength, carbon cycle parameters (carbon reservoir size, fluxes, etc.), and ^{14}C production rates (Stuiver and Polach, 1977; Stuiver et al., 1986). Thus, DOC molecules recently photosynthesized in the ocean should inherit the same relatively depleted Δ^{14}C values of surface DIC. In contrast, contemporaneously photosynthesized terrestrial organic matter will possess the higher Δ^{14}C value of atmospheric CO_2, and therefore appear artificially "younger" than its marine counterpart.

Second, the abundance of atmospheric ^{14}C exhibited natural variability for at least the past ~50,000 years, with a dramatic decline in Δ^{14}C values beginning ~30,000 years ago (Figure 6.1a). This is primarily due to natural variations in both the ^{14}C production rate and major fluxes of the global carbon cycle. More recently, Δ^{14}C values have been steadily declining since the late 1800s due to a combination of natural variations and human behavior collectively known as the "Suess Effect." This trend was primarily caused by fossil fuel combustion, which has diluted the atmosphere and sea with ^{14}C-free ($\Delta^{14}C = -1000$‰) CO_2 (Suess, 1953). The Suess Effect continues to this day, but the downward trend in Δ^{14}C values was interrupted

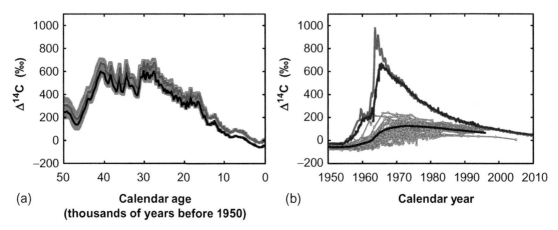

FIGURE 6.1   Temporal variations in global mean Δ^{14}C values. (a) Reconstructions of atmospheric CO_2 (IntCal09, red line) and surface ocean DIC (Marine09, black line) Δ^{14}C variations during the past 50,000 years (Reimer et al., 2009). Gray shaded areas correspond to ±1 standard deviation. Note that marine Δ^{14}C values are lower than contemporaneous atmospheric values due to the oceanic reservoir effect. (b) The "bomb spike" in Δ^{14}C values of mean northern (red line) and southern (blue line) hemispheric CO_2 (Hua et al., 2013, and references therein), individual records of surface ocean DIC (gray circles) and a model of surface DIC (heavy black line; Reimer et al., 2009). Surface ocean DIC records were primarily based on banded corals and shells from sites throughout the tropical Pacific, Indian, and Atlantic Oceans (Druffel, 1981, 1982, 1987, 1996; Druffel and Griffin, 1993, 1995, 1999; Druffel et al., 2001; Grumet et al., 2004; Guilderson et al., 2000, 2004; Hua et al., 2005; Schmidt et al., 2004), and from higher latitudes in the North Atlantic Ocean (Kilada et al., 2007; Scourse et al., 2012; Weidman and Jones, 1993; Witbaard et al., 1994).

in the mid-twentieth century by a more galvanizing human exploit. The concentration of ^{14}C in the atmosphere was nearly doubled by testing thermonuclear weapons in the atmosphere ("bomb spike," Figure 6.1b) during the 1950s and 1960s (Nydal, 1963; Nydal et al., 1980). Clearly, the steady-state ^{14}C abundance assumed in radiocarbon dating is not universally applicable. However, the bomb spike provides a unique method of tracing the origins and ages of molecules produced within the last *ca.* 60 years.

For example, the mean Northern Hemisphere Δ^{14}C value of atmospheric CO_2 peaked at nearly +1000‰ when the 1964 nuclear test ban treaty came into effect (Figure 6.1b). The atmospheric bomb spike then decreased faster than expected based solely on radioactive decay. This decrease was evidence for exchange between atmospheric CO_2 and the principle reservoirs of the carbon cycle, including marine DIC, which peaked in Δ^{14}C ~10 years later (Figure 6.1b) due to the long equilibration time for exchange between the atmosphere

and surface ocean (Broecker and Peng, 1982). The marine DIC bomb spike only reached *ca.* +200‰, however, because the bomb-^{14}C-labeled CO_2 mixed into the much larger DIC pool (Eq. 6.2). Since DOC inherits the Δ^{14}C values of its source material, compounds with Δ^{14}C values greater than ~0‰ must contain bomb-^{14}C and were therefore synthesized after ~1955. Furthermore, compounds with Δ^{14}C values exceeding ~300‰ were most likely synthesized on land from atmospheric CO_2 during this same period of time and subsequently delivered to the sea.

The bomb spike would serve as a simple Δ^{14}C-age calibration curve for recently produced molecules if it were not for two complications. First, the same high Δ^{14}C values appear on both sides of the bomb spike's peak and suggest two possible ages for an individual Δ^{14}C measurement. Without additional knowledge of the lifespan of a sample, it can only be said that it contains bomb-^{14}C. Second, radioactive decay would artificially place Δ^{14}C values measured today on a more recent portion of the bomb spike's tail and lead to a slight underestimation of the age. The latter complication can be overcome by expressing ^{14}C abundances in units of "fraction modern" rather than $\Delta^{14}C$ values. Since this alternative unit does not change with radioactive decay (Stuiver and Polach, 1977), bomb-era ages can be estimated by comparing measured fraction modern values with those of the bomb spike via calibration models (Reimer et al., 2009).

Given the above complications, the 4000-6000-year conventional ^{14}C ages of deep DOM (Bauer et al., 1992; Druffel et al., 1992; Williams and Druffel, 1987) are not equal to calendar years. Instead, they are a convenient measure of persistence that is frequently cited and should be interpreted with the following considerations. First, conventional ^{14}C ages are corrected for fractionation. Thus, ^{14}C ages are inherited by molecules of DOC from their source materials (e.g., DIC or organic precursor molecules) and subsequently increase as radioactive decay proceeds. Second, conventional ^{14}C ages assume atmospheric ^{14}C concentrations have remained constant and that ^{14}C decays with a "Libby" half-life of 5568 years (Stuiver and Polach, 1977). These two assumptions compensate for the natural variations in global ^{14}C abundances noted above, while simultaneously providing consistency with the scores of measurements that antedate the accepted "Cambridge" half-life of 5730 years. This distinction in half-lives is critical when modeling DOM turnover times with ^{14}C measurements. For example, molecules synthesized in the surface ocean more than ~4000 calendar years before 1950 will have older calendar ages than conventional ^{14}C ages would suggest (Figure 6.2). In contrast, molecules synthesized between 1950 AD and ~4000 calendar years ago will have shorter calendar ages than conventional ^{14}C ages. Finally, molecules synthesized since the 1950s would have negative conventional ^{14}C ages due to bomb-^{14}C (see above) and are therefore simply deemed "modern." In all three cases, we must consider that the conventional ^{14}C ages of

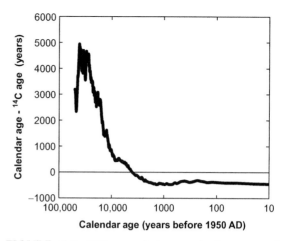

FIGURE 6.2 Differences between calendar ages and modeled global mean conventional ^{14}C ages of DIC in the surface ocean versus calendar age (plotted on a logarithmic scale), based on the Marine09 ^{14}C calibration (Reimer et al., 2009).

marine DOC are products of the marine reservoir effect (Reimer et al., 2009), variability in the ^{14}C content of DIC (Nydal, 1963; Nydal et al., 1980; Suess, 1953), and an operational DOC definition that implies a mixture of molecules that can, in principle, vary in age (e.g., Loh et al., 2004). As such, uncertainties reported with conventional ^{14}C age measurements correspond to measurement precisions rather than distributions of molecular ages.

III DOC ISOTOPE RATIO METHODS

High precision DOC $\delta^{13}C$ and $\Delta^{14}C$ values are presently quantified by isotope ratio mass spectrometry (IRMS) and accelerator mass spectrometry (AMS), respectively (e.g., see Faure (1986) and Tuniz et al. (1998) for reviews of instrumentation). AMS is necessarily more complex and sensitive than IRMS because it must quantify $^{14}C/^{12}C$ ratios (ca. 10^{-12} to 10^{-15}) that are 10 to 13 orders of magnitude smaller than most $^{13}C/^{12}C$ ratios (ca. 10^{-2}). Regardless, both instruments measure isotope ratios with uncertainties that largely depend on the number of atoms of detected. Additional uncertainty arises when correcting the isotope ratio measurements for contamination that is inherently introduced to all samples by all methods (i.e., the "blank"). Thus, the final measurement uncertainty of a sample depends on the magnitudes and uncertainties of the masses, mole fractions, and isotope ratios of both the sample and its blank (see Bevington and Robinson (2002) or Glover et al. (2011) for general discussions of error propagation). The importance of this relationship cannot be overstated, because measurement uncertainty is a quantitative measure of our ability to answer scientific questions. Therefore, the uncertainty's dependence on sample mass and purity dictates the scale of the methods used for isotopic analyses of DOC. For example, AMS is routinely used to measure $\Delta^{14}C$ values in samples containing 100-1000 µg C with a precision of a few permil, and is now capable of quantifying $\Delta^{14}C$ values in samples containing as little as 2-10 µg C with a precision of a few percent (Santos et al., 2007). The superior sensitivity of AMS has thus reduced the detection limit for bulk marine DOC $\Delta^{14}C$ measurements to <100 mL of seawater (Beaupré and Druffel, 2009) and, more impressively, permitted $\Delta^{14}C$ analyses on individual compounds within the DOM pool (Aluwihare, 1999).

Unfortunately, neither IRMS nor AMS are currently capable of directly measuring the carbon isotope ratios of raw DOM in seawater. Thus, the analytical challenge is to prepare samples for these instruments by first oxidizing the complex mixture of organic molecules into CO_2. This process homogenizes DOC into a single, volatile compound that can be injected directly into an IRMS. Indeed, several labs have already developed interfaces that directly couple DOC oxidizing instruments to IRMS for rapid $\delta^{13}C$ measurements (Bouillon et al., 2006; Lang et al., 2007, 2012; Osburn and St. Jean, 2007; Panetta et al., 2008). AMS requires the additional step of converting CO_2 gas into a solid graphite target

(Vogel et al., 1987), which is then efficiently sputtered into the instrument as a beam of negative carbon ions. An exciting new technological development is the gas-accepting ion source that circumvents graphitization and permits direct injection of CO_2 into an AMS (Roberts et al., 2011; Ruff et al., 2010). While this technology holds promise for expedited, lower-cost ^{14}C measurements (Roberts et al., 2013), analytical challenges such as detection limits, blanks, and flow rate matching, have so far precluded direct coupling of marine DOC oxidizing instruments to an AMS. Thus, all DOC ^{14}C measurements, and most $\delta^{13}C$ measurements, are still obtained via offline oxidation of DOM.

Early attempts to convert marine DOM into CO_2 were based on an established method of combusting the residues from evaporated freshwater samples (Krogh, 1930). This approach, however, was found to be "of no avail" when applied to seawater because of the overwhelming proportion of salts (Krogh and Keys, 1934). Alternatively, the community embraced the reactions of strong chemical oxidizers added directly to seawater (Atkins, 1922; Duursma, 1961; Krogh and Keys, 1934; Lieb and Krainick, 1931; Natterer, 1892; Wilson, 1961). Eventually, persulfate oxidation became the standard "wet chemical oxidation" method (Menzel and Vaccaro, 1964), but it was not amenable to large volumes of seawater necessary for isotopic analyses. Recognizing this limitation, Williams and Zirino (1964) attempted to recover larger masses of DOM by coprecipitation with hydrated metal oxides (Bader et al., 1960; Jeffrey and Hood, 1958; Park et al., 1962; Tatsumoto et al., 1961; Williams, 1961). They presented the following salient geochemical, rather than analytical, explanations for their unexpectedly low yields (16-54%) from deep samples. First, the most reactive DOC constituents must have been naturally removed from the deep sea in situ, leaving behind solutes that were more resistant to scavenging, oxidation, and biochemical utilization. Second, at least some fraction of the residual DOC must have been extremely persistent, and $\Delta^{14}C$ values could reveal its age (Williams and Zirino, 1964). Shortly thereafter, Armstrong et al. (1966) successfully oxidized marine DOC by irradiating 110 mL seawater samples with ultraviolet light (i.e., "UV oxidation") and suggested that their system could be scaled up for isotopic analyses. Using this approach, Pete Williams published the first bulk marine DOC δ^{13}C values in 1968, followed by the first Δ^{14}C values in 1969 (Williams, 1968; Williams et al., 1969).

Subsequent bulk marine DOC δ^{13}C and Δ^{14}C measurements were primarily obtained via UV oxidation because this approach can process large batches of seawater with reproducibly small blanks and high yields (Bauer et al., 1992; Beaupré and Druffel, 2009, 2012; Beaupré et al., 2007; Druffel et al., 1989; Williams and Druffel, 1987; Williams and Gordon, 1970). The principle disadvantages of UV oxidation are the need for custom quartz reactors, dedicated vacuum lines, and significant lengths of time to process each sample (Beaupré et al., 2007). Alternative methods have been devised to overcome these hindrances, such as repeat-injection high-temperature combustion (HTC) (Bauer et al., 1992; Druffel et al., 1992), super-critical fluid oxidation (Le Clercq et al., 1997), sealed-tube thermal sulfate reduction (Fry et al., 1996; Johnson and Komada, 2011), and combined UV-persulfate oxidation (Bauer et al., 1991). These alternatives are presently subject to larger and more variable blanks than the UV oxidation method (Figure 6.3), which ultimately increases the propagated uncertainties for their concentration, δ^{13}C, and Δ^{14}C measurements.

All of the above methods can generally be divided into a sequence of three steps: (1) removal of residual DIC, (2) oxidation of the organic matter, and (3) isolation of the resulting CO_2. DIC is typically removed from DOC samples by acidifying (pH ~2.5) and sparging seawater with purified gas (e.g., helium). Although volatile organic carbon (VOC) is co-sparged with DIC, the losses are assumed to be a negligible proportion of the DOC pool (~5%; MacKinnon, 1979; Sharp, 2002). Likewise, losses of carbon atoms

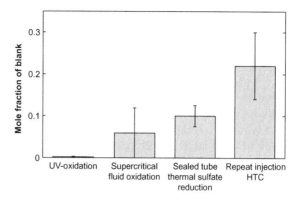

FIGURE 6.3 Average mole fractions of blank carbon from published methods for measuring bulk δ^{13}C or Δ^{14}C values of marine DOC. For comparison, the published blanks (Bauer et al., 1992; Beaupré et al., 2007; Fry et al., 1996; Le Clercq et al., 1997) were calculated as mole fractions of C in a hypothetical near-surface DOC sample (e.g., ~70 μM C), with error bars representing ±1 standard deviation. Blanks for the sealed tube thermal sulfate reduction method were originally published as a range spanning 5-15% of the sample mass (Fry et al., 1996). Therefore, the blank was plotted here as 10 ± 2.5% of the sample mass, assuming a normal distribution of blank masses in which the reported range spanned *ca.* ±2 standard deviations.

due to pH-dependent structural transformations (e.g., decarboxylation, hydrolysis, etc.) and corresponding changes in volatility or solubility are also assumed to be negligible. Therefore, implicit to the operational definition of DOC is the fact that the concentrations and isotope ratios are based on nonvolatile constituents that are also stable with respect to acid-base chemistry on the time scales of laboratory analysis. Given these complications and potential for additional contamination during acidification, why would we remove DIC prior to oxidation?

DIC must be removed or else it will serve as an overwhelming blank ($X_{blank} > ca.$ 0.96, Eqs. 6.2–6.4) that renders the corrected DOC isotope ratio virtually useless. For example, consider surface water from the central North Pacific Ocean with concentrations and Δ^{14}C values of DIC (2007 ± 2 μM, +137 ± 5‰) and DOC (82 ± 2 μM, −163 ± 5‰) that were measured by

Druffel et al. (1992). If the DIC (i.e., the "blank") had not been removed prior to UV oxidation, the ensuing mixture ($2089 \pm 3\,\mu M$ and $+125 \pm 5‰$) would have yielded a blank-corrected DOC $\Delta^{14}C$ value of $-161 \pm 177‰$, i.e., it would have inflated the uncertainty from $\pm 5‰$ to $\pm 177‰$. This uncertainty spans 344‰ on the $\Delta^{14}C$ scale and is nearly equal to the range of $\Delta^{14}C$ values in Druffel et al.'s (1992) corresponding DOC depth profile (from -546 to $-163‰$, a difference of 383‰). Thus, DIC removal is a critical step in obtaining meaningful DOC isotope ratio measurements.

Alternative methods have been employed to measure DOC isotope ratios without the challenges of removing DIC or oxidizing DOC in solution. These approaches involve measuring $\delta^{13}C$ and $\Delta^{14}C$ values of DOC constituents isolated in large quantities from seawater by ultrafiltration (UF) or solid phase adsorption onto polystyrene cross-linked resins (XAD). This includes, for example, analyses of "humic materials" (Druffel et al., 1992; Meyers-Schulte and Hedges, 1986), high molecular weight (HMW) and colloidal DOC (Guo et al., 1996; McCarthy et al., 2011; Santschi et al., 1995; Walker and McCarthy, 2012; Walker et al., 2011), specific compounds (Aluwihare, 1999; Repeta and Aluwihare, 2006), procedurally defined compound classes (Loh et al., 2004), and even black carbon (Ziolkowski and Druffel, 2010). These methods sufficiently concentrate organic matter for oxidation by methods such as sealed-tube combustion (see references above for details) that are much simpler than UV oxidation of bulk DOC in seawater. In addition to simplifying the oxidation method, these approaches permit a suite of chemical analyses that complement and enhance our understanding of DOC biogeochemistry (Aluwihare et al., 2002; Amon and Benner, 1994; Benner, 2002; Benner et al., 1992; McCarthy et al., 1996). While studying subsets of the total DOC pool presents several analytical and intellectual advantages, it also presents one geochemical limitation: ultra-filtered DOC

represents $\leq \sim 20\text{-}40\%$ of the mass of DOC, and the proportions of individual components are accordingly smaller.

Regardless of the method employed, DOC is an operationally defined mixture of compounds that, in principle, should vary in reactivity according to their structures. Fractionation during incomplete oxidation of individual compounds can, in principle, lead to erroneous $\delta^{13}C$ values. If the compounds also vary in isotopic signature, then incomplete oxidation will produce erroneous $\Delta^{14}C$ measurements that are not representative of the original bulk mixture. This sensitivity to yield was demonstrated by observing an asymptotic decrease in $\Delta^{14}C$ values of CO_2 released during UV oxidation of DOC samples from the eastern North Pacific Ocean (Beaupré and Druffel, 2012; Beaupré et al., 2007). Therefore, the practical metric for obtaining precise $\Delta^{14}C$ measurements is to continue oxidizing until DOC yields are within measurement uncertainty of 100%. A minimum of 4h of exposure to UV light was required for the system employed by Beaupré et al. (2007), but will likely differ for other systems. Similar relationships between yields and isotopic signatures may exist for alternative methods of oxidation, but have not yet been reported.

The development costs and painstaking implementation of the methods described above have been justified by the intellectual value that they continue to provide. However, there is considerable room for progress in efficiency, throughput, and detection limits. For example, a global survey of bulk DOC $\Delta^{14}C$ values on par with the very illuminating body of concentration measurements ($n > 23{,}000$) would require more than 159 years of daily UV oxidations and cost more than \$18,000,000 in lab fees using today's technology. Higher spatiotemporal resolution is absolutely essential to continued progress, but such costs do not justify an endeavor on this scale. Continued methodological innovations are needed to expand our knowledge on a more practical budget.

IV ISOTOPIC COMPOSITION OF BULK MARINE DOC

The isotopic composition of DOC is clearly a powerful tracer of both mass and time. It is also cumbersome to measure. Perhaps this is why so few measurements have populated the literature (Figure 6.4) and yet produced so many long-lasting paradigms. Two outstanding reviews have previously addressed many of these concepts (Bauer, 2002; McNichol and Aluwihare, 2007). While there is considerable overlap, the present review focuses on (i) open ocean DOC, (ii) what has been learned from several key studies, and (iii) recent advances since the previous reviews.

A The First Measurements

Much of our view of marine DOM was laid out prior to the first DOC $\delta^{13}C$ or $\Delta^{14}C$ measurements. For example, Duursma (1961) and Menzel (1964) demonstrated that DOC concentrations exhibited little spatiotemporal variability in the deep sea. Combined with low yields during co-precipitation experiments (Williams and Zirino,

1964) and biological incubations (Barber, 1968), the logical conclusion was that the majority of deep DOM consisted of material that was stable to chemical or biological degradation. This conclusion was only further supported by publication of the practically invariant bulk $\delta^{13}C$ values ($-22.9\pm0.2‰$ to $-22.5\pm0.2‰$) of deep DOC in offshore waters near Southern California (Williams, 1968). More interestingly, similarity between these $\delta^{13}C$ values and those of cellulose and lignin extracted from plankton (Degens et al., 1968a; Williams, 1968) hinted at its composition and source. At the same time, Calder and Parker (1968) published $\delta^{13}C$ values of bulk DOC and petrochemical effluents in the Gulf of Mexico and coastal waters of Galveston Bay, Texas. They applied one of the first isotopic mass balance calculations (e.g., Eqs. 6.2 and 6.3) to apportion sources of the comparatively $\delta^{13}C$-depleted pollutants to the bulk DOC pool. Combined, the first bulk $\delta^{13}C$ measurements provided independent clues as to the composition, origin, and longevity of DOM in two different systems, but they did not provide a tangible timescale upon which to evaluate DOM in the larger carbon cycle.

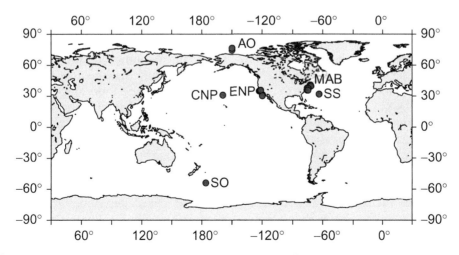

FIGURE 6.4 Map of all locations at which depth profiles of bulk DOC $\Delta^{14}C$ have been reported, with stations in the Arctic Ocean (AO; 75°00′N, 150°00′W and 78°00′N, 150°6′W), central North Pacific (CNP; 31°00′N, 159°00′W), eastern North Pacific (ENP: i.e., "Station M"; 34°50′N, 123°00′W), Mid-Atlantic Bight (MAB; multiple profiles), Sargasso Sea (SS; 31°50′N, 63°30′W), and Southern Ocean (SO; 54°00′S, 176°00′W).

About 1 year later, however, the first bulk $\Delta^{14}C$ measurements of deep (~1900 m) DOC were published: $-341 \pm 25\permil$ and $-351 \pm 27\permil$ (Williams et al., 1969). These measurements translated to conventional ^{14}C ages of 3350 ± 300 and 3470 ± 330 years, respectively, and were significantly older than the 1480 ± 80 to 2194 ± 70 year ages of DIC from 2000 m depth at a nearby site in the Pacific (Bien et al., 1965; Williams et al., 1969). Logically, the organic and inorganic constituents could not have followed identical paths from the surface ocean to the deep interior if they took dramatically different lengths of time to get there. Some of this age difference was attributed to an apparent rejuvenation of deep DIC by mineralization of rapidly sinking, ^{14}C-enriched particles. The most galvanizing explanation, however, recognized that a bulk ^{14}C age is simply the average age of all molecules in a sample (i.e., Eq. 6.2). That is, if some fraction of the DOC pool could be "younger" than the measured bulk ages, then there must be some fraction of molecules that are even "older" still. Thus, the authors posited that older molecules were "recycled" through the system.

Although somewhat vague, Williams et al.'s (1969) hypothesis of "old," recycled DOC has at least two possible interpretations. Some molecules could persist, intact, substantially longer than the timescale for ventilation (e.g., ~595 years in the Pacific, based on DIC ^{14}C; Stuiver et al., 1983) and eventually mix with recently produced DOC in the surface ocean to yield average ages exceeding those of ambient DIC. Alternatively, a fraction of the carbon atoms in these molecules could be reworked into progressively more resilient molecules during their residence in the sea. These interpretations may not be mutually exclusive because DOC is operationally defined in terms of carbon atoms, but its longevity (and hence, ^{14}C-ages) depends on the inherent reactivities of organic molecules. Under this construct, and assuming both a steady-state DOC inventory and negligible ^{14}C reservoir age, Williams et al. (1969) equated the

average of their duplicate ^{14}C-age measurements, 3400 years, to the residence time (τ) of DOC in the deep ocean. Further assuming a global burden of 630 Pg of deep DOC (i.e., the postulated mean concentration of 500 μg DOC/L multiplied by the 1.26×10^{21} L volume of the ocean between 300 and 3800 m depth), they were able to calculate a net input rate of 0.185 Pg DOC/year into the deep sea. The latter values are remarkably similar to our current best estimates of the 615 Pg global burden of deep DOC ($\geq$200 m depth, after Hansell et al., 2009) and the ~0.2 Pg of DOC exported to the deep ocean each year ($\geq$500 m depth; Hansell et al., 2009). Although neither set of calculations was published with formal uncertainties (Hansell et al., 2009; Williams et al., 1969), they may be considered statistically indistinguishable when constrained by their least significant digits. We should certainly place more stock in the recent estimates because they benefited from a comparatively massive data set of robust DOC concentration measurements and relied less upon the assumption of global uniformity. But, we may also be inspired by the ability of these pioneers to see the entire ocean within a few bottles of seawater.

Thus, with just one replicated measurement of ^{14}C-ages, Williams et al. (1969) placed marine DOC biogeochemistry on a globally averaged timescale. We now knew that the average organic molecule dissolved in the deep sea was "born" before the founding of Rome. And, although most of the organic molecules were born from living organisms (see review by Carlson, this volume), we did not know how these molecules "died"—Williams et al. (1969) did not speculate on loss mechanisms. And herein lies the genesis of just one of many interesting geochemical paradoxes. Since ^{14}C measurements suggest DOC is the longest-lived reservoir of carbon in seawater, we might expect it to have accumulated over geological time to also become the most abundant. But, it is not. It is, in fact, a comparatively dilute solute in the open ocean that exhibits a remarkably small range of concentrations (34 to *ca.* 80 μM; Hansell et al., 2009). After more than 40 years since the first ^{14}C-age determinations, the loss mechanisms and associated rates that maintain this globally unvarying distribution remain prominent targets for DOC research.

Although pioneering and prescient, Williams et al. (1969) based their residence time calculations on several implicit assumptions that warrant further discussion. First, their calculations implicitly assumed that bulk marine DOC contains a negligible proportion of allochthonous molecules with different initial Δ^{14}C values. For example, it did not consider organic matter synthesized from carbon stocks with different Δ^{14}C values or delivered to the sea after a sufficient period of aging. The ^{14}C ages of such molecules would not necessarily correspond to their residence times in the sea. If these molecules comprise a significant proportion of the total DOC pool, then their individual turnover times and the collective turnover time of DOC may be considerably different than expected from bulk ^{14}C ages (Trumbore and Druffel, 1995). By analogy, Florida residents have the 5th oldest median (i.e., bulk) age in the United States (Howden and Meyer, 2002) largely due to the net immigration of (i.e., allochthonous) people ≥60 years old (The Florida Legislature, 2011) rather than to inherent longevity. Although the methods for studying marine DOC dynamics are less routine than those of the U.S. census, we now recognize several potentially significant sources of isotopically distinct DOC to the open ocean water column (Bauer et al., 1995; McCarthy et al., 2011; Pohlman et al., 2010; Wang et al., 2001). Consider, for example, the isotopic composition of HMW DOC venting from the flanks of the Juan de Fuca Ridge (McCarthy et al., 2011). The Δ^{14}C values of HMW DOC (−834.9‰ to −771.5‰) and DIC (*ca.* −790‰) in the venting fluids were consistently lower than HMW DOC (−442‰) or DIC (*ca.* −250‰) in the overlying water column. Furthermore, the venting HMW

DOC's δ^{13}C values (−26.1‰ and −34.5‰) were also significantly lower than HMW DOC (*ca.* −21‰) in the overlying water column. These results suggested that the HMW DOC venting into the deep sea at these sites was produced chemosynthetically from DIC within the crust. That is, the venting HMW DOC was a source of pre-aged organic matter that did not originate in the surface ocean. Quantifying the influence of such sources on DOC's apparent longevity is complicated by several uncertainties that include, but are not limited to, (i) the number of possible mechanisms by which pre-aged organic matter can enter the DOC pool, (ii) the number of such systems that are active at any given time in history, their (iii) lifespans, (iv) fluid flow rates, (v) DOC concentrations, (vi) Δ^{14}C values, and (vii) the associated statistical distributions of the preceding terms. Given our limited knowledge of each of these terms, we can at best multiply their ranges of postulated values to constrain each flux. In this manner, McCarthy et al. (2011) identified mid-ocean ridge flanks as a source of 1.2 to 5×10^{12} g/year of HMW DOC with ^{14}C-ages spanning 11,800-14,400 years (i.e., ~1-5% of the steady-state input of DOC to the deep ocean).

The second implicit assumption by Williams et al (1969) lies in their use of a traditional residence time equation (i.e., $\tau =$ total mass/input rate) that is derived explicitly for first order (1°) kinetic processes (Li, 1977). While this assumption may be reasonable, it is unlikely that all DOC processes proceed exclusively with 1° order kinetics. For example, enzymatic reactions that follow Michaelis-Mentin kinetics can, in theory, proceed anywhere between 0° and 1° depending upon the relative concentration of available substrates. In addition, a recent study has shown that photomineralization can proceed with apparent second order (2°) kinetics when exposing DOC to the output of a high-energy UV lamp (Beaupré and Druffel, 2012). The consequences of higher order processes are nontrivial. Aside from confounding our notions of a traditional DOC residence time, the chemical half-life of residual DOC with respect to a higher order (e.g., 2°) process will increase as degradation proceeds. Thus, reaction order itself may be an import control on the longevity of molecules in the sea (Beaupré and Druffel, 2012). Our understanding of DOC turnover times could therefore benefit from additional process studies that couple kinetic and isotopic analyses.

Finally, Williams et al. (1969) naturally assumed that their isotopic measurements were, in fact, accurate. Nearly 20 years would pass before they could be compared to another bulk DOC $\Delta^{14}C$ measurement (Williams and Druffel, 1987). The new measurements were in the form of a detailed depth profile that suggested the original bulk DOC ^{14}C ages from the deep Pacific were underestimated by almost 3000 years.

B The First $\delta^{13}C$ and $\Delta^{14}C$ Depth Profiles

The most remarkable feature of the first detailed depth profiles of bulk marine DOC $\delta^{13}C$ values was, ironically, their lack of prominent features (Williams and Gordon, 1970). The profiles were nearly constant, averaging $-22.6 \pm 0.7\permil$ ($n = 18$) despite variations in depth, temperature, dissolved oxygen, nutrients, season, and sampling location in the northeast Pacific. This uniformity only galvanized the notion that DOC was stable to biochemical transformations after export to the deep ocean, and once again raised questions of composition, sources, and sinks. For example, these DOC $\delta^{13}C$ values most closely resembled bulk POC (average $= -23.2 \pm 0.8\permil$, $n = 5$) and the cellulose ($-22.4\permil$) and insoluble residues ($-23.1\permil$) of plankton extracts (Williams and Gordon, 1970, and references therein). If this similarity was evidence for a primarily photosynthetic origin, then the DOC should have contained multiple constituents with different reactivities, such as a large proportion of ^{13}C-enriched amino acids and sugars

that were likely lost en route to deeper waters. Another possible source of marine DOC was the large flux of riverine DOC (e.g., $-28.5\permil$ in the Amazon River), but the uniform and isotopically disparate $\delta^{13}C$ values of marine DOC ($-22.6\permil$) suggested that terrigenous material was also rapidly lost and did not accumulate to a significant proportion in the sea. In other words, the $\delta^{13}C$ signatures were consistent with a primarily marine origin. Finally, Williams and Gordon (1970) considered burial in marine sediments as a possible sink. The $\delta^{13}C$ values of marine DOC ($-22.6\permil$) and sediments near their study area ($-20.8\permil$) were similar, but not identical, and therefore could not unambiguously record coprecipitation from overlying waters. If burial was indeed a sink for marine DOC, then it must have involved either fractionation at the sediment water interface or mixing with ^{13}C-enriched matter from surface plankton (Williams and Gordon, 1970).

The most commonly cited feature of the first detailed depth profile of bulk DOC $\Delta^{14}C$ values (Figure 6.5, gray circles) was an average age of 6000 conventional ^{14}C years ($\Delta^{14}C = -525\permil$) of DOC in the deep North Pacific (Williams and Druffel, 1987). This ^{14}C age was nearly twice as old as the only prior measurements (~3400 years; Williams et al., 1969). The difference between these ages was attributed to lower yields (~70%) in the 1969 study and the *unstated* assumption that a partial UV oxidation would leave behind molecules that were ^{14}C-depleted (Williams and Druffel, 1987). This clandestine hypothesis related photochemical lability to ^{14}C age and is consistent with recently observed trends toward significantly lower $\Delta^{14}C$ values during UV oxidation of both surface and deep DOC (Beaupré and Druffel, 2012; Beaupré et al., 2007). Thus, the higher yields, improved measurement precision (from $\pm ca.$ 26‰ in 1969 to as low as $\pm 8\permil$ in 1987), and uniform $\Delta^{14}C$ values at depth lent greater credibility to the 6000 year age of deep DOC in the Pacific. As a result, the previously estimated steady-state input of DOC

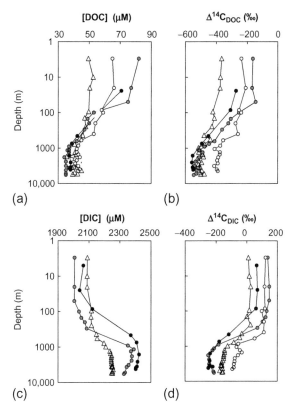

[DOC] (μM)

Δ¹⁴C_DOC (‰)

(a)

(b)

[DIC] (μM)

Δ¹⁴C_DIC (‰)

(c)

(d)

FIGURE 6.5 Depth profiles of (a) DOC concentration, (b) DOC Δ^{14}C, (c) DIC concentration, and (d) DIC Δ^{14}C from the Southern Ocean (triangles), Sargasso Sea (open circles), central North Pacific (gray circles), and eastern North Pacific (black circles) as illustrated in Figure 6.4. Profiles from the eastern North Pacific are presented as mean values at consistent depth horizons from 12 separate cruises to Station M (1991-2004). Depths are plotted on a logarithmic scale so that structures at shallower depths can be seen more clearly. *Reprinted from Beaupré, S.R., Aluwihare, L.I., (2010), with permission from Elsevier.*

to the deep ocean was revised from 0.185 PgC/year (Williams et al., 1969) down to 0.1 PgC/year (Williams and Druffel, 1987).

Three additional but no less important features of the DOC Δ^{14}C profile (Figure 6.5, gray circles) were articulated by Williams and Druffel (1987). First, the DOC Δ^{14}C profile was similar in shape to contemporaneous profiles of DOC concentrations and DIC Δ^{14}C values. All three profiles exhibited high surface ocean values that decreased monotonically until becoming practically invariant below 1000 m. This similarity argued strongly for a predominantly surface ocean source of marine DOC. Second, the DIC Δ^{14}C profile was enriched with bomb-^{14}C, but the corresponding DOC Δ^{14}C values were uniformly lower by *ca.* 300‰ at all depths. Therefore, surface ocean DOC must have contained a mixture (i.e., Eq. 6.2) of bomb-^{14}C enriched molecules that were recently synthesized from surface DIC and some very old, ^{14}C-depleted molecules that were driving the bulk DOC Δ^{14}C values down. Third, the difference (390‰) between Δ^{14}C values of surface and deep DOC was identical to that of DIC. Combined, these three features suggested that DOC produced in the surface ocean was redistributed to deeper waters through processes similar to those that control DIC (Williams and Druffel, 1987). In addition, the presence of bomb-^{14}C in mesopelagic DIC (<900 m) was a reminder of the speed with which surface processes transport carbon to the ocean interior. Thus, the hypothesized parallel migration of DOC and DIC based on ^{14}C analyses was considered the most definitive evidence at that time for export of recently produced DOC to deeper waters (Hedges, 1987). Moreover, similarity between DOC and DIC Δ^{14}C profiles also was taken as evidence for a fraction of DOC that was transported with DIC along the path of deep thermohaline circulation from the North Atlantic to the North Pacific by way of the Southern Ocean (Williams and Druffel, 1987). This implied that the global pattern of deep DOC Δ^{14}C values should be similar to that of deep DIC, but more profiles were needed to test this hypothesis.

C New Depth Profiles and Spatiotemporal Variability

Bulk DOC Δ^{14}C depth profiles have been reported for just six regions globally since the first measurements were published in 1969 (Figures 6.4

TABLE 6.1   Carbon Isotopic Signature of Bulk Marine DOC

Region	Depths (m)	Concentration (μM)	$\delta^{13}C$ (‰)	$\Delta^{14}C$ (‰)	References
Canada Basin, Arctic	3 to 350	53 to 77	−24.4 to −21.5	−375 to −216	Griffith et al. (2012)
	400 to 3800	35 to 51	−24.3 to −22.7	−494 to −335	Griffith et al. (2012)
Caribbean	30 to 1430	37 to 67	−22.2 to −22.1	n.d.	Jeffrey (1969) and Eadie et al. (1978)
Central North Pacific	3 to 900	37 to 82	−21.4 to −20.4	−477 to −163	Williams and Druffel (1987) and Druffel et al. (1992)
	21 to 915	40 to 73	−23.2 to −20.4	−479 to −246	Walker et al. (2011)
	1150 to 5720	34 to 37	−21.4 to −20.7	−546 to −489	Williams and Druffel (1987) and Druffel et al. (1992)
Eastern North Pacific	1900	31	−22.5	−346	Williams et al. (1969)
	5 to 800	38 to 77	−22.4 to −20.5	−526 to −238	Beaupré and Druffel (2009) and Bauer et al. (1998a,b)
	900 to 4100	33 to 43	−22.4 to −22.4	−584 to −468	Beaupré and Druffel (2009) and Bauer et al. (1998a,b)
Gulf of Mexico	0 to 3000	33 to 72	−23.6 to −19.6	n.d.	Jeffrey (1969) and Eadie et al. (1978)
North Atlantic	40 to 1870	37 to 67	−23.1 to −22.2	n.d.	Jeffrey (1969) and Eadie et al. (1978)
Sargasso Sea	Surface film	580	−28.4	−869	Druffel et al. (1992)
	0.1 to 600	45 to 75	−21.8 to −20.9	−356 to −210	Druffel et al. (1992)
	850 to 4500	40 to 44	−21.3 to −20.2	−414 to −375	Druffel et al. (1992)
Southern Ocean	3 to 770	41 to 53	−22.7 to −21.3	−475 to −366	Druffel and Bauer (2000)
	970 to 5440	36 to 45	−22.7 to −21.3	−529 to −476	Druffel and Bauer (2000)
South Pacific	0 to 2700	43 to 80	−22.2 to −19.9	n.d.	Eadie et al. (1978)

and 6.5, Table 6.1). Three were located at anchor points along the postulated route of global thermohaline circulation: the Sargasso Sea (SS), the Southern Ocean (SO), and the central North Pacific (CNP) (Bauer et al., 1992; Druffel and Bauer, 2000; Druffel et al., 1992). Two were located near the North American boundaries of adjacent ocean basins: ~200 km off the coast of Southern California at Station M, and throughout the Mid-Atlantic Bight (Bauer and Druffel, 1998; Bauer et al., 1998a,b, 2002; Beaupré and

Druffel, 2009). The most recent profiles were reported from the Canada Basin in the Arctic Ocean (Griffith et al., 2012). The DOC $\Delta^{14}C$ profiles of the Canada Basin were unique in that layers of Atlantic and Pacific source waters were readily discernable as discontinuities throughout the upper 1000 m (Griffith et al., 2012). The Southern Ocean DOC profiles were unique in exhibiting the lowest surface values (50 μM and $\Delta^{14}C = -366$‰) and smallest ranges of concentrations (35.5-52.6 μM) and $\Delta^{14}C$ values (−529‰

to −366‰), indicative of stronger vertical mixing at this site. Localized features aside, all of these more recent DOC $\Delta^{14}C$ profiles exhibited the same basic shape first identified by Williams and Druffel (1987): a surface maximum that decreased to nearly constant bathypelagic values. And, in all cases, the very old ^{14}C ages in deep waters still argued for a component of DOM that was resistant to biological or chemical degradation. However, the deep waters did not all possess the same ^{14}C ages or DOC concentrations.

As hypothesized by Williams and Druffel (1987), average DOC concentrations and $\Delta^{14}C$ values in deep (>1000 m) oligotrophic waters decreased along the path of thermohaline circulation from 43 µM and −396‰ (i.e., 4050 years) in the Sargasso Sea (Bauer et al., 1992; Druffel et al., 1992), to 41 µM and −502‰ (i.e., 5600 years) in the Southern Ocean (Druffel and Bauer, 2000), and finally to 35 µM and −524‰ (i.e., 6000 years) in the central North Pacific Ocean (Williams and Druffel, 1987). All of these deep DOC ages exceeded the *ca.* 500 year DIC ^{14}C-based globally averaged replacement time for deep waters, as well as the *ca.* 1650 year DIC ^{14}C-based transit time from the North Atlantic to the North Pacific (Stuiver et al., 1983). Since the ^{14}C content of marine DIC is generally considered a reliable constraint on the timescales of ocean mixing (Broecker and Peng, 1982; Stuiver et al., 1983), the comparatively excessive DOC ^{14}C-ages once again suggested that deep DOC was refractory, survived multiple cycles of ocean turnover, and aged quasi-conservatively during deep-water transport. A mass balance argument suggested that as much as 80% of deep DOC in both the North Atlantic and North Pacific (with $\Delta^{14}C = -490‰$ and −602‰, respectively) was recycled during each mixing cycle, with the remainder having pre-bomb $\Delta^{14}C$ values characteristic of each basin (Druffel et al., 1992).

Additionally, the apparent 1950 conventional ^{14}C year transit time of DOC between the N. Atlantic and N. Pacific (i.e., 6000-4050 years) was ~300 years longer than that of DIC (~1650 years). This difference in transit times implied that DOC was not merely a conservative tracer, but perhaps a more dynamic reservoir than previously thought. Druffel et al (1992) postulated that larger fluxes of ^{14}C-enriched organic matter to the deep ocean would have produced higher bulk $\Delta^{14}C$ values in the deep N. Atlantic than the N. Pacific, thus enhancing the age difference. They also considered the larger flux of riverine organic matter to the N. Atlantic compared to the N. Pacific (Meybeck, 1982), but noted the lack of chemical or isotopic evidence for a significant proportion of terrestrial organic matter in the deep sea. Finally, the deep N. Atlantic had evidence of bomb-^{14}C in the DIC pool, implying that the bulk DOC $\Delta^{14}C$ values may also have been elevated by post-bomb organic matter (Druffel et al., 1992). These explanations place new limits on our ability to estimate deep DOC degradation rates because they relied upon the N. Atlantic station having a distinctively ^{14}C-enriched DOM.

Inconsistencies between estimates of DOC aging and DIC ^{14}C-based water transit times are more apparent when considering the Southern Ocean profile, which lies approximately midway between the Atlantic and Pacific anchors of deep thermohaline circulation. Deep DOC appears to have aged by 1600 years and decreased by 2 µM in transit from the North Atlantic to Southern Ocean station while DIC aged just 800 years (Bauer and Druffel, 1998; Druffel and Bauer, 2000). The opposite is true as DOC continued its journey from the Southern Ocean to the North Pacific: DOC apparently aged just 400 years and lost ~6 µM while DIC aged 700 years (Druffel and Bauer, 2000; Williams and Druffel, 1987). As before, these discrepancies may be explained by local differences in sources and sinks of isotopically unique organic matter. Indeed, allochthonous ^{14}C-depleted organic matter has been shown to enter the sea (Masiello and Druffel, 1998; McCarthy et al., 2011; Pohlman et al., 2010), and selective utilization of modern carbon (83% preference over older carbon) has been observed in surface waters at Station M (Cherrier et al., 1999).

The question remains as to whether these or other mechanisms are capable of operating at rates expected throughout the deep ocean, and it underscores the difficulty of explicating global-scale biogeochemistry with so few observations. Resolving inconsistencies between DOC aging and ^{14}C-based transit times will require improved spatiotemporal DOC sampling resolution, comprehensive cataloguing of molecular constituents and their isotopic compositions, and process studies that provide independent estimates of reaction rates.

D Mass Balance Constraints on Bulk Δ^{14}C Values

The very old ^{14}C ages and uniform concentrations of deep DOC suggested that it must be well-mixed throughout the water column. Its presence in the surface ocean and mixing with newly photosynthesized DOC would explain why bulk DOC Δ^{14}C values are ~200-380‰ lower than DIC. This can be summarized by the following simplified equations for mass balance.

$$DOC_{surface} = DOC_{deep} + DOC_{new} \qquad (6.7)$$

$$\Delta^{14}C_{surface} = \frac{DOC_{deep}\Delta^{14}C_{deep} + DOC_{new}\Delta^{14}C_{new}}{DOC_{surface}} \quad (6.8)$$

For example, the observed mean bulk Δ^{14}C value (−146‰) of surface DOC (89 μM) in the central North Pacific was identical to the Δ^{14}C value calculated for a hypothetical mixture of old, deep DOC (38 μM, Δ^{14}C = −525‰) and newly produced DOC (89 − 38 μM = 49 μM, and Δ^{14}C = +150‰ = Δ^{14}C of surface DIC) (Williams and Druffel, 1987). Similar mass balance calculations predicted surface DOC Δ^{14}C values that were also identical, within measurement uncertainty, to observed values in profiles from the Sargasso Sea, Southern Ocean, and eastern North Pacific (Bauer and Druffel, 1998; Bauer et al., 1998a; Beaupré and Druffel, 2009; Druffel and Bauer, 2000; Druffel et al., 1992).

This "two-component model" was able to reproduce observations of surface DOM with remarkable simplicity. It did not consider the variety of molecular structures or reactivities that inherently characterize DOM. Nor did it consider that a steady-state DOM pool with a predominant surface source *requires* some proportion of these molecules to have ^{14}C ages between those of the surface and the deep (i.e., ~1270 and 6000 years, respectively, in the North Pacific). Furthermore, deep DOM is not necessarily a homogenous entity, but could also be composed of multiple pools spanning a range of Δ^{14}C values. Williams and Druffel (1987) recognized the latter possibility, calculating that up to 51% (~19 μM) of *deep* DOC could be radiocarbon-dead (Δ^{14}C = −1000‰, i.e., ≥57,000 years) with the remainder having a pre-bomb surface signature (*ca.* −40‰).

Despite these limitations, Mortazavi and Chanton (2004) successfully demonstrated that covariance in entire DOC concentration and Δ^{14}C depth profiles could also be explained by two-component "Keeling plots." These mixing models originally examined the diurnal variability of atmospheric CO_2 concentrations and δ^{13}C values in rural environments (Keeling, 1958). Although conceptually similar, Keeling plots of marine DOC depth profiles are based on spatial variability of a radioactive isotope (^{14}C) rather than temporal variability of a stable isotope (^{13}C). Since the ocean mixes on much longer timescales than the atmosphere, this approach must assume that variability due to ^{14}C decay is negligible compared to the timescale of mixing or the magnitude of measurement uncertainties. Accordingly, DOC at *any* depth (z) can be described as a highly persistent background (bg) component that is uniformly distributed throughout the water column, to which a second component of recent origin is added in excess (xs) (Figure 6.6a and b).

$$DOC_z = DOC_{bg} + DOC_{xs} \qquad (6.9)$$

$$\Delta^{14}C_z = \frac{DOC_{bg}\Delta^{14}C_{bg} + DOC_{xs}\Delta^{14}C_{xs}}{DOC_z} \quad (6.10)$$

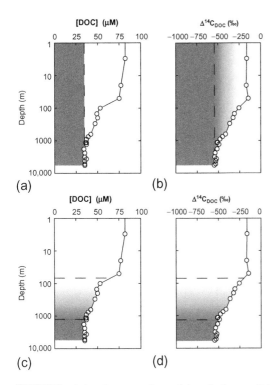

(a)   (b)

(c)   (d)

FIGURE 6.6 Conceptual models of the traditional Keeling plot (a and b) and solution-based two-component mixing (c and d) as applied to depth profiles of DOC concentration and $\Delta^{14}C$ measurements in the CNP (open circles; Druffel et al., 1989). (a) The traditional Keeling plot model with background (dark gray area) and excess (white area enclosed by data) DOC. The vertical dashed line represents a hypothetical depth profile of background DOC, while the solid line represents the total concentration of DOC at each depth. (b) The dashed vertical line represents a hypothetical profile of background DOC $\Delta^{14}C$ values while the solid line represents the mass-balanced bulk DOC $\Delta^{14}C$ throughout the water column. (c) and (d) The solution based mixing model with deep (dark gray area) and surface (white area) water components. Areas shaded with gray-scale gradients represent mass balance mixtures in various proportions of two components throughout the mesopelagic. *Reprinted from Beaupré, S.R., Aluwihare, L.I., (2010), with permission from Elsevier.*

Like the surface and deep-water components of Williams and Druffel's (1987) original two-component model, the background and excess components in DOC Keeling plots are nondescript mixtures of molecules. Substituting Eq. (6.9) into Eq. (6.10) and rearranging predicts a hyperbolic relationship between bulk $\Delta^{14}C$ values and DOC concentrations.

$$\Delta^{14}C_z = \frac{(\Delta^{14}C_{bg} - \Delta^{14}C_{xs})DOC_{bg}}{DOC_z} + \Delta^{14}C_{xs} \quad (6.11)$$

If only the excess component's concentration (DOC_{xs}) is allowed to vary (i.e., DOC_{bg}, $\Delta^{14}C_{bg}$, and $\Delta^{14}C_{xs}$ are constants), then a plot of $\Delta^{14}C_z$ versus $1/DOC_z$ will be linear. The slope of the line will be a function of DOC_{bg}, $\Delta^{14}C_{bg}$, and $\Delta^{14}C_{xs}$, and will require additional independent information for interpretation. However, the intercept of a geometric-mean linear regression will be equal to the isotope ratio of the excess component ($\Delta^{14}C_{xs}$). Therefore, Keeling plots provide a simple means for identifying two-component behavior as well as estimating the isotopic composition of the excess component. For example, the Keeling plot of central North Pacific DOC profiles ($r^2 = 0.92$, $n = 19$) yielded an intercept ($\Delta^{14}C_{xs} = +162 \pm 39‰$) that was within uncertainty of DIC ($\Delta^{14}C = +146‰$) from the upper 50 m (Mortazavi and Chanton, 2004; Williams and Druffel, 1987). In fact, Keeling plots of all the subtropical profiles exhibited strong correlations ($r^2 > 0.86-0.99$) and intercepts that were within uncertainty of surface DIC $\Delta^{14}C$ values at each site (Figure 6.7) (Mortazavi and Chanton, 2004). Interestingly, Keeling plots of near-shore profiles from the Mid-Atlantic Bight indicated that $\Delta^{14}C_{xs}$ values varied seasonally between $\Delta^{14}C$ values of riverine DOC ($\geq ca. +200‰$) and marine DOC ($+50$ to $+80‰$) (Bauer et al., 2001; Mortazavi and Chanton, 2004; Raymond and Bauer, 2001). In all cases, a single regression line elegantly summarized entire depth profiles, supported the two-component model with high coefficients of determination, and identified geochemically meaningful sources of marine DOC.

The depth profiles from polar seas were more complicated to interpret. The Southern Ocean

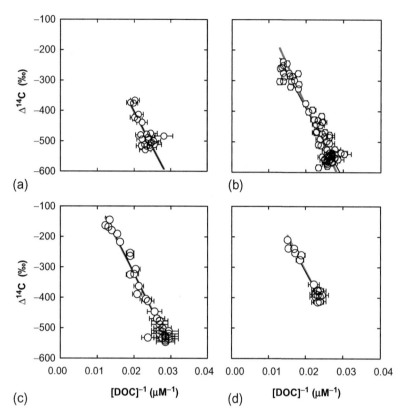

FIGURE 6.7 DOC Keeling plots, geometric mean best-fit lines (black lines), and solution mixing models (gray lines) from (a) the Southern Ocean, (b) eastern North Pacific, (c) central North Pacific, and (d) Sargasso Sea. Gray lines predicted by solution mixing are in close agreement with, and therefore partially obscured by, the black lines of the geometric mean linear regressions. *Reprinted from Beaupré, S.R., Aluwihare, L.I., (2010), with permission from Elsevier.*

Keeling plot was unique in yielding the lowest coefficient of determination ($r^2 = 0.54$, $n = 24$) and a highly uncertain intercept ($+60 \pm 130‰$) (Beaupré and Aluwihare, 2010). The poorer fit was partly due to uncharacteristic scatter in the deep DOC concentrations ($1\sigma = \pm 2.4\,\mu M$). It was also a statistical artifact of fitting a line to a very small range of concentrations (maximum − minimum = $17\,\mu M$) and $\Delta^{14}C$ values ($163‰$) compared to profiles such as the central North Pacific (spanning $48\,\mu M$ and $401‰$, respectively, with deep concentrations $1\sigma = \pm 1.1\,\mu M$) (McNichol and Aluwihare, 2007). Nevertheless, the Southern Ocean Keeling plot still followed the same trend as the other profiles and yielded an intercept that was within uncertainty of surface DIC (mean $\Delta^{14}C = +17 \pm 6‰$, $n = 4$, upper $100\,m$).

In contrast to the Southern Ocean, the two Arctic profiles (Griffith et al., 2012) yielded

Keeling plots with higher coefficients of determination ($r^2 = 0.72$ and 0.80) and more precise intercepts ($-26 \pm 55‰$ and $-10 \pm 30‰$) that were within uncertainty of surface DIC $\Delta^{14}C$ values at each site ($+31 \pm 4‰$ and $-1 \pm 5‰$) (Beaupré, unpublished). These summary statistics were consistent with two-component mixing, but the data undulated (rather than scattered) about the regression line in concordance with distinct water masses on the stations' temperature-salinity diagrams (Beaupre, unpublished; Griffith et al., 2012). Thus, the residuals remind us that Keeling plots are overly simplistic models of DOM, and that more realistic models are required to accurately capture the vertical structure. This is surely true for the other DOC depth profiles, but $\Delta^{14}C$ measurement uncertainties presently obscure finer hydrographic details on their Keeling plots.

While Keeling plots directly reveal the excess component's $\Delta^{14}C$ value, the background component's $\Delta^{14}C$ value and concentration must be constrained by additional information. For example, uniform *minimum* concentrations and $\Delta^{14}C$ values in profiles from the eastern North Pacific set an *upper* limit for background DOC concentrations and $\Delta^{14}C$ values equal to mean deep DOC ($38 \pm 2\,\mu M$ and $-548 \pm 20\permil$) at that site. A lower limit for the background component can obtained by extrapolating the Keeling plot regression line toward the lowest possible values of bulk DOC concentrations and $\Delta^{14}C$ values (i.e., $0\,\mu M$ and $-1000\permil$). In the eastern North Pacific, this extrapolated regression line intercepted the concentration axis, rather than the $\Delta^{14}C$ axis, indicating that up to $21 \pm 1\,\mu M$ of DOC could have an unquantifiable age ($\Delta^{14}C = -1000\permil$, i.e., $>ca.$ 57,000 years) (Beaupré and Druffel, 2009). Similar concentration limits were determined for a $-1000\permil$ background component in the Sargasso Sea, Southern Ocean, central North Pacific ($20 \pm 1\,\mu M$, $23 \pm 3\,\mu M$, and $21 \pm 1\,\mu M$, respectively; Beaupré and Aluwihare, 2010), and Arctic Ocean ($19 \pm 3\,\mu M$ and $17 \pm 2\,\mu M$; Beaupré, unpublished, based on data from Griffith et al., 2012). These limits were in close agreement with Williams and Druffel's (1987) original constraint of up to $\sim 19\,\mu M$ of $-1000\permil$ DOC in the deep central North Pacific, presenting the intriguing possibility of an ancient ($>57{,}000$ years) component of DOC that is present everywhere in the World Ocean. However, this ubiquitous background component can only exist if the entire deep World Ocean is also uniformly populated with a significant proportion ($ca.$ 50%) of modern DOC in order to return the observed bulk $\Delta^{14}C$ values. The latter constraint is much harder to rationalize and casts doubt on the existence of a large ($\sim 20\,\mu M$) ancient fraction. By similar reasoning (e.g., Eqs. 6.2–6.4), however, the likelihood of a smaller ancient DOC component will increase if future measurements can justify a model with additional components of intermediate ^{14}C ages.

Keeling plots do not explicitly address the underlying mechanisms by which DOC can be redistributed. This is evident from Eq. (6.9), which is a valid expression for conservation of mass only if DOC_{bg} and DOC_{xs} represent the concentrations of each component *after* mixing has already occurred (Beaupré and Aluwihare, 2010). Otherwise, changes in the abundance of DOC could only occur via in situ production or decomposition (e.g., photosynthesis, particle-solute interactions, egestion) to the exclusion of water mass transport and mixing. As an alternative, the two-component model may be derived for mixing volumes (v) of water (Figure 6.6c and d), in various proportions, from two different end-member solutions (subscripts 1 and 2) and the corresponding approximate equations for conservation of mass (Eqs. 6.12–6.14) (Beaupré and Aluwihare, 2010).

$$v_z = v_1 + v_2 \tag{6.12}$$

$$DOC_z = \frac{v_1 DOC_1 + v_2 DOC_2}{v_z} \tag{6.13}$$

$$\Delta^{14}C_z = \frac{v_1 DOC_1 \Delta^{14}C_1 + v_2 DOC_2 \Delta^{14}C_2}{v_z\, DOC_z} \tag{6.14}$$

Combining Eqs. (6.12)–(6.14) and rearranging predicts a linear relationship between bulk $\Delta^{14}C$ values and $1/DOC$ that is identical in form to traditional Keeling plots.

$$\Delta^{14}C_z = \left(\frac{DOC_1 DOC_2 \left(\Delta^{14}C_2 - \Delta^{14}C_1 \right)}{DOC_1 - DOC_2} \right) \frac{1}{DOC_z}$$
$$+ \left(\frac{DOC_1 \Delta^{14}C_1 - DOC_2 \Delta^{14}C_2}{DOC_1 - DOC_2} \right) \tag{6.15}$$

However, this relationship differs from traditional Keeling plots (Eq. 6.11) in that the slopes and intercepts are functions of the end-member DOC concentrations and $\Delta^{14}C$ values *prior* to mixing (i.e., they are constants). Furthermore, the intercept obtained under this model cannot be equal to, or lie between, the $\Delta^{14}C$ values of either end-member solution if they have different initial $\Delta^{14}C$ values. Therefore, care must be taken when interpreting the regression coefficients of a Keeling plot (Figure 6.7). For example, the slope and intercept of a Keeling plot from the central North Pacific ($-24,300 \pm 800\,\mu M\%_0$, and $+164 \pm 20\%_0$) were within uncertainty of the slope and intercept ($-23,700 \pm 2700\,\mu M\%_0$, and $+150 \pm 60\%_0$) calculated for mixtures of surface and deep water (Eq. 6.15). Similar agreement was reported for each profile from the North Atlantic, Southern Ocean, and eastern North Pacific (Beaupré and Aluwihare, 2010).

Additional information is needed to determine which mixing model applies to DOC (Eq. 6.11 or Eq. 6.15) and therefore how to interpret depth profiles. One approach is to extend the assumption of conservative mixing to other solutes. For example, if two solutes in each of two solutions mix conservatively (Eq. 6.15), then their isotope ratios (e.g., $\Delta^{14}C$) will be hyperbolically related. DIC is a good candidate as a second solute because it is a recognized tracer of water mass movement and its ^{14}C reservoir effect, by definition, is evidence for mixing with older material from deeper waters. Not surprisingly, Keeling plots of DIC fit the volume mixing relationship with high coefficients of determination ($r^2 = 0.85\text{-}0.99$) and intercepts ($-2000\%_0$ to $-2700\%_0$) that lie beyond the range of possible $\Delta^{14}C$ values (minimum $= -1000\%_0$) but are within uncertainty of $\Delta^{14}C$ values calculated from mixtures of surface and deep DIC (Eq. 6.15) (Beaupré and Aluwihare, 2010).

Since Keeling plots of both DOC and DIC were linear throughout the water column, it was surprising that DOC and DIC $\Delta^{14}C$ values were not uniformly correlated throughout the water column. Instead, the expected hyperbolic trends were confined to deeper waters and separated from the surface mixed layer by sharp cusps (Figure 6.8). This discontinuity was evidence

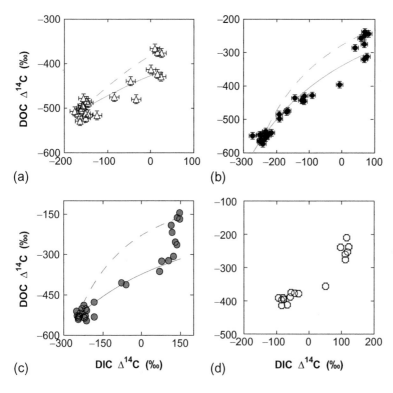

FIGURE 6.8 Correlation plots of DOC and DIC $\Delta^{14}C$ values from the (a) Southern Ocean, (b) eastern North Pacific, (c) central North Pacific, and (d) Sargasso Sea. Dashed lines indicate the relationship predicted for mixtures of surface water with mean deep (>1500m) water. Solid gray lines indicate the relationship predicted for mixtures of mean deep water with water near the cusp (~85-250m). Model predictions are not shown for the Sargasso Sea (d) because the equations rely upon DIC concentrations, which were not concurrently measured with DOC at that site. *Reprinted from Beaupré, S.R., Aluwihare, L.I., (2010), with permission from Elsevier.*

for three depth horizons that distribute DOC by different mechanisms. In the epipelagic, above ~85-250 m, Δ^{14}C gradients in DOC depth profiles were absent from DIC (Figure 6.5), suggesting that these two solutes were not jointly redistributed by water-mass mixing. Instead, DOC variability in this horizon was likely dominated by biogeochemistry (e.g., in situ production and loss) rather than advection. For example, the epipelagic pattern (Figure 6.8) may have resulted from the production of isotopically enriched, labile (i.e., "excess") DOC in the surface ocean that was consumed before export to great depths. In the mesopelagic (~100-1000 m), Δ^{14}C values of DOC and DIC decreased concordantly with the two-component solution model (Eq. 6.15, Figure 6.8). Lastly, bathypelagic (below ~1000 m) DOC and DIC Δ^{14}C values were uniform with depth (Figure 6.5), fell into clusters at one end of the solute-mixing hyperbola on Δ^{14}C correlation plots (Figure 6.8), and are therefore difficult to interpret without more information (see Section IV.C). Since diapycnal mixing on this scale is unlikely at the sub-polar sites, the mesopelagic gradients in DOC Δ^{14}C depth profiles (Figure 6.5) more likely resulted from horizontal gradients in the upper ocean transposed vertically by advection along isopycnal surfaces. These findings were consistent with observations of dissolved neutral sugars in the central North Pacific that possessed Δ^{14}C values (+47 to +67‰) slightly depleted relative to DIC (+72±7‰) in the surface ocean, and slightly enriched (−133 to −108‰) relative to ambient DIC (−155±7‰) at 600 m depth (Repeta and Aluwihare, 2006). The 600 m values suggested that the majority (~85%) of dissolved neutral sugars were advected to this depth while a smaller fraction (~15%) may have been derived from dissolution of Δ^{14}C-enriched sinking particles. Data scatter and limited sampling depths in the bulk Δ^{14}C profiles precluded additional estimates of the significance of particle-solute interactions at depth (Beaupré and Aluwihare, 2010).

In summary, the two-component model is a simple, convenient, and insightful description of DOC biogeochemistry. But, there is no a priori reason to assume that DOC should behave as just two groups of molecules with distinct properties. There should be at least as many components as there are dominant sources. For example, see the excellent review by Bauer (2002) for an application of δ^{13}C and Δ^{14}C in a three-component model that constrained terrestrial, local, and marine sources of DOM to the hydrographically complex mid-Atlantic Bight. In the open ocean, however, it is unlikely that additional nondescript "components" will significantly improve the mixing model's fit to the data. Instead, more advanced descriptions that consider the isotopic compositions of chemical constituents are required to fully capture the observed spatiotemporal variability of marine DOM (Bauer et al., 1998a; Beaupré and Druffel, 2009).

V ISOTOPIC COMPOSITION OF DOM CONSTITUENTS

The ideal characterization of marine DOM would identify the structures of all unique compounds and quantify their respective δ^{13}C and Δ^{14}C values. This is not presently feasible due to the very low natural abundance of ^{14}C, low concentrations of individual compounds in seawater, and the analytical challenges associated with isolating many of these compounds from complex mixtures. Instead, we can examine the isotopic composition of marine DOM by studying a few select compounds or fractions that share both analytically and geochemically meaningful properties, such as size, polarity, solubility, or functional groups. The earliest attempt to directly probe the isotopic spectrum within individual DOC samples involved isolation of "humic substances" by absorption onto polymer resins (XAD, Table 6.2) (Druffel et al., 1992). The different resins consistently isolated ~4% to ~20% w/w of the bulk DOC, with Δ^{14}C values in the Sargasso Sea ranging from a low of −587‰ (7100 ^{14}C years) at 3200 m to a high of −329‰ (3600 ^{14}C years) at 50 m. Similar results were reported for samples collected from the central North Pacific. None of these Δ^{14}C values were equal to local surface DIC Δ^{14}C or bulk DOC Δ^{14}C from the same samples. Instead, they were depleted in ^{14}C relative to bulk DOC at all depths studied. This was unequivocal evidence for the presence of dissolved organic molecules with unique Δ^{14}C values throughout the water column (Druffel et al., 1992). Additional evidence came through ^{14}C analyses of size fractions, compound classes, individual compounds, and a few novel approaches to isotopic characterization.

It is important to remember that all compound class and compound specific analyses require many individual molecules to measure useful Δ^{14}C values. For example, there are 8×10^{16} molecules of glucose ($C_6H_{12}O_6$) in the 10 µg C needed for a small sample AMS analysis, and very few of those molecules (~7 × 10^{-10} %) have any ^{14}C atoms whatsoever—i.e., there is but one ^{14}C atom for every trillion C atoms in modern samples. This does not necessarily mean that the glucose sample was created >50,000 years ago and subsequently contaminated with ~600,000 extraordinarily "hot" molecules. It does, however, underscore the fact that we cannot unambiguously assume one carbon source just because we have purified a single compound. Therefore, all carbon isotopic measurements, even compound specific radiocarbon analyses, must be interpreted within the context of mass balance (Eqs. 6.2–6.4) and the isotopic signatures of candidate source materials.

A Characterization by Size Fractions

More detailed age distributions have been obtained by measuring the ^{14}C content of groups of molecules defined by their physical dimensions (Table 6.2). These DOC "size fractions" are isolated by flowing seawater through a series of filters with progressively smaller pore sizes (e.g., 0.2 µm), with the smallest fractions captured by tangential flow UF (~1 nm). Naturally, the proportion of DOC isolated in each fraction depends upon the size distribution of molecules and the pore sizes of each filter. For example, tangential flow UF recovers molecules with a nominal molecular weight of at least 1000 Daltons (1 kiloDalton, or 1 kDa), representing ~20-40% of bulk DOC in the open ocean (Benner, 2002). This material is commonly referred to as the high molecular weight (HMW) fraction of DOM or ultrafiltered DOC (UDOC), and it can be collected in sufficiently large masses (>1 g) to perform compound class and compound specific ^{14}C analyses (Benner et al., 1992). These analyses are often aided by subsequent diafiltration to remove sea salts from the mass of organic matter (see Chapter 2). While simple in principle, intercomparison studies have shown that UF is sensitive to the type of filter used, how it is cleaned, and the length of time it is used to collect samples (Buesseler et al., 1996). These factors, and the risk of contamination, require strict adherence to sampling protocols and meticulous assessments of blanks in order to obtain reliable measurements of isotope ratios (e.g., Aluwihare, 1999; Guo and Santschi, 1996).

Guo et al. (1996) used UF to create a coarsely resolved, sized-based Δ^{14}C spectrum of organic matter (Figure 6.9) from surface waters in the Mid-Atlantic Bight. The Δ^{14}C values decreased monotonically with fraction size from "modern" in sinking particulate organic matter (POM) down to −257±8‰ in low molecular weight (LMW) DOM (deemed "ultrafiltered organic matter," UOM, by the authors). The latter value was inferred by mass balance because, by definition, the LMW fraction was not retained by the UF apparatus and could not be measured directly. The trend from larger, ^{14}C-enriched fractions to smaller, ^{14}C-depleted fractions was consistent with photosynthetically produced DOC that degraded into progressively smaller molecules as it aged. Further analyses of the

TABLE 6.2 Carbon Isotope Signatures of Size Fractions, Compound Classes, and Specific Compounds in Marine DOC

Region	Depths (m)	Component	Concentration (μM)	$\delta^{13}C$ (‰)	$\Delta^{14}C$ (‰)	References
Arctic	80	UDOC[a]	n.d.	n.d.	−90	Benner et al. (2004)
	3200	UDOC	n.d.	n.d.	−348	Benner et al. (2004)
	11 to 51	C18 extract	n.d.	n.d.	−262 to −199	Benner et al. (2004)
	100 to 173	C18 extract	n.d.	n.d.	−353 to −87	Benner et al. (2004)
	1971 to 2949	C18 extract	n.d.	n.d.	−379 to −378	Benner et al. (2004)
Central North Pacific	5 to 183	Humics-XAD extracts	3 to 5	−23.3 to −20.4	−410 to −310	Druffel et al. (1992) and Meyers-Schulte and Hedges (1986)
	21	UDOC[b]	10 to 23	−22.1 to −21.2	−131 to −6	Walker et al. (2011)
	670	UDOC[b]	3 to 9	−21.7 to −21.3	−424 to −306	Walker et al. (2011)
	915	UDOC[b]	3 to 9	−22.3 to −21.4	−552 to −345	Walker et al. (2011)
	21	DNA[c]	n.d.	−19.0	+60	Hansman et al. (2009)
	670	DNA[c]	n.d.	−20.1	−140	Hansman et al. (2009)
	915	DNA[c]	n.d.	−19.7	−87	Hansman et al. (2009)
	20	UDOC	n.d.	−21.8	−92	Loh et al. (2004)
	900	UDOC	n.d.	−21.5	−381	Loh et al. (2004)
	1800	UDOC	n.d.	−21.3	−434	Loh et al. (2004)
	20	Protein-like	n.d.	−21	−21	Loh et al. (2004)
	900	Protein-like	n.d.	−21.0	−279	Loh et al. (2004)
	1800	Protein-like	n.d.	−20.8	−332	Loh et al. (2004)
	20	Carbohydrate-like	n.d.	−21.4	+7	Loh et al. (2004)
	900	Carbohydrate-like	n.d.	−20.4	−302	Loh et al. (2004)
	1800	Carbohydrate-like	n.d.	−20.3	−406	Loh et al. (2004)
	20	Lipid extract	n.d.	−27.6	−551	Loh et al. (2004)
	900	Lipid extract	n.d.	−29.4	−865	Loh et al. (2004)
	1800	Lipid extract	n.d.	−28	−881	Loh et al. (2004)
North Pacific Subtropical Gyre	3	UDOC	n.d.	n.d.	+46	Repeta and Aluwihare (2006)
	15	UDOC	n.d.	−21.9	+10	Repeta and Aluwihare (2006)
	670	UDOC	n.d.	−20.9	−262 and −255	Repeta and Aluwihare (2006)

(Continued)

TABLE 6.2   Carbon Isotope Signatures of Size Fractions, Compound Classes, and Specific Compounds in Marine DOC—cont'd

Region	Depths (m)	Component	Concentration (µM)	$\delta^{13}C$ (‰)	$\Delta^{14}C$ (‰)	References
	3	Monosaccharides	n.d.	n.d.	+57 to +103	Repeta and Aluwihare (2006)
	15	Monosaccharides	n.d.	−19.9 to −15.6	+10 to +67	Repeta and Aluwihare (2006)
	670	Monosaccharides	n.d.	n.d.	−133 to −108	Repeta and Aluwihare (2006)
Eastern North Pacific	4	DNA[d]	0.009 to 0.017	−22.3 to −20.7	−61 to −13	Cherrier et al. (1999)
Mid Atlantic Bight, salinity >34	1 to 2	UDOC	28 to 63[e]	−28.2 to −21.8	−182 to −89	Guo et al. (1996)
	25 to 250	UDOC	16 to 28[e]	−26.8 to −20.2	−399 to −308	Guo et al. (1996)
	750 to 2600	UDOC	13 to 14[e]	−24.4 to −20.7	−403 to −376	Guo et al. (1996)
	1 to 2	UDOC-10[f]	4 to 19[e]	−27.6 to −22.7	−160 to −6	Guo et al. (1996)
	25 to 250	UDOC-10[f]	3 to 4[e]	−25.9 to −22.9	−611 to −132	Guo et al. (1996)
	750 to 2600	UDOC-10[f]	1 to 2[e]	−28.3 to −23.9	−709 to −442	Guo et al. (1996)
Mid Atlantic Bight, marine	2	UDOC	37[e]	n.d.	−10	Aluwihare et al. (2002)
	300	UDOC	4[e]	n.d.	−375	Aluwihare et al. (2002)
	750	UDOC	6[e]	n.d.	−255	Aluwihare et al. (2002)
	1 to 2	Monosaccharides	10[g]	n.d.	+49 to +92	Aluwihare et al. (2002)
	300	Monosaccharides	1[g]	n.d.	−120	Aluwihare et al. (2002)
	750	Monosaccharides	1[g]	n.d.	−59	Aluwihare et al. (2002)
	2	Polysaccharides	n.d.	−28.2	+26	Santschi et al. (1998)
	2600	Polysaccharides	n.d.	−20.7	−321	Santschi et al. (1998)
Sargasso Sea	50	Humics-XAD extracts	3 to 12	−23.1 to −20.8	−402 to −329	Druffel et al. (1992)
	850 to 3237	Humics-XAD extracts	1 to 10	−22.9 to −20.5	−587 to −381	Druffel et al. (1992)
	3	UDOC	n.d.	−21.8	−5	Loh et al. (2004)
	850	UDOC	n.d.	−21	−270	Loh et al. (2004)
	1500	UDOC	n.d.	−21.2	−262	Loh et al. (2004)
	3	Protein-like	n.d.	−21.2	+2	Loh et al. (2004)
	850	Protein-like	n.d.	−20.4	−190	Loh et al. (2004)
	1500	Protein-like	n.d.	−20.8	−215	Loh et al. (2004)

(Continued)

TABLE 6.2   Carbon Isotope Signatures of Size Fractions, Compound Classes, and Specific Compounds in Marine DOC—cont'd

Region	Depths (m)	Component	Concentration (μM)	$\delta^{13}C$ (‰)	$\Delta^{14}C$ (‰)	References
Sargasso Sea	3	Carbohydrate-like	n.d.	−21.5	+13	Loh et al. (2004)
	850	Carbohydrate-like	n.d.	−20.4	−228	Loh et al. (2004)
	1500	Carbohydrate-like	n.d.	−21	−309	Loh et al. (2004)
	3	Lipid extract	n.d.	−28	−637	Loh et al. (2004)
	850	Lipid extract	n.d.	−28.1	−730	Loh et al. (2004)
	1500	Lipid extract	n.d.	−28	−830	Loh et al. (2004)

[a]UDOC represents that fraction of DOC between 1 kDa and 0.2 μm.

[b]Values from testing a range of concentration factors and the effects of diafiltration.

[c]DNA isolated from the 0.2 to 0.5 μm size range.

[d]DNA isolated from the 0.2 to 0.8 μm size range. Concentrations estimated from reported water volumes and yields of combusted CO_2.

[e]Concentrations estimated from reported DOC concentrations and % HMW DOC.

[f]UDOC-10 represents that fraction of DOC between 10 kDa and 0.2 μm.

[g]Concentration calculated as the product of [DOC], % HMW, % APS of HMW, % carbohydrates of AP, and mole fraction of carbonates.

relationship between size and age were limited because neither the POM nor bulk DOC were sampled concurrently with Guo et al.'s (1996) UF fractions. Nevertheless, similar relationships between size and age were found by others (Table 6.2). For example, Aluwihare et al. (2002) measured HMW DOC $\Delta^{14}C$ (−10‰, −375‰, and −255‰) at three depths in the Mid-Atlantic Bight that were consistently higher than corresponding bulk DOC $\Delta^{14}C$ values (−32‰, −414‰, and −405‰, respectively). Although not stated explicitly by the authors, mass balance dictates that the LMW DOC must have had $\Delta^{14}C$ values that were lower than the bulk material. Similar arguments by Loh et al. (2004) showed that HMW DOC $\Delta^{14}C$ values measured at three different depths in the North Atlantic and North Pacific Oceans (e.g., −5‰ and −92‰, respectively, at 3 m) were significantly higher than $\Delta^{14}C$ values of LMW DOC inferred by mass balance (e.g., −280‰ and −210‰, respectively).

The relationship between DOC molecular size and ^{14}C age was clear, observed throughout the water column at several sites, and reproduced by multiple laboratories. However, this generalized relationship belies the fact that size fractions are operationally defined pools which, like bulk DOC, must contain a variety of molecules and unknown distributions of ^{14}C ages. One can easily argue that surface ocean HMW DOC contains a mixture of recently photosynthesized organic matter and older molecules because its $\Delta^{14}C$ values typically lie between those of DIC and bulk DOC. This was demonstrated with Keeling plots of size-fractionated DOC from the Gulf of Mexico and Mid-Atlantic Bight ($r^2 = 0.87$ and 0.99) that yielded the $\Delta^{14}C$ signature of surface DIC in the excess component ($\Delta^{14}C_{xs} = +108 \pm 35$‰ and $+141 \pm 34$‰) (Mortazavi and Chanton, 2004). The LMW fractions are also problematic because the size-age relationship does not consider small molecules that are metabolically labile (e.g., sugars) or potentially volatile (e.g., DMS). Thus, the existence of an age-distribution in the LMW fraction may reconcile the notion that the smallest molecules must also be the most persistent. Finally, Walker et al. (2011) showed that the degree to

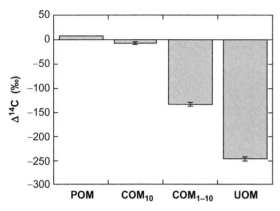

FIGURE 6.9   Mean $\Delta^{14}C$ values for various size fractions of organic matter from surface waters (2 m) of the
Mid-Atlantic Bight (stations 10, 12, and 13), as defined by
Guo et al. (1996): POM (particulate organic matter), COM_{10}
("colloidal" organic matter, 10 kDa-0.2 μm), COM_{1-10} ("colloidal" organic matter, 1-10 kDa; calculated as the difference between COM_{10} and a separate ultrafiltration fraction
ranging from 1 kDa to 0.2 μm), and UOM (ultrafiltered
organic matter, <1 kDa; determined by mass balance from
COM_{10}, COM_1, and measurements of bulk DOC $\Delta^{14}C$ provided by Bauer et al, unpublished). Error bars represent
±1 propagated standard deviation for the mean COM_{10},
COM_{1-10}, and UOM $\Delta^{14}C$ values. Error bars were not reported by Guo et al. (1996) for POM, and therefore do not
appear in this figure. *Reprinted after Guo, L., Santschi, P.H.,
Cifuentes, L.A., Trumbore, S.E., Southon, J., (1996). Copyright
2013 by the Association for the Sciences of Limnology and
Oceanography, Inc.*

which seawater is concentrated by UF affects the
relative proportions of ^{14}C-enriched HMW and
^{14}C-depleted LMW DOC in UDOC, and hence,
the UDOC's measured $\Delta^{14}C$ value (Table 6.2).
Ultimately, patterns in $\Delta^{14}C$ values of UDOC
will be most revealing when each sample is normalized for its isolation conditions.

B Characterization by Compounds and Compound Classes

Compound specific isotope analyses represent
the ultimate tool for testing hypotheses of DOC

provenance and reactivity on long time scales.
While the community would benefit from a
catalogue of chemical structures and their ^{14}C
ages, analytical challenges direct us toward a
few compounds that can be isolated from seawater with sufficient yields and manageable
blanks. As a consequence, much of this work
has focused on prominent biochemical classes of
compounds—such as carbohydrates, proteins,
and lipids (Table 6.2)—which can be isolated
with classical techniques. For example, Santschi
et al. (1998) reported elevated $\Delta^{14}C$ values (e.g.,
+26‰) for polysaccharides in the Mid Atlantic
Bight that were precipitated from surface HMW
DOC (−112‰) using a 70% solution of ethanol.
These authors found similar results at other
depths here and in the Gulf of Mexico.

Aluwihare et al. (2002) and Repeta and
Aluwihare (2006) also studied carbohydrates,
and reported the first compound specific ^{14}C
analyses of organic molecules dissolved in the
sea. Among many results, they showed that 7
unique monosaccharides from surface waters
at two sites in the North Pacific Ocean (average
$\Delta^{14}C = +57 \pm 6‰$ and $+89 \pm 13‰$) were isotopically similar to both surface DIC ($+72 \pm 7‰$ and
$+89 \pm 7‰$) and HMW DOC (10‰ and 46‰).
This was the most definitive evidence to date
that the vast majority of dissolved sugars in the
surface ocean were produced by photosynthesis. It also confirmed the presence of "modern"
organic molecules in the surface ocean that were
predicted by the two-component mixing model
of Williams and Druffel (1987). The monosaccharide $\Delta^{14}C$ signatures contained bomb-^{14}C,
arguing for their production within the last
~60 years. Assigning more specific ages to these
molecules was complicated by $\Delta^{14}C$ values that
could have originated from either the rising or
falling sides of the bomb spike. Therefore, the
carbohydrates in this study likely had surface
ocean residence times ranging from <3 years to
between 20 and 25 years. In contrast, deep-sea
sugars ($-123 \pm 10‰$ at 650 m) were, on average,
enriched in ^{14}C relative to both ambient DIC

(−155±7‰) and HMW DOC (−255‰). Unless these sugars were slowly traveling remnants from an elevated point on the bomb spike, the difference in $\Delta^{14}C$ values could be explained by the dissolution of rapidly sinking, ^{14}C-enriched particles at depth. Uncertainties in the data did not permit robust constraints on either mechanism, but the authors were able to estimate that up to 15% of the deep sugars could have been delivered by particles.

A separate study examined carbohydrate-like, protein-like, and lipid-like fractions of marine DOM that were defined by their respective extraction procedures (Loh et al., 2004; Table 6.2). The $\Delta^{14}C$ values of the carbohydrate-like fraction decreased with depth at sites in the central North Pacific and Sargasso Sea, but remained higher than the $\Delta^{14}C$ values of the HMW DOM from which they were derived. The protein-like fraction followed the same pattern, but the lipid extracts were universally ^{14}C-depleted relative to HMW DOM and bulk DOM, reaching values as low as −881‰ (i.e., 17,000 ^{14}C years). The very old ^{14}C ages of the lipid-like fraction were consistent with the geochemical resiliency of lipids in other environments (Gaines et al., 2008), and may have contributed significantly to the old ^{14}C ages of bulk DOM. However, it is difficult to know exactly what comprised the lipid-like fraction because it was isolated nonselectively by solvent extraction from HMW DOC.

Recently, Ziolkowski and Druffel (2010) reported the first $\Delta^{14}C$ values of black carbon isolated from HMW DOM, which ranged from −858±38‰ (15,700 years) in the deep North Atlantic to −918±31‰ (20,100 years) in the deep eastern North Pacific. Finding that 0.3 μM of this extremely old material was associated with UDOM, and assuming that black carbon comprises as much as 22% of bulk DOC (Masiello and Druffel, 1998), Ziolkowski and Druffel (2010) concluded that the majority of black carbon resided in the LMW fraction of DOC. The extreme $\Delta^{14}C$ values measured in this study provided more evidence for a very broad age spectrum in the organic matter of the sea.

VI SUMMARY AND CONCLUSIONS

The studies presented above have made a strong case for the persistence of many organic molecules in the sea. Multiple observations of bulk DOC, size fractions, and compound classes have demonstrated the existence of molecules with ^{14}C ages that are, on average, significantly longer than the timescales for ocean circulation. The corollary that both modern and relict molecules should exist in the surface ocean was also confirmed through compound class and compound specific analyses. Combined with mass balance mixing models, these results support the paradigm of a primarily photosynthetic origin for marine DOC in the surface ocean. However, this simplified picture does not address many unresolved questions: What happens to DOC in the dark ocean, and how do we reconcile the dramatic differences in apparent aging between DOC and DIC along the path of deep-water circulation? How significant are allochthonous sources of carbon to the marine DOC pool? What are the principle loss mechanisms for DOC, and how do they influence the observed isotope ratios? Ultimately, these and many other questions address our uncertainty on a fundamental characteristic of DOC that is common to any population: its "age" spectrum. We must continue to make progress in understanding the DOC age spectrum, its primary controls, how has it changed in the past, and how is it projected to change in the future.

Our understanding of the isotopic composition of marine DOC has largely relied upon a limited number of analyses and mass balance constraints. For example, at the time of this writing, there are no published bulk DOC $\Delta^{14}C$ profiles from the central gyres of the South Atlantic, South Pacific, or Indian Ocean (e.g., Figure 6.4). But we have the capacity to obtain additional measurements and to make more sophisticated isotopic analyses. Our long-standing uncertainties could be reduced with continued effort in several areas, including, but not limited to, (i) expanding our observations throughout the global ocean, (ii) developing a more comprehensive library of compound specific analyses, (iii) corroborating new isotopic analyses with complementary observations of the physical, chemical, and biological environment, (iv) monitoring molecular and isotopic transformations during physical, chemical, and biological processes, (v) pursuing higher resolution time-series studies of natural variability, (vi) constraining the fluxes of allochthonous material to the DOC pool, (vii) improving methods that facilitate all of the above endeavors on practical timescales, and (viii) fostering creative experiments and models that will answer our questions and, hopefully, reveal the transformative information that is needed most.

Acknowledgments

I graciously acknowledge Dennis Hansell and Craig Carlson for their encouragement, support, and patience in the preparation of this chapter; Lihini Aluwihare, Tim Eglinton, Sheila Griffin, Jeomshik Hwang, Bill Jenkins, David Kieber, Tomoko Komada, Matthew McCarthy, Ann McNichol, Georg Meyer, Eustace B. Nifkin, Mark Roberts, Guaciara dos Santos, John Southon, Karl von Reden, Brett Walker, Xiaomei Xu, Oliver Zafirio, Lori Ziolkowski, and the NOSAMS staff for their support. Most importantly, I thank Ellen Druffel for her inspiring work in this field, and for providing outstanding guidance, support, and friendship throughout my career. This work was supported under a Cooperative Agreement (OCE-2310753487) with the US National Science Foundation.

References

Aluwihare, L.I., 1999. High Molecular Weight (HMW) Dissolved Organic Matter (DOM) in Seawater: Chemical Structure, Sources and Cycling. Massachusetts Institute of Technology/Woods Hole Oceanographic Institution, Woods Hole.

Aluwihare, L.I., Repeta, D.J., Chen, R.F., 2002. Chemical composition and cycling of dissolved organic matter in the Mid-Atlantic Bight. Deep-Sea Res. II 49 (20), 4421–4437.

Amon, R.M.W., Benner, R., 1994. Rapid cycling of high-molecular-weight dissolved organic matter in the ocean. Nature 369, 549–552.

Anderson, E.C., Libby, W.F., Weinhouse, S., Reid, A.F., Kirshenbaum, A.D., Grosse, A.V., 1947. Radiocarbon from cosmic radiation. Science 105 (2735), 576–577.

Armstrong, F.A.J., Williams, P.M., Strickland, J.D.H., 1966. Photo-oxidation of organic matter in sea water by ultra-violet radiation, analytical and other applications. Nature 211 (5048), 481–483.

Atkins, W.R.G., 1922. The respirable organic matter of sea water. J. Mar. Biol. Assoc. UK 12 (4), 772–780.

Bader, R.G., Hood, D.W., Smith, J.B., 1960. Recovery of dissolved organic matter in sea-water and organic sorption by particulate material. Geochim. Cosmochim. Acta 19 (4), 236–243.

Barber, R.T., 1968. Dissolved organic carbon from deep water resists microbial oxidation. Nature 220 (5164), 274–275.

Bauer, J.E., 2002. Carbon isotopic composition of DOM. In: Hansell, D.A., Carlson, C.A. (Eds.), Biogeochemistry of Marine Dissolved Organic Matter. Academic Press, San Diego, CA, pp. 405–453.

Bauer, J.E., Druffel, E.R.M., 1998. Ocean margins as a significant source of organic matter to the deep ocean. Nature 392 (6675), 482–485.

Bauer, J.E., Haddad, R.I., Des Marais, D.J., 1991. Method for determining stable isotope ratios of dissolved organic carbon in interstitial and other natural marine waters. Mar. Chem. 33 (4), 335–351.

Bauer, J.E., Williams, P.M., Druffel, E.R.M., 1992. Carbon-14 activity of dissolved organic carbon fractions in the north-central Pacific and Sargasso Sea. Nature 357 (6380), 667–670.

Bauer, J.E., Reimers, C.E., Druffel, E.R.M., Williams, P.M., 1995. Isotopic constraints on carbon exchange between deep ocean sediments and sea water. Nature 373 (6516), 386–389.

Bauer, J.E., Druffel, E.R.M., Williams, P.M., Wolgast, D.M., Griffin, S., 1998a. Temporal variability in dissolved organic carbon and radiocarbon in the eastern North Pacific Ocean. J. Geophys. Res. 103 (C2), 2867–2881.

Bauer, J.E., Druffel, E.R.M., Wolgast, D.M., Griffin, S., Masiello, C.A., 1998b. Distributions of dissolved organic and inorganic carbon and radiocarbon in the eastern North Pacific continental margin. Deep-Sea Res. Part II—Topical Stud. Oceanogr. 45 (4–5), 689–713.

Bauer, J.E., Druffel, E.R.M., Wolgast, D.M., Griffin, S., 2001. Sources and cycling of dissolved and particulate organic radiocarbon in the northwest Atlantic continental margin. Global Biogeochem. Cycles 15 (3), 615–636.

Bauer, J.E., Druffel, E.R.M., Wolgast, D.M., Griffin, S., 2002. Temporal and regional variability in sources and cycling of DOC and POC in the northwest Atlantic continental shelf and slope. Deep-Sea Res. II 49 (20), 4387–4419.

Beaupré, S.R., Aluwihare, L.I., 2010. Constraining the two-component model of marine dissolved organic radiocarbon. Deep-Sea Res. II 57 (16), 1494–1503.

Beaupré, S.R., Druffel, E.R.M., 2009. Constraining the propagation of bomb-radiocarbon through the dissolved organic carbon (DOC) pool in the northeast Pacific Ocean. Deep-Sea Res. I 56 (10), 1717–1726.

Beaupré, S.R., Druffel, E.R.M., 2012. Photochemical reactivity of ancient marine dissolved organic carbon. Geophys. Res. Lett. 39 (L18602), 1–5.

Beaupré, S.R., Druffel, E.R.M., Griffin, S., 2007. A low-blank photochemical extraction system for concentration and isotopic analyses of marine dissolved organic carbon. Limnol. Oceanogr. Methods 5, 174–184.

Benner, R., 2002. Chemical composition and reactivity. In: Hansell, D.A., Carlson, C.A. (Eds.), Biogeochemistry of Marine Dissolved Organic Matter. Academic Press, San Diego, CA, pp. 59–90.

Benner, R., Pakulski, J.D., McCarthy, M., Hedges, J.I., Hatcher, P.G., 1992. Bulk chemical characteristics of dissolved organic matter in the ocean. Science 255 (5051), 1561–1564.

Benner, R., Benitez-Nelson, B., Kaiser, K., Amon, R.M.W., 2004. Export of young terrigenous dissolved organic carbon from rivers to the Arctic Ocean. Geophys. Res. Lett. 31, 1–4, L05305.

Bevington, P., Robinson, D.K., 2002. Data Reduction and Error Analysis for the Physical Sciences, third ed. McGraw-Hill Science/Engineering/Math.

Bien, J.P., Rakestraw, N.W., Suess, H.E., 1965. Radiocarbon in the Pacific and Indian Oceans and its relation to deepwater movements. Limnol. Oceanogr. 10 (5), R25–R37.

Bouillon, S., Korntheuer, M., Baeyens, W., Dehairs, F., 2006. A new automated setup for stable isotope analysis of dissolved organic carbon. Limnol. Oceanogr. Methods 4, 216–226.

Boutton, T.W., 1991. Stable carbon isotope ratios in natural materials. II. Atmospheric, terrestrial, marine and freshwater environments. In: Coleman, D.C., Fry, B. (Eds.), Carbon Isotope Techniques. Academic Press, New York, NY.

Broecker, W.S., Peng, T.H., 1982. Tracers in the Sea. Eldigio Press, Palisades, New York.

Buesseler, K.O., Bauer, J.E., Chen, R.F., Eglinton, T.I., Gustafsson, Ö., Landing, W., et al., 1996. An intercomparison of cross-flow filtration techniques used for sampling marine colloids: overview and organic carbon results. Mar. Chem. 55 (1–2), 1–31.

Calder, J.A., Parker, P.L., 1968. Stable carbon isotope ratios as indices of petrochemical pollution of aquatic systems. Environ. Sci. Technol. 2 (7), 535–539.

Calvin, M., Heidelberger, C., Reid, J.C., Tolbert, B.M., Yankwich, P.F., 1949. Isotopic Carbon: Techniques in its Measurement and Chemical Manipulation. John Wiley & Sons, New York.

Cherrier, J., Bauer, J.E., Druffel, E.R.M., Coffin, R.B., Chanton, J.P., 1999. Radiocarbon in marine bacteria: evidence for the ages of assimilated carbon. Limnol. Oceanogr. 44 (3), 730–736.

Criss, R., 1999. Principles of Stable Isotope Distribution. Oxford University Press, New York, NY.

Dalrymple, G.B., 1991. The Age of the Earth. Stanford University Press, Stanford, CA.

Degens, E.T., Behrendt, M., Gotthard, B., Reppmann, E., 1968a. Metabolic fractionation of carbon isotopes in marine plankton—II. Data on samples collected off the coasts of Peru and Ecuador. Deep-Sea Res. 15 (1), 11–20.

Degens, E.T., Guillard, R.R.L., Sackett, W.M., Hellebust, J.A., 1968b. Metabolic fractionation of carbon isotopes in marine plankton—I. Temperature and respiration experiments. Deep-Sea Res. 15 (1), 1–9.

Druffel, E.M., 1981. Radiocarbon in annual coral rings from the eastern Tropical Pacific Ocean. Geophys. Res. Lett. 8 (1), 59–62.

Druffel, E.M., 1982. Banded corals: changes in oceanic carbon-14 during the Little Ice Age. Science 218 (4567), 13–19.

Druffel, E.R.M., 1987. Bomb radiocarbon in the Pacific: annual and seasonal timescale variations. J. Mar. Res. 45, 667–698.

Druffel, E.R.M., 1996. Post-bomb radiocarbon records of surface corals from the tropical Atlantic Ocean. Radiocarbon 38 (3), 563–572.

Druffel, E.R.M., Bauer, J.E., 2000. Radiocarbon distributions in Southern Ocean dissolved and particulate organic matter. Geophys. Res. Lett. 27 (10), 1495–1498.

Druffel, E.R.M., Griffin, S., 1993. Large variations of surface ocean radiocarbon: evidence of circulation changes in the southwestern Pacific. J. Geophys. Res. Oceans 98 (C11), 20249–20259.

Druffel, E.R.M., Griffin, S., 1995. Regional variability of surface ocean radiocarbon from southern Great Barrier Reef corals. Radiocarbon 37 (2), 517–524.

Druffel, E.R.M., Griffin, S., 1999. Variability of surface ocean radiocarbon and stable isotopes in the southwestern Pacific. J. Geophys. Res. Oceans 104 (C10), 23607–23613.

Druffel, E.R.M., Williams, P.M., Robertson, K., Griffin, S., Jull, A.J.T., Donahue, D., et al., 1989. Radiocarbon in dissolved organic and inorganic carbon from the Central North Pacific. Radiocarbon 31 (3), 523–532.

Druffel, E.R.M., Williams, P.M., Bauer, J.E., Ertel, J.R., 1992. Cycling of dissolved and particulate organic matter in the open ocean. J. Geophys. Res. 97 (C10), 15639–15659.

Druffel, E.R.M., Griffin, S., Guilderson, T.P., Kashgarian, M., Southon, J., Schrag, D.P., 2001. Changes of subtropical north Pacific radiocarbon and correlation with climate variability. Radiocarbon 43 (1), 15–25.

Duursma, E.K., 1961. Dissolved organic carbon, nitrogen and phosphorous in the sea. Neth. J. Sea Res. 1 (1–2), 1–141.

Eadie, B.J., Jeffrey, L.M., Sackett, W.M., 1978. Some observations on the stable carbon isotope composition of dissolved and particulate organic carbon in the marine environment. Geochim. Cosmochim. Acta 42 (8), 1265–1269.

Falkowski, P.G., 1991. Species variability in the fractionation of ^{13}C and ^{12}C by marine phytoplankton. J. Plankton Res. 13 (Suppl. 1), 21–28.

Faure, G., 1986. Principles of Isotope Geology, second ed. John Wiley & Sons, New York.

Fry, B., Peltzer, E.T., Hopkinson Jr., C.S., Nolin, A., Redmond, L., 1996. Analysis of marine DOC using a dry combustion method. Mar. Chem. 54 (3–4), 191–201.

Gaines, S.M., Eglinton, G., Rullkotter, J., 2008. Echoes of Life: What Fossil Molecules Reveal About Earth History. Oxford University Press, New York, NY.

Glover, D.M., Jenkins, W.J., Doney, S.C., 2011. Modelling Methods for Marine Science. Cambridge University Press, Cambridge, UK.

Godwin, H., 1962. Half-life of radiocarbon. Nature 195 (4845), 984.

Griffith, D.R., McNichol, A.P., Xu, L., McKaughlin, F.A., Macdonald, R.W., Brown, K.A., et al., 2012. Carbon dynamics in the western Arctic Ocean: insights from full-depth carbon isotope profiles of DIC, DOC, and POC. Biogeosciences 9 (3), 1217–1224.

Grosse, A.V., 1934. An unknown radioactivity. J. Am. Chem. Soc. 56 (9), 1922–1923.

Grumet, N.S., Abram, N.J., Beck, J.W., Dunbar, R.B., Gagan, M.K., Guilderson, T.P., et al., 2004. Coral radiocarbon records of Indian Ocean water mass mixing and wind-induced upwelling along the coast of Sumatra, Indonesia. J. Geophys. Res. Oceans 109 (C5), C05003.

Guilderson, T.P., Schrag, D.P., Goddard, E., Kashgarian, M., Wellington, G.M., Linsley, B.K., 2000. Southwest subtropical Pacific surface water radiocarbon in a high-resolution coral record. Radiocarbon 42 (2), 249–256.

Guilderson, T.P., Schrag, D.P., Cane, M.A., 2004. Surface water mixing in the Solomon Sea as documented by a high-resolution coral ^{14}C record. J. Climate 17 (5), 1147–1156.

Guo, L., Santschi, P.H., 1996. A critical evaluation of the cross-flow ultrafiltration technique for sampling colloidal organic carbon in seawater. Mar. Chem. 55, 113–127.

Guo, L., Santschi, P.H., Cifuentes, L.A., Trumbore, S.E., Southon, J., 1996. Cycling of high-molecular-weight dissolved organic matter in the Middle Atlantic Bight as revealed by carbon isotopic (^{13}C and ^{14}C) signatures. Limnol. Oceanogr. 41 (6), 1242–1252.

Hansell, D.A., Carlson, C.A. (Eds.), 2002. Biogeochemistry of Marine Dissolved Organic Matter. Academic Press, San Diego, CA.

Hansell, D.A., Carlson, C.A., Repeta, D.J., Schlitzer, R., 2009. Dissolved organic matter in the ocean: a controversy stimulates new insights. Oceanography 22 (4), 202–211.

Hansman, R.L., Griffin, S., Watson, J.T., Druffel, E.R.M., Ingalls, A.E., Pearson, A., et al., 2009. The radiocarbon signature of microorganisms in the mesopelagic ocean. Proc. Natl. Acad. Sci. U. S. A. 106 (16), 6513–6518.

Hedges, J.I., 1987. Organic matter in sea water. Nature 330 (6145), 205–206.

Hoefs, J., 1997. Stable Isotope Geochemistry. Springer, Berlin.

Howden, L., Meyer, J.A., 2002. 2010 Census briefs: age and sex composition 2010. In: U.S. Census Bureau (Ed.), U.S. Department of Commerce, E.a.S.A, http://www.census.gov/prod/cen2010/briefs/c2010br-2003.pdf (accessed 2013).

Hua, Q., Woodroffe, C.D., Smithers, S.G., Barbetti, M., Fink, D., 2005. Radiocarbon in corals from the Cocos (Keeling) Islands and implications for Indian Ocean circulation. Geophys. Res. Lett. 32 (21), L21602.

Hua, Q., Barbetti, M., Rakowski, A.Z., 2013. Atmospheric radiocarbon for the period 1950-2010. Radiocarbon 55 (4), 2059–2072.

Jeffrey, L.M., 1969. Development of a Method for Isolation of Gram Quantities of Dissolved Organic Matter from Seawater and Some Chemical and Isotopic Characteristics of the Isolated Material. Texas A&M University, College Station, TX.

Jeffrey, L.M., Hood, D.W., 1958. Organic matter in sea water—an evaluation of various methods used for isolation. J. Mar. Res. 17, 247–271.

Johnson, L., Komada, T., 2011. Determination of radiocarbon in marine sediment porewater dissolved organic carbon by thermal sulfate reduction. Limnol. Oceanogr. Methods 9, 485–495.

Keeling, C.D., 1958. The concentration and isotopic abundances of atmospheric carbon dioxide in rural areas. Geochim. Cosmochim. Acta 13, 322–334.

Kilada, R.W., Campana, S.E., Roddick, D., 2007. Validated age, growth, and mortality estimates of the ocean quahog (*Arctica islandica*) in the western Atlantic. ICES J. Mar. Sci. 64 (1), 31–38.

Krogh, A., 1930. Eine mikromethode für die organische Verbrennungs-analyse, besonders von gelösten substanzen. Biochem. Zeitschr. 221, 247.

Krogh, A., Keys, A., 1934. Methods for the determination of dissolved organic carbon and nitrogen in sea water. Biol. Bull. 67 (1), 132–144.

Lang, S.Q., Lilley, M.D., Hedges, J.I., 2007. A method to measure the isotopic (^{13}C) composition of dissolved organic carbon using a high temperature combustion instrument. Mar. Chem. 103 (3–4), 318–326.

Lang, S.Q., Bernasconi, S.M., Früh-Green, G.L., 2012. Stable isotope analysis of organic carbon in small (μg C) samples and dissolved organic matter using a GasBench preparation device. Rapid Commun. Mass Spectrom. 26 (1), 9–16.

Le Clercq, M., Van der Plicht, J., Meijer, H.A.J., De Baar, H.J.W., 1997. Radiocarbon in marine dissolved organic carbon (DOC). Nucl. Instrum. Methods Phys. Res. B 123 (1–4), 443–446.

Li, Y.-H., 1977. Confusion of the mathematical notation for defining the residence time. Geochim. Cosmochim. Acta 41 (4), 555–556.

Libby, W.F., 1952. Radiocarbon Dating. The University of Chicago Press, Chicago, IL.

Libby, W.F., 1955. Radiocarbon Dating. The University of Chicago Press, Chicago, IL.

Lide, D.R. (Ed.), 2007. CRC Handbook of Chemistry and Physics, Internet Version 2007, 87th ed. Taylor and Francis, Boca Raton, FL.

Lieb, H., Krainick, H.G., 1931. Eine neuu mikrobestimmung des kohlenstoffs durch nasse verbrennung. Mickrochemie 3, 367.

Loh, A.N., Bauer, J.E., Druffel, E.R.M., 2004. Variable ageing and storage of dissolved organic components in the open ocean. Nature 430 (7002), 877–881.

MacKinnon, M.D., 1979. The measurement of the volatile organic fraction of the TOC in seawater. Mar. Chem. 8 (2), 143–162.

Masiello, C.A., Druffel, E.R.M., 1998. Black carbon in deep-sea sediments. Science 280 (5371), 1911–1913.

McCarthy, M., Hedges, J., Benner, R., 1996. Major biochemical composition of dissolved high molecular weight organic matter in seawater. Mar. Chem. 55 (3–4), 281–297.

McCarthy, M., Beaupré, S.R., Walker, B.D., Voparil, I., Guilderson, T.P., Druffel, E.R.M., 2011. Chemosynthetic origin of ^{14}C-depleted dissolved organic matter in a ridge-flank hydrothermal system. Nat. Geosci. 4, 32–36.

McNichol, A.P., Aluwihare, L.I., 2007. The power of radiocarbon in biogeochemical studies of the marine carbon cycle: insights from studies of dissolved and particulate organic carbon. Chem. Rev. 107 (2), 443–466.

Menzel, D.W., 1964. The distribution of dissolved organic carbon in the Western Indian Ocean. Deep-Sea Res. 11 (5), 757–765.

Menzel, D.W., Vaccaro, R.F., 1964. The measurement of dissolved organic and particulate carbon in seawater. Limnol. Oceanogr. 9 (1), 138–142.

Meybeck, M., 1982. Carbon, nitrogen, and phosphorus transport by world rivers. Am. J. Sci. 282 (4), 401–450.

Meyers-Schulte, K.J., Hedges, J.I., 1986. Molecular evidence for a terrestrial component of organic-matter dissolved in ocean water. Nature 321 (6065), 61–63.

Mortazavi, B., Chanton, J.P., 2004. Use of Keeling plots to determine sources of dissolved organic carbon in nearshore and open ocean systems. Limnol. Oceanogr. 49 (1), 102–108.

Natterer, K., 1892. Chemische untersuchengen im Oestlichen Mittelmeer. Denkschr. Akad. Wiss. Wien 59, 53–116.

Nydal, R., 1963. Increase in radiocarbon from the most recent series of thermonuclear tests. Nature 200 (4903), 212–214.

Nydal, R., Lövseth, K., Skogseth, F.H., 1980. Transfer of bomb ^{14}C to the ocean surface. Radiocarbon 22 (3), 626–635.

O'Leary, M.H., 1981. Carbon isotope fractionation in plants. Phytochemistry 20 (4), 553–567.

O'Leary, M.H., 1988. Carbon isotopes in photosynthesis. BioScience 38 (5), 328–336.

Osburn, C.L., St. Jean, G., 2007. The use of wet chemical oxidation with high-amplification isotope ratio mass spectrometry (WCO-IRMS) to measure stable isotope values of dissolved organic carbon in seawater. Limnol. Oceanogr. Methods 5, 296–308.

Panetta, R.J., Ibrahim, M., Gélinas, Y., 2008. Coupling a high-temperature catalytic oxidation total organic carbon analyzer to an isotope ratio mass spectrometer to measure natural-abundance δ^{13}C-dissolved organic carbon in marine and freshwater samples. Anal. Chem. 80 (13), 5232–5239.

Park, K., Prescott, J.M., Williams, W.T., Hood, D.W., 1962. Amino acids in deep-sea water. Science 138 (3539), 531.

Pohlman, J.W., Bauer, J.E., Waite, W.F., Osburn, C.L., Chapman, N.R., 2010. Methane hydrate-bearing seeps as a source of aged dissolved organic carbon to the oceans. Nat. Geosci. 4, 37–41.

Rau, G.H., Sweeney, R.E., Kaplan, I.R., 1982. Plankton ^{13}C:^{12}C ratio changes with latitude: differences between northern and southern oceans. Deep-Sea Res. 29 (8A), 1035–1039.

Rau, G.H., Takahashi, T., Des Marais, D.J., Repeta, D.J., Martin, J.H., 1992. The relationship between δ^{13}C of organic matter and [CO$_2$ (aq)] in ocean surface water: data from a JGOFS site in the northeast Atlantic Ocean and a model. Geochim. Cosmochim. Acta 56 (3), 1413–1419.

Raymond, P.A., Bauer, J.E., 2001. Riverine export of aged terrestrial organic matter to the North Atlantic Ocean. Nature 409 (6819), 497–500.

Reimer, P.J., Baillie, M.G.L., Bard, E., Bayliss, A., Beck, J.W., Blackwell, P.G., et al., 2009. IntCal09 and Marine09 radiocarbon age calibration curves, 0–50,000 years Cal BP. Radiocarbon 51 (4), 1111–1150.

Repeta, D.J., Aluwihare, L.I., 2006. Radiocarbon analysis of neutral sugars in high-molecular-weight dissolved organic carbon: Implications for organic carbon cycling. Limnol. Oceanogr. 51 (2), 1045–1053.

Roberts, M.L., von Reden, K.F., McIntyre, C.P., Burton, J.R., 2011. Progress with a gas-accepting ion source for Accelerator Mass Spectrometry. Nucl. Instrum. Methods Phys. Res. B 269 (24), 3192–3195.

Roberts, M.L., Beaupré, S.R., Burton, J.R., 2013. A high-throughput low-cost method for analysis of carbonate samples for ^{14}C. Radiocarbon 55 (2–3), 585–592.

Ruff, M., Fahrni, S., Gäggeler, H.W., Hajdas, I., Suter, M., Synal, H.-A., et al., 2010. On-line radiocarbon measurements of small samples using elemental analyzer and MICADAS gas ion source. Radiocarbon 52 (4), 1645–1656.

Santos, G.M., Southon, J.R., Griffin, S., Beaupré, S.R., Druffel, E.R.M., 2007. Ultra small-mass AMS ^{14}C sample preparation and analyses at KCCAMS/UCI facility. Nucl. Instrum. Methods Phys. Res. B 259 (1), 293–302.

Santschi, P.H., Guo, L., Baskaran, M., Trumbore, S., Southon, J., Bianchi, T.S., et al., 1995. Isotopic evidence for the contemporary origin of high-molecular weight organic matter in oceanic environments. Geochim. Cosmochim. Acta 59 (3), 625–631.

Santschi, P., Balnois, E., Wilkinson, K.J., Zhang, J., Buffle, J., Guo, L., 1998. Fibrillar polysaccharides in marine macromolecular organic matter as imaged by atomic force microscopy and transmission electron microscopy. Limnol. Oceanogr. 43 (5), 896–908.

Schlesinger, W.H., 1997. Biogeochemistry: An Analysis of Global Change. Academic Press, San Diego, CA.

Schmidt, A., Burr, G.S., Taylor, F.W., O'Malley, J., Beck, J.W., 2004. A semiannual radiocarbon record of a modern coral from the Solomon Islands. Nucl. Instrum. Methods Phys. Res., Sect. B 223, 420–427.

Scourse, J.D., Wanamaker, A.D.J., Weidman, C., Heinemeier, J., Reimer, P.J., Butler, P.G., et al., 2012. The marine radiocarbon bomb pulse across the temperate North Atlantic: a compilation of Δ^{14}C time histories from *Arctica islandica* growth increments. Radiocarbon 54 (2), 165–186.

Sharp, J.H., 2002. Analytical methods for total DOM pools. In: Hansell, D.A., Carlson, C.A. (Eds.), Biogeochemistry of Marine Dissolved Organic Matter. Academic Press, San Diego, CA, pp. 35–58.

Stuiver, M., Polach, H.A., 1977. Discussion: reporting of ^{14}C data. Radiocarbon 19 (3), 355–363.

Stuiver, M., Quay, P.D., Ostlund, H.G., 1983. Abyssal water carbon-14 distribution and the age of the world oceans. Science 219 (4586), 849–851.

Stuiver, M., Pearson, G.W., Braziunas, T., 1986. Radiocarbon age calibration of marine samples back to 9000 CAL YR BP. Radiocarbon 28 (2B), 980–1021.

Suess, H.E., 1953. Natural radiocarbon and the rate of exchange of carbon dioxide between the atmosphere and the sea. In: Proceedings of the Conference on Nuclear Processes in Geologic Settings. University of Chicago Press, Chicago, IL, pp. 52–56.

Tatsumoto, M., Williams, W.T., Prescott, J.M., Hood, D.W., Hood, D.W., 1961. Amino acids in samples of surface sea water. J. Mar. Res. 19 (2), 89.

The Florida Legislature, O.o.E.a.D.R., 2011. Florida: Demographics. http://edr.state.fl.us/Content/presentations/population-demographics/DemographicOverview_4-20-11.pdf.

Trumbore, S.E., Druffel, E.R.M., 1995. Carbon isotopes for characterizing sources and turnover of nonliving organic matter. In: Zepp, R.G., Sonntag, C. (Eds.), The Role of Nonliving Organic Matter in the Earth's Carbon Cycle. John Wiley & Sons, Chichester, West Sussex, England, pp. 7–22.

Tuniz, C., Bird, J.R., Herzog, G.F., Fink, D., 1998. Accelerator Mass Spectrometry: Ultrasensitive Analysis for Global Science. CRC Press, Boca Raton, FL.

Vogel, J.S., Nelson, D.E., Southon, J.R., 1987. ^{14}C background levels in an accelerator mass spectrometry system. Radiocarbon 29 (3), 323–333.

Walker, B.D., McCarthy, M.D., 2012. Elemental and isotopic characterization of dissolved and particulate organic matter in a unique California upwelling system: importance of size and composition in the export of labile material. Limnol. Oceanogr. 57 (6), 1757–1774.

Walker, B.D., Beaupré, S.R., Guilderson, T.P., Druffel, E.R.M., McCarthy, M.D., 2011. Large-volume ultrafiltration for the study of radiocarbon signatures and size vs. age relationships in marine dissolved organic matter. Geochim. Cosmochim. Acta 75 (18), 5187–5202.

Wang, X.-C., Chen, R.F., Whelan, J., Eglinton, L., 2001. Contribution of "old" carbon from natural marine hydrocarbon seeps to sedimentary and dissolved organic pools in the Gulf of Mexico. Geophys. Res. Lett. 28 (17), 3313–3316.

Weidman, C.R., Jones, G.A., 1993. A shell-derived time history of bomb ^{14}C on Georges Bank and its Labrador Sea implications. J. Geophys. Res. Oceans 98 (C8), 14577–14588.

Williams, P.M., 1961. Organic acids in Pacific Ocean waters. Nature 189 (476), 220.

Williams, P.M., 1968. Stable carbon isotopes in the dissolved organic matter of the sea. Nature 219 (5150), 152–153.

Williams, P.M., 1969. The determination of dissolved organic carbon in seawater: a comparison of two methods. Limnol. Oceanogr. 14 (2), 297–298.

Williams, P.M., Druffel, E.R.M., 1987. Radiocarbon in dissolved organic matter in the central North Pacific Ocean. Nature 330 (6145), 246–248.

Williams, P.M., Gordon, L.I., 1970. Carbon-13: carbon-12 ratios in dissolved and particulate organic matter in the sea. Deep-Sea Res. 17 (1), 19–27.

Williams, P.M., Zirino, A., 1964. Scavenging of 'dissolved' organic matter from sea-water with hydrated metal oxides. Nature 204 (4957), 462–464.

Williams, P.M., Oeschger, H., Kinney, P., 1969. Natural radiocarbon activity of dissolved organic carbon in North-East Pacific Ocean. Nature 224 (5216), 256–258.

Wilson, R.F., 1961. Measurement of organic carbon in sea water. Limnol. Oceanogr. 6 (3), 259–261.

Witbaard, R., Jenness, M.I., van der Borg, K., Ganssen, G., 1994. Verification of annual growth increments in *Arctica islandica* L. from the North Sea by means of oxygen and carbon isotopes. Neth. J. Sea Res. 33 (1), 91–101.

Ziolkowski, L.A., Druffel, E.R.M., 2010. Aged black carbon identified in marine dissolved organic carbon. Geophys. Res. Lett. 37, 1–4, L16601.

CHAPTER

7

Reasons Behind the Long-Term Stability of Dissolved Organic Matter

Thorsten Dittmar

Research Group for Marine Geochemistry (ICBM-MPI Bridging Group), Institute for Chemistry and Biology of the Marine Environment (ICBM), University of Oldenburg, Oldenburg, Germany

CONTENTS

I INTRODUCTION: THE PARADOX OF DOM PERSISTENCE

Marine dissolved organic matter (DOM) is, in many aspects, an enigmatic pool of carbon. Particularly puzzling is the enormous size of the global DOM pool and its great age. At about 660×10^{15} g carbon, DOM contains one thousand times more carbon than all living organisms in the oceans combined (Hansell et al., 2009), and its apparent radiocarbon age is 3000-6000 years

in the deep ocean (Bauer et al., 2002; Williams and Druffel, 1987; Williams et al., 1969). Both the apparent age and the inventory are surprising, because marine microheterotrophs are typically very efficient in decomposing DOM. Marine biota release DOM to the water while growing or via cell lysis (Azam and Malfatti, 2007; Jiao et al., 2011). Heterotrophic production and the microbial loop are largely driven by this freshly produced, highly labile DOM (Carlson et al., 2007; del Giorgio et al., 1997). About half of the

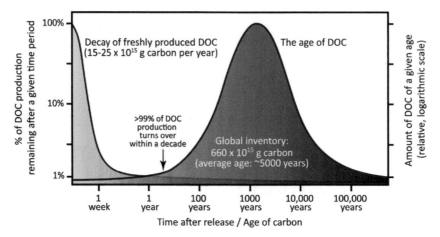

FIGURE 7.1   Conceptual scheme illustrating the age distribution of marine DOC (blue) and the decay of freshly released DOM (green). Most DOC is quickly consumed within days after production. After a decade >99% of the original DOC is consumed. A small fraction of DOC decomposes very slowly and accumulates in the ocean over thousands of years. Overall, there is an inverse relationship between DOC age and decomposition rate. Due to continuous slow degradation, at some point, this relationship inverts and only few molecules are preserved beyond hundreds of thousands of years. For references, see main text.

net primary production in the ocean is quickly catabolized in the form of DOM by bacteria and archaea (Azam and Malfatti, 2007; Carlson et al., 2007; del Giorgio et al., 1997). Within a decade after release into the water, the vast majority of newly produced DOM is respired (Figure 7.1) and decomposed into its inorganic constituents (Robinson and Williams, 2005) at a rate of ~10 µmol carbon kg^{-1}year^{-1} (Hansell, 2013). As a result, the bioavailable fraction of DOM occurs in very low concentrations in most parts of the ocean (Carlson, 2002).

However, a minuscule fraction of the annual DOM production escapes decomposition for thousands of years and has accumulated to become the largest organic carbon pool in marine waters (Hansell, 2013). Hansell and Carlson (2013) found that refractory DOM is conserved like salinity during much of its circulation in the deep ocean, and they postulated regional sinks for DOM in the deep ocean. As a consequence

of the conservation of a DOM fraction, the marine organisms that produce, transform, and decompose organic matter are surrounded by an enormous excess of residual DOM. This sets DOM apart from most other major pools of organic matter on Earth, where detrital organic carbon is preserved in distinctly different environments compared to the locations where primary production took place. Accumulation of nonliving organic matter occurs if essential ingredients for life are missing, such as electron acceptors (e.g., oxygen or sulfate), nutrients, or liquid water (Leahy and Colwell, 1990; Schmidt et al., 2011). If organic matter is locked away in such inhibiting environments, it is preserved until the environmental conditions change. Huge buildups of peat in permafrost, for example, can quickly decompose in a warming climate (Dorrepaal et al., 2009). Even petroleum that was trapped away from active cycling for millions of years is decomposed by microorganisms in the marine water column or coastal sediments on timescales of months to years (Swannell et al., 1996).

In the case of DOM, a paradoxical situation exists: phytoplankton that produce and bacterioplankton that decompose organic matter are surrounded by an enormous pool of organic matter that appears to remain untouched for thousands of years. DOM is not trapped in an uninhabitable environment like the other major organic carbon reservoirs on Earth. In fact, the accumulation of DOM over large spatial and temporal scales occurs under environmental conditions that are favorable to life. If bacteria are so efficient in decomposing freshly produced DOM, why does so much DOM accumulate? Bacteria would gain essential elements and energy from the oxidation of refractory DOM. Even in the subtropical gyres, the macronutrients residing in DOM are not used to satisfy the biological demand for those nutrients (Halm et al., 2012). This resilience is particularly surprising because the subtropical gyres are among the most oligotrophic areas of the surface ocean.

The long-term persistence of DOM appears enigmatic, but similarly puzzling is the fact that DOM does not accumulate for even longer periods in the ocean. Certain factors clearly inhibit the decay of refractory DOM, but why does DOM eventually vanish after a few thousand years? Any mechanism that causes the accumulation of DOM is obviously counterbalanced by removal processes. Otherwise, refractory DOM would accumulate over geological time spans in the ocean. The mechanisms controlling the size of the global DOM pool and its long-term turnover are unknown. This lack of knowledge is of concern because DOM is a major player in the global carbon cycle. Changes in the global pool of refractory DOM could cause major perturbations of atmospheric CO_2 and the radiation balance of Earth (Sexton et al., 2011). Analogous

to the biological carbon pump in the ocean, it was proposed that the formation of refractory DOM could constitute a microbial carbon pump through which atmospheric CO_2 is sequestered from active cycles for several thousand years (Jiao et al., 2011).

In this chapter, current hypotheses behind the millennium-scale stability of DOM are summarized and discussed. For more in-depth information on the chemical composition of DOM, and the definition of reactivity fractions of DOM, the reader is referred to the respective chapters of this book. This chapter focuses on the refractory and ultra-refractory fractions of DOM and the hypotheses proposed to mechanistically explain their millennium-scale stability. This is a very active and developing field of research. Our current understanding on this topic is largely hypothetical and the evidence is far from being conclusive. Explanations for stability are summarized under the umbrella of three major hypotheses (Figure 7.2). (1) The "environment hypothesis" relates the reactivity of DOM to particular environmental conditions prevailing in certain regions or during specific periods of Earth history. (2) The "intrinsic stability hypothesis" links the reactivity of DOM to its molecular structure. Refractory molecular structures may be biosynthesized by organisms, or they may result from secondary, abiotic molecular modifications. (3) The "molecular diversity hypothesis" proposes extremely dilute concentrations of substrate molecules in seawater as a reason for low turnover rates. Evidence for each of these hypotheses can be found in the literature, and it is likely that all proposed mechanisms (and possibly others) are contributing simultaneously to the long-term stability of specific fractions of DOM. Because the evidence on this topic is not conclusive, this chapter aims to stimulate discussions and promote future research rather than provide definite answers.

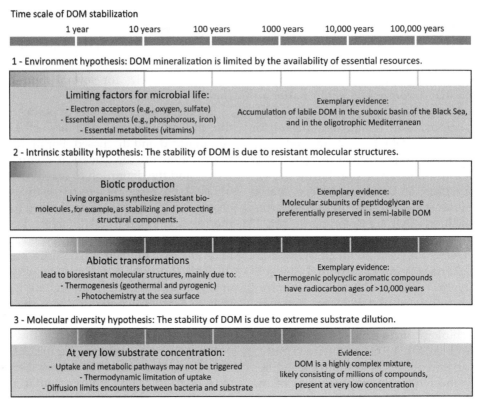

FIGURE 7.2   Conceptual summary of different hypotheses explaining the stability of dissolved organic matter in the contemporary ocean, and the respective timescales (color bars) on which the proposed mechanisms may operate.

II THE ENVIRONMENT HYPOTHESIS

Living cells rely on the availability of a certain set of chemical elements that are absolutely essential for life. Despite the enormous versatility of microorganisms and their remarkable capacity to perform molecular transformations, microorganisms are powerless in the face of the immutability of the elements (Merchant and Helmann, 2012). In the most nutrient-depleted regions of the upper ocean, photoautotrophic microorganisms have driven the concentrations of mineral nutrients to extremely low levels, and there is little supply of new nutrients from deeper layers of the ocean or from the continents or atmosphere

(Moore et al., 2013). As a consequence, microbial production and consumption of organic substrate may be hindered by the lack of essential minerals in the oligotrophic ocean.

Probably the best-studied system in this context is the oligotrophic Mediterranean Sea, where low phosphate availability limits the growth of phytoplankton and heterotrophic bacteria (Kritzberg et al., 2010; Thingstad et al., 1998; Zohary et al., 2005). The largest and most oligotrophic region of the world ocean, the South Pacific Gyre, however, is heavily understudied in this respect (Halm et al., 2012). For the Mediterranean, several experiments have impressively demonstrated that a highly labile organic substrate, such as glucose, cannot be used

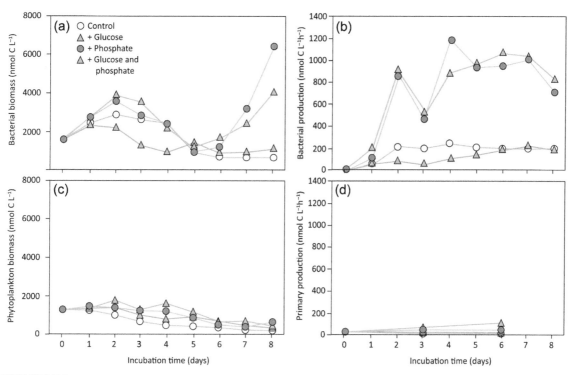

FIGURE 7.3   Results from an incubation experiment in phosphate-limited waters (Tanaka et al., 2009). Experimental addition of phosphate to surface water of the northwestern Mediterranean Sea strongly enhanced bacterial production and biomass (a and b). Addition of glucose alone did not stimulate bacterial production. Bacteria outcompeted phytoplankton (c and d) after addition of phosphate.

by heterotrophic microorganisms due to the lack of phosphate. In large-scale in situ experiments in the oligotrophic Cyprus Gyre (Eastern Mediterranean Sea), phosphate additions caused significant increases in bacterial production (Thingstad et al., 2005). Also, experimental addition of phosphate to surface water of the northwestern Mediterranean Sea strongly enhanced bacterial production (Tanaka et al., 2009). In the latter experiment, a heterotrophic bacterial community grew on natural DOM a few days after phosphate was added (Figure 7.3). Addition of glucose alone, on the other hand, did not stimulate bacterial production. These experiments (Tanaka et al., 2009; Thingstad et al., 2005) clearly demonstrated that there is a large component of DOM in Mediterranean surface waters that is

accessible to bacteria if the phosphate deficit is alleviated. Also in an Arctic pelagic ecosystem, Thingstad et al. (2008) observed organic carbon accumulating when bacterial growth rate was limited by mineral nutrients in the system. Another example for nutrient limitation is the vast "high nutrient low chlorophyll" (HNLC) region of the Southern Ocean off Antarctica, where phytoplankton are constrained by limited inputs of iron, while nitrate and phosphate abound. Different from the situation in the Mediterranean, heterotrophic bacteria in the HNLC region are constrained primarily by the availability of DOM (Church et al., 2000). However, simultaneous additions of organic substrate and iron revealed that bacterial growth efficiency and nitrogen utilization may be partly constrained by iron availability in the HNLC region of the Southern Ocean (Church et al., 2000). A similar observation was made in the northwestern Sargasso Sea, where simultaneous addition of mineral nutrients and labile DOM to the oligotrophic waters stimulated bacterial growth and utilization of seasonally accumulated semi-labile DOM. Addition of nitrate and phosphate alone, however, did not enhance DOM utilization and bacterial growth (Carlson et al., 2002).

The above studies illustrate that the availability of inorganic and organic nutrients can directly limit bacterial growth and the utilization of DOM. Furthermore, shifts in mineral nutrient limitation and stoichiometry may also cause significant and unexpected ecosystem responses. These responses can have major implications on the carbon transfer and the accumulation of DOM in marine systems (Thingstad et al., 2005). Interestingly, in the Mediterranean, the addition of phosphate did not stimulate phytoplankton. In the mesocosm experiment (Figure 7.3), bacteria outcompeted phytoplankton for nutrients after a few days (Tanaka et al., 2009). Similarly, in the large-scale in situ experiment in the Cyprus Gyre, phosphate addition increased bacterial production, while chlorophyll concentrations declined (Thingstad et al., 2005). In the Arctic, phytoplankton were outcompeted by bacteria when labile DOM was added under nutrient-replete conditions, and despite the addition of DOM, less organic carbon accumulated in the system (Thingstad et al., 2008).

Consequently, even outside distinct geographic features of nutrient limitation, bacterial utilization of DOM may be largely controlled by ecological competition. Thingstad et al. (1997) proposed a provocative hypothesis whereby bacterial consumption of DOM can be restricted because growth and biomass of bacteria are kept in check by food web mechanisms. Bacterial growth rate is lower than one would expect due to competition with phytoplankton for mineral nutrients. At the same time, predators keep

bacterial biomass low. With such a dual mechanism, otherwise labile DOM may accumulate in the upper ocean and may become subject to chemical transformation and vertical transport (Thingstad et al., 1997). Furthermore, possibly as a consequence of microbial biogeography (Azam and Malfatti, 2007; Martiny et al., 2006), there are also latitudinal gradients in the degradation of marine DOM (Arnosti et al., 2011). Such geographic gradients in the spectrum of substrates accessible by microbial communities cause regional differences in the production and decomposition rates of DOM. Regional accumulation of DOM may be a consequence.

Perhaps due to the above-described mechanisms, DOM accumulates over seasonal or annual timescales in parts of the upper ocean. Eventually, this DOM may enter the deep ocean by diffusion and advective downwelling. In the dark realm of the ocean, where most of the refractory DOM resides, heterotrophs do not compete with photoautotrophs for essential resources. In contrast to the surface ocean, there is wide consensus that carbon limits prokaryotic activity in the deep ocean (Aristegui et al., 2005, 2009). Most of the respiration in the interior of the ocean is fueled by sinking debris derived from production in the euphotic zone (Aristegui et al., 2002), indicating that prokaryotes obviously find favorable conditions in the deep ocean to efficiently oxidize sinking particulate organic matter. DOM, on the other hand, largely escapes consumption in the deep ocean. This can hardly be explained by the lack of essential nutrients in today's deep ocean. Also, electron acceptors, required for the oxidation of organic matter, abound in the deep ocean. The concentration of dissolved oxygen (O_2), the thermodynamically most favorable electron acceptor (Jørgensen, 1982), is usually higher than the concentration of dissolved organic carbon (DOC), that is, complete oxidation of DOC via oxygen would be possible. The concentration of sulfate and nitrate, which serve as electron acceptors usually under the absence of oxygen (Jørgensen, 1982), also surpasses the amounts required for the complete oxidation of DOC by several orders of magnitude. Overall, in most parts of the deep sea, there seems to be no lack of essential resources or significant competition for them that would explain the stability of refractory DOM.

On geological timescales, however, elemental availability on Earth has changed dramatically (Habicht et al., 2002; Konhauser et al., 2009; Quigg et al., 2003). Sexton et al. (2011) hypothesized that in the Eocene past, the global pool of DOC could have been three times larger than today, possibly because of the lack of oxygen related to reduced deep ocean ventilation. Cyclic changes in deep ocean ventilation could have resulted in large-scale DOC accumulation and oxidation events. The concomitant fluctuations of CO_2 exchange with the atmosphere could have affected the radiation balance of Earth and triggered the Eocene global warming events (Sexton et al., 2011). It remains speculative whether such profound changes in the global DOM pool are possible, and whether lack of oxygen can cause such extreme DOM accumulation as proposed by Sexton et al. (2011). Probably today's best example of a poorly ventilated deep ocean is the Black Sea. In the Black Sea, there is a continuous flux of organic debris from the productive surface to the deep water column. Microbial oxidation of sinking debris and the lack of deep ocean ventilation has driven oxygen levels in the deep Black Sea to very low levels, while sulfate still abounds (Albert et al., 1995). Consistently, the DOC concentrations in the deep basin of the Black Sea (Ducklow et al., 2007) are high ($120 \mu mol L^{-1}$) compared to those in the deep open ocean ($\sim 45 \mu mol L^{-1}$). This difference in concentration would be sufficient to explain the accumulation of DOC in Earth's past, as postulated by Sexton et al. (2011). Interestingly, a significant component of DOM in the Black Sea water column is lactate, acetate, and formate

(Albert et al., 1995). Around 10-20% of DOC is composed of these three low molecular-weight fatty acids at some depths/stations in the Black Sea (Albert et al., 1995; Ducklow et al., 2007). Given the extremely labile nature of these compounds, their high concentrations are surprising. Despite the concurrent presence of highly labile organic substrate and sulfate, sulfate reduction in the deep Black Sea is low (Albert et al., 1995). The latter authors concluded that the failure of the sulfate-reducing bacteria to utilize labile substrates implies that they were limited by something else. Though the hypoxic conditions in the Black Sea do not pose direct constraints on microbial life, the reducing conditions may cause a lack of essential elements, which would indirectly limit microbial life.

In conclusion, there is evidence that DOM can accumulate under nutrient-limiting conditions. The molecular composition of DOM probably plays a secondary role since highly labile substrates also accumulate. Competition with phytoplankton, combined with limited supply of new nutrients, is the main reason for the lack of mineral nutrients to heterotrophic organisms. An exception to this general observation is the apparent lack of some essential elements in the hypoxic deep Black Sea. Competition with phytoplankton can be ruled out in the darkness, although abiotic reactions in the reduced environment could play a role. All the discussed processes are regional or transient features in today's ocean. The widespread and long-term accumulation of DOM on timescales beyond a season or few years seems unlikely the direct result of environmental constraints. However, shorter and regional periods of DOM accumulation may be important on the long term, because exposure time to sunlight or other specific environmental conditions might be crucial for the abiotic formation of recalcitrant molecular structures in DOM. Hypotheses on the formation of such recalcitrant structure are discussed in the following section.

III THE INTRINSIC STABILITY HYPOTHESIS

In the deep ocean, heterotrophic microorganisms are limited by the availability of organic matter, while surrounded by large amounts of DOM. This apparent contradiction is generally explained by inferring an intrinsic stability of refractory DOM (Barber, 1968; Jiao et al., 2011). Evolution has equipped organisms with a finite number of enzymatic pathways to decompose organic molecules. If, for any reason, molecular structures are generated for which no decomposition pathway has evolved, these structures will accumulate in the environment. A well-known example of such an intrinsically recalcitrant molecule is graphite. Graphitic carbon is formed when organic matter is exposed to high temperature and pressure in Earth's crust. Organisms are not able to decompose this form of carbon, being returned to active cycles only via abiotic oxidation, in particular through volcanism (Dickens et al., 2004).

The search for recalcitrant molecular structures in marine DOM is challenging. Linking molecular structure to long-term stability requires knowledge both of the molecular structure of DOM and of the stability of these structures in the ocean. The molecular structure of DOM remains largely undetermined because only a few percent of deep-sea DOM is composed of known biochemicals (Dittmar and Paeng, 2009; Kaiser and Benner, 2009). For details on the molecular composition of DOM and the associated analytical challenges, I refer the reader to Chapter 2 in this book. Assessing the turnover rate of individual compounds in DOM has been approached from different angles. Incubation experiments were performed to directly study the stability of refractory DOM and the formation of these compounds from labile organic substrates (Brophy and Carlson, 1989; Gruber et al., 2006; Ogawa et al., 2001). The term "refractory" is loosely used in these studies to describe the DOM that persisted over the course of these

experiments, some of which lasted for more than 1 year. Despite the long duration of these studies, the extrapolation from a year to millennia, that is, the relevant timescale for the turnover of refractory DOM in the ocean, is highly speculative. Much of our knowledge also comes from water-column distribution measurements. This approach is based on the idea that the reactivity of DOM varies among water masses, and that the least reactive fraction predominates in the oldest water masses of the deep sea (Hansell, 2013). It is then commonly assumed that the compositional differences of DOM across the different water masses are related to the stage of decomposition and reactivity (Benner et al., 1992; Dittmar and Kattner, 2003; Hertkorn et al., 2006). The radiocarbon age of DOM varies systematically between water masses in support of this approach, but one must keep in mind that, at any location, DOM represents an isotopically heterogeneous mixture of compounds with a wide range of different ages (Flerus et al., 2012; Loh et al., 2004). Radiocarbon dating of defined compound groups is an elegant approach to circumvent this uncertainty. There are a few studies where operationally defined polarity or size fractions of DOM (Loh et al., 2004; Walker et al., 2011) or even structurally defined groups of molecules (Repeta and Aluwihare, 2006; Ziolkowski and Druffel, 2010) were radiocarbon dated. Possible input of pre-aged DOM, for example, from fossil deposits or hydrothermal circulation, introduces further uncertainty (McCarthy et al., 2011; Ziolkowski and Druffel, 2010).

Despite the inherent sources of uncertainty, these studies have tremendously advanced our knowledge on refractory DOM in the ocean. In general agreement, the component of DOM that can be molecularly characterized as common biomolecules, especially those that release amino acids, neutral sugars, or amino sugars upon acidic hydrolysis, is preferentially used by bacteria (Amon et al., 2001; Davis et al., 2009; Kaiser and Benner, 2012). Consequently, the concentration of hydrolyzable compounds sharply decreases in the course of microbial degradation. While hydrolyzable compounds are present at detectable levels throughout the water column (McCarthy et al., 1996), the bulk of DOM resists hydrolysis, and as such, it is largely outside the analytical window of established chromatographic analytical techniques. In contrast to the low level of hydrolyzable carbohydrates, nuclear magnetic resonance (NMR) analyses indicate an abundance of carbohydrates in at least the high molecular weight fraction of DOM (Benner et al., 1992). These non-hydrolyzable carbohydrates that persist after more labile DOM has been degraded are possibly biosynthesized and belong to the family of acylated polysaccharides (Aluwihare et al., 1997; Panagiotopoulos et al., 2007). The latter observations are of major relevance because they imply that biosynthesized compounds that are presumably not modified by secondary abiotic reactions resist degradation on the long term. This scenario is consistent with the finding that DOM throughout the water column has a clear molecular overprint of bacterial metabolism in the form of D-amino acids (Dittmar et al., 2001; McCarthy et al., 1998). D-amino acids are only produced by bacteria and archaea, not by algae, and the relative proportion of the different amino acid stereoisomers points clearly

toward a major bacterial source of DOM in the ocean. D-amino acids are mainly cell wall constituents and part of peptidoglycan. Because of their protective function, these and other structural polymers are more resistant to hydrolysis and enzymatic attack than other common biopolymers (Amon et al., 2001; McCarthy et al., 1998). D-amino acids can also be formed via abiotic racemization, but the preferential enrichment of only some distinct D-amino acids in the ocean is not consistent with an unselective abiotic racemization process (Dittmar et al., 2001; McCarthy et al., 1998). Besides D-amino acids, other bacterial biomarkers have been identified in marine DOM, namely muramic acid (Benner and Kaiser, 2003), an amino sugar found only in the bacterial cell wall polymer peptidoglycan, and short-chain beta-hydroxy fatty acids (Wakeham et al., 2003), a bacterial membrane component. The persistence of biosynthesized compounds in seawater was also demonstrated in incubation experiments (Brophy and Carlson, 1989; Gruber et al., 2006; Ogawa et al., 2001). Most remarkably, a natural bacterial community growing on simple monomeric substrates (glucose or glutamate, Figure 7.4) released a non-hydrolyzable form of DOM that persisted throughout the course of an incubation experiment (Ogawa et al., 2001).

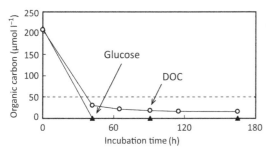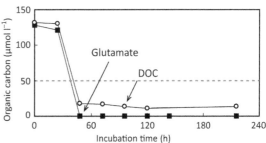

FIGURE 7.4 Results from incubation experiments illustrating the production of recalcitrant DOM by marine bacteria. A natural assembly of marine bacteria was grown on dissolved glucose and glutamate as single substrates (Ogawa et al., 2001). The DOC concentration quickly dropped during the first hours of the experiment, because the added substrates were consumed to undetectable levels within a few days. At the same time, however, DOM was released by bacteria. The concentration of the microbially derived DOC remained at almost constant levels well below the concentration in the deep North Atlantic of $50 \, \mu mol \, L^{-1}$ (dotted line). Only 10-15% of the bacterially derived DOM was identified as hydrolyzable amino acids and sugars, a feature consistent with marine DOM.

The observed transformation of labile DOM into more stable forms by microorganisms could constitute a mechanism through which carbon is sequestered from active cycles for many years (Jiao et al., 2011). However, the reason behind the stability of bacterial products in the ocean remains elusive. It is not known whether these compounds are intrinsically stable or whether other stabilization mechanisms are at work. Keil and Kirchman (1994) observed that highly labile proteins are protected from bacterial decomposition under the presence of natural marine DOM. They proposed reversible aggregation of molecules as the protection mechanism. DOM is rich in small amphiphilic molecules that have the potential to form micelles and hydrophobic hydrolysis-resistant coatings around other polar compounds (Dittmar and Kattner, 2003). This mechanism may at least temporarily protect certain compound groups in seawater. Whether these micelle-forming compounds are biosynthesized in the ocean is unknown.

Abiotic transformation steps likely contribute to the formation of recalcitrant molecular structures. There is evidence that light-induced reactions at the sea surface modify the molecular structure of biomolecules in a way that they become inaccessible to microorganisms. In experiments, photochemically modified DOM resisted microbial degradation over months (Benner and Biddanda, 1998). Photochemistry

at the sea surface contributes to the formation of refractory compounds, but at the same time, aged refractory compounds from the deep sea are decomposed and returned back into active cycles by photochemical reactions (Beaupre and Druffel, 2012; Obernosterer and Benner, 2004). This ambivalent role of photochemistry in the DOM cycle makes photochemistry a particularly interesting mechanism. However, refractory molecules that are unambiguous photochemical products of marine biomolecules have not been identified in the deep ocean. Finding such molecular proof will be a challenge. DOM contains an estimated 8% of carboxylic-rich alicyclic structures (Hertkorn et al., 2006). It was hypothesized that the formation of these structures (Figure 7.5a) might be related to photochemical reactions (Hertkorn et al., 2006). The abiotic cross-linking of polyunsaturated fatty acids may also occur in the photic zone of the ocean (Harvey et al., 1983), but such structures (Figure 7.5b) have not unambiguously been identified in DOM yet. A detailed discussion on photochemistry and other reactions at the sea surface is provided in Chapter 8 in this book.

Recently, heat was discovered as a quite unexpected abiotic factor contributing to the bioresistance of DOM in the ocean. Dittmar and Paeng (2009) estimated that about 2% of DOM, that is, 12×10^{15} g carbon, is composed of derivatives of polycyclic aromatic

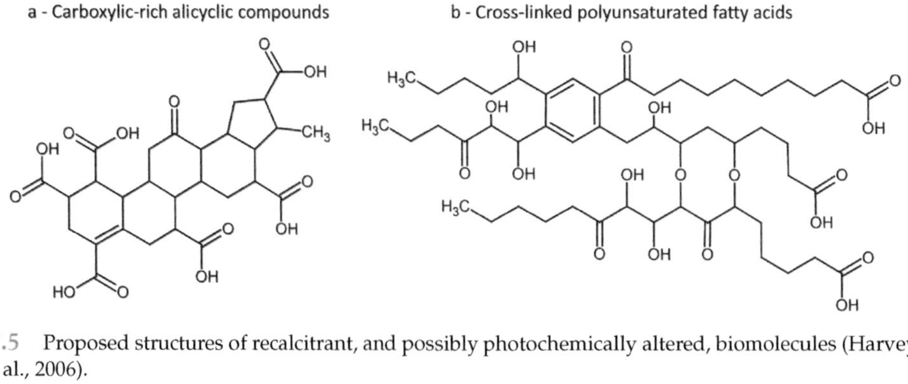

FIGURE 7.5 Proposed structures of recalcitrant, and possibly photochemically altered, biomolecules (Harvey et al., 1983; Hertkorn et al., 2006).

hydrocarbons. These compounds have a core structure consisting of five to eight fused benzene rings and carboxylic functional groups in their periphery. These functional groups make the polycyclic aromatic hydrocarbons more soluble in water (Dittmar and Koch, 2006; Kim et al., 2004). On a first view, it is most surprising to find this group of compounds in such abundance in the ocean because no organism is known to biosynthesize large polycyclic aromatic compounds. Only under conditions of excessive heat, either on land during fires or in geothermal settings in marine sediments and the crust, is organic matter modified to yield large polycyclic aromatic compounds. The presence of these compounds is therefore unequivocal evidence for a heat-related history of DOM (Kim et al., 2004).

There have been vegetation fires on Earth since plants evolved on land, and humans have used fire intensively to shape the landscape (Bowman et al., 2011). Part of the biomass does not completely burn during fire events but is charred at a wide range of temperatures (Forbes et al., 2006). Moderate dehydration occurs at low temperatures and some of the products are highly bioreactive (Norwood et al., 2013). At high temperature, large polycyclic aromatic structures, a main component of charcoal, are produced. These compounds are relatively resistant to microbial degradation and accumulate in soils and sediments over decades to thousands of years (Masiello and Druffel, 1998; Singh et al., 2012). During degradation in soils, charcoal is oxidized and partially solubilizes (Kim et al., 2004; Mannino and Harvey, 2004). Decades after a fire event, large amounts of dissolved polycyclic aromatic compounds, or dissolved black carbon (DBC), are then released from the watershed into rivers to be carried ultimately into the oceans (Dittmar et al., 2012a). Globally, the flux of dissolved charcoal from the continents to the ocean amounts to $26.5 \pm 1.8 \times 10^{12}$ g carbon per year, which is ~10% of the global riverine flux of DOC (Jaffé et al.,

2013). A fully unconstrained potential source of DBC to the deep ocean is hydrothermal circulation through deep marine sediments (Dittmar and Koch, 2006). Advective water transport exposes DOM to high temperature and pressure in marine sediments and the upper crust. The environmental conditions would facilitate the formation of polycyclic aromatic compounds from DOM or sedimentary organic matter. Concentration profiles of DBC in the deep ocean indicate that this pathway may be significant (Dittmar and Paeng, 2009), but at this point, the release of DBC from hydrothermal fluids has not been directly investigated.

In the deep ocean, DBC appears highly refractory. Large-scale distribution patterns of DBC concentrations resemble those of salinity (Figure 7.6). In the region between South Africa and Antarctica, salinity is largely affected by conservative mixing of the water masses, evaporation in the subtropics (north of the Antarctic front system), and ice melt in Antarctica (south of the Antarctic front system). DBC concentrations follow exactly the same pattern as salinity, which indicates salt-like stability of DBC in the ocean. The enormous resistance to biodegradation of DBC in the ocean is also reflected in its radiocarbon age. The colloidal fraction of DBC (>1 nm) exhibits a radiocarbon age of 15,000-20,000 years (Ziolkowski and Druffel, 2010). The radiocarbon age of the bulk of DBC is unknown, but it is very unlikely to be younger than the colloidal fraction.

The distribution of DBC in the ocean, as well as its radiocarbon age, provide evidence for millennium-scale stability of DBC in the ocean. If DBC is protected from microbial decomposition by its molecular structure, why is the deep ocean not further enriched in thermogenic DOM? The likely answer to this question is another abiotic process, removal by photodegradation. In large parts of the Southern Ocean, deep waters are transported to the sea surface, where they are exposed to sunlight during the summer months. In an experiment where Atlantic deep

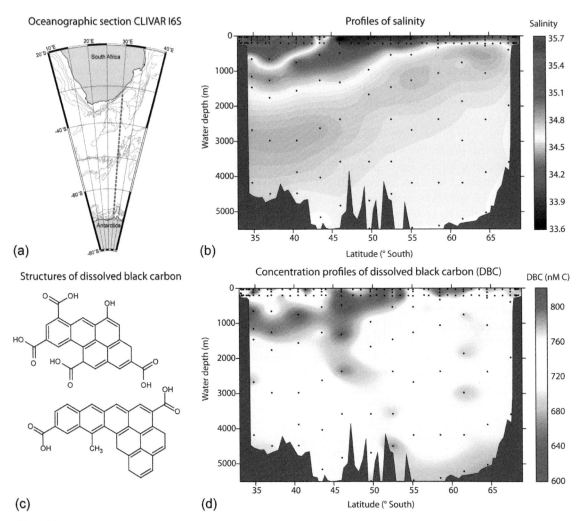

FIGURE 7.6   The distribution of dissolved black carbon (DBC) in the Southern Ocean (Dittmar and Paeng, 2009). The oceanographic section CLIVAR I6S (a) covers major global water masses that can be distinguished (e.g., via salinity, (b)). The deep water mass with elevated salinity is North Atlantic Deep Water formed from sinking surface waters near Greenland. The intermediate and bottom waters of low salinity are from sinking surface waters off Antarctica that are freshened by ice melt and precipitation. Subtropical surface water is enriched in salinity due to evaporation. Proposed structures for DBC (Dittmar and Koch, 2006) are shown in (c). The concentration profiles of DBC (d) resemble those of salinity, which is evidence for biogeochemically very stable, salt-like (conservative) properties.

water was exposed to simulated sunlight, DBC was very efficiently photodegraded (Figure 7.7), while the concentration of bulk DOC changed only marginally (Stubbins et al., 2012). From this experiment, it was estimated that between 20 and 490×10^{12} g DBC may photodegrade every year in the ocean, balancing the entire riverine input. Furthermore, this rough estimate leads to an estimated photochemical half-life for oceanic DBC of <800 years, more than an order of magnitude shorter than the radiocarbon age of DBC in the ocean. While this number is a preliminary estimate, it illustrates that photodegradation is probably the single most important removal

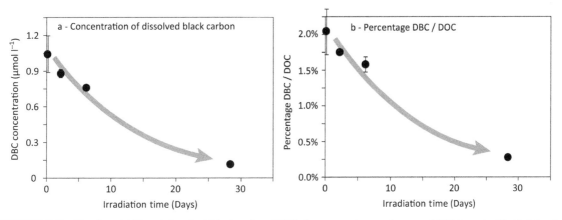

FIGURE 7.7 The photolability of dissolved black carbon (DBC). In an experiment, deep-sea DOM from the North Atlantic (Bermuda) was exposed to artificial sunlight (Stubbins et al., 2012). Within less than a month, most black carbon was preferentially decomposed, and the black carbon content of DOC was reduced from 2.1% to 0.2%.

process for DBC in the ocean. The apparent survival of DBC molecules in the oceans for millennia appears to be facilitated not only by their inherent inertness against microbial degradation but also by the rate at which they are cycled through the surface ocean's photic zone (Stubbins et al., 2012).

In the case of thermogenic DOM, an extraordinary piece of the global carbon cycle emerges (Figure 7.8): It seems that organic matter that has been stabilized by heat is then only remobilized by another abiotic process, that is, exposure to sunlight. Since thermogenesis and photodegradation are uncoupled processes and no direct feedback mechanisms are involved, the concentration of thermogenic DOM in the ocean seems largely a function of deep ocean ventilation.

IV THE MOLECULAR DIVERSITY HYPOTHESIS

The decomposition rate of an organic substrate can also depend on its concentration. Culture experiments indicate that transporter proteins and catabolic pathways are only expressed when a certain threshold concentration of substrate is reached, and organisms may not be able to consume a substrate below that threshold (Kovarova-Kovar and Egli, 1998). Under extreme oligotrophy, unusual metabolic rearrangements enable cells to substitute substrates that are present at too low concentration by others if those are more abundant (Carini et al., 2013). Under optimum growth conditions, on the other hand, substrate concentrations are high so that all uptake sites of a cell and the respective catabolic pathways are saturated. A further increase of substrate concentration will not enhance the uptake rate of an individual cell. The threshold concentration at which maximum uptake of an individual cell is reached (Figure 7.9) is determined by the species-specific substrate affinity (Kovarova-Kovar and Egli, 1998). In response to higher substrate concentrations, cells divide and their number increases so that the uptake rate of the microbial community as a whole increases. Under optimum conditions, there is a dynamic equilibrium between substrate supply, for example, from phytoplankton, and substrate consumption that is controlled by bacterial cell number and substrate affinity (Billen et al., 1980). When substrate supply ceases, substrate concentration drops due

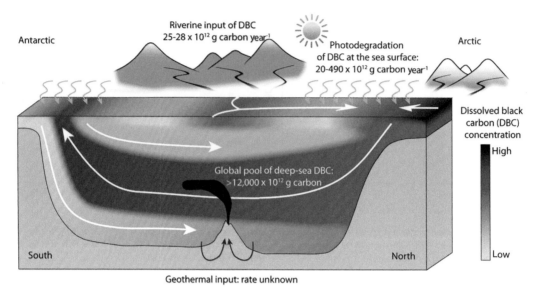

FIGURE 7.8 Conceptual scheme of dissolved black carbon (DBC) turnover in the ocean, illustrated for the Atlantic Ocean. Arrows indicate main ocean currents, the shades of blue concentrations of DBC. Rivers are the only constrained input (Jaffé et al., 2013). Groundwater discharge (Dittmar et al., 2012b) and geothermal and atmospheric inputs are other potentially important sources. Riverine input can only be injected in significant amounts into the deep ocean in the North Atlantic, the only site worldwide where major deep water formation occurs in a terrestrially influenced region. During transport from the river mouths to the North Atlantic an unknown fraction of terrigenous DBC is photodegraded. Once in the deep ocean, black carbon is preserved for many thousands of years, as indicated by deep ocean concentrations and radiocarbon age (Dittmar and Paeng, 2009; Ziolkowski and Druffel, 2010). Deep water masses are brought back to the surface only in the Southern Ocean, where dissolved black carbon partially photodegrades before it sinks back to the bottom or to intermediate depths off Antarctica.

to continuous consumption. At suboptimum concentrations, substrate utilization still yields energy, but the rate between energy consumption and gain is reduced. Under these conditions, uptake is thermodynamically limited (LaRowe et al., 2012). At an even lower concentration, the energy demand of protein biosynthesis exceeds the energy gained from substrate utilization. In this case, substrate utilization is thermodynamically inhibited (LaRowe et al., 2012) and cell number declines. Theoretically, substrate concentration and cell number continue declining until the very last remaining active cell has reached its physiological limit, and an absolute minimum concentration of substrate is reached. Due to the absence of active cells, this substrate could then persist virtually infinitely in the ocean.

According to this concept, the concentration of DOM in the ocean would be a function of the basal power requirements of microbial consumers. The physiological limit of a cell is constrained by its basal power requirements, which is the energy flux associated with the minimal complement of functions required to sustain a metabolically active state (Hoehler and Jørgensen, 2013). In theory, the higher the basal power requirement, the higher would be the concentration of substrate leftover. The basal power requirement of marine microorganisms is unknown, but studies of the deep biosphere have yielded valuable insights with respect to the basal power requirement of natural microbial communities. There is no supply of new substrate to the deep biosphere hundreds of

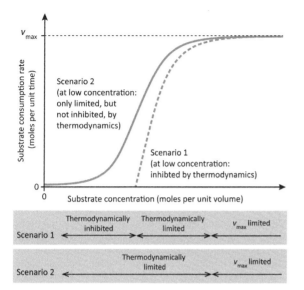

FIGURE 7.9 Schematic diagram on the dependency of the rate of an energy-yielding, microbially catalyzed reaction on the concentration of a substrate, after LaRowe et al. (2012). Above a certain threshold concentration, a maximum substrate consumption rate (V_{max}) is reached. Below that threshold, consumption rate is thermodynamically limited. At extremely low substrate concentrations, consumption may come to a complete halt (scenario 1) or continue at rates commensurate with concentration (scenario 2).

meters deep in the sediment. In this environment, there are active sulfate-reducing cells that respire on average only one sulfate ion (as the terminal electron acceptor) per cell per second (Hoehler and Jørgensen, 2013). Many known cellular functions are not possible at such a low metabolic rate, for example, to meet the energy demand of a single rotation of a bacterial flagellum, the energy gained from several minutes of sulfate reduction at the above low rate would be required (Hoehler and Jørgensen, 2013). These observations indicate that in natural environments, substrate utilization may not be thermodynamically inhibited at extremely low substrate concentrations (Figure 7.9), and that previous results from culture experiments (Kovarova-Kovar and Egli, 1998), may not be applicable to the natural marine environment. Without ther-

modynamic inhibition, all substrate molecules in the ocean could theoretically be utilized and DOM could reach zero concentration in seawater, if there was no input of new substrates. But, even if there was no thermodynamic inhibition at extremely low substrate concentration, the rate of uptake and substrate utilization is limited by physical and physiological constraints. As an absolute physical maximum, a cell cannot take up more molecules than it encounters via molecular diffusion (Stocker, 2012). The lower the substrate concentration, the higher the distance between a microbial cell and a molecule of interest, and the longer it takes to reencounter a substrate molecule.

Based on the above, the DOM decomposition rate could, in theory, be a function of its concentration (Barber, 1968; Kattner et al., 2011). On the first view, however, this seems a highly unlikely scenario. In the deep sea, the DOC concentration is ~40 µmol L^{-1} (Hansell et al., 2009). The average DOM molecule contains about 20 carbon atoms (Hertkorn et al., 2006; Koch et al., 2005). Consequently, there are about 2 µmol of DOM molecules per liter of seawater. At this concentration, the maximum rate of substrate consumption (V_{max}) is expected to be reached (Azam and Hodson, 1981). One has to consider, however, that DOM does not consist of a single substrate compound. A microbial cell can express only a limited number of transporter proteins and catabolic pathways, and can therefore only process a limited number of different substrates. There is experimental evidence that substrate limitation provokes the simultaneous expression of many catabolic enzyme systems, even if the appropriate carbon sources are absent, preparing cells to immediately utilize the substrates if they become available (Kovarova-Kovar and Egli, 1998). Under such conditions, some organisms are reported to grow on a mixture containing 45 different organic compounds (Van der Kooij et al., 1982). This number of compounds is dwarfed by the extraordinary molecular diversity of DOM. The number of different compounds in DOM is

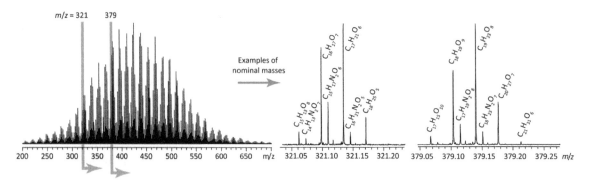

FIGURE 7.10   The molecular diversity of deep-sea DOM, illustrated by an ultrahigh-resolution mass spectrum of solid-phase extracted DOM from the deep North Pacific. The mass spectrum was obtained on a 15 Tesla Fourier-transform ion cyclotron resonance mass spectrometer (Bruker Solarix FT-ICR-MS). The precise masses of negatively singly charged ions of intact molecules are detected with this method. The whole mass spectrum and two exemplary nominal masses (321 and 379 Da) are shown. Due to the ultrahigh mass accuracy, molecular formulae can be assigned to >5000 detected masses. The molecular diversity of DOM is probably much higher than indicated by the number of molecular formulae because many structural isomers exist per molecular formula. *Modified from Dittmar and Stubbins (2014).*

unknown, but ultrahigh-resolution mass spectrometry (Figure 7.10) and multidimensional NMR spectroscopy (Figure 7.11) have revealed many thousand structural features in DOM. More than 5000 different molecular formulae of DOM compounds have already been identified via the Fourier-transform ion cyclotron resonance mass spectrometry (FT-ICR-MS) technique (Dittmar and Stubbins, 2014; Gonsior et al., 2011; Koch et al., 2005). The number of structural isomers behind each molecular formula is unknown, but recent multidimensional NMR analyses have revealed about 1500 different structural units in DOM (Hertkorn et al., 2013). Taking these results together, it seems possible that DOM is composed of millions of different compounds. At this diversity, the concentration of individual compounds is exceedingly low (roughly $2\,\mu mol\,L^{-1}$ divided by the number of different compounds). As a consequence, the turnover rate of individual DOM compounds could indeed be controlled by their concentration. However, as long as the number of different DOM compounds and their concentrations are unknown, a link between the observed decomposition rates of DOM and the concentration of individual DOM compounds remains hypothetical.

Though entirely speculative at this point, the molecular diversity hypothesis offers a strikingly different view on organic matter persistence in the ocean. Extreme substrate dilution could provide a mechanistic explanation for the persistence of DOM in the ocean and also an answer to the question of why DOM is eventually decomposed over the course of millennia and not infinitely accumulated. The concept of extreme substrate dilution also implies that most dissolved organic molecules are degradable when at high enough concentration. Degradation rates would be coupled to production rate, because compounds that are produced at a high rate encounter cells more frequently than rare compounds. As a consequence, DOM and its individual constituents could be in a dynamic steady state, where inputs and outputs are balanced. From the perspective of extreme substrate dilution, the deep ocean appears a hostile environment for free-living heterotrophic microorganisms. Once released from sinking particles, microorganisms face an extreme shortage of suitable substrate. Though there is enough organic material dissolved in the ocean, the substrate diversity might be so excessive that assimilation rates of free-living deep-sea bacteria are similar to those in the deep biosphere.

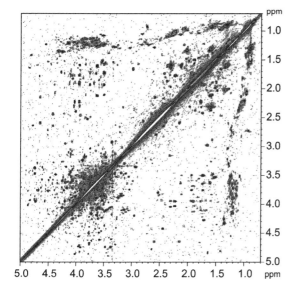

FIGURE 7.11   The molecular diversity of deep-sea DOM, illustrated by a two-dimensional nuclear magnetic resonance (NMR) spectrum of solid-phase extracted DOM from the surface Atlantic ocean (^{1}H, ^{1}H COSY NMR, obtained on a 800 MHz Bruker Avance III NMR spectrometer). Different from mass spectrometry (Figure 7.10), NMR does not detect the mass of a molecule and its elemental composition, but reveals information on the spatial arrangement of elements in a molecule, that is, structural features. Each dot in this spectrum represents a distinct structural feature in DOM. Hertkorn et al. (2013) were able to apply this technique to identify ~1500 molecular units and their structures within the DOM pool. *Modified from Hertkorn et al. (2013).*

V CONCLUDING REMARKS

Despite several decades of intense research, the reason behind the long-term stability of DOM remains unknown. It is most likely a complex interplay of different mechanisms that causes some organic molecules to persist in the ocean for several thousand years. Complex food web mechanisms and shortage of mineral nutrients or electron acceptors can cause an accumulation of DOM over the course of a season or up to decades. During this intermediate accumulation, DOM can be subject to secondary, stabilizing abiotic modifications. The resulting molecular structures may be inaccessible to the metabolic toolbox of microorganisms. Photochemical modifications and thermogenesis on land or in hydrothermal settings are likely modifying DOM to yield persistent structures. Photochemical reactions, on the other hand, can also return refractory molecules formed in the dark of the deep ocean or in the dark of soils, back into active cycles. The idea of extreme substrate dilution does not require any assumptions on molecular structure. It links the slow turnover of DOM to extremely low concentrations of individual compounds in DOM. In light of the size of the global DOM pool, and the perturbations in our climate system that changes in the DOM pool could cause, our lack of knowledge is of concern. The hypotheses discussed in this chapter provide a solid framework for future research, but finding definite answers will remain a major challenge for the coming decades.

Acknowledgments

I thank Dennis Hansell and Craig Carlson for their support and guidance. This manuscript also benefited from reviews of Jutta Niggemann, Aron Stubbins, and two anonymous colleagues.

References

Albert, D.B., Taylor, C., Martens, C.S., 1995. Sulfate reduction rates and low-molecular-weight fatty-acid concentrations in the water column and surficial sediments of the Black-Sea. Deep-Sea Res. Pt. I 42, 1239–1260.

Aluwihare, L.I., Repeta, D.J., Chen, R.F., 1997. A major biopolymeric component to dissolved organic carbon in surface sea water. Nature 387, 166–169.

Amon, R.M.W., Fitznar, H.P., Benner, R., 2001. Linkages among the bioreactivity, chemical composition, and diagenetic state of marine dissolved organic matter. Limnol. Oceanogr. 46, 287–297.

Aristegui, J., Duarte, C.M., Agusti, S., Doval, M., Alvarez-Salgado, X.A., Hansell, D.A., 2002. Dissolved organic carbon support of respiration in the dark ocean. Science 298, 1967.

Aristegui, J., Agusti, S., Middelburg, J.J., Duarte, C.M., 2005. Respiration in the mesopelagic and bathypelagic zones of the oceans. In: del Giorgio, P., Williams, P. (Eds.) Respiration in Aquatic Ecosystems. Oxford University Press, Oxford UK.

Aristegui, J., Gasol, J.M., Duarte, C.M., Herndl, G.J., 2009. Microbial oceanography of the dark ocean's pelagic realm. Limnol. Oceanogr. 54, 1501–1529.

Arnosti, C., Steen, A.D., Ziervogel, K., Ghobrial, S., Jeffrey, W.H., 2011. Latitudinal gradients in degradation of marine dissolved organic carbon. PLoS One 6, e28900, 1–6.

Azam, F., Hodson, R.E., 1981. Multiphasic kinetics for D-glucose uptake by assemblages of natural marine-bacteria. Mar. Ecol. Prog. Ser. 6, 213–222.

Azam, F., Malfatti, F., 2007. Microbial structuring of marine ecosystems. Nat. Rev. Microbiol. 5, 782–791.

Barber, R.T., 1968. Dissolved organic carbon from deep waters resists microbial oxidation. Nature 220, 274–275.

Bauer, J.E., Druffel, E.R.M., Wolgast, D.M., Griffin, S., 2002. Temporal and regional variability in sources and cycling of DOC and POC in the northwest Atlantic continental shelf and slope. Deep-Sea Res. Pt. II 49, 4387–4419.

Beaupre, S.R., Druffel, E.R.M., 2012. Photochemical reactivity of ancient marine dissolved organic carbon. Geophys. Res. Lett. 39, L18602, 1–5.

Benner, R., Biddanda, B., 1998. Photochemical transformations of surface and deep marine dissolved organic matter: effects on bacterial growth. Limnol. Oceanogr. 43, 1373–1378.

Benner, R., Kaiser, K., 2003. Abundance of amino sugars and peptidoglycan in marine particulate and dissolved organic matter. Limnol. Oceanogr. 48, 118–128.

Benner, R., Pakulski, J.D., Mccarthy, M., Hedges, J.I., Hatcher, P.G., 1992. Bulk chemical characteristics of dissolved organic-matter in the ocean. Science 255, 1561–1564.

Billen, G., Joiris, C., Wijnant, J., Gillain, G., 1980. Concentration and microbial utilization of small organic molecules in the Scheld Estuary, the Belgian coastal zone of the North Sea and the English Channel. Estuar. Coast. Mar. Sci. 11, 279–294.

Bowman, D., Balch, J., Artaxo, P., Bond, W.J., Cochrane, M.A., D'antonio, C.M., et al., 2011. The human dimension of fire regimes on Earth. J. Biogeogr. 38, 2223–2236.

Brophy, J.E., Carlson, D.J., 1989. Production of biologically refractory dissolved organic-carbon by natural seawater microbial-populations. Deep-Sea Res. 36, 497–507.

Carini, P., Steindler, L., Beszteri, S., Giovannoni, S.J., 2013. Nutrient requirements for growth of the extreme oligotroph 'Candidatus Pelagibacter ubique' HTCC1062 on a defined medium. ISME J. 7, 592–602.

Carlson, C.A., 2002. Production and removal processes. In: Hansell, D.A., Carlson, C.A. (Eds.), Biogeochemistry of Marine Dissolved Organic Matter. Academic Press, Boston, MA.

Carlson, C.A., Giovannoni, S.J., Hansell, D.A., Goldberg, S.J., Parsons, R., Otero, M.P., et al., 2002. Effect of nutrient amendments on bacterioplankton production, community structure, and DOC utilization in the northwestern Sargasso Sea. Aquat. Microb. Ecol. 30, 19–36.

Carlson, C.A., Del Giorgio, P.A., Herndl, G.J., 2007. Microbes and the dissipation of energy and respiration: from cells to ecosystems. Oceanography 20, 89–100.

Church, M.J., Hutchins, D.A., Ducklow, H.W., 2000. Limitation of bacterial growth by dissolved organic matter and iron in the Southern Ocean. Appl. Environ. Microbiol. 66, 455–466.

Davis, J., Kaiser, K., Benner, R., 2009. Amino acid and amino sugar yields and compositions as indicators of dissolved organic matter diagenesis. Org. Geochem. 40, 343–352.

Del Giorgio, P.A., Cole, J.J., Cimbleris, A., 1997. Respiration rates in bacteria exceed phytoplankton production in unproductive aquatic systems. Nature 385, 148–151.

Dickens, A.F., Gelinas, Y., Masiello, C.A., Wakeham, S., Hedges, J.I., 2004. Reburial of fossil organic carbon in marine sediments. Nature 427, 336–339.

Dittmar, T., Kattner, G., 2003. Recalcitrant dissolved organic matter in the ocean: major contribution of small amphiphilics. Mar. Chem. 82, 115–123.

Dittmar, T., Koch, B.P., 2006. Thermogenic organic matter dissolved in the abyssal ocean. Mar. Chem. 102, 208–217.

Dittmar, T., Paeng, J., 2009. A heat-induced molecular signature in marine dissolved organic matter. Nat. Geosci. 2, 175–179.

Dittmar, T., Stubbins, A., 2014. Dissolved organic matter in aquatic systems. In: Birrer, B., Falkowski, P., Freeman, K. (Eds.), Treatise of Geochemistry. second ed.. Elsevier, Oxford.

Dittmar, T., Fitznar, H.P., Kattner, G., 2001. Origin and biogeochemical cycling of organic nitrogen in the eastern Arctic Ocean as evident from D- and L-amino acids. Geochim. Cosmochim. Acta 65, 4103–4114.

Dittmar, T., De Rezende, C.E., Manecki, M., Niggemann, J., Coelho Ovalle, A.R., Stubbins, A., et al., 2012a. Continuous flux of dissolved black carbon from a vanished tropical forest biome. Nat. Geosci. 5, 618–622.

Dittmar, T., Paeng, J., Gihring, T.M., Suryaputra, I.G.N.A., Huettel, M., 2012b. Discharge of dissolved black carbon from a fire-affected intertidal system. Limnol. Oceanogr. 57, 1171–1181.

Dorrepaal, E., Toet, S., Van Logtestijn, R.S.P., Swart, E., Van De Weg, M.J., Callaghan, T.V., et al., 2009. Carbon respiration from subsurface peat accelerated by climate warming in the subarctic. Nature 460, 616–620.

Ducklow, H.W., Hansell, D.A., Morgan, J.A., 2007. Dissolved organic carbon and nitrogen in the Western Black Sea. Mar. Chem. 105, 140–150.

Flerus, R., Lechtenfeld, O.J., Koch, B.P., Mccallister, S.L., Schmitt-Kopplin, P., Benner, R., et al., 2012. A molecular perspective on the ageing of marine dissolved organic matter. Biogeosciences 9, 1935–1955.

Forbes, M.S., Raison, R.J., Skjemstad, J.O., 2006. Formation, transformation and transport of black carbon (charcoal) in terrestrial and aquatic ecosystems. Sci. Total Environ. 370, 190–206.

Gonsior, M., Peake, B.M., Cooper, W.T., Podgorski, D.C., D'andrilli, J., Dittmar, T., et al., 2011. Characterization of dissolved organic matter across the Subtropical Convergence off the South Island, New Zealand. Mar. Chem. 123, 99–110.

Gruber, D.F., Simjouw, J.P., Seitzinger, S.P., Taghon, G.L., 2006. Dynamics and characterization of refractory dissolved organic matter produced by a pure bacterial culture in an experimental predator-prey system. Appl. Environ. Microbiol. 72, 4184–4191.

Habicht, K.S., Gade, M., Thamdrup, B., Berg, P., Canfield, D.E., 2002. Calibration of sulfate levels in the Archean Ocean. Science 298, 2372–2374.

Halm, H., Lam, P., Ferdelman, T.G., Lavik, G., Dittmar, T., Laroche, J., et al., 2012. Heterotrophic organisms dominate nitrogen fixation in the South Pacific Gyre. ISME J. 6, 1238–1249.

Hansell, D.A., 2013. Recalcitrant dissolved organic carbon fractions. In: Carlson, C.A., Giovannoni, S.J. (Eds.), Annual Review of Marine Science, vol. 5, Annual Reviews, Palo Alto CA, USA.

Hansell, D.A., Carlson, C.A., 2013. Localized refractory dissolved organic carbon sinks in the deep ocean. Global Biogeochem. Cycles 27, 705–710.

Hansell, D.A., Carlson, C.A., Repeta, D.J., Schlitzer, R., 2009. Dissolved organic matter in the ocean: a controversy stimulates new insights. Oceanography 22, 202–211.

Harvey, G.R., Boran, D.A., Chesal, L.A., Tokar, J.M., 1983. The structure of marine fulvic and humic acids. Mar. Chem. 12, 119–132.

Hertkorn, N., Benner, R., Frommberger, M., Schmitt-Kopplin, P., Witt, M., Kaiser, K., et al., 2006. Characterization of a major refractory component of marine dissolved organic matter. Geochim. Cosmochim. Acta 70, 2990–3010.

Hertkorn, N., Harir, M., Koch, B.P., Michalke, B., Schmitt-Kopplin, P., 2013. High-field NMR spectroscopy and FTICR mass spectrometry: powerful discovery tools for the molecular level characterization of marine dissolved organic matter. Biogeosciences 10, 1583–1624.

Hoehler, T.M., Jørgensen, B.B., 2013. Microbial life under extreme energy limitation. Nat. Rev. Microbiol. 11, 83–94.

Jaffé, R., Ding, Y., Niggemann, J., Vahatalo, A.V., Stubbins, A., Spencer, R.G.M., et al., 2013. Global charcoal mobilization from soils via dissolution and riverine transport to the oceans. Science 340, 345–347.

Jiao, N., Herndl, G.J., Hansell, D.A., Benner, R., Kattner, G., Wilhelm, S.W., et al., 2011. Microbial production of recalcitrant dissolved organic matter: long-term carbon storage in the global ocean. Nat. Rev. Microbiol. 8, 593–599.

Jørgensen, B.B., 1982. Mineralization of organic-matter in the sea bed—the role of sulfate reduction. Nature 296, 643–645.

Kaiser, K., Benner, R., 2009. Biochemical composition and size distribution of organic matter at the Pacific and Atlantic time-series stations. Mar. Chem. 113, 63–77.

Kaiser, K., Benner, R., 2012. Organic matter transformations in the upper mesopelagic zone of the North Pacific:

chemical composition and linkages to microbial community structure. J. Geophys. Res.-Oc. 117, C01023.

Kattner, G., Simon, M., Koch, B.P., 2011. Molecular characterization of dissolved organic matter and constraints for prokaryotic utilization. In: Jiao, N., Azam, F., Sanders, S. (Eds.), Microbial Carbon Pump in the Ocean. American Association for the Advancement of Science, Washington, DC.

Keil, R.G., Kirchman, D.L., 1994. Abiotic transformation of labile protein to refractory protein in sea-water. Mar. Chem. 45, 187–196.

Kim, S.W., Kaplan, L.A., Benner, R., Hatcher, P.G., 2004. Hydrogen-deficient molecules in natural riverine water samples—evidence for the existence of black carbon in DOM. Mar. Chem. 92, 225–234.

Koch, B.P., Witt, M.R., Engbrodt, R., Dittmar, T., Kattner, G., 2005. Molecular formulae of marine and terrigenous dissolved organic matter detected by electrospray ionization Fourier transform ion cyclotron resonance mass spectrometry. Geochim. Cosmochim. Acta 69, 3299–3308.

Konhauser, K.O., Pecoits, E., Lalonde, S.V., Papineau, D., Nisbet, E.G., Barley, M.E., 2009. Oceanic nickel depletion and a methanogen famine before the Great Oxidation Event. Nature 458, 750–753.

Kovarova-Kovar, K., Egli, T., 1998. Growth kinetics of suspended microbial cells: from single-substrate-controlled growth to mixed-substrate kinetics. Microbiol. Mol. Biol. Rev. 62, 646–666.

Kritzberg, E.S., Arrieta, J.M., Duarte, C.M., 2010. Temperature and phosphorus regulating carbon flux through bacteria in a coastal marine system. Aquat. Microb. Ecol. 58, 141–151.

Larowe, D.E., Dale, A.W., Amend, J.P., Van Cappellen, P., 2012. Thermodynamic limitations on microbially catalyzed reaction rates. Geochim. Cosmochim. Acta 90, 96–109.

Leahy, J.G., Colwell, R.R., 1990. Microbial-degradation of hydrocarbons in the environment. Microbiol. Rev. 54, 305–315.

Loh, A.N., Bauer, J.E., Druffel, E.R.M., 2004. Variable ageing and storage of dissolved organic components in the open ocean. Nature 430, 877–881.

Mannino, A., Harvey, H.R., 2004. Black carbon in estuarine and coastal ocean dissolved organic matter. Limnol. Oceanogr. 49, 735–740.

Martiny, J.B.H., Bohannan, B.J.M., Brown, J.H., Colwell, R.K., Fuhrman, J.A., Green, J.L., et al., 2006. Microbial biogeography: putting microorganisms on the map. Nat. Rev. Microbiol. 4, 102–112.

Masiello, C.A., Druffel, E.R.M., 1998. Black carbon in deep-sea sediments. Science 280, 1911–1913.

Mccarthy, M., Hedges, J., Benner, R., 1996. Major biochemical composition of dissolved high molecular weight organic matter in seawater. Mar. Chem. 55, 281–297.

Mccarthy, M.D., Hedges, J.I., Benner, R., 1998. Major bacterial contribution to marine dissolved organic nitrogen. Science 281, 231–234.

Mccarthy, M.D., Beaupre, S.R., Walker, B.D., Voparil, I., Guilderson, T.P., Druffel, E.R.M., 2011. Chemosynthetic origin of C-14-depleted dissolved organic matter in a ridge-flank hydrothermal system. Nat. Geosci. 4, 32–36.

Merchant, S.S., Helmann, J.D., 2012. Elemental economy: microbial strategies for optimizing growth in the face of nutrient limitation. In: Poole, R.K. (Ed.), Advances in Microbial Physiology, 60, 91–210, Elsevier, Amsterdam, The Netherlands.

Moore, C.M., Mills, M.M., Arrigo, K.R., Berman-Frank, I., Bopp, L., Boyd, P.W., et al., 2013. Processes and patterns of oceanic nutrient limitation. Nat. Geosci. 6, 701–710.

Norwood, M.J., Louchouarn, P., Kuo, L.-J., Harvey, O.R., 2013. Characterization and biodegradation of water-soluble biomarkers and organic carbon extracted from low temperature chars. Org. Geochem. 56, 111–119.

Obernosterer, I., Benner, R., 2004. Competition between biological and photochemical processes in the mineralization of dissolved organic carbon. Limnol. Oceanogr. 49, 117–124.

Ogawa, H., Amagai, Y., Koike, I., Kaiser, K., Benner, R., 2001. Production of refractory dissolved organic matter by bacteria. Science 292, 917–920.

Panagiotopoulos, C., Repeta, D.J., Johnson, C.G., 2007. Characterization of methyl sugars, 3-deoxysugars and methyl deoxysugars in marine high molecular weight dissolved organic matter. Org. Geochem. 38, 884–896.

Quigg, A., Finkel, Z.V., Irwin, A.J., Rosenthal, Y., Ho, T.Y., Reinfelder, J.R., et al., 2003. The evolutionary inheritance of elemental stoichiometry in marine phytoplankton. Nature 425, 291–294.

Repeta, D.J., Aluwihare, L.I., 2006. Radiocarbon analysis of neutral sugars in high-molecular-weight dissolved organic carbon: implications for organic carbon cycling. Limnol. Oceanogr. 51, 1045–1053.

Robinson, C., Williams, P.J.L.B., 2005. Respiration and its measurement in surface marine waters. In: del Giorgio, P.A., Williams, P.J.le B. (Eds.), Respiration in Aquatic Ecosystems. Oxford University Press, Oxford.

Schmidt, M.W.I., Torn, M.S., Abiven, S., Dittmar, T., Guggenberger, G., Janssens, I.A., et al., 2011. Persistence of soil organic matter as an ecosystem property. Nature 478, 49–56.

Sexton, P.F., Norris, R.D., Wilson, P.A., Paelike, H., Westerhold, T., Roehl, U., et al., 2011. Eocene global warming events driven by ventilation of oceanic dissolved organic carbon. Nature 471, 349–352.

Singh, N., Abiven, S., Torn, M.S., Schmidt, M.W.I., 2012. Fire-derived organic carbon in soil turns over on a centennial scale. Biogeosciences 9, 2847–2857.

Stocker, R., 2012. Marine microbes see a Sea of Gradients. Science 338, 628–633.

Stubbins, A., Niggemann, J., Dittmar, T., 2012. Photo-lability of deep ocean dissolved black carbon. Biogeosciences 9, 1661–1670.

Swannell, R.P.J., Lee, K., Mcdonagh, M., 1996. Field evaluations of marine oil spill bioremediation. Microbiol. Rev. 60, 342–365.

Tanaka, T., Thingstad, T.F., Gasol, J.M., Cardelus, C., Jezbera, J., Sala, M.M., et al., 2009. Determining the availability of phosphate and glucose for bacteria in P-limited mesocosms of NW Mediterranean surface waters. Aquat. Microb. Ecol. 56, 81–91.

Thingstad, T.F., Hagstrom, A., Rassoulzadegan, F., 1997. Accumulation of degradable DOC in surface waters: is it caused by a malfunctioning microbial loop? Limnol. Oceanogr. 42, 398–404.

Thingstad, T.F., Zweifel, U.L., Rassoulzadegan, F., 1998. P limitation of heterotrophic bacteria and phytoplankton in the northwest Mediterranean. Limnol. Oceanogr. 43, 88–94.

Thingstad, T.F., Krom, M.D., Mantoura, R.F.C., Flaten, Ga.F., Groom, S., Herut, B., et al., 2005. Nature of phosphorus limitation in the ultraoligotrophic eastern Mediterranean. Science 309, 1068–1071.

Thingstad, T.F., Bellerby, R.G.J., Bratbak, G., Borsheim, K.Y., Egge, J.K., Heldal, M., et al., 2008. Counterintuitive carbon-to-nutrient coupling in an Arctic pelagic ecosystem. Nature 455, 387.

Van Der Kooij, D., Oranje, J.P., Hijnen, Wa.M., 1982. Growth of *Pseudomonas aeruginosa* in tap water in relation to utilization of substrates at concentrations of a few micrograms per liter. Appl. Environ. Microbiol. 44, 1086–1095.

Wakeham, S.G., Pease, T.K., Benner, R., 2003. Hydroxy fatty acids in marine dissolved organic matter as indicators of bacterial membrane material. Org. Geochem. 34, 857–868.

Walker, B.D., Beaupre, S.R., Guilderson, T.P., Druffel, E.R.M., Mccarthy, M.D., 2011. Large-volume ultrafiltration for the study of radiocarbon signatures and size vs. age relationships in marine dissolved organic matter. Geochim. Cosmochim. Acta 75, 5187–5202.

Williams, P.M., Druffel, E.R.M., 1987. Radiocarbon in dissolved organic mater in the central north Pacific Ocean. Nature 330, 246–248.

Williams, P.M., Oeschger, H., Kinney, P., 1969. Natural radiocarbon activity of dissolved organic carbon in north-east Pacific Ocean. Nature 224, 256–258.

Ziolkowski, L.A., Druffel, E.R.M., 2010. Aged black carbon identified in marine dissolved organic carbon. Geophys. Res. Lett. 37, L16601.

Zohary, T., Herut, B., Krom, M.D., Mantoura, R.F.C., Pitta, P., Psarra, S., et al., 2005. P-limited bacteria but N and P co-limited phytoplankton in the Eastern Mediterranean—a microcosm experiment. Deep-Sea Res. Pt. II 52, 3011–3023.

Marine Photochemistry of Organic Matter: Processes and Impacts

Kenneth Mopper*, David J. Kieber†, Aron Stubbins‡

*Department of Chemistry and Biochemistry, Old Dominion University, Norfolk, Virginia, USA

†College of Environmental Science and Forestry, Department of Chemistry,
State University of New York, Syracuse, New York, USA

‡Skidaway Institute of Oceanography, Department of Marine Sciences,
University of Georgia, Savannah, Georgia, USA

CONTENTS

Biogeochemistry of Marine Dissolved Organic Matter,
http://dx.doi.org/10.1016/B978-0-12-405940-5.00008-X

I INTRODUCTION

Dissolved organic matter (DOM) plays a dominant role in the absorption of ultraviolet (UV) and visible light in the open ocean. As DOM absorbance is regulated in part by photobleaching processes (see Chapter 10), light availability for photosynthesis and the penetration of UV radiation within the marine environment are influenced by the abiotic photochemical transformations that are the focus of this chapter. In addition to its control on UV light fields, DOM photochemistry strongly impacts the biogeochemical cycling of biologically important elements in surface seawater, including mineral nutrients and trace metals. Examples of important light-driven chemical reactions in the photic zone include reduction of trace metals like iron, manganese, and copper. The biogeochemistry of these reduced species is quite different from their oxidized counterparts. Photochemical oxidation of DOM produces a suite of free radicals and other short-lived species including the superoxide anion, the carbonate radical, singlet oxygen, the hydroxyl radical, dibromide radical anion, excited state triplets, and a number of poorly described organic radicals. These species are generally more reactive compared to the corresponding diamagnetic and ground state species and are expected to influence biological and chemical processes in sunlit surface waters. Photolysis of DOM is also an important source of a variety of atmospherically important gases that are emitted from the ocean into the atmosphere. These gases

affect the chemistry of the atmosphere and exert control over the Earth's radiation balance. Surface seawater concentrations, and hence emissions, of alkyl nitrates, carbon monoxide (CO), carbon dioxide (CO_2), carbonyl sulfide (OCS), and dimethylsulfide (DMS) are all partly regulated through photochemical processes involving DOM, which acts as a reactant, a photosensitizer, and a source of oxidative radical species.

Photochemistry results in the partial or complete remineralization of DOM in the oceans, thereby playing an important role in turning over carbon, nitrogen, sulfur, and phosphorus in the photic zone. Photoreactions also alter the bioavailability of refractory DOM. Incomplete oxidation of biologically recalcitrant DOM yields substrates (and nutrients such as ammonium) to marine heterotrophs thereby injecting extra "fuel" into the food web. However, this oceanic photochemical/biological coupling is poorly constrained, and it is complicated because photochemistry may destroy as well as form biological substrates. Nonetheless, this coupled pathway appears to be a major sink for riverine DOM (see Chapter 11) and an important control of the geochemical cycling of refractory marine DOM (see Chapter 7).

All abiotic photochemical transformations in seawater are ultimately controlled by the absorption of sunlight by inorganic matter such as nitrate or nitrite and dissolved and particulate organic matter (POM). Decay of this absorbed energy occurs through competing photophysical and photochemical pathways. Generally, only UV and blue light are energetic enough to promote photochemical transformations in natural waters; however, these transformations account for only a small fraction of the solar energy that is absorbed, since most of the absorbed energy is dissipated as heat. The efficiency of a photochemical process, known as the quantum yield, is a unitless ratio equal to the number of moles of species formed or photolyzed divided by the number of moles of photons absorbed at any one wavelength by the chromophore responsible for the photoprocess. In the case of DOM, the specific chromophore (or chromophores) giving rise to a phototransient or product is (are) not known for indirect photochemical processes and many direct photochemical reactions. Therefore, for environmental photochemical studies, the photoefficiency is often related to the total number of photons absorbed by DOM and the term apparent quantum yield (AQY) is applied. AQYs are always less (typically much less) than true quantum yields for pure compounds. Nevertheless, the AQY is a fundamental parameter needed to quantify environmental photochemical processes and to model these processes in seawater. This is because the rate of a photoreaction is equal to the product of the AQY for that reaction and the specific rate of light absorption.

The fundamental equations used to describe direct and indirect photochemical reactions, and their application to aquatic systems, are described elsewhere (e.g., Leifer, 1988) and will not be detailed here. In addition, evaluation of errors associated with AQY determinations and with the assumption that AQYs are constant with photon exposure (i.e., irradiance integrated over time of exposure) are discussed in detail elsewhere (Neale and Kieber, 2000; Reader and Miller, 2012). Understanding and evaluating these errors and assumptions are key to accurately modeling photochemical processes in seawater on a global scale (Fichot and Miller, 2010; Powers and Miller, 2014; see Section V in this chapter).

This review focuses on the impact of photochemistry on marine biogeochemical cycles of carbon, sulfur, nitrogen, and phosphorus,

highlighting and evaluating recent advances and areas for future work. Although photochemical reactions also impact trace metal cycling in seawater, this topic is discussed in detail elsewhere (Gledhill and Buck, 2012; Wells, 2002) and will not be reviewed here. Likewise, analytical techniques, marine optics, and photoinhibition, although important topics, are not covered here. The experimental details of many of the studies reviewed here (e.g., light source, pH, ionic strength, temperature) are generally not given. However, it is likely that some of the differences in results noted throughout this review may be due in part to differences in the experimental design, particularly with respect to the light source. Unless stated otherwise, for the studies discussed in this chapter, irradiations were performed using either natural sunlight or various artificial light sources (e.g., xenon arc, low pressure Hg, UV-A fluorescent bulbs) with a cutoff at ~300 nm, the approximate solar cutoff at the Earth's surface. The reader is referred to the original publications for experimental details and to Burns et al. (2012) for a review of recent photochemical experimental methods. Even though our emphasis is on marine photochemistry, many of the findings and conclusions from freshwater studies are applicable to marine waters, especially terrestrially impacted coastal and inland (e.g., estuarine) waters. Therefore, results of freshwater studies will be mentioned when appropriate.

II IMPACT OF PHOTOCHEMISTRY ON ELEMENTAL CYCLES

A Carbon

During the past two to three decades, there has been an exponential increase in the number of studies focused on the role of photochemical degradation of DOM in carbon cycling in freshwater and oceanic environments. Many of these studies through 2001 were reviewed in Mopper and Kieber (2002), with recent updates reviewed in Zepp et al. (2011). In the Mopper and Kieber (2002) review, two major photochemical oxidation/degradation pathways of DOM were discussed: *an abiotic pathway* involving the photochemical production of CO and CO_2 and *a coupled photochemical/biotic pathway* involving photodegradation of DOM to biolabile substrates, in particular low molecular weight (LMW) compounds, followed by microbial uptake and conversion of these substrates into biomass and CO_2. The production of substrates via the abiotic pathway occurs through direct or indirect photochemical mechanisms (Goldstone et al., 2002; Mopper and Kieber, 2000; Pullin et al., 2004; Tedetti et al., 2009). The two DOM degradation pathways (abiotic and coupled photochemical/biotic) contribute to the mineralization of dissolved organic carbon (DOC) (Obernosterer and Benner, 2004; Scully et al., 2003b) and together can result in the rapid removal of terrestrially derived DOM in coastal waters (Aarnos et al., 2012; Amon and Benner, 1996; Fichot and Benner, 2014; Kieber, 2000; Kieber et al., 1990; Miller and Moran, 1997; Miller et al., 2002; Miller and Zepp, 1995; Mopper and Kieber, 2000; Moran et al., 2000; Rossel et al., 2013) and in the geochemical turnover of marine biologically refractory DOM (Anderson and Williams, 1999; Gonsior et al., 2014; Mopper and Kieber, 2000; Mopper et al., 1991). Cherrier et al. (1999) reported the presence of isotopically old carbon (^{14}C depleted) in bacteria from open oceanic surface water, suggesting that photoreactions were responsible for increasing the bioavailability and cycling of old, bio-refractory DOC. Beaupré and Druffel (2012) analyzed the ^{14}C content of photochemically formed dissolved inorganic carbon (DIC) in sterile-filtered marine surface waters, concluding that abiotic photomineralization plays an important role in determining the residence time of ^{14}C-depleted DOM in the sea. Gonsior et al. (2014) found that irradiation of biologically refractory DOM from open ocean surface water gives rise to polyols, which are expected to be

rapidly consumed by biota. In addition, Stubbins et al. (2012) indicated that photodegradation is the primary sink for otherwise biologically refractory oceanic dissolved black carbon (DBC).

The quantitative importance of the two degradation pathways for the photoremineralization of terrestrially derived DOM or DOC (tDOM or tDOC) was demonstrated by Miller and Moran (1997). They determined that the percentage of tDOC in coastal waters utilized by heterotrophs (i.e., incorporated plus respired) as a result of DOM photodegradation was approximately equal to the abiotic photoproduction of CO and CO_2 in those samples; they also found that ~10-15% of DOC in various freshwater and coastal seawater samples was photooxidized to CO and CO_2. Combining these results suggests that ~20-30% of tDOC in coastal waters can be rapidly photodegraded and consumed. However, recent studies indicate that this estimate is probably conservative. Based on AQY spectra for the formation of biologically labile photoproducts in terrestrially impacted coastal waters, Miller et al. (2002) estimated that the annual global production of these photoproducts (0.21 Pg C) was approximately the same as the annual global input of riverine DOC (~0.25 Pg C; McNeil et al., 2003). Wang et al. (2004) estimated that the abiotic DOM photolysis pathway alone (based on CO_2 photoproduction rates) exceeded the rate of the input of riverine DOM to the sea, which is in agreement with Andrews et al. (2000), whose estimates were based on AQYs for photochemical oxygen consumption and photochemical modeling. Using AQY measurements for DIC photoformation and photochemical modeling, Aarnos et al. (2012) concluded that the annual loss of DOC in the Baltic Sea, including the direct photomineralization and bacterial mineralization of photoproduced labile DOM, exceeded the annual river loading of photolabile DOC (assuming that ~half of the total river DOC input to the Baltic Sea was photolabile, i.e., contained sunlight-absorbing chromophores). Fichot and Benner (2014) used a lignin-based optical proxy for tDOC and a shelf-wide mass balance approach to quantitatively assess the fate of tDOC discharged from the Mississippi-Atchafalaya River System to the Louisiana shelf. They estimated that during their ~1-year sampling period, 40% of the tDOC was mineralized to CO_2, mainly by bacterial mineralization of photoproduced labile DOM. In contrast, Reader and Miller (2012) estimated that abiotic, photochemical production of DIC and CO accounted for about 3% of the DOC delivered to the coastal waters of the South Atlantic Bight; however, this estimate was based on the residence time of DOC in a very confined area (inshore and estuarine) and did not apply to further oxidation during transit through the entire South Atlantic Bight. Furthermore, microbial mineralization of photoproduced labile DOM was not examined in that study.

1 Coupled Photochemical-Microbial DOC Degradation: Impact on Marine Food Web Dynamics

Strome and Miller (1978) hypothesized that the degradation of DOM in lakes by solar irradiation stimulates bacterial growth by the photochemical formation of substrates. This hypothesis was subsequently supported by Geller (1986), who performed experiments with extracted and irradiated lacustrine DOM that was inoculated with bacterial isolates. Mopper and Stahovec (1986) demonstrated that LMW carbonyl compounds, such as formaldehyde, were produced during the irradiation of coastal seawater and that their steady state concentrations underwent diurnal variations in response to interactions between photochemical production and microbial uptake. They hypothesized that these interactions could be important in marine food web dynamics. Kieber et al. (1989) supported this hypothesis by showing a strong correlation between the photochemical production of pyruvate and its microbial utilization in coastal and open ocean waters. In addition to providing LMW carbon substrates, e.g., formaldehyde, acetaldehyde, pyruvate, oxalate, and other LMW organic acids (Bertilsson and Tranvik, 1998, 2000; Goldstone et al., 2002; Pullin et al., 2004; Remington

et al., 2011), photochemical reactions can enhance microbial activity through a variety of pathways, for example, by increasing the availability of essential trace metals (e.g., iron, copper, and manganese) via production or destruction of organic ligands and altering trace metal redox state/solubility (Gledhill and Buck, 2012; Grabo et al., 2014; Wells, 2002), by release of macronutrients including inorganic nitrogen and phosphorus (Aarnos et al., 2012; Bushaw et al., 1996; Bushaw-Newton and Moran, 1999; Koopmans and Bronk, 2002; Moran and Covert, 2003; Rossel et al., 2013; Smith and Benner, 2005; Tarr et al., 2001; Tranvik et al., 2000; Vähätalo and Järvinen, 2007; Vähätalo et al., 2003, 2011; Xie et al., 2012), by reactivation of enzymes bound to humic substances, i.e., by photolytic cleavage of humic enzyme complexes (Wetzel et al., 1995), by depolymerization of biopolymers such as polysaccharides (Kovac et al., 1998), and by destruction of inhibitory substances such as acrylic acid (Bajt et al., 1997) and methyl mercury (Tai et al., 2014; Kim and Zoh, 2013). Thus, photochemical reactions involving DOM can be important in food web dynamics in seawater and freshwater, as pointed out in numerous reviews (Bauer and Bianchi, 2011; Bauer et al., 2013; Benner and Ziegler, 2000; Clark and Zika, 2000; Crosby, 1994; De Haan, 1992; Ehrhardt, 2005; Kieber, 2000; Kieber et al., 2003; Miller, 2000; Mopper and Kieber, 2000; Mopper and Kieber, 2002; Moran and Covert, 2003; Moran and Zepp, 2000; Mostofa et al., 2013; Osburn and Morris, 2003; Porcal et al., 2009; Sulzberger, 2000; Sulzberger and Durisch-Kaiser, 2009; Tranvik, 1992; Tranvik, 1998; Williams, 2000; Zepp, 1997; Zepp et al., 1998, 2003, 2007, 2011). In addition, photochemical reactions involving DOM can affect community structure (Abboudi et al., 2008; Langenheder et al., 2006; Lønborg et al., 2013; Pérez and Sommaruga, 2007; Piccini et al., 2009).

Experiments to evaluate the coupled photochemical-microbial degradation pathway are generally performed as follows (Mopper and Kieber, 2002): A natural water sample is sterile-filtered, transferred to irradiation vessels (usually quartz glass but, for a few studies, boro-silicate glass, UV transparent or semitransparent plastic), irradiated for several hours to days in natural sunlight or in a solar simulator, allowed to stand in the dark for ~24 h in order for the hydrogen peroxide and other reactive species to decay, inoculated with bacteria (usually isolated from the same sample or from water from a similar depth, but sometimes from other sources such as purified cultures) and incubated in the dark for several hours to several days or weeks. The net effect on microbial activity is then measured relative to controls (usually samples incubated in the dark with *non*irradiated filtered water).

In related experiments, specific DOM components are added to sterile-filtered water in order to determine the effect of irradiation on microbial utilization of the added component (Mopper and Kieber, 2002). These added components include extracted DOM and humic substances (Anesio et al., 2005; Goldstone et al., 2002; Judd et al., 2007; Miller and Moran, 1997; Nieto-Cid et al., 2006; Rossel et al., 2013), size-fractionated DOM (Kaiser and Sulzberger, 2004), algal extracts/exudates and higher plant leachates (aquatic and terrestrial) (Anesio et al., 1999; Tranvik and Kokalj, 1998; Wetzel et al., 1995), and specific biochemicals such as proteins, amino acids, pyruvate, and glucose (Hoikkala et al., 2009; Lignell et al., 2008; Obernosterer et al., 2001a,b). In these addition experiments, controls are usually unamended irradiated samples and amended samples stored in the dark.

In another variant, unfiltered or non-sterile filtered (e.g., ~0.7 μm-filtered) samples are irradiated with the microbial community present to determine the net effect (abiotic + biotic) of light-related processes, in particular direct UV inhibition (e.g., *in vivo* damage to DNA) and ambient photochemical reactions (e.g., production/destruction of substrates and production of mineral nutrients), on microbial activity (Lignell et al., 2008; Ortega-Retuerta et al., 2007; Remington et al., 2011; Rossel et al., 2013; Vähätalo and

Järvinen, 2007; Vähätalo et al., 2011). However, because of this experimental design, it is difficult or impossible to deconvolute the contribution of photochemical DOM degradation reactions to the overall change in microbial activity.

Various techniques have been employed to quantify the effect of DOM photochemistry on microbial activity. The most common approach has been to determine the change in bacterial carbon production based on the uptake of radio-labeled leucine or thymidine. Other approaches for assessing bacterial carbon production include measuring bacterial biomass, bacterial growth (cell number or density), and incorporation of specific radiolabeled substrates (e.g., pyruvate, glucose). Alternatively, respiration (e.g., O_2 consumption, DIC production, or DOC loss) has been used. However, using either carbon production or respiration alone leads to incomplete and hence erroneous conclusions regarding the effect of DOM photochemistry on microbial activity (i.e., respiration plus biomass production) because photochemistry induces changes in the bacterial growth efficiency (BGE), as discussed below. Experimental details of the approaches used to assess photochemically related changes in microbial activity are given in Mopper and Kieber (2002), who grouped studies according to whether positive, negative, or a mixed effect of photochemistry on microbial activity were reported. Results of more recent studies (and selected past studies) are given here.

Most irradiation studies reported positive effects on microbial activity (Amaral et al., 2013; Grzybowski, 2002a; Lignell et al., 2008; McCallister et al., 2005; Miller et al., 2002; Nieto-Cid et al., 2006; Remington et al., 2011; Smith and Benner, 2005; Vähätalo et al., 2003) and several studies showed that photochemically produced substrates can be transferred up the food chain, thus potentially enhancing productivity in higher organisms (Daniel et al., 2006; De Lange et al., 2003; Vähätalo and Järvinen, 2007; Vähätalo et al., 2011). Studies with positive effects on microbial activity can be separated into two types based on their experimental design. In the most common design, marine or fresh waters were irradiated and the effect on microbial activity (usually bacterial production and/or respiration) was measured (Amaral et al., 2013; Hoikkala et al., 2009; Lignell et al., 2008). In the second type of experiment, the positive effect on microbial activity was measured or inferred from the photochemical formation of specific substrates, e.g., organic acids and polyols (Gonsier et al., 2014) and, in some cases, their uptake using radiolabeled substrates (Kieber et al., 1989; Remington et al., 2011).

Although the majority of past studies yielded positive effects, several studies have shown that photodegradation of DOM can also result in a negative effect, a mixed effect (negative and positive), or no effect on microbial activity (Abboudi et al., 2008; Anesio et al., 2005; Biddanda and Cotner, 2003; Calza et al., 2008; Hoikkala et al., 2009; Judd et al., 2007; Kaiser and Sulzberger, 2004; Kramer and Herndl, 2004; Obernosterer and Benner, 2004; Ortega-Retuerta et al., 2007; Pérez and Sommaruga, 2007; Pringault et al., 2009; Reader and Miller, 2014; Tedetti et al., 2009); for earlier studies, see Mopper and Kieber (2002). A negative effect can be due to a number of factors. In particular, photochemistry can destroy (as well as form) biological substrates. Abiotic cross-linking, humification, and polymerization of labile biomolecules, such as proteins, unsaturated lipids, polysaccharides, fresh algal extracts or exudates, and fresh macrophyte leachates, can occur in the dark (Brophy and Carlson, 1989; Keil and Kirchman, 1994; Thomas and Lara, 1995; Tranvik, 1993), but are enhanced by irradiation (Kieber et al., 1997a; Obernosterer et al., 2001a; Scully et al., 2004) and high concentrations of DOM, as encountered in humic-rich fresh waters (Vähätalo et al., 2000), coastal marshes, and mangrove stands (Scully et al., 2004). Benner and Biddanda (1998) hypothesized that, in the ocean, recently produced algal-derived DOM (which mainly occurs in surface waters) becomes less available, while humic-rich DOM (dominant in upwelled deep waters) becomes more available

to bacteria upon irradiation. Therefore, the net effect of photochemistry on microbial growth may depend on the relative proportions of these two DOM types in the irradiated sample. Results of subsequent field studies are consistent with this hypothesis (Benner and Ziegler, 2000; Bertilsson, 1999; Herndl et al., 2000; Obernosterer et al., 1999a). Compelling evidence in support of this hypothesis comes from a study of over thirty lakes, in which Tranvik et al. (2000) and Tranvik and Bertilsson (2001) successfully modeled the degree of photochemical stimulation or inhibition to microbial activity on the basis of the relative proportions of fresh algal-derived DOM (based on Chl a) versus older humic DOM.

Interactions of photochemically produced reactive oxygen species (ROS) with fresh algal-derived DOM (based on Chl a) versus older humic DOM may be partly responsible for the stimulation or inhibition of microbial activity observed in the above studies. Some ROS, in particular the OH radical by reaction with DOM, can indirectly stimulate microbial activity by the production of biologically labile substrates, e.g., LMW organic acids, (Goldstone et al., 2002; Pullin et al., 2004; Tedetti et al., 2009). On the other hand, Scully et al. (2003a,b) speculated that ROS mainly destroy biologically labile photoproducts and thus compete with their microbial uptake. These results suggest that biological and photochemical processes compete in the mineralization of some common fraction of DOC (Obernosterer and Benner, 2004; Obernosterer et al., 1999a; Tedetti et al., 2009). However, Amado et al. (2006) concluded that the two DOC mineralization pathways (photochemical and microbial) in Amazonian waters were essentially independent and complementary, with allochthonous DOC (i.e., tDOM) degraded mainly by photochemical mineralization and autochthonous DOC by microbial processes. Clearly, the effect of photochemistry and, in particular ROS, on bioavailable substrates (photochemically and non-photochemically formed) needs to be further examined.

While changes in the relative importance of photoproduction versus photodestruction of substrates appear to be responsible for the contrasting results in some studies, other factors may also have been important. Factors resulting in a negative or mixed effect include photochemical production of inhibitory substances such as hydrogen peroxide (Angel et al., 1999; Baltar et al., 2013; Farjalla et al., 2001; Gjessing and Källqvist, 1991; Kaiser and Sulzberger, 2004; Leunert et al., 2014; Lund and Hongve, 1994; Morris et al., 2011; Scully et al., 2003a; Tranvik and Kokalj, 1998; Weinbauer and Suttle, 1999), release of toxic metals (e.g., Pb, Cu, Ni, Cd, and Hg) from DOM complexes (Haverstock et al., 2012; Tonietto et al., 2011; Winch and Lean, 2005), photolysis of DOM to form substances that are both biologically and photochemically refractory (Kieber, 2000; Stubbins et al., 2010), deactivation of enzymes (Scully et al., 2003b; Vähätalo et al., 2003), changes in the BGE (Abboudi et al., 2008; McCallister et al., 2005; Mopper and Kieber, 2002; Pullin et al., 2004; Smith and Benner, 2005), changes in microbial populations (i.e., community structure) in response to photoproduced substrates or toxic substances (Abboudi et al., 2008; Calza et al., 2008; Lønborg et al., 2013; Piccini et al., 2009), and prior photochemical history, i.e., photon dose-related bleaching (Reader and Miller, 2014). In addition to the above factors, a negative or mixed effect on biological activity can result from reactions of *biologically* produced ROS, e.g., H_2O_2 (Palenik and Morel, 1990; Diaz et al., 2013) with photochemically produced reduced metals, e.g., Fe(II), Mn(II), and Cu(I) (Barbeau, 2006; Brinkmann et al., 2003; Sunda and Huntsman, 1994) or other photoproducts that are reactive towards ROS (Garg et al., 2013; Rose and Waite, 2003). In particular, the reaction between photoproduced Fe(II) and biologically produced H_2O_2 produces the highly reactive OH radical via the photo-Fenton reaction (White et al., 2003; Zepp et al., 1992), which can destroy biologically labile substrates and inactivate membrane-bound enzymes (Jagger, 1985; Mee, 1987).

In the case where studies found no discernible effect on microbial activity, several factors may have been involved. The photoproduction of labile substrates and mineral nutrients from humic substances may have been offset by the photodestruction of algal-produced DOM (Benner and Biddanda, 1998) or balanced by the photoproduction of toxic substances (Calza et al., 2008; Diamond, 2003; Haverstock et al., 2012; Ortega-Retuerta et al., 2007; Tonietto et al., 2011; Winch and Lean, 2005). Alternatively, carbon limitation may have been important. For example, based on their results in the Amazon River basin, Amon and Benner (1996) observed that photoproduction of carbon substrates had a much greater impact on bacteria growing under carbon-limited conditions compared to carbon replete conditions. Similar results were obtained in the Baltic Sea (Hoikkala et al., 2009) and black water Amazonian systems (Amaral et al., 2013).

Finally, experimental design may have played a major role in determining whether a positive, negative, no, or mixed effect was observed. For example, results appear to be strongly influenced by the incubation time, i.e., the time that the microbes were allowed to grow after inoculation into the irradiated water sample (Biddanda and Cotner, 2003). Studies with short, post-irradiation dark incubation times (e.g., hours to days) sometimes exhibited a negative or no effect on microbial activity (Biddanda and Cotner, 2003; Kaiser and Sulzberger, 2004; Kramer and Herndl, 2004; Ortega-Retuerta et al., 2007; Pérez and Sommaruga, 2007) and enzymatic activity (Vähätalo et al., 2003). In contrast, studies with long-term incubation times (e.g., days to weeks or longer) often showed an initial lag period (with no or negative effect) followed by a positive effect on microbial activity (Anesio et al., 2005; Biddanda and Cotner, 2003; Judd et al., 2007; McCallister et al., 2005; Obernosterer and Benner, 2004; Pullin et al., 2004; Smith and Benner, 2005) or enzymatic activity (Vähätalo et al., 2003). The initial lag (or negative effect) may have been due to the photoproduction of inhibitory substances such as ROS or toxins (Anesio et al., 1999, 2005; Angel et al., 1999; Baltar et al., 2013; Calza et al., 2008; Diamond, 2003; Farjalla et al., 2001; Goldstone et al., 2002; Kaiser and Sulzberger, 2004; Leunert et al., 2014; Lund and Hongve, 1994; Morris et al., 2011; Scully et al., 2003b; Tranvik and Kokalj, 1998; Weinbauer and Suttle, 1999) that decayed over time (Anesio et al., 2005; Goldstone et al., 2002), or induced a change in the microbial community structure to species less affected by photoproduced toxins and/or better adapted to utilizing photoproduced products (Abboudi et al., 2008; Judd et al., 2007; Langenheder et al., 2006; Pérez and Sommaruga, 2007; Piccini et al., 2009; Pullin et al., 2004; Vähätalo et al., 2003). Alternatively, the lag period may be the time needed for microbes to biodegrade photoaltered high molecular weight (HMW) DOM to LMW membrane-transportable substrates (Grzybowski, 2002a).

The method by which microbial activity (or response) is quantified can also affect the conclusions regarding the effect of photochemistry on microbial activity. In particular, changes in the BGE may be more important than currently recognized (Abboudi et al., 2008; Amaral et al., 2013; del Giorgio and Cole, 2000; Jahnke and Craven, 1995; Mopper and Kieber, 2002; Pullin et al., 2004; Williams, 2000; see Chapter 3). The BGE is usually expressed as the fraction of carbon incorporated relative to total uptake (i.e., incorporated/ (incorporated + respired)). In many studies, the impact of photochemistry on bacterial activity (upon incubation with irradiated DOM) is evaluated by measuring only the change in bacterial carbon production, usually by incorporation of radiolabeled thymidine and/or leucine, or by direct counting of bacterial abundance and converting to biomass (Mopper and Kieber, 2002). However, bacterial production gives an incomplete picture of the amount of total carbon taken up because it only accounts for the fraction that is incorporated into biomass, which is usually a minor fraction of the total photochemically derived carbon taken up. Most of the assimilated

carbon is usually respired to CO_2. The percent respired varies greatly among different bacteria, but in the open ocean it is generally in the range of ~80 to >95% of the total carbon taken up and in coastal waters about 60 to >75% (del Giorgio and Cole, 2000). Miller et al. (2002) reported that bacterial carbon biomass accumulation accounted for <4% of total C utilization in irradiated coastal marine samples. Smith and Benner (2005) speculated that photodegradation of tDOM had an overall negative impact on food web dynamics in coastal marine waters due to the decrease in the BGE.

Factors controlling the BGE in marine bacterial assemblages are not well understood but appear to be mainly related to environmental and physiological conditions, substrate quality, and nutrient availability (del Giorgio and Cole, 2000; Kragh et al., 2008; Smith and Benner, 2005; Williams, 2000; also see Chapter 3). For example, photodegradation of DOM produces LMW substrates, such as organic acids, that are more oxidized than the precursor molecules (Pullin et al., 2004). Highly oxidized substrates are generally utilized much less efficiently by bacteria than more reduced, energy-rich substrates (del Giorgio and Cole, 2000). This factor alone may cause a significant decrease in the BGE, even if no shift in the actively growing bacterial population occurs. However, shifts in the BGE can also occur as a result of changes in the bacterial population because the latter is a complex assemblage of microorganisms in varying growth states and with different affinities for substrates. In fact, photochemical changes in the concentrations and types of substrates cause a shift in the actively growing bacterial population (Abboudi et al., 2008; Calza et al., 2008; Lønborg et al., 2013; Piccini et al., 2009), with a concomitant change in the net BGE (del Giorgio and Cole, 2000; Pullin et al., 2004). It is likely that both factors (i.e., substrate oxidation state and shift in assemblage composition) were important in yielding the wide range of responses in many published studies. However, since only bacterial production/biomass was measured in most studies, it is not possible to conclude if the *net* effect of DOM photodegradation was stimulatory or inhibitory to bacterial growth; the BGE was usually not monitored or simply assumed to be constant (e.g., Aarnos et al., 2012). In studies where the BGE was reported, it generally shifted to lower values in irradiated samples due to increased respiration relative to incorporation (Amado et al., 2006; Anesio et al., 2000; Farjalla et al., 2001; McCallister et al., 2005; Mopper and Kieber, 2002; Pullin et al., 2004; Smith and Benner, 2005; Vähätalo and Salonen, 1996). In contrast, Kragh et al. (2008) observed increases in the BGE when phosphorus was added to irradiated freshwater DOM extracts and concluded that the availability of phosphorus was important for the quantitative transfer of carbon in microbial food webs. Given the overall importance of respiration (del Giorgio and Cole, 2000; Williams, 2000), it appears to be inappropriate to use bacterial production measurements alone to assess the effect of DOM photodegradation on bacterial activity and certainly upon carbon cycling. Thus, conclusions from many past irradiation studies regarding the net photodestruction of substrates (or photochemical formation of biologically recalcitrant humic substances from substrates) need to be reevaluated.

In addition to the above factors, photochemical (i.e., abiotic) oxygen consumption (Andrews et al., 2000) can compete with microbial oxygen consumption, which is often used to assess bacterial respiration in the light (Pringault et al., 2009; Reader and Miller, 2014). This competition can result in an overestimation of respiration and an underestimation of phytoplankton photosynthetic carbon fixation (Kitidis et al., 2014), as the contribution from photochemical oxygen consumption has not been taken into account in past studies. Furthermore, Pringault et al. (2009) found that bacterial respiration in the light was enhanced relative to bacterial biomass production, resulting in a decrease in BGE during irradiation. Thus, BGEs

may have been significantly underestimated in studies examining the effect of photochemistry on bacterial activity in natural waters (Lignell et al., 2008; Ortega-Retuerta et al., 2007; Reitner et al., 1997; Remington et al., 2011; Vähätalo and Järvinen, 2007; Vähätalo et al., 2011).

A further complication with assessing photochemically induced shifts in the BGE is a possible interference with radiolabeled bacterial production measurements. This interference was revealed in glucose addition experiments that addressed microbial carbon limitation (Mopper and Kieber, 2002). Surprisingly, it was observed that ^{14}C-labeled glucose additions often caused a significant decrease in radiolabeled leucine and thymidine incorporation. Varying effects of glucose additions on the growth of bacteria due to varying degrees of carbon limitation have been observed (Carlson et al., 1999; Hoikkala et al., 2009; Kirchman, 1990; Lignell et al., 2008; Pomeroy et al., 1995). In the case of decreased radiolabeled leucine and thymidine incorporation, it is possible that the added glucose stimulated the internal production of leucine and thymidine thereby diluting the radiolabel, which would result in an apparent decrease in the uptake of the radiolabel (Falkowski, personal communication). A similar effect was found for pyruvate in Antarctic waters (Mopper and Kieber, unpublished results). Since pyruvate is a photochemically produced substrate (Kieber et al., 1989), the question arises whether the apparent "inhibition" effects reported in several studies (i.e., decreased microbial production after irradiation) may be partly due to the photochemical production of substrates that, after being taken up, stimulate the internal production of leucine or thymidine (and thus diluting the ^{14}C label). Like the oxygen consumption effect discussed above, the BGE would also be underestimated. In addition to interference effects, the added radiolabel can be partially respired. Hill et al. (2013) found that ^{3}H-leucine additions overestimated ambient leucine uptake at low uptake rates and underestimated uptake at high uptake rates, which, in turn, would affect BGE estimates. Clearly, these questions, as well as those involving the complex interactions between biological systems and abiotic photochemical reactions discussed above, need to be addressed in future studies designed to evaluate or model the impact of photochemistry on aquatic food web dynamics (Reader and Miller, 2014).

2 Photochemical DIC Formation and Oxygen Consumption

a DIC PHOTOPRODUCTION

While it is well known that photolysis of DOM yields LMW organic products, many of which are biologically labile (see Section II.A.1 and Kieber, 2000), the major carbon photoproducts are inorganic species, in particular CO (Section II.A.3) and DIC. Photochemical DIC formation may strongly impact carbon cycling in seawater and other natural waters (Aarnos et al., 2012; Amado et al., 2006; Anesio and Granéli, 2003, 2004; Bauer et al., 2013; Clark et al., 2004; Granéli et al., 1998; Johannessen and Miller, 2001; Johannessen et al., 2007; Kieber et al., 1999a, 2001; Lindell et al., 2000; Ma and Green, 2004; Miller and Moran, 1997; Miller and Zepp, 1995; Minor et al., 2006; Mopper and Kieber, 2000; Osburn et al., 2001; Pers et al., 2001; Porcal et al., 2013; Vähätalo et al., 2000; Wang et al., 2004; Xie et al., 2004, 2012). In these studies, DIC photoproduction was usually measured either by the increase in DIC concentration or by the loss of DOC upon irradiation. For marine studies, the former approach has generally been used because of the low precision of current DOC analyzers (typically 1-3%) and the relatively low DOC concentration in seawater. However, measuring DIC photoproduction in seawater is still challenging due to the analytical difficulties involved in measuring a small change in DIC concentration (sub-µM) over a large DIC background (~2 mM). Consequently, samples are typically acidified and purged to

remove DIC (Miller and Zepp, 1995) and the pH adjusted back to ~8 prior to irradiation (Johannessen and Miller, 2001) to minimize pH effects on the rates and yields (Miller and Zepp, 1995). However, due the sample manipulation involved, the potential for DOM and DIC/CO_2 contamination is high (White et al., 2010). In addition, the potential of the major pH changes employed to alter DOM optical (i.e., absorbance) and photochemical properties have been inadequately addressed. Even when DIC stripping is employed, analyses are challenging in the low DOM open ocean. Therefore, most marine studies have been conducted in DOM-rich coastal waters or estuarine systems (Aarnos et al., 2012; Guo et al., 2012; White et al., 2010; Xie et al., 2004, 2012) in order to obtain sufficiently large DIC production signals against the somewhat variable blank. Recently, the problems related to acidification, purging, and pH readjustment have been overcome using the pool isotope exchange (PIE) method. In this approach, most of the sample's $DI^{12}C$ is replaced with $DI^{13}C$ at the natural pH and temperature so that $^{12}CO_2$ from DOM photooxidation during the irradiation elevates the $^{12}CO_2/^{13}CO_2$ ratios, which are then measured by GC-IRMS (Wang et al., 2009). This approach was intercalibrated with the "traditional" acidification, purging, and pH readjustment method with excellent agreement (Wang et al., 2009; White et al., 2008). Alternatively, photochemical production of DIC in DOM-rich waters (fresh and coastal marine) can be estimated from careful mass balance measurements of pCO_2 sources and sinks (Clark et al., 2004).

Only a few studies have attempted to measure open ocean rates of DIC photoproduction (Johannessen and Miller, 2001; Johannessen et al., 2007; Wang et al., 2009; White et al., 2008). Interestingly, open ocean waters have been reported to have higher DIC photoproduction AQYs than coastal, inshore, and freshwaters (Johannessen and Miller, 2001). This difference was hypothesized to be due to autochthonous marine chromophoric DOM (CDOM) having higher DIC AYQs compared to terrestrial CDOM and/or to preferential photobleaching of less productive, terrestrially derived CDOM in the open ocean. In support of this hypothesis, Johannessen and Miller (2001) found that the DIC AQY for a coastal water sample increased upon irradiation, which is contrary to the photoproduction of other species, such as CO (Stubbins, 2002; Zhang et al., 2006). Interestingly, analogous to the open ocean DIC AQYs, Peterson et al. (2012) found that open Lake Superior water was more efficient at producing singlet oxygen than the CDOM-rich riverine and river-impacted waters and speculated that autochthonously produced open lake CDOM had a higher singlet oxygen production efficiency (i.e., AQYs) than terrestrial CDOM or that the AQYs of open-lake DOM increased by photochemically induced humification (i.e., cross-linking). An alternative explanation for the increased DIC (and singlet oxygen) AQYs in the above studies is that photodegradation (bleaching) selectively removed less (and non-) photoproductive chromophores while not significantly decreasing overall photoproduction rates of DIC (and singlet oxygen), which would result in higher AQYs. This explanation is supported by a number of recent studies (Mostafa and Rosario-Ortiz, 2013; Zhang et al., 2012) and may be in part due to LMW CDOM having higher photoproduction efficiencies than HMW CDOM (Dalrymple et al., 2010; Golanoski et al., 2012; Halladja et al., 2007; Richard et al., 2004; Sharpless, 2012), the latter being selectively photodegraded.

Apparent quantum yields for DIC photoproduction, which are critical for water column photochemical modeling to assess its impact on carbon cycling (Aarnos et al., 2012; Bélanger et al., 2006), vary nearly two orders of magnitude. For example, at 350 nm, AQYs for DIC photoproduction range from ~6 × 10^{-5} mol DIC (mol quanta)$^{-1}$ for inshore waters (Johannessen et al., 2007) to ~2 × 10^{-3} mol DIC (mol quanta)$^{-1}$ for DOM-rich coastal river

water (Gao and Zepp, 1998; Osburn et al., 2009) and ~5 × 10^{-3} mol DIC (mol quanta)$^{-1}$ on the Mackenzie Shelf, Beaufort Sea (Osburn et al., 2009). Within this wide range, there is no apparent relation of AQYs to DOM type (e.g., fresh vs. inshore vs. coastal vs. open ocean) or to CDOM absorbance, temperature, or salinity (Reader and Miller, 2012; White et al., 2010). For example, in a salinity transect in the Delaware Estuary (from salinity 3 to 21), AQYs at 350 nm were remarkably constant, at about 1 × 10^{-4} mol DIC (mol quanta)$^{-1}$ (±~10% RSD), while corresponding CO AQYs decreased by a factor of about 1.6 with increasing salinity (from ~1.6 × 10^{-5} mol CO (mol quanta)$^{-1}$ to ~1.0 × 10^{-5} mol CO (mol quanta)$^{-1}$), suggesting that DIC and CO are formed by different photochemical mechanisms (White et al., 2010). A similar decrease in CO AQYs down a temperate estuary was observed by Stubbins et al. (2011).

The AQYs for DIC photoproduction are about 4-30 (or more) times greater than for CO photoproduction (White et al., 2010), with a mean of about 23 for coastal waters (Reader and Miller, 2012). This variability appears to be due to differences in DOM/CDOM type, e.g., terrestrial vs. microbial, prior photobleaching history and irradiation wavelength (Boyle et al., 2009; Del Vecchio and Blough, 2004; Golanoski et al., 2012; Ma et al., 2010; Reader and Miller, 2012; Stubbins et al., 2011; White et al., 2010; see discussion of CDOM sources in Chapter 10). Similar variability was also observed for DIC photoproduction by broadband irradiations (Reader and Miller, 2012; White et al., 2010). Despite this variability, using this approximate relation between DIC and CO photoproduction, coupled with the relative ease in measuring the latter, global DIC photoproduction has been estimated from CO photoproduction (Mopper and Kieber, 2000; Stubbins et al., 2006a; Wang et al., 2009); see Section V for further discussion. These estimates are proposed to exceed the rate of input of riverine DOM to the sea,

~0.25 Pg C year^{-1} (Wang et al., 2009) and, therefore, should be incorporated into oceanic CO_2 cycle models.

b DIC PHOTOPRODUCTION AND PHOTOCHEMICAL OXYGEN CONSUMPTION, AND MECHANISMS OF DIC PHOTOFORMATION

In addition to its role in carbon cycling, DIC photoproduction is likely responsible for a major fraction of abiotic oxygen consumption in irradiated surface waters, e.g., in DOM-rich fresh waters (Andrews et al., 2000; Lindell and Rai, 1994) and open ocean waters (Obernosterer et al., 2001b). Rates of photochemical oxygen consumption can equal or exceed bacterial oxygen consumption rates (Amon and Benner, 1996; Obernosterer et al., 2001b, 2005; Reitner et al., 1997). In addition, abiotic (photochemical) oxygen consumption may be accompanied by oxygen isotope fractionation (Chomicki and Schiff, 2008) and may result in an underestimation of primary production based on the classic oxygen method (Kitidis et al., 2014).

Photochemical pathways of oxygen consumption and how they are linked to DIC formation are not well understood (Andrews et al., 2000; Pullin et al., 2004). DIC photoproduction from DOM is thought to occur via two main decarboxylation routes: one by reaction with molecular oxygen and/or ROS and the other by oxygen-independent pathways (e.g., direct decarboxylation). In support of the oxygen-dependent pathway, Miles and Brezonik (1981) proposed a decarboxylation mechanism that involved the consumption of molecular oxygen at a rate of one mole of molecular oxygen consumed per two moles of CO_2 produced (i.e., O_2 consumption to CO_2 production molar ratio of 0.5), corresponding to the stoichiometry: $RCOOH + \frac{1}{2}O_2 \rightarrow ROH + CO_2$. However, this simple mechanism requires that all photochemical oxygen consumption be accounted for by DIC photoproduction, i.e., oxygen is not consumed by any other photochemical oxidative processes.

In contrast, other studies (Amon and Benner, 1996; Andrews et al., 2000; Gao and Zepp, 1998; Lindell and Rai, 1994; Lindell et al., 2000; Xie et al., 2004) found a molar O_2 consumption to CO_2 evolution ratio close to 1 (range 0.8-1.2) for natural waters, while (Estapa and Mayer, 2010) obtained values around 1.3 for irradiated suspended marine sediments. This wide range (0.5-1.3) is probably due to two factors: (1) differences in the redox state of the original DOM, i.e., if the DOM was initially highly oxidized, then the resultant O_2 consumption to CO_2 production ratio would be low and (2) the degree to which O_2 is consumed by other photochemical processes and microbial contamination. For example, >50% of all photochemically consumed oxygen appears to be required for the production of H_2O_2 via dismutation of superoxide (Andrews et al., 2000; Blough, 1997; Blough and Caron, 1995; Blough and Zepp, 1995; Petasne and Zika, 1987; Zhang et al., 2012). Incorporation of molecular oxygen into DOM upon irradiation was demonstrated using $^{18}O_2$ (Cory et al., 2010), while increased oxygenation of DOM was demonstrated by mass spectroscopy (Gonsior et al., 2009; Kujawinski et al., 2004) and nuclear magnetic resonance (NMR) spectroscopy (Schmitt-Kopplin et al., 1998). However, in the latter studies, the source of the oxygen that was incorporated into DOM was not determined. Although it is usually assumed that molecular oxygen is the main source (Andrews et al., 2000; Xie et al., 2004), oxygen may have also been derived from water via reactions involving excited state quinones (Gan et al., 2008; Görner, 2003; Pochon et al., 2002). Mechanisms of DOM photooxygenation and sources of oxygen during DOM photooxidation, i.e., O_2 versus H_2O, have not been systematically examined.

The above studies suggest that molecular oxygen is required for DIC photoproduction, so its depletion may limit DIC photoproduction in natural waters (Lindell and Rai, 1994; Lindell et al., 2000; Xie et al., 2004). In support of the oxygen requirement, several studies observed increases in DIC photoproduction when pure oxygen was substituted for air (Gao and Zepp, 1998; Wang et al., 2009; Xie et al., 2004). However, the presence of dissolved oxygen is not absolutely necessary for DIC photoproduction. Wang et al. (2009) found that DIC photoproduction for Suwannee River humic acid decreased by only 36% when pure nitrogen was substituted for air, while Gao and Zepp (1998) observed an ~60% drop for sterile-filtered DOM-rich coastal river water (Satilla River, GA) when nitrogen was substituted for air. Schmitt-Kopplin et al. (1998) observed an ~60% decrease in DOM weight loss rate (i.e., DIC and CO photoproduction rate) for irradiated sterile-filtered solutions of extracted soil humic substances when nitrogen was substituted for air.

A variety of oxygen-independent and -dependent decarboxylation pathways have been identified using model compounds (Budac and Wan, 1992). Some potentially important photochemical mechanisms include: cleavage of a carboxyl group attached to an aromatic ring via an intramolecular or intermolecular charge transfer reaction (Budac and Wan, 1992); ligand-to-metal charge transfer (LMCT) excitation of metal complexes (or chelates) followed by cleavage of the carboxyl group (Brinkmann et al., 2003; Langford et al., 1973); photolysis of aromatic rings followed by decarboxylation of ring cleavage products (Brinkmann et al., 2003; Chen et al., 1978; Vähatalo et al., 1999); and photolysis of LMW α-keto acids (Bockman et al., 1996) such as pyruvate and glyoxylate (Guzmán et al., 2007; Kieber, 1988; Klementova and Wagnerova, 1990) and oxalate (Bertilsson and Tranvik, 1998, 2000), probably via photosensitized oxidation (Klementova and Wagnerova, 1990). In irradiated oxygen-free aqueous solutions of LMW organic acids, the radical O=C–OH is observed, which undergoes hydrogen atom abstraction to yield CO_2 (Bockman et al., 1996). In natural waters, the HCO_2 radical rapidly deprotonates yielding the carbon dioxide radical anion, which is a powerful one-electron reductant. Photoproduced radical cations of benzophenone carboxylic acids (Budac and Wan, 1992), which are typical fulvic acid proxy compounds (Canonica et al., 2000; Grebel et al., 2011; Stubbins et al., 2008), oxidize in aqueous solution by reaction with the nucleophilic OH^- ion followed by α-scission (with generation of aryl radicals) and CO_2 elimination (Säuberlich et al., 1996). While all the above oxygen-independent pathways are feasible, as shown using model compounds, their relevance to CO_2 photoproduction from natural water DOM have not been investigated to date.

If DIC photoproduction from DOM occurs mainly by direct decarboxylation, one might expect DIC photoproduction to be correlated with DOM carboxyl content. However, in a study of DIC photoproduction by twelve humic substances from diverse terrestrial and marine environments added to fresh water, Anesio et al. (2005) found that the only good correlation ($r^2 = 0.81$) was with aromaticity (as determined by solid state NMR) and not with total carboxyl carbon content. The lack of a significant correlation with total carboxyl carbon is likely due to regeneration of carboxyl groups during photodecarboxylation (Xie et al., 2004) and/or to the varying composition of molecules to which carboxyls are attached. Thus, the correlation of DIC photoproduction with aromaticity suggests that carboxyl groups attached to aromatic rings are preferentially cleaved off the ring to form CO_2, consistent with studies of model compounds (Budac and Wan, 1992). This pathway is also consistent with IR spectroscopic results for fulvic acids that showed the carboxyl signal was preferentially lost during UV irradiation of aqueous solutions, while the aliphatic signal increased and the fulvic acids became more oxidized and less aromatic (Chen et al., 1978; Helms et al., 2013a,b). Similar results were obtained by solid-phase ^{13}C and two-dimensional NMR analyses of photolyzed aqueous fulvic and humic substance solutions (Kulovaara et al., 1996; Schmitt-Kopplin et al., 1998).

ROS, such as the OH radical and superoxide radical anion (O_2^-), may be involved in DOM photooxygenation and DOC mineralization

(Cory et al., 2010; Mopper and Zhou, 1990; Schmitt-Kopplin et al., 1998; Scully et al., 2003a,b; Zafiriou et al., 1990). For example, the OH radical has been implicated in decarboxylation of amino acids (Steffen et al., 1991) and organic acids (Zafiriou et al., 1990). Goldstone et al. (2002) showed that the OH radical (generated by radiolysis) reacted with humic substances (Suwannee River fulvic and humic acids) to produce DIC with an efficiency of ~0.3 mol CO_2 (mol of OH radical)$^{-1}$. This ratio is close to the expected yield of DIC per mol OH of ~0.25 (i.e., four OH radicals are needed to produce one CO_2 molecule from DOM) assuming that all OH radicals produced were consumed during the production of DIC (Goldstone et al., 2002). The predicted value of 0.25 is based on the average oxidation state of carbon in their humic substance samples of approximately 0 (IHSS, 2014), while that of CO_2 is 4. In contrast, Pullin et al. (2004) obtained much higher values (~1.1) using sterile-filtered whole DOM-rich river and estuarine water samples (i.e., not DOM isolates), suggesting that other oxidants (in addition to the OH radical) oxidized DOM to CO_2 or that OH radical reactions with DOM were selective for more oxidized moieties (i.e., initial DOM oxidation state >0). In seawater, the OH radical reacts predominantly (>95%) with bromide (to form the Br_2^- radical anion); consequently, steady-state OH radical concentrations are typically very low, e.g., 10^{-16}-10^{-18} M, and thus its reaction with DOM is likely negligible (Mopper and Zhou, 1990), except in high iron, nitrate, and/or nitrite containing coastal and estuarine waters (Miller et al., 2013; White et al., 2003).

In contrast to the OH radical, singlet oxygen, hydrogen peroxide, and organic peroxides are insufficiently reactive to be important DOM oxidants (Andrews et al., 2000; Cooper et al., 1989; Cory et al., 2010), and thus probably do not significantly contribute to DIC photoproduction in natural waters. On the other hand, superoxide is relatively reactive (Cooper et al., 1989), with second-order rate constants of 10^7-10^9 M^{-1} s^{-1} for

reaction with many aromatic compounds, such as quinones (Bielski et al., 1985). Although ~60-75% of photoproduced superoxide in coastal seawater is consumed by hydrogen peroxide production (Petasne and Zika, 1987), the fate of the unaccounted 25-40% is unknown. Most of this unknown fraction is likely lost mainly via recombination and metal ion reactions (Garg et al., 2007; Rose and Waite, 2005, 2006; Zhang et al., 2012) and to a lesser degree by reaction with DOM to form stable organoperoxides (Garg et al., 2007). Thus, the role of superoxide in directly oxidizing DOM oxidation and producing DIC is unknown, but likely minor. Highly energetic, excited state ketone triplets $^3(R_1R_2\text{-C=O})^*$ are also known to oxidize DOM (Canonica et al., 2000; Cottrell et al., 2013; Golanoski et al., 2012), but like superoxide, their role in oxidizing DOM and producing DIC in natural waters is not known.

It is clear from the above studies that the major pathways of DIC photoproduction in seawater and other natural waters involve direct photolysis reactions and indirect photooxidation by ROS, with direct photolysis probably being more important. Direct photochemical reactions would result in electron transfer to O_2 and/or formation of peroxy (and cation) radical intermediates (Garg et al., 2007; Zhang et al., 2012), which would subsequently photolyze to yield CO_2 (Blough, personal communication). Both photochemical pathways result in oxygen consumption. Interestingly, de Bruyn et al. (2011) proposed a similar combination of mechanisms to explain the photoproduction of LMW carbonyl compounds (e.g., formaldehyde, acetaldehyde, and acetone) in seawater. Clearly, the nature and relative importance of these pathways to DOM photooxidation and oxygen consumption in seawater and other natural waters are not well understood and warrant systematic study.

c CARBON MONOXIDE PHOTOPRODUCTION AND TRANSFER TO THE ATMOSPHERE

This section will review studies focused on the production of CO from DOM photochemistry. The production of CO from marine particle and sea ice photochemistry will be reviewed in Sections IV.A and B, respectively.

Photochemistry was first hypothesized as a source of aquatic CO based on strong diurnal cycles in sea-surface concentrations, peaking in the mid-afternoon and reaching a minimum around dawn (Swinnerton et al., 1970; Figure 8.1a). Subsequent work has revealed that the diurnal CO cycle is driven by its production via sunlight photolysis of DOM (Conrad et al., 1982; Redden, 1982; Stubbins et al., 2006a; Valentine and Zepp, 1993; Wilson et al., 1970; Zafiriou et al., 2003; Zuo and Jones, 1995) and removal via microbial oxidation (Conrad and Seiler, 1980, 1982; Zafiriou et al., 2003), air-sea gas exchange (Bates et al., 1995; Conrad et al., 1982), and vertical mixing (Gnanadesikan, 1996; Johnson and Bates, 1996; Kettle, 2005; Najjar et al., 1995). Due to the strong photochemical source and microbial sink of CO, concentrations of CO are maximal in surface waters, leading to supersaturation (Bates et al., 1995; Stubbins et al., 2006b), and minimal at depth (Figure 8.1b).

As a major product of DOC photomineralization, CO photoproduction is a significant term in the global carbon cycle. In the open ocean, CO AQYs and CDOM levels are relatively constant, allowing CO photoproduction to be reasonably well constrained (30 to 90 Tg CO year⁻¹; Fichot and Miller, 2010; Stubbins et al., 2006a; Zafiriou et al., 2003). However, CO AQYs are higher in freshwaters (Gao and Zepp, 1998; Kettle, 1994; Valentine and Zepp, 1993) than in seawater (Zafiriou et al., 2003; Ziolkowski and Miller, 2007). Thus, CO AQYs decrease with increasing salinity across the freshwater-marine mixing zone (White et al., 2010; Zhang et al., 2006). The decrease in CO AQYs within estuaries exceeds that predicted by conservative mixing of high AQY river water and low AQY seawater (Stubbins et al., 2011; White et al., 2010). Photodegradation is known to decrease CO AQYs (Stubbins, 2002; Zhang et al., 2006). Therefore, in-estuary photodegradation may be a major process driving rapid decreases in CO AQYs across an estuarine salinity gradient.

The CO photoproduction efficiency from simple CDOM analogues (e.g., substituted phenolic compounds) decreases when electron-donating groups (e.g., alkyl, methoxyl, and hydroxyl) are substituted with electron-withdrawing groups (e.g., carboxyl and carbonyl) around the aromatic ring (Stubbins et al., 2008). However, it is not known whether similar changes in the

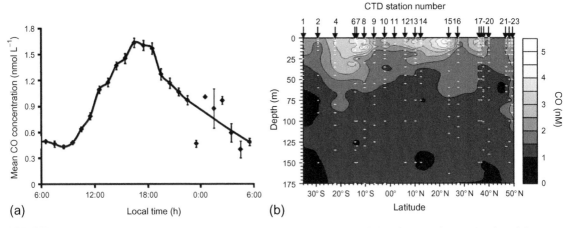

FIGURE 8.1 (a) Mean hourly surface-water CO concentration versus time of day during Atlantic Meridional Cruise 10 (AMT-10), which transited from 35°S:55°W to 54°N:0°W during April and May 2000 (adapted from Stubbins et al., 2006). (b) Dissolved CO concentrations in the Atlantic Ocean during AMT-10. White points represent sampling depths. *Adapted from Stubbins (2002).*

substituent chemistry of aromatic chromophores (e.g., by carboxylation/decarboxylation during photochemical oxygen consumption; Abdulla et al., 2010; Andrews et al., 2000; Xie et al., 2004) occurring across the freshwater-marine interface drive the observed trends in estuarine CO AQYs.

Sensitive analytical techniques and a low CO background in seawater facilitate precise and accurate quantification of CO photoproduction. Consequently, CO photochemistry is well studied and CO has been used as a proxy for other, less easily quantified photoreaction rates such as DIC photoproduction (e.g., Stubbins et al., 2006a). CO has also emerged as a key tracer for testing and tuning models of mixed-layer processes (Doney et al., 1995; Kettle, 2005; Najjar et al., 1995) and for the exploration of photochemical mechanisms (Stubbins et al., 2008).

As a result of photoproduction, ocean surface waters are supersaturated in CO (daily mean concentrations ~1 nmol L^{-1}; diurnal range ~0.1-5 nmol L^{-1}), resulting in atmospheric emissions of 2-11 Tg CO year^{-1} (Bates et al., 1995; Rhee, 2000; Stubbins et al., 2006b). Emissions of CO from the ocean are of interest as CO is one of the primary determinants of tropospheric OH radical concentrations (Warneck, 2000), and therefore oceanic CO emission indirectly affects the atmospheric residence times of greenhouse gases such as methane and halocarbons that are predominately removed by OH-initiated oxidation (IPCC, 2007; Thompson, 1992). CO oxidation, in conjunction with NO$_x$ (sum of NO$_2$ and NO$_3$), is also pivotal to the abundance of tropospheric ozone, a potent greenhouse gas and atmospheric oxidant (Dignon and Hameed, 1985). Although some CO is emitted to the atmosphere, the major sink for water column CO is microbial oxidation (32 ± 18 Tg CO year^{-1}; Zafiriou et al., 2003).

B Sulfur

Organic sulfur compounds are present in seawater at low pM to nM concentrations due to close coupling between production and removal processes that include biological uptake and release, thermal reactions, and photochemical reactions. The principal trace sulfur compounds detected in seawater are DMS, dimethylsulfoniopropionate (DMSP), OCS, and dimethylsulfoxide (DMSO). Other minor species include hydrogen sulfide (H$_2$S), dimethyldisulfide (DMDS), carbon disulfide (CS$_2$), methane thiol (MeS), cysteine, glutathione, phytochelatins, and methionine. Except for disulfides and metal-phytochelatin complexes, none of these sulfur compounds absorb solar radiation in natural waters. Therefore, if they photochemically degrade, it will be through indirect photochemical reactions. Sulfur compounds undergo a variety of indirect photochemical transformations in aqueous media, especially alkyl sulfides, disulfides, and thiols involving an array of reactants including the superoxide anion, hydroxyl radical, hydrogen peroxide, and singlet oxygen.

1 Dimethylsulphoniopropionate

Dissolved DMSP does not undergo direct photolysis in seawater at ambient concentrations (ca. 10^{-9} M) (del Valle et al., 2007) nor does it react with ROS at appreciable rates at ambient seawater concentrations of these oxidants (for details, see parallel discussion in the next section on DMS photoreactions). However, many marine algae across several classes have cellular DMSP concentrations that are quite high, in the μM (e.g., diatoms) to high mM range (e.g., prymnesiophytes) (Keller and Korjeff-Bellows, 1996). Under these conditions, particulate DMSP is expected to react with the OH radical and O$_2^-$. Liu et al. (2014) determined that the second-order rate constant for reaction of DMSP with the superoxide anion was 8.3 ± 0.3 M^{-1} s^{-1}, which is relatively slow due to the high activation energy for this reaction (93 ± 8 kJ mol^{-1}). The main sulfur product formed from this reaction is DMS, accounting for >95% of the DMSP loss. The conjugate acid, the HO$_2$ radical, does not react with DMSP, and as a result the second-order rate constant exhibits a strong pH dependence (Liu et al., 2014).

The average, second-order rate constant for the reaction between DMSP and the OH radical as determined by ion chromatography-mass spectrometry ($9 \times 10^8 M^{-1} s^{-1}$) is orders of magnitude greater than for the reaction between DMSP and O_2^- (Spiese, 2010). Additionally, the DMSP-OH radical rate constant is approximately threefold higher than reported using the base-cleavage method for DMSP analysis (Sunda et al., 2002), indicating that the primary reaction site for the OH radical, accounting for nearly 75% of the observed DMSP loss, must be located on the propionyl chain and not on either the sulfur atom or the sulfonium methyl groups. Interestingly, DMS and DMSO are formed during the initial loss of DMSP, and therefore this reaction may be an important source for these compounds to the dissolved phase in DMSP-containing algae that lack DMSP lyases (Spiese, 2010).

2 Dimethylsulfide

Dimethylsulfide loss in the photic zone is generally controlled by its microbial uptake and photochemical degradation (Dacey et al., 1998; del Valle et al., 2009; Galí and Simó, 2010; Kieber et al., 1996; Simó and Pedrós-Alió, 1999; Toole et al., 2004). The photochemical loss of nMDMS in seawater follows pseudo first-order kinetics (e.g., Brimblecombe and Shooter, 1986; Brugger et al., 1998; Kieber et al., 1996), with photochemical turnover times ranging from hours to a few days. However, at high concentrations of added DMS, observed reaction kinetics approach zero order with respect to DMS (Kieber et al., 1996). This saturation-type behavior suggests that the photochemical loss of DMS may occur through a binding (or catalytic) mechanism, presumably involving components of DOM and perhaps reactive species that are generated by DOM (Kieber and Jiao, 1995) or via complexation to Hg (Yang et al., 2007).

The mechanism for photosensitized DMS loss in seawater is poorly described as most studies have focused on the wavelength dependence for this reaction and determination of photolysis rates. Photochemically generated singlet oxygen ($^1O_2$) can be an important oxidant at μM levels of DMS (Brimblecombe and Shooter, 1986). However, it is a relatively minor oxidant at ambient DMS and $^1O_2$ concentrations, accounting for about 14% of total DMS photochemical loss, as determined from direct DMSO measurements and sodium azide (a singlet oxygen quencher) addition experiments (Kieber et al., 1996). This finding is consistent with calculations that yield a DMS removal rate (or DMSO production rate) of $2.1 \times 10^{-11} M h^{-1}$, given $1 \times 10^{-9} M$ DMS, $1 \times 10^{-13} M$ $^1O_2$, and a bimolecular rate constant of $5.8 \times 10^7 M^{-1} s^{-1}$ (Wilkinson et al., 1995). As with singlet oxygen, hydroxyl radical steady state concentrations (ca. $10^{-18} M$) are also too low to effectively remove DMS from seawater, even though the bimolecular rate constant for this reaction is near the diffusion-controlled limit ($k = 1.9 \times 10^{10} M^{-1} s^{-1}$; Buxton et al., 1988). Dimethylsulfide also reacts with hydrogen peroxide in seawater (ca. $k = 0.14 M^{-1} s^{-1}$; Shooter and Brimblecombe, 1989) but at ambient concentrations of DMS and H_2O_2 (ca. 10^{-9} and $10^{-7} M$, respectively), reaction rates are too slow (ca. $0.1 \times 10^{-12} M h^{-1}$) to be a significant removal mechanism for DMS in the photic zone. These calculations assume that DMS is dissolved in the bulk phase. However, if DMS is associated with DOM as seen for hydrophobic pollutants (e.g., the hydrated electron, Burns et al., 1997; singlet oxygen, Grandbois et al., 2008), then they would experience locally higher ROS concentrations relative to the bulk phase. Under this condition, DMS-ROS reactions may be important. Appreciable rates may also be expected when high concentrations of DMS and/or ROS are encountered such as may be observed inside an algal cell (Sunda et al., 2002) or possibly during a rain event (Cooper et al., 1987) or under bloom conditions when relatively high H_2O_2 and DMS concentrations may occur in the water column.

Based on results of laboratory studies in well-defined solutions, it has been suggested that DMS is photochemically degraded in seawater through reactions involving the carbonate or dibromide radical, the latter of which is produced from the photolysis of nitrate (Bouillon and Miller, 2005). However, this supposition is not consistent with the finding that addition of nitrate to Sargasso Sea seawater to the same level present in Ross Sea seawater (~30 µM) did not significantly increase the rate constant for DMS photolysis when both samples were exposed to solar radiation in the same water bath; the nitrate-amended Sargasso Sea seawater DMS photolysis rate constant was ~13-fold lower than that obtained with Antarctic seawater (Toole et al., 2004). Based on this finding, Toole et al. (2004) concluded that over 60% of the observed DMS photolysis in Antarctic waters involved DOM and not nitrate, and that Antarctic DOM was quite different from DOM in the Sargasso Sea with respect to DMS photolysis, even though CDOM absorption spectra were nearly identical for these two water samples.

Thus, although it is evident that DOM and ROS are involved in the photosensitized loss of DMS in seawater, none of the known reactions can account for the observed DMS loss when considered alone or together. Furthermore, the predominant products have not been identified. DMSO is the only product that has been detected in seawater at ambient DMS concentrations, and DMSO generally represents <35% of the observed DMS loss (Kieber and Jiao, 1995; Kieber et al., 1996; Toole et al., 2004; Yang et al., 2007); higher percentages of DMSO formation have been observed in the Ross Sea (41-61%, del Valle et al., 2009) and North Sea (21-99%, Hatton, 2002). Although DMSO is generally not the main photochemical product formed from DMS photolysis in seawater, no other products have been identified to date.

The wavelength-dependent photolysis of DMS has been determined in a range of seawater samples on several occasions both with monochromatic (Deal et al., 2005; Toole et al., 2003) and polychromatic irradiation systems (Bouillon and Miller, 2004; Toole et al., 2004). In all cases, an exponential decrease in AQYs for DMS photolysis has been observed with increasing wavelength in the UV (Bouillon and Miller, 2004; Deal et al., 2005; Toole et al., 2003). Unlike AQYs for compounds photochemically formed in seawater, wavelength-dependent AQYs for compounds that are photochemically degraded in natural waters through first-order kinetics (including DMS) are proportional to their initial concentration (Deal et al., 2005). Therefore, DMS AQYs have been scaled to 1 nM DMS (Bouillon and Miller, 2004; Deal et al., 2005; Toole et al., 2003), a typical background concentration in the open ocean surface mixed layer. Spectral slopes for DMS AQYs are remarkably similar in contrasting oceanic waters, ranging from 0.028 to $0.041 \, nm^{-1}$ in the northeastern Pacific (Bouillon and Miller, 2005), 0.043 to $0.05 \, nm^{-1}$ in the biologically productive Bering Sea (Deal et al., 2005), and $0.05 \, nm^{-1}$ in the oligotrophic Sargasso Sea (Toole et al., 2003). When sunlight-normalized rates were determined from wavelength-dependent DMS AQYs, it was seen that photolysis was dominated by UV-A solar radiation (ca. 60%) and UV-B (ca. 40%) (Deal et al., 2005; Toole et al., 2003, 2004). Except for one study in the equatorial Pacific (Kieber et al., 1996), these findings indicate that DMS photolysis in the open ocean is largely confined to the UV region of the solar spectrum.

Spectral slopes for DMS AQYs are comparable to other, well-studied species such as CO or hydrogen peroxide. However, published DMS AQYs, which range from $\sim 10^{-6}$ at 290 nm to $\sim 10^{-9}$ at 400 nm, are considerably lower than AQYs for other compounds including hydrogen peroxide, CO, and CO_2.

Even though DMS AQY spectral slopes are similar in contrasting oceanic waters, AQYs for these waters are quite different. For example, at the wavelength of peak DMS photolysis in seawater (~330 nm; Deal et al., 2005; Toole et al.,

2003, 2004), the DMS AQY for the Sargasso Sea sample is a factor of three to ten times larger than corresponding AQY at different stations in the Bering Sea; likewise, the AQY at 330 nm in the Bering Sea is approximately a factor of four lower than in the northeast Pacific. These differences may arise from differences among samples in the relative proportion of CDOM that photosensitize DMS photolysis (Deal et al., 2005) and not due to differences in nitrate concentrations (Toole et al., 2004).

3 Dimethylsulfoxide

DMSO is released into seawater through biological processes (Lee et al., 1999), and it may be produced in seawater through the reaction of DMS with singlet oxygen (Brimblecombe and Shooter, 1986; Kieber et al., 1996). DMSO may also be produced within algal cells (Spiese, 2010) or in the digestive systems in marine protists through reactions of cellular DMSP or DMS with ROS. Once formed, DMSO will readily diffuse out of the cell into the dissolved phase (Spiese, 2010). DMS may also react with other ROS such as the carbonate or dibromide radical (Bouillon and Miller, 2005), which are known to react with organic sulfides (Huang and Mabury, 2000), to form DMSO. DMSO does not undergo direct photolysis in seawater, since it does not absorb light at wavelengths present in solar radiation at the Earth's surface (ca. >290 nm). Therefore, if DMSO photochemical loss is observed, it must occur through indirect photochemical pathways. Reaction of DMSO with the OH radical is too slow to be quantitatively important. Given a bimolecular rate constant of $6.6 \times 10^9 \, M^{-1} s^{-1}$ (Buxton et al., 1988) and ambient seawater concentrations of DMSO and the OH radical of $25 \times 10^{-9} \, M$ and $1 \times 10^{-18} \, M$, respectively (Mopper and Zhou, 1990; Vaughan and Blough, 1998), the rate of DMSO loss in surface seawater is only $6 \times 10^{-13} \, M \, h^{-1}$. Significant rates are only expected at higher concentrations of reactants, which may be encountered, for example, in a marine algal cell containing high DMSP.

Inside the cell, DMSO may play a role in alleviating cellular OH radical stress (Lee and de Mora, 1999; Sunda et al., 2002). Based on rate calculations with known reactants and observations by Brimblecombe and Shooter (1986), it is unlikely that DMSO will be photochemically degraded in seawater through indirect photochemical reactions as suggested by Lee et al. (1999). Indeed, Kieber et al. (1996) and Toole et al. (2004) observed no DMSO loss in Pacific or Antarctic seawater, when samples were exposed to solar radiation for ~8 h. Further photochemical studies with ambient levels of radiolabeled DMSO are needed to confirm this observation. Given that the photochemical loss of DMSO in marine waters is unlikely, the main loss of this nonvolatile compound in the oceans should be through mixing and its microbial uptake (Tyssebotn et al., 2014).

4 Carbonyl Sulfide

OCS is the most atmospherically stable, naturally occurring sulfur species that is ventilated from the oceans into the atmosphere. It has a lifetime in the troposphere of more than 1 year (Khalil and Rasmussen, 1984) and diffuses into the lower stratosphere where it is photooxidized to sulfuric acid, giving rise to the Junge aerosol layer (Crutzen, 1976). In seawater, concentrations of OCS are low compared to DMS, DMSO, and DMSP, ranging from ~0.07-$1.2 \times 10^{-9} \, M$ in coastal waters to about $30 \times 10^{-12} \, M$ in the open ocean. Low concentrations of OCS partly reflect its low production rate in seawater and its rapid turnover in the water column (von Hobe et al., 2003). In particular, although gas-phase OCS is relatively long-lived in the troposphere, it is very reactive in seawater due to its hydrolysis to carbon dioxide and H_2S (Elliot et al., 1987, 1989; Flöck and Andreae, 1996). The half-life of OCS in seawater is ~2 days ($k = 3.8 \times 10^{-6} \, s^{-1}$) at 298 K. Photochemical loss of OCS is unlikely because it will not undergo direct photolysis in seawater and its reactions with known photochemically formed ROS will be slow at ambient ROS and OCS concentrations.

OCS is produced in the photic zone by both thermal and photochemical pathways. Non-photochemical (thermal) production of OCS may involve DOM, dissolved organosulfur species, polysulfides, and thiyl radicals (Flöck et al., 1997; Kamyshny et al., 2003; Pos et al., 1998) but the mechanism is poorly described. Nonetheless, thermal formation of OCS is an important source for this compound in seawater, comparable to photochemical production rates (Flöck and Andreae, 1996; Ulshöfer et al., 1996). Non-photochemical production of OCS is presumably the main source for this compound deeper in the water column.

Photochemical OCS production is an important source for this compound in sunlit surface waters (Cutter et al., 2004; Ferek and Andreae, 1984; von Hobe et al., 2003), with rates in the $10^{-12} M h^{-1}$ range (Flöck and Andreae, 1996). Photochemical production of OCS involves DOM both as a photosensitizer and as a source of sulfur (Andreae and Ferek, 1992). However, oxygen-dependent oxidants such as singlet oxygen and the OH radical are not involved in OCS formation (Zepp and Andreae, 1994). A number of model sulfur compounds generate OCS photochemically, including cysteine, 3-mercaptopropionic acid, glutathione, thiols, sulfides, and DOM itself, while DMSP, DMSO, and dimethylsulfone produced very little OCS (Flöck et al., 1997; Zepp and Andreae, 1994). Production rates increase when DOM is added as a photosensitizer (Flöck et al., 1997; Uher and Andreae, 1997); Cutter et al. (2004) determined that OCS photoproduction rates in the Sargasso Sea were dependent on dissolved sulfur and CDOM concentrations. Based on these and other published studies, the mechanism for the photochemical production of OCS is proposed to involve thiyl or sulfhydryl radicals (Flöck et al., 1997; Pos et al., 1998). The source(s) of these radicals is not known. Gun et al. (2000) studied polysulfide formation in Lake Kinneret and suggested that OCS formation resulted from the reaction of polysulfide radicals and carbon monoxide. This mechanism may explain the strong correlation between OCS photoproduction and CO concentrations observed in marine waters (Flöck et al., 1997; Pos et al., 1998). Perhaps the source of thiyl radicals is the photolysis of metal-sulfide complexes through a LMCT reaction. These complexes are known to occur in seawater (Luther and Tsamakis, 1989), but their stabilities, especially with respect to photolysis, are not known.

AQYs for OCS photoproduction range from $\sim 10^{-7}$ to 10^{-5} between 297 and 436 nm, and decrease exponentially with increasing wavelength in the UV and blue regions of the solar spectrum (Cutter et al., 2004; Weiss et al., 1995; Zepp and Andreae, 1994), yielding a broad solar response curve between 290 and 400 nm with a maximum response centered at ~ 340 nm (Weiss et al., 1995). This trend is similar to that observed for many other photochemically generated species in seawater (see Blough, 1997 for review). However, AQYs for OCS production (ca. 10^{-7}) are much lower than observed for other species (ca. 10^{-4}-10^{-3}) such as CO, CO_2, and hydrogen peroxide, indicating that, like DMS, specific (and minor) components of DOM are precursors to OCS. AQYs for OCS photoproduction are higher in the Zepp and Andreae (1994) study in coastal waters (e.g., 6.4×10^{-7} at 365 nm) and in the Cutter et al. (2004) study in the Sargasso Sea (e.g., $\sim 3.3 \times 10^{-6}$ at 365 nm) compared to those obtained in Pacific Ocean waters by Weiss et al. (1995) (e.g., 9.3×10^{-8} at 365 nm). Other than analytical artifacts, the only way to generate this large difference in AQYs is to invoke differences in the concentration (and possibly type) of CDOM and sulfur precursors present in these samples. If AQYs for OCS formation vary greatly in seawater, then it may be difficult to accurately model production rates on a global scale (von Hobe et al., 2003), especially when using remotely sensed data based on DOM fluorescence or absorbance (Neale and Kieber, 2000).

5 Minor Sulfur Species

CS_2, MeS, and H_2S are present in seawater at pM concentrations, and all are involved in photochemical transformations (*vide infra*).

Other reduced sulfur compounds that may be involved in photochemical transformations in oxic seawater include cysteine, cystine, and glutathione (Flöck et al., 1997). These studies were conducted using μM additions of these compounds to seawater; it is not known if these processes will be important at the low nM concentrations that are detected in seawater.

CS_2, an atmospherically important trace gas, is photochemically produced in surface waters. AQYs for CS_2 formation decrease exponentially with increasing wavelength from about 14×10^{-8} at 308 nm to 0.2×10^{-8} at 400 nm (Xie et al., 1998). AQYs for CS_2 formation are quite low relative to hydrogen peroxide, DIC, etc., and they are ~25% of corresponding AQYs for OCS formation (Weiss et al., 1995). Both cysteine and cystine are efficient precursors of CS_2 and the OH radical is likely an important reactant (Xie et al., 1998). Based on field and laboratory evidence, Xie et al. (1998) concluded that the mechanisms for CS_2 and OCS formation were similar and that photoproduction was the primary source of CS_2 in seawater.

MeS and H_2S are both removed from seawater through photochemical transformations. Average MeS removal rates in the northeast Atlantic and Aegean Sea are low, 6.7 and $36 \times 10^{-12} M h^{-1}$, respectively (Flöck and Andreae, 1996). Although the mechanism is not known, the photochemical loss of MeS likely involves binding to DOM, possibly through complexation or formation of a thioketal (Kiene et al., unpublished results).

The photochemical degradation of H_2S follows first-order kinetics in Biscayne Bay, FL seawater, with loss rates of 5.1 and $20.2 \times 10^{-11} M h^{-1}$. Sulfide photolysis is independent of dissolved oxygen and involved DOM and possibly a thiyl radical intermediate (Pos et al., 1997). The rate of sulfide loss does not decrease exponentially with increasing wavelength as observed for so many other species in seawater. Rather maximum HS^- loss rates are in the blue region of the solar spectrum (Pos et al., 1997) similar to that observed for DMS photolysis in the equatorial Pacific (Kieber et al., 1996).

It is clear from the above studies that photochemical transformations affect many reduced sulfur compounds in seawater, but there is still much we do not know. In particular, we do not know the role of iron or other metals in complexing sulfur compounds in seawater, especially H_2S and thiols; recent evidence suggests that complexation with Hg may be important in freshwater systems but not in seawater where chloro-Hg complexes predominate (Zhang and Hsu-Kim, 2010). Sulfur compounds complex to type B metals, such as iron and copper, to form very strong complexes. Thus, metallo-sulfur complexes are likely to be important in seawater, both in stabilizing sulfur to thermal oxidation and destabilizing sulfur to photochemical oxidation through LMCT transfer reactions. Other less-known reactants may be important in sulfur photodegradation in seawater, including the carbonate radical, the superoxide anion, and organic radicals. Systematic studies are needed to quantify and delineate these sulfur phototransformations.

C Nitrogen and Phosphorus

In comparison to carbon or sulfur, surprisingly little is known about photochemical transformations of nitrogen in natural waters. Studies have been conducted to determine photochemical degradation rates, product formation rates, and the impact of N-containing compounds on microbial and photochemical CDOM and FDOM formation rates (Biers et al., 2007). However, very little work has been done to understand the mechanisms involved in organic nitrogen photochemical transformations. In addition to N-containing DOM and POM, several N-containing compounds and compound classes have been studied, including ammonium, nitrate, nitrite, dissolved free amino acids (DFAA), peptides, dissolved primary amines (DPAs), and amino sugars. Of these, ammonium is the main photoproduct that is detected, with production rates generally in the 10^{-6}-$10^{-9} M h^{-1}$ range (Aarnos et al., 2012; Bronk, 2002; Bronk et al., 2010; Buffam and McGlathery,

2003; Bushaw et al., 1996; Bushaw-Newton and Moran, 1999; Gardner et al., 1998; Gao and Zepp, 1998; Kieber, 2000; Kitidis et al., 2006a, 2008; Koopmans and Bronk, 2002; Morell and Corredor, 2001; Stedmon et al., 2007a; Tarr et al., 2001; Vähätalo et al., 2003; Vähätalo and Zepp, 2005; Vähätalo and Jarvinen, 2007; Wang et al., 2000; Xie et al., 2012).

In one of the earliest studies, Bushaw et al. (1996) exposed whole water and fulvic acid extracts, isolated from a boreal pond and a series of high DOM rivers in Georgia (including the Satilla River), to sunlight (or artificial light) and measured rates of NH_4^+ production ranging from 40 to $370 \times 10^{-9} \, M \, h^{-1}$. Photochemical NH_4^+ production was also observed in solar-simulator-irradiated Satilla River water, employing specific detection of ammonium by HPLC; ammonium photoproduction was highly correlated to CDOM loss (Kieber, 2000). Gardner et al. (1998) conducted a $^{15}NH_4^+$ isotope dilution study in filter-sterilized lake water and found photoproduction rates for ammonium (ca. $2 \times 10^{-7} \, M \, h^{-1}$) comparable to those obtained by Bushaw et al. (1996). Most studies have been conducted with organic-rich fresh, estuarine, or coastal waters. However, two studies were conducted with oligotrophic seawater, one in the eastern Mediterranean Sea (Kitidis et al., 2006a) and the second in the eastern tropical Pacific Ocean (Bronk, 2002). Photochemical production rates of ammonium determined in the open ocean from the Pacific ($4.4 \, nM \, h^{-1}$, Bronk, 2002) and Mediterranean Sea (0.4-$3.0 \, nM \, h^{-1}$, Kitidis et al., 2006a) are lower than reported in terrestrial or coastal waters. An important finding of several studies is that photoproduction rates of ammonium increase with increasing salinity, a trend opposite to that observed for CO, DIC, or other carbon-compound photoproduction rates (Aarnos et al., 2012; Bronk et al., 2010). This finding suggests that, whereas terrestrial sources predominate in the photochemical formation of many carbon compounds,

marine sources are proportionately more important for photoammonification. However, this trend is not universal as Xie et al. (2012) observed rates decreasing with increasing salinity in the southeastern Beaufort Sea, in line with carbon photoproduction rate and CDOM absorption coefficient trends with salinity (Xie et al., 2012).

Several studies estimated the importance of photochemically produced ammonium as a source of nitrogen to the food web and found that its importance was variable, ranging from a few percent to >50% of the bioavailable nitrogen, depending on the relative magnitude of other nitrogen fluxes (Aarnos et al., 2012; Bronk, 2002; Buffam and McGlathery, 2003; Bushaw et al., 1996; Kitidis et al., 2006a; Morell and Corredor, 2001; Stedmon et al., 2007a; Vähätalo and Jarvinen, 2007; Vähätalo and Zepp, 2005; Xie et al., 2012). In general, these estimates are based on measured photochemical production rates and measurements or estimates of the depth dependence for ammonium photoproduction compared to nitrogen fluxes from other sources (e.g., atmospheric input). The implicit, but mostly untested, assumption of these estimates (except for Vähätalo and Zepp, 2005, and Vähätalo and Jarvinen, 2007) is that photochemical production of ammonium is from a biologically recalcitrant reservoir of dissolved organic nitrogen (DON) as opposed to an interconversion of one biologically available form of nitrogen to another.

Four studies cited above made fairly robust estimates of the importance of photochemistry to nitrogen cycling in the upper water column because they determined ammonium photoproduction AQYs; from these, they calculated the depth dependence for this process (Aarnos et al., 2012; Vähätalo and Jarvinen, 2007; Vähätalo and Zepp, 2005; Xie et al., 2012). All of these studies employed a polychromatic radiation system to determine AQYs, assumed *a priori* that AQYs decreased exponentially with increasing wavelength and fit photochemical rate data to an exponential equation, as has been done for other photochemical species (e.g., DMS, CO, DIC). However, the assumed exponential decrease in AQY with increasing wavelength has not been verified for ammonium photoproduction employing monochromatic radiation, which makes no assumptions about the spectral shape of the AQY wavelength dependence. The exponential decrease in AQY for photoammonification with increasing wavelength was primarily confined to the UV in Baltic Sea water, with AQYs ranging from $<1 \times 10^{-6}$ at 400 nm to $1.5\text{-}6 \times 10^{-5}$ at 300 nm (Aarnos et al., 2012; Vähätalo and Jarvinen, 2007; Vähätalo and Zepp, 2005). A notable exception is Xie et al. (2012) who reported lower AQYs at 300 nm (ca. 5×10^{-6}) and a much flatter spectral slope for wavelength-dependent AQYs, with a long tail in the visible for ammonium photoproduction in Arctic waters from the Mackenzie River shelf. Differences in AQY spectra may reflect differences in the nitrogen photolability of the source waters and/or differences in temperature-dependent AQYs, the latter of which are not known for ammonium photoproduction. Despite differences in AQY spectra, corresponding sunlight-normalized rates for all studies showed a peak spectral response at ~330 nm, with most of the observed ammonium photoproduction confined to the UV and more than 60% in the UV-A. Only in the Xie et al. (2012) study was there significant production in the visible (ca. 10-20% of the total observed production). All published photoammonification AQYs are from one to three orders of magnitude lower than AQY for the photochemical formation of DIC, but similar to AQY for CO, implying that a much smaller subset of DOM precursors are responsible for ammonium or CO photoproduction compared to that for DIC (Xie et al., 2012).

Despite its reported importance as a source of nitrogen to the aquatic ecosystem, ammonium photoproduction is not ubiquitous in natural waters. There are several reports that show neither photoproduction nor a loss of NH_4^+ when water samples are exposed to solar radiation or a solar simulator. Vähätalo and Salonen (1996),

Jørgensen et al. (1998), Bertilsson et al. (1999), Wiegner and Seitzinger (2001), Grzybowski (2002b), McCallister et al. (2005), and Southwell et al. (2009) all observed no change in ammonium concentrations when filter-sterilized water was exposed to sunlight, while Jørgensen et al. (1999) and Vähätalo et al. (2003) observed a loss (or no change) of NH_4^+ in irradiated Baltic Sea and Lake Valkea-Kotinen water samples, respectively. Likewise, Gardner et al. (1998) detected variable but consistent photochemical loss of ammonium ($2\text{-}130 \times 10^{-9}\,M\,h^{-1}$) in Lake Maracaibo, presumably due to incorporation into DOM since ammonium does not directly photooxidize in seawater (Hamilton, 1964). The Gardner et al. (1998) findings are consistent with results of a laboratory study showing that ammonium is incorporated into humic extracts during photolysis of filtered seawater (Kieber et al., 1997b).

Kitidis et al. (2008) irradiated several water samples from the North Sea, the Tyne Estuary, the River Tyne, and River Tay, and determined that ammonium photoproduction rates depended on the length of irradiation. Initially, there was no change in the ammonium concentration. This lag phase was followed by an increase and then decrease in the ammonium concentration. This trend was proposed to result from an interplay between photochemical production and loss of ammonium due to its incorporation into DOM. The nonlinearity observed by Kitidis et al. (2008) was not related to initial conditions including ammonium concentration and the CDOM absorbance. They suggested that this kinetic profile may explain differences that have been noted in the literature regarding the production or loss of NH_4^+ during exposure of natural waters or humic/fulvic extracts to solar radiation. Nonlinear kinetics have been reported elsewhere (e.g., Bushaw et al., 1996; Kieber, 2000; Wang et al., 2000; Xie et al., 2012) but often no lag period is observed and, in some studies, ammonium accumulation is linear with exposure time (e.g., Kitidis et al., 2006a). As suggested by Kitidis et al. (2006a), these differences likely result from differences in exposure time. Wang et al. (2000) found that ammonium accumulation in samples was initially linear with irradiation exposure but after an extended irradiation time the rate of accumulation decreased as ammonium precursors were depleted. Differences in kinetic results further point to fundamental differences in mechanisms for photoproduction or inherent differences in the DOM composition among the different water samples and differences in the relative importance of photoproduction versus photodestruction pathways for ammonium (Gardner et al., 1998). Contrasting results may also partly reflect differences and limitations in the light source used to irradiate samples, the analytical methodology used to quantify NH_4^+, contamination issues, and initial ammonium concentrations.

An important conclusion from published studies showing a cessation of ammonium accumulation in irradiated samples is that natural waters have a limited capacity for ammonium photoproduction, and therefore unless NH_4^+ photoprecursors are continually refreshed in the photochemically active upper ocean through mixing or other processes (e.g., biological production and release), photoammonification will have a finite effect on aquatic food webs, especially if sunlight-driven ammonium losses are important (e.g., Wang et al., 2000).

The mechanism for ammonium photoproduction in natural waters has been proposed to involve organonitrogen intermediates (e.g., imines) that are converted to ammonium by hydrolysis or biological pathways (e.g., Bushaw et al., 1996; Kitidis et al., 2008; Wang et al., 2000). Thus, the loss of DON observed by Letscher et al. (2013) may be initiated by organonitrogen photoreactions in the upper water column and finished by hydrolysis or biological uptake in deeper waters (Zepp, personal communication). Only one study to date has investigated the mechanism for the photochemical production of ammonium in some detail (Tarr et al., 2001). Tarr et al. used a solar simulator to irradiate

Suwannee River fulvic and humic acid isolates (SRFA and SRHA, respectively), and organic-rich water from the west Pearl River and Bayou Trepagnier containing 20-30 mg DOC L^{-1}. Samples were irradiated both in the presence and absence of an OH radical scavenger 1-proponal. They determined that ammonium production resulted through at least two pathways in these organic-rich samples, an OH-radical dependent and independent pathway. The OH-initiated pathway was proposed to involve hydrogen-peroxide-mediated production of the OH radical, since ammonium production proceeded in the dark. This pathway may involve the Fenton reaction, as it was inhibited by the addition of 1-propanol. However, addition of 1-propanol to irradiated samples had little effect on ammonium photoproduction, suggesting that light-mediated production was independent of the OH radical. This interpretation however does not consider internal OH radical production within humic-bound DOM and its direct attack of humic-bound DON that may proceed even in the presence of a water-soluble OH radical scavenger, since this scavenger may not access internally produced OH radicals (Burns et al., 1997; Grandbois et al., 2008; Hassett, 2006; Latch and McNeill, 2006).

Tarr et al. (2001) determined that added amino acids were photodegraded in solutions containing DOM through a photosensitized pathway, and this process yielded ammonium. However, addition of 1-propanol did not affect the rate of amino acid photorelease from DOM (with no added amino acids). Therefore amino acids were only a minor source of ammonium in Suwannee River humic and fulvic acids or Bayou Trepagnier samples, implying that other unknown functional groups (or DOM-bound amino acids/peptides) were responsible for ammonium production. Interestingly, no published study to date has conducted an experiment to determine if ammonium is photoreleased from DOM into the dissolved phase due to the photochemical oxidation and subsequent loss of ammonium-DOM complexes. The photochemical breakdown of DOM into smaller, oxidized molecules may no longer bind ammonium, causing its release into solution. This photochemical release of ammonium may partly account for the nonlinear NH_4^+ production that is often observed. It is also possible that the photooxidation of DOM may produce oxidized, highly functionalized DOM (e.g., with increased carboxyl content) that results in a net loss of ammonium due to its increased complexation to the oxidized DOM.

Photochemical production (or loss) of DFAA is not as well documented as photoammonification, and all published results to date have been obtained with carbon-rich waters and not open ocean seawater or even coastal seawater. Kieber (2000) found no chromatographic (HPLC) evidence for primary amine or amino acid production in a Satilla River fulvic acid isolate, even after 11 h of solar irradiation. Jørgensen et al. (1998) also employed HPLC, but in their study DFAA and carbohydrates increased when water from Lake Skärshult was exposed to sunlight. However, increases in DFAA were small (13-23%), and no information was given regarding changes in specific amino acids. In a follow-up study, Jørgensen et al. (1999) observed that, in some irradiated samples, DFAA concentrations increased, while there was a net decrease (or no change) in DFAA concentrations in other samples. Again, no specific compositional changes in the DFAA pool were reported. Rosenstock et al. (2005) had a similar finding in Antarctic waters; concentrations of DFAA increased from 50-100% in Polar Front samples when exposed to UV-B solar radiation, while a slight decrease in DFAA was observed for two Weddell Sea samples. No photoproduction was observed in any samples exposed to only PAR or UV-A and PAR. Tarr et al. (2001) determined that several DFAA were photoproduced in organic-rich isolates and bayou samples at rates ranging from 0.03-9.5 nM h^{-1}. DFAA that were photochemically produced included alanine, asparagine, citrulline, glutamic acid, histidine, norvaline, and serine. Buffam and McGlathery (2003) similarly observed photoproduction of alanine and glycine in the 1-nM-h^{-1} range in coastal lagoon water, but no photoproduction was observed for serine or other common DFAA. In both studies, DFAA photoproduction rates were low compared to the µM h^{-1} rates observed for NH_4^+ and are therefore not likely to be an important class of substrates for microbial growth in natural waters. Published DFAA photoproduction rates should be considered net rates (Tarr et al., 2001), since it has been shown that several DFAA photodegrade in natural waters, albeit at high µM concentrations through reactions with singlet O_2 and the OH radical (Boreen et al., 2008; Reitner et al., 2002). It is not clear whether these photodegradation reactions are significant at ambient low nMDFAA concentrations present in seawater.

In a study to assess the effect of sunlight on protein lability, irradiation of seawater with added protein made the protein less available to microorganisms, possibly due to the production of refractory DOM with a concomitant loss of labile proteinaceous nitrogen (Keil and Kirchman, 1994). In related studies, Amador et al. (1989; 1991) demonstrated that the photolysis of humic-bound organics such as amino acids and aromatic compounds greatly increased the rate and extent that microorganisms degraded these compounds relative to dark controls. They found that only 20% of the ^{14}C-labeled glycine in the humic sample was utilized over a 60-day period in the dark. Most of the glycine (ca. 80%) was bound to intermediate and HMW humics (>5000 Da) that could not be used by the microbes. After humic acid solutions were exposed to sunlight, the percentage of glycine that was mineralized by the heterotrophic bacteria in the dark increased significantly to 40-60% of the total. This increase paralleled the increase in the percentage of glycine associated with the LMW humic acid fraction, consistent with the finding that bacteria mineralize only the LMW fraction. However, not all humic-bound organics studied by Amador et al. were biologically labile after exposure to sunlight. This biologically unavailable fraction was attributed to the formation of photochemically and biologically resistant N-heterocycles in the humic acids (Amador et al., 1989, 1991).

Biers et al. (2007) determined the effects of microbial and photochemical organonitrogen transformations on CDOM formation and loss. For all photochemical experiments, they amended 0.2-μm filtered tidal creek water with a single nitrogen source to a final concentration of either 1 or 100 μM followed by irradiation for 8 h in a water bath employing a Xe lamp-based, simulated solar irradiation system. Several N-containing compounds were examined including amino acids (tryptophan, glutamic acid, and aspartic acid) and amino sugars (galactosamine, glucosamine, and mannosamine). Of all the compounds tested, only tryptophan affected CDOM photochemistry. Relative to the control that showed CDOM photobleaching, the tryptophan-amended sample increased CDOM absorption coefficients and decreased the spectral slope. Their results suggest that, other than tryptophan, none of these compounds were incorporated into or otherwise affected the photodynamics of CDOM. This finding is not surprising given that only tryptophan is known to undergo primary photolysis to form radical intermediates that could be incorporated into CDOM; none of the other compounds tested absorb solar radiation or undergo photolysis in solution at environmentally relevant wavelengths. This finding is similar to that of Reitner et al. (2002) who observed that, of the compounds they tested (alanine, tryptophan, and bovine serum albumin (BSA)), only tryptophan photoreacted to produce CDOM. In addition to the DFAA examined in these two studies, histidine, methionine, and tyrosine should also be considered because they are known to photochemically degrade in natural waters (Boreen et al., 2008) and therefore they may affect CDOM photodynamics. It is likely

that tryptophan or other photoreactive DFAA would not be nearly as effective if tied up in peptides or proteins. For example, BSA did not produce CDOM or accelerate its photolysis relative to unamended controls even though BSA contains several tryptophan residues (Reitner et al., 2002).

Koopmans and Bronk (2002), in one of the few studies that examined dissolved primary amines (DPAs), observed that DPA concentrations generally did not change when ground water or estuarine water was exposed to irradiation from a solar simulator. Likewise, in a 2010 study, Bronk (2002) found no production of DPA after a 24-h exposure of treated wastewater effluent to sunlight and only a slight increase when samples were irradiated for an additional 24 h. Using a batch fluorometric technique, Bushaw et al. (1996) and Bushaw-Newton and Moran (1999) reported photoproduction of primary amines in DOM concentrated from the Skidaway and Satilla River samples but not in lower-DOC concentrates from the Skidaway River. Ammonium and DPA accounted for approximately one-third of the total biologically labile N products photoproduced in these samples; the remaining photoproducts, nearly two-thirds of the total biolabile N photoproduced, consisted of unknown N-containing compounds.

The lack of agreement in DPA, DFAA, and NH_4^+ results is also seen for other nitrogen compounds. Photoproduction of urea and nitrate/nitrite have been observed in natural waters (Bronk et al., 2010; Gao and Zepp, 1998; Jankowski et al., 1999; Jørgensen et al., 1998, 1999; Kieber et al., 1999b; Koopmans and Bronk, 2002; Spokes and Liss, 1996; Vähätalo and Jarvinen, 2007; Wiegner and Seitzinger, 2001). Concentrations of these compounds were also shown to remain constant or decreased when filter-sterilized samples were exposed to sunlight or a solar simulator (Bronk et al., 2010; Buffam and McGlathery, 2003; Jankowski et al., 1999; Jørgensen et al., 1999; Kieber et al., 1999b; Koopmans and Bronk, 2002; Vähätalo and Salonen, 1996; Vähätalo and Zepp, 2005). Variable results may be expected for nitrite (i.e., net production, loss or no change), since it undergoes direct photolysis (Zafiriou and Bonneau, 1987; Zafiriou and True, 1979) and is photoproduced in seawater. In a coastal seawater study, the rate of direct photolysis of nitrite was $\sim 2.3 \times 10^{-8} \, M h^{-1}$ compared to an average photochemical production rate of $4 \times 10^{-9} \, M h^{-1}$ (Kieber et al., 1999b).

Photochemical formation of phosphate was first reported by Francko and Heath (1982), who observed that sunlight exposure of DOM-rich, acidic lake water (pH 5.2-5.8) caused the release of phosphate complexed to HMW DOM. The rate of release of bound phosphate was highly correlated to Fe(III) reduction to Fe(II). Phosphate photoproduction was also observed in a humic-rich lake (Vähätalo and Salonen, 1996), wastewater treatment plant effluent (Bronk et al., 2010), and coastal shelf sediment slurries (Southwell et al., 2009). However, the mechanisms underlying phosphate photoproduction are poorly understood (Francko, 1990). There also appears to be a seasonality in the extent of phosphate release (Cotner and Heath, 1990). Photochemically induced increases in dissolved phosphorus levels have not been observed in other systems, including a culture of the marine diatom *A. anophageffers* (Gobler et al., 1997), a freshwater lake (Jørgensen et al., 1998), and several other freshwater and marine water samples (McCallister et al., 2005; Southwell et al., 2009; Wiegner and Seitzinger, 2001).

Several studies by Wetzel and colleagues observed that UV increased alkaline phosphatase activity in the presence of high levels of DOM or humic substances isolated from decaying aquatic macrophytes; the alkaline phosphatase used in these studies was isolated from several sources including aquatic algae (Wetzel, 1993; Kim and Wetzel, 1993; Wetzel et al., 1995). Based on their findings, they proposed that alkaline phosphatase and other enzymes are complexed with DOM, resulting in partial or complete enzyme inactivation; solar photolysis, particularly in the UV, causes the partial breakdown of the enzyme-DOM complexes, thereby increasing the alkaline phosphatase activity. It is not clear, however, if this light-induced activation will affect the photoproduction of inorganic phosphate observed by Francko and Heath (1982) and others, especially in oligotrophic oceanic waters. It is also not known how UV-enhanced activation of other DOM-bound enzymes might affect the photoproduction or loss of other LMW compounds, especially at much lower DOM concentrations in open ocean seawater. Even if this process is not important in phosphate or DOM photodynamics in seawater, it may be important in POM since the latter will contain much higher concentrations of organic matter and, for some particles, enzymes.

It is somewhat surprising that the irradiation of natural waters shows such diverse photochemical responses with respect to many nitrogen- or phosphorus-containing compounds. It is difficult to reconcile the diverse responses (i.e., no production, production, or loss) obtained among published studies because, for the most part, we do not understand the photochemical mechanisms underlying these transformations. For example, what is the effect of dissolved oxygen, pH, and temperature on production rates? Are some compounds simply photoreleased from a DOM-bound state or are they produced through photochemical reactions involving photosensitizers and ROS? We know that ammonium photoproduction kinetics are complex and variable, but what about the other compounds and compound classes? Are production rates linearly dependent on the photon exposure for DFAA, DPA, urea, and phosphate? What is the observed reaction order? What are wavelength-dependent AQYs? Conflicting results may be due to fundamental differences in the DOM among these waters or to variations in the balance between rates of photoproduction and photodestruction. They also may reflect differences in experimental design (e.g., the light source) and analytical artifacts that result from a lack of understanding of the factors that control nitrogen and phosphorus phototransformations in natural waters. Given the wide range of photochemical results that have been obtained for nitrogen or phosphorus, it is difficult to evaluate the overall importance of these photochemical transformations on corresponding elemental biogeochemical cycles. This is especially true in oligotrophic seawater where very few studies have been performed, almost certainly due to analytical limitations.

III DOM PHOTOLABILITY SPECTRUM AND FATE OF TERRESTRIAL DOM IN THE SEA

In this section, the lability or, by contrast, resistance of different DOM fractions to photodegradation is presented. Labile here refers to DOM compounds or fractions that are lost during irradiations. Biologically labile DOM is addressed in Chapters 2 and 3. Lability and reactivity are not synonymous. DOM components that initiate and are modified by a photoreaction are both reactive and labile. Other components, for instance compounds that do not absorb light (e.g., DMS), are not photoreactive unless they are present in water containing photoreactive DOM compounds wherein they may be photolabile, presumably through indirect reactions initiated by the photoreactive DOM (e.g., for review, see Mopper and Kieber, 2000). A stable photocatalyst would define a third pool of photoreactive, yet resistant, DOM. Differentiating between these DOM types is experimentally challenging, and limited information exists about the relative contributions of each type of DOM towards net changes in DOM chemistry observed during irradiations of natural waters. For this reason, the current section focuses mainly on photolabile DOM and the above text acts to remind the reader that not all components of DOM that are lost during an irradiation are involved in direct photoreactions.

A large number of studies have examined the photolability of DOC and CDOM, usually by exposing waters to natural or simulated sunlight. Early work revealed that all the color in unfiltered natural waters measured using a Nessler cylinder was removed during prolonged exposure to sunlight, and the loss of color resulted from sunlight and not microbial activity, due to the sterilizing activity of the sun's UV rays (Boston Water Board, 1895). In contemporary work, waters are commonly sterile filtered (0.1 or 0.2 μm) prior to irradiation to isolate photochemical from biological processes. When exposed to solar radiation, DOC concentrations and CDOM absorption decrease exponentially with time, indicating that the photoreactivity of the DOM decreases with light exposure (photon dose). Fitting exponential decay curves to DOC and CDOM data from long-term irradiations show that CDOM would be almost completely photobleached after infinite irradiation leaving behind a weakly absorbing pool of photorefractory DOC (Helms et al. 2013b, 2014; Moran et al., 2000; Spencer et al., 2009; Stubbins et al., 2012). What absorbance remains is shifted to shorter UV wavelengths where solar radiation is unable to affect further direct photobleaching. The percentage of photorefractory DOC surviving irradiation (or formed during the irradiation) depends on several factors, including the initial sample composition and prior irradiation history, irradiation conditions (e.g., wavelengths and duration of light used), and the model used to fit the photobleaching kinetics. For aromatic-rich tDOM, 54-69% of the DOC has been characterized as photorefractory based upon exponential decay models (Moran et al., 2000; Obernosterer and Benner, 2004; Spencer et al., 2009; Vähätalo and Wetzel, 2004). By comparison, irradiation of North Atlantic Deep Water (NADW) removed ~90% of DOC in 28 days (Stubbins et al., 2012). It is not clear if the high % photolability of deep ocean DOC is intrinsic to the DOM itself or whether the high concentration of inorganics that can absorb light and initiate photoreactions (e.g., nitrate), forming radicals through indirect reactions (e.g., with chloride and bromide), increased DOC photolability. It is likely that DOC within the surface waters of the open ocean, particularly the stratified, sunbathed, CDOM-depleted waters of the subtropical gyres, would be less photolabile than that of the comparatively CDOM-rich NADW sample (Helms et al., 2013b).

The spectral dependence of bleaching produces marked changes in the CDOM spectral slope (S). The calculation and variation of S in natural waters are discussed in Chapter 10. At broadband scales (i.e., 250-500 or 700 nm), irradiation usually leads to steeper spectral slopes, although at this resolution S may also become shallower (Shank et al., 2010). Closer inspection reveals that slopes for different wavelength ranges have different responses. For example, the region between 275 and 295 nm ($S_{275-295}$) becomes consistently steeper, whereas the slope from 350 to 400 nm ($S_{350-400}$) shallows during photobleaching (Helms et al., 2008; Helms et al., 2013b). Using these restricted ranges provides more robust, sensitive indices for assessing CDOM quality than broader wavelength range slopes. In addition, the spectral slope ratio ($S_{275-295}$:$S_{350-400}$) is highly sensitive to changes in CDOM quality and increases during CDOM photodegradation (Helms et al., 2008, 2013b, 2014).

Fluorescent CDOM (FDOM) is also highly photolabile (Chen and Bada, 1992; Kouassi and Zika, 1990), with average percentage fluorescence losses exceeding both DOC and CDOM absorbance photobleaching (Moran et al., 2000). Thus, using the definitions applied here, the FDOM fraction represents the most photolabile DOM component. FDOM photobleaching follows reaction kinetics similar to DOC and CDOM absorbance photodegradation (Chen and Bada, 1992; Nieto-Cid et al., 2006) and exhibits spectral variations. Humic-like fluorophores that are generally representative of tDOM or remineralized deep ocean DOM (Coble, 2007; Jørgensen et al., 2011; McKnight et al., 2001) are more photolabile than amino acid-like fluorophores (Moran

et al., 2000; Stedmon et al., 2007a). As discussed in the Chapter 10, photodegradation is widely held to be the dominant sink for CDOM absorbance and humic-like fluorescence in the global ocean (Jørgensen et al., 2011; Kitidis et al., 2006b; Nelson et al., 2010; Swan et al., 2009, 2012).

In addition to absorbance losses, irradiation decreases the average molecular weight of terrestrial and estuarine DOM as determined by size exclusion chromatography and ultrafiltration (Helms et al., 2008; Lou and Xie, 2006; Thomson et al., 2004). In reference to these apparent changes in molecular weight, it is important to note that dissolved organic molecules appear to be considerably smaller than indicated by the above size classification techniques (Aiken and Malcolm, 1987; Simpson et al., 2002). For example, the DOM fraction recovered by ultrafiltration through a membrane with a 1-nm pore size (corresponding to ~1000 Da) is composed largely of small molecules below 1000 Da as revealed by Fourier transform ion cyclotron mass spectrometry (FT-ICR MS) (Hertkorn et al., 2006), suggesting that the physical sizes ascribed to DOM using techniques such as ultrafiltration and size exclusion chromatography pertain to the size of molecular aggregates within the sample rather than single, large macromolecules. As such, the drop in molecular weight observed during photodegradation experiments may be partly due to photodegradation resulting in a decrease in the average size of molecular aggregates (and thus a decrease in intermolecular connectivity) and to direct photodegradation of the sub-kilodalton molecules within DOM as observed when FT-ICR MS is applied to the study of photodegraded DOM (Stubbins et al., 2010). Photochemical molecular disaggregation could stem from the loss of organic structures and bonds or from the loss of inorganic ions required to stabilize the dissolved organic aggregates. Metal cations such as iron are thought to be important to the integrity of DOM aggregates (Chin et al., 1998; Simpson et al., 2002). As iron can be precipitated (Helms et al., 2013a) and the binding strength of organic ligands to iron decreases during photodegradation (Powell and Wilson-Finelli, 2003), it is plausible that the loss of this stabilizing cation plays a key role in the apparent drop in molecular weight observed during irradiation of DOM. Since the above experiments only describe photochemically induced DOM molecular weight shifts in terrestrial and estuarine DOM, it is not clear that photochemistry will also reduce the molecular weight of oceanic DOM. However, Helms et al. (2013b) observed a substantial increase in absorbance spectral slope upon simulated solar irradiation of deep sea DOM, which suggests a significant photo-induced reduction in molecular weight (Helms et al., 2008).

The absorption of sunlight by CDOM derives principally from aromatic moieties (Chin et al., 1994), which are also the main photoreactants within the DOM pool (Stubbins et al., 2008; Weishaar et al., 2003). A few studies have reported photochemically induced changes to DOM via NMR and Fourier transform infrared (FT-IR) spectroscopy, the two main tools biogeochemists use to assess DOM structure (see Chapter 2). These studies reveal that aromatic DOM is indeed the most photoreactive structural DOM component (Helms et al., 2014; Osburn et al., 2001). Recently, FT-ICR MS has been used by multiple investigators to assess the photodegradation of DOM. The extreme mass accuracy and precision of FT-ICR MS allows the exact elemental composition of ions to be calculated—a major breakthrough for the characterization of DOM and its molecular messages (Dittmar and Stubbins, 2013; Mopper et al., 2007). In each photochemical study employing FT-ICR MS, aromatic carbon or high double bond equivalent molecules constituted the most photolabile DOM fractions (Chen et al., 2014; Gonsior et al., 2009; Kujawinski et al., 2004; Stubbins et al., 2010). The average molecular weight of DOM also decreases during irradiations. This reduction in molecular weight is in broad agreement with the studies discussed earlier that used membranes or size exclusion columns to fractionate DOM by size. However,

it should not be taken as direct support for the large shifts in MW these earlier membrane and chromatography based techniques imply, as all the molecular masses identified by FT-ICR MS in initial and irradiated samples were below 1000 Da. For example, irradiation of Congo River water decreased average molecular weights for DOM within the analytical window of electrospray ionization FT-ICR MS from 424 to 409 Da (Stubbins et al., 2010).

Stubbins et al. (2010) reported FT-ICR MS data for Congo River water DOM that was irradiated under a solar simulator for 57 days and was not fractionated prior to analysis, thus providing a broad analytical window and revealing the presence of three photochemically defined pools of DOM: (1) photoresistant DOM—that survived irradiation and were present in both the initial and irradiated samples, (2) photolabile DOM—peaks that were lost during irradiation, and (3) photoproduced DOM—peaks that were formed during irradiation. Photoresistant DOM was heterogeneous with most molecular classes represented, although only a small number of aromatics and no condensed aromatics were identified. Aliphatic compounds dominated the photoproduced pool. Aromatic compounds were the most photolabile, with 90% being lost upon irradiation. Photochemistry also resulted in a significant drop in the number of molecules found and a decrease in overall stoichiometric (C:H:O) diversity. Solid-state ^{13}C-NMR data of Osburn et al. (2001) and Helms et al. (2014) for freshwater DOM samples revealed similar trends. In the Osburn et al. (2001) study, aromatic carbon (160-110 ppm) accounted for 58% of photolabile DOC and photoproduced DOM being dominated by C-alkyl carbon moieties (i.e., aliphatic, predominantly C-C; 60-0 ppm). O-alkyl (110-90 ppm) and aliphatic carbons associated with carbohydrates (90-60 ppm) also appeared to be photolabile, whereas carbonyl carbons (220-160 ppm) were largely unaltered, suggesting they were photorefractory (Table 8.1). In contrast, Helms et al. (2014) found carbohydrate-like

TABLE 8.1   Peak Areas Converted to Carbon Concentrations for Solid-State CP/MAS ^{13}C-NMR Spectra of Lacawac Sphagnum Bog DOM Samples

Carbon Type	NMR Peaks Areas				
	Carbonyl	Aromatic	O-Alkyl	Carbohydrates	C-Alkyl
Chemical shift (ppm)	220-160	160-110	110-90	90-60	60-0
Initial (μM-C) (% NMR signal × [DOC])	819	979	461	1037	1192
Post-irradiation (μM-C) (% NMR signal × [DOC])	788	561	323	807	1290
Photodegradation (μM-C) (initial—post-irradiation)	31	418	137	230	−98
% of photolabile DOC	4	58	19	32	−14

Adapted from Osburn et al. (2001) by calculating DOC-normalized changes in each NMR peak area.

and amide/peptide-like carbons were preserved during UV exposure. Thorn et al. (2010) performed liquid-state ^{13}C-NMR experiments for terrestrial humics, fulvics, and XAD-4 extracts of DOM photodegraded using a mercury lamp that emitted light at wavelengths significantly shorter than those present in solar radiation at the Earth's surface. Results confirmed the preferential photodegradation of aromatics, but also indicated carbonyls and O-alkyl carbons were photolabile. Further, ^{15}N-NMR analyses revealed that quinone/hydroquinone moieties (identified after reaction with hydroxylamine) were photodegraded more efficiently than ketone groups. The dominant photoproducts were LMW compounds such as formate, acetate, and succinate in accordance with a body of literature noting the photochemical production of LMW compounds (e.g., Bertilsson and Tranvik, 1998; Kieber et al., 1989; Mopper et al., 1991; Pullin et al., 2004). Finally, Thorn et al. (2010) reported that decarboxylated, predominantly C-alkyl and O-alkyl moieties appeared resistant even to extensive photodegradation, in agreement with FT-ICR MS results (Stubbins et al., 2010).

Further work has assessed the photolability of specific compounds and compound classes. For example, condensed aromatics, also known as dissolved black carbon (DBC), are generally thought to only be formed thermogenically, making them a specific tracer for thermally altered DOM (Dittmar, 2008). Quantified as the benzenepolycarboxylic acid (BPCA) oxidation products of condensed aromatics (Dittmar, 2008), DBC contributes ~10% to global riverine export of DOC to the oceans (Jaffé et al., 2013), contributes about 2% to the global store of oceanic DOC (Dittmar and Paeng, 2009), and represents the most refractory and most radiocarbon depleted class of organic molecules in the deep ocean (Dittmar and Paeng, 2009; Ziolkowski and Druffel, 2010). Despite being recalcitrant in the deep ocean, DBC is highly photolabile. In Congo River water, all DBC molecular signatures revealed by FT-ICR MS were lost during irradiation (Stubbins et al., 2010) and, in the open ocean, ~95% of DBC was photolabile when an unfiltered water sample of North Atlantic Deep Water was irradiated for 28 days (Stubbins et al., 2012). Further, the photolability of DBC components increased with their degree of aromatic condensation. These trends indicate that a continuum of compounds of varying photolability exists within the DOM reservoir. In this continuum, photolability scales with aromatic character such that photolability decreases from highly condensed DBC to less condensed DBC to bulk CDOM to bulk DOC. This result is akin to previous reports that the photolability of DOM molecular formulas identified by FT-ICR MS increased with the aromaticity index,

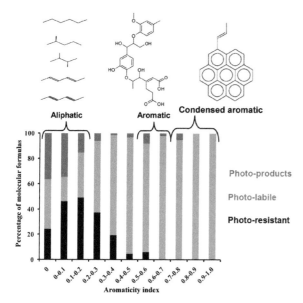

FIGURE 8.2 Bar chart depicting trends in DOM photoreactivity with aromaticity index (higher number = more double bonds). Blue indicates molecular formulae unique to initial Congo River water, red indicates those unique to photobleached Congo River water, and black indicates those common to both samples (Stubbins et al., 2010).

an indicator of molecular condensation (Stubbins et al., 2010; Figure 8.2).

Lignin is an aromatic biopolymer produced by vascular plants that is often used as a tracer of tDOM in the oceans (see Chapters 2 and 11). Several studies have determined the photolability of natural lignin-derived phenols (Hernes and Benner, 2003; Opsahl and Benner, 1998; Spencer et al., 2009), and others have studied the photochemistry of model lignin compounds (Lanzalunga and Bietti, 2000; McNally et al., 2005; Vähätalo et al., 1999). Irradiations of unfiltered natural waters indicate that lignin is highly photolabile, with 75% lost during 28-day irradiations of Mississippi River water (Opsahl and Benner, 1998) and more than 95% lost during 57-day irradiations of Congo River water (Spencer et al., 2009). In the only study to quantify lignin photolability in open ocean samples, Opsahl and Benner (1998) irradiated a DOM isolate obtained by ultrafiltration. Lignin within the low salt ultrafiltrate solution was not photodegraded after 4 days exposure to sunlight, suggesting that lignin-derived phenols within the ocean may not be as photolabile as those in freshwaters. However, the photolability of lignin in the natural seawater matrix is not known.

Photodegradation of DOM is driven by a mixture of direct photochemical reactions and indirect photoreactions usually involving ROS and energy transfer reactions. The most potent ROS are short-lived in natural waters and do not travel far from their source of production. Assuming reactive species are dominantly produced by aromatic moieties, which are the main DOM chromophores (Mopper and Kieber, 2000), the inherent proximity of these chromophores to the production source of degradative reactive species may be a secondary reason for the photolability of aromatics. One of the main ROS responsible for indirect photodegradation reactions is the hydroxyl radical (Mopper and Zhou, 1990; Zafiriou, 1974). In seawater, the OH radical is efficiently scavenged by halide ions (mainly bromide) (Zhou and Mopper, 1990), yielding reactive halide species, including reactive halide radicals (RX), and to a lesser extent the carbonate anion yielding the carbonate radical. The OH radical is a nonspecific reactant compared to RX. The latter targets electron-rich DOM chromophores, based on observations that the addition of chloride and bromide enhances CDOM absorbance photobleaching relative to both fluorescence and DOC photodegradation (Grebel et al., 2009). The importance of aromatic-seeking halide radicals to the biogeochemistry of marine DOM, the optics of natural waters, and the photolability of aromatic tracers such as lignin and DBC remains unclear.

There is still significant debate about the fate of tDOM in the oceans (see Chapters 2, 6, and 11). In order to track tDOM, tracers such as CDOM, CDOM spectral properties, humic-like florescence, lignin, and the $\delta^{13}C$ signature of DOC have been applied. As previously discussed, only 54-69% of tDOC is photolabile, whereas in the same or similar experiments, greater than 90% of terrestrial tracers, such as CDOM, humic-like fluorescence, and lignin compounds, are lost during comparable irradiations. Thus, the photochemically refractory tDOC pool is substantially depleted in these tracers. Due to the presence of aromatic-seeking halide radicals, this depletion of aromatic, terrestrial tracers may be even greater in marine waters. The photoresistant pool of terrestrial DOC is also enriched in ^{13}C relative to the starting material (Opsahl and Zepp, 2001; Osburn et al., 2001; Spencer et al., 2009), thereby increasing terrestrial DOC δ^{13}C enrichment (~−30 to −25‰) toward marine DOC values (~−23 to −18‰) (Hedges et al., 1997; Benner et al., 2005;see Chapter 6). These tDOM photofractionations transform terrestrial DOM to "look" more like marine DOM, complicating the task of tracking and quantifying terrigenous DOC in the oceans (Helms et al., 2014; Minor et al., 2006; Stubbins et al., 2010).

IV IMPACT OF PHOTOCHEMISTRY ON OTHER MARINE PROCESSES

A Particles and Photoflocculation

Photochemical processes occurring within or on the surface of living or nonliving POM or organic matter coating inorganic particles in the oceans have generally received very little attention because most of the organic matter in the oceans is present in the dissolved phase (POC is typically <2% of TOC). However, a particulate-phase photochemical process may be important if there is no corresponding dissolved phase photochemical reaction that occurs at an appreciable rate. For example, it has been known for some time that particles can play an important role in the photochemical transformations of compounds that are mainly associated with particulate matter, including hydrophobic, particle-bound pollutants (Miller and Zepp, 1979;

Zepp and Schlotzhauer, 1983) and phytoplankton pigments (Nelson, 1993; SooHoo and Kiefer, 1982a,b; Yentsch and Reichert, 1962). Likewise, if a particulate photoprocess proceeds by a static mechanism that does not involve diffusion of dissolved-phase reactants to the particle surface but instead only involves photochemical reactions of reactants that are sorbed onto particulate matter, then the particulate photoprocess may be important under the dilute conditions that exist in seawater. Many particulate-bound pollutants and pigments will photolyze via a static mechanism that will not involve other reactants either on the particle surface or in the dissolved phase. By contrast, particulate-phase reactions that proceed via a dynamic mechanism that may involve diffusion of dissolved-phase reactants or photosensitizer molecules to the particle surface will less likely be important in seawater, especially if the same photoprocess occurs in the dissolved phase.

Visible and UV-initiated pigment photodegradation in algal detritus and POM has been studied in some detail to examine specific transformations of several important chlorophyll compounds and their phytol side chains and to evaluate the possible use of the products as photochemical biomarkers (Cuny and Rontani, 1999; Cuny et al., 1999; Rontani and Marchand, 2000; Rontani et al., 2003). As with algal pigments, some other lipids readily undergo photooxidation during reaction with ROS, including singlet oxygen, superoxide, and the hydroxyl radical (for review, see Girotti, 2001). Several studies examined fatty acid and sterol photooxidation reactions in marine phytodetritus and senescent algal cells to gain mechanistic insights, identify photoproducts, and evaluate their use as photochemical biomarkers (Marchand and Rontani, 2001, 2003; Marchand et al., 2005; Rontani, 2001; Rontani and Marchand, 2000; Rontani et al., 2009). Mayer et al. (2009a) photolyzed phytodetritus to quantify pigment loss in the context of chromophoric POM loss and production of DON, DOM, and ammonium.

Given the diversity and high concentrations of POM in phytodetritus, photochemical degradation of phytodetrital POM may form crosslinked fluorescent products (Kikugawa and Beppu, 1987; Yentsch and Reichert, 1962) that transform labile organic matter to refractory DOM (Hansell, 2013) in the oceans.

A second important area of POM photochemistry research has focused on the photochemical conversion of POM to DOC or DON or on the formation (or release) of specific photoproducts. With few exceptions, these studies have involved exposure of particle-rich samples (200 mg-24 g sediment per liter of seawater or estuarine water) to a solar simulator or (in a few cases) sunlight from hours to days. Results obtained from these studies are similar to what has been obtained in the dissolved phase. For example, exposure of sediment slurries to natural or simulated solar radiation results in the release of nutrients, including DON, ammonium, soluble reactive phosphorus (SRP), and loss of dissolved iron (Estapa and Mayer, 2010; Mayer et al., 2009b, 2011; Southwell et al., 2009). However, several studies found no production of SRP, nitrate, or nitrite (Mayer et al., 2011; Riggsbee et al., 2008; Southwell et al., 2009). As may be expected, irradiation of particulate matter results in photomineralization (or heterotrophic respiration) of POC to DIC (Estapa and Mayer, 2010; Riggsbee et al., 2008), oxygen consumption (Estapa and Mayer, 2010), production of CDOM from CPOM (chromophoric particulate organic matter) (Pisani et al., 2011), CPOM photobleaching (Mayer et al., 2009a), and loss of the terrestrial lignin biomarkers from the particulate phase (Mayer et al., 2009b). Absorbance and EEM PARAFAC spectra indicate that the DOM produced from POM has humic-like optical properties (Shank et al., 2011). Several studies have shown that POM is partly converted to DOC (Estapa and Mayer, 2010; Kieber et al., 2006; Mayer et al., 2006, 2009a,b, 2011; Pisani et al., 2011; Riggsbee et al., 2008; Shank et al., 2011), with equal contributions from both the UV and visible regions of the solar spectrum (Kieber et al., 2006; Mayer et al., 2006). Despite these and other changes, the bioavailability of DOM produced from POM was only slightly greater than observed in the dark controls over a 60-day incubation of Gulf of Mexico coastal sediments (Mayer et al., 2011).

Several POM studies have involved examination of particles not derived from sediments. Based partly on preliminary evidence presented in Stubbins et al. (2006a), Xie and Zafiriou (2009) and Song et al. (2013) showed that photochemical production of CO from the dissolved and particulate phases was important not only in coastal waters but in the open ocean as well. Song et al. (2013) determined AQY for CO photoproduction from DOM and POM in waters from the Beaufort Sea and Mackenzie River estuary and shelf. Particulate and dissolved-phase CO AQY decreased exponentially with increasing wavelength, but whereas dissolved-phase AQY spectra were primarily confined to the UV, particulate AQY spectra were flatter and extended well into the visible. This difference was apparent in modeled depth profiles for CO photoproduction, which showed the increasing importance of POM CO photoproduction with increasing depth. Overall, POM contributed from ~10 to 30% of the observed CO photoproduction in these Arctic waters. In Antarctic waters, del Valle et al. (2007) determined that exposure of unfiltered Ross Sea seawater samples to solar radiation resulted in the production of dissolved DMSO at concentrations greater than observed in the 0.2 μm-filtered controls. For marine gels 0.3-0.7 μm in size, solar UV radiation strongly inhibited gel formation, degrading the gels to lower molecular weight nanogels (Orellana and Verdugo, 2003; Orellana et al., 2011; see Chapter 9).

Although substantial advances have been made in understanding marine POM photochemistry, there are many questions unanswered as well as inherent limitations to current methods. In particular, in nearly all the studies discussed above, it has not been demonstrated that particulate-phase photochemistry makes an important contribution to the overall cycling of organic matter in an estuarine or coastal setting. It has also not been possible to directly intercompare results from different studies. This lack of quantitative assessment and ability to cross compare results is due to several important limitations, including the lack of information regarding the grain size or surface area and composition of exposed sediment particles, the presence of microorganisms and algae, and the general lack of quantification of the *in situ* light exposure in all studies except for Estapa et al. (2012) and Song et al. (2013). In addition, except for Song et al. (2013), no one has directly compared or modeled POM photochemical rates, irrespective of whether they involve specific biomolecules such as pigments or lipids or ill-defined, particle-bound humic or fulvic acids, mainly because the actinic light field has not been quantified in unfiltered samples in most studies. Furthermore, particle absorption and scattering need to be accounted for, as done in Mayer et al. (2006), Estapa et al. (2012) and Song et al. (2013). Measurement of particle absorbance and the use of chemical actinometers or *in situ* fiber-optic spectroradiometer probes placed in unfiltered samples are needed to advance this area of study. However, appropriate visible actinometers are not available and so need to be developed. The second limitation inherent in nearly all POM photolysis studies is that the unfiltered samples used in these studies are not axenic. Therefore, it is not possible, for example, to assess whether DIC production is due to its photochemical production from POM, heterotrophic respiration, or a combination of both. In some cases, it may be possible to add a biological inhibitor (e.g., Tuominen et al., 1994), but inhibitor additions need to be carefully assessed with controls. Xie and Zafiriou (2009) used sodium azide to inhibit microbial consumption to assess particle-associated CO photoproduction.

Southwell et al. (2009) autoclaved selected samples, but autoclaving is not recommended because it will change the sample pH, cause precipitation of Mg and Ca (Jones, 1964), and likely produce microbial detritus and phyto detritus that may photochemically degrade in light-exposed samples to release a plethora of compounds that may not otherwise be observed. Despite these limitations, there is growing evidence that photochemical processes occurring in the particulate phase may be important in particle-rich environments, especially in estuarine settings, and for specific compounds (e.g., pigments, CO) particulate-phase photochemistry may be significant even in the open ocean.

While photodissolution of marine POM and resuspended sediments has been extensively examined, the opposite process, photoflocculation, has received relatively little attention. To date, photoflocculation has been observed only for DOM-rich freshwaters (Gao and Zepp, 1998; Helms et al., 2013; Maurice et al., 2002; Porcal et al., 2013; Scully et al., 2004; von Wachenfeldt et al., 2008). With the exception of Helms et al. (2013a), no studies have examined the chemical composition of the flocs or investigated the coagulation mechanisms. Helms et al. (2013a) observed that, after 30 days of simulated solar UV irradiation of 0.1 μm filtered Great Dismal Swamp (Virginia) water, 7.1% of the DOC was converted to POC while 75% was remineralized. Approximately 87% of the iron was removed from the dissolved phase after 30 days, but iron did not flocculate until a major fraction of DOM was removed by photochemical degradation and flocculation (>10 days); thus, during the initial 10 days, there were sufficient organic ligands present and/or the pH was low enough to keep iron in solution. Although photoflocculation of iron did eventually occur, it is not clear if iron is required for the initial flocculation of DOM. Using NMR and FTIR techniques, Helms et al. (2013a) found that photochemically flocculated POM was enriched in aliphatics and amide functionality relative to the residual non-flocculated

DOM, while carbohydrate-like material was neither photochemical degraded nor flocculated. Based on this spectroscopic evidence, Helms et al. (2013a) proposed several mechanisms for the formation of the flocs during irradiation. They also speculated that abiotic photochemical flocculation may remove a significant fraction of tDOM and iron from the upper water column between headwaters and the ocean, including estuaries (Helms, 2012). However, none of the previous studies, including those of Helms, determined if photoflocculation occurs at the basic pH values and higher ionic strengths found in estuaries, coastal waters, or the open ocean.

B Sea Ice

At its greatest seasonal extent, sea ice covers ~15% of the world's oceans and represents an extensive surface with high albedo and profound effects on deep water formation and climate. Likewise, sea ice is a dynamic interface in the polar seas serving as an important ecological and photochemical medium, with important impacts on gas exchange and tropospheric chemistry (Grannas et al., 2007a; Law et al., 2013). It is a highly heterogeneous medium pocketed with brine channels and pores that contain rich and diverse microbial and algal communities that thrive in this extreme environment (Thomas and Dieckmann, 2002). Considering the richness of the sea ice biological community, it is not surprising that polar sea ice also contains high concentrations of organic matter relative to seawater (Song et al., 2011), especially towards the base of sea ice, which often has a yellow-brown color. During sea ice formation, freezing also enriches chromophoric and fluorophoric DOM relative to salts (Müller et al., 2011). Not all sea ice is enriched in organic matter; Baltic Sea ice has DOC and DON concentrations lower than in the underlying seawater (Stedmon et al., 2007b). In polar waters, the organic matter associated with sea ice contains CDOM (Xie and Gosselin, 2005; Xu et al., 2012), particulate organic matter

(Xu et al., 2012), and other compounds (e.g., mycosporine amino acids, nitrate) that absorb and scatter actinic solar radiation (Uusikivi et al., 2010). This absorption is expected to result in a wide range of photochemical transformations in sea ice similar to those observed in seawater, and these transformations are expected to have important impacts on the ecology of the sea ice, its optical properties, and the production of atmospherically important trace gases. However, there is very little direct evidence for sea-ice photochemical transformations other than a handful of observations, including photoproduction of hydrogen peroxide and carbon monoxide (King et al., 2005; Klánová et al., 2003; Song et al., 2011; Xie and Gosselin, 2005) and photochemical loss of DMS (Asher et al., 2011; Hellmer et al., 2006).

There is also considerable interest in the snowpack covering the sea ice because it represents a rich photochemical medium with a decidedly atmospheric organic and inorganic signature (Grannas et al., 2007a). Unlike sea ice, snowpack photochemistry has been studied intensively; but nonetheless, neither medium is understood very well from a photochemical standpoint. Photolyses of snowpack constituents result in the production of several important atmospheric species including NO_x (NO, NO_2, and NO_3 radicals) (Boxe et al., 2006), HONO (Zhou et al., 2001), Hg (Bartels-Rausch et al., 2010), carbonyl compounds (Sumner and Shepson, 1999), alkenes (Swanson et al., 2002), and halocarbons (Adams et al., 2002; Swanson et al., 2002). In some cases, photochemical production in the snowpack is the main source of these constituents in the marine boundary layer.

One important finding in snowpack studies that should be relevant to sea ice is that the photochemistry within and on the surfaces of these media cannot necessarily be extrapolated based on results obtained in the associated aqueous phase (Grannas et al., 2007b). For example, rates may be much faster in ice than predicted based on studies in the aqueous phase (Kahan et al., 2010) or they can be nearly the same as observed for pyruvic acid photolysis (Guzmán et al., 2007). Salts have been shown to reduce photochemical rates in ice but not in water (Kahan et al., 2010), and therefore, it will be important to determine whether photochemical transformations occur within the ice or in brine pockets. One difficulty in studying sea ice photochemistry that has not been problematic in snowpack studies is how to best quantify the light field and light dose for photochemical (or photobiological) studies. In snowpack, chemical actinometers and radiometers have been used (for a review see Grannas et al., 2007a), but these methods have not been developed for sea ice studies and will require some unique capabilities. Another major difficulty is how to interpret results of sea ice photochemical experiments that include the microbial-algal sea ice community. Simple filtration used to remove organisms and particles from seawater will likely not translate well to ice studies because the freezing-thawing process will undoubtedly release DOM from POM.

C Siderophores and Toxins

Siderophores. Iron is a limiting nutrient throughout large regions of the oceans, and owing to its scarcity many heterotrophic bacteria and at least some cyanobacteria go to great lengths to acquire iron from the sea by producing high-affinity, iron-binding siderophores. Unlike prokaryotes, eukaryotic phytoplankton generally do not produce or use siderophores, other than possibly through an algal-bacterial mutualism or a cell surface reductase mechanism (Amin et al., 2009a; Maldonado and Price, 2001), even though they have a relatively high cellular requirement for iron. Given the essential biological requirement for iron and the commonplace occurrence of strong and weak iron-binding ligands in the open oceans (Boyd and Ellwood, 2010; Gledhill and Buck, 2012), many studies have been undertaken to isolate, identify, and determine the properties of marine siderophores including their photochemical lability.

The archetypical siderophores that contain only hydroxymate or catecholate moieties that bind iron with high affinity are not photochemically reactive when complexed to iron (Barbeau et al., 2003) even though they contain a LMCT absorption band in the UV-blue portion of the solar spectrum. The uncomplexed, hydroxymate siderophores are also photostable, but this is not the case for the free, uncomplexed catecholate-based siderophores that readily undergo photolysis in seawater (Barbeau et al., 2003). Many of the marine siderophores that have been produced in culture have hydroxymate or catecholate iron-binding ligands present but also have in common the presence of one or more α-hydroxy carboxylic acid ligands in the form of either citrate or β-hydroxyaspartic acid (Vraspir and Butler, 2009). All marine siderophores isolated to date that contain the α-hydroxy carboxylic acid moiety undergo photolysis when complexed to iron but are photostable when exposed to sunlight as the uncomplexed ligand (Barbeau, 2006).

The aquachelin-Fe complexes were the first marine mixed-ligand siderophores shown to undergo photolysis when exposed to sunlight (Barbeau et al., 2001). Photolysis results in decarboxylation of the siderophore and reduction of iron (Figure 8.3). Interestingly, the photolyzed siderophore retains its strong binding affinity for Fe(III), albeit not as strong as the non-photolyzed siderophore (Barbeau et al., 2001; Grabo et al., 2014; Kupper et al., 2006). Several studies have since demonstrated that photolysis occurs for a range of marine siderophores, including the synechobactins, the ochrobactins, aquachelin, aerobactin, and petrobactin, resulting in the oxidation of the siderophore and photochemical reduction of Fe(III) to Fe(II) (Amin et al., 2009a,b; Barbeau et al., 2002, 2003; Grabo et al., 2014; Ito and Butler, 2005; Kupper et al., 2006). Marine siderophores also cause the light-induced dissolution of colloidal iron hydroxides; however, they are not directly involved in surface photochemical reactions but rather function as the transfer agent for the iron from the particulate to dissolved phase (Borer et al., 2005).

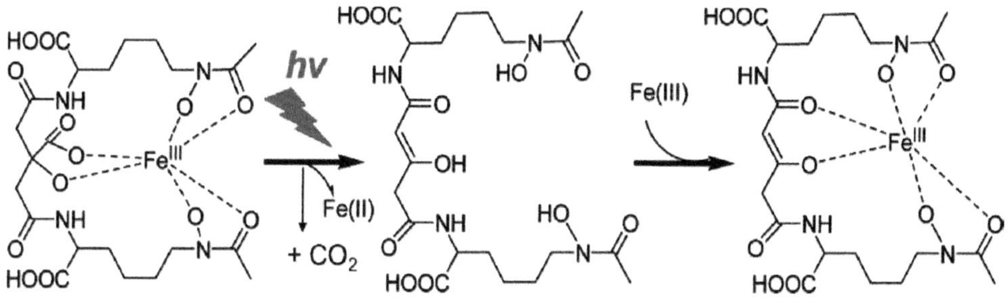

FIGURE 8.3 Photoreaction of Fe(III)-aerobactin and production of Fe(II) and CO_2. *Adapted from Kupper et al. (2006).*

Although siderophores are important iron-binding ligands, they have, in nearly all cases, only been produced in marine microbial cultures. It is not known if they represent the ubiquitous strong and weak iron-binding ligands that have been detected in seawater, and, even if they do, they may not be quantitatively important, as it has been suggested that carbohydrates and colloids may complex most of the iron in the oceans (Benner, 2011). Marine siderophores have not been observed in the open ocean and the genes that encode their production or uptake have not been detected. In fact, these genes are conspicuously absent in open ocean samples (Hopkinson and Morel, 2009). Likewise, unlike the marine siderophores that have been isolated from culture, the strong and weak ligands that have been detected in seawater generally do not photolyze (Barbeau, 2006; Moffett, 2001). However, it is interesting to note that when marine siderophores are photolyzed via decarboxylation (for example), they appear to be stable to further photochemical degradation. It is possible that the photochemically stable strong and weak ligands are in part derived from photochemical remnants of marine siderophore precursors produced by prokaryotes. Indeed, the stability constants of photolyzed siderophore-Fe(III) complexes are remarkably similar to binding constants determined for the L_2-type ligands found in seawater (Barbeau, 2006).

Toxins. The occurrence of harmful algal blooms (HABs) are increasing worldwide (Chretiennot-Dinet, 1998). Toxin levels of brevetoxins, ciguatoxin, saxitoxin, gonyaotoxin, nodularin, or domoic acid associated with HABs are also expected to increase (Richardson, 1997). Some bloom-forming cyanobacteria also form toxic HABs, although they are not common and are mostly associated with fresh waters. Several species of the marine diatom *Pseudo-nitzchia* produce a neurotoxin, domoic acid, a tricarboxylic amino acid that complexes at least two metals, Fe^{3+} and Cu^{2+} (Rue and Bruland, 2001).

A few toxins are known to undergo photolysis in seawater. For example, domoic acid photolyzes in coastal seawater, estuarine water, and high-purity laboratory water with and without added humic acids, copper, or iron (Bates et al., 2003; Bouillon et al., 2006). The observed first-order rate constant, k, at temperate latitudes during the summer is ~0.15 h^{-1}, resulting in a surface water turnover time of 6.7 h. The photolysis rate is nearly same under all conditions, and is independent of pH or dissolved oxygen concentration but has a small temperature dependence ($E_a = 13$ kJ mol^{-1}) (Bouillon et al., 2006). These results are consistent with a direct photochemical loss for domoic acid due to its weak absorption in the UV (Bouillon et al., 2006; Falk et al., 1989). Indirect photolysis through reaction with ROS or via an energy transfer does not occur at appreciable rates under solar radiation, although photosensitized reactions cannot be entirely ruled out (Bouillon et al., 2006). In contrast to the Bouillon et al. (2006) results, Fisher et al. (2006) observed that direct photolysis was a minor loss pathway for low μM levels of domoic acid compared to indirect pathways involving DOM or iron. They argued that the Fe-domoic acid complex was needed for domoic acid to undergo photolysis in solution; this reaction was inhibited by added phosphate presumably due to the phosphate binding the iron. The reason for this discrepancy between these studies is not known, although it may have resulted from the high domoic acid, metal, and sensitizer concentrations used by Fisher et al. (2006), which are orders-of-magnitude above ambient levels and, thus, may have favored secondary pathways that would not be competitive at the low nM concentrations used in the Bouillon et al. (2006) study.

Wavelength-dependent quantum yields for photolysis decrease rapidly in the UV from 0.2 at 280 nm to 0 at 404 nm in Monterey Bay seawater. The corresponding spectral dependence of the photolysis rate (i.e., the cross-product of the action spectrum and the solar spectral irradiance) has a peak response at ~330 nm, with >95% of the photolysis between 300 and 380 nm; modeling photolysis rates indicates that photolysis is important only in the upper few meters of the water column in coastal waters. Depths corresponding to 1% of surface rates are 0.8, 2.3, and 3.4 m for Prince Edward Island, Washington State coast, and Monterey Bay, respectively (Bouillon et al., 2006).

Brevetoxin PbTx-2 is one of a number of potent neurotoxins produced by the bloom-forming dinoflagellate *Karenia brevis*. Unlike domoic acid, PbTx-2 undergoes photolysis that is sensitized by CDOM and trace metals (Pitt, 2007). At least 18 photoproducts are formed, several that are novel and all of which appear to be photostable (Hardman et al., 2004). As seen with many compounds, PbTx-2 photolysis kinetics is first order. The average first-order rate constant and turnover time for PbTx-2 degradation in coastal seawater is $0.20\,h^{-1}$ and 6 h at 25°C. The rate constant decreases to nearly half when PbTx-2 is added to UV-irradiated seawater, consistent with DOM-sensitized photolysis (Pitt, 2007). Preliminary evidence also suggests that the other brevetoxins, including PbTx-1, -3, -6, and -9, undergo photolysis in seawater (Pitt, 2007).

Considering the plethora of toxins produced by HAB species (Plumley, 1997) and the many others yet to be discovered, it is likely that many of these toxins are photochemically reactive in seawater. However, other that domoic acid and to a lesser extent the brevetoxins, the importance of photolysis with respect to marine toxin cycling is poorly known. The photochemical stability and toxicity of the photoproducts is also virtually unknown. Since most HABs occur in coastal waters, it is also possible that surface photochemistry of toxins on particulate matter will be important. It is expected that this emerging area of research will receive considerable attention in the coming years owing to its importance in carbon cycling, marine ecology, and human health.

V MODELING PHOTOCHEMICAL RATES AND IMPACT ON MARINE CARBON CYCLING

Although photochemistry is known to play a significant role in the biogeochemistry of DOM, real-world rates of photochemical processes remain poorly constrained. For example, photochemical turnover time estimates range between 30 to 800 years in the case of dissolved black carbon (Stubbins et al., 2012) and 500 to 2100 years in the case of the refractory deep ocean DOC (Mopper et al., 1991). For site- and time-specific studies, the rate of a photoreaction at a given depth, for a given sample, on a given day can be empirically determined using drifter buoy experiments where a quartz flask filled with sample is irradiated at natural light levels by being suspended in the water column (e.g., del Valle et al., 2007; Kieber et al., 1997a, 2014; Qian et al., 2001; Toole et al., 2006; Yocis et al., 2000). Although this approach offers an empirical measure of photochemical rates at water column light levels, it is difficult to extrapolate such data to larger temporal and spatial scales. Therefore, for studies at large spatial and/or seasonal scales, most researchers couple reductionist laboratory quantification of photochemical rates to models of sunlight irradiance and light penetration within oceanic waters.

The first step in the latter approach is to determine the efficiency of the photoreaction being studied (i.e., AQY). For CDOM, the exact nature, much less the molar concentrations and molar absorption coefficients, of the chromophores are unknown. Therefore AQYs are reported. As the photons absorbed by CDOM are not all incident upon molecules capable of a given photoreaction, the calculated AQYs are significantly lower than true quantum yields. Additionally, quantum yields for pure compounds undergoing a direct photolysis are generally independent of wavelength (Calvert and Pitts, 1966). However, CDOM is a complex mixture of organic compounds dissolved in an aqueous matrix together with a suite of inorganic solutes, and empirical evidence reveals that AQYs involving CDOM generally decrease exponentially with increasing wavelength (e.g., Blough, 1997; Gao and Zepp, 1998; Miller et al., 2002; Powers and Miller, 2014; Valentine and Zepp, 1993; Weiss et al., 1995; Zepp and Andreae, 1994). Consequently, spectrally and temperature-dependent AQYs need to be determined when studying CDOM-driven photoreactions. To accomplish this, researchers have determined photoreaction efficiency at various wavelengths under narrow (typically <20nm; e.g., Valentine and Zepp, 1993) or broadband (using long pass filters e.g., Johannessen and Miller, 2001) irradiations.

The AQYs for CO production are among the most commonly measured. As for most CDOM photoreactions, CO AQY spectra decrease with increasing wavelength. In addition, CO AQYs at a given wavelength are higher for high CDOM terrestrial waters than for low CDOM marine waters (Gao and Zepp, 1998; Kettle, 1994; Valentine and Zepp, 1993; White et al., 2010; Zafiriou et al., 2003; Ziolkowski and Miller, 2007). CO AQYs at a given wavelength also fall rapidly with

increasing absorbed photon dose (i.e., irradiation time; Stubbins, 2002; Zhang et al., 2006) indicating that the more extensive photodegradation of marine CDOM (i.e., irradiation history) may be responsible for its lower CO AQYs. Although the mechanism causing lower CO AQYs in the ocean remains unclear, CDOM absorption coefficients can provide a good proxy for CO AQYs across the river-estuary-ocean transition (Stubbins et al., 2011). These empirical proxies may provide a means to remotely sense CO AQYs in natural waters and to estimate CO AQY spectra and CO photoproduction in estuarine systems from discrete measurements of CDOM absorption (Stubbins et al., 2011).

In the open ocean, CO wavelength-dependent AQYs vary minimally over large spatial scales so photochemical studies have used constant AQYs to estimate global open ocean CO photoproduction (Fichot and Miller, 2010; Stubbins et al., 2006a; Zafiriou et al., 2003). Once a CO AQY spectrum is defined, the next step in estimating a rate of CO photoproduction is to determine the photon flux absorbed by CDOM. In order to achieve this, researchers model solar irradiance reaching the surface of the ocean, the penetration of sunlight into the water column, and the proportion and spectral dependence of the water column light field absorbed by CDOM. The reader is referred to Fichot and Miller (2010), Miller et al. (2002), Moran and Zepp (1997), Powers and Miller, 2014; Smyth (2011), Stubbins et al. (2006a), and Zafiriou et al. (2003) for various models and assumptions used in determining the photon flux absorbed by CDOM in the ocean. In the open ocean, CDOM dominates light absorbance at photochemically relevant wavelengths between ~280 and 400nm (Stubbins et al., 2006a), whereas in turbid estuarine waters particles compete for available light (Stubbins et al., 2011). Global models of CO photoproduction have used average CDOM absorbance spectra to estimate the proportion of light absorbed by CDOM (Stubbins et al., 2006a; Zafiriou et al., 2003). However, variations in estimates of CDOM light absorption can have a significant influence on the proportion of light absorbed by CDOM and resultant calculations of photochemical rates (Powers and Miller, 2014; Reader and Miller, 2011). An improvement in this regard utilizes remote sensed CDOM levels to provide an estimated global ocean CO photoproduction rate ($\sim$41$\times$10^{12} g-C year^{-1}; Fichot and Miller, 2010). Importantly, use of regionally and temporally resolved solar irradiance and CDOM levels provide similarly resolved estimates of CO production and the ability to determine depth resolved rates of CO production (Fichot and Miller, 2010).

At present, our best estimates of the rate of global open ocean DOC photomineralization to DIC are based on simple and perhaps flawed extrapolations from CO photoproduction rates assuming average ratios of DIC:CO photoproduction, the latter ratios coming from broadband irradiations. Based on the reported DIC:CO photoproduction ratio of $\sim$14 (Gao and Zepp, 1998; Miller and Zepp, 1995), global open ocean DIC photoproduction would range between 0.4 and 1.2$\times$10^{15} g-C year^{-1} (Fichot and Miller, 2010; Stubbins et al., 2006a; Zafiriou et al., 2003). Recent work by Reader and Miller (2012) indicates the ratio of DIC:CO production is closer to 20, in which case CO photoproduction would suggest a global oceanic DIC photoproduction rate of 0.6-1.7$\times$10^{15} g-C year^{-1}. Although this rate is modest compared to net primary production in the open ocean ($\sim$50$\times$10^{15} g-C year^{-1}; Field et al., 1998; del Giorgio and Williams, 2005), it is significant when it is considered that nearly all of the carbon fixed by phytoplankton is respired back to DIC in surface waters such that only a tiny percentage accumulates as DOC in the ocean (Hansell and Carlson, 1998; Robinson and Williams, 2005) and that photochemistry appears to target aromatic organics that plankton struggle to mineralize. Furthermore, open ocean photomineralization rates are sufficient to mineralize all of the riverine DOC delivered to the oceans ($\sim$0.26$\times$10^{15} g-C year^{-1}; see Chapter 11), exceeds organic carbon burial in deep ocean sediments (0.01$\times$10^{15} g-C year^{-1}; Dunne et al.,

2007) by at least an order of magnitude, and is similar to rates of burial in coastal and shelf sediments ($\sim$0.7$\times$10^{15} g-C year^{-1}; Dunne et al., 2007).

In temperate estuaries, photomineralization removes minor amounts of the riverine DOC (White et al., 2010), with $\sim$1% of riverine DOC inputs to the Tyne estuary in northeast England photomineralized in the estuary (Cai and Wang, 1998; Stubbins et al., 2011). In the Arctic, Bélanger et al. (2006) report that $\sim$3% of Mackenzie River DOC export could be photomineralized in the Beaufort Sea, although this percentage may double if Arctic warming leads to ice-free waters. Other work in the temperate Mid-Atlantic Bight used a novel, remote-sensing approach to estimate DOC mineralization rates (Del Vecchio et al., 2009). First, levels of CDOM on the shelf were estimated between 1998 and 2007 using data from the satellite-based SeaWiFS ocean color sensor. A deficit in CDOM stocks on the shelf was noted in summer months, being attributed to the photobleaching of terrestrial DOC. This CDOM deficit was then converted to a DOC loss (CO_2 production) based on relationships between DOC and CDOM absorbance in the region, yielding an estimate of Mid-Atlantic Bight DOC mineralization (3-7$\times$10^{10} g-C year^{-1}; Del Vecchio et al., 2009). Such an approach could be readily expanded to other regions although it comes with a number of caveats, not least the assumption of a minimally varying CDOM:DOC ratio, a ratio known to vary strongly during photodegradation of DOM (see Section III and see Chapter 10). The CDOM-deficit approach may also be a useful tool for constraining photobleaching of CDOM and could be of great value if used in conjunction with more reductionist approaches for defining photochemical rates (e.g., Fichot and Miller, 2010).

VI FUTURE DIRECTIONS

A Mechanistic Studies

Relatively few comprehensive photochemical mechanistic studies involving DOM have been carried out in seawater, and those that have been conducted have concentrated on identifying reactive transients and on measuring their steady-state concentrations, rate laws, and production rates in seawater (e.g., Blough, 1997; Blough and Zepp, 1995; Faust, 1999; Kieber et al., 2003). Little is known about the major reactive sites (or chromophores) within marine DOM that are responsible for production of reactive transients and stable photoproducts in the upper ocean. In order to develop a predictive and mechanistic understanding of the rates of and controls on important photochemically produced compounds, it will be necessary to: (1) identify the DOM precursors and photosensitizers (autochthonous/microbial vs. terrestrial/higher plant derived) that affect specific photochemical reactions and determine how these precursors vary spatiotemporally in the oceans, (2) develop additional selective probes for reactive species suitable for a seawater medium, and (3) track element mass balances resulting from photochemical transformations.

B Photoproduction and Air-Sea Exchange of Important Atmospheric Trace Gases and Volatile Organic Compounds

Climate models are greatly affected by the degree to which air-sea gas exchanges of CO, OCS, DMS, alkyl nitrates gases, and other volatile organic compounds (VOCs) are influenced by photochemical reactions at the sea surface (Blough, 1997; Doney et al., 1995; de Bruyn et al., 2011; Erickson III et al., 2000; Fischer et al., 2012; Gnanadesikan, 1996; Méndez-Díaz et al., 2014; Najjar et al., 1995). Studies are needed to quantify the photoproduction (or loss) of CO_2 and volatile sulfur species in surface seawater and to determine its effect on their air-sea exchange, as has been done for other reactive species in the sea surface microlayer (Thompson and Zafiriou, 1983). Furthermore, the impact of changes in global UV radiation on air-sea fluxes of important trace gases will need to be assessed. Another uncertainty is whether photochemical reactions (production and destruction) on the surface microlayer affect the flux of volatile species at the air-sea interface (Blough, 1997). These reactions have not been taken into account in air-sea flux models as the microlayer has historically been assumed to be passive, but there is a growing body of evidence indicating that the microlayer is a very dynamic interface (Cunliffe et al., 2013). Even though the residence time of the material in the microlayer is relatively short (Blough, 1997), steady state concentrations of potential reactants there can be high relative to bulk seawater (Cunliffe et al., 2013).

C DOM Marine Food Web Dynamics and Trace Metals

In our original review (Mopper and Kieber, 2002), we suggested that the photochemical processes by which biorefractory DOM is made bioavailable, and by which bioavailable DOM is made unavailable, need to be identified. Although much work has been done in this direction, there remains a need to further resolve these processes to develop a predictive understanding of changes in DOM lability due to coupled photochemical-biological transformations. In this context, quantifying the effect of photochemistry on BGE is essential. In a related area, the role of extracellular release of DOM as a photoprotective mechanism in autotrophs and heterotrophs needs to be examined, i.e., what is the role or effect of externally (and internally) produced reactive phototransient species (e.g., free radicals) on the growth of autotrophs and heterotrophs? What factors determine whether photochemically altered DOM becomes more or less bioavailable? What potentially inhibitory substances are formed or destroyed, e.g., HAB toxins, methyl mercury?

Photochemistry can also affect trace element speciation, including complexation and redox state, by several processes (Blough and Zepp, 1995; Faust, 1999), which in turn impacts trace metal bioavailability/toxicity and food web dynamics. Redox-sensitive trace elements include iron, copper, manganese, chromium, mercury, bromine and iodine (Méndez-Díaz et al., 2014; Wells, 2002). For example, photochemical reactions alter the concentration and reactivity of organic ligands involved in the complexation/solubilization of trace metals (Moffett, 1995). The mechanisms by which these photoredox reactions occur, and the nature of the DOM ligands involved, including siderophores, are not known and are clearly areas for future research. While DOM photochemistry significantly impacts the chemistry of important trace metals in seawater, the reverse, i.e., the effects of trace metals on the photochemical reactivity and oxidative degradation of DOM, e.g., via LMCT reactions, is not well known. Is the trace metal complexing capacity, and thus bioavailability, of the resulting photooxidized DOM increased or decreased relative to the non-photodegraded DOM?

D Photodissolution and Photoflocculation of POM

A number of recent studies indicate that photochemical processes occurring in the particulate phase may be important in particle-rich environments, especially in estuarine settings and, for specific compounds, even in the open ocean (see Section IV.A). However, it has generally not been demonstrated that POM photochemistry makes a quantitatively important contribution to the cycling of organic matter in estuaries, coastal waters, or the open ocean. Methods are being developed to accurately define POM light fields in order to better quantify POM photoreaction rates and compare results (Estapa, 2011; Song et al., 2013). Separating photochemical and biological processes associated with POM remains a challenge.

While the opposite process, photoflocculation, has been demonstrated to be a significant process in DOM- and iron-rich freshwaters (Gao and Zepp, 1998; Helms et al., 2013), it is not known whether this process occurs at the alkaline pH and higher ionic strengths found in estuaries or

seawater. Furthermore, in open ocean surface waters, iron is generally at very low concentrations and DOM is also at much lower concentrations than in the DOM-rich samples irradiated by Helms et al. (2013a). In addition, tDOM has a different chemical character than marine DOM. Thus, it still needs to be established whether photoflocculation is a significant biogeochemical term within saline, basic estuaries, and further out, beyond the coastal zone.

E Improved Quantification of Photochemical Rates

As discussed in Section V, major advances have been made in the modeling and extrapolation of photochemical rates to regional and global scales. However, further work is required to determine variations in AQYs for a suite of biogeochemically relevant photoreactions, with region, temperature, CDOM quality, and other parameters known to influence photoreaction rates such as salinity, concentrations of nitrate and trace metals (particularly iron, copper, and manganese), and pH. For example, how will ocean acidification (and increasing UV-B fluxes from atmospheric ozone depletion) affect the rates of photochemical processes, e.g., DOM photooxidation and photobleaching of DOM optical properties (absorbance and fluorescence)? Although a small change in pH may not affect rates directly, ocean acidification may substantially change the ecology, which in turn would likely change the quantity and quality of the DOM and POM and, thus, alter DOM photoreactivity.

Without robust AQYs it is impossible to improve our assessment of the quantitative role of photochemistry in the cycling of DOM and other biogeochemical constituents (see Chapter 2). In some cases, measurements of reliable AQYs in oceanic waters are limited by analytical precision, most notably in the case of arguably the most important marine photochemical rate measurement: the photomineralization of DOC to DIC.

Further work should also use field or remote sensed data for variations in the photoproduct (e.g., CO) or photoreactant (e.g., CDOM) to assess the performance of empirically driven photochemical models. In this regard, *in situ* drifter studies where a photoreaction rate is determined for a sample suspended in a quartz flask in the water column would offer a valuable empirical check against rates predicted by photochemical models (Neale and Kieber, 2000). Although photomineralization of bulk DOC appears a modest flux when compared to gross primary production, photomineralization may be significant when compared to net ecosystem primary production. Furthermore, photochemistry more than likely plays a significant role in the removal of otherwise refractory organics (including DBC) from the ocean; AQYs for the losses of these materials have yet to be determined.

Acknowledgments

KM would like to thank Neil V. Blough for insightful comments on Section II.A.2 (Photochemical Dissolved Inorganic Carbon Formation and Oxygen Consumption). KM would also like to acknowledge comments by William L. Miller, Hussain Abdulla, and Hongmei Chen. The authors would also like to acknowledge discussions with Richard G. Zepp, Robert G. M. Spencer, Chris L. Osburn, Colin A. Stedmon, and Cedric Fichot and the thoughtful, thorough reviews that were provided by Larry M. Mayer and Richard G. Zepp. Financial support was provided by NSF Chemical Oceanography (OCE-0728634, OCE-0850635 and OCE-1235005 to KM; OCE-0961831 and OCE-1129896 to DJK; and OCE-1234704 to AS), NSF Biological Oceanography (OCE-1029569 to DJK), NSF Division of Environmental Biology (DEB-1146161 to AS), and NSF Office of Polar Programs (ANT-0944686 to DJK).

References

Aarnos, H., Ylöstalo, P., Vähätalo, A.V., 2012. Seasonal phototransformation of dissolved organic matter to ammonium, dissolved inorganic carbon, and labile substrates supporting bacterial biomass across the Baltic Sea. J. Geophys. Res. 117. http://dx.doi.org/10.1029/2010JG001633, G01004.

Abboudi, M., Jeffrey, W., Ghiglione, J.F., Pujo-Pay, M., Oriol, L., Sempéré, R., et al., 2008. Effects of photochemical transformations of dissolved organic matter on bacterial metabolism and diversity in three contrasting coastal sites in the northwestern Mediterranean Sea during summer. Microb. Ecol. 55, 344–357.

Abdulla, H.A.N., Minor, E.C., Dias, R.F., Hatcher, P.G., 2010. Changes in the compound classes of dissolved organic matter along an estuarine transect: a study using FTIR and ^{13}C NMR. Geochim. Cosmochim. Acta 74, 3815–3838.

Adams, J.W., Holmes, N.S., Crowley, J.N., 2002. Uptake and reaction of HOBr on frozen and dry salt surfaces. Atmos. Chem. Phys. 2, 79–91.

Aiken, G.R., Malcolm, R.L., 1987. Molecular weight of aquatic fulvic acids by vapor pressure osmometry. Geochim. Cosmochim. Acta 51, 2177–2184.

Amado, A.M., Farjalla, V.F., Esteves, de A.F., Bozelli, R.L., Roland, F., Enrich-Prast, A., 2006. Complementary pathways of dissolved organic carbon removal pathways in clear-water Amazonian ecosystems: photochemical degradation and bacterial uptake. FEMS Microbiol. Ecol. 56, 8–17.

Amador, J.A., Alexander, M., Zika, R.G., 1989. Sequential photochemical and microbial degradation of organic matter bound to humic acid. Appl. Environ. Microb. 55, 2843–2849.

Amador, J.A., Alexander, M., Zika, R.G., 1991. Degradation of aromatic compounds bound to humic acid by the combined action of sunlight and microorganisms. Environ. Toxicol. Chem. 10, 475–482.

Amaral, J.H.F., Suhet, A.L., Melo, S., Farjalla, V.F., 2013. Seasonal variation and interaction of photodegradation and microbial metabolism of DOC in black water Amazonian ecosystems. Aquat. Microb. Ecol. 70, 157–168.

Amin, S.A., Green, D.H., Hart, M.C., Kupper, F.C., Sunda, W.G., Carrano, C.J., 2009a. Photolysis of iron-siderophore chelates promotes bacterial-algal mutualism. Proc. Natl. Acad. Sci. U. S. A. 106, 17071–17076.

Amin, S.A., Green, D.H., Kupper, F.C., Carrano, C.J., 2009b. Vibrioferrin, an unusual marine siderophore: iron binding, photochemistry, and biological implications. Inorg. Chem. 48, 11451–11458.

Amon, R.M.W., Benner, R., 1996. Photochemical and microbial consumption of dissolved organic carbon and dissolved oxygen in the Amazon River system. Geochim. Cosmochim. Acta 60, 1783–1792.

Anderson, T.R., Williams, P.J.le B., 1999. A one-dimensional model of dissolved organic carbon cycling in the water column incorporating combined biological-photochemical decomposition. Global Biogeochem. Cycles 13, 337–349.

Andreae, M., Ferek, R., 1992. Photochemical production of carbonyl sulfide in seawater and its emission to the atmosphere. Global Biogeochem. Cycles 6, 175–183.

Andrews, S.S., Caron, S., Zafiriou, O.C., 2000. Photochemical oxygen consumption in marine waters. A major sink for colored dissolved organic matter? Limnol. Oceanogr. 45, 267–277.

Anesio, A.M., Granéli, W., 2003. Increased photoreactivity of DOC by acidification: implications for the carbon cycle in humic lakes. Limnol. Oceanogr. 48, 735–744.

Anesio, A.M., Granéli, W., 2004. Photochemical mineralization dissolved organic carbon in lakes of differing pH and humic content. Arch. Hydrobiol. 160, 105–116.

Anesio, A.M., Denward, C.M.T., Tranvik, L.J., Granéli, W., 1999. Decreased bacterial growth on vascular plant detritus due to photochemical modification. Aquat. Microb. Ecol. 17, 159–165.

Anesio, A.M., Theil-Nielsen, J., Granéli, W., 2000. Bacterial growth on photochemically transformed leachates from aquatic and terrestrial primary producers. Microb. Ecol. 40, 200–208.

Anesio, A.M., Granéli, W., Aiken, G.R., Kieber, D.J., Mopper, K., 2005. Effect of humic substance photodegradation on bacterial growth and respiration in lake water. Appl. Environ. Microbiol. 71, 6267–6275.

Angel, D., Fiedler, U., Eden, N., Kress, N., Adelung, D., Herut, B., 1999. Catalase activity in macro- and microorganisms as an indicator of biotic stress in coastal waters of the eastern Mediterranean Sea. Helgoland Mar. Res. 53, 209–218.

Asher, E.C., Dacey, J.W.H., Mills, M.M., Arrigo, K.R., Tortell, P.D., 2011. High concentrations and turnover rates of DMS, DMSP and DMSO in Antarctic sea ice. Geophys. Res. Lett. 38, 1–5.

Bajt, O., Sket, B., Faganeli, J., 1997. The aqueous photochemical transformation of acrylic acid. Mar. Chem. 58, 255–259.

Baltar, F., Reinthaler, T., Herndl, G.J., Pinhassi, J., 2013. Major effect of hydrogen peroxide on bacterioplankton metabolism in the Northeast Atlantic. PLoS One 8. http://dx.doi.org/10.1371/journal.pone.0061051.

Barbeau, K., 2006. Photochemistry of organic iron(III) complexing ligands in oceanic systems. Photochem. Photobiol. 82, 1505–1516.

Barbeau, K., Rue, E.L., Bruland, K.W., Butler, A., 2001. Photochemical cycling of iron in the surface ocean mediated by microbial iron(III)-binding ligands. Nature 413, 409–413.

Barbeau, K., Zhang, G.P., Live, D.H., Butler, A., 2002. Petrobactin, a photoreactive siderophore produced by the oil-degrading marine bacterium *Marinobacter hydrocarbonoclasticus*. J. Am. Chem. Soc. 124, 378–379.

Barbeau, K., Rue, E.L., Trick, C.G., Bruland, K.T., Butler, A., 2003. Photochemical reactivity of siderophores produced by marine heterotrophic bacteria and cyanobacteria based on characteristic Fe(III) binding groups. Limnol. Oceanogr. 48, 1069–1078.

Bartels-Rausch, T., Krysztofiak, G., Bernhard, A., Schlappi, M., Schwikowski, M., Ammann, M., 2010. Photoinduced reduction of divalent mercury in ice by organic matter. Chemosphere 82, 199–203.

Bates, T.S., Kelly, K.C., Johnson, J.E., Gammon, R.H., 1995. Regional and seasonal variations in the flux of oceanic carbon monoxide to the atmosphere. J. Geophys. Res. 100, 23093–23101.

Bates, S.S., Léger, C., Wells, M.L., Hardy, K., 2003. Photodegradation of domoic acid. In: Bates, S.S. (Ed.), Canadian technical report of fisheries and aquatic sciences 2498. Proceedings of the Eighth Canadian Workshop on Harmful Marine Algae. pp. 30–35.

Bauer, J.E., Bianchi, T.S., 2011. Dissolved organic carbon cycling and transformation. In: Wolanski, E., McLusky, D.S. (Eds.), Treatise on Estuarine and Coastal Science. Academic Press, Waltham, pp. 7–67.

Bauer, J.E., Cai, W.-C., Raymond, P.A., Bianchi, T.S., Hopkinson, C.S., Regnier, P.A.G., 2013. The changing carbon cycle of the coastal ocean. Nature 504, 61–70.

Beaupré, S.R., Druffel, E.R.M., 2012. Photochemical reactivity of ancient marine dissolved organic carbon. Geophys. Res. Lett. 39, http://dx.doi.org/10.1029/2012GL052974,L18602.

Bélanger, S., Xie, H., Krotkov, N., Larouche, P., Vincent, W.F., Marcel Babin, M., 2006. Photomineralization of terrigenous dissolved organic matter in Arctic coastal waters from 1979 to 2003: interannual variability and implications of climate change. Global Biogeochem. Cycles. 20, http://dx.doi.org/10.1029/2006GB002708, GB4005.

Benner, R., 2011. Loose ligands and available iron in the ocean. Proc. Natl. Acad. Sci. U. S. A. 108, 893–894.

Benner, R., Biddanda, B., 1998. Photochemical transformations of surface and deep marine dissolved organic matter: effects on bacterial growth. Limnol. Oceanogr. 43, 1373–1378.

Benner, R., Ziegler, S., 2000. Do photochemical transformations of dissolved organic matter produce biorefractory as well as bioreactive substrates? In: Bell, C.R., Brylinsky, M., Johnson-Green, P.C. (Eds.), Microbial Biosystems: New Frontiers. Proceedings of the 8th International Symposium on Microbial Ecology, 1998. Atlantic Canada Society for Microbial Ecology, Halifax, Canada, pp. 181–192.

Benner, R., Louchouarn, P., Amon, R.M., 2005. Terrigenous dissolved organic matter in the Arctic Ocean and its transport to surface and deep waters of the North Atlantic. Global Biogeochem. Cycles 19, http://dx.doi.org/10.1029/2004GB002398, GB2025.

Bertilsson, S., 1999. Photochemical Alterations of Dissolved Organic Matter - Impact on Heterotrophic Bacteria and Carbon Cycling in Lakes (Ph.D. thesis), Linköping University, Linköping, Sweden.

Bertilsson, S., Tranvik, L.J., 1998. Photochemically produced carboxylic acids as substrates for freshwater bacterioplankton. Limnol. Oceanogr. 43, 885–895.

Bertilsson, S., Tranvik, L.J., 2000. Photochemical transformation of dissolved organic matter in lakes. Limnol. Oceanogr. 45, 753–762.

Bertilsson, S., Stepanauskas, R., Cuadros-Hansson, R., Granéli, W., Wikner, J., Tranvik, L., 1999. Photochemically induced changes in bioavailable carbon and nitrogen pools in a boreal watershed. Aquat. Microb. Ecol. 19, 47–56.

Biddanda, B.A., Cotner, J.B., 2003. Enhancement of dissolved organic matter bioavailability by sunlight and its role in the carbon cycle of Lakes Superior and Michigan. J. Great Lakes Res. 29, 228–241.

Bielski, B.H.J., Cabelli, D.E., Arudi, R.L., 1985. Reactivity of HO_2/O_2^- radicals in aqueous solution. J. Phys. Chem. Ref. Data 14, 104–1100.

Biers, E.J., Zepp, R.G., Moran, M.A., 2007. The role of nitrogen in chromophoric and fluorescent dissolved organic matter formation. Mar. Chem. 103, 46–60.

Blough, N.V., 1997. Photochemistry in the sea-surface microlayer. In: Liss, P.S., Duce, R.A. (Eds.), The Sea Surface and Global Change. Cambridge University Press, Oxford, UK, pp. 383–424.

Blough, N.V., Caron, S., 1995. Photochemical production of superoxide and peroxy radicals in natural waters. In: American Chemical Society 210th National Meeting. pp. 390–393.

Blough, N.V., Zepp, R.G., 1995. Reactive oxygen species in natural waters. In: Foote, C.S., Pacentine, J.S., Greenberg, A., Liebman, J.F. (Eds.), Active Oxygen in Chemistry. Chapman and Hall, New York, NY, pp. 280–333.

Bockman, T.M., Hubig, S.M., Kochi, J.K., 1996. Direct observation of carbon-carbon bond cleavage in ultrafast decarboxylations. J. Am. Chem. Soc. 118, 4502–4503.

Boreen, A.L., Edhlund, B.L., Cotner, J.B., McNeill, K., 2008. Indirect photodegradation of dissolved free amino acids: the contribution of singlet oxygen and the differential reactivity of DOM from various sources. Environ. Sci. Technol. 42, 5492–5498.

Borer, P.M., Sulzberger, B., Reichard, P., Kraemer, S.M., 2005. Effects of siderophores on the light-induced dissolution of colloidal iron (III) (hydr)oxides. Mar. Chem. 93, 179–193.

Boston Water Board, 1895. Nineteenth annual report of the Boston Water Board for the year ending January 31, 1895. Rockwell and Churchill, City Printers, Boston, MA.

Bouillon, R.-C., Miller, W.L., 2004. Determination of apparent quantum yield spectra of DMS photo-degradation in an in situ iron-induced Northeast Pacific Ocean bloom. Geophys. Res. Lett. 31. http://dx.doi.org/10.1029/2004GL019536, L06310.

Bouillon, R.-C., Miller, W.L., 2005. Photodegradation of dimethyl sulfide (DMS) in natural waters: laboratory assessment of the nitrate-photolysis-induced DMS oxidation. Environ. Sci. Technol. 39, 9471–9477.

Bouillon, R.C., Knierim, T.L., Kieber, R.J., Skrabal, S.A., Wright, J.L.C., 2006. Photodegradation of the algal toxin domic acid in natural water matrices. Limnol. Oceanogr. 51, 321–330.

Boxe, C.S., Colussi, A.J., Hoffman, M.R., Perez, I.M., Murphy, J.G., Cohen, R.C., 2006. Kinetics of NO and NO_2 evolution from illuminated frozen nitrate solutions. J. Phys. Chem. 110, 3578–3583.

Boyd, P.W., Ellwood, M.J., 2010. The biogeochemical cycle of iron in the ocean. Nat. Geosci. 3, 675–682.

Boyle, E.S., Guerriero, N., Thiallet, A., Del Vecchio, R., Blough, N.V., 2009. Optical properties of humic substances and CDOM: relation to structure. Environ. Sci. Technol. 43, 2262–2268.

Brimblecombe, P., Shooter, D., 1986. Photo-oxidation of dimethylsulphide in aqueous solution. Mar. Chem. 19, 343–353.

Brinkmann, T., Hörsch, P., Sartorius, D., Frimmel, F.H., 2003. Photoformation of low-molecular-weight organic acids from brown water dissolved organic matter. Environ. Sci. Technol. 37, 4190–4198.

Bronk, D.A., 2002. Dynamics of DON. In: Hansell, D.A., Carlson, C.A. (Eds.), Biogeochemistry of Marine Dissolved Organic Matter. Academic Press, Elsevier, Amsterdam, pp. 153–247.

Bronk, D.A., Roberts, Q.N., Sanderson, M.P., Canuel, E.A., Hatcher, P.G., Mesfioui, R., et al., 2010. Effluent organic nitrogen (EON): bioavailability and photochemical and salinity-mediated release. Environ. Sci. Technol. 44, 5830–5835.

Brophy, J.E., Carlson, D.J., 1989. Production of biologically refractory dissolved organic carbon by natural seawater microbial populations. Deep-Sea Res. 36, 497–507.

Brugger, A., Slezak, D., Obernosterer, I., Herndl, G.J., 1998. Photolysis of dimethylsulfide in the northern Adriatic Sea: dependence on substrate concentration, irradiance and DOC concentration. Mar. Chem. 59, 321–331.

Budac, D., Wan, P., 1992. Photodecarboxylation: mechanism and synthetic utility. J. Photochem. Photobiol. A: Chem. 67, 135–166.

Buffam, I., McGlathery, K.J., 2003. Effect of ultraviolet light on dissolved nitrogen transformations in coastal lagoon water. Limnol. Oceanogr. 48, 723–734.

Burns, S.E., Hassett, J.P., Rossi, M.V., 1997. Mechanistic implications of the intrahumic dechlorination of mirex. Environ. Sci. Technol. 31, 1365–1371.

Burns, J.M., Cooper, W.J., Ferry, J.L., King, D.W., DiMento, B.P., McNeill, K., et al., 2012. Methods for reactive oxygen species (ROS) detection in aqueous environments. Aquat. Sci. 74, 683–734.

Bushaw, K.L., Zepp, R.G., Tarr, M.A., Schultz-Jander, D., Bourbonniere, R.A., Hodson, R.E., et al., 1996. Photochemical release of biologically available nitrogen from aquatic dissolved organic matter. Nature 381, 404–407.

Bushaw-Newton, K.L., Moran, M.A., 1999. Photochemical formation of biologically available nitrogen from dissolved humic substances in coastal marine systems. Aquat. Microb. Ecol. 18, 285–292.

Buxton, G.V., Greenstock, C.L., Helman, W.P., Ross, A.B., 1988. Critical review of rate constants for reactions of hydrated electrons, hydrogen atoms and hydroxyl radicals in aqueous solution. J. Phys. Chem. Ref. Data 17, 513–886.

Cai, W.J., Wang, Y., 1998. The chemistry, fluxes, and sources of carbon dioxide in the estuarine waters of the Satilla and Altamaha Rivers, Georgia. Limnol. Oceanogr. 43, 657–668.

Calvert, J.G., Pitts, J.N., 1966. Photochemistry. Wiley & Sons, New York, NY.

Calza, P., Massolino, C., Pelizzetti, E., Minero, C., 2008. Solar driven production of toxic halogenated and nitroaromatic compounds in natural seawater. Sci. Total Environ. 398, 196–202.

Canonica, S., Hellrung, B., Wirz, J., 2000. Oxidation of phenols by triplet aromatic ketones in aqueous solution. J. Phys. Chem. A 104, 1226–1232.

Carlson, C.A., Bates, N.R., Ducklow, H.W., Hansell, D.A., 1999. Estimation of bacterial respiration and growth efficiencies in the Ross Sea, Antarctica. Aquat. Microb. Ecol. 19, 229–244.

Chen, R.F., Bada, J.L., 1992. The fluorescence of dissolved organic matter in seawater. Mar. Chem. 37, 191–221.

Chen, Y., Khan, S.U., Schnitzer, M., 1978. Ultraviolet irradiation of dilute fulvic acid solutions. Soil Sci. Soc. Am. J. 42, 292–296.

Chen, H.-M., Stubbins, A., Perdue, E.M., Green, N., Helms, J.R., Mopper, K., Hatcher, P.G., 2014. Ultrahigh resolution mass spectrometric differentiation of dissolved organic matter isolated by coupled reverse osmosis-electrodialysis from various major oceanic water masses. Mar. Chem. http://dx.doi.org/10.1016/j.marchem.2014.06.002. In press.

Cherrier, J., Bauer, J.E., Druffel, E.R.M., Coffin, R.B., Chanton, J.P., 1999. Radiocarbon in marine bacteria: evidence for the ages of assimilated carbon. Limnol. Oceanogr. 44, 730–736.

Chin, Y.-P., Aiken, G., O'Loughlin, E., 1994. Molecular weight, polydispersity, and spectroscopic properties of aquatic humic substances. Environ. Sci. Technol. 28, 1853–1858.

Chin, Y.-P., Traina, S.J., Swank, C.R., Backhus, D., 1998. Abundance and properties of dissolved organic matter in pore waters of a freshwater wetland. Limnol. Oceanogr. 43, 1287–1296.

Chomicki, K.M., Schiff, S.L., 2008. Stable oxygen isotopic fractionation during photolytic O_2 consumption in stream waters. Sci. Total Environ. 404, 236–244.

Chretiennot-Dinet, M.J., 1998. Global increase of algal blooms, toxic events, casual species introductions and biodiversity. Oceans 24, 223–238.

Clark, C.D., Zika, R.G., 2000. Marine organic photochemistry: from the sea surface to the marine aerosols. In: Wangersky, P.J. (Ed.), Handbook of Environmental Chemistry, vol. 5. Springer, Berlin, pp. 1–33.

Clark, C.D., Hiscocka, W.T., Millero, F.J., Hitchcock, G., Brand, L., Miller, W.L., et al., 2004. CDOM distribution and CO_2 production on the Southwest Florida Shelf. Mar. Chem. 89, 145–167.

Coble, P.G., 2007. Marine optical biogeochemistry: the chemistry of ocean color. Chem. Rev. 107, 402–418.

Conrad, R., Seiler, W., 1980. Photooxidative production and microbial consumption of carbon monoxide in seawater. FEMS Microbiol. Lett. 9, 61–64.

Conrad, R., Seiler, W., 1982. Utilization of traces of carbon monoxide by aerobic oligotrophic microorganisms in ocean, lake and soil. Arch. Microbiol. 132, 41–46.

Conrad, R., Seiler, W., Bunse, G., Giehl, H., 1982. Carbon monoxide in seawater (Atlantic Ocean). J. Geophys. Res. 87, 8839–8852.

Cooper, W.J., Saltzman, E.S., Zika, R.G., 1987. The contribution of rainwater to variability of surface ocean hydrogen peroxide. J. Geophys. Res. 92, 2970–2980.

Cooper, W.J., Zika, R.G., Petasne, R.G., Fischer, A.M., 1989. Sunlight-induced photochemistry of humic substances in natural waters: major reactive species. In: Suffet, I.H., MacCarthy, P. (Eds.), Aquatic Humic Substances: Influence on Fate and Treatment of Pollutants. American Chemical Society, Washington, D.C., pp. 333–362.

Cory, R.M., McNeill, K., Cotner, J.P., Amado, A., Purcell, J., Marshall, A.G., 2010. Singlet oxygen in the coupled photochemical and biochemical oxidation of dissolved organic matter. Environ. Sci. Technol. 44, 3683–3689.

Cotner Jr., J.B., Heath, R.T., 1990. Iron redox effects on photosensitive phosphorus release from dissolved humic materials. Limnol. Oceanogr. 35, 1175–1181.

Cottrell, B.A., Timko, S.A., Devera, L., Robinson, A.K., Gonsior, M., Vizenor, A.E., et al., 2013. Photochemistry of excited-state species in natural waters: a role for particulate organic matter. Water Res. 47, 5189–5199.

Crosby, D.G., 1994. Photochemical aspects of bioavailability. In: Hamelink, J.L. (Ed.), Bioavailability Proceedings Pellston Workshop, 13th, Session 5, Dynamic Environmental Factors. Lewis, Boca Raton, FL, pp. 109–118.

Crutzen, J.P., 1976. The possible importance of COS for the sulfate layer of the stratosphere. Geophys. Res. Lett. 3, 73–75.

Cunliffe, M., Engel, A., Frka, S., Gašparović, B., Guitart, C., Murrell, J.C., et al., 2013. Sea surface microlayers: a unified physicochemical and biological perspective of the air–ocean interface. Prog. Oceanogr. 109, 104–116.

Cuny, P., Rontani, J.F., 1999. On the widespread occurrence of 3-methyldiene-7,11,15-trimethylhexadecan-1,2-diol in the marine environment: a specific isoprenoid marker of chlorophyll photodegradation. Mar. Chem. 65, 155–165.

Cuny, P., Romano, J.C., Beker, B., Rontani, J.F., 1999. Comparison of the photodegradation rates of chlorophyll chlorin ring and phytol side chain in phytodetritus: is the phytyldiol versus phytol ratio (CCPI) a new biogeochemical index. J. Exp. Mar. Biol. Ecol. 237, 271–290.

Cutter, G.A., Cutter, L.S., Filippino, K.C., 2004. Sources and cycling of carbonyl sulfide in the Sargasso Sea. Limnol. Oceanogr. 49, 555–565.

Dacey, J.W.H., Howse, F.A., Michaels, A.F., Wakeham, S.G., 1998. Temporal variability of dimethylsulfide and dimethylsulfoniopropionate in the Sargasso Sea. Deep-Sea Res. I 45, 2085–2104.

Dalrymple, R.M., Carfagno, A.K., Sharpless, C.M., 2010. Correlations between dissolved organic matter optical properties and quantum yields of singlet oxygen and hydrogen peroxide. Environ. Sci. Technol. 44, 5824–5829.

Daniel, C., Granéli, W., Kritzberg, E.S., Anesio, A.M., 2006. Stimulation of metazooplankton by photochemically modified dissolved organic matter. Limnol. Oceanogr. 51, 101–108.

de Bruyn, W.J., Clark, C.D., Pagel, L., Takehara, C., 2011. Photochemical production of formaldehyde, acetaldehyde and acetone from chromophoric dissolved organic matter in coastal waters. J. Photochem. Photobiol. A: Chem. 226, 16–22.

De Haan, H., 1992. Impacts of environmental changes on the biogeochemistry of aquatic humic substances. Hydrobiology 229, 59–71.

De Lange, H.J., Morris, D.P., Williamson, C.E., 2003. Solar ultraviolet photodegradation of DOC may stimulate freshwater food webs. J. Plankton Res. 25, 111–117.

Deal, C.J., Kieber, D.J., Toole, D.A., Stamnes, K., Jiang, S., Uzuka, N., 2005. Dimethylsulfide photolysis rates and apparent quantum yields in Bering Sea seawater. Cont. Shelf Res. 25, 1825–1835.

del Giorgio, P.A., Cole, J.J., 2000. Bacterial energetics and growth efficiency. In: Kirchman, D.L. (Ed.), Microbial Ecology of the Oceans. Wiley & Sons, New York, NY, pp. 289–325.

del Giorgio, P.A., Williams, P.J.le B. (Eds.), 2005. Respiration in Aquatic Ecosystems. Oxford University Press, USA, 326 pp.

del Valle, D.A., Kieber, D.J., Bisgrove, J., Kiene, R.P., 2007. Light-stimulated production of dissolved DMSO by a particle-associated process in the Ross Sea, Antarctica. Limnol. Oceanogr. 52, 2456–2466.

del Valle, D.A., Kieber, D.J., Toole, D.A., Bisgrove, J., Kiene, R.P., 2009. Dissolved DMSO production via biological and photochemical oxidation of dissolved DMS in the Ross Sea, Antarctica. Deep-Sea Res. I. 56, 166–177.

Del Vecchio, R., Blough, N.V., 2004. On the origin of the optical properties of humic substances. Environ. Sci. Technol. 38, 3885–3891.

Del Vecchio, R., Subramaniam, A., Uz, S.S., Ballabrera-Poy, J., Brown, C.W., Blough, N.V., 2009. Decadal time-series of SeaWiFS retrieved CDOM absorption and estimated CO_2 photoproduction on the continental shelf of the eastern United States. Geophys. Res. Lett. 36. http://dx.doi.org/10.1029/2008GL036169, L02602.

Diamond, S.A., 2003. Photoactivated toxicity in aquatic environments. In: Helbling, E.W., Zargarese, H. (Eds.), UV Effects in Aquatic Organisms and Ecosystems. RSC Publishing, Cambridge, UK, pp. 219–250.

Diaz, J.M., Hansel, C.M., Voelker, B.M., Mendes, C.M., Andeer, P.F., Zhang, T., 2013. Widespread production of extracellular superoxide by heterotrophic bacteria. Science 340, 1223–1226.

Dignon, J., Hameed, S., 1985. A model investigation of the impact of increases in anthropogenic NO_x emissions between 1967 and 1980 on tropospheric ozone. J. Atmos. Chem. 3, 491–506.

Dittmar, T., 2008. The molecular level determination of black carbon in marine dissolved organic matter. Org. Geochem. 39, 396–407.

Dittmar, T., Paeng, J., 2009. A heat-induced molecular signature in marine dissolved organic matter. Nat. Geosci. 2, 175–179.

Dittmar, T., Stubbins, A., 2013. Dissolved organic matter in aquatic systems. In: Holland, H.D., Turekian, K.K. (Eds.), Treatise on Geochemistry. Elsevier, Oxford, UK, pp. 125–156.

Doney, S.C., Najjar, R.G., Stewart, S., 1995. Photochemistry, mixing and diurnal cycles in the upper ocean. J. Mar. Res. 53, 341–369.

Dunne, J.P., Sarmiento, J.L., Gnanadesikan, A., 2007. A synthesis of global particle export from the surface ocean and cycling through the ocean interior and on the seafloor. Global Biogeochem. Cycles 21. http://dx.doi.org/10.1029/2006GB002907, GB4006.

Ehrhardt, M.G., 2005. Hydrocarbon breakdown in the sea-surface microlayer. In: Liss, P.S., Duce, R.A. (Eds.), Sea Surface and Global Change. Cambridge University Press, Cambridge, UK, pp. 425–444.

Elliot, S., Lu, E., Rowland, F.S., 1987. Carbonyl sulfide hydrolysis as a source of hydrogen sulfide in open ocean seawater. Geophys. Res. Lett. 14, 131–134.

Elliot, S., Lu, E., Rowland, F.S., 1989. Rates and mechanisms for hydrolysis of carbonyl sulfide in natural waters. Environ. Sci. Technol. 23, 458–461.

Erickson III, D.J., Zepp, R.G., Atlas, E., 2000. Ozone depletion and the air-sea exchange of greenhouse and chemically reactive trace gases. Chemosphere 2, 137–149.

Estapa, M.L., 2011. Photochemical Reactions of Marine Particulate Organic Matter (Ph.D. thesis), University of Maine, Orono, ME.

Estapa, M.L., Mayer, L.M., 2010. Photooxidation of particulate organic matter, carbon/oxygen stoichiometry, and related photoreactions. Mar. Chem. 122, 138–147.

Estapa, M.L., Mayer, L.M., Boss, E., 2012. Rate and apparent quantum yield of sedimentary organic matter. Limnol. Oceanogr. 57, 1743–1756.

Falk, M., Walter, J.A., Wiseman, P.W., 1989. Ultraviolet spectrum of domoic acid. Can. J. Chem. 67, 1421–1425.

Farjalla, V.F., Anesio, A.M., Bertilsson, S., Granéli, W., 2001. Photochemical reactivity of aquatic macrophyte leachates: abiotic transformations and bacterial response. Aquat. Microb. Ecol. 24, 187–195.

Faust, B.C., 1999. Aquatic photochemical reactions in atmospheric, surface, and marine waters. Influences on oxidant formation and pollutant degradation. In: Handb. Environ. Chem. Springe, Berlin, pp. 101–122.

Ferek, R.J., Andreae, M.O., 1984. Photochemical production of carbonyl sulfide in marine surface waters. Nature 307, 148–150.

Fichot, C.G., Benner, R., 2014. The fate of terrigenous dissolved organic carbon in a river-influenced ocean margin. Global Biogeochem. Cycles 28, 300–318. http://dx.doi.org/10.1002/2013GB004670.

Fichot, C.G., Miller, W.L., 2010. An approach to quantify depth-resolved marine photochemical fluxes using remote sensing: application to carbon monoxide (CO) photoproduction. Remote Sens. Environ. 114, 1363–1377.

Field, C.B., Behrenfeld, M.J., Randerson, J.T., Falkowski, P., 1998. Primary production of the biosphere: integrating terrestrial and oceanic components. Science 281, 237–240.

Fischer, E.V., Jacob, D.J., Millet, D.B., Yantosca, R.M., Mao, J., 2012. The role of the ocean in the global atmospheric budget of acetone. Geophys. Res. Lett. 39. http://dx.doi.org/10.1029/2011GL050086, L01807.

Fisher, J.M., Reese, J.G., Pellechia, P.J., Moeller, P.L., Ferry, J.L., 2006. Role of Fe(III), phosphate, dissolved organic matter, and nitrate during the photodegradation of domoic acid in the marine environment. Environ. Sci. Technol. 40, 2200–2205.

Flöck, O.R., Andreae, M.O., 1996. Photochemical and non-photochemical formation and destruction of carbonyl sulfide and methyl mercaptan in ocean waters. Mar. Chem. 54, 11–26.

Flöck, O.R., Andreae, M.O., Dräger, M., 1997. Environmentally relevant precursors of carbonyl sulfide in aquatic systems. Mar. Chem. 59, 71–85.

Francko, D.A., 1990. Alteration of bioavailability and toxicity by phototransformation of organic acids. In: Perdue, E.M., Gjessing, E.T. (Eds.), Organic Acids in Aquatic Ecosystems: Report of the Dahlem Workshop on Organic Acids in Aquatic Ecosystems. John Wiley & Sons, New York, NY, pp. 167–177.

Francko, D.A., Heath, R.T., 1982. UV-sensitive complex phosphorus: association with dissolved humic material and iron in a bog lake. Limnol. Oceanogr. 27, 564–569.

Galí, M., Simó, R., 2010. Occurrence and cycling of dimethylated sulfur compounds in the Arctic during summer receding of the ice edge. Mar. Chem. 122, 105–117.

Gan, D., Jia, M., Vaughan, P.P., Falvey, D.E., Blough, N.V., 2008. Aqueous photochemistry of methyl-benzoquinone. J. Phys. Chem. A 112, 2803–2812.

Gao, H., Zepp, R.G., 1998. Factors influencing photoreactions of dissolved organic matter in a coastal river of the southeastern United States. Environ. Sci. Technol. 32, 2940–2946.

Gardner, W.S., Cavaletto, J.F., Bootsma, H.A., Lavrentyev, P.J., Troncone, F., 1998. Nitrogen cycling rates and light effects in tropical Lake Maracaibo. Venezuela. Limnol. Oceanogr. 43, 1814–1825.

Garg, S., Rose, A.L., Waite, D.T., 2007. Superoxide mediated reduction of organically complexed iron(III): comparison of non-dissociative and dissociative reduction pathways. Environ. Sci. Technol. 41, 3205–3212.

Garg, S., Ito, H., Rose, A.L., Waite, T.D., 2013. Mechanism and kinetics of dark iron redox transformations in previously photolyzed acidic natural organic matter solutions. Environ. Sci. Technol. 47, 1861–1869.

Geller, A., 1986. Comparison of mechanisms enhancing biodegradability of refractory lake-water constituents. Limnol. Oceanogr. 31, 755–764.

Girotti, A.W., 2001. Photosensitized oxidation of membrane lipids: reaction pathways, cytotoxic effects, and cytoprotective mechanisms. J. Photochem. Photobiol. B: Biol. 63, 103–113.

Gjessing, E.T., Källqvist, T., 1991. Algicidal and chemical effect of UV-radiation of water containing humic substances. Water Res. 25, 491–494.

Gledhill, M., Buck, K.N., 2012. The organic complexation of iron in the marine environment: a review. Front. Microbiol. 3, Article 69, 1–17.

Gnanadesikan, A., 1996. Modeling the diurnal cycle of carbon monoxide: sensitivity to physics, chemistry, biology, and optics. J. Geophys. Res. 101, 12,177–12,191.

Gobler, C.J., Hutchins, D.A., Fisher, N.S., Cosper, E.M., Sanudo-Wilhelmy, S.A., 1997. Release and bioavailability of C, N, P, Se and Fe following viral release of a marine chrysophyte. Limnol. Oceanogr. 42, 1492–1504.

Golanoski, K.S., Fang, S., Del Vecchio, R., Blough, N.V., 2012. Investigating the mechanism of phenol photooxidation by humic substances. Environ. Sci. Technol. 46, 3912–3920.

Goldstone, J.V., Pullin, M.J., Bertilsson, S., Voelker, B.M., 2002. Reactions of hydroxyl radical with humic substances: bleaching, mineralization, and production of bioavailable carbon substrates. Environ. Sci. Technol. 36, 364–372.

Gonsior, M., Hertkorn, N., Conte, M.H., Cooper, W.J., Bastviken, D., Druffel, E., Schmitt-Kopplin, P., 2014. Photochemical production of polyols arising from significant photo-transformation of dissolved organic matter in the oligotrophic surface ocean. Mar. Chem. 163, 10–18.

Gonsior, M., Peake, B.M., Cooper, W.T., Podgorski, D., D'Andrilli, J., Cooper, W.J., 2009. Photochemically induced changes in dissolved organic matter identified by ultra-high resolution Fourier transform ion cyclotron resonance mass spectrometry. Environ. Sci. Technol. 43, 698–703.

Görner, H., 2003. Photoprocesses of p-benzoquinones in aqueous solution. J. Phys. Chem. A 107, 11587–11595.

Grabo, J.E., Chrisman, M.A., Webb, L.M., Baldwin, M.J., 2014. Photochemical reactivity of the iron(III) complex of a mixed-donor,α-hydroxy acid-containing chelate and its biological relevance to photoactive marine siderophores. Inorg. Chem. 53, 5781–5787. http://dx.doi.org/10.1021/ic500635q.

Grandbois, M., Latch, D.E., McNeill, K., 2008. Microheterogeneous concentrations of singlet oxygen in natural organic matter isolate solutions. Environ. Sci. Technol. 42, 9184–9190.

Granéli, W., Lindell, M., de Faria, B.M., de Assis Esteves, F., 1998. Photoproduction of dissolved inorganic carbon in temperate and tropical lakes—dependence on wavelength band and dissolved organic carbon concentration. Biogeochemistry 43, 175–195.

Grannas, A.M., Jones, A.E., Dibb, J., Ammann, M., Anastasio, C., Beine, H.J., et al., 2007a. An overview of snow photochemistry: evidence, mechanisms and impacts. Atmos. Chem. Phys. 7, 4329–4373.

Grannas, A.M., Bausch, A.R., Mahanna, K.M., 2007b. Enhanced aqueous photochemical reaction rates after freezing. J. Phys. Chem. A 111, 11043–11049.

Grebel, J.E., Pignatello, J.J., Song, W., Cooper, W.J., Mitch, W.A., 2009. Impact of halides on the photobleaching of dissolved organic matter. Mar. Chem. 115, 134–144.

Grebel, J.E., Pignatello, J.J., Mitch, W.A., 2011. Sorbic acid as a quantitative probe for the formation, scavenging and steady-state concentrations of the triplet-excited state of organic compounds. Water Res. 45, 6535–6544.

Grzybowski, W., 2000. Effect of short-term sunlight irradiation on absorbance spectra of chromophoric organic matter dissolved in coastal and riverine water. Chemosphere 40, 1313–1318.

Grzybowski, W., 2002a. Short-term sunlight irradiation of organic matter dissolved in lake water increased its susceptibility to subsequent biooxidation. Acta Hydrochim. Hydrobiol. 30, 285–292.

Grzybowski, W., 2002b. The significance of DOM photolysis on NH_4 production in natural waters. Oceanologia 44, 355–365.

Gun, J., Goifman, A., Shkrob, I., Kamyshny, A., Ginzburg, B., Hadas, O., et al., 2000. Formation of polysulfides in an oxygen rich freshwater lake and their role in the production of volatile sulfur compounds in aquatic systems. Environ. Sci. Technol. 34, 4741–4746.

Guo, W., Yang, L., Yu, X., Zhai, W., Hong, H., 2012. Photoproduction of dissolved inorganic carbon from dissolved organic matter in contrasting coastal waters in the southwestern Taiwan Strait China. J. Environ. Sci. 24, 1181–1188.

Guzmán, M.I., Hoffmann, M.R., Colussi, A.J., Guzmán, M.I., Hoffmann, M.R., Colussi, A.J., 2007. Photolysis of pyruvic acid in ice: possible relevance to CO and CO_2 ice core record anomalies. J. Geophys. Res. 112. http://dx.doi.org/10.1029/2006JD007886.

Halladja, S., Ter Halle, A., Aguer, J.P., Boulkamh, A., Richard, C., 2007. Inhibition of humic substances mediated photooxygenation of furfuryl alcohol by 2,4,6-trimethylphenol: evidence for reactivity of the phenol with humic triplet excited states. Environ. Sci. Technol. 41, 6066–6073.

Hamilton, R.D., 1964. Photochemical processes in the inorganic nitrogen cycle of the sea. Limnol. Oceanogr. 9, 107–111.

Hansell, D.A., 2013. Recalcitrant dissolved organic carbon fractions. Ann. Rev. Mar. Sci. 5, 421–445.

Hansell, D.A., Carlson, C.A., 1998. Net community production of dissolved organic carbon. Global Biogeochem. Cycles 12, 443–453.

Hardman, R.C., Cooper, W.J., Bourdelais, A.J., Gardinali, P., Baden, D.G., 2004. Brevetoxin degradation and by-product formation via natural sunlight. In: Steidinger, K.A., Landsberg, J.H., Tomas, C.R., Vargo, G.A. (Eds.), Harmful Algae. Florida Fish and Wildlife Conservation Commission, Florida Institute of Oceanography, Intergovernmental Oceanographic Commission of UNESCO, Florida, USA, pp. 153–154.

Hassett, J.P., 2006. Dissolved natural organic matter as a microreactor. Science 311, 1723–1724.

Hatton, A.D., 2002. Influence of photochemistry on the marine biogeochemical cycle of dimethylsulphide in the northern North Sea. Deep-Sea Res. II 49, 3039–3052.

Haverstock, S., Sizmur, T., Murimboh, J., O'Driscoll, N.J., 2012. Modeling the photo-oxidation of dissolved organic matter by ultraviolet radiation in freshwater lakes: implications for mercury bioavailability. Chemosphere 88, 1220–1226.

Hedges, J., Keil, R., Benner, R., 1997. What happens to terrestrial organic matter in the ocean? Org. Geochem. 27, 195–212.

Hellmer, H.H., Haas, C., Dieckmann, G.S., Schroder, M., 2006. Sea Ice feedbacks observed in the western Weddell Sea. Trans. Amer. Geophys. Un. 87, pp. 173 and 179..

Helms, J.R., 2012. Spectroscopic Characterization of Dissolved Organic Matter: Insights into Composition, Photochemical Transformation and Carbon Cycling (Ph.D. thesis), Old Dominion University, Norfolk, VA.

Helms, J.R., Stubbins, A., Ritchie, J.D., Minor, E.C., Kieber, D.J., Mopper, K., 2008. Absorption spectral slopes and slope ratios as indicators of molecular weight, source, and photobleaching of chromophoric dissolved organic matter. Limnol. Oceanogr. 53, 955–969.

Helms, J.R., Mao, J.-D., Schmidt-Rohr, K., Mopper, K., 2013a. Photochemical flocculation of terrestrial dissolved organic matter and iron. Geochim. Cosmochim. Acta 121, 398–413.

Helms, J.R., Stubbins, A., Perdue, E.M., Green, N.W., Chen, H.-M., Mopper, K., 2013b. Photochemical bleaching of oceanic dissolved organic matter and its effect on absorption spectral slope and fluorescence. Mar. Chem. 155, 81–91.

Helms, J.R., Mao, J.-D., Stubbins, A., Schmidt-Rohr, K., Mopper, K., 2014. Loss of optical and molecular indicators of terrigenous dissolved organic matter during long-term photobleaching. Aquat. Sci., 76, 353–373. http://dx.doi.org/10.1007/s00027-014-0340-0.

Herndl, G.J., Arrieta, J.M., Kaiser, E., Obernosterer, I., Pausz, C., Reitner, B., 2000. Role of ultraviolet radiation in aquatic systems: interaction between mixing processes, photochemistry, and microbial activity. In: Bell, C.R., Brylinsky, M., Johnson-Green, P.C. (Eds.), Microbial Biosystems: New Frontiers. Proceedings of the 8th International Symposium on Microbial Ecology, 1998.

Hernes, P.J., Benner, R., 2003. Photochemical and microbial degradation of dissolved lignin phenols: implications for the fate of terrigenous dissolved organic matter in marine environments. J. Geophys. Res. 108 (C9), 3291. http://dx.doi.org/10.1029/2002JC001421.

Hertkorn, N., Benner, R., Frommberger, M., Schmitt-Kopplin, P., Witt, M., Kaiser, K., et al., 2006. Characterization of a major refractory component of marine dissolved organic matter. Geochim. Cosmochim. Acta 70, 2990–3010.

Hill, P.G., Warwick, P.E., Zubkov, M.V., 2013. Low microbial respiration of leucine at ambient oceanic concentration in the mixed layer of the central Atlantic Ocean. Limnol. Oceanogr. 58, 1597–1604.

Hoikkala, L., Aarnos, H., Lignell, R., 2009. Changes in nutrient and carbon availability and temperature as factors controlling bacterial growth in the northern Baltic Sea. Estuar. Coast. Shelf Sci. 32, 720–733.

Hopkinson, B.M., Morel, F.M., 2009. The role of siderophores in iron acquisition by photosynthetic marine microorganisms. Biometals 22, 659–669.

Huang, J., Mabury, S.A., 2000. The role of carbonate radical in limiting the persistence of sulfur-containing chemicals in sunlit natural waters. Chemosphere 41, 1775–1782.

IHSS, 2014. Standard and Reference Collection, International Humic Substances Society, http://www.humicsubstances.org/elements.html.

IPCC, 2007. In: Solomon, S., Qin, D., Manning, M., Chen, Z., Marquis, M., Averyt, K.B., Tignor, M., Miller, H.L. (Eds.), Contribution of working group I to the fourth assessment report of the Intergovernmental Panel on Climate Change. Cambridge University Press, Cambridge, UK and New York, NY, pp. 19–91.

Ito, Y., Butler, A., 2005. Structure of synechobactins, new siderophores of the marine cyanobacterium *Synechococcus* sp. PCC 7002. Limnol. Oceanogr. 50, 1918–1923.

Jaffé, R., Ding, Y., Niggemann, J., Vähätalo, A.V., Stubbins, A., Spencer, R.G., et al., 2013. Global charcoal mobilization from soils via dissolution and riverine transport to the oceans. Science 340, 345–347.

Jagger, J., 1985. Solar-UV Actions on Living Cells. Praeger Publishing, New York, 202 pp.

Jahnke, R.A., Craven, D.B., 1995. Quantifying the role of heterotrophic bacteria in the carbon cycle: a need for respiration rate measurements. Limnol. Oceanogr. 40, 436–441.

Jankowski, J.J., Beaupre, S.R., Kieber, D.J., Mopper, K., Hofsetz, B.D., 1999. Concentration and photochemical production of nitrite in Antarctic seawater. EOS, Trans. Amer. Geophys. Un. OS271.

Johannessen, S.C., Miller, W.L., 2001. Quantum yield for the photochemical production of dissolved inorganic carbon in seawater. Mar. Chem. 76, 271–283.

Johannessen, S.C., Peña, M.A., Quenneville, M.L., 2007. Photochemical production of carbon dioxide during a coastal phytoplankton bloom. Estuar. Coast. Shelf Sci. 73, 236–242.

Johnson, J.E., Bates, T.S., 1996. Sources and sinks of carbon monoxide in the mixed layer of the tropical South Pacific Ocean. Global Biogeochem. Cycles 10, 347–359.

Jones, G.E., 1964. Precipitates from autoclaved seawater. Limnol. Oceanogr. 12, 165–167.

Jørgensen, N.O.G., Tranvik, L.J., Edling, H., Granéli, W., Lindell, M., 1998. Effects of sunlight on occurrence and bacterial turnover of specific carbon and nitrogen compounds in lake water. FEMS Microbiol. Ecol. 25, 217–227.

Jørgensen, N.O.G., Tranvik, L.J., Berg, G.M., 1999. Occurrence and bacterial cycling of dissolved nitrogen in the Gulf of Riga, the Baltic Sea. Mar. Ecol. Prog. Ser. 191, 1–18.

Jørgensen, L., Stedmon, C.A., Kragh, T., Markager, S., Middelboe, M., Søndergaard, M., 2011. Global trends in the fluorescence characteristics and distribution of marine dissolved organic matter. Mar. Chem. 126, 139–148.

Judd, K., Crump, B., Kling, G., 2007. Bacterial responses in activity and community composition to photo-oxidation of dissolved organic matter from soil and surface waters. Aquat. Sci. 69, 96–107.

Kahan, T.F., Kwamena, N.O.A., Donaldson, D.J., 2010. Different photolysis kinetics at the surface of frozen freshwater vs. frozen salt solutions. Atmos. Chem. Phys. 10, 10917–10922.

Kaiser, E., Sulzberger, B., 2004. Phototransformation of riverine dissolved organic matter (DOM) in the presence of abundant iron: effect on DOM bioavailability. Limnol. Oceanogr. 49, 540–554.

Kamyshny, A., Goifman, D., Rizkov, D., Lev, O., 2003. Formation of carbonyl sulfide by the reaction of carbon monoxide and inorganic polysulfides. Env. Sci. Technol. 37, 1865–1872.

Keil, R.G., Kirchman, D.L., 1994. Abiotic transformation of labile protein to refractory protein in seawater. Mar. Chem. 45, 187–196.

Keller, M.D., Korjeff-Bellows, W., 1996. Physiological aspects of the production of dimethylsulfoniopropionate (DMS) by marine phytoplankton. In: Kiene, R.P., Visscher, P.T., Keller, M.D., Kirst, G.O. (Eds.), Biological and Environmental Chemistry of DMSP and Related Sulfonium Compounds. Plenum Press, New York, NY, pp. 131–142.

Kettle, A.J., 1994. A Model of the Temporal and Spatial Distribution of Carbon Monoxide in the Mixed Layer (M.S. thesis), Massachusetts Institute of Technology, MA.

Kettle, A.J., 2005. Diurnal cycling of carbon monoxide (CO) in the upper ocean near Bermuda. Ocean Model. 8, 337–367.

Khalil, M.A.K., Rasmussen, R.A., 1984. Global sources, lifetimes and mass balances of carbonyl sulfide (OCS) and carbon disulfide (CS_2) in the earth's atmosphere. Atmos. Environ. 18, 1805–1813.

Kieber, D.J., 1988. Marine Biogeochemistry of α-Keto Acids (Ph.D. thesis), University of Miami, Rosenstiel School of Marine and Atmospheric Science, Miami, FL.

Kieber, D.J., 2000. Photochemical production of biological substrates. In: de Mora, S.J., Demers, S.J.S., Vernet, M. (Eds.), The Effects of UV Radiation in the Marine Environment. Cambridge University Press, New York, pp. 130–148.

Kieber, D.J., Jiao, J., 1995. Photochemistry of dimethyl sulfide in seawater. Div. Env. Chem., Amer. Chem. Soc. 35, 523–525, Preprint of Papers Presented at the 210th ACS National Meeting, August 22–24, 1995, Chicago, IL, pp. 523–525.

Kieber, D.J., McDaniel, J.A., Mopper, K., 1989. Photochemical source of biological substrates in seawater: implications for carbon cycling. Nature 341, 637–639.

Kieber, R.J., Zhou, X., Mopper, K., 1990. Formation of carbonyl compounds from UV-induced photodegradation of humic substances in natural waters: fate of riverine carbon in the sea. Limnol. Oceanogr. 35, 1503–1515.

Kieber, D.J., Jiao, J.F., Kiene, R.P., Bates, T.S., 1996. Impact of dimethylsulfide photochemistry on methyl sulfur cycling in the equatorial Pacific Ocean. J. Geophys. Res. C 101, 3715–3722.

Kieber, D.J., Yocis, B.H., Mopper, K., 1997b. Free-floating drifter for photochemical studies in the water column. Limnol. Oceanogr. 42, 1829–1833.

Kieber, R.J., Dydro, L.H., Seaton, P.J., 1997a. Photooxidation of triglycerides and fatty acids in seawater: implication toward the formation of marine humic substances. Limnol. Oceanogr. 42, 1454–1462.

Kieber, D.J., Mopper, K., Qian, J.G., 1999a. Photochemical formation of dissolved inorganic carbon in seawater and its impact on the marine carbon cycle. Trans. Am. Geophys. Un 80(49), 128.

Kieber, R.J., Li, A., Seaton, P.J., 1999b. Production of nitrite from the photodegradation of dissolved organic matter in natural waters. Environ. Sci. Technol. 33, 993–998.

Kieber, D.J., Mopper, K., Qian, J.G., Zafiriou, O.C., 2001. Photochemical formation of dissolved inorganic carbon in seawater and its impact on the marine carbon cycle. In: Book of Abstracts, 2001 ASLO Meeting, Albuquerque, NM, Feb. 11–16, pp. 80.

Kieber, D.J., Peake, B.M., Scully, N.M., 2003. Reactive oxygen species in aquatic ecosystems. In: Helbling, E.W., Zagarese, H. (Eds.), UV Effects on Aquatic Organisms and Ecosystems, Comprehensive Series in Photochemistry and Photobiology. The Royal Society of Chemistry, Cambridge, UK, pp. 253–288.

Kieber, R.J., Whitehead, R.F., Skrabal, S.A., 2006. Photochemical production of dissolved organic carbon from resuspended sediments. Limnol. Oceanogr. 51, 2187–2195.

Kieber, D.J., Miller, G.W., Neale, P.J., Mopper, K., 2014. Wavelength and temperature-dependent apparent quantum yields for photochemical formation of hydrogen peroxide in seawater. Env. Sci.: Processes Impacts. http://dx.doi.org/10.1039/C4EM00036F.

Kikugawa, K., Beppu, M., 1987. Involvement of lipid oxidation products in the formation of fluorescent and cross-linked proteins. Chem. Phys. Lipids 44, 277–296.

Kim, B., Wetzel, R.G., 1993. The effect of dissolved humic substances on the alkaline phosphatase and the growth of microalgae. Int. Ver. Theor. Angew. Limnol. Verh. 25, 129–132.

Kim, M.-K., Zoh, K.-D., 2013. Effects of natural water constituents on the photo-decomposition of methylmercury and the role of hydroxyl radical. Sci. Total Environ. 449, 95–101.

King, M.D., France, J.L., Fisher, F.N., Beine, H.J., 2005. Measurement and modelling of UV radiation penetration and photolysis rates of nitrate and hydrogen peroxide in Antarctic sea ice: an estimate of the production rate of hydroxyl radicals in first-year sea ice. J. Photochem. Photobiol. A: Chem. 176, 39–49.

Kirchman, D.L., 1990. Limitation of bacterial growth by dissolved organic matter in the subarctic Pacific. Mar. Ecol. Prog. Ser. 62, 47–54.

Kitidis, V., Stubbins, A., Uher, G., Upstill-Goddard, R.C., Law, C.S., Woodward, E.M.S., 2006a. Variability of chromophoric organic matter in surface waters of the Atlantic Ocean. Deep-Sea Res. II 53, 1666–1684.

Kitidis, V., Uher, G., Upstill-Goddard, R.C., Mantoura, R.F.C., Spyres, G., Woodward, E.M.S., 2006b. Photochemical production of ammonium in the oligotrophic Cyprus Gyre (Eastern Mediterranean). Biogeosciences 3, 439–449.

Kitidis, V., Uher, G., Woodward, E.M.S., Owens, N.J.P., Upstill-Goddard, R.C., 2008. Photochemical production and consumption of ammonium in a temperate river–sea system. Mar. Chem. 112, 118–127.

Kitidis, V., Tilstone, G.H., Serret, P., Smyth, T.J., Torres, R., Robinson, C., 2014. Oxygen photolysis in the Mauritanian upwelling: implications for net community production. Limnol. Oceanogr. 59, 299–310.

Klánová, J., Klán, P., Nosek, J., Holoubek, I., 2003. Environmental ice photochemistry: monochlorophenols. Environ. Sci. Technol. 37, 1568–1574.

Klementova, S., Wagnerova, D.M., 1990. Photoinitiated transformation of glyoxalic and glycolic acids in aqueous solution. Mar. Chem. 30, 89–103.

Koopmans, D.J., Bronk, D.A., 2002. Photochemical production of dissolved inorganic nitrogen and primary amines from dissolved organic nitrogen in waters of two estuaries and adjacent surficial groundwaters. Aquat. Microb. Ecol. 26, 295–304.

Kouassi, A.M., Zika, R.G., 1990. Light-induced alteration of the photophysical properties of dissolved organic matter in seawater Part I. Photoreversible properties of natural water fluorescence. Neth. J. Sea Res. 27, 25–32.

Kovac, N., Faganeli, J., Sket, B., Bajt, O., 1998. Characterization of macroaggregates and photodegradation of their water soluble fraction. Org. Geochem. 29, 1623–1634.

Kragh, T., Søndergaard, M., Tranvik, L., 2008. Effect of exposure to sunlight and phosphorus-limitation on bacterial degradation of coloured dissolved organic matter (CDOM) in freshwater. FEMS Microbiol. Ecol. 64, 230–239.

Kramer, G.D., Herndl, G.J., 2004. Photo- and bioreactivity of chromophoric dissolved organic matter produced by marine bacterioplankton. Aquat. Microb. Ecol. 36, 239–246.

Kujawinski, E.B., Del Vecchio, R., Blough, N.V., Klein, G.C., Marshall, A.G., 2004. Probing molecular-level transformations of dissolved organic matter: insights on photochemical degradation and protozoan modification of DOM from electrospray ionization Fourier transform ion cyclotron resonance mass spectrometry. Mar. Chem. 92, 23–37.

Kulovaara, M., Corin, N., Backlund, P., Tervo, J., 1996. Impact of UV-254-radiation on aquatic humic substances. Chemosphere 33, 783–790.

Kupper, F.C., Carrano, C.J., Kuhn, J.U., Butler, A., 2006. Photoreactivity of iron(III)-aerobactin: photoproduct structure and iron(III) coordination. Inorg. Chem. 45, 6028–6033.

Langenheder, S., Sobek, S., Tranvik, L.J., 2006. Changes bacterial community composition along solar radiation gradient in humic waters. Aquat. Sci. 68, 415–424.

Langford, C.H., Wingham, M., Sastri, V.S., 1973. Ligand photooxidation in copper (II) complexes of nitrilotriacetic acid. Environ. Sci. Technol. 7, 820–822.

Lanzalunga, O., Bietti, M., 2000. Photo-and radiation chemical induced degradation of lignin model compounds. J. Photochem. Photobiol. B: Biol. 56, 85–108.

Latch, D.E., McNeill, K., 2006. Microheterogeneity of singlet oxygen distributions in irradiated humic acid solutions. Science 311, 1743–1747.

Law, C.S., Brévière, E., de Leeuw, G., Garçon, V., Guieu, C., Kieber, D.J., et al., 2013. Evolving research directions in surface ocean–lower atmosphere (SOLAS) science. Environ. Chem. 10, 1–16.

Lee, P.A., de Mora, S.J., 1999. Intracellular dimethylsulfoxide (DMSO) in unicellular marine algae: speculations on its origin and possible biological role. J. Phycol. 35, 8–18.

Lee, P.A., de Mora, S.J., Levasseur, M., 1999. A review of dimethylsulfoxide in aquatic environments. Atmos. Oceans 37, 439–456.

Leifer, A., 1988. The Kinetics of Environmental Aquatic Photochemistry: Theory and Practice. American Chemical Society, Washington, DC.

Letscher, R.T., Hansell, D.A., Carlson, C.A., Lumpkin, R., Knapp, A.N., 2013. Dissolved organic nitrogen in the global surface ocean: distribution and fate. Global Biogeochem. Cycles 27, 141–153.

Leunert, F., Eckert, W., Paul, A., Gerhardt, V., Grossart, H.-P., 2014. Phytoplankton response to UV-generated hydrogen peroxide from natural organic matter. J. Plankton Res. 36, 185–197.

Lignell, R., Hoikkala, L., Lahtinen, T., 2008. Effects of inorganic nutrients, glucose and solar radiation on bacterial growth and exploitation of dissolved organic carbon and nitrogen in the northern Baltic Sea. Aquat. Microb. Ecol. 51, 209–221.

Lindell, M.J., Rai, H., 1994. Photochemical oxygen consumption in humic waters. Arch. Hydrobiol. Beih. Ergebn. Limnol. 43, 145–155.

Lindell, M.J., Granéli, H.W., Bertilsson, S., 2000. Seasonal photoreactivity of dissolved organic matter from lakes with contrasting humic content. Can. J. Fish. Aquat. Sci. 57, 875–885.

Liu, C., Spiese, C.E., Kinsey, J.D., Kieber, D.J., Kiene, R.P., 2014. Production of dimethylsulfide by superoxide reduction of dimethylsulfoniopropionate. In preparation.

Lønborg, C., Martínez-Garcíad, S., Eva Teirad, E., Álvarez-Salgadoa, X.A., 2013. Effects of photochemical transformation of dissolved organic matter on bacterial physiology and diversity in a coastal system. Estuar. Coast. Shelf Sci. 129, 11–18.

Lou, T., Xie, H., 2006. Photochemical alteration of the molecular weight of dissolved organic matter. Chemosphere 65, 2333–2342.

Lund, V., Hongve, D., 1994. Ultraviolet irradiated water containing humic substances inhibits bacterial metabolism. Water Res. 28, 1111–1116.

Luther, G.W., Tsamakis, E., 1989. Concentration and form of dissolved sulfide in the oxic water column of the ocean. Mar. Chem. 27, 165–177.

Ma, X., Green, S.A., 2004. Photochemical transformation of dissolved organic carbon in Lake Superior—an in-situ experiment. J. Great Lakes Res. 30, 97–112.

Ma, J.H., Del Vecchio, R., Golanoski, K.S., Boyle, E.S., Blough, N.V., 2010. Optical properties of humic substances and CDOM: effects of borohydride reduction. Environ. Sci. Technol. 44, 5395–5402.

Maldonado, M.T., Price, N.M., 2001. Reduction and transport of organically bound iron by *Thalassiosira oceanica* (Bacillariophyceae). J. Phycol. 37, 298–310.

Marchand, D., Rontani, J.-F., 2001. Characterisation of photo-oxidation and autoxidation products of phytoplanktonic monounsaturated fatty acids in marine particulate matter and recent sediments. Org. Geochem. 32, 287–304.

Marchand, D., Rontani, J.-F., 2003. Visible light-induced oxidation of lipid components of purple sulfur bacteria: a significant process in microbial mats. Org. Geochem. 34, 61–79.

Marchand, D., Marty, J.-C., Miquel, J.-C., Rontani, J.-F., 2005. Lipids and their oxidation products as biomarkers for carbon cycling in the northwestern Mediterranean Sea: results from a sediment trap study. Mar. Chem. 95, 129–147.

Maurice, P.A., Cabaniss, S.E., Drummond, J., Ito, E., 2002. Hydrogeochemical controls on the variations in chemical characteristics of natural organic matter at a small freshwater wetland. Chem. Geol. 187, 59–77.

Mayer, L.M., Schick, L.L., Skorko, K., Boss, E., 2006. Photodissolution of particulate organic matter from sediments. Limnol. Oceanogr. 51, 1064–1071.

Mayer, L.M., Schick, L.L., Bianchi, T.S., Wysocki, L.A., 2009a. Photochemical changes in chemical markers of sedimentary organic matter source and age. Mar. Chem. 113, 123–128.

Mayer, L.M., Schick, L.L., Hardy, K.R., Estapa, M.L., 2009b. Photodissolution and other photochemical changes upon irradiation of algal detritus. Limnol. Oceanogr. 54, 1688–1698.

Mayer, L.M., Thornton, K.H., Schick, L.L., 2011. Bioavailability of organic matter photodissolved from coastal sediments. Aquat. Microb. Ecol. 64, 275–284.

McCallister, S.L., Bauer, J.E., Kelly, J., Ducklow, H.W., 2005. Effects of sunlight on decomposition of estuarine dissolved organic C, N and P and bacterial metabolism. Aquat. Microb. Ecol. 40, 25–35.

McKnight, D.M., Boyer, E.W., Westerhoff, P.K., Doran, P.T., Kulbe, T., Andersen, D.T., 2001. Spectrofluorometric characterization of dissolved organic matter for indication of precursor organic material and aromaticity. Limnol. Oceanogr. 46, 38–48.

McNally, A.M., Moody, E.C., McNeill, K., 2005. Kinetics and mechanism of the sensitized photodegradation of lignin model compounds. Photochem. Photobiol. Sci. 4, 268–274.

McNeil, B.I.M., Key, R.J., Bulliste, R.M., Sarmiento, J.L., 2003. Anthropogenic CO_2 uptake by the ocean based on the global chlorofluorocarbon data set. Science 299, 235–239.

Mee, L.K., 1987. In: Radiation Chemistry of Biopolymers. VCH Publishers, New York, pp. 477–499.

Méndez-Díaz, J.D., Shimabuku, K.K., Ma, J., Enumah, Z.O., Pignatello, J.J., Mitch, W.A., Dodd, M.C., 2014. Sunlight-driven photochemical halogenation of dissolved organic matter in seawater: a natural abiotic source of organobromine and organoiodine. Environ. Sci. Technol. 48, 7418–7427. http://dx.doi.org/10.1021/es5016668.

Miles, C.J., Brezonik, P.L., 1981. Oxygen consumption in humic-colored waters by a photochemical ferrous-ferric catalytic cycle. Environ. Sci. Technol. 15, 1089–1095.

Miller, W.L., 2000. An overview of aquatic photochemistry as it relates to microbial production. In: Bell, C.R., Brylinsky, M., Johnson-Green, P.C. (Eds.), Microbial Biosystems: New Frontiers. Proceedings of the 8th International Symposium on Microbial Ecology, 1998. Atlantic Canada Society for Microbial Ecology, Halifax, Canada, pp. 201–207.

Miller, W.L., Moran, M.A., 1997. Interaction of photochemical and microbial processes in the degradation of refractory dissolved organic matter from a coastal marine environment. Limnol. Oceanogr. 42, 1317–1324.

Miller, G.C., Zepp, R.G., 1979. Effects of suspended sediments on photolysis rates of dissolved pollutants. Water Res. 13, 453–459.

Miller, W.L., Zepp, R.G., 1995. Photochemical production of dissolved inorganic carbon from terrestrial organic matter: significance to the oceanic organic carbon cycle. Geophys. Res. Lett. 22, 417–420.

Miller, W.L., Moran, M.A., Sheldon, W.M., Zepp, R.G., Opsahl, S., 2002. Determination of apparent quantum yield spectra for the formation of biologically labile photoproducts. Limnol. Oceanogr. 47, 343–352.

Miller, C.J., Rose, A.L., David Waite, T.D., 2013. Hydroxyl radical production by H_2O_2-mediated oxidation of Fe(II) complexed by Suwannee River fulvic acid under circumneutral freshwater conditions. Environ. Sci. Technol. 47, 829–835.

Minor, E.C., Pothen, J., Dalzell, B.J., Abdulla, H., Mopper, K., 2006. Effects of salinity changes on the photodegradation and ultraviolet-visible absorbance of terrestrial dissolved organic matter. Limnol. Oceanogr. 51, 2181–2186.

Moffett, J.W., 1995. Temporal and spatial variability of copper complexation by strong chelators in the Sargasso Sea. Deep-Sea Res. I 42, 1273–1295.

Moffett, J.W., 2001. Transformations among different forms of iron in the ocean. In: Turner, D.R., Hunter, K.A. (Eds.), The Biogeochemistry of Iron in Seawater. John Wiley & Sons, West Sussex, England, pp. 343–372.

Mopper, K., Kieber, D.J., 2000. Marine photochemistry and its impact on carbon cycling. In: de Mora, S.J., Demers, S., Vernet, M. (Eds.), The Effects of UV Radiation in the Marine Environment. Cambridge University Press, Cambridge, UK, pp. 101–129.

Mopper, K., Kieber, D.J., 2002. Photochemistry and the cycling of carbon, sulfur, nitrogen and phosphorus. In: Hansell, D.A., Carlson, C.A. (Eds.), Biogeochemistry of Marine Dissolved Organic Matter. Academic Press, San Diego, CA, pp. 455–507.

Mopper, K., Stahovec, W.L., 1986. Sources and sinks of low-molecular-weight organic carbonyl compounds in seawater. Mar. Chem. 19, 305–321.

Mopper, K., Zhou, X., 1990. Hydroxyl radical photoproduction in the sea and its potential impact on marine processes. Science 250, 661–664.

Mopper, K., Zhou, X., Kieber, R.J., Kieber, D.J., Sikorski, R.J., Jones, R.D., 1991. Photochemical degradation of dissolved organic carbon and its impact on the oceanic carbon cycle. Nature 353, 60–62.

Mopper, K., Stubbins, A., Ritchie, J.D., Bialk, H.M., Hatcher, P.G., 2007. Advanced instrumental approaches for characterization of marine dissolved organic matter: extraction techniques, mass spectrometry, and nuclear magnetic resonance spectroscopy. Chem. Rev. 107, 419–442.

Moran, M.A., Covert, J.S., 2003. Photochemically-mediated linkages between dissolved organic matter and bacterioplankton. In: Findlay, S.E.G., Sinsabaugh, R.L. (Eds.), Aquatic Ecosystems: Interactivity of Dissolved Organic Matter. Academic Press, New York, NY, pp. 243–262.

Moran, M.A., Zepp, R.G., 1997. Role of photoreactions in the formation of biologically labile compounds from dissolved organic matter. Limnol. Oceanogr. 42, 1307–1316.

Moran, M.A., Zepp, R.G., 2000. UV radiation effects on microbes and microbial processes. In: Kirchman, D.L. (Ed.), Microbial Ecology of the Oceans. Wiley-Liss, Inc., New York, NY, pp. 201–228.

Moran, M.A., Sheldon, W.M., Zepp, R.G., 2000. Carbon loss and optical property changes during long-term photochemical and biological degradation of estuarine dissolved organic matter. Limnol. Oceanogr. 45, 1254–1264.

Morell, J.M., Corredor, J.E., 2001. Photomineralization of fluorescent dissolved organic matter in the Orinoco River plume: estimation of ammonium release. J. Geophys. Res. C 8, 16,807–16,813.

Morris, J.J., Johnson, Z.I., Szul, M.J., Keller, M., Zinser, E.R., 2011. Dependence of the cyanobacterium *Prochlorococcus* on hydrogen peroxide scavenging microbes for growth at the ocean's surface. PLoS One 6, e16805. http://dx.doi.org/10.1371/journal.pone.0016805.

Mostafa, S., Rosario-Ortiz, F.L., 2013. Singlet oxygen formation from wastewater organic matter. Environ. Sci. Technol. 47, 8179–8186.

Mostofa, K.M.G., Liu, C.-Q., Minakata, D., Wu, F., Vione, D., Mottaleb, M.A., 2013. Photoinduced and microbial degradation of dissolved organic matter in natural waters. In: Mostofa, K.M.G., Yoshioka, T., Mottaleb, A., Vione, D. (Eds.), Photobiogeochemistry of Organic Matter. Springer-Verlag, Berlin Heidelberg, pp. 273–364.

Müller, S., Vähätalo, A.V., Granskog, M.A., Autio, R., Kaartokallio, H., 2011. Behaviour of dissolved organic matter during formation of natural and artificially grown Baltic Sea ice. Ann. Glaciol. 52, 233–241.

Najjar, R.G., Erickson III, D.J., Madronich, S., 1995. Modeling the air-sea fluxes of gases formed from the decomposition of dissolved organic matter: carbonyl sulfide and carbon monoxide. In: Zepp, R.G., Sonntag, C. (Eds.), Role of Nonliving Organic Matter in the Earth's Carbon Cycle. Wiley & Sons, Chichester, UK, pp. 107–132.

Neale, P.J., Kieber, D.J., 2000. Assessing biological and chemical effects of UV in the marine environment: spectral weighting functions. In: Hester, R.E., Harrison, R.M. (Eds.), Causes and Environmental Implications of Increased UV-B Radiation. The Royal Society of Chemistry, Cambridge, UK, pp. 61–83.

Nelson, J.R., 1993. Rates and possible mechanism of light-dependent degradation of pigments in detritus derived from phytoplankton. J. Mar. Res. 51, 155–179.

Nelson, N.B., Siegel, D.A., Carlson, C.A., Swan, C.M., 2010. Tracing global biogeochemical cycles and meridional overturning circulation using chromophoric dissolved organic matter. Geophys. Res. Lett. 37, L03610.

Nieto-Cid, M., Alvarez-Salgado, X.A., Perez, F.F., 2006. Microbial and photochemical reactivity of fluorescent dissolved organic matter in a coastal upwelling system. Limnol. Oceanogr. 51, 1391–1400.

Obernosterer, I., Benner, R., 2004. Competition between biological and photochemical processes in the mineralization of dissolved organic carbon. Limnol. Oceanogr. 49, 117–124.

Obernosterer, I., Kraay, G., DeRanitz, E., Herndl, G.J., 1999a. Concentrations of low molecular weight carboxylic acids and carbonyl compounds in the Aegean Sea (Eastern Mediterranean) and the turnover of pyruvate. Aquat. Microb. Ecol. 20, 147–156.

Obernosterer, I., Reitner, B., Herndl, G.J., 1999b. Contrasting effects of solar radiation on dissolved organic matter and its bioavailability to marine bacterioplankton. Limnol. Oceanogr. 44, 1645–1654.

Obernosterer, I., Sempére, R., Herndl, G.J., 2001a. Ultraviolet radiation induces a reversal of the bioavailability of DOM to marine bacterioplankton. Aquat. Microb. Ecol. 24, 61–68.

Obernosterer, I., Ruardu, P., Herndl, G.J., 2001b. DOM-photoreactivity across the subtropical Atlantic Ocean: spatial and temporal dynamics of DOM fluorescence and H_2O_2 and photochemical oxygen demand of surface water DOM. Limnol. Oceanogr. 46, 632–643.

Obernosterer, I., Catala, P., Reinthaler, T., Herndl, G.J., Lebaron, P., 2005. Enhanced heterotrophic activity in the surface microlayer of the Mediterranean Sea. Aquat. Microb. Ecol. 39, 293–302.

Opsahl, S., Benner, R., 1998. Photochemical reactivity of dissolved lignin in river and ocean waters. Limnol. Oceanogr. 43, 1297–1304.

Opsahl, S.P., Zepp, R.G., 2001. Photochemically-induced alteration of stable carbon isotope ratios ($\delta^{13}C$) in terrigenous dissolved organic carbon. Geophys. Res. Lett. 28, 2417–2420.

Orellana, M.V., Verdugo, P., 2003. Ultraviolet radiation blocks the organic carbon exchange between the dissolved phase and the gel phase in the ocean. Limnol. Oceanogr. 48, 1618–1623.

Orellana, M.V., Matrai, P.A., Leck, C., Rauschenberg, C.D., Lee, A.M., Coz, E., 2011. Marine microgels as a source of cloud condensation nuclei in the high Arctic. Proc. Natl. Acad. Sci. U. S. A. 108, 13,612–13,617.

Ortega-Retuerta, E., Pulido-Villena, E., Reche, I., 2007. Effects of dissolved organic matter photoproducts and mineral nutrient supply on bacterial growth in Mediterranean inland waters. Microb. Ecol. 54, 161–169.

Osburn, C.L., Morris, D.P., 2003. Photochemistry of chromophoric dissolved organic matter in natural waters. In: Hebling, E.W., Zargarese, H. (Eds.), UV Effects on Aquatic Organisms and Ecosystems, Comprehensive Series in Photochemistry and Photobiology. The Royal Society of Chemistry, Cambridge, UK, pp. 185–217.

Osburn, C.L., Morris, D.P., Thorn, K.A., Moeller, R.E., 2001. Chemical and optical changes in freshwater dissolved organic matter exposed to solar radiation. Biogeochemistry 54, 251–278.

Osburn, C.L., Retamal, L., Vincent, W.F., 2009. Photoreactivity of chromophoric dissolved organic matter transported by the Mackenzie River to the Beaufort Sea. Mar. Chem. 115, 10–20.

Palenik, B., Morel, F.M.M., 1990. Amine oxidases of marine phytoplankton. Appl. Environ. Microbiol. 57, 2440–2443.

Pérez, M.T., Sommaruga, R., 2007. Interactive effects of solar radiation and dissolved organic matter on bacterial activity and community structure. Environ. Microbiol. 9, 2200–2210.

Pers, C., Rahm, L., Jonsson, A., Bergstrom, A.-K., Jansson, M., 2001. Modelling dissolved organic carbon turnover in Humic Lake Örträsket. Sweden Environ. Model. Assess. 6, 159–172.

Petasne, R.G., Zika, R.G., 1987. Fate of superoxide in coastal sea water. Nature 325, 516–518.

Peterson, B.M., McNally, A.M., Cory, R.M., Thoemke, J.D., Cotner, J.B., McNeill, K., 2012. Spatial and temporal distribution of singlet oxygen in Lake Superior. Environ. Sci. Technol. 46, 7222–7229.

Piccini, C., Conde, D., Pernthaler, J., Sommaruga, R., 2009. Alteration of chromophoric dissolved organic matter by solar UV radiation causes rapid changes in bacterial community composition. Photochem. Photobiol. Sci. 8, 1321–1328.

Pisani, O., Yamashita, Y., Jaffé, R., 2011. Photo-dissolution of flocculent, detrital material in aquatic environments: contributions to the dissolved organic matter pool. Water Res. 45, 3836–3844.

Pitt, J., 2007. Photochemistry of Brevetoxin, PbTx-2, Produced by the Dinoflagellate, *Karenia brevis* (M.S. thesis), University of North Carolina at Wilmington, Wilmington, NC.

Plumley, F.G., 1997. Marine algal toxins: biochemistry, genetics, and molecular biology. Limnol. Oceanogr. 42, 1252–1264.

Pochon, A., Vaughan, P.P., Gan, D., Vath, P., Blough, N.V., Falvey, D.E., 2002. Photochemical oxidation of water by 2-methyl-1,4-benzoquinone: evidence against the formation of free hydroxyl radical. J. Phys. Chem. A 106, 2889–2894.

Pomeroy, L.R., Sheldon, J.E., Sheldon Jr., W.M., Peters, F., 1995. Limits to growth and respiration of bacterioplankton in the Gulf of Mexico. Mar. Ecol. Prog. Ser. 117, 259–268.

Porcal, P., Koprivnjak, J.F., Molot, L.A., Dillon, P.J., 2009. Humic substances-part 7: the biogeochemistry of dissolved organic carbon and its interactions with climate change. Environ. Sci. Pollut. Res. 16, 714–726.

Porcal, P., Dillon, P.J., Molot, L.A., 2013. Seasonal changes in photochemical properties of dissolved organic matter in small boreal streams. Biogeosciences 10, 5533–5543.

Pos, W.H., Milne, P.J., Riemer, D.D., Zika, R.G., 1997. Photoinduced oxidation of H_2S species: a sink for sulfide in seawater. J. Geophys. Res. D 102, 12,831–12,837.

Pos, W.H., Riemer, D.D., Zika, R.G., 1998. Carbonyl sulfide (OCS) and carbon monoxide (CO) in natural waters: evidence of a coupled production pathway. Mar. Chem. 62, 89–101.

Powell, R.T., Wilson-Finelli, A., 2003. Photochemical degradation of organic iron complexing ligands in seawater. Aquat. Sci. 65, 367–374.

Powers, L.C., Miller, W.L., 2014. Blending remote sensing data products to estimate photochemical production of hydrogen peroxide and superoxide in the surface ocean. Environ. Sci.: Processes Impacts 16, 792–806.

Pringault, O., Tesson, S., Rochelle-Newall, E., 2009. Respiration in the light and bacterio-phytoplankton coupling in a coastal environment. Microb. Ecol. 57, 321–334.

Pullin, M.J., Bertilsson, S., Goldstone, J.V., Voelker, B.M., 2004. Effects of sunlight and hydroxyl radical on dissolved organic matter: bacterial growth efficiency and production of carboxylic acids and other substrates. Limnol. Oceanogr. 49, 2011–2022.

Qian, J.G., Mopper, K., Kieber, D.J., 2001. Photochemical production of the hydroxyl radical in Antarctic waters. Deep-Sea Res. I 48, 741–759.

Reader, H.E., Miller, W.L., 2011. Effect of estimations of ultraviolet absorption spectra of chromophoric dissolved organic matter on the uncertainty of photochemical production calculations. J. Geophys. Res. 116. http://dx.doi.org/10.1029/2010JC006823, C08002.

Reader, H.E., Miller, W.L., 2012. Variability of carbon monoxide and carbon dioxide apparent quantum yield spectra in three coastal estuaries of the South Atlantic Bight. Biogeosciences 9, 4279–4294.

Reader, H.E., Miller, W.L., 2014. The efficiency and spectral photon dose dependence of photochemically induced changes to the bioavailability of dissolved organic carbon. Limnol. Oceanogr. 59, 182–194.

Redden, G.D., 1982. Characteristics of Photochemical Production of Carbon Monoxide in Seawater (M.S. thesis), Oregon State University, Corvallis, OR.

Reitner, B., Herndl, G.J., Herzig, A., 1997. Role of ultraviolet-B radiation on photochemical and microbial oxygen consumption in a humic-rich shallow lake. Limnol. Oceanogr. 42, 950–960.

Reitner, B., Herzig, A., Herndl, G.J., 2002. Photoreactivity and bacterioplankton availability of aliphatic versus aromatic amino acids and a protein. Aquat. Microb. Ecol. 26, 305–311.

Remington, S., Krusche, A., Richey, J., 2011. Effects of DOM photochemistry on bacterial metabolism and CO_2 evasion during falling water in a humic and a whitewater river in the Brazilian Amazon. Biogeochemistry 105, 185–200.

Rhee, T.S., 2000. The Process of Air-Water Exchange and its Application (Ph.D. thesis), Texas A and M University, College Station, TX.

Richard, C., Trubetskaya, O., Trubetskoj, O., Reznikova, O., Afanas'eva, G., Aguer, J.-P., et al., 2004. Key role of the low molecular size fraction of soil humic acids for fluorescence and photoinductive activity. Environ. Sci. Technol. 38, 2052–2057.

Richardson, K., 1997. Harmful or exceptional phytoplankton blooms in the marine ecosystem. Adv. Mar. Biol. 31, 301–385.

Riggsbee, J.A., Orr, C.H., Leech, D.M., Doyle, M.W., Wetzel, R.G., 2008. Suspended sediments in river ecosystems: photochemical sources of dissolved organic carbon, dissolved organic nitrogen, and adsorptive removal of dissolved iron. J. Geophys. Res. 113. http://dx.doi.org/10.1029/2007JG000654, G03019.

Robinson, C., Williams, P.J., 2005. Respiration and its measurement in surface marine waters. In: Respiration in Aquatic Ecosystems. Oxford University Press, New York, pp. 147–180.

Rontani, J.-F., 2001. Visible light-dependent degradation of lipidic phytoplanktonic components during senescence: a review. Phytochemistry 58, 187–202.

Rontani, J.-F., Marchand, D., 2000. Photoproducts of phytoplanktonic sterols: a potential source of hydroperoxides in marine sediments? Org. Geochem. 31, 169–180.

Rontani, J.-F., Rabourdin, A., Marchand, D., Aubert, C., 2003. Photochemical oxidation and autoxidation of chlorophyll phytyl side chain in senescent phytoplanktonic cells: potential sources of several acyclic isoprenoid compounds in the marine environment. Lipids 38, 241–254.

Rontani, J.-F., Zabeti, N., Wakeham, S.G., 2009. The fate of marine lipids: biotic vs. abiotic degradation of particulate sterols and alkenones in the Northwestern Mediterranean Sea. Mar. Chem. 113, 9–18.

Rose, A.L., Waite, T.D., 2003. Predicting iron speciation in coastal waters from the kinetics of sunlight-mediated iron redox cycling. Aquat. Sci. 65, 375–383.

Rose, A.L., Waite, D.T., 2005. Reduction of organically complexed ferric iron by superoxide in a simulated natural water. Environ. Sci. Technol. 39, 2645–2650.

Rose, A.L., Waite, D.T., 2006. Role of superoxide in the photochemical reduction of iron in seawater. Geochim. Cosmochim. Acta 70, 3869–3882.

Rosenstock, B., Zwisler, W., Simon, M., 2005. Bacterial consumption of humic and non-humic low and high molecular weight DOM and the effect of solar irradiation on the turnover of labile DOM in the Southern Ocean. Microb. Ecol. 50, 90–101.

Rossel, P.E., Vähätalo, A.V., Witt, M., Dittmar, T., 2013. Molecular composition of dissolved organic matter from a wetland plant (*Juncus effusus*) after photochemical and microbial decomposition (1.25 yr): common features with deep sea dissolved organic matter. Org. Geochem. 60, 62–71.

Rue, E., Bruland, K., 2001. Domoic acid binds iron and copper: a possible role for the toxin produced by the marine diatom *Pseudo-nitzschia*. Mar. Chem. 76, 127–134.

Säuberlich, J., Brede, O., Beckert, D., 1996. Photoionization of benzophenone carboxylic acids in aqueous solution. A FT EPR and optical spectroscopy study of radical cation decay. J. Phys. Chem. 100, 18101–18107.

Schmitt-Kopplin, P., Hertkorn, N., Schulten, H.-R., Kettrup, A., 1998. Structural changes in a dissolved soil humic acid during photochemical degradation processes under O_2 and N_2 atmosphere. Environ. Sci. Technol. 32, 2531–2541.

Scully, N.M., Cooper, W.J., Tranvik, L.J., 2003a. Photochemical effects on microbial activity in natural waters: the interaction of reactive oxygen species and dissolved organic matter. FEMS Microbiol. Ecol. 46, 353–357.

Scully, N.M., Tranvik, L.J., Cooper, W.J., 2003b. Photochemical effects on the interaction of enzymes and dissolved organic matter in natural waters. Limnol. Oceanogr. 48, 1818–1824.

Scully, N.M., Maie, N., Dailey, S.K., Boyer, J.N., Jones, R.D., Jaffé, R., 2004. Early diagenesis of plant-derived dissolved organic matter along a wetland, mangrove, estuary ecotone. Limnol. Oceanogr. 49, 1667–1678.

Shank, G.C., Zepp, R.G., Vahatalo, A., Lee, R., Bartels, E., 2010. Photobleaching kinetics of chromophoric dissolved organic matter derived from mangrove leaf litter and floating Sargassum colonies. Mar. Chem. 119, 162–171.

Shank, G.C., Evans, A., Yamashita, Y., Jaffé, R., 2011. Solar radiation-enhanced dissolution of particulate organic matter from coastal marine sediments. Limnol. Oceanogr. 56, 577–588.

Sharpless, C.M., 2012. Lifetimes of triplet dissolved natural organic matter (DOM) and the effect of $NaBH_4$ reduction on singlet oxygen quantum yields: implications for DOM photophysics. Environ. Sci. Technol. 46, 4466–4473.

Shooter, D., Brimblecombe, P., 1989. Dimethylsulphide oxidation in the ocean. Deep-Sea Res. 36, 577–585.

Simó, R., Pedrós-Alió, C., 1999. Short-term variability in the open ocean cycle of dimethylsulfide. Global Biogeochem. Cycles 13, 1173–1181.

Simpson, A.J., Kingery, W.L., Hayes, M.H., Spraul, M., Humpfer, E., Dvortsak, P., et al., 2002. Molecular structures and associations of humic substances in the terrestrial environment. Naturwissenschaften 89, 84–88.

Smith, E.M., Benner, R., 2005. Photochemical transformations of riverine dissolved organic matter: effects on estuarine bacterial metabolism and nutrient demand. Aquat. Microb. Ecol. 40, 37–50.

Smyth, T.J., 2011. Penetration of UV irradiance into the global ocean. J. Geophys. Res. 116. http://dx.doi.org/10.1029/2011JC007183, C11020.

Song, G., Xie, H., Aubry, C., Zhang, Y., Gosselin, M., Mundy, C.J., et al., 2011. Spatiotemporal variations of dissolved organic carbon and carbon monoxide in first-year sea ice in the western Canadian Arctic. J. Geophys. Res. 116, http://dx.doi.org/10.1029/2010JC006867.

Song, G., Xie, H., Bélanger, S., Leymarie, E., Babin, M., 2013. Spectrally resolved efficiencies of carbon monoxide (CO) photoproduction in the western Canadian Arctic: particles versus solutes. Biogeosciences 10, 3731–3748.

SooHoo, J.B., Kiefer, D.A., 1982a. Vertical distribution of phaeopigments—I. A simple grazing and photooxidative scheme for small particles. Deep-Sea Res. I 29, 1539–1551.

SooHoo, J.B., Kiefer, D.A., 1982b. Vertical distribution of phaeopigments—II. Rates of production and kinetics of photooxidation. Deep-Sea Res. I 29, 1553–1563.

Southwell, M.W., Kieber, R.J., Mead, R.N., Brooks Avery, G., Skrabal, S.A., 2009. Effects of sunlight on the production of dissolved organic and inorganic nutrients from resuspended sediments. Biogeochemistry 98, 115–126.

Spencer, R.G., Stubbins, A., Hernes, P.J., Baker, A., Mopper, K., Aufdenkampe, A.K., et al., 2009. Photochemical degradation of dissolved organic matter and dissolved lignin phenols from the Congo River. J. Geophys. Res. 114. http://dx.doi.org/10.1029/2009JG000968, G03010.

Spiese, C.E., 2010. Cellular Production and Losses of Dimethylsulfide in Marine Algae (Ph.D. thesis), State University of New York, College of Environmental Science and Forestry, Syracuse, NY.

Spokes, L.J., Liss, P.S., 1996. Photochemically induced redox reactions in seawater. II. Nitrogen and iodine. Mar. Chem. 54, 1–10.

Stedmon, C.A., Markager, S., Tranvik, L., Kronberg, L., Slätis, T., Martinsen, W., 2007a. Photochemical production of ammonium and transformation of dissolved organic matter in the Baltic Sea. Mar. Chem. 104, 227–240.

Stedmon, C.A., Thomas, D.N., Granskog, M., Kaartokallio, H., Papadimitriou, S., Kuosa, H., 2007b. Characteristics of dissolved organic matter in Baltic coastal sea ice: allochthonous or autochthonous origins? Environ. Sci. Technol. 41, 7273–7279.

Steffen, L.K., Glass, R.S., Sabahi, M., Wilson, G.S., Schoeneich, C., Mahling, S., et al., 1991. Hydroxyl radical induced decarboxylation of amino acids. Decarboxylation vs bond formation in radical intermediates. J. Am. Chem. Soc. 113, 2141–2145.

Strome, D.J., Miller, M.C., 1978. Photolytic changes in dissolved humic substances. Int. Theor. Angew. Limnol. Verh. 20, 1248–1254.

Stubbins, A., 2002. Aspects of Aquatic CO Photoproduction from CDOM (Ph.D. thesis), Newcastle University, Newcastle upon Tyne, UK.

Stubbins, A., Uher, G., Kitidis, V., Law, C.S., Upstill-Goddard, R.C., Woodward, E.M.S., 2006a. The open-ocean source of atmospheric carbon monoxide. Deep-Sea Res. II 53, 1685–1694.

Stubbins, A., Uher, G., Law, C.S., Mopper, K., Robinson, C., Upstill-Goddard, R.C., 2006b. Open-ocean carbon monoxide photoproduction. Deep-Sea Res. II 53, 1695–1705.

Stubbins, A., Hubbard, V., Uher, G., Law, C.S., Upstill-Goddard, R.C., Aiken, G.R., et al., 2008. Relating carbon monoxide photoproduction to dissolved organic matter functionality. Environ. Sci. Technol. 42, 3271–3276.

Stubbins, A., Spencer, R.G.M., Chen, H., Hatcher, P.G., Mopper, K., Hernes, P.J., et al., 2010. Illuminated darkness: molecular signatures of Congo River dissolved organic matter and its photochemical alteration as revealed by ultrahigh precision mass spectrometry. Limnol. Oceanogr. 55, 1467–1477.

Stubbins, A., Law, C., Uher, G., Upstill-Goddard, R., 2011. Carbon monoxide apparent quantum yields and photoproduction in the Tyne estuary. Biogeosciences 8, 703–713.

Stubbins, A., Niggemann, J., Dittmar, T., 2012. Photo-lability of deep ocean dissolved black carbon. Biogeosciences 9, 1661–1670.

Sulzberger, B., 2000. Photooxidation of dissolved organic matter: role for carbon bioavailability and for the penetration depth of solar UV-radiation. In: Gianguzza, A., Pelizzetti, E., Sammartano, S. (Eds.), Chemical Processes in the Marine Environment. Springer-Verlag, Berlin, pp. 75–90.

Sulzberger, B., Durisch-Kaiser, E., 2009. Chemical characterization of dissolved organic matter (DOM): a prerequisite for understanding UV-induced changes of DOM absorption properties and bioavailability. Aquat. Sci. 71, 104–126.

Sumner, A.L., Shepson, P.B., 1999. Snowpack production of formaldehyde and its effect on the Arctic troposphere. Nature 398, 230–233.

Sunda, W.G., Huntsman, S.A., 1994. Photoreduction of manganese oxides in seawater. Mar. Chem. 46, 133–152.

Sunda, W., Kieber, D.J., Kiene, R.P., Huntsman, S., 2002. DMSP dynamics in marine algae in relation to iron, photosynthesis, and oxidative stress. Nature 418, 317–320.

Swan, C.M., Siegel, D.A., Nelson, N.B., Carlson, C.A., Nasir, E., 2009. Biogeochemical and hydrographic controls on chromophoric dissolved organic matter distribution in the Pacific Ocean. Deep-Sea Res. I 56, 2175–2192.

Swan, C.M., Nelson, N.B., Siegel, D.A., Kostadinov, T.S., 2012. The effect of surface irradiance on the absorption spectrum of chromophoric dissolved organic matter in the global ocean. Deep-Sea Res. I 63, 52–64.

Swanson, A.L., Blake, N.J., Dibb, J.E., Albert, M.R., Blake, D.R., Rowland, S.F., 2002. Photochemically induced production of CH$_3$Br, CH$_3$I, C$_2$H$_5$I, ethene, and propene within surface snow at Summit. Greenland. Atmos. Environ. 36, 2671–2682.

Swinnerton, J.W., Linnenbo, V.J., Lamontag, R.A., 1970. Ocean: a natural source of carbon monoxide. Science 167, 984–986.

Tai, C., Li, Y., Yin, Y., Scinto, L.J., Jiang, G., Cai, Y., 2014. Methylmercury photodegradation in surface water of the Florida Everglades: importance of dissolved organic matter-methylmercury complexation. Environ. Sci. Technol. 48, 7333–7340.

Tarr, M.A., Wang, W., Bianchi, T.S., Engelhaupt, E., 2001. Mechanisms of ammonia and amino acid photoproduction from aquatic humic and colloidal matter. Water Res. 35, 3688–3696.

Tedetti, M., Joux, F., Charrière, B., Mopper, K., Sempéré, R., 2009. Contrasting effects of solar radiation and nitrates on the bioavailability of dissolved organic matter to marine bacteria. J. Photochem. Photobiol. A: Chem. 201, 243–247.

Thomas, D.N., Dieckmann, G.S., 2002. Antarctic Sea ice–a habitat for extremophiles. Science 295, 641–644.

Thomas, D.N., Lara, R.J., 1995. Photodegradation of algal derived dissolved organic carbon. Mar. Ecol. Progr. Ser. 116, 309–310.

Thompson, A.M., 1992. The oxidizing capacity of the Earths atmosphere—probable past and future changes. Science 256, 1157–1165.

Thompson, A.M., Zafiriou, O.C., 1983. Air-sea fluxes of transient atmospheric species. J. Geophys. Res. 88, 6696–6708.

Thomson, J., Parkinson, A., Roddick, F.A., 2004. Depolymerization of chromophoric natural organic matter. Environ. Sci. Technol. 38, 3360–3369.

Thorn, K.A., Younger, S.J., Cox, L.G., 2010. Order of functionality loss during photodegradation of aquatic humic substances. J. Environ. Qual. 39, 1416–1428.

Tonietto, A.E., Lombardia, A.T., Vieira, A.A.H., 2011. The effects of solar irradiation on copper speciation and organic complexation. J. Braz. Chem. Soc. 22, 1695–1700.

Toole, D.A., Kieber, D.J., Kiene, R.P., Siegel, D.A., Nelson, N.B., 2003. Photolysis and the dimethylsulfide (DMS) summer paradox in the Sargasso Sea. Limnol. Oceanogr. 48, 1088–1100.

Toole, D.A., Kieber, D.J., Kiene, R.P., White, E.M., Bisgrove, J., del Valle, D.A., et al., 2004. High dimethylsulfide photolysis rates in nitrate-rich Antarctic waters. Geophys. Res. Lett. 31. http://dx.doi.org/10.1029/2004GL019863, L11307.

Toole, D.A., Slezak, D., Kieber, D.J., Kiene, R.P., Siegel, D.A., 2006. Effects of solar radiation on dimethylsulfide cycling in the western Atlantic Ocean. Deep-Sea Res. 53, 136–153.

Tranvik, L.J., 1992. Allochthonous dissolved organic matter as an energy source for pelagic bacteria and the concept of the microbial loop. Hydrobiologia 229, 107–114.

Tranvik, L.J., 1993. Microbial transformation of labile dissolved organic matter into humic-like matter in seawater. FEMS Microbiol. Ecol. 12, 177–183.

Tranvik, L.J., 1998. Degradation of dissolved organic matter in humic waters by bacteria. In: Aquatic Humic Substances. Ecological Studies, vol. 133. Springer-Verlag, Berlin Heidelberg, pp. 259–283.

Tranvik, L.J., Bertilsson, S., 2001. Contrasting effects of solar UV radiation on dissolved organic sources for bacterial growth. Ecol. Lett. 4, 458–463.

Tranvik, L.J., Kokalj, S., 1998. Decreased biodegradability of algal DOC due to interactive effects of UV radiation and humic matter. Aquat. Microb. Ecol. 14, 301–307.

Tranvik, L.J., Olofsson, H., Bertilsson, S., 2000. Photochemical effects on bacterial degradation of dissolved organic matter in lake water. In: Bell, C.R., Brylinsky, M., Johnson-Green, P.C. (Eds.), Microbial Biosystems: New Frontiers. Proceedings of the 8th International Symposium on Microbial Ecology, 1998, Halifax, NS.

Tuominen, L., Kairesalo, T., Hartikainen, H., 1994. Comparison of methods for inhibiting bacterial activity in sediment. Appl. Environ. Microbiol. 60, 3454–3457.

Tyssebotn, I.M.B., Kinsey, J.D., Kieber, D.J., Kiene, R.P., Rellinger, A.N., Motard-Côté, J., et al., 2014. Acrylate and dimethylsulfoxide concentrations and turnover rates in the Gulf of Mexico. In preparation.

Uher, G., Andreae, M.O., 1997. Photochemical production fo carbonyl sulfide in North Sea water: a process study. Limnol. Oceanogr. 42, 432–442.

Ulshöfer, V.S., Flöck, O.R., Uher, G., Andreae, M.O., 1996. Photochemical production and air-sea exchange of carbonyl sulfide in the eastern Mediterranean Sea. Mar. Chem. 53, 25–39.

Uusikivi, J., Vähätalo, A.V., Granskog, M.A., Sommaruga, R., 2010. Contribution of mycosporine-like amino acids and colored dissolved and particulate matter to sea ice optical properties and ultraviolet attenuation. Limnol. Oceanogr. 55, 703–713.

Vähätalo, A.V., Järvinen, M., 2007. Photochemically produced bioavailable nitrogen from biologically recalcitrant dissolved organic matter stimulates production of a nitrogen-limited microbial food web in the Baltic Sea. Limnol. Oceanogr. 52, 132–143.

Vähätalo, A., Salonen, K., 1996. Enhanced bacterial metabolism after exposure of humic lake water to solar radiation. Nordic Humus Newsletter 3, 36–43.

Vähätalo, A.V., Wetzel, R.G., 2004. Photochemical and microbial decomposition of chromophoric dissolved organic matter during long (months–years) exposures. Mar. Chem. 89, 313–326.

Vähätalo, A.V., Zepp, R.G., 2005. Photochemical mineralization of dissolved organic nitrogen to ammonium in the Baltic Sea. Environ. Sci. Technol. 39, 6985–6992.

Vähätalo, A.V., Salonen, K., Salkinoja-Salonen, M., Hatakka, A., 1999. Photochemical mineralization of synthetic lignin in lake water indicates enhanced turnover of aromatic organic matter under solar radiation. Biodegradation 10, 415–420.

Vähätalo, A.V., Salkinoja-Salonen, M., Taalas, P., Salonen, K., 2000. Spectrum of the quantum yield for photochemical mineralization of dissolved organic carbon in a humic lake. Limnol. Oceanogr. 43, 664–676.

Vähätalo, A.V., Salonen, K., Münster, U., Järvinen, M., Wetzel, R.G., 2003. Photochemical transformation of allochthonous organic matter provides bioavailable nutrients in a humic lake. Arch. Hydrobiol. 156, 287–314.

Vähätalo, A.V., Aarnos, H., Hoikkala, L., Lignell, R., 2011. Photochemical transformation of terrestrial dissolved organic matter supports hetero- and autotrophic production in coastal waters. Mar. Ecol. Prog. Ser. 423, 1–14.

Valentine, R.L., Zepp, R.G., 1993. Formation of carbon monoxide from the photodegradation of terrestrial dissolved organic carbon in natural waters. Environ. Sci. Technol. 27, 409–412.

Vaughan, P.P., Blough, N.V., 1998. Photochemical formation of hydroxyl radical by constituents of natural waters. Environ. Sci. Technol. 32, 2947–2953.

von Hobe, M., Najjar, R.G., Kettle, A.J., Andreae, M.O., 2003. Photochemical and physical modeling of carbonyl sulfide in the ocean. J. Geophys. Res. C 108, 3229. http://dx.doi.org/10.1029/2000JC000712.

von Wachenfeldt, E., Sobek, S., Bastviken, D., Tranvik, L.J., 2008. Linking allochthonous dissolved organic matter and boreal lake sediment carbon sequestration: the role of light-mediated flocculation. Limnol. Oceanogr. 53, 2416–2426.

Vraspir, J.M., Butler, A., 2009. Chemistry of marine ligands and siderophores. Ann. Rev. Mar. Sci. 1, 43–63.

Wang, W., Tarr, M., Bianchi, T., Engelhaupt, E., 2000. Ammonium photoproduction from aquatic humic and colloidal matter. Aquat. Geochem. 6, 275–292.

Wang, X.-C., Chen, R.F., Gardner, G.B., 2004. Sources and transport of dissolved and particulate organic carbon in the Mississippi River estuary and adjacent coastal waters of the northern Gulf of Mexico. Mar. Chem. 89, 241–256.

Wang, W., Johnson, C.G., Takeda, K., Zafiriou, O., 2009. Measuring the photochemical production of carbon dioxide from marine dissolved organic matter by pool isotope exchange. Environ. Sci. Technol. 43, 8604–8609.

Warneck, P., 2000. Chemistry of the Natural Atmosphere, second ed. Academic Press, San Diego, CA, 757 pp.

Weinbauer, M., Suttle, C., 1999. Lysogeny and prophage induction in coastal and offshore bacterial communities. Aquat. Microb. Ecol. 18, 217–225.

Weishaar, J.L., Aiken, G.R., Bergamaschi, B.A., Fram, M.S., Fujii, R., Mopper, K., 2003. Evaluation of specific ultraviolet absorbance as an indicator of the chemical composition and reactivity of dissolved organic carbon. Environ. Sci. Technol. 37, 4702–4708.

Weiss, P.S., Andrews, S.S., Johnson, J.E., Zafiriou, O.C., 1995. Photoproduction of carbonyl sulfide in South Pacific Ocean waters as a function of irradiation wavelength. Geophys. Res. Lett. 22, 215–218.

Wells, M.L., 2002. Marine colloids and trace metals. In: Hansell, D.A., Carlson, C.A. (Eds.), Biogeochemistry of Marine Dissolved Organic Matter. Academic Press, San Diego, CA, pp. 367–404.

Wetzel, R.G., 1993. Humic compounds from wetlands: complexation, inactivation, and reactivation of surface-bound and extracellular enzymes. Int. Ver. Theor. Angew. Limnol. Verh. 25, 122–128.

Wetzel, R.G., Hatcher, P.G., Bianchi, T.S., 1995. Natural photolysis by ultraviolet irradiance of recalcitrant dissolved organic matter to simple substrates for rapid bacterial metabolism. Limnol. Oceanogr. 40, 1369–1380.

White, E.M., Vaughan, P.P., Zepp, R.G., 2003. Role of the photo-Fenton reaction in the production of hydroxyl radicals and photobleaching of colored dissolved organic matter in a coastal river of the southeastern United States. Aquat. Sci. 65, 402–414.

White, E.M., Kieber, D.J., Mopper, K., 2008. Determination of photochemically produced carbon dioxide in seawater. Limnol. Oceanogr.: Meth 6, 441–453.

White, E.M., Kieber, D.J., Sherrard, J., Miller, W.L., Mopper, K., 2010. Carbon dioxide and carbon monoxide photoproduction quantum yields in the Delaware Estuary. Mar. Chem. 118, 11–21.

Wiegner, T.N., Seitzinger, S.P., 2001. Photochemical and microbial degradation of external dissolved organic matter inputs to rivers. Aquat. Microb. Ecol. 24, 27–40.

Wilkinson, F., Helman, W.P., Ross, A.B., 1995. Rate constants for the decay and reactions of the lowest electronically excited singlet state of molecular oxygen in solution. An expanded and revised compilation. J. Phys. Chem. Ref. Data 24, 663–1021.

Williams, P.J.le B., 2000. Heterotrophic bacteria and the dynamics of dissolved organic matter. In: Kirchman, D.L. (Ed.), Microbial Ecology of the Oceans. Wiley-Liss, Inc., New York, NY, pp. 153–200.

Wilson, D., Swinnerton, J., Lamontagne, R., 1970. Production of carbon monoxide and gaseous hydrocarbons in seawater: relation to dissolved organic carbon. Science 168, 1577–1579.

Winch, S., Lean, D., 2005. Comparison of changes in metal toxicity following exposure of water with high dissolved organic carbon content to solar, UV-B and UV-A radiation. Photochem. Photobiol. 81, 1469–1480.

Xie, H., Gosselin, M., 2005. Photoproduction of carbon monoxide in first-year sea ice in Franklin Bay, southeastern Beaufort Sea. Geophys. Res. Lett. 32. http://dx.doi.org/10.1029/2005GL022803, L12606.

Xie, H., Zafiriou, O.C., 2009. Evidence for significant photochemical production of carbon monoxide by particles in coastal and oligotrophic marine waters. Geophys. Res. Lett. 36. http://dx.doi.org/10.1029/2009GL041158, L23606.

Xie, H., Moore, R.M., Miller, W.L., 1998. Photochemical production of carbon disulfide in seawater. J. Geophys. Res. 103, 5635–5644.

Xie, H., Zafiriou, O.C., Cai, W.-J., Zepp, R.G., Wang, Y., 2004. Photooxidation and its effects on the carboxyl content of dissolved organic matter in two coastal rivers in the southeastern United States. Environ. Sci. Technol. 38, 4113–4119.

Xie, H., Bélanger, S., Song, G., Benner, R., Taalba, A., Blais, M., et al., 2012. Photoproduction of ammonium in the southeastern Beaufort Sea and its biogeochemical implications. Biogeosciences 9, 3047–3061.

Xu, Z., Yang, Y., Wang, G., Cao, W., Li, Z., Sun, Z., 2012. Optical properties of sea ice in Liaodong Bay. China J. Geophys. Res. 117. http://dx.doi.org/10.1029/2010JC006756, C03007.

Yang, G., Chengxuan, L., Jialin, Q., Lige, H., Haijun, J., 2007. Photochemical oxidation of dimethylsulfide in seawater. Acta Oceanol. Sinica 26, 34–42.

Yentsch, C.S., Reichert, C.A., 1962. The interrelationship between water soluble yellow substances and chloroplastic pigments in marine algae. Botanica Marina 3, 65–74.

Yocis, B.H., Kieber, D.J., Mopper, K., 2000. Photochemical production of hydrogen peroxide in Antarctic Waters. Deep-Sea Res. I 47, 1077–1099.

Zafiriou, O.C., 1974. Photochemistry of halogens in the marine atmosphere. J. Geophys. Res. 79, 2730–2732.

Zafiriou, O.C., Bonneau, R., 1987. Wavelength-dependent quantum yield of OH radical formation from photolysis of nitrite ion in water. Photochem. Photobiol. 45, 723–727.

Zafiriou, O.C., True, M.B., 1979. Nitrate photolysis in seawater by sunlight. Mar. Chem. 8, 33–42.

Zafiriou, O.C., Blough, N.V., Micinski, E., Dister, B., Kieber, D.J., Moffett, J., 1990. Molecular probe systems for reactive transients in natural waters. Mar. Chem. 30, 45–70.

Zafiriou, O.C., Andrews, S.S., Wang, W., 2003. Concordant estimates of oceanic carbon monoxide source and sink processes in the Pacific yield a balanced global "bluewater" CO budget. Global Biogeochem. Cycles 17, 1015. http://dx.doi.org/10.1029/2001GB001638.

Zepp, R.G., 1997. Interactions of marine biogeochemical cycles and the photodegradation of dissolved organic carbon and dissolved organic nitrogen. In: Gianguzza, A., Pelizzetti, E., Sammartano, S. (Eds.), Marine Chemistry. An Environmental Analytical Chemistry Approach. Kluwer Academic Publishing, Netherlands, pp. 329–351.

Zepp, R.G., Andreae, M.O., 1994. Factors affecting the photochemical production of carbonyl sulfide in seawater. Geophys. Res. Lett. 21, 2813–2816.

Zepp, R.G., Schlotzhauer, P.F., 1983. Influence of algae on photolysis rates of chemicals in water. Environ. Sci. Technol. 17, 462–468.

Zepp, R.G., Faust, B.C., Hoigné, J., 1992. Hydroxyl radical formation in aqueous reactions (pH 3-8) of iron(II) with hydrogen peroxide: the photo-Fenton reaction. Environ. Sci. Technol. 26, 313–319.

Zepp, R.G., Callaghan, T.V., Erickson, D.J., 1998. Effects of enhanced solar ultraviolet radiation on biogeochemical cycles. J. Photochem. Photobiol. B: Biol. 46, 69–82.

Zepp, R.G., Callaghan, T.V., Erickson III, D.J., 2003. Interactive effects of ozone depletion and climate change on biogeochemical cycles. Photochem. Photobiol. Sci. 2, 51–61.

Zepp, R.G., Erickson III, D.J., Paul, N.D., Sulzberger, B., 2007. Interactive effects of solar UV radiation and climate change on biogeochemical cycling. Photochem. Photobiol. Sci. 6, 286–300.

Zepp, R.G., Erickson III, D.J., Paul, N.D., Sulzberger, B., 2011. Effects of solar UV radiation and climate change on biogeochemical cycling: interactions and feedbacks. Photochem. Photobiol. Sci. 10, 261–279.

Zhang, T., Hsu-Kim, H., 2010. Photolytic degradation of methylmercury enhanced by binding to natural organic ligands. Nat. Geosci. 3, 473–476.

Zhang, Y., Xie, H., 2012. The sources and sinks of carbon monoxide in the St. Lawrence estuarine system. Deep-Sea Res. II 81–84, 114–123.

Zhang, Y., Xie, H., Chen, G., 2006. Factors affecting the efficiency of carbon monoxide photoproduction in the St. Lawrence estuarine system (Canada). Environ. Sci. Technol. 40, 7771–7777.

Zhang, Y., Del Vecchio, R., Blough, N.V., 2012. Investigating the mechanism of hydrogen peroxide photoproduction by humic substances. Environ. Sci. Technol. 46, 11836–11843.

Zhou, X., Beine, H.J., Honrath, R.E., Fuentes, J.D., Simpson, W., Shepson, P.B., et al., 2001. Snowpack photochemical production of HONO: a major source of OH in the Arctic boundary layer in springtime. Geophys. Res. Lett. 28, 4087–4090.

Ziolkowski, L., Druffel, E., 2010. Aged black carbon identified in marine dissolved organic carbon. Geophys. Res. Lett. 37. http://dx.doi.org/10.1029/2010GL043963, L16601.

Ziolkowski, L.A., Miller, W.L., 2007. Variability of the apparent quantum efficiency of CO photoproduction in the Gulf of Maine and Northwest Atlantic. Mar. Chem. 105, 258–270.

Zuo, Y., Jones, R., 1995. Formation of carbon monoxide by photolysis of dissolved marine organic material and its significance in the carbon cycling of the oceans. Naturwissenschaften 82, 472–474.

Marine Microgels

Mónica V. Orellana*, Caroline Leck†

*Polar Science Center, University of Washington/Institute for Systems Biology, Seattle, Washington, USA
†Department of Meteorology, University of Stockholm, Stockholm, Sweden

CONTENTS

I INTRODUCTION

Marine gels are three-dimensional (3D) polymer networks proposed to play a pivotal role in regulating ocean-basin-scale biogeochemical dynamics. Microgels structurally link biological production and microbial degradative processes at the ocean's surface to biogeochemical dynamics at the ocean's interior, cloud properties, radiative balance, and global climate (Figure 9.1). Formed *in situ* in the ocean's water column, gels provide structure to the microbial loop by forming metabolic "hot spots." Assembling from dissolved biopolymers in the microlayer at the ocean-atmosphere interface,

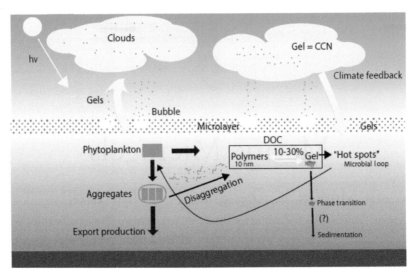

FIGURE 9.1　Marine polymer gel dynamics. DOM is mainly produced by phytoplankton and bacteria, which release free biopolymers and/or micron-size gels into the seawater by diverse mechanisms including disaggregation of phytoplankton cells (Table 9.1). Gel assembly is an important mechanism by which refractory short chain biopolymers (~1 nm or 1000 Da; Aluwihare et al., 1997; Chin et al., 1998) form "hot spots" for microbial activity (Azam, 1998; Azam and Malfatti, 2007; Baltar et al., 2010), which are particularly important in the dark ocean (Herndl and Reinthaler, 2013). Although microgels are non-sinking soft matter particles, they could also be exported to the deep ocean by undergoing volume phase transition which might increase their density and settling velocity (Chin et al., 1998; Orellana and Hansell, 2012). Furthermore, microgels can form cloud condensation nuclei (CCN), which have an important role in climate feedback (see text; Orellana et al., 2011b).

gels are also found in clouds, serving as cloud condensation nuclei (CCN). Problematically, the quantitative role of marine gels in these important processes is yet to be decided.

The dissolved organic carbon (DOC) pool is the largest bioreactive reservoir of organic carbon (C) in the world's oceans (662 ± 32 Gt; Hansell et al., 2009). This pool of C is similar in size to the atmospheric reservoir 750 Gt C (Sarmiento and Gruber, 2006). The sizes of these pools relative to the fluxes between them imply that changes in C flux through DOC can significantly influence the global C cycle on relatively short timescales (Hedges and Oades, 1997; Kwon et al., 2009). Scientists have made tremendous progress in understanding DOC dynamics (Hansell, 2013; Hansell and Carlson, 2002; Hansell et al., 2009); however, we still have only a fragmentary knowledge of the source and sink mechanisms, and specifically of the role played

by biopolymers and their dynamics. Marine polymer dynamics can be viewed in the context of soft matter physics, with clear benefits for developing accurate models of the response of biogeochemical cycles to environmental forcing.

Chin et al. (1998) applied the principles of soft matter physics to understand marine biopolymer dynamics, demonstrating that they assemble into 3D gel networks. By applying the tools and the conceptual framework of polymer physics, a mechanistic understanding of the structure of the DOC field and of the dynamics of DOC biopolymers emerged. This perspective brought new insights and a complementary "soft-material"-related understanding to DOC; assembly has thus emerged as a dominant concept to mechanistically explain the abiotic formation of particles in seawater, known since the time of Riley (1963), Alldredge and Jackson (1995), Chin et al. (1998), Sheldon et al. (1967), and Wells (1998).

Assembly of marine polymer gels occurs in the oceans when a chemically heterogeneous polydispersed mixture of biopolymers interact to form randomly tangled 3D cross-linked networks held together by ionic bonds, hydrophobic forces, and/or hydrogen bonds, depending on the nature of the polymers and the relation with the solvent (in this case seawater) (Chin et al., 1998; Ding et al., 2008; Orellana et al., 2011b; Radić et al., 2011). Marine polymer assembly is reversible, with an approximate thermodynamic yield at equilibrium of at least 10% in surface waters (maybe lower in deep waters); thus, 10% of the DOC remains in dynamic and reversible assembly equilibrium, forming porous networks (Chin et al., 1998; Orellana et al., 2011b). As a result, marine polymer gels may account for at least ~70GC of the global organic DOC pool (Verdugo, 2012). This colossal mass of polymer gel represents the largest pool of biodegradable organic C available to the microbial loop (Orellana and Verdugo, 2003; Verdugo and Santschi, 2010), which might be especially important in the dark ocean (Herndl and Reinthaler, 2013). Micron-sized gels can further aggregate to form macroscopic polymer gels (Radić et al., 2011; Svetličić et al., 2005). This dynamic conceptual framework constitutes a particle size continuum for dissolved-to-gel-to-particulate organic matter (POM) in seawater (Azam, 1998; Chin et al., 1998; Koike et al., 1990; Verdugo et al., 2004) that ranges from single dissolved monomers and oligomers, to nanometer-colloidal and micron-sized gel networks (nm to μm), to larger size (mm) macrogels and aggregates (Svetličić et al., 2011; Verdugo et al., 2004). The particle size continuum is a critical mechanism through which truly dissolved organic matter biopolymers (<10nm; Aluwihare et al., 2005; Chin et al., 1998; Jiao et al., 2010), characterized by being old, short-chain, and biologically refractory (Walker et al., 2014), are transferred to nutrient-rich gel networks, creating discrete "hot spots" accessible to microbial degradation and remineralization (Figure 9.1; Amon and Benner, 1994, 1996; Azam, 1998; Azam

and Malfatti, 2007; Orellana and Verdugo, 2003; Verdugo and Santschi, 2010; Verdugo et al., 2004).

Excellent reviews of marine polymer gels (Verdugo, 2012; Verdugo et al., 2004; Verdugo and Santschi, 2010) and books already explain soft matter physics (Edwards, 1986; Grosberg and Khokhlov, 1994; Poon and Andelman, 2006). This chapter has a different aim: to comprehensively summarize the salient aspects of recent discoveries about the role polymer gels play in DOC dynamics. Some fundamental aspects of DOC dynamics are well documented (Hansell, 2013; Hansell and Carlson, 2013; Hansell et al., 2009), but polymer gel theory and relevant mathematical and modeling tools (de Gennes and Leger, 1982; Edwards, 1986; Ohmine and Tanaka, 1982; Tanaka et al., 1980) have not been systematically applied to achieve a mechanistic understanding of the dynamics of marine biopolymers, their emergent physical properties that arise from their interactions with seawater, or their control by environmental stimuli. Nor do we fully understand their role in colloidal trace metal scavenging (Wells, 2002), the microbial loop, C cycling, and the microbial pump, trace metal complexation and size reactivity relationships, or biogeopolymer condensate formation or their role in cloud formation (Orellana et al., 2011b).

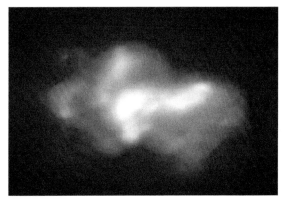

FIGURE 9.2 Microgel sorted from seawater by flow cytometry. *From Orellana et al., 2007 with modifications.*

II WHAT ARE POLYMER GELS?

Chin et al. (1998) offer this summary: Polymer gels are a distinctive form of supramolecular organization formed by a deformable "3D polymer network and a solvent which, in the case of marine hydrogels, is seawater (Figure 9.2). Although the solvent prevents the collapse of the network, the network entraps and holds the solvent, creating a microenvironment that is in thermodynamic equilibrium with the surrounding media. The polymer chains that form the 3D network are interconnected by chemical (covalent bonds) or physical cross-links. In physical gels, polymers are interconnected by tangles and/or low energy links" (hydrogen bonding, metal coordination, hydrophobic forces, van der Waals forces, pi-pi interactions and electrostatic effects) "are continuously being made and broken, thus these networks are in continuous assembly/dispersion equilibrium." "Depending on the characteristics of the polymer chains (such as polyelectrolytic properties, degree of hydrophobicity, length, and linear or branched chains) and the dielectric properties of the solvent, osmolarity, ionic composition, pH, and environmental conditions (temperature, pressure) the polymers in the gel's matrix can interact strongly with each other, with the solvent, or with smaller solutes. These interactions determine the gel's physical properties." This range of structural and ultra-structural features give polymer gels unique emergent physical properties (assembly, volume phase transition; de Gennes, 1992; de Gennes and Leger, 1982), as well as unique chemical and biological reactivities. These properties are very different from those of the dispersed polymeric components comprising the gels. The soft matter "polymer gel concept" offers a solid and powerful physicochemical predictive theory that mechanistically explains the dynamics of biopolymers in the DOC pool, thereby aiding our understanding of DOC biogeochemistry (Verdugo, 2012).

III STRUCTURE, PROPERTIES, AND DYNAMICS OF MARINE POLYMER GELS

A Composition of Marine Polymer Gels

Marine gels are tangled 3D networks that assemble from the complex and heterogeneous polydispersed mixture of biopolymers present in marine and freshwater dissolved organic matter (DOM) (Chin et al., 1998; Pace et al., 2012; Verdugo, 2012). These polymer gels range in size from colloidal (1-1000 nm) to several microns and even to meters, such as those reported for the Adriatic Sea (Svetličić et al., 2005; Verdugo et al., 2004). Fluorescent probes, histochemical stains, and fluorescently labeled antibody probes demonstrate that gels contain polyanionic polysaccharides, proteins (Table 9.2), nucleic acids, and other amphiphilic and hydrophobic moieties (Chin et al., 1998; Ding et al., 2008; Orellana et al., 2007, 2011b). This composition is not surprising given the diverse spectrum of biopolymers in DOM, though DOM itself has not been completely characterized (Aluwihare et al., 1997; Benner, 2002; Close et al., 2013; Hansman et al., 2009; Hertkorn et al., 2006; Kaiser and Benner, 2008; Kaiser and Benner, 2012; McCarthy et al., 1993, 1998; see Chapter 2). Phytoplankton and bacteria in the euphotic zone are known to produce a complex variety of polysaccharides and proteins, and their monomers (amino acids and sugars) have been isolated from the DOM pool (Aluwihare et al., 2005; Benner, 2002; Jiao et al., 2010; Kaiser and Benner, 2008; McCarthy et al., 1993). The mechanisms of production of these biopolymers include several processes (Table 9.1). Phytoplankton alone can release ~10-30% of their primary production, depending on the species (Wetz and Wheeler, 2007) and their physiological stage (Biddanda and Benner, 1997; Biersmith and Benner, 1998), into the DOC pool in the form of tangled polymers forming micron-sized gels and/or free biopolymers (Biller et al., 2014; Chin et al., 2004; Orellana et al., 2011a). While some species secrete biopolymers as

TABLE 9.1   Biopolymer Production Mechanisms by Phytoplankton, Bacteria, and Other Sources (the Relative Importance of Each is Unknown)

Process	Reference
Direct release	Decho (1990)
Viral lysis	Suttle (2007); Vardi et al. (2012)
Apoptosis and programmed cell death	Berman-Frank et al. (2004); Bidle and Falkowski (2004); Orellana et al. (2013)
Microbial degradation of particulate matter	Nagata and Kirchman (1997)
Grazing	Strom (2008); Strom et al. (1997)
Zooplankton sloppy feeding	Jumars et al., 1989
Particle dissolution	Azam and Long (2001); Carlson (2002); Kiørboe and Jackson (2001); Smith et al. (1992); Nagata et al. (2010)
Vesicle production and regulated exocytosis	Biller et al. (2014); Chin et al. (1998, 2004); Orellana et al. (2011a)

TABLE 9.2   Proteins Found in Marine Polymer Gels (Orellana and Hansell, 2012; Orellana et al., 2007; Powell et al., 2005)

ATP/GTP-binding site motif A (P-loop)

Cation channel, non-ligand-gated cation channel

Eukaryotic thiol (cysteine) protease cysteine-type endopeptidase

Peptidase activity, tail specific

Eukaryotic thiol (cysteine) protease cysteine-type endopeptidase activity

Enolase

Actin/actin-like protein

Aldose 1-epimerase/galactose metabolism Zn-finger (putative), N-recognin ubiquitin-protein ligase activity) (serine/threonine-protein kinase)

Transporting ATPase

Glucan 1,4-alpha-glucosidase

ATP/GTP-binding site motif A (P-loop)

RuBisCO (ribulose-1,5-bisphosphate carboxylase oxygenase)

polymer gels (e.g., *Phaeocystis* (Chin et al., 2004), *Fragilariopsis cylindrus* (Aslam et al., 2012)), the ratio of polymers released as gels to free polymers is unknown for most phytoplankton species. These 3D gel networks contain polysaccharides, in line with the molecular-level analysis of DOM (Aluwihare and Repeta, 1999; Aluwihare et al., 1997; Benner, 2002; Biddanda and Benner, 1997; Kaiser and Benner, 2009; see Chapter 2); peptides, protein, including ribulose-1,5-bisphosphate oxygenase (RuBisCO; Table 9.2; Chin et al., 1998; Orellana and Hansell, 2012; Orellana et al., 2007); hydrophobic and amphiphilic moieties including phospholipids (Orellana et al., 2013); nucleic acids (Chin et al., 1998); and other metabolites (Kujawinski et al., 2009).

Bacteria also release biopolymers, and ~50% of the bacterial production in the ocean is released into the DOM pool by viral burst and mortality (Suttle, 2007), probably releasing bacterial membrane porins containing D-amino acid enantiomers (D-alanine, D-glutamic, D-serine, D-aspartate, and D-glutamate; Benner and Kaiser, 2003; Benner et al., 1992; Kaiser and Benner, 2008; McCarthy et al., 1993; Ogawa et al., 2001; Tanoue et al., 1996; Tanoue et al., 1995), or by direct production of D-amino acids when bacteria detach from particles, similarly to the breakdown of biofilms (Kolodkin-Gal et al., 2010); hydrolyases and other exoenzymes (Arnosti, 2011; Smith et al., 1992); carbohydrates and amino sugars, such as glucosamine, galactosamine, and muramic acid (an amino sugar from peptidoglycan), a bacterial cell wall polymer (Benner and Kaiser, 2003; Kaiser and Benner, 2009; McCarthy et al., 1993); fatty acids, phospholipids, and lipopolysaccharides (Popendorf et al., 2011; Wakeham et al., 2003) perhaps produced by virus while infecting and lysing phytoplankton cells (Vardi et al., 2009). Bacteria can also release biologically refractory and chemically unknown molecules (Jiao et al., 2010; Nagata et al., 2010). This DOM pool of microbial origin, estimated at 165 PgC (Benner and Herndl, 2011), resists degradation and is typically described as high molecular weight (HMW) DOC (>1000 Da but <0.1-0.2 μm; 1 nm pore size; Benner et al., 1992). This pool is chemically defined as containing a high proportion of acetylated polysaccharides (APS) (Aluwihare et al., 1997), composed mainly of neutral monosaccharides and amino sugars that accumulate in the deep ocean (Aluwihare et al., 2005). It probably has a distinct ^{615}N-DON signature, indicating complex degradation (Calleja et al., 2013), and proves itself to form the bulk of the DOC pool; 10-30% of all DOM polymers assemble into forming gels.

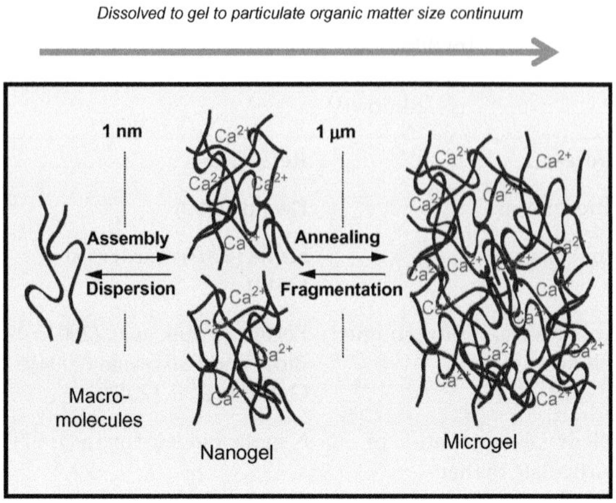

B Dynamics: Assembly, Size, and Stability of Gels in Seawater

Marine gels are held together by ionic, hydrophobic, or hydrogen interactions, with the most dominant being ionic interactions.

1 Ionic Interactions

Irrespective of the biopolymer chemistry, biopolymer interactions in the ocean's DOM pool spontaneously form 3D microscopic polymer hydrogel networks that continuously assemble and disperse at equilibrium (Figure 9.3; Chin et al., 1998). These micron-sized gels form hydrated Ca^{2+}-linked supramolecular networks whose interactions with the solvent (seawater) provide them with a gel-like texture (de Gennes and Leger, 1982). Using independent and parallel methods, including dynamic laser scattering (DLS) or photon correlation spectroscopy, flow cytometry, and environmental scanning electron microscopy, Chin et al. (1998) demonstrated that polyanionic polymers (proteins, nucleic acids, polysaccharides) associate to form microscopic polymer gels as a result of the local interactions of negatively charged polymers and multivalent cations (such as the metal ions present in sea water). When these authors chelated the metals in seawater with ethylenediaminetetraacetic

Dissolved to gel to particulate organic matter size continuum

FIGURE 9.3 Dynamics of self-assembling polyanionic marine polymer gels. Hydrogels consist of a three-dimensional polymer network. Polyanionic polymers found in the DOM pool (662 PgC) assemble spontaneously, forming nanometer-sized tangled networks that are stabilized by Ca^{2+} bonds. The tangled nature of these nanogels allows polymers to interpenetrate neighboring gels, annealing into larger microgels. Polymer gels exhibit emergent properties that are different from those of the dispersed polymers that make up these networks. Polymer chains inside the microgels interact to creating a microenvironment of high-substrate concentration serving as a rich source of substrate to microorganisms. *From Orellana and Verdugo (2003), with modifications.*

acid (EDTA, 10 mM) or removed them by dialysis using Ca^{2+}-free artificial seawater, the microscopic polymer gel networks did not form, indicating the cross-linking role of Ca^{2+}. Using x-ray energy dispersive spectroscopy, they further demonstrated that the polymer association was driven mainly by Ca^{2+} ionic bonds (binding sites on marine polymers include carboxyl, hydroxyl, phosphate, sulfate, amino, and sulfhydryl groups (Benner et al., 1992; Hung et al., 2003; Hung et al., 2001; Mopper et al., 1995; Zhou et al., 1998a). This finding is consistent with Ca^{2+} ions attaining high concentrations in seawater (10 mM). Furthermore, Ca^{2+} has a smaller Stokes radius than other abundant marine divalent metal ions (e.g., Mg^{2+}), thus increasing the probability of interaction and linking the polyanionic marine polymers. Calcium also acts as an important cross-linker in other natural gels such as pectin, alginates, and mucous (Pollack, 2001; Verdugo, 1994).

While the spontaneous assembly of polymers is common in seawater, its variability is determined by the mass and composition of the biopolymer backbone, the concentration and charge density of biopolymers (i.e., coastal vs. oceanic and temporal variability), and by the concentration, valence, size, and shape of the cross-linking counter anion (i.e., coastal vs. oceanic waters, surface vs. deep waters). For example, while Ca^{2+} cross-linking might dominate, trace metals, which reach extremely dilute concentrations in seawater (e.g., low nM), exhibit multivalent properties (i.e., Fe^{3+}, Al^{3+}), allowing the fast spontaneous assembly of very stable networks that have important consequences for biogeochemistry. In fact, Wells and Goldberg (1991, 1992) demonstrated that colloids (120 nm) in coastal California surface waters contain polycations such as Fe and Al (see below).

Chin et al. (1998) also demonstrated that spontaneous assembly is reversible, follows second-order kinetics, and, at equilibrium, has an average thermodynamic yield of 10% at room temperature. New measurements indicate that the equilibrium yield at *in situ* temperature can reach 30%, as in the case of Arctic biopolymers (see below, Orellana et al., 2011b). Reversibility refers to the fact that, at equilibrium, the electrostatic interactions between the polymers continuously break and form, assembling and dispersing, so that polymers that form a gel are in dynamic equilibrium with chains that remain in the bulk solution. Second-order kinetics characterizes the system's behavior over time as a two-step assembly. Free polymer chains with molecular dimensions (from angstroms to a few nanometers) first form nanometer-sized gel networks (100-200 nm), which then anneal and interpenetrate themselves by the mechanism of diffusional reptation (thermal motion of long linear, entangled macromolecules in concentrated polymer solutions) to form colloidal-sized gels (visualized by transmission atomic force microscopy), which subsequently anneal and

aggregate to form micron-sized gels that stabilize *in vitro* and at room temperature to about 4.5-5 µm (microgels) as measured by DLS and visualized by environmental electron microscope (Figure 9.3; Chin et al., 1998). In the oceans, with an enormous availability of free polymers, gels may assemble to yet greater sizes.

Physical aggregation and agglomeration have often been used to explain marine colloidal growth into larger particles, as well as the transfer of truly dissolved substances into micron-size particles (larger than 1.0 µm in size; Paerl, 1973; Riley, 1963; Wells, 2002; Wells and Goldberg, 1993). However, the mechanism and nature of the physical interaction of the polymers was not completely understood then (Paerl, 1973; Wells and Goldberg, 1993). Referring to the discovery of self-assembly, Wells (1998) commented that the "assembly processes mechanistically explains the colloidal annealing and aggregation behavior, including trace metal colloidal complexation in seawater" (Wells, 2002). The influence of attractive interactions on the aggregation of biopolymers and gel formation in colloidal systems is now well described (Poon and Andelman, 2006).

In fact, the spontaneous self-assembly of marine biopolymers is consistent with the time frames of colloidal aggregation of radioactive thorium species (e.g., [234]Th, [230]Th, [228]Th), a process known as "colloidal pumping" (Guo and Santschi, 1997; Guo et al., 2000; Honeyman and Santschi, 1988; Santschi et al., 1995). Th species have also been used as a tracer in a range of processes, including particle cycling (Clegg and Whitfield, 1990, 1991; Dunne et al., 1997), C export flux (e.g., Buesseler, 1998; Murray et al., 1989), boundary scavenging (Santschi et al., 1999), and paleo-circulation (Moran et al., 2002). Scavenging of dissolved iron (Fe^{3+}), a key limiting nutrient in the high-nitrate, low-chlorophyll (HNLC) regions of the Southern Ocean, as well as in the subarctic and equatorial Pacific (Martin et al., 1990, 1994; Moore and Doney, 2007), is consistent with the conceptual model developed for the scavenging of Th isotopes, whereby particle scavenging is a two-step process of scavenging by organic colloidal and small particulates followed by aggregation and removal on larger sinking particles. Removal of dissolved iron from subsurface waters (where iron concentrations are often well below 0.6 nM) occurs by "aggregation" and by sinking particles of Fe^{3+} bound to organic colloids (Moore et al., 2002). In this respect, according to polymer theory, the assembly of biopolymers is accelerated not only by the presence of metal-binding ligand groups and siderophores (Butler and Theisen, 2010) but also by the fact that electrostatic interactions in polyanionic polymers form cross-links that are proportional to the square of the valence of the counter ion (Fe^{3+}), conferring remarkable stability to the 3D gel architecture (Ohmine and Tanaka, 1982; Okajima et al., 2012; Verdugo, 2012). Chuang et al. (2013) demonstrated the role of proteins, polysaccharides, sugars such as uronic acids (containing carbonyl and carboxylic acid functional groups), and cathecolamines as major carriers of radionuclides and other metals in the Atlantic Ocean. Cathecolamines, siderophores, and uronic acids are known to assemble as gels as well (Menyo et al., 2013; Okajima et al., 2012). Additionally, sacran, a cyanobacterial sugar containing uronic acids, easily forms gels with trivalent metal ions (Fe^{3+}, Ga^{3+}, Al^{3+}) (Okajima et al., 2012). Measuring the kinetics of marine biopolymer assembly and iron complexation in conjunction with electron probe analysis would improve our mechanistic understanding of iron scavenging, as well as other ions in the oceans.

2 Hydrophobic Interactions

Polymer gels also form at hydrophobic interfaces, such as at the air-water interface and at bubble surfaces (Orellana et al., 2011b; Verdugo, 2012; Wheeler, 1975). Although the bulk of the polymers produced by phytoplankton and bacteria are polyanionic, they also produce amphiphilic polymers (Decho, 1990; Orellana et al., 2007, 2011b; Stoderegger and Herndl, 2004; Wingender et al., 1999). However, the assembly of marine amphiphilic polymers is less well understood. Ding et al. (2008) demonstrated that nanomolar concentrations of amphiphilic exopolymers ($20 \mu g L^{-1}$) released by the proteobacteria *Sagittula stellata* induced polymer network formation with very rapid rates of assembly. Likewise, the remarkable predominance of amphiphilic siderophores secreted by oceanic bacteria enables microbial iron acquisition, thus playing an important role in upper-ocean iron cycling (Butler and Theisen, 2010; Martinez et al., 2003). Polymers secreted by phytoplankton, such as the cyanobacteria *Synechococcus*, the prymnesiophyte *Emiliania huxleyi*, and the centric diatom *Skeletonema costatum*, contain proteins exhibiting hydrophobic domains (with 30% being hydrophobic amino acids) that self-assemble in Ca^{2+}-free seawater or in low Ca^{2+} concentrations (Ding et al., 2009). While the magnitude of the production of hydrophobic moieties in the world oceans is not well known, the presence of lipids in the North Pacific subtropical gyre indicates that lipid compositional signatures of colloidal-sized particles (0.2-0.5 μm) are depth specific (surface vs. mesopelagic) and that these particles account for an important percentage of the exported material in oligotrophic waters, making an important contribution to the biological pump (Close et al., 2013). Therefore, these moieties probably play an important role in self-assembly of colloidal-sized gels, as has also been demonstrated in the Arctic. In fact, in the high Arctic, amphiphilic and hydrophobic moieties play a significant role in accelerating the assembly of gels at the water-air interface, where self-assembly takes place in a matter of minutes (Figure 9.4).

Hydrophobic moieties form the core of Arctic gels, containing lipids and proteins that include hydrophobic amino acid residues such as leucine, isoleucine, phenylalanine, and cysteine (Figure 9.5; Orellana et al., 2011b). These amino acid residues create hydrophobic pockets capable of allowing the assembly and further aggregation of polymer gels (Maitra et al., 2001). Aromatic residues (tryptophan and phenylalanine) may also be important. Perhaps hydrophobic polymers are

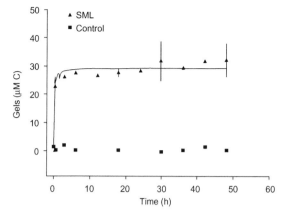

FIGURE 9.4 Polymer gel assembly as a function of time in high Arctic surface waters (87-88° N, 2-10° W). The assembly of polymer gels was monitored by measuring percent polymers assembled as microgels (Ding et al., 2007) at 4 °C (triangles). Control experiments in which Ca^{2+} was chelated from seawater with 10 mM EDTA showed no assembled gels, regardless of the time of observation (squares). Each point corresponds to the average of three replicates. Note that the assembly kinetics is very fast, with the concentration and size of assembled polymer gels reaching equilibrium in 6 h. An average yield of assembly equal to 32% of the polymers present in the DOM pool was measured for either subsurface (SSW) or surface microlayer (SML) water samples. *From Orellana et al., 2011b.*

more abundant in recently produced gels containing a higher concentration of proteins with amphiphilic side chains, and therefore having lower C:N ratios, during and at the end of phytoplankton blooms, as in the case of the Arctic gels. Polymers with higher nitrogen content biodegrade faster in the water column (Cherrier and Bauer, 2004; Cherrier et al., 1996; Davis and Benner, 2007) and are more abundant in the surface than deep oceans (Hopkinson and Vallino, 2005; Walker et al., 2014). Walker et al. (2014) demonstrated relations between age, C:N ratio, and size of the particles; older particles are smaller and nitrogen-poor. These relations may provide an explanation for the preferential storage of polyanionic molecules in the ocean and for their forming the bulk of the interacting polymers that spontaneously assemble into polymer gels.

3 Size and Stability of Gels

Polymer length, concentration, and charge density are key features conferring stability to marine 3D gel networks. The probability of assembly, the equilibrium size, and the stability of tangled

FIGURE 9.5 Hydrophobic moieties in polymer gels. (a) Several cloud microgels assembled by amphiphilic polymers and stained in red (Nile Red), indicating the presence of hydrophobic moieties, and hydrophilic moieties stained green with quinacrine. (b) Enrichment of dissolved hydrophobic amino acids (leucine, isoleucine, phenylalanine, and cysteine) in the surface microlayer (SML) with respect to subsurface (SSW) waters (Orellana et al., 2011b). *x*-axis is sampling day-of-year (DoY); *y*-axis is the observed enrichment of the amino acids (see symbol legend) in the surface microlayer relative to sea surface water.

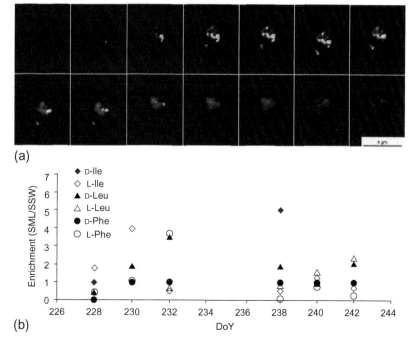

gel networks are determined by polymer length (Edwards, 1986). The assembly of polymers into polymer gels and their stability once assembled increases with the square of the polymer length (de Gennes and Leger, 1982; Edwards, 1986). Longer chains assemble into more stable and bigger equilibrium size polymer networks. When the length of the polymers increases, the interactions between the polymer chains (ionic, van der Waals, hydrophobic, etc.), tangles, and interpenetrations become stronger, requiring more energy to disrupt them. In contrast, short-length chains experience lower degrees of interactions, and therefore increased gel destabilization. Using ultraviolet (UV)-B irradiation to photocleave free marine polymers and assembled microgels, Orellana and Verdugo (2003) demonstrated that as the biopolymers fragmented with increasing exposure time to UV-B ($\lambda = 280\text{-}320\,\text{nm}$), the assembly kinetics of those short, fragmented polymer chains took more time to reach equilibrium and the polymer gel size at equilibrium decreased exponentially (Figure 9.6). Thus, depending on the extent of fragmentation, UV-B-irradiated polymers assembled into smaller, less stable submicron-sized gel networks; at the end (>12h of UV exposure), monomers and short oligomers completely failed to anneal and to form stable networks. Assembled gel networks irradiated with UV-B dispersed, and the resulting short-chain polymers failed to assemble into stable gels, drastically disrupting the exchange between the dissolved and the gel phases (Figure 9.6; Orellana and Verdugo, 2003).

A gel's network dependency on polymer size can explain the distribution of biopolymers in the oceans. For example, at the end of spring blooms, phytoplankton release a high percentage of their primary production as long polysaccharide chains and proteins that generally assemble to form large gel networks, as occurs with gels produced by *Phaeocystis* blooms in polar regions (Janse et al., 1996; Vernet et al., 1998) and macroscopic-sized, gel-like transparent exopolymer particles (TEP) (Alldredge et al., 1993; Mopper et al., 1995;

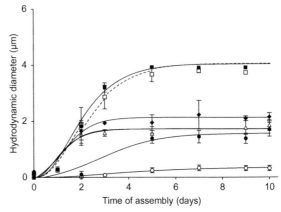

FIGURE 9.6   Effect of UV-A and UV-B radiation on spontaneous assembly of DOM polymers. Assembly of polymers follows characteristic second-order kinetics, forming polymer gels that grow from colloidal (nanometer) to multimicron size in ~60h. Filled squares represent the assembly of non-irradiated polymers (the control run). Assembly of polymers irradiated for 24h with UV-A ($10\,\text{W m}^{-2}$, $\lambda = 320\text{--}400\,\text{nm}$) are in open squares, showing no statistical difference between the assembly kinetics of controls and UV-A-irradiated samples. The assembly of DOC polymers in seawater samples exposed to UV-B ($0.5\,\text{W m}^{-2}$) for 30 min (filled diamonds), 1h (open triangles), 6h (filled circles), and 12h (open circles) follow a similar second-order kinetic profile; however, the time to reach equilibrium is longer and the equilibrium size of the polymer gel networks is smaller and less stable. Data points correspond to the mean $\pm$SD of 30 dynamic laser scattering measurements. *Adapted from Orellana and Verdugo (2003).*

Passow, 2002b), as well as protein-containing particles (Long and Azam, 1996). An extreme case of macroscopic accumulation (meters) of polymer networks held together by hydrogen bonds took place in the Adriatic Sea, with the polymers produced by the diatom *Cylindrotheca closterium* (Svetličić et al., 2005, 2011). Similarly, the high concentrations of longer younger polymers produced at the end of spring blooms perhaps explains the high yield of gels found in the high Arctic at the end of a bloom of the ubiquitous diatom *Melosira arctica* (30% of polymers present as DOM, assembled as gels; Orellana et al., 2011b).

Polymer gel theory can also explain why the distribution of marine DOM is highly skewed toward low molecular weight (LMW) components (Benner, 2002; see Chapter 2). Truly dissolved substances make up $70 \pm 5\%$ of the total organic C, and the HMW fraction and colloids essentially make up the remaining 25% (Benner, 2002). Because the probability of short-chain oligomers and monomers to interact and form stable gel networks is very low or nil, they accumulate and probably account for the short chain (~1 nm), biologically refractory DOC pool found in the world oceans (Druffel et al., 1998; Druffel and Williams, 1990; Walker et al., 2008, 2014). Macroscopic gels form sporadically, mainly at the end of spring blooms (Passow, 2002b) when newly long polymeric chains are released by phytoplankton (diatoms), aggregating to form big particles embedded with cells (Alldredge and Jackson, 1995), degrading and/ or sedimenting as POM relatively quickly in the water column (Figure 9.1; Smetacek, 2000). However, HMW colloids are the most abundant gels within this pool. Most importantly, some 10-30% of this pool of polymers anneal to form micron-sized 3D gel networks, representing the biggest and most important shunt of polymer material that transforms the dissolved DOC into a gel phase, and then to POM in seawater (Verdugo, 2012; Verdugo et al., 2004). This shunt provides a constant supply of C to heterotrophic microorganisms in a dilute ($40 \mu M$ C average) and highly viscous ocean, where the foraging distances are large and contain only a limiting nutrient supply to the microbial world (Fenchel, 1984; Jumars, 1993). Polymer gel networks are critical to providing a natural 3D matrix scaffolding capable of holding a heterotrophic microbial community together (Moon et al., 2007; Orellana et al., 2000), particularly in the deep ocean where non-sinking particles support microbial life (Baltar et al., 2010; Herndl and Reinthaler, 2013). Indeed, the presence of gels in the deep ocean may explain the discrepancy between bacterial C demand by the deep-water heterotrophic microbial community and the particulate organic C (POC) supply from the surface

ocean (Baltar et al., 2010; Herndl and Reinthaler, 2013). Polymer gels contain a rich nutritional microenvironment (polysaccharides, lipids, proteins, and nucleic acid chains; (Chin et al., 1998; Ding et al., 2008; Orellana et al., 2007), providing an optimal microenvironment of biodegradable substances—that is, a "hot spot" (Azam, 1998; Azam and Malfatti, 2007; Grossart et al., 2007; Malfatti and Azam, 2009) where heterotrophic bacteria/Archaea can interact and grow (Figure 9.2; Gram et al., 2002; Hmelo and Van Mooy, 2010; Malfatti and Azam, 2009). Bacterial growth is four to seven times faster when grown on gels than on free unassembled polymers (Orellana et al., 2000). Therefore, marine polymer gels form a fundamental structure and bacterial microenvironment (Malfatti and Azam, 2009).

What is the relation between gel networks, lability, age, and the size of the DOC polymers in seawater? Based on turnover time, five fractions of DOC have been conceptualized (Carlson, 2002; Hansell, 2013), with four relevant here: (a) a highly labile pool of young polymeric material with a fast turnover lasting from hours to a few days; (b) a semi-labile fraction with turnover of months to a few years; (c) a semi-refractory fraction with lifetimes of a decade, and (d) a refractory, nonbiodegradable fraction with a turnover lasting thousands of years, exceeding the timescale of deep water circulation (an average of about 5000 years; see Chapter 6 and Bauer (2002)). These fractions overlap in distribution in the upper water column, with the refractory fraction being the most abundant and, chemically, the least understood (Hansell, 2013; Hansell and Carlson, 2013). The bioreactivity of marine polymers depends on many factors, including (a) the nature of the biopolymers (C:N ratio, amino acid composition) and their size (Amon and Benner, 1996), (b) the abundance and taxonomic composition of the colonizing microbial community, bearing the genetic capacity toward attachment to particles (Bauer, 2006; Carlson et al., 2004; Cottrell and Kirchman, 2000; DeLong et al., 2006), and (c) the functional genomic fingerprint of the bacterial population colonizing and degrading the gels and/or particles (Fuhrman, 2009; McCarren et al., 2010; Moran et al., 2007; Palenik et al., 2006). LMW DOC (<1000 Da) exhibits the lowest bioreactivity, while HMW DOC (>1000 Da but <0.2 μm) and colloids support the bulk of the marine heterotrophic microbial production, thus LMW DOC accumulates (Amon and Benner, 1994, 1996; Simon et al., 2002; Verdugo, 2012). These LMW short-chain chemical moieties are too large to be incorporated into the bacterial cell (>600 Da) (Weiss et al., 1991) but too short to assemble into stable microgels and so remain dispersed in solution as part of the refractory DOC. Additionally, we now have a better understanding of bacterial ecophysiological strategies and bacterial interactions among themselves, and other taxa (Cuadrado-Silva et al., 2013; Dobretsov et al., 2009; Gram et al., 2002; Hmelo and Van Mooy, 2010; Orellana et al., 2013), as well as of the structure of the organic matter continuum network (Malfatti and Azam, 2009). These fundamental characteristics and the interplay between the DOC field and bacteria imply that microbial oxidation of DOC depends significantly on the structure of the organic matter field and the quaternary conformation of larger molecules and/or networks of smaller chains forming an organic matter continuum, rather than on smaller dilute single free monomers or oligomer chains (DOC ~40-70 μM) (Azam, 1998; Malfatti and Azam, 2009; Orellana et al., 2000; Simon et al., 2002; Verdugo, 2012; Verdugo et al., 2004).

Polymer gel networks, as hot spots, likely play a fundamental role in influencing the biogeochemical fate of C in the oceans, with consequences ranging from bacterial interactions in the microbial loop, especially in the deep ocean, to climate change (Aristegui et al., 2009; Azam, 1998; Baltar et al., 2010; Herndl and Reinthaler, 2013; Verdugo, 2012). When microbial heterotrophic assemblages colonize these 3D gel networks, they hydrolyze and crack the long-chain polymers with their ectoenzymes (Arnosti,

2011); this ongoing process affects the assembly/dispersion equilibrium of the marine gels. Physical fragmentation of biopolymers by UV photolysis (see Chapter 8; Mopper and Kieber, 2002; Orellana and Verdugo, 2003) or by bacterial enzymatic degradation (Benner and Kaiser, 2011) produce short-chain monomers and oligomers that do not interact to assemble into stable networks, and therefore they accumulate in the ocean. Thus, the natural process of spontaneous assembly of DOC polymers is disrupted and inhibited (Figure 9.4; Orellana and Verdugo, 2003; Orellana et al., 2000). Biological degradation will also disperse assembled microgels, driving short chain molecules to diffuse into a dilute DOC environment (Orellana and Verdugo, 2003). The molecular architecture of bacterial porins confirms that bacteria/Archaea can utilize only biopolymers <600 Da (Weiss et al., 1991), with some exceptions (Payne and Smith, 1994). Therefore, free polymer chains between 600 and 1000 Da are too large to pass through the microbial membrane porin channels for bacterial assimilation and metabolization and too short to assemble into stable networks that bacteria can colonize and degrade (Azam and Malfatti, 2007). Although we have no clear consensus on why old refractory DOC accumulates in the world oceans (Hansell, 2013; Jiao et al., 2010), soft matter polymer theory can explain the mechanisms by which short-chain polymers (<1000 Da) accumulate in the world oceans. Irrespective of their chemistry, monomers and oligomers remain dispersed (unassembled) in seawater due to their inability to form stable gels, since the probability of assembly of polymer gels depend on the square of the polymer length. Therefore, short-chain oligomers and probably complex molecules cycle slowly in the world oceans (Aluwihare et al., 1997; Bauer et al., 1992; Benner and Kaiser, 2011; Chin et al., 1998); unstable colloidal-size gels (100 and 200 nm) too small for microorganisms to colonize will have a similar character (Koike et al., 1990; Wells and Goldberg, 1992). The nanocolloids accumulate in the oceans, forming part of the refractory DOC (Benner, 2002). This refractory fraction also may include polymers that have undergone volume phase transition (see below) and/or are chemically complex (Vandenbroucke and Largeau, 2007).

4 Macroscopic Polymer Gels and TEP

Marine particles and their dynamics have long interested oceanographers (Paerl, 1973; Riley, 1963). Attention has grown as studies have brought new knowledge about colloid-sized particles—and their role in trace metal scavenging (Koike et al., 1990; Wells and Goldberg, 1992, 1993, 1994), reviewed by Wells (2002); marine snow and macroscopic particles such as TEP and gelatinous macro-aggregates—as well as their role in the biological pump (Alldredge et al., 1993; Burd and Jackson, 2009; Jackson and Burd, 1998; Jackson, 1990; Passow, 2002b). Today, scientific literature contains >1000 papers about TEP (Google Scholar). TEP (>0.7-500 μm particles) have been studied *in vitro* and *in situ*, from the Atlantic (Engel, 2004; Martin et al., 2011b) to the Pacific (Alldredge et al., 1993) to the Indian Oceans (Kumar et al., 1998), from tropical to temperate to polar (Hong et al., 1997) and from the sea surface microlayer (SML) to the deep ocean (Wurl et al., 2009). Abundant TEP also exist in freshwater lakes (Chateauvert et al., 2012; de Vicente et al., 2010; Pace et al., 2012).

Marine particles have been investigated in the context of polymer physics (Chin et al., 1998), bringing new insights and understanding about their dynamics. Macroscopic gels assemble from large polymer chains held together by metal ions (Ca^{2+} (Chin et al., 1998; Ding et al., 2007; Orellana et al., 2007; Orellana and Verdugo, 2003; Verdugo, 2012; Verdugo and Santschi, 2010), hydrophobic polymers (Ding et al., 2008), and/or hydrogen bonds (Radić et al., 2011). TEP commonly form the matrix for the aggregation of cell debris and detritus and small organisms, arising mainly on the demise of phytoplankton blooms (Alldredge and Jackson, 1995; Logan

et al., 1995; Passow, 2002b). TEP contain a subgroup of all polymers encountered in the ocean (only acidic polysaccharides) and of polymers that assemble into marine gels (Verdugo, 2012; Verdugo and Santschi, 2010). TEP are operationally defined as acidic polysaccharides—especially sulfate ester groups (Mopper et al., 1995; Zhou et al., 1998b), carboxyl, sulfate, and phosphate-containing particles (Hung et al., 2001, 2003; Passow, 2002a, b) that can be visualized when stained with Alcian blue (a cationic copper phthalocyanine dye that complexes carboxyl ($-COO^-$) and half-ester sulfate (OSO reactive groups of acidic polysaccharides) at pH 2.5). Stained exopolymer particles can be measured in semiquantitative assays, either microscopically or colorimetrically, using xanthan gum equivalents ($\mu g L^{-1}$), a proxy for marine polymers for calibration purposes, which makes the quantitative expression of these particles difficult to associate with measures of C export in the ocean and are rarely measured (Engel and Passow, 2001; Mari, 1999). Furthermore, as the C content of marine polymers per Alcian Blue binding site varies, the conversion between xanthan gum equivalent and C also varies; it needs to be ascertained for each analytical case (Engel and Passow, 2001; Mari, 1999; Mari et al., 2001).

TEP-forming acidic polysaccharides are secreted by phytoplankton and bacteria, especially diatoms (Alldredge and Gotschalk, 1989; Mopper et al., 1995); however, phytoplankton groups such as dinoflagellates, cyanobacteria, and prymnesiophytes can produce large amounts of TEP-forming polymers as well (Alldredge et al., 1998; Hung et al., 2003; Passow, 2002b). TEP can form fast, in time frames of hours (Wurl et al., 2011), and can be exported swiftly into the water column and sediments, reaching average rates equal to $20 g C m^2 day^{-1}$ (dry weight) (Logan et al., 1995) or 10-100 mg xanthum gum eq. $m^{-2} day^{-1}$ (Passow et al., 2001). Thus, while TEP might have an important role in the biological pump, it suffers from its semiquantitative nature and the difficulty of transforming xanthan gum equivalent into actual C concentrations (Alkire et al., 2012; Martin et al., 2011b).

TEP have been described as gels because of their gelatinous nature (Wurl and Holmes, 2008). While TEP might be gels, they have not been demonstrated to have the emergent physicochemical characteristics of polymer gel networks (spontaneous assembly, volume phase transition). Furthermore, the processes of aggregation of TEP are based on coagulation theory, collision, and stickiness (the probability of two particles adhering once they collide; Engel et al., 2004); absent is the mechanistic understanding of the intermolecular association, energies, and assembly processes of the polymers that form them (Verdugo, 2012). Molecular dynamic simulations are today used to model polymer aggregations, ranging from pure proteins to lipids to carbohydrates (Poon and Andelman, 2006). This approach could be used in the future to accurately simulate marine polymer dynamics and aggregation using the distributions of bonds, angles, and dihedrals (Lee et al., 2009). While these simulations have not been used in marine biogeochemistry, they might advance understanding of C dynamics, trace metal scavenging, and export into the deep ocean.

IV PHASE TRANSITION

All polymer gels have a remarkable feature: they undergo a reversible, mechanical deformation that does not affect the structure of the network. Polymer gels can undergo volume phase transition, from a swollen, hydrated phase to a condensed and compact phase. The gel volume can change several hundred times in response to infinitesimal changes and shifts in environmental conditions; such changes have been observed universally in synthetic and natural polymer gels (Pollack, 2001; Skubatz et al., 2013; Tanaka, 1992; Tanaka et al., 1980). Changes in volume occur in association with the internal dielectric properties of the gel polymer matrix in response

to environmental stimuli, on the nature of interactions between polymers and on the environment (Tanaka et al., 1980), such as temperature and UV light (Mamada et al., 1990); trace metals and pollutants (Jadhav et al., 2010; Rosen, 1993); and probably many other unknown chemicals. Volume phase transition in marine gels can be induced by changes in temperature (Verdugo, 2012; Verdugo and Santschi, 2010); pH (Chin et al., 1998; Orellana et al., 2011b); climate-relevant substances, such as dimethyl sulfide (DMS) and its precursor dimethylsulfoniopropionate (DMSP) (Orellana et al., 2011b); and polycations, such as polyamines (Nishibori et al., 2003). External stimuli induce marine microgel volume phase transitions (swelling and condensation) and the further collapse of the microgels by expulsion of water into a dense, compact polymeric network, increasing its specific weight and thus probably its sedimentation rate into the deep ocean; however, the settling rates of gels under these conditions have not been measured (Chin et al., 1998; Orellana and Hansell, 2012; Orellana et al., 2011b; Verdugo and Santschi, 2010). During the process of phase transition, molecules such as the photosynthetic protein RuBisCO (ribulose-1,5-bisphosphate carboxylase/oxygenase), DMSP, and DMS can be entrapped and packaged or caged at very high concentrations within the gel (Fernandez et al., 1991; Orellana and Hansell, 2012; Orellana et al., 2011a; Verdugo et al., 1995). These gels could then be carried to the deep ocean. In the Equatorial and North Pacific, for example, microgels containing immunologically recognizable RuBisCO were found at 3000m (see below, Orellana and Hansell, 2012).

The first demonstration of phase transition in marine gels (Chin et al., 1998) found marine microgels to collapse at pH 4.5; these gels exhibited a 20-fold volume collapse, from 5 μm to 250 nm. This pH corresponds with the pKa of carboxylic acid groups, consistent with measurements showing that one out of every six C residues in DOC corresponds to a carboxylic group (see

Chapter 2; Benner et al., 1992; Hertkorn et al., 2006; Lechtenfeld et al., 2014). Sulfuric acid (H_2SO_4), an oxidation product of the climate-relevant chemical DMS, also induces reversible volume phase transitions in microgels found at high latitudes in the high Arctic (Orellana et al., 2011b). In this case, the steep volume transition from the hydrated to the condensed phase took place as the pH decreased from 8 to 6. This pH transition inflection point is higher than observed (pH 4.5) for marine polymer gels from other latitudes (Chin et al., 1998), perhaps reflecting differences in polymer composition. The low transition pH measured by Chin et al. (1998) was used as an experimental demonstration that marine gels do indeed undergo discontinuous volume changes; but this value is not yet relevant in today's changing and acidifying ocean (pH 8.05), even in corrosive upwelling areas (pH 7.6; Feely et al. 2008, 2009). However, H_2SO_4 is germane to nanometer-sized atmospheric particles (Leck and Bigg, 2005a). Furthermore, recent research has demonstrated that microgels containing and entrapping RuBisCO in the Pacific Ocean (Orellana and Hansell, 2012) undergo a steep volume condensation and phase transition *in vitro*. These gels collapse from a swollen network at pH 8 to a nonporous, tight network at pH 7, where they reach a size of at least six times smaller than that at pH 8. The volume phase transition of the microgels containing RuBisCO resulted in nonporous polymeric networks analogous to a stone-like particle smaller than 300 nm, probably preventing further bacterial enzymatic degradation by inhibiting diffusion of the enzymes into the collapsed network (Verdugo, 2012). One may expect that RuBisCO, as a soluble enzyme within algal chloroplasts (pH 8), would be biodegraded rather quickly by microzooplankton and/or zooplankton in the water column. But the pH inside the vacuoles of protozoa and in zooplankton guts can decrease abruptly with the production of degrading enzymes (Laybourn-Parry, 1984; Pond et al., 1995), inducing physicochemical alterations and

a microgel volume phase transition from swollen and hydrated to condensed and collapsed at pH 7. While these networks should swell once back in seawater, entrapped and caged in a gel, RuBisCO would be protected from microbial degradation (Verdugo et al., 1995), explaining the presence of algal RuBisCO in the deep Pacific Ocean (Orellana and Hansell, 2012). The combined effect of ocean pressure and temperature on marine proteins is not known.

The process of phase transition can have important biogeochemical implications. Apparently, it prevents the degradation of proteins such as RuBisCO and probably other proteins present in the DOM pool by inhibiting microbial degradation caused by decreasing permeability of the collapsed network by exoenzymes (Orellana and Hansell, 2012; Verdugo, 2012). Phase transition could also explain how autotrophic biomolecules, specifically proteins, escape biodegradation in the water column, why the dissolved organic nitrogen pool, which exists mainly as amides in the interior of the ocean, resists decay, and why marine proteins are preserved in the deep ocean (Aluwihare et al., 2005; McCarthy et al., 2004; Orellana and Hansell, 2012); however, this protective mechanism needs to be demonstrated *in situ*.

Volume condensation, followed by volume collapse of these polymer microgel networks can also be induced by micromolar levels of DMS and its biogenic precursor DMSP (Figure 9.7). The concentrations of DMS and DMSP inducing microgel phase transitions are consistent with extracellular and intracellular concentrations of these climate-relevant chemical compounds produced by polar phytoplankton and ice algae (Matrai and Vernet, 1997), as well as by phytoplankton from the Sargasso (Gabric et al., 2009) and the tropical Pacific (Bates and Quinn, 1997). The results showing microgels undergoing phase transition with DMS and DMSP are also analogous to the finding that high concentrations of DMSP and DMS are stored in condensed state in the acidic secretory vesicles of

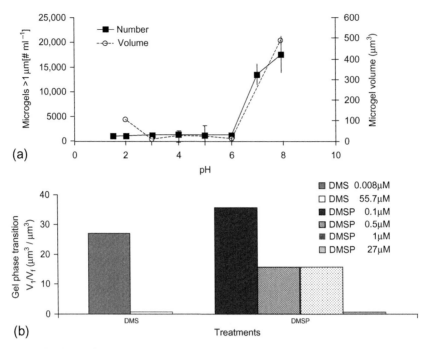

FIGURE 9.7   Microgel volume phase transition. (a) pH. Marine polymer gels undergo fast, reversible volume phase transition (<1 min) from a swollen or hydrated phase to a condensed and collapsed phase by changing the pH of the seawater with H_2SO_4. The sizes of the polymer gels were monitored by confocal microscopy and the number of gels by flow cytometry. The swelling/condensation transition is reversible and has a steep sigmoidal change in the volume of the gels. Each datum corresponds to the average and standard deviation of three samples. (b) DMS and DMSP. Marine polymer gels can undergo fast, reversible change from a swollen/hydrated phase to a condensed phase as a function of DMS and DMSP concentrations; data expressed as the ratio between initial (V_i) and final (V_f) microgel volume, before and after addition of the inducing compound, respectively, and measured with confocal microscopy (Orellana et al., 2011b).

the Prymnesiophyceae microalga *Phaeocystis* in condensed state, within the polyanionic gel matrix (Orellana et al., 2011a). The secretory vesicles are stimulated by blue light (and probably other unknown stimuli) to release their concentrated content into the environment by the process of exocytosis (Chin et al., 2004). Secretion of polymer gels by this process is accompanied by elevated DMS and DMSP concentrations, suggesting that these substances are released with the polymer gel matrix into the water column (Matrai and Vernet, 1997) as well as into the clouds (Orellana et al., 2011b).

Phase transition of freshwater gels has also been demonstrated at a critical point between pH 6.5 and 7, with serious consequences for the optical landscape of the organic matter field (Pace et al., 2012). At low pH, as these authors demonstrated, DOC polymers and colloids are condensed and compact. This volume condensation affects the optical characteristics of colored DOM (CDOM) by limiting the exposure of chromophores to light; conversely, at higher pH, polymers and colloids expand, exposing chromophores to light. This experimental dynamic change in volume of the freshwater gels resulted in changes of light absorption and photobleaching, which affects water transparency and ultimately may strongly influence C cycling, including the balance of autotrophy and heterotrophy in freshwater ecosystems (Pace et al., 2012). However, the effects of pH on the structure of the CDOM are different in marine waters. Marine waters do not show big changes in pH except in microenvironments or diurnal changes in pH related to photosynthetic processes (Cornwall et al., 2013; Schmalz and Swanson, 1969). Volume phase transitions of CDOM may also be induced by changes in salinity gradients along estuaries, as indicated by discontinuous transitions in maximal fluorescence emission by CDOM when small changes in salinity (33-35) occurred in samples taken from Puget Sound, the Orinoco River plume, and the west Florida shelf (Del Castillo et al., 2000). Additionally, volume phase transitions driven by hydrophobic interactions can be induced by temperature changes, with critical transition points at 5 °C and from 24 to 30 °C (Verdugo, 2012). Finally, volume phase changes may also be implicated in the production of biogeocondensates in sediments, where large changes in *in situ* pH and other chemical parameters are known (Fenchel et al., 2012), preventing microbial exoenzymatic degradation of kerogen and algaeans (Vandenbroucke and Largeau, 2007).

V MARINE GELS IN THE ATMOSPHERE AND THEIR RELEVANCE FOR CLOUD FORMATION

Clouds remain a weakness in our understanding of the climate system and consequently in climate modeling (IPCC, 2007). Clouds form when water vapor condenses. But, water vapor needs something to condense on—tiny airborne aerosol particles known as CCN. Typically, CCN fall within the submicron size fraction, about 100 nm in diameter. Depending on their properties and heights, clouds can either warm their surfaces by triggering a localized greenhouse effect or cool them by reflecting solar radiation. If CCN are scarce, the resulting clouds will contain fewer and larger droplets.

Such clouds will reflect little sunlight to space while blocking the escape of heat from Earth's surface, causing it to warm. However, if CCN are plentiful, many fine droplets form; the resultant clouds are better reflectors, thus cooling the surface below. Over remote marine areas such as in the high Arctic (>87°N), anthropogenic particles are virtually absent. Instead, biological sources of particles may dominate (Bigg and Leck, 2001a,b; Leck and Persson, 1996; Leck and Bigg, 2005a,b; Leck et al., 2002, 2013; Orellana et al., 2011b). This "clean" air, with few CCN, makes the low-level stratocumulus clouds optically thin, with fewer but larger droplets. But, if a warming climate spurs the activity of microbiota, organic sources of CCN might become more prominent and lead to increasingly radiation-reflective clouds. Over the last 15 years, articles have emphasized the presence and enrichment of organic matter particles of submicron sizes in airborne aerosols and cloud water (Bigg and Leck, 2001, 2008; Leck and Bigg, 2008 and Karl et al., 2013; Duce and Hoffman, 1976; Facchini et al., 2008; Gaston et al., 2011; Keene et al., 2007; Leck and Bigg, 2005a,b, 2010; Middlebrook et al., 1998; O'Dowd et al., 2004; Russell et al., 2010; Yoon et al., 2007). The detection of organic substances—specifically of exopolymer like particles—in the atmosphere was first discovered by Bigg and Leck (Bigg and Leck, 2001a,b; Leck and Bigg, 2008; Bigg and Leck, 2008; Leck and Bigg, 2005a; Leck et al., 2002). These authors recognized that these particles could bear the physicochemical characteristics of marine gels. This followed from their studies of a possible link between cloud formation and polymer gels in the SML (<100 μm thick at the air-sea interface) in the high Arctic sea ice (Bigg et al., 2004).

A Is the "Gel Theory of Marine CCN" Coupled to the Sulfur Cycle?

Charlson et al. (1987) reviewed existing evidence that implicated DMS (produced by the microbial food webs) in the production of CCN over remote marine areas. This provocative CLAW hypothesis (so named informally after the paper's authors Charlson, Lovelock, Andreae, and Warren) stated that, in the marine domain, DMS emissions and their subsequent oxidation products—methane sulfonic acid, sulfur dioxide, and sulfuric acid—trigger cloud formation, which cools the ocean surface. This cooling would, in turn, affect further emissions of DMS by changing the speciation/abundance of marine phytoplankton, establishing a feedback loop. Observations in the early 1990s from the Arctic did indeed show that the intermediate oxidation products provided most of the mass for the CCN-sized particles observed over pack ice (Leck and Persson, 1996). The source of most of the DMS, though, was found at the fringe of the central Arctic Ocean, at the hospitable edges of the pack ice. At that time, this distribution suggested that winds carried DMS-rich air toward the North Pole, and oxidation of this DMS created extremely small sulfuric acid particles. Theoretically, these particles would then grow slowly by further condensation of the acids until they were large enough to serve as CCN. Surprisingly, it turns out that sulfuric acid had nothing to do with the small precursors of CCN. Instead, observations from the Arctic in the mid-1990s showed that these small precursors were mostly particles resembling viruses and nanogels and microgels that were accompanied by other larger particles, such as bacteria and fragments of diatoms (Figure 9.8; Bigg and Leck, 2001a,b; Leck and Bigg, 1999; Leck and Bigg, 2005a, 2010; Leck et al., 2002). Subsequently, Bigg et al. (2004) detected large numbers (10^6-10^{14} mL^{-1}) of similar particles within the thin surface film at the water-air interface between ice floes (Figure 9.9).

We know that polymer gel networks assemble preferentially at the water-air interface (Verdugo et al., 2004). The SML has long been known as a source of gels (Sieburth, 1983), but the connection between the airborne particles and the SML was found much more recently.

Airborne microgels may have the chemical surfactant properties necessary to act as nuclei for clouds (CCN), but to behave as effective CCN,

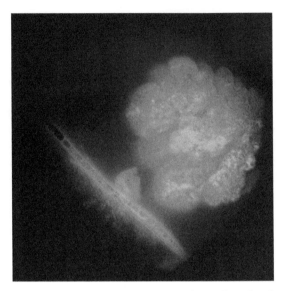

FIGURE 9.8   A polymer gel microcolloid collected in the air over the Arctic pack ice at 89°N. *Adapted from Leck and Bigg (2005a).*

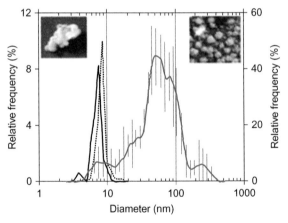

FIGURE 9.9   Red curve. With standard error bars: median particle number size distribution of five SML samples expressed as relative frequency of occurrence (right *y*-axis). Black curve, relative particle number size distribution of the individual components of the particles presented in the median size distribution (left *y*-axis). Black dotted curve not described. Inset, upper right. Microscopic image of particles from samples, corresponding to the red curve. Inset, upper left. Microscopic detail of an aggregated particle typical of those sized in the red curve. *Modified from Bigg et al. (2004).*

they must reach a critical size and meet other physicochemical properties and energy constraints of the system. Leck and Bigg (1999, 2010) and Karl et al. (2013) speculated that the primary marine gel would disintegrate under some circumstances, generating smaller particles, probably due to UV radiation cleavage (Orellana and Verdugo, 2003). However, it is not clear whether the polymeric material reached smaller sizes due to breakage caused by UV radiation, by reversible volume phase transition, or by a combination of both, as has been described for marine microgels (Chin et al., 1998; Orellana et al., 2011b; Orellana and Verdugo, 2003); thus, this remains an open question. Furthermore, the gels could also provide sites for condensation of the oxidation products of DMS. In 2005, when Leck and Bigg tested predominantly airborne sulfate particles for the presence of microgels, they detected marine microgel material in half or more of their samples coated with sulfuric acid. A specific fluorescently labeled antibody probe developed against *in situ* seawater and SML biopolymers confirmed for the first time that the particles found in the atmosphere (aerosol/fog/cloud) originated in the surface sea water, including the subsurface and SML (Orellana et al., 2011b). The particles were released by sea-ice algae, phytoplankton, and bacteria, and behaved as nanogels and microgels (Figure 9.10). The gel networks were held together by random entanglements and Ca^{2+} ionic bonds (Figure 9.3), as well as by hydrophobic moieties (Figure 9.5). The gels comprised as much as 50% of the total organic C in surface waters and the SLM, and they assembled faster than previously observed, probably due to the presence of hydrophobic moieties enhancing polymer assembly (Figure 9.4; Ding et al., 2008). The gels also underwent volume phase transitions induced by DMSP as well as DMS, another indication that those particles displayed the physicochemical characteristics of gels (Figure 9.7). Gels were abundant in the subsurface seawater and enriched in the SML. They also correlated with enrichment of proteins containing hydrophobic amino acids and DMSP in the SML.

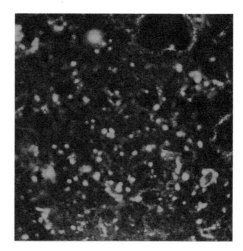

FIGURE 9.10 Cloud water: polymer gels immune-labeled with a fluorescently labeled antibody developed against seawater biopolymers (Orellana et al., 2011b).

The co-occurrence of atmospheric organic material and biologically active marine waters has, since the mid-1990s, been confirmed for the high Arctic waters, but it has also been documented for temperate waters (Facchini et al., 2008; Leck and Bigg, 2008, 2010; Russell et al., 2010). Observations from the Arctic called into question the key role given to DMS in the CLAW hypothesis (Leck and Bigg, 2007). In the emerging picture of the Arctic atmosphere, DMS concentrations will determine the mass of the particles by producing material for their growth. But, it is the number of airborne gels that will primarily influence the number of CCN and the resulting optical properties of the cloud droplets. Indeed, research during the past two decades—reviewed by Quinn and Bates (2011)—does not corroborate the CLAW hypothesis for other regions either.

B The Effect of Gels on Bubble Properties and Bursting

Leck and Bigg (Leck et al., 2004; Leck and Bigg, 2005b and Bigg and Leck, 2008; Leck and Bigg, 1999; Leck et al., 2002) had hypothesized that the source of gels found in clouds was the open water between ice floes and that those particles would be transferred to the atmosphere by the bursting of air bubbles at the air-sea interface. Transport of aerosols by bubble bursting under experimental conditions is well known (e.g., Aller et al., 2005; Fuentes et al., 2010a; Fuentes et al., 2010b; Fuentes et al., 2011; Keene et al., 2007; Kuznetsova et al., 2005). However, experimental setups and different conditions show enormous variability as to aerosol size and concentration. Factors that affect the variability include the presence of surfactants, salinity, viscosity, Langmuir circulation, turbulence, wave breaking, etc. (Fuentes et al., 2010a).

Bubbles usually result from entrainment of air induced by wind stress at the air-water interface (Blanchard, 1971; Blanchard and Syzdek, 1988); bubble bursting produces primary aerosol particles in CCN sizes. Over the summer pack ice, near-surface wind speeds are typically low ($<6\,\mathrm{m\,s^{-1}}$), and the extent of open water in the pack ice leads is usually modest (10-30%), shortening fetches and limiting the generation of waves. A recent study confirmed, in spite of the low winds, both the presence and the temporal variability of a population of bubbles within the open leads; proposed was a non-wave bubble source mechanism; the loss of heat to the atmosphere produced temperature differences at the surface layer and bulk water creating convective mixing and driving local fluctuations in gas saturation, that subsequently generated both film and jet droplets (Norris et al., 2011). Possible non-wind-related sources of bubbles include releases of bubbles trapped in melting sea ice, as well as those expelled by freezing water, or transported to the surface by increased turbulence caused by super-cooling conditions (Grammatika and Zimmerman, 2001). Further sources of bubbles include photosynthesis and respiration of phytoplankton (Johnson and Wangersky, 1987), and falling raindrops on the ocean surface (Lewis and Schwartz, 2004).

Bubbles scavenge DOM to the bubble film (Zhou et al., 1998b), particularly demonstrating the role of surface-active polysaccharides in the formation of large TEP by bubble adsorption in seawater. However, during bubble bursting over marine areas, bubbles scavenge not only debris and HMW soluble organic surface-active compounds but also sea salt, as they rise through the water prior to their injection into the atmosphere (Blanchard and Syzdek, 1988). It has generally been assumed that particles derived from bubble bursting would be composed of sea salt only, and would thus contribute a significant fraction of the CCN population (O'Dowd et al., 1999). However, transmission electron microscopy of individual particles (Bigg, 1980; Bigg and Leck, 2001, 2008; Gras and Ayers, 1983; Leck and Bigg, 2005a,b; Leck et al., 2002; Pósfai et al., 2003) over the pristine perennial arctic ice, and at remote marine locations at lower latitudes, have failed to find evidence of sea salt particles $<200\,\mathrm{nm}$ in diameter. To explain this, Bigg and Leck (2008) proposed a mechanism for transporting polymer-microgel-rich organic material from the bulk seawater into the open lead SML. They suggested that the highly surface-active polymer gels could concentrate on the surface of rising bubbles and then aggregate by interpenetration and polymer tangling. Consequently, rising bubbles can selectively carry polymer gels, as well as embedded solid particles such as bacteria and phytoplankton and their detritus, to the SML. Before bursting, bubbles rest in the microlayer; therefore, they likely have a film envelope composed largely of gels, and embedded particulate matter that may become points of weakness as the water drains from between their envelope. Following the burst, the film drops fragments, containing surfactant material, salt-free water, and any particle attached to the fragments. These suggestions are consistent with the fact that gels assemble preferentially at the microlayer interface (Verdugo et al., 2004) as well as on bubble films (Gao et al., 2012).

An alternative process for expelling polymer gels from the ocean surface involves charge repulsion by the negatively charged surface of the Earth, which acts as a spherical capacitor. The Earth has a net negative charge of about a million coulombs, and an equal positive charge resides in the atmosphere (Feynman et al., 1964). No observational measurement to explain this path of the aerosols in the atmosphere has been done.

C Is the Gel Theory for the Origin of Marine CCN Consistent with Primary Marine Aerosol Observations?

The Arctic studies of primary organic polymer gel aggregates derived from the SML between sea ice leads, performed over the last two decades, have been expanded to other marine areas (Facchini et al., 2008; Fuentes et al., 2010b; Leck and Bigg, 2008; Russell et al., 2010). These studies show strong similarities in the morphological and chemical characteristics of the particles both previously and recently described (Leck and Bigg, 2005b, 2007, 2008; Leck et al., 2013). Most of the studies have focused on size-segregated chemical characterization of marine aerosol particles (Cavalli, 2004; Keene et al., 2007; O'Dowd et al., 2004; Rinaldi et al., 2010; Russell et al., 2010; Sciare et al., 2009). They all have suggested marine DOM at the ocean surface as the source for these particles. However, they have not looked at gel's physical-chemical characteristics and the emergent properties (assembly, phase transition) of this sort of particulate matter and thus have not confirmed that they are marine gels.

Recently, Martin et al. (2011a) suggested that internally mixed amphiphilic biopolymers and gels do not have the ability to activate as CCN, instead CCN were dominated by the sulfate fraction of the particles, which could be trapped within the polyanionic matrix of the gel (Leck and Bigg, 2005a; Orellana et al., 2011a). However, we know that marine gels have the important property of being solvated in water, highly surface active and also of undergoing reversible volume phase transitions. Marine gels

can reversibly change their morphology from a swollen, hydrated phase to a condensed, collapsed phase. The volume phase transition may, for instance, be caused by changes of pH, DMS, and DMSP (Orellana et al., 2011b). Moreover, Arctic gels were shown to consist of hydrophilic and hydrophobic segments (Orellana et al. 2011b) in agreement with their chemical behavior. This behavior was recently confirmed by molecular dynamic modeling by (Xin et al., 2013). Hence, the colloidal gel shows only a partial wetting character below 100 percent relative humidity (RH) thus showing only weak hygroscopic growth but, at the same time, high CCN activation efficiency is shown at RH above 100 percent. (Orellana et al., 2011b). Experimental data reported by Ovadnevaite et al. (2011) suggest a dichotomous behavior for the primary marine organic aerosol; water vapor does not uniformly undergo condensation, since only part of the surface exhibits strong hydrophilicity. Hence, the colloidal gel shows only a partial wetting character (<100% relative humidity), thus showing only weak hygroscopic growth but, at the same time, high CCN activation efficiency. The 3D structure of marine microgels, with their hydrophilicity or surface-active properties and only partial wetting character that result from the hydrophobic characteristics of the polymers, could explain such dichotomous behavior. Our understanding of gels and their emergent behavior as CCN is still in its infancy; however, the gel conceptual framework provides a predictive theory (Edwards, 1986; Tanaka, 1981, 1992) for understanding soft matter colloidal processes and controlling factors. While, we have no observations of gels as CCN outside the high Arctic, these observations suggest this might be a universal process.

Acknowledgments

We thank the editors, D.A. Hansell and C.A. Carlson, for the invitation to prepare this chapter on marine microgels. The chapter benefited greatly from two anonymous reviewers for helpful comments, and the NSF Biological Oceanography

Program, the Arctic Natural Sciences Program, the Polar
Science Center (University of Washington), the Institute for
Systems Biology, the Swedish Research Council, mentoring
by Pedro Verdugo, and collaborations with Patricia Matrai,
Keith Bigg, Robert Moritz and many students.

References

Alkire, M.B., D'Asaro, E., Lee, C., Perry, M.J., Gray, A., Cetinić, I.,
et al., 2012. Estimates of net community production and
export using high-resolution, Lagrangian measurements
of O_2, NO_3^-, and POC through the evolution of a spring
diatom bloom in the North Atlantic. Deep-Sea Res. Pt. I
64, 157–174.

Alldredge, A.L., Gotschalk, C.C., 1989. Direct observation of
the mass flocculation of diatom blooms: characteristics,
settling velocities and formation of diatom aggregates.
Deep-Sea Res. 36, 159–171.

Alldredge, A.L., Jackson, G.A., 1995. Aggregation in marine
systems: preface. Deep-Sea Res. Pt. II 42, 1–7.

Alldredge, A.L., Passow, U., Logan, E.B., 1993. The abundance and significance of a class of large, transparent organic particles in the oceans. Deep-Sea Res. 40, 1131–1140.

Alldredge, A.L., Passow, U., Haddock, S.H.D., 1998. The
characteristics and transparent exopolymer particle
(TEP) content of marine snow formed from thecate dinoflagellates. J. Plankton Res. 20, 393–406.

Aller, J.Y., Kuznetsova, M.R., Jahns, C.J., Kemp, P.F., 2005. The
sea surface microlayer as a source of viral and bacterial
enrichment in marine aerosols. J. Aerosol. Sci. 36, 801–812.

Aluwihare, L.I., Repeta, D.J., 1999. A comparison of the
chemical characteristics of oceanic DOM and extracellular DOM produced by marine algae. Mar. Ecol. Prog. Ser.
186, 105–117.

Aluwihare, L.I., Repeta, D.J., Chen, R.F., 1997. A major bio-
polymeric component to dissolved organic carbon in surface sea water. Nature 387, 166–169.

Aluwihare, L.I., Repeta, D.J., Pantoja, S., Johnson, C.G., 2005.
Two chemically distinct pools of organic nitrogen accumulate in the ocean. Science 308, 1007–1010.

Amon, R.M.W., Benner, R., 1994. Rapid cycling of high-
molecular-weight dissolved organic matter in the ocean.
Nature 369, 549–552.

Amon, R.M.W., Benner, R., 1996. Bacterial utilization of different size classes of dissolved organic matter. Limnol.
Oceanogr. 41, 41–51.

Aristegui, J., Gasol, J.M., Duarte, C.M., Herndl, G.J., 2009.
Microbial oceanography of the dark ocean's pelagic
realm. Limnol. Oceanogr. 54, 1501–1529.

Arnosti, C., 2011. Microbial extracellular enzymes and the
marine carbon cycle. Ann. Rev. Mar. Sci. 3, 401–425.

Aslam, S.N., Cresswell-Maynard, T., Thomas, D.N.,
Underwood, G.J.C., 2012. Production and character-

ization of the intra- and extracellular carbohydrates
and polymeric substances (EPS) of three sea ice diatom
species, and evidence for a cryoprotective role for EPS.
J. Phycol. 48, 1494–1509. http://dx.doi.org/10.1111/
jpy.12004.

Azam, F., 1998. Oceanography: microbial control of oceanic
carbon flux: the plot thickens. Science 280, 694–696.

Azam, F., Long, R.A., 2001. Oceanography—sea snow microcosms. Nature 414, 495–498.

Azam, F., Malfatti, F., 2007. Microbial structuring of marine
ecosystems. Nat. Rev. Microbiol. 5, 782–791.

Baltar, F., Arístegui, J., Sintes, E., Gasol, J.M., Reinthaler, T.,
Herndl, G.J., 2010. Significance of non-sinking particulate
organic carbon and dark CO_2 fixation to heterotrophic
carbon demand in the mesopelagic northeast Atlantic.
Geophys. Res. Lett. 37, 1–6, L09602.

Bates, T.S., Quinn, P.K., 1997. Dimethylsulfide (DMS) in the
equatorial Pacific Ocean (1982 to1996): evidence of a
climate feedback? Geophys. Res. Lett. 24, 861–864.

Bauer, J.E., 2002. Carbon Isotopic Composition of DOM.
Academic Press, San Diego, California.

Bauer, M., 2006. Whole genome analysis of the marine
Bacteroidetes 'Gramella forseti' reveals adaptations
to degradation of polymeric organic matter. Environ.
Microbiol. 8, 2201–2213.

Bauer, J.E., Williams, P.M., Druffel, E.R.M., 1992. 14C activity
of dissolved organic carbon fractions in the north-central
Pacific and Sargasso Sea. Nature 357, 667–670.

Benner, R., 2002. Chemical composition and reactivity.
In: Hansell, D., Carlson, C. (Eds.), Biogeochemistry of
Marine Dissolved Organic Matter. Academic Press, USA,
pp. 59–90.

Benner, R., Herndl, G.J., 2011. Bacterially derived dissolved
organic matter in the microbial carbon pump. In: Jiao, N.,
Azam, F., Sanders, S. (Eds.), Microbial Carbon Pump in
the Ocean. Science/AAAS, Washington, DC, pp. 46–48.
http://dx.doi.org/10.1126/science.opms.sb0001.

Benner, R., Kaiser, K., 2003. Abundance of amino sugars and
peptidoglycan in marine particulate and dissolved organic matter. Limnol. Oceanogr. 48, 118–128.

Benner, R., Pakulski, J.D., McCarthy, M., Hedges, J.I., Hatcher,
P.G., 1992. Bulk chemical characteristics of dissolved organic matter in the ocean. Science 255, 1561–1564.

Berman-Frank, I., Bidle, K., Haramaty, L., Falkowski, P., 2004.
The demise of the marine cyanobacterium, *Trichodesmium*
spp., via an autocatalyzed cell death pathway. Limnol.
Oceanogr. 49, 997–1005.

Biddanda, B., Benner, R., 1997. Carbon, nitrogen, and carbohydrate fluxes during the production of particulate
and dissolved organic matter by marine phytoplankton.
Limnol. Oceanogr. 42, 506–518.

Bidle, K.D., Falkowski, P.G., 2004. Cell death in planktonic,
photosynthetic microorganisms. Nat. Rev. Microbiol. 2,
643–655.

Biersmith, A., Benner, R., 1998. Carbohydrates in phytoplankton and freshly produced dissolved organic matter. Mar. Chem. 63, 131–144.

Bigg, E.K., 1980. Comparison of aerosol at four baseline atmospheric monitoring stations. J. Appl. Meteorol. 19, 521–533.

Bigg, E.K., Leck, C., 2001a. Properties of the aerosol over the central Arctic Ocean. J. Geophys. Res. 106, 32101–32109.

Bigg, E.K., Leck, C., 2001b. Cloud-active particles over the central Arctic Ocean. J. Geophys. Res. 106, 32155–32166.

Bigg, E.K., Leck, C., 2008. The composition of fragments of bubbles bursting at the ocean surface. J. Geophys. Res. 113 (D1), 1–7, D11209.

Bigg, E.K., Leck, C., Tranvik, L., 2004. Particulates of the surface microlayer of open water in the central Arctic Ocean in summer. Mar. Chem. 91, 131–141.

Biller, S.J., Schubotz, F., Roggensack, S.E., Thompson, A.W., Summons, R.E., Chisholm, S.W., 2014. Bacterial vesicles in marine ecosystems. Science 343, 183–186.

Blanchard, D.C., 1971. The oceanic production of volatile cloud nuclei. J. Atmos. Sci. 28, 811–812.

Blanchard, D.C., Syzdek, L.D., 1988. Film drop production as a function of bubble size. J. Geophys. Res. 93, 3649–3654.

Buesseler, K.O., 1998. The de-coupling of production and particle export in the surface ocean. Global Biogeochem. Cycles 12, 297–310.

Burd, A.B., Jackson, G.A., 2009. Particle aggregation. Ann. Rev. Mar. Sci. 1, 65–90.

Butler, A., Theisen, R.M., 2010. Iron (III)–siderophore coordination chemistry: reactivity of marine siderophores. Coord. Chem. Rev. 254, 288–296.

Calleja, M.L., Batista, F., Peacock, M., Kudela, R., McCarthy, M.D., 2013. Changes in compound specific [615]N amino acid signatures and d/l ratios in marine dissolved organic matter induced by heterotrophic bacterial reworking. Mar. Chem. 149, 32–44.

Carlson, C.A., 2002. Production and removal processes. In: Hansell, D.A., Carlson, C. (Eds.), Biogeochemistry of Dissolved Organic Matter. Academic Press, San Diego, CA, pp. 91–151.

Carlson, C.A., Giovannoni, S.J., Hansell, D.A., Goldberg, S.J., Parsons, R.V., Vergin, K., 2004. Interactions between DOC, microbial processes, and community structure in the mesopelagic zone of the northwestern Sargasso Sea. Limnol. Oceanogr. 49, 1073–1083.

Cavalli, F., 2004. Advances in identification of organic matter in marine aerosol. J. Geophys. Res. 109, http://dx.doi.org/10.1029/2004JD005137, D24215.

Charlson, R.J., Lovelock, J.E., Andreae, M.O., Warren, S.G., 1987. Oceanic phytoplankton, atmospheric sulfur, cloud albedo and climate. Nature 326, 655–661.

Chateauvert, C.A., Lesack, L.F.W., Bothwell, M.L., 2012. Abundance and patterns of transparent exopolymer

particles (TEP) in Arctic floodplain lakes of the Mackenzie River Delta. J. Geophys. Res. 117, G04013.

Cherrier, J., Bauer, J.E., 2004. Bacterial utilization of transient plankton-derived dissolved organic carbon and nitrogen inputs in surface ocean waters. Aquat. Microb. Ecol. 35, 229–241.

Cherrier, J., Bauer, J.E., Druffel, E.R.M., 1996. Utilization and turnover of labile dissolved organic matter by bacterial heterotrophs in eastern North Pacific surface waters. Mar. Ecol. Prog. Ser. 139, 267–279.

Chin, W.-C., Orellana, M.V., Verdugo, P., 1998. Spontaneous assembly of marine dissolved organic matter into polymer gels. Nature 391, 568–572.

Chin, W.-C., Orellana, M.V., Quesada, I., Verdugo, P., 2004. Secretion in unicellular marine phytoplankton: demonstration of regulated exocytosis in *Phaeocystis globosa*. Plant Cell Physiol. 45, 535–542.

Chuang, C.-Y., Santschi, P.H., Ho, Y.-F., Conte, M.H., Guo, L., Schumann, D., Ayranov, M., Li, Y.H., 2013. Role of biopolymers as major carrier phases of Th, Pa, Pb, Po, and Be radionuclides in settling particles from the Atlantic Ocean. Mar. Chem. 157, 131–143.

Clegg, S.L., Whitfield, M., 1990. A generalized model for the scavenging of trace metals in the open ocean: I. Particle cycling. Deep-Sea Res. 37, 809–832.

Clegg, S.L., Whitfield, M., 1991. A generalized model for the scavenging of trace metals in the open ocean: II. Thorium scavenging. Deep-Sea Res. 38, 91–120.

Close, H.G., Shah, S.R., Ingalls, A.E., Diefendorf, A.F., Brodie, E.L., Hansman, R.L., et al., 2013. Export of submicron particulate organic matter to mesopelagic depth in an oligotrophic gyre. Proc. Natl. Acad. Sci. 110, 12565–12570.

Cornwall, C.E., Hepburn, C.D., McGraw, C.M., Currie, K.I., Pilditch, C.A., Hunter, K.A., et al., 2013. Diurnal fluctuations in seawater pH influence the response of a calcifying macroalga to ocean acidification. Proc. R. Soc. Lond. B Biol. Sci. 280, 1471–2954.

Cottrell, M.T., Kirchman, D.L., 2000. Natural assemblages of marine proteobacteria and members of the Cytophaga-Flovobacter cluster consuming low and high molecular weight dissolved organic matter. Appl. Environ. Microbiol. 66, 1692–1697.

Cuadrado-Silva, C.T., Castellanos, L., Arévalo-Ferro, C., Osorno, O.E., 2013. Detection of quorum sensing systems of bacteria isolated from fouled marine organisms. Biochem. Syst. Ecol. 46, 101–107.

Davis, J., Benner, R., 2007. Quantitative estimates of labile and semi-labile dissolved organic carbon in the western Arctic Ocean: a molecular approach. Limnol. Oceanogr. 52, 2434–2444.

de Gennes, P.G., 1992. Soft matter. Science 256, 495–497.

de Gennes, P.G., Leger, L., 1982. Dynamics of entangled polymer chains. Annu. Rev. Phys. Chem. 33, 49–61.

de Vicente, I., Ortega-Retuerta, E., Mazuecos, I.P., Pace, M.L., Cole, J.J., Reche, I., 2010. Variation in transparent exopolymer particles in relation to biological and chemical factors in two contrasting lake districts. Aquat. Sci. 72, 443–453.

Decho, A.W., 1990. Microbial exopolymer secretions in ocean environments: their role(s) in food webs and marine processes. Oceanogr. Mar. Biol. Annu. Rev. 28, 73–153.

Del Castillo, C.E., Gilbes, F., Coble, P.G., 2000. On the dispersal of riverine colored dissolved organic matter over the west Florida Shelf. Limnol. Oceanogr. 45, 1425–1432.

DeLong, E.F., Preston, C.M., Mincer, T., Rich, V., Hallam, S.J., Frigaard, N.U., 2006. Community genomics among stratified microbial assemblages in the ocean's interior. Science 311, 496–503.

Ding, Y.-X., Chin, W.-C., Verdugo, P., 2007. Development of a fluorescence quenching assay to measure the fraction of organic carbon present in self-assembled gels in seawater. Mar. Chem. 106, 456–462.

Ding, Y.-X., Chin, W.-C., Rodriguez, A., Hung, C.-C., Santschi, H.P., Verdugo, P., 2008. Amphiphilic exopolymers from *Sagittula stellata* induce DOM self-assembly and formation of marine microgels. Mar. Chem. 112, 11–19.

Ding, Y.-X., Hung, C.-C., Santschi, P.H., Verdugo, P., Chin, W.-C., 2009. Spontaneous assembly of exopolymers from phytoplankton. Terr. Atmos. Ocean Sci. 20, 741–747.

Dobretsov, S., Teplitski, M., Paul, V., 2009. Mini-review: quorum sensing in the marine environment and its relationship to biofouling. Biofouling 25, 413–427.

Druffel, E.R.M., Williams, P.M., 1990. Identification of a deep marine source of particulate organic carbon using bomb [14]C. Nature 347, 172–174.

Druffel, E.R.M., Griffin, S., Honjo, S., Manganini, S.J., 1998. Evidence of old carbon in the deep water column of the Panama Basin from natural radiocarbon measurements. Geophys. Res. Lett. 25, 1733–1736.

Duce, R.A., Hoffman, E.J., 1976. Chemical fractionation at the air/sea interface. Annu. Rev. Earth Planet Sci. 4, 187–228.

Dunne, J.P., Murray, J.W., Young, J., Balistrieri, L.S., Bishop, J., 1997. [234]Th and particle cycling in the central equatorial Pacific. Deep-Sea Res. Pt. II 44, 2049–2084.

Edwards, S.F., 1986. The theory of macromolecular networks. Biorheology 23, 589–603.

Engel, A., 2004. Distribution of transparent exopolymer particles (TEP) in the northeast Atlantic Ocean and their potential significance for aggregation processes. Deep-Sea Res. Pt. I 51, 83–92.

Engel, A., Passow, U., 2001. Carbon and nitrogen content of transparent exopolymer particles (TEP) in relation to their Alcian Blue adsorption. Mar. Ecol. Prog. Ser. 219, 1–10.

Engel, A., Thoms, S., Riebesell, U., Rochelle-Newall, E., Zondervan, I., 2004. Polysaccharide aggregation as a potential sink of marine dissolved organic carbon. Nature 428, 929–932.

Facchini, M.C., Rinaldi, M., Decesari, S., Carbone, C., Finessi, E., Mircea, M., et al., 2008. Primary submicron marine aerosol dominated by insoluble organic colloids and aggregates. Geophys. Res. Lett. 35, L17814.

Feely, R.A., Sabine, C.L., Hernandez-Ayon, J.M., Ianson, D., Hale, B., 2008. Evidence for upwelling of corrosive "acidified" water onto the Continental Shelf. Science 320, 1490–1492. http://dx.doi.org/10.1126/science.1155676.

Feely, R.A., Doney, S.C., Cooley, S.R., 2009. Ocean acidification: present conditions and future changes in a high-CO_2 world. Oceanography 22, 36–47.

Fenchel, T., 1984. Suspended bacteria as a food source. In: Fasham, M.J.R. (Ed.), Flows of Energy and Materials in Marine Ecosystems. Plenum, New York, pp. 301–315.

Fenchel, T., Blackburn, H., King, G.M., 2012. Bacterial Biogeochemistry: The Ecophysiology of Mineral Cycling. third ed. Elsevier Ltd, San Diego, CA.

Fernandez, J.M., Villalón, M., Verdugo, P., 1991. Reversible condensation of mast cell secretory products in vitro. Biophys. J. 59, 1022–1027.

Feynman, R.P., Leighton, R.B., Sands, M., 1964. In: The Feynman Lectures on Physics. Mainly Electromagnetism and Matter, vol. II. Basic Books, USA.

Fuentes, E., Coe, H., Green, D., de Leeuw, G., McFiggans, G., 2010a. Laboratory-generated primary marine aerosol via bubble-bursting and atomization. Atmos. Meas. Tech. 3, 141–162.

Fuentes, E., Coe, H., Green, D., de Leeuw, G., McFiggans, G., 2010b. On the impacts of phytoplankton-derived organic matter on the properties of the primary marine aerosol—Part 1: source fluxes. Atmos. Chem. Phys. 10, 9295–9317.

Fuentes, E., Coe, H., Green, D., McFiggans, G., 2011. On the impacts of phytoplankton-derived organic matter on the properties of the primary marine aerosol—Part 2: composition, hygroscopicity and cloud condensation activity. Atmos. Chem. Phys. 11, 2585–2602.

Fuhrman, J.A., 2009. Microbial community structure and its functional implications. Nature 459, 193–199.

Gabric, A.J., Matrai, P.A., Kiene, R.P., Cropp, R., Dacey, J.W.H., DiTullio, G.R., et al., 2009. Factors determining the vertical profile of dimethylsulfide in the Sargasso Sea during summer. Deep-Sea Res. Pt. II 55, 1505–1518.

Gao, Q., Leck, C., Rauschenberg, C., Matrai, P.A., 2012. On the chemical dynamics of extracellular polysaccharides in the high Arctic surface microlayer. Ocean Sci. 8, 401–418.

Gaston, C.J., Furutani, H., Guazzotti, S.A., Coffee, K.R., Bates, T.S., Quinn, P.K., et al., 2011. Unique ocean-derived particles serve as a proxy for changes in ocean chemistry. J. Geophys. Res. 116, http://dx.doi.org/10.1029/2010JD015289, D18310.

Gram, L., Grossart, H.-P., Schlingloff, A., Kiørboe, T., 2002. Possible quorum sensing in marine snow bacteria: production of acylated homoserine lactones by *Roseobacter* strains isolated from marine snow. Appl. Environ. Microbiol. 68, 4111–4116.

Grammatika, M., Zimmerman, W.B., 2001. Microhydrodynamics of flotation processes in the sea surface layer. Dyn. Atmos. Oceans 34, 327–348.

Gras, J.L., Ayers, G.P., 1983. Marine aerosol at southern mid-latitudes. J. Geophys. Res. 88, 10661–10666.

Grosberg, A.Y., Khokhlov, A.R., 1994. Statistical Physics of Macromolecules. AIP Press, Woodbury, NY.

Grossart, H.-P., Tang, K.W., Kiørboe, T., Ploug, H., 2007. Comparison of cell-specific activity between free-living and attached bacteria using isolates and natural assemblages. FEMS Microbiol. Lett. 266, 194–200.

Guo, L., Santschi, P.H., 1997. Composition and cycling of colloids in marine environments. Rev. Geophys. 35, 17–40.

Guo, L., Santschi, P.H., Warnken, K.W., 2000. Trace metal composition of colloidal organic material in marine environments. Mar. Chem. 70, 257–275.

Hansell, D.A., 2013. Recalcitrant dissolved organic carbon fractions. Ann. Rev. Mar. Sci. 5, 421–445.

Hansell, D.A., Carlson, C., 2002. Biogeochemistry of Marine Dissolved Organic Matter. Academic Press, San Diego.

Hansell, D.A., Carlson, C.A., 2013. Localized refractory dissolved organic carbon sinks in the deep ocean. Global Biogeochem. Cycles 27, 1–6.

Hansell, D.A., Carlson, C.A., Repeta, D.J., Schlitzer, R., 2009. Dissolved organic matter in the ocean. Oceanography 22, 52–61.

Hansman, R.L., Griffin, S., Watson, J.T., Druffel, E.R.M., Ingalls, A.E., Pearson, A., et al., 2009. The radiocarbon signature of microorganisms in the mesopelagic ocean. Proc. Natl. Acad. Sci. 106, 6513–6518.

Hedges, J.I., Oades, J.M., 1997. Comparative organic geochemistries of soils and marine sediments. Org. Geochem. 27, 319–361.

Herndl, G.J., Reinthaler, T., 2013. Microbial control of the dark end of the biological pump. Nat. Geosci. 6, 718–724.

Hertkorn, N., Benner, R., Frommberger, M., Schmitt-Kopplin, P., Witt, M., Kaiser, K., et al., 2006. Characterization of a major refractory component of marine dissolved organic matter. Geochim. Cosmochim. Acta 70, 2990–3010.

Hmelo, L., Van Mooy, B., 2010. Kinetic constraints on acylated homoserine lactone-based quorum sensing in marine environments. Aquat. Microb. Ecol. 54, 127–133.

Honeyman, B.D., Santschi, H.P., 1988. A Brownian-pumping model for oceanic trace metal scavenging: evidence from Th isotopes. J. Mar. Res. 47, 951–992.

Hong, Y., Smith, W.O., White, A.-M., 1997. Studies on transparent exopolymer particles (TEP) in the Ross Sea (Antarctica) and by Phaeocystis antarctica (Prymnesiophyceae). J. Phycol. 33, 368–376.

Hopkinson, C.S., Vallino, J.J., 2005. Efficient export of carbon to the deep ocean through dissolved organic matter. Nature 433, 142–145.

Hung, C.-C., Tang, D., Warnken, K., Santschi, P.H., 2001. Distributions of carbohydrates, including uronic acids, in estuarine waters of Galveston Bay. Mar. Chem. 73, 305–318.

Hung, C.-C., Guo, L., Schultz, G.E., Pinckney, J.L., Santschi, P.H., 2003. Production and flux of carbohydrate species in the Gulf of Mexico. Global Biogeochem. Cycles 17, 1055.

IPCC, 2007. The Physical Science Basis. Contribution of Working Group I to the Fourth Assessment Report of the Intergovernmental Panel on Climate Change, Cambridge University Press, Cambridge.

Jackson, G.A., 1990. A model of the formation of marine algal flocs by physical coagulation processes. Deep-Sea Res. 37, 1197–1211.

Jackson, G., Burd, A., 1998. Aggregation in the marine environment. Environ. Sci. Tech. 32, 2805–2814.

Jadhav, S.R., Vemula, P.K., Kumar, R., Raghavan, S.R., John, G., 2010. Sugar-derived phase-selective molecular gelators as model solidifiers for oil spills. Angew. Chem. Int. Ed. 49, 7695–7698.

Janse, I., van Rijssel, M., Gottschall, J.C., Lancelot, C., Gieskes, W.W.C., 1996. Carbohydrates in the North Sea during spring blooms of Phaeocystis: a specific fingerprint. Aquat. Microb. Ecol. 10, 97–103.

Jiao, N., Herndl, G.J., Hansell, D.A., Benner, R., Kattner, G., Wilhelm, S.W., et al., 2010. Microbial production of recalcitrant dissolved organic matter: long-term carbon storage in the global ocean. Nat. Rev. Microbiol. 8, 593–599.

Johnson, B.D., Wangersky, P.J., 1987. Microbubbles: stabilization by monolayers of adsorbed particles. J. Geophys. Res. 92, 14641–14647.

Jumars, P.A., 1993. Concepts in Biological Oceanography. Oxford University Press, New York.

Jumars, P.A., Penry, D., Barros, J.A., Perry, M.J., Frost, B.W., 1989. Closing the microbial loop: dissolved carbon pathway to heterotrophic bacteria from incomplete ingestion, digestion and absorption in animals. Deep-Sea Res. 36, 483–485.

Kaiser, K., Benner, R., 2008. Major bacterial contribution to the ocean reservoir of detrital organic carbon and nitrogen. Limnol. Oceanogr. 53, 99–112.

Kaiser, K., Benner, R., 2009. Biochemical composition and size distribution of organic matter at the Pacific and Atlantic time-series stations. Mar. Chem. 113, 63–77.

Kaiser, K., Benner, R., 2012. Organic matter transformations in the upper mesopelagic zone of the North Pacific: chemical composition and linkages to microbial community structure. J. Geophys. Res. Oceans 117, 1–12, C01023.

Karl, M., Leck, C., Coz, E., Heintzenberg, J., 2013. Marine nanogels as a source of atmospheric nanoparticles in the high Arctic. Geophys. Res. Lett. 40, 3738–3743.

Keene, W.C., Maring, H., Maben, J.R., Kieber, D.J., Pszenny, A.A.P., Dahl, E.E., et al., 2007. Chemical and physical characteristics of nascent aerosols produced by bursting bubbles at a model air-sea interface. J. Geophys. Res. 112, 1–16, D21202.

Kiørboe, T., Jackson, G.A., 2001. Marine snow, organic solute plumes and optimal chemosensory behavior of bacteria. Limnol. Oceanogr. 46, 1309–1318.

Koike, I., Hara, S.I., Terauchi, K., Kogure, K., 1990. Role of sub-micrometre particles in the ocean. Nature 345, 242–244.

Kolodkin-Gal, I., Romero, D., Cao, S., Clardy, J., Kolter, R., Losick, R., 2010. D-amino acids trigger biofilm disassembly. Science 328, 627–629.

Kujawinski, E.B., Longnecker, K., Blough, N.V., Vecchio, R.D., Finlay, L., Kitner, J.B., et al., 2009. Identification of possible source markers in marine dissolved organic matter using ultrahigh resolution mass spectrometry. Geochim. Cosmochim. Acta 73, 4384–4399.

Kumar, M.D., Sarma, V.V.S.S., Ramaiah, N., Gauns, M., de Sousa, S.N., 1998. Biogeochemical significance of transport exopolymer particles in the Indian Ocean. Geophys. Res. Lett. 25, 81–84.

Kuznetsova, M., Lee, C., Aller, J., 2005. Characterization of the proteinaceous matter in marine aerosols. Mar. Chem. 96, 359–377.

Kwon, E.Y., Primeau, F., Sarmiento, J.L., 2009. The impact of remineralization depth on the air-sea carbon balance. Nat. Geosci. 2, 630–635.

Laybourn-Parry, J., 1984. Protozoan Plankton Ecology. Chapman and Hall, London.

Lechtenfeld, O.J., Kattner, G., Flerus, R., McCallister, S.L., Schmitt-Kopplin, P., Koch, B.P., 2014. Molecular transformation and degradation of refractory dissolved organic matter in the Atlantic and Southern Ocean. Geochim. Cosmochim. Acta 126, 321–337.

Leck, C., Bigg, E., 1999. Aerosol production over remote marine areas—a new route. Geophys. Res. Lett. 23, 23. http://dx.doi.org/10.1029/1999GL010807.

Leck, C., Bigg, E.K., 2005a. Biogenic particles in the surface microlayer and overlaying atmosphere in the central Arctic Ocean during summer. Tellus B 57, 305–316.

Leck, C., Bigg, E.K., 2005b. Source and evolution of the marine aerosol—a new perspective. Geophys. Res. Lett. 32, L19803.

Leck, C., Bigg, E.K., 2007. A modified aerosol–cloud–climate feedback hypothesis. Envir. Chem. 4, 400–403.

Leck, C., Bigg, E.K., 2008. Comparison of sources and nature of the tropical aerosol with the summer high Arctic aerosol. Tellus B 60, 118–126.

Leck, C., Bigg, E.K., 2010. New particle formation of marine biological origin. Aerosol Sci. Tech. 44, 570–577.

Leck, C., Persson, C., 1996. The central Arctic as a source of dimethyl sulfide-Seasonal variability in relation to biological activity. Tellus B 48, 156–177.

Leck, C., Norman, M., Bigg, E.K., Hillamo, R., 2002. Chemical composition and sources of the high Arctic aerosol relevant for cloud formation. J. Geophys. Res. 107, 4135.

Leck, C., Tjernström, M., Matrai, P., Swietlicki, E., 2004. Microbes, clouds and climate: Can marine microorganisms influence the melting of the Arctic pack ice? EOS Transactions, 85, 25–36.

Leck, C., Gao, G., Rad, M.F., Nilsson, U., 2013. Size resolved airborne particulate polysaccharides in summer high. Atmos. Chem. Phys. 13, 9801–9847.

Lee, H., de Vries, A.H., Marrink, S.-J., Pastor, R.W., 2009. A coarse-grained model for polyethylene oxide and polyethylene glycol: conformation and hydrodynamics. J. Phys. Chem. B 113, 13186–13194.

Lewis, E.R., Schwartz, S.E. (Eds.), 2004. Sea Salt Aerosol Production: Mechanisms, Methods, Measurements, and Models. AGU, Washington, DC.

Logan, B.E., Passow, U., Alldredge, A.L., Grossartt, H.-P., Simont, M., 1995. Rapid formation and sedimentation of large aggregates is predictable from coagulation rates (half-lives) of transparent exopolymer particles (TEP). Deep-Sea Res. Pt. II 42, 203–214.

Long, R.A., Azam, F., 1996. Abundant protein-containing particles in the sea. Aquat. Microb. Ecol. 10, 213–221.

Maitra, U., Mukhpadhyay, S., Sarkar, A., Rao, P., Indi, S.S., 2001. Hydrophobic pockets in a nonpolymeric aqueous gel: observation of such a gelation process by color change. Angew. Chem. Int. Ed. 40, 2281–2283.

Malfatti, F., Azam, F., 2009. Atomic force microscopy reveals microscale networks and possible symbioses among pelagic marine bacteria. Aquat. Microb. Ecol. 58, 1–14.

Mamada, A., Tanaka, T., Kungwatchakun, D., Irie, M., 1990. Photoinduced phase transition of gels. Macromolecules 23, 1517–1519.

Mari, X., 1999. Carbon content and C:N ratio of transparent exopolymeric particles (TEP) produced by bubbling exudates of diatoms. Mar. Ecol. Prog. Ser. 183, 59–71.

Mari, X., Beauvais, S., Lemee, R., Pedrotti, M.L., 2001. Non-Redfield C:N ratio of transparent exopolymeric particles in the northwestern Mediterranean Sea. Limnol. Oceanogr. 46, 1831–1836.

Martin, J.H., Fitzwater, S.E., Gordon, R.M., 1990. Iron deficiency limits phytoplankton growth in Antarctic waters. Global Biogeochem. Cycles 4, 5–12.

Martin, J.H., Coale, K.H., Johnson, K.S., Fitzwater, S.E., Gordon, R.M., Tanner, S.J., et al., 1994. Testing the iron hypothesis in ecosystems of the equatorial Pacific Ocean. Nature 371, 123–129.

Martin, M., Chang, R.Y.W., Sierau, B., Sjogren, S., Swietlicki, E., Abbatt, J.P.D., et al., 2011a. Cloud condensation nuclei closure study on summer arctic aerosol. Atmos. Chem. Phys. 11, 11335–11350.

Martin, P., Lampitt, R.S., Perry, M.J., Sanders, R., Lee, C., D'Asaro, E., 2011b. Export and mesopelagic particle flux during a North Atlantic spring diatom bloom. Deep-Sea Res. Pt. I 58, 338–349.

Martinez, J.S., Carter-Franklin, J.N., Mann, E.L., Martin, J.D., Haygood, M.G., Butler, A., 2003. Structure and membrane affinity of a suite of amphiphilic siderophores produced by a marine bacterium. Proc. Natl. Acad. Sci. 100, 3754–3759.

Matrai, P.A., Vernet, M., 1997. Dynamics of the vernal bloom in the marginal ice zone of the Barents Sea: dimethyl sulfide and dimethylsulfoniopropionate budgets. J. Geophys. Res. 102, 22965–22979.

McCarren, J., Becker, J.W., Repeta, D.J., Shi, Y., Young, C.R., Malmstrom, R.R., et al., 2010. Microbial community transcriptomes reveal microbes and metabolic pathways associated with dissolved organic matter turnover in the sea. Proc. Natl. Acad. Sci. 107, 16420–16427.

McCarthy, M.D., Hedges, J.I., Benner, R., 1993. The chemical composition of dissolved organic matter in seawater. Chem. Geol. 107, 503–507.

McCarthy, M.D., Hedges, J.I., Benner, R., 1998. Major bacterial contribution to marine dissolved organic nitrogen. Science 281, 231–234.

McCarthy, M.D., Benner, R., Lee, C., Hedges, J.I., Fogel, M.L., 2004. Amino acid carbon isotopic fractionation patterns in oceanic dissolved organic matter: an unaltered photoautotrophic source for dissolved organic nitrogen in the ocean? Mar. Chem. 92, 123–134.

Menyo, M.S., Hawker, C.J., Waite, J.H., 2013. Versatile tuning of supramolecular hydrogels through metal complexation of oxidation-resistant catechol-inspired ligands. Soft Matter 9, 10314–10323.

Middlebrook, A.M., Murphy, D.M., Thomson, D.S., 1998. Observations of organic material in individual marine particles at Cape Grim during the First Aerosol Characterization Experiment (ACE 1). J. Geophys. Res. 103, 16475–16483.

Moon, A.A.O., Ng, C., Turhill, J., Dimitrijeva, J., Verdugo, P., 2007. Bacterial colonization on marine gel. In: ASLO Aquatic Sci Meeting, Feb 4-9, Santa Fe, NW.

Moore, J.K., Doney, S.C., 2007. Iron availability limits the ocean nitrogen inventory stabilizing feedbacks between marine denitrification and nitrogen fixation. Global Biogeochem. Cycles 21, 1–12, GB2001.

Moore, J.K., Doney, S.C., Glover, D.M., Fung, I.Y., 2002. Iron cycling and nutrient-limitation patterns in surface waters of the World Ocean. Deep-Sea Res. Pt. II 49, 463–507.

Mopper, K., Kieber, D.J., 2002. Photochemistry and the cycling of carbon, sulfur, nitrogen and phosphorus. In: Hansell, D.A., Carlson, C.A. (Eds.), Biogeochemistry of Dissolved Organic Matter. Academic Press, San Diego, pp. 455–507.

Mopper, K., Zhou, J., Sri Ramana, K., Passow, U., Dam, H.G., Drapeau, D.T., 1995. The role of surface-active carbohydrates in the flocculation of a diatom bloom in a mesocosm. Deep-Sea Res. Pt. II 42, 47–73.

Moran, S.B., Shen, C.-C., Edmonds, H.N., Weinstein, S.E., Smith, J.N., Edwards, R.L., 2002. Dissolved and particulate Pa-231 and Th-230 in the Atlantic Ocean: constraints on intermediate/deepwater age, boundary scavenging, and Pa-231/Th-230 fractionation. Earth Planet. Sci. Lett. 203, 999–1014.

Moran, M.A., Belas, R., Schell, M.A., González, J.M., Sun, F., Sun, S., et al., 2007. Ecological genomics of marine Roseobacters. Appl. Environ. Microbiol. 73, 4559–4569.

Murray, J.W., Downs, J., Strom, S., Wei, C.-L., Jannasch, H., 1989. Nutrient assimilation, export production and ^{234}Th scavenging in the Eastern Equatorial Pacific. Deep-Sea Res. 36, 1471–1489.

Nagata, T., Kirchman, D., 1997. Role of submicron particles and colloids in microbial food webs and biogeochemical cycles within marine environments. Adv. Microb. Ecol. 15, 81–103.

Nagata, T., Tamburini, C., Arístegui, J., Baltar, F., Bochdansky, A.B., Fonda-Umani, S., et al., 2010. Emerging concepts on microbial processes in the bathypelagic ocean - ecology, biogeochemistry, and genomics. Deep-Sea Res. Pt. II 57, 1519–1536.

Nishibori, N., Matuyama, Y., Uchida, T., Moriyama, T., Ogita, Y., Oda, M., et al., 2003. Spatial and temporal variations in free polyamine distributions in Uranouchi Inlet, Japan. Mar. Chem. 82, 307–314.

Norris, S.J., Brooks, I.M., de Leeuw, G., Sirevaag, A., Leck, C., Brooks, B.J., et al., 2011. Measurements of bubble size spectra within leads in the Arctic summer pack ice. Ocean Sci. 7, 129–139.

O'Dowd, C.D., Lowe, J.A., Smith, M.H., Kaye, A.D., 1999. The relative importance of non-sea-salt sulphate and sea-salt aerosol to the marine cloud condensation nuclei population: an improved multi-component aerosol-cloud droplet parametrization. Q. J. Roy. Meteorol. Soc. 125, 1295–1313.

O'Dowd, C.D., Facchini, M.C., Cavalli, F., Ceburnis, D., Mircea, M., Decesari, S., et al., 2004. Biogenically driven organic contribution to marine aerosol. Nature 431, 676–680.

Ogawa, H., Amagai, Y., Koike, I., Kaiser, K., Benner, R., 2001. Production of refractory dissolved organic matter by bacteria. Science 292, 917–920.

Ohmine, I., Tanaka, T., 1982. Salt effects on phase transition of ionic gels. J. Chem. Phys. 77, 5725–5729.

Okajima, M.K., Nguyen, Q.T.l., Tateyama, S., Masuyama, H., Tanaka, T., Mitsumata, T., et al., 2012. Photoshrinkage in polysaccharide gels with trivalent metal ions. Biomacromolecules 13, 4158–4163.

Orellana, M.V., Hansell, D., 2012. Ribulose-1,5-bisphosphate carboxylase/oxygenase (RuBisCO): a long-lived protein in the deep ocean. Limnol. Oceanogr. 57, 826–834.

Orellana, M.V., Verdugo, P., 2003. Ultraviolet radiation blocks the organic carbon exchange between the dissolved phase and the gel phase in the ocean. Limnol. Oceanogr. 48, 1618–1623.

Orellana, M.V., Vetter, Y.A., Verdugo, P., 2000. The assembly of DOM polymers into POM microgels enhances their suceptibility to bacterial degradation. In: Aquat Sci Meeting, Jan 24-28, San Antonio, Texas (Am. Geophysical. Union.& Am Soc.Limnol.Oceanogr.).

Orellana, M.V., Petersen, T.W., Diercks, A.H., Donohoe, S., Verdugo, P., van den Engh, G., 2007. Marine microgels: optical and proteomic fingerprints. Mar. Chem. 105, 229–239.

Orellana, M.V., Matrai, P.A., Janer, M., Rauschenberg, C., 2011a. DMSP storage in *Phaeocystis* secretory vesicles. J. Phycol. 47, 112–117.

Orellana, M.V., Matrai, P.A., Leck, C., Rauschenberg, C.D., Lee, A.M., Coz, E., 2011b. Marine microgels as a source of cloud condensation nuclei in the high Arctic. Proc. Natl. Acad. Sci. 108, 13612–13617.

Orellana, M.V., Pang, W.L., Durand, P.M., Whitehead, K., Baliga, N.S., 2013. A role for programmed cell death in the microbial loop. PLoS One 8, e62595.

Ovadnevaite, J., Ceburnis, D., Martucci, G., Bialek, J., Monahan, C., Rinaldi, M., et al., 2011. Primary marine organic aerosol: a dichotomy of low hygroscopicity and high CCN activity. Geophys. Res. Lett. 38, http://dx.doi.org/10.1029/2011GL048869.

Pace, M., Reche, I., Cole, J., Fernández-Barbero, A., Mazuecos, I., Prairie, Y., 2012. pH change induces shifts in the size and light absorption of dissolved organic matter. Biogeochemistry 108, 109–118.

Paerl, H.W., 1973. Bacterial uptake of dissolved organic matter in relation to detrital aggregation in marine and freshwater systems. Limnol. Oceanogr. 19, 966–972.

Palenik, B., Ren, Q., Dupont, C.L., Myers, G.S., Heidelberg, J.F., Badger, J.H., 2006. Genome sequence of *Synechococcus* CC9311: insights into adaptation to a coastal environment. Proc. Natl. Acad. Sci. 103, 13555–13559.

Passow, U., 2002a. Production of transparent exopolymer particles (TEP) by phyto-and bacterioplankton. Mar. Ecol. Prog. 236, 1–12.

Passow, U., 2002b. Transparent exopolymer particles (TEP) in the marine environment. Prog. Oceanogr. 55, 287–333.

Passow, U., Shipe, R.F., Murray, A., Pak, D.K., Brzezinski, M.A., Alldredge, A.L., 2001. The origin of transparent exopolymer particles (TEP) and their role in the sedimentation of particulate matter. Cont. Shelf Res. 21, 327–346.

Payne, J.W., Smith, M.W., 1994. Peptide transport by micro-organisms. Adv. Microb. Physiol. 36, 1–80.

Pollack, G.H., 2001. Cell, Gels and the Engines of Life: A New Unifying Approach to Cell Function. first ed. Ebner and Sons Publishers, Korea.

Pond, D.W., Harris, R.P., Brownlee, C., 1995. A microinjection technique using a pH-sensitive dye to determine the gut pH of *Calanus helgolandicus*. Mar. Biol. 123, 75–79.

Poon, W.C.K., Andelman, D., 2006. In: first ed.. Soft Matter Physics in Molecular and Cell Biology, vol. 59. Taylor & Francis, Boca Raton, FL.

Popendorf, K.J., Lomas, M.W., Van Mooy, B.A.S., 2011. Microbial sources of intact polar diacylglycerolipids in the western North Atlantic Ocean. Org. Geochem. 42, 803–811.

Pósfai, M., Simonics, R., Li, J., Hobbs, P.V., Buseck, P.R., 2003. Individual aerosol particles from biomass burning in southern Africa: 1. Compositions and size distributions of carbonaceous particles. J. Geophys. Res. 108, 1–13, 8483.

Powell, M.J., Sutton, J.N., Del Castillo, C.E., Timperman, A.T., 2005. Marine proteomics: generation of sequence tags for dissolved proteins in seawater using tandem mass spectrometry. Mar. Chem. 95, 183–198.

Quinn, P.K., Bates, T.S., 2011. The case against climate regulation via oceanic phytoplankton sulphur emissions. Nature 480, 51–56.

Radić, T.M., Svetličić, V., Žutić, V., Boulgaropoulos, B., 2011. Seawater at the nanoscale: marine gel imaged by atomic force microscopy. J. Mol. Recognit. 24, 397–405.

Riley, G.A., 1963. Organic aggregates in seawater and the dynamics of their formation and utilization. Limnol. Oceanogr. 8, 372–381.

Rinaldi, M., Decesari, S., Finessi, E., Giulianelli, L., Carbone, C., Fuzzi, S., et al., 2010. Primary and secondary organic marine aerosol and oceanic biological activity: recent results and new perspectives for future studies. Adv. Meteor. 2010, 10.

Rosen, S.L., 1993. Fundamental Principles of Polymeric Materials. second ed. John Wiley and Sons, New York.

Russell, L.M., Hawkins, L.N., Frossard, A.A., Quinn, P.K., Bates, T.S., 2010. Carbohydrate-like composition of submicron atmospheric particles and their production from ocean bubble bursting. Proc. Natl. Acad. Sci. 107, 6652–6657.

Santschi, P.H., Guo, L., Baskaran, M., Trumbore, S., Southon, J., Bianchi, T., et al., 1995. Isotopic evidence for the contemporary origin of high-molecular weight organic matter in oceanic environments. Geochim. Cosmochim. Acta 59, 625–631.

Santschi, P.H., Guo, L., Walsh, I.D., Quigley, M.S., Baskaran, M., 1999. Boundary exchange and scavenging of radionuclides in continental margin waters of the Middle Atlantic Bight. Implications for organic carbon fluxes. Cont. Shelf Res. 19, 609–636.

Sarmiento, J.L., Gruber, N., 2006. Ocean Biogeochemical Dynamics. first ed. Princeton University Press, Princeton.

Schmalz, R.F., Swanson, F.J., 1969. Diurnal variations in the carbonate saturation of seawater. J. Sediment. Res. 39, 255–267.

Sciare, J., Favez, O., Sarda-Esteve, R., Oikonomou, K., Kazan, V., 2009. Long-term observations of carbonaceous aerosols in the Austral Ocean atmosphere: evidence of a biogenic marine organic source. J. Geophys. Res. 114, D15302.

Sheldon, R.W., Evelyn, T.P.T., Parsons, T.R., 1967. On the occurrence and formation of small particles in seawater. Limnol. Oceanogr. 12, 367–375.

Sieburth, J.M., 1983. Microbiological and Organic-Chemical Processes in the Surface and Mixed Layers. Reidel Publishers Co., Hingham, MA.

Simon, M., Grossart, H.-P., Schweitzer, B., Ploug, H., 2002. Microbial ecology of organic aggregates in aquatic eco-system. Aquat. Microb. Ecol. 28 (175–211), 175–211.

Skubatz, H., Orellana, M.V., Howald, W.N., 2013. A NAD (P) reductase like protein is the salicylic acid receptor in the appendix of the *Sauromatum guttatum* inflorescence. Int. Disord. Proteins 1, e26372.

Smetacek, V., 2000. The giant diatom dump. Nature 406, 574–575.

Smith, D.C., Simon, M., Alldredge, A.L., Azam, F., 1992. Intense hydrolytic enzyme activity on marine aggregates and implications for rapid particle dissolution. Nature 359, 139–142.

Stoderegger, K.E., Herndl, G.J., 2004. Dynamics in bacterial cell surface properties assessed by fluorescent stains and confocal laser scanning microscopy. Aquat. Microb. Ecol. 36, 29–40.

Strom, S.L., 2008. Microbial ecology of ocean biogeochem-istry: a community perspective. Science 320, 1043–1045.

Strom, S.L., Benner, Ronald, Ziegler, S., Dagg, M.J., 1997. Planktonic grazers are a potentially important source of marine dissolved organic carbon. Limnol. Oceangr. 42, 1364–1374.

Suttle, C.A., 2007. Marine viruses—major players in the global ecosystem. Nat. Rev. Microbiol. 5, 801–812.

Svetličić, V., Žutic, V., Zimmermann, A.H., 2005. Biophysical scenario of giant gel formation in the Northern Adriatic Sea. Ann. N. Y. Acad. Sci. 1048, 524–527.

Svetličić, V., Žutić, V., Urbani, R., 2011. Polymer networks produced by marine diatoms in the Northern Adriatic Sea. Mar. Drugs 9, 666–679.

Tanaka, T., 1981. Gels. Sci. Am. 244, 124–138.

Tanaka, T., 1992. Phase transitions of gels. In: Polyelectrolyte Gels. American Chemical Society, Washington, DC, pp. 1–21.

Tanaka, T., Fillmore, D., Sun, S.-T., Nishio, I., Swislow, G., Shah, A., 1980. Phase transitions in ionic gels. Phys. Rev. Lett. 45, 1636–1639.

Tanoue, E., Nishiyama, S., Kamo, M., Tsugita, A., 1995. Bacterial membranes: possible source of dissolved protein in seawater. Geochim. Cosmochim. Acta 59, 2643–2648.

Tanoue, E., Ischii, M., Midorikawa, T., 1996. Discrete dis-solved and particulate proteins in oceanic waters. Limnol. Oceanogr. 41, 1334–1343.

Vandenbroucke, M., Largeau, C., 2007. Kerogen origin, evo-lution and structure. Org. Geochem. 38, 719–833.

Vardi, A., Van Mooy, B.A.S., Fredricks, H.F., Popendorf, K.J., Ossolinski, J.E., Haramaty, L., et al., 2009. Viral glyco-sphingolipids induce lytic infection and cell death in ma-rine phytoplankton. Science 326, 861–865.

Vardi, A., Haramaty, L., Van Mooy, B.A.S., Fredricks, H.F., Kimmance, S.A., Larsen, A., et al., 2012. Host–virus

dynamics and subcellular controls of cell fate in a natural coccolithophore population. Proc. Natl. Acad. Sci. 109, 19327–19332.

Verdugo, P., 1994. Polymer gel phase transition in condension-decondensation of secretory products. Adv. Polymer. Sci. 110, 145–156.

Verdugo, P., 2012. Marine microgels. Ann. Rev. Mar. Sci. 4, 9.1–9.25.

Verdugo, P., Santschi, P.H., 2010. Polymer dynamics of DOC networks and gel formation in seawater. Deep-Sea Res. Pt. II 57, 1486–1493.

Verdugo, P., Orellana, M.V., Freitag, C., 1995. The secretory granule as a biomimetic model for drug delivery. In: Proc 22nd Int Symp on Controlled Release of Bioactive Materials 22. pp. 25.

Verdugo, P.A.L., Alldredge, F., Azam, D.L., Kirchman, P.U., Santschi, P., 2004. The oceanic gel phase: a bridge in the DOM–POM continuum. Mar. Chem. 92, 67–85.

Vernet, M., Matrai, P.A., Andreassen, I., 1998. Synthesis of particulate and extracellular carbon by phytoplankton at the marginal ice zone in the Barents Sea. J. Geophys. Res. 103, 1023–1037.

Wakeham, S.G., Pease, T.K., Benner, R., 2003. Hydroxy fatty acids in marine dissolved organic matter as indicators of bacterial membrane material. Org. Geochem. 34, 857–868.

Walker, B.D., McCarthy, M.D., Fisher, A.T., Guilderson, T.P., 2008. Dissolved inorganic carbon isotopic composition of low-temperature axial and ridge-flank hydrother-mal fluids of the Juan de Fuca Ridge. Mar. Chem. 108, 123–136.

Walker, B.D., Guilderson, T.P., Okimura, K.M., Peacock, M.B., McCarthy, M.D., 2014. Radiocarbon signatures and size–age–composition relationships of major organic matter pools within a unique California upwelling sys-tem. Geochim. Cosmochim. Acta 126, 1–17.

Weiss, M., Abele, U., Weckesser, J., Welte, W., Schiltz, E., Schulz, G.E., 1991. Molecular architecture and elec-trostatic propertiesof a bacterial porin. Science 254, 1627–1630.

Wells, M.L., 1998. Marine colloids: a neglected dimension. Nature 391, 530–531.

Wells, M.L., 2002. Marine colloids in seawater. In: Hansell, D., Carlson, C. (Eds.), Biogeochemistry of Marine Dissolved Organic Matter. Academic Press, San Diego, CA, pp. 367–404.

Wells, M.L., Goldberg, E.D., 1991. Occurrence of small col-loids in seawater. Nature 353, 342–344.

Wells, M.L., Goldberg, E., 1992. Occurrence of small colloids in seawater. Nature 353, 342–344.

Wells, M.L., Goldberg, E.D., 1993. Colloid aggregation in seawater. Mar. Chem. 41, 353–358.

Wells, M.L., Goldberg, E.D., 1994. The distribution of colloids in the North Atlantic and Southern Oceans. Limnol. Oceanogr. 39, 286–302.

Wetz, M.S., Wheeler, A.P., 2007. Release of dissolved organic matter by coastal diatoms. Limnol. Oceanogr. 52, 798–807.

Wheeler, J.R., 1975. Formation and collapse of surface films. Limnol. Oceanogr. 20, 338–3342.

Wingender, D.J., Neu, T.R., Fleming, H.C., 1999. Microbial Extracellular Substances. Springer-Verlag, Berlin.

Wurl, O., Holmes, M., 2008. The gelatinous nature of the sea-surface microlayer. Mar. Chem. 110, 89–97.

Wurl, O., Miller, L., Röttgers, R., Vagle, S., 2009. The distribution and fate of surface-active substances in the sea-surface microlayer and water column. Mar. Chem. 115, 1–9.

Wurl, O., Miller, L., Vagle, S., 2011. Production and fate of transparent exopolymer particles in the ocean. J. Geophys. Res. 116, C00H13.

Xin Li, Leck, C., Sun, L., Hede, T., Tu, Y., Ågren, H., 2013. Cross-Linked Polysaccharide Assemblies in Marine Gels: An Atomistic Simulation, Journal of Physical Chemistry Letters, 4, 2637–2642. http://dx.doi.org/10.1021/jz401276r.

Yoon, Y.J., Ceburnis, D., Cavalli, F., Jourdan, O., Putaud, J.P., Facchini, M.C., et al., 2007. Seasonal characteristics of the physicochemical properties of North Atlantic marine atmospheric aerosols. J. Geophys. Res. 112, 1–14, D04206.

Zhou, J., Mopper, K., Passow, U., 1998. The role of surface-active carbohydrates in the formation of transparent exopolymer particles by bubble adsorption of seawater. Limnol. Oceanogr. 43, 1860–1871.

CHAPTER

10

The Optical Properties of DOM in the Ocean

Colin A. Stedmon[*], *Norman B. Nelson*[†]

[*]National Institute of Aquatic Resources, Technical University of Denmark,
Charlottenlund, Denmark
[†]Earth Research Institute, University of California Santa Barbara, California, USA

CONTENTS

I INTRODUCTION

In all natural waters, a portion of the dissolved organic compounds is colored, or chromophoric, resulting in the term chromophoric (or colored) dissolved organic matter (CDOM). At high concentrations, CDOM causes a yellow or brown water color, so in early research (beginning of the 1900s) it was termed "yellow substance" ("Gelbstoff" in German) (Kalle, 1938). It was later

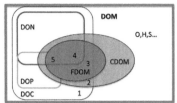

These approaches are also applicable to the hydrography of the Arctic Ocean, which receives considerable amounts of terrestrial CDOM from rivers (Granskog et al., 2012; see Chapter 14).

In estuarine and coastal waters in particular, high concentrations of CDOM have a considerable influence on water color, light penetration, and spectral quality. Even in the open ocean, CDOM can dominate the absorption spectra in the blue and ultraviolet wavebands (Figure 10.2),

FIGURE 10.1 Schematic of the dissolved organic matter (DOM) pool and how it is primarily composed of carbon, oxygen, hydrogen, nitrogen, phosphorus, and sulfur. The boxes represent different subsets of compounds (numbered 1-4) that together make up DOM. Some compounds present are chromophores (i.e., absorb light, CDOM, 2) and a fraction of these fluoresce (3). Examples of types of organic compounds that may exist in each of the subsets are listed on the right.

found that a fraction of CDOM also exhibited a blue fluorescence (Duursma, 1965; Kalle, 1966), referred to here as fluorescent DOM (FDOM). The linkages between DOM, CDOM, and FDOM are depicted in Figure 10.1. The absorbance and fluorescence properties of DOM are "optical markers" comparable to traditional biomarkers used in geochemistry (e.g., lignin). However, while biomarkers represent specific chemical compounds that can be linked to a specific source (synthesis process), optical signatures of DOM are supposedly the product of a complex mixture of compounds. The word "supposedly" is used here as little is currently known about the responsible chromophores and fluorophores.

Several fields of marine research have fueled the study of CDOM during the last 50 years. On the whole, these can be grouped into three categories: hydrography, optics, and biogeochemistry. Due to the leaching of organic matter from soils, rivers often have high concentrations of CDOM, such that estuarine and coastal mixing can be traced by combining CDOM and salinity measurements (Laane and Kramer, 1990). This use is particularly valuable when there is more than one source of freshwater but with differing CDOM concentrations. If the end members are adequately constrained, mixing equations can be solved and water samples fractionated into the respective contributions from each source (Granskog et al., 2007; Højerslev et al., 1996; Stedmon et al., 2010).

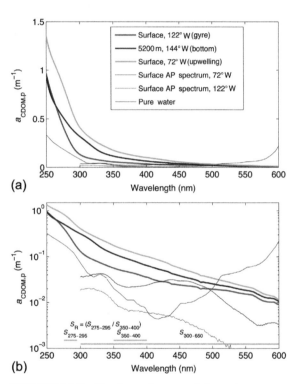

FIGURE 10.2 The dynamic range of absorption spectra in the ocean from a section along 32°S in the Pacific Ocean. (a) Surface (5 m) CDOM absorption spectrum from the coastal upwelling zone off Chile (green) contrasted with a surface (8 m) spectrum from the subtropical gyre (red) and a deep-ocean (5200 m) sample from the central South Pacific. (b) The same spectra, in a semi-logarithmic plot. Horizontal gray lines show parts of the spectral range used to compute slope parameters and the slope ratio, S_R. Also shown for comparison are surface particulate absorption spectra (thin black and thin purple lines) from the gyre and upwelling stations and the absorption spectrum of pure water (from Smith and Baker, 1981; Pope and Fry, 1997) (thin blue line).

indicating that variability in CDOM abundance is the primary factor controlling the penetration of light energy into the oceans (Siegel et al., 2002, 2005). This control has ecological implications (Urtizberea et al., 2013), as CDOM can limit light available for photosynthesis (Arrigo and Brown, 1996) by absorbing light in the blue wavebands where chlorophylls and photosynthetic carotenoids have absorption peaks (Figure 10.2). Conversely, CDOM absorbs potentially harmful UVB (280-320 nm) or UVA (320-400 nm) radiation, limiting the negative impacts of high UV exposure on plankton populations (Arrigo and Brown, 1996). Because of its influence on aquatic light absorption, CDOM is also an important component of the visible wavelength remote sensing signal collected by satellites. Proper quantification and compensation for CDOM are necessary preconditions for estimating chlorophyll concentrations from space (Siegel et al., 2005).

Finally, CDOM research has also been driven by marine biogeochemical studies. As a result of its UV light-absorbing properties, CDOM takes part in a suite of photochemical processes contributing to the direct remineralization of organic carbon (Miller and Zepp, 1995; Mopper et al., 1991), photolysis of dimethyl sulfide (DMS) (Toole et al., 2003; see Chapter 8), and production of bioavailable DOM (Kieber et al., 1989). The properties of CDOM and FDOM have also been used as an indicator for quantitative and qualitative changes occurring to marine DOM as a whole. These comparatively rapid and simple optical measurements are well suited to intensive sampling programs and development of in situ sensing instrumentation. These measurements potentially offer insight into the dynamics of quantitative and qualitative changes in DOM at a much higher temporal and spatial resolution than other chemical techniques. In a recent review, Hansell indicated advances in the current understanding of the role of DOM in the global carbon cycle are hampered by "the inability to quantify the production of specific fractions or to detect their presence in the water column via specific chemical markers" (Hansell,

2013). Measurements of the optical properties of DOM are potentially capable of providing a means to resolve this issue; however, a better understanding of what drives the changes in these signatures is required, a topic currently the basis of much research across a wide range of aquatic systems, both natural and industrial.

The first edition of this book included two chapters extensively covering the coastal and open ocean dynamics of CDOM (Blough and Del Vecchio, 2002; Nelson and Siegel, 2002), comprehensively reviewing the progress to date. Since then, the number of publications on DOM absorption and fluorescence has increased substantially from ~25 publications per year to 110 publications annually in 2009-2011. A comprehensive review of the 700+ studies published since 2002 would be difficult; our aim here is to present an introduction, general overview, and synthesis of the current status with emphasis on introducing readers new to the field. Other important reviews of marine CDOM research that are useful and recommended for additional reading are Coble (2007) and Nelson and Siegel (2013).

II UV-VISIBLE SPECTROSCOPY OF DOM

In the short wavelength region of the electromagnetic spectrum, photons have enough energy to promote bond electrons in organic molecules to higher energy levels. These wavelengths correspond to the ultra violet and visible region of the electromagnetic spectrum (UV-Vis), 200-400 and 400-800 nm, respectively. The part of an organic compound capable of absorbing light (>200 nm, Braslavsky, 2007) is called a chromophore (from the Greek "bearing color").

A Absorption Properties of CDOM

Light absorption by a chromophore is characterized by its intensity and nature. The intensity is a product of the molar absorptivity (also known as extinction coefficient) and concentration of the chromophore, according to the Beer-Lambert law. The nature of the light absorption (the wavelengths across which it occurs) depends on the electronic transitions involved. For the UV-Vis region, these transitions primarily involve lone pairs or π-electrons. Increasing conjugation or the presence of electron donors on aromatic rings increases both the wavelength of absorption and the molar absorptivity of the chromophore. As a result, the absorption spectra of relatively simple aromatic compounds are limited to wavelengths <300 nm. As the degree of conjugation and substitution increases, the wavelength of maximum absorption increases and the spectra

broaden. An example is seen in Figure 10.3, where the molar absorptivity of three aromatic acids are plotted. Syringic acid has an additional "–OR" group substitution compared to vanillic acid and as a result one would predict from the Woodward-Fieser rule a 5-10-nm-higher wavelength of maximum absorbance, as observed (Figure 10.3). Note that the molar absorptivities of these compounds are, however, comparable. The effect on increasing conjugation can be seen by comparing the vanillic acid and ferulic acid spectra, where there is an increase in both the wavelength of maximum absorbance and the molar absorptivity (Figure 10.3).

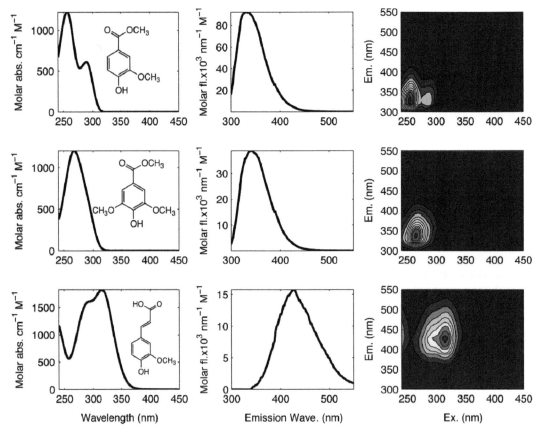

FIGURE 10.3 The UV-visible spectral properties of vanillic (top), syringic (middle), and ferulic (bottom) acid. The left panel shows the molar absorptivity, middle panel molar fluorescence (vanillic acid, $Ex = 260$; syringic acid, $Ex = 275$; ferulic acid, $Ex = 320$ nm), and right panel the fluorescence excitation and emission matrix.

The absorption spectrum of CDOM is comparatively featureless (Figure 10.2). Across much of the UV and visible spectra, the majority of the measured absorbance in a filtered seawater sample is due to CDOM and water (if the latter is not already subtracted during measurement). At wavelengths <260 nm, inorganic constituents of seawater also absorb (e.g., nitrate, nitrite, and sulfide) (see Figure 10.1 in Johnson and Coletti (2002)). In oxic waters (devoid of sulfide), however, most absorption by dissolved material at wavelengths >240 nm is due to CDOM. In CDOM spectra, there is often an exponential decrease in absorbance from 240 to 600 nm (Figure 10.2). Between 260 and 270 nm, there is often a slight shoulder in the absorption spectrum. From 300 nm upwards, the absorption spectrum very closely follows an exponential function.

There are two explanations for the absorption spectrum character of CDOM. To some extent, the absorption spectrum represents the sum of the overlapping spectra of the chromophores present (so-called linear supposition). This explanation is rather simplistic, inferring that the chromophores are essentially electronically isolated from each other (Del Vecchio and Blough, 2004a). In order to reproduce the featureless shape common to CDOM spectra (Figure 10.2), there would have to be a very complex mixture of chromophores. Adding the absorption spectra of a simple mixture of a series of benzoic acid-based derivatives and amino acids replicates the lower part of the CDOM spectrum (<350 nm), but not the gradual tailing in absorption at longer wavelengths. The presence of more complex conjugated structures such as flavins, melanins, and tannins could explain some of the long wavelength absorption (Seritti et al., 1994). They often have several absorption maxima in the UV-Vis region; in principle, a complex mixture of derivatives could replicate the gradual tail in the CDOM absorption spectrum.

An alternate theory proposes that there are a limited number of chromophores in CDOM and that electronic interactions between them are responsible for the smooth featureless shape of the absorption spectrum at >350 nm (Andrew et al., 2013; Boyle et al., 2009; Del Vecchio and Blough, 2004a). Charge transfer from an electron donor to an acceptor results in additional low-intensity and low-energy (high-wavelength) charge transfer absorption bands that otherwise do not exist in either the donor or acceptor alone. This scenario has been elegantly exemplified in coastal and ocean samples by the addition of a reductant (sodium borohydride), which inhibits charge transfer, resulting in a reduction in the long wavelength absorption (Andrew et al., 2013; Ma et al., 2010).

The absorption properties of CDOM are measured on a spectrophotometer as absorbance (A_λ), a unitless ratio of spectral radiant power transmitted through the sample across the pathlength (l) (Braslavsky, 2007). A_λ is then converted to the (Napierian) absorption coefficient (a_λ, m^{-1}) according to Eq. (10.1):

$$a_\lambda = 2.303 A_\lambda / l \qquad (10.1)$$

Typically, measurements are performed using a long-pathlength quartz cuvette (e.g., 0.1 m) or a liquid core waveguide (0.5-5 m) with submicron-filtered (0.2 or ~0.7 μm) seawater; this measurement includes contributions (both absorption and scattering) from colloids (Floge and Wells, 2007; Zhang et al., 2013), essentially neutrally buoyant microparticles. The colloid contribution varies with distance from coast and with productivity (Floge and Wells, 2007; Stolpe et al., 2010). The colloid fraction consists of organic and inorganic particles, polymeric gels, and biomass (microbes and viruses). Large colloids (0.4-1 μm) have been found to contribute significantly to light attenuation, having scattering coefficients greater than pure water in the visible wavelength range (Stramski and Wozniak, 2005; see Chapter 9).

In low-CDOM waters, measurement of the absorption spectrum of CDOM using a conventional spectrophotometer is challenging. Due to the exponential decline of absorption with increasing wavelength, absorbance values can fall below the detection threshold of the detectors used, typically 0.03-0.06 m^{-1} (Nelson et al., 1998). Devices with longer effective pathlengths, such as reflective tube absorption meters (up to 25 cm, Zaneveld et al., 1992), liquid waveguide cells (up to 200 cm; Miller et al., 2002; Nelson et al., 2007), and integrating cavity meters (up to 25 m; Röttgers and Doerffer, 2007) attempt to resolve this problem. In many cases, the use of these devices is still experimental but procedures are being developed that will allow their routine use for measuring absorption spectra of CDOM in the visible. An overview of the principal methodologies for CDOM absorption (and fluorescence) is available in Nelson and Coble (2012).

The absorption spectrum provides both quantitative and qualitative information on CDOM. The intensity of a_λ at a specific wavelength (λ) is used as an expression of the concentration of CDOM. The shape of the absorption spectrum indicates changes in CDOM composition. The dilution of a concentrated CDOM sample results in a decrease in a_λ but no significant changes to the shape of the absorption spectrum (Stedmon and Markager, 2003). Currently, there are three commonly used approaches to characterizing the shape of the CDOM absorption spectrum. The most common is to fit an exponential model similar to that in Eq. (10.2):

$$a_\lambda = a_{\lambda_0} e^{-S(\lambda - \lambda_0)} \qquad (10.2)$$

For most samples, this model is appropriate within the wavelength range 300-700 nm, especially if fitted with a nonlinear regression routine (Stedmon et al., 2000). S is termed the spectral slope coefficient and is a descriptor of CDOM quality (i.e., characteristics). Values normally range between 5 and 30 μm^{-1} (0.005-0.030 nm^{-1}). The value of S is in particular dependent on the lower bounds of the wavelength range across which the model is fitted, so direct comparisons between S values calculated for different wavelength ranges are not always possible (Blough and Del Vecchio, 2002; Stedmon and Markager, 2003; Stedmon et al., 2000). However, if the lower limit is common (e.g., 300 nm) and a nonlinear fitting routine is used, the effect of different upper

wavelength limits (between 500 and 700 nm) on S values derived will be minor. When comparing across a wide range of contrasting aquatic systems, there is a negative correlation between S and weighted average molecular weight of DOM (Figure 10.4), which suggests that S can be used as a simple proxy for average molecular weight (Floge and Wells, 2007; Green and Blough, 1994; Guéguen and Cuss, 2011).

The second approach for characterizing CDOM absorption spectra was introduced by Helms et al. (2008). They re-emphasized that UVB wavelengths contain considerable information on CDOM, as it is this region where most aromatic compounds have their absorption maxima (see earlier). The S value calculated for a small UVB wavelength range (275-295 nm) and also the ratio of S values from UVB and UVA (350-400 nm), termed the slope ratio, S_R, are correlated with molecular weight and exposure

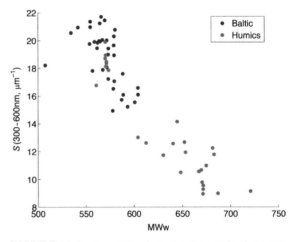

FIGURE 10.4 Correlation between the weighted average molecular weight determined using ESI-QTOF-MS (electrospray ionization quadrapole time-of-flight mass spectrometer) and CDOM spectral slope calculated between 300 and 600 nm. The blue data are from the Baltic Sea (styrene divinyl benzene polymer (Varian PPL) extracts) and the green data are fulvic acid XAD isolates from contrasting environments (including: Pony Lake, Antarctica; Missouri River; Ohio River; Pacific Ocean; Arctic coastal water; Alaskan stream). Samples provided by E. Kritzberg, G. Aiken, D. McKnight and R. Cory. Analyses carried out by Christensen and Nielsen; Stedmon et al. (in prep).

to photochemical degradation (Guéguen and Cuss, 2011; Helms et al., 2008). S_R values for terrestrial CDOM typically are <1 whereas oceanic CDOM and extensively photodegraded terrestrial CDOM are typically >1.5.

The third approach is to calculate either derivative spectra or continuous spectral slopes across, for example, 20-nm wavebands (Loiselle et al., 2009). The advantage of these approaches is that subtle and contrasting differences or responses in CDOM spectra can be resolved. In particular, Loiselle et al. (2009) demonstrated that while photodegradation causes the slopes calculated at UVA wavelengths to increase, the slopes in the visible region decrease. This result supports the trends reported for the S_R by Helms et al. (2008).

An additional qualitative measure of CDOM is the carbon-specific absorption, which is absorption normalized to the DOC, with units of $m^2 g^{-1} C$. (Note: this unit is a simplification of $m^{-1} mg^{-1} C \, L$.) There has been some confusion in the literature as some studies have used the Napierian absorption and others the decadic absorption, resulting in values that differ by a factor of 2.303 (Murphy et al., 2010). The majority of earlier oceanographic studies have normalized the Napierian absorption coefficient to the DOC concentration and termed this the specific absorption coefficient (Blough and Del Vecchio, 2002; Carder et al., 1989; Stedmon and Markager, 2001), which is referred to here as the carbon-specific absorption. In freshwaters and industrial water treatment applications, the specific UV absorbance (*SUVA*) has proved to be a valuable indicator of DOM quality (Chin et al., 1994; Weishaar et al., 2003); it is now often reported in marine studies. It is calculated by normalizing the decadic absorption at 254 nm (A_{254}/l in Eq. 10.1) to the DOC concentration, with units of $m^2 g^{-1} C$. Both these parameters provide a measure of the color intensity of CDOM and are positively correlated to the molecular weight and aromaticity (Chin et al., 1994). As a rule of thumb, *SUVA* should range between 0.5 and $5 m^2 g^{-1} C$, with values >5 indicating possible interference by dissolved iron (Aiken, pers. comm.).

B Fluorescence Properties of FDOM

After the absorption of light (energy), a fraction of the organic compounds in CDOM will emit light as fluorescence. This process returns molecules from electronically excited state to ground state after first losing energy to molecular vibration and internal conversion. Although organic chromophores can have several absorption bands corresponding to the transition from ground state to several excited states, the fluorescence (emission) spectrum of a given fluorophore generally has only one peak. This singularity is due to the fact fluorescence only occurs from transition for the lowest excited state to the ground state. Rigid structures such as aromatic compounds have fewer vibrational degrees of freedom and therefore often exhibit fluorescence. Conversely, in aliphatic compounds relaxation occurs without fluorescence. As with absorption (excitation), substitution and conjugation of aromatic compounds influence a compound's fluorescence spectra. As the size of the aromatic compound increases, the energy difference between ground and excited states decreases, resulting in a longer wavelength of fluorescence. This lengthening can be seen when comparing the emission spectra of two of the compounds presented earlier, vanillic acid and ferulic acid (Figure 10.3), which have maxima that differ in position by ~100 nm.

The fluorescence properties of FDOM are largely limited to excitation wavelengths 240-500 nm and emission wavelengths 300-600 nm (Figure 10.5). A combined plot of the fluorescence properties is referred to as an excitation-emission matrix (EEM). At wavelengths <240 nm, the excitation intensity in standard fluorometers is too low and inner filter effects in standard cuvettes are very high due to the high absorbance by CDOM. These limitations result in a barely detectable fluorescence signal, with substantially lower signal to noise ratio. Before fluorescence measurements can be used to study FDOM dynamics, a series of instrument and sample biases (e.g., spectral correction, inner filter corrections, and intensity calibration) must be carried out.

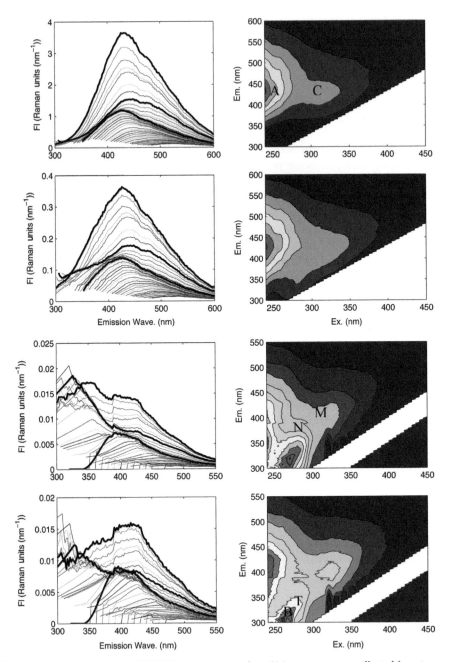

FIGURE 10.5   Fluorescence properties of DOM. Top row: average from 214 measurements collected from two major streams flowing into Horsen estuary in Denmark (Stations 7 and 9, Stedmon and Markager, 2005a). Second row: average from 500 samples collected in Horsen estuary (Stedmon and Markager, 2005a). Third row: average from 500 surface water (0-100 m) ocean samples (Jørgensen et al., 2011). Fourth row average from 500 deep water (100-4000 m) ocean samples (Jørgensen et al., 2011). Left panels: the emission spectra from different excitation wavelengths are plotted on top of each other. The bold black spectra represent the emission spectra from 240, 275, and 320 nm excitation. Right panels: the same data but plotted as a contour plot.

An inter-laboratory comparison and calibration has resulted in a standardized approach to the procedure with a freely available MATLAB toolbox to implement it (Murphy et al., 2010, 2013). In order for FDOM measurements to be a useful optical marker for studying DOM dynamics tracer, this standard approach has to be applied, otherwise much of the variability and trends observed across studies may be due to measurement artifacts, which are often substantial.

Typically, FDOM EEMs are characterized by two signals: protein-like and humic-like fluorescence (Coble, 2007). Despite the limited informative value of these terms, they persist largely as a matter of consistency. For these two signals, wavelength regions of DOM fluorescence have been classified (Table 10.1; Figure 10.5) (Coble, 2007). Protein-like fluorescence is so termed because it is similar to that of three fluorescent amino acids that occur in proteins: tryptophan, tyrosine, and phenylalanine (Determann et al., 1998). In their free dissolved forms in water, they have fluorescence maxima at approximately Ex. 275/Em. 350 nm, Ex. 275/Em. 300, and Ex. 255/Em. <300 nm, respectively. The fluorescence (efficiency and peak position) of these aromatic amino acids depends greatly on the environment they are in, i.e., free dissolved or bound in peptides, pH, and solvent polarity. When bound in proteins, the fluorescence of tryptophan often

shifts to shorter wavelengths due to shielding from water, with the extent of the shift depending on the type of the protein (Lakowicz, 2006; Wolfbeis, 1985). The fluorescence of tyrosine when bound in proteins is difficult to detect due to energy transfer to tryptophan and quenching by neighboring groups. A good example is seen in the fluorescence of human serum albumin (HSA), which resembles protein-bound tryptophan (emission peak at shorter wavelength), despite containing 18 times more tyrosine than tryptophan and both compounds having comparable fluorescence efficiencies (Lakowicz, 2006). When proteins containing tyrosine are denatured/degraded, there is an increase in the observed fluorescence by tyrosine (Determann et al., 1998), as detectable in marine samples (Stedmon and Markager, 2005a). Adding to the complexity of interpreting the fluorescence in this region, many other naturally occurring aromatic compounds exhibit fluorescence in this wavelength region (Figure 10.3). This occurrence has also been shown in natural samples using size exclusion chromatography (Maie et al., 2007) and calls into question the appropriateness of the current terminology "protein-like fluorescence." Despite this, a study in Japanese coastal waters found strong correlations between the concentrations of tyrosine and tryptophan in total hydrolysable amino acids and the fluorescence in the UVA region (Yamashita and Tanoue, 2003a). This study was unique in combining bulk fluorescence measurements with actual measurements of amino acids; similar studies from contrasting environments would be valuable. To avoid some confusion from here onwards, this fluorescence will be referred to as *UVA fluorescence*, reflecting the wavelength region of emission.

The humic-like fluorescence is characterized as having broad emission spectra generally 400-600 nm. The term humic-like refers to the fact that similar fluorescence was reported originally in soil organic matter and XAD extracts. From here onwards, this fluorescence will be referred to as the visible fluorescence, akin to that

TABLE 10.1   Common regions of fluorescence maxima

Type	Label	Ex. Region (nm)	Em. Region (nm)
Visible	A	260	400-460
Visible	M	290-310	370-410
Visible	C	320-360	420-460
UVA	N	280	370
UVA	B	275	305
UVA	T	275	340

Modified from Coble (2007).

of UVA fluorescence described in the previous paragraph. In contrast to UVA fluorescence, visible fluorescence is ubiquitous in all aquatic environments. The fluorescence in this region has been classified into different peaks, commonly referred to as A, C, and M, which are apparent from visual inspection of the EEMs (Figure 10.5). However, applications in mathematical chromatography techniques, such as parallel factor analysis (PARAFAC), have revealed that there is often a suite of different fluorescent signals associated with visible fluorescence (Murphy et al., 2013; Stedmon and Markager, 2003), but the actual fluorophores potentially responsible remain unknown.

The fluorescence properties of DOM are also thought to represent a combination of the superposition of overlapping spectra and the effects of charge transfer (Del Vecchio and Blough, 2002). As has been shown with absorption, reduction of the organic matter with sodium borohydride and the resultant suppression of charge transfer causes changes in the fluorescence characteristics, greatly enhancing the UVA fluorescence and slightly reducing the visible fluorescence (Andrew et al., 2013). In particular, the visible fluorescence from excitation at wavelengths >350 nm is affected. This region in the fluorescence landscape is characterized by a gradually increasing emission maximum with increasing excitation wavelength (Figure 10.5). It has been hypothesized that polyphenolic compounds such as those found in lignin are responsible for this trend. The apparent shifts in emission maxima with excitation wavelength is not only due to FDOM being composed of a mixture of fluorophores but, at longer excitation wavelengths in particular, also due to charge transfer (Boyle et al., 2009; Ma et al., 2010).

EEMs represent a detailed map of the fluorescence properties of DOM, containing a considerable amount of information that until a decade ago was largely disregarded, defeating the objective of the measurement approach. In general, the data were characterized by visual identification of peaks (peak picking), integration of the signal in different wavelength regions, or developing indices from ratios of fluorescence at specific wavelengths. While these approaches should not be disregarded as they offer a necessary, and often sufficient, data analysis, there are now additional complementary multivariate approaches that utilize more of the information available (Boehme et al., 2004; Hall and Kenny, 2007; Persson and Wedborg, 2001; Stedmon et al., 2003). A review of these techniques, their applicability, and suitability is reported in Murphy et al. (2014). One approach that has become popular is PARAFAC, which decomposes the combined fluorescence signal into underlying individual, independently variable fluorescent signals (Bro, 1997; Murphy et al., 2013, 2014; Stedmon and Bro, 2008; Stedmon et al., 2003). PARAFAC is simpler to comprehend than other approaches, such as principle component analysis (PCA), as the algorithm at its core is well suited to fluorescence. PARAFAC provides both a quantitative and qualitative model of the data resolving the excitation and emission spectra of individual components that are directly correlated to their specific absorption (excitation) and quantum yield of fluorescence (emission). The reader is referred to Stedmon and Bro (2008) and Murphy et al. (2013, 2014) for a further introduction and description of the analysis. Although, when applied correctly, PARAFAC analysis of fluorescence data can provide a wealth of information on the dynamics of different DOM fractions otherwise not apparent due to the overlapping nature of the spectra (e.g., Stedmon and Markager 2005a), it is not always suitable, never a necessity, and does not replace other approaches.

Another approach to characterize DOM fluorescence is to calculate the quantum yield (ϕ) of fluorescence, which represents the fraction of photons emitted relative to those absorbed (Green and Blough, 1994; Vodacek et al., 1995, 1997). ϕ can be calculated by referencing to the fluorescence of a well-characterized standard fluorophore such as quinine sulfate according to Eq. (10.3):

$$\phi_\lambda = \frac{F_\lambda^m a_{350}^{QS} \phi_r}{a_\lambda^m F_{350}^{QS}}, \qquad (10.3)$$

where ϕ_λ is the quantum yield at excitation wavelength λ, F_λ^m is the integrated fluorescence (emission) from excitation wavelength λ for the sample, a_λ^m is the absorption coefficient at wavelength λ, F_{350}^{QS} is the integrated fluorescence of quinine sulfate from an excitation at 350 nm, a_{350}^{QS} is the absorption coefficient of quinine sulfate at 350 nm, and ϕ_{350} is the quantum yield of quinine sulfate at 350 nm. If the composition of FDOM does not vary within a set of samples, ϕ_λ will remain constant.

C Coupling CDOM and DOC

As the carbon within CDOM represents a fraction of dissolved organic carbon (DOC) (Figure 10.1), many studies have investigated whether CDOM measurements can be used to estimate DOC concentrations. The main motivations for developing CDOM to DOC relationships are estimation of the spatial distribution of DOC in surface waters and quantification of land-to-ocean carbon fluxes (Mannino et al., 2008). The major approaches to estimating DOC using CDOM optical properties use remote sensing of ocean color (Mannino et al., 2008), lidar measurements (Vodacek et al., 1995), or in situ FDOM measurements made with submersible fluorometers (Amon et al., 2003; Guay et al., 1999). Remote sensing or in situ approaches would considerably expand the current capabilities to study the seasonal and spatial dynamics of DOC, across scales ranging from centimeters to kilometers. The major assumptions for the existence of such a relationship is that the proportion of uncolored dissolved organic matter is invariant and the proportion of different CDOM fractions, each with their own respective contribution to DOC, is also constant. With regard to a universal relationship, both these assumptions

are essentially flawed as DOM consists of very complex mixtures of both colored and uncolored compounds and the relative composition of these vary depending on mixing, photochemical oxidation, and microbial degradation. The differential impacts these processes have on DOC and CDOM result in a decoupling and nonlinear correlation (Del Vecchio and Blough, 2004b; Nelson and Siegel, 2013; Skoog et al., 2011). In the open ocean (away from continental shelves) in particular, there is no correlation between DOC and CDOM as quantified by the absorption coefficient (Nelson and Siegel, 2002, 2013; Nelson et al., 1998).

However, under certain conditions where mixing is simplified and the degradation processes are small compared to the ambient concentrations, relationships can be found. This is especially true for coastal waters that are influenced by runoff. The vast majority of the correlations between CDOM and DOC have been reported in such waters (Coble, 2007; Ferrari et al., 1996; Mannino et al., 2008; Stedmon et al., 2000; Vodacek et al., 1995, 1997), typically characterized by simple and rapid mixing of a high-DOM concentration freshwater source with a low-DOM marine component. The addition of more DOM sources or alterations during mixing deteriorates this relationship. The slope and intercept of the CDOM to DOC relationship is dependent on the concentrations in the two end-members and are therefore often site- and season-specific. Despite this, with a documented understanding of local conditions (mixing and seasonality), these empirical relationships can be valuable for extrapolating high-resolution optical measurements to estimate DOC concentrations (Mannino et al., 2008).

Alternative empirical approaches to linking CDOM and DOC concentrations in coastal and shelf waters exist (Fichot and Benner, 2011, 2012; Fichot et al., 2013). The first is based on a multiple linear regression of CDOM absorption at two wavelengths (275 and 295 nm), providing a good site-specific estimate of DOC in northern Gulf of

Mexico shelf waters. Although, care should be taken when using two estimator variables that are highly intercorrelated (Asmala et al., 2012). The second approach utilizes the fact that a negative correlation exists between the spectral slope coefficient and the specific absorption coefficient (Anderson and Stedmon, 2007; Blough and Del Vecchio, 2002; Norman et al., 2011). After regressing a nonlinear model to this relationship, absorption measurements (a_λ and S) alone can estimate DOC concentrations (Fichot et al., 2013). This method provides a better estimate of DOC as it allows the CDOM-DOC relationship to vary as a function of mixing of CDOM end-members, as depicted by the S values.

The question of how much organic carbon is bound as colored (or fluorescent) substances is difficult to resolve. Most estimates are from measurements of the humic content of DOM using solid-phase extraction and assuming all the isolated humic material is colored. Alternatively, CDOM absorption or fluorescence has been linearly regressed against DOC concentrations and the intercept of the DOC axis inferred to represent the uncolored DOC fraction (e.g., Ferrari et al., 1996). These relationships exist primarily where mixing is strong and there are large gradients in DOC and CDOM, such as in estuaries and coastal waters where terrestrial DOM supplied by rivers is mixed with coastal DOM. The slopes and intercepts of the relationships contain no chemical information and are solely determined by the properties of the end-members. Instrumentation with the potential to quantify the chemical composition of CDOM is the LC-OC-TN analyzer (liquid chromatography-organic carbon-total nitrogen analyzer), which measures organic carbon and absorbance at 254 nm continuously across a size fractionation chromatogram (Huber and Frimmel, 1994; Huber et al., 2011). The instrument is suboptimal for saltwater samples and a suitable data analysis technique needs to be developed to mathematically fractionate and link the chromatogram from the organic carbon detector to CDOM's (A_{254}) chromatogram.

Incomplete combustion of biomass and fossil fuels on land produces a suite of polycyclic aromatic hydrocarbons collectively referred to as black carbon. The soluble fraction (dissolved black carbon, DBC) is leached from soils and is transported to the ocean (Jaffé et al., 2013; Mannino and Harvey, 2004) or is supplied to surface waters via atmospheric deposition (Jurado et al., 2008). The molecular structure of DBC indicates that it likely contributes to CDOM (see Chapter 7). The UV absorption properties of the oxidation products are used to quantify and characterize DBC (Dittmar, 2008). Although DBC constitutes 2% of oceanic DOC (Dittmar and Paeng, 2009) it is still unclear to what extent CDOM's optical properties originate from DBC. Stubbins et al. (2012) have shown that the photochemical degradation of DBC and CDOM are closely coupled and it is likely that the riverine supplies of both CDOM and DBC are also coupled.

III SOURCES OF CDOM TO THE MARINE ENVIRONMENT

The major sources of DOM/CDOM in the marine environment are allochthonous or autochthonous inputs. Allochthonous CDOM is that formed outside the system being studied while autochthonous CDOM is formed within the system. DOM in the allochthonous fraction was originally synthesized by terrestrial or freshwater plants, subsequently processed and modified in limnic systems and eventually exported to coastal waters. Autochthonous material in the marine environment is originally fixed by marine plants or phytoplankton and processed within the marine food web. As estuaries and coastal waters are often very productive systems, they represent very dynamic sites with high external loading of allochthonous material and high local (autochthonous) production of DOM (see Chapter 11), which are processed and exported to shelf seas. Inputs of CDOM from sediments are less well understood, although their character may reflect the nature of the particular organic matter (POM) from which they are derived (allochthonous terrestrial POM transported by rivers versus autochthonous POM and larger aquatic plant detritus) (Osburn et al., 2012). For this section, a description of the qualitative differences in the CDOM optical properties between different sources will be given. There are also considerable differences in the concentrations (intensities) of CDOM from different sources and these are addressed afterwards.

A Terrestrially Derived CDOM

There are fundamental differences in the chemical characteristics of DOM derived from terrestrial material compared to that generated by aquatic organisms, and this is reflected in the optical properties of CDOM. In general, the spectral slopes (S) of terrestrial CDOM absorption spectra (Figure 10.4) are lower reflecting a greater relative absorption at longer wavelengths, although the absolute value depends on both the wavelength range it is calculated across and the extent of previous exposure to degradation, as mentioned earlier. Coastal CDOM S values calculated across a broad wavelength range spanning the UVA and visible, 300-600 nm, typically range between 15 and 19 μm^{-1} (Del Vecchio and Blough, 2004b; Granskog, 2012; Stedmon et al., 2000) (Figure 10.6). The UV slope values ($S_{275-295}$) for terrestrial material in coastal waters typically range between 10 and 20 μm^{-1} (Asmala et al., 2012; Fichot and Benner, 2012; Granskog et al., 2012; Helms et al., 2008; Mizubayashi et al., 2013).

The absorption intensity of CDOM (a_λ) in coastal waters varies greatly, being largely dependent on intensity of absorption in the freshwater supply and the flux of freshwater. It can be difficult to compare between studies as often absorption values at different wavelengths are reported. Before 2008, the majority of studies

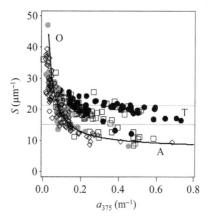

FIGURE 10.6 Spectral slope calculated across 300-650 nm plotted against the absorption coefficient at 375 nm for samples from the Arctic Ocean (Stedmon et al., 2011). The labels T, A, and O refer to three CDOM endmembers identified as terrestrial (riverine), autochthonous (marine productivity), and oceanic (background pool of CDOM in ocean waters), respectively. Symbols change with different depth layers: black dots are samples from the polar mixed layer (0-50 m), squares are samples from the halocline (50-300 m), gray dots are samples from the Atlantic Water (300-900 m), and diamonds are deep water (>900 m). *Reprinted from Stedmon et al. (2011), with permission from Elsevier.*

reported S values from the UVA-visible range; these could be used to estimate the absorption coefficients at any other wavelength in the range using Eq. (10.2). Now there is a tendency to only report the UV slopes specific to the 275-295-nm range, making it impossible to convert reported absorption data to other wavelengths. Despite this limitation in the published data, some generalizations can be made. Rivers tend to have a_{375} between 5 and 25 m^{-1} depending on the catchment they are draining. In estuaries and coastal waters, a_{375} is typically 0.5-5 m^{-1}. These ranges can be converted to other wavelengths assuming S values in the ranges reported above.

Terrestrial FDOM supplied to coastal waters by rivers is often characterized by emission spectra that are broad and extend above 450 nm (Figure 10.5). The EEMs are typically dominated by the visible fluorescence signal. Often the fluorescence in this region can be separated into three or more components, either by eye or by PARAFAC analysis with emission maxima between 400 and 500 nm. The relative contribution of each of these signals varies between systems and across the year depending on the catchment type and season (Fellman et al., 2009; Stedmon and Markager, 2005b). Longer wavelength emission maxima and broad emission spectra are characteristic of higher molecular weight material (Boehme and Wells, 2006) and the prevalence of these fluorescence signals in freshwaters reflects the export of soil organic matter into aquatic systems (Guéguen and Cuss, 2011). In estuarine waters, terrestrial FDOM is diluted with its marine counterpart and in many systems the visible fluorescence signals have been observed to mix relatively conservatively (Coble, 2007; Guo et al., 2011; Stedmon et al., 2003). However, this conservation is highly dependent on the balance between mixing and freshwater supply rates relative to the internal production and removal rates of FDOM. The visible fluorescence intensities (concentrations) and supply of terrestrial FDOM are often much greater than that produced by aquatic organisms so the signal is overwhelmed and dominated by the overall dilution of the terrestrial material. The major processes influencing fluorescence at these wavelengths are for the most part only important under weak estuarine mixing conditions. For example, such processes include sediment pore water release after sediment resuspension (Osburn et al., 2012), photochemical degradation, or production by aquatic organisms (Stedmon and Markager, 2005a), each covered in the following section.

B Oceanic CDOM

Away from the influence of terrestrial inputs, absorption measurements distinguish two dominant CDOM pools in the ocean (Kitidis et al., 2006; Nelson et al., 2010; Stedmon and Markager, 2001). The two pools are best depicted in a plot of the UVA-visible spectral slope against a_{375} (Castillo and Coble, 2000; Stedmon and Markager, 2001; Stedmon et al., 2011; Swan et al., 2013). For oceanic water, a_{375} values vary between 0 and 0.5 m^{-1} and S values between 10 and 30 µm^{-1} (Figure 10.7). In general, higher a_{375} values, indicating recently produced CDOM, are associated with low S values (10-15 µm^{-1}) (Kitidis et al., 2006; Stedmon and Markager, 2001). The background oceanic CDOM pool has low a_{375} values and more variable S values although distinctly higher (20-30 µm^{-1}). As these two pools mix in the oceans, a mixing curve joins these two end-members (e.g., Figure 10.6). In general, photobleaching in the surface ocean causes the a_{375} values to be lower there than in deeper waters and corresponding S values are higher (Figure 10.7). The UV spectral slope reveals a similar pattern with the highest slopes in surfaces waters, varying between 20 and 60 µm^{-1}. The decrease in the slopes, and corresponding increases in a_{375}, indicate that, in general, the deep aphotic ocean is a source of CDOM. Superimposed on this vertical pattern, there are also often local CDOM maxima associated with, or just below, phytoplankton biomass maxima (deep chlorophyll maxima) in the lower photic zone (Kitidis et al., 2006; Kowalczuk et al., 2013; Stedmon and Markager, 2001).

A variety of heterotrophic and autotrophic organisms produce CDOM. Steinberg et al. (2004) reported that many organisms release a suite of different chromophores with very distinct absorption spectra. Similar peaks are produced by sea ice microbial communities (Norman et al., 2011; Uusikivi et al., 2010). Despite the existence of these distinct absorption peaks, the majority of measured ocean CDOM spectra are relatively featureless (Figure 10.2) so it is likely that these colored metabolites are rapidly processed by bacteria, either transformed to more persistent CDOM or remineralized. This supposition is supported by the experiments of Ortega-Retuerta et al. (2009) with Antarctic krill. During short-term incubations (hours), CDOM with peaks at ~275 and 340 nm

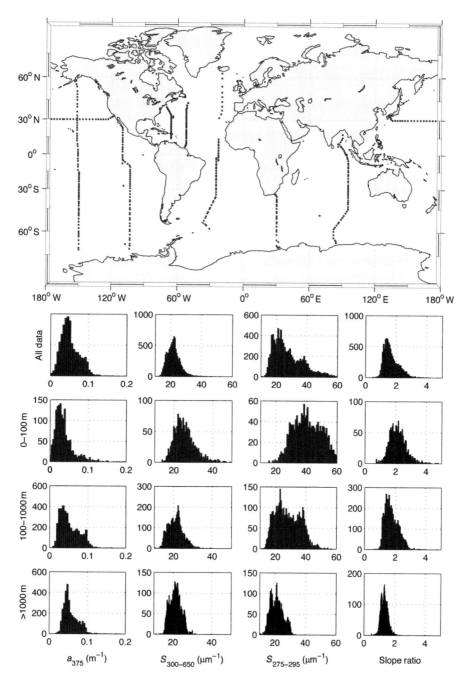

FIGURE 10.7 Histograms of CDOM absorption in oceanic waters originating on U.S. CLIVAR cruises shown on the map. The top row summarizes all data (9067 samples). The next three rows plot data by depth intervals shown in the y-axis label. The columns represent a_{375}, spectral slope (300-650 nm), UV slope (275-295 nm), and the Slope Ratio. *Data originally published in Nelson et al. (2010).*

was produced. For longer-term incubations (days), the spectral signature was both reduced and altered to generate broader, featureless spectra, typical for oceanic CDOM. Similarly, CDOM absorption spectra in surface ocean waters systematically have a greater absorption at wavelengths <300 nm relative to longer wavelengths, likely representing the production of CDOM from organisms and the photochemical degradation of CDOM absorbing at longer wavelengths (Yamashita and Tanoue, 2009). Experiments with axenic phytoplankton cultures have also indicated that phytoplankton in isolation release CDOM with low S values (Chari et al., 2013; Zhang et al., 2013).

CDOM can also be formed photochemically, although the effect of longer-term exposure undoubtedly results in net removal. Kieber et al. (1997) reported that photooxidation of colorless fatty acids and triglycerides produced CDOM. Additionally, the photochemical exposure of free dissolved tryptophan mixed with natural DOM produces CDOM (Biers et al., 2007; Reitner et al., 2002). In solar-simulator experiments, Swan et al. (2012) documented a nitrate-dependent photoproduction of CDOM absorbing between 400 and 500 nm in selected oceanic water samples, mostly from high-nutrient-low-chlorophyll areas.

Figure 10.5 shows the fluorescence properties of surface and deep-ocean FDOM. The spectra are an average of several hundred EEMs collected on a global survey (Jørgensen et al., 2011). The major contrast to terrestrial FDOM is that the fluorescence is predominantly at lower wavelengths and characterized by the UVA fluorescence peaks and a peak with an emission around 400 nm. By comparing the surface and deep-ocean spectra, one sees a greater visible fluorescence signal in deep waters. This signal is linked to differences in the dominant processes controlling the production and removal of FDOM; the latter is discussed in the next section.

The visible fluorescence signal in aphotic ocean waters has two components: one emitting in the region of 400 nm and another at 440 nm (Heller et al., 2013; Jørgensen et al., 2011; Kowalczuk et al., 2013; Yamashita et al., 2010). Fluorescence at these wavelengths represents material that is produced by heterotrophic bacteria, as strong correlations have been found with oxygen utilization and nutrient mineralization (Jørgensen et al., 2011; Nieto-Cid et al., 2006; Yamashita and Tanoue, 2008). In surface waters, both UVA and visible fluorescence signals are produced (Lønborg et al., 2010; Rochelle-Newall and Fisher, 2002; Romera-castillo et al., 2010; Shimotori et al., 2009; Stedmon and Markager, 2005a; Suksomjit et al., 2009) (Figure 10.8). Most studies concluded that the visible fluorescence produced was, as in aphotic waters, bacterially derived given as absence of correlation to phytoplankton biomass and its production subsequent to a phytoplankton bloom. However, Romera-Castillo et al. (2010) and Chari et al. (2013) found that visible fluorescence can be directly produced by phytoplankton. Both UVA and visible fluorescence was produced in four different axenic phytoplankton cultures. Urban-Rich et al. (2006) reported the excretion of visible fluorescent material by zooplankton.

There is evidence for abiotic pathways of visible fluorescence production as well. Parallel to that mentioned above for CDOM, photochemical reactions can result in the initial formation of FDOM (Biers et al., 2007; Kieber et al., 1997). Additionally, cross-linking of unsaturated plankton lipids is a viable formation pathway of FDOM (Harvey et al., 1983, 1984).

UVA fluorescence is created by many planktonic organisms (Shimotori et al., 2009, 2012; Stedmon and Markager, 2005a; Urban-Rich et al., 2006). In marine waters, these signals are correlated to the concentrations of free or bound amino acids (Nieto-Cid et al., 2005; Seritti et al., 1994; Yamashita and Tanoue, 2003b). The vertical distribution of UVA fluorescence differs from that of visible fluorescence by being predominantly associated with surface waters (Figure 10.8).

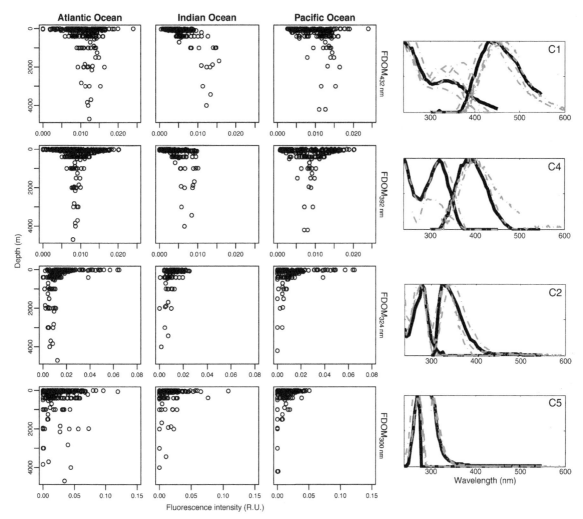

FIGURE 10.8 Fluorescence intensities (three left columns) of four components (C1, C4, C2, and C5) identified by Jørgensen et al. (2011) in the Atlantic, Pacific, and Indian Oceans. The spectra in the right-hand column show the spectral properties of each component. The excitation and emission spectra are plotted in black. The excitation and emission spectra of components identified in earlier studies (Murphy et al., 2006; Stedmon et al., 2007a,b; Walker et al., 2009; Yamashita et al., 2010) are overlain in gray.

C Sediments as Sources of CDOM

As for DOC (see Chapter 12), sediments are sources of CDOM to the marine environment, in particular in coastal waters and shelf seas where sediment resuspension and hypoxia events occur (Osburn et al., 2012; Skoog et al., 1996).

Hydrothermal vents are sources of CDOM too (Yang et al., 2012). Despite the known existence of benthic CDOM sources, there has been comparatively little research quantifying the benthic environment as a CDOM source and characterizing the CDOM released. Lübben et al. (2009) show that tidal flat pore waters of the Wadden

Sea release both CDOM and FDOM. In sediment pore waters, the visible fluorescence increases with depth and is correlated to DOC (Burdige et al., 2004; Chen and Bada, 1994). This fluorescence was interpreted in the context of a pore water reactivity model as representing refractory low molecular weight material generated during diagenesis of sediment organic matter. This interpretation supports the findings of Skoog et al. (1996) that visible fluorescence is generated in and released from sediment pore waters under anoxic conditions. Under oxic conditions, this fluorescent signal is thought to be retained in sediments by association with iron oxides. In contrast, UVA fluorescence exhibits little trend with depth in surface sediments (Burdige et al., 2004).

The impact of DOM generated and subsequently released from sediments on oxygen and nutrient concentrations in the overlying waters is poorly resolved. If it is bioavailable, it can sustain oxygen depletion events and reintroduce nutrients into coastal waters. The fact that this material has such a high-intensity optical signal, which is strongly correlated to DOC and nutrients (Burdige et al., 2004; Skoog et al., 1996), presents the opportunity to develop in situ sensors to improve flux estimates.

IV REMOVAL OF CDOM IN THE MARINE ENVIRONMENT

A Photochemistry

Taking into account for CDOM and FDOM: (i) their riverine export to the coast (Blough and Del Vecchio, 2002), (ii) their widespread formation in the deep ocean (Jørgensen et al., 2011; Nelson et al., 2010; Yamashita and Tanoue, 2008), and (iii) their vertical and geographical distributions in the ocean (Figures 10.9 and 10.10), there must be a significant removal process to balance inputs. This process is photochemical degradation (Miller and Zepp, 1995). The absorption of high-energy (low wavelength) light by CDOM can result in a photochemical reaction to form either a new stable product or a suite of reactive species (inorganic and organic radicals), which subsequently initiate secondary reactions (Blough and Zepp, 1995; O'Sullivan et al., 2005; Qian et al., 2001; see also Chapter 8). Here, it is important to note that photons in the UVA and B region (280-400 nm) have sufficient energy to directly cleave C–H, C–O, C–C, C–N, and C–P bonds (bond dissociation energies ranging between approximately 280 and 420 kJ/mol). While primary reactions only influence CDOM, the secondary reactions driven by reactive

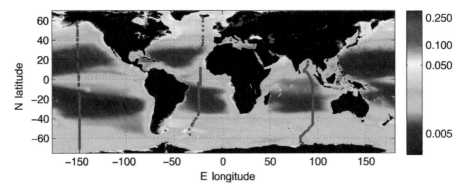

FIGURE 10.9 Global distribution of CDOM plus detrital particulate materials inferred from the Aqua spacecraft's MODIS ocean color imagery (2002-2013 mission average). The color scale is the sum of the absorption coefficient at 443 nm (m^{-1}) of CDOM and of non-phytoplankton (detrital) particulate materials. Red dots show the location of stations comprising the sections shown in Figure 10.10. *Data courtesy of Siegel, Maritorena, and Fields; http://wiki.eri.ucsb.edu/measures.*

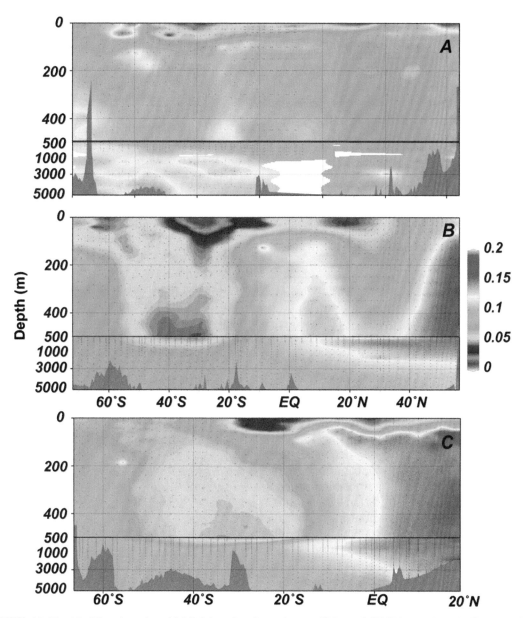

FIGURE 10.10 Meridional sections highlighting the absorption coefficient of CDOM at 340 nm (m⁻¹) across the Atlantic (a), Pacific (b), and Indian (c) Ocean basins. Note depth scale change at 500 m in each plot.

photoproducts affect all compounds (colored and noncolored) and microorganisms. The impact of photodegradation is both biogeochemical via direct mineralization of carbon (Johannessen and Miller, 2001) and nitrogen (Aarnos et al., 2012; Stedmon et al., 2007a; Vähätalo and Zepp, 2005; Xie et al., 2012) and ecological through the resulting increased light penetration in surface waters and the production of biolabile substrates for microbes (Nieto-Cid et al., 2006).

The role of photodegradation is evident from the global ocean distributions of CDOM, with lowest CDOM abundance in the subtropical gyres where solar exposure is high and mixed layer depths are shallow (Heller et al., 2013; Kowalczuk et al., 2013; Nelson et al., 2010; Siegel et al., 2002; Yamashita and Tanoue, 2009). Seasonally, the establishment of stratification and increasing solar radiation leads to subsequent decreases in CDOM abundance (e.g., Nelson et al., 1998; Vodacek et al., 1997). The overall effect of photodegradation on the CDOM spectrum is to cause the absorption to decrease and the slope to increase (Del Vecchio and Blough, 2002; Helms et al., 2013; Swan et al., 2012; Vähätalo and Wetzel, 2004). These results can be seen in the surface layer values shown in Figure 10.7 and in the spectra in Figure 10.2.

FDOM is generally more sensitive to photodegradation than CDOM. The visible fluorescence signal decreases rapidly, while the UV fluorescence decreases to a lesser extent (Helms et al., 2013; Nieto-Cid et al., 2006; Stedmon and Markager, 2005a). Ocean sections of visible fluorescence reveal depletion in surface waters (Heller et al., 2013; Kowalczuk et al., 2013; Yamashita and Tanoue, 2009) similar to that seen for CDOM (Figure 10.10). However, in coastal and shelf seas, where contributions from terrestrial FDOM are stronger, a visible fluorescence signal with a UVC excitation can be photochemically produced (Stedmon et al., 2007a) and persist (Murphy et al., 2008). The overall result of photodegradation is that FDOM EEMs change dramatically with a shift in the fluorescence maxima to shorter excitation and emission wavelengths (Coble, 2007; Helms et al., 2013).

The responses of CDOM, FDOM, and DOC to photodegradation, measured as losses of color, fluorescence, and organic carbon, respectively, are not proportional. They each differ in their sensitivity. The loss of fluorescence is always greater than the loss of absorption, which in turn is greater than carbon remineralization (Helms et al., 2013; Nelson et al., 1998; Tzortziou et al., 2007; Vähätalo and Wetzel, 2004).

B Microbial Activity

As with photochemistry, the turnover of DOM by heterotrophic microbes alters its spectral properties. Although microbes either incorporate or respire organic matter they also, in the process, modify DOM. With respect to CDOM and FDOM, this modification is often observed as transient peaks or a more general microbial "humification," with an increase in the longer wavelength absorption and fluorescence. In dark laboratory microbial culture experiments, Nelson et al. (2004) observed rapid production of CDOM that was subsequently degraded (Figure 10.11). Parallel to DOM as a whole, it is likely that CDOM consists of a mixture of compounds spanning a continuum of labilities. It is often difficult to detect the short-lived metabolites in field (ocean) samples. They are commonly only observed in laboratory experiments (Nelson et al., 2004; Ortega-retuerta et al., 2009) or immediately after phytoplankton blooms (Sasaki et al., 2005), where production rates can be high enough to

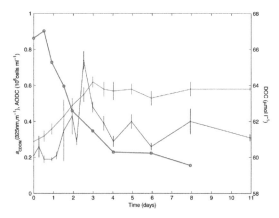

FIGURE 10.11　Time course of bacterial counts (blue, as acridine orange direct counts; AODC) CDOM absorption coefficient at 325 nm (black) and DOC (green) in a dark culture experiment conducted on water from the western Sargasso Sea in the spring of 1996 (Nelson et al., 2004). Error bars on the cell counts and CDOM are standard deviation of replicate samples. In this experiment, net CDOM production occurred during growth of the bacterial culture as the DOC declined. When the bacteria reached stationary phase the CDOM declined rapidly, suggesting consumption of labile CDOM by microbes (Nelson et al., 2004).

exceed removal/transformation rates. The often featureless CDOM absorption spectra observed in ocean samples most likely reflect a more persistent fraction that resists microbial degradation. However, the persistent fraction is still distinguishable from the overall background (and photobleached) CDOM pool as recently produced CDOM has systematically lower spectral slopes and higher absorption values (see section earlier).

The multidimensional characteristics and higher sensitivity of fluorescence measurements shed light on the underlying changes occurring to DOM as a result of microbial activity. Parallel to CDOM, visible fluorescence is ubiquitous, reduced by photodegradation and produced during the microbial turnover of recently produced DOM (Lønborg et al., 2010; Nieto-Cid et al., 2006; Rochelle-Newall and Fisher, 2002; Stedmon and Markager, 2005a; Suksomjit et al., 2009). Production rates of visible fluorescence in microbial incubations are up to five times slower than the DOC consumption rate (Lønborg et al., 2010), indicating that the material responsible for this fluorescence is not generated from the turnover of very labile DOM but more likely a semi-labile fraction (Jørgensen et al., 2014). UVA fluorescence, in contrast, represents a labile fraction that persists long enough to be measured across a wide range of systems, but restricted mostly to surface waters (Jørgensen et al., 2011; Lønborg et al., 2010; Nieto-Cid et al., 2006; Stedmon and Markager, 2005a). This fraction is degraded faster than the bulk DOC (Lønborg et al., 2010).

V DISTRIBUTION

A Oceanic

The global distribution of CDOM has been assessed by measurements of absorption spectra (field observations) or inference from ocean color reflectance spectra (satellite observations). The distribution of CDOM in surface waters (Figure 10.9) exhibits features superficially similar to the distribution of chlorophyll a (Siegel et al., 2002). There is a decline in CDOM abundance with distance from

the coast representing the dilution of CDOM originating from rivers, productive coastal and shelf waters, and sediments. The highest abundances of CDOM are found in coastal regions, on the continental shelf, restricted seas, and river outflows (Figure 10.9), suggesting that in these systems terrestrial input dominates the CDOM source. In the open ocean, CDOM is more abundant at high latitudes and in upwelling areas (coastal and equatorial), reflecting a connection to both productivity and CDOM-rich deep waters. The lowest concentrations of CDOM are in the subtropical gyres because of low productivity, high stratification, and high photochemical removal.

Meridional sections of CDOM in the Atlantic, Pacific, and Indian oceans are shown in Figure 10.10. Oceanic distributions of CDOM and FDOM represent a balance between surface photochemical removal, mesopelagic microbial formation and water mass mixing (Álvarez-Salgado et al., 2013; Nelson et al., 2010; Yamashita and Tanoue, 2009). The fact that the meridional sections differ between the ocean basins largely reflects the origins and time since subduction of the water masses. The North Atlantic demonstrates the influence of Arctic terrestrial DOM entrained into bottom waters, resulting in higher deep-ocean values relative to the general deep-ocean CDOM and FDOM produced by microbial respiration (Álvarez-Salgado et al., 2013; Jørgensen et al., 2011).

In the North Pacific and North Indian basins (Figure 10.10b and c), the deep-water CDOM absorption coefficient (and fluorescence) is higher than in the South Pacific, Southern Ocean, or Atlantic basins (Nelson et al., 2010; Swan et al., 2012; Yamashita and Tanoue, 2008, 2009). The higher deep-water CDOM reflects the accumulation of CDOM from the remineralization of sinking particles (Yamashita and Tanoue, 2009) and comparatively slower overturning circulation (Nelson et al., 2010). As a result, a strong correlation between CDOM or FDOM with apparent oxygen utilization (AOU) exists. In the Atlantic (Figure 10.10a), rapid mixing of high-CDOM/ low-AOU waters to depth obscures the remineralization-related signal (Álvarez-Salgado et al., 2013; Jørgensen et al., 2011; Nelson and Siegel, 2013). Nevertheless, correlations between CDOM abundance and CFC-12 as a proxy for ventilation age in the North Atlantic (Nelson et al., 2007) suggest that CDOM accumulates slowly in subsurface water masses.

B Sea Ice

When ice forms from seawater, the dissolved constituents are rejected, accumulating in a network of brine channels within the ice. The concentrations of these solutes can be very high. For example, salinities are often 50-150 and DOC concentrations can reach millimolar levels (Norman et al., 2011). Despite the extreme nature of these environments, with temperatures typically between −15 and 0 °C, there are active autotrophic and heterotrophic organisms influencing the biogeochemistry of carbon and nutrients in the brines (Thomas and Dieckmann, 2010). CDOM in sea ice brines either primarily originates from the parent seawater and subsequently concentrated or is produced by microbial communities inhabiting the brine channels. Measurements of the spectral characteristics of CDOM make it possible to differentiate these sources, thereby tracing the impact that abiotic and biotic processes have on DOM concentrations and characteristics (Stedmon et al., 2007b, 2011).

CDOM is an important optical component of sea ice, contributing significantly to the seasonal and spatial variability in ice transmittance (Belzile et al., 2002; Perovich et al., 1998). This impact on light has a consequence for sea ice organisms in a similar fashion to open seawater, with CDOM influencing the penetration of both harmful UV radiation and photosynthetically available radiation (PAR). The UV-absorbing properties of CDOM make it an important player in the photochemistry of sea ice. CDOM can be directly mineralized, transformed to volatile organics, or fuel the production of reactive oxygen species (Beine et al., 2012). These species can subsequently react with other constituents such as bromide, influencing atmospheric ozone and mercury levels via the resulting bromine (Beine et al., 2012).

VI CONCLUSIONS AND FUTURE RESEARCH NEEDS

Considerable progress has been made in CDOM research during the last two decades. The increased sensitivity, speed, and availability of instrumentation is largely responsible. The field has evolved from mostly focusing on single wavelength measurements at coastal sites to spectral characterization on a global scale. In concert with this growth, data treatment methods for characterizing absorption and fluorescence spectra have been developed. These methods are capable of reducing and summarizing the vast amounts of spectral data. Global survey and experimental results reveal that CDOM measurements have the potential to provide optical markers to trace turnover of DOM that is not apparent from elemental analyses alone. However, the key challenge that frustratingly remains is to improve our understanding of the chemical compounds responsible for the optical signals measured and to assess how much of the carbon bound in DOM is associated with CDOM. The distribution and variability in CDOM intensity and characteristics have been relatively well mapped. If this measurement approach is to continue to contribute to the study of DOM biogeochemistry beyond photochemistry and bio-optics, it is vital that clearer links be made between optical measurements and the ever-expanding battery of analytical chemical techniques.

Acknowledgments

Linda Jørgensen for help with Figure 10.8. CS was funded by the Danish Strategic Research Council NAACOS project Grant 10-093903 and the Danish Council for Independent Research-Natural Sciences Grant DFF—1323-00336. NBN would like to acknowledge the support of the U.S. National Aeronautics and Space Administration and the National Science Foundation for the research described in this review.

References

Aarnos, H., Ylöstalo, P., Vähätalo, A.V., 2012. Seasonal phototransformation of dissolved organic matter to ammonium, dissolved inorganic carbon, and labile substrates supporting bacterial biomass across the Baltic Sea. J. Geophys. Res. 117, 1–14.

Álvarez-Salgado, X.A., Nieto-Cid, M., Álvarez, M., Pérez, F.F., Morin, P., Mercier, H., 2013. New insights on the mineralization of dissolved organic matter in central, intermediate, and deep water masses of the northeast North Atlantic. Limnol. Oceanogr. 58, 681–696.

Amon, R.M.W., Budeus, G., Meon, B., 2003. Dissolved organic carbon distribution and origin in the Nordic Seas: Exchanges with the Arctic Ocean and the North Atlantic. J. Geophys. Res. 108, 3221.

Anderson, N.J., Stedmon, C.A., 2007. The effect of evapoconcentration on dissolved organic carbon concentration and quality in lakes of SW Greenland. Freshw. Biol. 52, 280–289.

Andrew, A.A., Del Vecchio, R., Subramaniam, A., Blough, N.V., 2013. Chromophoric dissolved organic matter (CDOM) in the Equatorial Atlantic Ocean: optical properties and their relation to CDOM structure and source. Mar. Chem. 148, 33–43.

Arrigo, K., Brown, C., 1996. Impact of chromophoric dissolved organic matter on UV inhibition of primary productivity in the sea. Mar. Ecol. Prog. Ser. 140, 207–216.

Asmala, E., Stedmon, C.A., Thomas, D.N., 2012. Linking CDOM spectral absorption to dissolved organic carbon concentrations and loadings in boreal estuaries. Estuar. Coast. Shelf Sci. 111, 107–117.

Beine, H., Anastasio, C., Domine, F., Douglas, T., Barret, M., France, J., et al., 2012. Soluble chromophores in marine snow, seawater, sea ice and frost flowers near Barrow, Alaska. J. Geophys. Res. 117, D00R15.

Belzile, C., Gibson, J.A.E., Vincent, W.F., 2002. Colored dissolved organic matter and dissolved organic carbon exclusion from lake ice: implications for irradiance transmission and carbon cycling. Limnol. Oceanogr. 47, 1283–1293.

Biers, E.J., Zepp, R.G., Moran, M.A., 2007. The role of nitrogen in chromophoric and fluorescent dissolved organic matter formation. Mar. Chem. 103, 46–60.

Blough, N.V., Del Vecchio, R., 2002. Chromophoric DOM in the coastal environment. In: Biogeochemistry of Marine Dissolved Organic Matter. Hansell, D., Carlson, C. (Eds.), Academic Press, San Diego, pp. 509–546.

Blough, N.V., Zepp, R.G., 1995. Reactive oxygen species in natural waters. In: Foote, C.S., et al. (Eds.), Active Oxygen in Chemistry. Chapman & Hall, Glasgow, UK, pp. 280–333.

Boehme, J., Wells, M., 2006. Fluorescence variability of marine and terrestrial colloids: examining size fractions of chromophoric dissolved organic matter in the Damariscotta River estuary. Mar. Chem. 101, 95–103.

Boehme, J., Coble, P., Conmy, R., Stovall-Leonard, A., 2004. Examining CDOM fluorescence variability using principal component analysis: seasonal and regional modeling of three-dimensional fluorescence in the Gulf of Mexico. Mar. Chem. 89, 3–14.

Boyle, E.S., Guerriero, N., Thiallet, A., Del Vecchio, R., Blough, N.V., 2009. Optical properties of humic substances and CDOM: relation to structure. Environ. Sci. Technol. 43, 2262–2268.

Braslavsky, S.E., 2007. Glossary of terms used in photochemistry, 3rd edition (IUPAC Recommendations 2006). Pure Appl. Chem. 79, 293–465.

Bro, R., 1997. PARAFAC. Tutorial and applications. Chemom. Intell. Lab. Syst. 38, 149–171.

Burdige, D.J., Kline, S.W., Chen, W., 2004. Fluorescent dissolved organic matter in marine sediment pore waters. Mar. Chem. 89, 289–311.

Carder, K.L., Steward, R.G., Harvey, G.R., Ortner, P.B., 1989. Marine humic and fulvic acids: their effects on remote sensing of ocean chlorophyll. Limnol. Oceanogr. 34, 68–81.

Castillo, C.E. Del, Coble, P.G., 2000. Seasonal variability of the colored dissolved organic matter during the 1994-95 NE and SW Monsoons in the Arabian Sea. Deep-Sea Res. 47, 1563–1579.

Chari, N.V.H.K., Keerthi, S., Sarma, N.S., Pandi, S.R., Chiranjeevulu, G., Kiran, R., et al., 2013. Fluorescence and absorption characteristics of dissolved organic matter excreted by phytoplankton species of western Bay of Bengal under axenic laboratory condition. J. Exp. Mar. Biol. Ecol. 445, 148–155.

Chen, R.F., Bada, J.L., 1994. The fluorescence of dissolved organic matter in porewaters of marine sediments. Mar. Chem. 45, 31–42.

Chin, Y.-P., Aiken, G., O'Loughlin, E., 1994. Molecular weight, polydispersity, and spectroscopic properties of aquatic humic substances. Environ. Sci. Technol. 28, 1853–1858.

Coble, P.G., 2007. Marine optical biogeochemistry: the chemistry of ocean color. Chem. Rev. 107, 402–418.

Del Vecchio, R., Blough, N.V., 2002. Photobleaching of chromophoric dissolved organic matter in natural waters: kinetics and modeling. Mar. Chem. 78, 231–253.

Del Vecchio, R., Blough, N.V., 2004a. On the origin of the optical properties of humic substances. Environ. Sci. Technol. 38, 3885–3891.

Del Vecchio, R., Blough, N.V., 2004b. Spatial and seasonal distribution of chromophoric dissolved organic matter and dissolved organic carbon in the Middle Atlantic Bight. Mar. Chem. 89, 169–187.

Determann, S., Lobbes, M., Reuter, R., 1998. Ultraviolet fluorescence excitation and emission spectroscopy of marine algae and bacteria. Mar. Chem. 62, 137–156.

Dittmar, T., 2008. The molecular level determination of black carbon in marine dissolved organic matter. Org. Geochem. 39, 396–407.

Dittmar, T., Paeng, J., 2009. A heat-induced molecular signature in marine dissolved organic matter. Nat. Geosci. 2, 175–179.

Duursma, E.K., 1965. Dissolved organic constituents of sea water. In: Riley, J.P., Skirrow, G. (Eds.), Chemical Oceanography, vol. 1. Academic Press, London, pp. 433–477.

Fellman, J.B., Hood, E., D'Amore, D.V., Edwards, R.T., White, D., 2009. Seasonal changes in the chemical quality and biodegradability of dissolved organic matter exported from soils to streams in coastal temperate rainforest watersheds. Biogeochemistry 95, 277–293.

Ferrari, G.M., Dowel, M.D., Grossi, S., Targa, C., 1996. Relationship between the optical properties of chromophoric dissolved organic matter and total concentration of dissolved organic carbon in the southern Baltic Sea region. Mar. Chem. 55, 299–316.

Fichot, C.G., Benner, R., 2011. A novel method to estimate DOC concentrations from CDOM absorption coefficients in coastal waters. Geophys. Res. Lett. 38, L03610.

Fichot, C.G., Benner, R., 2012. The spectral slope coefficient of chromophoric dissolved organic matter (S275-295) as a tracer of terrigenous dissolved organic carbon in river-influenced ocean margins. Limnol. Oceanogr. 57, 1453–1466.

Fichot, C.G., Kaiser, K., Hooker, S.B., Amon, R.M.W., Babin, M., Bélanger, S., et al., 2013. Pan-Arctic distributions of continental runoff in the Arctic Ocean. Sci. Rep. 3, 1053.

Floge, S.A., Wells, M.L., 2007. Variation in colloidal chromophoric dissolved organic matter in the Damariscotta Estuary. Limnol. Oceanogr. 52, 32–45.

Granskog, M.A., 2012. Changes in spectral slopes of colored dissolved organic matter absorption with mixing and removal in a terrestrially dominated marine system (Hudson Bay, Canada). Mar. Chem. 134–135, 10–17.

Granskog, M.A., Macdonald, R.W., Mundy, C.-J., Barber, D.G., 2007. Distribution, characteristics and potential impacts of chromophoric dissolved organic matter (CDOM) in Hudson Strait and Hudson Bay, Canada. Cont. Shelf Res. 27, 2032–2050.

Granskog, M.A., Stedmon, C.A., Dodd, P.A., Amon, R.M.W., Pavlov, A.K., de Steur, L., et al., 2012. Characteristics of colored dissolved organic matter (CDOM) in the Arctic outflow in the Fram Strait: assessing the changes and fate of terrigenous CDOM in the Arctic Ocean. J. Geophys. Res. 117, C12021.

Green, S.A., Blough, N.V., 1994. Optical absorption and fluorescence properties of chromophoric dissolved organic matter in natural waters. Limnol. Oceanogr. 39, 1903–1916.

Guay, K., Klinkhammer, G.P., Kenison, K.K., Benner, R., Coble, P.G., Whitledge, T.E., et al., 1999. High-resolution measurements of dissolved organic carbon in the Arctic Ocean by in situ fiber-optic spectrometry. Geophys. Res. Lett. 26, 1007–1010.

Guéguen, C., Cuss, C.W., 2011. Characterization of aquatic dissolved organic matter by asymmetrical flow field-flow fractionation coupled to UV-visible diode array and excitation emission matrix fluorescence. J. Chromatogr. A 1218, 4188–4198.

Guo, W., Yang, L., Hong, H., Stedmon, C.A., Wang, F., Xu, J., et al., 2011. Assessing the dynamics of chromophoric dissolved organic matter in a subtropical estuary using parallel factor analysis. Mar. Chem. 124, 125–133.

Hall, G.J., Kenny, J.E., 2007. Estuarine water classification using EEM spectroscopy and PARAFAC-SIMCA. Anal. Chim. Acta 581, 118–124.

Hansell, D.A., 2013. Recalcitrant dissolved organic carbon fractions. Ann. Rev. Mar. Sci. 5, 421–445.

Harvey, G.R., Boran, D.A., Chesal, L.A., Tokar, J.M., 1983. The structure of marine fulvic and humic acids. Mar. Chem. 12, 119–132.

Harvey, G.R., Boran, D.A., Piotrowicz, S.R., Weisel, C.P., 1984. Synthesis of marine humic substances from unsaturated lipids. Nature 309, 244–246.

Heller, M.I., Gaiero, D.M., Croot, P.L., 2013. Basin scale survey of marine humic fluorescence in the Atlantic: relationship to iron solubility and H_2O_2. Global Biogeochem. Cycles 27, 88–100.

Helms, J.R., Stubbins, A., Ritchie, J.D., Minor, E.C., Kieber, D.J., Mopper, K., 2008. Absorption spectral slopes and slope ratios as indicators of molecular weight, source, and photobleaching of chromophoric dissolved organic matter. Limnol. Oceanogr. 53, 955–969.

Helms, J.R., Stubbins, A., Perdue, E.M., Green, N.W., Chen, H., Mopper, K., 2013. Photochemical bleaching of oceanic dissolved organic matter and its effect on absorption spectral slope and fluorescence. Mar. Chem. 155, 81–91.

Højerslev, N.K., Holt, N., Aarup, T., 1996. Optical measurements in the North Sea-Baltic Sea transition zone. I. On the origin of the deep water in the Kattegat. Cont. Shelf Res. 16, 1329–1342.

Huber, S.A., Frimmel, F.H., 1994. Direct gel chromatographic characterization and quantification of marine dissolved organic carbon using high-sensitivity DOC detection. Environ. Sci. Technol. 28, 1194–1197.

Huber, S., Balz, A., Abert, M., Pronk, W., 2011. Characterisation of aquatic humic and non-humic matter with size-exclusion chromatography—organic carbon detection—organic nitrogen detection (LC-OCD-OND). Water Res. 45, 879–885.

Jaffé, R., Ding, Y., Niggemann, J., Vähätalo, A.V., Stubbins, A., Spencer, R.G.M., et al., 2013. Global charcoal mobilization from soils via dissolution and riverine transport to the oceans. Science 340, 345–347.

Johannessen, S.C., Miller, W.L., 2001. Quantum yield for the photochemical production of dissolved inorganic carbon in seawater. Mar. Chem. 76, 271–283.

Johnson, K.S., Coletti, L.J., 2002. In situ ultraviolet spectrophotometry for high resolution and long-term monitoring of nitrate, bromide and bisulfide in the ocean. Deep-Sea Res. Part I 49, 1291–1305.

Jørgensen, L., Stedmon, C.A., Kragh, T., Markager, S., Middelboe, M., Søndergaard, M., 2011. Global trends in the fluorescence characteristics and distribution of marine dissolved organic matter. Mar. Chem. 126, 139–148.

Jørgensen, L., Stedmon, C.A., Granskog, M., Middelboe, M., 2014. Tracing the long-term microbial production of recalcitrant fluorescent dissolved organic matter in seawater. Geophys. Res. Lett. doi:10.1002/2014GL059428.

Jurado, E., Dachs, J., Duarte, C.M., Simó, R., 2008. Atmospheric deposition of organic and black carbon to the global oceans. Atmos. Environ. 42, 7931–7939.

Kalle, K., 1938. Zum problem der meerwasserfarbe. Ann. Hydrol. Mar. mitteilungen 66, 1–13.

Kalle, K., 1966. The problem of the Gelbstoff in the sea. Oceanagr. Mar. Biol. Ann. Rev. 4, 91–104.

Kieber, D., McDaniel, J., Mopper, K., 1989. Photochemical source of biological substrates in sea water: implications for carbon cycling. Nature 353, 60–62.

Kieber, R.J., Hydra, L.H., Seaton, P.J., 1997. Photooxidation of triglycerides and fatty acids in seawater: implication toward the formation of marine humic substances. Limnol. Oceanogr. 42, 1454–1462.

Kitidis, V., Stubbins, A.P., Uher, G., Upstill Goddard, R.C., Law, C.S., Woodward, E.M.S., 2006. Variability of chromophoric organic matter in surface waters of the Atlantic Ocean. Deep Sea Res. Part II Top. Stud. Oceanogr. 53, 1666–1684.

Kowalczuk, P., Tilstone, G.H., Zabłocka, M., Röttgers, R., Thomas, R., 2013. Composition of dissolved organic matter along an Atlantic Meridional Transect from fluorescence spectroscopy and Parallel Factor Analysis. Mar. Chem. 157, 170–184.

Laane, R.W.P.M., Kramer, K.J.M., 1990. Natural fluorescence in the North Sea and its major estuaries. Neth. J. Sea Res. 26, 1–9.

Lakowicz, J.R., 2006. Principles of Fluorescence Spectroscopy. third ed. Springer, New York.

Loiselle, S.A., Bracchini, L., Dattilo, A.M., Ricci, M., Tognazzi, A., Rossi, C., 2009. Optical characterization of chromophoric dissolved organic matter using wavelength distribution of absorption spectral slopes. Limnol. Oceanogr. 54, 590–597.

Lønborg, C., Álvarez-Salgado, X.A., Davidson, K., Martínez-García, S., Teira, E., 2010. Assessing the microbial bioavailability and degradation rate constants of dissolved organic matter by fluorescence spectroscopy in the coastal upwelling system of the Ría de Vigo. Mar. Chem. 119, 121–129.

Lübben, A., Dellwig, O., Koch, S., Beck, M., Badewien, T.H., Fischer, S., et al., 2009. Distributions and characteristics of dissolved organic matter in temperate coastal waters (Southern North Sea). Ocean Dyn. 59, 263–275.

Ma, J., Del Vecchio, R., Golanoski, K.S., Boyle, E.S., Blough, N.V., 2010. Optical properties of humic substances and CDOM: effects of borohydride reduction. Environ. Sci. Technol. 44, 5395–5402.

Maie, N., Scully, N.M., Pisani, O., Jaffé, R., 2007. Composition of a protein-like fluorophore of dissolved organic matter in coastal wetland and estuarine ecosystems. Water Res. 41, 563–570.

Mannino, A., Harvey, H.R., 2004. Black carbon in estuarine and coastal ocean dissolved organic matter. Limnol. Oceanogr. 49, 735–740.

Mannino, A., Russ, M.E., Hooker, S.B., 2008. Algorithm development and validation for satellite-derived distributions of DOC and CDOM in the U.S. Middle Atlantic Bight. J. Geophys. Res. 113, C07051.

Miller, W.L., Zepp, R.G., 1995. Photochemical production of dissolved inorganic carbon from terrestrial organic matter: significance to the oceanic organic carbon cycle. Geophys. Res. Lett. 22, 417–420.

Miller, R.L., Belz, M., Castillo, C. Del, Trzaska, R., 2002. Determining CDOM absorption spectra in diverse coastal environments using a multiple pathlength, liquid core waveguide system. Cont. Shelf Res. 22, 1301–1310.

Mizubayashi, K., Kuwahara, V.S., Segaran, T.C., Zaleha, K., Effendy, A.W.M., Kushairi, M.R.M., et al., 2013. Monsoon variability of ultraviolet radiation (UVR) attenuation and bio-optical factors in the Asian tropical coral-reef waters. Estuar. Coast. Shelf Sci. 126, 34–43.

Mopper, K., Zhou, X., Kieber, R.J., Kieber, D.J., Sikorski, R.J., Jones, R.D., 1991. Photochemical degradation of dissolved organic carbon and its impact on the oceanic carbon cycle. Nature 353, 60–62.

Murphy, K.R., Ruiz, G.M., Dunsmuir, W.T.M., Waite, T.D., 2006. Optimized parameters for fluorescence-based verification of ballast water exchange by ships. Environ. Sci. Tech. 40 (7), 2357–2362.

Murphy, K.R., Stedmon, C.A., Waite, T.D., Ruiz, G.M., 2008. Distinguishing between terrestrial and autochthonous organic matter sources in marine environments using fluorescence spectroscopy. Mar. Chem. 108, 40–58.

Murphy, K.R., Butler, K.D., Spencer, R.G.M., Stedmon, C.A., Boehme, J.R., Aiken, G.R., 2010. Measurement of dissolved organic matter fluorescence in aquatic environments: an interlaboratory comparison. Environ. Sci. Technol. 44, 9405–9412.

Murphy, K.R., Stedmon, C.A., Graeber, D., Bro, R., 2013. Fluorescence spectroscopy and multi-way techniques. PARAFAC. Anal. Methods 5, 6557–6566.

Murphy, K.R., Stedmon, C.A., Bro, R., 2014. Chemometric analysis of organic matter fluorescence. In: Coble, P., Lead, J.R., Baker, A., Reynolds, D., Spencer, R.G.M. (Eds.), Aquatic Organic Matter Fluorescence. Cambridge University Press.

Nelson, N.B., Coble, P.G., 2012. Optical analysis of chromophoric dissolved organic matter. In: Wurl, O. (Ed.), Practical Guidelines for the Analysis of Seawater. CRC, Boca Raton, pp. 79–96.

Nelson, N.B., Siegel, D.A., 2002. Chromophoric DOM in the open ocean. In: Biogeochemistry of Marine Dissolved Organic Matter. Hansell, D., Carlson, C. (Eds.), Academic Press, San Diego, pp. 547–578.

Nelson, N.B., Siegel, D.a., 2013. The global distribution and dynamics of chromophoric dissolved organic matter. Ann. Rev. Mar. Sci. 5, 447–476.

Nelson, N.B., Siegel, D.A., Michaels, A.F., 1998. Seasonal dynamics of colored dissolved material in the Sargasso Sea. Deep Sea Res. I 45, 931–957.

Nelson, N.B., Carlson, C.A., Steinberg, D.K., 2004. Production of chromophoric dissolved organic matter by Sargasso Sea microbes. Mar. Chem. 89, 273–287.

Nelson, N.B., Siegel, D.A., Carlson, C.A., Swan, C., Smethie, W.M., Khatiwala, S., 2007. Hydrography of chromophoric dissolved organic matter in the North Atlantic. Deep Sea Res. Part I Oceanogr. Res. Pap. 54, 710–731.

Nelson, N.B., Siegel, D.A., Carlson, C.A., Swan, C.M., 2010. Tracing global biogeochemical cycles and meridional overturning circulation using chromophoric dissolved organic matter. Geophys. Res. Lett. 37, .

Nieto-Cid, M., Álvarez-Salgado, X.A., Gago, J., Pérez, F.F., 2005. DOM fluorescence, a tracer for biogeochemical processes in a coastal upwelling system (NW Iberian Peninsula). Mar. Ecol. Prog. Ser. 297, 33–50.

Nieto-Cid, M., Álvarez-Salgado, X., Pérez, F., 2006. Microbial and photochemical reactivity of fluorescent dissolved organic matter in a coastal upwelling system. Limnol. Oceanogr., 51, 1391–1400.

Norman, L., Thomas, D.N., Stedmon, C.A., Granskog, M.A., Papadimitriou, S., Krapp, R.H., et al., 2011. The characteristics of dissolved organic matter (DOM) and chromophoric dissolved organic matter (CDOM) in Antarctic sea ice. Deep Sea Res. Part II Top. Stud. Oceanogr. 58, 1075–1091.

O'Sullivan, D.W., Neale, P.J., Coffin, R.B., Boyd, T.J., Osburn, C.L., 2005. Photochemical production of hydrogen peroxide and methylhydroperoxide in coastal waters. Mar. Chem. 97, 14–33.

Ortega-retuerta, E., Frazer, T.K., Duarte, C.M., Ruiz-halpern, S., Tovar-sa, A., Arrieta, J.M., et al., 2009. Biogeneration of chromophoric dissolved organic matter by bacteria and krill in the Southern Ocean. Limnol. Oceanogr. 54, 1941–1950.

Osburn, C.L., Handsel, L.T., Mikan, M.P., Paerl, H.W., Montgomery, M.T., 2012. Fluorescence tracking of dissolved and particulate organic matter quality in a river-dominated estuary. Environ. Sci. Technol. 46, 8628–8636.

Perovich, D.K., Roesler, S., Pegau, W.S., 1998. Variability in Arctic sea ice optical properties. J. Geophys. Res. 103, 1193–1208.

Persson, T., Wedborg, M., 2001. Multivariate evaluation of the fluorescence of aquatic organic matter. Anal. Chim. Acta 434, 179–192.

Pope, R.M., Fry, E.S., 1997. Absorption spectrum (380–700 nm) of pure water. II. Integrating cavity measurements. Appl. Opt. 36, 8710–8723.

Qian, J., Mopper, K., Kieber, D.J., 2001. Photochemical production of the hydroxyl radical in Antarctic waters. Deep Sea Res. 48, 741–759.

Reitner, B., Herzig, A., Herndl, G.J., 2002. Photoreactivity and bacterioplankton availability of aliphatic versus aromatic amino acids and a protein. Aquat. Microb. Ecol. 26, 305–311.

Rochelle-Newall, E.J., Fisher, T.R., 2002. Production of chromophoric dissolved organic matter fluorescence in marine and estuarine environments: an investigation into the role of phytoplankton. Mar. Chem. 77, 7–21.

Romera-castillo, C., Sarmento, H., Álvarez-Salgado, X.A., Gasol, J.M., Marrase, C., 2010. Production of chromophoric dissolved organic matter by marine phytoplankton. Limnol. Oceanogr. 55, 446–454.

Röttgers, R., Doerffer, R., 2007. Measurements of optical absorption by chromophoric dissolved organic matter using a point-source integrating-cavity absorption meter. Limnol. Oceanogr. Methods 5, 126–135.

Sasaki, H., Miyamura, T., Saitoh, S., Ishizaka, J., 2005. Seasonal variation of absorption by particles and colored dissolved organic matter (CDOM) in Funka Bay, southwestern Hokkaido, Japan. Estuar. Coast. Shelf Sci. 64, 447–458.

Seritti, A., Morelli, E., Nannicini, L., Vecchio, R. Del, 1994. Production of hydrophobic fluorescent organic matter by the marine diatom *Phaeodactylum tricornutum*. Chemosphere 28, 117–129.

Shimotori, K., Omori, Y., Hama, T., 2009. Bacterial production of marine humic-like fluorescent dissolved organic matter and its biogeochemical importance. Aquat. Microb. Ecol. 58, 55–66.

Shimotori, K., Watanabe, K., Hama, T., 2012. Fluorescence characteristics of humic-like fluorescent dissolved organic matter produced by various taxa of marine bacteria. Aquat. Microb. Ecol. 65, 249–260.

Siegel, D.A., Maritorena, S., Nelson, N.B., Hansell, D.A., Lorenzi-Kayser, M., 2002. Global distribution and dynamics of colored dissolved and detrital organic materials. J. Geophys. Res. 107, 3228.

Siegel, D.A., Maritorena, S., Nelson, N.B., Behrenfeld, M.J., McClain, C.R., 2005. Colored dissolved organic matter and its influence on the satellite-based characterization of the ocean biosphere. Geophys. Res. Lett. 32, L20605.

Skoog, A., Hall, P.O.J., Hulth, S., Paxéus, N., Rutgers van der Loeff, M., Westerlund, S., 1996. Early diagenetic production and sediment-water exchange of fluorescent dissolved organic matter in the coastal environment. Geochim. Cosmochim. Acta 60, 3619–3629.

Skoog, A., Wedborg, M., Fogelqvist, E., 2011. Decoupling of total organic carbon concentrations and humic substance fluorescence in a an extended temperate estuary. Mar. Chem. 124, 68–77.

Smith, R.C., Baker, K.S., 1981. Optical properties of the clearest natural waters (200–800 nm). Appl. Opt. 20, 177–184.

Stedmon, C.A., Bro, R., 2008. Characterizing dissolved organic matter fluorescence with parallel factor analysis: a tutorial Appendix 1. Limnol. Oceanogr. Methods 6, 1–6.

Stedmon, C.A., Markager, S., 2001. The optics of chromophoric dissolved organic matter (CDOM) in the Greenland Sea: an algorithm for differentiation between marine and terrestrially derived organic matter. Limnol. Oceanogr. 46, 2087–2093.

Stedmon, C.A., Markager, S., 2003. Behaviour of the optical properties of coloured dissolved organic matter under conservative mixing. Estuar. Coast. Shelf Sci. 57, 973–979.

Stedmon, C.A., Markager, S., 2005a. Tracing the production and degradation of autochthonous fractions of dissolved organic matter by fluorescence analysis. Limnol. Oceanogr. 50, 1415–1426.

Stedmon, C.A., Markager, S., 2005b. Resolving the variability in dissolved organic matter fluorescence in a temperate estuary and its catchment using PARAFAC analysis. Limnol. Oceanogr. 50, 686–697.

Stedmon, C.A., Markager, S., Kaas, H., 2000. Optical properties and signatures of chromophoric dissolved organic matter (CDOM) in Danish coastal waters. Estuar. Coast. Shelf Sci. 51, 267–278.

Stedmon, C.A., Markager, S., Bro, R., 2003. Tracing dissolved organic matter in aquatic environments using a new approach to fluorescence spectroscopy. Mar. Chem. 82, 239–254.

Stedmon, C.A., Markager, S., Tranvik, L., Kronberg, L., Slätis, T., Martinsen, W., 2007a. Photochemical production of ammonium and transformation of dissolved organic matter in the Baltic Sea. Mar. Chem. 104, 227–240.

Stedmon, C.A., Thomas, D.N., Granskog, M., Kaartokallio, H., Papadimitriou, S., Kuosa, H., 2007b. Characteristics of dissolved organic matter in Baltic coastal sea ice: allochthonous or autochthonous origins? Environ. Sci. Technol. 41, 7273–7279.

Stedmon, C.A., Osburn, C.L., Kragh, T., 2010. Tracing water mass mixing in the Baltic–North Sea transition zone using the optical properties of coloured dissolved organic matter. Estuar. Coast. Shelf Sci. 87, 156–162.

Stedmon, C.A., Thomas, D.N., Papadimitriou, S., Granskog, M.A., Dieckmann, G.S., 2011. Using fluorescence to characterize dissolved organic matter in Antarctic sea ice brines. J. Geophys. Res. 116, G03027.

Steinberg, D.K., Nelson, N.B., Carlson, C.A., Prusak, A.C., 2004. Production of chromophoric dissolved organic matter (CDOM) in the open ocean by zooplankton and the colonial cyanobacterium Trichodesmium spp. Mar. Ecol. Prog. Ser. 267, 45–56.

Stolpe, B., Guo, L., Shiller, A.M., Hassellöv, M., 2010. Size and composition of colloidal organic matter and trace elements in the Mississippi River, Pearl River and the northern Gulf of Mexico, as characterized by flow field-flow fractionation. Mar. Chem. 118, 119–128.

Stramski, D., Wozniak, S.B., 2005. On the role of colloidal particles in light scattering in the ocean. Limnol. Oceanogr. 50, 1581–1591.

Stubbins, A., Niggemann, J., Dittmar, T., 2012. Photo-lability of deep ocean dissolved black carbon. Biogeosciences 9, 1661–1670.

Suksomjit, M., Nagao, S., Ichimi, K., Yamada, T., Tada, K., 2009. Variation of dissolved organic matter and fluorescence characteristics before, during and after phytoplankton. J. Oceanogr. 65, 835–846.

Swan, C.M., Nelson, N.B., Siegel, D.A., Kostadinov, T.S., 2012. The effect of surface irradiance on the absorption spectrum of chromophoric dissolved organic matter in the global ocean. Deep Sea Res. Part I Oceanogr. Res. Pap. 63, 52–64.

Swan, C.M., Nelson, N.B., Siegel, D.A., Fields, E.A., 2013. A model for remote estimation of ultraviolet absorption by chromophoric dissolved organic matter based on the global distribution of spectral slope. Remote Sens. Environ. 136, 277–285.

Thomas, D.N., Dieckmann, G.S., 2010. Sea Ice, second ed. Wiley-Blackwell, Oxford.

Toole, D.A., Kieber, D.J., Kiene, R.P., Siegel, D.A., Nelson, N.B., 2003. Photolysis and the dimethylsulfide (DMS) summer paradox in the Sargasso Sea. Limnol. Oceanogr. 48, 1088–1100.

Tzortziou, M., Osburn, C.L., Neale, P.J., 2007. Photobleaching of dissolved organic material from a tidal marsh-estuarine system of the Chesapeake Bay. Photochem. Photobiol. 83, 782–792.

Urban-Rich, J., McCarty, J.T., Fernández, D., Acuña, J.L., 2006. Larvaceans and copepods excrete fluorescent dissolved organic matter (FDOM). J. Exp. Mar. Biol. Ecol. 332, 96–105.

Urtizberea, A., Dupont, N., Rosland, R., Aksnes, D.L., 2013. Sensitivity of euphotic zone properties to CDOM variations in marine ecosystem models. Ecol. Model. 256, 16–22.

Uusikivi, J., Vähätalo, A.V., Granskog, M.A., Sommaruga, R., 2010. Contribution of mycosporine-like amino acids and colored dissolved and particulate matter to sea ice optical properties and ultraviolet attenuation. Limnol. Oceanogr. 55, 703–713.

Vähätalo, A.V., Wetzel, R.G., 2004. Photochemical and microbial decomposition of chromophoric dissolved organic matter during long (months–years) exposures. Mar. Chem. 89, 313–326.

Vähätalo, A., Zepp, R., 2005. Photochemical mineralization of dissolved organic nitrogen to ammonium in the Baltic Sea. Environ. Sci. Technol. 39, 6985–6992.

Vodacek, A., Hoge, F.E., Swift, R.N., Yungel, J.K., Peltzer, E.T., Blough, N.V., 1995. The use of in situ and airborne fluorescence measurements to determine UV absorption coefficients and DOC concentrations in surface waters. Limnol. Oceanogr. 40, 411–415.

Vodacek, A., Blough, N.V., DeGrandpre, M.D., Peltzer, E.T., Nelson, R.K., 1997. Seasonal variation of CDOM and DOC in the Middle Atlantic Bight: terrestrial inputs and photooxidation. Limnol. Oceanogr. 42, 674–686.

Walker, S.A., Amon, R.M.W., Stedmon, C., Duan, S., Louchouarn, P., 2009. The use of PARAFAC modeling to trace terrestrial dissolved organic matter and fingerprint water masses in coastal Canadian Arctic sur- face waters. J. Geophys. Res. 114, G00F06. doi:10.1029/2009JG000990.

Weishaar, J.L., Aiken, G.R., Bergamaschi, B.A., Fram, M.S., Fujii, R., Mopper, K., 2003. Evaluation of specific ultraviolet absorbance as an indicator of the chemical composition and reactivity of dissolved organic carbon. Environ. Sci. Technol. 37, 4702–4708.

Wolfbeis, O.S., 1985. Fluorescence optical sensors in analytical chemistry. Trends Anal. Chem. 4, 184–188.

Xie, H., Bélanger, S., Song, G., Benner, R., Taalba, A., Blais, M., et al., 2012. Photoproduction of ammonium in the southeastern Beaufort Sea and its biogeochemical implications. Biogeosciences 9, 3047–3061.

Yamashita, Y., Tanoue, E., 2003. Chemical characterization of protein-like fluorophores in DOM in relation to aromatic amino acids. Mar. Chem. 82, 255–271.

Yamashita, Y., Tanoue, E., 2008. Production of bio-refractory fluorescent dissolved organic matter in the ocean interior. Nat. Geosci. 1, 579–582.

Yamashita, Y., Tanoue, E., 2009. Basin scale distribution of chromophoric dissolved organic matter in the Pacific Ocean. Limnol. Oceanogr. 54, 598–609.

Yamashita, Y., Cory, R.M., Nishioka, J., Kuma, K., Tanoue, E., Jaffé, R., 2010. Fluorescence characteristics of dissolved organic matter in the deep waters of the Okhotsk Sea and the northwestern North Pacific Ocean. Deep Sea Res. Part II Top. Stud. Oceanogr. 57, 1478–1485.

Yang, L., Hong, H., Guo, W., Chen, C.A., Pan, P., Feng, C., 2012. Absorption and fluorescence of dissolved organic matter in submarine hydrothermal vents off NE Taiwan. Mar. Chem. 128–129, 64–71.

Zaneveld, J.R.V., Kitchen, J.C., Bricaud, A., Moore, C.C., 1992. Analysis of in-situ spectral absorption meter data. In: Gilbert, G.D. (Ed.), Ocean Optics XI, International Society for Optics and Photonics, San Diego, pp. 187–200.

Zhang, Y., Liu, X., Wang, M., Qin, B., 2013. Compositional differences of chromophoric dissolved organic matter derived from phytoplankton and macrophytes. Org. Geochem. 55, 26–37.

CHAPTER

11

Riverine DOM

Peter A. Raymond*, Robert G.M. Spencer†

*Yale School of Forestry and Environmental Studies, New Haven, Connecticut, USA
†Department of Earth, Ocean and Atmospheric Science, Tallahassee, Florida, USA

CONTENTS

I INTRODUCTION

A Rivers and Oceanic DOM

The steady state concentration of many dissolved chemical constituents in seawater is dependent on the amount delivered by rivers (Goldschmidt, 1932; Mackenzi and Garrels, 1966). In some cases, since rivers are the major input term, the removal rate of a constituent in the ocean can be indirectly assessed if the riverine term is known. Pioneering work on the radiocarbon age of oceanic dissolved

organic matter (DOM) suggested that the riverine input of DOM would be large enough to maintain the steady state age and concentration of oceanic DOM (Bauer et al., 1992; Druffel et al., 1989; Williams et al., 1969). Thus in this case the input rate provided by rivers matched an independent check on the rate of removal, suggesting the source of oceanic DOM could be terrestrial material that aged in the ocean. The importance of rivers to steady state concentrations and chemical character of oceanic DOM motivated oceanographer's early interest in riverine DOM, which due to the complexity of the composition of DOM continues to engage researchers.

Early methods for measuring DOM in inland waters were rather crude, often involving measuring the loss on ignition of a vacuum dried sample, attempting to correct for dissolved salts (Hutchinson, 1957). These methods were employed in lacustrine environments more than fluvial systems (Hutchinson, 1957), and were adopted by other studies looking at average river water concentrations globally (Livingstone, 1963). Method improvements in the 1960s involved oxidizing samples to CO_2 and detecting with infrared analyzers (Menzel and Vaccaro, 1964; Vanhall et al., 1963). An early study using these methods on the Amazon River reported concentrations of dissolved organic carbon (DOC) of ~3.5 mg L^{-1}, or 3-4x greater than surface seawater (Williams, 1968).

B Importance of Riverine DOM to Estuarine and Coastal Processes

The flux of riverine DOM from land to ocean is a major source of reduced carbon to marine environments, with biogeochemical cycling in coastal margins near riverine outflows dominated by the influx of terrestrial organic matter and nutrients. DOM can undergo a variety of biogeochemical reactions in river plumes, estuaries, and at the land-ocean interface that ultimately determines both the concentration of DOC and composition of DOM reaching the ocean. Unsurprisingly, the behavior of terrestrial DOM during estuarine mixing is highly variable. Conservative behavior of DOC in numerous estuaries (Abril et al., 2002; Alvarez-Salgado and Miller, 1998; Mantoura and Woodward, 1983) reflects comparatively minor estuarine modification of the terrestrial DOM. In other estuaries, nonconservative behavior at low salinities has been attributed to the removal of specific components of the DOM pool (Benner and Opsahl, 2001; Hernes and Benner, 2003), and DOM composition has been shown to exert a control on the

degree of DOC removal (Spencer et al., 2007a). Nonconservative behavior of DOM within estuaries can also result from inputs from sources such as anthropogenic pollution (e.g., the Scheldt and the Tyne), phytoplankton (e.g., Chesapeake Bay and the Mississippi), intertidal areas (e.g., the Sado and the Ems), salt marshes, tributaries (e.g., the Pearl), and desorption from sediments and the flushing of porewaters (e.g., the Tamar and the Gironde) (Abril et al., 1999, 2002; Benner and Opsahl, 2001; Chen and Gardner, 2004; Raymond and Hopkinson, 2003; Spencer et al., 2007b). A number of factors ultimately play into determining how much DOC, and the composition of DOM, exported to coastal waters, including flocculation, adsorption onto suspended sediments, and microbial and photochemical degradation (Hernes and Benner, 2003; Shank et al., 2005; Uher et al., 2001).

Coastal and estuarine microbial communities may be better adapted at consuming terrestrial DOM than riverine microbial communities (Fellman et al., 2010; Stepanauskas et al., 1999; Wikner et al., 1999). Increasing ionic strength also causes biochemical and compositional changes of DOM that can lead to greater bacterial utilization (Dehaan et al., 1987; Kerner et al., 2003). However, Fellman et al. (2010) found no significant difference in DOC utilization using a riverine microbial inoculation added to both river water and river water made up to different salinities with an artificial salt mixture. Marine microbes have the potential for more intensive synthesis of extracellular enzymes with greater diversity, thus allowing them to more effectively metabolize terrestrial DOM (Stepanauskas et al., 1999). Riverine and estuarine microbial communities also show different preferences with respect to degrading various components of the DOM pool (Fellman et al., 2010; Sondergaard et al., 2003). However, some studies have found no significant difference in DOC utilization between marine and freshwater microbial communities (Langenheder et al., 2003; Sondergaard et al., 2003). Coastal regions characterized as receiving elevated riverine DOM and nutrient inputs with microbial communities that are highly adaptable to shifting physico-chemical gradients (e.g., light and salinity) have been deemed hot spots for priming effects on terrestrial DOM (Bianchi, 2011).

Naturally, riverine inputs of DOM impact ocean DOM where inputs are greatest, as is apparent in dissolved lignin phenol (a vascular plant biomarker) concentrations in surface waters. For example, lignin phenol concentrations in Arctic surface waters, a major marine system exhibiting the highest river input to volume ratio, are ~10 times greater than reported for the Atlantic, which are in turn 40% greater than those in the Pacific (Hernes and Benner, 2006). The Arctic Ocean receives ~11% of global riverine discharge into ~1% of the global ocean volume, thus imparting estuarine gradients throughout the Arctic Ocean (McClelland et al., 2012). Similarly, concentrations of dissolved black carbon originating from combustion sources (e.g., wildfires, fossil fuel burning) are elevated in coastal waters likely due to continental runoff (Dittmar et al., 2012b). Riverine DOM has a strong impact on the bio-optical properties of the coastal ocean due to its chromophoric character (CDOM), which in coastal waters dominates the inherent light absorption at ultraviolet and blue wavelengths (e.g., 20-70% at 440 nm; Del Vecchio and Subramaniam, 2004; Mannino et al., 2008). Utilizing satellite imagery, and with appropriate algorithm development, CDOM can trace terrestrial inputs of DOC in coastal regions (Fichot and Benner, 2012; Mannino et al., 2008).

II LAND TRANSPORT

With respect to lateral transport of terrestrial material, there are two important fluxes to consider: the mobilization of terrestrial DOM to inland waters and the export of terrestrial and riverine DOM to coastal waters. The latter most interests oceanographers, but the former is important to understanding the amount and character of DOM

that ultimately reaches the ocean. In small headwater streams, almost all of the DOC exported is of terrestrial origin (Hynes, 1963; Royer and David, 2005). During transport, however, this material can be removed by heterotrophs (Fisher and Likens, 1972; Sondergaard and Middelboe, 1995), by photo-oxidation (Gjessing and Gjerdahl, 1970; Moran and Hodson, 1994), and by flocculation (Sholkovitz, 1976). Furthermore, autochthonous DOM can be added by algae (Allen, 1956; Baines and Pace, 1991a), wetlands and submerged vegetation (Mulholland and Kuenzler, 1979; Raymond and Hopkinson, 2003), and anthropogenic point sources such as sewage (Griffith and Raymond, 2011).

Wetlands are an important control on the transport of DOM off the terrestrial landscape (Wetzel, 1992). The constant contact of wetland and emergent vegetation with water, and the low oxygen content of soils, result in the transfer of large amounts of OM as DOM into inland waters. River floodplains can also simulate wetlands during flood periods, exporting appreciable DOM (Shen et al., 2012; Tockner et al., 1999). This material is generally modern in age (Mayorga et al., 2005; Raymond and Hopkinson, 2003) and contains a biolabile component (Findlay et al., 1992). It is generally less microbially available than phytoplankton exudates and other riverine DOM sources (Mann and Wetzel, 1995; Moran and Hodson, 1994; Wetzel, 1992) but highly photo-reactive (Franke et al., 2012).

Upland ecosystems are increasingly seen as important to drainage network DOM delivery. Fluxes off the landscape during spring runoff in temperate and high-latitude systems can be high when hydrologic flow paths do not interact with wetlands (Laudon et al., 2004). Furthermore the export of DOM from upland systems during large precipitation events is high, and studies which attempt to determine the importance of events are demonstrating a larger lateral transport than previously estimated by studies that have not explicitly sampled events (Raymond and Saiers, 2010).

The largest storms can be responsible for upwards of half of long-term annual average DOC fluxes (Schiff et al., 1998; Yoon and Raymond, 2012), and due to the bypassing of watershed filters, have impacts on inland and coastal carbon fluxes (Bianchi et al., 2013; Klug et al., 2012). The predicted increased proportion of annual rainfall that will be delivered with large events forced by climate change (IPCC, 2012) should therefore significantly impact these lateral fluxes (Jeong et al., 2012; Sebestyen et al., 2009) and coastal ocean processes.

The fraction of terrestrial DOM removed during transport to the ocean is not currently well constrained. DOM from small headwater systems demonstrates a significant biolabile component (Buffam et al., 2001; Fellman et al., 2009a; Holmes et al., 2008; Volk et al., 1997; Wilson et al., 2013b). DOM exported from headwater streams during high-flow events has a high degree of aromaticity and therefore is highly photo-reactive (Fellman et al., 2009a; Vidon et al., 2008); laboratory experiments have confirmed a high degree of photoreactions of riverine DOM (Franke et al., 2012; Gao and Zepp, 1998). The light-absorbing properties of this DOM can influence phytoplankton structure and production by both changing the depth of the photic zone and fractionating the wavelengths of available light (Frenette et al., 2012). Photoreactions are also responsible for the direct loss and transformation of DOM. Most work to date on DOM lability, both from bacteria and sunlight, has been done using laboratory experiments where DOM is incubated over time under controlled conditions. A few studies have directly injected DOM into streams demonstrating a highly biolabile pool and a much larger pool that is conservatively transported (Kaplan et al., 2008). It is, however, difficult to undertake these experiments with DOM that is representative of stream DOM during high-flow periods or in major riverine systems. Lauerwald et al. (2012) assessed the total amount of DOM exported to the oceans from large watersheds and the amount exported from small sub-watersheds, finding

that 75% of DOM exported from land makes it to the ocean. The smallest headwater streams are not commonly monitored and therefore were excluded from this analysis. The 25% loss is net, however, due to the addition of autochthonous DOM during transport; the total amount of terrestrial DOM removed has to be >25%.

It is difficult to determine how much autochthonous DOM is added during transport to the coast. Within rivers there can be large additions from submerged and emergent wetlands (Tzortziou et al., 2011; Wetzel, 1992). Phytoplankton also exude a significant percentage of gross primary production as DOM (Baines and Pace, 1991b), which is often detected in river waters (Massicotte and Frenette, 2011; Raymond and Bauer, 2001a). The light-limited nature of many rivers and short transport times, however, often reduces the importance of authochthonous inputs to low-flow summer periods (Goni et al., 2003; Helie and Hillaire-Marcel, 2006; Roach, 2013), resulting in a low contribution of autochthonous DOM to annual riverine DOM fluxes (Stepanauskas et al., 2005). This outcome might not hold true for large rivers that have high nutrient loadings and large reservoirs (Bianchi et al., 2004). In low-flow rivers with large water withdrawals for drinking water or irrigation, sewage DOM can be detected (Butman et al., 2012; He et al., 2010), making important contributions to heterotrophic processes (Griffith and Raymond, 2011).

A Global Fluxes

Several estimates of global DOC flux from land to ocean via rivers have been made. Early studies grouped DOC and particulate organic carbon (POC) together for a flux of ~0.4 PgC year^{-1} (Schlesinger and Melack, 1981). Meybeck (1982) grouped a limited number of river DOC measurements into four river classes to estimate a global flux of 0.22 PgC year^{-1}. Meybeck (1993) later improved upon this estimate with measurements from ~40 rivers, reporting a flux of 0.20 PgC year^{-1}. Ludwig et al. (1996) improved upon this using a similar set of riverine DOC measurements but employing an empirical relationship between

DOC flux and discharge, slope and soil C to obtain a flux of 0.21 PgC year^{-1}. Cauwet (2002) used a similar typology to Meybeck but updated the DOC concentrations to obtain a global flux of 0.24 PgC year^{-1}. Dai et al. (2012) used a similar approach, binning measurements from 118 rivers to different typologies to report a flux of 0.20-0.21 PgC year^{-1}. Aitkenhead and McDowell (2000) binned river DOC fluxes from 164 rivers into biomes to describe a relationship between river DOC flux and soil C:N, then used this relationship to obtain a flux of 0.36 PgC year^{-1}. GlobalNEWS has estimated DOC fluxes based on discharge and wetland extent (Seitzinger et al., 2005). Early estimates from GlobalNEWS, calibrated using measurements from 68 large river systems, had fluxes of ~0.17 PgC year^{-1} (Mayorga et al., 2010; Seitzinger et al., 2005).

Although the estimates of global river DOC export are in a narrow range, they are still rather coarse. Estimates generally share the same training data taken from a limited number of systems with insufficient temporal sampling (i.e., absent event sampling) required for accurate estimates of DOC fluxes. Some regions, such as the Arctic, are not dealt with well in GlobalNEWS, resulting in inaccurately low fluxes from this region. The estimates also share a similar global discharge data set. Annual variations in discharge, mostly driven by changes in precipitation, are 32,000-41,000 km^3 year^{-1} or ~25% (Wisser et al., 2010), thus annual variations in riverine carbon fluxes may vary by ~20-30%. Dai et al. (2012) reported a large uncertainty in the annual flux of 30%. Thus advances in global/regional export estimates are still possible.

The top 30 rivers with respect to discharge contribute 51% of the total annual global discharge, with the Amazon providing >18%. These rivers drain 36% of the Earth's exorheic land surface, providing vast quantities of terrestrially derived DOM to the ocean. In total, these rivers export 90.2 Tg DOC-C year^{-1} as DOC (Table 11.1; Figure 11.1a), or 36% of global DOC flux to the ocean (250 Tg DOC-C year^{-1}; Hedges et al., 1997). The Amazon River alone exports DOC

at ~27 Tg DOC-C year^{-1} at Obidos (Table 11.1; Moreira-Turcq et al., 2003) or 11% of the total DOC flux to the oceans. However, if downstream major lowland tributaries such as the Tapajos and Xingu Rivers are included (1.50 and 0.95 Tg DOC-C year^{-1}, respectively) the total flux for the Amazon is 29.35 Tg DOC-C year^{-1} (Coynel et al., 2005; Moreira-Turcq et al., 2003), 11.7% of the global DOC flux. A number of other major tropical rivers are central to DOC flux as well, with the Congo the second largest exporter (12.4 Tg DOC-C year^{-1}; 5% of global total) and the Parana and Orinoco important as well (Table 11.1; Figure 11.1a). The major Russian Arctic rivers, such as the Lena (5.68 Tg DOC-C year^{-1}), Yenisey (4.65 Tg DOC-C year^{-1}), and Ob' (4.12 Tg DOC-C year^{-1}), support large fluxes of DOC to the Arctic Ocean (Raymond et al., 2007); these fluxes are considerably greater than the next principal rivers (e.g., Amur at 2.50 Tg DOC-C year^{-1} and Mississippi at 2.10 Tg DOC-C year^{-1}; Table 11.1; Figure 11.1a).

A number of these major rivers are important point or individual sources (e.g., the Congo to the Atlantic Ocean) while others merge with other substantial rivers at their outlets to the ocean, particularly in deltaic systems, thus representing combined DOM sources to the ocean. For example, the Ganges (1.70 Tg DOC-C year^{-1}) joins the Brahmaputra (1.90 Tg DOC-C year^{-1}) and subsequently the Meghna River (annual discharge = 111 km^3 year^{-1}; Meybeck and Ragu, 1996) before together discharging through the Ganges-Brahmaputra-Meghna delta into the Bay of Bengal. Similarly, the Irrawaddy (0.89 Tg DOC-C year^{-1}) and the Salween (0.23 Tg DOC-C year^{-1}) discharge into the Gulf of Martaban in the eastern Indian Ocean along a similar length of deltaic coast to the Ganges-Brahmaputra (Bird et al., 2008). At the mouth of the Amazon, the Tocantins River also joins the Amazon delta, exporting DOC at 30.47 Tg DOC-C year^{-1} (Table 11.1). Naturally, rivers in the top 30 globally also share deltaic systems with other smaller rivers, for example, the Mississippi (2.10 Tg DOC-C year^{-1}) and the Atchafalya (1.19 Tg DOC-C year^{-1}; Spencer et al., 2013) before entering the Gulf of Mexico.

TABLE 11.1 Discharge, watershed area and DOC fluxes and yields for the top 30 rivers ranked by discharge globally. Discharge and watershed area estimates are from Meybeck and Ragu (1996) if not specified.

River Rank by Discharge	River Name	Discharge (km³/yr)	Area (Mkm²)	DOC Flux (TgC/yr)	Global DOC Flux (%)	DOC Yield (gC/m²/yr)
1	Amazon	6590	6.112	26.90[d]	10.8	4.4
2	Congo	1325[a]	3.698	12.40[a]	5.0	3.4
3	Orinoco	1135	1.100	4.98[e]	2.0	4.5
4	Changjiang (Yangtze)	928	1.808	1.58[f]	0.6	0.9
5	Yenisey	673[b]	2.540[b]	4.65[b]	1.9	1.8
6	Lena	588[b]	2.460[b]	5.68[b]	2.3	2.3
7	Mississippi	580	2.980	2.10[g]	0.8	0.7
8	Parana	568	2.783	5.92[h]	2.4	2.1
9	Brahmaputra	510	0.580	1.90[h]	0.8	3.3
10	Ganges	493	1.050	1.70[h]	0.7	1.6
11	Irrawaddy (Ayeyarwady)	486	0.410	0.89[i]	0.4	2.2
12	Mekong	467	0.795	1.11*	0.4	1.4
13	Ob'	427[b]	2.990[b]	4.12[b]	1.6	1.4
14	Tocantins	372	0.757	1.12*	0.4	1.5
15	Amur	344	1.855	2.50[j]	1.0	1.3
16	St. Lawrence	337	1.780	1.55[h]	0.6	0.9
17	Mackenzie	316[b]	1.780[b]	1.38[b]	0.6	0.8
18	Zhujiang (Pearl)	280[c]	0.437	0.40[c]	0.2	0.9
19	Magdalena	237	0.235	0.47*	0.2	2.0
20	Columbia	236	0.669	0.40[g]	0.2	0.6
21	Salween (Thanlwin)	211	0.325	0.23[i]	0.1	0.7
22	Yukon	208[b]	0.830[b]	1.47[b]	0.6	1.8
23	Danube	207	0.817	0.59[k]	0.2	0.7
24	Essequibo	178	0.164	0.89*	0.4	5.4
25	Niger	154	1.200	0.53[h]	0.2	0.4
26	Ogooue	150	0.205	1.25*	0.5	6.1
27	Uruguay	145	0.240	0.50[h]	0.2	2.1

River Rank by Discharge	River Name	Discharge (km³/yr)	Area (Mkm²)	DOC Flux (TgC/yr)	Global DOC Flux (%)	DOC Yield (gC/m²/yr)
28	Fly	141	0.064	0.55*	0.2	8.6
29	Kolyma	136[b]	0.650[b]	0.82[b]	0.3	1.3
30	Pechora	131	0.324	1.66[l]	0.7	5.1
Sum		18553	41.6	90.2	36.1	–
Total		36000[m]	116.0[n]	250.0[o]	100.0	–
Average		–	–	–	–	2.3

[a]Coynel et al., (2005)

[b]Holmes et al., (2012)

[c]Ni et al., (2008) estimates for the whole Pearl Delta

[d]Moreira-Turcq et al., (2003) estimate for Obidos see section 2.1, for discussion

[e]Lewis and Saunders (1989);

[f]Wang et al., (2012)

[g]Spencer et al., (2013)

[h]Degens et al., (1991)

[i]Bird et al., (2008)

[j]Nakatsuka et al., (2004)

[k]Cauwet (2002)

[l]Gordeev et al., (1996)

[m]Milliman and Farnsworth (2011)

[n]Vorosmarty et al., (2000) represents the exorheic land mass (87% of the total nonglacierized land mass)

[o]Hedges et al., (1997)

*Estimated DOC flux: Mekong mean value for DOC concentration data from across the annual hydrograph (2.38 mg/L; range = 1.75 to 3.19 mg/L; Spencer unpublished data) multiplied by annual river discharge data; Tocantins assumed an average DOC concentration of 3 mg/L multiplied by annual river discharge data as no DOC concentration data are currently available but it is described as a clear-water river similar to the Tapajos and Xingu Rivers (Goulding et al., 2003) that have DOC concentrations typically in the 3 to 5 mg/L range (Moreira-Turcq et al., 2003); Magdalena assumed an average DOC concentration of 2 mg/L multiplied by annual river discharge data as no DOC concentration data are currently available but it is an extremely sediment-rich river that typically exhibits low DOC concentrations and estimates for total dissolved solids that are approximately eight-fold higher than the estimated DOC flux (30×10⁶ t/yr; Restrepo et al., 2006); Essequibo assumed an average DOC concentration of 5 mg/L multiplied by annual river discharge data as no DOC concentration data are currently available but it is described as a black-water river (Hammond, 2005); Ogooue mean value for DOC concentration of 8.33 mg/L (Cadee et al., 1984) multiplied by annual river discharge data; Fly mean value for DOC concentration of 3.85 mg/L (Alin et al., 2008) multiplied by annual river discharge data.

DOC yields (i.e., normalization of the carbon load to the watershed area) for the 30 major rivers ranges from $0.4\,gC\,m^{-2}\,year^{-1}$ in the Niger to $8.6\,gC\,m^{-2}\,year^{-1}$ in the Fly (Table 11.1; Figure 11.1b). At a total DOC flux to the ocean of $250\,Tg\,DOC\text{-}C\,year^{-1}$ (Hedges et al., 1997), the average DOC yield from the remaining rivers is $2.15\,gC\,m^{-2}\,year^{-1}$, whereas the average DOC yield from the top 30 rivers is $2.3\,gC\,m^{-2}\,year^{-1}$. Thus, the major rivers have on average slightly elevated yields in relation to the remaining rivers or, alternatively, the global DOC flux to the oceans is underestimated. If the average DOC yield from the top 30 rivers is applied to the remaining exorheic land surface area that they do not drain, the total DOC flux to the ocean is $\sim260\,Tg\,DOC\text{-}C\,year^{-1}$. A number of recent studies have highlighted DOC fluxes from major global rivers that are elevated in comparison to historic estimates, particularly the major Arctic rivers (Holmes et al., 2012; Raymond et al., 2007; Spencer et al., 2009a). For example, Holmes et al. (2012) note that recent estimates of pan-Arctic riverine flux to the ocean range from 34 to $38\,Tg\,DOC\text{-}C\,year^{-1}$, much higher

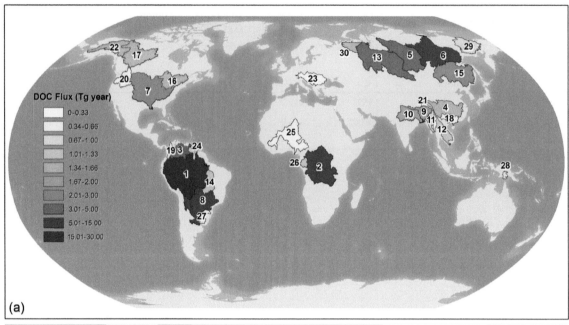

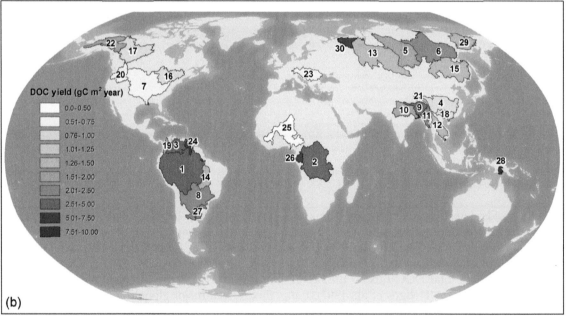

FIGURE 11.1   DOC fluxes (a) and yields (b) for the top 30 global rivers ranked by discharge. *Data from Table 1.*

than previous estimates of 18-26 Tg DOC-C year^{-1}. These improved estimates result from higher temporal resolution sampling of the major Arctic rivers (Lena, Yenisey, Ob', Mackenzie, Yukon, and Kolyma), including sampling over the spring freshet. Even these improved DOC flux estimates may be underestimated as the limited data that exist for other northern high-latitude rivers often show higher DOC yields (e.g., the Pechora River; Table 11.1; Gordeev et al., 1996; Lobbes et al., 2000).

Relative to DOC yields from the top 30 rivers, smaller tropical watersheds exhibit higher yields, such as the Fly (8.6 gC m^{-2} year^{-1}), Ogooue (6.1 gC m^{-2} year^{-1}), and the Essequibo (5.4 gC m^{-2} year^{-1}; Table 11.1; Figure 11.1b). Major tropical rivers such as the Amazon, Congo, and Orinoco exhibit high yields for DOC (4.4, 3.4, and 4.5 gC m^{-2} year^{-1}, respectively). The Pechora River displays a high yield (5.13 gC m^{-2} year^{-1}; Table 11.1; Figure 11.1b), attributable to vast tracts of forest and swamp, including organic carbon-rich soils in the watershed (Gordeev et al., 1996; Tarnocai et al., 2009). Watersheds characterized by elevated DOC yields are important areas for future study in order to improve DOC flux estimates to the ocean.

B Coastal Vegetation Inputs

The coastal ocean also receives substantial inputs of DOM from fringing wetlands and mangroves. The global areas of salt marshes, mangroves, and seagrasses are estimated at ~400,000, 200,000, and 400,000 km^2, respectively (Jennerjahn and Ittekkot, 2002; McLeod et al., 2011). Although relatively small areas, these systems export a large amount of DOM per unit area due to consistently flooded root systems and direct contact between plants, their detritus, and water. Direct export of DOM from these systems is ~10-100 gC m^{-2} year^{-1} (Adame and Lovelock, 2011; Barron and Duarte, 2009; Childers et al., 2000; Dame et al., 1991; Dittmar et al., 2006; Gonneea et al., 2004; Happ et al., 1977; Ziegler and Benner, 1999). Using a surface area of 1,000,000 km^2, a global flux of

10-100 Tg DOC-C year^{-1} is 4-40% of global riverine input. However, a large portion of exported detrital material is transformed to DOM in coastal waters (Dittmar et al., 2006). Remembering that the productivities of these systems are among the highest in the world (Jennerjahn and Ittekkot, 2002), this indirect DOC production in coastal waters could be important. For mangrove forests, the export of litter is ~100-300 gC m^{-2} year^{-1} (Gong and Ong, 1990; Twilley et al., 1997; Woodroffe et al., 1988), with ~50% of this exported detritus converted to DOC in coastal waters (Adame and Lovelock, 2011; Dittmar et al., 2006). Data on litter export from other coastal vegetated ecosystems are sparse, but assuming the range for mangroves (50-150 g indirect DOC production m^{-2} year^{-1}) holds for these other sites suggests an additional flux of 50-150 Tg DOC-C year^{-1}. Combining the direct and indirect input of DOC from these ecosystems results in a flux that is 24-100% of riverine input. Material exported from salt marshes and mangroves may be particularly important to CDOM input to coastal waters (Clark et al., 2008).

III RIVERINE DOM COMPOSITION

DOM composition in riverine systems strongly influences DOM's role in the environment. For example, DOM is a freshwater quality constituent of concern, impacting the formation of carcinogenic and mutagenic disinfection by-products (Chow et al., 2007; Weishaar et al., 2003) as well as the transport and reactivity of toxic substances such as mercury (Aiken et al., 2011; Bergamaschi et al., 2011). Furthermore, riverine DOM composition influences bacterioplankton community structure and function (Crump et al., 2009), so composition data for rivers and streams is of interest to a diverse range of scientists and engineers. To characterize DOM in riverine ecosystems, researchers have focused on bulk properties such as C:N ratios, stable (δ^{13}C, δ^{15}N) and radiocarbon isotopes (Δ^{14}C), optical properties (CDOM absorbance and fluorescence), as well as biomarkers (e.g., amino acids, carbohydrates, black carbon, and lignin phenols). High-resolution analytical techniques, such as Fourier transform ion cyclotron mass spectrometry (FT-ICR-MS) and advanced nuclear magnetic resonance (NMR) spectroscopy, are now also routinely used to investigate riverine DOM.

Riverine DOM was typically perceived as relatively stable from a geochemical standpoint, representing the degraded remains of vascular plant materials aged in soils (Hedges et al., 1994, 1997). A number of studies have shown however that DOC is predominantly modern and younger than POC and therefore derived from recent plant production (Butman et al., 2012; Mayorga et al., 2005; Raymond and Bauer, 2001b; Raymond et al., 2007; Spencer et al., 2012). Furthermore, the importance of phase history has been highlighted by a number of studies when utilizing biomarkers to assess DOM reactivity (Aufdenkampe et al., 2001; Hernes et al., 2007). For example, the vascular plant biomarker lignin has been used extensively in studies of riverine DOM, and studies have established that degradation of plant tissues leads to increased proportions of oxidized (i.e., acidic) lignin phenols (Opsahl and Benner, 1998), making acid:aldehyde ratios of lignin phenols relative indictors of diagenetic state. In the Amazon River Basin, elevated acid:aldehyde ratios in riverine DOM relative to the ratios in POM has been cited as evidence that DOM is more highly biologically degraded than POM (Ertel et al., 1986; Hedges et al., 2000). However, Hernes et al. (2007) showed that elevated acid:aldehyde ratios can be produced in DOM through the abiotic processes of dissolution and sorption. Similarly, Aufdenkampe et al. (2001) demonstrated the same process with amino acids, finding that young DOM gains a degraded signature from dissolution and sorption. The apparent degraded biochemical signature of riverine DOM, along with modern radiocarbon ages of DOC, has been reported in systems from the Arctic to the equator (Raymond et al., 2007; Spencer et al., 2008, 2012). As such, physical processes need to be accounted for when interpreting

biomarkers in riverine DOM; these findings reconcile earlier reports that apparently stable riverine DOM had little presence in the ocean (Hedges et al., 1997).

A ^{14}C-Age of River DOM

The age of riverine DOM provides information on its sources as well as residence times in the Earth's carbon cycle. Riverine organic matter pools are made up of complex mixtures of compounds of varying ages. In some instances, age is linked to its reactivity, with young pools being comprised of OM of high nutritional content that is removed quickly, while older pools are highly oxidized and resistant to further degradation (Loh et al., 2008). There are mechanisms such as soil freezing (Goulden et al., 1998), high water levels in wetlands (Laiho, 2006; Oechel et al., 1993), and mineral layer physical protection of OM bound to soils (Keil et al., 1994; Mayer, 1994) that can remove biolabile pools from active cycling. However, these stored pools can be released through disturbance or natural changes in physical conditions, altering the bulk age of the riverine OM (Evans et al., 2007). Finally, there are petrochemical, "^{14}C-dead" sources of DOM that can have a large impact on the river's DOC ages (Griffith et al., 2009), potentially fueling riverine metabolic processes (Griffith and Raymond, 2011).

There are a growing number of ^{14}C measurements for river DOC. One of the first studies reported ages varying from 1384 years B.P. to modern, with most samples being enriched in bomb carbon (Raymond and Bauer, 2001b). This general trend, a range of riverine DOC ages with a predominance of young ages, has held in subsequent compilations and surveys (Butman et al., 2012; Mayorga et al., 2005; Raymond et al., 2004; Sickman et al., 2010), underscoring the complex nature and multiple sources of river DOM.

These studies provide clues as to the major controls on DOC age. Relatively undisturbed small watersheds demonstrate the propensity for DOM to be young (Longworth et al., 2007). Even in watersheds underlain by shale (Longworth et al.,

2007) and peat (Evans et al., 2007), where one might expect to encounter older DOC, ^{14}C-enrichment dominates. Although there are reports of older DOC with baseflow (Neff et al., 2006; Raymond et al., 2007; Schiff et al., 1997), low-flow periods are inherently less important with respect to fluxes and, therefore, to the average age of DOC exported to the ocean. The dominance of young age is consistent with DOM exported from wetlands and terrestrial systems originating in upper soil profiles (Mayorga et al., 2005). The exception to young DOC can be found in rivers that are rapidly eroding old soil profiles (Masiello and Druffel, 2001) or systems with thawed permafrost or glacier inputs (Hood et al., 2009; Vonk et al., 2013).

Land use and disturbance results in export of aged components. Agricultural watersheds have been shown to export aged DOC (Longworth et al., 2007; Sickman et al., 2010), presumably due to soil disruption and the use of petroleum-based agrochemicals. Disturbance of peatlands also releases older DOC (Moore et al., 2013). The legacy of land use on DOM export can remain for decades (Dittmar et al., 2012a).

Petroleum byproducts make important contributions to riverine DOM. As mentioned, agricultural systems have demonstrated old DOC likely due to agrochemicals (Sickman et al., 2010). One of the first ^{14}C-DOC studies in rivers demonstrated the input of petroleum DOM to rivers from cities (Spiker and Rubin, 1975). Twenty-five percent of DOM in wastewater treatment plants appears to be from petroleum-based household products (Griffith et al., 2009). Glacial streams in both Alaska (Hood et al., 2009) and Europe (Singer et al., 2012) hold old DOC, a fraction of which may be from anthropogenic aerosols (Stubbins et al., 2012) that are bioavailable (Hood et al., 2009; Singer et al., 2012). The potential for human activities to add aged DOC to rivers has been demonstrated in a recent survey of large U.S. rivers (Butman et al., 2012). It appears that the majority of DOM exported to the global ocean from natural ecosystems is young, while human activities inject ancient DOC into riverine systems and the ocean.

B Microbially Reactive Fraction

Rivers and streams are major sources of CO_2 to the atmosphere (Aufdenkampe et al., 2011; Butman and Raymond, 2011; Cole et al., 2007; Striegl et al., 2012). The degree to which riverine DOM fuels this efflux is debatable but recent work suggests that biolabile terrestrial DOM is an important source. Firstly, riverine DOM has been shown to be more biolabile in Arctic Rivers in conjunction with an elevated lignin carbon-normalized yield (Λ_8) and a modern radiocarbon age (Holmes et al., 2008; Raymond and Oh, 2007; Spencer et al., 2008). Secondly, lignin phenols and a host of other phenolic compounds, largely derived from terrestrial macromolecules, are quickly mineralized in Amazon River water (Ward et al., 2013). The breakdown of these terrestrially derived macromolecules appears to be the primary driver for river-to-air CO_2 fluxes in the Amazon (Mayorga et al., 2005; Ward et al., 2013). Finally, the emerging view of the utility of biomarkers such as lignin with respect to the information these biomarkers provide is supported from recent studies from the soil organic matter community that highlight environmental factors can mediate molecular structure effects with respect to the long-term persistence of organic matter (Kleber and Johnson, 2010; Schmidt et al., 2011). The inherent ability of specific molecular structures to resist microbial degradation may constrain the use of some historically utilized indicators of chemical stability.

The radiocarbon age of DOC is also a poor predictor of riverine DOM biolability. The most striking example comes from DOC mobilized from Siberian Yedoma permafrost (DOC > 21,000 ^{14}C year), which was highly biolabile (34 ± 0.8% decomposed during 14 day incubations under dark, oxygenated conditions at ambient river temperatures) (Vonk et al., 2013). In a similar vein, glacial DOM is both aged and biologically labile (Hood et al., 2009; Singer et al., 2012). Anthropogenic aerosols have been suggested as the source of this aged organic matter, given low levels of vascular plant-derived compounds and the presence of combustions products commonly found in anthropogenic aerosols (Stubbins et al., 2012). These combustion products were enriched in condensed aromatics, aliphatic compounds, and particularly fatty acids with chain lengths 30+ carbons, which is compatible with a water-soluble organic carbon aerosol source originating from the incomplete combustion of fossil fuels (Stubbins et al., 2012).

These examples do not mean that biochemical composition is not important to DOM biolability but that the influence of composition depends on environmental controls (Schmidt et al., 2011). A number of studies have linked DOM composition and shifts in radiocarbon age to its biolability (Butman et al., 2007; Holmes et al., 2008; Raymond et al., 2007). For example, biodegradation has been shown to preferentially remove specific components of the DOM pool such as amino acids and carbohydrates (Benner and Kaiser, 2011; Volk et al., 1997; Weiss and Simon, 1999). Lignin, for example, is more resistant to biodegradation than other components in plant material such as carbohydrates (Benner et al., 1987). Elevated protein-like fluorescence in DOM, determined in optical assessments, has been linked to elevated biolability (Balcarczyk et al., 2009; Fellman et al., 2009b; Hood et al., 2009; Wickland et al., 2012). Finally, the conventional wisdom suggesting that younger, relatively unaltered DOM should be more easily metabolized by microbial communities has been demonstrated in the Amazon and Arctic rivers (Holmes et al., 2008; Mayorga et al., 2005; Raymond et al., 2007).

DOM composition impacts its biolability but other environmental factors are also at play. For example, the assimilation capability of the ambient microbial community must be taken into account; physicochemical gradients and the potential of priming effects will all impact DOM biolability (Bianchi, 2011). Linking land to ocean, rivers and their receiving estuaries, and coastal waters provide numerous environmental hot spots for DOM biodegradation (e.g., soil pore-waters, hyporheic zones, the confluence of rivers with contrasting chemistries, the freshwater-seawater interface, and coastal zones) (McClain et al., 2003; Morris et al., 1978).

C Photochemically Labile Fraction

Although photochemical mineralization of DOM in rivers and estuaries is typically minimal (Stubbins et al., 2011; White et al., 2010), the material is highly susceptible upon dilution with ocean waters in plumes and coastal waters (Moran et al., 2000; Vodacek and Blough, 1997). Photochemical degradation preferentially removes components commonly utilized to track riverine inputs in the ocean, such as CDOM and lignin phenols, and also enriches DOM δ^{13}C (Benner and Kaiser, 2011; Hernes and Benner, 2003; Opsahl and Benner, 1998; Opsahl and Zepp, 2001; Spencer et al., 2009b).

A typical response of photochemical degradation is observed in a 57-day irradiation experiment of Congo River water (Spencer et al., 2009b; Stubbins et al., 2010; Figure 11.2). Photochemical loss of DOC was slower than of CDOM by photobleaching (Figure 11.2a), consistent with previous studies (Moran et al., 2000; Vähätalo et al., 2010). Lignin phenol concentrations showed a >95% decrease, highlighting the susceptibility of dissolved lignin to photochemical degradation as with past studies (Hernes and Benner, 2003; Opsahl and Benner, 1998; Opsahl and Zepp, 2001). The preferential photodegradation of aromatic compounds such as lignin results in an order of magnitude drop in the DOC-specific absorption coefficient of CDOM at α_{325} (α_{CDOM}^{*} [$m^2\,g^{-1}$] $=\alpha_{325}\,m^{-1}/DOC\,g\,m^3$) from $1.90\,m^2\,g^{-1}$ in the initial sample to $0.19\,m^2\,g^{-1}$ following irradiation (Stubbins et al., 2010). This latter value is within the range reported previously for Atlantic waters (<0.1 to $0.3\,m^2\,g^{-1}$; Nelson et al., 2007). Similarly, the increase in lignin acid:aldehyde ratios by the end of the irradiation was comparable to previously reported values for marine samples (Hernes and Benner, 2003; 2006; Spencer et al., 2009b). FT-ICR-MS

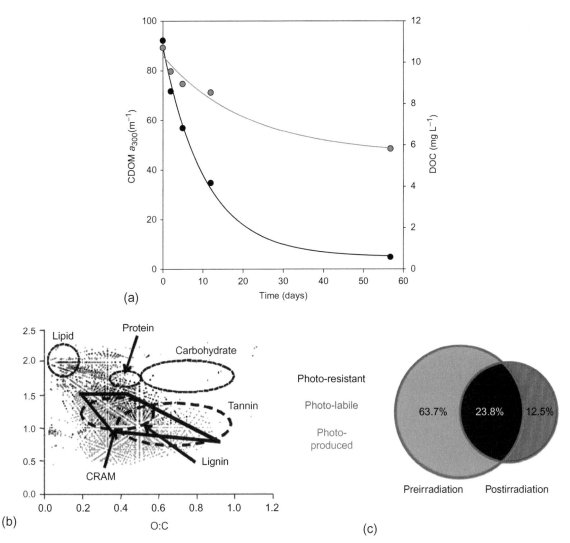

FIGURE 11.2 Congo River DOC and DOM compositional shifts during a 57-day irradiation. (a) Loss of CDOM absorbance (a_{300}, black circles, black line) and DOC (gray circles, gray line). (b) van Krevelen diagram (blue, red, and black data points indicate molecular formulae unique to initial, photo-bleached, and those common to both, respectively). Areas that are assigned as carboxylic-rich alicyclic molecules (CRAM), lignin, tannin, lipid, and protein are delineated. (c) Venn diagram of initial and photo-bleached Congo River water. Area of overlap in black indicates the photo-resistant molecular formulae present in both samples. The blue area represents the unique photo-labile molecular formulae and the red area the unique photo-produced molecular formulae. *Modified from Spencer et al. (2009b) and Stubbins et al. (2010).*

reveals three fractions based on photoreactivity: (1) photo-resistant, (2) photo-labile, and (3) photo-produced (Figure 11.2b and c; Stubbins et al., 2010). The photo-resistant fraction was heterogeneous, with most molecular classes represented although no condensed aromatics persisted and only a small number of aromatics were present. Overall, irradiation caused a significant drop in the number of molecules identified, a decrease in their structural diversity, and a 90% loss in aromatic compounds. (Stubbins et al., 2010).

As photochemical degradation preferentially removes lignin (Hernes and Benner, 2003; Opsahl and Benner, 1998; Spencer et al., 2009b) and other classes typically defined as stable, such as black carbon, the process likely represents a major sink for these compounds (Stubbins et al., 2010; 2012). Black carbon molecules are particularly resistant to biodegradation (Kim et al., 2006), comprising ~10% of the global riverine flux of DOC to the oceans (Jaffe et al., 2013). Thus photochemistry is exceptionally important in its preferential removal of common tracers of terrestrial DOM. The preferential elimination of the terrestrial signature of riverine DOM via photochemistry is clearly exhibited via FT-ICR-MS, shifting the molecular signature of riverine DOM toward that of marine DOM, thus complicating the tracking of terrestrial DOM in the ocean (Gonsior et al., 2009; Kujawinski et al., 2002; Stubbins et al., 2010).

IV ANTHROPOGENIC INFLUENCES

A Fluxes

Across and within watersheds, DOM fluxes scale with discharge (Meybeck, 1993; Raymond and Oh, 2007). Thus, it is generally accepted that increasing water discharge, due to climate change or land management, will result in higher fluxes. Globally, precipitation is predicted to increase, particularly in wet regions. Many studies are also now demonstrating that the increase in DOC concentration with large discharge events (Schiff et al., 1998) is a common feature in watersheds (Raymond and Saiers, 2010). This is true even for the largest events. Tropical storm Irene in New England, a 500-year storm event, caused stream DOM concentrations to increase such that export of 40% of average annual DOC export occurred in just a few days (Figure 11.3; Yoon and Raymond, 2012). These large events are obviously important to coastal carbon budgets (Bianchi et al., 2013). Thus, it is not only the amount of precipitation, but also the number and size of precipitation

events that are important to DOC transfers from land to ocean. The projected increase in the proportion of high rainfall events will likely strengthen the land-ocean DOC connection.

Temperature is also important to DOC fluxes (Figure 11.4). DOC production normalized to discharge is generally higher in warmer months (Wilson et al., 2013a), presumably partly due to temperature controls on dissolution. Many other ecosystem processes important to DOC production, such as microbial degradation, are controlled by temperature, creating indirect links between temperature and DOC export. Increases in DOC production can be offset by temperature modulated increases in evapotranspiration and therefore decreases in discharge, although the interannual variation in evapotranspiration due to temperature is secondary to interannual variation in precipitation in regulating DOC fluxes (Raymond and Oh, 2007). A recent study has argued for an important role between temperature and DOC concentration for watersheds above 43.5° N latitude (Laudon et al., 2012).

Little is known about the influence of land use change on the mass of DOC export. It is well demonstrated that wetlands are hot spots for DOM export off the landscape. This is true for both inland (Junk et al., 2013; Kortelainen, 1993; Mulholland and Kuenzler, 1979; Raymond and Hopkinson, 2003) and coastal wetlands (Jiang et al., 2013; Raymond and Hopkinson, 2003; Tzortziou et al., 2011). Both classes of wetlands continue to be lost to other land uses globally (Dahl, 1990; Junk et al., 2013; McLeod et al., 2011; Waycott et al., 2009). For watersheds of the mid-Atlantic Bight, for instance, wetland coverage has reduced from ~12% to 6% over the past 200 years (Dahl, 1990), presumably resulting in a 20-30% reduction in DOC export to that system (Raymond et al., 2004). The impact of losses of inland marshes and coastal vegetated habitats on lateral transports of DOC is probably significant, but not currently known.

The thawing of permafrost will undoubtedly influence carbon fluxes, although its impact on

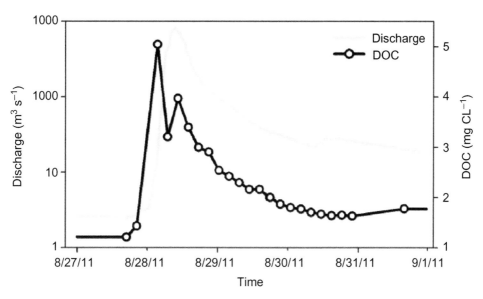

FIGURE 11.3 DOC concentration and discharge for Esopus Creek in New York during Tropical Storm Irene. *Adapted from Yoon and Raymond (2012).*

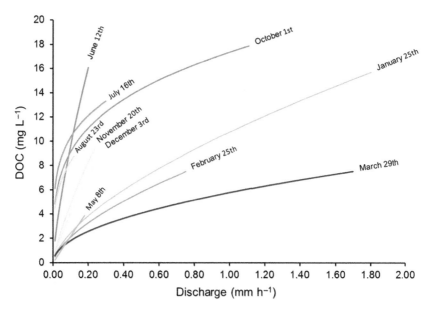

FIGURE 11.4 The temporally variable relationships between discharge and DOC concentrations during precipitation or snow melt events in Bigelow Brook, Harvard Forest. *Adapted from Wilson et al., 2013a,b.*

DOC fluxes is unclear. Authors have argued for an increase in DOC export from high-latitude watersheds based on both mechanistic (Freeman et al., 2001) and space for time substitution water-shed studies (Frey and Smith, 2005). Conversely, the Yukon River demonstrated a decrease in discharge-normalized DOC export over time per-haps due to increases in flow path, residence time,

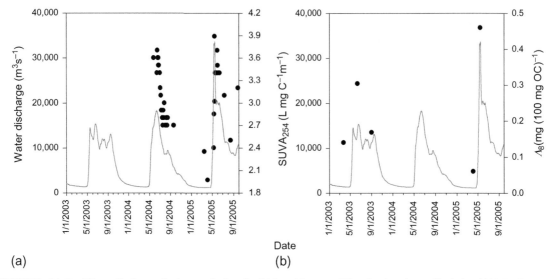

FIGURE 11.5   Water discharge hydrograph for the Yukon River at Pilot Station (gray line) for 2003-2005 versus: (a) SUVA$_{254}$ and (b) lignin carbon-normalized yield (Λ_8) (black circles). *Modified from Spencer et al., 2008, 2009a.*

and terrestrial microbial utilization (Striegl et al., 2005). Tropical peatlands demonstrated a ~30% increase in the flux of DOC, much of it aged, with disturbance (Moore et al., 2013). Thus, ongoing monitoring is required in both Arctic and tropical regions to track the DOC yield responses of these globally significant carbon pools to climate change and anthropogenic impacts such as deforestation and land-use conversion. Numerous rivers in northeastern North America and northwestern Europe are demonstrating changes in DOC concentration (Evans et al., 2005; Findlay, 2005; Monteith et al., 2007). Causes have been suggested to include nitrogen deposition, altered hydrology, and reduced acid rain. Though unmeasured, fluxes have likely increased.

The retreat of glaciers impacts DOM fluxes. Relic glacial water contains low concentrations of DOM, but this material is a net addition to the oceans. Over the 2003-2009 period ~840 km^3 year^{-1} of relic water was added to the world's oceans (Gardner, 2013). If this flow is considered a river, it is the fifth largest in the world (Table 11.1). Concentrations of DOC in glacial melt water are low (Hood et al., 2009) and therefore

the direct contribution to DOC fluxes are low at ~1 Tg DOC-C year^{-1}. Land revegetated after glacier decline, however, exports many times more DOC than the ice-covered landscape (Hood and Scott, 2008); thus, glacial retreat may be resulting in a larger indirect input of DOC. DOM exported from glaciers is also compositionally unique and reactive so coastal regions that receive direct inputs may be impacted (Hood et al., 2009; Singer et al., 2012; Stubbins et al., 2012).

B Composition

Riverine DOM composition varies with hydrologic drivers in watersheds from the Arctic to the equator (Amon et al., 2012; Bouillon et al., 2012; Butman et al., 2012; Spencer et al., 2008). Typically at higher discharge, riverine DOM takes on a more terrestrial character due to a shift in sources (i.e., greater surface runoff and leaching of organic-rich soil horizons and surface litter layers), while at lower discharge the increased flow path, residence time, and therefore greater potential for microbial mineralization and physical protection shifts DOM composition to a less terrestrial nature (Boyer et al., 1997; McGlynn and McDonnell, 2003; Striegl et al., 2005). These shifts are particularly evident in Arctic rivers which change from very low discharge under-ice to peak spring freshet in a very short time period, leading to extensive leaching of surface organic-rich layers. This thaw and extensive leaching imparts a strong terrestrial signature on riverine DOM during the freshet (e.g., elevated $SUVA_{254}$ values, lignin carbon-normalized yields; Figure 11.5). Therefore, the spring flush is a period of high vascular plant-derived and aromatic DOM export in Arctic rivers. After the freshet, $SUVA_{254}$ and lignin carbon-normalized yields typically decline due to increased residence time of DOM in contact with subsurface microbial communities and soils; during the summer-autumn period rainfall can increase terrestrial character again alongside increasing discharge due to mobilization of DOM from surface organic-rich layers (Figure 11.5; Spencer et al., 2008, 2009a). With respect to DOM composition in rivers, hysteresis has been reported, with a more terrestrial signature imparted on the rising limb of the hydrograph (e.g., Bouillon et al., 2012; Hood et al., 2006).

Relative inputs from different land cover types are also readily apparent in stream and riverine DOM composition. For example, carbon-normalized vanillyl phenol yields (V) in the Amazon River $(0.67 \, (\mathrm{mg} \, (100 \, \mathrm{mg} \, \mathrm{OC})^{-1}))$ (Hedges et al., 2001) are elevated compared to the Mississippi $(0.44 \, (\mathrm{mg} \, (100 \, \mathrm{mg} \, \mathrm{OC})^{-1}))$ (Benner and Opsahl, 2001; Hernes and Benner, 2003) and Arctic rivers $(0.13\text{-}0.15 \, (\mathrm{mg} \, (100 \, \mathrm{mg} \, \mathrm{OC})^{-1}))$ (Lobbes et al., 2000; Spencer et al., 2008), reflecting greater vascular plant inputs derived from the tropical rainforest within the Amazon Basin. The relatively low V in Arctic Rivers is not likely a result of dilution by autochthonous production within these systems as that has been shown to be typically low (Meon and Amon, 2004). Low V in Arctic rivers largely reflects the large inputs from nonvascular sources within these watersheds, such as bryophytes that contain little if any

lignin (Spencer et al., 2008). Similarly, the presence of wetlands impacts the terrestrial character of riverine DOM as rivers draining wetland-dominated systems typically contain DOM of greater terrestrial character (e.g., relatively high DOC-specific absorption; Fellman et al., 2009b; Tzortziou et al., 2008).

Riverine DOM composition is impacted by a host of anthropogenic land use and land management practices (Bernardes et al., 2004; Williams et al., 2010; Wilson and Xenopoulos, 2009; Yamashita et al., 2011). Agricultural watersheds export DOM with reduced structural complexity and increased microbial contributions (Wilson and Xenopoulos, 2009), as well as a shifted $\delta^{13}C$ signature (Bernardes et al., 2004). One of the most striking impacts of agriculture on DOM composition is the legacy effect of long ceased agricultural practices. Dissolved black carbon continues to be mobilized and exported from Brazil's Paraiba do Sul watershed despite widespread forest burning having ceased in 1973 (Dittmar et al., 2012b). As major tropical watersheds such as the Amazon and Congo continue to face deforestation and agricultural expansion, much work needs to be undertaken to understand what this means for exported riverine DOM composition and the role this material plays in downstream receiving ecosystems.

Acknowledgments

We thank the U.S. National Science Foundation for supporting our research. Specifically R.G.M.S. was supported by ETBC-0851101, DEB-1145932, OPP-1107774, and ANT-1203885 which funded research discussed in this chapter. We also thank anonymous reviewers and Aron Stubbins for providing helpful comments on this chapter.

References

Abril, G., Etcheber, H., Le Hir, P., Bassoullet, P., Boutier, B., Frankignoulle, M., 1999. Oxic/anoxic oscillations and organic carbon mineralization in an estuarine maximum turbidity zone (the Gironde, France). Limnol. Oceanogr. 44, 1304–1315.

Abril, G., Nogueira, M., Etcheber, H., Cabeçadas, G., Lemaire, E., Brogueira, M.J., 2002. Behaviour of organic carbon in nine contrasting European estuaries. Estuar. Coast. Shelf Sci. 54, 241–262.

Adame, M.F., Lovelock, C.E., 2011. Carbon and nutrient exchange of mangrove forests with the coastal ocean. Hydrobiologia 663, 23–50.

Aiken, G.R., Hsu-Kim, H., Ryan, J.N., 2011. Influence of dissolved organic matter on the environmental fate of metals, nanoparticles, and colloids. Environ. Sci. Technol. 45, 3196–3201.

Aitkenhead, J.A., McDowell, W.H., 2000. Soil C:N ratio as a predictor of annual riverine doc flux at local and global scales. Global Biogeochem. Cycles 14, 127–138.

Allen, M.B., 1956. Excretion of organic compounds by *Chlamydomonas*. Arch. Mikrobiol. 24, 163–168.

Alvarez-Salgado, X.A., Miller, A.E.J., 1998. Dissolved organic carbon in a large macrotidal estuary (the Humber, UK): Behaviour during estuarine mixing. Mar. Pollut. Bull. 37, 216–224.

Amon, R.M.W., Rinehart, A.J., Duan, S., Louchouarn, P., Prokushkin, A., Guggenberger, G., et al., 2012. Dissolved organic matter sources in large Arctic rivers. Geochim. Cosmochim. Acta 94, 217–237.

Aufdenkampe, A.K., Hedges, J.I., Richey, J.E., Krusche, A.V., Llerena, C.A., 2001. Sorptive fractionation of dissolved organic nitrogen and amino acids onto fine sediments within the Amazon basin. Limnol. Oceanogr. 46, 1921–1935.

Aufdenkampe, A.K., Mayorga, E., Raymond, P.A., Melack, J.M., Doney, S.C., Alin, S.R., et al., 2011. Riverine coupling of biogeochemical cycles between land, oceans, and atmosphere. Front. Ecol. Environ. 9, 53–60.

Baines, S.B., Pace, M.L., 1991a. The production of dissolved organic-matter by phytoplankton and its importance to bacteria—patterns across marine and fresh-water systems. Limnol. Oceanogr. 36, 1078–1090.

Baines, S.B., Pace, M.L., 1991b. The production of dissolved organic matter by phytoplantkon and its importance to bacteria: patterns across marine and freshwater systems. Limnol. Oceanogr. 36, 1078–1090.

Balcarczyk, K.L., Jones, J.B., Jaffe, R., Maie, N., 2009. Stream dissolved organic matter bioavailability and composition in watersheds underlain with discontinuous permafrost. Biogeochemistry 94, 255–270.

Barron, C., Duarte, C.M., 2009. Dissolved organic matter release in a posidonia oceanica meadow. Mar. Ecol. Prog. Ser. 374, 75–84.

Bauer, J.E., Williams, P.M., Druffel, E.R.M., 1992. C-14 activity of dissolved organic-carbon fractions in the north-central Pacific and Sargasso Sea. Nature 357, 667–670.

Benner, R., Kaiser, K., 2011. Biological and photochemical transformations of amino acids and lignin phenols in riverine dissolved organic matter. Biogeochemistry 102, 209–222.

Benner, R., Opsahl, S., 2001. Molecular indicators of the sources and transformations of dissolved organic matter in the Mississippi River plume, Org. Geochem. 32, 597–611.

Benner, R., Fogel, M.L., Sprague, E.K., Hodson, R.E., 1987. Depletion of C-13 in lignin and its implications for stable carbon isotope studies. Nature 329, 708–710.

Bergamaschi, B.A., Fleck, J.A., Downing, B.D., Boss, E., Pellerin, B., Ganju, N.K., et al., 2011. Methyl mercury dynamics in a tidal wetland quantified using in situ optical measurements. Limnol. Oceanogr. 56, 1355–1371.

Bernardes, M.C., Martinelli, L.A., Krusche, A.V., Gudeman, J., Moreira, M., Victoria, R.L., et al., 2004. Riverine organic matter composition as a function of land use changes, southwest Amazon. Ecol. Appl. 14, S263–S279.

Bianchi, T.S., 2011. The role of terrestrially derived organic carbon in the coastal ocean: a changing paradigm and the priming effect. Proc. Natl. Acad. Sci. U. S. A. 108, 19473–19481.

Bianchi, T.S., Filley, T., Dria, K., Hatcher, P.G., 2004. Temporal variability in sources of dissolved organic carbon in the lower Mississippi River. Geochim. Cosmochim. Acta 68, 959–967.

Bianchi, T.S., Garcia-Tigreros, F., Yvon-Lewis, S.A., Shields, M., Mills, H.J., Butman, D., 2013. Enhanced transfer of terrestrially derived carbon to the atmosphere in a flooding event. Geophys. Res. Lett. 40, 116–122.

Bird, M.I., Robinson, R.A.J., Win Oo, N., Maung Aye, M., Lu, X.X., Higgitt, D.L., et al., 2008. A preliminary estimate of organic carbon transport by the Ayeyarwady (Irrawaddy) and Thanlwin (Salween) rivers of Myanmar. Quat. Int. 186, 113–122.

Bouillon, S., Yambélé, A., Spencer, R.G.M., Gillikin, D.P., Hernes, P.J., Six, J., et al., 2012. Organic matter sources, fluxes and greenhouse gas exchange in the Oubangui River (Congo River Basin). Biogeosciences 9, 2045–2062.

Boyer, E.W., Hornberger, G.M., Bencala, K.E., McKnight, D.M., 1997. Response characteristics of DOC flushing in an alpine catchment. Hydrol. Process. 11, 1635–1647.

Buffam, I., Galloway, J.N., Blum, L.K., McGlathery, K.J., 2001. A stormflow/baseflow comparison of dissolved organic matter concentrations and bioavailability in an Appalachian stream. Biogeochemistry 53, 269–306.

Butman, D., Raymond, P.A., 2011. Significant efflux of carbon dioxide from streams and rivers in the United States. Nat. Geosci. 4, 839–842.

Butman, D., Raymond, P., Oh, N.-H., Mull, K., 2007. Quantity, ^{14}C age and lability of desorbed soil organic carbon in fresh water and seawater. Org. Geochem. 38, 1547–1557.

Butman, D., Raymond, P.A., Butler, K., Aiken, G., 2012. Relationships between delta C-14 and the molecular quality of dissolved organic carbon in rivers draining to the coast from the conterminous United States. Global Biogeochem. Cycles 26.

Cauwet, G., 2002. Dom in the coastal zone. In: Hansell, D.A., Carlson, C.A. (Eds.), Dom in the Coastal Zone. Academic Press, San Diego.

Chen, R.F., Gardner, G.B., 2004. High-resolution measurements of chromophoric dissolved organic matter in the Mississippi and Atchafalaya River plume regions. Mar. Chem. 89, 103–125.

Childers, D.L., Day, J.W., McKellar, H.N., 2000. Twenty more Years of Marsh and Estuarine Flux Studies: Revisiting Nixon (1980).

Chow, A.T., Dahlgren, R.A., Harrison, J.A., 2007. Watershed sources of disinfection byproduct precursors in the Sacramento and San Joaquin Rivers, California. Environ. Sci. Technol. 41, 7645–7652.

Clark, C.D., Litz, L.P., Grant, S.B., 2008. Salt marshes as a source of chromophoric dissolved organic matter (CDOM) to Southern California coastal waters. Limnol. Oceanogr. 53, 1923–1933.

Cole, J.J., Prairie, Y.T., Caraco, N.F., McDowell, W.H., Tranvik, L.J., Striegl, R.G., et al., 2007. Plumbing the global carbon cycle: integrating inland waters into the terrestrial carbon budget. Ecosystems 10, 171–184.

Coynel, A., Seyler, P., Etcheber, H., Meybeck, M., Orange, D., 2005. Spatial and seasonal dynamics of total suspended sediment and organic carbon species in the Congo River. Global Biogeochem. Cycles 19.

Crump, B.C., Peterson, B.J., Raymond, P.A., Amon, R.M.W., Rinehart, A., McClelland, J.W., et al., 2009. Circumpolar synchrony in big river bacterioplankton. Proc. Natl. Acad. Sci. U. S. A. 106, 21208–21212.

Dahl, T.E., 1990. Wetlands losses in the United States 1780's to 1980's. http://www.npwrc.usgs.gov/resource/othrdata/wetloss/wetloss.htm.

Dai, M.H., Yin, Z.Q., Meng, F.F., Liu, Q., Cai, W.J., 2012. Spatial distribution of riverine doc inputs to the ocean: an updated global synthesis. Curr. Opin. Environ. Sustain. 4, 170–178.

Dame, R.F., Spurrier, J.D., Williams, T.M., Kjerfve, B., Zingmark, R.G., Wolaver, T.G., et al., 1991. Annual material processing by a salt-marsh estuarine-basin in South-Carolina, USA. Mar. Ecol. Prog. Ser. 72, 153–166.

Dehaan, H., Jones, R.I., Salonen, K., 1987. Does ionic-strength affect the configuration of aquatic humic substances, as indicated by gel-filtration. Freshw. Biol. 17, 453–459.

Del Vecchio, R., Subramaniam, A., 2004. Influence of the Amazon River on the surface optical properties of the western tropical north Atlantic ocean. J. Geophys. Res. Ocean 109.

Dittmar, T., Hertkorn, N., Kattner, G., Lara, R.J., 2006. Mangroves, a major source of dissolved organic carbon to the oceans. Global Biogeochem. Cycles 20.

Dittmar, T., de Rezende, C.E., Manecki, M., Niggemann, J., Ovalle, A.R.C., Stubbins, A., et al., 2012a. Continuous flux of dissolved black carbon from a vanished tropical forest biome. Nat. Geosci. 5, 618–622.

Dittmar, T., Paeng, J., Gihring, T.M., Suryaputra, I., Huettel, M., 2012b. Discharge of dissolved black carbon from a fire-affected intertidal system. Limnol. Oceanogr. 57, 1171–1181.

Druffel, E.R.M., Williams, P.M., Robertson, K., Griffin, S., Jull, A.J.T., Donahue, D., et al., 1989. Radiocarbon in dissolved organic and inorganic carbon from the central North Pacific. Radiocarbon 31, 523–532.

Ertel, J.R., Hedges, J.I., Devol, A.H., Richey, J.E., Ribeiro, M.D.G., 1986. Dissolved humic substances of the Amazon river system. Limnol. Oceanogr. 31, 739–754.

Evans, C.D., Monteith, D.T., Cooper, D.M., 2005. Long-term increases in surface water dissolved organic carbon: observations, possible causes and environmental impacts. Environ. Pollut. 137, 55–71.

Evans, C.D., Freeman, C., Cork, L.G., Thomas, D.N., Reynolds, B., Billett, M.F., et al., 2007. Evidence against recent climate-induced destabilisation of soil carbon from ^{14}C analysis of riverine dissolved organic matter. Geophys. Res. Lett. 34, L07407.

Fellman, J.B., Hood, E., Edwards, R.T., D'Amore, D.V., 2009a. Changes in the concentration, biodegradability, and fluorescent properties of dissolved organic matter during stormflows in coastal temperate watersheds. J. Geophys. Res.—Biogeosci. 114.

Fellman, J.B., Hood, E., Edwards, R.T., D'Amore, D.V., 2009b. Changes in the concentration, biodegradability, and fluorescent properties of dissolved organic matter during stormflows in coastal temperate watersheds. J. Geophys. Res. 114, G01021.

Fellman, J.B., Hood, E., Spencer, R.G.M., 2010. Fluorescence spectroscopy opens new windows into dissolved organic matter dynamics in freshwater ecosystems: a review. Limnol. Oceanogr. 55, 2452–2462.

Fichot, C.G., Benner, R., 2012. The spectral slope coefficient of chromophoric dissolved organic matter (s275–295) as a tracer of terrigenous dissolved organic carbon in river-influenced ocean margins. Limnol. Oceanogr. 57, 1453–1466.

Findlay, S.E.G., 2005. Increased carbon transport in the Hudson river: unexpected consequence of nitrogen deposition? Front. Ecol. Environ. 3, 133–137.

Findlay, S., Pace, M.L., Lints, D., Howe, K., 1992. Bacterial metabolism of organic-carbon in the tidal fresh-water Hudson estuary. Mar. Ecol. Prog. Ser. 89, 147–153.

Fisher, S.G., Likens, G.E., 1972. Stream ecosystem—organic energy budget. Bioscience 22, 33.

Franke, D., Hamilton, M.W., Ziegler, S.E., 2012. Variation in the photochemical lability of dissolved organic matter in a large boreal watershed. Aquat. Sci. 74, 751–768.

Freeman, C., Evans, C.D., Monteith, D.T., Reynolds, B., Fenner, N., 2001. Export of organic carbon from peat soils. Nature 412, 785.

Frenette, J.J., Massicotte, P., Lapierre, J.F., 2012. Colorful niches of phytoplankton shaped by the spatial connectivity in a large river ecosystem: a riverscape perspective. PLoS One 7.

Frey, K.E., Smith, L.C., 2005. Amplified carbon release from vast West Siberian peatlands by 2100. Geophys. Res. Lett. 32.

Gao, H.Z., Zepp, R.G., 1998. Factors influencing photoreactions of dissolved organic matter in a coastal river of the southeastern United States. Environ. Sci. Technol. 32, 2940–2946.

Gardner, A.S., 2013. A reconciled estimate of glacier contributions to sea level rise: 2003 to 2009. Science 340, 1168 (vol. 340, pg 852, 2013).

Gjessing, E.T., Gjerdahl, T., 1970. Influence of ultra-violet radiation on aquatic humis. Vatten 26, 144–145.

Goldschmidt, V.M., 1932. Grunderlagen der quantitativen geochemie. Forsschr. d. Min. 17, 112–156.

Gong, W.K., Ong, J.E., 1990. Plant biomass and nutrient flux in a managed mangrove forest in Malaysia. Estuar. Coast. Shelf Sci. 31, 519–530.

Goni, M.A., Teixeira, M.J., Perkey, D.W., 2003. Sources and distribution of organic matter in a river-dominated estuary (Winyah Bay, SC, USA). Estuar. Coast. Shelf Sci. 57, 1023–1048.

Gonneea, M.E., Paytan, A., Herrera-Silveira, J.A., 2004. Tracing organic matter sources and carbon burial in mangrove sediments over the past 160 years. Estuar. Coast. Shelf Sci. 61, 211–227.

Gonsior, M., Peake, B.M., Cooper, W.T., Podgorski, D., D'Andrilli, J., Cooper, W.J., 2009. Photochemically induced changes in dissolved organic matter identified by ultra-high resolution Fourier transform ion cyclotron resonance mass spectrometry. Environ. Sci. Technol. 43, 698–703.

Gordeev, V.V., Martin, J.M., Sidorov, I.S., Sidorova, M.V., 1996. A reassessment of the Eurasian river input of water, sediment, major elements, and nutrients to the Arctic Ocean. Am. J. Sci. 296, 664–691.

Goulden, M.L., Wofsy, S.C., Harden, J.W., Trumbore, S.E., Crill, P.M., Gower, S.T., et al., 1998. Sensitivity of boreal forest carbon balance to soil thaw. Science 279, 214–217.

Griffith, D.R., Raymond, P.A., 2011. Multiple-source heterotrophy fueled by aged organic carbon in an urbanized estuary. Mar. Chem. 124, 14–22.

Griffith, D.R., Barnes, R.T., Raymond, P.A., 2009. Inputs of fossil carbon from wastewater treatment plants to U.S. rivers and oceans. Environ. Sci. Technol. http://dx.doi.org/10.1021/es9004043.

Happ, G., Gosselink, J.G., Day, J.W., 1977. Seasonal distribution of organic-carbon in a Louisiana estuary. Estuar. Coast. Mar. Sci. 5, 695–705.

He, B.Y., Dai, M.H., Zhai, W.D., Wang, L.F., Wang, K.J., Chen, J.H., et al., 2010. Distribution, degradation and dynamics of dissolved organic carbon and its major compound classes in the Pearl River estuary, China. Mar. Chem. 119, 52–64.

Hedges, J.I., Cowie, G.L., Richey, J.E., Quay, P.D., Benner, R., Strom, M., et al., 1994. Origins and processing of organic-matter in the Amazon river as indicated by carbohydrates and amino-acids. Limnol. Oceanogr. 39, 743–761.

Hedges, J.I., Keil, R.G., Benner, R., 1997. What happens to terrestrial organic matter in the ocean? Org. Geochem. 27, 195–212.

Hedges, J.I., Mayorga, E., Tsamakis, E., McClain, M.E., Aufdenkampe, A., Quay, P., et al., 2000. Organic matter in Bolivian tributaries of the Amazon River: a comparison to the lower mainstream. Limnol. Oceanogr. 45, 1449–1466.

Helie, J.F., Hillaire-Marcel, C., 2006. Sources of particulate and dissolved organic carbon in the St Lawrence river: isotopic approach. Hydrol. Process. 20, 1945–1959.

Hernes, P.J., Benner, R., 2003. Photochemical and microbial degradation of dissolved lignin phenols: implications for the fate of terrigenous dissolved organic matter in marine environments. J. Geophys. Res. Ocean 108.

Hernes, P.J., Benner, R., 2006. Terrigenous organic matter sources and reactivity in the North Atlantic Ocean and a comparison to the Arctic and Pacific oceans. Mar. Chem. 100, 66–79.

Hernes, P.J., Robinson, A.C., Aufdenkampe, A.K., 2007. Fractionation of lignin during leaching and sorption and implications for organic matter "freshness". Geophys. Res. Lett. 34, L17401.

Holmes, R.M., McClelland, J.W., Raymond, P.A., Frazer, B.B., Peterson, B.J., Stieglitz, M., 2008. Lability of doc transported by Alaskan rivers to the Arctic Ocean. Geophys. Res. Lett. 35.

Holmes, R.M., McClelland, J.W., Peterson, B.J., Tank, S.E., Bulygina, E., Eglinton, T.I., et al., 2012. Seasonal and annual fluxes of nutrients and organic matter from large rivers to the Arctic Ocean and surrounding seas. Estuar. Coast. Shelf Sci. 35, 369–382.

Hood, E., Gooseff, M.N., Johnson, S.L., 2006. Changes in the character of stream water dissolved organic carbon during flushing in three small watersheds. Oregon. J. Geophys. Res.-Biogeosci. 111, 8.

Hood, E., Scott, D., 2008. Riverine organic matter and nutrients in southeast Alaska affected by glacial coverage. Nat. Geosci. 1, 583–587.

Hood, E., Fellman, J., Spencer, R.G.M., Hernes, P.J., Edwards, R., D'Amore, D., et al., 2009. Glaciers as a source of ancient and labile organic matter to the marine environment. Nature 462, 1044-U100.

Hutchinson, G.E., 1957. A Treatise on Limnology. Volume 1. Geography, Physics, and Chemistry. John Wiley and Sons, New York.

Hynes, H.B.N., 1963. Imported organic matter and secondary productivity in streams. In: Moore, J.A. (Ed.), Proceedings of the XVI International Congress of Zoology. XVI International Congress of Zoology, Washington, DC, pp. 324–329.

IPCC, 2012. Managing the risks of extreme events and disasters to advance climate change adaptation. In: Field, C.B., Barros, V., Stocker, T.F., Qin, D., Dokken, D.J., Ebi, K.L., et al. (Eds.), A special report of working groups I and II of the Intergovermental Panel on Climate Change. Cambridge University Press, Cambridge, UK.

Jaffe, R., Ding, Y., Niggemann, J., Vahatalo, A.V., Stubbins, A., Spencer, R.G.M., et al., 2013. Global charcoal mobilization from soils via dissolution and riverine transport to the oceans. Science 340, 345–347.

Jennerjahn, T.C., Ittekkot, V., 2002. Relevance of mangroves for the production and deposition of organic matter along tropical continental margins. Naturwissenschaften 89, 23–30.

Jeong, J.J., Bartsch, S., Fleckenstein, J.H., Matzner, E., Tenhunen, J.D., Lee, S.D., et al., 2012. Differential storm responses of dissolved and particulate organic carbon in a mountainous headwater stream, investigated by high-frequency, in situ optical measurements. J. Geophys. Res.—Biogeosci. 117.

Jiang, L.Q., Cai, W.J., Wang, Y., Bauer, J.E., 2013. Influence of terrestrial inputs on continental shelf carbon dioxide. Biogeosciences 10, 839–849.

Junk, W.J., An, S.Q., Finlayson, C.M., Gopal, B., Kvet, J., Mitchell, S.A., et al., 2013. Current state of knowledge regarding the world's wetlands and their future under global climate change: a synthesis. Aquat. Sci. 75, 151–167.

Kaplan, L.A., Wiegner, T.N., Newbold, J.D., Ostrom, P.H., Gandhi, H., 2008. Untangling the complex issue of dissolved organic carbon uptake: a stable isotope approach. Freshw. Biol. 53, 855–864.

Keil, R.G., Montlucon, D.B., Prahl, F.G., Hedges, J.I., 1994. Sorptive preservation of labile organic-matter in marine-sediments. Nature 370, 549–552.

Kerner, M., Hohenberg, H., Ertl, S., Reckermann, M., Spitzy, A., 2003. Self-organization of dissolved organic matter to micelle-like microparticles in river water. Nature 422, 150–154.

Kim, S., Kaplan, L.A., Hatcher, P.G., 2006. Biodegradable dissolved organic matter in a temperate and a tropical stream determined from ultra-high resolution mass spectrometry. Limnol. Oceanogr. 51, 1054–1063.

Kleber, M., Johnson, M.G., 2010. Advances in understanding the molecular structure of soil organic matter: implications for interactions in the environment. In: Sparks, D.L. (Ed.), Advances in Agronomy, vol. 106.

Klug, J.L., Richardson, D.C., Ewing, H.A., Hargreaves, B.R., Samal, N.R., Vachon, D., et al., 2012. Ecosystem effects of a tropical cyclone on a network of lakes in northeastern North America. Environ. Sci. Technol. 46, 11693–11701.

Kortelainen, P., 1993. Content of total organic-carbon in Finnish lakes and its relationship to catchment characteristics. Can. J. Fish. Aquat. Sci. 50, 1447–1483.

Kujawinski, E.B., Hatcher, P.G., Freitas, M.A., 2002. High-resolution Fourier transform ion cyclotron resonance mass spectrometry of humic and fulvic acids: improvements and comparisons. Anal. Chem. 74, 413–419.

Laiho, R., 2006. Decomposition in peatlands: reconciling seemingly contrasting results on the impacts of lowered water levels. Soil Biol. Biochem. 38, 2011–2024.

Langenheder, S., Kisand, V., Wikner, J., Tranvik, L.J., 2003. Salinity as a structuring factor for the composition and performance of bacterioplankton degrading riverine doc. FEMS Microbiol. Ecol. 45, 189–202.

Laudon, H., Kohler, S., Buffam, I., 2004. Seasonal TOC export from seven boreal catchments in Northern Sweden. Aquat. Sci. 66, 223–230.

Laudon, H., Buttle, J., Carey, S.K., McDonnell, J., McGuire, K., Seibert, J., et al., 2012. Cross-regional prediction of long-term trajectory of stream water doc response to climate change. Geophys. Res. Lett. 39.

Lauerwald, R., Hartmann, J., Ludwig, W., Moosdorf, N., 2012. Assessing the nonconservative fluvial fluxes of dissolved organic carbon in North America. J. Geophys. Res.—Biogeosci. 117.

Livingstone, D.A., 1963. Chemical composition of rivers and lakes. U.S.G.S. Professional Paper.

Lobbes, J.M., Fitznar, H.P., Kattner, G., 2000. Biogeochemical characteristics of dissolved and particulate organic matter in Russian rivers entering the Arctic Ocean. Geochim. Cosmochim. Acta 64, 2973–2983.

Loh, A.N., Canuel, E.A., Bauer, J.E., 2008. Potential source and diagenetic signatures of oceanic dissolved and particulate organic matter as distinguished by lipid biomarker distributions. Mar. Chem. 112, 189–202.

Longworth, B.E., Petsch, S.T., Raymond, P.A., Bauer, J.E., 2007. Linking lithology and land use to sources of dissolved and particulate organic matter in headwaters of a temperate, passive-margin river system. Geochim. Cosmochim. Acta 71, 4233–4250.

Ludwig, W., Probst, J.L., Kempe, S., 1996. Predicting the oceanic input of organic carbon by continental erosion. Glob. Biogeochem. Cycle 10, 23–41.

Mackenzi, F.T., Garrels, R.M., 1966. Chemical mass balance between rivers and oceans. Am. J. Sci. 264, 507.

Mann, C.J., Wetzel, R.G., 1995. Dissolved organic carbon and its utilization in a riverine wetland ecosystem. Biogeochemistry 31, 99–120.

Mannino, A., Russ, M.E., Hooker, S.B., 2008. Algorithm development and validation for satellite-derived distributions of doc and CDOM in the U.S. Middle Atlantic Bight. J. Geophys. Res. Ocean 113.

Mantoura, R.F.C., Woodward, E.M.S., 1983. Conservative behavior of riverine dissolved organic-carbon in the Severn estuary - chemical and geochemical implications. Geochim. Cosmochim. Ac. 47, 1293–1309.

Masiello, C.A., Druffel, E.R.M., 2001. Carbon isotope geochemistry of the Santa Clara River. Global Biogeochem. Cycles 15, 407–416.

Massicotte, P., Frenette, J.J., 2011. Spatial connectivity in a large river system: resolving the sources and fate of dissolved organic matter. Ecol. Appl. 21, 2600–2617.

Mayer, L.M., 1994. Relationships between mineral surfaces and organic-carbon concentrations in soils and sediments. Chem. Geol. 114, 347–363.

Mayorga, E., Aufdenkampe, A.K., Masiello, C.A., Krusche, A.V., Hedges, J.I., Quay, P.D., et al., 2005. Young organic matter as a source of carbon dioxide outgassing from Amazonian rivers. Nature 436, 538–541.

Mayorga, E., Seitzinger, S.P., Harrison, J.A., Dumont, E., Beusen, A.H.W., Bouwman, A.F., et al., 2010. Global nutrient export from watersheds 2 (news 2): model development and implementation. Environ. Model. Software 25, 837–853.

McClain, M.E., Boyer, E.W., Dent, C.L., Gergel, S.E., Grimm, N.B., Groffman, P.M., et al., 2003. Biogeochemical hot spots and hot moments at the interface of terrestrial and aquatic ecosystems. Ecosystems 6, 301–312.

McClelland, J.W., Holmes, R.M., Dunton, K.H., Macdonald, R.W., 2012. The arctic ocean estuary. Estuar. Coast. Shelf Sci. 35, 353–368.

McGlynn, B.L., McDonnell, J.J., 2003. Role of discrete landscape units in controlling catchment dissolved organic carbon dynamics. Water Resour. Res. 39.

McLeod, E., Chmura, G.L., Bouillon, S., Salm, R., Bjork, M., Duarte, C.M., et al., 2011. A blueprint for blue carbon: toward an improved understanding of the role of vegetated coastal habitats in sequestering CO_2. Front. Ecol. Environ. 9, 552–560.

Menzel, D.W., Vaccaro, R.F., 1964. The measurement of dissolved organic and particulate carbon in seawater. Limnol. Oceanogr. 9, 138–142.

Meon, B., Amon, R.M.W., 2004. Heterotrophic bacterial activity and fluxes of dissolved free amino acids and glucose in the Arctic rivers Ob, Yenisei and the adjacent Kara Sea. Aquat. Microb. Ecol. 37, 121–135.

Meybeck, M., 1982. Carbon, nitrogen, and phosphorus transport by world rivers. Am. J. Sci. 282, 401–450.

Meybeck, M., 1993. Riverine transport of atmospheric carbon—sources, global typology and budget. Water Air Soil Pollut. 70, 443–463.

Meybeck, M., Ragu, A., 1996. River discharges to the oceans: an assessment of suspended solids and major ions and nutrients. In: UNEP (ed.).

Monteith, D.T., Stoddard, J.L., Evans, C.D., de Wit, H.A., Forsius, M., Hogasen, T., et al., 2007. Dissolved organic carbon trends resulting from changes in atmospheric deposition chemistry. Nature 450, 537-540.

Moore, S., Evans, C.D., Page, S.E., Garnett, M.H., Jones, T.G., Freeman, C., et al., 2013. Deep instability of deforested tropical peatlands revealed by fluvial organic carbon fluxes. Nature 493, 660.

Moran, M.A., Hodson, R.E., 1994. Dissolved humic substances of vascular plant-origin in a coastal marine-environment. Limnol. Oceanogr. 39, 762–771.

Moran, M.A., Sheldon, W.M., Zepp, R.G., 2000. Carbon loss and optical property changes during long-term photochemical and biological degradation of estuarine dissolved organic matter. Limnol. Oceanogr. 45, 1254–1264.

Moreira-Turcq, P., Seyler, P., Guyot, J.L., Etcheber, H., 2003. Exportation of organic carbon from the Amazon river and its main tributaries. Hydrol. Process. 17, 1329–1344.

Morris, A.W., Mantoura, R.F.C., Bale, A.J., Howland, R.J.M., 1978. Very low salinity regions of estuaries—important sites for chemical and biological reactions. Nature 274, 678–680.

Mulholland, P.J., Kuenzler, E.J., 1979. Organic carbon export from upland and forested wetland watershed. Limnol. Oceanogr. 24, 960–966.

Neff, J.C., Finlay, J.C., Zimov, S.A., Davydov, S.P., Carrasco, J.J., Schuur, E.A.G., et al., 2006. Seasonal changes in the age and structure of dissolved organic carbon in Siberian rivers and streams. Geophys. Res. Lett. 33.

Nelson, N.B., Siegel, D.A., Carlson, C.A., Swan, C., Smethie, W.M., Khatiwala, S., 2007. Hydrography of chromophoric dissolved organic matter in the north Atlantic. Deep-Sea Res. Part I—Oceanogr. Res. Pap. 54, 710–731.

Oechel, W.C., Hastings, S.J., Vourlitis, G., Jenkins, M., Riechers, G., Grulke, N., 1993. Recent change of Arctic tundra ecosystems from a net carbon-dioxide sink to a source. Nature 361, 520–523.

Opsahl, S., Benner, R., 1998. Photochemical reactivity of dissolved lignin in river and ocean waters. Limnol. Oceanogr. 43, 1297–1304.

Opsahl, S.P., Zepp, R.G., 2001. Photochemically-induced alteration of stable carbon isotope ratios (delta C-13) in terrigenous dissolved organic carbon. Geophys. Res. Lett. 28, 2417–2420.

Raymond, P.A., Bauer, J.E., 2001a. Doc cycling in a temperate estuary: a mass balance approach using natural C-14 and C-13 isotopes. Limnol. Oceanogr. 46, 655–667.

Raymond, P.A., Bauer, J.E., 2001b. Riverine export of aged terrestrial organic matter to the North Atlantic Ocean. Nature 409, 497–500.

Raymond, P.A., Hopkinson, C.S., 2003. Ecosystem modulation of dissolved carbon age in a temperate marsh-dominated estuary. Ecosystems 6, 694–705.

Raymond, P.A., Oh, N.-H., 2007. An empirical study of climatic controls on riverine c export from three major U.S. watersheds. Global Biogeochem. Cycles 21.

Raymond, P.A., Saiers, J.E., 2010. Event controlled doc export from forested watersheds. Biogeochemistry 100, 197–209.

Raymond, P.A., Bauer, J.E., Caraco, N.F., Cole, J.J., Longworth, B., Petsch, S.T., 2004. Controls on the variability of organic matter and dissolved inorganic carbon ages in northeast US rivers. Mar. Chem. 92, 353–366.

Raymond, P.A., McClelland, J.W., Holmes, R.M., Zhulidov, A.V., Mull, K., Peterson, B.J., et al., 2007. Flux and age of dissolved organic carbon exported to the arctic ocean: a carbon isotopic study of the five largest arctic rivers. Global Biogeochem. Cycles 21.

Roach, K.A., 2013. Environmental factors affecting incorporation of terrestrial material into large river food webs. Freshw. Sci. 32, 283–298.

Royer, T.V., David, M.B., 2005. Export of dissolved organic carbon from agricultural streams in Illinois, USA. Aquat. Sci. 67, 465–471.

Schiff, S.L., Aravena, R., Trumbore, S.E., Hinton, M.J., Elgood, R., Dillon, P.J., 1997. Export of doc from forested catchments on the Precambrain Shield of Central Ontario: clues from 13C and 14C. Biogeochemistry 36, 43–65.

Schiff, S., Aravena, R., Mewhinney, E., Elgood, R., Warner, B., Dillon, P., et al., 1998. Precambrian shield wetlands: hydrologic control of the sources and export of dissolved organic matter. Clim. Change 40, 167–188.

Schlesinger, W.H., Melack, J.M., 1981. Transport of organic-carbon in the worlds rivers. Tellus 33, 172–187.

Schmidt, M.W.I., Torn, M.S., Abiven, S., Dittmar, T., Guggenberger, G., Janssens, I.A., et al., 2011. Persistence of soil organic matter as an ecosystem property. Nature 478, 49–56.

Sebestyen, S.D., Boyer, E.W., Shanley, J.B., 2009. Responses of stream nitrate and doc loadings to hydrological forcing and climate change in an upland forest of the northeastern United States. J. Geophys. Res.—Biogeosci. 114, 11.

Seitzinger, S.P., Harrison, J.A., Dumont, E., Beusen, A.H.W., Bouwman, A.F., 2005. Sources and delivery of carbon, nitrogen, and phosphorus to the coastal zone: an overview of Global Nutrient Export from watersheds (news) models and their application. Global Biogeochem. Cycles 19.

Shank, G.C., Zepp, R.G., Whitehead, R.F., Moran, M.A., 2005. Variations in the spectral properties of freshwater and estuarine CDOM caused by partitioning onto river and estuarine sediments. Estuar. Coast. Shelf Sci. 65, 289–301.

Shen, Y., Fichot, C.G., Benner, R., 2012. Floodplain influence on dissolved organic matter composition and export from the Mississippi-Atchafalaya River system to the Gulf of Mexico. Limnol. Oceanogr. 57, 1149–1160.

Sholkovitz, E.R., 1976. Flocculation of dissolved organic and inorganic matter during mixing of river water and seawater. Geochim. Cosmochim. Acta 40, 831–845.

Sickman, J.O., DiGiorgio, C.L., Davisson, M.L., Lucero, D.M., Bergamaschi, B., 2010. Identifying sources of dissolved organic carbon in agriculturally dominated rivers using radiocarbon age dating: Sacramento-San Joaquin river basin, California. Biogeochemistry 99, 79–96.

Singer, G.A., Fasching, C., Wilhelm, L., Niggemann, J., Steier, P., Dittmar, T., et al., 2012. Biogeochemically diverse organic matter in Alpine glaciers and its downstream fate. Nat. Geosci. 5, 710–714.

Sondergaard, M., Middelboe, M., 1995. A cross-system analysis of labile dissolved organic-carbon. Mar. Ecol. Prog. Ser. 118, 283–294.

Sondergaard, M., Stedmon, C.A., Borch, N.H., 2003. Fate of terrigenous dissolved organic matter (DOM) in estuaries: aggregation and bioavailability. Ophelia 57, 161–176.

Spencer, R.G.M., Ahad, J.M.E., Baker, A., Cowie, G.L., Ganeshram, R., Upstill-Goddard, R.C., et al., 2007a. The estuarine mixing behaviour of peatland derived dissolved organic carbon and its relationship to chromophoric dissolved organic matter in two North Sea Estuaries (UK). Estuar. Coast. Shelf Sci. 74, 131–144.

Spencer, R.G.M., Baker, A., Ahad, J.M.E., Cowie, G.L., Ganeshram, R., Upstill-Goddard, R.C., et al., 2007b. Discriminatory classification of natural and anthropogenic waters in two U.K. estuaries. Sci. Total Environ. 373, 305–323.

Spencer, R.G.M., Aiken, G.R., Wickland, K.P., Striegl, R.G., Hernes, P.J., 2008. Seasonal and spatial variability in dissolved organic matter quantity and composition from the Yukon River basin, Alaska. Global Biogeochem. Cycles 22.

Spencer, R.G.M., Aiken, G.R., Butler, K.D., Dornblaser, M.M., Striegl, R.G., Hernes, P.J., 2009a. Utilizing chromophoric dissolved organic matter measurements to derive export and reactivity of dissolved organic carbon exported to the Arctic ocean: a case study of the Yukon river, Alaska. Geophys. Res. Lett. 36, L06401.

Spencer, R.G.M., Stubbins, A., Hernes, P.J., Baker, A., Mopper, K., Aufdenkampe, A.K., et al., 2009b. Photochemical degradation of dissolved organic matter and dissolved lignin phenols from the Congo River. J. Geophys. Res. 114, G03010.

Spencer, R.G.M., Hernes, P.J., Aufdenkampe, A.K., Baker, A., Gulliver, P., Stubbins, A., et al., 2012. An initial investigation into the organic matter biogeochemistry of the Congo River. Geochim. Cosmochim. Acta 84, 614–627.

Spencer, R.G.M., Aiken, G.R., Dornblaser, M.M., Butler, K.D., Holmes, R.M., Fiske, G., et al., 2013. Chromophoric dissolved organic matter export from U.S. Rivers. Geophys. Res. Lett. 40, 1575–1579.

Spiker, E.C., Rubin, M., 1975. Petroleum pollutants in surface and groundwater as indicated by the carbon-14 activity of dissolved organic carbon. Science 187, 61–64.

Stepanauskas, R., Leonardson, L., Tranvik, L.J., 1999. Bioavailability of wetland-derived DON to freshwater and marine bacterioplankton. Limnol. Oceanogr. 44, 1477–1485.

Stepanauskas, R., Moran, M.A., Bergamasch, B.A., Hollibaugh, J.T., 2005. Sources, bioavailability, and photoreactivity of dissolved organic carbon in the Sacramento-San Joaquin River Delta. Biogeochemistry 74, 131–149.

Striegl, R.G., Aiken, G.R., Dornblaser, M.M., Raymond, P.A., Wickland, K.P., 2005. A decrease in discharge-normalized doc export by the Yukon river during summer through autumn. Geophys. Res. Lett. 32.

Striegl, R.G., Dornblaser, M.M., McDonald, C.P., Rover, J.R., Stets, E.G., 2012. Carbon dioxide and methane emissions from the yukon river system. Global Biogeochem. Cycles 26, GB0E05.

Stubbins, A., Spencer, R.G.M., Chen, H., Hatcher, P.G., Mopper, K., Hernes, P.J., et al., 2010. Illuminated darkness: molecular signatures of Congo river dissolved organic matter and its photochemical alteration as revealed by ultrahigh precision mass spectrometry. Limnol. Oceanogr. 55, 1467–1477.

Stubbins, A., Law, C.S., Uher, G., Upstill-Goddard, R.C., 2011. Carbon monoxide apparent quantum yields and photoproduction in the Tyne estuary. Biogeosciences 8, 703–713.

Stubbins, A., Hood, E., Raymond, P.A., Aiken, G.R., Sleighter, R.L., Hernes, P.J., et al., 2012. Anthropogenic aerosols as a source of ancient dissolved organic matter in glaciers. Nat. Geosci. 5, 198–201.

Tarnocai, C., Canadell, J.G., Schuur, E.A.G., Kuhry, P., Mazhitova, G., Zimov, S., 2009. Soil organic carbon pools in the northern circumpolar permafrost region. Global Biogeochem. Cycles 23.

Tockner, K., Pennetzdorfer, D., Reiner, N., Schiemer, F., Ward, J.V., 1999. Hydrological connectivity, and the exchange of organic matter and nutrients in a dynamic river-floodplain system (Danube, Austria). Freshw. Biol. 41, 521–535.

Twilley, R.R., Pozo, M., Garcia, V.H., RiveraMonroy, V.H., Bodero, R.Z.A., 1997. Litter dynamics in riverine mangrove forests in the Guayas River estuary, Ecuador. Oecologia 111, 109–122.

Tzortziou, M., Neale, P.J., Osburn, C.L., Megonigal, J.P., Maie, N., Jaffe, R., 2008. Tidal marshes as a source of optically and chemically distinctive colored dissolved organic matter in the Chesapeake Bay. Limnol. Oceanogr. 53, 148–159.

Tzortziou, M., Neale, P.J., Megonigal, J.P., Pow, C.L., Butterworth, M., 2011. Spatial gradients in dissolved carbon due to tidal marsh outwelling into a Chesapeake Bay estuary. Mar. Ecol. Prog. Ser. 426, 41–56.

Uher, G., Hughes, C., Henry, G., Upstill-Goddard, R.C., 2001. Non-conservative mixing behavior of colored dissolved organic matter in a humic-rich, turbid estuary. Geophys. Res. Lett. 28, 3309–3312.

Vähätalo, A.V., Aarnos, H., Mäntyniemi, S., 2010. Biodegradability continuum and biodegradation kinetics of natural organic matter described by the beta distribution. Biogeochemistry 100, 227–240.

Vanhall, C.E., Stenger, V.A., Safranko, J., 1963. Rapid combustion method for determination of organic substances in aqueous solutions. Anal. Chem. 35, 315.

Vidon, P., Wagner, L.E., Soyeux, E., 2008. Changes in the character of doc in streams during storms in two midwestern watersheds with contrasting land uses. Biogeochemistry 88, 257–270.

Vodacek, A., Blough, N.V., 1997. Seasonal Variation of CDOM in the Middle Atlantic Bight: Terrestrial Inputs and Photooxidation.

Volk, C.J., Volk, C.B., Kaplan, L.A., 1997. Chemical composition of biodegradable dissolved organic matter in streamwater. Limnol. Oceanogr. 42, 39–44.

Vonk, J.E., Mann, P.J., Davydov, S., Davydova, A., Spencer, R.G.M., Schade, J., et al., 2013. High biolability of ancient permafrost carbon upon thaw. Geophys. Res. Lett. 40, 2689–2693.

Ward, N.D., Keil, R.G., Medeiros, P.M., Brito, D.C., Cunha, A.C., Dittmar, T., et al., 2013. Degradation of terrestrially derived macromolecules in the Amazon river. Nat. Geosci. 6, 530–533.

Waycott, M., Duarte, C.M., Carruthers, T.J.B., Orth, R.J., Dennison, W.C., Olyarnik, S., et al., 2009. Accelerating loss of seagrasses across the globe threatens coastal ecosystems. Proc. Natl. Acad. Sci. U. S. A. 106, 12377–12381.

Weishaar, J.L., Aiken, G.R., Bergamaschi, B.A., Fram, M.S., Fujii, R., Mopper, K., 2003. Evaluation of specific ultraviolet absorbance as an indicator of the chemical composition and reactivity of dissolved organic carbon. Environ. Sci. Technol. 37, 4702–4708.

Weiss, M., Simon, M., 1999. Consumption of labile dissolved organic matter by limnetic bacterioplankton: the relative significance of amino acids and carbohydrates. Aquat. Microb. Ecol. 17, 1–12.

Wetzel, R.G., 1992. Gradient-dominated ecosystems—sources and regulatory functions of dissolved organic-matter in fresh-water ecosystems. Hydrobiologia 229, 181–198.

White, E.M., Kieber, D.J., Sherrard, J., Miller, W.L., Mopper, K., 2010. Carbon dioxide and carbon monoxide photoproduction quantum yields in the Delaware estuary. Mar. Chem. 118, 11–21.

Wickland, K.P., Aiken, G.R., Butler, K., Dornblaser, M.M., Spencer, R.G.M., Striegl, R.G., 2012. Biodegradability of dissolved organic carbon in the Yukon River and its tributaries: seasonality and importance of inorganic nitrogen. Global Biogeochem. Cycles 26, GB0E06.

Wikner, J., Cuadros, R., Jansson, M., 1999. Differences in consumption of allochthonous DOC under limnic and estuarine conditions in a watershed. Aquat. Microb. Ecol. 17, 289–299.

Williams, P.M., 1968. Organic and inorganic constituents of Amazon river. Nature 218, 937.

Williams, P.M., Oeschger, H., Kinney, P., 1969. Natural radiocarbon activity of dissolved organic carbon in north-east Pacific Ocean. Nature 224, 256.

Williams, C.J., Yamashita, Y., Wilson, H.F., Jaffe, R., Xenopoulos, M.A., 2010. Unraveling the role of land use and microbial activity in shaping dissolved organic matter characteristics in stream ecosystems. Limnol. Oceanogr. 55, 1159–1171.

Wilson, H.F., Xenopoulos, M.A., 2009. Effects of agricultural land use on the composition of fluvial dissolved organic matter. Nat. Geosci. 2, 37–41.

Wilson, H.F., Saiers, J.E., Raymond, P.A., Sobczak, W.V., 2013a. Hydrologic drivers and seasonality of dissolved organic carbon concentration, nitrogen content, bioavailability, and export in a forested New England stream. Ecosystems.

Wilson, H.F., Saiers, J.E., Raymond, P.A., Sobczak, W.V., 2013b. Hydrologic drivers and seasonality of dissolved organic carbon concentration, nitrogen content, bioavailability, and export in a forested New England stream. Ecosystems 16, 604–616.

Wisser, D., Fekete, B.M., Vorosmarty, C.J., Schumann, A.H., 2010. Reconstructing 20th century global hydrography: a contribution to the Global Terrestrial Network-Hydrology (GTN-H). Hydrol. Earth Sys. Sci. 14, 1–24.

Woodroffe, C.D., Bardsley, K.N., Ward, P.J., Hanley, J.R., 1988. Production of mangrove litter in a macrotidal embayment, Darwin Harbor, N.T., Australia. Estuar. Coast. Shelf Sci. 26, 581–598.

Yamashita, Y., Kloeppel, B.D., Knoepp, J., Zausen, G.L., Jaffe, R., 2011. Effects of watershed history on dissolved organic matter characteristics in headwater streams. Ecosystems 14, 1110–1122.

Yoon, B., Raymond, P.A., 2012. Dissolved organic matter export from a forested watershed during Hurricane Irene. Geophys. Res. Lett. 39.

Ziegler, S., Benner, R., 1999. Dissolved organic carbon cycling in a subtropical seagrass-dominated lagoon. Mar. Ecol. Prog. Ser. 180, 149–160.

Sediment Pore Waters

David J. Burdige*, Tomoko Komada†

*Department of Ocean, Earth and Atmospheric Sciences, Old Dominion University, Noroflk, Virginia, USA

†Romberg Tiburon Center, San Francisco State University, Tiburon, California, USA

CONTENTS

Biogeochemistry of Marine Dissolved Organic Matter,
http://dx.doi.org/10.1016/B978-0-12-405940-5.00012-1

I PREFACE

Dissolved organic matter (DOM) in marine sediment pore waters plays an important role in sediment carbon and nitrogen remineralization and may also be involved in sediment carbon preservation. It also plays a role in pore water metal complexation, affecting dissolved metal and metal-complexing ligand fluxes from sediments (e.g., Shank et al., 2004).

Since the publication of this chapter in the first edition of this book (Burdige, 2002), a number of key advances have been made in this field. Despite these advances, the following introductory questions from the original chapter still represent an appropriate basis for starting the discussion here:

1. What do we know about the composition and reactivity of pore water DOM, with particular reference to its role in sediment organic matter remineralization?
2. What do we know about the controls on pore water DOM concentrations?
3. What is the role of benthic DOM fluxes in the global ocean cycles of carbon and nitrogen?
4. What is the role of pore water DOM in sediment carbon preservation?

In an effort to be comprehensive, key information from the earlier version of this chapter is presented here along with newer results. However, in a number of places, reference is simply made to this earlier chapter for details that, for brevity and conciseness, we felt could be omitted.

II INTRODUCTION

Dissolved organic matter (DOM) in sediment pore waters is a heterogeneous collection of organic compounds, ranging in size from relatively large macromolecules (e.g., dissolved proteins or humic substances) to smaller molecules such as individual amino acids or short-chain organic acids. As is also the case in the water column (e.g., Benner, 2002; see Chapter 2), much of the pore water DOM remains uncharacterized at the compound-class or molecular-level (see Section IV and Burdige, 2001, 2002 for details).

In many sediments, pore water concentrations of DOM—both dissolved organic carbon (DOC) and dissolved organic nitrogen (DON)—are elevated by up to an order of magnitude over bottom water values (Figure 12.1). This implies that there is net production of DOM in sediments as a result of organic matter degradation processes. Based on diffusive arguments alone, sediments are a potential source of both DOC and DON to the overlying waters (see Section VII). Much of the total pore water DOC and DON is of relatively low-molecular weight (LMW; see Section III.A) and appears to be recalcitrant, at least in a bulk sense. Such observations are consistent with the results of water column studies showing that high-molecular weight (HMW) DOC represents a more reactive and less diagenetically altered fraction of the total DOC than the more abundant, and presumably more recalcitrant, LMW-DOC (e.g., Amon and Benner, 1994; Benner, 2002; Santschi et al., 1995).

Several lines of evidence suggest that the DOM accumulating in sediments is indeed recalcitrant. The first is simply that relatively high

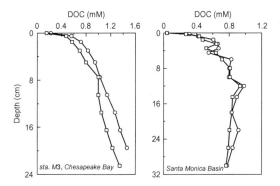

FIGURE 12.1   Pore water DOC concentration profiles in anoxic marine sediments. Symbols on the upper *x*-axes represent concentrations in bottom water samples obtained by hydrocasts. (Left) Cores collected at site M3 in the mesohaline Chesapeake Bay (July 1995, open square; October 1995, open circle); data from Burdige and Zheng (1998). (Right) Replicate cores collected in Santa Monica Basin in March 1994 (Burdige, 2002). Also see Jahnke (1990) and Komada et al. (2013) for general information on the geochemistry of these sediments. See Figure 12.9 for additional DOC profiles in anoxic sediments.

concentrations of this material can be found in sediments and that DOM accumulates with depth in most sediments as rates of particulate organic matter (POM) remineralization and the reactivity of sediment POM both decrease (e.g., Burdige, 1991a; Middelburg, 1989; Westrich and Berner, 1984). Such trends are seen across a wide range of time and space (sediment depth) scales, ranging from ~1 m or less in coastal and nearshore sediments to 100s of meters or more in deeply buried marine sediments (see Burdige, 2002, and discussions in Section VI for details). If the majority of pore water DOM had a high degree of reactivity, then one might expect its concentration to eventually decrease with depth due to microbial remineralization of the material. Second, in many coastal marine sediments, humic-like fluorescence of pore water DOM is strongly correlated with total DOC concentrations, suggesting that much of the pore water DOM may be considered dissolved humic substances (see Section III.B). Third, a number of

incubation experiments support the suggestion that what appears to be recalcitrant DOC can be produced on time scales comparable to those over which POC remineralization and inorganic nutrient regeneration occur (Brüchert and Arnosti, 2003; Chipman et al., 2010; Hee et al., 2001; Komada et al., 2012; Robador et al., 2010; Weston and Joye, 2005).

As a starting point for our discussions, Figure 12.2 shows a conceptual model of DOM cycling in sediments based on the classic anaerobic food chain model (e.g., Fenchel et al., 1998; Megonigal et al., 2003) and the pore water size/reactivity (PWSR) model of Burdige and Gardner (1998). The model in this figure also incorporates DOM production pathway(s) inferred from more recent work (Komada et al., 2012; Robador et al., 2010; Weston and Joye, 2005). In anaerobic settings, the majority of these processes are mediated by bacteria or archaea, with many catalyzed by microbial exoenzymes that must break down organic molecules to sufficiently small sizes to pass into microbial cells for further degradation (e.g., Arnosti, 2004).

In the model in Figure 12.2, we think of the degradation of sediment POM to inorganic end products as occurring by a series of hydrolytic (or oxidative), fermentative, and eventually respiratory processes that produce and consume pore water DOM intermediates with increasingly smaller molecular weights. Although this process leads to a continuum of DOM compounds (in terms of molecular weights and reactivities), the model assumes that there is an initial class of HMW-DOM (box "A" in Figure 12.2) containing biological polymers such as dissolved proteins and polysaccharides resulting from the initial hydrolysis or oxidative cleavage (depolymerization) of sediment POM. Most HMW-DOM is further hydrolyzed and fermented, producing and consuming labile DOM compounds of decreasingly smaller molecular weights (box "B" in Figure 12.2). Eventually, this results in the production of monomeric LMW-DOM compounds

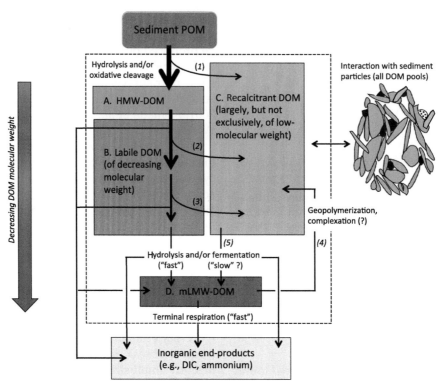

FIGURE 12.2   A conceptual model for DOM cycling in sediments. The processes illustrated here (including pathways (1)-(4)) are discussed in more detail in Section II. The terms "fast" and "slow" are used in a relative sense to imply that the turnover of either labile DOM (box "B") or mLMW-DOM (box "D") occurs much more rapidly than that of recalcitrant DOM (box "C"). This figure is based on information from several sources (e.g., Burdige and Gardner, 1998; Fenchel et al., 1998; Megonigal et al., 2003).

such as acetate, other small organic acids, and individual amino acids (mLMW-DOM; box "D" in Figure 12.2), which are then utilized in terminal respiratory processes such as iron reduction, sulfate reduction, or methanogenesis. A recent food web model for aerobic deep-sea sediments (Rowe and Deming, 2011) also supports the basic aspects of the model in Figure 12.2, although this study notes that the initial breakdown of POM in these sediments occurs by microbial exoenzymes as well as by metazoan feeding (which may in part be mediated by microbes living in the digestive tracts of these higher organisms). In addition, viral lysis of living bacterial cells may be important in adding DOM compounds to these sediment pore waters (Rowe and Deming, 2011).

While boxes "A" and "B" in Figure 12.2 represent labile compounds that generally turn over rapidly, not all carbon flow follows the vertical path along the left side of this figure. Some of the carbon end products of reactions along this pathway may actually be DOM compounds of lower reactivity that appear recalcitrant on the overall time scales of remineralization or the production of inorganic end products (i.e., this DOM falls into box "C" and following pathways (1), (2), or (3) in Figure 12.2; also see discussion in Section V.A). More recent sediment incubation studies point to the importance of direct production of inherently recalcitrant DOM through processes that may be similar to those discussed here (Komada et al., 2012, 2013; Robador et al., 2010; Weston and Joye, 2005).

For example, consider the following hypothetical fermentation reaction (modified after Yao and Conrad, 2000),

$$DOM_1 + aH_2O \rightarrow DOM_2 + DOM_3 + bCH_3COOH + cH_2 + dCO_2, \quad (12.1)$$

where a, b, c, and d are arbitrary coefficients that determine the relative production of acetic acid, H_2, and CO_2 in this process based on the chemical formulae of the three DOM molecules. Here both DOM_2 and DOM_3 have lower molecular weights than DOM_1. If DOM_2 is a labile molecule then this part of the carbon flow in this reaction will follow the vertical path along the left side of Figure 12.2 while if DOM_3 is recalcitrant, then this part of the carbon flow will go into box "C."

In the original PWSR model (Burdige, 2002; Burdige and Gardner, 1998), this recalcitrant DOM was referred to as polymeric LMW-DOM (pLMW-DOM), and in addition to production as discussed above, it was suggested that it might be produced through internal transformations of more reactive precursors (pathway (4) in Figure 12.2). Such processes include the possibility of forming recalcitrant DOM from mLMW-DOM compounds through geopolymerization processes such as the melanoidin or "browning" reaction (an abiotic sugar amino acid condensation reaction; Hedges, 1988) and complexation reactions (Christensen and Blackburn, 1982; Finke et al., 2007; Michelson et al., 1989).

Other recently proposed humification models suggest a slightly different pathway by which recalcitrant or reactive LMW-DOM compounds may serve as precursors for the formation of recalcitrant humic substances (Piccolo, 2001; Sutton and Sposito, 2005). Here, it is suggested that humic substances consist of a supramolecular cluster of relatively LMW-DOM compounds linked together by hydrogen bonds and hydrophobic interactions, as opposed to macromolecules in which covalent bonds formed by geopolymerization-type reactions link LMW-DOM reactants. In the context of the model in Figure 12.2, pathways (1)-(3) could contribute to the formation of recalcitrant dissolved humic substances. However, if such processes do occur on early diagenetic time scales in sediment pore waters, their products apparently still have relatively low molecular weights (Burdige and Gardner, 1998).

In the first edition of this chapter (Burdige, 2002), a reactive-transport (advection-diffusion-reaction) model based on the original PWSR model was presented for both strictly anoxic sediments (the ANS model) and mixed redox (bioturbated and/or bioirrigated) sediments (the BBS model). Carbon flow in the model equations is illustrated in Figure 12.3A while one set of model results, using the ANS model, is shown in Figure 12.3b. A key finding of this modeling effort was that pore water gradients of HMW-DOC and pLMW-DOC near the sediment surface were similar in magnitude, despite the fact that model-derived HMW-DOC concentrations were significantly lower than those of pLMW-DOC throughout most of the sediment column. As a result, in model calculations for both anoxic and mixed redox sediments, benthic fluxes of HMW-DOM were ~50-80% of the total benthic DOC flux (see Burdige, 2002 for details). Thus, in both types of sediments, model results suggest that benthic fluxes of both recalcitrant and reactive DOM can be similar in magnitude. The significance of this result will be discussed in later sections of this chapter.

A DOC and DON in Sediment Pore Waters: General Observations

Pore water DOC and DON profiles have been published from a wide range of surficial marine sediments, in environments ranging from the deep-sea to shallow water estuaries, salt marshes, and seagrass environments (for reviews of the earlier literature, see Burdige, 2002; Krom and Westrich, 1981; more recent results include Alkhatib et al., 2013; Chipman et al., 2012;

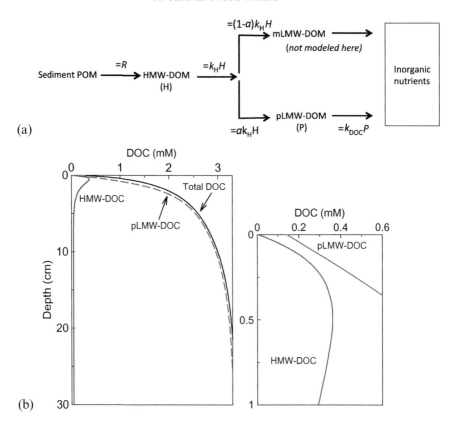

FIGURE 12.3 (a) A schematic representation of DOM remineralization based on the original PWSR model. In this model, production of HMW-DOM from sediment POM is given by Equation (12.2) in the text (see Section VI.A), while the remineralization of either HMW-DOM or pLMW-DOM is assumed to be a first-order process. The parameter a represents the fraction of HMW-DOM remineralization that occurs through the lower pathway; in the context of the model illustrated in Figure 12.2, this represents material that is remineralized through the recalcitrant DOM pool on the right side of the figure. The remaining HMW-DOM is then remineralized along the far left side of the figure. Since mLMW-DOM compounds are assumed to be a small fraction of the total DOM pool that are remineralized rapidly to inorganic end products (see Section IV), they are not explicitly modeled here (see Burdige, 2002, for details). This model also does not directly examine production of inorganic end products (i.e., DIC). (b) Model results for strictly anoxic sediments obtained using a reactive-transport model based on the carbon flow illustrated in part A. This model (the ANS model) is described in detail in Burdige (2002), where the specific parameters are also listed. Note that in spite of the fact that throughout most of the sediment column model-derived concentrations of more reactive HMW-DOC are significantly lower than those of more recalcitrant pLMW-DOC (which also makes up the bulk of the pore water DOC pool), the pore water gradients near the sediment surface of these two types of DOC are quite similar in magnitude.

Hall et al., 2007; Heuer et al., 2009; Komada et al., 2004; Lahajnar et al., 2005; Papadimitriou et al., 2002; Pohlman et al., 2010; Ståhl et al., 2004). Profile depths range from 10s of centimeters to several hundred meters.

In early attempts to describe some of the general controls on pore water DOC depth profiles (Krom and Sholkovitz, 1977; Starikova, 1970), it was suggested that these profiles fall into two general categories. In anoxic sediments where benthic macrofaunal processes (bioturbation and/or bioirrigation) are insignificant, pore water DOC (and DON) concentrations generally increase with depth, often approaching "asymptotic"

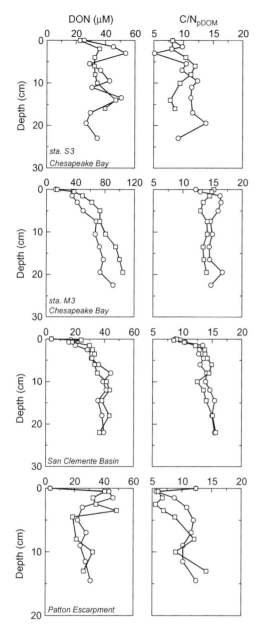

concentrations (Figures 12.1 and 12.4). In contrast, in what were then termed "oxic" or "oxidizing" sediments, DOC and DON concentrations are elevated above bottom water values, but are also much more constant with depth in the upper portions of the sediments (Figures 12.4 and 12.5). While some of these sediments may indeed be oxic, it may be more appropriate to recognize that such sediments are often bioturbated and/ or bioirrigated and have mixed (or oscillating) redox conditions (sensu Aller, 1994). The three-dimensional, heterogeneous geometry of such sediments leads to a situation in which organic matter remineralization can be thought of as oscillating between oxic and sub-oxic/anoxic processes (also see discussions in Burdige, 2006b).

We also note that among the published DOC and DON depth profiles, one will find some profiles that do not appear to be consistent with the discussions presented here. Our incomplete understanding of DOM cycling in sediments is likely a part of the reason for these differences. At the same time though, several studies have discussed potential sampling artifacts that can affect pore water DOM depth profiles, suggesting another possible explanation for these differences. In particular, in sediments containing abundant benthic macrofaunal populations, pore water collection via sediment centrifugation can lead to elevated DOC concentrations (and potentially elevated

FIGURE 12.4 Pore water DON concentrations (left) and the C/N ratios of pore water DOM (=[DOC]/[DON] and defined here as C/N_{pDOM}) (right) versus depth in contrasting marine sediments. Symbols on the upper x-axes represent values in bottom water samples obtained by hydrocasts. (First row) Cores collected at site S3 in the southern Chesapeake Bay (October 1996, open circle; August 1997, open square); data

FIGURE 12.4, CONT'D from Burdige and Zheng (1998) and Burdige (2001). (Second row) Cores collected at site M3 in the mesohaline Chesapeake Bay (July 1995, open circle; October 1995, open square); data from Burdige and Zheng (1998). (Third row) Replicate cores collected in San Clemente Basin (March 1994) (Burdige, 2002). (Fourth row) Replicate cores collected at the Patton Escarpment (March 1994) (Burdige, 2002). See Burdige et al. (1999) and McManus et al. (1997) for details on the geochemistry of the San Clemente Basin and Patton Escarpment sediments (also see Table 12.1). Note that DOC data from these Chesapeake Bay site M3 cores are shown in Figure 12.1, and those from the site S3 cores are shown in Figure 12.5.

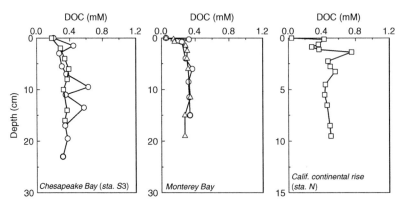

FIGURE 12.5 Pore water DOC profiles in bioturbated and/or bioirrigated (i.e., mixed redox) marine sediments. Symbols on the upper *x*-axes represent concentrations in bottom water samples obtained by hydrocasts. (Left) Cores collected at site S3 in the southern Chespapeake Bay (October 1996, open circle; August 1997, open square); data from Burdige and Zheng (1998) and Burdige (2001). (Center) Replicate cores collected at a site (95-m water depth) in Monterey Bay (Burdige, 2002). See Berelson et al. (2003) for details on the geochemistry of these sediments. (Right) A core collected at station N on the California continental rise (4100-m water depth) at the base of the Monterey deep sea fan (DOC data from Bauer et al., 1995). Also see Cai et al. (1995) for details on the geochemistry of the sediments at this site.

concentrations of individual components of the total DOC pool) as compared to concentrations in pore waters collected by more "gentle" techniques such as with sediment sippers (Alperin et al., 1999; Burdige and Gardner, 1998; Burdige and Martens, 1990; Holcombe et al., 2001; Jørgensen et al., 1981; Martin and McCorkle, 1994). These differences may be due to animal rupture or simple DOM release by benthic organisms during core processing. In deep-sea sediments, distinct maxima in DOC, DON, and dissolved carbohydrates just below the sediment-water interface also appear to be artifacts related to lysis of sediment bacteria due to decompression and/or warming during sediment core collection and recovery (Brunnegård et al., 2004; Hall et al., 2007).

One set of environments where DOM profiles may not follow many of the depth trends and general behaviors discussed here are highly permeable sandy coastal and shelf sediments (e.g., Boudreau et al., 2001; Chipman et al., 2012). These sediments, often described as behaving like a sand filter in a sewage treatment or water purification plant, may act as a DOC sink because of the complex interplay between microbial activity and advective flow through the sediments (see Chipman et al., 2010, and references therein). However, results from Avery et al. (2012) showed that sands from a high-energy beach act as a DOC source, which is more typical of most marine sediments. While the overall significance of net DOC remineralization by permeable sediments is currently unclear (e.g., source vs. sink to the water column), the process could be of large-scale importance because such highly permeable sands represent a major fraction of the continental shelf (Emery, 1968) and because continental shelf sediments are, in general, important sites of organic carbon preservation and remineralization (Burdige, 2007; Hedges and Keil, 1995).

III COMPOSITION AND DYNAMICS OF BULK PORE WATER DOM

A Molecular Weight Distribution

Using ultrafiltration techniques, Burdige and Gardner (1998) and Burdige and Zheng (1998) showed that the vast majority of the DOC and DON in Chesapeake Bay (estuarine) and Mid-Atlantic Bight (continental margin) sediment pore waters has molecular weights less than ~3 kDa (~80-90% of total DOM in estuarine sediments and ~60-70% of total DOM in continental margin sediments). Studies of DOM from other nearshore marine and freshwater sediments using ultrafiltration and size exclusion chromatography also generally support these conclusions (Chin and Gschwend, 1991; Chin et al., 1994, 1998; Nissenbaum et al., 1972; Ziegelgruber et al., 2013).

In contrast, earlier studies suggested that HMW-DOM (rather than LMW-DOM) accumulates with depth in sediment pore waters (Krom and Sholkovitz, 1977; Orem et al., 1986). However, these studies used filters with smaller nominal molecular weight cut-offs (0.5 and 1 kDa respectively) than those used in the ultrafiltration studies cited above, making a direct comparison of the results difficult. Marine water column studies generally show that the majority of oceanic DOC passes through a membrane with a nominal 1-kDa molecular weight cut-off (Benner, 2002), with LMW-DOC representing 60-75% of the DOC in the surface ocean and 75-80% of the DOC in the deep ocean. Given the bulk radiocarbon age and other chemical characteristics of marine DOM (e.g., Hansell et al., 2009), this provides additional evidence for the recalcitrant nature of LMW marine DOM in general.

B Fluorescence Spectroscopy

Fluorescence spectroscopy has been applied to pore water DOM in a variety of environments including estuarine and marine sediments (Benamou et al., 1994; Burdige et al., 2004; Chen and Bada, 1989; Chen and Bada, 1994; Chen et al., 1993; Coble, 1996; Komada et al., 2004; Komada et al., 2002; Seretti et al., 1997; Sierra et al., 2001; Skoog et al., 1996), intertidal sand flats and subterranean estuaries (Kim et al., 2012; Lübben et al., 2009), mangrove and salt marsh sediments (Marchand et al., 2006; Otero et al.,

2007), and deep-sea coral mounds (Larmagnat and Neuweiler, 2011). These studies examined properties of fluorescent DOM (FDOM) through the determination of fluorescence intensity at fixed excitation and emission wavelengths, emission spectra at a limited number of excitation wavelengths, synchronous scan spectra, or full three-dimensional excitation-emission matrix (EEM) spectra.

Regardless of the technique used, pore water FDOM invariably shows broad fluorescence in the spectral region thought to represent fluorescence of humic-like substances. Individual humic-like peaks have also been observed in EEM spectra of pore waters, and possible relationships between these specific humic-like peaks are discussed in Boehme and Coble (2000) and Burdige et al. (2004). Strong fluorescence consistent with emission bands of the aromatic proteins tyrosine and tryptophan is also evident in EEM and synchronous scan spectra of sediment pore waters (see references at the end of this section). The specific nomenclature and characteristic excitation and emission wavelengths of humic-like and protein-like fluorescence peaks seen in EEM spectra are discussed in Coble (1996, 2007), Burdige et al. (2004), and Chapter 10.

Fluorescence intensity of sediment pore waters is in many cases markedly higher than that of the overlying water column and typically increases with sediment depth (Chen and Bada, 1994; Chen et al., 1993; Komada et al., 2004; Lübben et al., 2009; Marchand et al., 2006; Sierra et al., 2001). This increase is due in part to the higher concentration of DOC (and hence FDOM) in sediments relative to bottom water. However, results from organic-rich, anoxic sediments also show considerably higher DOC-normalized fluorescence intensities in pore waters relative to the overlying bottom water (Burdige et al., 2004; Chen et al., 1993), indicating that pore water FDOM fluoresces more intensely than its water column counterpart, at least for these types of sediments. Combined with the fact that marine sediments are net sources of DOC to the water column (Section VII), sediments are also net sources of FDOM to the water column (Boss et al., 2001; Burdige et al., 2004; Chen et al., 1993; Lübben et al., 2009; Skoog et al., 1996).

A positive correlation between humic-like fluorescence intensity and DOC appears to be the norm in pore waters (e.g., Chen et al., 1993; Seretti et al., 1997; Sierra et al., 2001; Skoog et al., 1996). This is in contrast to water column FDOM, where this relationship can be of an inverse nature depending on the area of study (Coble, 2007). The intercept of the FDOM-DOC regression line further suggests that the overwhelming majority of pore water DOC is fluorescent (Burdige et al., 2004; Chen et al., 1993; Komada et al., 2004). These observations again support the hypothesis that humic-like fluorescence arises from molecularly complex (uncharacterized), poorly reactive DOC that comprises the bulk of the pore water DOC pool (Section II).

There are several lines of evidence supporting a direct link between FDOM and the molecularly uncharacterized, poorly reactive component of bulk DOC. First, slopes of humic-like fluorescence versus DOC concentration have been found to vary with redox potential, with lower slopes observed under more oxidizing (or mixed redox) conditions, both across a redox gradient within a given core (Komada et al., 2004) and across different sedimentary settings (Burdige et al., 2004). In a laboratory incubation experiment using coastal sediments, Skoog et al. (1996) also observed higher benthic FDOM fluxes under anoxic versus oxic conditions. As is discussed in Section VI.C, greater exposure to stronger oxidants such as O_2 appears to result in greater loss of more recalcitrant organic matter (e.g., Blair and Aller, 2012; Burdige, 2007; Zonneveld et al., 2010).

At present, it is not possible to link humic-like fluorescence to specific compounds or fluorophores because the molecular basis for natural DOM fluorescence is unclear (Boehme and Coble, 2000; Del Vecchio and Blough, 2004).

Nonetheless, the presence of specific peaks in EEM spectra should hold important clues regarding the composition of FDOM (e.g., Burdige et al., 2004; Coble, 2007; Komada et al., 2002; also see Chapter 10).

Pore water EEM spectra also contain peaks that coincide with those observed in spectra of the aromatic amino acids tryptophan and tyrosine (Burdige et al., 2004; Coble, 1996; Kim et al., 2012; Komada et al., 2002). These protein-like peaks have been associated with high biological activity in the water column (Coble, 2007) and with degradation of fresh biomass (Parlanti et al., 2000). Some studies have observed these peaks to be most pronounced near the sediment-water interface (e.g., Coble, 1996; Seretti et al., 1997), although Burdige et al. (2004) observed protein-like fluorescence to co-vary with bulk DOC, and proposed that protein-like fluorescence is associated with both labile DON that escapes the sediments as a benthic flux and DON that is part of the recalcitrant bulk DOM that accumulates in the pore waters (see Section III.F).

C Carbon Isotope Ratios

Abundances of natural ^{13}C and ^{14}C provide insight into the composition and dynamics of pore water DOC cycling that complement molecular and other compositional studies (e.g., McNichol and Aluwihare, 2007; Raymond and Bauer, 2001b). To the best of the authors' knowledge, only a handful of $\Delta^{14}C$ and $\delta^{13}C$ values for marine pore water DOC ($\Delta^{14}C_{DOC}$ and $\delta^{13}C_{DOC}$, respectively) have been published (Bauer et al., 1995; Heuer et al., 2009; Ijiri et al., 2012; Komada et al., 2013; Valentine et al., 2005). The scarcity of data may be due, in part, to the analytical challenges associated with determining natural C isotope ratios in marine DOC, especially $\Delta^{14}C$ (Bauer, 2002; Johnson and Komada, 2011; also see Chapter 6).

Pore water $\Delta^{14}C_{DOC}$ profiles have been reported for two sites in the northeastern Pacific

Ocean: station N on the continental rise at the base of the Monterey deep-sea fan (Bauer et al., 1995) and Santa Monica Basin (SMB) in the California Borderland (Bauer et al., 1995; Komada et al., 2013). At both sites, $\Delta^{14}C_{DOC}$ values are invariably higher than the $\Delta^{14}C$ value of the overlying bottom water DOC by >100‰ (e.g., see Section V.B), and pore water $\delta^{13}C_{DOC}$ values generally fall between −24‰ and −20‰. These results clearly indicate that DOC flux out of these sediments is dominated by material that is much younger than the DOC found in the deep ocean and that this material appears to be of marine origin. However, as discussed in Section VII.C, these observations belie the more complex role of sediments as a possible source of ^{14}C-depleted (and recalcitrant) DOC to the oceans.

Strongly ^{14}C-depleted DOC has been reported for waters associated with hydrothermal systems and methane seeps. McCarthy et al. (2011) determined $\Delta^{14}C$ and $\delta^{13}C$ signatures of DOC and DIC in ridge-flank and on-axis hydrothermal fluids from the Juan de Fuca Ridge, concluding that DOC in these fluids is synthesized from ^{14}C-depleted DIC by chemosynthetic bacteria. They further suggested that ridge-flank systems might be sources of new, yet ^{14}C-depleted, DOC to the deep ocean. Pohlman et al. (2010) observed that pore water DOC in northern Cascadia margin seep sediments was strongly depleted in both ^{14}C and ^{13}C, concluding that a significant amount of this DOC is derived from fossil methane.

$\delta^{13}C$ values of pore water DOC have been used to gain insight into carbon cycling and metabolic pathways in methane-bearing environments. Heuer et al. (2009) reported $\delta^{13}C$ values of acetate, lactate, and bulk DOC in the upper 190 m of sediments in the northern Cascadia Margin. $\delta^{13}C$ signatures of DOC and lactate ranged from ~−20‰ to −24‰ and were similar to, or slightly higher than, the $\delta^{13}C$ value of bulk sedimentary POC, suggesting little to no ^{13}C-fractionation during fermentation. In contrast, $\delta^{13}C$ signatures

of acetate varied widely with depth, which was attributed to the interplay among pathways, and the associated isotope fractionation, of acetate production (fermentation, acetogenesis) and consumption (sulfate reduction, acetoclastic methanogenesis). Ijiri et al. (2012) drew similar conclusions for the uppermost 14 m of sediments of the Bering Sea shelf break.

Valentine et al. (2005) reported $\delta^{13}C$ values for pore water DOC in the uppermost 25 cm of sediments of Hydrate Ridge, a site of intense methane seepage off the coast of Oregon (USA). In contrast to the findings of Heuer et al. (2009) and Ijiri et al. (2012), pore water DOC at this site had $\delta^{13}C$ signatures that deviated from those of bulk sedimentary POC by as much as ±10‰. Also, $\delta^{13}C$ signatures of DOC were as low as −38.3‰. Through mass balance calculations, these authors concluded that pore water DOC was an important carbon source for both heterotrophy and organic carbon accumulation at this site.

D Fourier Transform Ion Cyclotron Resonance Mass Spectrometry (FTICR-MS)

Fourier transform ion cyclotron resonance mass spectrometry (FTICR-MS) is a relatively new analytical technique with very high mass resolution and accuracy, capable of providing exact chemical formulas of LMW (<~1 kDa) DOM compounds (Sleighter and Hatcher, 2007). With this technique, many thousands of DOM compounds can be identified in a single sample, although quantification of the concentrations of these individual compounds is currently not possible (e.g., Kujawinski et al., 2004). Nevertheless, results to date provide important insights into the nature of recalcitrant LMW-DOM in natural systems.

Only a relatively small number of pore water samples have been analyzed by FTICR-MS (Koch et al., 2005; McKee, 2011; Schmidt et al., 2009, 2011; Tremblay et al., 2007), and so few conclusions can be drawn. A wide range of aliphatic and aromatic compounds have been observed in these studies, with suggested marine and terrestrial sources. Based on elemental ratios and inferred reaction pathways determined using a van Krevelen diagram (e.g., Sleighter and Hatcher, 2007) many of the compounds detected by FTICR-MS appear to represent recalcitrant compounds produced by microbial and/or photochemical degradation processes (also see related discussions in Chipman et al., 2010).

More specifically, weighted average H/C (~1.2-1.3) and O/C (~0.4-0.5) values for these pore water samples (Koch et al., 2005; Schmidt et al., 2009) plot within the region on a van Krevelen diagram that Hertkorn et al. (2006) identify as representing carboxyl-rich alicyclic molecules (CRAM). CRAM is thought to represent a major component of recalcitrant DOM in the oceans. When these results are examined in the context of Figure 12.2, we see that they support the notion of relatively LMW recalcitrant DOM being produced during the course of organic matter remineralization in sediments. FTICR-MS studies focusing on N-containing compounds in Black Sea sediment pore waters (Schmidt et al., 2011) similarly suggest that when these DON compounds are plotted on a van Krevelen diagram they too fall within the region discussed above for recalcitrant DOM.

In two contrasting organic-rich sediments (Mangrove Lake sediments and sediments on the northwest Iberian continental margin off Spain), a relatively large number of sulfur-containing DOM compounds were observed in pore waters by FTICR-MS (McKee, 2011; Schmidt et al., 2011). In both studies, it was suggested that organic matter sulfurization is responsible for these compounds. However, it is not clear whether this sulfur addition occurs to DOM precursors or whether sulfur addition to POM itself is the precursor to S-containing DOM compounds. Regardless of their mode of formation, S-containing DOM molecules are likely to be fairly recalcitrant (e.g., see related discussions in Kohnen et al., 1992; Tegelaar et al., 1989).

E Nuclear Magnetic Resonance (NMR)

Only a small number of pore water samples have been analyzed by nuclear magnetic resonance (NMR), and in all cases only HMW extracts (>0.5 or 1 kDa, depending on the study) were examined (Orem and Hatcher, 1987; Orem et al., 1986; Repeta et al., 2002). In the suite of samples studied by Orem and Hatcher (1987), they observed that samples from predominantly anoxic settings were dominated by carbohydrate resonances, while DOM from more aerobic settings had NMR spectra with diminished carbohydrate character and greater aromatic character. They suggested that these differences might be the result of more effective decomposition of carbohydrates under oxidizing conditions plus significantly lower rates of lignin degradation (as the source of the observed aromatic resonances) under anoxic conditions.

The HMW pore water DOM extracts analyzed by Repeta et al. (2002), all from settings that could be considered anoxic in the context of the definitions from the Orem and Hatcher (1987) study, also had major contributions to their NMR spectra from carbohydrates, as well as bound acetate, and lipids. Acetate and carbohydrate resonances, when coupled with monosaccharide analyses of acid-hydrolyzed extracts, were interpreted as being indicative of relatively high concentrations of acylated polysaccharides in these pore waters (see Section V.B for details).

F DON and the C/N Ratio of Pore Water DOM

Much of what has been said about DOC in sediment pore waters applies equally well to pore water DON, in that nitrogen-containing DOM is also a heterogeneous class of organic compounds that range from well-defined biochemicals such as urea or amino acids, to larger dissolved proteins and peptides, to more complex (and poorly characterized) N-containing humic and fulvic acids (see recent reviews in Aluwihare and

Meador, 2008; Worsfold et al., 2008). Similarly, much of the DON in pore waters is uncharacterized at the compound level. Lomstein et al. (1998) were able to quantify ~40% of the DON in Danish coastal sediment pore waters as dissolved free and combined amino acids, although as discussed in Section IV.C the details about what these combined amino acids actually represent is uncertain.

Nitrogen found in marine DOM can have several types of functionality, although in HMW-DOM in the water column much of this functionality appears to be in the amide ($-CONH_2$) form (e.g., Aluwihare and Meador, 2008). This issue has not yet been examined in sediment pore waters, although given the important role of proteins (in general) and amino acids (in particular) in sedimentary carbon and nitrogen remineralization (Burdige and Martens, 1988; Cowie et al., 1992; Henrichs and Farrington, 1987) and preservation (e.g., Moore et al., 2012), amide or amine ($-NH_2$) functionality for much of the pore water DON seems likely. The factors controlling the preservation of nitrogen in marine sediments is an important topic of research (Derenne et al., 1998; McKee and Hatcher, 2010) and reactions involving DON intermediates might be expected to play some role in the preservation process (see discussions in Burdige, 2001; Schmidt et al., 2011).

Despite the lack of detailed information regarding the specific components that make up DOM in pore waters, the simultaneous determination of pore water DOC and DON does allow for the examination of the C/N ratio of pore water DOM (=C/N_{pDOM}), which can yield insights into the composition, reactivity, and cycling of pore water DOM. Pore water profiles of DON and C/N_{pDOM} from selected marine sediments are shown in Figure 12.4 and Table 12.1 contains a more detailed summary of C/N_{pDOM} values and depth trends from a wide range of marine sediments.

At least four general conclusions can be made from these observations. The first is that when C/N_{pDOM} values in estuarine Chesapeake Bay sediments are compared with the C/N ratio of DOM benthic fluxes, we see that DOM accumulating in sediment pore waters is depleted in nitrogen as compared to that which escapes the sediments as a benthic flux (Figure 12.6). This uncoupling is consistent with other field observations (Blackburn et al., 1996; Landén-Hillmeyr, 1998) and model results (Burdige, 2002; and Figure 12.3) and will be discussed in further detail in Section VII.C.

A second observation is that in sediments where there is a significant input of terrestrial organic matter (e.g., site N3 in the Chesapeake Bay and ODP core 1075 from the Southwest African Margin), C/N_{pDOM} values tend to be higher. This is consistent with the fact that terrestrially derived organic matter is generally depleted in nitrogen as compared to marine organic matter (C/N values of ~20-80 vs. ~6-8; e.g., Burdige, 2006b).

A third observation is that in oxic or mixed redox sediments, C/N_{pDOM} values tend to be relatively low as compared to more strict anoxic sediments. This result can be seen in Chesapeake Bay site S3 sediments, in Patton Escarpment sediments, and in oxic, pelagic sediments in the southwest Pacific. One possible explanation for these observations involves benthic macrofaunal processes that may produce low C/N ratio organic compounds, such as urea (C/N = 0.5) (Burdige and Zheng, 1998; Lomstein et al., 1989). However, in the bioturbated sediments at site S3 in Chesapeake Bay, urea is not a significant component of the pore water DOM pool (Burdige, 2001). Another possible source of low C/N ratio pore water DOM compounds in mixed redox sediments are bacteria. Bacteria have C/N ratios that range from ~3 to 5 (Fenchel et al., 1998), so grazing of bacteria by higher organisms in oxic or mixed redox sediments (Kemp, 1990; Lee, 1992) could lead to the production of DOM with a low C/N ratio (e.g., certain amino acids; see Section IV.C for further details). The absence of bacterial grazing in anoxic sediments would

TABLE 12.1   C/N Ratios of Pore Water DOM (C/N$_{pDOM}$) in Marine Sediments

Site	Sediment Depth (max.)	C/N$_{pDOM}$ (range)	C/N$_{pDOM}$ (Depth Variations)	Sediment Characteristics
Coastal, estuarine sediments				
Chesapeake Bay site M3[a]	30 cm	~12-18	No coherent trend with depth, but depth-weighted averages show seasonal variations and are highest when remineralization rates are lowest	Anoxic, non-bioturbated; sulfate reduction and methanogenesis dominate organic matter remineralization
Chesapeake Bay site S3[a]	30 cm	~8-14	No coherent depth trends	Mixed redox conditions; sediments are bioturbated and bioirrigated
Chesapeake Bay site N3[a]	30 cm	~10-30	Increase with depth	Sediment organic matter is largely terrestrially derived; some bioturbation in the upper ~5 cm of sediment
Danish coastal sediments[b]	5 cm	~10-20	~20 (upper 1 cm), constant below this depth (=~10-11)	Shallow water depth (4 m); upper ~10 cm of sediment is bioturbated
Lower St. Lawrence estuary (CA)/Gulf of St. Lawrence[c]	~2-3 cm	~5-14	Sediment depth variations not reported, although average C/N$_{pDOM}$ values increase as one moves from the lower estuary into the Gulf	As one moves from the lower estuary into the Gulf sediment organic matter reactivity decreases, sediment oxygen exposure increases, and the relative importance of terrestrial organic matter in the sediments decreases; salinity is high (>30) in the bottom waters, with little variation along this transect
Continental margin and deep sea sediments				
Mid-Atlantic shelf/ slope break (400-750 m water depths)[d]	30 cm	~7-17	Increase with depth in most cores	Sub-oxic sediments with minimal bioturbation in the upper 20-30 cm of sediments; nitrate becomes undetectable in the upper 1-4 cm of sediment; linear sulfate gradients in the upper 25 cm of sediment
Santa Monica Basin (California Borderlands; 900 m water depth)[e]	30 cm	~7-18	Increase with depth (7-12 near sediment surface, 7-18 at depth)	Anoxic, sulfidic sediments with no bioturbation or bioirrigation
San Clemente Basin (California Borderlands; ~2000 m water depth)[e]	30 cm	~7-17	Increase with depth (7-10 near sediment surface, ~17 at depth)	Sub-oxic sediments; pore water O_2 depleted in the upper 1 cm of sediment, nitrate depleted by ~2-5 cm sediment depth

TABLE 12.1   C/N Ratios of Pore Water DOM (C/NpDOM) in Marine Sediments—cont'd

Site	Sediment Depth (max.)	C/N_{pDOM} (range)	C/N_{pDOM} (Depth Variations)	Sediment Characteristics
Patton Escarpment (eastern North Pacific; ~3700 m water depth)[e]	30 cm	~5-15	Minimum value observed ~1-2 cm below the sediment surface; increase with depth below	Pore water O_2 depleted at a sediment depth of ~2.5 cm, nitrate depleted by ~4-10 cm sediment depth; some sediment bioturbation and bioirrigation
Hatteras continental rise (northwest Atlantic; ~4200 m)[f]	30 cm	~6-11	No obvious depth trends	Sub-oxic sediments; nitrate depleted by ~8 cm sediment depth after an initial increase in the upper 1-2 cm of sediment
Southwest Pacific pelagic sediments (~2800-5400 m water depths)[g]	30-50 cm	~2-7	No obvious depth trends	Oxic sediments; nitrate increases with depth in an exponential-like fashion
"Deep" sediment cores (Southwest African Margin)[h]				
Lower Congo Basin (ODP Leg 175, core 1075; ~3000 m water depth)	60 m	20 (±7)	No obvious depth trends	Anoxic sediments; sulfate is depleted by ~30 m sediment depth; marine and terrestrial organic matter sources
Walvis Basin (ODP Leg 175, core 1082; ~1300 m water depth)	360 m	11.3 (±3.2)	No obvious depth trends	Anoxic sediments; sulfate is depleted by ~20 m sediment depth; predominant marine organic matter sources

[a]Data from Burdige and Zheng (1998) and Burdige (2001).

[b]Data from Lomstein et al. (1998).

[c]Data from Alkhatib et al. (2012, 2013).

[d]Data from Burdige and Gardner (1998) and Burdige et al. (2000, unpub. data).

[e]DOC and DON data from Burdige et al. (1999, unpub. data). Other data from Shaw et al. (1990), Berelson et al. (1996), and W. Berelson (pers. comm.).

[f]Data from Heggie et al. (1987).

[g]Data from Suess et al. (1980).

[h]DOC and DON data from Burdige et al. (unpub. data); other data from Wefer et al. (1998).

then minimize the production of low C/N ratio and bacterially derived DOM (also see discussions in Burdige, 2001).

However, studies conducted in the lower St. Lawrence estuary, Canada contrast with the second and third general observations discussed above. As one moves down this estuary into the Gulf of St. Lawrence, there is an increase in C/N_{pDOM} from ~5 to ~14 in surface (upper ~2-3 cm) sediment pore waters as the relative input of terrestrial organic matter to the sediments decreases and sediment oxygen exposure time increases (Alkhatib et al., 2013). These authors attribute these trends in C/N_{pDOM} to a decrease in sediment organic matter reactivity along with the increase in sediment oxygen exposure time (also see Alkhatib et al., 2012). Additional work is needed to examine the reasons for contrasting trends among these environments.

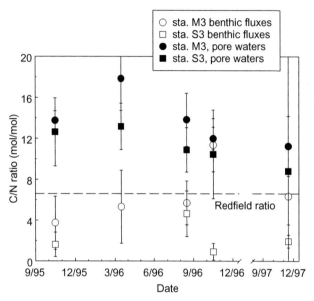

FIGURE 12.6 Temporal changes in the depth-weighted average value of C/N$_{pDOM}$ (e.g., see Figure 12.4) in the pore waters (closed symbols) and the C/N ratio of the DOM escaping the sediments (i.e., the ratio of the DOC benthic flux to the DON benthic flux, open symbols) at stations S3 and M3 in the Chesapeake Bay. *Data from Burdige and Zheng (1998) and Burdige et al. (2004).*

The final observation based on the results in Table 12.1 is that in most of the continental margin sediments examined to date, C/N$_{pDOM}$ values increase with sediment depth (also see the San Clemente Basin profiles in Figure 12.4). Since there is evidence for the occurrence of terrestrially derived organic matter in such continental margin sediments (see reviews in Blair and Aller, 2012; Burdige, 2006b), one explanation of these C/N$_{pDOM}$ values is that the increased remineralization of terrestrially derived organic matter becomes increasingly important with depth, leading to the production of DOM with higher C/N ratios. Implicit in this assumption is that N-depleted, terrestrially derived organic matter deposited in these sediments is less reactive than marine-derived POM (e.g., see Burdige, 1991a).

In contrast, C/N$_{pDOM}$ values in Chesapeake Bay estuarine sediments do not increase with sediment depth (Burdige and Zheng, 1998), despite decomposition studies suggesting that terrestrially derived POM undergoes remineralization with depth in these sediments (Burdige, 1991a). The reasons for these differences are not understood, although the results suggest that there may not be a tight coupling between the C/N ratio of sediment POM undergoing remineralization and that of its DOM intermediates (or its recalcitrant "end products," e.g., DOM in box "C" in Figure 12.2; also see discussions in Alkhatib et al., 2013). Schmidt et al. (2011) similarly suggested that FTICR-MS analyses of N-containing DOM compounds in sediment pore waters can be explained by proteins and peptides undergoing reactions that reduce the molecular size, nitrogen content, and potentially the reactivity of the products.

IV COMPOSITION AND DYNAMICS OF DOM AT THE COMPOUND AND COMPOUND-CLASS LEVELS

When looking at studies that have characterized pore water DOM at the compound or compound-class level, we note that most efforts have focused on examining concentrations and cycling of compounds that fall into the mLMW-DOM category (box "D" in Figure 12.2; also see Henrichs, 1993 for an earlier summary). While some work has been carried out examining the dynamics of the HMW-DOM pool (Arnosti, 2000; Boschker et al., 1995; Mayer and Rice, 1992; Pantoja and Lee, 1999; Robador et al., 2010), few studies have examined its chemical composition.

A Short Chain Organic Acids (SCOAs)

Interest in the study of short-chain organic acids (SCOAs) such acetate, lactate, formate, propionate, and butyrate in marine sediments stems from the observation that these compounds are important in situ substrates/electron donors (along with H_2) for terminal anaerobic remineralization processes such as iron reduction, sulfate reduction, and methanogenesis (e.g., Finke et al., 2007; Lovley and Phillips, 1987; Parkes et al., 1989; Sansone and Martens, 1982; Sørensen et al., 1981; Thamdrup, 2000; Valdemarsen and Kristensen, 2010). These organic acids are largely produced by fermentation reactions that fall along the left side of Figure 12.2 (e.g., Thauer et al., 1977). However, acetate can also be produced by CO_2 reduction using H_2 in a microbial process referred to as acetogenesis (Ragsdale and Pierce, 2008).

In the uppermost ~30cm of most anoxic sediments, acetate concentrations range from <1 to up to ~100µM, generally increasing with sediment depth (Albert and Martens, 1997; Barcelona, 1980; Christensen and Blackburn, 1982; Heuer et al., 2009; Hines et al., 1994; Knab et al., 2008; Novelli et al., 1988; also see Burdige, 2002 for a more detailed list of earlier publications). In many of these studies, total DOC concentrations were generally not determined along with acetate, although where both measurements were made (or where DOC concentrations can be obtained from other published studies), acetate usually accounts for at most ~5%, and often times <1%, of the total pore water DOC.

Concentrations of other SCOAs such as formate or propionate have not been determined as frequently as acetate although when determined, their concentrations can be comparable to those of acetate (Albert and Martens, 1997; Barcelona, 1980). Previous studies (summarized in Henrichs, 1993) have also concluded that much of the acetate and other SCOAs that can be chemically measured may not be biologically available, possibly due to complexation in the pore waters; Finke et al. (2007) support these earlier suggestions. Given the importance of SCOAs as substrates for terminal anaerobic remineralization processes, all of these observations reinforce the notion that material in the mLMW-DOM pool (Figure 12.2) represents a small fraction of the total sediment pore water DOM pool whose concentration is held at relatively low levels due to rapid microbial utilization.

It is generally thought that competition for key substrates (electron donors) such as acetate or H_2 plays a major role in regulating the biogeochemical zonation of different anaerobic terminal remineralization processes (Canfield et al., 2005; Hoehler et al., 1998; Lovley and Goodwin, 1988; Lovley and Phillips, 1987). Furthermore, concentrations of these key substrates increase with decreasing energy yield of the remineralization process. For example, iron reducers appear able to maintain H_2 concentrations at levels that are too low for sulfate reducers to utilize H_2, as sulfate reducers similarly do to out-compete methanogens for this substrate (Hoehler et al., 1998; Lovley and Goodwin, 1988).

Incubations of freshwater sediments by Lovley and Phillips (1987) show an overall order of magnitude increase in the acetate concentration when the dominant terminal remineralization process changes from iron reduction to sulfate reduction to methanogenesis. In contrast, similar competitive inhibition of acetate uptake did not affect sulfate reduction and iron reduction in surface (0-2cm) Arctic marine sediments (Finke et al., 2007). However, field data, as well as other sediment incubation studies and bioenergetics model calculations, generally do show that the transition in marine sediments (either spatially or temporally) from sulfate reduction to methanogenesis results in an increase in acetate concentrations (Alperin et al., 1994; Dale et al., 2006; Heuer et al., 2009; Hoehler et al., 1998; Knab et al., 2008).

Slightly different trends have been observed in deeply buried sediments (the so-called deep marine biosphere; i.e., depths beginning at 10s to 100s of meters below the seafloor, mbsf). For example, in gas hydrate-containing sediments on the Blake Ridge on the southeast US continental margin, acetate increases to a maximum concentration of ~15 mM (~15% of the total DOC) in methanogenic sediments at depths of ~750 mbsf (Egeberg and Dickens, 1999) while complete sulfate depletion in the pore waters occurs at a significantly shallower sediment depth, that is, ~20 mbsf. This acetate buildup is attributed to a temperature increase with depth (due to the natural geothermal gradient) that stimulates acetate production over acetate consumption by methanogens (Wellsbury et al., 1997). Of equal importance is the fact that acetate consumption appears to occur several hundred meters shallower in the sediment column relative to its depth of production (Egeberg and Dickens, 1999). Similar depth trends were also observed in hydrate-containing sediments on the Peru margin (site 1230, Leg 201; D'Hondt et al., 2003). Here acetate concentrations increase from <20 μM to ~60 μM at ~10 mbsf, where sulfate concentrations go to zero, and then increase again dramatically to ~200 μM (~2% of total DOC) at ~140 mbsf. This spatial separation between acetate production and consumption in the deep marine biosphere is very different than that seen in anoxic surficial sediments (see references above) where acetate production and consumption do not show such dramatic spatial uncoupling. Further study is required to assess the significance of these observations in terms of general biogeochemical dynamics of the deep marine biosphere.

B Carbohydrates

Total dissolved carbohydrates (TDCHOs) have been determined in a limited number of coastal and continental margin sediments, with concentrations that generally range from ~10 to 400 μM C (Arnosti and Holmer, 1999; Burdige et al., 2000; Jensen et al., 2005; Lyons et al., 1979; Robador et al., 2010). In most cases TDCHO concentrations increase with depth in the upper ~20-30 cm of sediment and represent ~10-40% of the total DOC. Such relative TDCHO concentrations generally decrease in this sediment depth range although the magnitude of these changes vary among the few sites that have been examined (see Burdige et al., 2000, for details).

There have been few studies of individual aldoses (monomeric neutral sugars) in sediment pore waters. In selected pore water samples from Chesapeake Bay and mid-Atlantic shelf/slope break sediments, after acid hydrolysis ~30-50% of the TDCHOs could be identified as individual aldoses (Burdige et al., 2000), although lower percentages were observed in Danish continental margin sediments of the Skagerrak, Kattegat, and Belt Seas (Jensen et al., 2005). In the Burdige et al. (2000) study, total aldose yields (total individual aldose concentrations as a percentage of DOC) are higher in continental margin sediment pore waters (~9%) than they are in the estuarine sediment pore waters (<5%), while values in the Danish continental margin sediments range from 0.05% to 4% (Jensen et al., 2005). In both studies, dissolved glucose is the predominant aldose.

These results suggest that dissolved carbohydrate concentrations in sediment pore waters are not strongly tied to particulate (sediment) carbohydrate concentrations, and TDCHO concentrations may be more strongly controlled by sediment remineralization processes (Burdige et al., 2000). Early studies also suggested that these TDCHOs likely represent some of the initial HMW intermediates produced and consumed during sediment POC remineralization (Arnosti and Holmer, 1999; Burdige et al., 2000). However, several recent studies indicate that this may not necessarily be the case.

In long-term (24 months) sediment incubations with Arctic and temperate sediments, Robador et al. (2010) observed an absolute and relative increase with time in TDCHO concentrations. They interpret this as being the result of the accumulation of recalcitrant carbohydrates, although the absence of molecular weight data does not allow us to determine whether these are HMW or LMW entities. At the same time, in pore waters collected from sediment depths of ~100-300 mbsf in the equatorial Pacific and Peru margin (ODP Leg 201; Burdige, 2006a), TDCHOs represent up to three quarters of the total DOC (0.76 ± 0.46, at the three open ocean sites; Figure 12.7). Furthermore, in these deeply buried sediments, as is also seen in shallow estuarine and continental margin sediments, there appears to be an inverse relationship between the relative concentrations of pore water TDCHOs and rates of sediment carbon oxidation (Figure 12.7).

One interpretation of these observations is that recalcitrant carbohydrates may be directly produced through the decomposition of more recalcitrant components of the POC pool; this has also been discussed in terms of a "decoupling"

of initial hydrolysis of POC and downstream fermentative and/or terminal metabolism (Robador et al., 2010). This observation is discussed in a more general sense in Section V. However, little is known about these recalcitrant carbohydrates. One suggestion regarding their occurrence is that structural changes (e.g., methylation, sulfurization) may render reactive carbohydrates less susceptible to further degradation. Such changes may also make individual aldoses in these carbohydrates difficult to identify after acid hydrolysis, yet may not impact their detection by the colorimetric procedures used in many of these studies to determine total dissolved carbohydrates (Burdige et al., 2000; Robador et al., 2010). Studies of the HMW isolates of three pore water samples from coastal and organic-rich continental margin sediments further indicate relatively high concentrations of acylated polysaccharides (APS) in pore waters (Repeta et al., 2002). APS represents a group of compounds that appear to be produced by marine phytoplankton and are

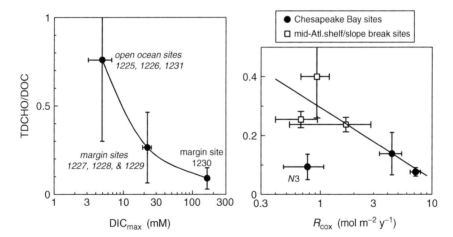

FIGURE 12.7 Apparent inverse relationship between the relative concentration of pore water TDCHOs and the rate of sediment carbon oxidation. (left panel) The relative concentrations of total dissolved carbohydrates (TDCHO/DOC) versus the maximum concentrations of DIC in the pore waters (DIC_{max}) at open-ocean and Peru margin sites from ODP Leg 201; data from Burdige (2006a). Note that here DIC_{max} is used as a proxy for the sediment carbon oxidation rate. (right panel) TDCHO/DOC versus the depth-integrated rate of sediment carbon oxidation at Chesapeake Bay (closed circles) and mid-Atlantic shelf/slope break sites (open squares); data from Burdige et al. (2000). Also shown here is the best fit line through this semi-log plot of the data ($y = -10.54 \ln(x) + 29.48$; $r^2 = 0.73$), excluding the data point from sta. N3. In contrast to the other sites in this panel, terrestrial organic matter predominates over marine organic matter in sta. N3 sediments (Burdige et al., 2000; Marvin-DiPasquale and Capone, 1998), which may explain why its values do not fall on the best fit line shown here.

defined as being "semi-reactive" in the surface ocean, i.e., having turnover times on the order of years to decades (Repeta and Aluwihare, 2006). While APS production might explain the results from the incubation studies of Robador et al. (2010), it is difficult to see how it could explain high relative concentrations of recalcitrant carbohydrates in the pore waters of deeply buried sediments where the time scales of remineralization are clearly significantly longer than decades (Burdige, 2006a).

C Amino Acids

Amino acids comprise significant amounts of the carbon and nitrogen that are remineralized in marine sediments (e.g., Burdige and Martens, 1988; Cowie and Hedges, 1992; Henrichs and Farrington, 1987) and pore water dissolved amino acids are likely important intermediates in this process (also see references cited below in this section). Dissolved amino acids in sediment pore waters may be "free" monomeric amino acids (DFAAs) as well as combined amino acids (DCAAs); the latter generally either represent dissolved peptides (originally produced by the hydrolysis of larger proteins, e.g., Roth and Harvey, 2006) or may be incorporated into humic-like substances. Combined amino acids (at least those found in dissolved peptides) can be further hydrolyzed and deaminated to produce, among other end products, smaller peptides, ammonium, and DFAAs (e.g., Jacobsen et al., 1987; Pantoja and Lee, 1999). DFAAs can be used in a wide range of fermentation reactions, generally producing H_2 and short-chain organic acids such as acetate (Barker, 1981; Thauer et al., 1977). DFAAs can also be used directly by sulfate-reducing bacteria (e.g., Takii et al., 2008).

Studies to date in anoxic sediments suggest that most remineralization of DFAAs occurs by fermentation rather than by sulfate reduction (Burdige, 1991b; Hansen and Blackburn, 1995; Valdemarsen and Kristensen, 2010), although

there appear to be exceptions (Hansen et al., 1993; Wang and Lee, 1995). In any event, when compared to H_2 or short-chain organic acids such as acetate, free amino acids are likely minor electron donors, at best, for total sulfate reduction in sediments (Burdige, 1989; Parkes et al., 1989; Valdemarsen and Kristensen, 2010)

DFAAs in pore waters have been examined in a wide range of sediments, including those in salt marsh, estuarine, coastal, continental margin, and deep-sea settings (e.g., Burdige and Martens, 1990; Haberstroh and Karl, 1989; Henrichs et al., 1984; Landén and Hall, 1998, 2000; Lomstein et al., 1998). Several of these studies have examined amino acid adsorption to sediments as well as DFAA turnover rates (also see Christensen and Blackburn, 1980; Ding and Henrichs, 2002; Henrichs and Sugai, 1993; Liu and Lee, 2007; Rosenfeld, 1979; Wang and Lee, 1993). Concentrations of total dissolved free amino acids (TDFAAs) generally decrease with sediment depth, with surface pore water concentrations usually ranging from ~20 to 200 μM and concentrations below 10-20 cm being <5-10 μM. In the few studies where total DOC and DON have been examined along with amino acids, TDFAAs represent 1-13% of the DON and <4% of the DOC (Henrichs and Farrington, 1987; Landén and Hall, 2000; Lomstein et al., 1998).

The predominant amino acids in the DFAA pool include glutamic acid, alanine, glycine, aspartic acid, and, in some sediments, the nonprotein amino acid β-aminoglutaric acid (β-aga), an isomer of glutamic acid. Among the protein amino acids, the DFAA pool is generally enriched in glutamic acid relative to the sediment hydrolyzable amino acid pool. Other nonprotein amino acids in addition to β-aga, such as β-alanine, can also be enriched in pore waters relative to the sediments. These and other compositional differences between pore water and sediment amino acids are likely related to both biological and physical (e.g., adsorption) processes that affect amino acids as they undergo remineralization (Burdige and Martens, 1990; Henrichs and Farrington, 1987).

In anoxic sediments, the mole percentage of β-aga generally increases with depth (upper ~30 cm), with β-aga often becoming the predominant DFAA in pore waters (> ~40 mol%) (Burdige and Martens, 1990; Henrichs and Farrington, 1979; Henrichs et al., 1984). In contrast, β-aga appears to be a much less important component of the pore water DFAA pool in oxic or mixed redox sediments (Caughey, 1982; Henrichs and Farrington, 1980; Landén and Hall, 2000). The source(s) of β-aga in sediment pore waters are not well characterized (Burdige, 1989; Henrichs and Cuhel, 1985) and its relative accumulation with depth in anoxic sediments may occur because it is more recalcitrant than other amino acids (Henrichs and Farrington, 1979). The proposed recalcitrant nature of β-aga may also be a function of sediment redox conditions, similar to that observed for other recalcitrant components of the pore water DOM pool (Burdige, 2001; also see Section VI.C).

Along similar lines, there could be broader compositional differences in the dissolved amino acid pool in anoxic versus oxic/mixed redox sediments that might help explain the low C/N_{pDOM} values in these latter sediments (see Section III.F). Specifically, Burdige (2002) suggested that glycine ($C/N = 2$) may be preferentially enriched in the pore waters of mixed redox versus strictly anoxic sediments. However, glycine is also an abundant amino acid in many benthic invertebrates (Henrichs and Farrington, 1980), and so elevated glycine levels may simply result from the release of this amino acid by benthic organisms during core collection and/or pore water processing (Burdige and Martens, 1990; Jørgensen et al., 1981). Further studies will be needed to critically examine these observations (also see discussions in Burdige, 2001).

In contrast to studies of DFAAs, there have been far fewer studies of DCAAs in marine sediment pore waters (Caughey, 1982; Kawahata and Ishizuka, 2000; Lomstein et al., 1998; Ogasawara et al., 2001; Pantoja and Lee, 1999). In general, concentrations of total DCAAs are ~1.5-4 times

that of total DFAAs. Given the small data set on pore water DCAAs, it is difficult to determine whether the composition of the DCAA pool is more similar to that of DFAAs or hydrolyzable sediment amino acids. Such information could be important in determining the extent to which the DCAA pool represents dissolved peptides or proteins (i.e., "reactive" HMW intermediates of sediment organic matter remineralization) or perhaps abiotic condensation products of, for example, melanoidin-type reactions. In the former case, the DCAA pool might be expected to be more similar to that of sediment (hydrolysable) amino acids, while in the latter case the amino acid distribution in DCAAs might be expected to look more like DFAAs (for further discussions, see Burdige, 2002).

V MODELING DOC CYCLING IN MARINE SEDIMENTS

A Production of Recalcitrant DOC: General Observations

Consistent with Figure 12.2 and discussions in Section II, a number of incubation studies show production of DOC that appears recalcitrant on time scales comparable to those over which organic matter remineralization and inorganic nutrient regeneration occur (Brüchert and Arnosti, 2003; Chipman et al., 2010; Hee et al., 2001; Komada et al., 2012; Robador et al., 2010; Weston and Joye, 2005). Production of recalcitrant DOC was observed in flow through column reactor studies run for only several hours (Chipman et al., 2010) to closed system sediment incubations lasting months (Komada et al., 2012) to years (Robador et al., 2010). In at least one of these studies (Chipman et al., 2010), FTICR-MS analyses of the produced DOM indicated the presence of lignin- and tannin-like compounds. These compounds are presumed to be fairly recalcitrant, in part because of their high degree of aromaticity (e.g., Bianchi and Canuel, 2011).

In cases where such compositional information is not available, accumulation of "recalcitrant" DOC in incubation studies could also be due to a decoupling between DOC production and degradation, as opposed to production of inherently recalcitrant DOC (Robador et al., 2010; Weston and Joye, 2005). Looking at Figure 12.2, a decrease in carbon flow from box "A" to "B" or box "B" to "D" relative to the initial organic matter breakdown (sediment POM to box "A") could, for example, result in the accumulation of DOM in boxes "A" or "B." This suggestion also reinforces the point that describing the DOC accumulating in these experiments as "recalcitrant" is somewhat subjective and in part may be a function of the time scale of the study.

In considering the production of recalcitrant DOC during overall POC remineralization, it may also be difficult to distinguish between "direct" production of recalcitrant DOC from POC or HWM intermediates (i.e., pathways 1-3 in Figure 12.2) versus pathways by which more labile LMW intermediates are transformed into recalcitrant components (i.e., pathway 4 in Figure 12.2). While there is evidence that complexation reactions can decrease the reactivity, or bioavailability, of simple monomers such as acetate (Christensen and Blackburn, 1982; Finke et al., 2007; Michelson et al., 1989), the long-term fate of these complexes is uncertain. At the same time, there is little direct evidence for the occurrence of aqueous phase geopolymerization reactions in nature (Hedges, 1988; Hedges and Oades, 1997; Henrichs, 1992), and abiotic condensation reactions involving LMW monomeric reactants are likely to be quite slow in comparison to their biological uptake or remineralization to inorganic end products (Alperin et al., 1994). If these internal conversion processes are indeed slow, it is unlikely that they would have been detected in incubation experiments such as those discussed here (also see discussions in Section VI.B).

Another important constraint on in situ production of recalcitrant DOC from labile intermediates comes from radiocarbon (^{14}C) analysis of the DOC produced in the incubation study of Komada et al. (2012). They saw net production of DOC depleted in ^{14}C by at least 200‰ relative to the DIC produced by fermentation and/or terminal respiration of POC, itself enriched in ^{14}C and containing at least some bomb-^{14}C. Given the extremely short time period of this incubation study (~4 months) relative to the half-life of ^{14}C (5730 year), in situ production and ageing of recalcitrant DOC is not possible and its production from pre-aged (^{14}C-depleted) POC is therefore a plausible explanation. Related discussions of this point are also presented in the next section.

B The Multi-G + DOC Model

In applying the model in Figure 12.2 to the incubation studies discussed above, an important point to recall is that different types of POC undergo remineralization with time, or in the case of sediments in situ, with depth in the sediment column. Remineralization therefore continually fractionates POC by preferentially using the most reactive material available and thus decreases the overall reactivity of that remaining (e.g., Cowie and Hedges, 1994; Middelburg, 1989). One approach to parameterizing these changes assumes that POC can be "quantized" into discrete classes with different reactivities and chemical properties, as in the multi-G model (Burdige, 1991a; Westrich and Berner, 1984). Other formulations allow for changes (decreases) in POC reactivity to be viewed as a continuous function of the age of the material, and therefore depth in a sediment column (Boudreau and Ruddick, 1991; Middelburg, 1989). There are some advantages to this latter approach to describe sediment organic matter reactivity. However, in the context of modeling DOC cycling in sediments (or in sediment incubation studies), this approach has distinct disadvantages with regard to parameterizing the changing properties of sediment POC undergoing remineralization beyond its bulk reactivity and therefore linking POC remineralization to DOC cycling. Thus, in the following discussions, we will use a multi-G approach to link DOC cycling to POC remineralization.

Referring to Figure 12.2, the specific DOC transformation pathways, their rates, and the types of recalcitrant DOC that may be produced, could all depend on the source POC. Further complicating this discussion is that there are few field data with which to parameterize the types of DOC cycling illustrated in Figure 12.2 in reactive-transport (early diagenesis) models. The model illustrated in Figure 12.3 (Burdige, 2002) was one such attempt (albeit highly simplified), as was the work of Dale et al. (2008). However, this latter study focused on DOC cycling as it primarily relates to mLMW-DOC compounds (box "D" in Figure 12.2), that is, substrates such as acetate and H_2 that are directly utilized by sulfate-reducing and methanogenic bacteria.

In our most recent work (Komada et al., 2013), we approached this problem using what we refer to as the multi-G+DOC model (Figure 12.8). In this model, it is assumed that isotopically and kinetically heterogeneous DOC is produced and consumed during remineralization of multiple pools of metabolizable organic matter (G_i), each of which is characterized by distinct $\Delta^{14}C$ and $\delta^{13}C$ signatures (Δ_i and δ_i in Figure 12.8). Different DOC fractions (DOC_i) are produced by first-order degradation of G_i with degradation

rate constant k_i, and it is assumed that DOC_i has the same isotopic values as its parent. This DOC_i is then oxidized to DIC without isotopic fractionation (Boehme et al., 1996; Heuer et al., 2009; Penning and Conrad, 2006) with rate constant k_{DOCi}. In the context of the conceptual model in Figure 12.2, the multi-G+DOC model takes the complexity of DOC cycling in this figure and simplifies it such that the k_{DOCi} value for each DOC_i pool effectively integrates (or averages) over the reactivity of all DOC intermediates produced and consumed downstream of the parent, particulate G_i material.

The multi-G+DOC model was used in the interpretation of pore water $\Delta^{14}C_{DOC}$ and $\delta^{13}C_{DOC}$ profiles obtained from the Santa Monica Basin (Komada et al., 2013; Figure 12.9). These kinetics were implemented in a steady-state, variable-porosity, reactive-transport model for carbon species in the uppermost 45 cm of the sediment column. The values of $\Delta^{14}C_{DOC}$ are about the same as, or slightly higher than, $\Delta^{14}C$ values of bulk POC, while $\Delta^{14}C_{DIC}$ is higher than $\Delta^{14}C$ values of bulk POC and DOC, indicating that POC remineralization is a selective process in which net oxidation of younger components of the POC pool to DIC occurs at a faster rate than older counterparts. Furthermore, the $\Delta^{14}C_{DOC}$ profile shows a large drop (~200‰) from the maximum value observed in the uppermost 2 cm to that observed at 30-cm sediment depth (Figure 12.9); this drop is too large

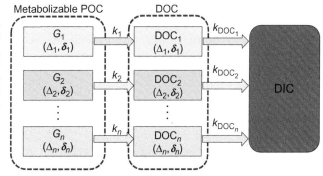

FIGURE 12.8 The multi-G+DOC model. Metabolizable POC is divided into n components, each with its own unique first-order degradation rate constant (k_i) and $\delta^{13}C$ and $\Delta^{14}C$ signatures (δ_i and Δ_i respectively). Each POC fraction (G_i) is metabolized to DOC_i without isotopic fractionation and each DOC_i fraction is oxidized to DIC (again without isotopic fractionation) with first-order rate constant k_{DOCi}. *Figure modified after Komada et al. (2013).*

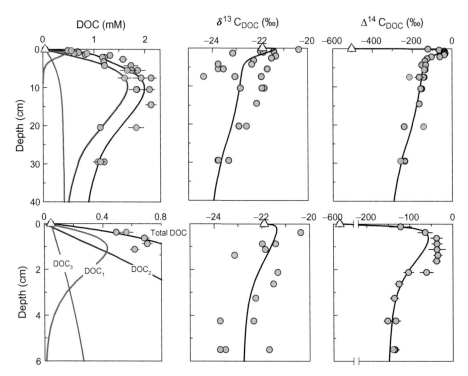

FIGURE 12.9 Pore water DOC concentration and isotopic signature profiles from Santa Monica Basin sediments (symbols) and best fit model results (lines) using the multi-G + DOC model (Figure 12.8) in a steady-state reactive-transport model (for details see Komada et al., 2013). Upper panels show the complete profiles over 40 cm, while the lower panels highlight the upper 6 cm of sediment. Shown in the concentration plots (far left) are the model results for the three DOC sub-fractions along with modeled bulk DOC (i.e., the sum of the three fractions). Yellow filled triangles are bottom water values. Where error bars are not visible, they are smaller than the symbols. *Figure modified after Komada et al. (2013).*

to be attributed to radioactive decay during diffusive transport over this distance. While it is possible that this gradient is caused by upward diffusion of ^{14}C-depleted DOC produced in deeper sediments, the relatively small DOC concentration gradient with depth (Figure 12.9) suggests this to be of minor importance. A more plausible explanation is that the isotopic composition of DIC and bulk DOC in the pore waters changes with depth as a result of the differential reactivities of the different G_i and DOC$_i$ pools, and their associated Δ^{14}C and δ^{13}C signatures.

The isotopic composition and k_{DOCi} values for the three DOC fractions obtained by the

model are presented in Table 12.2 and a subset of the model curves are shown in Figure 12.9. DOC$_1$ dominates the bulk DOC pool in the uppermost ~1 cm of the sediments, is highly reactive, and contains bomb-^{14}C. DOC$_2$ dominates the bulk DOC pool throughout much of the sediment column, exhibits intermediate reactivity, and has a modern (but prebomb) Δ^{14}C signature. DOC$_3$, which steadily increases with depth, is virtually nonreactive and has a Δ^{14}C signature of ~−500‰. DOC$_1$ and DOC$_2$ appear to be of marine origin, while DOC$_3$ appears slightly depleted in ^{13}C. These model results are discussed below in more detail in Sections VI.A and VII.C.

TABLE 12.2   Model-derived Rate Constants (k_{DOC}) and Turnover Times ($\tau_{DOC} = 1/k_{DOC}$) for Sediment DOC Consumption and Estimated Isotopic Values for Different Pore Water DOC Pools

Site	Depth Interval of Fit	k_{DOC} (year^{-1})	τ_{DOC} (year)	$\delta^{13}C$ (‰) [a]	$\Delta^{14}C$ (‰)[a]
Chesapeake Bay (site M3)[b]	0-25 cm	6.4 ± 2.1	~0.1-0.3		
North Carolina continental slope[c]	25-225 cm	$1.7 \pm 0.7 \times 10^{-3}$	~300-800		
Southwest African Margin (Walvis Bay, site 1082)[d]	1.5-115 m 1.5-370 m	$4.6 \pm 1.2 \times 10^{-5}$ $9.3 \pm 2.9 \times 10^{-5}$	~16,000-30,000 ~7000-14,000		
Santa Monica Basin (SMB)[e]	0-45 cm				
DOC_1		33-80	~0.01-0.03	−17 to -20.6	−58 to +68
DOC_2		0.16-0.23	~4-6	−22 to −23	−45 to −66
DOC_3		~1×10^{-4}	~10^4	−26 to −27	−480 to −520

[a]These isotopic values are derived from fitting the depth profiles of pore water DOC and DIC concentration and isotopic signatures to the multi-G + DOC model discussed in Section V.B and illustrated in Figures 12.8 and 12.9 (see Komada et al., 2013, for details). The values listed here are the δ_i and Δ_i values shown in Figure 12.8. Similar isotopic values are not available for pore water DOC at the three other sites discussed in this table.

[b]Data from Burdige and Zheng (1998). Model results from Burdige (2002). See Figure 12.11 for best fit profiles.

[c]Data from Alperin et al. (1999), model results from Burdige (2002). See Figure 12.11 for best fit profiles.

[d]Data and model results from Burdige (2002). See Figure 12.11 for best fit profiles.

[e]Data and model results from Komada et al. (2013). The ranges reported here for k_{DOC}, $\delta^{13}C$ and $\Delta^{14}C$ are based on different assumptions made in the model fits to the data.

VI CONTROLS ON DOC CONCENTRATIONS IN SEDIMENTS

A Controls on DOC Concentrations in Surficial Sediments

DOC (and DON) accumulate with depth in sediment pore waters due to a slight imbalance between production and consumption (e.g., Alperin et al., 1999; Burdige and Gardner, 1998). Because this accumulation occurs predominantly in the form of recalcitrant DOM, understanding the controls on DOC concentration at depth (or the asymptotic concentration, in cases where the DOC concentration gradient diminishes to zero) may provide useful information about the origin and dynamics of recalcitrant DOC. We start this discussion by noting that in Figure 12.10 there is a positive relationship between maximum pore water DOC concentrations in anoxic surficial sediments ($[DOC]_\infty$) and depth-integrated rates of sediment carbon oxidation (R_{Cox}). Bioturbated/bioirrigated sediments do not appear to fall on the trend line for anoxic sediments, for reasons discussed below in Section VI.C.

Two possibilities may explain the observations in Figure 12.10 for anoxic sediments. The first is that a balance occurs at depth between DOC production and DOC consumption (Alperin et al., 1994; Burdige and Gardner, 1998). This explanation was implicitly incorporated into the DOC reactive-transport model presented in Burdige (2002) and illustrated in Figure 12.3, in which it was assumed that the rate of DOC production from POC (R_{DOC}) could be expressed by an equation of the form,

$$R_{DOC} = \left(R_o - R_\infty\right)e^{-\alpha z} + R_\infty, \qquad (12.2)$$

where R_i is the rate of DOC production at the sediment surface ($i=0$) or at depth ($i=\infty$) and α is the attenuation constant for this rate expression. It was also shown in Burdige (2002) that model equations containing Equation (12.2) are consistent with the observations in Figure 12.10. Interestingly, Alperin et al. (1999) used a very different modeling approach to examine sediment DOC cycling and obtained the following relationship,

$$[DOC]_\infty \propto R_{cox} z^*, \qquad (12.3)$$

where z^* is the e-folding depth for remineralization. Assuming that z^* is roughly constant in all of the anoxic sediments shown in Figure 12.10, this equation is similarly consistent with the observations in this figure.

The second explanation for asymptotic DOC concentrations with depth is that DOC production rates go to zero and biotic or abiotic changes in the composition of the pore water DOC pool continually decreases its bulk reactivity. This scenario would eventually lead to a situation in which pore water DOC found at depth is effectively nonreactive on early diagenetic time scales and is therefore selectively preserved (see Burdige, 2006b). In this case, one might think of this DOC at depth much like one thinks of "inert" inorganic remineralization end products such as phosphate, ammonium, or DIC, which also show similar exponential-like profiles in anoxic sediments (e.g., Berner, 1980). This analogy would then predict that greater amounts of DOC would accumulate with depth in sediment pore waters as rates of sediment carbon oxidation increase (e.g., Krom and Westrich, 1981), as is seen in Figure 12.10. However, several lines of evidence (see Section V.A and references cited therein) argue against some aspects of this second suggestion, at least over the time and depth scales of surficial sediments.

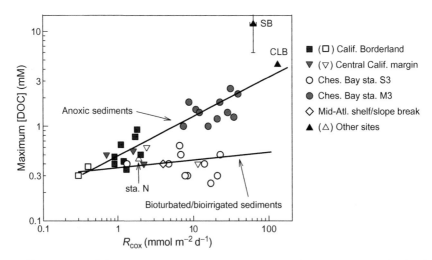

FIGURE 12.10   The maximum DOC concentration in the upper ~20-30 cm of sediment versus the depth-integrated sediment carbon oxidation rate. Open symbols represent bioturbated/bioirrigated sediments while closed symbols represent strictly anoxic sediments. The two lines highlight the general trends in the data sets but do not imply any functional relationships. Data sources: Chesapeake Bay sites M3 and S3—Burdige and Homstead (1994), Burdige and Zheng (1998), and Burdige (2001); California Borderlands and central California margin sites—Berelson et al. (1996) and Burdige et al. (1999, unpub. data); mid-Atlantic-shelf/slope break (site WC4)—Burdige et al. (2000, unpub. data); Cape Lookout Bight, NC (CLB)—Martens et al. (1992) and Alperin et al. (1994); Skan Bay (SB), Alaska—Alperin et al. (1992); station N (see Figure 12.2)—Bauer et al. (1995).

The multi-G+DOC model provides some insight into reconciling the two suggestions discussed above for explaining the accumulation of recalcitrant DOC with depth in anoxic sediments. Regarding the second suggestion, this model provides an explanation for how DOC reactivity decreases with depth in the absence of in situ processes that specifically produce recalcitrant material from more reactive constituents. Here, less reactive DOC is produced from less reactive POC, versus it being rendered less reactive as a result of in situ ageing or chemical transformation. This DOC also inherits it radiocarbon signature directly from its parent POC, again as opposed to acquiring it via in situ ageing.

Regarding the first suggestion that asymptotic DOC concentrations at depth represent a balance between production and consumption, we start by recognizing that in the multi-G+DOC model we express total DOC production as $\sum_i k_i G_i$ (Komada et al., 2013). We also assume here, for the sake of this argument, that the concentration of reactive POC in each fraction decreases exponentially with depth (e.g., Berner, 1980; Burdige, 2006b). In this case, the total rate of DOC production can be expressed as:

$$\sum k_i G_i \approx k_1 G^o, e^{-\alpha_1 z} + k_2 G_2^o e^{-\alpha_2 z} + k_3 G_3^o e^{-\alpha_3 z} + \dots \text{(12.4)}$$

where $\alpha_i = k_i/\omega$, ω is the sediment accumulation rate, and G_i^o is the initial concentration of organic matter in the ith fraction. However, in a given sedimentary setting (i.e., over specific time and depth scales), data analysis methods generally used to extract information about sediment organic matter reactivity can only determine, at most, two or three fractions of material, regardless of the number of fractions that may actually exist (e.g., Middelburg, 1989). Furthermore, since the various k_i values decrease for higher-order organic matter fractions, this implies that the e-folding depth for the remineralization of higher-order G_i fractions ($=1/\alpha_i = \omega/k_i$) will increase and eventually become large relative to the depth scale of the sediment system. Higher-order exponential terms in Equation (12.4) will

therefore be roughly constant with depth in this sediment (or roughly equal to zero as k_i values continually decrease) and as a result, this equation will simplify to either a single or double exponential function of depth plus a constant term, as in Equation. (12.2).

B Controls on DOC Concentrations in Deeply Buried Sediments

In contrast to the numerous DOC profiles that have been collected in surficial sediments (see Section II and earlier reviews in Burdige, 2002 and Krom and Westrich, 1981), far fewer studies have examined pore water DOC concentrations over larger depth and time scales (Burdige, 2002; Egeberg and Barth, 1998; Heuer et al., 2009; Smith et al., 2005). Additional DOC pore water data from deeply buried sediments are available in the scientific reports of selected ODP (Ocean Drilling Program) and IODP (International Ocean Drilling Program) cruises (see, most recently, Fisher et al., 2005; Tréhu et al., 2003; Wilson et al., 2003). Interestingly, when such DOC profiles are compared with results from surficial sediments (Figure 12.11) one often observes a general similarity in the exponential-like shape of the profiles, in spite of significant differences in both the depth and concentration scales for the profiles. Examining these observations in the context of Equation (12.3) leads to the conclusion that z^* (i.e., the increasing depth scale over which organic matter remineralization and its associated DOC production and consumption occurs) rather than R_{cox} plays a major role in explaining the shapes of these DOC profiles. Specifically, between site M3 in Chesapeake Bay and site C on the North Carolina (NC) continental slope, R_{cox} values actually decrease from ~20 to ~5 mmol m^{-2} d^{-1} while [DOC]$_\infty$ increases (Alperin et al., 1999; Burdige and Zheng, 1998). In addition, as one moves to deeply buried sediments (such as the Walvis Bay sediments), values of R_{cox} also presumably continue to decrease as [DOC]$_\infty$ increases.

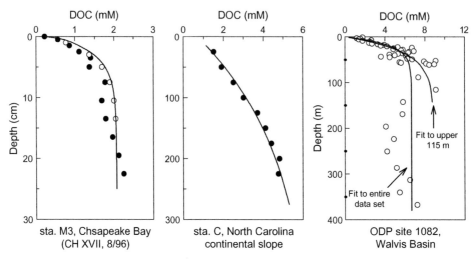

Pore water DOC concentration profiles from three contrasting anoxic marine sediments fit to the ANS model in Burdige (2002). The best fit rate constants for DOC consumption (k_{DOC}) are listed in Table 12.2, while the remaining fitting parameters are listed in the original reference. Note the factor of 4 difference in concentration scales and factor of >1000 difference in depth scales as one moves from the estuarine Chesapeake Bay sediments to the deep sediment (ODP) cores collected in Walvis Basin (also note that the depth scale for the Walvis Basin core is in units of meters while the depth scales for the other two sites are in units of centimeters). The different symbols for the Chesapeake Bay plot (left) represent replicate cores collected on this date at this site. At site C on the North Carolina continental slope (center) the data were fit starting at a sediment depth of 25cm based on the observation that the upper portion of these sediments are extensively bioturbated (Alperin et al., 1999). Finally, for the Walvis Basin sediments (right) the entire data set and the upper 115m of sediment were fit separately to the model. Although there are factor of ~2 differences in the resulting fitting parameters of each fit (see Table 12.2), both sets of results are consistent with the general trends discussed in the text regarding the comparison of the fitting parameters from all three sediments. Data sources: Chesapeake Bay—Burdige and Zheng (1998); North Carolina continental slope—Alperin et al. (1999); Walvis Bay—DOC data from Burdige et al. (unpubl. data), other model input parameters from Wefer et al. (1998).

The significance of z^* in describing the shape of DOC profiles in deeply buried sediments also has implications for POC remineralization over these longer depth and time scales as expressed, for example, in Equation (12.4). Here, we see that lower-order organic matter fractions are "rapidly" remineralized near the sediment surface (i.e., on early diagenetic time scales) while higher-order, less reactive fractions that appear recalcitrant (or even inert) on these short time scales become increasingly more important for fueling remineralization in much deeper sediments (see, e.g., recent discussions in Hoehler and Jørgensen, 2013 and Lomstein et al., 2012).

The results in Figure 12.11 along with those in Figure 12.9 from SMB sediments can be used to further examine the questions posed in the previous section regarding the controls on pore water DOC concentrations in sediments. We begin by comparing in Table 12.2 model-derived DOC remineralization rate constants (k_{DOC}) and the DOC turnover times ($\tau_{DOC} = 1/k_{DOC}$) for the four sites. For the three sites shown in Figure 12.11, these rate constants were determined using the steady-state reactive-transport model described in Burdige (2002), assuming that only one type of DOC is produced from POC according to Equation (12.2). In contrast, DOC consumption in SMB sediment pore waters was modeled using the multi-G+DOC model illustrated in Figure 12.10 with three reactive fractions of organic matter and therefore three DOC pools undergoing decomposition.

As a first observation, we note that even the largest k_{DOC} values listed here are, for the most part, much smaller (by up to two orders of magnitude) than analogous first-order rate constants for the decomposition of monomeric DOM such as acetate or amino acids (see summaries in Finke et al., 2007 and Henrichs, 1993). This observation is consistent with previous discussions that the bulk of the pore water DOC pool represents relatively recalcitrant material that turns over much more slowly than components of the mLMW-DOC pool. At the same time, rate constants for the degradation of synthetic melanoidins determined in studies with Alaska coastal sediments (Henrichs and Doyle, 1986) ranged from 0.7 to <0.09 year^{-1} (τ_{DOC} from ~1 to >10 year) and are more comparable to some of the k_{DOC} and τ_{DOC} values in Table 12.2. Since melanoidins have been proposed as models for recalcitrant marine humic materials (Hedges, 1988; Krom and Sholkovitz, 1977; Nissenbaum et al., 1972), this observation is perhaps not surprising.

A comparison of the SMB results with those from the other three sites indicates that the Chesapeake Bay rate constant is intermediate between the SMB k_{DOC1} and k_{DOC2} values while the NC continental slope value is intermediate between k_{DOC2} and k_{DOC3}. Given the broad types of POC deposited in these sediments, and their assumed reactivity, these observations suggest that rate constants derived from models assuming only one type of pore water DOC undergoing remineralization likely "average" over the rate constants one obtains by assuming the existence of multiple DOC pools with differing reactivities (see similar discussions in Arndt et al., 2013, regarding 1–G vs. multi-G POC-only degradation models).

Overall, these observations suggest that POC degradation through DOC intermediates can be minimally described as a two-step process as defined by the multi-G+DOC model in Figure 12.9. However, as noted above, this approach is clearly an oversimplification of the dynamics of DOC production and consumption illustrated in Figure 12.2, despite the success of the multi-G+DOC model in linking bulk POC reactivity to its isotopic signatures (δ^{13}C and Δ^{14}C) and those of its downstream DOC intermediates and remineralization end products (i.e., DIC). Future work should therefore be aimed at better incorporating aspects of the conceptual model in Figure 12.2 into kinetic models like the multi-G+DOC model, and eventually into DOC reactive-transport models.

Also of equal interest here is the almost seven orders-of-magnitude decrease in k_{DOC} values (or increase in τ_{DOC} values) as one moves from the most reactive DOC fraction in surficial SMB sediments to deep Walvis Basin sediments. When looked at in the context of the discussion above, these trends are most easily explained by the remineralization of increasingly recalcitrant sediment POC leading to the net production of increasingly recalcitrant pore water DOC.

Within SMB sediments (Komada et al., 2013), the depth distribution of Δ^{14}C$_{DOC}$ (Figure 12.9) places important constraints on possible sources of recalcitrant DOC in these sediments and suggests that it is highly unlikely that this material forms within the pore waters from labile moieties (see Section V.B for details). However, when this problem is examined over the much longer time and depth scales of deeply buried sediments, the possibility exists that some amount of the pore water DOC also undergoes internal transformations (by any number of possible mechanisms discussed in Section II) that lead to a decrease in its reactivity (also see discussions in Section V.A and in Burdige, 2002). With time, such processes could play a role in the accumulation of higher concentrations of more recalcitrant DOC and ultimately be involved in overall sediment carbon preservation (see Section VI.D for further discussions). More detailed compositional and/or structural studies of pore water DOC from a wide range of sedimentary settings may be able to provide additional insights into this problem.

C Possible Redox Controls on Pore Water DOC Concentrations

Results in Figure 12.10 illustrate that asymptotic DOC concentrations in mixed redox sediments are in general lower than those in strictly anoxic sediments. As noted in Section II, similar observations date back to studies of DOC cycling as early as the 1970s. Model calculations presented in Burdige (2002) illustrate that changes in sediment redox conditions alter carbon flow through DOM intermediates and lead to the buildup of recalcitrant DOM under anoxic conditions (also see discussions in Burdige, 2001). This accumulation appears to occur as a result of some combination, under mixed redox conditions, of either enhanced consumption of recalcitrant DOM (pathway 5 in Figure 12.2 or the lower pathway in Figure 12.3) or preferential carbon flow through mLMW-DOM intermediates (i.e., carbon flow via the vertical arrows on the left side of Figure 12.2 or the upper pathway in Figure 12.3 (a)).

Evidence in support of the former suggestion comes from observations that the ratio of humic-like fluorescence to total DOC concentration is higher in pore waters of strictly anoxic versus mixed redox sediments (Burdige et al., 2004; Komada et al., 2004; also see Section III.B). One possible explanation for this is that by analogy with discussions of the factors controlling the preservations of POM in sediments (see reviews in Blair and Aller, 2012; Burdige, 2007; Hedges and Keil, 1995; Zonneveld et al., 2010), the presence of O_2 and/or the occurrence of mixed redox conditions may similarly enhance the remineralization of certain types of recalcitrant DOM in sediment pore waters. Such "oxygen" effects may express themselves in a number of different direct and indirect ways that are described in greater detail in the reviews cited above.

The cycling of DOC in sediment pore waters may also be impacted by iron redox cycling, given the strong affinity of DOC compounds to adsorb to iron oxide surfaces (see discussions in Chin et al., 1998; Lalonde et al., 2012; Skoog and Arias-Esquivel, 2009). However, the broader-scale significance that this may have on redox controls of pore water DOC concentrations requires further study.

D Interactions Between DOM and Sediment Particles and the Possible Role of DOM in Sediment Carbon Preservation

In addition to biological processes that affect DOM, physical interactions with sediment particles, for example, adsorption, are also of some importance (Figure 12.2). If we assume adsorption is a reversible, equilibrium process, then a simple linear adsorption isotherm can be used to describe the process (Stumm and Morgan, 1996),

$$\overline{C} = K^{*}C, \qquad (12.5)$$

where $\overline{C}$ is the concentration of the adsorbed compound (in units of, e.g., $mmol\,kg^{-1}$, where kg is kilograms of dry sediment), C is the concentration of the compound in solution (in units of, e.g., mM), and K^{*} is the adsorption coefficient with units of $L\,kg^{-1}$.

Adsorption can decrease the availability of DOM to microbial degradation (e.g., Sugai and Henrichs, 1992) and may play a role in sediment carbon preservation (see the discussion at the end of this section). Ionic interactions or weaker Van der Waals interactions appear to be the predominant mechanisms by which reversible adsorption occurs (Arnarson and Keil, 2001; Henrichs and Sugai, 1993; Liu and Lee, 2007; Wang and Lee, 1993). For simple LMW-DOM such as glucose, amino acids, and short-chain organic acids, adsorption coefficients range from ~0.1 to ~300 $L\,kg^{-1}$ (e.g., Henrichs, 1992, 1995; Liu and Lee, 2007; Sansone et al., 1987), while larger molecules such as proteins or synthetic melanoidins have K^{*} values that range from 50 to 600 $L\,kg^{-1}$ (Ding and Henrichs, 2002; Henrichs, 1995; Henrichs and Doyle, 1986). Adsorption studies using natural assemblages of pore water DOM yield K^{*} values that range from 1 to 3200 $L\,kg^{-1}$ (Arnarson and Keil, 2001; Thimsen and Keil, 1998).

Positively charged organic molecules, for example, basic amino acids such as lysine, are more strongly adsorbed than neutral or negatively charged organic molecules; this has historically been attributed to the fact that most natural sediments have a negative surface charge (see Henrichs, 1992 for a summary). However, recent studies also suggest that in some marine sediments adsorption of LMW-DOM to sediment organic matter itself may be more important than adsorption to mineral surfaces (Liu and Lee, 2007). Furthermore, organic matter architecture as well as diagenetic history (also sometimes referred to as diagenetic maturity; e.g., Cowie and Hedges, 1994) appears to influence not only the overall extent of adsorption but also changes in the extent of adsorption as a function of DOM characteristics such as hydrophobicity (Liu and Lee, 2006, 2007; Wang and Lee, 1993). Increasing molecular weight and increasing hydrophobicity may enhance adsorption (Henrichs, 1992, 1995), although in the context of the recent studies discussed above, this generalization may require reexamination.

Several studies have also shown that much of this DOM adsorption is only partially reversible and in some cases is effectively irreversible (Ding and Henrichs, 2002; Henrichs and Sugai, 1993; Liu and Lee, 2007; Wang and Lee, 1993). The causes of this behavior may include: attachment of molecules to multiple binding sites that slow down desorption (Collins et al., 1995; Henrichs and Sugai, 1993), chemical reactions between adsorbed molecules and sedimentary organic matter (Henrichs and Sugai, 1993), and enzyme-type adsorption sites with high desorption activation energies (Liu and Lee, 2007).

Sorption of DOC to sediment particles can affect pore water DOC concentrations (Hedges and Keil, 1995; Henrichs, 1995; Papadimitriou et al., 2002), and it has been suggested that pore water DOC concentrations may be "buffered" by reversibly sorbed DOC in equilibrium with the pore waters (Thimsen and Keil, 1998). However, Alperin et al. (1999) concluded that buffering of pore water DOC concentrations by reversible

sorption is not an important controlling factor in explaining pore water DOC concentrations in NC continental slope sediments. Furthermore, in situations where rates of sediment accumulation and bioturbation are sufficiently low, reversible adsorption can be neglected in reactive-transport models (Berner, 1980; Komada et al., 2004), further minimizing the importance of reversible adsorption in impacting pore water DOC concentrations.

DOM-particle interactions such as those discussed above have also been examined in terms of broader, and more general, questions of how these interactions may impact sediment carbon preservation. Many aspects of this have been discussed previously (Burdige, 2007; Hedges and Keil, 1995; Henrichs, 1995) and for details the interested reader is urged to consult these publications. A more recent study examining the role of reactive iron in promoting organic carbon preservation in marine sediments also discusses the role pore water DOC may play in this type of sediment carbon preservation (Lalonde et al., 2012).

Similarly, given other recent studies examining the ways in which humic substances may form (Piccolo, 2001; Sutton and Sposito, 2005), many properties of pore water DOM (LMW, low degree of "bulk" reactivity) also argue for pore water DOM playing a potentially important role in the formation of presumably recalcitrant humic materials (see Section II for details). However, more work is needed to examine the role this may ultimately play in sediment carbon preservation.

VII THE ROLE OF BENTHIC DOM FLUXES IN THE OCEAN CARBON AND NITROGEN CYCLES

A Benthic DOC Fluxes

The fact that concentrations of both DOC and DON in sediment pore waters are often elevated over bottom water concentrations implies that sediments can be a potential source of DOM to overlying waters. The occurrence of these benthic fluxes was discussed in the literature beginning in the 1980s (e.g., Hedges, 1992; Heggie et al., 1987; Williams and Druffel, 1987) where it was suggested that sediments might represent a significant source of DOC to the deep ocean and that benthic DOC fluxes might provide an explanation for the apparent discrepancy between the "old" (~6000 ybp) ^{14}C age of deep water DOC and the average oceanic mixing time of ~1000 year (for details also see Burdige, 2002). The direct measurement of these fluxes was undertaken in studies beginning in the 1990s using either core incubation techniques or in situ benthic landers or chambers (see Burdige et al., 1999, for a summary). Benthic DOC fluxes determined in this manner in a number of estuarine, coastal, and continental margin sediments range from ~0.1 to 3 mmol m^{-2} d^{-1}.

In a compilation and synthesis of these results, Burdige et al. (1999) observed the following positive, but nonlinear, relationship between benthic DOC fluxes (BDF) and depth-integrated sediment carbon oxidation rates (R_{cox}):

$$BDF = 0.36 R_{cox}^{0.29} \qquad (12.6)$$

(BDF and R_{cox} both expressed here in units of mmol m^{-2} d^{-1}). The relationship implies that the ratio of BDF to R_{cox} increases with decreasing R_{cox}. Using Equation (12.6) Burdige et al. (1999) estimated that the integrated DOC flux from coastal and continental margin sediments (0-2000 m water depth) was ~180 Tg C year^{-1}. Three other estimates of this quantity for this sedimentary region, determined using different approaches, have also been presented; in one case, a near-identical value was obtained (Dunne et al., 2007), while in the other two cases, smaller values of 40 Tg C year^{-1} (Alperin et al., 1999) and 90 Tg C year^{-1} were obtained (Maher and Eyre, 2010). However, the Maher and Eyre (2010) value is based on a definition of the continental shelf that is significantly smaller in surface area (factor of ~10) than that used in the other studies.

In addition to coastal and continental margin sediments, Maher and Eyre (2010) estimated the integrated benthic DOC flux from intertidal and vegetated (seagrass, macroalgae, salt marsh, and mangrove) sediments not explicitly considered in the other estimates discussed above. For these sediments they obtained a value of ~170 Tg C year^{-1}, with vegetated sediments accounting for >90% of this flux. Their work (also see Maher and Eyre, 2011) further shows while these sites are net sources of DOC over diel cycles, this is often a balance between DOC uptake in the dark and release during daylight. These observations further highlight the role that higher plants may play in such environments in mediating DOC fluxes to the water column (also see discussions in Section VII.B).

Dunne et al. (2007) also estimated an integrated DOC flux of ~100 Tg C year^{-1} from sediments in water depths >2000 m. This estimate is extrapolated from calculated fluxes determined with DOC pore water profiles from cores collected at a single site at a water depth of 3500 m (Papadimitriou et al., 2002). Based on a comparison of these pore water profiles with those presented by Hall et al. (2007), and the discussion in Section II.A of possible artifacts associated with deep-sea pore water DOC determinations, it seems possible that this estimate of the integrated benthic DOC flux from deep-sea sediments may be an overestimate (also see discussions in Brunnegård et al., 2004). More work will be needed to evaluate deep-sea benthic DOC fluxes and noninvasive in situ techniques for measuring benthic fluxes may hold promise here (e.g., Swett, 2010).

Returning to the examination of benthic DOC fluxes from non-vegetated coastal and continental margin sediments, we note that these fluxes are generally less than ~10% of sediment carbon oxidation rates (for details see Burdige et al., 1999). Thus, these sediments are quite efficient in oxidizing DOM produced during remineralization processes, consistent with past discussions in this chapter. Similar trends also appear to be the case for sediment DON cycling (see next section) and both observations imply that net sediment DOM production is small in comparison to gross sediment DOM production. These trends are consistent with prior discussions regarding carbon and nitrogen flow through DOM intermediates during sediment POM remineralization and the role of DOM as an intermediate in the overall remineralization process.

A second implication of these results is that the integrated benthic DOC flux from coastal and continental margin sediments including vegetated sediments (~350 Tg C year^{-1}) is comparable to (or even larger than) estimates of the riverine DOC input of ~210 Tg C year^{-1} (Ludwig and Probst, 1996). Thus, marine sediments may be an important net source of DOC to the oceans (Burdige et al., 1999). However, the actual impact these fluxes have on the oceanic carbon cycle ultimately depends on the extent to which sediment-derived DOM is reactive in the water column. This point will be discussed in greater detail in Section VII.C (also see discussions in Alperin et al., 1999 and Maher and Eyre, 2010).

B Benthic DON Fluxes

Interest in benthic DON fluxes and their role in the marine nitrogen cycle is similar to that discussed above for benthic DOC fluxes and the marine carbon cycle. However, because nitrogen is a limiting nutrient in marine ecosystems (Gruber, 2008) and because marine phytoplankton can use DON as their nitrogen source (Mulholland and Lomas, 2008), there is additional interest in understanding the role of sediments as a source of DON to the water column.

A number of studies have examined benthic DON fluxes from coastal, estuarine, and continental margin sediments (see reviews in Bronk and Steinberg, 2008 and Burdige, 2002). Benthic DON fluxes show a tremendous range in both absolute magnitude and direction (into and out of the sediments). However, at estuarine or

coastal sites where repeated (or seasonal) studies have been carried out, mean or annual averages generally suggest that benthic DON fluxes are small, usually out of the sediments, and are only a small percentage of the benthic dissolved inorganic nitrogen (DIN) flux (e.g., see Burdige and Zheng, 1998 for more details; DIN = the sum of ammonium, nitrate, and nitrite). For example, in Chesapeake Bay sediments, benthic DON fluxes range from ~0.08 to 0.2 mmol m^{-2} d^{-1} in the anoxic sediments at site M3 and 0-0.4 mmol m^{-2} d^{-1} in the bioturbated and bioirrigated sediments at site S3 (Burdige, 2001; Burdige and Zheng, 1998). At both sites, benthic DON fluxes were only ~3-4% of benthic DIN fluxes.

In contrast, at some sites, benthic DON fluxes are comparable to, or even exceed, benthic DIN fluxes or integrated rates of sediment denitrification. These include high latitude (Arctic) sediments (Blackburn et al., 1996), North Sea continental margin sediments (Landén-Hillmeyr, 1998), estuarine mudflat sediments (Cook et al., 2004), microtidal fjord sediments (Sundbäck et al., 2004), and subtropical estuarine euphotic sediments (Ferguson et al., 2004).

The reasons for widely varying differences in the relative magnitude as well as the direction of benthic DON fluxes are not well understood. In some cases, they may be the result of transient (non-steady-state) events; for example, Blackburn et al. (1996) suggested that the large DON fluxes observed may have been a temporary phenomena associated with the recent sedimentation of fresh detrital material. Similarly, as is the case for benthic DOC fluxes (see the previous section), the presence of macroalgae, microphytobenthos, and benthic macrofauna can impact the magnitude and direction of benthic DON fluxes over a range of time scales (see discussions in Bronk and Steinberg, 2008; Ferguson et al., 2004; Maher and Eyre, 2011; Sundbäck et al., 2004; Tyler et al., 2003). Additional discussions on the controls of benthic DON fluxes can be found in Burdige and Zheng (1998) and Alkhatib et al. (2013).

C The Impact of Benthic DOM Fluxes on the Composition and Reactivity of Oceanic DOM

Interest in DOM fluxes from marine sediments stems in part from a recognition of the need to better understand the sources and sinks of DOM in the oceans (Hansell et al., 2009, 2012). Although benthic fluxes of DOC (Burdige et al., 1999; Dunne et al., 2007; Maher and Eyre, 2010) and DON (Alkhatib et al., 2013; Brunnegård et al., 2004; Burdige and Zheng, 1998) appear to be of similar magnitude to their riverine inputs, the impact of these fluxes on water column DOM concentrations and properties depends on the reactivity of sediment-derived DOM in the water column. If sediment-derived DOC (or DON) is reactive in the water column and undergoes remineralization on time scales shorter than that required for transport out of the benthic boundary region, or deep water residence times, then these fluxes will have a minimal impact on deep water DOM properties. Conversely, if this material is sufficiently recalcitrant, then these fluxes could represent an important source of DOM to the deep ocean and might also help explain, for example, the ^{14}C content of deep water DOC.

Several lines of evidence from contrasting marine sediments suggest that not all of the DOM escaping from sediments as a benthic flux is recalcitrant, and that it has the potential to be reactive in the water column. In estuarine Chesapeake Bay sediments, a comparison of measured benthic DOM fluxes versus calculated, diffusive DOM fluxes suggests that there is enhanced production of N-rich DOM at or near the sediment-water interface relative to the DOM accumulating in these sediment pore waters (Figure 12.6). Similar trends in DOM elemental ratios have been observed in other sediments (Blackburn et al., 1996; Landén-Hillmeyr, 1998) and were explained as being due to the diffusional loss of low C/N ratio DOM produced during the initial hydrolysis of fresh detrital organic matter near the sediment surface. This explanation is consistent with discussions in Burdige and Gardner (1998) regarding the spatial separation between sediment processes that produce the initial HMW intermediates of sediment POM remineralization and those responsible for the production of bulk recalcitrant pore water DOM. Such spatial separation of these processes can also be inferred from model results illustrated here in Figure 12.3 (Burdige, 2002).

Isotope studies of pore water DOC provide another approach to examining this problem. At two sites in the eastern North Pacific examined by Bauer et al. (1995), pore water DOC near the sediment surface was greatly enriched in ^{14}C as compared to bottom water DOC (also see Figure 12.9 and Section III.C), and based on these observations these authors concluded that sediments are not a major source of pre-aged (^{14}C-depleted) DOC to the oceans. However, it is important to remember that the ^{14}C signature of bulk DOC is actually a weighted average value based on the distribution of ^{14}C signatures of all of the molecules present in the sample (e.g., Loh et al., 2004). Recent work by Komada et al. (2013) supports this observation for sediment pore waters (Table 12.2 and Figure 12.9), suggesting that pore water DOC at any given depth is a mixture of components of varying reactivities and isotopic signatures. Pore water DOC near the surface of SMB sediments appears ^{14}C-young and labile (as was also noted by Bauer et al., 1995), and upon escaping the sediments as a benthic flux, the majority of this material is indeed likely oxidized in the bottom waters. However, modeling these pore water data also indicates the presence of an aged, recalcitrant DOC component. This recalcitrant (i.e., DOC_3) material represents ~3-8% of the total benthic DOC flux from these sediments (Komada et al., 2013), and as is shown in Table 12.2, is also depleted in ^{14}C. The significance of this is that the input of ^{14}C-depleted DOC to the ocean complicates the linkage between the apparent radiocarbon "age" of DOC in the deep ocean and its true deep ocean turnover time (Bauer and Druffel, 1998; Guo and

Santschi, 2000; McCarthy et al., 2011; Raymond and Bauer, 2001a)

On a global scale, the benthic DOC flux from non-vegetated coastal and continental margin sediments has been estimated at ~180 Tg C year^{-1} (see Burdige et al., 1999, and Section VII.A). If, based on the discussion above, we now assume that 3-8% of this material survives remineralization in the near-bottom waters, then 5-14 Tg C year^{-1} of recalcitrant DOC from coastal and continental margin sediments is added to the oceans (Komada et al., 2013). This flux represents 12-33% of the turnover rate of recalcitrant DOC in the deep ocean (=43 Tg C year^{-1}), the latter determined by modeling DOC distributions in the global oceans (Hansell et al., 2012). These results argue strongly for the important role of sediment processes and benthic DOC fluxes in adding recalcitrant and ^{14}C-depleted DOC to the deep oceans, although further work is needed to better constrain the production rate of recalcitrant pore water DOC in sediments and the role it plays in determining the Δ^{14}C value (and turnover time) of deep ocean DOC.

Finally, sediment pore water DOC interactions (Section VI.D) could play a role in a slightly different fashion in adding ^{14}C-depleted DOC to the deep ocean. This possibility is based on results from Guo and Santschi (2000), who observed that simple desorption of colloidal (>1 kDa) organic matter from continental margin sediments yields DOC that has a substantially greater ^{14}C-age than the bulk sediment organic matter (~3000 vs. 700 year, respectively). While the details of how this material ages in the sediments is somewhat uncertain, desorption of this material from sediments in the benthic nepheloid layer, coupled with its off-shore transport (e.g., Bauer and Druffel, 1998), could add pre-aged DOC to the deep ocean DOC.

VIII CONCLUDING THOUGHTS

Throughout this chapter, we have attempted to summarize and synthesize the existing data on DOM in marine sediment pore waters and present explanations that (at least to us) appear to best explain the observations. Addressing many of the important problems and questions that still remain will involve studies linking DOM composition, structure, and reactivity with a better understanding of the pathways and processes that are illustrated in Figure 12.2. Studies of the chemical composition and structure of the parent POM should also be useful here in understanding DOM cycling. As was noted in the first edition of this chapter (Burdige, 2002), many of these questions can be (and will continue to be) addressed in carefully conducted sediment incubation experiments, such as those discussed in Section V.A.

The general role of sediment redox conditions in DOM cycling, and the ways it may link DOM cycling and overall sediment carbon preservation, remain a continuing area of interest and inquiry. Finally, in terms of DOM cycling in specific sedimentary environments, a number of important and interesting questions remain about these processes in the very different realms of shallow, permeable sediments and deeply buried sediments (deep marine biosphere).

Acknowledgments

Over the years, our work on pore water DOM has been funded by the US National Science Foundation, the Office of Naval Research, and the Petroleum Research Fund of the American Chemical Society. We thank them for their financial support. We thank Dennis Hansell and Craig Carlson for taking the initiative to organize and put together this second edition of the "DOM" book. We also thank Bente Lomstein for her review of an earlier version of this chapter. Finally, we thank all of our current and former students, technicians, postdocs, and scientific colleagues for all they contributed over the years to our work on sediment DOM cycling.

References

Albert, D.B., Martens, C.S., 1997. Determination of low-molecular weight organic acid concentrations in seawater and pore-water samples via HPLC. Mar. Chem. 56, 27–37.

Alkhatib, M., Schubert, C.J., del Giorgio, P.A., Gelinas, Y., Lehmann, M.F., 2012. Organic matter reactivity indicators in sediments of the St. Lawrence Estuary. Estuar. Coast. Shelf Sci. 102–103, 36–47.

Alkhatib, M., del Giorgio, P.A., Gelinas, Y., Lehmann, M.F., 2013. Benthic fluxes of dissolved organic nitrogen in the lower St. Lawrence estuary and implications for selective organic matter degradation. Biogeosciences 10, 7609–7622.

Aller, R.C., 1994. Bioturbation and remineralization of sedimentary organic matter: effects of redox oscillation. Chem. Geol. 114, 331–345.

Alperin, M.J., Reeburgh, W.S., Devol, A.H., 1992. Organic carbon remineralization and preservation in sediments of Skan Bay, Alaska. In: Whelan, J.K., Farrington, J.W. (Eds.), Productivity, Accumulation, and Preservation of Organic Matter in Recent and Ancient Sediments. Columbia Univ. Press, pp. 99–122.

Alperin, M.J., Albert, D.B., Martens, C.S., 1994. Seasonal variations in production and consumption rates of dissolved organic carbon in an organic-rich coastal sediment. Geochim. Cosmochim. Acta 58, 4909–4929.

Alperin, M.J., Martens, C.S., Albert, D.B., Suayah, I.B., Benninger, L.K., Blair, N.E., et al., 1999. Benthic fluxes and porewater concentration profiles of dissolved organic carbon in sediments from the North Carolina continental slope. Geochim. Cosmochim. Acta 63, 427–448.

Aluwihare, L.I., Meador, T., 2008. Chemical composition of marine dissolved organic nitrogen. In: Capone, D.G., Bronk, D.A., Mulholland, M.R., Carpenter, E.J. (Eds.), Nitrogen in the Marine Environment, second ed. Elsevier, Amsterdam.

Amon, R.M.W., Benner, R., 1994. Rapid cycling of high-molecular-weight dissolved organic matter in the ocean. Nature 369, 549–552.

Arnarson, T.S., Keil, R.G., 2001. Organic-mineral interactions in marine sediments studied using density fractionation and X-ray photoelectron spectroscopy. Org. Geochem. 32, 1401–1415.

Arndt, S., Jørgensen, B.B., LaRowe, D.E., Middelburg, J.J., Pancost, R., Regnier, P., 2013. Quantifying the degradation of organic matter in marine sediments: a review and synthesis. Earth-Sci. Rev. 123, 53–86.

Arnosti, C., 2000. Substrate specificity in polysaccharide hydrolysis: contrasts between bottom water and sediments. Limnol. Oceanogr. 45, 1112–1119.

Arnosti, C., 2004. Speed bumps in the carbon cycle: substrate structural effects on carbon cycling. Mar. Chem. 92, 263–273.

Arnosti, C., Holmer, M., 1999. Carbohydrate dynamics and contributions to the carbon budget of an organic-rich coastal sediments. Geochim. Cosmochim. Acta 63, 353–403.

Avery Jr., G.B., Kieber, R.J., Taylor, K.J., Dixon, J.L., 2012. Dissolved organic carbon release from surface sand of a high energy beach along the Southeastern Coast of North Carolina, USA. Mar. Chem. 132–133, 23–27.

Barcelona, M.J., 1980. Dissolved organic carbon and volatile fatty acids in marine sediment pore waters. Geochim. Cosmochim. Acta 44, 1977–1984.

Barker, H.A., 1981. Amino acid degradation by anaerobic bacteria. Annu. Rev. Biochem. 50, 23–40.

Bauer, J.E., 2002. Carbon isotopic composition of DOM. In: Hansell, D.A., Carlson, C. (Eds.), Biogeochemistry of Marine Dissolved Organic Matter. Academic Press, San Diego.

Bauer, J.E., Druffel, E.R.M., 1998. Ocean margins as a significant source of organic matter to the deep open ocean. Nature 392, 482–485.

Bauer, J.E., Reimers, C.E., Druffel, E.R.M., Williams, P.M., 1995. Isotopic constraints on carbon exchange between deep ocean sediments and sea water. Nature 373, 686–689.

Benamou, C., Richou, M., Benaïm, J.Y., Loussert, A., Bartholin, F., Richou, J., 1994. Laser-induced fluorescence of marine sedimentary interstitial dissolved organic matter. Mar. Chem. 46, 7–23.

Benner, R., 2002. Chemical composition and reactivity. In: Hansell, D.A., Carlson, C.D. (Eds.), Biogeochemistry of Marine Dissolved Organic Matter. Academic Press, San Diego.

Berelson, W.M., McManus, J., Kilgore, T., Coale, K., Johnson, K.S., Burdige, D., et al., 1996. Biogenic matter diagenesis on the sea floor: a comparison between two continental margin transects. J. Mar. Res. 54, 731–762.

Berelson, W., McManus, J., Coale, K., Johnson, K., Burdige, D., Kilgore, T., Colodner, D., Chavez, F., Kudela, R., Boucher, J., 2003. A time series of benthic flux measurements from Monterey Bay, CA. Cont. Shelf Res. 23, 457–481.

Berner, R.A., 1980. Early Diagenesis: A Theoretical Approach. Princeton University Press, Princeton, N.J.

Bianchi, T.S., Canuel, E.A., 2011. Chemical Biomarkers in Aquatic Ecosystems. Princeton University Press, Princeton N.J.

Blackburn, T.H., Hall, P.O.J., Hulth, S., Landén, A., 1996. Organic-N loss by efflux and burial associated with a low efflux of inorganic N and with nitrate assimilation in Arctic sediments (Svalbard, Norway). Mar. Ecol. Prog. Ser. 141, 283–293.

Blair, N.E., Aller, R.C., 2012. The fate of terrestrial organic carbon in the marine environment. Ann. Rev. Mar. Sci. 4, 401–423.

Boehme, J.R., Coble, P.G., 2000. Characterization of colored dissolved organic matter using high-energy laser fragmentation. Environ. Sci. Technol. 34, 3283–3290.

Boehme, S.E., Blair, N.E., Chanton, J.P., Martens, C.S., 1996. A mass balance of ^{13}C and ^{12}C in an organic-rich methane-producing marine sediment. Geochim. Cosmochim. Acta 60, 3835–3848.

Boschker, H.T.S., Bertilsson, S.A., Dekkers, E.M.J., Cappenberg, T.E., 1995. An inhibitor-based method to measure initial decomposition of naturally occurring polysaccharides in sediments. Appl. Environ. Microbiol. 61, 2186–2192.

Boss, E., Pegau, W.S., Zaneveld, J.R.V., Barnard, A.H., 2001. Spatial and temporal variability of absorption by dissolved material at a continental shelf. J. Geophys. Res. 106, 9499–9507.

Boudreau, B.P., Ruddick, B.R., 1991. On a reactive continuum representation of organic matter diagenesis. Am. J. Sci. 291, 507–538.

Boudreau, B.P., Huettel, M., Forster, S., Jahnke, R.A., McLachlan, A., Middelburg, J.J., et al., 2001. Permeable marine sediments, overturning an old paradigm. Eos Trans. Am. Geophys. Union 82, 133.

Bronk, D.A., Steinberg, D.K., 2008. Nitrogen regeneration. In: Capone, D.G., Bronk, D.A., Mulholland, M.R., Carpenter, E.J. (Eds.), Nitrogen in the Marine Environment, second ed. Academic Press, San Diego.

Brüchert, V., Arnosti, C., 2003. Anaerobic carbon transformation: experimental studies with flow-through cells. Mar. Chem. 80, 171–183.

Brunnegård, J., Grandel, S., Ståhl, H., Tengberg, A., Hall, P.O.J., 2004. Nitrogen cycling in deep-sea sediments of the Porcupine Abyssal Plain, NE Atlantic. Prog. Oceanogr. 63, 159–181.

Burdige, D.J., 1989. The effects of sediment slurrying on microbial processes, and the role of amino acids as substrates for sulfate reduction in anoxic marine sediments. Biogeochemistry 8, 1–23.

Burdige, D.J., 1991a. The kinetics of organic matter mineralization in anoxic marine sediments. J. Mar. Res. 49, 727–761.

Burdige, D.J., 1991b. Microbial processes affecting alanine and glutamic acid in anoxic marine sediments. FEMS Microbiol. Ecol. 85, 211–231.

Burdige, D.J., Homstead, J., 1994. Fluxes of dissolved organic carbon from Chesapeake Bay sediments. Geochim. Cosmochim. Acta 58, 3407–3424.

Burdige, D.J., 2001. Dissolved organic matter in Chesapeake Bay sediment pore waters. Org. Geochem. 32, 487–505.

Burdige, D.J., 2002. Sediment pore waters. In: Hansell, D.A., Carlson, C.D. (Eds.), Biogeochemistry of Marine Dissolved Organic Matter. Academic Press, San Diego.

Burdige, D.J., 2006a. Data report: dissolved carbohydrates in interstitial waters from the Equatorial Pacific and Peru Margin, ODP Leg 201. In: Jørgensen, B.B., D'Hondt, S.L., Miller, D.J. (Eds.), Proceedings of the ODP, Scientific Results 201, 1–10.

Burdige, D.J., 2006b. Geochemistry of Marine Sediments. Princeton Univ. Press, Princeton N.J.

Burdige, D.J., 2007. The preservation of organic matter in marine sediments: controls, mechanisms and an imbalance in sediment organic carbon budgets? Chem. Rev. 107, 467–485.

Burdige, D.J., Gardner, K.G., 1998. Molecular weight distribution of dissolved organic carbon in marine sediment pore waters. Mar. Chem. 62, 45–64.

Burdige, D.J., Martens, C.S., 1988. Biogeochemical cycling in an organic-rich marine basin -10. The role of amino acids in sedimentary carbon and nitrogen cycling. Geochim. Cosmochim. Acta 52, 1571–1584.

Burdige, D.J., Martens, C.S., 1990. Biogeochemical cycling in an organic-rich marine basin -11. The sedimentary cycling of dissolved free amino acids. Geochim. Cosmochim. Acta 54, 3033–3052.

Burdige, D.J., Zheng, S., 1998. The biogeochemical cycling of dissolved organic nitrogen in estuarine sediments. Limnol. Oceanogr. 43, 1796–1813.

Burdige, D.J., Berelson, W.M., Coale, K.H., McManus, J., Johnson, K.S., 1999. Fluxes of dissolved organic carbon from California continental margin sediments. Geochim. Cosmochim. Acta 63, 1507–1515.

Burdige, D.J., Gardner, K.G., Skoog, A., 2000. Dissolved and particulate carbohydrates in contrasting marine sediments. Geochim. Cosmochim. Acta 64, 1029–1041.

Burdige, D.J., Kline, S.W., Chen, W., 2004. Fluorescent dissolved organic matter in marine sediment pore waters. Mar. Chem. 89, 289–311.

Cai, W.-J., Reimers, C.E., Shaw, T.J., 1995. Microelectrode studies of organic carbon degradation and calcite dissolution at a California continental rise site. Geochim. Cosmochim. Acta 59, 497–511.

Canfield, D., Thamdrup, B., Kristensen, E., 2005. Aquatic Geomicrobiology. Elsevier, San Diego.

Caughey, M. E., 1982. A Study of the Dissolved Organic Matter in the Pore Waters of Carbonate-Rich Sediment Cores from Florida Bay (M.S. thesis). Univ. of Texas at Dallas.

Chen, R.F., Bada, J.L., 1989. Seawater and porewater fluorescence in the Santa Barbara Basin. Geophys. Res. Lett. 16, 687–690.

Chen, R.F., Bada, J.L., 1994. The fluorescence of dissolved organic matter in porewaters of marine sediments. Mar. Chem. 45, 31–42.

Chen, R.F., Bada, J.L., Suzuki, Y., 1993. The relationship between dissolved organic carbon (DOC) and fluorescence in anoxic marine porewaters: implications for estimating benthic DOC fluxes. Geochim. Cosmochim. Acta 57, 2149–2153.

Chin, Y.-P., Gschwend, P.M., 1991. The abundance, distribution, and configuration of porewater organic colloids in recent sediments. Geochim. Cosmochim. Acta 55, 1309–1317.

Chin, Y.-P., Aiken, G., O'Loughlin, E., 1994. Molecular weight, polydispersity, and spectroscopic properties of aquatic humic substances. Environ. Sci. Technol. 28, 1853–1858.

Chin, Y.P., Traina, S.J., Swank, C.R., Backhus, D., 1998. Abundance and properties of dissolved organic matter in pore waters of a freshwater wetland. Limnol. Oceanogr. 43, 1287–1296.

Chipman, L., Podgorski, D., Green, S., Kostka, J., Cooper, W., Huettel, M., 2010. Decomposition of plankton-derived dissolved organic matter in permeable coastal sediments. Limnol. Oceanogr. 55, 857.

Chipman, L., Huettel, M., Laschet, M., 2012. Effect of benthic-pelagic coupling on dissolved organic carbon concentrations in permeable sediments and water column in the northeastern Gulf of Mexico. Cont. Shelf Res. 45, 116–125.

Christensen, D., Blackburn, T.H., 1980. Turnover of tracer (^{14}C, ^{3}H labelled) alanine in inshore marine sediments. Mar. Biol. 58, 97–103.

Christensen, D., Blackburn, T.H., 1982. Turnover of ^{14}C-labelled acetate in marine sediments. Mar. Biol. 71, 113–119.

Coble, P.G., 1996. Characterization of marine and terrestrial DOM in seawater using excitation-emission matrix spectroscopy. Mar. Chem. 51, 325–346.

Coble, P.G., 2007. Marine optical biogeochemistry: the chemistry of ocean color. Chem. Rev. 107, 402–418.

Collins, M.J., Bishop, A.N., Farrimond, P., 1995. Sorption by mineral surfaces: rebirth of the classical condensation pathway for kerogen formation? Geochim. Cosmochim. Acta 59, 2387–2391.

Cook, P.L.M., Revill, A.T., Butler, E.C.V., Eyre, B.D., 2004. Carbon and nitrogen cycling on intertidal mudflats of a temperate Australian estuary. II. Nitrogen cycling. Mar. Ecol. Prog. Ser. 280, 39–54.

Cowie, G.L., Hedges, J.I., 1992. Sources and reactivities of amino acids in a coastal marine environment. Limnol. Oceanogr. 37, 703–724.

Cowie, G.L., Hedges, J.I., 1994. Biochemical indicators of diagenetic alteration in natural organic matter mixtures. Nature 369, 304–307.

Cowie, G.L., Hedges, J.I., Calvert, S.E., 1992. Sources and reactivity of amino acids, neutral sugars, and lignin in an intermittently anoxic marine environment. Geochim. Cosmochim. Acta 56, 1963–1978.

Dale, A., Regnier, P., Van Cappellen, P., 2006. Bioenergetic controls on anaerobic oxidation of methane (AOM) in coastal marine sediments: a theoretical analysis. Am. J. Sci. 306, 246–294.

Dale, A., Regnier, P., Knab, N., Jorgensen, B., Van Cappellen, P., 2008. Anaerobic oxidation of methane (AOM) in marine sediments from the Skagerrak (Denmark): II. Reaction-transport modeling. Geochim. Cosmochim. Acta 72, 2880–2894.

Del Vecchio, R., Blough, N.V., 2004. On the origin of the optical properties of humic substances. Environ. Sci. Technol. 38, 3885–3891.

Derenne, S., Knicker, H., Largeau, C., Hatcher, P., 1998. Timing and mechanisms of changes in nitrogen functionality during biomass fossilization. In: Stankiewicz, B.A., van Bergen, P.F. (Eds.), Nitrogen-Containing Macromolecules in the Bio- and Geosphere. American Chemical Society, pp.243–253, Washington.

D'Hondt, S.L., Jørgensen, B.B., Miller, D.J., 2003. Proceedings of the Ocean Drilling Program, Initial Reports, Leg 201. Ocean Drilling Program, TX.

Ding, X., Henrichs, S.M., 2002. Adsorption and desorption of proteins and polyamino acids by clay minerals and marine sediments. Mar. Chem. 77, 225–237.

Dunne, J.P., Sarmiento, J.L., Gnanadesikan, A., 2007. A synthesis of global particle export from the surface ocean and cycling through the ocean interior and on the seafloor. Global Biogeochem. Cycles 21, GB4006.

Egeberg, P.K., Barth, T., 1998. Contribution of dissolved organic species to the carbon and energy budgets of hydrate bearing deep sea sediments (Ocean Drilling Program Site 997 Blake Ridge). Chem. Geol. 149, 25–35.

Egeberg, P.K., Dickens, G.R., 1999. Thermodynamic and pore water halogen constraints on gas hydrate accumulation at ODP site 997 (Blake Ridge). Chem. Geol. 153, 53–79.

Emery, K.O., 1968. Relict sediments on continental shelves of the world. AAPG Bull. 52, 445–464.

Fenchel, T., King, G.M., Blackburn, T.H., 1998. Bacterial Biogeochemistry: The Ecophysiology of Mineral Cycling. Academic Press, San Diego.

Ferguson, A.J., Eyre, B.D., Gay, J.M., 2004. Benthic nutrient fluxes in euphotic sediments along shallow sub-tropical estuaries, northern New South Wales, Australia. Aquat. Microb. Ecol. 37, 219–235.

Finke, N., Vandieken, V., Jørgensen, B.B., 2007. Acetate, lactate, propionate, and isobutyrate as electron donors for iron and sulfate reduction in Arctic marine sediments, Svalbard. FEMS Microbiol. Ecol. 59, 10–22.

Fisher, A.T., Urabe, T., Klaus, A., 2005. Proceedings of the IODP, 301. Integrated Ocean Drilling Program Management International, Inc., College Station, TX.

Gruber, N., 2008. The marine nitrogen cycle: overview and challenges. In: Capone, D.G., Bronk, D.A., Mulholland, M.R., Carpenter, E.J. (Eds.), Nitrogen in the Marine Environment, second ed. Academic Press, San Diego.

Guo, L., Santschi, P.H., 2000. Sedimentary sources of old high molecular weight dissolved organic carbon from the ocean margin nepheloid layer. Geochim. Cosmochim. Acta 64, 600–650.

Haberstroh, P.R., Karl, D.M., 1989. Dissolved free amino acids in hydrothermal vent habitats of the Guaymas Basin. Geochim. Cosmochim. Acta 53, 2937–2945.

Hall, P.O.J., Brunnegård, J., Hulthe, G., Martin, W.R., Stahl, H., Tengberg, A., 2007. Dissolved organic matter in abyssal sediments: core recovery artifacts. Limnol. Oceanogr. 52, 19–31.

Hansell, D.A., Carlson, C.A., Repeta, D.J., Schlitzer, R., 2009. Dissolved organic matter in the ocean: a controversy stimulates new insights. Oceanography 22, 202–211.

Hansell, D.A., Carlson, C.A., Schlitzer, R., 2012. Net removal of major marine dissolved organic carbon fractions in the subsurface ocean. Global Biogeochem. Cycles 26, GB1016. http://dx.doi.org/10.1029/2011GB004069.

Hansen, L.S., Blackburn, T.H., 1995. Amino acid degradation by sulfate-reducing bacteria: evaluation of four methods. Limnol. Oceanogr. 40, 502–510.

Hansen, L., Holmer, M., Blackburn, T., 1993. Mineralization of organic nitrogen and carbon (fish food) added to anoxic sediment microcosms: role of sulphate reduction. Mar. Ecol. Prog. Ser. 102, 199–204.

Hedges, J.I., 1988. Polymerization of humic substances in natural environments. In: Frimmel, F.C., Christman, R.C. (Eds.), Humic Substances and Their Role in the Environment. J. Wiley and Sons, Chichester.

Hedges, J.I., 1992. Global biogeochemical cycles: progress and problems. Mar. Chem. 39, 67–93.

Hedges, J.I., Keil, R.G., 1995. Sedimentary organic matter preservation: an assessment and speculative synthesis. Mar. Chem. 49, 81–115.

Hedges, J.I., Oades, J.M., 1997. Comparative organic geochemistries of soils and marine sediments. Org. Geochem. 27, 319–361.

Hee, C.A., Pease, T.K., Alperin, M.J., Martens, C.S., 2001. Dissolved organic carbon production and consumption in anoxic marine sediments: a pulsed-tracer experiment. Limnol. Oceanogr. 46, 1908–1920.

Heggie, D., Maris, C., Hudson, A., Dymond, J., Beach, R., Cullen, J., 1987. Organic carbon oxidation and preservation in NW Atlantic continental margin sediments. In: Weaver, P.P.E., Thomson, J. (Eds.), Geology and Geochemistry of Abyssal Plains. Blackwell Sci. Publ., London.

Henrichs, S.M., 1992. Early diagenesis of organic matter in marine sediments: progress and perplexity. Mar. Chem. 39, 119–149.

Henrichs, S.M., 1993. Early diagenesis of organic matter: the dynamics (rates) of cycling of organic compounds. In: Engel, M., Macko, S. (Eds.), Organic Geochemistry. Plenum Press, New York.

Henrichs, S.M., 1995. Sedimentary organic matter preservation: an assessment and speculative synthesis—a comment. Mar. Chem. 49, 127–136.

Henrichs, S.M., Cuhel, R., 1985. Occurrence of ß-aminoglutaric acid in marine bacteria. Appl. Environ. Microbiol. 50, 543–545.

Henrichs, S.M., Doyle, A.P., 1986. Decomposition of 14C-labelled organic substrates in marine sediments. Limnol. Oceanogr. 31, 765–778.

Henrichs, S.M., Farrington, J.W., 1979. Amino acids in interstitial waters of marine sediments. Nature 279, 319–322.

Henrichs, S.M., Farrington, J.W., 1980. Amino acids in interstitial waters of marine sediments: a comparison of results from varied sedimentary environments. In: Douglass, A.G., Maxwell, J.R. (Eds.), Advances in Organic Geochemistry. Pergamon Press, Oxford.

Henrichs, S.M., Farrington, J.W., 1987. Early diagenesis of amino acids and organic matter in two coastal marine sediments. Geochim. Cosmochim. Acta 51, 1–15.

Henrichs, S.M., Sugai, S.F., 1993. Adsorption of amino acids and glucose by sediments of Resurrection Bay, Alaska, USA: functional group effects. Geochim. Cosmochim. Acta 57, 823–835.

Henrichs, S.M., Farrington, J.W., Lee, C., 1984. Peru upwelling region sediments near 15°S. 2. Dissolved free and total hydrolyzable amino acids. Limnol. Oceanogr. 29, 20–34.

Hertkorn, N., Benner, R., Frommberger, M., Schmitt-Kopplin, P., Witt, M., Kaiser, K., et al., 2006. Characterization of a major refractory component of marine dissolved organic matter. Geochim. Cosmochim. Acta 70, 2990–3010.

Heuer, V.B., Pohlman, J.W., Torres, M.E., Elvert, M., Hinrichs, K.-U., 2009. The stable carbon isotope biogeochemistry of acetate and other dissolved carbon species in deep subseafloor sediments at the northern Cascadia Margin. Geochim. Cosmochim. Acta 73, 3323–3336.

Hines, M.E., Banta, G.T., Giblin, A.E., Hobbie, J.E., Tugel, J.B., 1994. Acetate concentrations and oxidation in salt marsh sediments. Limnol. Oceanogr. 39, 140–148.

Hoehler, T.M., Jørgensen, B.B., 2013. Microbial life under extreme energy limitation. Nat. Rev. Microbiol. 11, 83–94.

Hoehler, T.M., Alperin, M.J., Albert, D.B., Martens, C.S., 1998. Thermodynamic control on hydrogen concentrations in anoxic sediments. Geochim. Cosmochim. Acta 62, 1745–1756.

Holcombe, B.L., Keil, R., Devol, A.H., 2001. Determination of pore-water dissolved organic carbon fluxes from Mexican margin sediments. Limnol. Oceanogr. 46, 298–308.

Ijiri, A., Harada, N., Hirota, A., Tsunogai, U., Ogawa, N.O., Itaki, T., et al., 2012. Biogeochemical processes involving acetate in sub-seafloor sediments from the Bering Sea shelf break. Org. Geochem. 48, 47–55.

Jacobsen, M.E., Mackin, J.E., Capone, D.G., 1987. Ammonium production in sediments inhibited with molybdate: implications for the sources of ammonium in anoxic marine sediments. Appl. Environ. Microbiol. 53, 2435–2439.

Jahnke, R.A., 1990. Early diagenesis and recycling of biogenic debris at the seafloor, Santa Monica Basin, California. J. Mar. Res. 48, 413–436.

Jensen, S.I., Kühl, M., Glud, R.N., Jørgensen, L.B., Priemé, A., 2005. Oxic microzones and radial oxygen loss from roots of Zostera marina. Mar. Ecol. Prog. Ser. 293, 49–58.

Johnson, L., Komada, T., 2011. Determination of radiocarbon in marine sediment porewater dissolved organic carbon by thermal sulfate reduction. Limnol. Oceanogr.: Methods 9, 485–498.

Jørgensen, N.O.G., Lindroth, P., Mopper, K.H., 1981. Extraction and distribution of free amino acids and ammonium in sediment interstitial waters from the Limfjord, Denmark. Oceanol. Acta 4, 465–474.

Kawahata, H., Ishizuka, T., 2000. Amino acids in interstitial waters from ODP Sites 689 and 690 on the Maud Rise, Antarctic Ocean. Geochem. J. 34, 247–261.

Kemp, P.F., 1990. The fate of benthic bacterial production. Rev. Aquat. Sci. 2, 109–124.

Kim, T.-H., Waska, H., Kwon, E., Suryaputra, I.G.N., Kim, G., 2012. Production, degradation, and flux of dissolved organic matter in the subterranean estuary of a large tidal flat. Mar. Chem. 142-144, 1–10.

Knab, N.J., Cragg, B.A., Borowski, C., Parkes, R.J., Pancost, R., Jørgensen, B.B., 2008. Anaerobic oxidation of methane (AOM) in marine sediments from the Skagerrak (Denmark): I. Geochemical and microbiological analyses. Geochim. Cosmochim. Acta 72, 2868–2879.

Koch, B.P., Witt, M., Engbrodt, R., Dittmar, T., Kattner, G., 2005. Molecular formulae of marine and terrigenous dissolved organic matter detected by electrospray ionization Fourier transform ion cyclotron resonance mass spectrometry. Geochim. Cosmochim. Acta 69, 3299–3308.

Kohnen, M.E.L., Schouten, S., Damste, J.S.S., de Leeuw, J.W., Merrit, D., Hayes, J.M., 1992. The combined application of organic sulphur and isotope geochemistry to assess multiple sources of palaeobiochemicals with identical carbon skeletons. Org. Geochem. 19, 403–419.

Komada, T., Schofield, O.M.E., Reimers, C.E., 2002. Fluorescence characteristics of organic matter released from coastal sediments during resuspension. Mar. Chem. 79, 81–97.

Komada, T., Reimers, C.E., Luther III, G.W., Burdige, D.J., 2004. Factors affecting dissolved organic matter dynamics in mixed-redox to anoxic coastal sediments. Geochim. Cosmochim. Acta 68, 4099–4111.

Komada, T., Polly, J.A., Johnson, L., 2012. Transformations of carbon in anoxic marine sediments: implications from ^{14}C and ^{13}C signatures. Limnol. Oceanogr. 57, 567–581.

Komada, T., Burdige, D.J., Crispo, S.M., Druffel, E.R.M., Griffin, S., Johnson, L., et al., 2013. Dissolved organic carbon dynamics in anaerobic sediments of the Santa Monica Basin. Geochim. Cosmochim. Acta 110, 253–273.

Krom, M.D., Sholkovitz, E.R., 1977. Nature and reactions of dissolved organic matter in the interstitial waters of marine sediments. Geochim. Cosmochim. Acta 41, 1565–1573.

Krom, M. D., Westrich, J. T., 1981. Dissolved organic matter in the pore waters of recent marine sediments; a review. Biogéochemie de la matière organique à l'interface eau-sédiment marin. Colloques Internationaux du C.N.R.S.

Kujawinski, E.B., Del Vecchio, R., Blough, N.V., Klein, G.C., Marshall, A.G., 2004. Probing molecular-level transformations of dissolved organic matter: insights on photochemical degradation and protozoan modification of DOM from electrospray ionization Fourier transform-ion cyclotron resonance mass spectrometry. Limnol. Oceanogr. 92, 23–37.

Lahajnar, N., Rixen, T., Gate-Haake, B., Schafer, P., Ittekkot, V., 2005. Dissolved organic carbon (DOC) fluxes of deep-sea sediments from the Arabian Sea and NE Atlantic. Deep-Sea Res. Pt. II, 1947–1964.

Lalonde, K., Mucci, A., Ouellet, A., Gélinas, Y., 2012. Preservation of organic matter in sediments promoted by iron. Nature 483, 198–200.

Landén, A., Hall, P.O.J., 1998. Seasonal variation of dissolved and adsorbed amino acids and ammonium in a near-shore marine sediment. Mar. Ecol. Prog. Ser. 170, 67–84.

Landén, A., Hall, P.O.J., 2000. Benthic fluxes and pore water distributions of dissolved free amino acids in the open Skagerrak. Mar. Chem. 71, 53–68.

Landén-Hillmeyr, A., 1998. Nitrogen Cycling in Continental Margin Sediments with Emphasis on Dissolved Organic Nitrogen and Amino Acids (Ph.D. dissertation). Göteborg University.

Larmagnat, S., Neuweiler, F., 2011. Exploring a link between Atlantic coral mounds and Phanerozoic carbonate mud-mounds: insights from pore water fluorescent dissolved organic matter (FDOM), Pen Duick mounds, offshore Morocco. Mar. Geol. 282, 149–159.

Lee, C., 1992. Controls on organic carbon preservation: the use of stratified water bodies to compare intrinsic rates of decompositin in oxic and anoxic systems. Geochim. Cosmochim. Acta 56, 3233–3335.

Liu, Z., Lee, C., 2006. Drying effects on sorption capacity of coastal sediment: the importance of architecture and polarity of organic matter. Geochim. Cosmochim. Acta 70, 3313–3324.

Liu, Z., Lee, C., 2007. The role of organic matter in the sorption capacity of marine sediments. Mar. Chem. 105, 240–257.

Loh, A.N., Bauer, J.E., Druffel, E.R., 2004. Variable ageing and storage of dissolved organic components in the open ocean. Nature 430, 877–881.

Lomstein, B.A., Blackburn, T.H., Henricksen, K., 1989. Aspects of nitrogen and carbon cycling inthe northern Bering Shelf sediments. I. The significance of urea turnover in the mineralization of NH_4^+. Mar. Ecol. Prog. Ser. 57, 237–247.

Lomstein, B.A., Jensen, A.-G.U., Hansen, J.W., Andreasen, J.B., Hansen, L.S., Berntsen, J., et al., 1998. Budgets of sediment nitrogen and carbon cycling in the shallow water of Knebel Vig, Denmark. Aquat. Microb. Ecol. 14, 69–80.

Lomstein, B.A., Langerhuus, A.T., D'Hondt,, S., Jørgensen, B.B., Spivack, A.J., 2012. Endospore abundance, microbial growth and necromass turnover in deep sub-seafloor sediment. Nature 484, 101–104.

Lovley, D.R., Goodwin, S., 1988. Hydrogen concentrations as an indicator of the predominant terminal electron-accepting reactions in aquatic sediments. Geochim. Cosmochim. Acta 52, 2993–3003.

Lovley, D.R., Phillips, E.J.P., 1987. Competitive mechanisms for inhibition of sulfate reduction and methane production in the zone of ferric iron reduction in sediments. Appl. Environ. Microbiol. 53, 2636–2641.

Lübben, A., Dellwig, O., Koch, S., Beck, M., Badewien, T.H., Fischer, S., et al., 2009. Distributions and characteristics of dissolved organic matter in temperate coastal waters (Southern North Sea). Ocean Dynam. 59, 263–275.

Ludwig, W., Probst, J.-L., 1996. Predicting the input of organic carbon by continental erosion. Global Biogeochem. Cycles 10, 23–41.

Lyons, W.B., Gaudette, H.E., Hewitt, A.D., 1979. Dissolved organic matter in pore waters of carbonate sediments from Bermuda. Geochim. Cosmochim. Acta 43, 433–437.

Maher, D.T., Eyre, B.D., 2010. Benthic fluxes of dissolved organic carbon in three temperate Australian estuaries: implications for global estimates of benthic DOC fluxes. Biogeosciences 115, G04039.

Maher, D., Eyre, B.D., 2011. Insights into estuarine benthic dissolved organic carbon (DOC) dynamics using δ^{13}C-DOC values, phospholipid fatty acids and dissolved organic nutrient fluxes. Geochim. Cosmochim. Acta 75, 1889–1902.

Marchand, C., Albéric, P., Lallier-Vergès, E., Baltzer, F., 2006. Distribution and characteristics of dissolved organic matter in mangrove sediment pore waters along the coastline of French Guiana. Biogeochemistry 81, 59–75.

Martens, C.S., Haddad, R.I., Chanton, J.P., 1992. Organic matter accumulation, remineralization and burial in an anoxic marine sediment. In: Whelan, J.K., Farrington, J.W. (Eds.), Productivity, Accumulation, and Preservation of Organic Matter in Recent and Ancient Sediments. Columbia Univ. Press, New York, pp. 82–98.

Martin, W.R., McCorkle, D.C., 1994. Dissolved organic carbon in continental margin sediments: an assessment of the relative rates of cabon oxidation to CO_2 and net DOC production. Eos Trans. Am. Geophys. Union 75, 323.

Marvin-DiPasquale, M.C., Capone, D.G., 1998. Benthic sulfate reduction along the Chesapeake Bay central channel. I. Spatial trends and controls. Mar. Ecol. Prog. Ser. 168, 213–228.

Mayer, L.M., Rice, D.L., 1992. Early diagenesis of proteins: a seasonal study. Limnol. Oceanogr. 37, 280–295.

McCarthy, M.D., Beaupre, S.R., Walker, B.D., Voparil, I., Guilderson, T.P., Druffel, E.R.M., 2011. Chemosynthetic origin of ^{14}C-depleted dissolved organic matter in a ridge-flank hydrothermal system. Nat. Geosci. 4, 32–36.

McKee, G.A., 2011. The Nature, Origin and Preservation of Amide Organic Nitrogen in Organic Matter (Ph.D. dissertation). Old Dominion Univ., USA.

McKee, G.A., Hatcher, P.G., 2010. Alkyl amides in two organic-rich anoxic sediments: a possible new abiotic route for N sequestration. Geochim. Cosmochim. Acta 74, 6436–6450.

McManus, J., Berelson, W.M., Coale, K., Johnson, K., Kilgore, T.E., 1997. Phosphorus regeneration in continental margin sediments. Geochim. Cosmochim. Acta 61, 2891–2907.

McNichol, A.P., Aluwihare, L.I., 2007. The power of radiocarbon in biogeochemical studies of the marine carbon cycle: insights from studies of dissolved and particulate organic carbon (DOC and POC). Chem. Rev. 107, 443.

Megonigal, J.P., Hines, M.E., Visscher, P.T., 2003. 8.08—Anaerobic metabolism: linkages to trace gases and aerobic processes. In: Holland, H.D., Turekian, K.K. (Eds.), Treatise on Geochemistry. Pergamon, Oxford.

Michelson, A.R., Jacobsen, M.E., Scranton, M.I., Mackin, J.E., 1989. Modeling the distribution of acetate in anoxic estuarine sediments. Limnol. Oceanogr. 34, 747–757.

Middelburg, J.J., 1989. A simple rate model for organic matter decomposition in marine sediments. Geochim. Cosmochim. Acta 53, 1577–1581.

Moore, E.K., Nunn, B.L., Goodlett, D.R., Harvey, H.R., 2012. Identifying and tracking proteins through the marine water column: insights into the inputs and preservation mechanisms of protein in sediments. Geochim. Cosmochim. Acta 83, 324–359.

Mulholland, M.R., Lomas, M.W., 2008. Nitrogen uptake and assimilation. In: Capone, D.G., Bronk, D.A., Mulholland, M.R., Carpenter, E.J. (Eds.), Nitrogen in the Marine Environment, Second ed. Academic Press, San Diego.

Nissenbaum, A., Baedecker, M.J., Kaplan, I.R., 1972. Studies on dissolved organic matter from interstitial waters of a reducing marine fjord. In: von Gaerter, H.R., Wehner, H. (Eds.), Advances in Organic Geochemistry 1971. Pergamon Press, Oxford.

Novelli, P.C., Michelson, A.R., Scranton, M.I., Banta, G.T., Hobbie, J.E., Howarth, R.W., 1988. Hydrogen and acetate cycling in two sulfate-reducing sediments: Buzzards Bay and Town Cove, Mass. Geochim. Cosmochim. Acta 52, 2477–2486.

Ogasawara, R., Ishiwatari, R., Shimoyama, A., 2001. Detection of water extractable dipeptides and their characteristics in recent sediments of Tokyo Bay. Geochem. J. 35, 439–450.

Orem, W.H., Hatcher, P.G., 1987. Solid-state ^{13}C NMR studies of dissolved organic matter in pore waters from different depositional environments. Org. Geochem. 11, 73–82.

Orem, W.E., Hatcher, P.G., Spiker, E.C., Szeverenyi, N.M., Machel, G.E., 1986. Dissolved organic matter in anoxic pore waters from Mangrove Lake, Bermuda. Geochim. Cosmochim. Acta 50, 609–618.

Otero, M., Mendonça, A., Válega, M., Santos, E.B.H., Pereira, E., Esteves, V.I., et al., 2007. Fluorescence and DOC contents of estuarine pore waters from colonized and non-colonized sediments: effects of sampling preservation. Chemosphere 67, 211–220.

Pantoja, S., Lee, C., 1999. Peptide decomposition by extracellular hydrolysis in coastal seawater and salt marsh sediment. Mar. Chem. 63, 273–291.

Papadimitriou, S., Kennedy, H., Bentaleb, I., Thomas, D.N., 2002. Dissolved organic carbon in sediments from the eastern North Atlantic. Mar. Chem. 79, 37–47.

Parkes, R.J., Gibson, G.R., Mueller-Harvey, I., Buckingham, W.J., Herbert, R.A., 1989. Determination of the substrates for sulfate-reducing bacteria within marine and estuarine sediments with different rates of sulfate reduction. J. Gen. Microbiol. 135, 175–187.

Parlanti, E., Wörz, K., Geoffrey, L., Lamotte, M., 2000. Dissolved organic matter fluorescence spectroscopy as a tool to estimate biological activity in a coastal zone submitted to anthropogenic inputs. Org. Geochem. 31, 1765–1781.

Penning, H., Conrad, R., 2006. Carbon isotope effects associated with mixed-acid fermentation of saccharides by *Clostridium papyrosolvens*. Geochim. Cosmochim. Acta 70, 2283–2297.

Piccolo, A., 2001. The supramolecular structure of humic substances. Soil Sci. 166, 810–832.

Pohlman, J.W., Bauer, J.E., Waite, W.F., Osburn, C.L., Chapman, N.R., 2010. Methane hydrate-bearing seeps as a source of aged dissolved organic carbon to the oceans. Nat. Geosci. 4, 37–41.

Ragsdale, S.W., Pierce, E., 2008. Acetogenesis and the Wood-Ljungdahl pathway of CO_2 fixation. Biochim. Biophys. Acta 1784, 1873–1898.

Raymond, P.A., Bauer, J.E., 2001a. Riverine export of aged terrestrial organic matter to the North Atlantic Ocean. Nature 409, 497–500.

Raymond, P.A., Bauer, J.E., 2001b. Use of ^{14}C and ^{13}C natural abundances for evaluating riverine, estuarine, and coastal DOC and POC sources and cycling: a review and synthesis. Org. Geochem. 32, 469–485.

Repeta, D.J., Aluwihare, L.I., 2006. Radiocarbon analysis of neutral sugars in high-molecular-weight dissolved organic carbon: implications for organic carbon cycling. Limnol. Oceanogr. 51, 1045–1053.

Repeta, D.J., Quan, T.M., Aluwihare, L.I., Accardi, A., 2002. Chemical characterization of high molecular weight dissolved organic matter in fresh and marine waters. Geochim. Cosmochim. Acta 66, 955–962.

Robador, A., Brüchert, V., Steen, A.D., Arnosti, C., 2010. Temperature induced decoupling of enzymatic hydrolysis and carbon remineralization in long-term incubations of Arctic and temperate sediments. Geochim. Cosmochim. Acta 74, 2316–2326.

Rosenfeld, J.K., 1979. Amino acid diagenesis and adsorption in nearshore anoxic marine sediments. Limnol. Oceanogr. 24, 1014–1021.

Roth, L.C., Harvey, H.R., 2006. Intact protein modification and degradation in estuarine environments. Mar. Chem. 102, 33–45.

Rowe, G.T., Deming, J.W., 2011. An alternative view of the role of heterotrophic microbes in the cycling of organic matter in deep-sea sediments. Mar. Biol. Res. 7, 629–636.

Sansone, F.J., Martens, C.S., 1982. Volatile fatty acid cycling in organic-rich marine sediments. Geochim. Cosmochim. Acta 45, 101–121.

Sansone, F.J., Andrews, C.A., Okamoto, M.Y., 1987. Adsorption of short-chain organic acids onto nearshore sediments. Geochim. Cosmochim. Acta 51, 1889–1896.

Santschi, P.H., Guo, L., Baskaran, M., Trumbore, S., Southon, J., Bianchi, T.S., et al., 1995. Isotopic evidence for the contemporary origin of high-molecular-weight organic matter in oceanic environments. Geochim. Cosmochim. Acta 59, 625–631.

Schmidt, F., Elvert, M., Koch, B.P., Witt, M., Hinrichs, K.U., 2009. Molecular characterization of dissolved organic matter in pore water of continental shelf sediments. Geochim. Cosmochim. Acta 73, 3337–3358.

Schmidt, F., Koch, B.P., Elvert, M., Schmidt, G., Witt, M., Hinrichs, K.U., 2011. Diagenetic transformation of dissolved organic nitrogen compounds under contrasting sedimentary redox conditions in the Black Sea. Environ. Sci. Technol. 45, 5223–5229.

Seretti, A., Nannicini, L., Del Vecchio, R., Giordani, P., Balboni, V., Miserocchi, S., 1997. Optical properties of sediment pore waters of the Adriatic Sea. Toxicol. Environ. Chem. 61, 195–209.

Shank, G.C., Skrabal, S.A., Whitehead, R.F., Kieber, R.J., 2004. Fluxes of strong Cu-complexing ligands from sediments of an organic-rich estuary. Estuar. Coast. Shelf Sci. 60, 349–358.

Shaw, T.J., Gieskes, J.M., Jahnke, R.A., 1990. Early diagenesis in differing depositional environments: the response of transition metals in pore waters. Geochim. Cosmochim. Acta 54, 1233–1246.

Sierra, M.M.D., Donard, O.F.X., Etcheber, H., Soriano-Sierra, E.J., Ewald, M., 2001. Fluorescence and DOC contents of pore waters from coastal and deep-sea sediments in the Gulf of Biscay. Org. Geochem. 32, 1319–1328.

Skoog, A.C., Arias-Esquivel, V.A., 2009. The effect of induced anoxia and reoxygenation on benthic fluxes of organic carbon, phosphate, iron, and manganese. Sci. Total Environ. 407, 6085–6092.

Skoog, A., Hall, P.O.J., Hukth, S., Paxéus, N., Rutgers van der Loef, M., 1996. Early diagenetic production and sediment-water exchange of fluorescent dissolved organic matter in the coastal environment. Geochim. Cosmochim. Acta 60, 3619–3629.

Sleighter, R.L., Hatcher, P.G., 2007. The application of electrospray ionization coupled to ultrahigh resolution mass spectrometry for the molecular characterization of natural organic matter. J. Mass Spectrosc. 42, 559–574.

Smith, D.C., Jørgensen, B.B., D'Hondt, S.L., Miller, D.J., 2005. Data report: dissolved organic carbon in interstitial waters, equatorial Pacific and Peru margin. In: Proceedings of the ODP, Scientific Results Volume 201. Ocean Drilling Program, College Station, TX.

Sørensen, J., Christensen, D., Jørgensen, B.B., 1981. Volatile fatty acids and hydrogen as substrates for sulfate-reducing bacteria in anaerobic marine sediments. Appl. Environ. Microbiol. 42, 5–11.

Ståhl, H., Hall, P.O.J., Tengberg, A., Josefson, A.B., Streftaris, N., Zenetos, A., et al., 2004. Respiration and sequestering of organic carbon in shelf sediments of the oligotrophic northern Aegean Sea. Mar. Ecol. Prog. Ser. 269, 33–48.

Starikova, N.D., 1970. Vertical distribution patterns of dissolved organic carbon in sea water and interstitial solutions. Oceanology 10, 796–807.

Stumm, W., Morgan, J.J., 1996. Aquatic Chemistry: Chemical Equilibria and Rates in Natural Waters. Wiley-Interscience, New York.

Suess, E., Müller, P.J., Reimers, C.E., 1980. A closer look at nitrification in pelagic sediments. Geophys. J. Roy. Astron. Soc. 14, 129–137.

Sugai, S.F., Henrichs, S.M., 1992. Rates of amino acid uptake and remineralization in Resurrection Bay (Alaska) sediments. Mar. Ecol. Prog. Ser. 88, 129–141.

Sundbäck, K., Linares, F., Larson, F., Wulff, A., Engelsen, A., 2004. Benthic nitrogen fluxes along a depth gradient in a microtidal fjord: the role of denitrification and microphytobenthos. Limnol. Oceanogr. 49, 1095–1107.

Sutton, R., Sposito, G., 2005. Molecular structure in soil humic substances: the new view. Environ. Sci. Technol. 39, 9009–9015.

Swett, M. P., 2010. Assessment of Benthic Flux of Dissolved Organic Carbon in Estuaries Using the Eddy-Correlation Technique (M.S. thesis). The University of Maine.

Takii, S., Hanada, S., Hase, Y., Tamaki, H., Uyeno, Y., Sekiguchi, Y., et al., 2008. *Desulfovibrio marinisediminis* sp. nov., a novel sulfate-reducing bacterium isolated from coastal marine sediment via enrichment with Casamino acids. Int. J. Syst. Evol. Microbiol. 58, 2433–2438.

Tegelaar, E.W., de Leeuw, J.W., Derenne, S., Largeau, C., 1989. A reappraisal of kerogen formation. Geochim. Cosmochim. Acta 53, 3103–3106.

Thamdrup, B., 2000. Bacterial manganese and iron reduction in aquatic sediments. Adv. Microbiol. Ecol. 16, 41–83.

Thauer, R.K., Jungermann, K., Decker, K., 1977. Energy conservation in chemotrophic anaerobic bacteria. Microbiol. Mol. Biol. Rev. 41, 100.

Thimsen, C.A., Keil, R.G., 1998. Potential interactions between sedimentary dissolved organic matter and mineral surfaces. Mar. Chem. 62, 65–76.

Tréhu, A.M., Bohrmann, G., Rack, F.R., Torres, M.E., 2003. Proceedings of the ODP, Initial Reports, 204. Ocean Drilling Program, College Station, TX.

Tremblay, L.B., Dittmar, T., Marshall, A.G., Cooper, W.J., Cooper, W.T., 2007. Molecular characterization of dissolved organic matter in a North Brazilian mangrove porewater and mangrove-fringed estuaries by ultrahigh resolution Fourier transform-ion cyclotron resonance mass spectrometry and excitation/emission spectroscopy. Mar. Chem. 105, 15–29.

Tyler, A.C., McGlathery, K.J., Anderson, I.C., 2003. Benthic algae control sediment-water column fluxes of organic and inorganic nitrogen compounds in a temperate lagoon. Limnol. Oceanogr. 48, 2125–2137.

Valdemarsen, T., Kristensen, E., 2010. Degradation of dissolved organic monomers and short-chain fatty acids in sandy marine sediment by fermentation and sulfate reduction. Geochim. Cosmochim. Acta 74, 1593–1605.

Valentine, D.L., Kastner, M., Wardlaw, G.D., Wang, X., Purdy, A., Bartlett, D.H., 2005. Biogeochemical investigations of marine methane seeps, Hydrate Ridge, Oregon. J. Geophys. Res. 110, G02005.

Wang, X.-C., Lee, C., 1993. Adsorption and desorption of aliphatic amines, amino acids and acetate by clay minerals in marine sediments. Mar. Chem. 44, 1–23.

Wang, X.-C., Lee, C., 1995. Decomposition of aliphatic amines and amino acids in anoxic salt marsh sediments. Geochim. Cosmochim. Acta 59, 1787–1797.

Wefer, G., Berger, W.H., Richter, C., et al., 1998. Proceedings of the Ocean Drilling Program, Init. Repts. vol. 175. Ocean Drilling Program, College Station, TX.

Wellsbury, P., Goodman, K., Barth, T., Cragg, B.A., Barnes, S.P., Parkes, R.J., 1997. Deep marine biosphere fuelled by increasing organic matter availability during burial and heating. Nature 388, 573–576.

Weston, N.B., Joye, S.B., 2005. Temperature-driven decoupling of key phases of organic matter degradation in marine sediments. Proc. Natl. Acad. Sci. U. S. A. 102, 17036–17040.

Westrich, J.T., Berner, R.A., 1984. The role of sedimentary organic matter in bacterial sulfate reduction: the G model tested. Limnol. Oceanogr. 29, 236–249.

Williams, P.M., Druffel, E.R.M., 1987. Radiocarbon in dissolved organic matter in the central North Pacific Ocean. Nature 330, 246–248.

Wilson, D.S., Teagle, D.A.H., Acton, G.D., 2003. Proceedings of the ODP, Initial Reports, 206. Ocean Drilling Program, College Station, TX.

Worsfold, P.J., Monbet, P., Tappin, A.D., Fitzsimons, M.F., Stiles, D.A., McKelvie, I.D., 2008. Characterisation and quantification of organic phosphorus and organic nitrogen components in aquatic systems: a review. Anal. Chim. Acta 624, 37–58.

Yao, H., Conrad, R., 2000. Electron balance during steady-state production of CH_4 and CO_2 in anoxic rice soil. Eur. J. Soil Sci. 51, 369–378.

Ziegelgruber, K.L., Zeng, T., Arnold, W.A., Chin, Y.-P., 2013. Sources and composition of sediment pore-water dissolved organic matter in prairie pothole lakes. Limnol. Oceanogr. 58, 1136–1146.

Zonneveld, K.A.F., Versteegh, G.J.M., Kasten, S., Eglinton, T.I., Emeis, K.C., Huguet, C., et al., 2010. Selective preservation of organic matter in marine environments; processes and impact on the sedimentary record. Biogeosciences 7, 483–511.

13

DOC in the Mediterranean Sea

Chiara Santinelli

Istituto di Biofisica, Pisa, Italy

CONTENTS

The size of the sea is of a human scale. No long days of navigation are spent between harbor and harbor, and the sailors can wait in safety for acceptable sea conditions. The trips of fishermen rarely take them away for more than one day. The sea keeps them together, as neighbors around the back fence. (**R. Margalef**)

I INTRODUCTION

A Main Features of the Mediterranean Sea

The Mediterranean Sea (Med Sea) is the largest semi-enclosed basin on the Earth. It has a surface area of $2.5 \times 10^6 \, km^2$ and $46 \times 10^3 \, km$ of coastline.

Biogeochemistry of Marine Dissolved Organic Matter,
http://dx.doi.org/10.1016/B978-0-12-405940-5.00013-3

In spite of its limited size (0.7% surface, 0.25% volume of the global ocean), the basin is considered one of the most complex marine environments. It is surrounded by three continents (Europe, Africa, and Asia) as suggested by its name: "Mediterranean" derives from the Latin word "Mediterraneus," meaning "in the middle of the land" or "between lands" (from medius, "middle, between" and terra, "land, earth"). The Med Sea was formed ~150 million years ago, and during the Messinian salinity crisis (~5.60 million years ago), it became mostly desiccated by evaporation with closing of the Gibraltar Strait. Roughly 5.33 million years ago, Atlantic waters rapidly refilled the basin (Garcia-Castellanos et al., 2009) to form the Med Sea as it is today.

The average depth is 1500 m with the deepest recorded point of 5267 m located in the Ionian Sea. The bottom relief shows irregularities, having many deep basins and troughs as well as several seamounts (Figure 13.1). The Tyrrhenian Sea hosts two main seamounts: the *Marsili*, a large undersea volcano, with its crater 450 m below the sea's surface and the *Vavilov*, an extinct volcano 2700 m high with its peak 800 m below the sea's surface. Two deep hypersaline anoxic basins were discovered at the bottom of the Eastern Med Sea (East Med) (Jongsma et al., 1983), where a wide diversity of prokaryotes was observed, including a new, abundant, deeply branching order within the Euryarchaeota (Van der Wielen et al., 2005) (Figure 13.1).

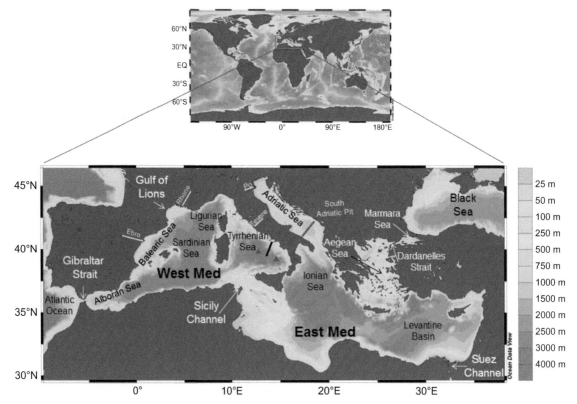

FIGURE 13.1   The Mediterranean Sea, with key features and main rivers identified. Locations of the *Marsili* and *Vavilov* volcanoes (red triangles; east and west, respectively) and of the deep hypersaline anoxic basins *Tyro* and *Bannock* (orange circles; east and west, respectively) are shown. Solid lines represent the sections where DOC temporal variability was studied.

Many aspects make the Med Sea an ideal basin to study the main physical and biogeochemical processes occurring in the oceans, with particular regards to dissolved organic matter (DOM) dynamics. It features: (1) unique thermohaline circulation including deep-water formation, despite being located at temperate latitudes; (2) meso and sub-mesocale activity that strongly affects dissolved organic carbon (DOC) and nutrient distributions; (3) fast ventilation rates and residence times for deep waters (20-126 years) (Andrie and Merlivat, 1988; Schlitzer et al., 1991); (4) deep-water temperatures ~10 °C higher than in the open ocean; (5) enhanced respiration (Christensen et al., 1989) and DOC mineralization (Santinelli et al., 2010) rates at depth; and (6) diverse patterns of stratification in areas located within close proximity (e.g., southern Tyrrhenian and Adriatic Seas).

From a biogeochemical perspective, the Med Sea is defined as a low-nutrient, low-chlorophyll system (MerMex group, 2011 and references therein). It shows a high N:P ratio of nutrients relative to Redfield expectations (Redfield, 1934) (N:P > 25 in the East Med; N:P ~ 20 in the West Med; Krom et al., 2005; Ribera d'Alcalà et al., 2003), and it is characterized by decreasing primary production, particulate carbon export, and nutrient availability eastward (Moutin and Raimbault, 2002; Van Wambeke et al., 2002).

The occurrence of distinct bio-provinces with distinct behaviors in terms of phytoplankton blooms has been recently described (D'Ortenzio and Ribera d'Alcalà, 2009). In particular, the authors identified: (i) *"non-blooming"* areas (60% of the basin), exhibiting a subtropical regime and (ii) *"blooming"* and *"intermittently blooming"* areas where, under particular conditions (both atmospheric and hydrographic), North Atlantic bloom-like events occur.

Finally, because of its modest dimensions, the basin responds rapidly to environmental change. Processes can be studied here on a smaller spatial and shorter temporal scale than in the open ocean.

B Water Masses and Thermohaline Circulation: Main Patterns

The Mediterranean thermohaline circulation is strongly affected by two key circulatory constrictions: (1) the Gibraltar Strait, 14 km wide and 300 m deep and the only connection between the Med Sea and the oceans and (2) the Sicily Channel, 140 km wide and 500 m deep, separating the Med Sea as two subbasins: the Western Med Sea (West Med) and the East Med (Figure 13.1).

The Med Sea is a concentration basin: evaporation is higher than the input of freshwater by precipitation and runoff; Atlantic water (AW) entering the basin through the Gibraltar Strait compensates the imbalance. The inflow of AW, coupled with the outflow of the saltier intermediate waters, keeps the total salt content almost constant. During transit into the basin (Figure 13.2), AW becomes saltier and cooler. Driven by surface buoyancy loss, AW sinks to form intermediate waters during convective events. Intermediate water formation mainly occurs in cyclonic gyres, favored by reduced stratification in their core, and mostly within the Rhodes Gyre, where the Levantine Intermediate Water (LIW) is formed.

LIW flows west along a complex route, exiting through the Gibraltar Strait. This water mass can be easily recognized in the whole basin by its high salinity.

AW and the LIW are the only two water masses flowing through the whole basin. The sill at the Sicily Channel hinders the exchange of waters at >500 m. As a consequence, deep waters found in the subbasins do not mix directly (Iudicone et al., 2003; Sparnocchia et al., 1999). Finally, Black Sea waters enter through the Dardanelles Strait in the northeastern part of the basin (Figure 13.1).

Winter formation of deep waters takes place in a few key areas (Figure 13.2). These sites are characterized by intense heat fluxes driven by winter storms and by reduced water column stability, derived from preexisting cyclonic circulation or by shallow bathymetry.

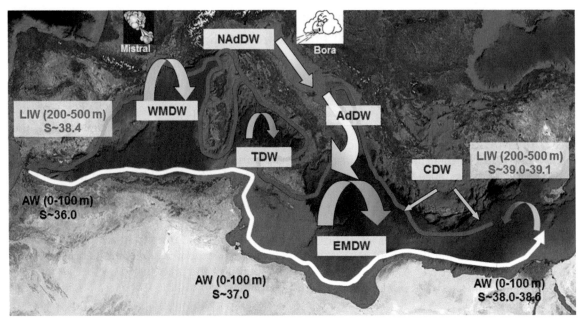

FIGURE 13.2   Schematic representation of the thermohaline circulation along with the strong, cold, and usually dry regional winds contributing to deep-water formation: the Mistral and Bora. The basic scheme of circulation reported by Millot and Taupier-Letage (2005) is strongly simplified and mesoscale activity is not reported. The main flows of the surface Atlantic Water (AW) and the Levantine Intermediate Water (LIW), along with characteristic salinities (S) and depths, are shown in white and red, respectively. AW return flow is not indicated. LIW formation is indicated by the white-red arrow. Deep-water formation is also indicated, with light blue arrows for water masses formed at the surface and green arrows for those formed by mixing at depth: WMDW, Western Mediterranean Deep Water; TDW, Tyrrhenian Deep Water; EMDW, Eastern Mediterranean Deep Water; NAdDW, North Adriatic Dense Water; AdDW, Adriatic Deep Water; and CDW, Cretan Deep Water. Yellow arrows indicate CDW outflows from the Aegean Sea.

The main deep-water formation sites are:

(i) The Gulf of Lions, where a large cyclonic gyre can be easily seen on satellite images due to its high chlorophyll levels. In winter, convection chimneys are clearly observed by isopycnal distribution and the surface water is mixed with LIW, increasing its salinity (e.g., Schott et al., 1996). The loss of heat, due to a dry and cold wind, named the Mistral (Figure 13.2), makes the surface water dense enough to sink to the bottom, creating Western Mediterranean Deep Water (WMDW) ($S = 38.45$-38.47; $\theta = 12.85$-$12.90\,°C$; dissolved oxygen (DO) = 200-$220\,\mu M$). Formation rates of 0.14-$1.2\,Sv$ (Sverdrups = $10^6\,m^3\,s^{-1}$, Rhein, 1995) occur over periods of weeks to months. Very high

WMDW formation rates were reported in the winter of 2005 ($2.4\,Sv$, Schroeder et al., 2008), when particularly strong deep-water cascading was observed (Canals et al., 2006).

(ii) The Tyrrhenian Sea, where Tyrrhenian Deep Water (TDW) ($S = 38.54$-38.62; $\theta = 13.1$-$13.4\,°C$; DO = 175-$188\,\mu M$) is formed mostly by the mixing between WMDW and LIW. TDW generally displays low oxygen, indicating that it has not been in contact with the atmosphere recently. TDW might result also from occasional local deep-water formation events occurring within the Tyrrhenian Sea itself (Buongiorno Nardelli and Salusti, 2000; Fuda et al., 2002).

(iii) The Adriatic Sea, where deep water is formed by two processes: (1) open ocean convection, in the South Adriatic Pit, which mainly affects the upper 800 m. This process is strongly preconditioned by the presence of a permanent cyclonic gyre and by an intense dry and cold wind, blowing from the northeast in winter, named the Bora (Figure 13.2) and (2) extreme densification over the shallow northern Adriatic Sea due to Bora, in winter. This process produces the North Adriatic Dense Water (NAdDW, $S < 38.3$; $\theta < 11.35\,°C$) that flows to the bottom of the South Adriatic Pit (depth >1000 m), where it can contribute to the formation of the Adriatic Deep Water (AdDW, $S < 38.74$; $\theta < 13.1\,°C$; DO = 222-236 µM). The AdDW, exiting the Otranto Straits, sinks to depth >3000 m in the Ionian Sea and represents the main source of Eastern Mediterranean Deep Water (EMDW) (Bignami et al., 1990; Cardin et al., 2011; Malanotte-Rizzoli, 1991; Manca et al., 2002; Vilibić and Supić, 2005) (Figure 13.2).

(iv) The Aegean Sea that can sporadically be a source of dense water for the East Med, depending on the meteorological conditions over the East Med (Theocharis et al., 1999, 2002; Zervakis et al., 2000, 2004).

The North Ionian Gyre alternates between cyclonic and anticyclonic states on decadal timescales; the East Med circulation changes according to the direction of this gyre (Gačič et al., 2010), with an impact on the ecosystem functioning of the Adriatic and Ionian Seas (Civitarese et al., 2010). This feedback mechanism between variations in the thermohaline properties of waters formed in the Southern Adriatic and the Ionian circulation has been named Adriatic-Ionian Bimodal Oscillating System (BiOS). BiOS is probably related to the big change in the basic circulation pattern of the East Med observed during the 1990s and termed the Eastern Mediterranean Transient (EMT) (Malanotte-Rizzoli et al., 1999; Roether et al., 1996). During the EMT, the Adriatic Sea stopped the production of deep water, leaving the source of deep water for the East Med the

Aegean Sea, with some consequences for the biogeochemistry of the basin (Klein et al., 2003; Seritti et al., 2003).

The high evaporation and intense air-sea interactions occurring in the Med Sea play an important role not only in Mediterranean deep-water formation and thermohaline circulation but also on an oceanic global scale. Mediterranean outflow is a key player in preconditioning the deep convection cells of the polar Atlantic, where it arrives through direct pathways or indirect mixing processes (Artale et al., 2006).

II DOC DISTRIBUTION AT BASIN SCALE

A DOC Spatial Variability

In the past decade, extensive surveys of DOC have been carried out in the Med Sea, illuminating DOM spatial distributions and dynamics on the basin scale. The availability of a consensus reference material has strongly improved the data quality (Hansell, 2005).

The concentrations of DOC in the Med Sea are similar to those observed in the open ocean. DOC ranges between 31 and 128 µM, with values higher than 128 µM observed in coastal areas impacted by rivers (e.g., the Gulf of Lions and northern Adriatic Sea) and in the Marmara Sea (Table 13.1). Vertical profiles show the highest concentrations in the mixed layer, with a gradual decrease to a minimum of 40-48 µM between 200 and 500 m, in the core of LIW. Below 500 m DOC shows almost constant values (34-43 µM) with a slight increase close to the bottom (Figure 13.3). There are exceptions to this general pattern, mainly linked to the physical processes that occur in the sampling areas.

1 Surface Layer

In the surface layer (0-100 m), where the largest variability is observed, mean DOC concentrations range between 57 and 101 µM with a marked influence of (i) terrestrial inputs, (ii) mesoscale

TABLE 13.1   DOC Concentrations Reported in the Literature for Different Areas of the Mediterranean Sea and Main Rivers

Study Area	Sampling Period	Depth (m)	DOC (µM)		DOC Reference Material[a]	Sample Treatment	Reference
			Range	Mean ± std			
Gibraltar Strait	September 1997	0-125	82-115	91 ± 11	DAW	No filtration + HgCl$_2$	Dafner et al. (2001a)
		180-822	61-69	66 ± 3			
	April 1998	0-67	54-66	60 ± 4	DPW	No filtration + HgCl$_2$	Dafner et al. (2001b)
		71-831	39-47	44 ± 2			
Alboran Sea (Almerian-Oran front)	November-December 1997	0-100	63-83[b]	72 ± 5[b]	DAW	No filtration + HgCl$_2$	Sempéré et al. (2003)
		200-2000	45-67				
Balearic Sea	June 1995	0-100	52-95	68 ± 4	n.a.	GF/F filter	Doval et al. (1999)
		200-2000	44-63	51 ± 2			
	September 1996	0-100	50-120		DW	GF/F filter + HCl at 4°C	Lucea et al. (2003)
		200-500	40-60				
		600-1000	50-80				
Gulf of Lions	June 1987	0-100	70-220[c]		n.a.	GF/F filter + HCl or + HgCl$_2$	Cauwet et al. (1997)
		100-2000	60-80[c]				
	November 1991	0-100	85-250				
		100-2000	60-80				
	November 1994	0-100	75-114	101 ± 12	n.a.	GF/F filter + HgCl$_2$	Yoro et al. (1997)
		100-1000	65-96	77 ± 10			
	September-October 1997	0-89	54-396[d]	162 ± 70[d]	n.a.	0.22 µm filter + H$_3$PO$_4$	Ferrari (2000)
	September 1984	Mixed layer	67-69		DW	GF/F filter at -20°C	Aminot and Kérouel (2004)
		200-600	46-48				
		800-1500	46				

Rhone River	June 1994-May 1995	Surface	133-386	217±60	n.a.	GF/F filter + HCl	Sempéré et al., (2000)
	May 2007-June 2009	Surface	80-199	134±31	CRM	GF/F and 0.2 filter + H$_3$PO$_4$; at 4°C	Panagiotopoulos et al. (2012)
Ligurian Sea (Dyfamed station)	1991-1992 monthly	0-100	62-92		n.a.	GF/F filter	Copin-Montégut and Avril (1993)
		150-2000	50-58				
	1991-1994 monthly	2-200	55-87	82±5	IE	GF/F filter + HgCl$_2$	Avril (2002)
		200-1000	50-65	51±2			
		1000-2000	49-56				
Ligurian Sea	May 2003	0-100	50-100	68±10	CRM	0.2-μm filter at 4°C	Santinelli et al. (2010)
		200-500	40-58	50±4			
		2000-2800	44-60	52±4			
Sardinian Sea	March 2001	0-25	54-79	67±4	CRM	0.2-μm filter at 4°C	Santinelli et al. (2008)
		300-700	38-49	45±1			
		900-2500	45-56	51±2			
	September 2001	0-25	64-80	68±5			
		300-600	38-49	45±1			
		900-2500	37-53	45±3			
Algerian basin	August 2000	0-100	65-88	75±7	CRM	0.2-μm filter at 4°C	Santinelli et al. (2002)
		300-600	49-57	54±3			
		900-1500	51-59	55±2			
		1700-3000	49-60	56±4			

(Continued)

TABLE 13.1 DOC Concentrations Reported in the Literature for Different Areas of the Mediterranean Sea and Main Rivers—cont'd

Study Area	Sampling Period	Depth (m)	DOC (μM)		DOC Reference Material[a]	Sample Treatment	Reference
			Range	Mean ± std			
West Med	September 1999	Mixed Layer	57-67[b]	62±4[b]	CRM	No filtration + H_3PO_4	Santinelli et al. (2012a)
		200-500	36-46				
		1000-3500	37-42				
	May 2005	0-100	48-95	62±9	CRM	0.2-μm filter at 4°C	Santinelli et al. (2010)
		200-500	35-58	48±5			
		2000-3000	45-76	56±9			
	March 2008	0-100	46-84	57±7			
		200-500	38-49	44±3			
		2000-3000	38-49	42±3			
	June-July 2008	0-150	50-69		CRM	GF/F filter + H_3PO_4	Pujo-Pay et al. (2011)
		200-800	38-44				
		1000-2900	38-42	40±1			
Northern Tyrrhenian Sea	January 1999	0-100	56-76		CRM	0.2 μm filter at 4°C	Vignudelli et al. (2004)
		200-400	44-69				
	September 1997	Surface	74-115	92±9	n.a.	0.2-μm filter at 4°C	Seritti et al. (1998)
Tyrrhenian Sea	August 2000	0-100	58-88	71±11	CRM	0.2-μm filter at 4°C	Santinelli et al. (2002)
		300-600	43-46	44±1			
		900-1500	51-58	55±3			
		1700-3500	55-63	59±3			
	May 2005, November 2006	0-100	47-73	58±7	CRM	0.2-μm filter at 4°C	Santinelli et al. (2010)
		200-500	39-54	44±4			
		2000-3000	37-48	42±3			

Location	Period	Depth	Range	Mean ± SD	CRM	Filter/storage	Reference
Northern Adriatic Sea	June 1996	0-50	74-281	161 ± 43	n.a.	GF/F filter at −20°C	Pettine et al. (1999)
	February 1997	0-50	53-123	81 ± 16	n.a.		
	February–May 2003	Surface	60-210	114 ± 36	n.a.	GF/F filter + HgCl$_2$ at −20°C	Berto et al. (2010)
	February–June 2004	Surface	60-260	155 ± 31			
	1999-2002	0-70		79 ± 12[e] / 157 ± 69[e]	n.a.	GF/F filter + HgCl$_2$ at −20°C	Giani et al. (2005)
	1999-2003 monthly	0-26	50-194	87 ± 16[f] / 103 ± 23[f]	n.a.	GF/F filter at −20°C	De Vittor et al. (2008)
Po River	February 1995–March 1996	Surface	108-308[d]	175 ± 50[d]	n.a.	GF/F filter at −20°C	Pettine et al. (1998)
Southern Adriatic Sea	November 2006	0-100	46-91	60 ± 9	CRM	0.2-μm filter at 4°C	Santinelli et al. (2010)
		200-500	39-54	46 ± 4			
		800-1200	46-66	54 ± 6			
	September 2007	0-200	57-79	70 ± 7	CRM	0.2-μm filter at 4°C	Santinelli et al. (2012b)
		200-600	45-54	49 ± 2			
		800-1200	57-56	51 ± 2			
	January 2008	0-200	49-59	53 ± 3			
		200-600	45-54	50 ± 2			
		800-1200	50-60	54 ± 3			

(Continued)

TABLE 13.1   DOC Concentrations Reported in the Literature for Different Areas of the Mediterranean Sea and Main Rivers—cont'd

Study Area	Sampling Period	Depth (m)	DOC (μM) Range	DOC (μM) Mean ± std	DOC Reference Material[a]	Sample Treatment	Reference
Ionian Sea	January 1999	0-100	50-73	62 ± 5	CRM	0.2-μm filter at 4 °C	Seritti et al. (2003)
		200-800	34-62	46 ± 8			
		1000-5000	31-48	42 ± 5			
	April 2002	0-100	54-82	66 ± 6	CRM	0.2-μm filter at 4 °C	Santinelli et al. (2010)
		200-500	47-63	53 ± 4			
		3000-5000	52-72	63 ± 5			
	March 2008	0-100	50-73	58 ± 6			
		200-500	37-51	45 ± 4			
		3000-4000	37-50	45 ± 4			
Marmara Sea	August 2008	0-20	116-217	172 ± 37	CRM	0.45-μm filter + HCl; at 4 °C	Zeri et al. (2014)
		20-1200	52-73	61 ± 5			
North Aegean Sea	September 1997	0-100	52-128	82 ± 11[b]	DAW	No filtration + HgCl$_2$	Sempéré et al. (2002)
		100-970	47-56				
		20-1200	58-66	63 ± 4			
South Aegean Sea	September 1997	0-100	55-87	70 ± 3[b]	DAW	No filtration + HgCl$_2$	Sempéré et al. (2002)
		100-1500	48-66				
South East Levantine Basin	May 2002	Photic zone	60-110		CRM	GF/F filter + H$_3$PO$_4$ at 4 °C	Krom et al. (2005)
		500-3000	40-60	45 ± 5			

East Med	September 1999	Mixed layer	62-74[b]	66 ± 5[b]	CRM	No filtration + H_3PO_4	Santinelli et al. (2012a)
		200-500	39-55				
		1000-2500	36-41				
	June 2007	0-100	54-75	64 ± 5	CRM	0.2-μm filter at 4°C	Santinelli et al. (2010)
		200-500	40-52	46 ± 4			
		2750-4000	34-47	41 ± 4			
	June-July 2008	0-100	56-72		CRM	GF/F filter + H_3PO_4	Pujo-Pay et al. (2011)
		200-800	37-59				
		1000-3000	38-42	41 ± 1			

DOC range and/or mean concentration ± std are shown. DOC analyses were carried out by using a Shimadzu TOC analyzer, except when indicated.

[a]*The use of DOC reference material is indicated: n.a., not available; DW, deep oceanic water, DAW, deep Atlantic Water, DPW, deep Pacific Water supplied by J. Sharp; IE, Inter-comparison Exercise (Sharp et al., 2002); CRM, Consensus Reference Material supplied by D.A. Hansell (Hansell, 2005).*

[b]*Values calculated from 0 to 100-m DOC stocks.*

[c]*Values obtained using a homemade automatic UV/persulfate analyzer.*

[d]*Values obtained using a Carlo Erba 480 analyzer.*

[e]*Minimum and maximum of the monthly averages.*

[f]*Minimum and maximum of the annual averages.*

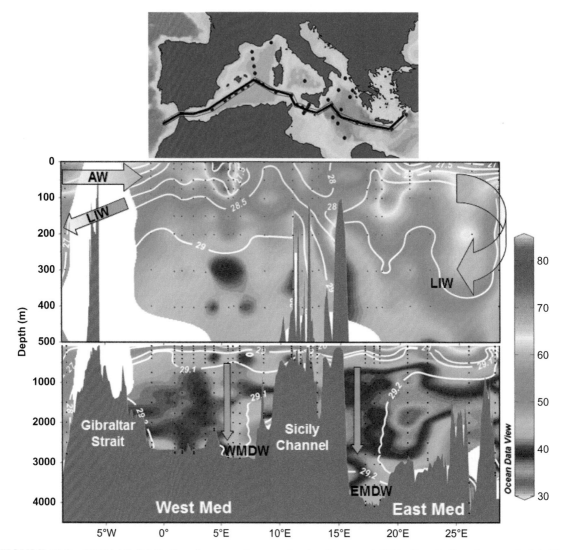

FIGURE 13.3　DOC (μM) distribution along a southern section (see inset map). White lines indicate isopycnals (σ_θ). The highest values (65-85 μM) are observed in the core of the anticyclonic eddy at 5° E and in the eastern (>18° E) basin. Data collected in May/June 2007 and March/April 2008. *Images created using Ocean Data View (R. Schlitzer, http://odv.awi.de).*

activity, and (iii) stratification/destratification pattern. DOC shows high values in the riverine plumes (Cauwet et al., 1997; De Vittor et al., 2008; Seritti et al., 1998), in the mixed layer of high stratified waters (Avril, 2002; Santinelli et al., 2013), and in the core of anticyclonic eddies (Santinelli et al., 2008) (Table 13.1). The lowest surface values are observed where the water column is completely mixed due to winter

convection (e.g., Gulf of Lions, Ligurian Sea, southern Adriatic Sea) (Avril, 2002; Santinelli et al., 2013) or to the occurrence of cyclonic eddies. In the areas characterized by an evident cycle of stratification, seasonal variations in DOC values and distributions have been observed (Avril, 2002; Santinelli et al., 2013). Where stratification is strong throughout the year, variations in surface DOC are mainly due to the mesoscale activity (Santinelli et al., 2013). It is worth noting that DOC always accumulates above the pycnocline, and the amount of accumulated DOC is a function of the strength of stratification. Finally, an eastward increase in DOC is usually observed, with values >65 µM in the easternmost part of the basin (Figure 13.3) (Pujo-Pay et al., 2011; Santinelli et al., 2012a).

2 Intermediate Layer

The water mass circulating at intermediate depths is the LIW, easily recognizable throughout the basin by its high salinity. The basin-scale DOC and DO distributions result from the merger of various data sets collected in the salinity maximum (Figure 13.4) (Santinelli et al., 2010, 2012a).

Following the flow of LIW, both DOC and DO concentrations decreased from 64-67 µM C and 230-255 µM O_2 in the east, at its formation site, to 41-45 µM C and 180-200 µM O_2 at the Sicily Channel. In the West Med, LIW shows minima in both DOC (36-44 µM) and DO (155-200 µM) (Figures 13.3 and 13.4). The same pattern has been observed using DOC data collected in the whole basin in September 1999 (Santinelli et al., 2012a) as well as DOC data collected over 8 years (2001-2008) during ten oceanographic cruises (Santinelli et al., 2010). The good relationship between DOC and apparent oxygen utilization (AOU = oxygen at saturation minus observed oxygen concentration) expressed as C equivalent (AOU-C_{eq} = AOU × 0.72, Doval and Hansell, 2000) suggests that DOC

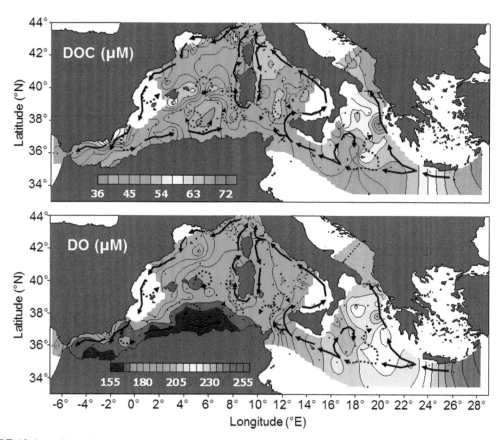

FIGURE 13.4   DOC and dissolved oxygen (DO) in the core of the LIW. Solid arrows indicate the main LIW circulation while the dashed arrows indicate mesoscale activity. Data from 10 cruises between 2001 and 2008 were merged to create these maps (Santinelli et al., 2010).

removal in the core of LIW could be due to microbial mineralization. The multiple regression between DOC, AOU_Ceq, and salinity allows for the removal of the effect of mixing from the relationship. It shows a slope of −0.38, −0.49, indicating that during LIW route from the Levantine Basin to the Tyrrhenian Sea, 38-49% of oxygen consumption is due to DOC mineralization (Santinelli et al., 2010, 2012a). These values are slightly higher than those reported for the North Atlantic Ocean (5-14%, Carlson et al., 2010), the South Pacific (15%, 34%), and the Indian Ocean (13%, 31%) (Doval and Hansell, 2000). If the DOC removal is in fact biotic, then it may be inferred that the semi-labile fraction of DOC is exported and mineralized while the LIW travels from the Levantine Basin to the Sicily Channel, with mineralization rates of 2.2-2.3 μM DOC year^{-1} (Santinelli et al., 2010, 2012a).

3 Deep Layer

DOC in the deep Med Sea shows some intriguing features. Between 500 m and the bottom, very low DOC concentrations (34-44 μM) are observed in both the East and West Med (Figure 13.3). These values are the lowest observed in the open ocean, thus consisting of the refractory DOC fraction, with a lifetime of 16,000 years (Hansell, 2013). The DOC lifetime is significantly higher than the residence time of WMDW (10-20 years; Andrie and Merlivat, 1988) and EMDW (126 years; Schlitzer et al., 1991). Another peculiar aspect of the Med Sea is that DOC is somewhat elevated in the near-bottom waters (Figure 13.3), with highly variable concentrations in the areas subject to deep-water formation (Santinelli et al., 2010); where, an increase in DO concentrations can also be observed (Figure 13.5).

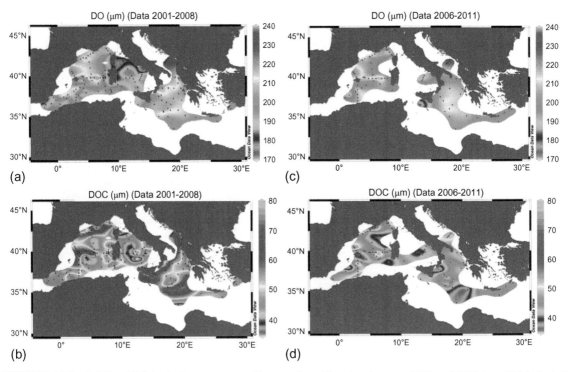

FIGURE 13.5   DOC and DO in the bottom waters. Data are from 10 cruises between 2001 and 2008 (a and b) (adapted from Santinelli et al., 2010) and from 6 cruises between 2006 and 2011 (c and d). High DOC concentrations, due to anomalous deep-water formation events in April 2002 in the Adriatic Sea and in May 2005 in the West Med, are evident. *Images created using Ocean Data View (R. Schlitzer, http://odv.awi.de).*

Bottom-water DOC values range between 35 and 60 µM, with the highest concentrations in the southern Adriatic Sea, as well as the Ligurian Sea and Gulf of Lions, and the lowest ones in the Tyrrhenian and Ionian Seas (Figure 13.5d). Bottom-water DOC is also highly variable. Particularly strong deep-water formation events export an amount of DOC markedly higher than in normal winters. These events may explain the exceptionally high DOC (>60 µM) and DO (220-240 µM) concentrations observed in April 2002 in the southern Adriatic and Ionian Sea (Manca et al., 2006), and in April 2005 in the Gulf of Lions (Canals et al., 2006) (Figure 13.5b). Bottom-water DOC values were lower after 2006 (Figure 13.5d), indicating its fast removal by mixing and/or consumption.

DOC mineralization rates are difficult to estimate because of the complex deep-water circulation and because of mixing between the recently ventilated deep waters and surrounding, older waters. However, it has been possible to estimate DOC mineralization rates in the South Adriatic Pit (Figure 13.1). Bottom topography prevents the waters below 1000 m to flow out of the pit unless there is an input of the NAdDW, mainly occurring in winter. This area represents a natural laboratory to study deep-DOC mineralization rates in natural conditions. DOC variability, observed during repeated cruises, suggest a DOC mineralization rate of 14.4 µM year^{-1} (Santinelli et al., 2010).

The high DOC concentrations together with the high DOC removal rates suggest that the fraction of DOC exported to depth is semi-labile. This observation is further confirmed by deep heterotrophic metabolism (Azzaro et al., 2012; Boutrif et al., 2011). It is noteworthy that the lowest DOC values are located in areas affected by volcanic activity (the Tyrrhenian Sea) or by the occurrence of hyperhaline anoxic basins (Ionian Sea) (Figure 13.1). Even though more data and an in-depth analysis are necessary, the high temperature, the high prokaryotic diversity, and the peculiar chemical conditions of these areas could play a role in DOC removal as also hypothesized for hydrothermal vents (Hansell and Carlson, 2013).

B DOC Temporal Variability

Observations of DOC temporal variability in the Med Sea are scarce and limited to a few areas (Avril, 2002; Copin-Montégut and Avril, 1993; Santinelli et al., 2013). Though the basin is located at temperate latitudes and it is characterized by an evident seasonal cycle, DOC temporal variability shows some interesting differences depending on the area taken into consideration. There are areas (e.g., the Ligurian and the Adriatic Seas) where DOC demonstrates seasonality, similar to that observed in temperate regions of the oceans (Hansell, 2002); and other areas, where stratification never breaks, winter mixing rarely exceeds 150 m (e.g., the Tyrrhenian Sea) and DOC does not show a clear seasonality.

At the Dyfamed station (43° 25′ N, 07° 52′ E, Ligurian Sea), where the oldest DOC time series is available (Avril, 2002; Copin-Montégut and Avril, 1993), DOC vertical profiles, collected monthly from January 1991 to November 1994, show seasonality. In the upper 200 m, DOC accumulates from spring to autumn reaching concentrations of 80-95 µM, while DOC export can be observed in winter, when values higher than 60 µM occur down to 1200 m (Avril, 2002).

In the Adriatic Sea, the highest concentrations are observed in the mixed layer at the end of summer and in autumn. A surface minimum can be observed in winter, together with an increase of DOC values in the deep waters, because of the vertical mixing (Figure 13.6). This seasonal cycle is linked to the stratification pattern. A 7-13 µM difference is observed in the 0-50-m layer between winter (52-58 µM) and autumn (65 µM). A marked interannual variability can also be observed, perhaps linked to the stratification pattern. As an example, the exceptionally high temperatures that affected Europe in fall 2006 and winter 2007 (Luterbacher et al., 2007) could have been responsible for weaker winter mixing, resulting

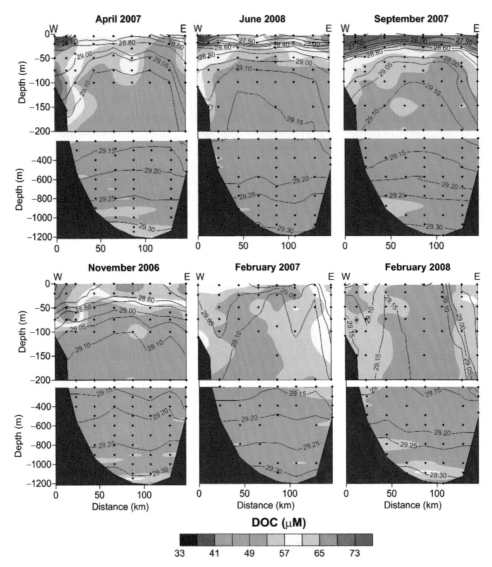

FIGURE 13.6   Temporal variability of DOC in the southern Adriatic Sea (red line in Figure 13.1) during 6 cruises, superimposed by isopycnals. The graphs are ordered by time of year in order to highlight seasonal variability (Santinelli et al., 2013).

in surface DOC values 6 μM higher (58 μM) in February 2007 than the following winter (52 μM) (Figure 13.6) (Santinelli et al., 2013).

In the Tyrrhenian Sea, the surface layer seems isolated from underlying water (absence of strong vertical mixing), with DOC elevated at the surface throughout the year. Variations of 3-5 μM are observed between winter (57-59 μM)

and summer (62 μM), while below the pycnocline DOC concentration remains low (36-40 μM) (Santinelli et al., 2013) (Figure 13.7). This behavior is similar to that observed in subtropical areas (Hansell, 2002). Interannual variability is observed as well, but the DOC temporal variability is likely due to the mesoscale activity. In April 2007, when the occurrence of an anticyclonic

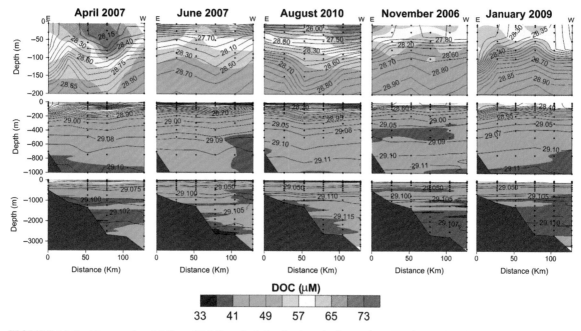

FIGURE 13.7 Temporal variability of DOC vertical distributions in the southern Tyrrhenian Sea (black line in Figure 13.1) during 5 cruises, superimposed by isopycnals. The graphs are ordered by time of year in order to highlight seasonal variability (Santinelli et al., 2013).

eddy was inferred by depressed isopycnals, DOC showed high concentrations (65 μM) down to the depth influenced by the eddy (Figure 13.7).

III THE ROLE OF DOC IN CARBON EXPORT

Since the 1970s (McCave, 1975), particulate organic carbon (POC) has been considered the main player in C export. POC fluxes occur throughout the year, with strong spatial and temporal variability. However, in oligotrophic areas, POC represents only a small fraction of the total organic carbon and most of the sinking POC is transformed to DOC and mineralized by microbes in the upper 1000 m of the water column (Buesseler and Boyd, 2009; Martin et al., 1987).

The importance of DOC in C export was recognized later (Carlson et al., 1994). It plays an important role in carbon export down to meso- and bathypelagic layers by two main processes:

(i) convective, downward mixing of the DOC accumulated in the mixed layer (Carlson et al., 1994; Copin-Montégut and Avril, 1993; Hansell, 2002; Santinelli et al., 2013; Sohrin and Sempéré, 2005) and (ii) transport with deep-water formation (Carlson et al., 2010; Santinelli et al., 2010).

These processes can be easily studied in the Med Sea, which hosts both areas with seasonal cycles of vertical stratification and areas where deep-water formation occurs, with a subsequent increase of DOC in deep waters.

A Winter Mixing

The seasonal cycle of DOC in areas characterized by clear stratification/destratification cycles shows DOC accumulation in the mixed layer during high stratification periods and low concentrations in winter (Figure 13.6). DOC accumulation is usually due to a decoupling between production and removal processes. The DOC produced is not consumed on a short temporal scale (weeks-months) by heterotrophic prokaryotes so its concentration increases. Currently, it is unknown if DOC accumulates because of its molecular characteristics, the inability of the above-pycnocline layer microbial community to consume the accumulated DOC, and/or the lack of micronutrients, vitamins and/or other substances necessary for bacterial growth (Carlson, 2002). Once exported to the mesopelagic layer, DOC can be quickly consumed by mesopelagic prokaryotes (mineralization rates of 1.8-13.2 μM C year^{-1}) (Santinelli et al., 2013). In agreement with oceanic observations (Hansell, 2013), we assume that the accumulated DOC is semi-labile (Figure 13.8). The export of semi-labile DOC by winter mixing sequesters carbon in the mesopelagic layer for short periods (months-years). The efficiency of this process depends on the amount of semi-labile DOC that accumulates above the pycnocline during stratification periods and by the intensity and depth of winter mixing.

Temporal data sets in fixed locations are necessary to calculate the DOC export by winter mixing. The first estimate of 14.8 gC m^{-2} year^{-1} was obtained at the Dyfamed station (Copin-Montégut and Avril, 1993). A more recent estimate indicates a DOC flux to >100 m of 10.9 gC m^{-2} year^{-1} (Avril, 2002). In the southern Tyrrhenian and Adriatic Seas, winter mixing may export below 50 m 3.24 and 15.36 gC m^{-2} year^{-1}, respectively. These fluxes are similar to those observed in the Sargasso Sea (4.8-16.8 gC m^{-2} year^{-1}) (Carlson et al., 1994; Hansell and Carlson, 2001). It is noteworthy that they are greater or similar than the mean export of POC measured in the same areas: Gulf of Lions, 16 gC m^{-2} year^{-1}; southern Tyrrhenian Sea, 2.4 gC m^{-2} year^{-1} (Moutin and Raimbault, 2002); and southern Adriatic Sea, 3.6 and 4.8 gC m^{-2} year^{-1} in 1997-1998 and 2006-2007, respectively (Boldrin et al., 2002; Turchetto et al., 2012).

B Deep-Water Formation

The large spatial and temporal variability of deep-water DOC concentrations, the high DOC and oxygen concentrations in the areas impacted by deep-water formation (Figure 13.5) (Santinelli et al., 2010) and the respiration rates higher in the WMDW than in open ocean waters (Christensen et al., 1989) are evidence that deep-water formation is capable of carrying a great amount of DOC to depths >1000 m. This geographically localized process determines long-term carbon sequestration. Deep-water formation occurs almost each winter and it lasts over several weeks. The fast transport of DOC to depth and its quick removal suggest that semi-labile DOC is exported (Santinelli et al., 2010).

The amount of DOC exported by deep-water formation is 0.85-1.19 × 10^{12} gC year^{-1} in the Adriatic Sea (Santinelli et al., 2013) and 0.53-18.16 × 10^{12} gC year^{-1} in the Gulf of Lions (Santinelli et al., 2010). These values are estimated by multiplying the deep-water formation rates (0.07 Sv for the NAdDW and 0.14-1.2 Sv for the WMDW) by the difference between DOC concentration at the surface and in the deep waters (32-45 μM in the Adriatic Sea and 10-40 μM in the Gulf of Lions). By the same process, the LIW formation may carry 8.3-12.5 × 10^{12} gC year^{-1} from the surface to intermediate layer. These values are calculated by multiplying LIW formation rates (1.0-1.5 Sv) by the difference between DOC concentration at the surface and in the intermediate layer (22 μM). DOC exported to 200-500 m is responsible for short-term C sequestration, but when LIW arrives in the Atlantic Ocean, it sinks to ~1000 m and contributes to North Atlantic deep-water formation; as a consequence, the DOC occurring in this water mass can be sequestered on a long temporal scale.

IV DOC INVENTORY AND FLUXES

In the past decade, the number of DOC data available for the Med Sea has strongly increased, though important gaps exist in areas such as the Levantine Basin, the northern Tyrrhenian Sea, the Gibraltar Strait, and the Aegean Sea. The large data set available allows re-assessment of DOC stocks in the different depth layers of two subbasins and of its fluxes at the straits (Copin-Montégut, 1993; Sempéré et al., 2000). Even though more data would have been necessary to have robust estimates of external inputs, in particular from the atmosphere, a rough estimate of DOC fluxes due to river runoff and atmospheric depositions is reported here.

A DOC Stocks

DOC stocks are calculated multiplying the mean DOC concentration in the surface (0-100 m), intermediate (100-500 m) and deep layers (between 500 m and the bottom) by the volume of the three layers in each subbasin considering an average depth of 1500 m (Table 13.2). The mean concentrations were calculated using data collected in June 2007 and April 2008 (Figure 13.3). These values are assumed to be representative of DOC concentrations in the Med Sea since they are comparable to other recently published data (Pujo-Pay et al., 2011; Santinelli et al., 2012a,b). However, surface DOC stocks could be underestimated due to the sampling periods, and DOC deep stocks could be slightly underestimated since they are calculated by multiplying the average DOC below 500 m by the volume of the two subbasins. In this way, they do not take into account the slightly higher DOC concentrations in bottom waters.

The mean concentration observed in the 100-500 m layer is higher in the East Med (48 µM) than in the West Med (45 µM) because of DOC import with LIW formation (Figure 13.3). The total Mediterranean DOC stock is 1980×10^{12} g C (671×10^{12} g C in the West Med and 1309×10^{12} g C in the East Med) (Table 13.2), about 1.5 times lower than that reported by Sempéré et al. (2000). The difference is explained by the high mean DOC concentration in water below 200 m (54 µM) reported in the past as well as by the larger volume of Med Sea taken into consideration in that paper.

B DOC Fluxes at the Mediterranean Straits

DOC fluxes at the Gibraltar Strait and Sicily Channel are calculated by multiplying the water fluxes by their DOC concentrations. In order to obtain a more accurate estimate, the range of both water flows and DOC concentrations are taken into account for each water mass (Table 13.3).

The net imports (west to east) are 7.7-9.7×10^{12} g C year^{-1} at the Gibraltar Strait and 4.6-7.3×10^{12} g C year^{-1} at the Sicily Channel. DOC input at the Gibraltar Strait is lower than that estimated by Copin-Montégut (1993) (15×10^{12} g C year^{-1}), explained by the very high DOC concentrations used in that paper. The net fluxes are to the east, from the Atlantic Ocean into

TABLE 13.2   DOC Inventory in the West and East Med

Depth Layer (m)	Western Mediterranean				Eastern Mediterranean			
	Volume (10^5 km³)	Mean ± std DOC (µM)	Sample (n.)	Stock (10^{12} g)	Volume (10^5 km³)	Mean ± std DOC (µM)	Sample (n.)	Stock (10^{12} g)
0-100	0.85	58 ± 7	85	59	1.65	59 ± 6	112	117
100-500	3.40	45 ± 3	76	184	6.60	48 ± 5	87	380
500-1500	8.50	42 ± 3	146	428	16.5	41 ± 4	177	812
Total DOC stock (10^{12} g)	671				1309			

TABLE 13.3 DOC Fluxes at the Gibraltar Strait and Sicily Channel

		Gibraltar Strait			Sicily Channel		
		DOC (μM)	Water Flux ($10^6\,m^3\,s^{-1}$)	DOC Flux ($10^{12}\,gC\,year^{-1}$)	DOC (μM)	Water Flux ($10^6\,m^3\,s^{-1}$)	DOC Flux ($10^{12}\,gC\,year^{-1}$)
Inflow (AW)	Low	59[a]	1.28[b,c]	28.6	54[d]	1.00[e]	20.4
	High	62[f]	1.68[b,c]	39.4	59[g]	1.40[e]	31.3
Outflow (LIW)	Low	46[a]	1.20[b,c]	20.9	44[g]	0.95[e]	15.8
	High	49[f]	1.60[b,c]	29.7	47[d]	1.35[e]	24.0
Net DOC inflow ($10^{12}\,gC\,year^{-1}$)		7.7-9.7			4.6-7.3		

[a]Mean DOC concentrations in the Atlantic inflow and Mediterranean outflow from Dafner et al. (2001b, Figure 9).

[b]Hopkins (1999).

[c]Bethoux and Gentili (1999).

[d]Mean DOC concentrations in the AW and LIW calculated using data collected in March 2008 (Santinelli, unpublished).

[e]Astraldi et al. (1996).

[f]Mean DOC concentrations in the Atlantic inflow and Mediterranean outflow calculated by using data collected in October 2004 (Santinelli, unpublished).

[g]Mean DOC concentrations in the AW and LIW calculated using data collected in September 2008 (Santinelli, unpublished).

the Med Sea and from the West to the East Med. The 27-40% reduction of DOC flux from Gibraltar Strait to Sicily Channel is mainly due to a reduction of water transport although DOC concentrations also diminish (Table 13.3). From these data, net DOC input into the West Med (inflow at the Gibraltar Strait in excess of outflow at the Sicily Channel) is 2.4-3.1 × 10^{12} gC year^{-1}. Input from the Black Sea through the Dardanelles Straits may account for 1.32-1.44 × 10^{12} g DOC year^{-1}, but its influence is limited to the Aegean Sea (Polat and Tugrul, 1996).

C River Runoff

Because of the extensive coastline, the high human population density and the modest dimensions of the basin, terrestrial inputs strongly influence DOC distributions in the Med Sea. Estimates of river discharges (Ludwig et al., 2009) combined with the DOC concentrations available for some rivers allow estimates of the riverine DOC input (Table 13.4). DOC concentrations found in river mouths are high and highly variable, ranging between 92 and 366 μM.

The major rivers (with a mean discharge higher than 10 km^3 year^{-1}) are the Rhone, Po, and Ebro, with DOC concentrations of 134-218 μM. DOC concentrations in the Rhone and Po rivers are from multiyear observations (Table 13.1) (Panagiotopoulos et al., 2012; Pettine et al., 1998; Sempéré et al., 2000), leading greater confidence in the estimates. In contrast, DOC data are taken sporadically in some minor rivers (Table 13.4). We overcome the lack of DOC data in some rivers by using the mean DOC concentration computed by all the available data and multiplying this value by the discharge of the rivers without DOC data. From this calculation, the total river input accounts for 0.644-0.712 × 10^{12} gC year^{-1} (0.235-0.303 × 10^{12} gC year^{-1} into the West Med and 0.409 × 10^{12} gC year^{-1} into the East Med). This summed value is roughly 21-24% of that reported by Copin-Montégut (1993). The discrepancy is due to the higher discharge value used in that paper and to the inclusion of POC in the calculation.

TABLE 13.4 DOC Fluxes from Rivers into the Mediterranean Sea (Discharge and Drainage Basin Area from Ludwing et al., 2009)

River	Discharge (km³ year⁻¹)	Drainage Basin Area (10³ km²)	DOC (µM)	DOC Flux (10⁹ gC year⁻¹)
Western Mediterranean Sea				
Rhone	54.3	95.6	134[a]-217[b]	87.28-141.33
Ebro	13.1	84.2	218[c]	34.32
Tevere	6.8	16.5	227[d]	18.62
Garigliano	3.8	5.0	183[d]	8.31
Arno	2.6	8.2	319[e]	10.09
Volturno	2.6	5.5	366[d]	11.36
Sele	2.2	1.2	287[d]	7.49
Aude	1.4	5.0	275[f]	4.56
Herault	1.3	2.6	208[f]	3.18
Var	1.2	1.8	133[f]	1.90
Orb	0.8	1.3	225[f]	2.29
Argens	0.4	2.6	200[f]	0.91
Tet	0.3	1.4	292[f]	1.15
Total	91	231	176-225[g]	191-246
Rivers without DOC data	21	-	176-225[g]	44-57[h]
Eastern Mediterranean Sea				
Po	49.5	70.1	175[i]	80.77
Evros	6.8	55.0	250[f]	20.40
Nile	6.0	2870.0	292[f]	21.00
Axios	4.9	24.7	117[f]	6.86
Strymon	2.6	16.5	175[f]	5.44
Acheloos	1.6	1.3	92[f]	1.78
Aliakmon	1.6	5.0	100[f]	1.89
Nestos	1.2	4.4	142[f]	2.12
Total	74	3047	157[g]	140
Rivers without DOC data	143	-	157[g]	269[h]
Mediterranean Sea				
Total	329			644-712

[a]Panagiotopoulos et al. (2012) (data refer to May 2007-June 2009).

[b]Sempéré et al. (2000) (data refer to June 1994-May 1995).

[c]Gómez-Gutiérrez et al. (2006).

[d]Data collected in November 2010 (Santinelli, unpublished).

[e]Mean DOC concentration calculated by using (i) data seasonally collected in 2012 (Santinelli, unpublished), (ii) data from Vignudelli et al. (2004), and (iii) data from Seritti et al. (1998).

[f]UNEP/MAP/MED POL, 2003 technical report.

[g]The mean concentration is computed as (DOC flux)/[(discharge) $\times 12 \times 10^6$].

[h]Computed by using the DOC mean concentration of the rivers where DOC data are available.

[i]Pettine et al. (1998) (data refer to February 1995-March 1996).

The total DOC input from the rivers divided by the entire volume of the basin ($3.75 \times 10^6 \, km^3$) suggests a total flux of 0.014-$0.016 \, \mu M \, C \, year^{-1}$. This value seems low, but if we assume no removal for terrestrial DOC, river input would have increased DOC concentration of 14-$16 \, \mu M$ every 1000 years. This simple calculation indicates that removal mechanisms of riverine DOC occur.

D Atmospheric Input

An important role could be played by atmospheric input of DOM to the Med Sea, but data on such are scarce. A single paper reports direct measurements of DOC atmospheric fluxes at Cap Ferrat during 2006 (0.04-$1.20 \, mmol \, C \, m^{-2} \, day^{-1}$) (Pulido-Villena et al., 2008); the highest values are associated with Saharan dust deposition events. Assuming this range valid for the whole basin, a total input of 0.4-$13.1 \times 10^{12} \, gC \, year^{-1}$ can be estimated. Similar values (1.97-$4.60 \times 10^{12} \, gC \, year^{-1}$) are obtained assuming that the input of DOC

from the atmosphere is roughly the same than that measured in the southern Spain (0.18-$0.42 \, mmol \, C \, m^{-2} \, day^{-1}$) (De Vicente et al., 2012). Direct measurements of TOC in rainwater, at the Crete Island (Eastern Mediterranean), indicate that wet deposition can account for an input of $1.5 \times 10^{12} \, gC \, year^{-1}$ (Economou and Mihalopoulos, 2002). Recently, new information about dissolved organic nitrogen (DON) and phosphorus (DOP) deposition to the Med Sea have been reported (Markaki et al., 2010). Using the available DON data and assuming a DOC/DON ratio in atmospheric DOM ranging between 10 and 18, typical ranges of Mediterranean DOM (Table 13.5), we obtain an estimate of atmospheric DOC input in the same range (5-$9 \times 10^{12} \, gC \, year^{-1}$). These rough estimates suggest that DOC input from the atmosphere could be up to 18 times higher than river inputs. This source was underestimated in previous papers (Copin-Montégut, 1993); because of its importance it should be quantified by direct measurements.

TABLE 13.5   DON and DOP Values and DOC:DON:DOP Ratios Reported for the Mediterranean Sea

Area	Sampling Period	Depth (m)	DON (µM)	DOP (µM)	DOC:DON	DOC:DOP	DON:DOP	Reference
Open Ocean (Eastern North Pacific; Southern Ocean)		0-100	2.9-4.5	0.15-0.23	11-17	222-338	17-27	Loh and Bauer (2000)
		> 1000	1.7-4.4	0.06-0.17	9-16	237-688	16-44	
Balearic Sea	June 1995	0-100	3.3-6.2 (4.5 ± 0.3)		15-16			Doval et al. (1999)
		150-2000	2.8-4.7 (4.0 ± 0.2)		13-16			
Gulf of Lions	September 1984	Mixed layer	4-4.2	0.08	16-17	920-970	50-52	Aminot and Kérouel (2004)
		200-600 m	3.0	0.04	15-16	1100-1200	75	
		800-1500 m	2.7	0.03-0.04	17	1100-1800	67-90	
West Med	May-June 1996	0-100 m	4.6 ± 0.6	0.08 ± 0.02			57	Moutin and Raimbault (2002)
		300-500 m	3.1 ± 0.3	0.06 ± 0.01			52	
		>1000 m	2.8 ± 0.1	0.03 ± 0.02			93	

TABLE 13.5   DON and DOP Values and DOC:DON:DOP Ratios Reported for the Mediterranean Sea—cont'd

Area	Sampling Period	Depth (m)	DON (µM)	DOP (µM)	DOC:DON	DOC:DOP	DON:DOP	Reference
	June-July 2008	0-100 m	4.1-5.5	0.03-0.09	12.5	1050	84	Pujo-Pay et al. (2011)
		200-800 m	2.5-4.2		12.4	3100	250	
		1000-2900 m	2.9-4.3	0-0.03	12.5	5000	400	
Southern Adriatic Sea	September 2007	0-200 m	2.5-6.9	0.03-0.07	16 ± 3	1411 ± 343	86 ± 16	Santinelli et al. (2012b)
		200-600 m	2.8-5.2	0.02-0.06	13 ± 2	1279 ± 396	97 ± 32	
		800-bottom	3.4-6.2	0.03-0.08	11 ± 2	993 ± 326	85 ± 22	
	January 2008	0-200 m	2.3-7.2	0.02-0.08	14 ± 3	1189 ± 333	88 ± 26	
		200-600 m	1.8-5.3	0.03-0.06	14 ± 4	1107 ± 265	83 ± 15	
		800-bottom	2.9-4.6	0.02-0.04	15 ± 2	1693 ± 566	108 ± 46	
Otranto Strait		0-200	4-6[a]	0.02-0.06[a]			100-200	Civitarese et al. (1998)
		200-1200	5-7[a]	0.04-0.08[a]			87-125	
East Med	May-June 1996	0-100 m	4.5 ± 0.6	0.06 ± 0.03			75	Moutin and Raimbault, (2002)
		300-500 m	3.7 ± 0.4	0.05 ± 0.02			74	
		>1000 m	3.3 ± 0.3	0.05 ± 0.02			66	
	June-July 2008	0-100 m	3.5-6.3	0.02-0.10[a]	13.0	1560	120	Pujo-Pay et al. (2011)
		200-800 m	2.1-4.5	-	12.4	3100	250	
		1000-3000 m	2.1-4.0	0-0.07	12.1	3150	260	
North Aegean Sea	August 2008	0-20	3.1-5.7	-	17-23	-	-	Zeri et al. (2014)
		20-1200	2.7-4.5	-	15-25	-	-	
Marmara Sea	August 2008	0-20	6.3-9.3	-	14-27	-	-	Zeri et al. (2014)
		20-1200	1.0-5.1	-	13-58	-	-	
South East Levantine Basin	May 2002	Photic zone	3-11	0.05	10-20	1200-2200	60-220	Krom et al. (2005)
		500-1200	1-2	0.04	40-30	1000-1500	25-50	

The range and/or mean ± std is reported.

[a]*Values were calculated subtracting DIN and DIP by TDN and TDP data reported by the authors.*

V DOM STOICHIOMETRY

An intriguing aspect of the Med Sea is related to DOM stoichiometry and to inorganic nutrient dynamics. New DON and DOP data have been published, leading to an improvement in the knowledge of DOM stoichiometry in the Med Sea (Pujo-Pay et al., 2011; Santinelli et al., 2012b). Mediterranean DOC and DON concentrations as well DOC:DON ratios are similar to those observed in the oceans (Table 13.5). Surface C:N:P ratios (1050-1560:84-120:1) show that Mediterranean DOM is depleted in DOP with respect to DOC and DON. In the "old" (not recently ventilated) Mediterranean deep water, DOM stoichiometry shows ratios similar to the highest values reported for the open ocean. In contrast, lower ratios have been reported for areas of deep-water formation such as the southern Adriatic Sea (Table 13.5).

Despite the similarity in DON and DOP values and ratios, Mediterranean deep-water inorganic nutrient concentrations are about 2.5-10 times lower than the deep-ocean ones (Ribera d'Alcalà et al., 2003). This observation is difficult to explain and leads to DON/TDN (Total Dissolved Nitrogen) and DOP/TDP (Total Dissolved Phosphorous) percentages higher than in deep oceans, in particular in recently ventilated deep waters (e.g., Adriatic Sea), where DON can represent 40-60% of TDN and DOP 20-40% of TDP. These data suggest that a large fraction of both N and P is organic in recently ventilated deep waters and that in the whole Med Sea a larger fraction of N is organic than in deep oceanic waters.

VI DOC DYNAMICS IN THE MED SEA, A COMPARISON WITH THE OCEANS

Recently, the major fractions in oceanic DOC pool have been quantitatively and qualitatively characterized and a new nomenclature has been proposed (Hansell, 2013). Most of DOC in the oceans is recalcitrant (i.e., it can accumulate), representing a dynamically stable reservoir of carbon that under circumstances, not completely understood, may be removed by mineralization and/or abiotic processes. The labile DOC (LDOC) is the fraction that fuels the microbial loop and does not, by definition, accumulate. At least four fractions, defined by their lifetimes, occur in the recalcitrant DOC: semi-labile DOC (SLDOC, lifetime ~1.5 years), semi-refractory DOC (SRDOC, lifetime ~20 years), refractory DOC (RDOC, lifetime 16,000 years), and ultra-refractory DOC

(URDOC, lifetime ~40,000 years). With the knowledge available today, the removal of a fraction is the only way to demonstrate its presence and to quantify it (Hansell, 2013).

Based on the characteristics of these geochemically defined fractions, we can determine their contributions to Mediterranean DOC pool, as done for the Sargasso and Ross Seas (Figure 5 of Hansell, 2013). DOC vertical profiles were studied in the southern Tyrrhenian and Adriatic Seas during three periods, characterized by different extents of vertical stratification (Figure 13.8). A strong similarity is observed between the Tyrrhenian and Sargasso Sea.

SLDOC is the fraction responsible for seasonal variations. We assume that this fraction may account for the difference in DOC concentration between summer (high stratification) and winter (minimum of stratification) (13 and $23\,\mu M$ in the southern Tyrrhenian and Adriatic Seas, respectively) (Figure 13.8). This is the fraction that can be exported in other areas by horizontal advection and to depths $>1000\,m$ by deep-water formation. The occurrence of $\sim 10\,\mu M$ SLDOC at $>800\,m$ in the Adriatic Sea (Figure 13.8) represents a peculiarity of the Med Sea because of the frequency (every winter) and the duration (some weeks) of deep-water formation. The high and highly variable deep DOC concentrations as well as the high mineralization rates in the AdDW ($14.4\,\mu M\,C\,year^{-1}$, Santinelli et al., 2010) suggest the export of SLDOC to great depth. Given this mineralization rate, the SLDOC in deep waters has a lifetime lower than 1 year and it represents a significant source of energy for deep metabolism. This great depth of export is the biggest difference in DOC dynamics between the Med Sea and the open oceans.

DOC concentration below $1000\,m$ in the Tyrrhenian Sea is almost constant ($\sim 40\,\mu M$), likely representing RDOC. The Tyrrhenian Sea is stratified also in winter (Figure 13.8, January 2009), as a consequence SRDOC may accumulate and occur down to $1000\,m$, as in the Sargasso Sea. To explain

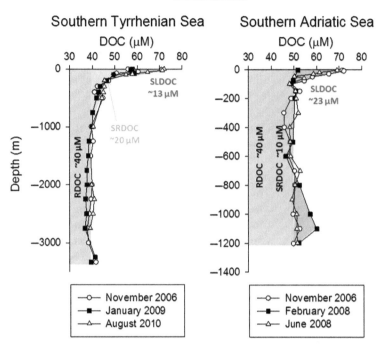

FIGURE 13.8   DOC vertical profiles in the southern Adriatic and Tyrrhenian Seas in three periods characterized by different extents of stratification. The hypothetical contribution of the DOC recalcitrant fractions is shown with different colors according to Hansell (2013). SLDOC, semi-labile DOC; SRDOC, semi-refractory DOC; RDOC, refractory DOC.

DOC vertical profiles in the Adriatic Sea, we hypothesize that the RDOC has the same concentration in the whole Med Sea, so SRDOC may explain the difference between the minimum (50 μM) observed in both surface (0-200 m) and deep water (800-1200 m) and the RDOC (40 μM), giving a SRDOC concentration of 10 μM (Figure 13.8). Since almost every winter, the water column is completely mixed down to 500-800 m, we assume that SRDOC is homogeneously distributed along the water column in the southern Adriatic Sea. The DOC values lower than 50 μM observed between 200 and 800 m as well as DOM stoichiometry (low C:N:P ratio, Table 13.5) support that not all the DOC below 100 m is refractory.

Focusing on RDOC, the less understood but most important fraction for long-term C storage, its dynamics suggests that some removal mechanisms are more efficient in the Med Sea than in the greater open ocean, as hypothesized by Hansell and Carlson (2013). In some areas of the Med Sea, concentrations <40 μM (34-39 μM) have been observed (Figures 13.3, 13.4, 13.7; Pujo-Pay et al., 2011; Santinelli et al., 2010). These concentrations are lower than the RDOC values in North Atlantic source waters, suggesting that some mechanisms of RDOC removal occurs on a time scale shorter than 100 years (the maximum estimated renewal time for Mediterranean waters).

Assuming that 40 μM represents the RDOC in the Med Sea as well as in the AW entering the Med Sea, a net inflow of RDOC at the Gibraltar Strait of 0.027 μM C year^{-1} can be estimated multiplying the net water inflow (0.08 Sv, Table 13.3) by

40 μM and then dividing it by the water volume of the Med Sea (3.75×10^6 km³). Assuming no removal or removal rates similar to those estimated for the oceans (0.003 μM C year⁻¹) (Hansell, 2013), it would take a period of ~1500 years to reach a Mediterranean RDOC concentration of 40 μM. Since the Med Sea was refilled by the AW about 5.33 million years ago (Garcia-Castellanos et al., 2009), RDOC removal mechanisms are probably occurring in the Med Sea (Hansell and Carlson, 2013), otherwise the RDOC concentration would be markedly higher.

RDOC might have been sequestered in the sediments, accounting for long-term C storage. On the other hand, it might have been mineralized, representing an important source of CO_2 to the atmosphere. The fast turnover of Mediterranean waters, determining a more frequent light exposure of the deep-water RDOC, might facilitate its mineralization.

VII SUMMARY

This chapter summarizes the present knowledge of DOC dynamics in the Med Sea, including a re-assessment of its stocks and fluxes on a basin scale.

1. The Med Sea is a good model for DOC dynamics on the global ocean scale. Mediterranean DOC not only shows concentrations and distributions similar to the oceanic ones, but its spatial and temporal dynamics are driven by the same processes and the lateral transports can be easily quantified.

2. DOC concentrations range between 31 and 128 μM in open seawaters, with values up to 366 μM in coastal areas impacted by river inputs. Its distribution is influenced by (i) terrestrial inputs, mesoscale activity, and stratification/destratification patterns in the surface layer and (ii) water mass circulation and deep-water formation in the intermediate and deep layers.

3. DOC plays a key role in carbon export to depth by winter mixing (between 3.24 and 15.36 gC m⁻² year⁻¹, depending on the region) and deep-water (1.38-19.35×10^{12} gC year⁻¹) or LIW (8.3-12.5×10^{12} gC year⁻¹) formation.

4. The total Mediterranean DOC stock is about 1980×10^{12} gC. Net flux into the Med Sea from the Atlantic Ocean is roughly 7.7-9.7×10^{12} gC year⁻¹, while total river and atmospheric input account for 0.6-0.7×10^{12} gC year⁻¹ and 0.4-13.1×10^{12} gC year⁻¹, respectively.

5. DOM stoichiometry is similar to that observed in the oceans. A large fraction of both N and P is organic in recently ventilated mediterranean deep waters.

6. The study of vertical profiles allows for an assessment of the occurrence of different recalcitrant DOC fractions. RDOC accounts for 40 μM in the entire basin and the occurrence of SLDOC in the deep water, because of deep-water formation, may explain the high mineralization rates estimated for this basin.

VIII OPEN QUESTIONS

7. What is the age of DOC in the different areas and water masses of the Med Sea?
8. Why does DOC accumulate in stratified waters and in the East Med?
9. What is the role of atmospheric DOC deposition in DOC cycle?
10. What is the impact of undersea volcanoes and hypersaline anoxic basins in DOC dynamics?
11. What are the mechanisms of RDOC removal in the Med Sea? And why are they more efficient than in the open ocean?
12. How will climate change (changes in stratification patterns, thermohaline circulation, deep-water formation rates, terrestrial inputs, light radiation, frequency, of extreme events) affect DOM dynamics in the Med Sea?

Acknowledgments

I would like to dedicate this chapter to my teacher and guide Alfredo Seritti, who taught me chemical oceanography, handing down to me the passion for DOC dynamics. I particularly thank Luciano Nannicini, who made most of the DOC analysis, and transmitted to me the laboratory skills and the passion for the work at sea. Without them all this work would have not been possible. I express my gratitude to Dennis Hansell and Craig Carlson for inviting me to write this chapter. I also thank Richard Sempéré and another anonymous reviewer for their helpful suggestions and comments that improved the earlier version of the chapter. This work was supported by the European project SESAME (GOCE-036949) and Italian Ministry of Research and University (VECTOR Project).

References

Aminot, A., Kérouel, R., 2004. Dissolved organic carbon, nitrogen and phosphorus in the N-E Atlantic and the N-W Mediterranean with particular reference to non-refractory fractions and degradation. Deep-Sea Res. I 51, 1975–1999.

Andrie, C., Merlivat, L., 1988. Tritium in the western Mediterranean Sea during 1981 Phycemed cruise. Deep-Sea Res. I 35, 247–267.

Artale, V., Calmanti, S., Malanotte-Rizzoli, P., Pisacane, G., Rupolo, V., Tsimplis, M., 2006. The Atlantic and Mediterranean Sea as Connected Systems. Mediterranean Climate Variability. Elsevier, Amsterdam, pp. 283–323.

Astraldi, M., Gasparini, G., Sparnocchia, S., Moretti, M., Sansone, E., 1996. The characteristics of the water masses and the water transport in the Sicily strait at long time scales. Bull. Inst. Oceanogr. 17, 95–116.

Avril, B., 2002. DOC dynamics in the northwestern Mediterranean Sea (DYFAMED site). Deep-Sea Res. II 49, 2163–2182.

Azzaro, M., La Ferla, R., Maimone, G., Monticelli, L.S., Zaccone, R., Civitarese, G., 2012. Prokaryotic dynamics and heterotrophic metabolism in a deep convection site of Eastern Mediterranean Sea (the Southern Adriatic Pit). Cont. Shelf Res. 44, 106–118.

Berto, D., Giani, M., Savelli, F., Centanni, E., Ferrari, C.R., Pavoni, B., 2010. Winter to spring variations of chromophoric dissolved organic matter in a temperate estuary (Po River, northern Adriatic Sea). Mar. Environ. Res. 70, 73–81.

Bethoux, J.P., Gentili, B., 1999. Functioning of the Mediterranean Sea: past and present changes related to fresh water input and climatic changes. J. Mar. Syst. 20, 33–47.

Bignami, F., Salusti, E., Schiarini, S., 1990. Observations on a bottom vein of dense water in the Southern Adriatic and Ionian Seas. J. Geophys. Res. 95N (C5), 7249–7259.

Boldrin, A., Miserocchi, S., Rabitti, S., Turchetto, M.M., Balboni, V., Socal, G., 2002. Particulate matter in the Southern Adriatic and Ionian Sea: characterisation and downward fluxes. J. Mar. Syst. 33–34, 389–410.

Boutrif, M., Garel, M., Cottrell, M.T., Tamburini, C., 2011. Assimilation of marine extracellular polymeric substances by deep-sea prokaryotes in the NW Mediterranean Sea. Environ. Microbiol. Rep. 3, 705–709. http://dx.doi.org/10.1111/j.1758-2229.2011.00285.x.

Buesseler, K.O., Boyd, P.W., 2009. Shedding light on processes that control particle export and flux attenuation in the, twilight zone of the open ocean. Limnol. Oceanogr. 54 (4), 1210–1232.

Buongiorno Nardelli, B., Salusti, E., 2000. On dense water formation criteria and their application to the Mediterranean Sea. Deep-Sea Res. I 47, 193–221.

Canals, M., Puig, P., Durrieu de Madron, X., Heussner, S., Palanques, A., Fabres, J., 2006. Flushing submarine canyons. Nature 444, 354–357.

Cardin, V., Bensi, M., Pacciaroni, M., 2011. Variability of water mass properties in the last two decades in the South Adriatic Sea with emphasis on the period 2006-2009. Cont. Shelf Res. 31, 951–965. http://dx.doi.org/10.1016/j.csr.2011.03.002.

Carlson, C.A., 2002. Production and removal processes. In: Hansell, D.A., Carlson, C.A. (Eds.), Biogeochemistry of Marine Dissolved Organic Matter. Elsevier, San Diego, CA, pp. 91–151.

Carlson, C.A., Ducklow, H.W., Michaels, A.F., 1994. Annual flux of dissolved organic carbon from the euphotic zone in the northwestern Sargasso Sea. Nature 371, 405–408.

Carlson, C.A., Hansell, D.A., Nelson, N.B., Siegel, D.A., Smethie, W.M., Khatiwala, S., et al., 2010. Dissolved organic carbon export and subsequent remineralization in the mesopelagic and bathypelagic realms of the North Atlantic basin. Deep-Sea Res. II 57, 1433–1445.

Cauwet, G., Miller, A., Brasse, S., Fengler, G., Mantoura, R.F.C., Spitzy, A., 1997. Dissolved and particulate organic carbon in the western Mediterranean Sea. Deep-Sea Res. II 44, 769–779.

Christensen, J.P., Packard, T.T., Dortch, F.Q., Minas, H.J., Gascard, J.C., Richez, C., et al., 1989. Carbon oxidation in the deep Mediterranean Sea: evidence for dissolved organic carbon source. Global Biogeochem. Cycles 3, 315–335.

Civitarese, G., Gačič, M., Vetrano, A., Boldrin, A., Bregant, D., Rabitti, S., et al., 1998. Biogeochemical fluxes through the Strait of Otranto (Eastern Mediterranean). Cont. Shelf Res. 18, 773–789.

Civitarese, G., Gačič, M., Lipizer, M., Borzelli, G.L.E., 2010. On the impact of the Bimodal Oscillating System (BIOS) on the biogeochemistry and biology of the Adriatic and Ionian Seas (Eastern Mediterranean). Biogeosciences 7, 3987–3997. http://dx.doi.org/10.5194/bg-7-3987-2010.

Copin-Montégut, G., 1993. Alkalinity and carbon budgets in the Mediterranean. Global Biogeochem. Cycles 7, 915–925.

Copin-Montégut, G., Avril, B., 1993. Vertical distribution and temporal variation of dissolved organic carbon in the north-western Mediterranean Sea. Deep-Sea Res. I 40, 1963–1972.

D'Ortenzio, F., Ribera d'Alcalà, M., 2009. On the trophic regimes of the Mediterranean Sea: a satellite analysis. Biogeosciences 6, 139–148.

Dafner, E., Gonzalez-Davila, M., Santana-Casiano, J.M., Sempéré, R., 2001a. Total organic and inorganic carbon exchange through the Strait of Gibraltar in September 1997. Deep-Sea Res. I 48, 1217–1235.

Dafner, E.V., Sempéré, R., Bryden, H.L., 2001b. Total organic carbon distribution and budget through the Strait of Gibraltar in April 1998. Mar. Chem. 73, 233–252.

De Vicente, I., Ortega-Retuerta, E., Morales-Baquero, R., Reche, I., 2012. Contribution of dust inputs to dissolved organic carbon and water transparency in Mediterranean reservoirs. Biogeosciences 9, 5049–5060. http://dx.doi.org/10.5194/bg-9-5049-2012.

De Vittor, C., Paoli, A., Fonda Umani, S., 2008. Dissolved organic carbon variability in a shallow coastal marine system (Gulf of Trieste, northern Adriatic Sea). Estuar. Coast. Shelf Sci. 78, 280–290.

Doval, M.D., Hansell, D.A., 2000. Organic carbon and apparent oxygen utilization in the western South pacific and central Indian Oceans. Mar. Chem. 68, 249–264.

Doval, M.D., Pérez, F.F., Berdalet, E., 1999. Dissolved and particulate organic carbon and nitrogen in the northwestern Mediterranean. Deep-Sea Res. I 46, 511–527.

Economou, C., Mihalopoulos, N., 2002. Formaldehyde in the rainwater in the eastern Mediterranean: occurrence, deposition and contribution to organic carbon budget. Atmos. Environ. 36, 1337–1347.

Ferrari, G.M., 2000. The relationship between chromophoric dissolved organic matter and dissolved organic carbon in the European Atlantic coastal area and in the West Mediterranean Sea (Gulf of Lions). Mar. Chem. 70, 339–357.

Fuda, J.L., Etiope, G., Millot, C., Favali, P., Calcara, M., Smriglio, G., et al., 2002. Warming, salting and origin of the Tyrrhenian Deep Water. Geophys. Res. Lett. 29 (19), http://dx.doi.org/10.1029/2001GL014072.

Gačič, M., Borzelli, G.L.E., Civitarese, G., Cardin, V., Yari, S., 2010. Can internal processes sustain reversals of the ocean upper circulation? The Ionian Sea example. Geophys. Res. Lett. 37, http://dx.doi.org/10.1029/2010GL043216, L09608.

Garcia-Castellanos, D., Estrada, F., Jiménez-Munt, I., Gorini, C., Fernàndez, M., Vergés, J., et al., 2009. Catastrophic flood of the Mediterranean after the Messinian salinity crisis. Nature 462, 778–782. http://dx.doi.org/10.1038/nature08555 778-782.

Giani, M., Savelli, F., Berto, D., Zangrando, V., Cosović, B., Vojvodić, V., 2005. Temporal dynamic of dissolved and particulate organic carbon in the northern Adriatic Sea in relation to the mucilage events. Sci. Total Environ. 353, 126–138.

Gómez-Gutiérrez, A.I., Jovera, E., Bodineau, L., Albaigés, J., Bayona, J.M., 2006. Organic contaminant loads into the Western Mediterranean Sea: estimate of Ebro River inputs. Chemosphere 65, 224–236.

Hansell, D.A., 2002. DOC in the global ocean carbon cycle. In: Hansell, D.A., Carlson, C.A. (Eds.), Biogeochemistry of Marine Dissolved Organic Matter. Elsevier, San Diego, CA, pp. 685–715.

Hansell, D.A., 2005. Dissolved organic carbon reference material program. EOS Trans. Amer. Geophys. Union 86 (35), 318.

Hansell, D.A., 2013. Recalcitrant dissolved organic carbon fractions. Ann. Rev. Mar. Sci. 5, 3.1–3.25. http://dx.doi.org/10.1146/annurev-marine-120710-100757.

Hansell, D.A., Carlson, C.A., 2001. Biogeochemistry of total organic carbon and nitrogen in the Sargasso Sea: control by convective overturn. Deep-Sea Res. II 48 (8–9), 1649–1667.

Hansell, D.A., Carlson, C.A., 2013. Localized refractory dissolved organic carbon sinks in the deep ocean. Global Biogeochem. Cycles 27, 705–710. http://dx.doi.org/10.1002/gbx.20067.

Hopkins, T.S., 1999. The thermohaline forcing of the Gibraltar exchange. J. Mar. Syst. 20, 1–31.

Iudicone, D., Buongiorno Nardelli, B., Santoleri, R., Marullo, S., 2003. Distribution and mixing of intermediate water masses in the Channel of Sicily (Mediterranean Sea). J. Geophys. Res. 108 (9), 8105. http://dx.doi.org/10.1029/2002JC001647.

Jongsma, D., Fortuin, A.R., Huson, W., Troelstra, S.R., Klaver, G.T., Peters, J.M., et al., 1983. Discovery of an anoxic basin within the Strabo Trench, eastern Mediterranean. Nature 305, 795–797. http://dx.doi.org/10.1038/305795a0.

Klein, B., Roether, W., Kress, N., Manca, B.B., Ribera d'Alcalà, M., Souvermezoglou, E., et al., 2003. Accelerated oxygen consumption in eastern Mediterranean deep waters following the recent changes in thermohaline circulation. J. Geophys. Res. 108 (C9), 8107. http://dx.doi.org/10.1029/2002JC001371.

Krom, M.D., Woodward, E.M.S., Herut, B., Kress, N., Carbo, P., Mantoura, R.F.C., et al., 2005. Nutrient cycling in the south east Levantine basin of the eastern Mediterranean: results from a phosphorus starved system. Deep-Sea Res. II 52, 2879–2896.

Loh, A.N., Bauer, J.E., 2000. Distribution, partitioning and fluxes of dissolved and particulate organic C, N and P in the eastern North Pacific and Southern Oceans. Deep-Sea Res. I 47, 2287–2316.

Lucea, A., Duarte, M., Agusti, S., Sondergaard, M., 2003. Nutrient (N, P and Si) partitioning in the NW Mediterranean. J. Sea Res. 49, 157–170.

Ludwig, W., Dumont, E., Meybeck, M., Heussner, S., 2009. River discharges of water and nutrients to the Mediterranean and Black Sea: major drivers for ecosystem changes during past and future decades? Prog. Oceanogr. 80, 199–217.

Luterbacher, J., Liniger, M.A., Menzel, A., Estrella, N., Della-Marta, P.M., Pfister, C., et al., 2007. Exceptional European warmth of autumn 2006 and winter 2007: historical context, the underlying dynamics, and its phenological impacts. Geophys. Res. Lett. 34, http://dx.doi.org/10.1029/2007GL029951, L12704.

Malanotte-Rizzoli, P., 1991. The Northern Adriatic Sea as a prototype of convection and water mass formation on the continental shelf. In: Chu, P.C., Gascard, J.P. (Eds.), Deep Convection and Deep Water Formation in the Oceans. Elsevier Oceanography Series, Elsevier, vol. 57. pp. 229–239.

Malanotte-Rizzoli, P., Manca, B.B., Ribera D'alcalà, M., Theocharis, A., Brenner, S., Budillon, G., et al., 1999. The Eastern Mediterranean in the 80s and in the 90s: the big transition in the intermediate and deep circulations. Dynam. Atmos. Ocean 29, 365–395.

Manca, B.B., Kovačević, V., Gačić, M., Viezzoli, D., 2002. Dense water formation in the Southern Adriatic Sea and spreading into the Ionian Sea in the period 1997-1999. J. Mar. Syst. 33-34, 133–154.

Manca, B.B., Ibello, V., Pacciaroni, M., Scarazzato, P., Giorgetti, A., 2006. Ventilation of deep waters in the Adriatic and Ionian Seas following changes in thermohaline circulation of the Eastern Mediterranean. Clim. Res. 31, 239–256.

Markaki, Z., Loÿe-Pilot, M.D., Violaki, K., Benyahya, L., Mihalopoulos, N., 2010. Variability of atmospheric deposition of dissolved nitrogen and phosphorus in the Mediterranean and possible link to the anomalous seawater N/P ratio. Mar. Chem. 120, 187–194.

Martin, J.H., Knauer, G.A., Karl, D.M., Broenkow, W.W., 1987. VERTEX: carbon cycling in the northeast Pacific. Deep-Sea Res. 34, 267–285.

McCave, I.N., 1975. Vertical flux of particles in the ocean. Deep-Sea Res. 22, 491–502.

Mermex group, 2011. Marine ecosystems' responses to climatic and anthropogenic forcings in the Mediterranean. Prog. Oceanogr. 91, 97–166.

Millot, C., Taupier-Letage, I., 2005. Circulation in the Mediterranean Sea. In: Saliot, A. (Ed.), The Handbook of Environmental Chemistry. Springer, Berlin, Heidelberg, pp. 29–66. http://dx.doi.org/10.1007/b107143.

Moutin, T., Raimbault, P., 2002. Primary production, carbon export and nutrients availability in western and eastern Mediterranean Sea in early summer 1996 (MINOS cruise). J. Mar. Syst. 33-34, 273–288.

Panagiotopoulos, C., Sempéré, R., Para, J., Raimbault, P., Rabouille, C., Charrière, B., 2012. The composition and flux of particulate and dissolved carbohydrates from the Rhone River into the Mediterranean Sea. Biogeosciences 9, 1827–1844. http://dx.doi.org/10.5194/bg-9-1827-2012.

Pettine, M., Patrolecco, L., Camusso, M., Crescenzio, S., 1998. Transport of carbon and nitrogen to the northern Adriatic Sea by the Po River. Estuar. Coast. Shelf Sci. 46, 127–142.

Pettine, M., Patrolecco, L., Manganelli, M., Capri, S., Farrace, M.G., 1999. Seasonal variations of dissolved organic matter in the northern Adriatic Sea. Mar. Chem. 64, 153–169.

Polat, C., Tugrul, S., 1996. Chemical exchange between the Mediterranean and the Black Sea via the Turkish Straits. Bull. Inst. Oceanogr. 17, 167–186.

Pujo-Pay, M., Conan, P., Oriol, L., Cornet-Barthaux, V., Falco, C., Ghiglione, J.-F., et al., 2011. Integrated survey of elemental stoichiometry (C, N, P) from the Western to Eastern Mediterranean Sea. Biogeosciences 8, 883–899.

Pulido-Villena, E., Wagener, T., Guieu, C., 2008. Bacterial response to dust pulses in the western Mediterranean: implications for carbon cycling in the oligotrophic ocean. Global Biogeochem. Cycles 22, http://dx.doi.org/10.1029/2007GB003091 GB1020.

Redfield, A.C., 1934. On the proportions of organic derivatives in sea water and their relation to the composition of plankton. In: Daniel, R.J. (Ed.), James Johnstone Memorial Volume. University Press of Liverpool, Liverpool, pp. 177–192.

Rhein, M., 1995. Deep water formation in the western Mediterranean. J. Geophys. Res. 100 (C4), 6943–6959.

Ribera d'Alcalà, M., Civitarese, G., Conversano, F., Lavezza, R., 2003. Nutrient ratios and fluxes hint at overlooked processes in the Mediterranean Sea. J. Geophys. Res. 108 (C9), 8106. http://dx.doi.org/10.1029/2002JC001650.

Roether, W., Manca, B.B., Klein, B., Bregant, D., Georgopoulos, D., Beitzel, V., et al., 1996. Recent changes in eastern Mediterranean deep waters. Science 271, 333–335.

Santinelli, C., Gasparini, G.P., Nannicini, L., Seritti, A., 2002. Vertical distribution of dissolved organic carbon (DOC) in the western Mediterranean Sea in relation to the hydrological characteristics. Deep-Sea Res. I 49, 2203–2219.

Santinelli, C., Ribotti, A., Sorgente, R., Gasparini, G.P., Nannicini, L., Vignudelli, S., et al., 2008. Coastal dynamics and dissolved organic carbon in the western Sardinian shelf (Western Mediterranean). J. Mar. Syst. 74, 167–188.

Santinelli, C., Nannicini, L., Seritti, A., 2010. DOC dynamics in the meso and bathypelagic layers of the Mediterranean Sea. Deep-Sea Res. II 57, 1446–1459.

Santinelli, C., Sempéré, R., Van Wambeke, F., Charriere, B., Seritti, A., 2012a. Organic carbon dynamics in the Mediterranean Sea: an integrated study.

Global Biogeochem. Cycles 26, http://dx.doi.org/10.1029/2011GB004151 GB4004.

Santinelli, C., Ibello, V., Lavezza, R., Civitarese, G., Seritti, A., 2012b. New insights into C, N and P stoichiometry in the Mediterranean Sea: the Adriatic Sea case. Cont. Shelf Res. 44, 83–93. http://dx.doi.org/10.1016/j.csr.2012.02.015.

Santinelli, C., Hansell, D.A., Ribera d'Alcalà, M., 2013. Influence of stratification on marine dissolved organic carbon (DOC) dynamics: the Mediterranean Sea case. Prog. Oceanogr. http://dx.doi.org/10.1016/j.pocean.2013.06.001.

Schlitzer, R., Roether, W., Oster, H., Junghansh, H., Hausmann, M., Johannsen, H., et al., 1991. Chlorofluoromenthane and oxygen in the Eastern Mediterranean. Deep-Sea Res. 38, 1531–1551.

Schott, F.M., Visbeck, M., Send, U., Fischer, J., Stramma, L., Desaubies, Y., 1996. Observations of deep convection in the Gulf of Lions, northern Mediterranean, during the winter of 1991/92. J. Phys. Oceanogr. 26, 505–524.

Schroeder, K., Ribotti, A., Borghini, M., Sorgente, R., Perilli, A., Gasparini, G.P., 2008. An extensive western Mediterranean deep water renewal between 2004 and 2006. Geophys. Res. Lett. 35, L18605. http://dx.doi.org/10.1029/2008GL035146.

Sempéré, R., Charrière, B., Van Wambeke, F., Cauwet, G., 2000. Carbon inputs of the Rhône River to the Mediterranean Sea: biogeochemical implications. Global Biogeochem. Cycles 14, 669–681.

Sempéré, R., Panagiotopoulos, C., Lafont, R., Marroni, B., Van Wambeke, F., 2002. Total organic carbon dynamics in the Aegean Sea. J. Mar. Syst. 33-34, 355–364.

Sempéré, R., Dafner, E., Van Wambeke, F., Lefèvre, D., Magen, C., Allègre, S., et al., 2003. Distribution and cycling of total organic carbon across the Almeria-Oran Front in the Mediterranean Sea: implications for carbon cycling in the western basin. J. Geophys. Res. 108, C11, 3361. http://dx.doi.org/10.1029/2002JC001475.

Seritti, A., Russo, D., Nannicini, L., Del Vecchio, R., 1998. DOC, absorption and fluorescence properties of estuarine and coastal waters of the Northern Tyrrhenian Sea. Chem. Spec. Bioavail. 10, 95–106.

Seritti, A., Manca, B.B., Santinelli, C., Murru, E., Boldrin, A., Nannicini, L., 2003. Relationships between dissolved organic carbon (DOC) and water mass structures in the Ionian Sea (winter 1999). J. Geophys. Res. 108, C9, 8112. http://dx.doi.org/10.1029/2002JC001345.

Sharp, J.H., Carlson, C.A., Peltzer, E.T., Castle-Ward, D.M., Savidge, K.B., Rinker, K.R., 2002. Final dissolved organic carbon broad community intercalibration and preliminary use of DOC reference materials. Mar. Chem. 77, 239–253.

Sohrin, R., Sempéré, R., 2005. Seasonal variation in total organic carbon in the Northeast Atlantic in

2000-2001. J. Geophys. Res. 110, C10S90. http://dx.doi.org/10.1029/2004JC002731.

Sparnocchia, S., Gasparini, G.P., Astraldi, M., Borghini, M., Pistek, P., 1999. Dynamics and mixing of the Eastern Mediterranean outflow in the Tyrrhenian basin. J. Mar. Syst. 20, 301–317.

Theocharis, A., Balopoulos, E., Kioroglou, S., Kontoyiannis, H., Iona, A., 1999. A synthesis of the circulation and hydrography of the south Aegean Sea and the Straits of the Cretan Arc (March 1994–January 1995). Prog. Oceanogr. 44, 469–509.

Theocharis, A., Klein, B., Nittis, K., Roether, W., 2002. Evolution and status of the Eastern Mediterranean Transient (1997–1999). J. Mar. Syst. 33-34, 91–116.

Turchetto, M., Boldrin, A., Langone, L., Miserocchi, S., 2012. Physical and biogeochemical processes controlling particle fluxes variability and carbon export in the Southern Adriatic. Cont. Shelf Res. 44, 72–82. http://dx.doi.org/10.1016/j.csr.2011.05.005.

van der Wielen, P.W.J.J., Bolhuis, H., Borin, S., Daffonchio, D., Corselli, C., et al., 2005. The enigma of prokaryotic life in deep hypersaline anoxic basins. Science 307, 121–123.

Van Wambeke, F., Christaki, U., Giannakourou, A., Moutin, T., Souvemerzoglou, K., 2002. Longitudinal and vertical trends of bacterial limitation by phosphorus and carbon in the Mediterranean Sea. Microb. Ecol. 43, 119–133.

Vignudelli, S., Santinelli, C., Murru, E., Nannicini, L., Seritti, A., 2004. Distributions of dissolved organic carbon (DOC) and chromophoric dissolved organic matter (CDOM) in coastal waters of the northern Tyrrhenian Sea (Italy). Estuar. Coast. Shelf Sci. 60, 133–149.

Vilibić, I., Supić, N., 2005. Dense water generation on a shelf: the case of the Adriatic Sea. Ocean Dynam. 55, 403–415.

Yoro, S.C., Sempéré, R., Turley, C., Unanue, M.A., durrieu de Madron, X., Bianchi, M., 1997. Cross-slope variations of organic carbon and bacteria in the Gulf of Lions in relation to water dynamics (northwestern Mediterranean). Mar. Ecol. Prog. Ser. 161, 255–264.

Zeri, C., Besiktepe, S., Giannakourou, A., Krasakopoulou, E., Tzortziouc, M., Tsoliakos, D., et al., 2014. Chemical properties and fluorescence of DOM in relation to biodegradation in the interconnected Marmara-North Aegean Seas during August 2008. J. Mar. Syst. 135, 124–136. http://dx.doi.org/10.1016/j.jmarsys.2013.11.019.

Zervakis, V., Georgopoulos, D., Drakopoulos, P.G., 2000. The role of the north Aegean in triggering the recent eastern Mediterranean climatic changes. J. Geophys. Res. 105 (C11) (26), 103–126.

Zervakis, V., Georgopoulos, D., Karageorgis, A.P., Theocharis, A., 2004. On the response of the Aegean Sea to climatic variability: a review. Int. J. Climatol. 24, 1845–1858. http://dx.doi.org/10.1002/joc.1108.

CHAPTER

14

DOM in the Arctic Ocean

Leif G. Anderson*, Rainer M.W. Amon[†]

*Department of Chemistry and Molecular Biology, University of Gothenburg, Gothenburg, Sweden
[†]Department of Marine Sciences and Oceanography, Texas A&M University at
Galveston, Galveston, Texas, USA

CONTENTS

I INTRODUCTION

The objective of this chapter is to summarize the present knowledge on the Arctic Ocean sources and sinks of DOC as well as the distribution of DOC within the Arctic Ocean. The Arctic Ocean, together with the Greenland, Iceland, and Labrador Seas, is a major area of deep water formation in the Northern Hemisphere. As this deep water contributes to the global thermohaline circulation it also adds DOC to the deep waters of all global oceans. In order to address the sources, sinks and distribution of DOC it is essential to consider water mass formation and circulation within the Arctic Ocean.

The precipitation that falls over Siberia and boreal North America largely drains into the Arctic Ocean through several major rivers. The six largest rivers in terms of discharge and watershed are Ob, Yenisey, Lena, and Kolyma from

Biogeochemistry of Marine Dissolved Organic Matter,
http://dx.doi.org/10.1016/B978-0-12-405940-5.00014-5

Siberia and Mackenzie and Yukon from North America. The Yukon River reaches the Arctic Ocean after entering the Bering Sea and flowing north through the Bering Strait. These six rivers contribute ~65% of the total annual discharge, while the other 35% are contributed by numerous mid-size and small rivers (Holmes et al., 2012). Consequently, the Arctic Ocean receives a disproportionately large fraction of the global river discharge, about 10% (Aagaard and Carmack, 1989), while only constituting about 1% of the global ocean volume (Menard and Smith, 1966; Opsahl et al., 1999). The runoff, especially that from Siberia, adds large amounts of terrigenous dissolved organic matter (DOM) to the upper Arctic Ocean (e.g., Amon et al., 2012; Gordeev et al., 1996; Holmes et al., 2012). The large terrestrial component of DOM distinguishes the Arctic Ocean from the Southern Ocean.

A Water Masses and Circulation

Seawater enters the Arctic Ocean from the Atlantic *via* the eastern Fram Strait and the Barents Sea, and from the Pacific through Bering Strait (Figure 14.1). Fresh water, in the form of runoff and sea-ice melt, mixes with seawater in the upper layers and exits the Arctic basin through the Canadian Arctic Archipelago and western Fram Strait (Jones et al., 1998, 2003). Some of the upper waters are entrained and transported to deeper layers by shelf plumes (e.g., Schauer and Fahrbach, 1999; Schauer et al., 1997). Intermediate-depth water (in the depth range of 500-2000 m) largely follows the topography, resulting in several large loops within the central Arctic Ocean (e.g., Rudels et al., 1994). Deep water both enters and exits the Arctic Ocean through Fram Strait over a sill at a depth of about 2200 m.

The inflowing Pacific water is relatively fresh, contributing significantly only to the upper water masses of the Arctic Ocean. However, a small amount of high-salinity water formed during sea-ice production in the Pacific (western) sector of the Arctic Ocean penetrates to the deepest parts of the Canada Basin (Jones et al., 1995). The Atlantic water, on the other hand, has salinity

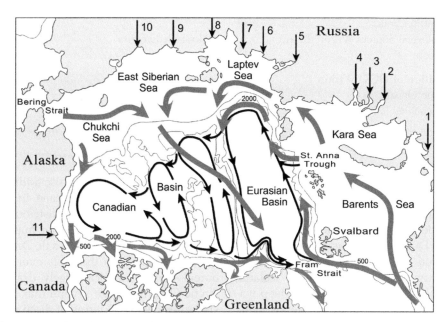

FIGURE 14.1 Geographic information and schematic circulation of surface water (gray arrows) and intermediate water (black arrows). The straight arrows locate the mouths of the rivers presented in Table 14.1.

close to 35 when entering the Fram Strait and the Barents Sea and the temperature is relatively high (around 4 °C). During transit through the Barents Sea, heat is lost to the atmosphere and, together with brine release from sea-ice production, the density increases to form waters that penetrate to several thousand meters in the deep Arctic basins (e.g., Schauer et al., 1997). Most of this high-density water enters the Arctic Ocean through the St. Anna Trough, though some water of Atlantic origin passes through the Kara Sea into the Laptev Sea, with a fraction continuing into the East Siberian Sea before meeting water of Pacific origin (Jones et al., 1998). The Atlantic water that flows through Fram Strait meets sea-ice, with the upper part of this warm water melting that ice, forming an ~100-m-thick surface water layer of low-salinity ($S \sim 34.2$) and with temperatures close to the freezing point (Rudels et al., 1996). This process constitutes the formation of the lower halocline water (LHW); it prevents deep water formation within the central Arctic Ocean by contributing to stratification. In addition, LHW hampers the penetration of heat from the Atlantic Layer water (the water of Atlantic origin with $T > 0$ inside the Arctic Ocean) to the overlying sea-ice cover.

Pacific water, and to a lesser extent Atlantic water, transport significant amounts of nutrients to Arctic shelf seas. The high nutrient supply and hydrographic conditions stabilizing the water column result in primary production (PP) rates that are high even in a global perspective (e.g., Codispoti et al., 2013; Hill et al., 2013; Matrai et al., 2013). In the Bering-Chukchi Seas region, new productivity has been estimated to be $288 \, g \, C \, m^{-2} \, year^{-1}$ (Hansell et al., 1993) of which much was attributed to the Bering Sea. Substantially lower values (5-$160 \, g \, C \, m^{-2} \, year^{-1}$) have been reported for the Chukchi Sea (e.g., Sakshaug, 2004). The new productivity in the Barents Sea is also considered high, being still higher within the marginal ice zone (Sakshaug and Skjoldal, 1989), varying from <8 to $100 \, g \, C \, m^{-2} \, year^{-1}$ (Sakshaug, 2004). This large range is mainly a result of the patchy productivity that makes it difficult to estimate a mean productivity rate. However, the vertical carbon flux at 75 m, as simulated by a 3D model, generally varied between 10 and $40 \, g \, C \, m^{-2} \, year^{-1}$, depending on the forcing conditions (Slagstad and Wassmann, 1996). Furthermore, the range of annual gross PP in the Barents Sea was estimated to 106-$134 \, g \, C \, m^{-2} \, year^{-1}$, while seasonally ice-covered areas had lower productivity (54-$67 \, g \, C \, m^{-2} \, year^{-1}$), based on a physically-biologically-coupled, nested 3D model with 4 km grid size (Reigstad et al., 2011; Wassmann et al., 2010). Recent studies have indicated that PP is rapidly changing due to a warmer and increasingly ice-free summer; it is suggested to increase in the future for the pan-Arctic region, especially on the continental shelves (Arrigo and van Dijken, 2011; Arrigo et al., 2012), while a recent review indicates a decline in PP in the Bering-Chukchi region over the last decade (Mathis et al., 2014).

B Sources of DOC to the Arctic Ocean

The highest concentration of DOC in the source waters to the Arctic Ocean is river runoff (mean annual discharge of 0.12 Sv), varying from ~600 to 990 μM in the big Siberian rivers while the Mackenzie River has a concentration ~350 μM, based on annual discharge-weighted means (Amon et al., 2012; Raymond et al., 2007; Stedmon et al., 2011). These concentrations are an order of magnitude higher than in the inflowing Atlantic (60 μM) and Pacific waters (70 μM), but the volume flux of the latter is about 60 times larger (~8 Sv) than that of continental runoff. Nevertheless, there is a clear signature of terrigenous DOC visible in the surface water over the central Arctic Ocean (Amon, 2004; Letscher et al., 2011; Mathis et al., 2014; Opsahl et al., 1999). A smaller source is the melting of sea-ice (basin-wide annual melt ~0.45 Sv), which can introduce DOC despite its relatively low concentrations in sea-ice (typically ~100 μM), except in zones within the ice of high biological activity where the concentration can be up to many hundreds µM (Thomas et al., 1995). These zones are typically found in the bottom layers of the ice and at the end of the winter-spring season. A major internal source for DOC in the Arctic Ocean is PP, for which the net PP has recently been estimated to average $101\,g\,C\,m^{-2}\,year^{-1}$ (Arrigo and van Dijken, 2011). Many of the investigations of organic carbon in the Arctic Ocean have reported data on unfiltered samples and are therefore total organic carbon (TOC) concentrations. The waters of the central Arctic Ocean are often very low in particles, which makes the difference between DOC and TOC concentrations small.

1 River Runoff Sources

Numerous rivers enter the Arctic Ocean. They drain enormous areas (total drainage basin area $\sim 17 \times 10^6\,km^2$) with variable vegetation and soil conditions. Consequently, DOC, as well as TOC, concentrations vary significantly between rivers (Table 14.1). There are also significant seasonal variations within rivers, as reported for the Lena by Cauwet and Sidorov (1996) and by Raymond et al. (2007) and Amon et al. (2012) for all the major rivers, based on the data from the PARTNERS and the Arctic great rivers observatory (Arctic-GRO) projects (http://arcticgreatrivers.org/data.html). In Figure 14.2, this variability is illustrated for Ob, Yenisey, Lena, and Mackenzie Rivers. The rivers with the largest variability of DOC concentrations (Yenisey and Lena) have a significant correlation between discharge and concentration, with maximum concentration ($\sim 1200\,µM$) observed during freshet, when the maximum discharge occurs (Figure 14.3). In autumn, the average concentration is around half of the maximum ($\sim 500\,µM$), with the values being even lower in the early season before the spring flood. The Arctic-GRO data were collected during all seasons, including open water and ice-covered conditions, thus amply illustrating the seasonal variability.

The Yenisey and Lena Rivers show a correlation between DOC and discharge (Figure 14.3b). Even so, these data were collected during several

seasons they still do not represent the seasonal variability well. The high short-term variability still represents a challenge to compute the annual input of DOC to the Arctic Ocean. Raymond et al. (2007) estimated an annual combined input of $15 \times 10^{12}\,g\,C$ for the Ob, Yenisey, Lena, Yukon, and Mackenzie Rivers. They scaled the fluxes of these rivers to the pan-Arctic watershed, assuming that areal yields in the unmonitored regions were equivalent to those in the monitored regions, and achieved a total flux to the Arctic Ocean of $25 \times 10^{12}\,g\,C$. Later, Holmes et al. (2012) confirmed the pan-Arctic estimate using a more extensive data set also including the Kolyma river.

The average discharge-weighted DOC concentration computed from the discharge and DOC input from the six rivers (Holmes et al., 2012) equals $667\,µMC$. Uncertainty in this estimate as a mean for the Arctic Ocean is mainly due to the fact that only data from the largest six rivers are included and they make up just over 50% of the total annual DOC flux (Holmes et al., 2012). Amon et al. (2012) demonstrated that Siberian rivers dominate terrestrial DOC input, contributing close to 90% of the total. They also showed that the vegetative cover of the watershed governs DOC concentrations in Arctic rivers; more plant biomass translates to higher DOC concentrations. Upscaling from large watersheds to smaller northern watersheds (mainly tundra) with less plant biomass, then, cannot be based on a linear relationship. Unfortunately, biomass estimates for the different watersheds are incomplete, but a general relationship between watershed vegetation and lignin export is suggested in Figure 14.4.

Earlier studies (Amon and Meon, 2004; Cauwet and Sidorov, 1996; Kattner et al., 1999; Köhler et al., 2003) estimated the average discharge-weighted concentration of DOC for rivers entering a given coastal area by regressing DOC *versus* salinity using data collected in the estuary and surrounding sea; this method yielded a theoretical end member. However,

TABLE 14.1   Concentrations of DOC and TOC in Arctic Rivers

No.	River	DOC (μM)	TOC (μM)	References	Shelf Seas
1	Pechora		1083	Gordeev et al. (1996)	Barents
2	Ob		592-733	Gordeev et al. (1996) and Telang et al. (1991)	Kara
		875		Amon et al. (2012) and Stedmon et al. (2011)	
3	Pyr		558	Gordeev et al. (1996)	Kara
4	Yenisey	711	617	Gordeev et al. (1996), Lobbes et al. (2000), and Telang et al. (1991)	Kara
		754		Stedmon et al. (2011)	
		733		Amon et al. (2012)	
5	Katanga		525	Gordeev et al. (1996)	Laptev
6	Olenek	850	600	Gordeev et al. (1996) and Lobbes et al. (2000)	Laptev
7	Lena	538-558	792-842	Cauwet and Sidorov (1996), Fitznar (1999), Gordeev et al. (1996), Lobbes et al. (2000), and Telang et al. (1991)	Laptev
		948		Amon et al. (2012) and Stedmon et al. (2011)	
8	Yana	232-264	558-611	Fitznar (1999), Gordeev et al. (1996), Lobbes et al. (2000), and Telang et al. (1991)	Laptev
9	Indigirka	404	642-754	Fitznar (1999), Gordeev et al. (1996), and Telang et al. (1991)	East Siberian
10	Kolyma	387	389-675	Fitznar (1999), Gordeev et al. (1996), and Telang et al. (1991)	East Siberian
		594		Stedmon et al. (2011)	
		547		Amon et al. (2012)	
11	Mackenzie	375-863	642-1050	Degens et al. (1991), Gordeev et al. (1996), Pocklington (1987), and Telang et al. (1991)	Beaufort
		385			
		363		Amon et al. (2012) and Stedmon et al. (2011)	
B.S.	Yukon	357-733	476-833	Degens et al. (1991), Pocklington (1987), and Telang et al. (1991)	Bering
		674		Stedmon et al. (2011)	
		637		Amon et al. (2012)	

Note. The Yukon River enters the Bering Sea (B.S.), outside the range of Figure 14.1, but its water enters the Arctic Ocean through Bering Strait.

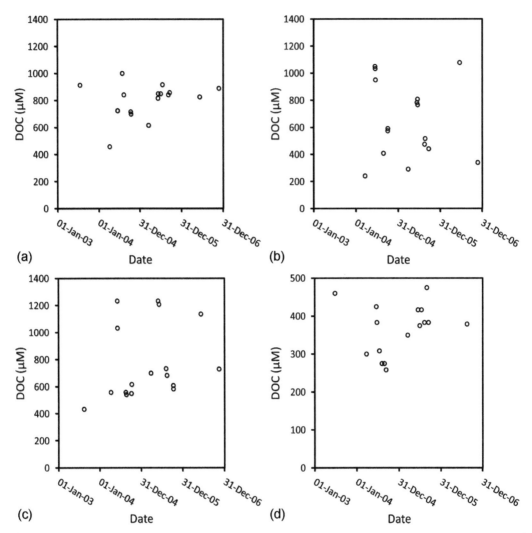

FIGURE 14.2   DOC in the four largest rivers (a Ob, b Yenisey, c Lena, d Mackensie) entering the Arctic Ocean during different seasons in the years 2003-2006. *Data from Arctic-GRO (http://arcticgreatrivers.org/data.html).*

this approach assumed that river DOC is largely refractory, which might not always be the case (Alling et al., 2010; Cooper et al., 2005; Hansell et al., 2004; Letscher et al., 2011). The amount of DOC delivered by rivers is a function of sources and sinks within the watersheds and river channels. The earlier studies were conducted in the late summer season after freshet, leading to the conclusion that DOC in large Arctic rivers is mostly refractory based on its apparently conservative mixing behavior as well as long-term decomposition studies (Köhler et al., 2003). But in fact, the seasonal fluctuation of DOC concentrations in the rivers is large (Raymond et al., 2007) and a significant portion (~30%) of the DOC during

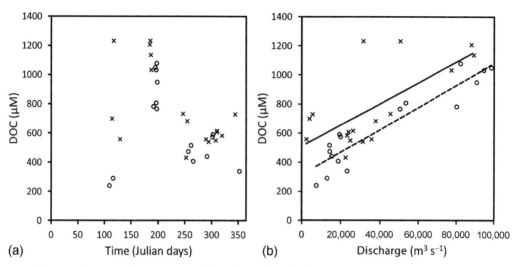

FIGURE 14.3 DOC concentrations *versus* (a) Julian day and (b) discharge, where o are for Yenisey and x for Lena. In (b), the strait line regressions are $y = 0.0081x + 444.04$, $R^2 = 0.7507$ and $y = 0.0076x + 315.24$, $R^2 = 0.8726$ for the Lena (dotted line) and Yenisey (interrupted line) Rivers, respectively. DOC data are the same as in Figure 14.2 with the discharge data from Arctic RIMS (http://rims.unh.edu/).

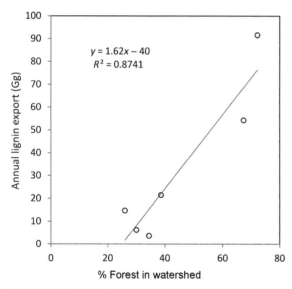

$$y = 1.62x - 40$$
$$R^2 = 0.8741$$

FIGURE 14.4 Relationship between the relative watershed area covered by forest and the annual export of lignin phenols.

freshet is labile on time scales of days to weeks (Holmes et al., 2008). Alling et al. (2010) used DOC measurements from the Laptev and East Siberian Seas with varying salinities to derive the zero salinity end member, comparing this value with mean observations from the major rivers of the region (see Figure 14.5 for a schematic illustration). Based on this comparison and a reported residence time of 3.5 year (Schlosser et al., 1994), a first-order removal rate of ~0.3 year^{-1} was computed. The main uncertainty lies in the residence time as the data were collected at different locations of the shelf and thus also had different residence times since leaving the river mouths. This uncertainty was partly overcome by Letscher et al. (2011) as they did a similar comparison, but based on data at the shelf break as well as at the transpolar drift area of the Makarov and Eurasian Basins. They achieved a first-order

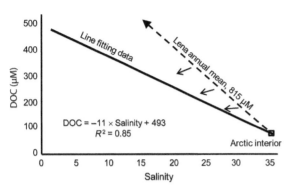

FIGURE 14.5　Schematic illustration of DOC distribution in the Lena River plume as observed in 2009 (Alling et al., 2010). The deviation between the DOC annual mean in the Lena River (815 μM) and that at zero salinity of the fitted line (493 μM) represents degradation during the terrigenous DOM's time over the Laptev Sea. The small arrows illustrate how sea-ice melt, having a DOC concentration of around 100 μM, will change the observed concentrations. The position of the arrow depends on the salinity of the water into which the sea ice-melt mixes.

decay constant of 0.24 ± 0.07 year^{-1}, meaning that more than half of the terrestrial DOC is removed while over the shelf.

There are several potential weaknesses with this approach. The first is the choice of the river end member. DOC data from the large Arctic rivers are based on samples collected several hundred km upstream of the river mouths (~800 km for the Lena River), such that DOC removal could have happened in the river or on the shelf; DOC losses due to bacterial degradation should happen sooner rather than later. Degradation in the river and the estuary would be consistent with earlier DOC concentration and degradation surveys in the Kara Sea (Amon and Meon, 2004; Köhler et al., 2003), which did not find elevated levels of DOC on the shelf even though this study covered 140,000 km^2 in the open Kara Sea. It would take about 5 months for the combined discharges of the Ob and Yenisei Rivers to pass the region covered in the study; sampling was conducted in August and September, about 3 months after freshet. Combined with the very

small DOC losses during decomposition experiments with Ob and Yenisey waters (Köhler et al., 2003; Meon and Amon, 2004), it seems that the bulk of riverine DOC losses happen before the freshet plume passes the estuaries. Significant losses of DOC within the rivers will have to be taken into account when calculating end members and estimating riverine DOC input to the Arctic Ocean.

The second potential weakness with salinity-DOC relationships is ice melt as a second freshwater source. The problem was directly addressed by Amon et al. (2012) showing that, when DOC concentrations are corrected for sea-ice formation and melt, theoretical river end members become larger. An insightful review of these processes in the Mackenzie estuary was given by Macdonald and Yu (2005). They suggested that river inflow into an estuary during winter and spring can result in freezing of river water in the form of subsurface frazil ice when the fresh river water gets in contact with seawater of subzero temperatures. These processes of freezing and melting near the river mouth result in DOM fractionation (concentration and dilution), causing a significant change in the salinity-DOC relationships within estuaries. Elevated levels of terrestrial DOC components at the bottom of the Laptev Sea indicate brine formation (resulting in concentration of terrestrial DOM) close to the river mouth (Amon, unpublished data). This process might even be more critical in Arctic Ocean surface waters where sea-ice melt is more prominent. DOC concentrations in Arctic surface waters reflect the net result of mixing background levels (Atlantic or Pacific source waters), river input, in situ production, in situ degradation, and sea-ice melt. The formation of frazil ice due to mixing of subsurface layers, as indicated by Macdonald and Yu (2005), has also been observed in the central Arctic Ocean (Gascard, pers. communication) where low-salinity surface waters are in contact with colder, more saline subsurface water, resulting in the formation of frazil ice that floats to the top and melts. The quantitative importance of this subsurface process is however not known at present.

If sea-ice processes are not considered when using DOC-salinity relationships, the theoretical river DOC end member can be severely underestimated because the slope of the regression line would be steeper after correcting for sea-ice processes. This point is best demonstrated by the relationship of lignin phenols (a tracer for terrigenous DOM) and salinity in Eurasian Basin surface waters (Figure 14.6). There is no apparent relationship between terrestrial DOC (represented by lignin phenols) and salinity, and the trendline would even suggest decreasing terrestrial DOC at lower salinity. Absence of the expected relationship indicates a significant dilution of lignin phenols by sea-ice melt, which is essentially free of lignin. A third potential problem for using DOC-salinity relationships to estimate river end members is the contribution of phytoplankton DOC at the high-salinity end. A significant increase in DOC due to phytoplankton activity will increase the slope resulting in an underestimate of the river end member.

Uncertainty due to sea-ice processes and phytoplankton contributions is greatest at high salinities where the runoff fraction is the lowest. On the other hand, uncertainty due to terrestrial DOC decomposition is greatest at zero salinity (in the rivers).

All things considered, it seems that freshet DOC has a significant biolabile portion that is degraded on the time scales of river transport and estuarine mixing. The labile fraction seems to be ~30% of the freshet DOC (e.g., Alling et al., 2010; Letscher et al., 2011). Using the basin-wide DOC export value of 25×10^{12} g C year^{-1} from the PARTNERS and Arctic-GRO projects (Holmes et al., 2012) and accounting for the 30% of short-term loss of labile freshet DOC, gives an annual terrestrial DOC input to the open Arctic Ocean of 20×10^{12} g C, with almost 90% of it coming from Siberia. For lignin phenols, the ultimate tracer for terrestrial DOC, the losses during river transport and estuarine mixing are possibly larger, as indicated by a rapid drop of lignin concentrations after the freshet period (Amon et al., 2012). Lignin concentrations drop by 60-80% between the freshet values and typical summer values. In addition, lignin yields, a measure for the contribution of lignin to total DOC, dropped by a factor of 2-3 in the large Siberian rivers. A drop in lignin yields indicates that lignin phenols were preferentially removed from the bulk DOM pool, or in other words, lignin was degraded faster than bulk DOC. It is conceivable that 50% of the freshet lignin is lost to degradation on a similar timescale as DOC. Such a loss would change the annual lignin input (given in Amon et al., 2012) to ~80 Gg of lignin phenols per year.

2 Seawater Sources

DOC concentrations in the inflowing Atlantic water vary between 52 and 108 µM (Børsheim and Myklestad, 1997; Engbroadt and Kattner, 2005; Fransson et al., 2000; Opsahl et al., 1999; Wheeler et al., 1997). A comprehensive investigation of the northern Nordic Seas in the summers of 1997 and 1998 found a mean

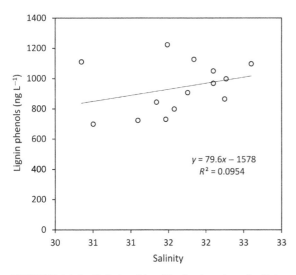

FIGURE 14.6 Relationship of lignin phenols and salinity in the upper 30 m of the Eurasian Basin.

DOC concentration of $58 \pm 5 \mu M$ ($n = 54$) in the Atlantic Water (50-600m depth) (Amon et al., 2003). This water, entering both through Fram Strait and Barents Sea, constitutes the source of the top ~500m of the central Arctic Ocean, except for in the Canadian Basin where the top 50-100m is of Pacific origin. Intermediate and deep waters also enter the Arctic Ocean through Fram Strait; these have DOC concentrations close to $50 \mu M$ (± 3, $n > 40$) (Amon et al., 2003). This water, even though it originates in the Atlantic, has been modified between the Greenland-Scotland Ridge and the Arctic Ocean. The Norwegian Sea deep water, flowing north through Fram Strait, is a mixture of Eurasian Basin deep water and Greenland Sea deep water (Swift et al., 1983). Some of the Eurasian Basin deep water that exits through Fram Strait flows around the Greenland Sea, mixing with Greenland Sea deep water before it reenters the deep Arctic Ocean.

It is difficult to assign a specific DOC concentration to the water flowing north through Bering Strait as the inflow consists of several water masses that have undergone considerable modifications during transit through the Bering Sea. The Arctic Ocean Section expedition in 1994 showed a TOC concentration span of 50-110μM in the waters with Pacific-derived characteristics (Wheeler et al., 1997). The mean DOC concentration, that is, when the particulate fraction was subtracted, computed for the samples at the continental slope off the Chukchi Sea stations equals ~70 ± 15μM. This value agrees with the range observed in the same region at the salinity range of the Bering Strait inflow ($S = 32$-34) (Hansell et al., 2004).

Shin and Tanaka (2004) showed a close DOC-salinity relationship in the Bering Strait inflow water with the linear correlation: $DOC = -15.17 \times S + 556$. However, this relationship is only valid up to a salinity of about 32.5, which is on the low end of the mean salinity of the inflow. Their data in the salinity range of 32-34, collected along the Chukchi shelf slope, give a mean of about $65 \pm 15 \mu M$ also being in the same range

as other observations in the Chukchi Sea (e.g., Hansell et al., 2004; Wheeler et al., 1997).

3 Biological Sources Within the Arctic Ocean

An additional source of DOM is through biological processes within the Arctic Ocean and its shelves. Primary productivity over the continental shelves is substantial and results in significant seasonal production of marine DOM. This seasonal signal can be observed in the surface waters (e.g., Mathis et al., 2007), and is one reason for the large concentration range in the waters flowing in from the Pacific Ocean. The outflow from the Barents Sea into the Arctic Ocean through the St. Anna Trough (containing a minor fraction of river runoff) also carries elevated surface DOC concentrations relative to waters > 150 m (Fransson et al., 2000). At depths < 150 m, the nutrient distribution indicated that PP had occurred during transit over the Barents Sea. The locally accumulated DOC can be estimated by subtracting the deep water DOC concentration (average of $52 \mu M$) from surface DOC values (c.f. Hansell and Carlson, 1998). The accumulated DOC amounts to $\sim 1.4 \, mol \, C \, m^{-2}$ integrated over the top 150 m (Fransson et al., 2000), which will be exported to central Arctic Ocean.

The Chukchi Sea also has a high biological productivity. In analogy with the above estimate of exported DOC, the difference between the highest (134μM) and lowest (67μM) monthly depth average, DOC concentration at the continental break of the north-western Chukchi Sea (Walsh et al., 1997) should reflect the marine DOC exported into the central Arctic Ocean from this area. However, DOC concentration in the Chukchi Sea is also heavily impacted by river runoff, calling for correction to this contribution. Mathis et al. (2007) computed the marine DOC concentration by correcting for ice melt and river water contribution using salinity and $\delta^{18}O$, showing that summer values in 2002 were up to ~15μM higher than the spring values.

The productivity of the central Arctic Ocean is small compared to the shelf seas. Based on the computed deficit of phosphate, Anderson et al. (2003) estimated an average export production of $0.04 \, mol \, m^{-2} \, year^{-1}$ there. In contrast, an *in situ* DOC production of more than $0.5 \, mol \, m^{-2} \, year^{-1}$ (assuming a 120-day productive season) was computed for the central Arctic Ocean in 1994 (Wheeler et al., 1997). If this DOC production is distributed over the top $50 \, m$, it will result in a concentration increase of $10 \, \mu M$. This increase is in agreement with what Mathis et al. (2007) estimated as the DOC production from spring to summer at the eastern shelf break of the Chukchi Sea. Such an annual DOC production cannot be sustained over the residence time of the Arctic Ocean surface water, 5-10 year, without building up unrealistic DOC concentrations, indicating extensive recycling. Hence, it is essential to consider the seasonal production and degradation of Arctic marine DOC.

DOC in sea-ice has been attributed to ice algae production (e.g., Smith et al., 1997). In shelf seas receiving much runoff, the sea-ice produced will include some terrigenous DOC. In spring (beginning of April to end of May) when sea-ice algae develops, Smith et al. (1997) found a good correlation between Chl *a* and DOC in the bottom ice of Resolute Passage in the Canadian Archipelago. DOC concentrations were much higher in the ice than in underlying water, especially in the ice covered with only a thin snow layer. The highest DOC concentration ($>3000 \, \mu M$) was measured in the bulk of the bottom ice on May 14 (Smith et al., 1997). However, the volume with such high concentration is limited and thus the integrated contribution of DOC from ice to the underlying water mass is small. That these high concentrations are mainly a result of biological activity can be inferred from the observations in sea-ice collected in the spring when the DOC concentration in 10-20 cm depth slices ranged from 19 to $25 \, \mu M$ (Mathis et al., 2007). Measurements of DOC release rates by ice algae were performed

by Gosselin et al. (1997) along the track of the Arctic Ocean Section in 1994, varying from <25 to $1600 \pm 1500 \, \mu mol \, C \, m^{-2} \, day^{-1}$, with the highest rates in the Chukchi Sea. Thomas et al. (1995) collected three ice cores of more than $2 \, m$ length in the Fram Strait. In two of these the DOC concentration was mostly $<100 \, \mu M$ throughout the core. In the third, the concentration was close to $100 \, \mu M$ in the top $\sim 1.9 \, m$, increasing to $\sim 700 \, \mu M$ $10 \, cm$ from the bottom. This increase was explained by a combination of DOM excretion by biota and decomposition of organisms (Thomas et al., 1995). The bulk concentration of DOC in sea-ice from the central Arctic Ocean varies substantially from about $300 \, \mu M$ (Melnikov, 1997) to normally much lower concentration (Opsahl et al., 1999; Thomas et al., 1995) with the averaging being $\sim 100 \, \mu M$. Hence, the melting of one meter of sea-ice will only marginally change the DOC concentration in the top $50 \, m$ (typical winter surface mixed layer; Rudels et al., 1996).

A biological source of DOC is release from the sediments by regeneration of organic material. Hulth et al. (1996) measured DOC concentrations in the range of 500-$8000 \, \mu M$ in pore water in the Svalbard area. The lowest concentrations were found at stations east of Svalbard, where a significant inverse linear correlation ($R^2 = 9.849$) of DOC concentrations with a sediment reactivity index (defined as the sediment oxygen consumption rate normalized to organic content) was found. This correlation suggests a coupling between reactivity of sediment organic matter and DOC lability in pore water. In a study of the eastern Eurasian Basin and adjacent shelves (Hulthe and Hall, 1997), DOC fluxes out of the sediment reached $3.6 \, mmol \, m^{-2} \, day^{-1}$. The highest fluxes were found on the shelves and the lowest in the deep basins and on the slopes. A positive correlation between DOC and dissolved inorganic carbon fluxes was observed, with DOC constituting up to 50% of the total benthic carbon flux at stations that had the highest benthic carbon fluxes. This high value indicates that the fraction of DOC oxidized to inorganic carbon decreases with increasing decomposition rates.

High DOC concentrations have been found in parts of the upper halocline north of the Chukchi Sea (Shin and Tanaka, 2004), coinciding with high nutrient and low oxygen concentrations, signatures of organic matter decay at the shelf sediment surface. This excess DOC likely is supplied from the sediment surface as a result of organic matter mineralization, with the organic matter potentially having both terrestrial and marine sources. Furthermore, Nakayama et al. (2011) found a maximum in humic-type fluorescent DOM associated with the nutrient maximum of the upper halocline at the shelf break and within the deep Canada Basin. However, not all nutrient-rich upper halocline water has elevated DOC concentrations (e.g., Hansell et al., 2004; Mathis et al., 2005; Wang et al., 2006), indicating that either the release of DOC from the sediment is not a continuous process, or that this fraction of DOC is quite labile for further decomposition within the water column. The latter scenario is supported by findings of elevated labile DOC in halocline waters of the Canada Basin (Davis and Benner, 2007). Based on recent estimates of pan-Arctic net PP of $440 \times 10^{12} \, gC$ $year^{-1}$ (Arrigo and van Dijken, 2011) and the assumption that 10% of net PP accumulates as DOC (Mathis et al., 2007), algae could introduce $44 \times 10^{12} \, gC$ of DOC annually. However, primary productivity estimates for the Arctic Ocean, including shelves, are wide ranging and most for the central Arctic Ocean are smaller than that by Arrigo and van Dijken (2011). We also need to distinguish between labile, semi-labile, and refractory DOM, as these influence the fate and shelf-basin transport of phytoplankton DOM. Recent studies have addressed the issue of bioavailability (Amon and Benner, 2003; Amon et al., 2001; Davis and Benner, 2005, 2007; Hansell, 2013). Labile DOM is usually utilized on time scales of hours to weeks, semi-labile DOM can last for months to several years, and DOM

that survives more than a few decades is considered refractory. Amon and Benner (2003) based their estimates of labile DOM on carbohydrate biomarkers, finding that for the central Arctic Ocean about 2% of surface water DOM is labile. Davis and Benner (2005, 2007) found higher proportions of labile DOM on the Chukchi and Beaufort shelves (up to 50%), with average values for the entire Arctic of 1-5% labile, 26-30% semi-labile, and roughly 70% refractory DOM. Interestingly, substantial amounts of labile and semi-labile DOM were found in subsurface halocline layers of the Canada Basin, indicating a relatively fast export from the shallow shelves to subsurface layers in the open Arctic Ocean.

II COMPOSITION OF DOC WITHIN THE ARCTIC OCEAN

Before the fluxes of DOC to and from of the Arctic Ocean are discussed, the quality of the terrigenous dissolved organic matter has to be considered. Will it flow with the water as a passive tracer or is it susceptible to diagenetic alteration or photochemical decomposition? Earlier conclusions that terrigenous fraction of DOC is refractory in the Arctic Ocean surface waters have been challenged (Alling et al., 2010; Hansell et al., 2004; Holmes et al., 2008; Letscher et al., 2011). Some studies reported a DOC half-life over the Siberian shelf of <3 year (Alling et al., 2010; Letscher et al., 2011) while a half-life of about 7 years was reported for the Beaufort Gyre (Hansell et a,l., 2004). This range reflects DOM components of varying labilities, with the key to understanding these lying in DOM's chemical composition (Amon and Benner, 1996).

A Elemental and Isotopic Composition

DOM's elemental stoichiometry traces its origin. The C/N molar ratio is generally high in terrigenous and low in marine DOM. An average C/N ratio of 20.5 ± 2.6 was reported for seven

Siberian rivers (Gordeev et al., 1996; recalculated by Wheeler et al., 1997). Cauwet and Sidorov (1996) found a similar ratio (22) for the Lena River, while significantly higher ratios (30-58) were reported for the Lena River by Lara et al. (1998). Lobbes et al. (2000) reported the mean C/N ratio for Yenisey, Olenek, Lena, Yana and Indigirka Rivers to be 47 ± 10. The variability in the reported ratios is mainly a result of variable DON concentrations. More recent studies typically place the C/N ratios of Arctic river DOM above 40 (Amon and Meon, 2004; Benner et al., 2005; Holmes et al., 2012). The high C/N ratios of DOM in runoff are characteristic of riverine fulvic and humic acids (Thurman, 1985).

The C/N ratio of marine DOM is dependent on biological activity in the water mass. When C/N ratios are plotted *versus* salinity for samples collected in the outer Laptev Sea, at the continental margin and in the eastern part of the Eurasian Basin, two regimes can be identified (Figure 14.7). At salinities <34.5 (depth <100m) C/N ratios increase with decreasing salinity, while at salinities >34.5 (depth >100m) the ratios vary from 12 to 30. The trend at $S < 34.5$ is mainly a result of water of Atlantic origin mixing

with runoff, as evidenced by the intercept 49.7 at $S = 0$. No trend, but a large C/N span, can be seen in the waters of $S > 34.5$, likely a result of these samples being deep waters and thus affected by decay of sinking particulate organic matter. A large variability in the C/N ratio of DOM, ranging from 10 to 40 with a peak around 15, was also observed in Fram Strait (Lara et al., 1998). Such large variability is not evident in more recent data sets, with ratios ranging from 16 to 20 depending on water mass with a standard deviation typically <2 (Benner et al., 2005; Davis and Benner, 2005). The lower ranges in recent investigations may result from methodological improvements of the DON determination. Changes of the C/N ratio due to the variable contribution of terrestrial DOM are small, typically less than the standard deviation, making the C/N ratio less useful for quantitative computations. It remains valuable as a qualitative tracer of terrigenous DOM, as demonstrated for the East Greenland Current (Benner et al., 2005).

More powerful tools to distinguish terrestrial from marine DOM in the Arctic are the stable isotopes of carbon (ratio of ^{13}C to ^{12}C, expressed as $\delta^{13}C$) and nitrogen (ratio of ^{15}N to ^{14}N, expressed as $\delta^{15}N$) (see Chapter 6). Vegetation in Arctic watersheds is dominated by plants that follow the so called C_3-pathway of carbon fixation, with strong fractionation of stable carbon isotopes resulting in highly depleted $\delta^{13}C$ (−24 to −33‰) values in the biomass. Accordingly, average $\delta^{13}C$ values for Arctic river DOM are typically between −25 and −28‰ (Amon and Meon, 2004; Opsahl et al., 1999; Raymond et al., 2007), while marine DOM is typically around −21‰ (Opsahl et al., 1999). Arctic Ocean DOM $\delta^{13}C$ values reported in the literature range from −21‰ to −23‰ with more depleted values in Arctic surface waters and the East Greenland Current (Benner et al., 2005). Stable nitrogen isotopes are much less explored in the Arctic region, but allow a similar distinction between terrestrial and marine DOM because terrestrial vegetation is characterized by nitrogen isotope

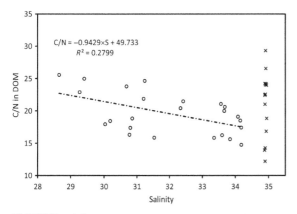

FIGURE 14.7 C/N molar ratios in dissolved organic matter (DOC/DON) in the eastern Eurasian Basin. The linear regression line is fitted to the upper 100m data (open circles) while the deeper water data are represented by ×. *Data from Fitznar (1999).*

values close to 0‰ as a result of nitrogen fixation, while marine DOM based on nitrate assimilation is more enriched (δ^{15}N in the range +4‰ to +7‰). Even with few data available from the Arctic region they reflect this general pattern, with Arctic river δ^{15}N values ranging from 1.8‰ to 3.3‰ (Amon and Meon, 2004) and Arctic Ocean DOM δ^{15}N ranging from 4.4‰ to 4.8‰ (Benner et al., 2005). Interestingly, based on stable N isotopes of DOM, surface waters from the Canada Basin (more enriched) were distinctly different from Eurasian Basin surface waters (more depleted) and within the Canada Basin LHW (with its Atlantic origin) DOM was distinct (more depleted) from surface water DOM and upper halocline water DOM (Benner et al., 2005). More work is required to understand the differences in stable nitrogen isotopes of DOM.

Radiocarbon has recently received more attention because the soil that supplies terrestrial DOC to the Arctic Ocean is a major global reservoir of old organic carbon, and its fate is hence a concern for the global carbon budget in the climate perspective (McGuire et al., 2010; Tarnocai et al., 2009). The old organic carbon currently locked in the vast permafrost regions of the Arctic watersheds will potentially be released under a warmer climate and part of that aged organic carbon is expected to be transported to the Arctic Ocean by rivers draining these watersheds. Such a mobilization would be reflected in the average ^{14}C-age of river DOC. Early studies in the Ob and Yenisei rivers have indicated that DOC flowing into the Arctic Ocean is of recent origin (Amon and Meon, 2004; Benner et al., 2004); this observation was later extended to the five largest Arctic rivers (Raymond et al., 2007). This result means that at present there is no sign of mobilization of old permafrost derived DOC into these rivers. Benner et al. (2004) reported the Δ^{14}C in DOC of Arctic Ocean DOM and related the ^{14}C-age to the lignin phenol concentrations in the same water samples. There was a strong positive relationship between these two constituents, showing that the terrestrial component

of DOC in the surface water was young relative to the marine component. A recent study (Vonk et al., 2012) has shown extensive loss of old organic carbon from ice complex deposits along the Siberian coast. Applying the δ^{13}C and Δ^{14}C isotopes and a mixing model, they estimated an annual input of old particulate carbon to the Eastern Siberian Shelf Seas equal to 44 ($\pm$ 10) $\times 10^{12}$ g C, with about two thirds being mineralized to carbon dioxide. No estimate on the magnitude of DOC production from this eroded POC was given.

B Molecular Level Composition

The most powerful tracers for the origin and diagenetic state of DOM are biomarkers. Quantitative tracers of terrestrial organic matter are lignin phenols, which have been determined in runoff to and in the surface waters of the Arctic Ocean (Amon et al., 2012; Benner et al., 2004; Kattner et al., 1999; Lobbes et al., 2000; Opsahl et al., 1999).

Kattner et al. (1999) determined lignin in the "humic" fraction of DOM and used this as a tracer for terrigenous influence, finding that the riverine-derived freshwater DOM contribution to the Laptev Sea is 8-30%. Combining this proportion with DOC concentrations in the Lena River and Laptev Sea suggested that 60% of the DOC in the surface layer of the Laptev Sea and adjacent Eurasian Basin is of terrigenous origin. In contrast, terrigenous dissolved organic nitrogen (DON) accounted for only 20-30% of the total DON (Kattner et al., 1999). This outcome agrees well with estimates of the terrigenous DON accounting for $28\pm13\%$ of the DON in the Laptev Sea as evaluated by principal-component analysis of amino acids (Dittmar, 2004). Low fractions were found in the deep Arctic Ocean, except for along the Laptev Sea continental margin down to about 2000-m depth where the terrigenous proportion was elevated (Dittmar, 2004).

The fraction of terrigenous DOM in the central Arctic Ocean was estimated from the carbon-normalized yields of lignin oxidation products (Λ_6) and $\delta^{13}C$ in ultra-filtered dissolved organic matter (UDOM) (Opsahl et al., 1999), resulting in 5-22% and 16-33%, respectively. UDOM is the high-molecular-weight fraction of DOM (>1 kDa), which is about 20-30% of total DOM. In Figure 14.8 the mean values (±variability) of samples from the Kara Sea (low $\delta^{13}C$), the surface water (medium $\delta^{13}C$) and deep Fram Strait and Greenland Sea (high $\delta^{13}C$) (Opsahl et al., 1999) are plotted *versus* Λ_6. The surface water ($32.0 < S < 34.5$) samples were collected at depths of 38-165 m from submarines operating within the SICEX program (e.g., Morison et al., 1998). Hence, these data do not include low-salinity surface waters. The contribution of terrigenous DOM in the surface water was computed from the mixing line of Figure 14.8 to be $15 \pm 6\%$, where the error represents the extreme variability of the data. This computation is based on the same hypothesis as that of Opsahl et al. (1999) that the deep Fram Strait and Greenland Sea data represent marine-derived organic matter and the Kara Sea data represent terrigenous-derived organic matter. Based on samples from

much of the Arctic Ocean as well as the Nordic Seas, Benner et al. (2005) also utilized Λ_6 and $\delta^{13}C$ data to compute the terrestrial-derived UDOM. Their results, 14-24% terrigenous UDOC in surface water (including the East Greenland Current) and about 2-3% in the deep waters of the Arctic and the Greenland and Norwegian Seas, are consistent with those by Opsahl et al. (1999) and Amon et al. (2003). More recent surveys with higher resolution sampling, however, indicate that the lignin concentrations and their contributions to DOC might have been underestimated (Amon et al., unpublished data), especially in the central Arctic Ocean.

The relative composition of lignin monomers can also be used as a diagnostic tool. The ratio of syringyl to vanillyl phenols (S/V) indicates oxidative changes. S/V ratios are reduced by diagenetic alterations in the Atlantic and Pacific Oceans (Opsahl and Benner, 1997) as well as by photodegradation of terrestrial DOC (Benner and Kaiser, 2011). Opsahl et al. (1999) and Lobbes et al. (2000) showed that the ratio distinguishes between DOM originating from angiosperm (high S/V ratios) and gymnosperms plants (low S/V ratios) in Russian rivers entering the Arctic Ocean. Opsahl et al. (1999) determined the S/V ratio in UDOM for different regions of the Arctic Ocean, finding values not too different in the Kara Sea (0.3-0.5; $n = 9$) and central Arctic Ocean (0.12-0.31; $n = 13$). The relatively constant S/V ratio within the Arctic Ocean indicates limited alteration of terrigenous DOM. A recent study on lignin phenol discharge from the six largest Arctic rivers (Amon et al., 2012) indicates that the dominant source of DOM is gymnosperm vegetation (50-70%), with smaller contributions by angiosperms and mosses. However, different rivers displayed different dominances of sources as well as diagenetic states. The Ob, for example, had a significantly larger contribution of angiosperm vegetation than Lena and Yenisei. DOM composition during base flow was distinct from the composition of freshet DOM in all rivers, indicating the need to use discharge-weighted

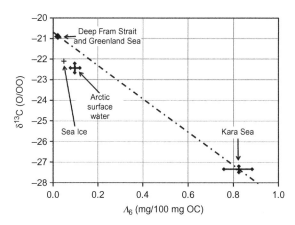

FIGURE 14.8 Mean (±standard deviation) carbon-normalized yields of lignin oxidation products (Λ_6) *versus* stable carbon isotopic composition ($\delta^{13}C$) of DOM from the Kara Sea, the Arctic surface water, deep Fram Strait, and Greenland Sea, as well as one sea-ice sample. *Data from Opsahl et al. (1999).*

means when calculating the input of DOM into the Arctic Ocean.

Another set of biomarkers that have been used are amino acids, neutral sugars, and amino sugars. These tracers are not necessarily useful for source identification but they determine the diagenetic history or lability of DOM. Both amino acids and neutral sugars contribute more to the bulk DOM in fresh and labile material and are preferentially lost during degradation; these biomarkers trace the amount of labile DOM (Amon and Benner, 2003; Amon et al., 2001). Davis and Benner (2007) used amino acids to distinguish between labile, semi-labile and refractory DOM in the western Arctic Ocean, determining that there is a small labile pool (1-5%), substantial semi-labile DOM (26-30%) and about 70% refractory DOM. Concentrations of dissolved amino acids and amino sugars were strongly positively correlated with chlorophyll-a in the waters of the Chukchi Sea and its slope (Davis and Benner, 2005) illustration the linkage between plankton productivity and labile DOM production.

Semi-labile DOM is less variable and not necessarily tied to local PP, indicating that a number of processes might contribute to the production of labile and semi-labile DOM. This complexity makes it difficult to estimate the production of labile and semi-labile DOM on an annual basin within the Arctic Ocean. The combined use of various biomarkers has the most potential to address this question (Davis et al., 2009).

C Optical Properties

Much progress was made investigating the optical properties of DOM in the Arctic Ocean over the last decade. The major advantage of optical properties, relative to biomarkers, is the high vertical resolution that can be obtained by *in situ* optical sensors and the real time data capabilities. While biomarkers are more specific than optical properties they require discrete sample volumes and are much more time consuming

in terms of sample preparation and analyses, reducing the vertical and horizontal resolution tremendously. Chromophoric dissolved organic matter (CDOM; see Chapter 10), the optically active portion of DOM, can be of marine as well as of terrestrial origin, with the distinction not always being clear. Generally, terrigenous CDOM is pronounced in the Arctic Ocean because of the large river inputs (Amon, 2004). The earliest study on Arctic CDOM fluorescence was published by Guay et al. (1999) who reported fluorescence data from a cross section of the Arctic Ocean along the 50-m isobath. From that early study we know that CDOM is highest in these waters of the Makarov Basin and decreases toward the Amundsen and Canada Basins. Although the excitation and emission wavelengths of that fluorometer were set to 320 and 420 nm, respectively, the strong signal over the Makarov Basin was interpreted to be mostly of terrestrial origin. Studies in the Fram Strait, using an *in situ* fluorometer with excitation wavelengths of 350-460 nm and an emission wavelength of 550 nm, revealed that the East Greenland Current transports a large amount of Arctic river CDOM to the North Atlantic (Amon et al., 2003). Interestingly, the maximum fluorescence along the EGC was found in subsurface layers (>50 m) and not at the lowest salinity. The strong positive relationship of *in situ* fluorescence and the lignin phenol concentration indicates that terrestrial DOM was responsible for most of the CDOM signal in this region.

In order to better understand the nature of the general CDOM signal in the Arctic Ocean, studies have looked more closely into absorbance characteristics from which new optical parameters were derived that can be related to CDOM origin, diagenetic history, and to some degree chemical composition. Such absorbance-derived parameters include absorption coefficients, specific UV absorbance at 254 nm (SUVA), and the spectral slope (S). Together, these indicators carry information about aromaticity (SUVA), source (S), molecular weight

(SUVA, S), and degradation state (both photodegradation and microbial alteration) (S) (Helms et al., 2008; Stedmon et al., 2011; Chapter 10). Another new approach applied to fluorescence data (excitation-emission matrices) in the Arctic Ocean is parallel factor analyses (PARAFAC), which is a statistical method to decompose the total fluorescence signal into subcomponents that contribute to the overall CDOM signal (Guéguen et al., 2012; Walker et al., 2009). For the Arctic Ocean these indictors have revealed that CDOM routinely measured with *in situ* backscatter fluorometers consists of terrigenous and marine sources and that the relative contribution of these two sources is not evenly distributed. While the Eurasian Basin is dominated by terrestrial CDOM, Canada Basin surface (<300 m) waters are dominated by marine CDOM (Stedmon et al., 2011). Absorbance-derived parameters, like absorption coefficients and spectral slope, identified three main CDOM components in the Arctic Ocean: a background marine CDOM pool, a recently produced marine CDOM pool, and the often dominant terrestrial CDOM pool (Stedmon et al., 2011). This study also indicated a relatively minor impact of photodegradation on the CDOM in surface waters.

Studies based on PARAFAC indicated 5-6 CDOM components with several being of terrestrial origin and two being marine (Guéguen et al., 2012; Walker et al., 2009). All studies indicate a subsurface (>30 m) CDOM maximum in the Arctic Ocean, particularly in the Canada Basin, and a surprising lack of CDOM in the top 50 m of the Canada Basin. Guéguen et al. (2012) argued for photodegradation of CDOM in Canada Basin surface waters, while absorbance based indicators and lignin S/V ratios do not indicate significant photochemical removal of CDOM in this region (Opsahl et al., 1999; Stedmon et al., 2011). It is more likely that dilution with sea-ice melt having low CDOM concentrations causes the surface water minima. CDOM measurements are semi-quantitative but

correlations to biomarkers will improve their quantitative use (Walker et al., 2013).

III DISTRIBUTION AND MASS BALANCE OF DOM

Because of the high vertical sampling resolution and satellite remote sensing potential, CDOM is a powerful tool to study the horizontal and vertical distribution of DOM in the Arctic Ocean. First attempts to use satellites to trace CDOM in the Arctic region have been published (Griffin et al., 2011; Fichot et al., 2013). Satellite remote sensing is most helpful in coastal waters (absent sea-ice cover during summer) but will become more applicable for the central Arctic Ocean as sea-ice retreat continues under a warmer climate. Fichot et al. (2013) demonstrated that satellite derived data can be used to develop an algorithm that traces lignin phenols in Arctic Ocean surface waters (<10 m). According to their study, export of Mackenzie River discharge to the central Beaufort Gyre happens more frequently than previously thought (Macdonald et al., 2002). The other interesting pattern revealed by the satellite data is a considerable amount of terrestrial CDOM on the East Siberian shelf, indicating that a significant portion of the Eurasian river discharge is not entrained directly into the Transpolar Drift, but rather continues east before entering the central basins. For central Arctic Ocean surface waters, maximum CDOM values were found over the Makarov basin, between the Lomonosov Ridge and the Alpha-Mendeleyev Ridge, and associated with the Transpolar Drift, a feature observed in 1998, 1999, 2005, and 2007 (Amon et al., unpublished data; Guay et al., 1999). This pattern is not reflected in the salinity distribution in the surface Arctic Ocean, which indicates lowest salinities in the surface waters of the Canadian Basin (Carmack et al., 2008). It is consistent with the freshwater transport pathway from the East Siberian Shelf to the central Canada Basin recently suggested by Morison et al. (2012). A significant export of CDOM (terrigenous and marine) has been observed in the EGC (Amon et al., 2003; Granskog et al., 2012) as well as in the Canadian Archipelago (Walker et al., 2009). CDOM levels in surface waters are lowest over the Nansen Basin and the central Canada Basin (Amon et al., unpublished data; Guéguen et al., 2012).

The vertical distribution of CDOM is even more intriguing because, with the exception of the Makarov Basin, CDOM is most abundant in subsurface layers (Amon et al., 2003; Cooper et al., 2005; Guéguen et al., 2012; Nakayama et al., 2011). The subsurface CDOM maximum in the Eurasian Basin is between 10 and 70 m, but it is deeper in the Canada Basin (120-220 m; Amon et al., unpublished data) where it can be found in both, the upper and the lower halocline. While Eurasian CDOM is mostly of terrigenous origin (Amon, 2004; Amon et al., 2003), the subsurface CDOM signal in the Canada Basin is a mix of terrestrial and marine-derived CDOM

(Amon et al., unpublished data; Guéguen et al., 2012; Stedmon et al., 2011). The Pacific-derived upper halocline is dominated by marine CDOM produced by heterotrophic processes on the Chukchi Shelf (Nakayama et al., 2011) while the lower, Atlantic-derived halocline is dominated by terrestrial CDOM coming from the Siberian rivers (Amon et al., unpublished data). Relating CDOM to salinity shows that for the majority of the existing observations, maximum CDOM values are found at intermediate salinity and low temperature, regardless of CDOM origin (Figure 14.9) and consistent with previous studies in Fram Strait (Amon et al., 2003) and the Canada Basin (Cooper et al., 2005; Guéguen et al., 2012; Matsuoka et al., 2012). The low CDOM level at the surface (<15 m in the Eurasian Basin and <50 m in the Canadian Basin) is due to sea-ice melt (Amon et al., unpublished data). The fact that CDOM is strongly related to hydrographic parameters makes it a useful tool not only for biogeochemists but also

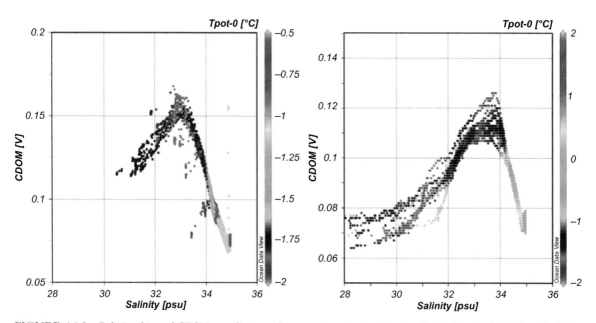

FIGURE 14.9 Relationships of CDOM to salinity and temperature in the Eurasian Basin (left) and the Canada Basin (right). *Figure drawn using Ocean Data View (Schlitzer, 2011).*

for physical oceanographers. The similarity of the CDOM distribution within the different regions of the Arctic Ocean argues for a consistent hydrographic control rather than a biological or photochemical one.

The abundance of CDOM in the halocline layers is consistent with previous studies (Aagaard and Carmack, 1989) identifying sea-ice formation in the shelf regions as important to sustaining the halocline layers by brine injection. Because DOM behaves much like salt during sea-ice formation (Amon, 2004), it is logical that the distribution of CDOM will mirror the distribution of brine. Sea-ice formation and melt partitions the CDOM (including the bulk of terrestrial DOM) by enriching CDOM in the halocline layers (Matsuoka et al., 2012) and depleting it in the surface layers, which are affected by sea-ice melt. The distribution of CDOM to some degree contradicts the distribution of freshwater fractions in the Canada Basin presented by Carmack et al. (2008); the lack of sea-ice melt and the abundance of meteoritic water in surface waters cannot be reconciled with the lack of CDOM signal (Amon et al., unpublished data).

Seasonal production of DOC by biological processes in surface waters increases the concentrations. In the central Arctic Ocean there is little biological activity, so little autochthonous DOC should accumulate. The annual amount of marine-produced DOC in the shelf seas is however lower compared to the amount added by river runoff, except for the Barents Sea with its low runoff contribution and the Bering-Chukchi Sea with its high PP. Even if the terrestrial DOC is not fully conservative, see Section I.B.1, it still dominates the observed distribution within large parts of the Arctic Ocean. However, as has been observed (e.g., Hansell et al., 2004; Letscher et al., 2011) degradation of DOC in waters of longer residence times, for example, in the Beaufort Gyre, results in lower DOC concentrations, which is further reinforced by the dilution with sea-ice melt water. The impact of two different fresh waters result in substantial scatter in DOC

versus salinity, a feature that can be substantially improved by plotting the DOC *versus* river runoff fraction. The latter can be computed from salinity and $\delta^{18}O$ values (e.g., Östlund and Hut, 1984). Cooper et al. (2005) found a significant linear correlation between DOC and river runoff fraction in the waters of the northern Chukchi Sea and southern Canadian Basin generated during the 2002 Shelf Basin Interaction cruises:

$$DOC = 149.9 \times \text{runoff fraction} + 55;$$
$$r^2 = 0.84; n = 274.$$

Theoretically, the distribution of terrestrial DOC in the central Arctic Ocean surface waters will follow the pattern of river runoff fraction, with highest concentrations in the Canadian and Makarov Basins and lowest in the Nansen Basin. However, this is not always the case, indicating a need to improve our understanding of DOC distributions in the surface, central Arctic Ocean.

DOC concentrations in the Arctic Ocean deep waters are lower than in the inflowing Atlantic water (Bussmann and Kattner, 2000; Opsahl et al., 1999). Opsahl et al. (1999) found $61 \, \mu M$ in the inflowing Atlantic water of Fram Strait and $65 \, \mu M$ in that recirculating in Fram Strait, while Bussmann and Kattner (2000) found $59 \, \mu M$ ($n = 37$) in the Atlantic Layer water of the central Arctic Ocean. These values can be compared to those of Amon et al. (2003), where Atlantic water entering through the Fram Strait had an average concentration of $61 \pm 5 \, \mu M$ in 1997. This value is slightly higher than found in the recirculating Atlantic Layer water, $59 \pm 5 \, \mu M$, while the outflowing deep water from the Eurasian Basin was significantly lower with a mean of $50 \pm 2 \, \mu M$ ($n = 20$) (Amon et al., 2003). Concentrations in the deep central Arctic Ocean are similar, with the mean of $50 \, \mu M$ in the Nansen Basin ($n = 53$), $54 \, \mu M$ in the Amundsen Basin ($n = 67$), and $56 \, \mu M$ in the Makarov Basin (Bussmann and Kattner, 2000). The range is $6 \, \mu M$, which is close to but significantly above the analytical range. The cause for these lower DOC concentrations likely is the "age" difference of the deep waters. Sedimentation of particulate organic matter is very low in the central Arctic basins and thus the supply of "new" DOC is limited; at the same time, the available DOC is degraded. In the Nordic Seas the deep water is renewed by convection and particulate organic matter is constantly falling down from the productive surface waters, both adding to the DOC supply. Even with these lower concentrations of DOC in the outflowing deep water there

is a small contribution of terrigenous DOC, on the order of 2-3% (Benner et al., 2005).

Fluxes of DOC to and from the Arctic Ocean are given in Table 14.2, based on measured concentrations and reported volume transports of the different waters. This budget does not distinguish between terrigenous and marine DOC. Generally the terrigenous DOC is high in the surface waters and low in the deep waters (e.g., Benner et al., 2005). It should be

TABLE 14.2 Budget of DOC fluxes to and from the arctic ocean

Water Mass	Volume Transport (Sv)	DOC (μM)	DOC Flux (10^{12} g C year^{-1})
In			
Runoff	0.12	667 ± 50[a]	25 ± 2
Pacific water	0.8[b]	70 ± 15[c]	18 ± 4
Atlantic water	7.5[d]	58 ± 5[e]	137 ± 12
Deep water	4.6[d]	50 ± 4[e]	73 ± 6
Total in	**13.0**		**253 ± 14**
Out			
Surface water through CAA	1.5[b]	80 ± 20[f]	38 ± 9
Sea ice	0.1[d]	100 ± 50[g]	3 ± 3
Surface water through Fram Strait	1.2	79 ± 19[e]	30 ± 7
Atlantic Layer water	3.3[d]	64 ± 5[h]	67 ± 5
Deep water	6.9[d]	49 ± 2[a]	107 ± 4
Total out	**13.0**		**244 ± 14**
Net outflow			**−9 ± 20**

Note. River runoff includes all continental freshwater input, and all outflows are through Fram Strait except for the polar water that also exits through the Canadian Arctic Archipelago (CAA). As discussed in the text, errors in the organic carbon fluxes do not include errors in the volume transports.

[a]Mean of Walsh et al. (1997), Wheeler et al. (1997), and Guay et al. (1999).

[b]The average discharge-weighted concentration computed for the six largest rivers.

[c]Schauer et al. (2004; 2008) and Schauer and Beszczynska-Möller (2009).

[d]Beszczynska-Möller et al. (2011).

[e]Amon et al. (2003).

[f]Walker et al. (2009).

[g]Thomas et al. (1995) and Opsahl et al. (1999).

[h]Benner et al. (2005).

noted that the total DOC *fluxes* of Table 14.2 are about twice as large as reported earlier (e.g., Anderson et al., 1998), mainly a result of substantially higher volume fluxes observed during the last decade.

Uncertainties in Table 14.2 reflect variability in DOC concentrations reported for the water masses. No consideration of uncertainties in the volume fluxes is included. The largest uncertainties are in the Atlantic and deep water volume fluxes but these waters have fairly constant DOC concentrations. In order to conserve mass an error in the volume transport into the Arctic Ocean has to be compensated by a comparable error in the volume transport out of the Arctic Ocean. As this is largest in the deep waters it result in very close to the same error in the in- and outfluxes of DOC, and hence have a small impact on the net DOC flux out of the Arctic Ocean. Summing the in- and outfluxes of Table 14.2 gives $-9 (\pm 20) \times 10^{12} \, \text{g C}$ year^{-1}; considering the uncertainty in the estimate, the Arctic Ocean is neither a sink nor a source of DOC.

The *in situ* productivity of marine DOC within the central Arctic Ocean has been estimated to $6.1 \, \text{g C m}^{-2}$ year^{-1} and the *in situ* respiration to be $8.8 \, \text{g C m}^{-2}$ year^{-1} (Wheeler et al., 1997). Combining these numbers with the area of the deep central Arctic Ocean ($5.8 \times 10^{12} \, \text{m}^2$) gives a total *in situ* productivity of $35 \times 10^{12} \, \text{g C}$ year^{-1} and a total *in situ* respiration of $51 \times 10^{12} \, \text{g C}$ year^{-1}. These numbers are based on a single summer investigation in a limited area so the uncertainties must be significant when applying them to a whole year and the whole central Arctic Ocean. Recent estimates of net community production given above are around $440 \times 10^{12} \, \text{g C}$ year^{-1} and $44 \times 10^{12} \, \text{g DOC}$ year^{-1} (Arrigo and van Dijken, 2011). It is interesting to note that the *in situ* respiration of DOC exceeds that of *in situ* production of marine DOC, while the latter is on the same order as the added terrigenous DOC ($35 \times 10^{12} \, \text{g C}$ year^{-1} relative to $20 \times 10^{12} \, \text{g C}$ year^{-1}). These results indicate that the *in situ*

respiration of DOC in the central Arctic Ocean will quantitatively consume all the marine DOC produced in the central Arctic Ocean and some of that added by river runoff.

Even if the Arctic Ocean itself is neither a sink nor a source of DOC there is significant export to the North Atlantic. This flux ($27 \times 10^{12} \, \text{g C}$ year^{-1}) results from inflow of Pacific water ($18 (\pm 5) \times 10^{12} \, \text{g C}$ year^{-1}), river runoff ($25 (\pm 2) \times 10^{12} \, \text{g C}$ year^{-1}), and the difference between *in situ* production ($35 \times 10^{12} \, \text{g C}$ year^{-1}) and respiration ($51 \times 10^{12} \, \text{g C}$ year^{-1}) within the Arctic Ocean.

The finding that the outflowing deep water DOC concentrations are lower than (or very similar to) the inflowing concentrations indicates that (i) small amounts of terrigenous DOM are exported to deep layers (as was also concluded by Benner et al. (2005) on the basis of lignin and δ^{13}C analysis) and (ii) limited net export of marine DOM occurs to deep layers. The latter statement is supported by the arguments above that most marine DOC produced in the central Arctic Ocean is respired in the surface layers. The fact that little terrigenous DOC is exported to the deep waters of the Arctic Ocean, through dense plumes originating on the selves where they are initiated by brine drainage from sea-ice production, is an important finding as it puts constraints on the global DOC budget.

With regard to carbon transport in and out of the Arctic Ocean, DOC fluxes are about 5% of total carbon fluxes, calculated as the sum of dissolved inorganic and organic carbon (Anderson et al., 1998). However, while the dissolved inorganic carbon concentration largely has a positive correlation with salinity over the range from fresh water to seawater, the DOC concentration has a negative one. Consequently, *in situ* production and respiration of DOC plays a relatively more important role for the inorganic carbon cycle in the low-salinity surface waters, relative to deeper layers, and it is the surface water that is in contact with the atmosphere linking the marine carbon cycle to climate.

630

630

Acknowledgments

Financial support from the Swedish Research Council, the US National Science Foundation, and Texas A&M University at Galveston are greatly acknowledged. The initial version of the chapter was improved by the comments of Gerhard Kattner.

References

Aagaard, K., Carmack, E.C., 1989. The role of sea ice and other fresh water in the Arctic circulation. J. Geophys. Res. 94, 14,485–14,498.

Alling, V., Sanchez-Garcia, L., Porcelli, D., Pugach, S., Vonk, J., van Dongen, B., et al., 2010. Non-conservative behavior of dissolved organic carbon across the Laptev and East Siberian Seas. Global Biogeochem. Cycles 24, http://dx.doi.org/10.1029/2010GB003834, GB4033.

Amon, R.M.W., 2004. The role of dissolved organic matter for the Arctic Ocean carbon cycle. In: Stein, R., Macdonald, R.W. (Eds.), The Arctic Ocean Organic Carbon Cycle: Present and Past. Springer, New York, pp. 83–99.

Amon, R.M.W., Benner, R., 1996. Bacterial utilization of different size classes of dissolved organic matter. Limnol. Oceanogr. 41, 41–51.

Amon, R.M.W., Benner, R., 2003. Combined neutral sugars as indicators of the diagenetic state of dissolved organic matter in the Arctic Ocean. Deep-Sea Res. Pt. I 50, 151–169.

Amon, R.M.W., Meon, B., 2004. The biogeochemistry of dissolved organic matter and nutrients in two large Arctic estuaries and potential implications for our understanding of the Arctic Ocean system. Mar. Chem. 92, 311–330.

Amon, R.M.W., Fitznar, H.-P., Benner, R., 2001. Linkages among bio reactivity, chemical composition, and diagenetic state of marine dissolved organic matter. Limnol. Oceanogr. 46, 287–297.

Amon, R.M.W., Budeus, G., Meon, B., 2003. Dissolved organic carbon distribution and origin in the Nordic seas: exchanges with the Arctic Ocean and North Atlantic. J. Geophys. Res. 108, 3221. http://dx.doi.org/10.1029/2002JC001594.

Amon, R., Rinehart, A.J., Duan, S., Louchouarn, P., Prokushkin, A., Guggenberger, G., et al., 2012. Dissolved organic matter sources in large Arctic rivers. Geochim. Cosmochim. Acta 94, 217–237.

Anderson, L.G., Olsson, K., Chierici, M., 1998. A carbon budget for the Arctic Ocean. Global Biogeochem. Cycles 12, 455–465.

Anderson, L.G., Jones, E.P., Swift, J.H., 2003. Export production in the central Arctic Ocean as evaluated from phosphate deficit. J. Geophys. Res. 108, 3199. http://dx.doi.org/10.1029/2001JC001057.

Arrigo, K.R., van Dijken, G.L., 2011. Secular trends in Arctic Ocean net primary production. J. Geophys. Res. 116, http://dx.doi.org/10.1029/2011JC007151, C09011.

Arrigo, K.R., Perovich, D.K., Pickart, R.S., Brown, Z.W., van Dijken, G.L., Lowry, K.E., et al., 2012. Massive phytoplankton blooms under Arctic sea ice. Science 15, http://dx.doi.org/10.1126/science.1215065.

Benner, R., Kaiser, K., 2011. Biological and photochemical transformations of amino acids and lignin phenols in riverine dissolved organic matter. Biogeochemistry 102, 209–222. http://dx.doi.org/10.1007/s10533-010-9435-4.

Benner, R., Benitez-Nelson, B., Kaiser, K., Amon, R.M.W., 2004. Export of young terrigenous dissolved organic carbon from rivers to the Arctic Ocean. Geophys. Res. Lett. 31, http://dx.doi.org/10.1029/2003GL019251, L05305.

Benner, R., Louchouarn, P., Amon, R.M.W., 2005. Terrigenous dissolved organic matter in the Arctic Ocean and its transport to surface and deep waters of the North Atlantic. Global Biogeochem. Cycles 19, http://dx.doi.org/10.1029/2004GB002398, GB2025.

Beszczynska-Möller, A., Woodgate, R.A., Lee, C., Melling, H., Karcher, M., 2011. A synthesis of exchanges through the main oceanic gateways to the Arctic Ocean. Oceanography 24 (3), 82–99. http://dx.doi.org/10.5670/oceanog.2011.59.

Børsheim, K.Y., Myklestad, S.M., 1997. Dynamics of DOC in the Norwegian Sea inferred from monthly profiles collected during 3 years at 66°N, 2°E. Deep-Sea Res. 44 (4), 593–601.

Bussmann, I., Kattner, G., 2000. Distribution of dissolved organic carbon in the central Arctic Ocean: the influence of physical and biological properties. J. Mar. Syst. 27, 209–219.

Carmack, E.C., McLaughlin, F.A., Yamamoto-Kawai, M., Itoh, M., Shimada, K., Krishfield, R., et al., 2008. Freshwater storage in the northern ocean and the special role of the Beaufort Gyre. In: Dickson, R.R., Meincke, J., Rhines, P. (Eds.), Arctic–Subarctic Ocean Fluxes. Springer, Netherlands, pp. 145–169.

Cauwet, G., Sidorov, I., 1996. The biogeochemistry of Lena River: organic carbon and nutrients distribution. Mar. Chem. 53, 211–227.

Codispoti, L.A., Kelly, V., Thessen, A., Matrai, P., Suttles, S., Hill, V., et al., 2013. Synthesis of primary production in the Arctic Ocean: III. Nitrate and phosphate based estimates of net community production. Prog. Oceanogr. 110, 126–150.

Cooper, L.W., Benner, R., McClelland, J.W., Peterson, B.J., Holmes, R.M., Raymond, P.A., 2005. Linkages among runoff, dissolved organic carbon, and the stable oxygen isotope composition of seawater and other water mass indicators in the Arctic Ocean. J. Geophys. Res. 110, http://dx.doi.org/10.1029/2005JG000031, G02013.

Davis, J., Benner, R., 2005. Seasonal trends in the abundance, composition and bioavailability of particulate and dissolved organic matter in the Chukchi/Beaufort Seas and western Canada Basin. Deep-Sea Res. Pt. II 52, 3396–3410.

Davis, J., Benner, R., 2007. Quantitative estimates of labile and semi-labile DOC in the western Arctic Ocean: a molecular approach. Limnol. Oceanogr. 5, 2434–2444.

Davis, J., Kaiser, K., Benner, R., 2009. Amino acid and amino sugar yields and compositions as indicators of dissolved organic matter diagenesis. Org. Geochem. 40, 343–352.

Degens, E.T., Kempe, S., Richey, J.E., 1991. Summary: biogeochemistry of major world rivers. In: Degens, E.T., Kempe, S., Richey, J.E. (Eds.), Biogeochemistry of Major World Rivers. John Wiley, New York, pp. 323–347.

Dittmar, T., 2004. Evidence for terrigenous dissolved organic nitrogen in the Arctic deep sea. Limnol. Oceanogr. 49 (1), 148–156.

Engbroadt, R., Kattner, G., 2005. On the biogeochemistry of dissolved carbohydrates in the Greenland Sea (Arctic). Org. Geochem. 36, 937–948.

Fichot, C.G., Kaiser, K., Hooker, S.B., Amon, R.M.W., Babin, M., Bélanger, S., et al., 2013. Pan-Arctic views of continental runoff in the Arctic Ocean. Sci. Rep. 3, doi:10.1038/srep01053, Article number: 1053.

Fitznar, H.P., 1999. D-Amino acids as tracers for biogeochemical processes in the river-shelf-ocean-system of the Arctic. Polar Report No. 334, Alfred Wegener Institute for Polar and Marine Research.

Fransson, A., Chierici, M., Anderson, L.G., Bussman, I., Kattner, G., Jones, E.P., et al., 2000. The importance of shelf processes for the modification of chemical constituents in the waters of the eastern Arctic Ocean. Cont. Shelf Res. 21, 225–242.

Gordeev, V.V., Martin, J.M., Sidorov, I.S., Sidorova, M.V., 1996. A reassessment of the eurasian river input of water, sediment, major elements, and nutrients to the Arctic Ocean. Am. J. Sci. 296, 664–691.

Gosselin, M., Levasseur, M., Wheeler, P.A., Horner, R.A., Booth, B.C., 1997. New measurements of phytoplankton and ice algal production in the Arctic Ocean. Deep-Sea Res.Pt. II 44 (8), 1623–1644.

Granskog, M.A., Stedmon, C.A., Dodd, P.A., Amon, R.M.W., Pavlov, A.K., de Steur, L., et al., 2012. Characteristics of colored dissolved organic matter (CDOM) in the Arctic outflow in the Fram Strait: assessing the changes and fate of terrigenous CDOM in the Arctic Ocean. J. Geophys. Res. 117, http://dx.doi.org/10.1029/2012JC008075, C12021.

Griffin, C.G., Frey, K.E., Rogan, J., Holmes, R.M., 2011. Spatial and interannual variability of dissolved organic matter in the Kolyma River, East Siberia, observed using satellite imagery. J. Geophys. Res. 116, http://dx.doi.org/10.1029/2010JG001634, G03018.

Guay, C.K., Klinghammer, G.P., Falkner, K.K., Benner, R., Coble, P.G., Whitledge, T.E., et al., 1999. High-resolution measurements of dissolved organic carbon in the Arctic Ocean by in situ fiber-optic spectrometer. Geophys. Res. Lett. 26, 1007–1010.

Guéguen, C., McLaughlin, F.A., Carmack, E.C., Itoh, M., Narita, H., Nishino, S., 2012. The nature of colored dissolved organic matter in the southern Canada Basin and East Siberian Sea. Deep-Sea Res. Pt. II 81–84, 102–113. http://dx.doi.org/10.1016/j.dsr2.2011.05.004.

Hansell, D.A., 2013. Recalcitrant dissolved organic carbon fractions. Ann. Rev. Mar. Sci. 5, 421–445.

Hansell, D.A., Carlson, C.A., 1998. Net community production of dissolved organic carbon. Global Biogeochem. Cycles 12, 443–453.

Hansell, D.A., Whitledge, T.E., Goering, J.J., 1993. Patterns of nitrate utilization and new production over the Bering-Chukchi shelf. Cont. Shelf Res. 13, 601–628.

Hansell, D.A., Kadko, D., Bates, N.R., 2004. Degradation of terrigenous dissolved organic carbon in the western Arctic Ocean. Science 304, 858–861.

Helms, J.R., Stubbins, A., Ritchie, J.D., Minor, E.C., Kieber, D.J., Mopper, K., 2008. Absorption spectral slopes and slope ratios as indicators of molecular weight, source, and photobleaching of chromophoric dissolved organic matter. Limnol. Oceanogr. 53 (3), 955–969. http://dx.doi.org/10.4319/lo.2008.53.3.0955.

Hill, V.J., Matrai, P.A., Olson, E., Suttles, S., Steele, M., Codispoti, L.A., et al., 2013. Synthesis of integrated primary production in the Arctic Ocean: II. In situ and remotely sensed estimates. Prog. Oceanogr. 110, 107–125.

Holmes, R.M., McClelland, J.W., Raymond, P.A., Frazer, B.B., Peterson, B.J., Stieglitz, M., 2008. Lability of DOC transported by Alaskan rivers to the Arctic Ocean. Geophys. Res. Lett. 35, http://dx.doi.org/10.1029/2007GL032837, L03402.

Holmes, R.M., McClelland, J.W., Peterson, B.J., Tank, S.E., Bulygina, E., Eglinton, T.I., et al., 2012. Seasonal and annual fluxes of nutrients and organic matter from large rivers to the Arctic Ocean and surrounding seas. Estuar. Coast. Shelf Sci. 35, 369–382. http://dx.doi.org/10.1007/s12237-011-9386-6.

Hulthe, G., Hall, P., 1997. Benthic carbon fluxes—DOC versus ΣCO₂ in shelf, slope and deep-sea environments, and relation to oxygen fluxes. Rep. Polar Res. 226, 115–116.

Hulth, S., Hall, P.O.J., Blackburn, T.H., Landen, A., 1996. Arctic sediments (Svalbard): pore water and solid phase distributions of C, N, P and Si. Polar Biol. 16, 447–462.

Jones, E.P., Rudels, B., Anderson, L.G., 1995. Deep waters of the Arctic Ocean: origin and circulation. Deep-Sea Res. 42, 737–760.

Jones, E.P., Anderson, L.G., Swift, J.H., 1998. Distribution of Atlantic and Pacific waters in the upper Arctic Ocean: implications for circulation. Geophys. Res. Lett. 25, 765–768.

Jones, E.P., Swift, J.H., Anderson, L.G., Lipizer, M., Civitarese, G., Falkner, K.K., et al., 2003. Tracing Pacific water in the North Atlantic Ocean. J. Geophys. Res. 108 (C4), 3116. http://dx.doi.org/10.1029/2001JC001141.

Kattner, G., Lobbes, J.M., Fitznar, H.P., Engbrodt, R., Nöthig, E.-M., Lara, R.J., 1999. Tracing dissolved organic substances and nutrients from the Lena River through Laptev Sea (Arctic). Mar. Chem. 65, 25–39.

Köhler, H., Meon, B., Gordeev, V.V., Spitzy, A., Amon, R.M.W., 2003. Dissolved organic matter (DOM) in the estuaries of Ob and Yenisei and the adjacent Kara-Sea, Russia. Proc. Mar. Sci. 6, 281–309.

Lara, R.J., Rachold, V., Kattner, G., Hubberten, H.W., Guggenberger, G., Skoog, A., et al., 1998. Dissolved organic matter and nutrients in the Lena River, Siberian Arctic: characteristics and distribution. Mar. Chem. 59, 301–309.

Letscher, R.T., Hansell, D.A., Kadko, D., 2011. Rapid removal of terrigenous dissolved organic carbon over the Eurasian shelves of the Arctic Ocean. Mar. Chem. 123, 78–87. http://dx.doi.org/10.1016/j.marchem.2010.10.002.

Lobbes, J.M., Fitznar, H.P., Kattner, G., 2000. Biogeochemical characteristics of dissolved and particulate organic matter in Russian rivers entering the Arctic Ocean. Geochim. Cosmochim. Acta 64 (17), 2973–2983.

Macdonald, R.W., Yu, Y., 2005. The Mackenzie estuary of the Arctic Ocean. In: Wangersky, P.J. (Ed.), Water Pollution: Estuaries. Springer-Verlag, Heidelberg, pp. 91–120. http://dx.doi.org/10.1007/698_5_027.

Macdonald, R.W., McLaughlin, F.A., Carmack, E.C., 2002. Freshwater and its sources during the SHEBA drift in the Canada Basin of the Arctic Ocean. Deep-Sea Res.Pt. I 49 (10), 1769–1785.

Mathis, J.T., Hansell, D.A., Bates, N.R., 2005. Strong hydrographic controls on spatial and seasonal variability of dissolved organic carbon in the Chukchi Sea. Deep-Sea Res.Pt. II 52, 3245–3258.

Mathis, J.T., Hansell, D.A., Kadko, D., Bates, N.R., Cooper, L.W., 2007. Determining net dissolved organic carbon production in the hydrographically complex western Arctic Ocean. Limnol. Oceanogr. 52, 1789–1799. http://dx.doi.org/10.4319/lo.2007.52.5.1789.

Mathis, J.T., Grebmeier, J.M., Hansell, D.A., Hopcroft, R.R., Kirchman, D.L., Lee, S.H., et al., 2014. Carbon biogeochemistry of the western Arctic: primary production, carbon export and the controls on ocean acidification. In: Grebmeier, J.M., Maslowski, W., Zhao, J.-P. (Eds.), Biogeochemistry of the Pacific Arctic Region". Springer, New York, in press.

Matrai, P.A., Olson, E., Suttles, S., Hill, V., Codispoti, L.A., Light, B., et al., 2013. Synthesis of primary production in the Arctic Ocean: I. Surface waters, 1954-2007. Prog. Oceanogr. 110, 93–106.

Matsuoka, A., Bricaud, A., Benner, R., Para, J., Sempéré, R., Prieur, L., et al., 2012. Tracing the transport of colored dissolved organic matter in water masses of the southern Beaufort Sea: relationship with hydrographic characteristics. Biogeosciences 9, 925–940. http://dx.doi.org/10.5194/bg-9-925-2012.

McGuire, A.D., Macdonald, R.W., Schuur, E.A.G., Harden, J.W., Kuhry, P., Hayes, D.J., et al., 2010. The carbon budget of the northern cryospere region. Curr. Opin. Environ. Sustainability 2, 231–236.

Melnikov, I.A., 1997. The Arctic Ice Ecosystem. Gordon and Breach Science Publisher, The Netherlands, 204 pp.

Menard, H.W., Smith, S.M., 1966. Hypsometry of ocean basin provinces. J. Geophys. Res. 71, 4305–4325.

Meon, B., Amon, R.M.W., 2004. Heterotrophic bacterial activity and fluxes of dissolved free amino acids (DFAA) and glucose in the Arctic rivers Ob, Yenisei and the adjacent Kara Sea. Aquat. Microb. Ecol. 37, 121–135.

Morison, J.H., Steele, M., Andersen, R., 1998. Hydrography of the upper Arctic Ocean measured from the nuclear submarine USS Pargo. Deep-Sea Res. 45, 15–38.

Morison, J., Kwok, R., Peralta-Ferriz, C., Alkire, M., Rigor, I., Andersen, R., et al., 2012. Changing Arctic Ocean freshwater pathways. Nature 481, 66–70.

Nakayama, Y., Fujita, S., Kuma, K., Shimada, K., 2011. Iron and humic-type fluorescent dissolved organic matter in the Chukchi Sea and Canada Basin of the western Arctic Ocean. J. Geophys. Res. 116, http://dx.doi.org/10.1029/2010JC006779, C07031.

Opsahl, S., Benner, R., 1997. Distribution and cycling of terrigenous dissolved organic matter in the ocean. Nature 386, 480–482.

Opsahl, S., Benner, R., Amon, R.M.W., 1999. Major flux of terrigenous dissolved organic matter through the Arctic Ocean. Limnol. Oceanogr. 44, 2017–2023.

Östlund, H.G., Hut, G., 1984. Arctic Ocean water mass balance from isotope data. J. Geophys. Res. 89 (C4), 6373–6381.

Pocklington, R., 1987. Arctic rivers and their discharge. Mitt. Geol.-Paläontol. Inst. Univ. Hamburg, SCOPE-UNEP. Sonderband 64, 261–268.

Raymond, P.A., McClelland, J.W., Holmes, R.M., Zhulidov, A.V., Mull, K., Peterson, B.J., et al., 2007. Flux and age of dissolved organic carbon exported to the Arctic Ocean: a carbon isotopic study of the five largest arctic rivers. Global Biogeochem. Cycles 21, http://dx.doi.org/10.1029/2007GB002934.

Reigstad, M., Carroll, J., Slagstad, D., Ellingsen, I., Wassmann, P., 2011. Intra-regional comparison of productivity, carbon flux and ecosystem composition within the northern Barents Sea. Prog. Oceanogr. 90, 33–46. http://dx.doi.org/10.1016/j.pocean.2011.02.005.

Rudels, B., Jones, E.P., Anderson, L.G., Kattner, G., 1994. On the intermediate depth waters of the Arctic Ocean. In: Johannessen, O.M., Muench, R., Overland, J.E. (Eds.), The Polar Oceans and Their Role in Shaping the Global Environment. American Geophysical Union, Washington, D.C., pp. 33–46.

Rudels, B., Anderson, L.G., Jones, E.P., 1996. Formation and evolution of the surface mixed layer and halocline of the Arctic Ocean. J. Geophys. Res. 101, 8807–8821.

Sakshaug, E., 2004. Primary and secondary production in the Arctic seas. In: Stein, R., Macdonald, R.W. (Eds.), The Organic Carbon Cycle in the Arctic Ocean. Springer, New York, pp. 57–81.

Sakshaug, E., Skjoldal, H.R., 1989. Life at the ice edge. Ambio 18, 60–67.

Schauer, U., Beszczynska-Möller, A., 2009. Problems with estimation and interpretation of oceanic heat transport: conceptual remarks for the case of Fram Strait in the Arctic Ocean. Ocean Sci. 5, 487–494. http://dx.doi.org/10.5194/os-5-487-2009.

Schauer, U., Fahrbach, E., 1999. A dense bottom water plume in the western Barents Sea: downstream modification and interannual variability. Deep-Sea Res. 46, 2095–2108.

Schauer, U., Muench, R., Rudels, B., Timokhov, L., 1997. The impact of eastern Arctic shelf waters on the Nansen Basin intermediate layers. J. Geophys. Res. 102, 3371–3382.

Schauer, U., Fahrbach, E., Osterhus, S., Rohardt, G., 2004. Arctic warming through the Fram Strait: oceanic heat transport from 3 years of measurements. J. Geophys. Res. 109, http://dx.doi.org/10.1029/2003JC001823, C06026.

Schauer, U., Beszczynska-Möller, A., Walczowski, W., Fahrbach, E., Piechura, J., Hansen, E., 2008. Variation of measured heat flow through the Fram Strait between 1997 and 2006. In: Dickson, R., Meincke, J., Rhines, P. (Eds.), Arctic-Subarctic Ocean Fluxes. Springer, Dordrecht, pp. 65–85.

Schlitzer, R., 2011. Ocean Data View, http://odv.awi.de.

Schlosser, P., Bauch, D., Fairbanks, R., Bönisch, G., 1994. Arctic river-runoff: mean residence time on the shelves and in the halocline. Deep-Sea Res. 41, 1053–1068.

Shen, Y., Fichot, C.G., Benner, R., 2012. Dissolved organic matter composition and bioavailability reflect ecosystem productivity in the western Arctic Ocean. Biogeosciences 9, 4993–5005. http://dx.doi.org/10.5194/bg-9-4993-2012.

Shin, K.H., Tanaka, N., 2004. Distribution of dissolved organic matter in the eastern Bering Sea, Chukchi Sea (Barrow Canyon) and Beaufort Sea. Geophys. Res. Lett. 31, http://dx.doi.org/10.1029/2004GL021039, L24304.

Slagstad, D., Wassmann, P., 1996. Climate change and carbon flux in the Barents Sea: 3-D simulations of ice-distribution, primary production and vertical export of particulate organic carbon. Mem. Natl. Inst. Polar Res. 51, 119–141.

Smith, R.E.H., Gosselin, M., Kudoh, S., Robineau, B., Taguchi, S., 1997. DOC and its relationship to algae in bottom ice communities. J. Mar. Syst. 11, 71–80.

Stedmon, C.A., Amon, R.M.W., Rhinehart, A.J., Walker, S.A., 2011. The supply and characteristics of colored dissolved organic matter (CDOM) in the Arctic Ocean. Mar. Chem. 124, 108–118.

Swift, J.H., Takahashi, T., Livingstone, H.D., 1983. The contribution of the Greenland and Barents Seas to the deep water of the Arctic Ocean. J. Geophys. Res. 88, 5981–5986.

Tarnocai, C., Canadell, J.G., Schuur, E.A.G., Kuhry, P., Mazhitova, G., Zimov, S., 2009. Soil organic carbon pools in the northern circumpolar permafrost region. Global Biogeochem. Cycles 23, http://dx.doi.org/10.1029/2008GB003327, GB2023.

Telang, S.A., Pocklington, R., Naidu, A.S., Romankevich, E.A., Gitelson, I.I., Gladyshev, M.I., 1991. Carbon and mineral transport in major North American, Russian Arctic, and Siberian rivers: the St. Lawrence, the Mackenzie, the Yukon, the Arctic Alaskan rivers, the Arctic basin rivers in the Soviet Union, and the Yenisey. In: Degens, E.T., Kempe, S., Richey, J.E. (Eds.), Biogeochemistry of Major World Rivers. John Wiley, New York, pp. 75–104.

Thomas, D.T., Lara, R.J., Eicken, H., Kattner, G., Skoog, A., 1995. Dissolved organic matter in Arctic multi-year sea ice during winter: major components and relationship to ice characteristics. Polar Biol. 15, 477–483.

Thurman, E.M., 1985. Aquatic humic substances. In: Thurman, E.M. (Ed.), Organic geochemistry of natural waters". Nijhoff/Junk Publishers, Dordrecht, pp. 273–361.

Vonk, J.E., Sanchez-Garcia, L., van Dongen, B.E., Alling, V., Kosmach, D., Charkin, A., et al., 2012. Activation of old carbon by erosion of coastal and subsea permafrost in Arctic Siberia. Nature 489 (7414), 137–140. http://dx.doi.org/10.1038/nature11392.

Walker, S.A., Amon, R.M.W., Stedmon, C., Duan, S.W., Louchouarn, P., 2009. The use of PARAFAC modeling to trace terrestrial dissolved organic matter and fingerprint water masses in coastal Canadian Arctic surface waters. J. Geophys. Res. 114, doi:10.1029/2009jg000990, G00F06.

Walker, S.A., Amon, R.M.W., Stedmon, C.A., 2013. Variations in high-latitude riverine fluorescent dissolved organic matter: a comparison of large Arctic rivers. J. Geophys. Res. Biogeosci. 118, 1689–1702. http://dx.doi.org/10.1002/2013JG002320.

Walsh, J.J., Dieterle, D.A., Muller-Karger, F.E., Aagaard, K., Roach, A.T., Whitledge, T.E., et al., 1997. CO_2 cycling in the costal ocean II. Seasonal organic loading of the Arctic Ocean from source waters in the Bering Sea. Cont. Shelf Res. 17, 1–36.

Wang, D., Henrichs, S.M., Guo, L., 2006. Distributions of nutrients, dissolved organic carbon and carbohydrates in the western Arctic Ocean. Cont. Shelf Res. 26, 1654–1667.

Wassmann, P., Slagstad, D., Ellingsen, I., 2010. Primary production and climatic variability in the European sector of the Arctic Ocean prior to 2007: preliminary results. Polar Biol. 33, 1641–1650. http://dx.doi.org/10.1007/s00300-010-0839-3.

Wheeler, P.A., Watkins, J.M., Hansing, R.L., 1997. Nutrients, organic carbon and organic nitrogen in the upper water column of the Arctic Ocean: implications for the sources of dissolved organic carbon. Deep-Sea Res. Pt. II 44, 1571–1592.

Modeling DOM Biogeochemistry

Thomas R. Anderson*, *James R. Christian*†, *Kevin J. Flynn*‡

*National Oceanography Centre Southampton, Southampton, UK
†Fisheries and Oceans Canada, Canadian Centre for Climate Modelling and Analysis,
Victoria, British Columbia, Canada
‡Centre for Sustainable Aquatic Research, Swansea University, Swansea, UK

CONTENTS

I INTRODUCTION

Up until the 1950s, the prevailing view of microbial ecology in the ocean was summed up by Keys et al. (1935): "It appears that in the sea there is generally an equilibrium such that only minimal bacterial activity at the expense of dissolved organic matter (DOM) takes place." Bacterial numbers, enumerated by plate counts, were typically a mere 100 per cubic centimeter (e.g., Reuszer, 1933). As Krogh (1934) put it, "...the number of bacteria present in ocean water is extremely small," continuing "... dissolved organic matter is not a suitable food." In similar vein, Harvey (1950) remarked that the numbers of bacteria in the sea are "limited by the great dilution of DOM serving as food in the water" and that attachment to particles provided a more favorable substrate for microbes. Changes in outlook came about in the following two decades with the realization that, based on direct microscope counts, numbers of bacteria were orders of magnitude higher than previously thought (e.g., Jannasch and Jones, 1959), along with the introduction of ^{14}C methods (Parsons and Strickland, 1961) and wet oxidation techniques (Menzel and Vaccaro, 1964) that demonstrated the dynamic nature of DOM in marine environments (e.g., Banoub and Williams, 1973). A fuller appreciation of the importance of the microbial loop was well on the way with Pomeroy's (1974) key paper. Further important publications followed which elaborated the ideas in more detail (e.g., Azam et al., 1983; Sorokin, 1977; Williams, 1981). Yet, there remained a considerable reluctance among biologists to reject the notion that food webs in the ocean are more complex than copepods feeding on phytoplankton (Sieburth, 1977).

This new paradigm regarding the importance of microbial processes and DOM dynamics was embraced slowly by biologists, and even more slowly by the modeling community. The first models of the marine ecosystem were developed by Richard Fleming and Gordon Riley in the 1940s and were followed, during the next three decades, by the nutrient-phytoplankton-zooplankton (NPZ) model of John Steele (e.g., Steele and Henderson, 1981). By the 1980s, computer simulations were becoming widespread. Most models continued to focus on the classical food chain of phytoplankton and zooplankton, as presented by Steele, without including DOM or bacteria (e.g., Franks et al., 1986; Hofmann and Ambler, 1988). There were, however, exceptions. Vézina and Platt (1988) conducted an inverse analysis of food web dynamics at stations in the English Channel and the Celtic Sea, concluding that microbial grazers were a major link in the transfer of carbon to mesozooplankton. Ducklow et al (1989) used a similar approach to assess food web dynamics in warm core rings. Other researchers took the first steps in creating dynamical, differential equation models of DOM and bacteria (Moloney et al., 1986; Pace et al., 1984).

The model of Fasham et al. (1990) provided an important step forward, incorporating bacteria and DOM, as well as an improved representation of recycling of inorganic nutrients by dividing N between nitrate and ammonium. Parameterizing DOM in models was, however, a major challenge and remains so today. Four years later, Berman and Stone (1994) wrote: "Theoretical ecologists have hardly begun to explore the implications of Pomeroy's 'new paradigm' of aquatic food webs. An explanation for this lies in the difficulties of modeling systems that are of a truly complex nature." Since then, ecosystem models have continued to proliferate, but Berman's statement is, we suggest, relevant even today. The NPZ models have been replaced by plankton-functional-type (PFT) models in which phytoplankton are divided into separate state variables for simulating groups such as diatoms, coccolithophores, and N_2-fixers (e.g., Gregg et al., 2003; Le Quéré et al., 2005). Most recently, the basic paradigm of the eukaryote plankton community has been called into question by a new emphasis on mixotrophy, with estimates that more than half of protistan primary producers are actively mixotrophic (Flynn et al., 2013). These mixotrophs use bacteria as a source of N, P, and Fe in support of photosynthetic carbon fixation, which substantially alters our view of the roles of DOM and bacteria in energy flow in pelagic food webs (Hartmann et al., 2012; Mitra et al., 2014).

Over the decades, the representation of DOM in models has also been progressively elaborated, including fractions of different lability and a variety of parameterizations of the production and fate of DOM in ocean ecosystems. Pragmatically, DOM is often associated with a filter pore size of <0.2-$0.7\,\mu m$, rather than a state of dissolution. Using this operational definition of DOM presents various challenges from a modeling perspective. Many models today, including those used in Earth System models (ESMs), do not even include DOM or any explicit representation of the microbial loop. Others do, but they also use apparently rudimentary assumptions about a very complex system. There is no right or wrong approach in this regard, although it is generally the case that increased mathematical complexity in models must be justified by specific conceptual objectives and not simply a desire for increased realism. Here, we examine the current state of the art, as well as the historical background, of modeling DOM in marine systems. The ongoing development of new schemes and associated parameterizations for representing DOM in models is discussed in context of the general trend toward increasing complexity in marine ecosystem models.

II MODELING APPROACHES

As we will show, there is an enormous range of approaches to modeling DOM in use, with the perennial issue of model complexity providing an ongoing subject of debate. At its crux is that models should be constructed with a level of complexity appropriate for the subject or hypothesis of interest (Allen and Fulton, 2010). Many studies have focused on biogeochemical cycling at basin or global scales, often using relatively simple representations of the marine ecosystem. This is not surprising given, in the words of Gordon Riley, the great pioneer in marine ecosystem modeling (Anderson and Gentleman, 2012), the "difficulty of deriving a system of mathematical equations subtle enough to meet the demands of widely varying environments and at the same time simple enough to be usable for practical application" (Riley et al., 1949). Alternatively, because marine microbial ecology, and the associated cycling of DOM, is multifaceted, complex models have also been favored in many instances.

A Models Without an Explicit Ecosystem

The use of modeling studies to examine the distribution of nutrients and carbon in the ocean need not necessarily involve the use of explicit ecosystem models. Early studies involving ocean general circulation models (OGCMs) calculated export by, for example, restoring nutrients to observed fields (Najjar et al., 1992) or using a simple Michaelis-Menten-type function of nutrient concentration (Bacastow and Maier-Reimer, 1990). Similar approaches have proved useful for studying the transport and fate of DOM in the ocean. These models specified production of DOM in surface waters and then examined its distribution in the ocean under the influences of circulation, mixing, and remineralization.

The first task facing modelers is to compartmentalize DOM into discrete entities that can be subject to simulation and represent the DOM pool as a whole. An immediate problem is the fact that the bulk DOM pool is largely uncharacterized (Chapter 2). Given the difficulty in characterizing DOM biochemically, the concept of lability may appear as a useful means of classifying DOM based on how readily different compounds are utilized by heterotrophic microbes. Conveniently for modelers, the bulk DOM pool has often been categorized into three fractions on the basis of turnover rates: labile, semi-labile, and refractory (Carlson and Ducklow, 1995; Cherrier et al., 1996; Kirchman et al., 1993). The early work of Ogura (1975), for example, showed that DOM decomposition in coastal seawater during bottle incubations occurred in two distinct phases with rates of 0.1-1 day^{-1} and 0.007 day^{-1}, with a third fraction remaining unutilized. Thus, labile material degrades rapidly on time scales of hours to days, semi-labile DOM turns over on seasonal timescales while the refractory fraction remains for centuries and may be treated as biologically inert on the timescales of most ocean model simulations. The concept of lability, which depends both on biochemical characteristics of DOM and the ecophysiology of microbial communities, is addressed in detail in Section IV. Suffice to say at this point that fractionation according to lability is challenging. For example, low molecular weight (LMW) DOM can be highly labile and its dynamics described within the time step used in most ecosystem models. Alternatively semi-labile/refractory LMW DOM is likely the remnants of high molecular weight (HMW) material after the more labile fractions have been successively stripped away (Amon and Benner, 1996) and may turnover on timescales appropriate for biogeochemical models. This raises an important question of whether models describing DOM in marine ecosystems should be categorized as those that attempt to describe biological processes (requiring suitable biological descriptions and appropriately small time steps) or biogeochemical processes (which for the most part will inevitably be concerned with longer lived DOM and less-labile remnants of DOM that would be outputs from detailed biological process models)? Pragmatically, Earth system scale models must fall into the latter camp (if only because their integration time step is far too long to do otherwise). However, as we shall see the history of the topic is rather schizophrenic in this regard, with constructions and parameterizations that arguably give acceptable results but not necessarily from realistic model structures.

We start by describing the model of Anderson and Williams (1999) as a characteristic early example of the methodology of modeling DOM in the ocean. This model, in combination with the complementary study of Anderson and Williams (1998; see next section), provided the basis for many of the models that followed, including models in use today (e.g., Keller and Hood, 2011, 2013). The model (Figure 15.1) has carbon as its base currency and contains three DOM fractions, labile, semi-labile, and refractory, as well as explicit heterotrophic bacteria. Primary production is specified as a fixed rate input (there is no explicit ecosystem), the fate of which is allocated between the labile and semi-labile DOC pools, exported particulate organic carbon (POC) and CO_2 with fractions of 14%, 41%, 10%, and 35%, respectively. These DOC fractions are then mixed downwards through a one-dimensional water

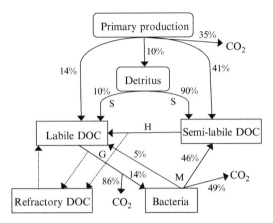

FIGURE 15.1 Model flow diagram showing the relative allocations of C between different pathways. S, solubilization; H, enzymatic hydrolysis; G, growth; M, mortality. Dotted arrows are quantitatively minor fluxes associated with the refractory pool. *Based on the model of Anderson and Williams (1999).*

column based on a prescribed profile of eddy diffusivity. Organic carbon is also exported to depth *via* sinking particles (10% of primary production). Losses from sinking detritus, which are distributed through the water column according to the hyperbolic function of Martin et al. (1987), are partitioned to DOC rather than direct remineralization to CO_2 with allocations of 10% and 90% to the labile and semi-labile pools, respectively. Both of these pools are utilized for growth by bacteria. It is assumed that the semi-labile pool is not taken up directly but instead enters the labile pool *via* enzymatic hydrolysis. Uptake of labile DOC is then calculated according to Michaelis-Menten kinetics (as is hydrolysis of semi-labile DOC) and is then used for growth with a fixed efficiency of 14% (the remainder released as CO_2). Bacteria are subject to mortality as a nonlinear function of biomass, with 51% allocated to DOC. Finally, a small fraction (0.35%) of the DOC processed by bacteria is allocated to the refractory pool, the sole loss term for which is photooxidation by ultraviolet radiation at the ocean surface.

The modeled vertical profile of DOC shows good agreement with data collected in the Atlantic

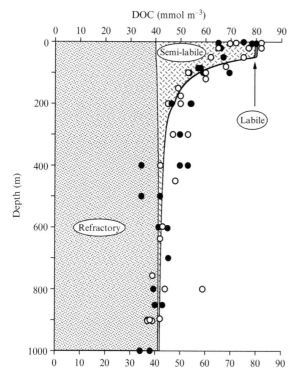

DOC (mmol m^{-3})

FIGURE 15.2 Vertical profile of DOC (mmolC m^{-3}) as simulated by the model of Anderson and Williams (1999), split into its component parts: labile, semi-labile, and refractory. Data are for various profiles in the Atlantic (solid points) and Pacific oceans (open points). *Taken from Christian and Anderson (2002).*

and Pacific oceans (Figure 15.2). The vertical gradient is maintained almost entirely by the semi-labile pool and results both from downward mixing of semi-labile DOC produced in the euphotic zone and DOC derived from solubilization of sinking detritus. Most of the organic C mixed downward is consumed by bacteria and remineralized to CO$_2$ within the upper 250 m of the water column, with associated turnover rates of between 0.01 and 0.001 day^{-1} in this depth range. It should be noted, however, that Anderson and Williams (1999) commented that their model may have underestimated downward transport due to lack of representation of physical processes such as subduction. Below

250 m, most of the semi-labile DOC in the water column arose from turnover of sinking particles. A large pool of refractory DOC, ~40 mmol m^{-3}, is maintained throughout the water column with concentrations decreasing slightly near the ocean surface due to photooxidation. In reality, it may be that refractory DOM is slowly remineralized, as shown by DOC gradients along the deep-ocean conveyor in the deep ocean (Hansell and Carlson, 1998). Recent work has also suggested that localized sinks of refractory DOM may control some of the patterns observed in the deep sea (Hansell and Carlson, 2013). There is little information available about the processes responsible for this remineralization (Hansell, 2013).

The model of Anderson and Williams (1999) illustrates many of the key challenges associated with representing DOM in ocean biogeochemical models: (1) the choice of which DOM pools to include and their associated definition; (2) choice of model currency and associated stoichiometry for models with multiple currencies; (3) parameterization of the turnover for each DOM pool; (4) choice of DOM source terms to include such as those *via* primary production in the euphotic zone, *via* detritus turnover, and *via* bacterial mortality (lysis), and their allocation among the different DOM pools.

DOM modeling studies that lack an explicit ecosystem (Table 15.1) simulate distributions of carbon and nutrients in the ocean, in both organic and inorganic form, and how these are influenced by circulation, vertical mixing, and DOM turnover rates. In this way the potential for C storage in the deep ocean can be examined along with the possible response of C storage to changing climate. The main focus of these models has therefore been the semi-labile pool, although in some instances slow-turnover biodegradable DOM is divided between semi-labile and semi-refractory pools (Hansell et al, 2012; Roussenov et al., 2006). A few models also include the refractory pool in order to directly compare with measured DOM. As a word of caution, however, different authors

TABLE 15.1   Model Characteristics: Models Without Ecosystem

Reference	Type	DOM Pools	Lifetime (years)	DOM Sources	Currencies
Yamanaka and Tajika (1997)	3D	SL, R	2, 16[a]	EuZ, D_{rem}	Redfield C:P
Anderson and Williams (1999)	1D	L, SL, R	B, B, 1.8[b]	EuZ, D_{rem}, B_{mort}	C only
Bendtsen et al. (2002)	1D[c]	SL, R	B, 150[a]	D_{rem}, B_{mort}	C only
Schlitzer (2002)	3D	SL	2	EuZ	C only
Roussenov et al. (2006)	3D	SL, SR	0.5, 6-12	EuZ	N, (P)
Kwon and Primeau (2006)	3D	SL	1	EuZ	Redfield C:P
Najjar et al. (2007)	3D	SL	2	EuZ	Redfield C:P
Kwon and Primeau (2008)	3D	SL	1.7	EuZ	Redfield C:P
Williams et al. (2011)	3D	SL, R	0.5, 6-12	EuZ	N only
Hansell et al. (2012)	3D	SL, SR, R	1.5, 20, 16 k	EuZ	C only

DOM sources, EuZ: euphotic zone; D_{rem}, detritus remineralization; B_{mort}, bacteria mortality.
Lifetime: B: variable, mediated by explicit bacteria state variable.
[a]EuZ only.
[b]At ocean surface, decreasing exponentially with depth within EuZ.
[c]Bendtsen et al. (2002) model is of aphotic zone only.

use different definitions for DOM pools. Efforts have been made to better constrain and unify these definitions based on data and improved estimates of turnover (Hansell et al., 2012, and see Chapter 3). The semi-labile pool, for example, has been variously described as both labile (e.g., Six and Maier-Reimer, 1996) and refractory (e.g., Levy et al., 1998; Walsh and Dieterle, 1994). Regarding currencies, most models include C for the obvious reason that DOM is a potentially important determinant of the ocean influence on atmospheric CO_2. Other elemental fluxes can be inferred by invoking the canonical Redfield ratio (Redfield et al., 1963), although DOM generally does not follow Redfield stoichiometry (Hopkinson and Vallino, 2005; Williams, 1995). DOM may also be important as a means of transporting nutrients into the oligotrophic gyres (see Section III) in which case N or P provide the main focus (e.g., Williams et al., 2011). The fundamental problem is, however, that DOM represents a suite of literally many 1000s of different chemicals about whose identities, concentrations, and

dynamics we know virtually nothing. And that is before one includes the microparticulates that are mixed in with the <0.2 μm size fraction, and perhaps excluding TEP and other material that arguably comprises "dissolved" material but does not pass the size fractionation.

The primary loss mechanism for DOM in the ocean is uptake by heterotrophic bacteria. All eukaryotic microorganisms, the protists, (Marchant and Scott, 1993; Sherr, 1988) and perhaps some metazoa (Wright and Manahan, 1989) have the potential to take up dissolved or colloidal organic matter, but it is not known how widespread or quantitatively significant this process is. Flynn and Berry (1999) sought to explain the apparent ubiquity of dissolved free amino acids (DFAA) uptake by phytoplankton through modeling a leak-recovery process. Through this, it can be seen that the leakage of metabolites such as DFAA (and sugars) across membranes with gradients running from high (mM) internal concentrations to low (nM) external concentrations presents a risk to the growth of the cell.

The recovery of these leaked metabolites can explain the "critical inoculum" problem often seen in algal culturing, where a sufficient load of microalgal cells must be introduced for the leak-recovery process to work to the advantage of the individual cell; if this does not occur then these organisms effectively bleed DFAA from their internal high-concentration metabolite pools to the external water (Flynn and Berry, 1999).

Setting aside the leak-recovery mechanism (which for biogeochemical modeling can be ignored as only the net loss is consequential), the first decision to make regarding the modeling of DOM turnover is whether or not to explicitly include heterotrophic bacteria. This represents one of the examples of schizophrenia mentioned earlier; because an explicit description of the DOM-linked biological C pump involving microbes must surely operate with time-steps far below those in ESMs. Indeed, it is difficult to reliably model bacteria in marine systems, a topic we discuss in detail when describing ecosystem-based models (Section II.B). In many instances, therefore, modelers have chosen to omit bacteria and use fixed (specified) lifetimes for DOM pools. The chosen lifetimes (turnover = 1/lifetime) typically vary between 0.5 and 2 years (Table 15.1), a fourfold range (Table 15.1). In reality, lability and associated turnover are variable because DOM represents a spectrum of compounds and as a result of physiological aspects of the bacterial community (Section IV). The models of Roussenov et al. (2006) and Hansell et al. (2012) therefore chose to have separate semi-labile and semi-refractory pools. Nevertheless, despite this apparently simplified approach, modeled distributions of carbon and nutrients are generally reasonable. Of the models listed, only that of Anderson and Williams (1999) included a labile pool. The refractory pool is rarely included in models because, at least according to conventional wisdom, it is biologically inert over time scales of hundreds to tens of thousands of years. The global inventory of refractory DOM, despite its large size, may be relatively insensitive to

climate change over the next 200 years if photo-oxidation is the sole loss term. Nevertheless, a new and upcoming challenge for biogeochemists and modelers is that of the microbial carbon pump (MCP), as proposed by Jiao et al. (2010). In the MCP framework, repetitive processing of DOM by microbes transforms reactive organic carbon into refractory DOM, potentially building up a reservoir for carbon storage. The processes involved include viral lysis and associated release of microbial cell wall material, the selective action of ecto- and exo-enzymes and losses associated with protist grazing (see Chapter 3).

The models listed in Table 15.1 each use simple methods to prescribe DOM input into the euphotic zone (in one case, Bendtsen et al. (2002), the euphotic zone was ignored altogether with the aphotic ocean as the sole focus). The most common approach is to derive either new or export production (the two are theoretically equivalent, at least at steady state), either by restoring nutrients to observed surface concentrations (e.g., Kwon and Primeau, 2006, 2008; Najjar et al., 1992) or as a simple function of surface nutrient concentrations (e.g., Roussenov et al., 2006; Schlitzer, 2002; Yamanaka and Tajika, 1997). The resulting export flux is divided between particulate and dissolved organic matter, with a perhaps surprisingly high allocation to the latter in most instances, for example, 50% (Schlitzer, 2002), 50-65% (Roussenov et al., 2006), 67% (Najjar et al., 2007 and Yamanaka and Tajika, 1997), or 74% (Kwon and Primeau, 2006). The model of Hansell et al. (2012) focused only on the production of DOM described as the square root of primary production and did represent sinking particulate organic matter (POM) in their analyses.

The early models listed in Table 15.1 exhibit a key difference compared to those published later, namely they include detritus turnover as a source of DOM in the deep ocean. As detritus sinks through the water column both attached bacteria and detritivorous zooplankton attenuate its flux. The former solubilize particulate matter using hydrolytic enzymes (Smith et al, 1992) and much of this solubilized material may be lost to surrounding waters thereby fuelling growth of free-living bacteria (Cho and Azam, 1988; Unanue et al., 1998). Hansman et al. (2009), for example, noted that at 950m in the meso-pelagic zone of the North Pacific subtropical gyre, free-living bacteria utilize carbon derived primarily from sinking particles. Zooplankton also release DOC as a result of breakup of food while grazing ("sloppy feeding"), excretion, and voiding of fecal material that includes dissolved substrates (Jumars et al., 1989). Both bacteria and zooplankton may play important roles in the processing and remineralization of organic C within the mesopelagic zone. Giering et al. (2014) used a simple steady state model to an-alyze data collected at the Porcupine Abyssal Plain site in the North Atlantic (49 °N, 16° 30′W). The model indicates that, although most remin-eralization can be attributed to microbes, detri-tivorous zooplankton may consume half of the sinking POC flux. Much of this POC consumed by zooplankton is however released as sus-pended detritus and DOM, thereby stimulating the microbial loop. Thus, the reality would ap-pear to be that a significant fraction of detritus turnover gives rise to the production of DOM.

B Models With an Explicit Ecosystem

Models with an explicit ecosystem (Tables 15.2 and 15.3) are intended to directly address in some detail the production and cycling of DOM in the euphotic zone, as well as its export to the ocean interior. As with models that do not incorporate an explicit ecosystem, it is neces-sary to choose which DOM pools to include and whether or not to opt for multiple curren-cies. Additional considerations are (1) whether to explicitly include bacteria and, if so, how to model their consumption and remineralisation of DOM and (2) the definition of DOM source terms and which pools (in terms of lability and stoichiometry) to allocate them to.

Models that have no explicit treatment of bac-teria (Table 15.2) are similar to those described in Section II.A above in that DOM fractions are assigned fixed (or temperature-dependent) life-times, but differ in that the production of DOM is related explicitly to ecosystem dynamics. Again, it is possible in these models to omit the labile pool and focus solely on the semi-labile DOM fraction. The chosen lifetimes for semi-labile pool are, however, often relatively short (<0.5 year) in these models, perhaps indicating a shift in emphasis from modeling carbon distri-butions in the ocean to modeling the dynamics of marine ecosystems. An interesting approach adopted by Pahlow et al. (Pahlow and Vézina, 2003; Pahlow et al., 2008) modeled DOC and DON labilities dynamically as the fractions of total DOM available for bacterial utilization. Freshly produced DOM was assumed to be en-tirely labile, with lability then decreasing over time according to first order decay coefficients. Some of the models, including more recent additions to the literature (Llebot et al., 2010; Shigemitsu et al., 2012), include only a single labile pool thereby providing a (relatively sim-plistic) representation of the microbial loop. It is notable that many of the models with an explicit ecosystem incorporate considerable complex-ity in the ecosystem representation (e.g., multi-ple phytoplankton and zooplankton functional types), yet choose to use simple, empirical ap-proaches to modeling DOM (Table 15.2).

Models with explicit descriptions of bacteria (Table 15.3) invariably include a state variable for labile DOM as it is this pool that is used directly for growth. A number of contrasting approaches have been used for modeling substrate uptake and utilization by bacteria. A popular choice is the Monod model (Monod, 1942) which was de-veloped by a French biologist, Jacques Monod, who showed that the growth rate of bacteria in culture, μ, can be described by:

$$\mu = \frac{\mu_{max}S}{K_S + S},$$ (15.1)

TABLE 15.2   Model Characteristics: Models with Ecosystem but Without Explicit Bacteria

Reference	Type	Food Web Structure[a]	DOM Pools[b]	DOM Currencies[c]	DOM Sources[d]	DOM Lifetime (years)[e]
Kawamiya et al. (1995)	1D	NPZD	S	N	P*, D#	0.09♦
Six and Maier-Reimer (1996)	3D	NPZD	S	P:(C)	P#, Z#	0.11♀
Aumont et al. (2003)	3D	3N2P2ZD	S	C	P#, Z#, D#	0.27
Moore et al. (2004)	3D	4N3PZ2D	(L), S	C:N:P	P#, Z†,#	0.27
Christian (2005)	1D	NPZ	L, S	C:N:P	P#, Z†	0.007, 2.7
Schmittner et al. (2005)	3D	NPZD	S	N	P#, Z†,#	5.8
Dutkiewicz et al. (2005)	3D	3N2PZD	S	P:Fe	P#, Z†,#	0.14
Huret et al. (2005)	3D	NPZD	S	N	P*, Z#,†, D#	0.05
Schartau et al. (2007)	0D	NP(Z+B)D	L	C:N	P#, Z#, D#	0.011-0.015
Salihoglu et al. (2008)	1D	3N3P2Z2D	S	N:P:(C)	P*, Z#, D#	0.15
Druon et al. (2010)	3D	NPZ2D	(L), S	C:N	P*, Z†, D#	0.18-0.36♦
Llebot et al. (2010)	0D	2N2PZD	L	N:P	P#, Z#, D#	0.027
Shigemitsu et al. (2012)	1D	3N2P3Z2D	L	N	P*, D#	0.018♦

[a]State variables: nutrient (N), phytoplankton (P), zooplankton (Z), bacteria (B), detritus (D), bacteria (B) are listed if present. Others not listed.
[b]DOM pools: labile (L, lifetime hours to days), semi-labile (S, weeks to months), refractory (R, decades and longer); terminology may differ in original texts.
[c]Nitrogen (N), phosphorus (P), carbon (C), iron (Fe); parentheses indicate fixed C/N or C/P ratios. Ammonium and nitrate not considered to be separate currencies.
[d]DOM sources: * via production; # via mortality (turnover) terms; † as fraction of grazing.
[e]DOM lifetime: ♦ value for 0 °C, with $Q_{10} = 2$; ♀ maximum rate as function of nutrients.

where S is substrate concentration in the surrounding environment, μ_{max} is the maximal growth rate and K_S is the substrate concentration at which $\mu = \mu_{max}/2$. Growth and uptake are assumed to operate in tandem, which is strictly only the case under steady-state conditions (i.e., Monod's cultures of bacteria). The model also assumes that growth is limited by only one factor (substrate); a most unlikely event in nature and certainly inappropriate for varying sources of DOM, noting that the single half saturation constant used is surrogate for perhaps many 10s of half saturation values for different components of DOM. Fasham et al. (1990) extended the basic formulation of Monod to address the simultaneous use of organic and inorganic nitrogen by defining the ratio, η, of the two which occurs for balanced growth:

$$\eta = \frac{\omega_C \theta_{DOM}}{\omega_N \theta_B} - 1, \tag{15.2}$$

where ω_x is the growth efficiency for C or N and θ_{DOM} and θ_B are the C:N ratios of DOM and bacteria, respectively. This formulation assumes that if there is sufficient NH_4^+, dissolved inorganic nitrogen (DIN), and DON are taken up in fixed ratio ($\eta = 0.6$). If not, DIN and DON jointly limit the bacterial growth rate. Respiration was parameterized as a constant biomass-specific loss of ammonium. A potential problem with the Fasham approach is that bacteria are unlikely to use ammonium if sufficient labile DON is available, depending on the concentration of the types of DON (DFAA could repress consumption of NH_4^+). If DON availability is used in this N-based model as a surrogate for DOC availability, then DON availability could indirectly support NH_4^+ use. An additional problem in models that have N as their sole currency is that in reality the consumption of ammonium requires a concurrent assimilation of DOC (*i.e.*, DOM with a sufficiently high C:N that there remains a demand for N even after respiratory and other considerations). Fasham's approach was nevertheless popular during the 1990s but, with increasing emphasis on the carbon cycle, was superseded by stoichiometric approaches. Experimental work by Goldman et al (1987) showed that excretion of N by bacteria decreases when DOM has a high C:N ratio indicating sparing of nutrient elements for growth while preferentially meeting the costs of respiration using C-rich substrates (Anderson, 1992). Based on these studies stoichiometric approaches were developed for modeling zooplankton (e.g., Anderson and Hessen, 1995; Hessen, 1992) and mineralization of soil organic matter (e.g., DeRuiter et al., 1993). Depending on θ_B and θ_{DOM}, either N or C can limit growth. If N is limiting, N excretion is zero. Indeed, NH_4^+ may be taken up to supplement DON for growth. Conversely, N is excreted under C-limiting conditions, *that is*, when N-rich substrates are utilized. The net excretion of nitrogen, E_B, is then:

$$E_B = U_C \left(\frac{1}{\theta_{DOM}} - \frac{\omega_C}{\theta_B} \right), \qquad (15.3)$$

where U_C is DOC uptake. A review of the literature by del Giorgio and Cole (1998) indicated a mean value for bacterial growth efficiency (BGE) of 0.15, with values in models varying between 0.15 and 0.3 (Table 15.3). In some cases, BGE is not a fixed parameter in models but instead varies as a result of separate basal and activity-related respiration terms (Luo et al., 2010; Polimene et al., 2006). Only a few models adopt a cell quota approach to bacterial stoichiometry (Blackford et al, 2004; Luo et al., 2010; Polimene et al., 2006). Nutrient uptake is calculated according to Michaelis-Menten kinetics and growth is a function of the internal quotas of elements such

as N and P. In contrast, the model of Flynn (2005) assumes fixed C:N:P but handles the fate of labile and semi-labile components by priority, giving a compact model that also describes different respiration rates and hence different growth efficiency depending on which nutrient is limiting and the degree of limitation.

Models that have a semi-labile pool often use a Monod approach whereby semi-labile DOM is hydrolyzed to form labile DOM, which is then suitable for uptake by bacteria. Values for the kinetic parameters, namely the maximum rate of hydrolysis and the half saturation constant, are not easily obtained and represent an amalgam of many different processes. Connolly et al. (1992) and Connolly and Coffin (1995) provided estimates for a variety of coastal and freshwater environments from which values for Santa Rosa Sound (FL, USA) were applied to the English Channel by Anderson and Williams (1998) and subsequently also used by some of the models which followed. Another approach is to allow bacteria to simultaneously take up both labile and semi-labile DOM directly for growth, but assuming that only a small fraction of the latter, for example, 1%, is bioavailable at any one time (e.g., Luo et al., 2010).

DOM is produced within marine ecosystems from a variety of sources (Figure 15.3; see Chapter 3), many of which are not well understood or quantified. Four main categories can be identified: phytoplankton exudation and mortality, grazer-associated losses, bacteria lysis, and mortality and detrital turnover. A variety of different combinations of these terms have been used in models, along with different methods of parameterization (Table 15.3). An immediate problem is that these processes are themselves poorly understood and further confounded with all the challenges over the description of what constitutes DOM in the first instance. Detrital turnover is, for example, often mediated by grazers together with bacterial activity, while most grazing-associated losses and mortality through cell lysis will liberate material as microparticulates or colloids, not

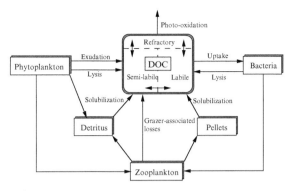

FIGURE 15.3 Idealized food web illustrating the four main DOC production terms (phytoplankton exudation, grazer-associated processes, lysis, detrital solubilization) and the two principal loss routes (bacterial uptake, photooxidation). *Taken from Christian and Anderson (2002).*

as true DOM (noting that the definition of DOM is operationally defined by filter pore size, traditionally <0.2μm diameter, and not with a state of dissolution).

Nearly all models with an explicit ecosystem include phytoplankton exudation, which involves both passive "leakage" and active release. The former results from the permeability of the plasma membrane to LMW compounds (Bjørnsen, 1988) whereas the active release of "overflow" carbon as DOC may occur as a result of metabolic instabilities (Williams, 1990). Accordingly, a key modeling choice has been as to whether to represent exudation as a "property tax" (some fraction of biomass per unit time), as suggested by Bjørnsen (1988), or as an "income tax" (a fraction of photosynthesis). Both approaches have been widely used in models (Table 15.3). In reality, release of all forms of DOM from an intact living cell represents an imbalance between internal *versus* external concentration gradients and whether the DOM molecular weight is small enough to pass through the plasma membrane unaided (ca. < 300 Daltons, which would include the bulk of primary metabolites which one would expect to also be highly labile). Such metabolites may often show peak internal concentrations during

unbalanced growth, such as during N-refeeding (Flynn, 1990). One may also expect (see Flynn and Berry, 1999) organisms to have active uptake systems to recover such metabolites as these may not only represent a loss of valuable material and energy but also provide an unwelcome attractant to grazers (e.g., Martel, 2006). If the molecular weight of the organic matter is larger than that permitted by passive diffusion then some form of reverse pinocytosis is required, or a specific structural feature such as a mucocyst. It appears that all protists can engage in osmotrophy, and that the bulk of organisms traditionally labeled as "phytoplankton" are mixotrophic (Flynn et al., 2013) and hence have phagocytic and pinocytic capabilities. As the rejection of partly digested material from a living cell cannot be passive, the release of HMW "DOM" should be viewed as active. Metabolic "instabilities" may thus apply to either or neither categories and is not a useful differentiator. Both the property tax and income tax approaches are part of a common explanation that sees leakage as a function of the concentration of the internal metabolite pool. Thus, from the models described by Flynn and Berry (1999) and Flynn et al. (2008), there is leakage (mainly of DOC) that increases (transiently) during nutrient stress and during N-refeeding (DON), while there will always be some level of background leakage of metabolites from living cells.

High (net) release rates may be a common feature of oligotrophic waters (Christian, 2005; Teira et al., 2001), possibly as a result of continued use of photosynthetic machinery after nutrient exhaustion (Wood and Van Valen, 1990). This observation is consistent also with the leak-recovery mechanism described in Flynn and Berry (1999) where thin suspensions of organisms are more likely to show a net "bleed" of organics. Exuded polymers may serve a number of fitness-promoting functions such as structural defences, storage, virus repellents, and protective metabolites (toxins) (Hessen and Anderson, 2008), few if any of which are well understood.

TABLE 15.3   Model Characteristics: Models with Ecosystem Including Explicit Bacteria

Reference	Type	Food Web Structure	DOM Pools	DOM Currencies	DOM Sources	Bacteria Model[a]
Fasham et al. (1990)	0D	NPZDB	L	N	P*, Z#, D#	Monod(L), FDM
Anderson and Williams (1998)	0D	NPZDB	L, S	C:N	P*,#, Z†, D#, B#	Monod(L), SL^H, AW98 ($\omega = 0.27$)
Levy et al. (1998)	1D	NPZDB	L, S	N	P*, Z#, D#	Monod(L), SL^H, FDM
Bissett et al. (1999)	1D	N4PB	L, R	C:N	P#, Z†	Monod(L), FDM
Tian et al. (2000)	1D	N2P2Z2DB	L	C:(N)	P#, Z†	Monod(L), $\omega = 0.15$
Vallino (2000)	0D	NPZDB	L, S	C:N	P*, D#	Monod(L), SL^H, $\omega_{max} = 0.8042^b$
Anderson and Pondaven (2003)	1D	NPZDB	L, S	C:N	P*,#, Z†,#, D#, B#	Monod(L), SL^H, AW98 ($\omega = 0.17$)
Pahlow and Vézina (2003), Pahlow et al. (2008)	1D	NPZDB	(L+S+R)	C:N	P*, Z#,†, D#	M-P(L+S+R), AW98 ($\omega = 0.25$)
Blackford et al. (2004)	1D	3N4P3ZDB	L, S	C:N:P	P*,#, Z†, D#, B#	M-M(L), SL^H, $\omega = $fn(active, basal resp)
Polimene et al. (2006)	0D	3N4P3ZDB	L, 2S	C:N:P	P*,#, Z†, B#	M-M(L+2SL), $\omega = $fn(active, basal resp)
Grégoire et al. (2008)	3D	2N3P2ZDB	L, S	C:N	P*,#, Z†, D#, B#	Monod(L), SL^H, AW98 ($\omega = 0.17$)
Luo et al. (2010)	1D	2N3P2ZDB	L, S, R	C:N:P	P#, Z†,#, D#, B#	M-M(L+SL), $\omega = $fn(active, basal resp)
Keller and Hood (2011)	0D	N2P2ZDB	L, S, R	C:N	P*,#, Z†,#, D#, B#	Monod(L), SL^H, AW98 ($\omega = 0.30$)

Nomenclature and symbols as for Table 15.2: food web structure and DOM pools, currencies, and sources.
[a]*Bacteria models: FDM (approach of Fasham et al. (1990): Equation 15.2); AW98 (method of stoichiometric balancing of Anderson and Williams, 1998); SL^H (semi-labile material not consumed directly by bacteria, but hydrolyzed to labile); ω (BGE), M-M (Michaelis-Menten uptake, used with cell quota model); M-P (Mayzaud-Poulet uptake, used with cell quota model); resp (respiration).*
[b]*BGE declines as N availability declines.*

The model of Polimene et al. (2006) explicitly assigns a fraction of DOC to represent capsular material, extracellular polysaccharides that produce a mucilaginous protective envelope and are assumed to be relatively refractory to hydrolysis and decomposition (Stoderegger and Herndl, 1998).

Most models also include grazer-associated losses to DOM, either *via* grazing or as mortality of the grazers themselves. A fixed fraction of the ingested material, typically between 20% and 40%, may be allocated to so-called sloppy feeding. Strictly, this term refers only to the release of dissolved compounds when prey cells are broken by mouthparts of crustacean zooplankton. However, grazer released DOM also includes the direct exudation of DOM by grazers, as well as DOM resulting from egestion and dissolution of fecal material (Jumars et al., 1989). Strom et al. (1997) estimated that between 16% and 37% of algal C is released as DOC during grazing by phagotrophic protozoa. Parameterizing DOM production *via* "sloppy feeding," egestion, excretion, and dissolution of fecal pellets in not straightforward. Microzooplankton, for example, use a variety of feeding mechanisms including pallium feeding tube feeding and direct engulfment (Hansen and Calado,1999) which may be expected to entail negligible sloppy feeding losses. Grazing by protists may actually result in the voiding of partly consumed prey and secondary metabolites (likely semi-labile DOM), material which may subsequently be consumed again if other food material is unavailable (Flynn and Davidson, 1993). Grazing in marine systems is often dominated by the microzooplankton, except in areas where large algal blooms occur (e.g., Verity et al., 1993). Release of DOM from fecal pellets may also be an important source, with as much as 50% of C in pellets being rapidly solubilized (Urban-Rich, 1999), although just how much of this material is really dissolved, as opposed to being fine suspended particulates, remains unknown.

When it comes to zooplankton non-grazing loss terms, namely excretion and mortality (including model closure that represents all higher trophic levels in the food chain), there is once again considerable variation among models. The model of Fasham et al. (1990), for example, included a zooplankton excretion term ($0.1\,day^{-1}$) in which 75% was released as ammonium and 25% as DON. The same model allocated zooplankton mortality, the closure term, as 33% to sinking detritus and the remainder to ammonium representing the cumulative losses associated with an infinite chain of predators. The same approach was taken by Anderson and Pondaven (2003) but with allocations of 29%, 33%, and 38% to detritus, inorganic nutrients, and DOM respectively. Thus, some models have chosen to include a fraction of zooplankton mortality as DOM (e.g., Blackford et al, 2004) whereas others have not (e.g., Grégoire et al, 2008; Tian et al, 2000). The justification of all these values is not clear and often represents little more than guestimates or values derived from tuning models to data.

Some recently published models include non-grazing bacterial mortality as a source of DOM. It is usually parameterized as a constant biomass-specific rate (e.g., Anderson and Williams, 1998; Grégoire et al., 2008) and represents viral infection and subsequent lysis. This process may account for 10-50% of bacterial mortality (Fuhrman, 1999) with the resulting lysis products fuelling microbial production (Noble and Fuhrman, 1999). Anderson and Ducklow (2001) included viral lysis of both phytoplankton and bacteria in their simple steady-state model of the microbial loop, concluding that production of DOC by this pathway is relatively small. Keller and Hood (2011) explicitly included viruses as a state variable in a complex model of DOM cycling, with lysis of bacteria contributing to both DOM and detritus pools. There is the continual problem of gauging just what of a lysed microbe is truly dissolved, *versus* being micro-particulate.

Finally, there is the decision whether to allocate detritus breakdown to DOM, which has already been addressed in some detail in Section II.A. above. It is interesting to note that many of the recently published models with an explicit ecosystem (Table 15.3) include detritus as a source of DOM, in contrast to models without an explicit ecosystem that do not (Table 15.2).

Fluxes to DOM produced *via* the various processes described above have to be divided between pools of different lability. "Passive" exudation (though likely often allied to active recovery) is often assumed to consist solely of small, labile molecules such as sugars and amino acids (e.g., Anderson and Williams, 1998; Billen and Becquevort, 1991). This can be distinguished from "active" exudation of DOC that can also include polymeric such as extracellular polysaccharides and may be allocated to the semi-labile pool because it is devoid of nutrients to support bacterial growth (e.g., Polimene et al., 2006). Christian (2005) found that under high-light, low nutrient conditions this DOC source must be primarily labile to prevent excessive accumulation of DOC. There is marked variation in the partitioning between labile and semi-labile pools between models, even more so than with allocation of the other DOM source terms. For example, Anderson and Williams (1998) used a labile fraction of 10% for non-phytoplankton processes compared to Anderson and Pondaven (2003) and Grégoire et al. (2008) who used labile fractions of 70% and 65%, respectively. Keller and Hood (2011) partitioned just 25% of zooplankton sloppy feeding and bacterial non-grazing mortality, but 40% of phytoplankton mortality, to labile DOM. Inevitably, these poorly known parameters are set by tuning to data, giving rise to at least some of the variability seen between models. Ultimately, as we shall turn to, lability depends on chemical structure and the need of organisms to utilize the structure. The latter is often linked to the presence of other nutrients (glucose is only "labile" when there is sufficient N, P (etc.) available to support its consumption and assimilation). However, a complicating factor affecting the longevity of DOM is the variable assimilation efficiency that is linked to nutrient sufficiency. Thus, P-limited bacterial growth can be far less efficient at assimilating DOC (Flynn, 2005; Pirt, 1982).

III MODELING THE ROLE OF DOM IN OCEAN BIOGEOCHEMISTRY

There have been several important developments in global-scale modeling of the role of DOM in ocean biogeochemical cycles since the first edition of this book (Christian and Anderson, 2002). First is the completion of the Ocean Carbon Cycle Model intercomparison project phase II (OCMIP-2, Najjar et al., 2007) which used a common biogeochemistry model with the key objective to diagnose the effects of intermodel differences in ocean circulation on modeled distributions of dissolved inorganic carbon (DIC) and oxygen. The experiment featured global ocean circulation models with a common, rather simplistic, scheme for biogeochemistry. The OCMIP-2 "biology" (Table 15.1) does not include explicit biology but calculated new production by nutrient restoring (Section II.A). A large fraction (two thirds) of this new production was allocated to DOM while the remainder, POM, was remineralized at various depths according to a parameterized ("Martin curve") representation of particle flux; the basic scheme is the same as used by Najjar et al. (1992). It is important to note that an allocation of two thirds of new production to DOM does not entail an equivalent 2:1 ratio of DOM to POM in the export flux. While 100% of the POM is exported, DOM produced in the surface layer is advected laterally by the model currents such that total carbon export from the euphotic zone in OCMIP-2 models is around 80% particulate, consistent with observations (Hansell et al., 2009; Najjar et al., 2007). The remineralization lifetime of DOM was set at 2 years, which is relatively high compared to other similar modeling studies (Table 15.1). The intercomparison study concluded that the semi-labile DOC distribution is "highly sensitive to ocean circulation, particularly the exchange of water between the mixed layer and the thermocline" and that there is need for better observational constraints, particularly "quantifying the DOM pool" and "constrain(ing) C:N and C:P ratios of semi-labile DOM" (Najjar et al., 2007). The chosen model structure may in fact have made the modeled distributions of organic and inorganic carbon more sensitive to ventilation processes than they would otherwise have been in a model more heavily weighted toward particulate export. Najjar et al. (2007) noted that models that include mixed-layer physics have greater relative export of DOM as a result of seasonal convection. The exact time scale chosen for remineralization of DOC also affects the sensitivity of ocean tracer distributions to modeled circulation processes.

A second important development has been the incorporation of ocean biogeochemistry, in some cases including DOM, into coupled ocean-atmosphere models. This entails a huge computational burden and so has led to the development of Earth System Models of Intermediate Complexity (EMICs). Decreased complexity is included in one or both components of the coupled ocean-atmosphere system, making EMICs suitable for experiments of greater duration than is generally possible with full dynamic coupled models (see Chapter 1). Representations of ocean and terrestrial biogeochemistry have also been included in recent years. While early EMICs used extremely simplified ocean models (e.g., Stocker et al., 1994), many of those now in use include a three-dimensional OGCM (albeit with coarse resolution) along with a highly simplified atmosphere model.

In recent years, several experiments have been published using EMICs that include DOM in the ocean biogeochemistry model (Matsumoto et al., 2008; Ridgwell et al, 2007; Schmittner et al., 2005). In one instance, a sophisticated data assimilation scheme was applied to estimate the parameter values that would allow the model to best reproduce the present-day global ocean distributions of phosphate and alkalinity, as well as other tracers of physical circulation (Ridgwell et al., 2007). The OCMIP-2 approach to representing DOM was used, with DOM as 66% of production and a turnover time of 2 years, but without optimization of these parameters. Optimization was carried out for the POC flux parameters, although changes from their a priori values were small (~10% for both the remineralization length scale and the fraction of export subject to remineralization). The C:P ratio was assumed to be in Redfield proportion, 106:1. The assumptions of a high production of DOM and a low C:P ratio are not as artificial as they may appear at first glance (see Najjar et al., 2007), although they were not tested in the parameter optimization experiment. Matsumoto et al. (2008) reevaluated some of the physical parameters of this model with respect to their effect on ocean ventilation and subsurface tracer distributions.

Most recently, an ensemble of model simulations was carried out that comprised the 5th Coupled Model Intercomparison Experiment (CMIP5). The aims of this comparison were to evaluate the performance of models in simulating the recent past, provide projections of future climate change and to understand the impact of key feedbacks such as clouds and the carbon cycle in those projections. Output from climate modeling groups around the world is being deposited into a globally accessible public-domain archive (Taylor et al., 2012). The models in question are not EMICs but, rather, fully dynamic coupled models. This is the first experiment of its kind that includes ocean biogeochemical fields. To date, output from more than 15 models have been submitted, of which at least 12 include DOC. This data resource is new and, as such, there is little or no published literature to draw upon. Hence, we include here a brief synthesis of these model results.

The global mean DOC profiles (for waters >200 m deep; semi-labile DOC only) of CMIP5 models can generally be grouped according to the attenuation length scale (Figure 15.4). Several models have elevated concentrations of semi-labile DOC within the upper 100 m, which rapidly drop off to near zero in the waters below, while others have appreciable concentrations throughout the upper kilometer that attenuate gradually through the thermocline. Most of the models have surface concentrations of semi-labile DOC of 10-25 μM, although a few are higher and one is substantially lower. Najjar et al. (2007) estimated the "observed" global mean surface concentration of semi-labile DOC as 29 μM; this is within the range of CMIP5 models although, in general, the models seem to have a slight low bias as compared to the OCMIP-2 models, most of which exhibit rapid attenuation of DOC with depth (Najjar et al, 2007). The relatively deeper penetration of DOC in the CMIP5 models may reflect greater ventilation of the intermediate and deep ocean due to higher resolution and improved mixing parameterizations, as well as

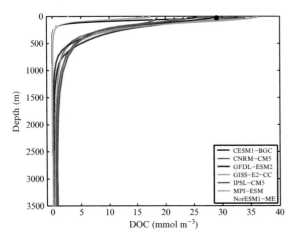

FIGURE 15.4 Global mean profiles of DOC concentration from CMIP5 models (mean for 1986-2005). The black dot and horizontal line represent an observation-based estimate of the global mean surface concentration of 29 ± 5 μM. Where a modeling group submitted DOC data from more than one model, all models are shown in the same color and the names in the legend are approximate (not official CMIP5 names). Refractory DOC was subtracted from the GFDL models (the only ones that were posted as total DOC rather than semi-labile) as the global mean for depths >3000 m.

more widespread inclusion of dynamical mixed-layer models.

The annual mean geographic distribution of surface DOC is shown in Figure 15.5, along with the seasonal cycles of zonally averaged surface DOC in Figure 15.6 (as in Figure 15.4, shelf waters with depths <200 m are excluded). The models show striking differences, with only half reproducing the generally observed trend of high concentrations in the tropics, decreasing towards the poles (Hansell et al., 2009). Others have high but transient accumulations in the mid-latitudes in summer, while some models have high concentrations both in the tropics and in summer in the mid-latitudes. Of all the models, only GISS-E2-R-CC reproduces the high concentrations of DOC in the stratified waters of the Arctic as a result of terrigenous inputs *via* rivers (Raymond et al., 2007). It is possible that the other models are underestimating the global mean due to poor representation of the Arctic,

neglecting fluvial DOC sources (see Chapter 14 regarding Arctic DOM). Many of the models show DOC concentrations (excluding the refractory baseline) near zero in the surface waters of the Southern Ocean for most or all of the year (Figure 15.5), a consequence of upwelling of deep water. Winter observations in the Ross Sea by Carlson et al. (2000) show uniform concentrations (at ~41 μM) down to 800 m, the data corresponding to "background" (*i.e.*, refractory) concentrations. Significant accumulations did occur in spring and summer in the upper 200 m although it should be noted that the Ross Sea, even well beyond the continental shelf, is generally more productive than most open ocean regions of the Southern Ocean. Nonetheless, several models do show significant accumulations in the Southern Ocean during austral summer (Figure 15.6). There are insufficient observations to know whether these modeled seasonal accumulations of DOC are realistic or not. There are few data available that allocate DOC to different classes of lability in a manner that is consistent with model formulation.

Several groups submitted more than one model, with identical biogeochemistry but different ocean circulation models (or the same ocean but different atmospheres) and in most cases they are sufficiently similar that it was not considered useful to display both in Figures 15.5 and 15.6. The differences between two or three variants of the same model are generally much smaller than the differences among different biogeochemical models (Figure 15.4). Thus, while the effects of circulation biases on subduction and subsurface remineralization of DOM may be important in terms of its role in biogeochemical cycling, modeled global distributions of DOC are mostly a function of the biogeochemical model used. The difference in resolution between the two versions of the MPI-ESM is quite large (about $1.4 \times 0.8°$ vs. $0.5 \times 0.5°$) but the difference in the mean DOC profiles is very small (Figure 15.4). In both GFDL-ESM2 and GISS-E2, the attenuation length scale (*i.e.*, the penetration depth of surface

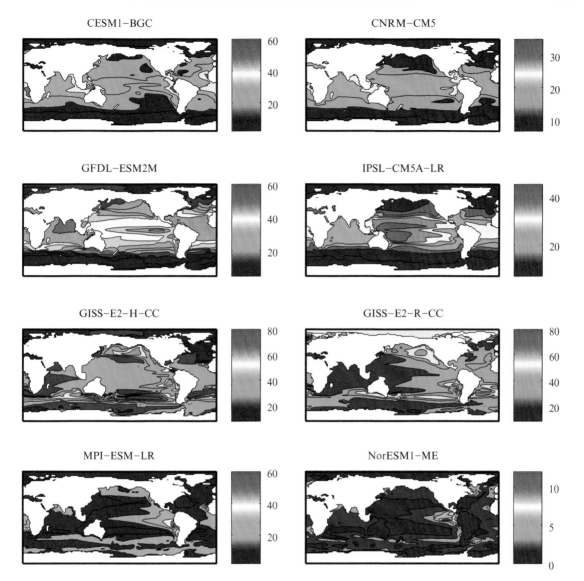

FIGURE 15.5 Global maps of annual mean surface semi-labile DOC concentration (mmol m⁻³) from eight CMIP5 models (mean for 1986-2005). Where a modeling group submitted DOC data from more than one model, only one is shown here, except for GISS-E2, because the two versions differ quite markedly. The models selected to represent multimodel groups are GFDL-ESM2M, IPSL-CM5A-LR, and MPI-ESM-LR.

produced DOC) is greater in the z-level (GFDL-ESM2M, GISS-E2-R-CC) than in the isopycnic (GFDL-ESM2G, GISS-E2-H-CC) models.

Ocean biogeochemical models have historically had difficulty simulating physical processes that introduce nutrients into the surface layer of the oligotrophic subtropical gyres, and may significantly underestimate new production in stratified regions (Hood and Christian, 2008). Proposed resolutions to this paradox have included mesoscale

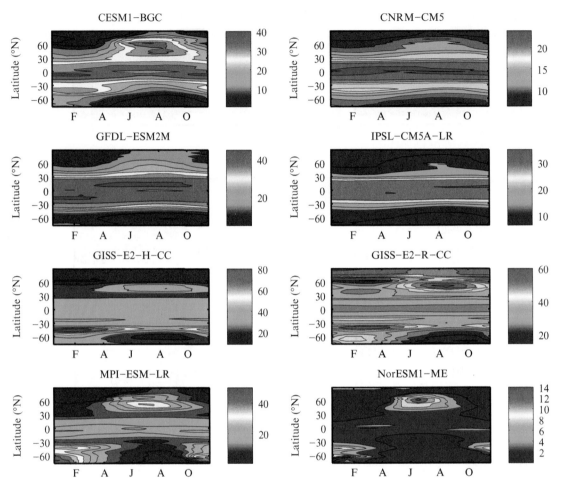

FIGURE 15.6   Annual cycle of zonal mean surface semi-labile DOC concentration (mmol m^{-3}) from eight CMIP5 models (same models as in Figure 15.5).

upwelling (e.g, McGillicuddy and Robinson, 1997) and N$_2$ fixation (Karl et al., 1997). The latter can only support a limited amount of additional production before phosphorus becomes limiting, even given a relatively flexible N/P stoichiometry (Christian, 2005). Horizontal advection of DOM has at times been suggested as a mechanism for transporting additional "new" N or P to the centers of the gyres (e.g., Abell et al, 2000; Hayward, 1991; Peltzer and Hayward, 1996), but strong evidence to support this hypothesis has been lacking. Modeling studies in the last few years have

provided a more quantitative underpinning of this conjecture, at least for the Atlantic Ocean (e.g., Roussenov et al., 2006; Torres-Valdés et al., 2009; Williams et al., 2011). As noted by Hood and Christian (2008), "models can provide estimates of quantities that are not directly observable, such as fluxes of nutrients by different physical processes." The results of these experiments show that in at least some regions, horizontal advection of DOM from outside the gyres is a source of "new" nutrients comparable to vertical transport of inorganic nutrients, although these tend to be the regions with the lowest total production (Torres-Valdés et al., 2009). The main process involved is Ekman transport from the tropics to the subtropics (Roussenov et al., 2006), and there is strong asymmetry across the basin with the transport and remineralization of DOM concentrated on the western side (Torres-Valdés et al., 2009). The largest gradients (*i.e.*, divergence) in meridional transport of DON and DOP are found between about 20°S and 30°N latitude, but particularly for P this transport can provide a source supporting more than 50% of particulate export in low-productivity regions up to 40-45°N (Torres-Valdés et al., 2009). Williams et al. (2011) illustrate the pathways by which DOM is transported into the gyres, which in the North Atlantic are strongly associated with the Gulf Stream and its associated mesoscale circulation. The supply of DOP appears to exceed (relative to phytoplankton requirements) the supply of DON in some regions, which could potentially help to sustain additional inputs of new N from N_2 fixation (Abell et al, 2000; Mahaffey et al, 2004; Roussenov et al., 2006).

IV LABILITY IN FOCUS: CONCEPTS AND DEFINITIONS

A Physiological Considerations

The usual approach taken by modelers is to divide bulk DOM into distinct pools, each with a defined lability (e.g., labile, semi-labile, semi-refractory, refractory; see Hansell, 2013 or Chapter 3) specified either as a fixed turnover rate or fixed kinetic parameters. It may then be assumed, for example, that semi-labile material is converted to labile by the action of microbial hydrolytic enzymes (Section II). A well known example is the activity of alkaline phosphatase, an enzyme common in marine prokaryote and eukaryote microbes, which cleaves P from DOP, leaving the phosphate to be taken up and the remaining organic component outside of the cell. Just how realistic is this type of approach? In order to answer this question, it is worth

considering the reasons why DOM accumulates in marine environments. The most obvious hypothesis is that the biochemical structure of the material prevents rapid degradation. An alternative hypothesis, proposed by Thingstad et al. (1997), is that of the "malfunctioning microbial loop" whereby bacteria are unable to consume otherwise degradable DOC released by the food web. They proposed that bacterial growth rate (and thereby consumption of DOM) is kept low by competition with phytoplankton for limiting nutrients. It may be that bacteria in some marine environments are limited by phosphorus, especially in areas such as the Mediterranean (Zweifel et al., 1993; Zohary and Robarts,1998). Another explanation is simply that the concentration of the individual compounds is too low to support net uptake against the concentration gradients (e.g., the leak-recovery argument of Flynn and Berry, 1999). While the concentration of bulk DOM may be high, that of the individual components, which are the subjects of transport kinetics, are several orders of magnitude lower (Flynn and Butler, 1986). Indeed, the concentrations of individual DFAA appear to reflect most closely the inverse of their "desirability" as substrates for consumption. This causes problems for experimental studies. Does one study the flux of a component of DOM that one can measure (perhaps because its flux is low) or one present at vanishingly low values (in which instance which substrate is released, and thence consumed, most)? Factors controlling DOM accumulation are reviewed in the Chapters 3 and 7 and will be described briefly here.

In order to appreciate the differences between the various hypotheses, we distinguish between potential and functional lability. Potential lability refers to whether a compound could ever be readily taken up and used by an organism. Functional lability of a compound refers to whether there is the means to readily use it at a point in space and time, and depends on a range of factors including biomass of consuming organisms, substrate concentration, expression of uptake pathways, and competing nutrient sources.

The potential lability of a compound depends on its biochemical composition and biophysical structure which dictates how readily it may be catabolized if nothing were to interfere with that consumption. A distinction can be made here between substrates that can be taken up directly *versus* those that require prior hydrolysis by extracellular enzymes. Only molecules below a certain size (ca. a few hundred Daltons) can be transported directly although small size does not in itself guarantee high potential lability. Histidine, for example, is not so readily catabolized (laying at the end of a unique biochemical pathway), as the other basic amino acids (Flynn, unpublished; but see Flynn and Butler, 1986); this protein amino acid is thus functionally semi-labile. Of the forms of DOM that cannot be directly taken up, most contain components that can be utilized after partial enzymatic hydrolysis and so molecular weight is not therefore indicative of potential lability although such compounds could be considered as semi-labile depending on the energetic and metabolic constraints on the degradation step.

Recalcitrance is typically related to chemical structures. An example is the tertiary bond, which presents a steric hindrance to terminal oxidation, and is perhaps a universally difficult structure to break (Alexander, 1994). The substitution of sulfate, nitrate, and the halogens chlorine and bromine as xenophores may similarly impede catabolism (Alexander, 1994), as may toxins that have complex multi-bond ring configurations. Finally, whereas many of the compounds that accumulate are in principle biologically degradable, the cost of doing so may in practice outweigh the benefits in terms of energy and/or nutrients obtained (e.g., Floodgate, 1995), especially when considering the availability of alternatives. In large measure this explanation is another facet of the low-concentration issue mentioned above; if concentrations were high enough it would be likely that some organisms would evolve mechanisms to extract elements and energy from these compounds.

The consumption and utilization of any one component of DOM by microbes is thus influenced by various factors associated with potential and functional lability: supply sufficient to support use, suitability for use, the specific need for particular compounds and the absence of competing alternative compounds. Intracellular concentrations of substrates in microbes tend to be in the low mM range in order to efficiently support enzymatic processes (e.g., Bjørnsen, 1988; Flynn, 1990). Regarding substrate concentrations in the surrounding environment, it is interesting to compare marine systems with research carried out on non-marine (especially medically important) bacteria (e.g., Egli et al., 1993; Harder and Dijkhuizen, 1982). External substrate concentrations in typical microbiology laboratory studies tend also to be in the mM range, which is orders of magnitude higher than concentrations of individual substrates observed in natural aquatic environments (typically nM, or lower). Thus, while Egli et al. (1993) detected "no apparent limiting nutrient concentration" for the utilization of sugars by *E. coli*, substrate concentrations recorded as "below detection" in biomedical studies are typically still far in excess of those in the ocean. The point to note here is that it is the concentration of substrates individually that counts, rather than in the total, because each is taken up using specific transporters (Driessen et al., 2000; Flynn and Butler, 1986). It may also be that in some instances individual substrates are present in concentrations that are too low for net consumption (see molecular diversity hypothesis in Chapter 7). This situation may be exacerbated by temperature in the deep ocean because substrate affinity may decrease at low temperatures (Aksnes and Egge, 1991; Nedwell, 1999).

Different sources of nutrition are favored, *via* investment in appropriate transporters and through (de)repression of biochemical pathways, by different microbial communities, so what is semi-labile for one may be labile for another (Carlson et al., 2011). A good example is provided by observations of DOC dynamics at the Bermuda Atlantic Time Series site where semi-labile DOC in surface waters is resistant to bacterial degradation whereas it is more readily utilized by bacteria from 250 m depth (Carlson et al., 2004). Bacteria can be divided into several different groups (Cottrell and Kirchman, 2003; Giovannoni and Stingl, 2005), each of which has its own characteristic pattern of substrate utilization (Elifantz et al., 2005). They invest in transport systems that selectively target different resources and so will vary for communities inhabiting, for example, different parts of the water column. Some bacteria are specialists operating on specific C compounds, with others having a general strategy (Gómez-Consarnau et al., 2012). Rapid shifts are seen in bacterial community structure in response to changing environmental conditions such as altered nutrient status (Teira et al., 2010).

The presence of preferred carbon sources prevents the expression of catabolic systems that enable the use of secondary substrates. Enzymes should only be synthesized for substrates that are present in sufficient quantities to warrant the cost of synthesis (Floodgate, 1995; Lengeler, 1993). Regulation is achieved by various mechanisms including transcription activation and repression and control of translation by an RNA-binding protein, in different bacteria (Görke and Stülke, 2008). A good example is the acquisition of phosphorus by marine bacteria. The preferred source of P for growth is inorganic phosphate but this element can also be obtained from DOP, which is dominated by two classes of compounds, phosphorus esters, and phosphonates. Phosphonates contain a highly stable C–P bond, in contrast to the more labile C–O–P phosphate ester bond. DOP in surface water exhibits a high proportion of phosphonates relative to phosphate esters, indicating that utilization by bacteria is a selective process with phosphate esters as the preferred source (Clark et al., 1998). Conditions where inorganic P (and presumably phosphate esters) is scarce lead to the introduction of genes controlling phosphonate uptake and proteins that cleave the C–P bond (McGrath et al., 1997), an

adaptation that appears widespread in marine bacteria (Martinez et al., 2010). The derepression of these genes and associated enzymes thus renders phosphonates functionally labile, a course of action that would be energetically wasteful under conditions when other sources of P are in plentiful supply. The same mechanism, induction of phosphonate transport under P-stressed conditions, is also seen in the diazotroph *Trichodesmium* (Dyrham et al., 2006). In similar fashion, the synthesis of hydrolytic enzymes is repressed when concentrations of readily utilizable substrates fulfill bacterial requirements for metabolism and growth (Gajewski and Chróst, 1995; Unanue et al., 1999). Under variable conditions, repression, and derepression of metabolic pathways leads to the exhibition of diauxic selectivity, where a substrate is only used when the preferred source is exhausted (Egli et al., 1993). The activity of ecto- and exo-enzymes (like the leak-recovery mechanism of Flynn and Berry, 1999) may also have a biomass-dependancy interaction because each enzyme molecule benefits more than just the individual cell, while the enzymes themselves are substrates for degradation.

B Modeling Implications

As described above, the division of the bulk DOM pool into a limited number of state variables (labile, semi-labile, etc.) of predefined lability is simplistic because (1) lability is dynamic, depending on the availability of different substrates and bacteria physiology and (2) substrates are not utilized in bulk but, rather, individual compounds may be taken up selectively to meet bacterial requirements.

The use of fixed kinetics for the labile pool may be reasonable in some cases, especially where this represents primary metabolites such as sugars, amino sugars, protein amino acids, nucleotides, ATP, etc. As used in models, the labile pool usually also includes HMW compounds (combined primary metabolites) that require extracellular hydrolysis with no account taken of possible repression of its use by early products of nutrient assimilation and other primary metabolites (Flynn, 1991; Magasanik, 1988). Semi-labile DOM is often the primary focus for modelers because of its roles in seasonal accumulation in surface waters, vertical distribution of DOM in the water column, and export of carbon to the deep ocean. In models where this pool is assigned a fixed lifetime, ranging from months to years (Section II), semi-labile DOM is effectively defined as being comprised exclusively of substrates that are of low potential lability from the outset. It is easy to envisage DOM lifetimes being maintained or increasing as material is exported below the euphotic zone into the deep ocean as compounds of low potential lability remain unused by bacteria. Such a picture is simplistic because it is functional lability that may define much of the semi-labile pool. Substrates that are functionally semi-labile in the euphotic zone may become functionally labile in deeper waters as transporters for their uptake by bacteria are derepressed. Other models simulate the degradation of semi-labile material to labile, usually invoking the action of hydrolytic enzymes, which seems entirely reasonable as combined primary metabolites will be acted on in this way. However, the conversion of semi-labile to labile DOM is parameterized as a fixed rate or by applying a fixed kinetic parameters that does not account for aspects of functional lability relating to competing substrates. In some instances induction of mechanisms to enable consumption of newly available substrates may occur, while in other instances a lack of repression signals from within the cell may result in expression of an ability to exploit an alternative nutrient source should such a source become available.

Most current models assume that dissolved organic nutrient elements (N, P, Fe, etc.) and carbon are taken up in proportion to their bulk availability, in the same way as different food particles are (incorrectly assumed to be) consumed by zooplankton. This assumption ignores the importance of the concentration of individual chemical moieties and also of selectivity between those chemicals.

The advantage of this assumption, however, is that the fate of the material consumed can then be calculated in accord with stoichiometric formulations (e.g., Anderson, 1992; Touratier et al., 1999). When nutrient elements such as N or P are limiting, excess carbon is left unutilized and may be consumed *via* the increased respiration that accompanies nutrient stress (Neijssel and Tempest, 1975; Pirt, 1982; Flynn, 2005) or release of C-rich polymers, notably polysaccharides (Decho, 1990). Laboratory studies have shown that bacteria have low C growth efficiency when confronted by large amounts of primary metabolites such as glucose (e.g., Goldman et al., 1987; Tezuka, 1990). Nevertheless, it is likely the case that when bacteria experience a heterogeneous resource environment, as is usually the case in the ocean, they strip out selected substrates. If this stripping is done to obtain nutrient elements, it would lead to an accumulation of DOC (of decreasing lability) that would not be predicted using conventional stoichiometric models. This could be seen as a facet of so-called malfunctioning microbial loop of Thingstad et al. (1997). Thus, the assumption of stoichiometric linkage between C and other elements, without the potential for nutrient stripping, is likely untenable for N, and is certainly so for P.

V DISCUSSION

Our survey of the literature has highlighted both similarity and diversity in the approaches taken for modeling DOM in marine systems. The similarity lies in the convenient approach of dividing DOM into labile, semi-labile, and refractory pools, although it should be remarked that differences in terminology exist between studies. The diversity, which is enormous, is in the chosen formulations for DOM production and consumption. How, one may ask, is it possible to reliably simulate the dynamics of DOM in marine systems given this diversity in formulations and associated parameterization? More fundamentally, we lack the chemical data (concentrations and fluxes) against which to configure and test such models. Unsurprisingly then, where common approaches exist, these formulations and associated parameterizations remain wide open to question. As we have shown, for example, lability is a dynamic concept and yet, with little in the way of alternatives, modelers today continue to use "labile," "semi-labile," and "refractory" pools as they were first conceptualized in the 1990s. Indeed, despite the current emphasis on increasing complexity and realism in models, it is interesting to note that some of the models in use today have chosen not only this basic construction, but also adopt some of the associated early parameterizations. For example the model of Anderson and Williams (1998), which includes a fixed C growth efficiency for bacteria, Michaelis-Menten consumption of labile and semi-labile pools and stoichiometric regulation of consumption/remineralization of ammonium, remains a popular choice (e.g., Grégoire et al., 2008; Keller and Hood, 2011).

To some extent, the differences seen in DOM parameterization between models can be attributed to objectives and focus. Simple approaches have often, for example, been used in OGCMs (Section III). Multiple DOM source terms, the whole issue of lability, a mixed bacterial community, and a heterogeneous environment all serve to make this a very complex and challenging system to model. Our understanding of the web of interactions is incomplete and so the parameterization of flows and the compartmentalization of DOM into meaningful structures in models are far from straightforward. Before addressing these difficulties, it is worth emphasizing that models have been relatively successful at simulating DOM in the ocean. On what basis have they apparently been so effective? There is a parallel here, we believe, with the nutrient-phytoplankton-zooplankton-detritus (NPZD) ecosystem models. Despite their simplicity, NPZD models have been largely successful as simulating bulk system properties

such as chlorophyll and primary production, which are constrained by nutrient availability, light, and grazing (Anderson, 2005; Franks, 2002). In the case of DOM, production is constrained largely by primary production, grazing, and detritus turnover although modelers have a free hand to allocate between the labile and semi-labile pools. Turnover rates are in reality variable but, nevertheless, mean lifetimes of months to years for semi-labile material are considered robust and reliable estimates. Simple models using DOM production terms as fractions of primary production, grazing and detritus, with a fixed turnover rates, may therefore do remarkably well.

The future development of marine biogeochemical models of DOM does, however, require assessment of the considerable difficulties of modeling this complex system. For example, there is the issue of just what constitutes DOM in the first place. Modelers have steered clear of the difficulties of characterizing DOM biochemically, adopting stoichiometric approaches to describing DOM instead. Processes such as phytoplankton exudation and release *via* messy feeding may therefore be assigned, for example, different C:N ratios thereby allowing simulation of the accumulation of C-rich DOM that is characteristic of many systems (e.g., Williams, 1995). The simulation of DOM consumption by bacteria is simplistic in that uptake is calculated as a function of bulk concentration whereas in reality DOM constitutes many compounds each with its own kinetic constraints. Bacteria selectively remove certain compounds; with time there will be an inevitable change in composition of DOM, with an increasing proportion of less energetically and less nutritionally favored chemicals remaining in solution.

Perhaps the most thorny issue for modelers is that of DOM lability and how to compartmentalize DOM into meaningful state variables in this regard. If one considers the traditional definitions, then exclusion of labile DOM in some models may be argued as reasonable on the basis that this pool has a fast turnover (hours to days) and so contributes little to C export *via* mixing or subduction. However, if one considers the role of DOM and bacteria in supporting primary production through the growth of the mixotrophic protists (Mitra et al., 2014), then describing the dynamics of bacteria growth over hours-days-weeks certainly warrants consideration; primary production supported by labile DOM could certainly contribute to C export and to the MCP. The new paradigm for planktonic primary production proposed by Mitra et al. (2014) places bacteria and DOM in a new light, and should act as a stimulus for DOM research and modeling. At the other end of the spectrum, the question of whether to include refractory DOM depends on the timescales of interest. Its turnover rate is so slow that it appears unnecessary to include it dynamically in models addressing climate change within the next 200 years, although recent work that points to localized removal of otherwise defined refractory DOC in the deep ocean (Hansell and Carlson, 2013) and emphasis on the MCP merits attention (Jiao et al., 2010). It is the semi-labile pool that is usually of greatest interest because of its role in carbon export and transport of nutrients within oligotrophic gyres. Yet, as we have discussed (Section IV), the factors that render DOM semi-labile are multifaceted such that there is a need to separate potential and functional lability. The latter depends on various factors including supply sufficient to support use, suitability for use, the specific need for particular compounds and the absence of competing alternative compounds. The way forward is, however, by no means clear. Matters are only complicated further by the fact that microbes constitute a mixed assemblage with varying growth requirements and, moreover, the factors controlling growth are not completely understood. Bacteria may be separated into several major groups (Cottrell and Kirchman, 2003; Giovannoni and Stingl, 2005), each with its own physiological attributes (Pinhassi and Hagström, 2000) and even so the individual organisms biochemistry changes in response to external factors.

An understanding of the relationship between bacteria community structure and biogeochemical function is therefore crucial for model development (Fuhrman and Steele, 2008; Höfle et al., 2008; Mitra et al., 2014).

The greatest diversity in approach by modelers is seen in the specification of source terms and turnover rates of the chosen DOM pools, as well as the associated representation of bacteria. Perhaps the most surprising difference seen between models is in the specification of the fate of detritus. In some models detritus turnover goes to DOM whereas in others remineralization is directly to inorganic nutrient and CO_2. Modeling studies by Anderson and Williams (1999) and Bendtsen et al. (2002) suggest that the majority of DOM present at depth is derived from turnover of sinking detritus rather than from DOM exported *via* mixing from the surface ocean. At first sight, it seems surprising that many modeling studies, especially those using OGCMs, have chosen not to include detritus turnover as a source term of DOM. Of the models listed in Table 15.1, for example, all of those published in the last 10 years chose this option. Depending on objectives, it may be that this choice is not particularly significant. If the aim is to model distributions of inorganic nutrients and carbon, including those in the deep ocean, then it may not matter much if detritus is remineralized directly to nutrient and CO_2. DOM is merely an intermediary. If, on the other hand, the aim is to model DOM distributions (e.g., Hansell et al, 2012), then we would argue that remineralization of detritus to DOM should be included. Further, detritus turnover is a significant source of nutrients and carbon in the euphotic zone. For example, Yool et al. (2011) simulated global biogeochemistry using an intermediate-complexity ecosystem model in which the turnover of detrital-N was allocated to DIN. Their model suggested that detritus turnover accounted for 35% of the source terms of DIN (0-100 m), with phytoplankton mortality, and zooplankton grazing, excretion, and mortality accounting for 15%, 23%, 13%, and 13% respectively. Further work is needed to assess the sensitivity of predicted biogeochemical fields to the choice of whether to allocate detritus turnover to DON or DIN. Similar arguments can be made for carbon. The accumulation of DOC in surface waters is potentially significant in, for example, the seasonal drawdown of DIC and pCO_2 (Anderson and Pondaven, 2003).

The reality is that detritus turnover is divided between DOM and inorganic nutrients, rather than 100% either way, the fractionation depending on a number of processes, and critically upon the time scale in question. Both bacteria and zooplankton are likely responsible for this turnover and organic matter is recycled *via* various pathways including DOM, suspended POM, and remineralization to inorganic forms (Giering et al., 2014). Particle-attached bacteria release large amounts of DOM to the surrounding water as the solubilization products of enzymatic hydrolysis (Cho and Azam, 1988; Smith et al., 1992). Zooplankton consume detritus, recycling it both as fecal material and DOM (Lampitt et al., 1990). Improved understanding and quantification of these processes is needed in order to formulate and parameterize the production and fate of DOM in the ocean.

Another point of departure in DOM models is whether or not bacteria are explicitly represented. In some cases, modelers have chosen to circumvent the difficulties of representing bacteria by not including them as a state variable, parameterizing DOM turnover using fixed rates (e.g., Llebot et al., 2010; Salihoglu et al., 2008). There is little evidence to suggest that these models are performing any worse than those that do include explicit bacteria. Nevertheless, it is unsurprising that models aimed at understanding the dynamics of the microbial loop and DOM cycling in marine systems usually favor the explicit representation of bacteria. Certainly descriptions of primary production, by protists other than by diatoms, which involve bactivory

(Flynn et al., 2013; Mitra et al., 2014) warrant a renewed consideration of the whole issue of bacteria PFT models. New parameterizations have been developed, for example, by dividing bacterial respiration into basal and active fractions giving rise to variable BGE depending on substrate availability (e.g., Flynn, 2005; Polimene et al., 2006; Luo et al., 2010). Our understanding of BGE, and notably the stoichiometric control thereof, nevertheless remains rudimentary despite excellent reviews (e.g., del Giorgio and Cole, 1998). The use of Michaelis-Menten kinetics and Blackman-style limitation of growth are also open to question. Although there have been calls for a more considered development of models for the bacterial consumption of multiple substrates (Button, 1993; Egli et al., 1993; van Dam et al., 1993), Michaelis-Menten remains the norm and may be criticized in the same way as equivalent models of phytoplankton growth (Flynn, 2003). Quota models may help (e.g., Luo et al., 2010) although other factors may also be important, such as the fact that bacteria are a mixed assemblage and simple responses to nutrient pulses may not be expected in the natural environment. The modeling of substrate uptake, bacterial nutrition, and resulting growth efficiency is probably more challenging than that for any other member of the plankton because the range of potential substrates and nutrient interactions is so great. Complex models have been developed (*i.e.*, Vallino et al., 1996) that describe bacterial utilization of DOM based on growth rate optimisation subject to constraints on energy, redox reactions, substrate uptake kinetics, and the C:N of bacteria. The C:N:P model of Flynn (2005) enables considerations of changes in growth efficiency with C, N, and/or P stress. The considerable variability in substrates experienced by bacteria in their natural environment makes models such as these difficult to employ in an ecosystem context.

Despite all the above, several reasons for the inclusion of DOM can be proposed. DOC plays a significant role in export to the deep ocean

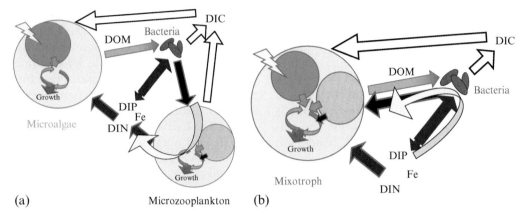

FIGURE 15.7 Schematic showing the detailed involvement of bacteria and DOM for the supply of nutrients to support primary production (yellow arrows) in the traditional paradigm for the linkage between marine primary production, DOM, and bacteria (a) versus that in the new paradigm suggested in (b) by Mitra et al. (2014). In the former, bacteria compete with primary producers for DIN, DIP, and Fe while their growth is supported by DOM released with primary production. Nutrients are then regenerated via the activity of microzooplankton. In the new paradigm, there is a direct support of primary production via nutrient acquisition in bactivory, with the traditional functionality of microalgae and microzooplankton merged in protist mixotrophs (see also Hartmann et al., 2012). Black arrows indicate predatory links.

(Hansell et al., 2009). The paradigm shift suggested by Mitra et al. (2014) sees an overturn of the concept of competition between the microbial loop and the "traditional food chain." The new paradigm (Figure 15.7) sees an intermediate relationship, not a competition, between primary production and bacteria mediated through DOM, a concept put forward by Flynn (1988). The seasonal accumulation of DOM alters nutrient inventories and pCO_2 in surface waters. Lateral transport of dissolved organic nutrients may affect primary production and nutrient budgets in the oligotrophic gyres. Most marine ecosystem models today do therefore incorporate DOM, although there are exceptions (e.g., Liu and Chai, 2009; Yool et al., 2011). The performance of two versions of a global ocean box model, with and without DOM, was compared by Popova and Anderson (2002). Export fluxes were relatively insensitive to the choice of whether or not to include DOM because they were dominated by sinking particles. Significant changes were nevertheless noted in primary production and the f-ratio, giving rise to changes in alkalinity that could be important when calculating pCO_2 and

the resulting air-sea flux of CO_2. Similar findings were made by Schmittner et al. (2005). The case for modeling DOM and the microbial loop is clear. The problem is that it is, as we have shown, it is a complex business. When explaining why DOM was excluded from their global biogeochemical model, Yool et al. (2011) cited the previous edition of this book (Christian and Anderson, 2002), summarizing our conclusions as "Assigning equations and parameter values for DOM cycling is fraught with difficulty given our limited understanding of interactions within microbial communities and of the physiology of heterotrophic bacteria."

Models have played a key role in the quantification and understanding of the biogeochemical cycling of DOM in the ocean. Yet, as we have shown, published equations and parameterizations are disparate and, if one were feeling uncharitable, can easily be criticized as being simplistic or unrealistic. Referring to this lack of consensus, Christian and Anderson (2002) commented that "the mathematical formulation of these terms remains tentative and speculative." Although the description of DOM cycling has become considerably elaborated in the last 10 years, we would argue that this statement still applies today. Modelers should not be satisfied with the current state of the art. Modeling DOM is challenging. Future progress requires a coordinated approach, with strong interlinking between modelers, experimentalists, and field researchers. It will be interesting to see how the field progresses in the next 10 years.

Acknowledgments

TRA and KJF acknowledge support from the Natural Environment Research Council (UK). KJF also acknowledges support from EURO-BASIN (Ref. 264933, 7FP, European Union). The authors thank the CMIP5 data originators for their contributions.

References

Abell, J., Emerson, S., Renaud, P., 2000. Distributions of TOP, TON and TOC in the North Pacific subtropical gyre: implications for nutrient supply in the surface ocean and remineralization in the upper thermocline. J. Mar. Res. 58, 203–222.

Aksnes, D.L., Egge, J.K., 1991. A theoretical model for nutrient uptake in phytoplankton. Mar. Ecol. Prog. Ser. 70, 65–72.

Alexander, M., 1994. Biodegradation and Bioremediation. Academic Press, San Diego, 302 pp.

Allen, J.I., Fulton, E.A., 2010. Top-down, bottom-up or middle-out? Avoiding extraneous detail and over-generality in marine ecosystem models. Prog. Oceanogr. 84, 129–133.

Amon, R.M.W., Benner, R., 1996. Bacterial utilization of different size classes of dissolved organic matter. Limnol. Oceanogr. 41, 41–51.

Anderson, T.R., 1992. Modelling the influence of food C:N ratio, and respiration on growth and nitrogen excretion in marine zooplankton and bacteria. J. Plankton Res. 14, 1645–1671.

Anderson, T.R., 2005. Plankton functional type modelling: running before we can walk? J. Plankton Res. 27, 1073–1081.

Anderson, T.R., Ducklow, H.W., 2001. Microbial loop carbon cycling in ocean environments studied using a simple steady-state model. Aquat. Microb. Ecol. 26, 37–49.

Anderson, T.R., Gentleman, W.C., 2012. The legacy of Gordon Arthur Riley (1911–1985) and the development of mathematical models in biological oceanography. J. Mar. Res. 70, 1–30.

Anderson, T.R., Hessen, D.O., 1995. Carbon or nitrogen limitation in marine copepods? J. Plankton Res. 17, 317–331.

Anderson, T.R., Pondaven, P., 2003. Non-Redfield carbon and nitrogen cycling in the Sargasso Sea: pelagic imbalances and export flux. Deep-Sea Res. I 50, 573–591.

Anderson, T.R., Williams, P.J. le B., 1998. Modelling the seasonal cycle of dissolved organic carbon at Station E1 in the English Channel. Estuar. Coast. Shelf Sci. 46, 93–109.

Anderson, T.R., Williams, P.J. le B., 1999. A one-dimensional model of dissolved organic carbon cycling in the water column incorporating combined biological-photochemical decomposition. Global Biogeochem. Cycles 13, 337–349.

Aumont, O., Maier-Reimer, E., Blain, S., Monfray, P., 2003. An ecosystem model of the global ocean including Fe, Si, P colimitations. Global Biogeochem. Cycles 17, 1060. http://dx.doi.org/10.1029/2001GB001745.

Azam, F., Fenchel, T., Field, J.G., Gray, J.S., Meyer-Reil, L.A., Thingstad, F., 1983. The ecological role of water-column microbes in the sea. Mar. Ecol. Prog. Ser. 10, 257–263.

Bacastow, R., Maier-Reimer, E., 1990. Ocean-circulation model of the carbon cycle. Climate Dynam. 4, 95–125.

Banoub, M.W., Williams, P.J.leB., 1973. Seasonal changes in the organic forms of carbon, nitrogen and phosphorus in sea water at E_1 in the English Channel During 1968. J. Mar. Biol. Assoc. U.K. 53, 695–703.

Bendtsen, J., Lundsgaard, C., Middelboe, M., Archer, D., 2002. Influence of bacterial uptake on deep-ocean dissolved organic carbon. Global Biogeochem. Cycles 16, 1127. http://dx.doi.org/10.1029/2002GB001947.

Berman, T., Stone, L., 1994. Musings on the microbial loop: twenty years after. Microb. Ecol. 28, 251–253.

Billen, G., Becquevort, S., 1991. Phytoplankton-bacteria relationship in the Antarctic marine ecosystem. Polar Res. 10, 245–253.

Bissett, P.W., Walsh, J.J., Dieterle, D.A., Carder, K.L., 1999. Carbon cycling in the upper waters of the Sargasso Sea: I. Numerical simulation of differential carbon and nitrogen fluxes. Deep Sea Res. Oceanogr. Res. Paper 46, 205–269.

Bjørnsen, P.K., 1988. Phytoplankton exudation of organic matter: why do healthy cells do it? Limnol. Oceanogr. 33, 151–154.

Blackford, J.C., Allen, J.I., Gilbert, F.J., 2004. Ecosystem dynamics at six contrasting sites: a generic modelling study. J. Mar. Syst. 52, 191–215.

Button, D.K., 1993. Nutrient-limited microbial growth kinetics: overview and recent advances. Antonie Van Leeuwenhoek 63, 225–235.

Carlson, C.A., Ducklow, H.W., 1995. Dissolved organic carbon in the upper ocean of the central equatorial Pacific Ocean, 1992: daily and finescale vertical variations. Deep-Sea Res. II 42, 639–656.

Carlson, C.A., Hansell, D.A., Peltzer, E.T., Smith Jr., W.O., 2000. Stocks and dynamics of dissolved and particulate organic matter in the southern Ross Sea, Antartica. Deep-Sea Res. II 47, 3201–3225.

Carlson, C.A., Giovannoni, S.J., Hansell, D.A., Goldberg, S.J., Parsons, R., Vergin, K., 2004. Interactions among dissolved organic carbon, microbial processes, and community structure in the mesopelagic zone of the northwestern Sargasso Sea. Limnol. Oceanogr. 49, 1073–1083.

Carlson, C.A., Hansell, D.A., Tamburini, C., 2011. DOC persistence and its fate after export within the ocean interior. In: Jiao, N., Azam, F., Sanders, S. (Eds.), Microbial Carbon Pump in the Ocean. Science/AAAS Washington, DC, pp. 57–59.

Cherrier, J., Bauer, J.E., Druffel, E.R.M., 1996. Utilization and turnover of labile dissolved organic matter by bacterial heterotrophs in eastern North Pacific surface waters. Mar. Ecol. Prog. Ser. 139, 267–279.

Cho, B.C., Azam, F., 1988. Major role of bacteria in biogeochemical fluxes in the ocean's interior. Nature 332, 441–443.

Christian, J.R., 2005. Biogeochemical cycling in the oligotrophic ocean: Redfield and non-Redfield models. Limnol. Oceanogr. 50, 646–657.

Christian, J.R., Anderson, T.R., 2002. Modeling DOM biogeochemistry. In: Hansell, D.A., Carlson, C.A. (Eds.), Biogeochemistry of marine dissolved organic matter. Academic Press, N.Y, pp. 717–755.

Clark, L.L., Ingall, E.D., Benner, R., 1998. Marine phosphorus is selectively remineralized. Nature 393, 426.

Connolly, J.P., Coffin, R.B., 1995. Model of carbon cycling in planktonic food webs. J. Environ. Eng. 121, 682–690.

Connolly, J.P., Coffin, R.B., Landeck, R.E., 1992. Modeling carbon utilization by bacteria in natural water systems. In: Hurst, C.J. (Ed.), Modelling the Metabolic and Physiologic Activities of Microorganisms. Wiley, New York, pp. 249–276.

Cottrell, M.T., Kirchman, D.L., 2003. Contribution of major bacterial groups to bacterial biomass production (thymidine and leucine incorporation) in the Delaware estuary. Limnol. Oceanogr. 48, 168–178.

De Ruiter, P.C., Van Veen, J.A., Moore, J.C., Brussaard, L., Hunt, H.W., 1993. Calculation of nitrogen mineralization in soil food webs. Plant & Soil 157, 263–273.

Decho, A.W., 1990. Microbial exopolymer secretions in ocean environments—their role(s) in food webs and marine processes. Oceanogr. Mar. Biol. Annu. Rev. 28, 73–153.

del Giorgio, P.A., Cole, J.J., 1998. Bacterial growth efficiency in natural aquatic systems. Annu. Rev. Ecol. Syst. 29, 503–541.

Driessen, A.J.M., Rosen, B.P., Konings, W.N., 2000. Diversity of transport mechanisms: common structural principles. Trends Biochem. Sci. 25, 397–401.

Druon, J.N., Mannino, A., Signorini, S., McClain, C., Friedrichs, M., Wilkin, J., et al., 2010. Modeling the dynamics and export of dissolved organic matter in the Northeastern U.S. continental shelf. Estuar. Coast. Shelf Sci. 88, 488–507.

Ducklow, H.W., Fasham, M.J.R., Vézina, A.F., 1989. Derivation and analysis of flow networks for open ocean plankton systems. In: Wulff, F., Field, J.G., Mann, K.H. (Eds.), Network Analysis in Marine ecosystems. Springer-Verlag, Berlin, pp. 159–205.

Dutkiewicz, S., Follows, M., Parekh, P., 2005. Interactions of the iron and phosphorus cycles: a three-dimensional model study. Global Biogeochem. Cycles 19, http://dx.doi.org/10.1029/2004GB002342, GB1021.

Dyrham, S.T., Chappell, P.D., Haley, S.T., Moffett, J.W., Orchard, E.D., Waterbury, J.B., et al., 2006. Phosphonate utilization by the globally important marine diazotroph *Trichodesmium*. Nature 439, 68–71.

Egli, T., Lendenmann, U., Snozzi, M., 1993. Kinetics of microbial growth with mixtures of carbon sources. Antonie Van Leeuwenhoek 63, 289–298.

Elifantz, H., Malmstrom, R.R., Cottrell, M.T., Kirchman, D.L., 2005. Assimilation of polysaccharides and glucose by major bacterial groups in the Delaware Estuary. Appl. Environ. Microbiol. 71, 7799–7805.

Fasham, M.J.R., Ducklow, H.W., McKelvie, S.M., 1990. A nitrogen-based model of plankton dynamics in the oceanic mixed layer. J. Mar. Res. 48, 591–639.

Floodgate, G.D., 1995. Some environmental aspects of marine hydrocarbon bacteriology. Aquat. Microb. Ecol. 9, 3–11.

Flynn, K.J., 1988. The concept of "primary production" in aquatic ecology. Limnol. Oceanogr. 33, 1215–1216.

Flynn, K.J., 1990. Composition of intracellular and extracellular pools of amino acids, and amino acid utilization of microalgae of different sizes. J. Exp. Biol. Ecol. 139, 151–166.

Flynn, K.J., 1991. Algal carbon-nitrogen metabolism: a biochemical basis for modelling the interactions between nitrate and ammonium uptake. J. Plankton Res. 13, 373–387.

Flynn, K.J., 2003. Modelling multi-nutrient interactions in phytoplankton: balancing simplicity and realism. Prog. Oceanogr. 56, 249–279.

Flynn, K.J., 2005. Incorporating plankton respiration in models of aquatic ecosystem function. In: del Giorgio, P.A., Williams, P.J. le B. (Eds.), Respiration in Aquatic Ecosystems. Oxford University Press, Oxford, pp. 248–266.

Flynn, K.J., Berry, L.S., 1999. The loss of organic nitrogen during marine primary production may be significantly overestimated when using ^{15}N substrates. Proc. R. Soc. Lond. B 266, 641–647.

Flynn, K.J., Butler, I., 1986. Nitrogen sources for the growth of marine microalgae: role of dissolved free amino acids. Mar. Ecol. Prog. Ser. 34, 281–304.

Flynn, K.J., Davidson, K., 1993. Predator-prey interactions between *Isochrysis galbana* and *Oxyrrhis marina*. II. Release of non-protein amines and faeces during predation of Isochrysis. J. Plankton Res. 15, 893–905.

Flynn, K.J., Clark, D.R., Xue, Y., 2008. Modelling the release of dissolved organic matter by phytoplankton. J. Phycol. 44, 1171–1187.

Flynn, K.J., Stoecker, D.K., Mitra, A., Raven, J.A., Gilbert, P.M., Hansen, P.J., et al., 2013. Misuse of the phytoplankton-zooplankton dichotomy: the need to assign organisms as mixotrophs within plankton functional types. J. Plankton Res. 35, 3–11.

Franks, P.J.S., 2002. NPZ models of plankton dynamics: their construction, coupling to physics, and application. J. Oceanogr. 58, 379–387.

Franks, P.J.S., Wroblewski, J.S., Flierl, G.R., 1986. Behavior of a simple plankton model with food-level acclimation by herbivores. Mar. Biol. 91, 121–129.

Fuhrman, J.A., 1999. Marine viruses and their biogeochemical and ecological effects. Nature 399, 541–548.

Fuhrman, J.A., Steele, J.A., 2008. Community structure of marine bacterioplankton: patterns, networks, and relationships to function. Aquat. Microb. Ecol. 53, 69–81.

Gajewski, A.J., Chróst, R.J., 1995. Production and enzymatic decomposition of organic matter by microplankton in a eutrophic lake. J. Plankton Res. 17, 709–728.

Giering, S.L.C., Sanders, R., Lampitt, R.S., Anderson, T.R., Tamburini, C., Boutrif, M., et al., 2014. Reconciliation of the carbon budget in the ocean's twilight zone. Nature 507, 480–483.

Giovannoni, S.J., Stingl, U., 2005. Molecular diversity and ecology of microbial plankton. Nature 437, 343–348.

Goldman, J.C., Caron, D.A., Dennett, M.R., 1987. Regulation of gross growth efficiency and ammonium regeneration in bacteria by substrate C:N ratio. Limnol. Oceanogr. 32, 1239–1252.

Gómez-Consarnau, L., Lindh, M.V., Gasol, J.M., Pinhassi, J., 2012. Structuring of bacterioplankton communities by specific dissolved organic carbon compounds. Environ. Microbiol. 14, 2361–2378.

Görke, B., Stülke, J., 2008. Carbon catabolite repression in bacteria: many ways to make the most out of nutrients. Nat. Rev. Microbiol. 6, 613–624.

Gregg, W.W., Ginoux, P., Schopf, P.S., Casey, N.W., 2003. Phytoplankton and iron: validation of a global three-dimensional ocean biogeochemical model. Deep-Sea Res. II 50, 3143–3169.

Grégoire, M., Raick, C. And, Soetaert, K., 2008. Numerical modeling of the central Black Sea ecosystem functioning during the eutrophication phase. Prog. Oceanogr. 76, 286–333.

Hansell, D.A., 2013. Recalcitrant dissolved organic carbon fractions. Ann. Rev. Mar. Sci. 5, 421–445.

Hansell, D.A., Carlson, C.A., 1998. Deep-ocean gradients in the concentration of dissolved organic carbon. Nature 395, 263–266.

Hansell, D.A., Carlson, C.A., 2013. Localized refractory dissolved organic carbon sinks in the deep ocean. Global Biogeochem. Cycles 27, 705–710.

Hansell, D.A., Carlson, C.A., Repeta, D.J., Schlitzer, R., 2009. Dissolved organic matter in the ocean. Oceanography 22 (4), 202–211.

Hansell, D.A., Carlson, D.A., Schlitzer, R., 2012. Net removal of major marine dissolved organic carbon fractions in the subsurface ocean. Global Biogeochem. Cycles 26, http://dx.doi.org/10.1029/2011GB004069, GB1016.

Hansen, P.J., Calado, A.J., 1999. Phagotrophic mechanisms and prey selection in free-living dinoflagellates. J. Eukaryotic Microbiol. 46, 382–389.

Hansman, R.L., Griffin, S., Watson, J.T., Druffel, E.R.M., Ingalls, A.E., Pearson, A., et al., 2009. The radiocarbon signature of microorganisms in the mesopelagic ocean. Proc. Natl. Acad. Sci. U. S. A. 106, 6513–6518.

Harder, W., Dijkhuizen, L., 1982. Strategies of mixed substrate utilization in microorganisms. Phil. Trans. R. Soc. Lond. B 297, 459–480.

Hartmann, M., Grob, C., Tarran, G.A., Martin, A.P., Burkill, P.H., Scanlan, D.J., et al., 2012. Mixotrophic basis of Atlantic oligotrophic ecosystems. Proc. Natl. Acad. Sci. U. S. A. 109, 5756–5760.

Harvey, H.W., 1950. On the production of living matter in the sea off Plymouth. J. Mar. Biol. Assoc. U.K. 29, 97–137.

Hayward, T.L., 1991. Primary production in the north Pacific central gyre: a controversy with important implications. TREE 6, 281–284.

Hessen, D.O., 1992. Nutrient element limitation of zooplankton production. Am. Nat. 140, 799–814.

Hessen, D.O., Anderson, T.R., 2008. Excess carbon in aquatic organisms and ecosystems: physiological, ecological, and evolutionary implications. Limnol. Oceanogr. 53, 1685–1696.

Höfle, M.G., Kirchman, D.L., Christen, R., Brettar, I., 2008. Molecular diversity of bacterioplankton: link to a predictive biogeochemistry of pelagic ecosystems. Aquat. Microb. Ecol. 53, 39–58.

Hofmann, E.E., Ambler, J.W., 1988. Plankton dynamics on the outer southeastern U.S. continental shelf. Part II: a time-dependent biological model. J. Mar. Res. 46, 883–917.

Hood, R.R., Christian, J.R., 2008. Ocean nitrogen cycle modeling. In: Capone, D.G., Bronk, D.A., Mulholland, M.R., Carpenter, E.J. (Eds.), Nitrogen in the Marine Environment. second ed. Elsevier, San Diego, pp. 1445–1495.

Hopkinson, C.S., Vallino, J.J., 2005. Efficient export of carbon to the deep ocean through dissolved organic matter. Nature 433, 142–145.

Huret, M., Dadou, I., Dumas, F., Lazure, P., Garçon, V., 2005. Coupling physical and biogeochemical processes in the Río de la Plata plume. Cont. Shelf Res. 25, 629–653.

Jannasch, H.W., Jones, G.E., 1959. Bacterial populations in sea water as determined by different methods of enumeration. Limnol. Oceanogr. 4, 128–139.

Jiao, N., Herndl, G.J., Hansell, D.A., Benner, R., Kattner, G., Wilhelm, S.W., et al., 2010. Microbial production of recalcitrant dissolved organic matter: long-term carbon storage in the global ocean. Nat. Rev. Microbiol. 8, 593–599.

Jumars, P.A., Penry, D.L., Baross, J.A., Perry, M.J., Frost, B.W., 1989. Closing the microbial loop: dissolved carbon pathway to heterotrophic bacteria from incomplete ingestion, digestion and absorption in animals. Deep-Sea Res. 36, 483–495.

Karl, D., Letelier, R., Tupas, L., Dore, J., Christian, J., Hebel, D., 1997. The role of nitrogen fixation in biogeochemical cycling in the subtropical North Pacific Ocean. Nature 388, 533–538.

Kawamiya, M., Kishi, M.J., Yamanaka, Y., Suginohara, N., 1995. An ecological-physical coupled model applied to Station Papa. J. Oceanogr. 51, 635–664.

Keller, D.P., Hood, R.R., 2011. Modeling the seasonal autochthonous sources of dissolved organic carbon and nitrogen in the upper Chesapeake Bay. Ecol. Model. 222, 1139–1162.

Keller, D.P., Hood, R.R., 2013. Comparative simulations of dissolved organic matter cycling in idealized oceanic, coastal, and estuarine surface waters. J. Mar. Syst. 109, 109–128.

Keys, A., Christensen, E.H., Krogh, A., 1935. The organic metabolism of sea-water with special reference to the ultimate food cycle in the sea. J. Mar. Biol. Assoc. U.K. 20, 181–196.

Kirchman, D.L., Lancelot, C., Fasham, M.J.R., Legendre, L., Radach, G., Scott, M., 1993. Dissolved organic matter in biogeochemical models of the ocean. In: Evans, G.T., Fasham, M.J.R. (Eds.), Towards a Model of Ocean Biogeochemical Processes. Springer-Verlag, Berlin, pp. 209–225.

Krogh, A., 1934. Conditions of life in the ocean. Ecol. Monogr. 4, 421–429.

Kwon, E.Y., Primeau, F., 2006. Optimization and sensitivity study of a biogeochemistry ocean model using an implicit solver and in situ phosphate data. Global Biogeochem. Cycles 20, http://dx.doi.org/10.1029/2005GB002631, GB4009.

Kwon, E.Y., Primeau, F., 2008. Optimization and sensitivity of a global biogeochemistry ocean model using combined in situ DIC, alkalinity, and phosphate data. J. Geophys. Res. 113, http://dx.doi.org/10.1029/2007JC004520, C08011.

Lampitt, R.S., Noji, T., von Bodungen, B., 1990. What happens to zooplankton faecal pellets? Implications for material flux. Mar. Biol. 104, 15–23.

Le Quéré, C., Harrison, S.P., Prentice, I.C., et al., 2005. Ecosystem dynamics based on plankton functional types for global ocean biogeochemistry models. Global Change Biol. 11, 2016–2040.

Lengeler, J.W., 1993. Carbohydrate transport in bacteria under environmental conditions, a black box? Antonie van Leeuwenhoek 63, 275–288.

Levy, M., Memery, L., Andre, J.-M., 1998. Simulation of primary production and export fluxes in the Northwestern Mediterranean Sea. J. Mar. Res. 56, 197–238.

Liu, G., Chai, F., 2009. Seasonal and interannual variation of physical and biological processes during 1994–2001 in the Sea of Japan/East Sea: a three-dimensional physical-biogeochemical modeling study. J. Mar. Syst. 78, 265–277.

Llebot, C., Spitz, Y.H., Solé, J., Estrada, M., 2010. The role of inorganic nutrients and dissolved organic phosphorus in the phytoplankton dynamics of a Mediterranean bay. A modeling study. J. Mar. Syst. 83, 192–209.

Luo, Y.-W., Friedrichs, M.A.M., Doney, S.C., Church, M.J., Ducklow, H.W., 2010. Oceanic heterotrophic bacterial nutrition by semilabile DOM as revealed by data assimilative modeling. Aquat. Microb. Ecol. 60, 273–287.

Magasanik, B., 1988. Reversible phosphylation of an inhancer binding protein regulates the transcription of bacteria nitrogen utilization genes. Trends Biochem. Sci. 13, 475–479.

Mahaffey, C., Williams, R.G., Wolff, G.A., Anderson, W.T., 2004. Physical supply of nitrogen to phytoplankton in the Atlantic Ocean. Global Biogeochem. Cycles 18, http://dx.doi.org/10.1029/2003GB002129, GB1034.

Marchant, H.J., Scott, F.J., 1993. Uptake of sub-micrometre particles and dissolved organic material by Antarctic choanoflagellates. Mar. Ecol. Prog. Ser. 92, 59–64.

Martel, C.M., 2006. Prey location, recognition and ingestion by the phagotrophic marine dinoflagellate Oxyrrhis marina. J. Exp. Mar. Biol. Ecol. 335, 210–220. http://dx.doi.org/10.1016/j.jembe.2006.03.006.

Martin, J.H., Knauer, G.A., Karl, D.M., Broenkow, W.W., 1987. VERTEX: carbon cycling in the Northeast Pacific. Deep-Sea Res. 34, 267–285.

Martinez, A., Tyson, G.W., DeLong, E.F., 2010. Widespread known and novel phosphonate utilization pathways in marine bacteria revealed by functional screening and metagenomic analyses. Environ. Microbiol. 12, 222–238.

Matsumoto, K., Tokos, K.S., Price, A.R., Cox, S.J., 2008. First description of the Minnesota Earth System Model for Ocean biogeochemistry (MESMO 1.0). Geosci. Model. Dev. 1, 1–15.

McGillicuddy, D.J., Robinson, A.R., 1997. Eddy induced nutrient supply and new production in the Sargasso Sea. Deep Sea Res. Oceanogr. Res. Paper 44, 1427–1450.

McGrath, J.W., Ternan, N.G., Quinn, J.P., 1997. Utilization of organophosphonate by environmental microorganisms. Lett. Appl. Microbiol. 24, 69–73.

Menzel, D.W., Vaccaro, R.F., 1964. The measurement of dissolved organic and particulate carbon in seawater. Limnol. Oceanogr. 9, 138–142.

Mitra, A., Flynn, K.J., Burkholder, J.M., Berge, T., Calbet, A., Raven, J.A., et al., 2014. The role of mixotrophic protists in the biological carbon pump. Biogeosciences 11, 1–11. http://dx.doi.org/10.5194/bg-11-1-2014.

Moloney, C.L., Bergh, M.O., Field, J.G., Newell, R.C., 1986. The effect of sedimentation and microbial nitrogen regeneration in a plankton community: a simulation investigation. J. Plankton Res. 8, 427–445.

Monod, J., 1942. Recherches sur la croissance des cultures bactériennes. Hermann, Paris.

Moore, K.J., Doney, S.C., Lindsay, K., 2004. Upper ocean ecosystem dynamics and iron cycling in a global three-dimensional model. Global Biogeochem. Cycles 18, 4028. http://dx.doi.org/10.1029/2004GB002220.

Najjar, R.G., Sarmiento, J.L., Toggweiler, J.R., 1992. Downward transport and fate of organic matter in the ocean: simulations with a general circulation model. Global Biogeochem. Cycles 6, 45–76.

Najjar, R.G., Jin, X., Louanchi, F., 2007. Impact of circulation on export production, dissolved organic matter, and dissolved oxygen in the ocean: results from Phase II of the Ocean Carbon-cycle Model Intercomparison Project (OCMIP-2). Global Biogeochem. Cycles 21, http://dx.doi.org/10.1029/2006GB002857, GB3007.

Nedwell, D.B., 1999. Effect of low temperature on microbial growth: lowered affinity for substrates limits growth at low temperature. FEMS Microbiol. Ecol. 30, 101–111.

Neijssel, O.M., Tempest, D.W., 1975. Regulation of carbohydrate-metabolism in Klebsiella aerogenes NCTC-418 organisms growing in chemostat culture. Arch. Microbiol. 106, 251–258.

Noble, R.T., Fuhrman, J.A., 1999. Breakdown and microbial uptake of marine viruses and other lysis products. Aquat. Microb. Ecol. 20, 1–11.

Ogura, N., 1975. Further studies on the decomposition of dissolved organic matter in coastal seawater. Mar. Biol. 31, 101–111.

Pace, M.L., Glasser, J.E., Pomeroy, L.R., 1984. A simulation analysis of continental shelf food webs. Mar. Biol. 82, 47–63.

Pahlow, M., Vézina, A.F., 2003. Adaptive model of DOM dynamics in the surface ocean. J. Mar. Res. 61, 127–146.

Pahlow, M., Vézina, A.F., Casault, B., Haass, H., Malloch, L., Wright, D.G., et al., 2008. Adaptive model of plankton dynamics for the North Atlantic. Prog. Oceanogr. 76, 151–191.

Parsons, T.R., Strickland, J.D.H., 1961. On the production of particulate organic carbon by heterotrophic processes in sea water. Deep-Sea Res. 8, 211–222.

Peltzer, E.T., Hayward, N.A., 1996. Spatial distribution and temporal variability of total organic carbon along 140 W in the equatorial Pacific Ocean in 1992. Deep Sea Res. II 43, 1155–1180.

Pinhassi, J., Hagström, A., 2000. Seasonal succession in marine bacterioplankton. Aquat. Microb. Ecol. 21, 245–256.

Pirt, S.J., 1982. Maintenance energy: a general model for energy-limited and energy-sufficient growth. Arch. Microbiol. 133, 300–302.

Polimene, L., Allen, J.I., Zavatarelli, M., 2006. Model of interactions between dissolved organic carbon and bacteria in marine systems. Aquat. Microb. Ecol. 43, 127–138.

Pomeroy, L.R., 1974. The ocean's food web, a changing paradigm. BioScience 24, 499–504.

Popova, E.E., Anderson, T.R., 2002. Impact of including dissolved organic matter in a global ocean box model on simulated distributions and fluxes of carbon and nitrogen. Geophys. Res. Lett. 29, 1303, http://dx.doi.10.1029/2001GL014274.

Raymond, P.A., McClelland, J.W., Holmes, R.M., Zhulidov, A.V., Mull, K., Peterson, B.J., et al., 2007. Flux and age of dissolved organic carbon exported to the Arctic Ocean: a carbon isotopic study of the five largest arctic rivers. Global Biogeochem. Cycles 21, GB4011. http://dx.doi.org/10.1029/2007GB002934.

Redfield, A.C., Ketchum, B.H., Richards, F.A., 1963. The influence of organisms on the composition of sea water. In: Hill, M.N. (Ed.), The Sea, 2. Interscience, New York, pp. 26–77.

Reuszer, H.W., 1933. Marine bacteria and their role in the cycle of life in the sea. III. The distribution of bacteria in the ocean waters and muds about Cape Cod. Biol. Bull. 65, 480–497.

Ridgwell, A., Hargreaves, J.C., Edwards, N.R., et al., 2007. Marine geochemical data assimilation in an efficient Earth System Model of global biogeochemical cycling. Biogeosciences 4, 87–104.

Riley, G.A., Stommel, H., Bumpus, D.F., 1949. Quantitative ecology of the plankton of the western North Atlantic. Bull. Bingham Oceanogr. Collect 12, 1–169.

Roussenov, V., Williams, R.G., Mahaffey, C., Wolff, G.A., 2006. Does the transport of dissolved organic nitrogen affect export production in the Atlantic Ocean? Global Biogeochem. Cycles 20, http://dx.doi.org/10.1029/2005GB002510, GB3002.

Salihoglu, B., Garçon, V., Oschlies, A., Lomas, M.W., 2008. Influence of nutrient utilization and remineralization stoichiometry on phytoplankton species and carbon export: a modeling study at BATS. Deep-Sea Res. I 55, 73–107.

Schartau, M., Engel, A., Schröter, J., Thoms, S., Völker, C. And, Wolf-Gladrow, D., 2007. Modelling carbon overconsumption and the formation of extracellular particulate organic carbon. Biogeosciences 4, 433–454.

Schlitzer, R., 2002. Carbon export fluxes in the Southern Ocean: results from inverse modeling and comparison with satellite-based estimates. Deep-Sea Res. II 49, 1623–1644.

Schmittner, A., Oschlies, A., Giraud, X., Eby, M., Simmons, H.L., 2005. A global model of the marine ecosystem for long-term simulations: sensitivity to ocean mixing, buoyancy forcing, particle sinking, and dissolved organic matter cycling. Global Biogeochem. Cycles 19, http://dx.doi.org/10.1029/2004GB002283, GB3004.

Sherr, E.B., 1988. Direct use of high molecular weight polysaccharide by heterotrophic flagellates. Nature 335, 348–351.

Shigemitsu, M., Okunishi, T., Nishioka, J., et al., 2012. Development of a one-dimensional ecosystem model including the iron cycle applied to the Oyashio region, western subarctic *Pacific*. J. Geophys. Res. 117, http://dx.doi.org/10.1029/2011JC007689 C06021.

Sieburth, J. McN, 1977. International Helgoland Symposium: Convenor's report on the informal session on biomass and productivity of microorganisms in planktonic ecosystems. Helgoländer wiss. Meeresunters 30, 697–704.

Six, K.D., Maier-Reimer, E., 1996. Effects of plankton dynamics on seasonal carbon fluxes in an ocean general circulation model. Global Biogeochem. Cycles 10, 559–583.

Smith, D.C., Simon, M., Alldredge, A.L., Azam, F., 1992. Intensive hydrolytic activity on marine aggregates and implications for rapid particle dissolution. Nature 359, 139–141.

Sorokin, Y.I., 1977. The heterotrophic phase of plankton succession in the Japan Sea. Mar. Biol. 41, 107–117.

Steele, J.H., Henderson, E.W., 1981. A simple plankton model. Am. Nat. 117, 676–691.

Stocker, T.F., Broecker, W.S., Wright, D.G., 1994. Carbon uptake experiments with a zonally-averaged global ocean circulation model. Tellus 46B, 103–122.

Stoderegger, K., Herndl, G.J., 1998. Production and release of bacterial capsular material and its subsequent utilization by marine bacterioplankton. Limnol. Oceanogr. 43, 877–884.

Strom, S.L., Benner, R., Ziegler, S., Dagg, M.J., 1997. Planktonic grazers are a potentially important source of marine dissolved organic carbon. Limnol. Oceanogr. 42, 1364–1374.

Taylor, K.E., Stouffer, R.J., Meehl, G.A., 2012. An overview of CMIP5 and the experiment design. Bull. Am. Meteorol. Soc. 93, 485–498.

Teira, E., Pazó, M.J., Serret, P., Fernández, E., 2001. Dissolved organic carbon (DOC) production by microbial populations in the Atlantic Ocean. Limnol. Oceanogr. 46, 1370–1377.

Teira, E., Martínez-García, S., Calvo-Díaz, A., Morán, X.A.G., 2010. Effects of inorganic and organic nutrient inputs on bacterioplankton community composition along a latitudinal transect in the Atlantic Ocean. Aquat. Microb. Ecol. 60, 299–313.

Tezuka, Y., 1990. Bacterial regeneration of ammonium and phosphate as affected by the carbon: nitrogen: phosphorus ratio of organic substrates. Microb. Ecol. 19, 227–238.

Thingstad, T.F., Hagström, Å., Rassoulzadegan, F., 1997. Accumulation of degradable DOC in surface waters: is it caused by a malfunctioning microbial loop? Limnol. Oceanogr. 42, 398–404.

Tian, R.C., Vézina, A.F., Legendre, L., Ingram, R.G., Klein, B., Packard, T., et al., 2000. Effects of pelagic food-web interactions and nutrient remineralization on the biogeochemical cycling of carbon: a modeling approach. Deep-Sea Res. 47, 637–662.

Torres-Valdés, S., Roussenov, V., Sanders, R., Reynolds, S., Pan, X., Mather, R., et al., 2009. Distribution of dissolved organic nutrients and their effect on export production over the Atlantic Ocean. Global Biogeochem. Cycles 23, GB4019. http://dx.doi.org/10.1029/2008GB003389.

Touratier, F., Legendre, L., Vézina, A., 1999. Model of bacterial growth influenced by substrate C:N ratio and concentration. Aquat. Microb. Ecol. 19, 105–118.

Unanue, M., Azúa, I., Arrieta, J.M., Labirua-Iturburu, A., Egeaand, L., Iriberri, J., 1998. Bacterial colonization and ectoenzymatic activity in phytoplankton-derived model particles: cleavage of peptides and uptake of amino acids. Microb. Ecol. 35, 136–146.

Unanue, M., Ayo, B., Agis, M., Slezak, D., Herndl, G.J., Iriberri, J., 1999. Ectoenzymatic activity and uptake of monomers in marine bacterioplankton described by a biphasic kinetic model. Microb. Ecol. 37, 36–48.

Urban-Rich, J., 1999. Release of dissolved organic carbon from copepod fecal pellets in the Greenland Sea. J. Exp. Mar. Biol. Ecol. 232, 107–124.

Vallino, J.J., Hopkinson, C.S., Hobbie, J.E., 1996. Modeling bacterial utilization of dissolved organic matter: optimization replaces Monod growth kinetics. Limnol. Oceanogr. 41, 1591–1609.

Vallino, J.J., 2000. Improving marine ecosystem models: use of data assimilation and mesocosm experiments. J. Mar. Res. 58, 117–164.

Van Dam, K., Jansen, N., Postma, P., Richard, P., Ruijter, G., Rutgers, M., et al., 1993. Control and regulation of metabolic fluxes in microbes by substrates and enzymes. Antonie Van Leeuwenhoek 63, 315–321.

Verity, P.G., Stoecker, D.K., Sieracki, M.E., Nelson, J.R., 1993. Grazing, growth and mortality of microzooplankton during the 1989 North Atlantic spring bloom at 47°N, 18°W. Deep-Sea Res. I 40, 1793–1814.

Vézina, A.F., Platt, T., 1988. Food web dynamics in the ocean. I. Best-estimates of flow networks using inverse methods. Mar. Ecol. Prog. Ser. 42, 269–287.

Walsh, J.J., Dieterle, D.A., 1994. CO_2 cycling in the coastal ocean. I—A numerical analysis of the southeastern Bering Sea with applications to the Chukchi Sea and the northern Gulf of Mexico. Prog. Oceanogr. 34, 335–392.

Williams, P.J. le B., 1981. Incorporation of microheterotrophic processes into the classical paradigm of the planktonic food web. Kieler Meeresforsch. Sonderh. 5, 1–28.

Williams, P.J. le B., 1990. The importance of losses during microbial growth: commentary on the physiology, measurement and ecology of the release of dissolved organic material. Mar. Microb. Food Webs 4, 175–206.

Williams, P.J. le B., 1995. Evidence for the seasonal accumulation of carbon-rich dissolved organic material, its scale in comparison with changes in particulate material and the consequential effect on net C/N assimilation ratios. Mar. Chem. 51, 17–29.

Williams, R.G., McDonagh, E., Roussenov, V.M., Torres-Valdes, S., King, B., Sanders, R., et al., 2011. Nutrient streams in the North Atlantic: advective pathways of inorganic and dissolved organic nutrients. Global Biogeochem. Cycles 25, http://dx.doi.org/10.1029/2010GB003853, GB4008.

Wood, A.W., Van Valen, L.M., 1990. Paradox Lost? On the release of energy-rich compounds by phytoplankton. Mar. Microb. Food Webs 4, 103–116.

Wright, S.H., Manahan, D.T., 1989. Integumental nutrient-uptake by aquatic organisms. Annu. Rev. Physiol. 51, 585–600.

Yamanaka, Y., Tajika, E., 1997. Role of dissolved organic matter in the marine biogeochemical cycle: studies using an ocean biogeochemical general circulation model. Global Biogeochem. Cycles 11, 599–612.

Yool, A., Popova, E.E., Anderson, T.R., 2011. MEDUSA-1.0: a new intermediate complexity plankton ecosystem model for the global domain. Geosci. Model. Dev. 4, 381–417.

Zohary, T., Robarts, R.D., 1998. Experimental study of microbial P limitation in the eastern Mediterranean. Limnol. Oceanogr. 43, 387–395.

Zweifel, U.L., Norrman, B., Hagström, Å., 1993. Consumption of dissolved organic carbon by marine bacteria and demand for inorganic nutrients. Mar. Ecol. Prog. Ser. 101, 23–32.

Index

Note: Page numbers followed by *f* indicate figures, and *t* indicate tables.

Printed and bound by CPI Group (UK) Ltd, Croydon, CR0 4YY

08/05/2025

01864866-0002